Geography

Human Population on the Earth at A.D. 2000

- □ The world's largest 25 cities in 2000 (with rank position shown)
- • The world's next largest 75 cities in 2000
- Generalized areas of concentrated human occupancy
- Generalized areas which are largely devoid of human population

Geographers over many centuries have struggled with the problem of displaying a three-dimensional spherical globe as a two-dimensional planar map. The various solutions proposed are called *map projections* and these are discussed in Chapter 21 of this book. The projection used here is a *Modified Breisemeister Projection*. It was developed in 1950 by the chief cartographer of the American Geographical Society, William Breisemeister. It shows the earth as an ellipse with areas preserved as the same proportionate area as on the globe. This makes it exceptionally well suited to showing the human population distribution. It does this at the expense of directions being distorted (so that the shape of some continents is skewed) and the Antarctic continent is split into two halves. These two hemispheres are split along the meridian of 10dgs. east so that the left-hand map is largely the western hemisphere and the right-hand map the eastern hemisphere.

Empty and Crowded Areas. The map shows in a generalized way how the human population of just over six billions is spread around the earth's surface at the start of the twenty-first century. Note that only some 27 percent of the earth's surface is land covered. The grey shading on the map shows the over a third of the land surface which is unpopulated or supports only very small populations. These include the polar climates of the Arctic and sub-Arctic (too cold for agriculture) and the world's desert areas (too dry for agriculture without irrigation). Over 98 percent of the world's population lives outside these hostile environments. The blue shading on the map shows the concentrated pockets of dense population. The great alluvial river valleys of China and India, the islands of Japan, the European peninsula, the eastern seaboard areas of both North and South America stand out. For the world as a whole, areas proximate to the sea (within 100km of the ocean or sea-navigable highway) have population densities over four times higher than inland areas.

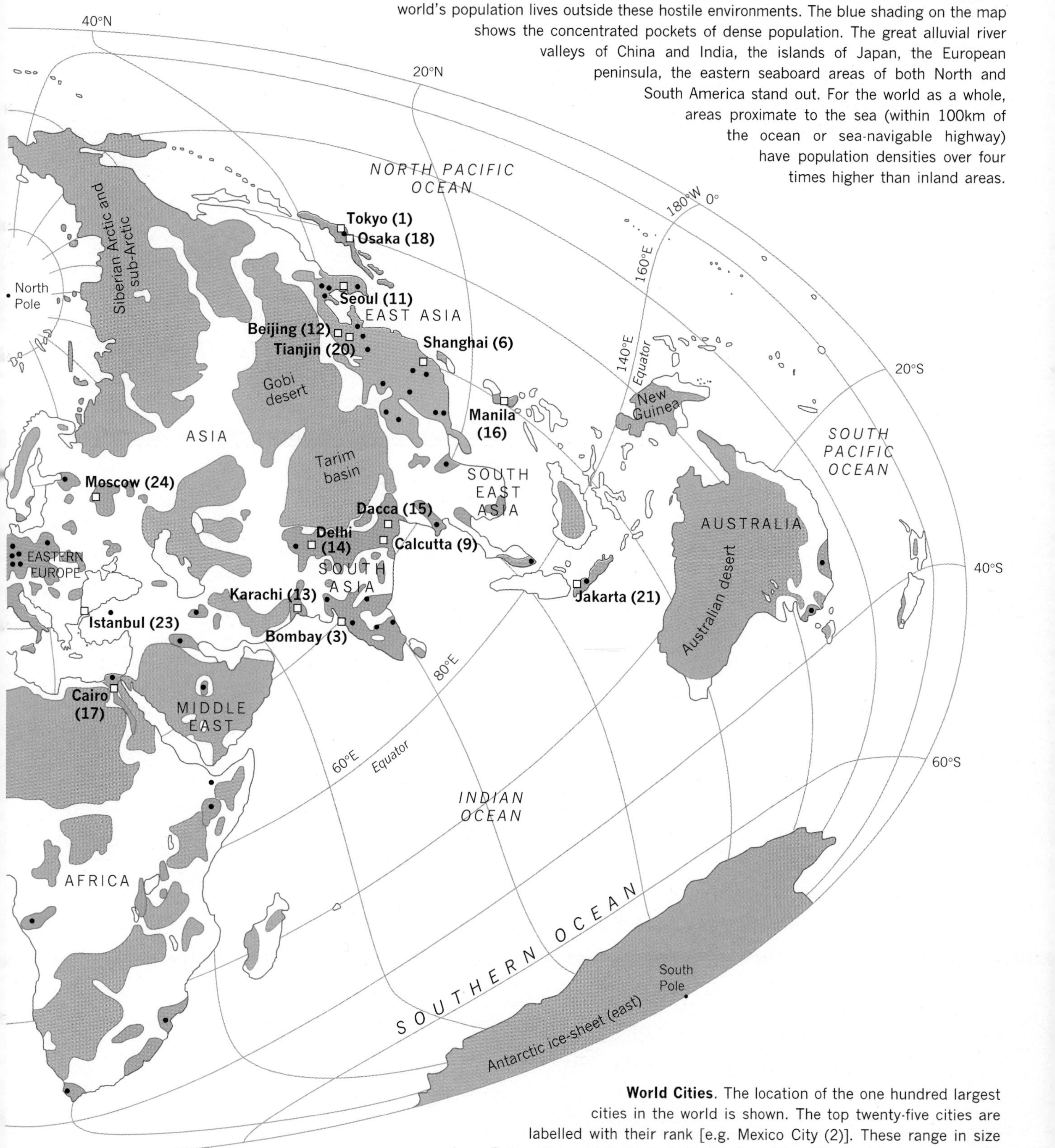

World Cities. The location of the one hundred largest cities in the world is shown. The top twenty-five cities are labelled with their rank [e.g. Mexico City (2)]. These range in size from Tokyo, the world's largest city, at over 28 million to London at rank 25 with 7.6 million. Together these top-ranking cities make up some five percent of the world's total population, i.e. about one person in every twenty live in these huge metropolitan areas. The next 75 cities are shown but not named. Together the top 100 cities make up one tenth of the world's total population and the proportion is rising. All figures are based on the latest estimates from the United Nations, the U.S. Census Bureau and the Population Reference Bureau. The problems of defining large cities is discussed in Chapter 8 of this book.

Pearson
Education

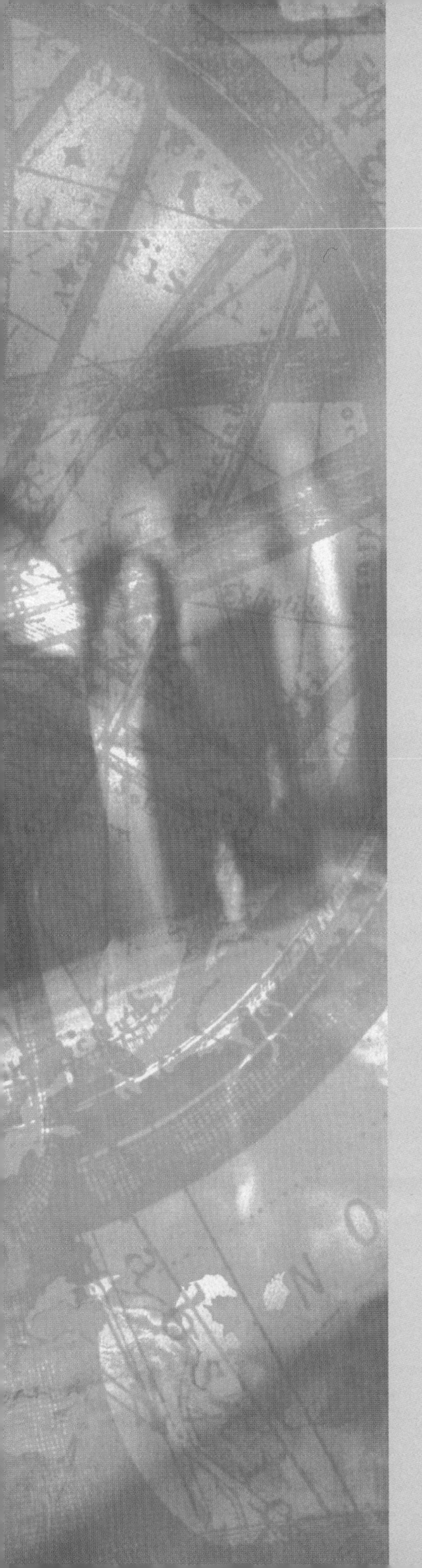

Geography

A Global Synthesis

PETER HAGGETT
University of Bristol

An imprint of Pearson Education
Harlow, England · London · New York · Reading, Massachusetts · San Francisco
Toronto · Don Mills, Ontario · Sydney · Tokyo · Singapore · Hong Kong · Seoul
Taipei · Cape Town · Madrid · Mexico City · Amsterdam · Munich · Paris · Milan

Pearson Education Limited
Edinburgh Gate
Harlow
Essex CM20 2JE
England

and Associated Companies throughout the world

Visit us on the World Wide Web at:
http://www.pearsoneduc.com

First published 2001

British Library Cataloguing-in-Publication Data
A catalogue record for this book is available from the British Library

Library of Congress Cataloging-in-Publication Data
Haggett, Peter.
Geography: a global synthesis/Peter Haggett.
p. cm.
Includes bibliographical references (p.).
ISBN 0-582-32030-5
1. Geography. I. Title

G128 .H29 2001
910--dc21 00-066898

ISBN 0 582 32030 5

10 9 8 7 6 5 4 3 2 1
05 04 03 02 01

Typeset by 3 in 10pt Sabon
Printed and bound in Italy by G. Canale & C.S.p.A

CONTENTS

PREFACE

> 'If I ONLY KNEW geography!' Chico Pacheco kept repeating the phrase between clenched teeth, lamenting the wasted days of his youth; He had been a notorious cutter of classes. And all the time he had lost during his life, frittering it away on nonsense, when he could have devoted himself, body and soul, to the intensive study of geography, a science whose utility he had only come to realize! . . . 'I'll have to send for Bahia for some textbooks'.
>
> JORGE AMADO 'Of the Drawbacks of Not Knowing Geography, and the Deplorable Tendency to Bluff at Poker', in *Os Velhos Marinheiros* (1963)

I start to write this preface somewhere high over the Bay of Bengal. The 747 appears to be motionless and below great tropical tower clouds boil up on the mid afternoon heat. In reality the aircraft is moving at approaching one thousand kilometres per hour, the outside air temperature is down to −46 °C (at the height of 11,000 metres), and I wouldn't survive a few seconds outside the protective cradle of the aircraft hull. I'm on my way to visit grandchildren in Australia and I reflect on the changes between my childhood, growing up in an English village, and theirs today.

Then, in the early 1930s, writing the above paragraph would have been dismissed as science fiction. Electricity had just arrived in our home but a refrigerator was a distant dream, the rural charabanc was the only reliable road transport from the village, our wireless set was still at the crystal stage. Then a journey to Australia by ship would have taken six weeks (and many months' earnings), air transport links still didn't exist, though one Bert Hinkler had just broken the record for a sixteen-day trip from London to Darwin by light aircraft. Most striking of all has been the change in the planet Earth. In the year of my birth the world's population had just reached two billion: today it is three times larger at over six billion. In those seven decades the annual rate of resource extraction has grown nearly a hundredfold and international travel a thousandfold.

Whether my rural roots and the love of the varied West Country landscape played any part in my decision to read geography at Cambridge I shall never know. World War II and petrol bans meant that I had to obey Carl Sauer's dictum that 'locomotion should be slow, the slower the better'. At first a battered bicycle allowed an ever-widening range of landscapes in England and postwar France to be explored. My first continent outside Europe was South America and it was there that I had the good fortune to meet the Brazilian novelist, Jorge Amado, learnt to love his perceptive writing on the Brazilian scene, and eventually stumbled on the geography-hungry Chico Pacheco whose words open this preface.

Whether the book now in your hands would have met Chico's demands I must leave to him. It was started a generation ago when I was visiting Northwestern University in Chicago and talked with a young economist, then editor for Mark Twain's old American publishing house. He encouraged me to write a textbook aimed not at those who were already well versed in the subject, but at those – like Chico Pacheco – who were coming to it for the first time and who would probably be going on to other careers. The book that eventually emerged was called *Geography: A Modern Synthesis* (borrowing the subtitle from the biologist Julian Huxley). It was first published in 1972, ran through three further revisions, was translated into six languages, and continued in print to the present.

Geography: A Global Synthesis is its successor. It was written for a new century, with a new title and with new features but continues to strive for the same goals as its predecessors. *Geography* is an attempt to present the whole spectrum of geography in a contemporary context and within a single volume. It tries to synthesize at two levels: first, by bringing together the different traditions and themes within the field; second, by stressing the synthesizing role of geography as a whole in relation to neighbouring fields. The book is designed to introduce the student with no previous geographic training to a field of rapidly expanding horizons and increasing consequence both as an academic subject and as an applied science. Lying athwart both the physical and social sciences, geography challenges students to abandon familiar and comfortable 'straightjackets' and to focus directly on relationships between people and the environment, their spatial consequences, and the resulting regional structures that have emerged on the earth's surface. Geography is uniquely relevant to current concerns both with the environment and ecology and with regional contrasts and imbalances in human welfare.

No single exponent of geography or any other academic field of enquiry can write about the whole of it in detail. Past efforts to do so seem naïve in retrospect. The problems that face the beginning student in geography, however, are now so complex that the challenge must be met. It is too easy to take the view that all one can or should do is to have a student take introductory courses in various easily identifiable subfields – physical geography, cultural geography, and so on – and hope that somehow these will produce an integrated view of geography as a whole. The osmosis, however, by which this is supposed to take place is rarely clearly defined.

To begin with the parts of a field of enquiry and go on to the whole seems to me to be a tactic of convenience, forced on us by the rising tide of research and continuous fission of new subdisciplines. Surely, to confine larger questions of the nature of geography to postgraduate seminars is an inversion of desirable sequence of scholarship. We owe it to those who are starting in a field, some of whom we hope will follow us, to look around and ahead just as far as we can. Therefore, in this book I have turned away for a while from my own research patch (a small corner of medical geography) and forced myself to put the various parts of geography together into what seems to me at this time to be an integrated form. I expand further on the structure of the book and the ways in which it can be used in Appendix C of the book.

With each succeeding decade a writer's debts grow. My original debt was to Cambridge where I studied as an undergraduate and graduate student and taught for many years. For the last 35 years my base has been in England's West Country, at the University of Bristol. My debt to successive generations of students and colleagues there is incalculable. Both Cambridge and Bristol allowed me free rein to roam widely across the academic world in pursuing my research. In the United States, I first taught at Berkeley and have subsequently held posts at Pennsylvania State, Wisconsin, Minnesota. Canada provided bases at Toronto, McMaster, and York; Australia at ANU and Monash; and New Zealand at Canterbury. Kuala Lumpur, Singapore, and Bandar provided experience in the growing universities of south and east Asia. Prolonged periods in Sweden and Iceland, and shorter visits to Denmark and Finland have provided rewarding insights into Scandinavian geography. Outside geography departments I've had the privilege to work in the leading epidemiological centres, notably the Centers for Disease Control (CDC), Atlanta, and the World Health Organization (WHO) at Geneva. All these years have left marks on the form and structure of the book.

To mention individual scholars is to open up an unending stream of debts. Don Meinig of Syracuse University edited the initial volumes from which this book has grown and I'm grateful to his outstanding example of scholarship and his wise counsel over the decades. I've been fortunate in my academic climbing companions over many years: Richard Chorley and Andy Cliff (Cambridge) and Matt Smallman Raynor (Nottingham) will recognize many sections stemming from our joint research. Geographers never meet but to talk and debate issues and Ron Abler (AAG), Antoinne Bailey (Geneva), Brian Berry (Texas), Sofus Christiansen (Copenhagen), Robert Geipel (Munich), Torsten Hägerstrand (Lund), David Harvey (Johns Hopkins), Ron Johnston (Bristol), Les King (McMaster), David Rhind (London), Ian Simmons (Durham), David Stoddart (Berkeley), and Alan Wilson (Leeds) have had more effect on this book than they will know. Sadly, the turning globe brings losses and I mourn the passing of three good friends, Jim Parsons and Jay Vance (Berkeley) and Peter Gould (Penn State), who shaped my thinking.

One learns most of all from one's undergraduate and graduate students. I count it a special privilege to have taught and learnt from Kevin Cox (Ohio State), Pip Forer (Auckland), Tony Gatrell (Lancaster), Ray Harris (Durham), Bill Macmillan (Oxford), Glen Norcliffe (York), Nigel Thrift (Bristol), and Roly Tinline (Queen's) among many others. There are too those anonymous referees and commentators who raised questions and clarified meanings and to them I both give thanks and invite readers of this new volume to join their ranks.

Complex books are increasingly the result of a team effort and I'm indebted to the talented team assembled by Matthew Smith at Pearson Education. It has been a pleasure to work with Tina Cadle, Ros Woodward, Faith Perkins, and the behind-the-scenes army who designed the book and drove the programme forward. The book as you see it is very much their creation.

Prefaces are too public a place to express the personal indebtedness of an author to family – they will know just how much I have to thank them for. Oskar Spate, the dean of Australian geographers and a member of my old Cambridge college, put the public and private life of a geographer better than my words can. His aim, he said, was simply 'to make what one can of this our earth, and if the cosmic order holds nothing for us either of despair or hope, to find our happiness in social duty and private love.' I have been blessed in both. I dedicated my first book a generation ago to my wife Brenda; we met as students a half century ago. This book (likely to be my last) I dedicate to the youngest of our four children, wishing God Speed to Andy, Janice, and Isobel, as they journey on into the new century.

PETER HAGGETT
Chew Magna

Special conventions used in this book

This book follows the normal rules of scientific writing but two special conventions used should be noted.

Geographic Scale For maps showing specific parts of the earth's surface, the normal linear scale in kilometres (or miles) is replaced by a unique star system which gives the magnitude on a scale that runs from the global (*) to the local (*****). Readers are advised to look at *Box 1.C: Orders of Geographic Magnitude* (pages 21–22) which describes the use of the scale in detail. Note that the scale is logarithmic so that the size difference between ** and *** is ten-fold, ** and **** a hundredfold, and so on.

Geographic Terms The language of geography is a complex one drawing from both the natural and behavioural sciences. In some cases commonplace terms are used in geographic writing in an unusual way. Terms of special interest in each chapter are therefore highlighted in *blue italics* in the text and ***black italics*** in boxes and figure captions. Such terms are also gathered together in *Appendix A: Glossary* together with a selection of other key words likely to be helpful to readers of this book.

COMPANION WEB SITE

GEOGRAPHY: A Global Synthesis

by Professor Peter Haggett

Visit the ***Geography: A Global Synthesis*** Companion Web Site at *www.booksites.net/haggett* to find valuable teaching and learning material including:

For students:

- Study material designed to help you improve your results
- Extensive links to valuable resources on the web
- Downloadable Study Guide
- Search for specific information on the site

For lecturers:

- A secure, password protected site with teaching material
- Downloadable Instructor's Manual
- A syllabus manager that will build and host a course web page

ACKNOWLEDGEMENTS

The publishers wish to thank the following for permission to reproduce the material:

Fig. 1.2 (b) from Heffer, *Orford Ness* (1966), Fig. 1.15 (a, b and c) from N.R. Hanson, *Patterns of Discovery: An Inquiry into the Conceptual Foundation of Science*, (1958), Fig. 4.15(b) 'Structure of the tropical rain forest' – profile diagram redrawn from P.W. Richards, *The Tropical Rain Forest* (1964), Fig. 11.4 from H.C. Darby, *The Domesday Geography of Eastern England* (1952), Fig. 16.2 drawn from date from August Lösch in A.D. Cliff, P. Haggett, J.K. Ord and R. Versey, *Spatial Diffusion* (1981) and Fig. 19.3(a) from A.D. Cliff, P. Haggett and M. Smallman-Raynor, *Deciphering Global Epidemics* (1999) by permission of Cambridge University Press; Fig. 1.13(c) Population map from R. Warwick Armstrong, Ed., *Atlas of Hawaii* (1973) by permission of The University of Hawai'i Press, Honolulu; Fig. 2.7(b) adapted from a diagram in *Geology* and later in S. Helmfrid (ed.), *The Geography of Sweden*, by permission of The National Atlas of Sweden, Vallingby; Fig. 2.17 adapted from L.W. Swan, in W.H. Osburn and H.E. Wright, Jnr., (Eds.), *Arctic and Alpine Environments* (1968) published by Indiana University Press, Bloomington, Indiana; Fig. 2.19a 'Earth's Hydrologic Cycle' modified from R.J. More, in R.J. Chorley and P. Haggett (eds.), *Models in Geography* (Methuen, London, 1967) and Fig. 4.17 from V. Olgay, from R.G. Barry and R.J. Chorley, *Atmosphere, Weather and Climate* (Methuen, London, 1968) by permission of Taylor and Francis; Fig. 3.19 (a, b and c) 'Mill Creek Ecology' from R.A. Bryson and D.A. Barreis, *Journal of the Iowa Archeological Society*, **15** (1960) pp 290–291 reproduced with the permission of the Journal of the Iowa Archeological Society; Fig. 3.23 from D.M. Herschfield and M.A. Kohler, *Journal of Geophysical Research* Vol. 75, page 1728 (1960) copyright by The American Geophysical Union; Figs. 4.2, 5.2, 5.3 (a and b), 6.19 and 19.18 from A.G. Fischer, in I. Douglas et al (eds.), *Companion Encyclopaedia of Geography* (Routledge, London, 1996), by permission of Taylor and Francis and Fig. 4.17 from V. Olgay, from R.G. Barry and R.J. Chorley, *Atmosphere, Weather and Climate* (1968) published by Methuen, London by permission of Taylor and Francis Ltd; Fig. 4.12 from F.K. Hare, *Geographical Review* **40** (1950), Fig. 4.16 from S. Haden-Guest et al. (eds), *World Geography of Forest Resources* (American Geographical Society, New York, 1956), Box 5B(b) Reproduction of the cover of *Agricultural Origins and Dispersals* by Carl Sauer published by The American Geographical Society, New York, Fig. 7.16(b) Map from F. Kniffen, *Geographical Review* **41** (1951), Fig. 5.7 from *Geography of Domestication* by Isaac, E. © 1970, Figs. 7.2 and 12.5 from *Cultural Geography of the United States* by Zelinsky, Wilbur, © 1973, Fig. 14.16 *The Merchant's World* by Vance, © 1971 and Fig. 19.11 from *The Geography of Economic Systems* by Berry/Cunkling, © 1975 reprinted by permission of Prentice Hall, Inc., Upper Saddle River, NJ; Fig. 5.12 Map showing intercontinental migration reprinted from *World Population and Production* by W.S. and E.S. Woytinsky, Copyright © 1953, with permission from The Century Foundation, Inc., formerly the Twentieth Century Fund; Fig. 6.3 Map showing natural components in population change, Population Reference Bureau, *Population Bulletin* **18**, 5 (1962), Fig. 1 by permission of the Population Reference Bureau, Washington, D.C., Fig. 6.7 from M.L. Levy, *Population et Sociétès*, Vol. 333 (Fig. 1, p.2) by permission of Institut National d'Etudes Demographiques (IINED), Paris; Fig. 6.20 Graph from T. Frejka, *Population Studies* **22** (1986) and data from Population Reference Bureau by permission of The Population Investigation Committee, London School of Economics; Fig. 7.7 (a and b) reprinted from The Canadian Geographer, Vol. 11, 1958, article by J.R. Mackay; Fig. 7.10 from Yi-Fu Tuan, Space and Place, (1977) and Fig. 16.18 from M. Levison, R.G. Ward and J.W. Webb, *The Settlement of Polynesia: A Computer Simulation* (1973) by permission of the University of Minnesota Press, Minneapolis; Fig. 7.14 from T. Friberg, *Everyday Life: Women's Adaptive Strategies in Time and Space* (1993) reprinted with permission of Tora Friberg, c/o Linköpings University, Sweden; Figs. 8.9 and 19.3(b) from R.E.G. Davis, *History of the World's Airlines* (1964) by permission of Oxford University Press; Fig. 8.18 from R.D. MacKinnon, in L.S. Bourne et al (eds.), *Urban Futures for Central Canada* (1974) and Fig. 8.22(a) adapted from J.W. Simmons and L.S. Bourne in L. Gentilcore (Ed.), *Ontario* (1972) by permission of University of Toronto Press, Canada; Fig. 9.17 Chart showing lead content of the Greenland ice cap reprinted from *Geochimica Cosmochimica Acta,* Vol. 33, M. Muroxumi et al, p. 1247 (1969) with permission from Elsevier Science; Fig. 10.3 Royal Air Force photograph of Grimes Graves in eastern England and Fig. 22.11(b) Infrared image of the River Axe lowlands, southwest England. (Royal Signal and Radar Establishment, Malvern): Crown copyright material is reproduced with the permission of the Controller of Her Majesty's Stationery Office; Fig. 10.16(b) Map from R.C. Lucas, *Natural Resources Journal*, Vol. 3, No. 3 (1962), Figure 3, pages 394–411 by permission of The Natural Resources Journal, University of New Mexico; Fig. 11.3 Maps from P.E. James, *Geographical Review*, **43** (1953) by permission of the American Geographical Society, New York; Box 11C: Australian (used) stamps with portrait of Griffith Taylor, geographer. Copyright © Australia Post. National Philatelic Collection, Australia Post; Fig. 11.5 (a, b, c and d) from H.C. Darby, Fig. 11.7 from A.N. Strahler, Fig. 11.10 (a, b, c and d) from J.T. Curtis, Fig. 13.8 from E.L. Ullman in W.L. Thomas, Jnr. (Ed.), *Man's Role in Changing the Face of the Earth* (1956) and Box 11A from the cover of by permission of the University of Chicago Press; Fig. 12.8 Map of India from A.T.A. Learmouth and O.H.K. Spate, *India and Pakistan, 3rd edition* (Methuen, London, 1967), Figure 13.1, page 408 by permission of Taylor and Francis; Fig. 12.14 (a, b and c) from data by R. Dannenbrink, Los Angeles City Planning Commission. From National Academy of Sciences, *Publication No. 1498* (1967), Figs. 2–4, pp. 107–112 courtesy of The National Academy of Sciences, Washington, D.C.; Fig. 12.16(a) from K. Lynch, *The Image of the City* (1960) by permission of MIT Press; Fig. 12.19: Data from Dieter Steiner, *Tijdschrift van het Koninklijk Nederlandsch Aardrijkskundig Genootschap Tweede Reeks*, **82** (1965) pp. 329–47 by permission of The Royal Dutch Geographical Society; Fig. 13.3 from J. Holmes and R.F. Pullinger, *Australian Geographer* (1973) by permission of Assoc. Professor J. Forrest, Editor of *Australian Geographer*, Macquarie University; Fig. 13.5 from H.C. Brookfield and D. Hart, *Melanesia* (Barnes & Noble, New York and Methuen, London) Figure 14.9, page 357 by permission of Taylor and Francis; Fig. 13.14 Old map of Kõnigsberg, Prussia from the New York Public Library (The Astor, Lennox and Tilden Foundations) by permission of The New York Public Library; Fig. 14.15(a) from G.W. Skinner, *Journal of Asian Studies* **34** (1964) reprinted with the permission of the Association for Asian Studies, Inc., Ann Arbor, Michigan; Fig. 14.19 from C. Madden, *Economic Development and Cultural Change* (1956) Vol. 4, p. 239, Fig. 1, Fig. 16.15 from L.W. Bowden, University of Chicago, Department of Geography, *Research Papers*, No. 97 (1965) and Fig. 18.5 adapted from N. Ginsburg, *Atlas of Economic Development* (1961) by permission of The University of Chicago Press; Fig. 15.9 from B.J.L. Berry et al., *Geographical Review*, **53** (1963) by permission of the American Geographical Society, New York; Fig. 15.14 from M.D.I. Chisholm, *Rural Settlement and Land use* (1966) by permission of Routledge; Fig. 15.15 Map of rural land use banding in the African tropics, from Directorate of Overseas Surveys, Gambia, Land Use Sheet 3/111, 1:25,000 (1958). Maps reproduced from Ordnance Survey mapping with the permission of the Controller of Her Majesty's Stationery Office, © Crown Copyright; Fig. 15.16

'Spatial expansion of land use zones'. Data from J.R. Peat, *Economic Geography*, **45** (1969), p. 295, Table 1; Fig. 15.19 from O. Lindberg, *Geografiska Annaler* **35** Series B (1952), Fig. 20, p. 39 by permission of *Geografiska Annaler* (Swedish Society for Anthropology and Geography); Fig. 16.3(b) Press cutting from *The Daily Telegraph*, (autumn 1977) by permission of the Telegraph Group Limited 2000; Fig. 16.4 'Types of spatial diffusion', data by A.G. Arthur in M.C.R. Edgell, *Monash Publications in Geography and Environmental Science*, No. 5 (1973) Fig. 2, p.7 with permission of M.C.R. Edgell and School of Geography and Environmental Science, Monash University; Fig. 16.7 from R.L. Morrill, *Economic Geography*, **46** (1970), p. 265, Fig. 12; Fig. 16.11 'Resistance to Change' after G.E. Jones, *Journal of Agricultural Economics* **15** (1963), Figs. 6 and 7(a), pp. 489–90 by permission of Agricultural Economic Society, Nottingham; Fig. 17.3 from *International Boundaries* by S.W. Boggs. © 1940 Columbia University Press. Reprinted by permission of the publisher; Fig. 17.5 from R.J. Davies, *Tropical Africa: An Atlas for Rural Development* (University of Wales Press, Cardiff, 1970), Plate 10, page 23 by permission of University of Wales Press; Figs. 17.9 and 17.26 from N.J.G. Pounds, *Political Geography* (1963), and Fig. 18.7 as adapted from J.O.M. Broek and J.W. Webb, *A Geography of Mankind* (1968) are reproduced with permission of The McGraw-Hill Companies; Fig. 17.12(b) from R. Muir, *Modern Political Geography* (Macmillan, London, 1975) and Figs. 6.4 (a, b and c) from C. Clark, *Population Growth and Land Use* (Macmillan, London, 1967), Figs reproduced with permission of Palgrave; Figs. 17.22 and 17.23(b) from L.M. Alexander, *Offshore Geography of Northwestern Europe* (1963) by permission of John Murray (Publishers) Ltd; Fig. 18.14(a) South African data for 1967 from D.M. Smith, *An Introduction to Welfare Geography*, Occasional Paper No. 11 reproduced by permission of Professor D.M. Smith, Queen Mary College, London; Fig. 18.17 'Intercity comparisons of social welfare'. Data from M.V. Jones and M.J. Flax, *The Quality of Life in Metropolitan Washington, D.C.: Some Statistical Benchmarks* (The Urban Institute, Washington, D.C., 1970) by permission of the authors and the Urban Institute; Fig. 19.4 from D. Bradley, in R. Steffen et al (eds.), *Travel Medicine* (Springer-Verlag, Berlin, 1988) Figure 1.4, pages 2–3 reproduced with permission of Springer-Verlag GmbH; Figs. 19.7(a), 19.12 and 19.14 from *World Development Report* 1999/2000 by World Bank, Copyright © 2000 by the International Bank for Reconstruction and Development/The World Bank. Used by permission of Oxford University Press, Inc.; Figs. 20.3 and 20.19 from *The World Health Report, 1998* (1999), Fig. 20.6 from *The World Health Report, 1995* (1995), Fig. 20.8 redrawn from maps in F. Fenner et al, *Smallpox and its Eradication* (1988), Fig. 20.12 from pamphlet *La poliomelite sera eradiquée* (7th April 1995) and Fig. 20.13 from *Geographical Distribution of Arthropod-Borne Diseases and their Principal Vectors*, (1989) by permission of The World Health Organization, Geneva; Fig. 20.16, redrawn from Peter R. Gould, *On Becoming a Geographer* (Syracuse University Press, Syracuse, 1999) by permission of the author's widow, Johanna S. Gould; Fig. 21.17 from J.R. Mackay, *Geographical Review*, **59** (1969) by permission of the American Geographical Society, New York; Fig. 21.21 from P. Haggett, 'The Edge of Space' in R.J. Bennett (Ed.), *European Progress in Spatial Analysis* (1981), p.62 by permission of Pion Limited, London; Fig. 21.22 from T. Hägerstrand, *Lund Studies in Geography*, **B**, No. 13 (1957), Fig. 38, p.73 with permission of Torsten Hägerstrand, Lund, Sweden; Fig. 22.14 (b) from Consultant's Report to the Food & Agriculture Organization of the United Nations on Assessment of Rainfall in North Eastern Oman (E.C. Barrett, Bristol, January 1977); Fig. 22.15 (a, b and c) from C.J. Willmott et al, *International Journal of Climatology*, Vol. 14 (1994), reproduced by permission of John Wiley and Sons Limited; Box 22B(a) Portrait of O.G.S. Crawford, deceased from his obituary notice by permission of Antiquity Publications Ltd., Cambridge and Irwin Scollar, Germany; Fig. 23.7 (a and b) from P.A. Burrough and R.A. McDonnell, *Principles of Geographical Information Systems* (1998) by permission of Oxford University Press; Fig. 24.4 (b) Portrait of George Perkins Marsh, 1801–1882, a leading American scholar and geographer by permission of the Trustees of Dartmouth College, Hanover, New Hampshire; Fig. 24.4(c) Portrait of Paul

Vidaldela Blanche, 1845–1918, a French scholar and geographer by permission of the American Geographic Society, New York.

Maps based on Ordnance Survey mapping with the permission of the Controller of Her Majesty's Stationery Office, © Crown copyright.

We would like to thank Cambridge University Press as the Crown's Patentee for an extract from *THE BOOK OF COMMON PRAYER*, the rights in which are vested in the Crown; Faber & Faber Ltd for extracts from the poems 'Little Gidding' from *FOUR QUARTETS* and 'Ash Wednesday' from *COLLECTED POEMS* 1909–1962 by T.S. Eliot.

The publishers would like to thank the following for supplying photographs: Aerofilms; Air Flight Service; Bryan & Cherry Alexander; J Allan Cash; Canadian Museum of Contemporary Photography, Ottawa; Chicago Department of Aviation; Cincinnati Enquirer; Corbis-Bettmann; Colorific!; Mary Evans Picture Library; John Frost Historical Newspaper Library; Colin Garratt; Dean & Chapter of Hereford Cathedral Mappa Mundi Trust; WHO; Michael Holford; Hulton Getty Picture Library; The Image Works; Integraph; Katz Pictures; Kew Enterprises; Knudsens Fotosenter; Frank Lane Picture Agency; MAGNUM; Matrix, New York; Milestone 92 1/2; Minnesota Historical Society; National Geographic Society; National Maritime Museum, Greenwich; Peter Newark's American Pictures; PANOS Pictures; Popperfoto; PUNCH; Quadrant Picture Library; Roger-Viollett; Royal Geographical Society, London; Selectron; Science Photograph Library; South American Pictures; Still Pictures; Tony Stone Images; Telegraph Colour Library; Topham Picturepoint; TRIP; Ullstein; USDA; Dr Tony Waltham. Detailed acknowledgements appear within the figure captions.

Whilst every effort has been made to trace the owners of copyright material, in a few cases this has proved impossible and we take this opportunity to offer our apologies to any copyright holders whose rights we may have unwittingly infringed.

TO THE STUDENT

> To teach I would build a trap such that, to escape, my students must learn.
>
> ROBERT M. CHUTE, *Environmental Insight* (1971)

Starting a course in a new subject at college is like driving into an unfamiliar city. We see sprawling new suburbs, the bustling freeways, the pockets of decay, but find it hard to get an overall impression of the structure or to know where we are. Geography is a Los Angeles among academic cities in that it sprawls over a very large area and merges with its neighbours. It is also hard to be sure which is the central business district.

This book has been written specifically for 'newcomers to the city' who have not previously taken courses in geography at college. It attempts to introduce some of the basic concepts geographers use as well as some of the essential environmental facts that form their background. The emphasis of the book is on ideas, or concepts. But these cannot be applied in a vacuum. Certain technical material has therefore been placed in separate discussions that are set outside the main text, and you may skip or explore them depending on the amount of time at your disposal and your taste.

The approach is essentially non-mathematical, and the book can be understood without any training in mathematics. On the other hand, geographers are using mathematics increasingly in their research, and certain aspects of a topic can be stated more explicitly in mathematical terms. These aspects, too, are presented outside the main text in separate discussions. You may wish to return to this material on a second reading.

For those of you who may be going on to further work in geography, each chapter makes some suggestions for further reading in the section entitled 'One Step Further ...'. These suggestions are largely confined to a handful of books that, in turn, open up other aspects of a subject. The final chapter points out some of the areas in which further training can be obtained and the kinds of courses you may wish to take.

Each of you may have your own favourite method for studying a textbook. Certainly no author can tell you which way is best for you, though many students find it useful to skip through a whole chapter quickly to get the general story. Figures have been designed to be self-contained wherever possible and so have been given somewhat fuller captions than normal. When, after a more lengthy reading of the chapter, you feel confident that you've understood it, you can turn to the concepts listed for review in the 'Reflections' section to check yourself.

For those of you whose formal study of geography takes you no further than this book, I hope the brief acquaintance will have been a provoking one. If you take with you even some small part of the concern and fascination geographers experience in their exploring of the earth's environment and our place in it, then I will feel that my job is done.

P.H.

NDRED
Sudbourn
Orford
Marsh House
Kings Marshes
Lights
Stone Ditch
rford Ness
Havergate
Beech

PROLOGUE

On the Beach

CHAPTER 1

On the Beach

■ denotes case studies

Figure 1.1 The beach environment Aerial view of a beach on the eastern coast of England. Orford Ness ('ness' is an Old English word for nose) is a major shingle structure created by the action of dominant waves and currents in the North Sea. They carry debris southwards, bending the mouth of the river Ore (the rivers Alde, right, and Butley, left, come together to form the river) ever further towards the south. Note the detailed curves of the spits in the foreground. North is at the top of the photo, the east–west distance across the middle of the photograph is around 5 km (3 miles). The small town of Orford is located towards the top of the photo.
[Source: Aerofilms.]

> We shall not cease from exploration
> And the end of all our exploring
> Will be to arrive where we started
> And know the place for the first time
>
> T.S. ELIOT *Little Gidding* (1942)

In Nevil Shute's compelling novel *On the Beach* the end of human occupation of the earth is forecast. If you have read the book or seen the old movie, you may recall the small group of survivors in Australia, waiting for radiation clouds to drift over the southern hemisphere – to complete the annihilation already accomplished in the north. Or you may have seen Leonardo Di Caprio starring as a young American backpacker in *The Beach* (based on Alex Garland's best-selling novel). He leaves his cell-phone and Internet world to search on a Thailand beach for a different and more realistic world.

Whatever the merits of Shute's grim forecast, or Garland's disconnected world, what justifies borrowing their titles for the opening chapter of a textbook? Well, there are good reasons for a geographer selecting this title. Early peoples were historically creatures of the strandline between water and land. They moved (as we still do today) like crabs in the denser bottom layer of gas on the surface of the earth, not occupying the water itself, but never far from it. In prehistoric times, the beaches were used as a highway for easy travel; in the Renaissance they were used as a springboard for colonization and conquest. Even at the start of the twenty-first century, the biggest cities are on the strandline. Three-quarters of the world's largest urban centres – those with over five million inhabitants – are on the ocean or lake shore. Most of the remainder are on major rivers.

Today's urban dweller remains in an ecological relationship with the earth's resources that is less intimate, but no less fundamental, than that in prehistoric times. This relationship has always been a finely balanced one, in which quite small swings could bring about discomfort or disaster. But the hazards for early people were essentially local, and new and empty lands could always be found over the horizon. For today's city dwellers the hazards are regional or global, and most of the new or empty lands have long since been filled, some with their resources squandered, or abandoned.

For over 2000 years geographers have been describing and analyzing the ways in which humanity has come to terms – or failed to come to terms – with the planetary environment. In this book we shall try to see what kind of insights into that environment have been achieved. So we begin with an ordinary beach (Section 1.1) and see what lessons it can teach us in both space and time (Section 1.2) and in environmental relations (Section 1.3). We then draw back and look at our beach in global focus (Section 1.4) before going on to sample the structure followed in the rest of the book (Section 1.5).

1.1 The Crowded Beach

Beach scenes are ordinary enough events, so what makes them of special interest to geographers? Let us answer that question in a round-about way.

If we give three similar rocks to three different people, they may respond in quite different ways: a sculptor may shape the rock into new and more interesting forms, a mineralogist may start to break his up to examine its chemical structure, and a protester may hurl hers through the nearest window. The trains of thought started and the actions taken are determined not by the object at hand (the rock), but by the attitudes of the three individuals towards it.

In the same way, a familiar beach may provoke different reactions among geographers with different interests. A physical geographer may head for the sand particles and study the fluid dynamics of the breaking waves. A human geographer may study the behaviour of the groups using the beach and the optimal location for the ice-cream sellers. A regional geographer

Box 1.A Space, Location, and Place

Three words that geographers use a lot are 'space', 'location', and 'place'. Since they are also words used in everyday language, we need to be sure just how they are being used in this book.

Space means extent or area, usually expressed in terms of the earth's surface. It does not mean space in the sense of outer space (e.g. NASA, the National Aeronautical and Space Administration) or space in the sense of to arrange things in tidy rows.

Location means a particular position within space, usually a position on the earth's surface. Like the word space, it is rather abstract in meaning when compared with the third word of the trio.

Place also means a particular position on the earth's surface; but, in contrast to location, it is not used in an abstract sense but confined to an identifiable location on which we load certain values. So a location becomes a place once it is identified with a certain content of information. Sometimes the content is a physical fact. For example, latitude 27° 59′N, longitude 86° 56′E is an abstract location which we only recognize as a place once we know it describes the position of Mount Everest, the highest point on the earth's land surface. In other cases, the information content is a human experience. What gives a place its particular identity was a question that occurred to physicists Niels Bohr and Werner Heisenberg when they visited Kronberg Castle in Denmark. Bohr said to Heisenberg:

> Isn't it strange how this castle changes as soon as one imagines that Hamlet lived here? As scientists we believe that a castle consists only of stones, and admire the way the architect put them together. The stones, the green roof with its patina, the wood carvings in the church, constitute the whole castle. None of this should be changed by the fact that Hamlet lived here, and yet it is changed completely. Suddenly, the walls and ramparts speak a quite different language. The courtyard becomes an entire world, a dark corner reminds us of the darkness in the human soul, we hear Hamlet's 'To be or not to be'. Yet all we really know about Hamlet is that his name appears in a thirteenth-century chronicle. No one can prove that he really lived, let alone that he lived here. But everyone knows the questions Shakespeare had him ask, the human depth he was made to reveal, and so he, too, had to be found in a place on earth, here in Kronberg. And once we know that, Kronberg becomes quite a different castle for us. [Werner Heisenberg, *Physics and Beyond: Encounters and Conversations* (Harper & Row, New York, 1972), p. 51, cited by Yi-Fu Tuan, *Space and Place*, University of Minnesota Press, Minneapolis, 1977), p. 4.]

might be concerned with all the variations among the different parts of the beach. But common to all the geographers in their reactions to the beach is trying to pin down exactly where events were occurring in space. A photograph taken from high ground or, better still, an aircraft or satellite allows a much more accurate assessment than a photograph taken on the ground. It is for this reason that most of the photographs you will find in this book are aerial photographs. Concern with locations in space is a characteristic of geographers' curiosity; specifying location accurately is one of the prime rules of the geographic game. An inaccurate description of location causes a geographer to wince in the same way as a linguist would over a mispronunciation, or a historian over an inaccurate date. (See Box 1.A on 'Space, location, and place'.)

The Physical Geography of the Beach

Our opening photograph (Figure 1.1 p. 2) shows an aerial view of a beach on the eastern coast of England. We shall look at this beach but I invite readers to substitute a beach they know and love: much the same lessons can be drawn from it.

If we ask how this great structure was formed we can find some clues from looking closely at the photograph. You can spot the parallel ridges where a succession of great storms have built up the ridges of shingle. If we look at old maps, these show that the spit has been pushed southward. Figure 1.2 shows the slow evolution of the beach over historical times. In (a) we see a map drawn two and a half centuries ago and in (b) the picture built up by the Cambridge geographer Alfred Steers of the intricate sequence of ridges that make up the beach structure. If we stand on the beach today we can see, and feel around our ankles, the drag of material continuing that building and rebuilding. Figure 1.3 shows the way in which this process of drifting (called 'longshore drift') occurs. Waves usually approach a beach at an oblique angle. As a wave breaks it sends material up the beach roughly in the same direction of wave approach. Much of that water quickly percolates down into the pebbles or sand, but the remainder

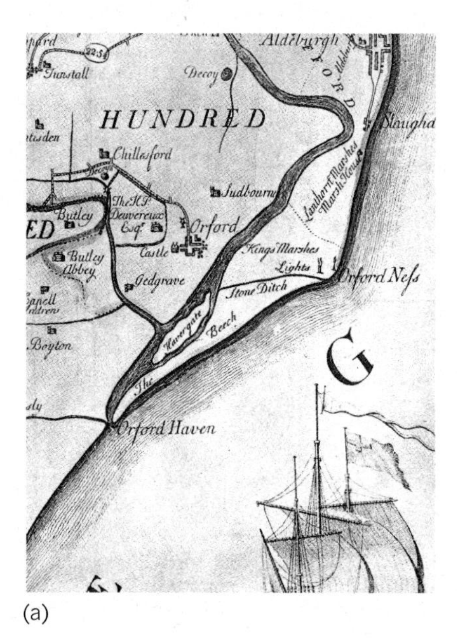

(a)

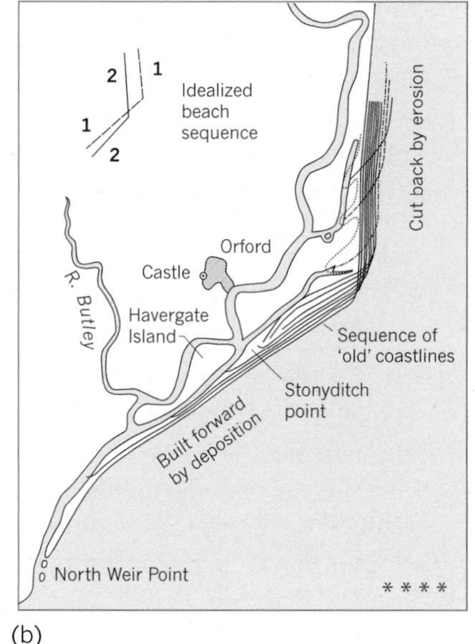

(b)

Figure 1.2 Evolution of a beach over historical time The slow evolution of a beach over the centuries can sometimes be tracked by using early maps and comparing them with the present. (a) John Kirkby's map of Orford Ness, 1736. (b) Generalized picture of the evolution of Orford Ness developed by the Cambridge geographer Alfred Steers. Note the erosion in the north being succeeded by deposition in the south. The inset gives two time periods (1 and 2) in an idealized evolutionary sequence.

[Source: *Orford Ness*, Heffer, Cambridge, 1966), p. 18. Plates reproduced with the kind permission of the Trustees of the British Museum.]

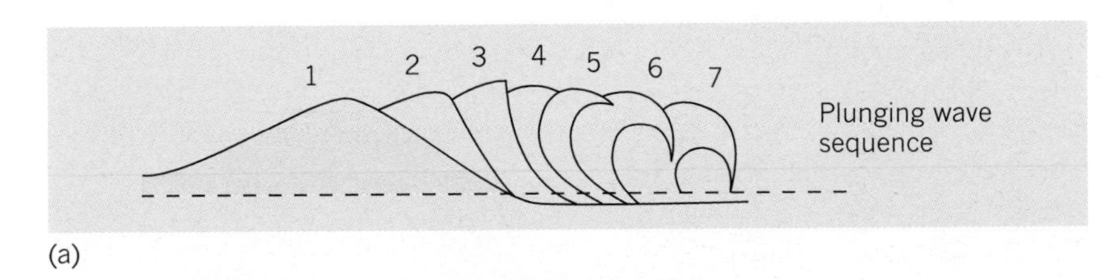

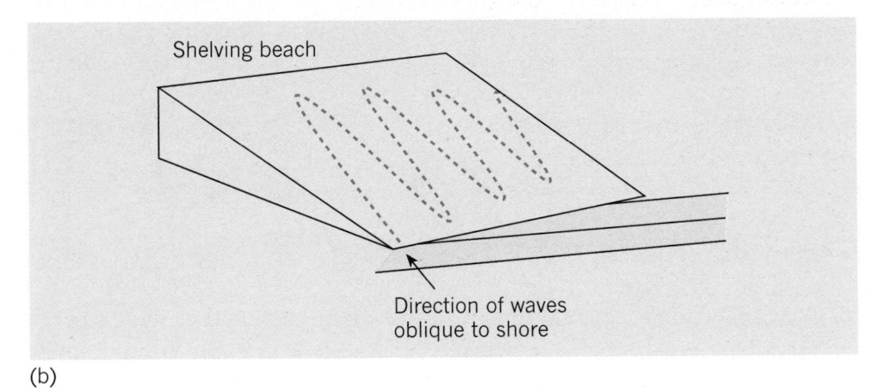

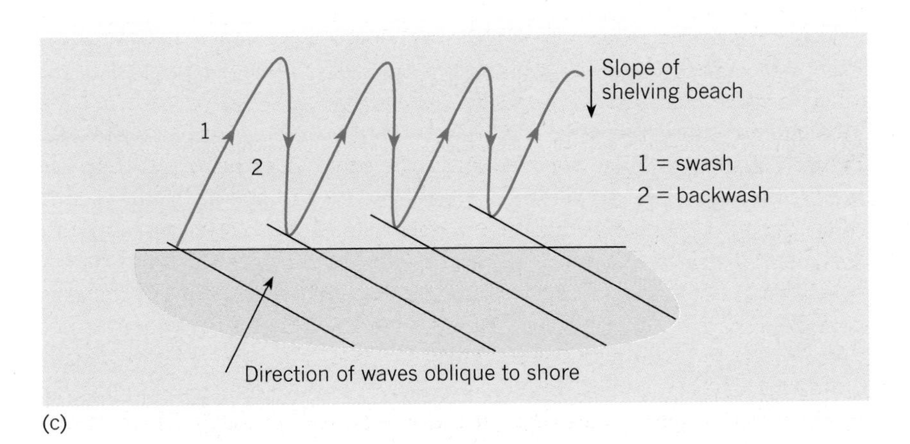

Figure 1.3 Evolution of a beach over a single day (a) Changing shape of a wave in profile as it nears the beach. (b), (c) Processes for wave sorting of beach material. The forward rush (swash) of the wave carries material up the sloping beach in the wave direction; the return flow (backwash) follows the contours of the beach back to the sea.

returns seaward following the steepest path back to the sea as *backwash*. Within less than a minute the next wave arrives and the process is repeated. Note that in building the beach we are concerned with varying periods of time: the ages over which a major beach has been built up (measured in decades and centuries) and the immediate process we observe on the beach (measured in minutes and seconds).

The Human Geography of the Beach

Figure 1.4 moves away from eastern England and shows two photographs of people on a beach in two different environments, in Norway and in China. Again, these photographs were taken from an elevated level to try to produce a maplike picture of the scene below.

From this concern with space come questions of spatial patterns and organization. So a second reaction to the beach scene in Figure 1.4 would be to try explaining the spatial variations that are observed. Why are some parts of the beach packed with people while others are deserted? How

(a)

(b)

Figure 1.4 People on the beach Two contrasting beach scenes in different environmental contexts. (a) Beach scene, Norway, July. (b) Crowded beach on a Chinese beach in late summer. Both photographs show the way in which human populations arrange themselves in the space provided by the beach. See the discussion of interpersonal distances (Figure 1.7) later in the chapter.

[Source: (a) Knudsens Fotosenters. (b) © MAGNUM/Eve Arnold.]

much does this circumstance relate to differences in the quality of the beach? Questions of this kind lead to a general concern with the relationship of people to their environment.

The Regional Geography of the Beach

A third geographic reaction to the beach scene shown here would be to try to sort the various elements in the photographs into some kind of regional order. ***Regions*** are a shorthand way of describing the variable character of an area in an efficient manner. One of the simplest ways of establishing sets of regions is to divide an area into several zones, each of which having cer-

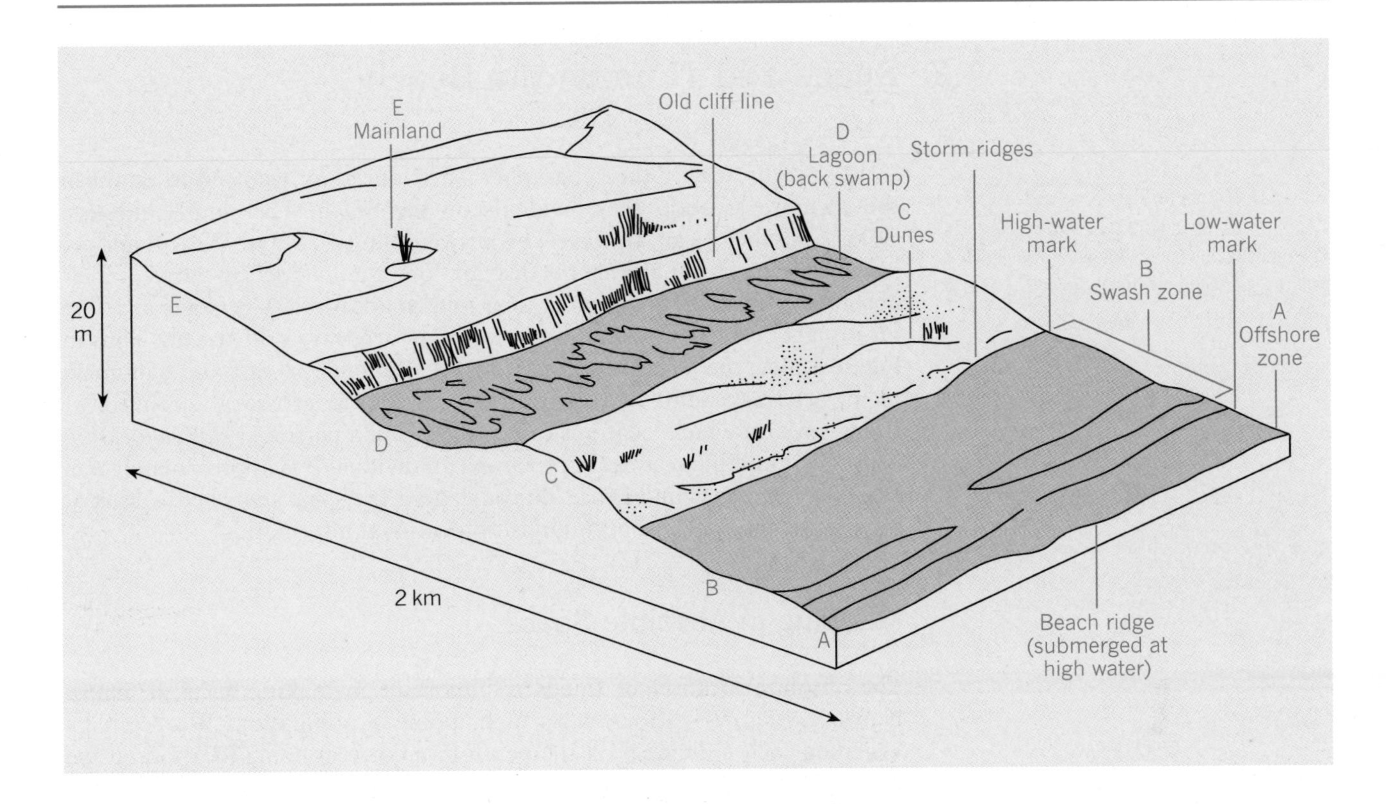

Figure 1.5 Beach environments Regional division of the beach environment into a series of five distinctive zones, A to E. The principal factors controlling the character of each zone are the tidal range (areas below the high water mark are shaded) and the surface material (shingle, sand, mud, or rock). The vertical scale has been exaggerated to emphasize the effect of height above sea level.

tain characteristics that give it a particular character. For example, we could divide the beach into three zones: a swash zone below the regular high-water mark, an upper zone above the regular high-water mark, and a belt of sand dunes behind. Figure 1.5 shows a cross-section of a typical beach zoned in this way. Geographers use this process of division, called dissection, to establish sets of regions. From this geographers go on to try to relate their findings on the beach to others around the world. That is, they try to place their regions in some kind of world focus.

To sum up: a geographer is concerned with three different but interlocking questions – (1) the question of location, in which the concern is to establish the precise spatial position of things within a particular area of the earth's surface; (2) the question of human–environment relations within the area; (3) the question of regions and the identification of the distinctive character of particular spatial subdivisions of the area.

All three questions focus on the particular character of the beach as a place, and we shall need later to ask a fourth and more general question. How do our findings on the beach relate to other regions, and to the world picture as a whole?

Perhaps at this point we should follow the conventional textbooks and attempt a definition of geography itself based on these central questions. If you like formal definitions, then please turn to the last chapter and look at those given in Table 24.1. These definitions have been deliberately placed near the end of the book to encourage you to draw your own portrait of a geographer as you work through the chapters; this should give you a better likeness of geography than would looking up a dictionary entry right away. So let us agree for the moment that 'geography is what geographers do' and go on to see them at work on the beach. Later in the book we shall see them working on wider, more important questions and in a global context.

1.2 Space and Time on the Beach

If we wish to answer the geographer's first question, we need to establish the accurate location of individuals on the beach. The simple question 'Where are they?' can, however, be answered in two ways. We can answer it, for example, by analyzing the distribution of individuals in terms of their absolute location, or in terms of their relative location. A person's *absolute location* is his or her position in terms of an arbitrary grid system. Thus in Figure 1.6(a), the location of individual A is about 9 m east and 6 m north of an arbitrary point of origin at zero. The grid provides a convenient framework on which locations can be fixed. But a person's *relative location* is usually more interesting. For example, individual B is approximately 6 m from A, whereas C and D are almost sharing the same spot. Let us look at the use we can make of both kinds of locational information.

Mapping in Absolute Space

The absolute location of things is important in making accurate maps. Figure 1.6 shows various ways of mapping a population. We begin by assigning each individual on the beach to a corresponding location on the

Figure 1.6 Population mapping
Diagrams (a) to (e) show various ways of mapping population density. (a) A simple dot map of a 25 m × 25 m segment of the beach. In (b) the area covered by the dot map is divided into 5 m × 5 m cells, and the number of persons in each cell is summed. (c) A choropleth map of the beach population. (d) An isopleth map. The values on each isopleth (2.5 to 5.5) describe the average number of people in each cell. (e) A three-dimensional isopleth map. All these maps are simply different ways of describing the same distribution of population.

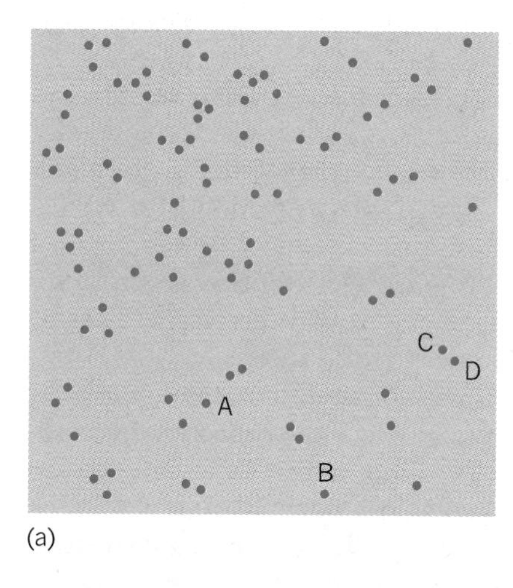

(a)

6	8	5	4	3
5	6	5	4	2
5	6	5	2	0
4	2	2	1	2
4	3	2	2	1

(b)

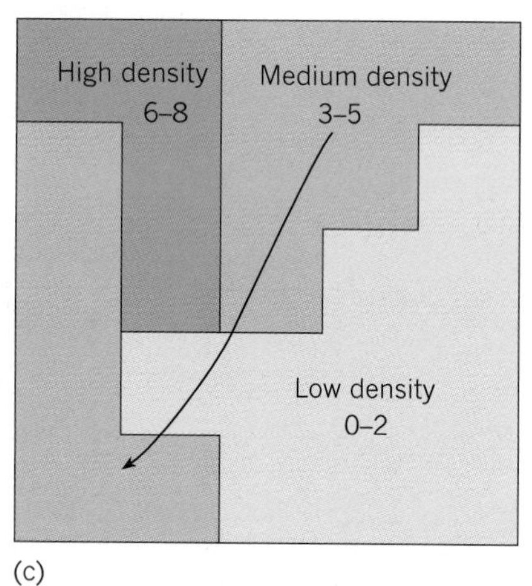

(c)

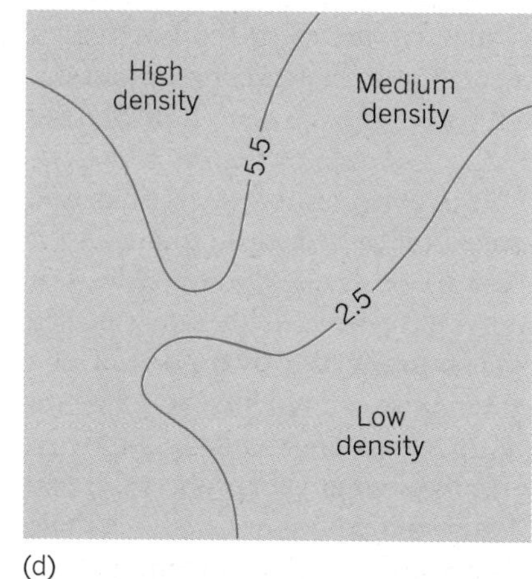

(d)

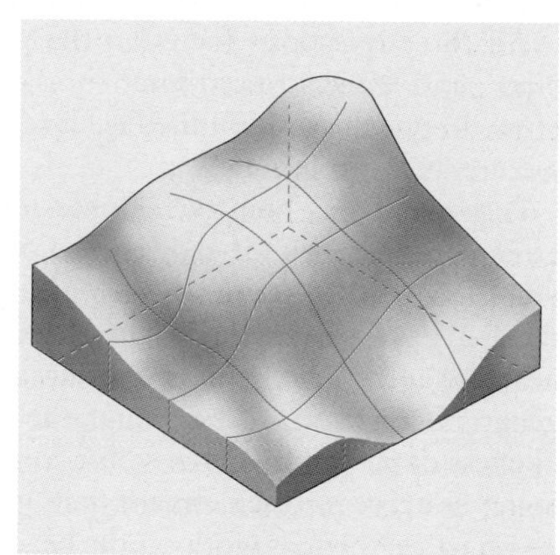

(e)

map, representing each person by a single dot. We then place a grid over the dots. By counting the number of dots that falls within each square cell of the grid, we can translate the distribution of dots, or people, into an array of numbers that reflect the density of the population (Figure 1.6(b)). Crowded parts of the beach are represented by cells with high values and sparser parts by cells with low values. Empty stretches of the beach have cells with zero values.

Two kinds of maps can be drawn from these cell values. We can make *choropleth maps* (from the Greek choros, area, and plethos, fullness or quantity) by assigning different shades of colour to each of the cells. By linking cells with similar colours, we can create a general picture of the distribution of population on the beach (Figure 1.6(c)). In doing this, we lose some information. Instead of nine different values (between 0 and 8), we now have only three. However, we have conveniently simplified our map, or grid, by reducing the spatial pattern from 25 cells to 4 areas. The second and more common way of mapping population distributions using a grid is to draw lines between all points having the same quality or value (Figure 1.6(d)). Such lines are known as isopleths (from the Greek *isos*, equal), and maps of this type are known as *isopleth maps*. These maps also give an accurate picture of variations in population density. Adding a third dimension to an isopleth map would show areas of denser population as peaks and sparser areas as hollows (Figure 1.6(e)).

Isopleth maps are one of the most common ways in which geographers show spatial distributions. Some of the isopleths are given distinctive names. Thus *isochrone* maps show lines of equal time; *isohyet* maps show lines of equal rainfall; *isotherm* maps show lines of equal temperature; *isotim* maps show lines of equal transport costs. One of the commonest isopleth maps you will encounter is the *contour* map, which shows lines of equal height of land above sea level. Contours are also used by geographers as a general term for any type of isopleth.

Organization in Relative Space

The relative location of the members of a population is also of great interest to geographers. It helps them to understand why a population is organized, or distributed, in a particular way. Animal behaviourists such as Konrad Lorenz in *On Aggression* and social psychologists such as Edward T. Hall in *The Hidden Dimension* have discussed how, like other primates, human beings have a strongly developed sense of territoriality. They surround themselves with visible or invisible 'space bubbles' that are sensitive to crowding. Hall distinguishes between four 'action zones' based on the distance between people: intimate space, personal space, social space, and public space. *Intimate space* is reserved for physical interactions such as loving or fighting, while *personal space* is used for soft talk and friendly interaction. The boundary between these first two zones is about 0.5 m (1½ ft). Beyond a distance of 1.5 m (about 4 ft), personal space gives way to the *social space* used for formal business and social contacts. At around 4 m (about 12 ft), the outer zone of public space, in which the preacher and the ice-cream-cone vendor on the beach may operate, begins. Clearly, these zones may vary from person to person and from culture to culture – the personal conversational space of, say, the French, may be seen by the more reserved English as an intrusion into their intimate space!

Because of the sense of territoriality, it is possible to argue that individuals

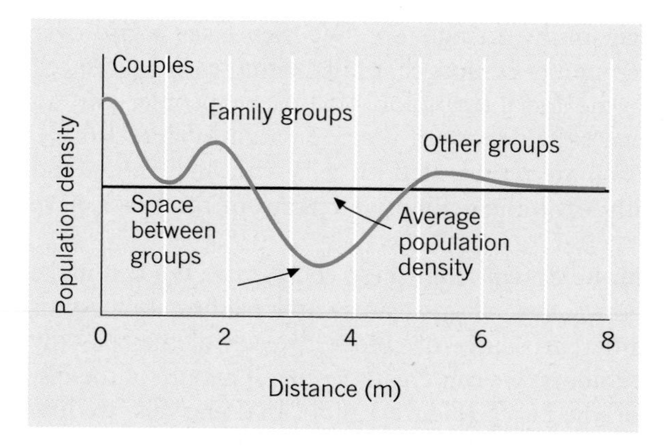

Figure 1.7 Interpersonal distances This is an idealized profile of the density of population on the beach in terms of the distance between individuals. Note the three characteristic peaks.

arrange themselves on a beach to achieve a certain distance from each other. Their object may be to be as near as possible to those they love, as far as possible from those they hate, and at a convenient distance from those about whom they feel less strongly. When the distribution of people in an area is regarded in this way, it is measured in terms of interpersonal space, the linear distance separating an individual from neighbours. To determine interpersonal space, we ask, 'How many metres is a given person from his nearest neighbour, his second nearest neighbour, his third nearest neighbour, and so on?' Geographers frequently use this approach in the study of human distributions and settlements. (See Section 14.1.)

The result of analyzing the relative location of a beach population is given in Figure 1.7. The three peaks, near zero, at 2 m, and at 5 m, may be interpreted as related to the space between couples less concerned with the beach (let alone geography!) than with each other, to family groups, and to strangers outside the family groups and keeping a respectful distance from them.

As relative distance increases, so the contacts between human groups change. The limited power of the human voice and eye mean that a large lecture theatre is about the limit of direct face-to-face communication. Beyond that the media and the communication system take over. Also as the size of space increases, so the number of people who can be involved gets larger. Thus the implications of the space seen on the beach extend far beyond it, underlying human organization right up to the level of the globe itself. We shall return to it again and again in this book.

Time Geography and Spatial Diffusion

The scenes in the photographs so far are unreal insofar as they freeze at one click of the camera's shutter a constantly changing pattern. The photograph is static; the real beach scene is dynamic. A similar photograph taken at another time would show a different picture: just how different would depend on the timing of the second shot. A picture taken a few seconds later would reveal little change in the population but would catch the movement of the breakers. One taken a few hours later would show an empty beach and a different tide level. One taken some months later in winter might show a change in some of the physical structures. One taken some years later might reveal a substantial change in the form of the beach itself due to erosion.

All geographers work within a specific time context. And as Figure 1.8 makes clear, this context vitally affects the conclusions they draw. In each

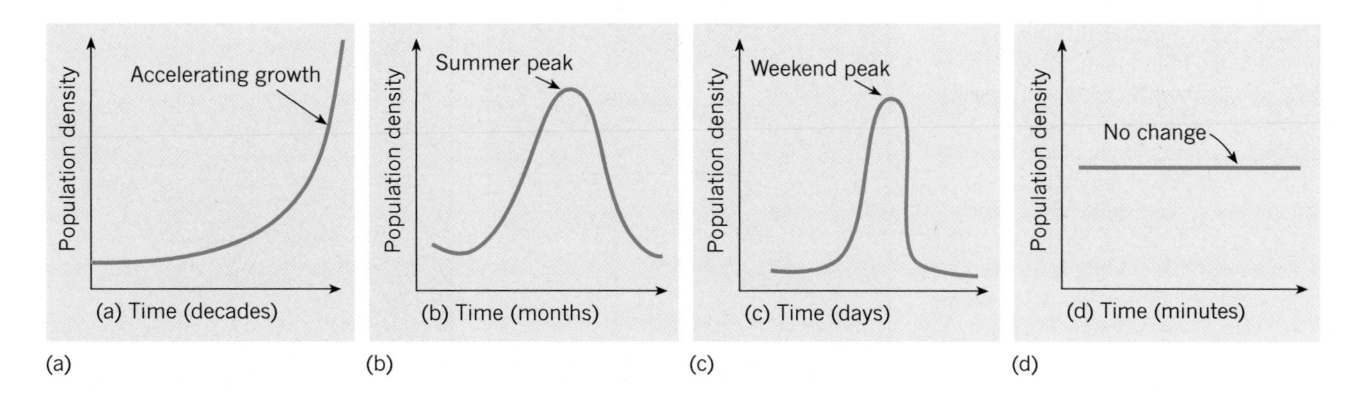

Figure 1.8 Changing population densities These graphs show the impact of the observation period on the trends detected in population density. Graph (a) shows the historical trend, (b) the seasonal cycle, (c) the weekly fluctuation, and (d) the short-term equilibrium.

graph the number of people on a beach is related to the passage of time. Over a period of 100 years (e.g. 1900–2000) the most noticeable change is the increasing use of the beach. On the average, more people were using the beach at the end of the period than at the beginning. This is not surprising when we think of the substantial increases in the populations of all the areas shown over this period and the changes in social attitudes towards leisure time.

If we reduce the period of observation to a single year, we find a wavelike trend with a peak in late summer and a trough in late winter. If we reduce it to a single week, the waveform narrows to a sharp peak on the weekend. But over a much shorter period (for example, half an hour), the number of people on the beach remains constant and the trend line is horizontal. These general trends – of accelerating growth, wavelike cycles, or stability – are functions of the period over which observations are made.

Consider what we would see if we observed the beach at regular intervals from daybreak. The first arrivals might well occupy what seemed to be the 'best' sites – what is best being determined by the requirements of the group: say, near the surfline for the youngsters, and near the parked car for the old folks. As the best sites were taken, new arrivals would have to occupy less attractive areas or crowd into the already occupied ones, reducing the amount of interpersonal space and producing the population pattern shown in the photograph.

Figure 1.9 traces the evolution of this pattern over two hours. This is a simple example of the spatial *diffusion* of a population, in which the location of an individual is related to the time he or she arrives. In geography, diffusion is the process of spreading out or scattering over an area of the earth's surface. It should not be confused with the physicist's use of the term to describe the slow mixing of gases or liquids with one another by molecular interpenetration. We shall look at more complex examples of spatial diffusion in later chapters (see especially Chapter 16). Observation of this process over time not only allows us to consider the present spatial pattern in the light of its past development; sometimes, it makes it possible to predict spatial diffusion in the future. Given the first two maps in Figure 1.9 (and knowing what has happened on the beach on similar days in the past), could you make a reasonably accurate prediction of the noontime pattern?

Spatial diffusion is one aspect of human geography in which Swedish geographers in general and Torsten Hägerstrand of Lund University, in particular have made massive contribution. Box 1.B on 'Hägerstrand and time–space geography' provides an introduction; specific models called Hägerstrand diffusion models are discussed later in the book in Chapter 16.

Figure 1.9 Spatial diffusion
Stages in the diffusion of people arriving on an empty beach at different stages of the morning. Note how the filling-up process is related to the point of access to the beach (top of the map) and to differences in the quality of the beach environment. Emptying the beach is not a simple reversal of filling it up. The homeward drift in late afternoon, shown in (d) below, is less orderly in spatial terms. Note that 'environmental quality' can include elements made by people (such as lifeguard surveillance) as well as natural features (e.g. sand *v.* shingle).

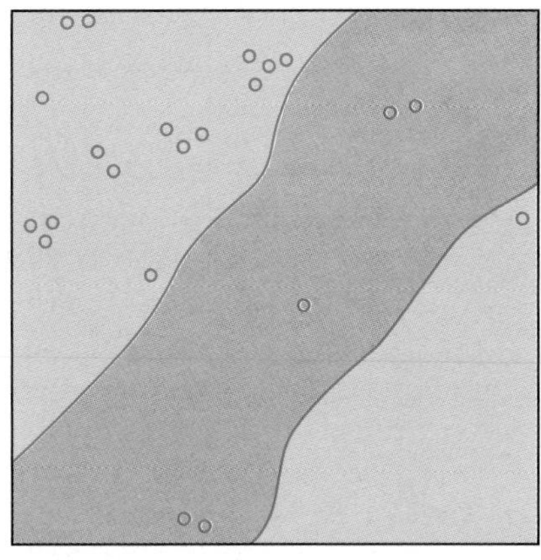

(a) 10 A.M.

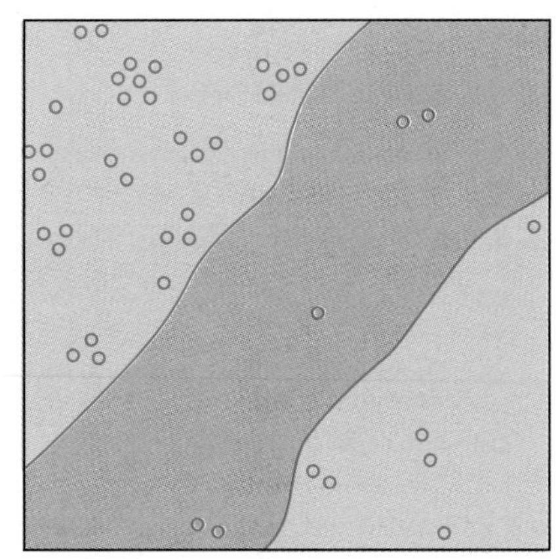

(b) 11 A.M.

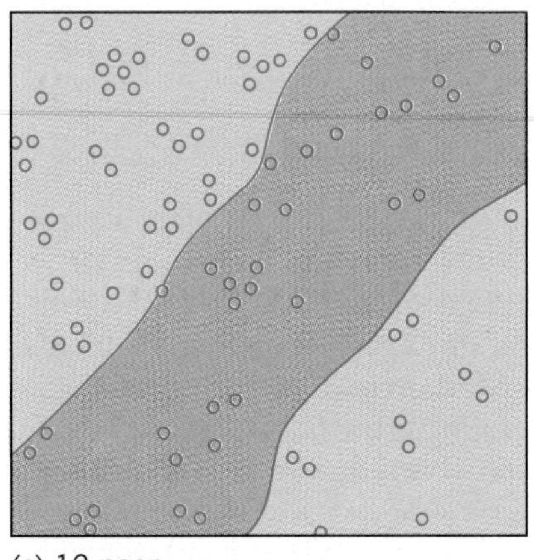

(c) 12 noon

Environmental quality

High Low

(d)

Box 1.B Hägerstrand and Time–Space Geography

Together, space and time form the framework of the cage within which human life unfolds. Torsten Hägerstrand (born Moheda, Sweden, 1916) is a Swedish geographer who has made major contributions to two research areas in human geography: ***time–space geography*** and spatial diffusion (see Chapter 16).

Hägerstrand studied at Lund under Helge Nelson and was strongly influenced by the mathematical approaches of Edgar Kant. The first introduced him to the meticulous tracking of human migration movements in small areas of rural Sweden, and the second to probabilistic ideas on settlement patterns. The result was his 1953 doctoral thesis on innovation waves (Swedish *innovationsförloppet*) in which the adoption of agricultural innovations by farmers in central Sweden was conceived as a series of diffusion waves whose passage could be mapped, modelled, and simulated. (The thesis was translated into English in 1967 as *Spatial Diffusion as an Innovation Process*.) The ideas there were elaborated and extended in *Migration in Sweden* (1957). This early work pioneered new methods in cartography and Monte Carlo simulation, which were rapidly taken up by US colleagues in the late 1950s and widely adapted and extended, particularly by computer modelling of diffusion processes. Over more recent decades Hägerstrand has moved from the study of aggregate time–space studies to the detailed dissection of individuals' movements over very short time periods. Hägerstrand and his Lund colleagues have been able to show how an understanding of changes in such trajectories at a micro-scale allows a clearer understanding of the changes in the aggregate pattern of human population distribution. His work has increasingly drawn him into government advisory work on future population changes at national and European levels.

The diagrams (a) to (c) show how his ideas on time–space geography can be simply illustrated and translated into our beach study. Imagine you are living in a small town near the coast (a). On a given Saturday in summer, you decide to go to the beach. There is a choice of three beaches: first, the local town beach at A to which you can cycle in half an hour; second, the ocean beach at B which means borrowing the family car and driving for an hour; finally, the best beach of all at C, with fine surf but lying so far to the north that it takes about 5 hours of hard driving to get there. Which are you going to choose?

You can plot your own choice as a path running from home to the beach and back in terms of a space–time box, looking like an aquarium. Diagram (b) shows what happens if you choose B. Notice that the vertical scale is that of the clock running from midnight to midnight over a single day. All three choices are shown in diagram (c).

Notice how your paths through time and space are limited in three different ways. There is the familiar rhythm of night and day, with the biological clock within each of us

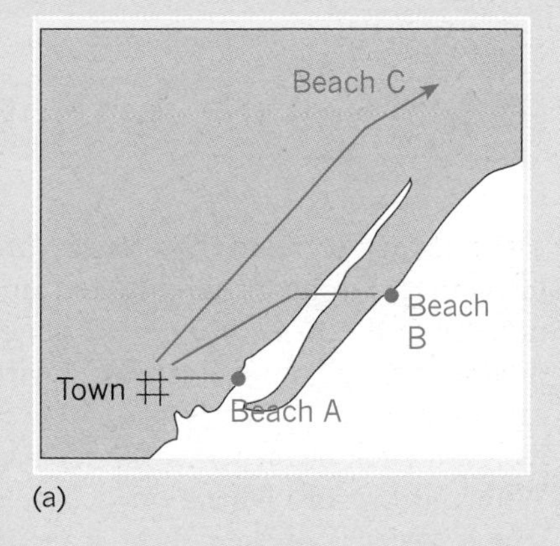

(a)

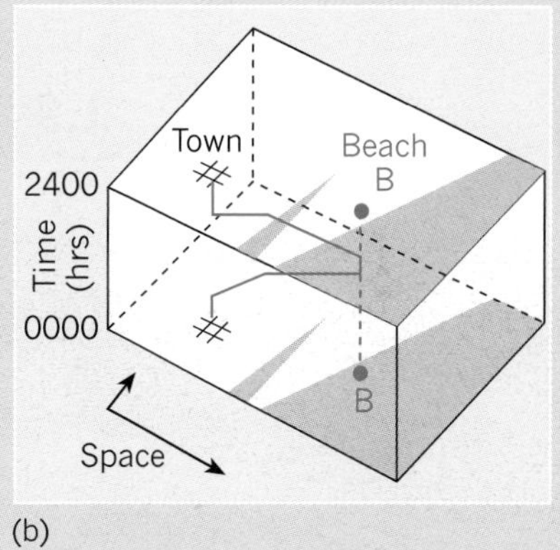

(b)

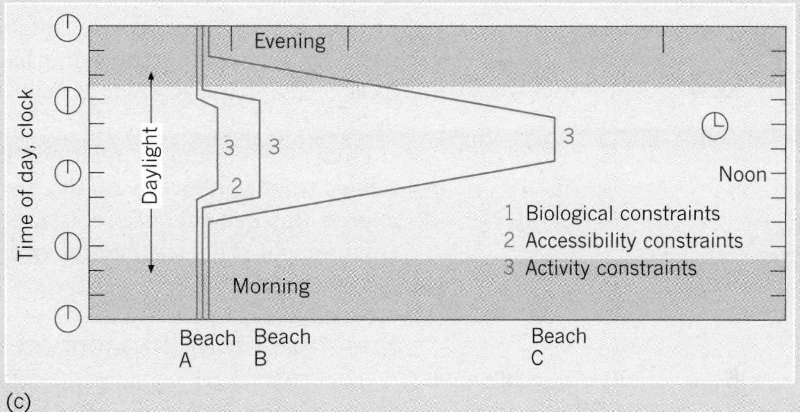

(c)

Box continued

demanding sleep at more or less regular intervals. So, we want to be home at a certain hour. Then there is the accessibility constraint, as shown by the diagonal lines in the diagram. Can you get hold of a car? Last, there is the activity constraint. You need at least an hour or two on the beach to make the trip worthwhile.

If you continue the three constraints, then you will see how they 'box in' your set of locational choices. If you cannot get the car, then constraint 2 means that only beach A is open to you. Even given the car, beach C lies so far away that it is hardly worth the long drive for the limited time there (constraint 3). So it looks like you will have to settle for beach B.

Geographers at Lund University in southwest Sweden have made a special study of the way these time and space constraints box in human activity. They have been able to show how the growth of cities, the range of jobs we can take, even the partners we meet and marry, are all related to the kind of constraints we have just discussed. If we are very poor or sick, our locational choice may be severely limited to the local area. If we are well off, then a helicopter or a private jet may greatly expand our spatial horizons. But, rich or poor, the basic biological constraint of the human being as a circadian (meaning 'almost daily') animal remains. None of us can climb out of that cage.

[Source: Photograph from Bristol University Department of Geography Collection.]

1.3 People and the Beach Environment

The concept of relations between the individual and the environment is basic to geographic thinking and underlies the second of our three basic questions. By an ***environment***, geographers mean the sum total of conditions that surround (literally, environ) a person at any one point on the earth's surface. For early people these conditions were largely natural and included such elements as the local climate, terrain, vegetation, and soils. With the rise of civilization people surrounded themselves with artifacts which, because of their sheer scale and longevity, became an integral part of their environment. For today's urban dweller the environment is dominated by the fixed stuctures of urban life (freeways, city blocks, asphalt surfaces). The natural environment has been either replaced or radically modified.

Human–environment relations have two sides to them. The first side relates to the influence of the environment on human activity. We can express this in symbols as E→H. Second, human activity may alter a given environment. This reverses the order to H→E.

Environmental Impacts on Human Beings (E→H)

We can bring the relationship between people and their environment into focus by going back to the beach. The population density on a beach is partly a function of environmental quality. Good beaches (i.e. beaches with

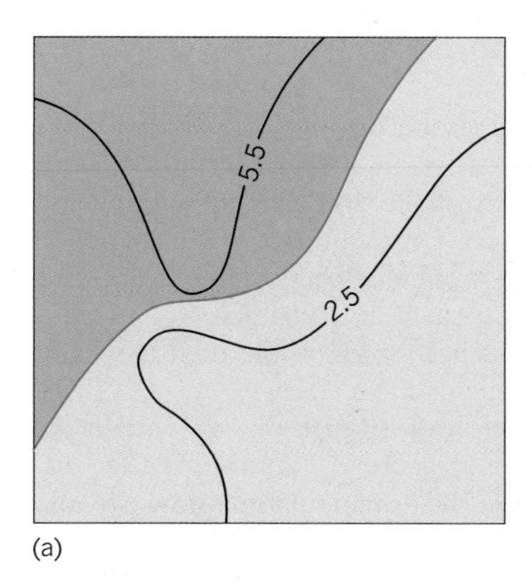

(a)

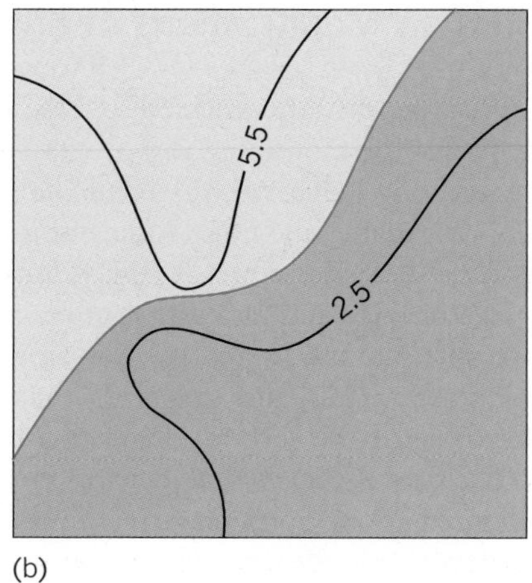

(b)

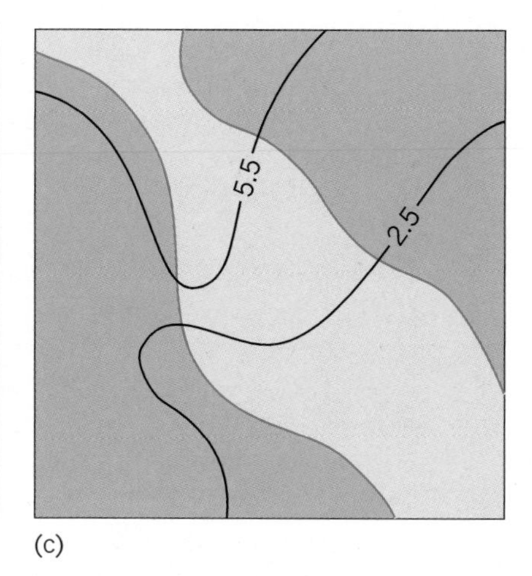

(c)

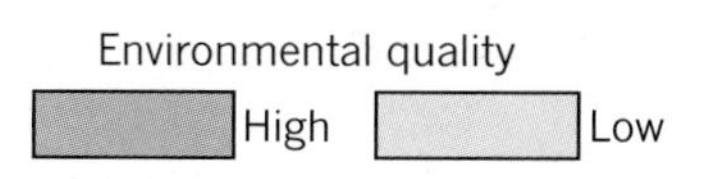

Figure 1.10 Spatial covariation Here we have three hypothetical distributions of population density on a beach. Density is shown by the numbered contours and environmental quality by shading. In (a) there is a strong positive correspondence between the distribution of population and the quality of the beach environment. In (b) there is a strong negative correspondence, and in (c) there is little correspondence.

fine sand or good surf) tend to attract more users, while poor beaches (those that are, say, polluted by oil or by the local dog population) are shunned. Other things such as weather and accessibility being equal, we can relate the capacity of a beach to attract a population to its environmental quality. Even within the area of our photograph, local variations in environmental quality can lead to variations in population density. Figure 1.10(a) shows the distribution of both the beach population and the beach environment as maps.

The study of two or more geographic distributions varying over the same area is a study of *spatial covariation*, an idea we shall meet repeatedly in this book. When the two maps look alike and the two distributions 'fit' one another closely, we say that the two phenomena are associated by area; that is, high values for population density in one area correspond with high values for environmental quality in the same area, and vice versa. Other hypothetical cases with little covariation are also illustrated in Figure 1.10. Comparing pairs of maps in this manner often tells us a great deal about the spatial covariation of different phenomena. Distributions can also be compared by statistical methods, but a discussion of these lies outside the scope of this introductory text.

To view the environmental quality of a beach in terms of, say, its surfing potential is clearly only one possible viewpoint – one strongly influenced by age and nationality of the users. For most of our history our questions would relate more to mundane matters of safe anchorage or shellfish yield than to questions of recreational use. We need then to assume some lens or filter which is placed between people and the environment. This means that what we *see* in a beach may be determined by our age, our interests, our income, our ethnic background, and so on. The environment provides a range of choices, only some of which have ever been seen. For example, a beach backed by very high dunes may suddenly become an attraction for hang-gliding enthusiasts. The environment (i.e. the dunes) has not changed: but what we choose to see and do in that environment has.

In this book, we shall spend much of Part I describing the earth's environment, and much of Part II discussing our reaction to it. We shall see there how strongly interpretation of a given environment is influenced by social, cultural, and technological factors.

Human Modifications of Environments (H→E)

While people are affected by their environment, they also have some capacity for modifying it. For example, they can alter the form of a beach by erecting defensive walls and change its quality by fouling it with oil and debris. Again time must enter our analysis, for the impact of human action is often *lagged* in time. A lagged impact is one that occurs later at the same place or later and at a different place. An example of the first type of lag is where toxic industrial waste is slowly concentrated in the food tissues of marine animals and the birds that feed on them. (See Figure 9.17.) An example of the second is where protection of a beach on one part of the coast may mean increased wave erosion at another. Likewise, sewage discharged at one point on a river may affect the fish populations downstream.

Beyond the beach most of the world's environments have been strongly influenced by human action. Chapters 9 to 11 will describe in detail how massive that change has been.

Human–Environment Systems (H/E)

Whether the effect of people on environment, or the effect of environment on people is more important is a chicken-and-egg question. Geographers find it more helpful to think of both relationships as part of a human–environment system. A *system* may be defined as a group of things or parts (called elements) that work together through a regular set of relations (called links) within defined limits (called the system boundary). Thus we can regard the beach as a system in which its various parts – shingle, sand, and mudbanks – are each linked together through a set of relations involving the energy of waves, tides, and winds.

Geographers are particularly interested in systems that link together human beings and environment. Figure 1.11 shows such a system, linking the users of a particular beach with its environment. Here a human system (H/H relations) and an environmental system (E/E relations) are linked into a human–environment system (H/E relations). The links are of various kinds – information, energy, and material. Together they form the flows that bind the five elements together. Note in particular the circular loop (called a *feedback* loop) connecting the character of the beach and the density of those using it. Higher levels of use may lead to two effects: first, to growing litter and pollution, which may lead people to shun the beach and

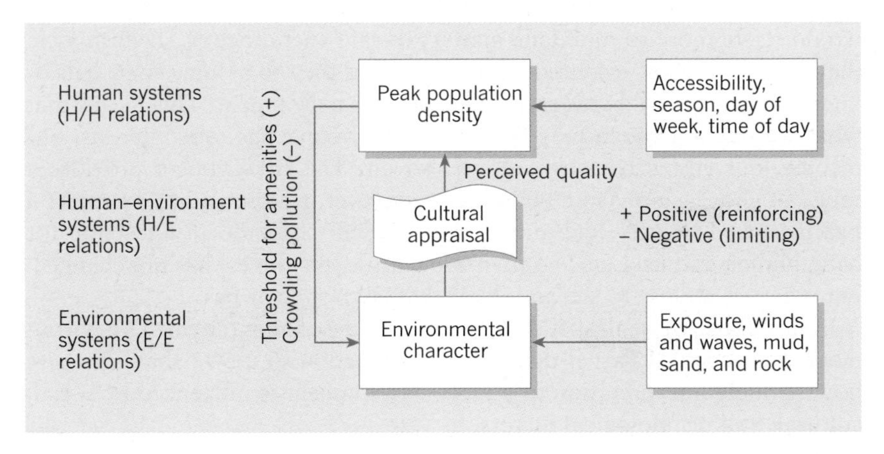

Figure 1.11 Human–environment relations as a simple feedback system Note the loop by which beach character affects the numbers who use it, but this beach-using population itself may feed back on perceived beach quality. Beneficial changes such as the installation of amenities will improve the beach character and attract more people (+), while hostile changes will have the reverse effect (−).

move elsewhere. Since this effect lessens the use of the beach, this kind of feedback is termed a *negative feedback*. A second and opposite effect may be for more people to increase the attractiveness of the beach. For example, on a popular beach it may be worthwhile to provide a lifeguard. This second effect is termed a *positive feedback*, since it tends to increase the level of use.

So our beach forms a very simple example of feedback in a human–environment system. Its study is worthwhile insofar as the same principles are incorporated into much more complex models, such as the world models discussed near the end of the book. (See Section *24.3.)

The Beach in World Focus 1.4

Beaches are fine and pleasant places, and we might all wish to spend more time there. But it would be misleading to suggest that geographers spend more time at the seashore than anyone else. We have chosen to concentrate on it in this chapter because it represents a microcosm, but only a microcosm, of the kinds of phenomena that geographers study.

Levels of Resolution

In other parts of this book are examples of humanity's relationship with the environment at very different scales. In Chapter 5 we see how the human population came to be distributed throughout the world, and in Chapter 17 there are examples of political influence distributed within nation-states. The world, and the nations of the world, are certainly more usual arenas for the geographer's work than a beach, which is only a tiny section of the human environment. The modern geographer deals with a continuum of environmental regions of increasing size. These range from the micro-environment of the individual in local surroundings to the macroenvironment of humanity as a whole.

Figure 1.12 shows what happens when we change our focus. On Figure 1.12(a) the picture in the camera viewfinder is clearly that of just two individuals on the beach. As we pull our camera back, the picture changes. By (c) the individuals have been lost in the crowd and by (d) even the crowd

Figure 1.12 Changing spatial focus on the beach continued overleaf

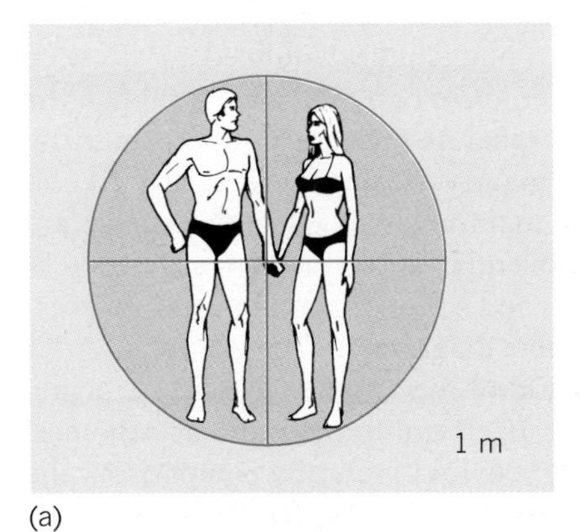

(a)

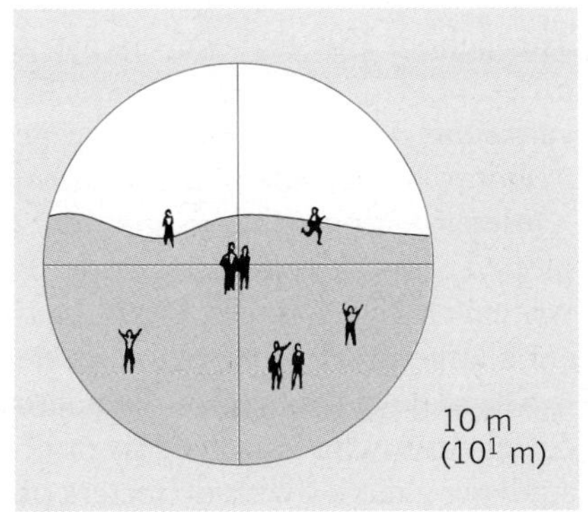

(b)

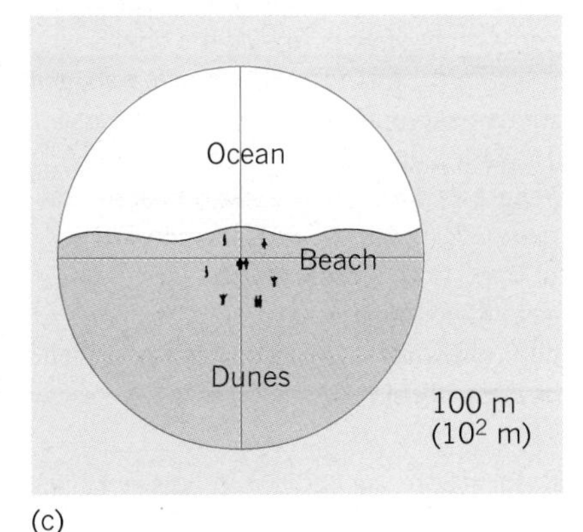

(c)

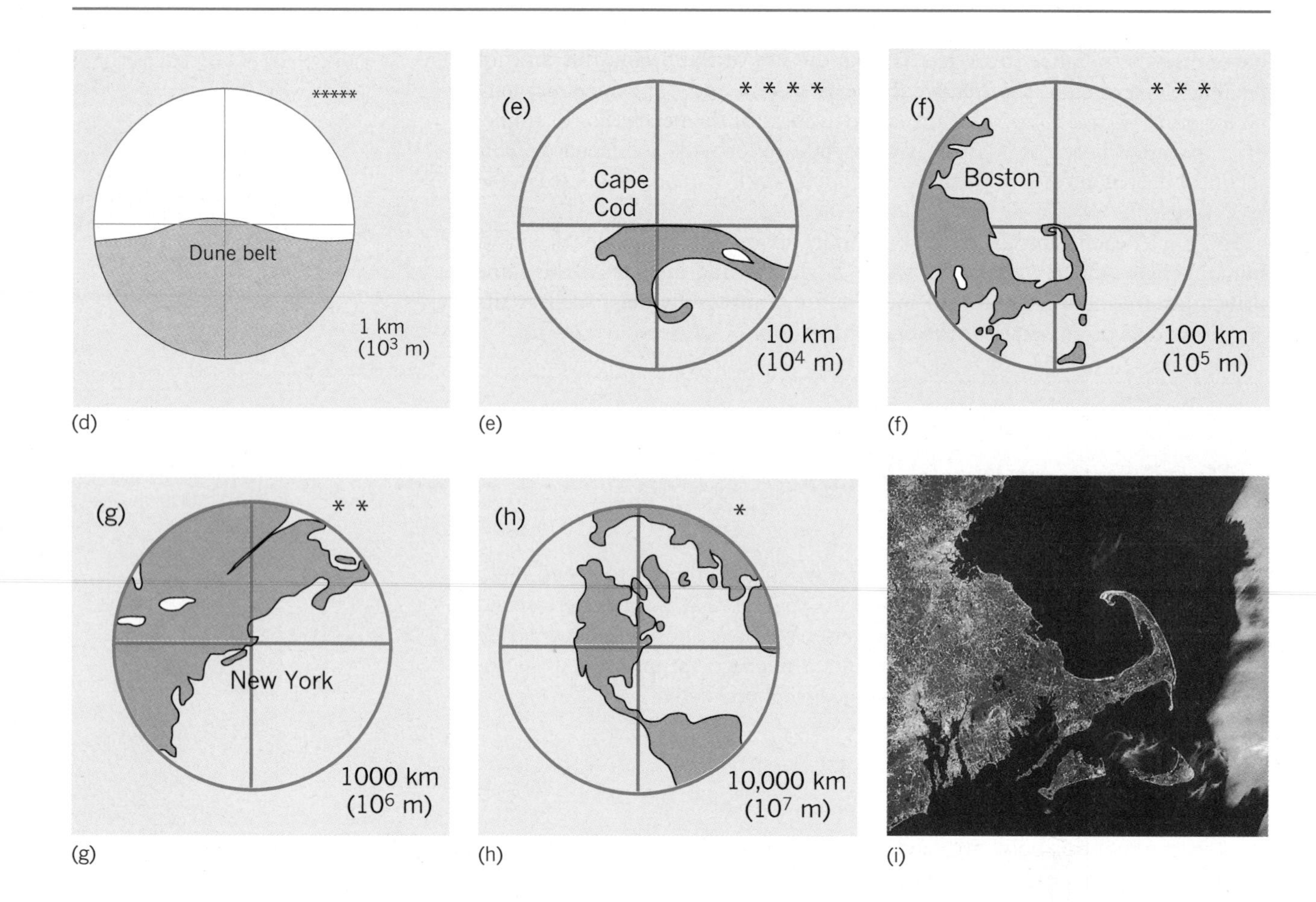

Figure 1.12 Changing spatial focus on the beach The maps show a couple of sunbathers on a Cape Cod beach viewed from higher and higher elevations. The radius of the resulting picture caught in the zoom lens is given in metres (m) or kilometres (km) (1000 m = 1 km). Note that each disk has a radius exactly ten times greater than or smaller than its neighbour, and that each disk *remains* centred on the two sunbathers. Because of the small size of the drawing the human figures quickly disappear and a new feature, Cape Cod, comes into view in (e). By the final drawing (h) this too has disappeared, but we can see one side of the whole earth. The stars, indicating orders of geographic magnitude, are discussed on page 22. (i) A LANDSAT infrared image of Cape Cod, Massachusetts. Areas of vegetation show up as dark shading. The city of Boston is the light area at top left of the picture.

Source: (i) Earth Satellite Corporation/Science Photo Library.]

is no longer visible. But notice that by (e) a new feature, the hook shape of Cape Cod on New England's coastline, has come into view. Finally this too disappears (g), and a new feature, the globe itself, comes into sharp focus (h).

So, as geographers stand back from the beach, new spatial realities emerge. Just as with a painting, you have to be the right distance away to see its meaning. For this depends on the level of detail, the spatial resolution, that can be seen. Go too close and the painting becomes a jumble of individual brush marks. Stand too far away and all you can see is a blurred rectangle of canvas on the distant wall.

Orders of Geographic Magnitude

Throughout this text we discuss geographic studies confined to sizes, or *orders of geographic magnitude*, between these two extremes. Although the variations are considerable, the geographer is concerned with a narrow 'window' within the scale of scientific inquiry. In the figure in Box 1.C on 'Orders of geographic magnitude' the main areas of scientific inquiry are plotted along a centimetre scale. Exponential notation (in which 1000 is written as 10^3, 1 as 10^0, 0.001 as 10^{-3}, and so on) is used in order to present a large range of variation on the same diagram.

The scale extends from the microworld of the atomic physicist, studying cosmic rays with wavelengths of about 10^{-15} cm, to the radio astronomer, studying galaxies with diameters of 10^{23} cm and more. By comparison, the geographer's world is confined to a narrow band along the middle range of

Box 1.C Orders of Geographic Magnitude

Geographers study a continuum of environments of varying size. However, even this continuum is rather limited when one considers the continuum of environments that constitutes the universe. This extends outwards from our familiar human space to the furthest edge of the galaxies. It also extends inwards, to the deep structure of atoms. This larger continuum and the zone of interest to geographers is represented by the scale in diagram (a). (Notice that this scale is logarithmic, not linear.) The enlargement of the zone of interest in (b)

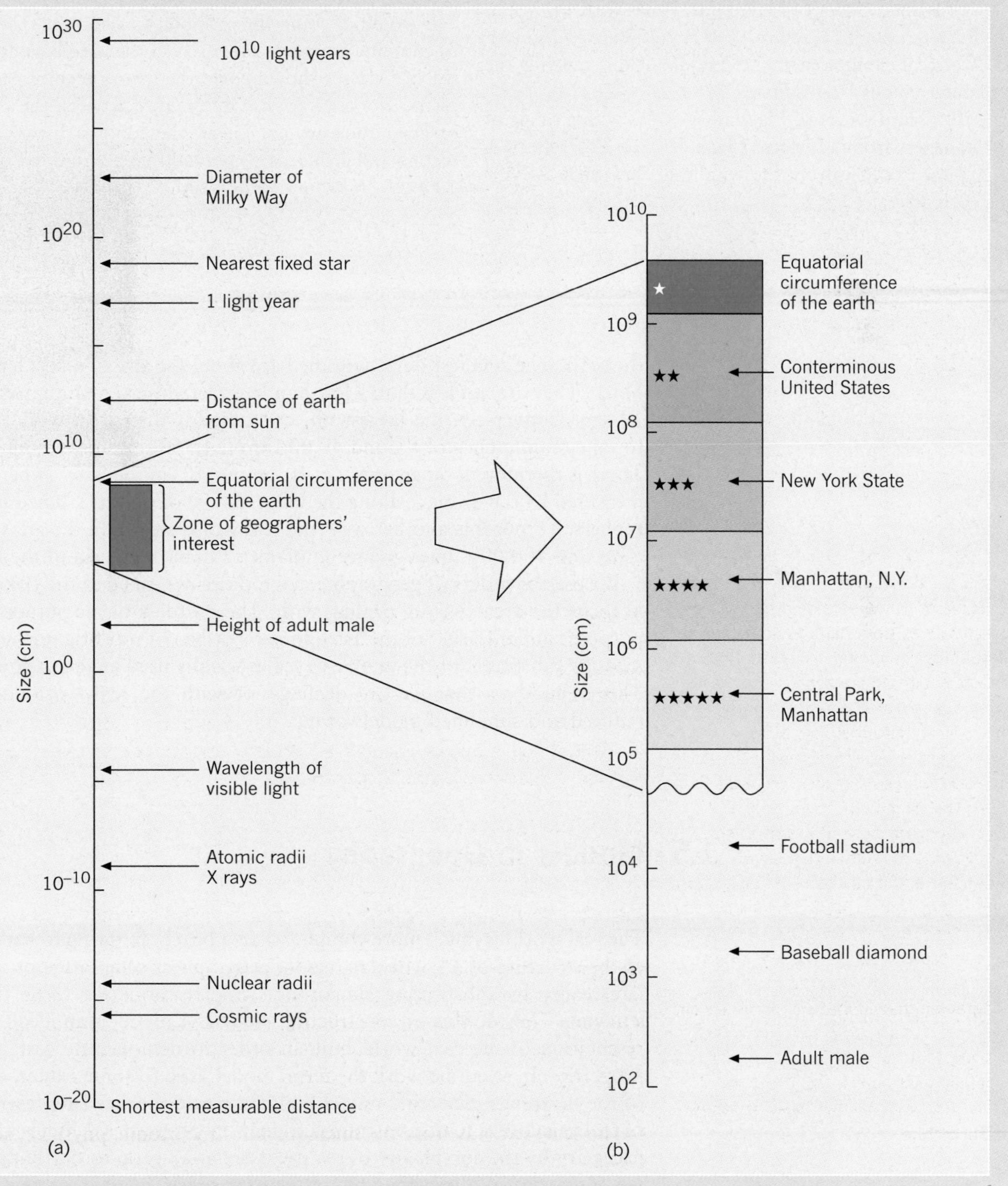

Box continued

enables us to divide it into zones that are roughly parallel to **the orders of geographic magnitude** described below. In this book we indicate the size in terms of orders of magnitude.

*First Order of Magnitude. Areas with a range of diameters from that of the surface of the earth itself (with an equatorial circumference of 40,000 km, or 24,860 miles) to 12,500 km (7700 miles).

**Second Order of Magnitude. Areas with a range of diameters from 12,500 to 1250 km (7770 to 777 miles). A typical example from the middle of this range is the conterminous United States.

***Third Order of Magnitude. Areas with a range of diameters from 1250 to 125 km (777 to 77.7 miles). A typical example from the middle of this range is New York State.

****Fourth Order of Magnitude. Areas with a range of diameters from 125 to 12.5 km (77.7 to 7.77 miles). A typical example from the middle of this range is New York City.

*****Fifth Order of Magnitude. Areas with a range of diameters from 12.5 to 1.25 km (7.77 to 0.777 miles). A typical example from the middle of this range is Central Park in New York City.

It would of course be possible, as diagram (a) shows, to continue downward to a sixth order and beyond, but the five classes shown cover the main range of geographers' work. Note that differences between the orders of magnitude are not linear: the contrast between the second and fifth order is not a difference of 3 but one of $10 \times 10 \times 10$ (i.e. 10^3, or 1000).

the scale. The smallest objects studied are about the size of a beach or a city block. They are not less than a few hundred metres across or approximately 10^4 cm. Conversely, the largest object studied is the earth, with an equatorial circumference of around 40,000 km (24,860 miles) or about 10^9 cm. There is therefore a range of 5, i.e. 10^9 minus 10^4 in the 'size' (where size is measured by the distance along the longest axis) of objects studied by geographers. To put this another way, the real world shown in the atlas section is around 100,000 times greater in diameter than the world of the beach.

We use the orders of geographic magnitudes described in the box to keep in focus the areas we are dealing with. These orders of magnitude serve a purpose similar to that of the astronomer's orders of star brightness and are a happy substitute for the jumble of scales usually used in geography books. They remind us that we are dealing not with the real world but with reduced and simplified models of it.

1.5 Models in Geography

The real world is much more complex than a beach. In trying to make sense of the structure of a particular region, geographers often attempt to simulate reality by substituting similar but simpler forms for those they are studying. They do this by constructing *models*. A model is an idealized representation of the real world built in order to demonstrate certain of its properties. In scientific work the term 'model' has, to some extent, all three of the meanings. Scientific model-builders create idealized representations of reality in order to demonstrate certain of its properties. Models are made necessary by the complexity of reality. They are a prop to our understanding and a source of working hypotheses for research. They convey not the whole truth, but a useful and apparently comprehensible part of it.

In everyday language the term model is used in at least three different ways. As a noun it signifies a representation; as an adjective it implies an ideal; as a verb, it means to demonstrate. Thus we are aware that when we refer to a model railway or a model husband we are using the same term in different senses.

Maps and Other Models

We have already seen a simple example of model-building in our study of the crowd on a beach. Figure 1.4 illustrates a first stage of abstraction. It represents the properties of the people on the beach faithfully, but in a distant perspective and on a different scale. It is common knowledge that most scientists make things appear larger in order to study them. The optical microscope, the electron microscope, and the radio telescope were scientific breakthroughs because they permitted increasingly powerful magnifications of reality. Geographers are curious folk in that they follow a reverse process. They bring reality down in size until it can be represented by a map. To shrink the universe to manageable size, they use various standard *linear scales*, which determine the ratio of the length of a line segment on a map to the true length of the line on the earth's surface. (See Table 1.1.) For example, 1 cm on a map may represent 1 km (100,000 cm) on the ground. Usually, the scale of a map is given as a representative fraction, as 1/100,000 or 1 : 100,000. This ratio applies equally to metric and non-metric measurement. Thus on a 1 : 100,000 map, 1 inch is equal to 100,000 inches (about 1.6 miles) on the ground, and 1 cm is equal to 100,000 cm (1 km) on the ground. We look further at maps and their scale in Chapter 21.

All the maps in this volume are very selective and partial models of the real world, with all the advantages – and the drawbacks – that simplification brings in its train. Simple scaled-down representations of reality are called iconic models. In Figure 1.4(a) we showed models of sunbathers' bodies at linear scales from around 1 : 50 (foreground) to 1 : 500 (background). Figure 1.6(a) constitutes the second stage of abstraction – an *analogue*

Table 1.1 Standard lengths and areas on ten commonly used map scales

		Equivalent on the earth's surface of standard measures on the map	
Class and scale	**Countries using the scale for major map series**	**1 cm on the map**	**1 inch on the map**
Large scales			
1:10,000	European	0.100 km	0.158 miles
1:10,560	British and Commonwealth	0.106 km	0.167 miles
1:24,000	United States	0.240 km	0.379 miles
1:25,000	British and Commonwealth	0.250 km	0.395 miles
Medium scales			
1:50,000	European	0.500 km	0.789 miles
1:62,500	United States	0.625 km	0.986 miles
1:63,360	British and Commonwealth	0.634 km	1.000 miles
1:100,000	European	1.000 km	1.578 miles
Small scales			
1:250,000	International	2.500 km	3.946 miles
1:1,000,000	International	10.000 km	15.783 miles

model. Here, real people have become points on a map. Clusters of people on the beach have become point clusters on the map. Abstraction is pushed still further in a third type of model, the *symbolic model*, in which real-world phenomena are represented by abstract mathematical expressions, such as that for the population density of a beach. Figure 1.6(d) is a symbolic model. It takes us a step further from reality than either a photograph or a map for a second example of progressive abstraction. We take the same idea further in Figure 1.13 which shows part of the Hawaiian island of

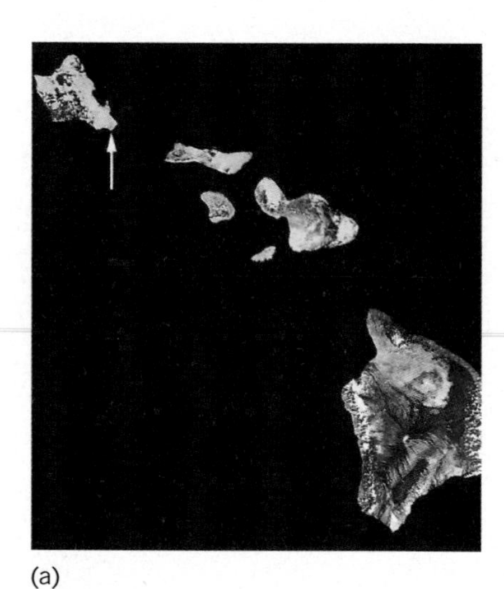

(a)

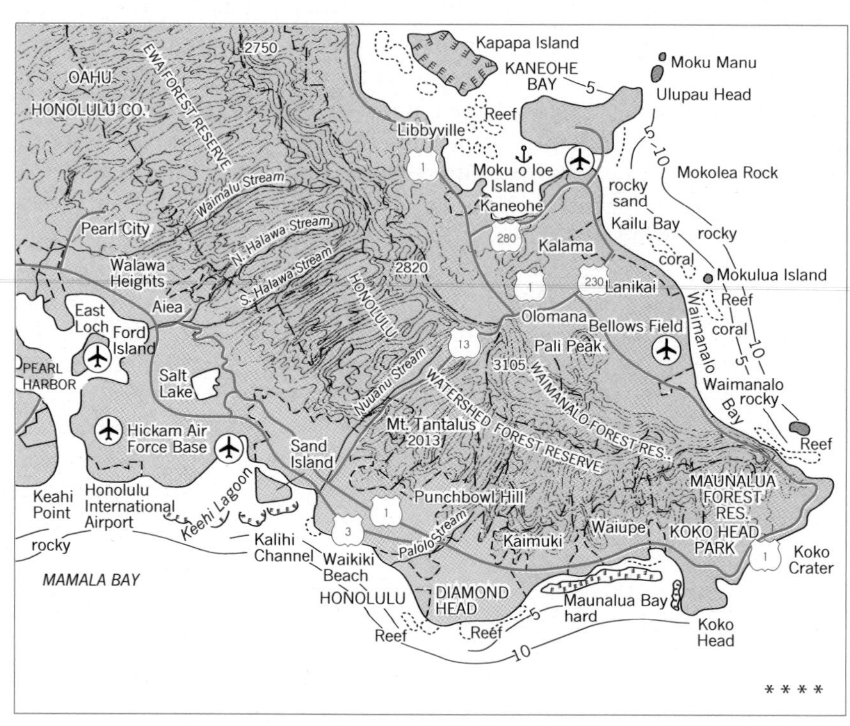

(b)

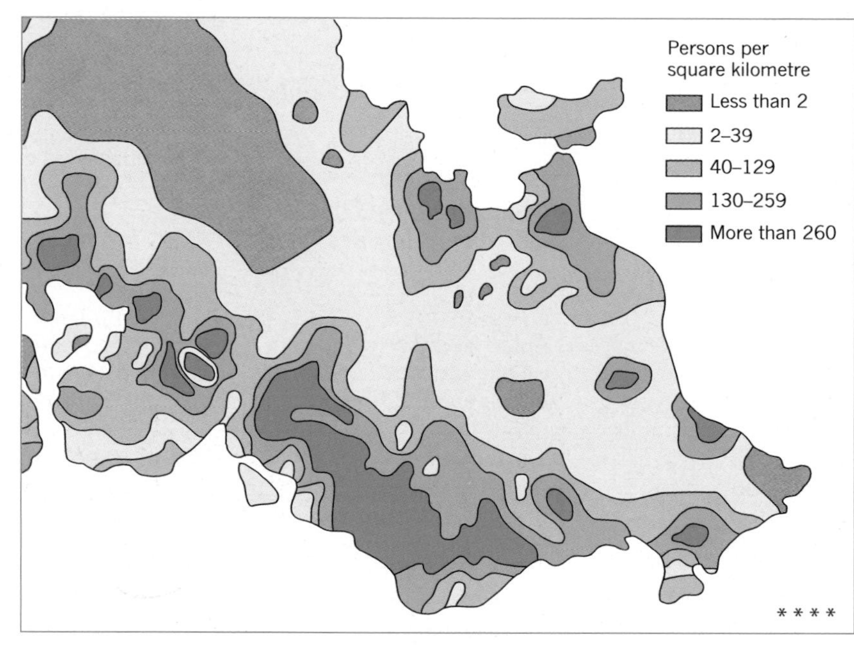

(c)

Figure 1.13 Maps as models
Three views of the southeastern corner of Oahu, one of the Hawaiian islands, consisting of the city of Honolulu and surrounding areas. (a) A satellite photograph of the entire island chain with superfluous ocean and clouds dropped out in order to accent the islands' configurations (Oahu is the island arrowed). (b) A reproduced section from a US Geological Survey topographic sheet. (c) A population density map of the area shown in (b), drawn from 1970 census records. The scale in (b) and (c) is 1:250,000. The scale in (a) approaches 1:6,000,000.

[Source: (a) Photograph courtesy of NASA. (b) Courtesy of US Geological Survey. (c) From R. Warwick Armstrong (ed.), *Atlas of Hawaii* (University Press of Hawaii, Honolulu, 1973), p. 118.]

Oahu (a) as seen from an earth satellite, (b) as shown on a topographic map on the same scale, and (c) as a population density map. Each represents increasing levels of abstraction.

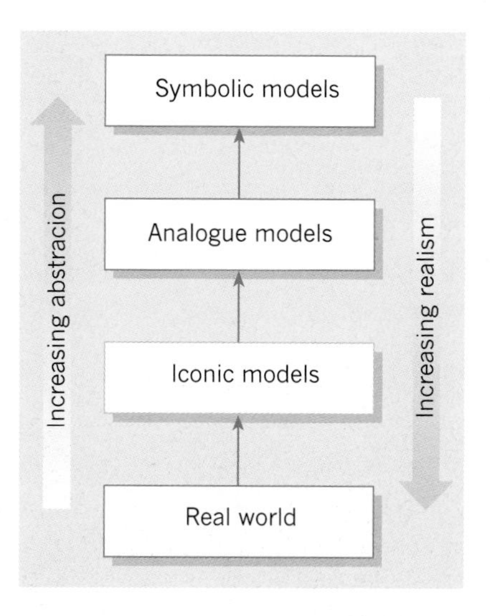

Figure 1.14 Model building
Model building is shown here as a three-stage process of increasing abstraction.

From Models to Paradigms

We may conveniently think of model-building as a three-stage process in which each stage represents a higher degree of abstraction than the last (see Figure 1.14). At each stage, information is lost and the model becomes less realistic but more general. Throughout this book we shall return to the ideal of models and look at their actual use. We shall encounter many models that are considerably more intricate than the three simple ones in the preceding section, but we will postpone discussion of them until we meet specific examples. It is useful at this point, however, to bring forward the idea of a *paradigm*. A paradigm is a kind of supermodel. It provides intuitive or inductive rules about the kinds of phenomena scientists should investigate and the best methods of investigation. This chapter – like the whole of this book – represents a paradigm, a particular view of geography.

Research in geography, like research in most fields, is based on a shared paradigm; that is, those who pursue this research are committed to probing the same problems, observing the same rules, and maintaining the same standards. Tedious methodological debates, and concern over what constitutes legitimate research or appropriate methods of analysis, are symptoms of transitional periods in the evolution of a science. Once a paradigm is fully established, debate languishes.

In his provoking history of modern science *The Structure of Scientific Revolutions*, the philosopher Thomas Kuhn contended that the origin, continuance, and eventual obsolescence of paradigms is the prime factor in the evolution of science. Modern geography has witnessed major shifts in emphasis from descriptive geography toward more analytic work. In the 1960s this focused on mathematical models of how regions grow and interact. In the 1970s there was a renewed stress on human behaviour and our response to environmental change. In the 1990s there has been great emphasis on Geographical Information Systems (GIS). The present decade is too close yet to discern how research will develop. In this book we try to illustrate both the traditional paradigm and the new ones. We regard the current emphases as the most recent phases in a long history of change, in which ways of looking at people on the beach and in the world have been refined, but the essential questions that geographers ask in the new century remain unaltered.

From Paradigms to Real-world Problems

How can we use the idea of spatial organization and humanity's relations with the environment in ways that are helpful to people? The answer depends partly on the paradigm within which geographers work, and there is a noticeable contrast between the traditional and modern views of the field.

The traditional role of the geographer has been the provision of two types of essential information: locational information on the exact position of events, and environmental information on the quality of particular areas. In response to demands for this kind of information the great geographic works of the Greeks were written, the exploration societies of the early nineteenth century were formed, and the universal geographies of the Victorian period were assembled.

Today, however, geographers are more concerned with optimization – with finding the 'best' location for things and making the 'best' use of areas. Where should a new model city be located? What is the best site for a hospital within a city? What is the best dividing line between two hostile communities? What is the best use for the more remote Appalachian areas? They are also interested in forecasting projected trends into the future and monitoring the likely effects of policy decisions in a wide variety of situations.

Geographers work in agencies that range from the World Health Organization or the World Bank on the international level to local city halls and county agencies. They dominate the regional planning sections of several countries (notably Britain) and form a significant element in government agencies from Washington to Moscow or Beijing. In the world of business or in private agencies, geographers perform essentially the same role: advising on locational priorities, watching for environmental feedbacks, and providing a geographic view of some part of the world.

On the Search for Spatial Order

A critical question in all scientific investigation is just what we are prepared to accept as a *signal* or a *pattern*, and what we dismiss as background *noise*. If we ask of a given world region whether its settlements are arranged in some predictable sequence, or its land-use zones are concentric, or its growth cyclical, and then the answer largely depends on what we are prepared to look for and what we accept as *order*. Order and chaos are not part of nature but part of the human mind: that there is more order in the world than appears at first sight is not discovered until the order is looked for. Cambridge geographer Richard Chorley has given a lively illustration of this problem as it afflicted Newton, newly struck on the head by an apple:

> Had he asked himself the obvious question: why did that particular apple choose that unrepeatable instance to fall on that unique head, he might have written the history of the apple. Instead of which he asked himself why apples fell and produced the theory of gravitation. The decision was not the apple's but Newton's.

We do not have look far to demonstrate that order depends not on the spatial geometry of the object we see but on the organizational framework in which we place it. Figure 1.15 gives three familiar examples of observational 'flips', which remain well worth repeating. Diagram (a) shows to some observers an old Parisienne lady, to others a young woman (drawn after the style of the French painter Toulouse-Lautrec). All normal retinas 'take' the same picture; and our sense-datum pictures must be the same for even if you see the *vielle femme* and I the *jeune fille*, the pictures we draw of what we see may turn out to be geometrically indistinguishable. Organization is not itself given by the lines or shapes; were there no preconceived model in our minds we would be left with nothing but an unintelligible pattern of lines. Consider the second diagram, (b). There some may see birds (each looking to the left), others may see antelopes (each looking to the right). But would people who had never seen an antelope, but only birds, see antelopes in this drawing?

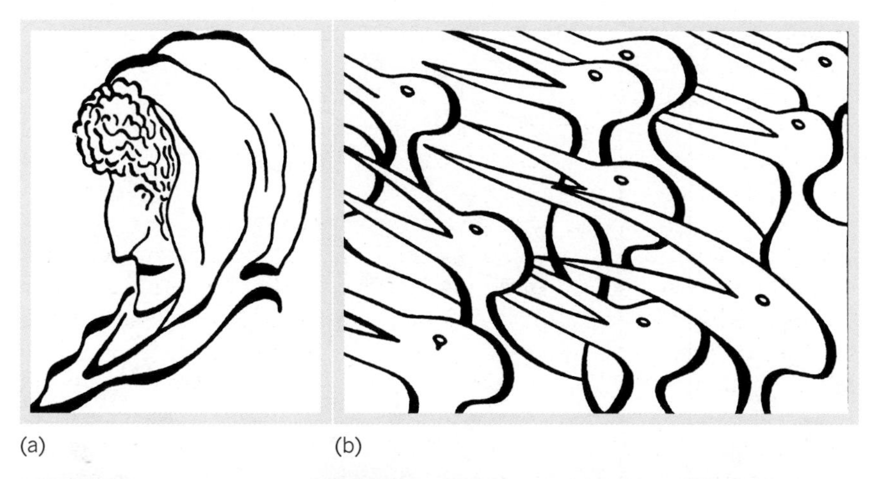

(a) (b)

(c)

Figure 1.15 Spatial order and meaning Patterns of lines with alternative (a) and (b) or hidden (c) interpretations. Look carefully at the three pictures before reading the interpretations suggested in the text.

[Source: Original diagrams in E.G. Boring, *American Journal of Psychology*, **42** (1930), p. 100. and P.B. Porter, *American Journal of Psychology*, **67** (1954), p. 550. Reproduced from N.R. Hanson, *Patterns of Discovery: An Inquiry into the Conceptual Foundation of Science* (Cambridge University Press, Cambridge, 1958), Figs 2, 5, and 7, pp. 11, 13 and 14.]

A third example is given in Figure 1.15(c). This pattern of black and white shapes is simply one of melting snow on a grass surface. But it also has a human interpretation: a late medieval representation of Jesus Christ. To see this you have to look very carefully. The upper margin of the picture cuts the brow, thus the top of the head is not shown. The point of the jaw, brightly illuminated, is just above the geometric centre of the picture. A white mantle covers the right shoulder with the right upper sleeve exposed as the rather black area at the lower left. The hair and beard are after the manner of a late medieval representation of the face of Christ. Once that face has been seen, then we live with it for life. I find it impossible to 'unsee' that pattern (once I have seen it) and thus may miss other and equally arguable interpretations.

The examples given in these figures will be familiar to many readers. But we often fail to appreciate how much the lessons they teach have implications for geographic observation. Consider the three maps in Figure 1.16. The upper diagram (a) shows a small 12 × 16 metre (40 × 52 feet) section of Salisbury Plain in southern England, with the blackened areas representing depressions in the old soil surface of this chalk downland (see the discussion in Box 22.B). The field archaeologists interpret these depression as the remnants of post holes dug to support the main timbers of a hut.

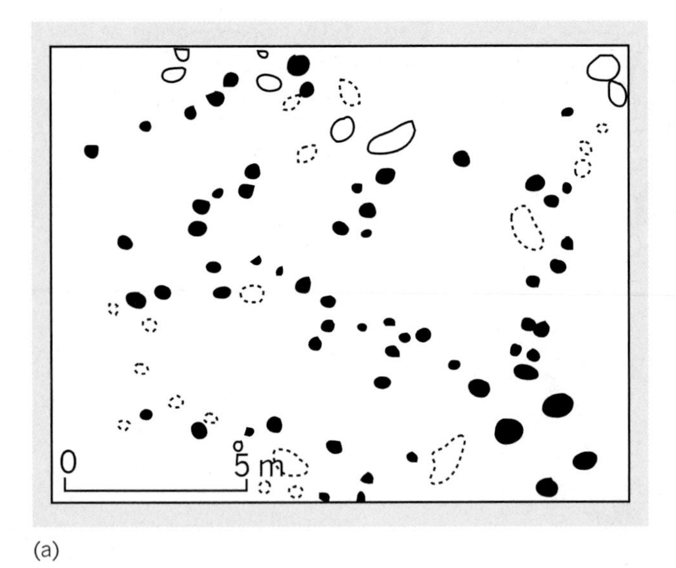

(a)

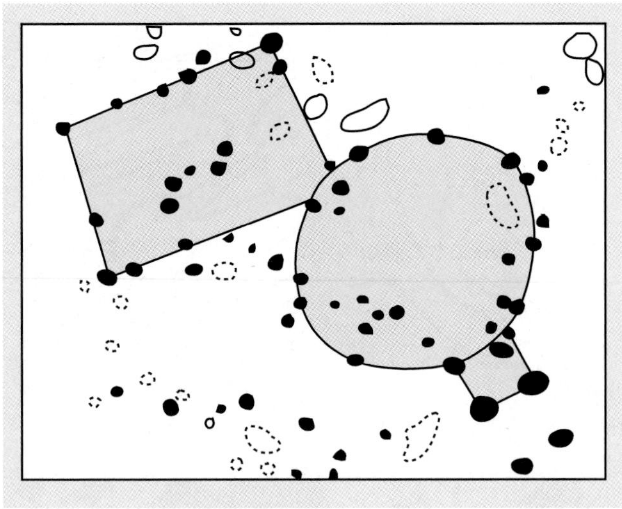

(b)

Figure 1.16 Alternative interpretations of a map (a) Map from an archaeological excavation showing holes marked in black. Some of these may be post holes for the timbers used in building a hut, others random excavation by burrowing animals. One group of archaeologists selects a set of holes as post holes and suggest a circular hut form (b), while another group favours a larger rectangular hut form (c). Both interpretations bring pre-existing hypotheses to bear on which pieces of the spatial evidence may be regarded as 'signals' and which are irrelevant 'noise'.

[Source: Richard H. Gregory and E.H. Gombrich (eds), *Illusion in Nature and Art* (Duckworth, London, 1973), Fig. 32.]

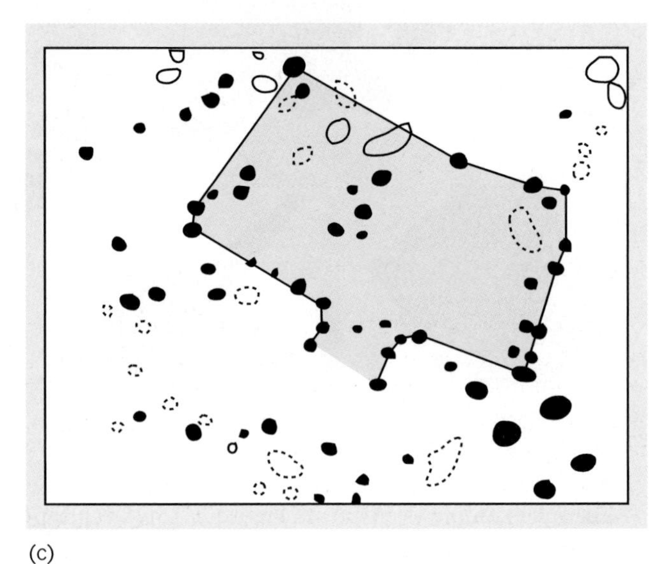

(c)

Questions now arise as to the age of the settlement and which tribal groups were the likely builders. To try to answer these, the original map was sent to two archaeologists in different universities for an independent opinion. Maps (b) and (c) show the two interpretations. The data have been linked to show a hut with a rectangular structure type of one period of settlement; the other a circular hut from a more primitive period. Exactly the same data have led to different models; the difference lies not in the evidence but what is seen in the data.

For geography the danger of model imposition is acute. For of all sciences it has traditionally placed greatest emphasis on 'seeing'. In how many field classes have we asked geography students to 'see' the clues of glaciation in the landscape? The 'seeing eye' beloved of physical geographers such as the late S.W. Wooldridge is a necessary part of our scientific equipment. But pattern and order exist in knowing what to look for, as well as how to look. And the order of one generation may be the chaos of the next.

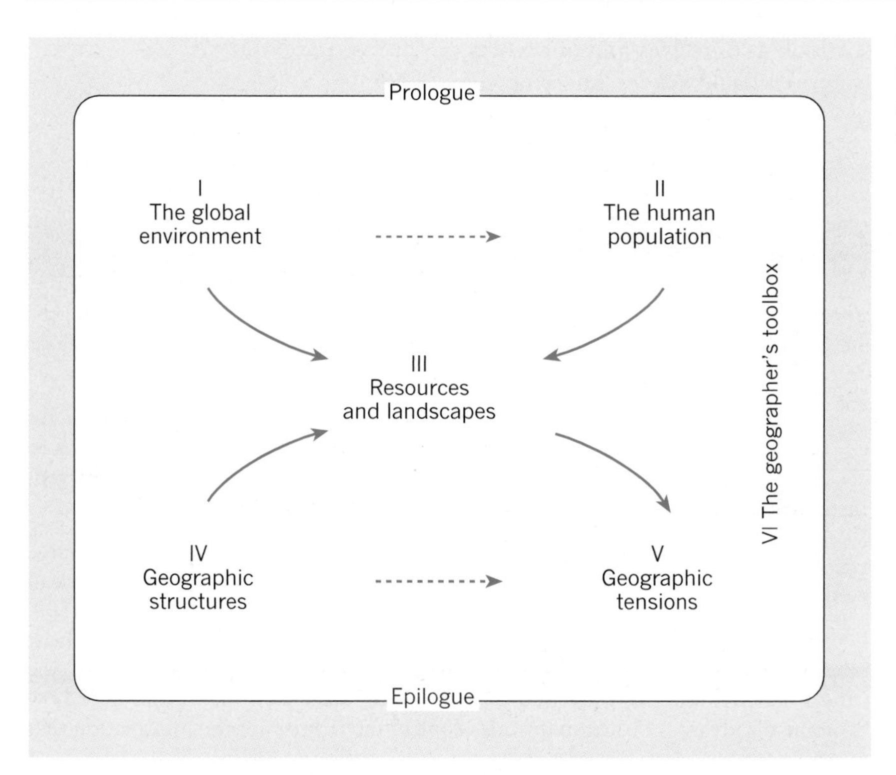

Figure 1.17 Organization of the book Ways in which the following chapters (2–24) are ordered in the remainder of the book.

Organization of the Book

We need to bear the lesson of Figure 1.16 in mind when we look at the rest of this book. Its organization as set out in Figure 1.17 is one way, I hope a logical and interesting way, of introducing geography. But other geographers would use different approaches and join up the 'dots on the map' in different and equally valid ways.

How important that geographic view is will depend on the problem being examined. In many instances, the importance of spatial and environmental considerations may be negligible. In such cases the contribution of the engineer, the economist, or the educator will surely outweigh that of the geographer. When large-scale environmental issues are discussed, however, a geographer's viewpoint will certainly be needed.

We do not intend to argue in this book for a purely geocentric view of the world's problems. At a time when the walls between academic subjects are crumbling, the isolationist subject makes about as little sense as an isolationist state. Geography has always been heavily dependent on its academic neighbours such as mathematics, the earth sciences, and the behavioural sciences, and it has everything to gain by remaining so.

Geography is interesting today not because of its solution of past problems, but because of its potential contribution to resolving future difficulties. Geographers as a group have been a little embarrassed to discover that locational and environmental questions, so long a part of their classroom discussion, are now a daily topic in the news media, in congressional and parliamentary committee rooms, and on the campus. For over two thousand years, geographers have been studying the world and the place of human beings in it. Suddenly, at the start of the third millennium, these seemingly academic preoccupations are being regarded as relevant – to us and to our children. We are taking a new look at ourselves and our world, and, like T.S. Eliot's explorer, or Di Caprio's backpacker, 'knowing the place for the first time.'

Reflections

1. Geography for you may be a new subject; or, perhaps, one you have already studied at school. Note down in a few sentences what you think you are likely to get out of a course in geography. Keep this statement in your file so that you can look back at the end of the term and compare what you expected to gain with what you actually gained.
2. It is sometimes said that disciplines are distinguished from one another by the questions they ask. We illustrate this in Section 1.1 with a very simple environment, the beach. What special questions do geographers ask? What questions should they ask?
3. How much can geographers studying the behaviour of humans learn from biologists studying the behaviour of animals? List (a) some advantages and (b) some disadvantages of such cross-borrowing between disciplines.
4. Human–environment relations may be usefully seen in system terms (H/E). Try to construct a diagram with arrows and boxes showing how (a) how the beach environment affects human behaviour and (b) how human behaviour (e.g. building sea defences) affects the beach.
5. Most geographic distributions are variable over time. Consider the graphs of population density over time in Figure 1.7. Try to construct similar graphs for (a) the campus restaurant, (b) the local shopping mall, and (c) the local airport. Comment on the similarities and differences in these graphs.
6. The scale of a geographer's interest ranges from the microgeographic to the macrogeographic (see Box 1.C). Try to place your own country and your own city on that geographic scale.
7. Use locally available maps with small, medium, and large scales to examine a beach of your choice. (For American students the maps will be at scales of 1 : 24,000, 1 : 62,500, and 1 : 250,000; students in other countries should check with Table 1.1 for guidance.) Familiarize yourself with the conventional symbols used on such maps, and note how the map scales control the richness of the information that can be conveyed.
8. Review your understanding of the following concepts using the text and the glossary in Appendix A of this book:

choropleth maps	orders of geographic magnitude
feedbacks	region
isopleth maps	scale
location	space
model	time–space geography

One Step Further . . .

Two useful general reviews of geography as an integrated discipline are given in Tim UNWIN, *The Place of Geography* (Longman, Harlow, 1992) and a book by a Norwegian geographer, Arild HOLT-JENSEN, *Geography – History and Concepts*, second edition (Paul Chapman, London, 1998). Personal views of geography are given by the author in Peter HAGGETT, *The Geographer's Art* (Blackwell, Oxford, 1990) while Peter GOULD in *Becoming a Geographer* (Syracuse University Press, Syracuse, 1999) and Anne BUTTIMER, *Geography and the Human Spirit* (Johns Hopkins University Press, Baltimore, 1993) provide engaging accounts of geography's attractions as a lifelong pursuit.

For some of the specific topics touched on in this chapter which are worth taking further, see Yi-Fu TUAN, *Space and Place* (University of Minnesota Press, Minneapolis, 1977) on interpersonal space and territoriality. On time geography, Torsten Hägerstrand's essay in T. CARLSTEIN, D. PARKES, and N.J. THRIFT, *Timing Space and Spacing Time* (Arnold, London, 1978), Vol. 2, pp.122–45 is a classic statement. An excellent review of time geography is also provided by Derek Gregory in Ron JOHNSTON *et al.*, *The Dictionary of Human Geography*, fourth edition (Blackwell, Oxford, 2000), pp. 830–3. The Johnston dictionary is an excellent companion to human geography and is well complemented for physical geography by David THOMAS and Andrew GOUDIE (eds), *Dictionary of Physical Geography*, third edition (Blackwell, Oxford, 2000). A briefer dictionary which covers both physical and human geography is R.J. SMALL and M. WITHERICK, *A Modern Dictionary of Geography*, third edition (Arnold, London, 1995).

Students of geography will find it helpful to look through recent numbers of geographic serials to see what kinds of work geographers are currently doing. *Geography Review* (quarterly) is specifically designed for students. Popular journals include one of the world's oldest (founded 1888) and largest selling American periodicals, the *National Geographic* (monthly). The *Geographical Magazine* (monthly) plays a similar role in the UK. Up-to-date reviews of work in geography at a more advanced level is provided in *Progress in Human Geography* (quarterly) and *Progress in Physical Geography* (quarterly). Most countries have several geographical journals and they are valuable to monitor during your course. For example, Australia (*Australian Geographical Studies*), Britain (*Geographical Journal*), Canada (*Canadian Geographer*), Netherlands (*Tijdschrift voor Economische en Geographische Geografie*), Sweden (*Geografiska Annaler*), United States (*Annals of the Association of American Geographers*). Finally, it will be useful to have at hand a standard world atlas such as *Goode's World Atlas*, 20th edition (Rand McNally, Skokie, Ill., 2000) or *Philip's Atlas of the World*, sixth edition (Reed Books, London, 1996). See the sites recommended in Appendix B at the end of this book for topics relevant to this chapter.

PART I

The Global Environment

CHAPTER 2

The Earth as a Planet

■ denotes case studies

Figure 2.1 **The earth as a planet** This classic satellite view from Apollo II emphasized the isolation of the beautiful but lonely planet on which the human population lives. It shows an 'Earthrise', with the planet rising over the surface of the moon. The lunar surface features are near the eastern limb of the moon as seen from the Earth. Some historians have dated the worldwide acceptance of the environmental movement to the impact of this image. [Source: NASA/Science Photo Library.]

> When the great markets by the sea shut fast
> All that calm Sunday that goes on and on:
> When even lovers, find their peace at last,
> And Earth is but a star that once had shone.
>
> JAMES ELROY FLECKER *The Golden Journey to Samarkand* (1913)

Flecker's verse and our opening photograph have something in common. Both speak to the immensity of space and of time. The famous photograph (Figure 2.1 p. 34), taken from outer space across the moon's barren surface, highlights the fragile planet with all its lonely beauty. The poet's lines, despite their technical licence (earth is a planet, not a dying star), stress the aeons of time in which the earth is set and the fact that everything must come to an end.

As we move outward into deep space, the intimate world of the beach and our small-scale ecosystems disappears. Viewed from this remote distance, the earth is simply a pale blue planet, much of it wreathed in clouds, with the outlines of tawny continents dimly visible. But even though the detail is invisible, relations between people and the environment are still observable. The scale of relations is vastly different though; our range of vision now includes the whole of humanity's six billions on the earth's half billion square kilometres of surface. In place of individual sunbathers on the beach in Chapter 1, we now have great clusters and nebulae of population set against a continental background, in place of wet sand and dry soil, we have an environmental range that runs from the Amazonian forest to Arctic ice caps.

In this chapter we shall look at the earth from a global perspective and assess some of its qualities as an environment for humanity. We first look at the earth's origins as a planet and the main environmental systems into which it can be divided (Section 2.1). Each of those divisions is then taken in turn. In Section 2.2 we consider the continental building blocks that make up the earth's surface and the framework of *landforms* that give it its distinctive skeleton. In Section 2.3 we turn to the thin but crucial envelope of the earth's atmosphere and in Section 2.4 to the waters that cover part of the earth. In both cases we ask how this sphere originated, how it is shaped, and how it is evolving. In the final part of the chapter (Section 2.5) we look at global change in the most recent period and its immediate effect on the earth as the home for humanity.

In raising such fundamental questions in the geography of the earth, we are entering the field of *physical geography*. This part of geography analyzes the physical structure of the earth's environment, such as its landforms, climate, vegetation, and soils. Partly because physical geography is closely linked to parallel natural sciences (geophysics, geology, meteorology, botany, etc.) and partly because of its longer history of scientific study, it currently forms the strongest and fully developed parts of geography (cf. Section 24.2, especially Figure 24.8). It has the most highly developed theoretical models, and its predictive capability has already achieved levels that are unlikely to be reached in human geography for some decades. Any attempt to summarize its contents within the next three chapters can only give a broad-brush picture. Those of you who have already taken introductory earth science courses or an elementary course in physical geography may wish to move quickly through these chapters to the further reading. If the concepts are new to you, we stress that they are the first steps in a field of geography that you will be able to study in much greater depth in advanced courses.

The Lonely Planet 2.1

When and how did planet Earth come to be? What processes led to its formation? How is the earth structured? And how do geographers go about studying it?

Origins of the Earth

There is no single, generally accepted view as to when and how the planets of the solar system were formed. The oldest rocks on the earth (the fifth largest of the planets in that system) are around 3.5 billion years old and scientists believe that our own planet was formed around a billion years earlier. So 4.5 billion years is the generally accepted age.

The mechanism of the earth's formation is equally a matter for debate. In 1755 the German philosopher, Immanuel Kant, proposed a nebular theory for the origin of the solar system. He argued that a huge *nebula* (a cloud of gas and dust) once swirled around the sun. As it swirled it slowly flattened out, causing eddies within the cloud that began to spin like whirlpools, drawing in gas and dust to collect at the centres of the whirlpools. This accretion of cosmic dust allowed gravitational attraction and the differential growth of the larger bodies (the nucleus of the planets) at the expense of the smaller bodies. Later astronomers have refined the theory with various tidal and double-star hypotheses being invoked.

The temperature at which this formation occurred is also a matter for competing theories. If it was an accretion of very hot particles, then this primitive heat would have been lost very early in the earth's history; if of cold particles, then the heat would have come solely from the radioactive decay of the materials that form its core. The earth's present internal heat continues to be generated in this way.

Figure 2.2 Main elements in the structure of the environment The four environmental 'spheres'. Note the way in which the biosphere is concentrated along the interfaces between the three abiotic environments – the atmosphere, hydrosphere, and lithosphere.

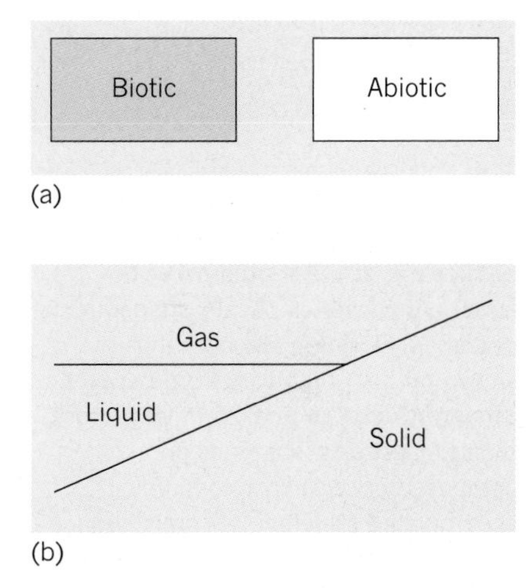

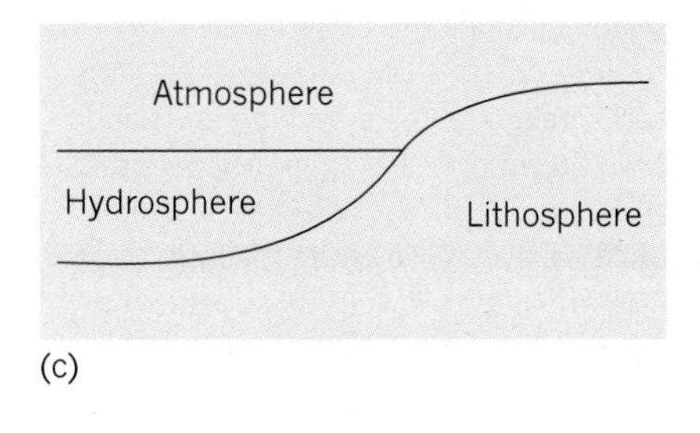

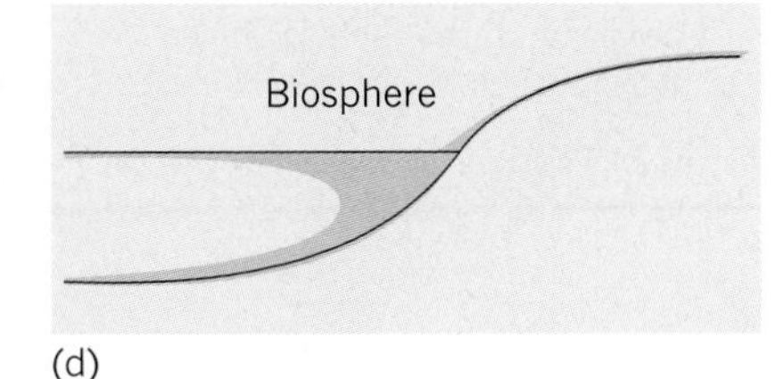

The Structure of the Environment

The most fundamental divisions of the environment are shown in Figure 2.2(a). Note there the primary and fundamental difference between the non-living world and the living world. The living world is termed the *biosphere*, from the Greek word *bios* meaning life. In environmental terms, the former is the abiotic environment and the latter is the biotic environment.

The abiotic environment can be subdivided in terms of its physical state as a solid, liquid, or gas. The solid earth is termed the *lithosphere* (from the Greek word, *lithos*, meaning rock). It refers to the earth's crust, a layer of rigid rocks some 70 km (43 miles) thick, lying above deeper rock layers that are plastic and can flow. These extend down to the earth's core itself.

The lithosphere is surrounded by two shells each up to 11 km (7 miles) thick: a discontinuous layer of liquid termed the *hydrosphere* (from the Greek *hudor*, meaning water) and a continuous layer of gas termed the *atmosphere* (Greek *atmos*, vapour). In comparison with the size of planet Earth with its diameter of 12,700 km (7900 miles), then these two layers are very thin indeed, much thinner in proportion than the skin on a good apple.

The living world of the biosphere (see Figure 2.2(d)) is even thinner. If we define the biosphere as the biotic environment in which living things are found, then it extends through the full depth of the oceans but is confined to the lower layers of the atmosphere (birds, flying insects, micro-organisms). On the land surface, the biosphere extends upward to over 110 m (360 ft) (the giant redwood trees of northwest California) and down many metres in the micro-organisms in the soil or in deep caves and rock fractures. The biosphere is separately described in Chapter 4.

The Nature of Earth Systems

A system is defined here as 'a set of components and the relationships between them.' As we saw in the opening chapter, a system consists of three essential ingredients, the components, the links between them, and the boundary that separates the system from the rest of the world.

Geographers are usually concerned with systems in which relations between individual components are built up by observed association to produce positive or negative bonds. Changes in the level of one component cause associated changes in other components. Such systems vary in the

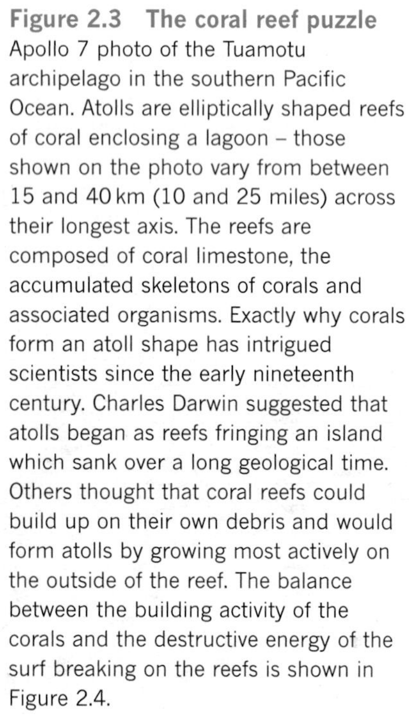

Figure 2.3 The coral reef puzzle Apollo 7 photo of the Tuamotu archipelago in the southern Pacific Ocean. Atolls are elliptically shaped reefs of coral enclosing a lagoon – those shown on the photo vary from between 15 and 40 km (10 and 25 miles) across their longest axis. The reefs are composed of coral limestone, the accumulated skeletons of corals and associated organisms. Exactly why corals form an atoll shape has intrigued scientists since the early nineteenth century. Charles Darwin suggested that atolls began as reefs fringing an island which sank over a long geological time. Others thought that coral reefs could build up on their own debris and would form atolls by growing most actively on the outside of the reef. The balance between the building activity of the corals and the destructive energy of the surf breaking on the reefs is shown in Figure 2.4.

[Source: Photograph courtesy of NASA.]

number of components they have, the strength of the links between them, and the arrangement of the links into negative or positive feedback loops.

We can illustrate the links and loops in an environmental system by considering the way in which one distinctive element on the earth's surface, coral reefs, are found (see Figure 2.3).

Corals are minute marine animals that live in huge colonies in shallow tropical seas where their combined limy skeletons form reefs and atolls. They have attracted great interest from geographers since Charles Darwin's Pacific voyages in the *Beagle* in the 1830s, and they are still one of the most fascinating of marine ecosystems. Many of the organisms that secrete calcium carbonate to make *coral reefs* are sensitive to the depth of water. As the depth decreases, sunlight becomes more abundant and the rate of growth of the reefs increases. This accelerated growth further decreases the depth of the water, increases the light, accelerates the growth of algae, and so on, in a positive feedback relationship. However, the inability of the organisms that make up the reefs to grow above sea level and the breakup of the reefs by pounding waves introduces an effective negative feedback that limits growth. Figure 2.4 presents a flow chart in which the elevation of the reef with respect to the sea level acts to regulate its growth, initiating either positive or negative feedback loops.

This example was chosen since its origin links together all four of the great environmental divisions: (1) the lithosphere that shapes the seamounts on which the corals grow; (2) the hydrosphere in which the reefs form; (3) the atmosphere that warms the tropical oceans; and (4) the biosphere in which corals evolved as organisms. We now look at the first three systems of the four in turn.

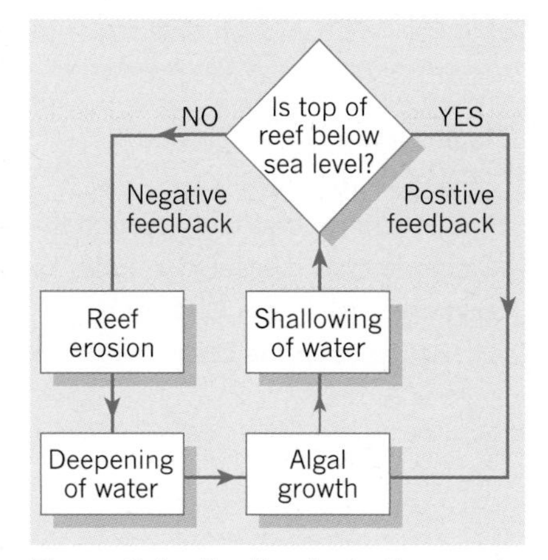

Figure 2.4 Feedbacks in the coral reef system The role of the sea level in switching on and off the growth of algae on a hypothetical coral reef is shown via a flow diagram. Notice that in this example each component (box) in the system contains a physical or biological process.

Lithosphere: The Continental Building Blocks 2.2

The surface of the earth cannot be properly understood without looking below that surface and trying to understand its structure. We look here at the fundamental division between continent and ocean and see how this provides the key to the pattern of global terrain.

Continents and Oceans

If we look at the world around us, the visual evidence of strong environmental contrasts is so compelling that we should begin by reminding ourselves that, compared with other wanderers in the solar system, the earth is a uniform planet. It is almost spherical, with a radius of 6365 km (3955 miles) that varies by less than 0.02 per cent. In height, no two places on its surface vary by more than 20 km (about 12 miles), or the length of Manhattan Island in their vertical distance (see Table 2.1).

The earth's crust consists of two fundamental components, continental crust and oceanic crust. *Continental crust* covers about 30 per cent of the earth's surface. Essentially the continents consist of granite (*sial*, dominated by minerals rich in the the two chemical elements silica and aluminium) about 20 km thick, with a thin cover of sedimentary rocks.

The *oceanic crust* consists of about 70 per cent of the earth's surface. The

Table 2.1 The basic dimensions of the earth

Total surface area	510,056,000 km^2	(196,934,000 square miles)
Area of land surface	149,137,000 km^2	(57,582,000 square miles)
Circumference measured around the poles	40,003 km	(24,857 miles)
Circumference measured around the equator	40,074 km	(24,901 miles)
Highest point on the land surface	+8.85 km	(+5.45 miles)
Lowest point on the ocean floor	−11.03 km	(−6.85 miles)

hydrosphere, being liquid, fills the gaps between the continental fragments, giving the oceans that we know today. But the oceanic crust is not an accidental distribution of water on a solid but irregular earth. The actual shorelines are rather incidental, depending on the height of sea level on the sloping shelves. Ocean waters extend onto continental rocks at continental shelves, and the true edges of the continents are the steeper continental slopes. The oceanic crust is made up of a layer of basaltic material (*sima*, dominated by the minerals rich in chemical elements silica and magnesium) about 7 km thick. The sima layer is thought to extend under the continents. The really striking thing about the sea floors is their young age. No sea floor is older than 200 million years, whereas some continental rocks date back to 3.8 billion years. How did that remarkable contrast come into being?

Pangaea and the Wandering Continents

To answer the question we have to go back nearly a century. The present and familiar patterns of continents and oceans have a comfortable solidity about them. In 1915 a German scientist, Alfred Wegener (see Figure 2.5(a)), disturbed this notion of tranquil stability by proposing that the continents were wandering about the earth's surface. Dismissed in the 1920s, the so-called *continental drift* theory was not considered valid until new evidence emerged after another 40 years.

Alfred Wegener was born in Berlin in 1880 and took his doctorate in astronomy. His controversial book, *The Origin of Continents and Oceans*, suggested that all the world's land areas had once been united in a single, primordial continent incorporating all the world's land masses. He called this supercontinent *Pangaea* after the Greek words for all (*pan*) and the earth (*Gaia*).

What led Wegener to this unusual reconstruction? The German scholar's evidence for the existence of Pangaea was of three main kinds. First, the contemporary world map presented features that could only be considered remarkable coincidences. The most obvious of these was the jigsaw-like fit between the Atlantic coasts of Africa and South America. The fit is closest when we take not the coastline itself but the line marking the edge of the continental shelf. (See the discussion of the continental shelf in Chapter 17.) This shelf is the shallow water area immediately adjacent to the continental land areas. Wegener also noted from the map that the world's major mountain ranges, such as the Alps or Andes, were mainly confined to narrow arc-like belts occurring in only a few areas. But this did not fit in with then prevalent ideas that mountains were the product of a shrinking, cooling planet. Surely, Wegener argued, a shrinking globe would have created a less localized pattern, more like the wrinkling all over a dried apple?

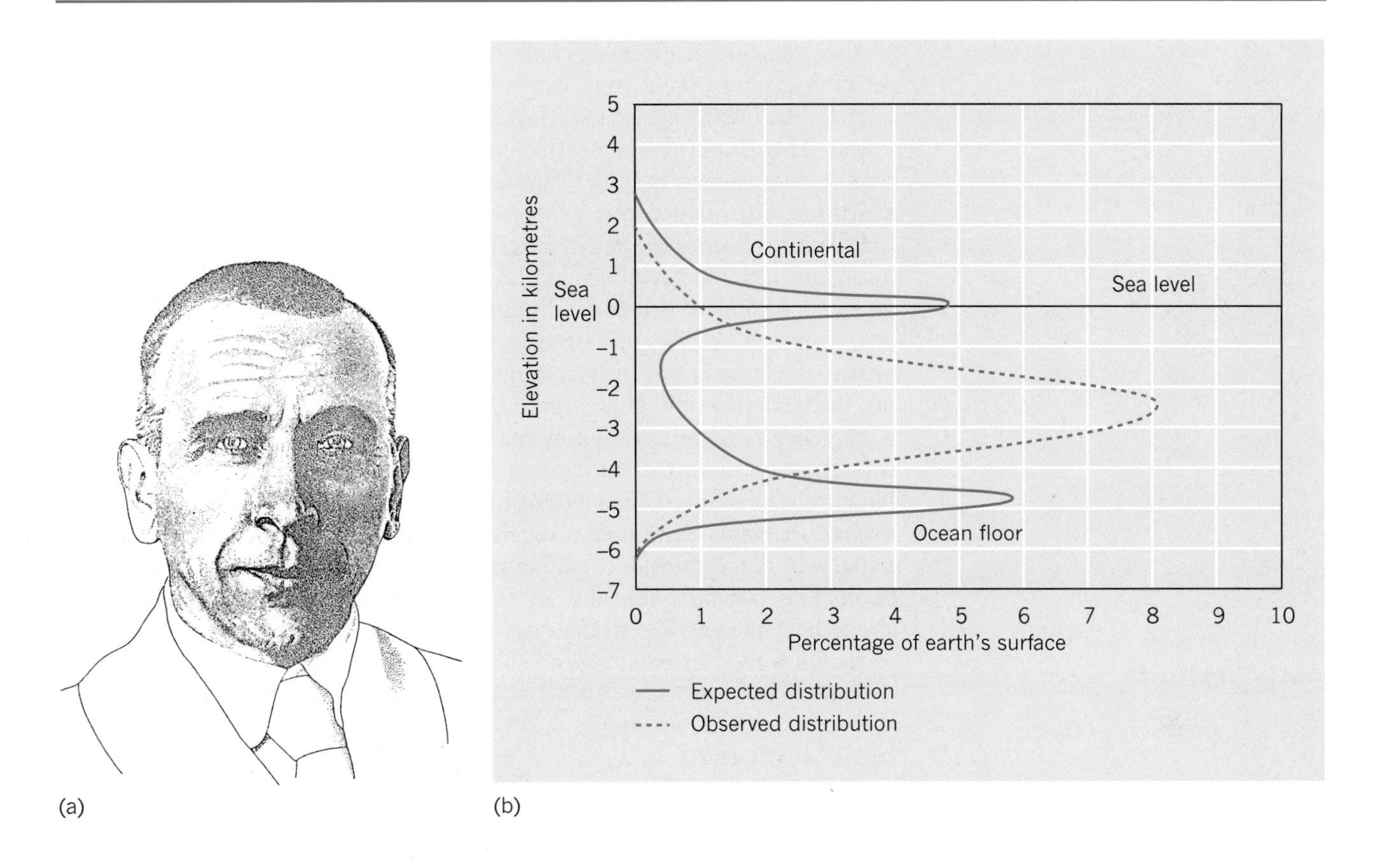

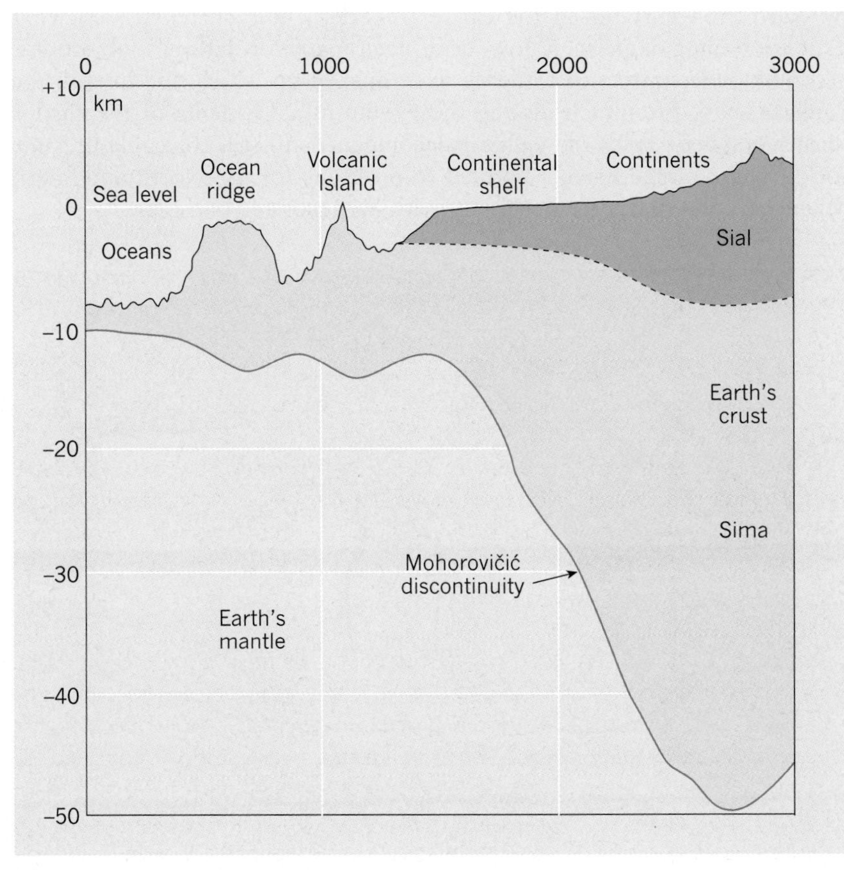

Figure 2.5 Continental drift (a) Alfred Wegener (1880–1930) was a German explorer and geophysicist who originated the theory of continental drift. First met with much hostility, it eventually led to the idea of plate tectonics (see next figure) which was established 30 years after his death. (b) One of the factors that affected Wegener's thinking was that the earth's elevation does not form a normal curve but occur in two distinct peaks, a continental platform and an abyssal sea floor. (c) Current views of the separate structure of continents and ocean basins shown in this idealized cross-section through the earth's crust. Note that the vertical scale is greatly exaggerated in relation to the horizontal scale.

[Source: (a) *Scientific American* (May 1972).]

A second line of evidence came from a statistical analysis of the earth's relief. A study of the actual distribution (see Figure 2.5(b)) shows two distinct levels: one at around sea level and a second at around −5 km (−3.1 miles). This disparity suggested that the continents and ocean floors were two distinct elements, rather than part of a continuous surface. Wegener's third line of evidence was geological: large blocks of very old rocks are found on both sides of the Atlantic. Putting South America and Africa back into place into Pangaea lines up the rock formations, notably those created under the extensive desert conditions of the Permian period, about 230 million years ago. As Wegener observed: 'It is just as if we were to refit the torn pieces of a newspaper by matching their edges, then checking whether or not the lines of print run smoothly across'.

Wegener's ideas received a mixed reception from his fellow scientists. He was regarded as an outsider and his 'evidence' was dismissed as being largely circumstantial. The weakest part of his arguments was his failure to find a force capable of achieving the movements he proposed. His fiercest critic was the Cambridge mathematician Sir Harold Jeffreys who in his book *The Earth*, published in the mid-1920s, demonstrated that the strength of the earth's surface was too great and the power of Wegener's proposed forces far too weak to achieve continental separation.

Plate Tectonics

Fresh light was not to come for another 30 years. Then an English geologist, Arthur Holmes, put forward the idea that continents could be moved by convection currents in the earth's mantle. These currents, somewhat akin to boiling jam on a low heat, met many of Jeffrey's objections. Secondly, new sources of evidence were opened up which tied in with the Pangaea map. Prominent among these were measurements of the earth's former magnetic fields (so-called palaeomagnetism) still traceable in some rocks. These measurements pointed to positions for the North and South Magnetic Poles that were consistent with Wegener's hypotheses.

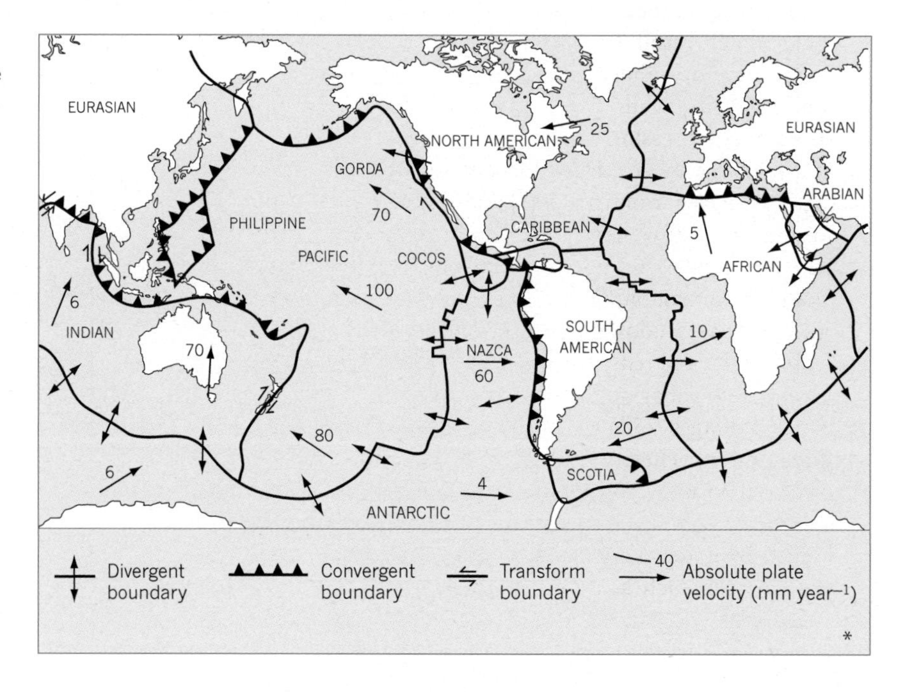

Figure 2.6 Main tectonic plates of the world Map of current ideas on the major tectonic plates which make up the earth's surface. The different types of boundaries for the plates are shown and the directions and estimated rates of movement (in mm per year).

The new conceptual framework is one of a dynamic earth model whose continents are on the move and whose ocean basins are opening and closing. Although this more dynamic model owes something to Wegener's original ideas, the dominant theoretical contribution is that of *plate tectonics*. This theory, largely developed during the 1960s, holds that the outer rind of the earth, the lithosphere, consists of a mosaic of rigid plates. A map of the main plates is shown in Figure 2.6.

The six main plates are supra-continental in scale. Indeed, the continents now appear as relatively secondary features, passengers on great moving plates approximately 100 km (60 miles) thick. The boundaries of the plates appear to be of two kinds. One type, called a convergent plate boundary, is where the two adjacent plates move together and either collide or one plate is forced downwards. A typical example is where the oceanic Nazca plate bends down into the Peru–Chile trench, and slides under the South American plate at about 6 cm (2.5 inches) a year. (A speed not dissimilar to the rate of growth of our fingernails.) A second type, called a divergent plate boundary, is where two plates move apart. The Mid-Atlantic ridge is a typical case. Parallel plate boundaries are where two adjacent plate boundaries move face-to-face along common junctions. The formation of rift valleys is discussed in Box 2.A.

Box 2.A Rift Valleys

A rift valley is a linear trough formed by subsidence or downthrusting in areas of continental crust within the interiors of tectonic plates. They occur where tensional stresses predominate in the lithosphere. The classic interpretation of rift valley morphology is that of a graben, with the rift floor being seen as a downthrown block bounded by normal faults, which create steep bounding escarpments. Diagram (a) shows a classic symmetrical graben. Faults are cracks or fissures in rocks which result from tectonic movements. *Normal* faults develop under a pattern of tensional stress (pulling apart) while *reverse* faults are associated with zones of compression.

One of the most extensive rift valley systems on the earth's surface runs for 6400 km (4000 miles) from Jordan in southwestern Asia southwards through the Red Sea into eastern Africa and reaching the Indian Ocean in Mozambique. It is marked by a series of lakes such as Lake Rudolph and Lake Nyasa. The plateaux adjacent to the rift valleys generally slope upward towards the valley and provide a distinctive sharp drop which averages 750 m (2500 ft) but in some places is three times this height. The rift is of great antiquity and has been forming for some 30 million years as Africa and the Arabian Peninsula separated. It has been accompanied by extensive volcanism along part of its length (e.g. massifs such as Kilimanjaro and Mount Kenya).

Seismic data suggest that over much of the great rift valley, the graben model is an over-simplification. The structure of many rifts appears to be asymmetric, with much of the downthrow occurring along a major boundary fault on one side only. These major faults tend to be discontinuous and may alternate along the rift, separated by transfer faults.

The infrared satellite image (see (b)) shows the northern end of the great rift valley system. The Dead Sea at centre lies between Israel (at left) and Jordan (at right). North is at the top. In the infrared range water appears

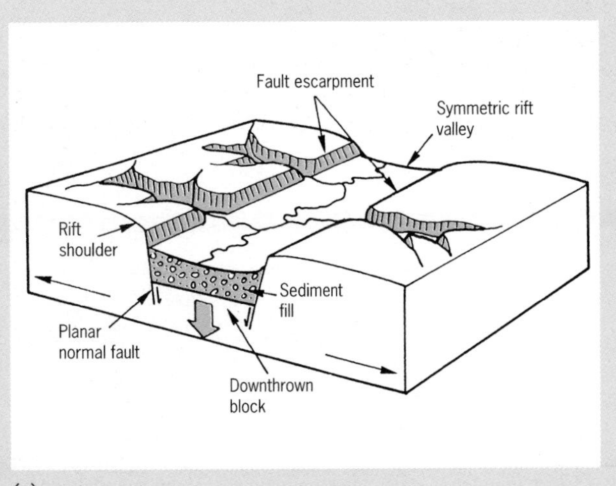

(a)

Box continued

(b)

black, vegetation is dark and barren ground is lighter. The River Jordan runs into the Sea from the north but the lake has no outlet and is highly saline. The Dead Sea is 400 m (1300 ft) below sea level, making it the lowest body of water on earth. A good discussion of rift valleys is given by M.A. Summerfield, *Global Geomorphology* (Longman, Harlow, 1991).

[Source: (b) Earth Satellite Corporation/Science Photo Library.]

The implications of this huge global instability in both the make-up and location of the continents and the position of the poles are shown in Figure 2.7. This shows (a) the present position of the Swedish capital city, Stockholm, located today on the Fennoscandian block at latitude 59°N.

In terms of the earth's history, this fixed location is an illusion. Chart (b) shows how Stockholm's position has varied over the past 2.7 billion years. During that period it has spent about half the period in the southern (not the northern) hemisphere and has been under the sea for long periods. Its probable climate has varied from tropical rainforest through aridity to antarctic icefields.

Instabilities in the Lithosphere

Although continental movements occur over huge time-spans outside direct human experience, they have direct implications on humankind today through two contemporary processes: earthquakes and volcanoes.

Earthquakes Earthquakes are sudden movements or fractures deep within the lithosphere measured as a series of shock waves on the earth's surface. About 150 thousand earth tremors are detected every year. These are measured on a Richter scale which ranges from zero to ten (although 8.9 is the largest value recorded so far). On this scale a value of 2.0 can be just felt as a tremor, with building damage occurring at values of 6.0 and over. One or two earthquakes each year cause very severe damage.

Although the specific timing of earthquakes is unpredictable and no effective counter-measures (other than appropriate building restrictions) are possible, the location of earthquake-prone areas is well established. Figure 2.8 shows that earthquakes occur principally in two elongated belts. The first passes around the Pacific Ocean and includes in its North American

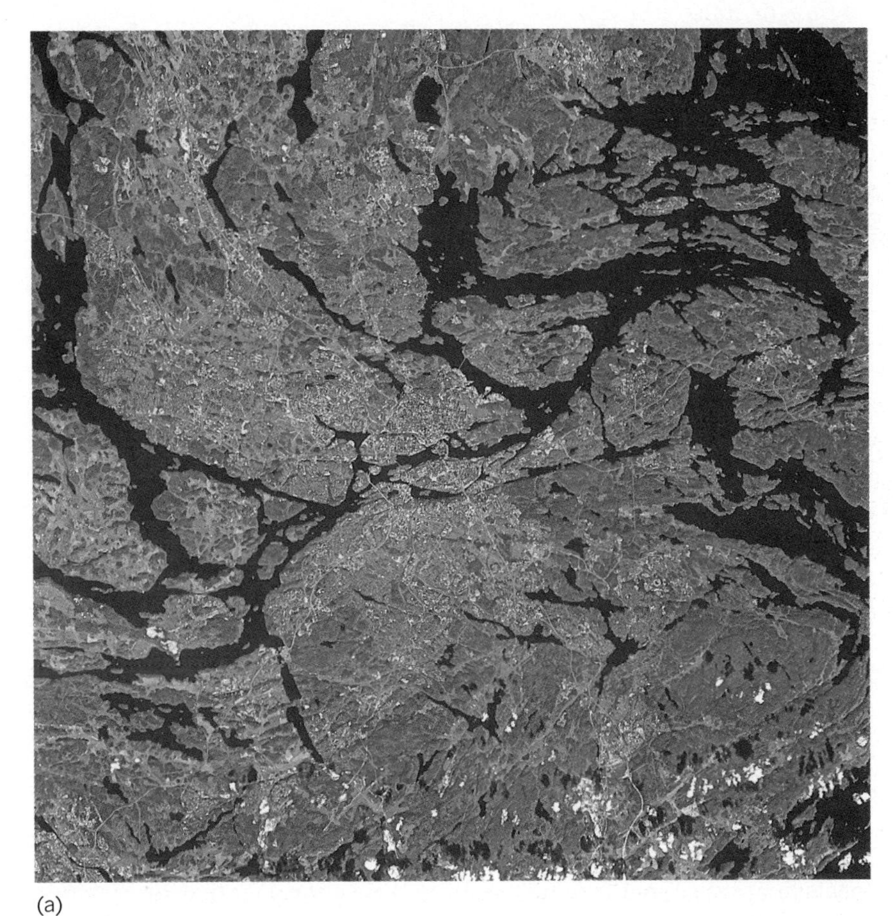

(a)

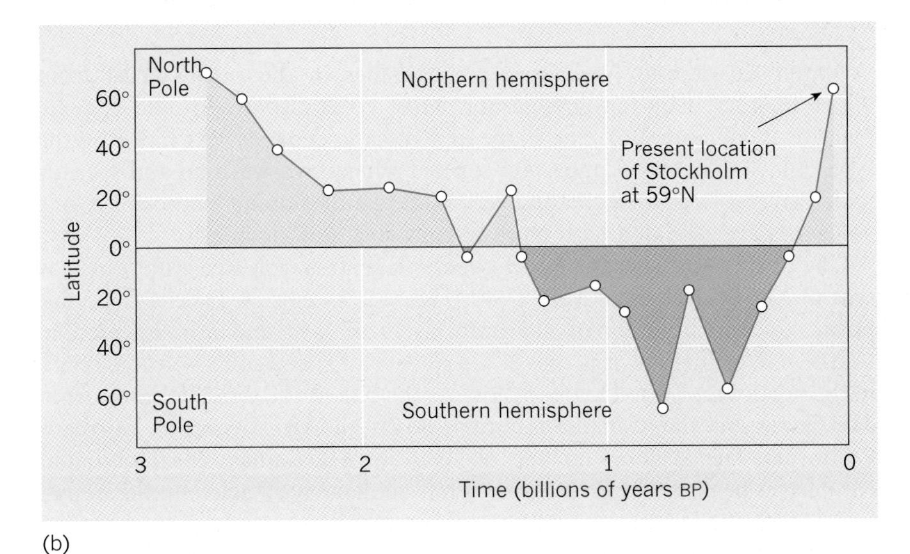

(b)

Figure 2.7 Wandering locations on the earth's surface Over long geological periods, locations on the earth's surface have shifted dramatically in relation to the poles and the equator. For example, present Stockholm was once near the South Pole. (a) Satellite image of Stockholm, the capital city of Sweden, located at centre of photo on islands between the Malaren Lake (far left) and the Baltic Sea (right). Its current latitude is 59°N. (b) Variations in the latitudinal position of the Stockholm area over the last 2.97 billion years based on studies of rock magnetism.

[Source: (a) CNES/Science Photo Library, 1995 Distribution Spot Image. (b) S. Helmfrid (ed.), *The Geography of Sweden* (National Atlas of Sweden, Stockholm), p. 19.]

segment the Aleutian Islands, southern Alaska, and the Pacific coast of Canada and the United States. This *circum-Pacific belt* is estimated to account for some 80 per cent of all the earthquake energy released on the planet. A second major belt runs from Portugal through the Mediterranean, the Middle East, and the Himalayas and meets the circum-Pacific belt in the Indonesian islands.

As more oceanographic exploration is completed, the presence of other

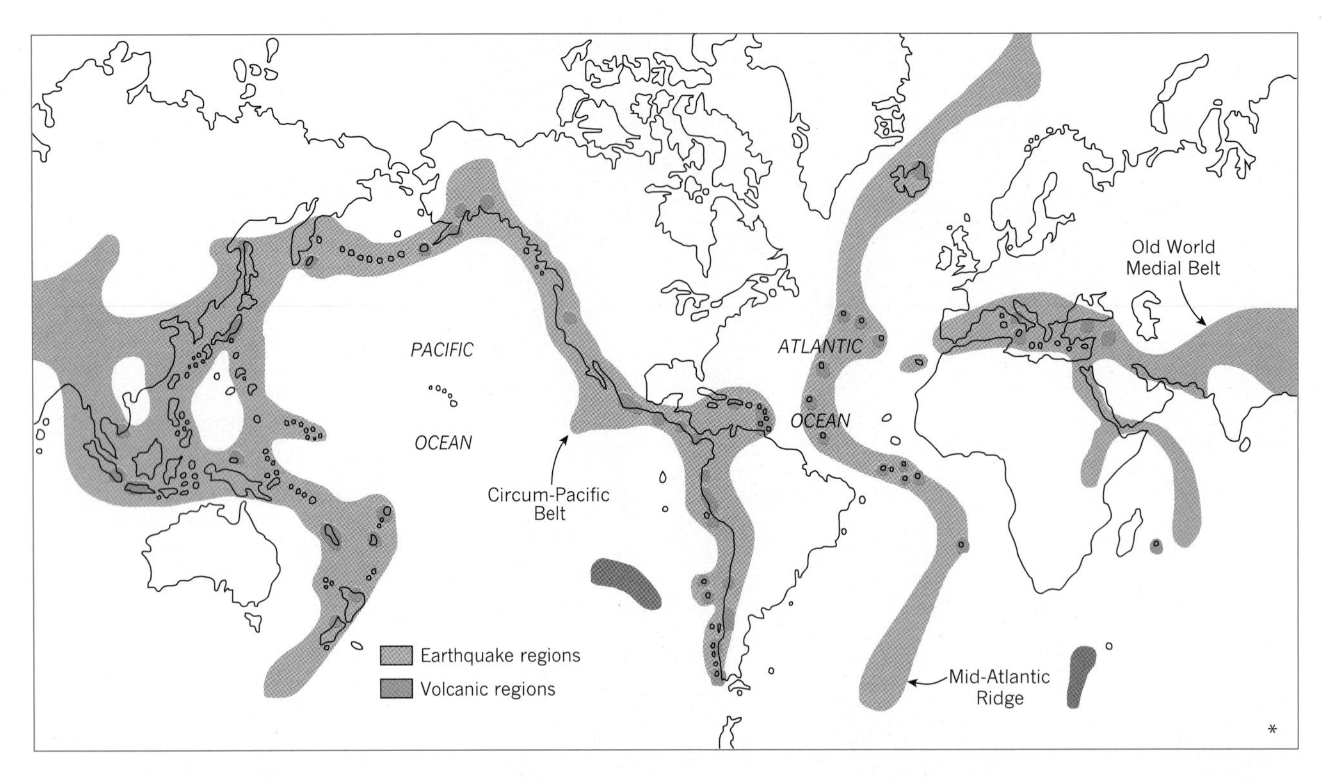

Figure 2.8 Earthquake–volcanic hazard zones Schematic map of main areas of earthquakes and volcanic activity during the most recent geological period (i.e. that most relevant to human beings). Note the concentration into three main zones. The most important is the circum-Pacific zone or girdle which accounts for about 80 per cent of all earthquake activity. The Old World medial belt running from the Mediterranean to Indonesia accounts for most of the remaining activity with the mid-Atlantic ridge forming a third and less active region.

earthquake belts associated with mid-ocean ridges is being established. These areas form the boundaries between a series of great structural plates rather like the sutures of a human skull. *Tectonic movements* and adjustments in pressure appear to take place in these critical tension zones within the earth's crust.

Volcanic Eruptions Volcanoes are openings in the earth's crust from which magma, ash, and gases erupt. Most volcanoes are located at plate margins (see Figure 2.6). The shape of a volcano depends very much on the type of lava. Shield volcanoes are formed where lava wells up and spreads widely over a large area creating very large, gently sloping landforms. Cone volcanoes are assciated with more viscous lava and much ash.

One of the most spectacular of twentieth-century volcanic eruptions was that of Mt. St. Helens in May 1980 (Figure 2.9). Mt. St. Helens is a composite volcano built up of alternate layers of lava and ash. Located in southern Washington, it is one of a sequence of spectacular volcanic cones in the northwest United States following the line of the Cascade range from Mt. Baker on the Canadian border down to Mt. Lassen in northern California. The sequence includes the majestic Mt. Rainier, Mt. Hood, and Mt. Shasta peaks and Crater Lake. While six have been active in the period since 1800, the last major eruption was in 1917.

The main Mt. St. Helen's eruption began with an earthquake of modest severity on 20 March 1980. It was followed by minor volcanic activity (steam and ash) for the next two months. On the morning of 18 May another earthquake was followed by one of the most massive volcanic explosions of recent history. It effectively blew the top off the mountain, leaving a massive gap where the cone had been. The blast of hot air charged with ash and stone particles flattened and seared the surrounding forest up to 30 km (18 miles) from the peak. A surge of ash, mud, logs, and debris flowed out, fingerlike, down the main valleys, choking and burying streams

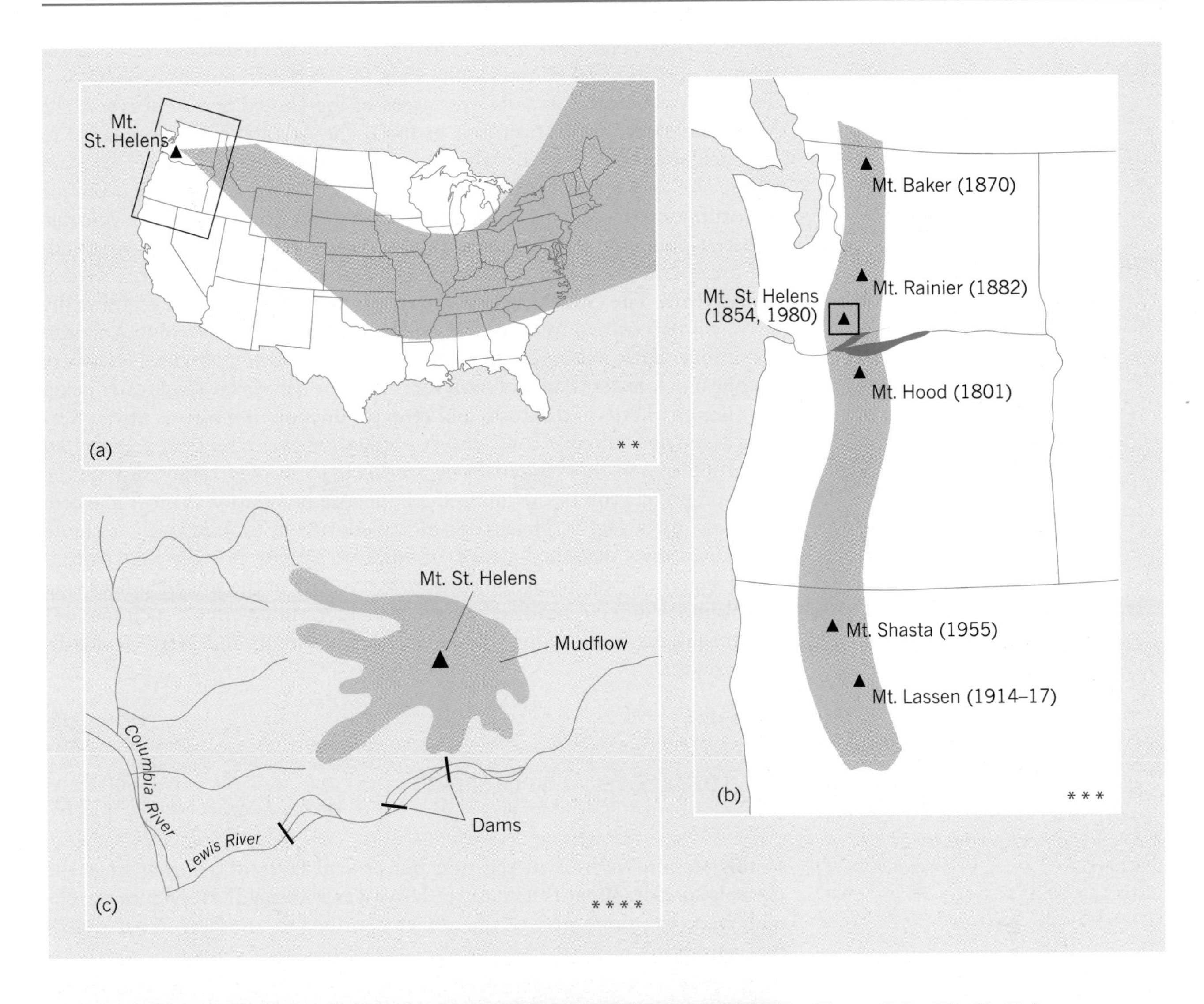

Figure 2.9 Mt. St. Helens (a) Dustfall from Mt. St. Helens over the United States, 1980. (b) Volcanoes in the Cascade range that have erupted since 1800. (c) Local impact of Mt. St. Helens eruption, May–June 1980. (d) Volcanic dust plume of Mt. St. Helens on 22 July 1980. This explosive event, some two months after the main explosion, threw ash up to a height of 18 km (11 miles) in the space of eight minutes.

[Source: (d) David Weintraub/Science Photo Library.]

up to 20 km (12 miles) away. Finally, a vertical column of dust mushroomed up within 10 minutes some 6 km (4 miles) into the atmosphere producing a swathe of dust falls over areas of the United States, heavy to the Montana border and measurable as far as the Atlantic coast. It also led to spectacular sunsets over the whole northern hemisphere.

The size and location of the Mt. St. Helens eruption have given unique opportunities to see how the earth's ecosystems recover from a volcanic disaster. The eruption destroyed all plant and animal life in the surrounding areas and deposited a choking ash and mud layer, in places several metres deep. The coincidence of the eruption with a spell of exceptionally heavy rainfall helped to keep dust and ash more localized than under drier conditions. Early studies suggest that the rates of plant and animal recovery may be much faster than had been forecast. The ash surface is already being colonized by herbs and shrubs, and crop production in the areas affected by dust is starting to come back. Insect populations were severely reduced by the eruption and their populations are likely to recover rather slowly; the effect on pollination of the fall in the honeybee population is most marked.

We can place Mt. St. Helens in a global context by looking back at Figure 2.8. This shows that the location of volcanic activity broadly follows that of earthquakes. The Pacific Northwest forms part of the so-called fiery ring of volcanoes in a circum-Pacific belt. It includes in its circuit such massive volcanic peaks as Ecuador's Cotapaxi, Japan's Fuji, and New Zealand's Mt. Egmont.

2.3 **Atmosphere:** The Ocean of Air

In this section we look at the thin but critical layer of gas that hugs the planet's surface. What is its nature? How was it formed? How does the climate vary from one part of the planet's surface to another? What causes that variation?

The Nature of the Atmosphere

The Origin of the Atmosphere We noted earlier in this chapter the uncertain ideas about how the earth might have been formed. However the earth might have originated, the early planet Earth had no atmosphere or a hydrosphere. If it originated at high temperatures, then the gases (which might have formed a primeval atmosphere) would be lost to space. If it started as a cold body, by accretion of cold rock fragments, it would have no atmosphere until gases escaped from the rock (mainly by so-called outgassing from early volcanic activity), and no oceans until water in the atmosphere condensed. Small losses to space continue to occur today but are now mainly restricted to the light gas helium (which must be replenished by outgassing) and some hydrogen.

Modern volcanoes erupt gases into the atmosphere: these including water vapour (70 per cent), carbon dioxide (15 per cent), and nitrogen (about 5 per cent). There is no oxygen, because any that was erupted would instantly react with hydrogen to form water, or with carbon to make carbon dioxide or carbon monoxide. The earth's atmosphere is made very different from those of other planets (and from the gas mixture erupted by volcanoes) by

Table 2.2 Composition of planetary atmospheres

Gas	Venus	Mars	Earth (without life)	Earth (with life)
Carbon dioxide	96.5%	95%	98%	0.03%
Nitrogen	3.5%	2.7%	1.9%	79%
Oxygen	trace	0.13%	0.0%	21%
Surface temperatures	459 °C	−53 °C	240–340 °C	13 °C

Source: Calculations by Lovelock, see C.D. Ollier, in I. Douglas *et al.* (eds) *Companion Encyclopaedia of Geography* (Routledge, London, 1996), Table 1.4, p. 37.

biological activity (see Table 2.2). The critical difference was provided by the evolution of photosynthesis, which allowed carbon dioxide, a major constituent of the early atmosphere, to be converted to oxygen. That story is told further in Chapter 4 on The Biosphere.

Earth's Atmospheric Filter The planet earth is surrounded by a thin but critically important shell of gases – its *atmosphere*. This shell is held to the earth's surface by gravitational attraction and, as we should expect, is densest at the bottom, thinning rapidly as we move upward. The zone of greatest importance to humans is the *troposphere*, which gets its name from the Greek phrase for a 'turbulent realm.' The upper boundary of the troposphere, the *tropopause*, varies seasonally in height but on the average is 9 km (5½ miles) above the poles and 17 km (10½ miles) above the equator. Above the tropopause lie the stratified layers of the *stratosphere*.

The troposphere is so thin, relative to the size of the earth, that it would be barely detectable on the regular classroom globe. Why then is it so important in any analysis of our planet as a home for humans? There are four overriding reasons:

1. It contains the invisible and odourless gas, oxygen (about 20 per cent of its volume), which is essential to human life. Climbers above about 6 km (about 20,000 ft) find that they need additional oxygen as the concentration of this element in the air diminishes. High-flying aircraft (and of course spacecraft) must carry with them their own oxygenated atmospheres.
2. Carbon dioxide, which is critical to plant life, is also present in the atmosphere, although in minute quantities (less than half of 1 per cent). The role of carbon dioxide in plant growth and the food chains dependent on it are discussed in Chapter 4.
3. Water vapour is drawn up from ocean and sea surfaces into the troposphere and is circulated and redistributed over the earth's surface as precipitation. Like oxygen and carbon dioxide, this water vapour is essential to the life and growth of organisms.
4. The gases in the troposphere and the layers above it act as a filter and a blanket. Harmful short-wave radiation from the sun is absorbed and reflected, while long-wave radiation from the earth itself is retained.

A detailed account of what happens to *solar energy* when it strikes the earth's atmospheric shell is given in Figure 2.10. A small amount (6 per cent) is reflected by the atmosphere and a larger amount (27 per cent) by

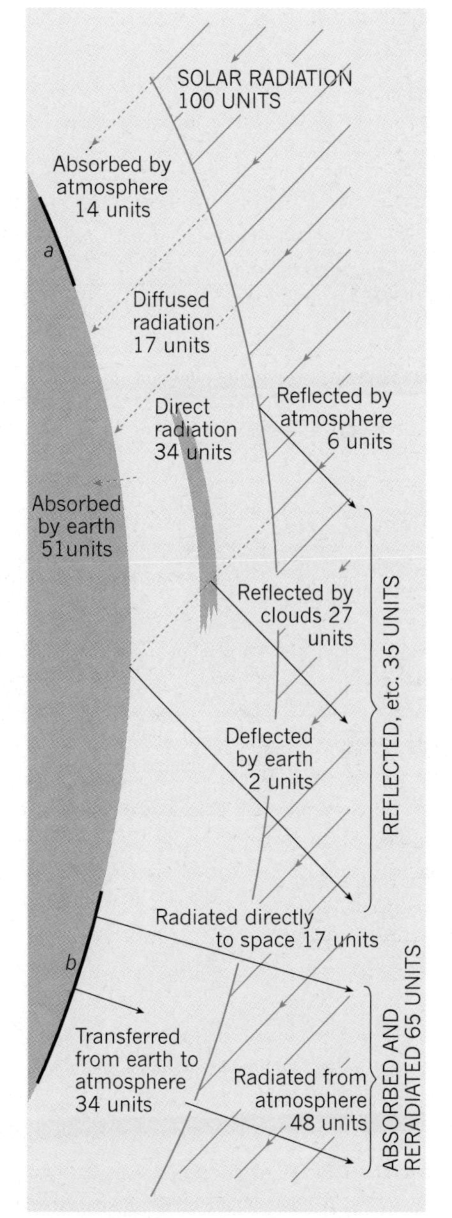

Figure 2.10 The earth's solar radiation budget The diagram shows how 100 units of solar radiation are affected by the thin layer of atmosphere surrounding the planet. (For demonstration purposes, the height of the atmosphere in relation to the earth's curvature is exaggerated.)

the world's cloud cover. The remaining two-thirds is absorbed by the earth and its atmosphere, and reradiated.

If you follow the arrows around the diagram you will find that the amount of energy received (100 units) is exactly balanced by the amount of energy reflected and reradiated. This *global energy balance* is the engine that powers not only the circulation of air and water (ocean currents), but the food chains of which humanity itself is a part.

Factors Shaping World Climate

Latitudinal Factors The variation between the torrid, temperate, and frigid zones recognized by the Greeks (and shown in Figure 2.11) is also related to this cycle of energy. Latitudinal variations in temperature indicate

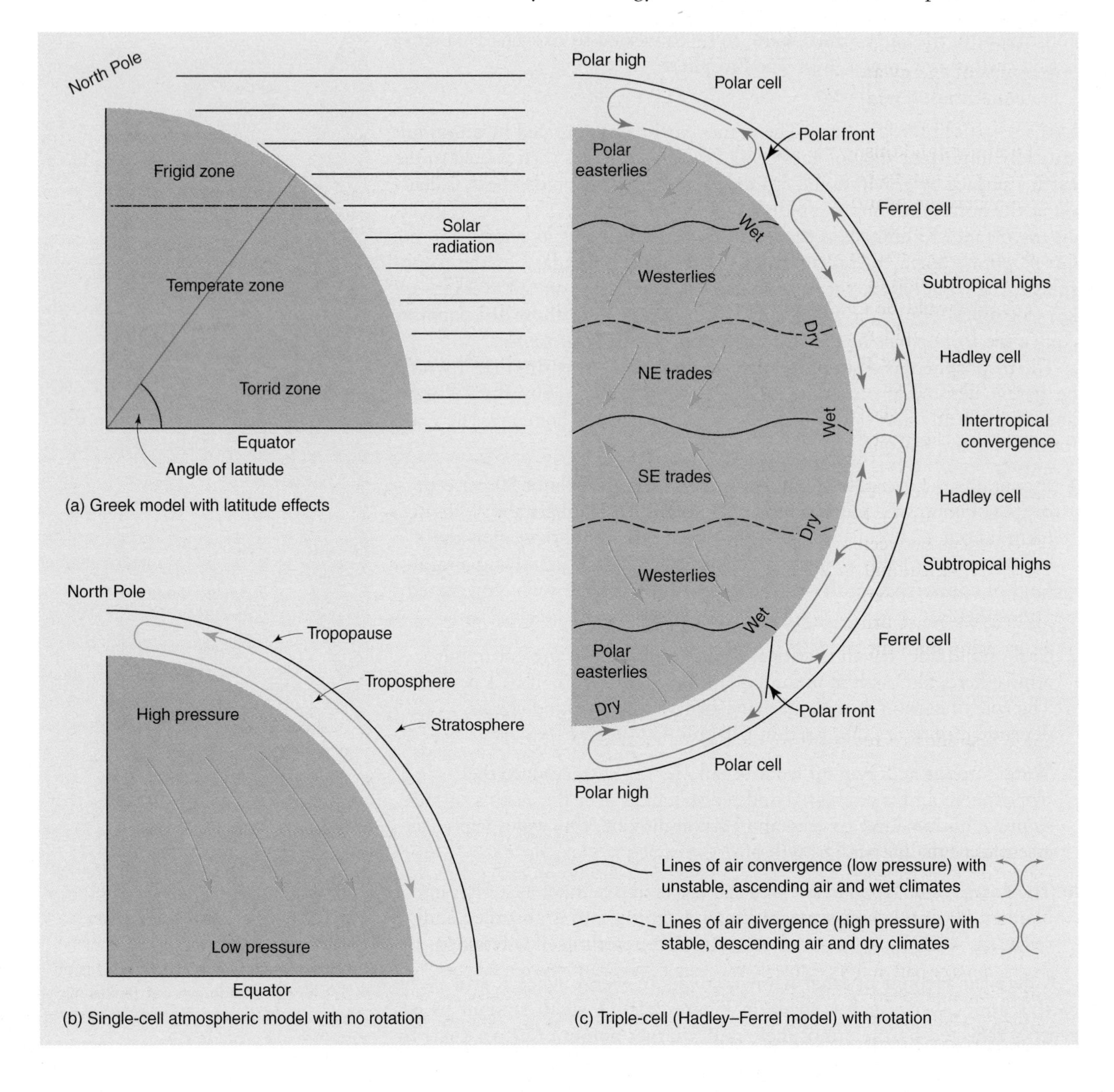

the average angle of incidence of the sun's rays, which is at a maximum in the tropics but decreases toward the North and South Poles. The energy mean from the sun must pass obliquely through the earth's atmosphere and falls on a wider surface area. Look back at Figure 2.10 and compare areas *a* and *b* to see the effect of these combined factors on the amount of energy received in the second area.

Geographers in ancient Greece regarded the climatic environment of any part of the earth's surface as largely a result of its latitude. The Earth was thought to slope (the Greek word *klima*, used in climatology, means slope) *away* from the sun north of the latitude of Greece and the Mediterranean Sea, making the northerly climate increasingly colder. The part of the planet to the south was thought to slope *towards* the sun, eventually becoming a torrid zone too hot for human life.

Figure 2.11(a) is a simplified picture of the Greek world view. The earth lies in a beam of solar radiation whose angle of incidence is directly related to latitude. As we move from the equator towards the poles, equivalent amounts of energy are spread over ever wider areas of the globe, producing an equatorial *torrid* zone, a mid-latitude *temperate* zone, and a high-latitude *frigid* zone.

To understand this low-latitude to high-latitude gradient we must look at the energy relations between the earth and the sun. Each day the earth intercepts a massive beam of solar energy (estimated at 17×10^{13} kilowatts). This energy is emitted on different wavelengths; the peak intensity is the visible (ultraviolet) and longer wave (infrared) emissions. If the earth were like the moon and had no atmosphere, the impact of these latitudinal variations in energy would be catastropic for biotic life. Daytime temperatures near the equator would rise to some hundred of degrees Celsius, while in the winter night at the poles, temperatures would fall almost to the absolute zero of outer space. The actual distribution of warm and cold areas on the earth's surface is shown in Figure 2.12.

Actually, the most extreme shade temperatures ever recorded at the earth's surface (in 1933 and 1960) differ by less than 150 °C (270 °F), and the average temperatures of the hottest and coldest locations on the earth vary by less than 90 °C (160 °F). Moreover, these are extremes (see Table 2.3). The average contrasts between any pairs of locations are much more subdued.

Clearly, latitudinal variations in temperature are dampened and modified in some way. To explain this softening of thermal contrasts on the earth's surface, we must bring a thin coating of air into our model of the earth's relationship with the sun.

General Circulation of the Atmosphere Like any gas or fluid, the atmosphere responds to temperature changes – increasing in density when it is

Figure 2.11 General circulation of the earth's atmosphere Here we see three stages of the historical evolution of concepts of atmospheric circulation and climatic zones. The Hadley–Ferrel model (c) names the three main wind belts of the world – the *polar easterlies*, the *westerlies* of the middle latitudes, and the *trades* of the tropical zone. (The word 'trades' comes from a Latin word indicating that the winds blow in a constant direction, and has nothing to do with trade in the sense of commerce.) Two important lines of air convergence are also shown. The *polar front* marks the junction of the polar easterlies and the westerlies in both the northern and southern hemispheres. Storms marked by low pressure move west along this front, playing a major part in the weather of the United States and Western Europe. The ***intertropical convergence zone (ITCZ)*** is an area of low atmospheric pressure separating the two trade wind systems. It lies more or less along the equator but moves slightly north and south with the seasons. Areas of air divergence with high pressures occur at both poles (the *polar highs*) and again at about 30°N and 30°S (the *subtropical highs).*

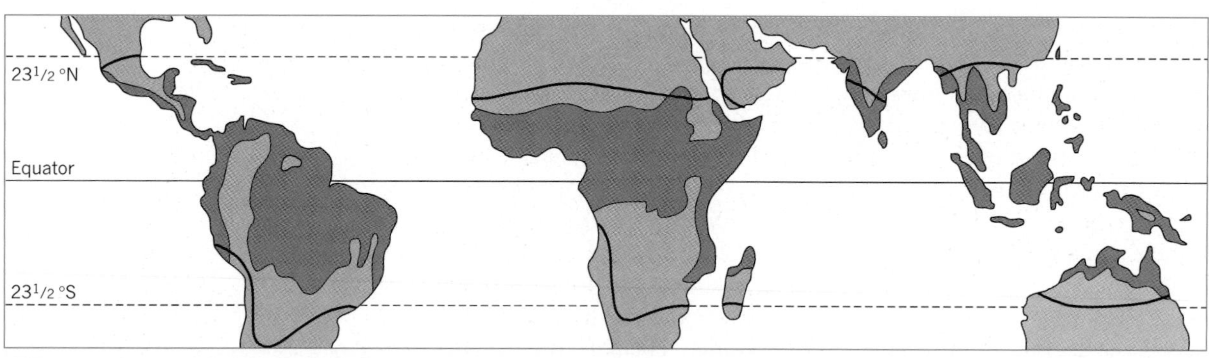

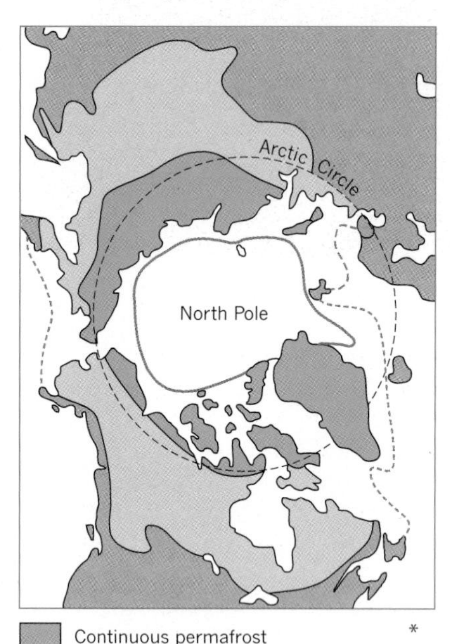

Figure 2.12 Global contrasts in warmth The ***tropics*** are formally defined as the areas of the earth between the Tropic of Cancer (latitude 23½°N) and the Tropic of Capricorn (latitude 23½°S). Thus the tropics lie in those latitudes where the sun is vertically overhead at some time of the year. These limits correspond very roughly with the two measures of warmth shown. At the other extreme the *polar* zones lie north of the Arctic Circle (66½°N) and south of the Arctic Circle (66½°S). Two thermal measures of the Arctic polar zone, ***permafrost*** (subsoil that remains frozen all the year round) and *pack ice* (large areas of ice covering the entire sea surface) are shown. Note that these extend south of the Arctic Circle in Siberia but lie well to the north over the sea areas north of Scandinavia.

[Source: R. Common, in G.H. Dury (ed.), *Essays in Geomorphology* (Heinemann, London, 1966), Fig. 7, p. 68.]

cooled, decreasing in density when it is warmed. As we might expect, therefore, the warmer (and lighter) air near the equator rises and flows poleward, to be replaced by cold (and dense) air near the poles moving equatorward along the surface. These two flows together make up a *convective circuit*, illustrated in Figure 2.11(b). The unequal heating of the air by the sun sets up compensating currents that act like escalators, redistributing heat across various latitudes, and tempering the simple Greek pattern of hot and cold zones.

As early as 1686, English astronomers had outlined a circulational model to explain why the lower latitudes, despite receiving more heat, did not become progressively hotter. By incorporating the effects of the earth's rotation into the model, it was possible to give a reasonable explanation of the ***trade winds***, those steady air currents blowing toward the equator from the northeast in the northern hemisphere and from the southeast in the southern hemisphere.

Table 2.3 Global extremes of temperature

Thermal characteristic	Temperature °C	Temperature °F	Location
Individual readings			
Highest	58.0	136.4	San Luis Potosí, Mexico (1933)
Lowest	−88.3	−126.9	Vostok, Antarctica (1960)
Range of readings			
Widest	88.9	160.0	Verkhoyansk, Siberia
Narrowest	13.4	24.1	Fernando de Noronha, South Atlantic
Annual averages			
Hottest	31.1	88.0	Lugh Ganane, Somalia
Coldest	−57.8	−72.0	78°S, 96°E, Antarctica

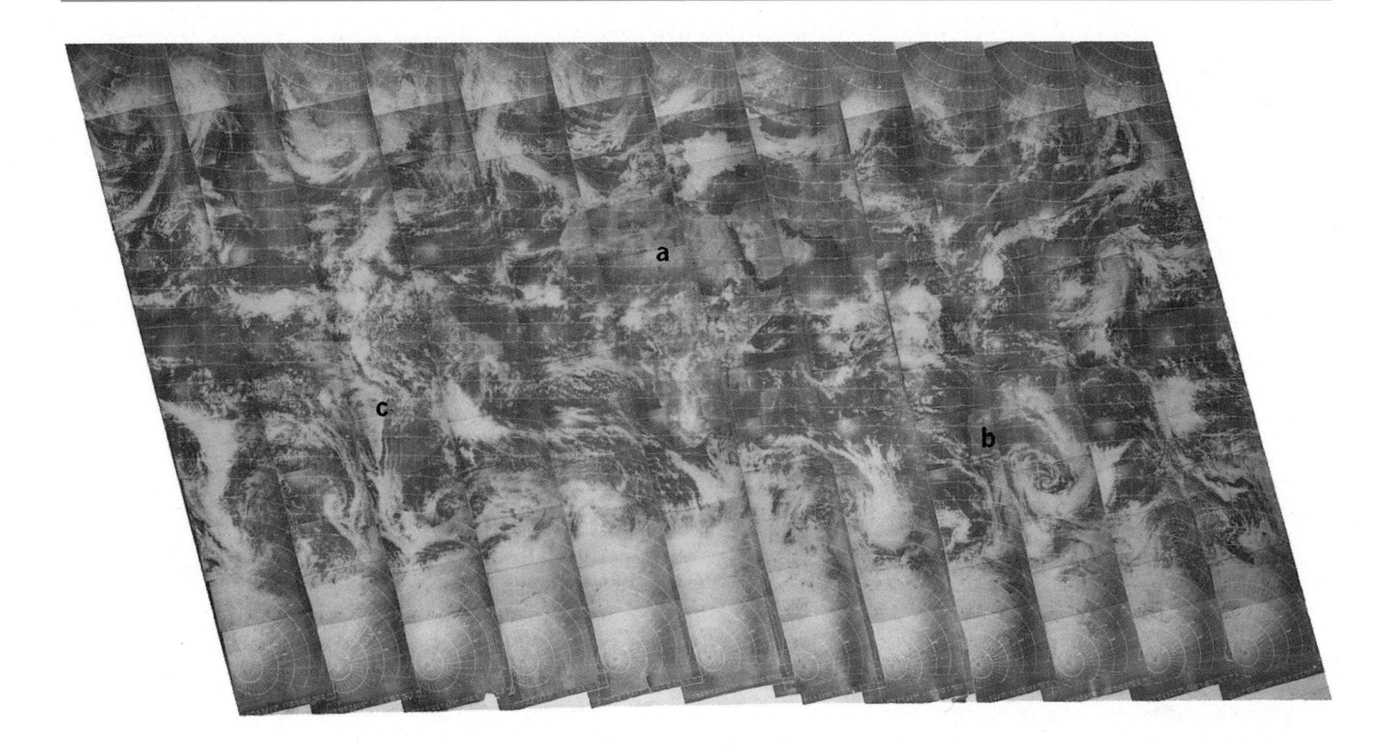

Figure 2.13 Earth's cloud cover Major climatic zones are suggested in this picture of the entire earth's cloud cover on a single day (13 February 1965). It is made up from 450 photographs received from the weather satellite Tiros 9 on its pole-to-pole orbit of the earth. Note the relative absence of cloud cover from the desert areas of North Africa and Arabia (a), central Australia (b), and the middle section of the west coast of South America (c).

[Source: NASA photograph.]

These early models were far from satisfactory, however. If we look at the motions of the earth's atmosphere as revealed by telltale cloud patterns in satellite photos (Figure 2.13), we can see two important features left out. First, there are extensive cloudless areas of very dry descending air at latitudes of about 30°N and 30°S (above the Sahara Desert, for example). Second, there are belts of strong westerly winds in the middle latitudes of both hemispheres. The westerly drift of clouds across the North Atlantic illustrates this circulatory force.

The missing elements in the explanation of atmospheric movements were provided by English scientist George Hadley in 1735 and American meteorologist William Ferrel in the following century. Their model (Figure 2.11(c)) replaced the single pole-to-equator convective circuit by a series of *three* convective circuits in each hemisphere. The *Hadley–Ferrel model* (*Hadley cell*) – although much modified in detail – still forms the basis of modern concepts of atmospheric circulation. It not only accounts for the puzzling feature of the *westerlies* (now seen as caused by the westward rotational deflection of poleward-moving air), but throws light on global patterns of precipitation. Each of the main belts of winds is defined in the caption of Figure 2.12.

Global Precipitation Patterns Though a number of factors affect precipitation, its general cause is the cooling of air that contains water vapour. Since water vapour is present throughout the troposphere, it is the location of cooling and drying processes at the global level that explains the wet and dry areas of the earth. Such processes are indicated in the Hadley–Ferrel model by areas of upward and downward motion in the general atmosphere. As Figure 2.14 shows, there are three main zones of upward motion: (a) an *intertropical convergence zone*, where the two trade-wind systems meet, and (b) two *polar front* zones at around 60°N and 60°S, where warm tropical westerly air overrides the colder air flowing outward from the two poles.

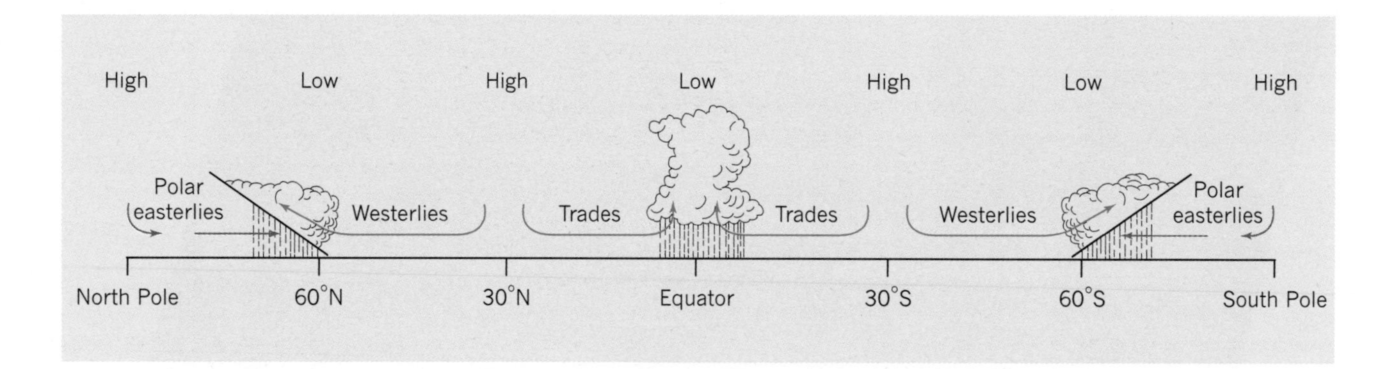

Figure 2.14 Circulation and precipitation patterns A simplified cross-section of the earth's atmosphere shows the general pattern of circulation. Note the association of the three zones of upward air motion (low pressures) with belts of rain. The four high-pressure areas are marked by descending air motion and very low precipitation. The sharp discontinuities at latitudes of 60°N and S mark the polar fronts.

A map of world precipitation (Figure 2.15) confirms that these are precisely the locations where high values are oberved.

Using the reverse argument, we should expect zones of low precipitation in areas of downward-moving air. As Figure 2.14 shows, such zones coincide with the two poles and with two zones of divergence at 30°N and 30°S; again, we find that these are zones of low precipitation on the world map. Note on Figure 2.15 the location of the world's main desert zones in relation to the Hadley–Ferrel model.

Smaller-scale Determinants of Climate Temperatures vary more on land than they do over the sea. But less than a third of the earth's surface is land, and this land is distributed in an asymmetric way in relation to latitude. Most of it is in the northern hemisphere. The climatic implications of the division of the planet into areas of land and water are dramatic. The thermal conductivity of the surface layers of land and sea is quite different. Land heats and cools much more quickly than the sea; bodies of water act as heat stores whose temperatures fluctuate much less than those of adjacent bodies of land.

Thus, there are considerable global variations in the seasonal range of temperatures. The *average* annual range is lowest at the equator and increases with latitude, from about 3 °C (5 °F) to 60 °C (110 °F) at the South Pole. The smallest ranges are in oceanic islands near the equator; the largest ranges are in mid-continental locations in high latitudes. On Saipan in the Mariana Islands of the western Pacific, the highest and lowest temperatures ever recorded, the *extreme* range, differ by only 12 °C (22 °F). By contrast, Olekminsk in central Siberia has an extreme range ten times longer, from −60 °C (−76 °F) to 45 °C (113 °F). The conventional distinction between continents and smaller land masses is an arbitrary one. All can be regarded as 'islands' of different sizes, and within each an increased range is detectable, though it weakens as the area becomes smaller.

A broad distinction can be drawn between the climates of continental interiors and areas near the sea. *Continental climates* are characterized by great ranges of temperature (both between day and night and between winter and summer), low humidities, and very variable precipitation. This variability in precipitation shows as a strong seasonal contrast, but also in year-to-year irregularity. *Maritime* climates have the reverse characteristics: smaller temperature ranges, higher humidities, and more uniform precipitation. These contrasts are not symmetrical on a continent, but are related to the latitudinal location of the land area with respect to the circulation pattern of both the atmosphere and the oceans. We can combine the effects of latitude and continentality and superimpose them on an idealized

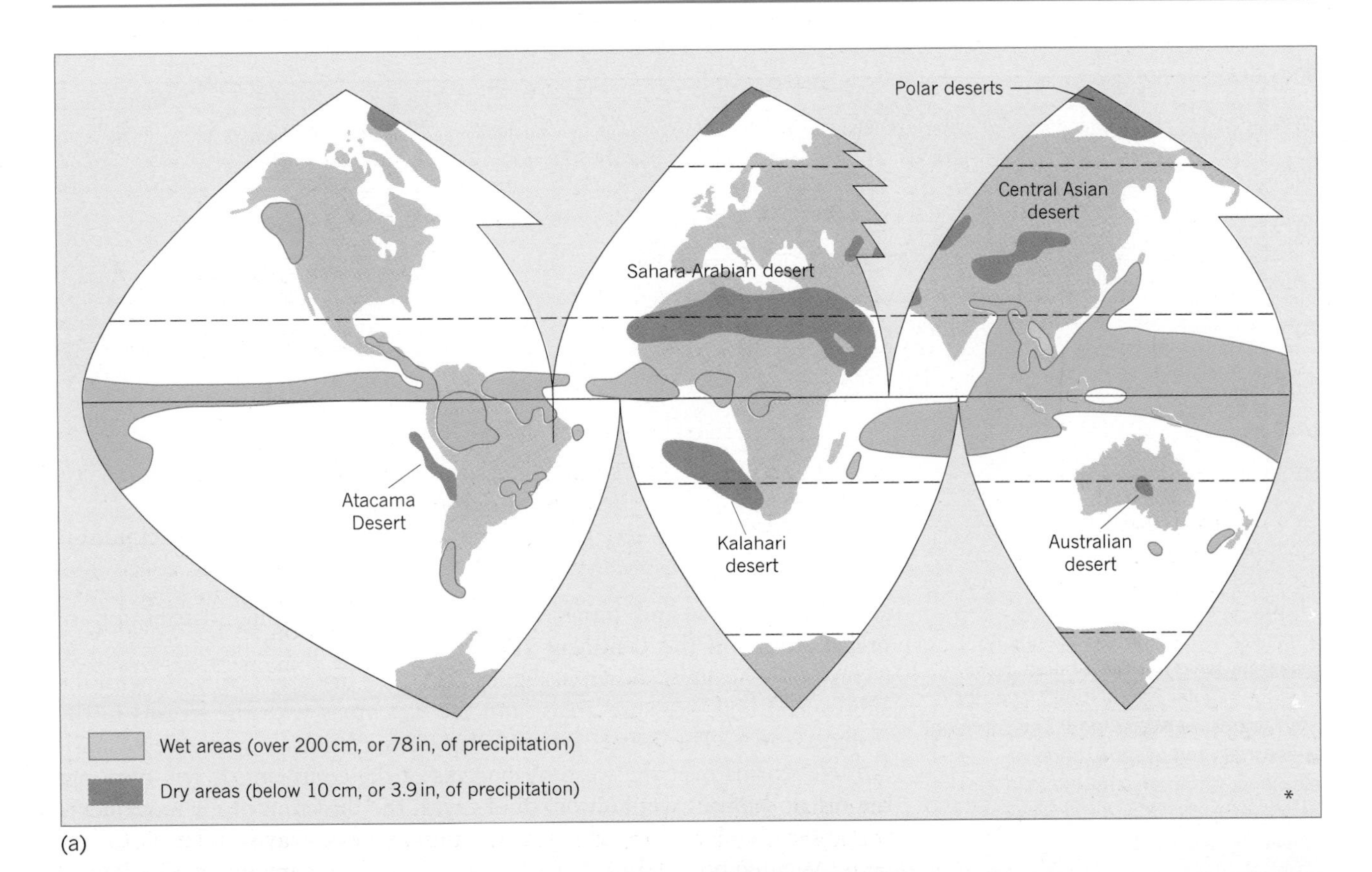

(a)

(b)

(c)

Figure 2.15 The world's wettest and driest areas
(a) The world's wettest areas straddle the equator and the windward margins of continents. The location of the driest areas is related to tropical high-pressure cells. (b) Desert dunes: shadows on the Coral Dunes, Colorado, USA. (c) Sahara desert: satellite image of the Great Erg desert on the Libya/Algeria border.

[Source: (b) FLPA/Dembinsky. (c) Earth Satellite Corporation/Science Photo Library.]

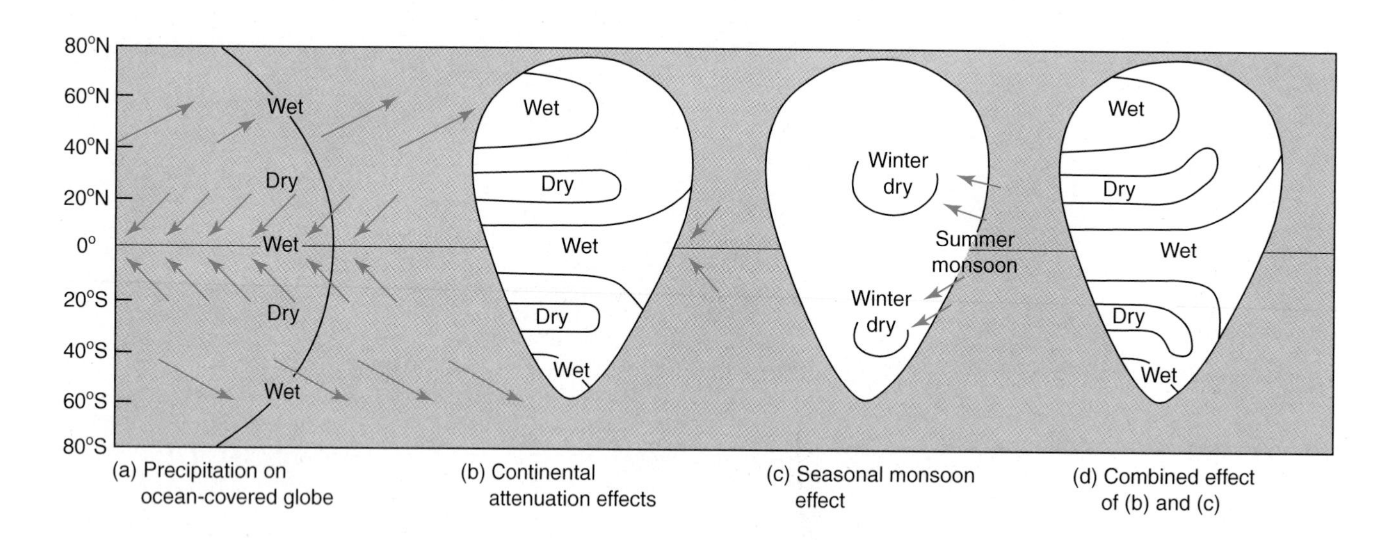

Figure 2.16 Continentality The impact of a hypothetical continent on global patterns of precipitation. Note how the regular bands of wet and dry are progressively distorted. The continent is assumed to have low, uniform elevation. Compare with the actual distribution of wet and dry areas as shown in Figure 2.15.

continent of low and uniform elevation. The resulting distribution of precipitation on the continent (Figure 2.16(b)) should be interpreted in terms of the airflows shown in Figure 2.11. The dry areas coincide with the subtropical high-pressure areas, and the wet areas with the storms in the tropical zones. The westerly circulation of air brings a sequence of mid-latitude storms over the western margins of the continent. If you compare the distribution of wet and dry zones with the direction of air movements worldwide, you will see why precipitation decreases away from the continent's oceanic boundary.

Because of the great variations in air temperature and pressure over continents, we must add a further seasonal effect to our model of precipitation patterns (Figure 2.16(c)). The differential heating of the continental air masses causes maritime air to be drawn inward in summer to replace the warm, light, and rising air over the continents. In winter, the colder, heavier, and descending air above the world's land masses flows outward toward the seas. The impact of these flows on the world's largest land mass, where the most striking seasonal reversals occur, is discussed in the next chapter. (See Section 3.3 on monsoon India.)

Mountains and Plains Within the overall climatic framework formed by the effects of latitude and continentality, we can detect a hierarchy of smaller-scale factors. The geographic scale of their impact is minor, but they may play a decisive part in determining the use that can be made of an area.

We have seen that the earth is almost a perfect sphere and that variations in its surface elevation are equal to only a small fraction of its radius. Land above sea level ranges up to 8.9 km (5.53 miles) in height, but over two-thirds of it is at elevations below 1 km (.62 miles), and less than one-tenth is above 2 km (1.24 miles). The city of Denver, the highest located major city in the United States, is at an elevation of 1.6 km, or almost 1 mile.

The direct effect of these relatively small differences in elevation on the characteristics of lowland and highland environments is striking. At an altitude of 8 km (roughly 5 miles), the density of the atmosphere is less than one-half its density at sea level. High elevations have a thinner shell of atmosphere above them and receive considerably more direct solar radi-

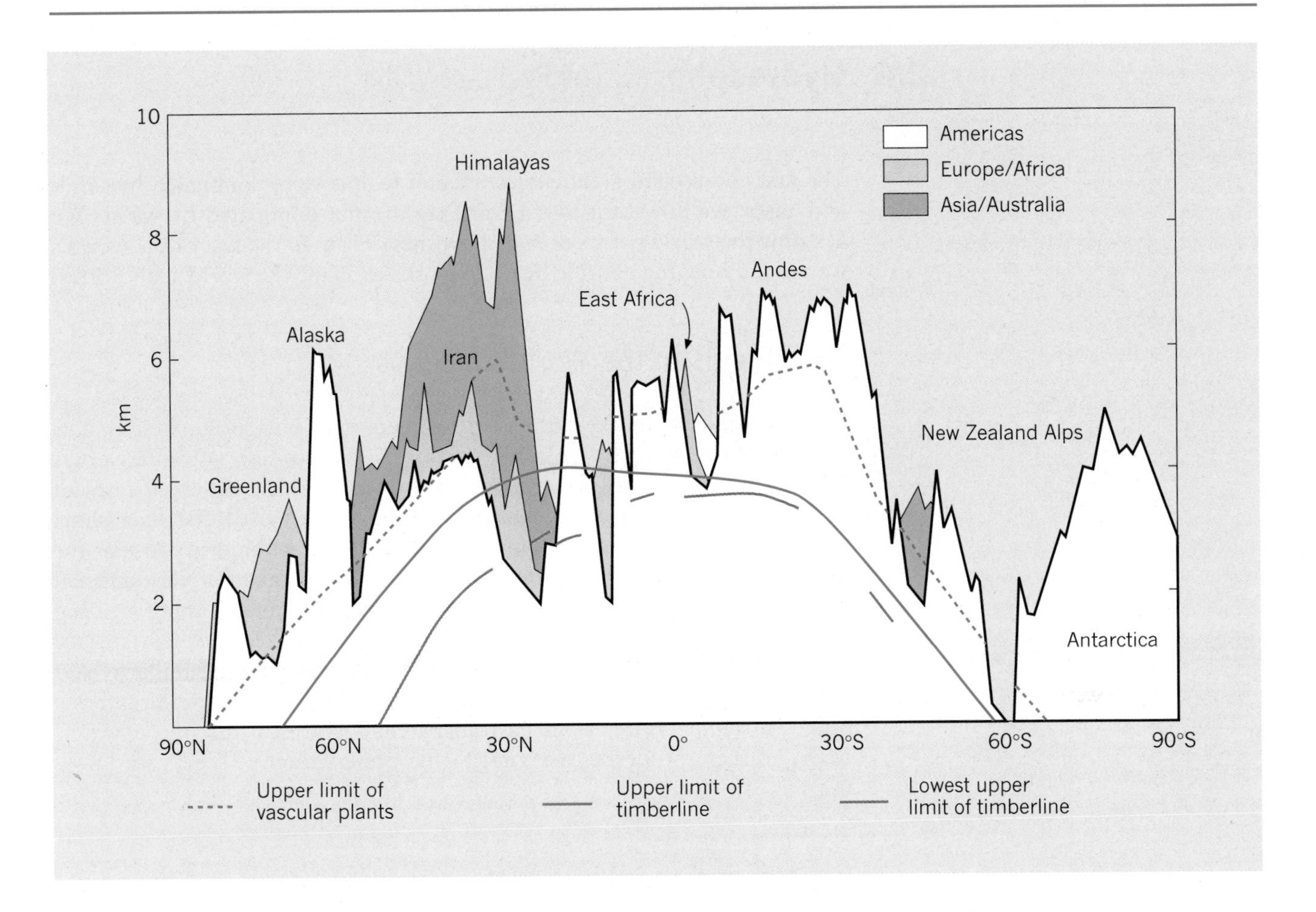

Figure 2.17 World timberline variations The altitude at which plants will grow in mountain areas changes at different latitudes, as also does the timberline. This pole to pole cross-section is drawn for the Americas (shown in white in the foreground). Peaks for the other continents are shown only when they exceed the height of the American ranges.

[Source: adapted from L.W. Swan, in W.H. Osburn and H.E. Wright, Jr. (eds), *Arctic and Alpine Environments* (Indiana University Press, Bloomington, Ind., 1968), Fig. 1, p. 32.]

ation than sea-level locations, but they lose much more heat by radiation from the ground surface. Within the lower layers of the atmosphere, temperature decreases with elevation at an average rate of about 6–4 °C per kilometre. This rate of loss in temperature with height is termed the *thermal lapse rate*. It is generally the same throughout the troposphere.

The effects of elevation on temperature are twofold: as elevation increases, the average temperature of an area decreases, and the daily range in temperature increases. Both effects are due to clearer, more rarified air, which allows more solar radiation to reach the ground surface (raising the midday temperature) and also permits a more rapid heat loss from the ground at night. The net effect is to reproduce the changes in temperature we may encounter as we move from one latitude to another over a short vertical distance. If we live on the equator and wish to see some snow, we can either travel 8000 km (5000 miles) poleward to find snowbanks at sea level or climb 4.5 km (15,000 ft) up!

The symmetry between changes in climate due to latitude and changes due to elevation is shown in Figure 2.17. Because snowlines change with the seasons, the upper limits of plant growth have been used as a substitute measure of climatic differences.

Timberlines reach their greatest sea-level extent at a latitude of about 72° in the northern hemisphere and 56° in the southern hemisphere. They are at higher altitudes as one moves from these latitudes toward the tropics. A mean temperature of 10 °C (50 °F) for the warmest part of the year appears to be a prominent factor limiting forest growth, and this limit is met at varying elevations in different parts of the world.

2.4 Hydrosphere: The Realm of Water

The first and second great environmental realms were dominated by solids and gases: we now turn to the third realm, that dominated by water. We ask how the world got its oceans, how these relate to the rest of the world's water, and how the hydrologic cycle works to shape the world's landforms.

The Origins of the World's Oceans

Like the atmosphere, the earth's hydrosphere must have originated by outgassing from rocks, by volcanic eruptions. Present-day volcanoes erupt volatiles which are 70 per cent water. At the currently observed rates of production, all the world's water could have been provided in a short geological time (as little as 55 million years). Water's ubiquity conceals the fact that it is a remarkable substance with a whole range of very unusual properties. The fact that fresh water has a maximum density above freezing point effects the icing and vertical circulation in lakes. Its high surface tension (highest of all liquids except mercury) is important in capillarity and in the physiology of plant cells. Its very high heat capacity prevents extreme ranges in temperature. Even its transparency is critical for life in water. Many of these properties are vital to the preservation of life as we know it.

Oceans and the World Water Cycle

Water vapour was noted earlier as one of the critical components of the earth's atmosphere. Water as a gas or small droplets in the air forms only 1 part in 100 thousand of the planet's overall water resources. Over 97 per cent is concentrated in the great water sheets of the world's oceans. Not only do the oceans cover over 70 per cent of the earth surface, but their lowest point (Marianas Trench in the Western Pacific, 11.03 km, or 6.85 miles deep) greatly exceeds the highest point on the land surface (Mount Everest in the Asian Himalayas, 8.85 km, or 5.5 miles high). Perhaps the simplest way to remember the oceans' size is to recall that if the earth had a smooth surface it would be covered everywhere, to a depth of about 3 km (1.86 miles), by water.

Water vapour is carried and distributed over the earth's sea and land surface by the global wind belts described in the Hadley–Ferrel model of the atmosphere. Each year an estimated 33,600 km^3 (8100 $miles^3$) of water evaporates from the ocean surface. Figure 2.18 shows what happens to that water.

About 89 per cent of it is returned directly to the ocean by precipitation. The remaining 11 per cent moves over the earth's land surfaces before precipitating out. Precipitation falling on land may be either returned directly to the atmosphere by evaporation and transpiration, or temporarily stored (in lakes, ice caps, in the upper soil layers or more deeply as groundwater). Box 2.B on 'Lake Baikal' describes one of the earth's great freshwater stores. Eventually all the water is returned to the oceans by flowing streams and melting glaciers. Hence, the balance between the moisture leaving the oceans as water vapour and returning as a liquid is maintained.

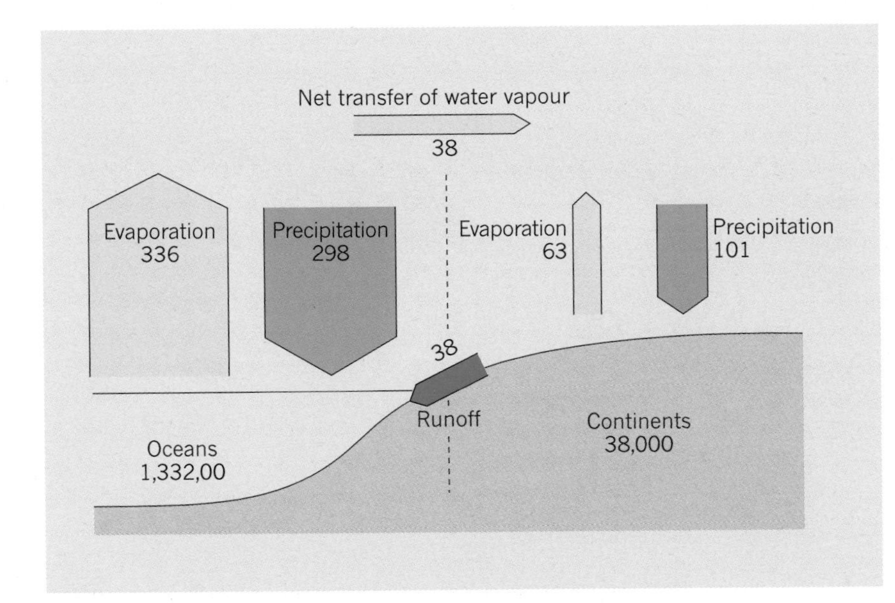

Figure 2.18 The world water balance The diagram shows major water flows between the oceans and the continents. The figures indicate flows in 100 tonne^3 of water.

[Source: A.N. Strahler, *The Earth Sciences*, 2nd edn (Harper & Row, New York, 1971), Fig. 33.2, p. 586.]

This global circulation of water is termed the *global hydrologic cycle*. Figure 2.19 shows the earth's main stores of water, as well as the direction of transfers. Outside the world's oceans most of the water is locked up in glaciers and ice caps. Only 0.001 per cent of the total is in the atmosphere. To understand the varying pattern of precipitation around the world, we must understand something of how that small atmospheric fraction is shunted from one location to another by the general circulation of the atmosphere.

This is made up of the oceans (which contain 97 per cent of all the world's water) plus five other stores or regulators – the atmosphere, the soil, the rocks, glaciers, and lakes and rivers. Note how the water circulates as inputs and outputs from one store to another, sometimes moving as a liquid and sometimes as water particles in a gas.

Box 2.B Lake Baikal

Set in the heart of the Asian land mass in Siberia, Lake Baikal is the world's oldest and deepest freshwater lake. It was formed over 25 million years ago within a rift valley and has a depth of 1600 m (5300 ft). It contains a water volume of some 23,000 km^3 (5500 cubic miles), equivalent to one-fifth of all the world's fresh water and four-fifths of Russia's fresh water.

Because of its central continental location, the climate is very severe. Temperatures remain below freezing in the surrounding mountains for up to two-thirds of the year and for five months of the year the lake is covered by a 2.5 m (8 ft) layer of ice. Despite this many of the hundreds of rivers that flow into the lake are warmed by thermal springs and remain unfrozen.

Because of its longevity and isolation, the lake hosts a wide range of animal and plant species with many hundreds of species not found anywhere else in the world. Despite the generally low population density around the lake, there are pockets of development linked to pulp-and-paper works and chemical and petrochemical plants. Tourism has also brought intense pressures in limited areas. Alarm signs from declining fish yields brought a vigorous response from the former Soviet government in the 1980s and specially protected zones have been set up around the lakeshore to protect a unique global resource. It has long been revered by Buddhists as a holy sea and a spiritual entity.

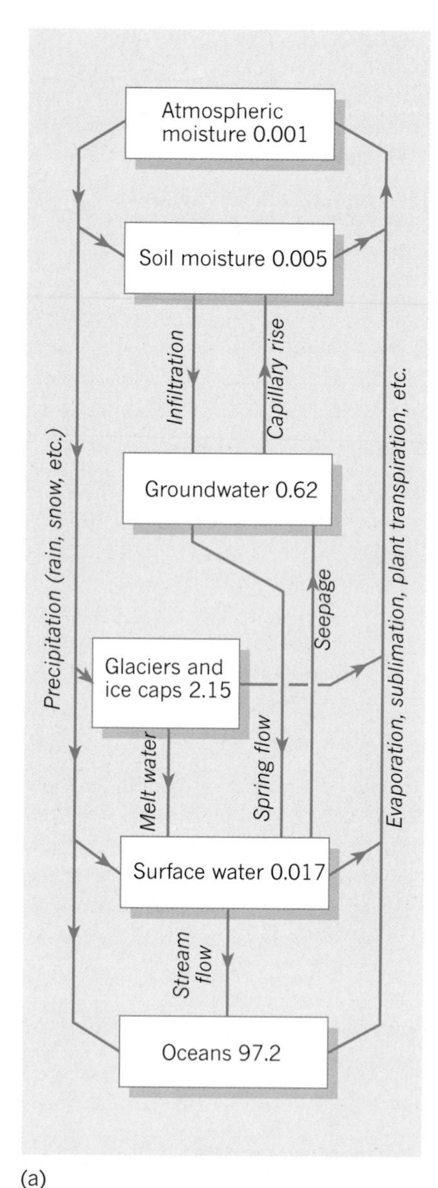

(a)

(b)

Figure 2.19 Earth's hydrologic cycle (a) The diagram shows the main components in the global hydrologic cycle. The figures refer to the percentage of all terrestrial water in each of the six main stores. (b) Water as solid and liquid stores. Summer sea beyond the Greenland ice-cap with icebergs calved from Glacier Cape York in northwest Greenland.

[Source: (a) Modified from R.J. More, in R.J. Chorley and P. Haggett (eds), *Models in Geography* (Methuen, London, 1967), p. 146, Fig. 5.1. (b) Bryan & Cherry Alexander.]

Water and Continental Erosion

We noted above how millions of km^3 of water evaporate from the oceans, move over the continents as water vapour, precipitate out, and return to the seas as rivers flow and ice melts. (Refer to Figure 2.18 and Table 2.4 for the details of this transfer.) But the rivers do not return unloaded to the sea. On the average, the Mississippi River brings back 1 tonne of sediment in every 1200 tonnes of water; in floods this figure may rise to 1 tonne in 400 tonnes of water. We can see one dramatic result of this process in the Mississippi delta, which looks like a growing bird's foot (see Figure 2.20).

Much less dramatic, but equally inexorable, is the general wearing away of the land surface within the area drained by the river (i.e. within its *catchment*, or *watershed*). In one person's lifetime the effect is miniscule, perhaps 3.6 mm is worn away in a 70-year span. But over a million years this adds up to an average lowering of the land of over 51 m (167 ft).

We should be careful not to extend our calculations too far. The slow reduction of the continental surfaces by this return flow of the hydrologic cycle does not necessarily mean that the elevation of the continents is

Table 2.4 Transfers of mineral matter from continents to oceans

Transfers	Million tonnes per year
Eroded from continents	
By streams	9.3
By wind	0.06 to 0.36
By glaciers	0.1
Total	**0.46 to 9.76**
Deposited in oceans	
Shallow waters (less than 3 km deep)	5 to 10
Deep waters (more than 3 km deep)	1.2
Total	**6.2 to 11.2**

Source: Data from S. Judson, *American Scientist*, **56**, No. 4 (1968), p. 371.

(a)

(b)

Figure 2.20 Erosional and sedimentational environments
(a) Erosion: the Grand Canyon of the Colorado River, Arizona, USA. North is at the top. In the infrared range green vegetation is a very dark shade, bare rocks appear as shades of grey and water is black. Images of the upper reaches of the Canyon with the Kaibab Plateau at upper left and the Painted Desert at right. The Canyon was formed during the last six million years due to erosion of the Colorado River which flows from to centre to lower left. The Canyon is 6–29 km (4–18 miles) wide and 1.6 km (1 mile) deep, exposing many layers of faulted sedimentary rock. Described by explorer John Wesley Powell as 'the most sublime spectacle on the earth', the Grand Canyon is still at an early stage in the Davisian cycle (see Box 2.C). Image taken by the French SPOT.3 satellite in June. (b) Sediment deposition: the Mississippi Delta. Shuttle *Challenger* view of the Mississippi delta on 6 November 1985. North is at the top of the photo with an east–west distance of around 80 km (50 miles). The detailed form of the delta shows lobes being formed at the mouth of the several distributaries to give a 'bird's foot' form.

[Source: (a) CNES/Science Photo Library 1996 Distribution Spot Image. (b) NASA/Science Photo Library.]

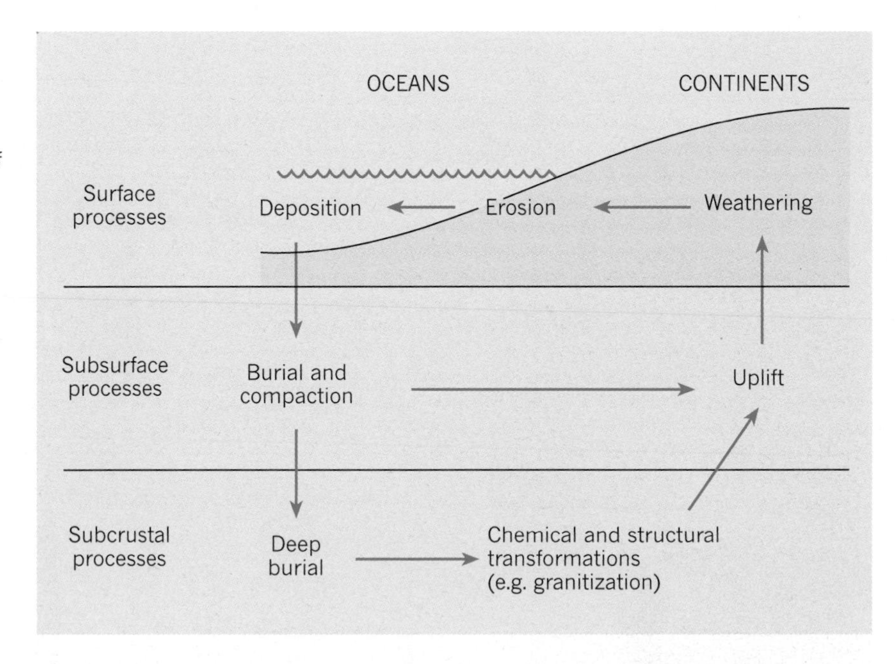

Figure 2.21 Erosion cycles This idealized cross-section shows major elements in the cycle of erosion, sedimentation, and uplift. Geographers are concerned with the surface phases of the cycle; geologists and geophysicists are mainly interested in the subsurface phases.

decreasing. There are two reasons for this. To understand the first, we must look more carefully at the earth's crust. As we noted earlier in this chapter, this crust is made up of two layers: an upper layer of lighter granite-like rocks with a heavier layer of basaltic rocks extending under the continents and the ocean basins. Thus, the lighter continents appear to float on the heavier rocks of the earth's mantle. Slow reductions in the continental mass by erosion are compensated for by upward movements of the land surface. But the balance is not achieved instantaneously, and the stresses caused by these adjustments are among the many reasons for earth tremors and earthquakes.

A second reason the elevation of the continents tends to remain the same is that the erosional forces set in motion by the hydrologic cycle are self-limiting. High regions are more easily eroded than lower ones, so as the elevation of the land is reduced the rate of lowering also declines. The interactions among the processes of erosion, sedimentation, and compensating uplift can be seen as part of yet another basic environmental cycle (Figure 2.21).

Physical geographers have been particularly interested in the erosion phases of the cycle. The Davisian school, named after the American geomorphologist, W.M. Davis, stressed the role of declining rates of erosion over time and developed a series of *geomorphic cycles*, each related to the shaping of land masses under different conditions. (See the discussion in Box 2.C on the 'The *Davisian cycle*'.)

Box 2.C The Davisian Cycle

William Morris Davis (born Philadelphia, Pennsylvania, USA, 1850; died Pasadena, California, USA, 1934) was probably the world's most influential physical geographer. In his lifetime he published over 600 papers and books which contained his evolving ideas about the way the world's terrain had developed. His voluminous writings included *The Rivers and Valleys of Pennsylvania* (1889), *Physical Geography* (1898), *Geographical Essays* (1909), and *The Coral Reef Problem* (1928). He spent most of his academic life at Harvard but travelled extensively both within North America and the other continents. He continued very active work for twenty years after his retirement in 1912.

The Davisian model of the way in which landforms change can be simply stated as an equation:

Landforms = function (Structure + Process + Stage)

By *landforms* Davis meant the physical shape of the earth's surface terrain; *structure* was the geological composition of the land, and its original elevation above sea level as the result of mountain-building forces. By *process* he covered all the decay and removal processes – such as chemical rotting, river erosion, downhill shuffling of material – by which the earth's surface was slowly reduced in height. *Stage*, the most important element in the Davisian equation, recognized that landforms evolve over time.

Davis found it was useful to divide the time period into three distinct and recognizable stages – youth, maturity, and old age. Together these formed a life cycle of landforms, generally known as the Davisian cycle. (See figure.)

Evolution of landforms under the action of running water finding its way back to the sea as part of the hydrologic cycle was regarded by Davis as the norm. It formed a standard cycle which could then be modified by climatic change (what Davis called *climatic accidents*). For instance, the sequence of landform stages under a very dry climate would give more importance to windblown material movements. The arid cycle also had to recognize that streams from occasional storms did not reach the sea but petered out in debris fans or salt lakes. In contrast, under glacial conditions ice takes the place of water as the major erosive element in the cycle. Second, the normal cycle could be modified by changes in the relative position of sea level. This could be due to an absolute drop or rise in its level as happened during the main ice ages when more or less volumes of ocean water were locked up as ice. It could also be caused by shifts due to earth-building movements. These changes in elevation were termed by Davis *interruptions* to the cycle. Such interruptions meant that a landscape that had reached an old-age stage could be rejuvenated and go back to a youthful stage.

William Morris Davis travelled very widely, wrote profusely, and fitted ever more complex landforms into his scheme. Although modern work has given different interpretations to some of his findings, notably by emphasizing the greater mobility of the earth's crust, Davis's contribution to understanding the evolution of the earth's surface form remains monumental. He was the first geographer to develop a cohesive scientific theory that allowed the natural world of terrain to be seen as an evolving living landscape. Throughout his life Davis, a geologist by training, stressed the importance of the physical earth in geographic studies. He played a leading part in the development of the Association of American Geographers (founded in 1905) and has left a permanent mark on the form and function of American geography.

For a sympathetic but critical biography of Davis with fascinating insights into both his academic and his private life see *The Life and Work of William Morris Davis* by Richard Chorley, R.P. Beckinsale, and A.J. Dunn (Methuen, London, 1973).

[Source: Davis's photo is from p. 438 of Chorley *et al.*; the diagram from W.M. Davis, *Physical Geography* (Ginn, Boston, 1898), Fig. 152.]

Most erosion was found to be due to the action of water rather than glacial action or wind (refer back to Table 2.4). Although these processes are, in the main, very slow, they can be punctuated by rapid changes that severely affect us. The shifting of the Chinese Hwang Ho River from an outlet north of Shantung in the years 1192 and 1938, and its reversal in 1852, resulted not only in many deaths but in long-term changes in patterns of settlement and land use. Similarly, compensating movements of the earth's crust may cause dramatic surface effects such as earthquakes and volcanic activity.

2.5 The Human Impact of Planetary Change

For most of the earth's long history, the three great cycles described in this chapter have run their complicated courses without human observation. Only in the last few seconds of our planet's history has human population been significant (see the discussion of human origins in Chapter 5). But today our huge population is affected by environmental change. We give here only some examples from lithosphere changes: climatic impacts are discussed in Chapter 3 and the impact of global warming in Chapter 19.

Impact of 'Normal' Tectonic Events

In terms of environmental hazards, a basic distinction can be drawn between changes such as rising sea levels that come on over centuries and those in which the speed of onset is very fast. There are a number of climatic conditions – blizzards, hurricanes, floods – that are felt by the human population within hours. Similarly there may be events within the slow evolution of the lithosphere or the oceans that are of short duration but of high impact (e.g. earthquakes, volcanic eruptions, avalanches, or tidal surges). These sudden changes in the environment form a class of *extreme* geophysical events.

Major environmental changes that affect human life are illustrated by earthquakes and volcanoes. Those produced by earthquakes probably stem from volcanic activity. The largest such explosion in historic times was probably the Krakatau explosion of 1883, which blew away two-thirds of an island and triggered a tidal wave estimated at 45 m in height (almost 150 ft) that broke upon the adjacent coast of Java with great destructive force. However, volcanic activity can also have beneficial effects. Slow accumulations of volcanic lava and dust may lead to the creation of entirely new land areas. The Hawaiian Islands, for example, were formed in this way. And while some ejected materials remain rocklike and sterile for centuries, other decompose into exceptionally fertile soils. Parts of Java, the Japanese island of Kyushu, and south India typify volcanic areas with fertile soil structures that support high populations.

We have noted earlier the St. Helens eruption (recall Figure 2.9). In human terms the eruption led to the loss of 33 lives killed by blast or trapped by ensuing mudflows. One was a US Geological Society (USGS) geologist observing the eruption from a post 8 km (5 miles) away. In economic terms the loss is estimated to be well above $2 billion.

If we measure earthquake tremors by the area over which their effects were felt, then the largest in recent times was probably the Assam earthquake of 1897, which affected an area of 4.2 million km^2 (over 1.6 million square miles, half the continental United States). The largest earthquake in the conterminous United States was the San Francisco earthquake of 1906. The Tangshan quake in northern China in 1976 may well have killed 400,000 people. In addition to the loss of life, we must add damage by fire and destruction of buildings caused by such events. There are also longer-term environmental changes resulting from the displacement of sediments and the redirection of river courses that may cause secondary hazards. Two recent examples of earthquake activity are shown in Figure 2.22.

Potential Impact of 'Super' Events

Earth scientists measure the magnitude of volanoes on an eight-point scale. As with the Richter scale for earthquakes, each point on the scale is ten times larger than the next point (so that point 8.0 on the scale is not twice as large as point 4.0, but 10^4 or 10,000 times as large). As a general rule (Figure 2.23(a)) very small eruptions or earthquakes are rather common, very large events are very rare.

To give some idea of the volcanic scale, the Mt. St. Helens eruption (1980) described earlier in this chapter measured point 5.0. This was one tenth smaller than Mt. Pinatubo, Philippines (1991). Another point 6 eruption was Santarini, a Greek island, some 3500 years ago.

Historical records suggest that big eruptions are powerful enough to have a global effect on climate. The massive point 7 eruption of an Indonesian volcano, Tambora, in 1815 was followed by one or two years of unusually cold climate around the world. In Europe, annual mean temperatures were 1–2.5 °C lower than normal, harvests were late or failed altogether, grain prices were at their highest and famine was widespread. (See Figure 2.23 (b).)

Special interest now attaches to the supervolcanoes, those that lie at the very top of the magnitude scale. These tend to be different in surface form and origin, their location marked today not by a cone but by a huge

Figure 2.22 Earthquakes
(a) Land movements in Anchorage, Alaska, 27 March 1964. (b) Turkish earthquake of 19 August 1999, with devastation to the town of Golcuk in northwest Turkey.

[Source: (a) FLPA/Steve McCutcheon/Alaska Pictorial Service. (b) Associated Press.]

(a)

(b)

collapsed crater called a caldera. One potentially dangerous supervolcano lies below Yellowstone National Park in the western United States. Detected in the 1960s by infrared satellite photographs, a magma-filled caldera 70 km × 30 km (43 × 19 square miles) was revealed. The Yellowstone supervolcano has erupted regularly over the last two million years at intervals of about 600,000 years. The last time was about 640,000 years ago, making another eruption 'imminent' in geological terms.

Still larger is the caldera at Toba on the Indonesian island of Sumatra. This last erupted only 74,000 years ago at point 8 on the scale. This massive eruption is thought to have emitted enough gas, dust, and debris into the atmosphere to block out sunlight and cause a temporary global 'volcanic winter' lasting several years. Geneticists studying human DNA hypothesize that this event may have reduced the small human population to only a few thousand and pushed humans to the brink of extinction. They argue that the genetic bottleneck in the mitochondrial DNA can be observed in our genes today.

Rather than being events so distant in time as to be of interest only to geologists and geophysicists, earth history is crucial to our understanding of human life, both its past and its future. Our toehold on this planet is recent in time (see Chapter 5) and tenuous. Not only are our lives brief in duration but the whole human story is of very limited span and, to return to the poet's words which opened the chapter, will be '. . . but a star that once had shone'.

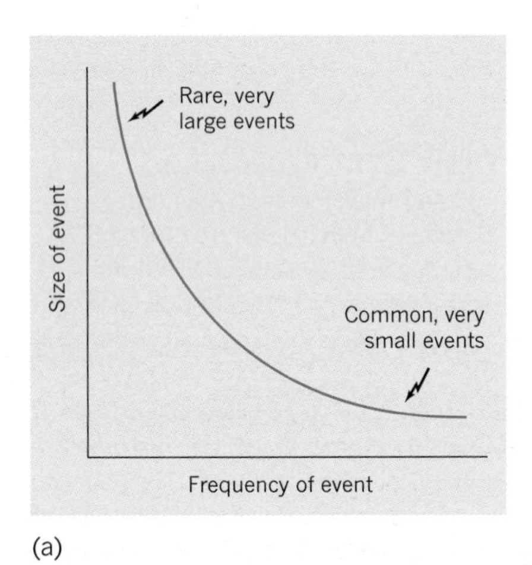

(a)

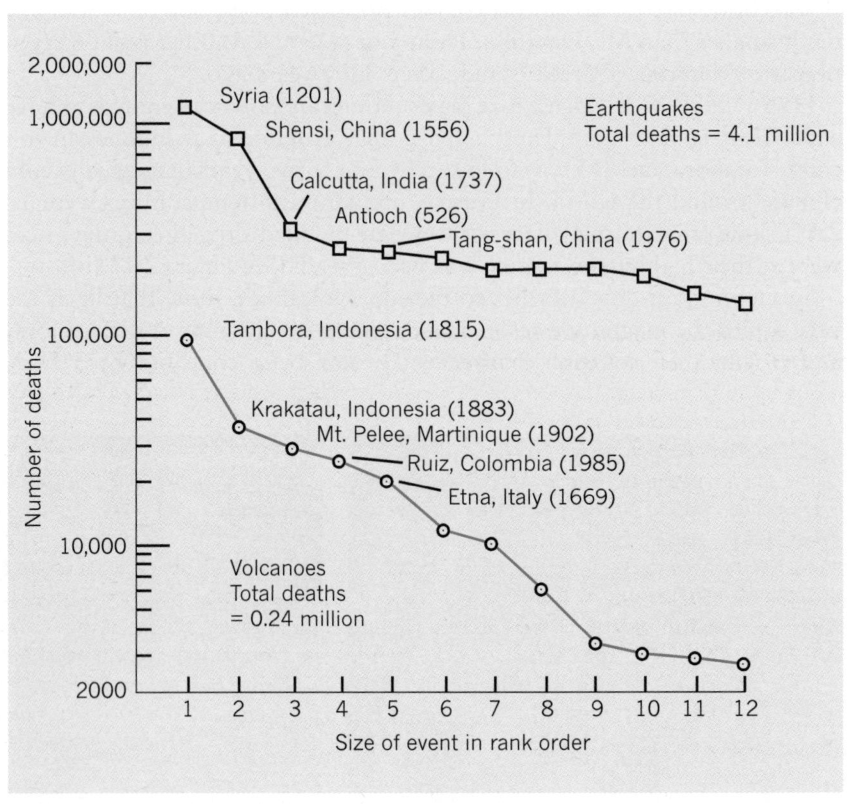

(b)

Figure 2.23 Frequency and impact (a) Inverse relation between the size and frequency of environmental events. (b) Estimated effects of volcanoes on human populations.

Reflections

1 The main elements of the planet's structure (lithosphere, atmosphere, hydrosphere, and biosphere) are shown in Figure 2.2. But remember that this is a diagram only, with an accentuated vertical scale. What would the same diagram look like if the various heights were drawn to scale? (You will need data on the highest mountain, deepest ocean, and the height of the atmosphere to do this.)

2 The structure of the earth is now thought to be made up of distinctive sets of plates (Figure 2.6) which are highly mobile. Yet a half century ago, the idea of 'continental drift' was seen as preposterous. List some of the factors that brought about this change of view.

3 We take our local environment as fixed in terms of latitude, but we know from geological evidence that land masses have been highly mobile. What kind of climate would the United States have if it were shifted several thousand miles south, so that the equator passed through Washington, D.C.? Would the country be more or less productive?

4 The distribution of wet and dry areas around the world follows a distinctive and logical pattern (see Section 2.3). What are the main wet and dry areas of the continent in which your country is located? Why are these areas located where they are?

5 Look at some atlas maps of January and June temperatures for North America. Why do the thermal contours bend equator-ward in winter and pole-ward in summer?

6 Human populations have occupied an earth that can, from time to time, produce environmental conditions that are very hazardous to human life. Which of the geological hazards do you rate as the most dangerous and why? Do you consider that you live in a 'very safe' or 'very unsafe' environment so far as natural hazards go? Explain your reasoning. Can you suggest any measures that individuals, corporations, or government might take to make conditions safer?

7 Using newspapers and recent magazines, make a list of any extreme geophysical events (floods, earthquakes, eruptions, etc.) reported. Then look at the spatial and temporal distribution of these events. Is there any pattern to them? How would you explain such patterns?

8 Review your understanding of the following concepts using the text and the glossary in Appendix A of this book:

circum-Pacific belt
Hadley–Ferrel model
hydrologic cycle
intertropical convergence
permafrost
plate tectonics
solar radiation budget
trade winds
tropopause
westerlies

One Step Further ...

A useful starting point is to look at the essay by Cliff Ollier in I. DOUGLAS *et al.* (eds), *Companion Encyclopaedia of Geography* (Routledge, London, 1996), pp. 15–43 and then dip into selected topics using A. GOUDIE (ed.), *Encyclopaedia of Global Change* (Oxford University Press, Oxford, 2000). There are a number of excellent texts covering those aspects of physical geography considered in this chapter. See for example D. BRIGGS and P. SMITHSON, *Fundamentals of Physical Geography* (Routledge, London, 1985); A.M. MANNION, *Global Environmental Change: A Natural and Cultural Environmental History*, second edition (Longman, Harlow, 1997); A. GOUDIE, *The Nature of the Environment*, fourth edition (Blackwell, Oxford, 2001) and H.O. SLAYMAKER and T. SPENCER, *Physical Geography and Global Environmental Change* (Longman, Harlow, 1998).

The earth's physical structure is clearly described in M.A. SUMMERFIELD, *Global Geomorphology: An Introduction to the Study of Landforms* (Longman, Harlow, 1991) and in A.N. STRAHLER and A.H. STRAHLER, *Elements of Physical Geography* (Wiley, New York, 1989). A useful introduction to volcanoes is given in P. FRANCIS, *Volcanoes: A Planetary Perspective* (Clarendon Press, Oxford, 1993).

Research on the global environment is reported in all the major geographical journals. The 'progress reports' section of *Progress in Physical Geography* (quarterly) will prove helpful. Look also at *Scientific American* (a monthly) and *New Scientist* (weekly) for regular reports on current work by environmental scientists. For readers with access to the World-Wide Web see also the sites recommended in Appendix B at the end of this book for topics relevant to this chapter.

CHAPTER 3

The Ever-changing Climate

■ denotes case studies

Figure 3.1 The fragile atmosphere Storm clouds are forming over Argentina in eastern South America, looking west towards the curve of the earth's horizon. The view shows a line of thunderclouds between the Parana river basin (foreground) and the Andes Mountains (background). The form of the mature cumulonimbus clouds with flat 'anvil' tops indicate the prospect of strong storm activity. This view was taken by the Shuttle *Endeavour* on 16 January 1993; west is at the top of the photo. [Source: NASA/Science Photo Library.]

> The world's a scene of changes, and to be Constant, in Nature were inconstancy.
>
> ABRAHAM COWLEY *Inconstancy* (1647)

The environmental patterns we see around us today are but the most recent frame in a feature-length film. The earth is probably over 4.5 billion years old, and in that time all the major environmental boundaries – of land and sea, mountain and lowland, tropic and pole – have never ceased to change. Environmental change, like death and taxes, is one of the few certainties in life. Many of these changes occurred so far back in time that they are of interest only to geologists. But others, which have been taking place during the last few millennia, are of direct interest to us all. They continue, slowly but perceptibly, to transform our environment.

Of all the elements in the environmental package, it is changes in the earth's atmosphere that are the most fickle, the most constantly changing. Our opening photograph catches catches one such moment (Figure 3.1 p. 68): in all the long history of the earth we shall never see *exactly* that scene again. The world's climate is forever changing and such changes impact in a multitude of ways upon human life.

In this chapter we follow the timescale of changes from the very short to the very long. We begin with the short-term changes of fluctuations over a single day or a single years (Section 3.1). These rapid, short-term changes form the familiar calendar of environmental change. The daily sequence of dark and light over most of the globe (note that conditions near the two poles are different) is one such cycle; the sequence of the seasons from high summer through autumn, winter, and spring is another. Again, as we shall see, the seasonal pattern varies in strength around the globe but is everywhere present in some degree.

We then contrast this with changes over the very long run, with special emphasis on the most recent geological period (the Quaternary Era) which is of direct relevance to human occupation (Section 3.2). This brings us logically to the time period between these two extremes: the middle-term. Here we look at changes that take place over 5 to 500 years. Explanations here are more difficult and we look separately at both recent fluctuations (in Section 3.3) and over past historical periods (in Section 3.4). Finally we turn to the question of climatic change as a hazard and try to see both the scale and the impact of those hazards (Section 3.5).

3.1 Short-term Fluctuations

Cambridge University philosopher Ludwig Wittgenstein liked to illustrate the motions of the earth on an afternoon walk with a friend. He would spin himself around, while at the same time circling around his companion. In the meantime, the friend was supposed to walk, following a leisurely, curving path, across the field. His biographers do not record how long this giddy game was kept up, but it had a relevant purpose.

The earth has three motions. First, it moves with the sun as it orbits the centre of the Milky Way once every 200 million years. Second, it travels around the sun once every 365.26 days. Third, it spins like a top around its own axis once every 23.94 hours. The planet's motion along the solar orbit is largely of astronomical interest, but its second and third motions are of direct and vital significance to us.

The Daily Round

The regular sequence of darkness and light that accompanies the daily rotation of the earth is so familiar that we ignore it. Yet we have evolved biologically in phase with this regular cycle, and our heartbeats, our blood pressure, our urine flow, even our sexual awareness, all have a distinct daily rhythm. Recall that in Chapter 1 a person on the beach was affected by this cycle; in Chapter 8 we go on to look at the basic rhythms of human communal activity and find that all our settlements are adapted to this same 24-hour beat.

From a strictly environmental viewpoint, the main effect of the earth's turning away from the sun is to cut off its darkened areas from massive inputs of solar energy. Thus the night is a period of energy loss by radiation from the land surface and falling temperatures. From dawn onwards, the average amount of incoming solar energy increases; it reaches a noontime peak, then declines again as evening approaches. Average air temperatures follow the same pattern, but peak in the early afternoon.

On some warm summer days we can observe this daily build-up and decline in temperatures by watching cycles of cloud formation. Figure 3.2, a *Gemini V* satellite photograph, shows the view looking south along Florida's Atlantic shoreline.

Note especially (a) that the clouds consist of thousands of isolated cells, giving the whole expanse a mottled appearance, and (b) that the clouds stop at the ocean's edge and are apparently absent from the main lake areas within the peninsula. This cloud pattern illustrates the noontime stage in the development of *tower* clouds, deep, rapidly developing clouds with small bases but considerable vertical depth. Such clouds are formed from the cooling of vertical columns of moist air that develop as the land surface heats up rapidly on a summer day. As some of the vertical columns grow stronger, larger and taller clouds predominate, some giving heavy rain showers. As the land cools toward evening, the tower clouds flatten and decay, so that the sky is mostly clear by nightfall. At dawn a new cycle of cloud formation will begin. Clear areas between the clouds over land are related to countervailing downward movements of air that compensate for the rising cloud columns. Clear areas over the sea and lakes are related to the different rates of warming of the land and sea. A further view of cloud formation is given in Box 3.A on 'Interpreting cloudscapes'.

The Seasonal Round: Temperature

Night and day are related to the rotation of the planet earth; winter and summer are related to its revolution around the sun. Figure 3.3 shows that the planet's path around the sun lies on an imaginary flat surface (termed the *orbital plane*) that cuts through the sun. The earth's axis, around which it spins (shown as a line connecting the two poles), does not stick up vertically into the orbital plane but is tilted at an angle of 23½°. It is this tilt,

Box 3.A Interpreting Cloudscapes

Landscape interpretation forms one of the central purposes of geography. But the most distinctive feature of the earth, whether viewed from space or by a ground observer, is that of clouds. Although these are a transient elements in the climate system, at any one time half the global surface is cloud covered. Clouds are composed of water droplets or ice crystals and occur at altitudes throughout the troposphere. Clouds are formed when air is cooled below its saturation point. The cooling is dominantly caused by (i) the ascent of air through vertical motion, (ii) an air mass being forced to lift over a topographic feature, or (iii) air being placed in contact with a cold surface. Diagram (a) show typical cloud formations from the middle-latitudes with characteristic forms and heights.

The World Meteorological Organization in its *International Cloud Atlas* group clouds into four basic categories and uses the Latin names originally given to them by an English chemist, Luke Howard, in 1803. *Cumulus* clouds are named from the latin for heap or pile; *stratus* clouds from a layer; *nimbus* from rain; and *cirrus* from a filament of hair. Other Latin terms can also be added so that *altocumulus* (describing a distinctive 'mackerel' sky) comes from adding the term *alto* meaning middle-level. Three examples of cloud types are shown. In (b) cirrus clouds occur in a jet stream. Cirrus is a high-altitude type of cloud found at heights of 6 to 13 km (4 to 8 miles) and formed of ice crystals. Photograph (c) shows cumulus tower clouds. These show vertical development as a steep-sided tower, extending from a few hundred metres up to 6 km (4 miles). They are associated with strong convectional activity. Finally (d) shows cumulonimbus cloud with its distinctive anvil-shaped top. The thunderhead of the anvil forms at a temperature inversion layer in the atmosphere. Clouds of this type can be up to 15 km (9 miles) tall with highly turbulent interiors with lightning, hail, and torrential rain.

[Source: (b) and (c) Pekka Parviainen/Science Photo Library, (d) Geoff Tompkinson Science Photo Library.]

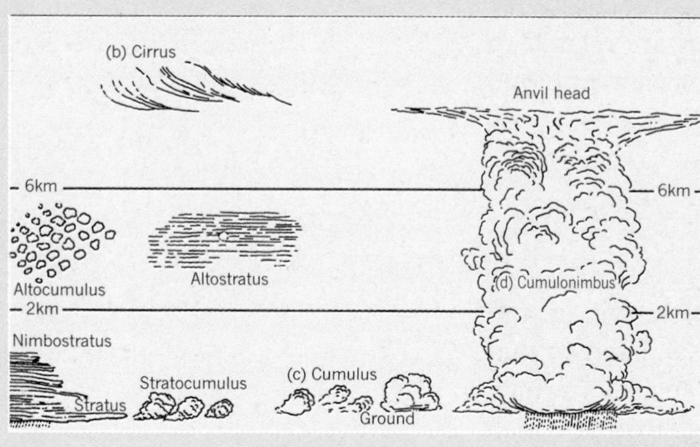

(a)

(b)

(c)

(d)

(a)

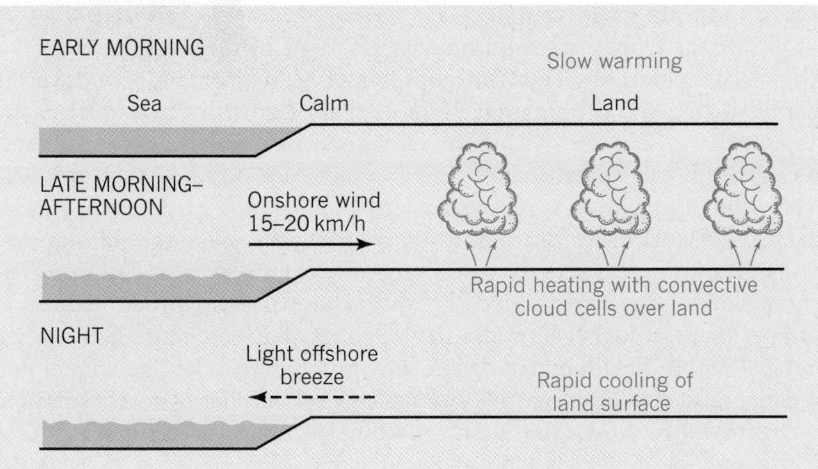

(b)

Figure 3.2 Daily cycles of environmental change (a) This view looking south over Florida's Atlantic coast with Cape Canaveral in mid-photo was taken from *Gemini V*. The distinctive shape of the Florida peninsula can be seen. The mottling over the land occurs as tower clouds form in response to rising warm-air currents. When the temperature falls and these thermal currents die away, the clouds will dissipate. Note that clouds are absent both from the sea and from the main lake areas. (b) Onshore and offshore winds are generated by the relative differences in the temperatures of land and sea areas. The cycle is at its peak in late morning and afternoon with strong convective currents giving tower clouds and strong onshore winds.

[Source: Photograph courtesy of NASA.]

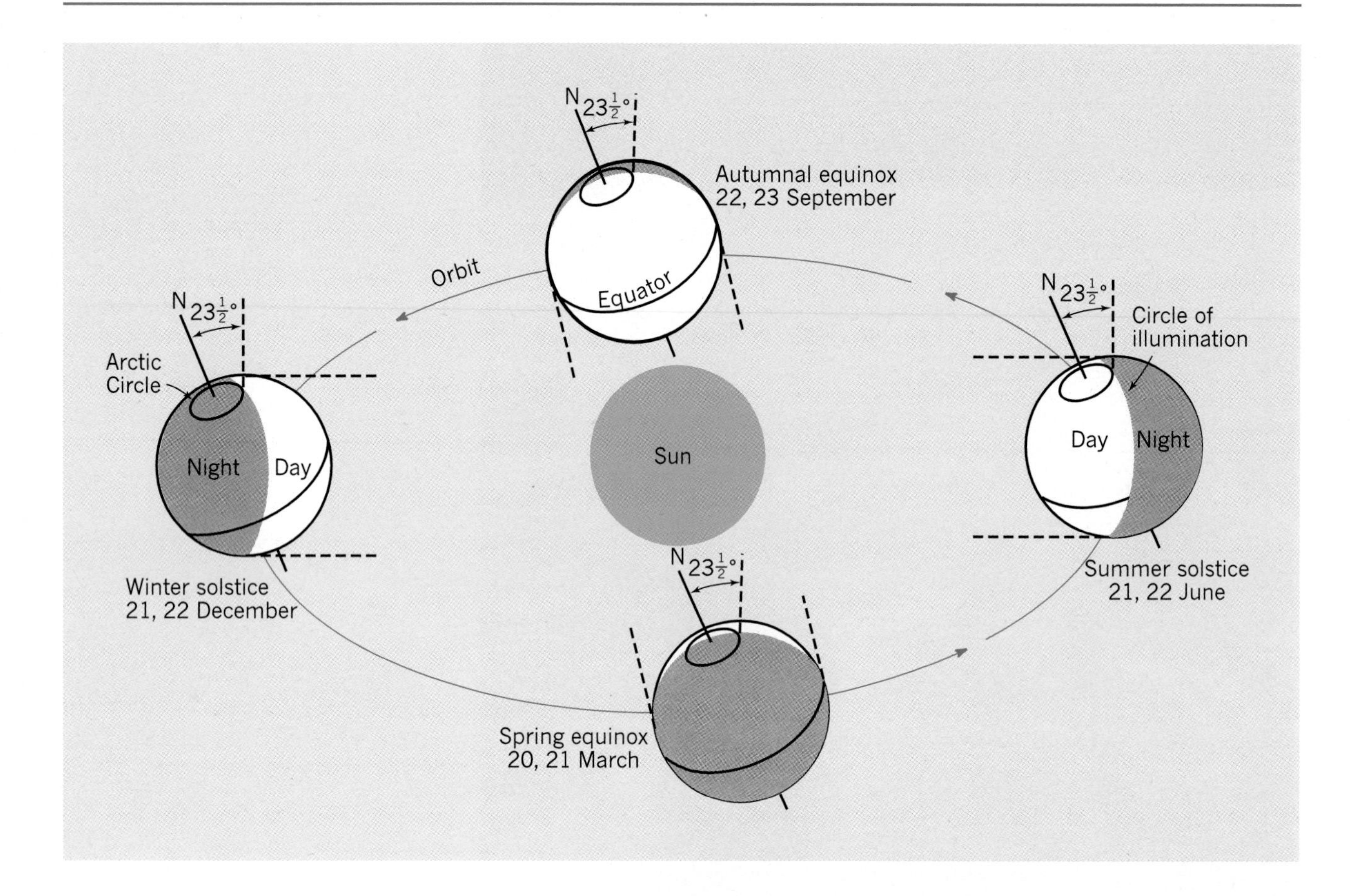

Figure 3.3 Seasonal rhythms This simplified diagram of the earth's annual orbit around the sun shows how the fixed orientation of the earth's axis in relation to its orbital plane gives rise to the familiar sequence of seasons. Note the regular seasonal sequence of ***solstices*** (December and June) and ***equinoxes*** (March and September).

together with the earth's motion around the sun, that produces our seasons. In late December the northern hemisphere is tilting away from the sun, so on each revolution it receives solar radiation for less than half a day. On 21 and 22 December the noon sun is vertically overhead at a latitude of 23½°S (the Tropic of Capricorn), but locations north of the Arctic Circle (latitude 66½ °N, i.e. 90°−23½°) receive no direct sunlight. If you follow the diagram around, you can work out the seasonal alteration through the northern spring, summer, and autumn, and trace the reverse patterns in the southern hemisphere. Thus the traditional spring months of March, April, and May in the northern hemisphere herald colder weather in the southern hemisphere and form its autumn.

Outside the tropics the essential characteristic of the seasonal cycle is a swing in temperatures. Thus for crops winter is the dormant period, spring that of sowing and germination, summer that of growth and maturity, and autumn that of harvest. As Figure 3.4 shows, this familiar cycle is related to changes in solar radiation.

How much radiation is received at any point on the earth's atmospheric surface is related to its location in terms of latitude. Note the shifting position of the latitude where the sun is vertically overhead in the diagram. The amount of solar energy received is not greatest at this latitude because in summer areas in higher latitudes have a longer day (i.e. more hours when they fall within the illuminated area shown in Figure 3.3). The annual flux of temperatures over the earth's surface lags behind that of solar radiation. As Figure 3.4(b) shows, this lag is caused by air temperatures that continue to rise so long as the incoming energy from solar radiation exceeds the energy reradiated from the earth. Thus, New York has its highest average

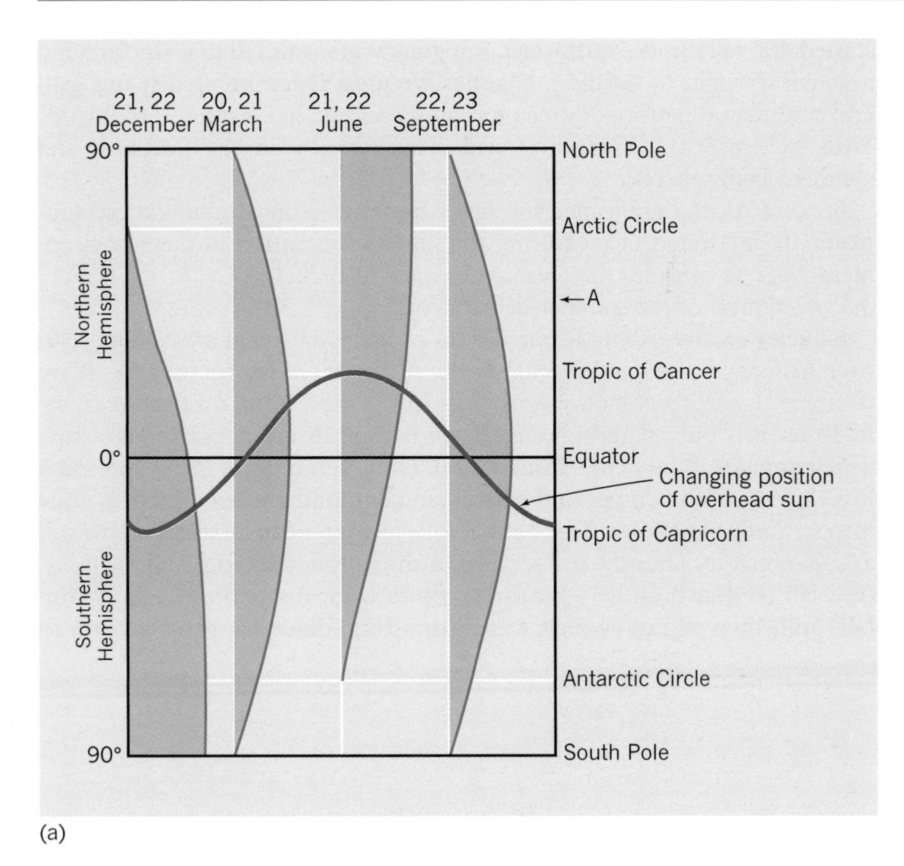

(a)

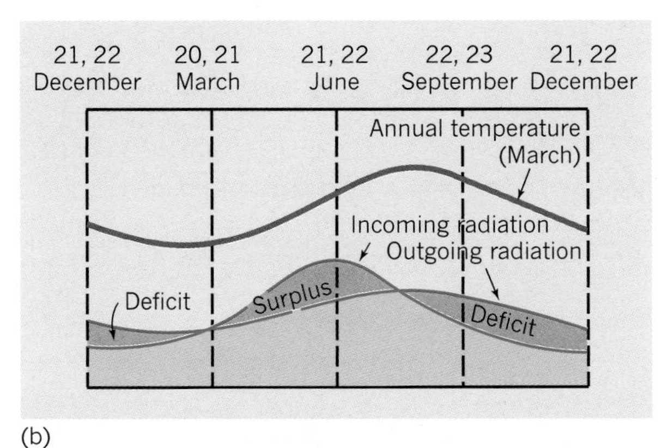

(b)

Figure 3.4 Seasonal variations in solar energy received (a) Variations with latitude. Note the changing position of the sun and the variations in solar energy received at different latitudes in the northern and southern hemispheres at different times of the year. (b) Seasonal variations for a single middle-latitude location in the northern hemisphere at location A. Note the lagged pattern of temperatures – the late January minimum and the late July maximum.

air temperatures not late in June but in the middle of August. Offshore, the lag in sea temperatures is still longer.

The Seasonal Round: Water Deficits

The regular swings of peak solar radiation cause a continuous north-to-south, south-to-north shift of the atmospheric circulatory system described in Figure 2.14. In late June, the northern summer, the whole zonal sequence is shifted north by up to 20° of latitude. This shift brings subtropical high-pressure areas with dry, warm, stable air over areas such as California and the Mediterranean, but sends the trade wind belt, with damp, unstable air, into northern Nigeria and Venezuela. By late December the system has been shifted 40° of latitude southward, bringing winter rainfall to California but a winter drought to northern Nigeria. We must therefore modify our general continental pattern of precipitation (described in Chapter 2) to include areas of seasonal deficits arranged symmetrically in the northern and southern hemispheres.

Because of its significance for vegetation and crop production, geographers are interested in measuring the seasonal variation in these environments. Let us consider the seasonal balance in Berkeley, California (Figure 3.5), as typical of the summer-deficit areas.

Berkeley receives about 63 cm (25 in) of precipitation in an average year, over half of it, as Figure 3.5(a) shows, in the three winter months. If we compute the area's potential water loss from evaporation over the year, we find that it is only slightly more, 70 cm (about 28 in); hence the moisture deficit appears to be 7 cm (close to 3 in). However, most of the evaporation loss (Figure 3.5(b)) comes in the hot, summer months when rainfall is at its lowest. Some of the precipitation that falls in winter can be stored in the soil as soil moisture. Thus the soil serves as a short-term reservoir, and the moisture can be drawn on by a growing crop to compensate for a lack of rainfall. Still, there is not enough water from this source for plant growth to

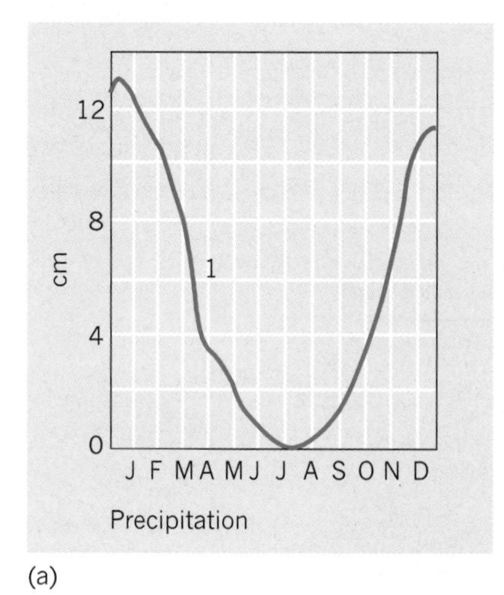

(a)

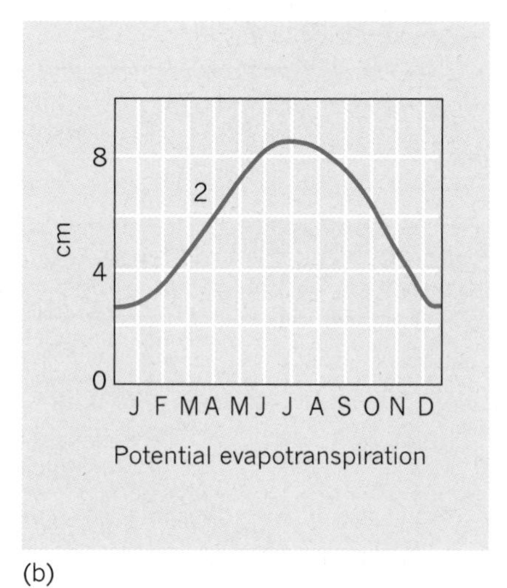

(b)

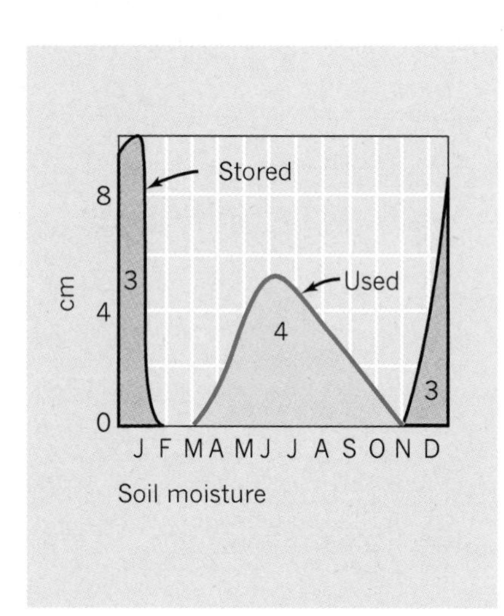

(c)

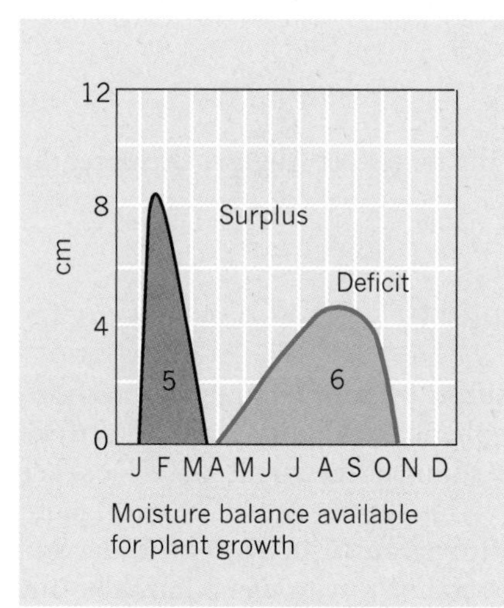

(d)

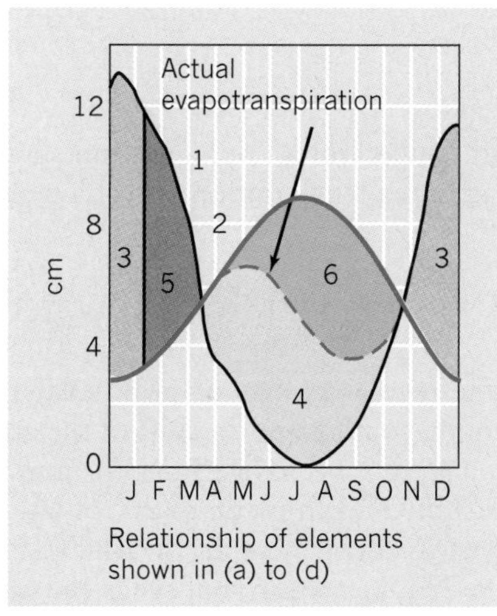

(e)

Figure 3.5 Seasonal moisture balances This series of graphs shows monthly changes in the relationships between precipitation, potential evapotranspiration, and moisture stored in the soil for Berkeley, California, on the west coast of the United States. 'Evapotranspiration' is the term used to describe the return of water to the atmosphere from the soil surface and from transpiration from plants. If you look closely at (e) you will see that it is made up from all the information shown in graphs (a) to (d). The same colours and numbers are used in all five charts. The most important chart is (d), which shows the severe late-summer drought in Berkeley in an average year, typical of a Mediterranean climate.

reach its full potential. From April on growth is inhibited, until by August the monthly deficit reaches 5 cm (2 in), as shown in Figure 3.5(d). These monthly deficits add to a yearly sum of 18 cm (17 in). This large moisture deficit is caused not by a lack of rainfall, but by its seasonal concentration. Excess winter rain cannot all be stored. Once the soil is sodden, additional rain runs off into streams.

Information on the moisture deficit of a particular environment is useful in assessing its irrigation needs. If we extend our calculations, we can show other significant deficit areas over the western part of the United States.

In tropical latitudes more complex patterns of seasonal moisture deficits are encountered. Annual swings of temperature are less significant, and the daily range in values is often greater than the seasonal range. Important seasonal changes are related more closely to periods of rain and drought than to variations in temperature. Rainy seasons are directly related to the weather at the zone of convergence between the trade winds of the two hemispheres as it swings first northward and then southward in its annual cycle. Figure 3.6 shows an ideal cycle of shifts with distinctive two-season peaks of precipitation at the equator, the first in March–April and a second in October–November. North and south of the equator the two peaks merge into a single rainy season.

In interpreting Figure 3.6 we should recall that it shows an ideal situation. In reality irregularities of air flow may blur the picture and convert a regular seasonal cycle into a tragically uncertain pattern of precipitation. This has been seen most recently in the prolonged drought in the *Sahel* region of Africa. This lies between latitude 10° and 15°N on the southern border of the Sahara Desert. As Figure 3.6(c) shows, rains should fall during the June–July hot season. Their failure has led to periodic famine conditions on an unprecedented scale. We turn to the reasons for such failures considering mid-term shifts in Section 3.3.

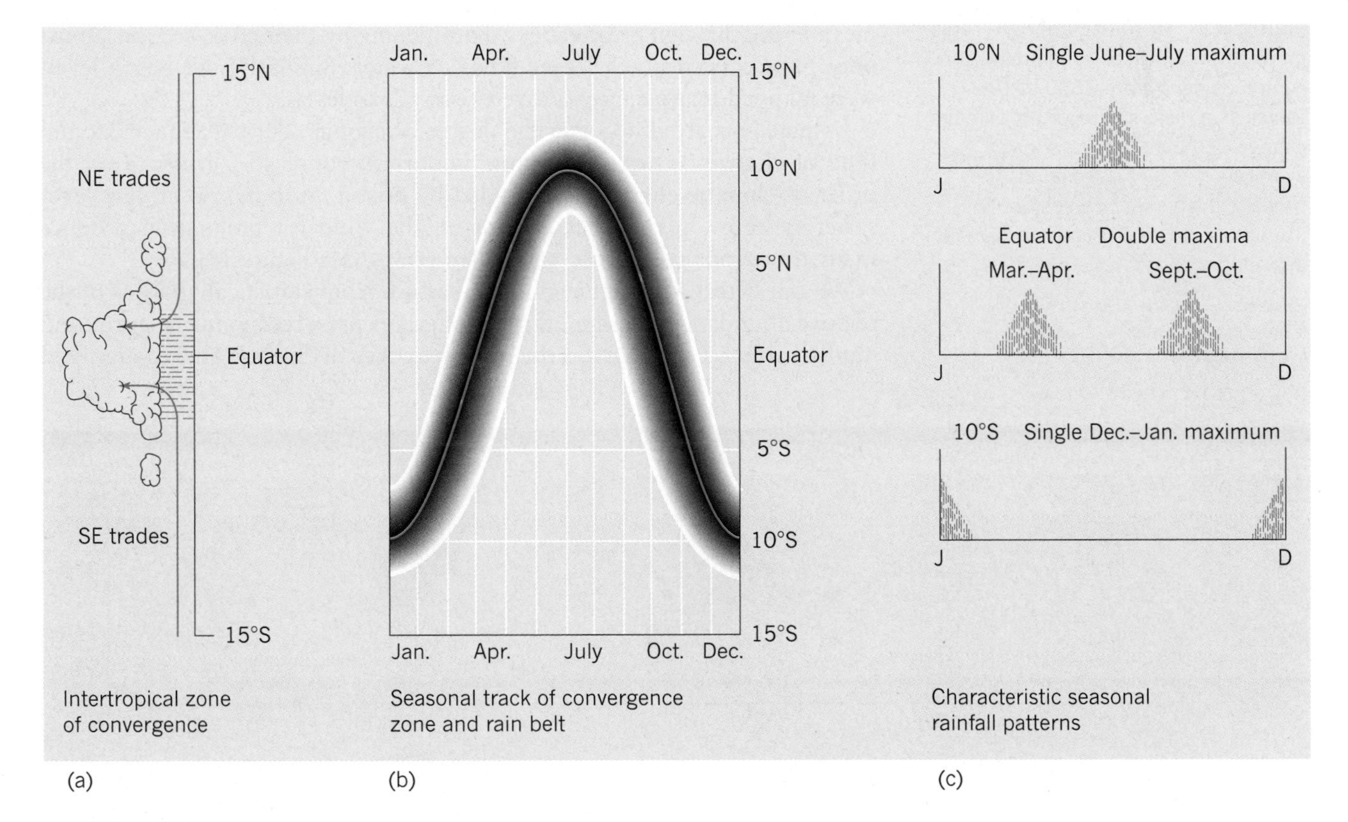

Figure 3.6 Seasonal patterns of precipitation in the tropics Tropical rainy seasons (c) are associated with the rhythmic northward and southward oscillation (b) of the high-rainfall belt associated with the intertropical zone of convergence where the trade winds from the two hemispheres meet (a). However, the uneven distribution of land and sea in the tropical zone, as well as monsoon effects, tend to blur the simple seasonal patterns shown here.

3.2 Long-term Swings

How do we know what environmental changes have been taking place since humans emerged on this planet? Dim memories of change have been passed down through the ages in oral traditions and written records. The account of the flood in Genesis may not be literally true, but there is no doubt from related evidence that major floods did occur in appropriate parts of Asia Minor in Biblical times. However, very precise evidence of climatic change, in the form of written records, is available only for a very short period. By including early occasional records of heavy storms or extended cold periods, we can construct long runs of meteorological observations for a few parts of Europe. Figure 3.7 presents a remarkable 280-year record of winter temperatures for central England compiled by climatologist Gordon Manley. But even with a record of this length, finding a pattern is difficult. It is not unlike trying to make sense of changes in Wall Street share prices.

Reading the Environmental Record

In practice, therefore, geographers find that they have to build up a picture of environmental change from a wide variety of indirect sources. Let us examine some of these sources. Early researchers were aware of significant climatic changes primarily from macroscopic organic remains. All sorts of evidence, from the excavation of elephant and rhinoceros skeletons on the edge of the tundra to the discovery of warm-water shells in cold-water streams, pointed to a substantial climatic change in the recent past. In the early nineteenth century, Victorian scientists reported the growing size of the Alpine glaciers in Central Europe. They also described small streams meandering through great valleys that, judging by their cross-sections, must once have carried much larger flows of water. Similarly, old beach levels were mapped many metres above existing lake levels.

Disputes occurred not over the degree of change, but over the order and time of changes. One of the most important sources of evidence on the order of climatic changes is provided by *pollen analysis*. As all hay fever sufferers know, plants that depend on the wind for pollination produce many thousands of microscopic pollen grains (see Figure 3.8).

We can detect a changing climatic pattern from statistical analysis of the relative abundance of different types of grains preserved in lakes, peats, and muds. Table 3.1 summarizes the main sequence of climatic and vegetational

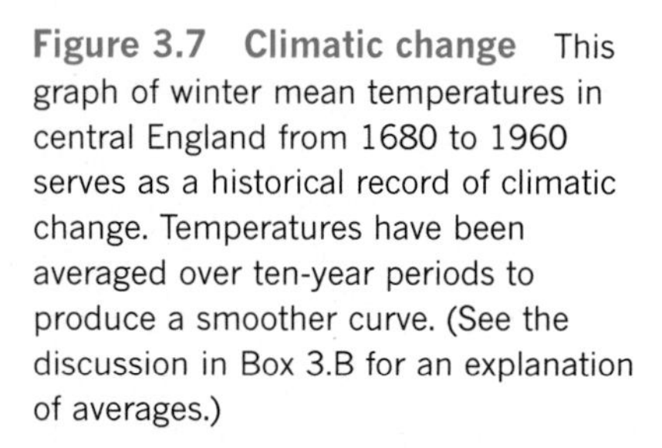

Figure 3.7 Climatic change This graph of winter mean temperatures in central England from 1680 to 1960 serves as a historical record of climatic change. Temperatures have been averaged over ten-year periods to produce a smoother curve. (See the discussion in Box 3.B for an explanation of averages.)

[Source: G. Manley, *Archiv for Meteorologie, Geophysik und Bioklimatologie* **9** (1959).]

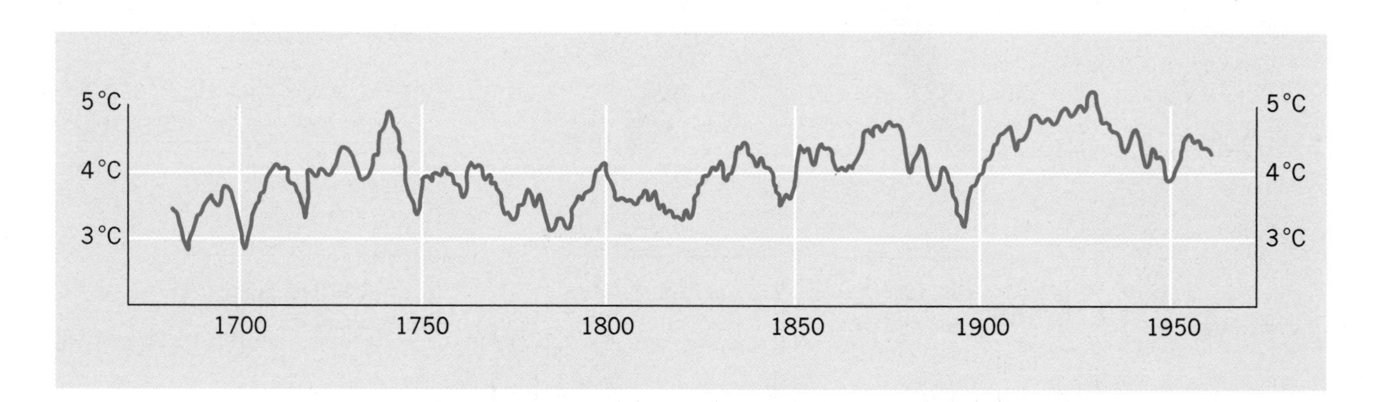

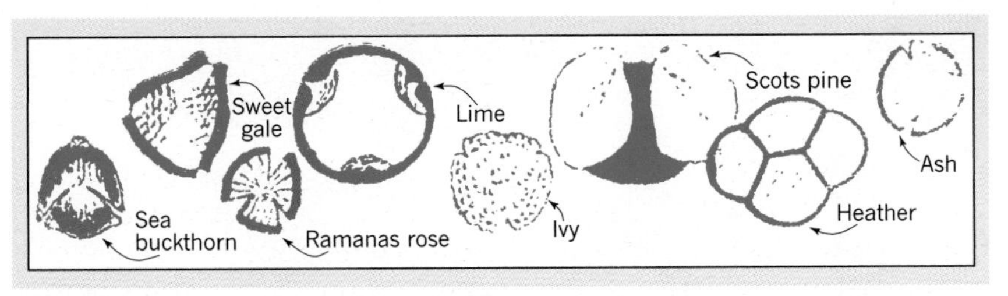

Figure 3.8 Pollen types The analysis of different types of pollen provides vegetational evidence of recent climatic shifts. The statistical frequency of pollen grains (magnified about 200 times in the diagram) preserved in peat deposits allows us to reconstruct the probable plant cover during different time periods.

[Source: H. Vedal and J. Lange, *Traer og Buske* (Politikens Forlag, Copenhagen, 1958), p. 208.]

conditions in Western Europe, indicated by pollen analysis, since the end of the last major expansion of the ice caps. The present cool rainy climate, which began about 400 BC, is the ninth phase in a series of postglacial oscillations. About 500 BC, this part of Europe had a rather warm continental climate conducive to extensive pine and hazel forests.

Pollen evidence of recent shifts in belts of vegetation has been studied mainly since the 1920s. It helps us to sort out the order of environmental changes, but it leaves unsolved the problem of exactly when they occurred. That we can give actual dates to within a few decades is due to the remarkable advances made in the early 1950s by physicist F. Willard Libby at the Institute for Nuclear Studies in the University of Chicago. By 1947, carbon-14, a radioactive form of carbon that loses half its radioactivity in the first 5750 years of its existence and half the remainder in each 5750 years that follow, had been discovered in nature. Its constant rate of decay enabled Libby to devise a method of dating organic material. This technique, called ***radio-carbon dating***, lets scientists correlate organic evidence of how the world looked in the past with other biological, geological, or archaeological evidence. Carbon dating has proved to be extremely accurate for a period of 1000 years, but evidence from tree-ring counts from the very old bristlecone pine suggests that corrections are needed for period beyond 2000 years (see Figure 3.9).

Radio-carbon dating has now been supplemented by other radioactive dating methods. Most new dating methods have been used to study the top few centimetres of deep-sea sediments because they hold the greatest promise as an archive of environmental change. Microscopic analysis of these sediments confirms the postglacial warming indicated by pollen analysis: a rise of about 8 °C in mean ocean temperatures in the North Atlantic during the last 15,000 years, and an even greater warming (12 °C) over a similar

Table 3.1 Main postglacial ecological changes

Period	Time (BC)	Climate	Dominant cover[a] (main species)
Sub-Atlantic	Since 400	Cool maritime	Woodland: beech and hornbeam
Cultivation and clearings			
Subboreal	3000–400	Continental (cold winters, warm summers)	Woodland: oak and ash
Cultivation and clearings			
Atlantic	5500–3000	Warm maritime	Woodland; oak and elm
Boreal	8000–5500	Warm continental	Woodland: pine and hazel
Later Dryas	9000–8000	Arctic	Tundra
Allerod	10,000–9000	Cool subarctic	Scrub: birch and aspen
Early Dryas	15,000–10,000	Arctic	Tundra: large barren areas

[a] General sequence typical of lowland areas of northwest Europe.

Figure 3.9 Dating world climatic changes The bristlecone pine (*Pinus longaevia*) is the earth's most long-lived inhabitant. In the White Mountains of California, a count of the annual tree rings suggests these trees may reach ages of nearly five thousand. Study of ***dendrochronology*** (from the Greek words for 'tree' and 'time') allows estimates of climatic conditions to be gauged from the width of the annual growth rings. The evidence of the bristlecone pine has proved particularly useful since it inhabits a semi-arid area and its growth is very sensitive to rainfall amounts. Rings are wide in wetter years, narrow in drier years. Sometimes parts of individual rings are missing, but multiple borings around the tree circumference allow a complete picture to be built up. By matching the rings of living and dead trees, the record has been extended back to 6200 BC. This long record has helped to check the dates derived from radio-carbon dating.

[Source: Photo FLPA/L. West.]

period in the Mediterranean. Most recent work is on the magnetic orientation of old sediments, which shows the position of the earth's magnetic field at the time they were laid down and allows us to date material as much as 20,000 years old.

Patterns of Change: The Pleistocene Epoch

What kind of environmental changes have these techniques revealed? In discussing this, it is helpful to use geologist's terms and confine ourselves to the Quaternary Period, the fourth and the last of the four major geological divisions of earth history. This period is conventionally divided into two epochs: the longer *Pleistocene epoch* of some 3.5 million years and the *Recent epoch*, covering the last 25,000 years. As we trace in Chapter 5, human beings probably developed from primate forebears sometime in the second half of the Pleistocene epoch. The human species emerged in a period that, from the standpoint of previous geological periods, was one of intense environmental contrasts and rapid changes. Global differences in elevation, climate, and vegetation, for example, were sharper than they had been for the last 250 million years.

The earth's climate, which had been cooling slowly for the last 65 million years, grew much colder about 2 million years ago. The effect of this cooling was to lock up more of the world's water in the form of ice. In North America ice caps formed over central Canada and Labrador and spread as far south as Missouri and southern Illinois. In Europe ice caps formed over Scandinavia and advanced south into England and west almost to Moscow. In the southern hemisphere and in the tropical highlands the evidence of glaciers is less clear, but a large expansion of the ice fields is indicated. This expansion did not occur in a single surge, but entailed several slow advances and retreats, separated by mild and sometimes long *interglacial periods* (see Figure 3.10(c)). In Europe there were four main periods of glacial expansion: the Gunz, Mindel, Riss, and Würm glaciations. These correspond timewise to the Nebraskan, Kansan, Illinoian, and Wisconsin glaciations in North America.

The impact of the ice sheets on the planet was threefold. The first effect we can predict from our knowledge of the earth's hydrologic cycle. (Refer back to Figure 2.19 for an outline of this cycle.) As more water was stored as ice, the return flow to the oceans lessened and sea levels fell. At their largest the expanded ice caps lowered the ocean levels by approximately 100–125 m (328–410 ft). Although the vertical drop in water levels was relatively small, the horizontal effects were striking. The shallow continental shelf fringing the main land masses was exposed. For example, the shoreline of the northeast United States was extended by 100–200 km (62–125 miles). As a result, new routes between continents and islands were opened. Although the archaeological evidence on this point is conflicting, it seems probable that the entry of humans into the New World by way of a land bridge with East Asia (now the Bering Straits) was in this period.

A second environmental change produced by the ice sheets was the compression of broad climatic and vegetation belts towards the equator. The productivity zones mapped in Chapter 4 were sharpened and realigned. For example, the Sahara Desert in North Africa (zone F) may have moved as far south as latitude 10°–15°N, compressing the savannah and equatorial zones into a narrow area. Figure 3.11 shows evidence of *desertification* in the sand dunes of this Pleistocene desert, now covered with vegetation and lying well south of the present desert margin.

Thirdly, expanding ice also bulldozed and realigned the river system of much of North America and Europe. The Great Lakes system emerged and was shaped as a frieze of vast seas on the edge of the retreating ice field. Northern Canada and Finland have a landscape fretted with millions of small lakes and pools left among the jumbled debris at the base of the ice sheets. Debris, ranging in size from a few centimetres to over 50 m (164 ft), was scraped and gouged from one area to be smoothed and plastered on another. It reshaped the terrain of the northern halves of both Europe and North America.

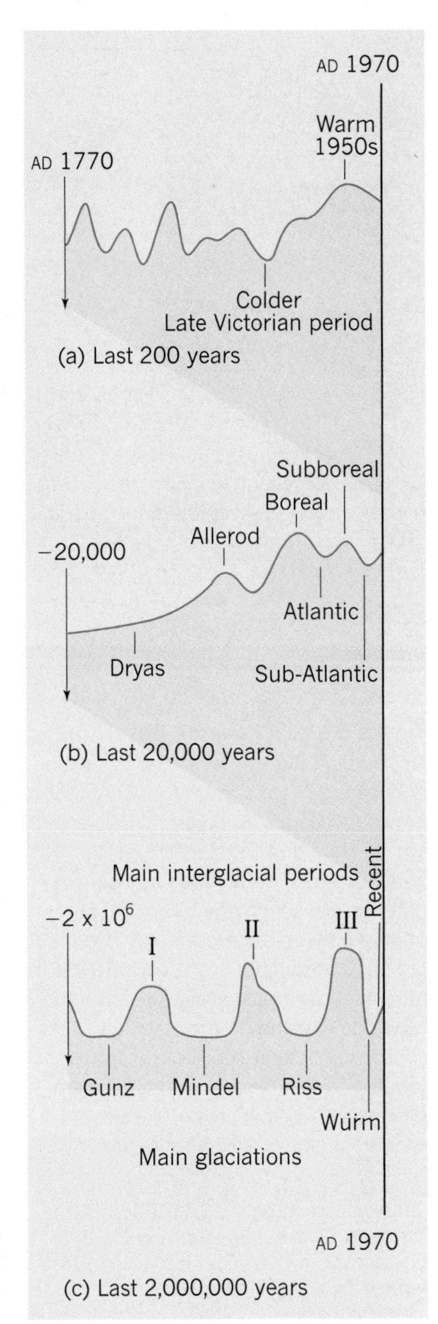

Figure 3.10 Continuities in climatic change The graphs show the general pattern of temperature changes over three time periods of different lengths: (a) the last 200 years, (b) the Recent epoch, and (c) the Pleistocene epoch. Note that each period is 100 times longer than the period immediately above it.

Patterns of Change: The Recent Epoch

Almost 10,000 years ago (about 8000 BC), the latest of the poleward shifts of climatic zones began. The continental ice caps contracted, and the glacial tundra climates of lowland North America and Europe gave way slowly to the present middle-latitude climates. The general warming continued until about 6000 BC, when the *Atlantic* climatic stage, characterized by temperatures 2.5 °C higher than the temperatures we experience today, began. This stage lasted until about 3000 BC (see Figure 3.10(b)). From this climatic optimum there has been a general but irregular deterioration. The *subboreal* stage (3000 BC to 400 BC) was colder than now, and sea levels were generally what they are today. The evidence of climatic changes in the last 2000 years is more detailed, and we can detect continuing swings of temperature. A low point in the record was reached in the northern hemisphere in the middle of the eighteenth century, and temperatures remained low into the nineteenth century. Whether the Recent epoch is a separate and warmer phase of earth's history is a matter of dispute. The present warmer conditions may be merely a prolonged interglacial stage.

These later stages of retreating ice caps have encompassed two opposing

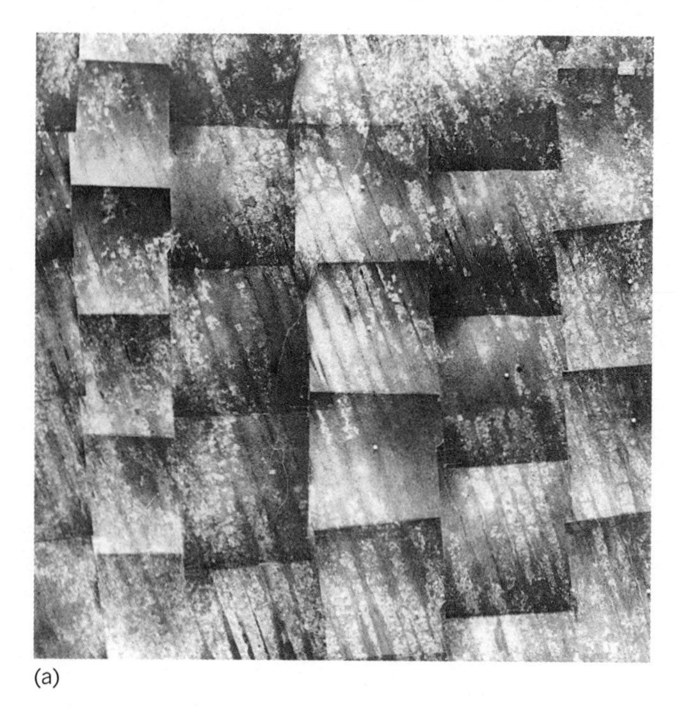

(a)

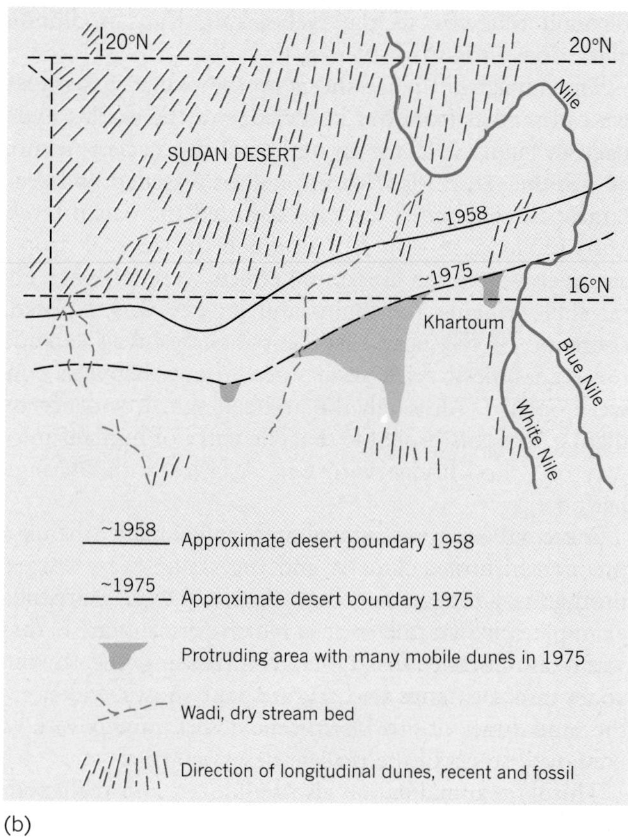

(b)

Figure 3.11 Changing desert margins (a) Ancient longitudinal dunes now under cultivation in the Savannah belt of northern Nigeria, were formed under desert conditions and provide evidence of major climatic shifts in the Pleistocene epoch. Note that the dunes form the parallel narrow strips; the large square patterns are simply the individual air photographs in the mosaic. (b) The prolonged droughts in the early 1970s in the Sahel region led some scientists to conclude that the desert margin is continuing to shift south but a cyclic variation remains more likely. The map shows Rapp's estimates of desert encroachment in a 17-year period.

[Source: (a) A.T. Grove and A. Warren, *Geographical Journal*, **134** (1968), Plate 1, p. 244. Crown copyright photos by Directorate of Overseas Surveys. (b) Map from A. Goudie, *The Human Impact* (Blackwell, Oxford, 1981), Fig. 2.6.]

phenomena: landward and seaward movements of shorelines. As the ice caps have melted, there has been a general worldwide rise in ocean levels of about 30 cm (1 ft) a century. This has resulted in a considerable net loss of land (see Figure 3.12), particularly because the coastal plains that emerged during the glacial maxima (the most intense periods of glaciation when the greatest volume of water was backed up in ice) provided some of the most attractive sites for early human settlement and communication. The rate of loss is trivial, however. For example, human populations in the Gulf of Mexico have been driven inland only 15 km (9.3 miles) every 1000 years.

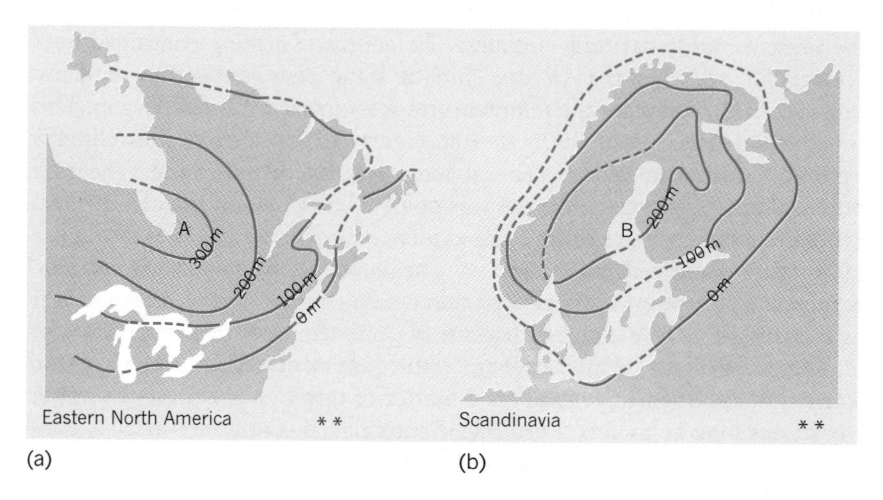

(a) (b)

Figure 3.12 Changing beach levels Contour lines show the height of old beach lines above the *present* sea level. They indicate how the melting of two huge ice caps (centred over points A and B) was followed by an upward warping of the land surface after the weight of ice had been released. Since the greatest weight was near the centres of the caps, it is here that the upward adjustment of the land surface has been greatest.

[Source: B.W. Sparks, *Geomorphology* (Longman, London, 1960), Figs 190, 192, pp. 329, 332.]

Around centres of former ice caps, shorelines have moved seaward and land has risen. At their maximum, the centres of the Labrador and Scandinavian ice caps may have been up to 3 km (1.87 miles) thick. This enormous weight caused a compensating downward displacement of the earth's crust. As the ice has melted, so the crust has recovered, but slowly and haltingly. Today the land around the former ice cap areas (e.g. the Hudson Bay and Scandinavia) is still slowly rising, and the sea is still retreating, leaving lines of old marine beaches inland from the present coast.

Mid-term Shifts: Present Evidence 3.3

Long environmental swings such as postglacial cooling are too remote to worry us; short daily and seasonal rhythms are so repetitive that we have learned to adapt to them. Even year-to-year fluctuations can be coped with if the surplus of one year can be carried over to the next. Irregular and uncertain fluctuations are what hit us hardest. But they are also one of the most difficult things for geographers to interpret. The records are too short for statistical patterns to appear, and the links to physical theory are too tenuous to provide reliable guides for forecasting. Here we shall illustrate with two regional examples the difficulties caused by irregular geographic phenomena.

The Great Plains: Mid-latitude Uncertainties

In the middle latitudes the boundaries between the major wind systems are continually shifting. Thus the polar front we described in Chapter 2 may wander considerably about its average location at about 60°N and 60°S. Fluctuations in the westerly circulation of the atmosphere in the middle latitudes are associated with waveforms that follow a four- to six-week cycle. The cycle begins with a zonal latitudinal flow in which waves of increasing amplitude build up to produce poleward and equatorward movements of air. Circulation then breaks these air movements into cellular patterns before the zonal flow is slowly re-established (Figure 3.13).

During the wave maximum (Figure 3.13(c)), strong incursions of freezing air from the north and warm, topical air from the south may greatly distort 'normal' climatic conditions. Past records of the world's climates reveal that these cycles are part of much larger swings which can last for several years. These cycles, in turn, are part of longer-term climatic shifts in the Recent epoch. Thus, the drought typical of arid zones may extend well outside the normal desert boundaries in one year, and precipitation characteristic of wetter regions may make incursions into an arid zone in the next year. A map of the world climate based on the meteorological records for 1981 would not be the same as one based on similar records for 1980, even if exactly the same classification criteria were used.

These shifts in climate become critical in areas where agriculture is carried on at the margins of humidity. For example, we will see in Chapter 4 that rainfall in the Great Plains of North America decreases from around 125 cm (49 in) in the humid east to 25 cm (10 in) in the dry west. But rainfall

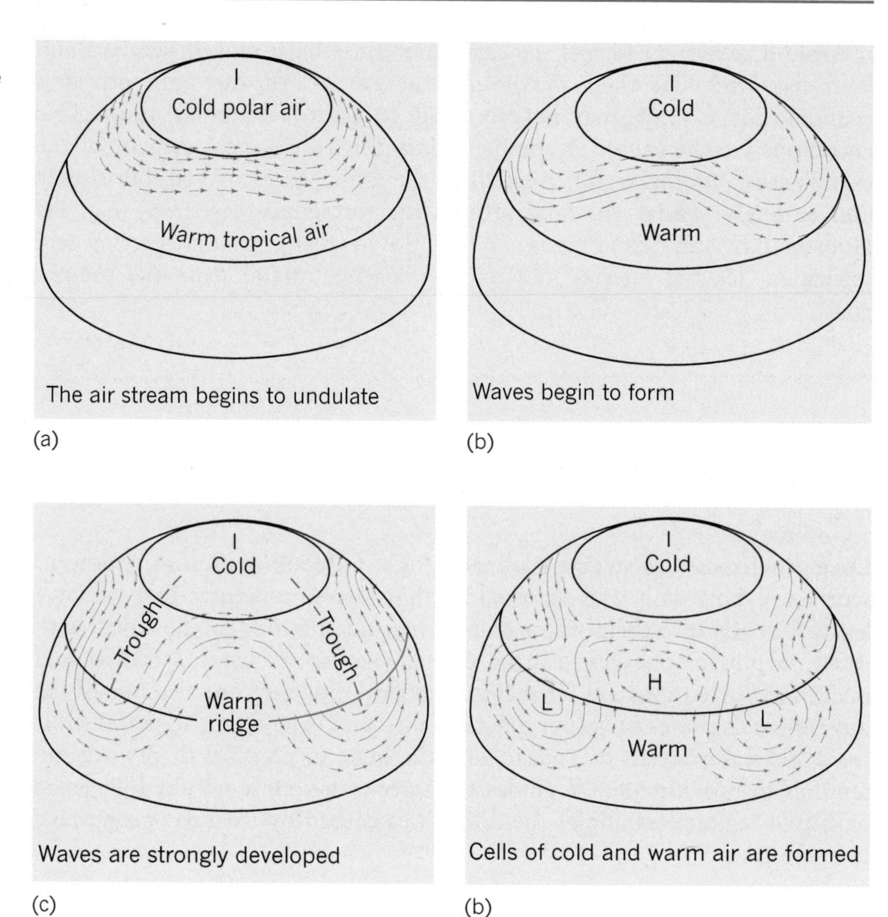

Figure 3.13 Changes in the westerlies This six-week cycle of wave formation and dissipation in westerly air flow (***Rossby waves***) is one of the more regular of the short- and long-term fluctuations that lead to the characteristic instability of mid-latitude climates. The general pattern of the westerlies was shown in Figure 2.11. Longer-term patterns of change are discussed later in this chapter (see Figure 3.18).

[Source: After J. Namias, from A.N. Strahler, *The Earth Sciences*, 2nd edn (Harper & Row, New York, 1971), Fig. 15.26, p. 247.]

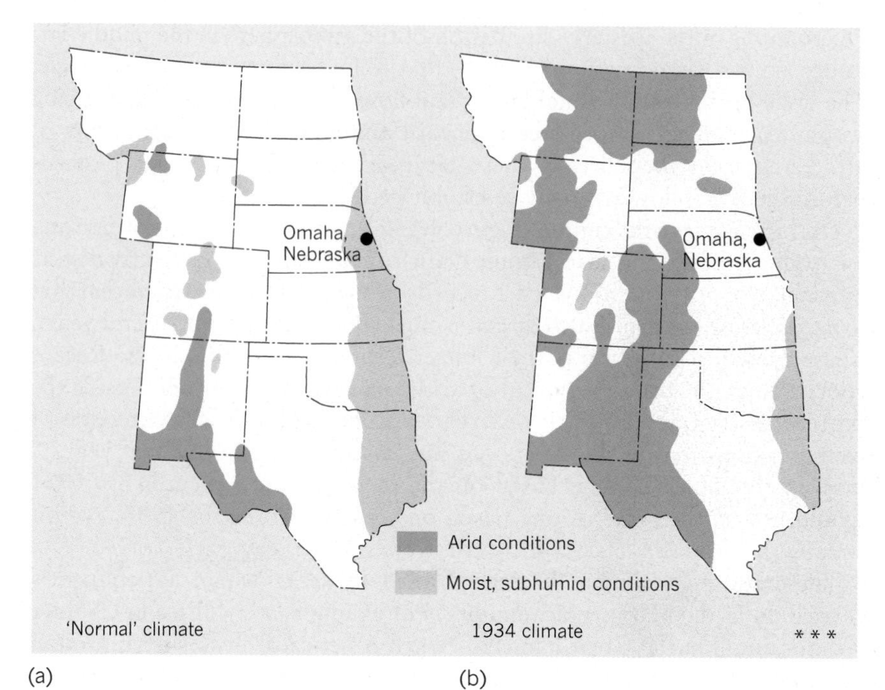

Figure 3.14 Climatic variations in the Great Plains The maps show the stark contrasts between (a) the 'normal' climatic pattern with that experienced in (b) the drought year of 1934. The second map shows the kind of conditions that were described so vividly in the opening chapters of John Steinbeck's *The Grapes of Wrath.* Note that arid conditions extended as far north as the Canadian border and covered an area five times greater than normal. (c) The photo shows dust clouds over the town of Springfield, Colorado, in May 1937 following the great droughts earlier in the 1930s.

[Sources: (a) and (b) C.W. Thornthwaite, in the US Department of Agriculture's *Climate and Man* (Government Printing Office, Washington, D.C., 1941), Fig. 2, p. 182. (c) Peter Newark's American Pictures.]

may fluctuate not only from year to year but from decade to decade. The 1930s saw a disastrous run of dry years in the plains; conversely, the 1940s were generally wetter there than average (see Figure 3.14). In the 1950s the regional pattern varied still further, with little rainfall in the southern plains but average conditions in the north.

What do we mean when we talk of 'average' conditions? One way of interpreting records is to filter out small variations and leave only the principal swings. Figure 3.15 displays a 70 year record of precipitation for a Great Plains location – Omaha, Nebraska. Graphing the original yearly values produces an irregular pattern, but by selecting 5, 10, or 20 year means we can smooth it out to emphasize the general decline in rainfall over the period. Smoothing out data can be useful in checking trends; but, as investors in stocks often find, the curves that result are of limited value in predicting what will happen next. Even wholly random data can have deceptively plausible rhythms and trends. (See Box 3.B on 'Climate variability'.)

The problem of variability in rainfall in the Great Plains has its counterpart in the other mid-latitude grasslands of the world – the South American Pampas, the South African Veld, the Australian Murray-Darling Plains, and so on. In humid regions the annual range of precipitation is small and poses few problems for agriculturalists; in the desert the drought is expected and plans are made accordingly. It is in semi-arid areas such as the Great Plains that settlers have frequently been fooled, because these locations are sometimes desert, sometimes humid, and sometimes a hybrid of the two. Good years draw optimistic individuals into very marginal areas where a run of bad years has caused failure and tragedy in the past. A knowledge of the climatic environment's intrinsic variability may prevent future misfortunes.

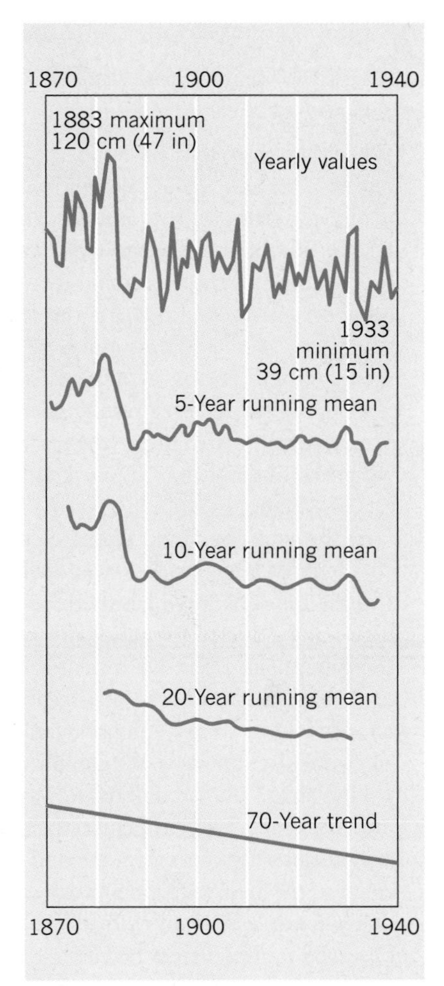

Figure 3.15 Rainfall trends in the Great Plains The graph shows the smoothing effect of running means on the record of annual precipitation in Omaha, Nebraska, from 1871 to 1940.

[Source: E.E. Foster, *Rainfall and Runoff* (Macmillan, New York, 1949).]

(c)

Box 3.B Climate Variability

When we wish to determine the average of a distribution of values, we usually find the *arithmetic mean*. If we have a string of five rainfall values of 57 cm, 69 cm, 85 cm, 96 cm, and 116 cm, we obtain the average value by summing all the values and dividing the result by the number of observations. In this case the sum of the values is 423 cm and the number of observations is 5, so the arithmetic mean, or average, is 84.6 cm. This is a reasonably satisfactory result, as it is very close to the *median* of the distribution (the middle value in the series when the observations are ranked from lowest to highest) of 85 cm.

In studying the Great Plains or monsoon rainfall we characteristically get less well-behaved sets of observations (i.e. sets in which there are some exceptionally high values due to exceptionally wet years). The same thing happens when we measure flood levels in rivers or, for that matter, the income of individuals. What happens when we try and use averages to describe these 'skewed' distributions? We can illustrate the problem by going back to our rainfall records and making one value much bigger – say, 216 cm rather than 116 cm. If we now recalculate our mean, we find it has risen to 104.6 cm (i.e. the new sum of 523 cm divided by 5). In this case we are less happy with the mean. Four of the five observed rainfall values lie *below* this average; 104.6 cm seems unrepresentative either of the four 'normal' years or of the one 'abnormal'. Note, however, that the median of the distribution is unaffected by the change in the final value; it remains at 85 cm. Clearly, the arithmetic mean is poorly representative of the many natural events that tend to have a few very high values and a large number of low ones. In these conditions the median is probably a preferable proxy.

The intense variability of the Asian monsoon environment is illustrated by the figure, which shows 40 years of July rainfall records for Anuradhapura in the dry zone of Sri Lanka. (The location of Anuradhapura is in north-central Sri Lanka.) Note how misleading is the average July rainfall of 3 cm (1.9 in). In fifteen years no rainfall at all was recorded in this month, while in one year nearly 20 cm (7.9 in) fell. The high values for a few years distort the average upward, so that the midpoint of the distribution, the median, is a better indicator of the probable rainfall in future Julys.

Another type of average, one we met in Figure 3.15, is the ***moving average***. Moving averages are used in the study of environmental trends and may be calculated from either arithmetic means or medians, depending on the skewness of the observations. Moving averages are a simple means of smoothing time series by adding the values at regular intervals over a period and dividing the result by the number of observations. If we have a set of yearly values for rainfalls (y), the five year moving average for a particular year (t) is

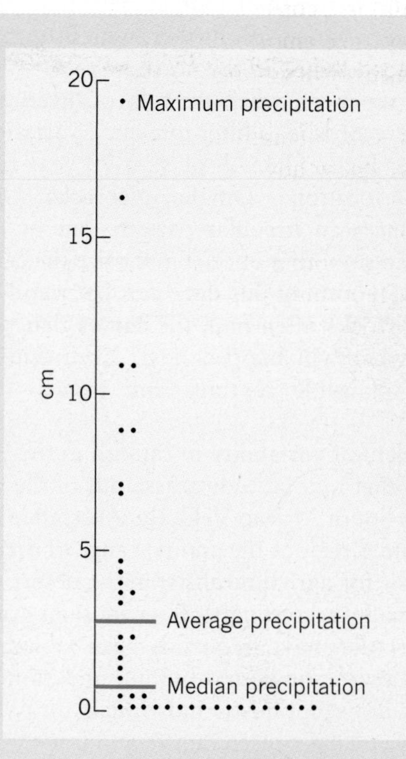

$$\frac{(yt-2) + (yt-1) + (yt) + (yt+1) + (yt+2)}{5}$$

Thus, if our rainfall values in cm for the first seven years are 57, 69, 85, 96, 116, 141, and 124, the corresponding five year moving averages are –, –, 84.6, 101.4, 112.4, –, and –. Note that moving averages cannot be computed for the end values of the series. They can, however, be calculated for any length of time, depending on the length of the series available and the degree of smoothing required. As Figure 3.15 shows, the longer the period of the moving average, the greater the amount of smoothing. It is preferable to use an odd number of years in calculating this kind of average, so that the midpoint of the period to which the average refers will be an actual year. Moving averages can also be extended to two dimensions to smooth map series.

[Source: Data from B.H. Farmer, in R.W. Steel and C.A. Fisher (eds), *Geographical Essays on British Tropical Lands* (George Philip, London, 1958), Fig. 4, p. 238.

Monsoon India: Tropical Uncertainties

Figure 3.16(a) is a conventional rainfall map of the Indian subcontinent. It shows the average amount of rain that may be expected to fall in any one year and distinguishes between very wet areas (southwest India, the eastern Himalayas and Assam, and the Burma coast) and very dry ones such as the Thar Desert. Like all maps based on averages, it should be viewed with caution until we know how much variability is concealed by the averages. We

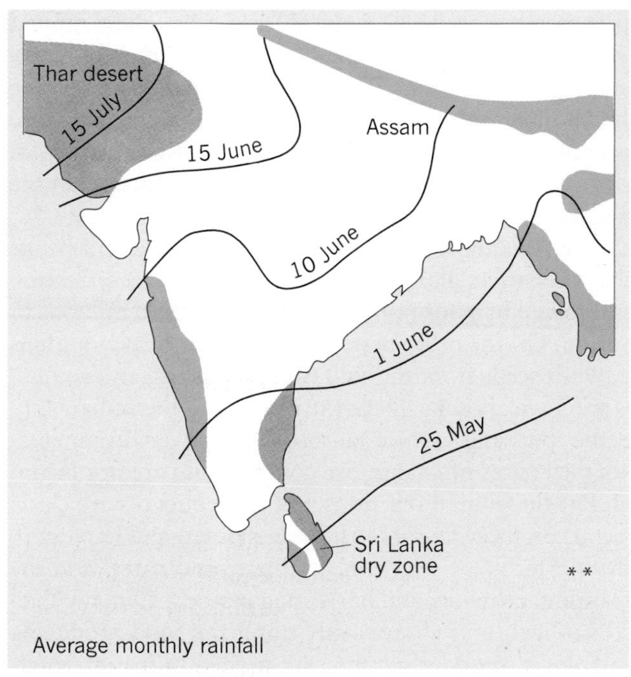

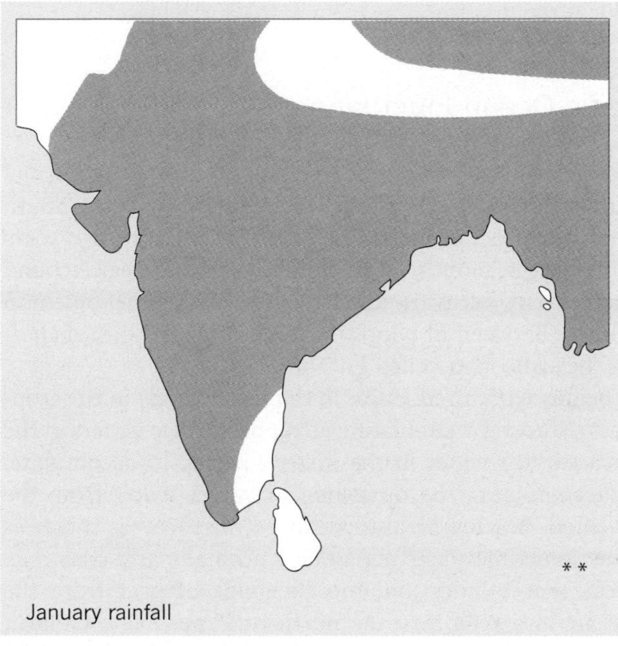

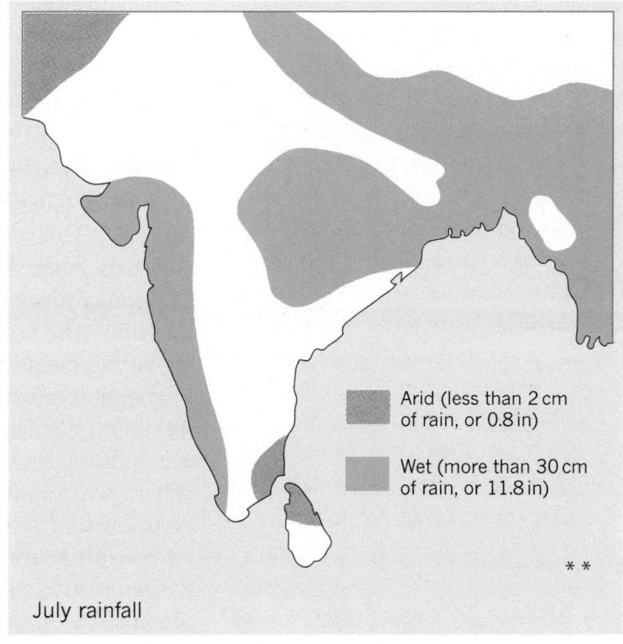

Figure 3.16 The Indian monsoon The map shows contrasts between (a) average monthly conditions, (b) dry-season conditions, and (c) wet-season conditions. Black lines in (a) show typical dates for the 'burst' of the monsoon. Dates are averages; major delays in the 'burst' occur in some years.

can illustrate the difficulty of working with averages by looking at maps for the two months of January and July (Figure 3.16(b) and (c)). The January map shows the situation during the *winter monsoon* period, when India is dominated by dry, colder air moving from the cold high-pressure cell over central Asia. (See the discussion of 'continentality' in Section 2.3.) The June map shows the contrasting period at the height of the *summer monsoon*, with moist, warm, tropical air moving from the Indian Ocean as southwesterly winds in response to the low-pressure cell developing over central Asia.

This regular seasonal reversal of wind directions and precipitation patterns lies at the heart of the Indian agricultural system. The rainy summer season from June to September provides some 90 per cent of the annual water supply of the subcontinent and is especially critical for crops like rice, which depend on waterlogged conditions. The end of the dry season and the sudden 'burst' of the summer monsoon is awaited with anxiety. Figure 3.16(a) shows the average dates for the onset of the monsoon rains. Sri Lanka in the south begins its wet season some two months earlier than the Indus Valley in the northwest.

The special anxiety over the monsoon relates to (a) its timing and (b) its character. Delays in its onset affect planting conditions, jeopardize irrigation regimes and may – if followed by poor rains – lead to famine and millions of deaths through starvation. On the other hand, exceptionally heavy rainfalls may lead to flooding, wash seeds from the soil, cause landslips, and so on.

Both the examples we have chosen, the Great Plains and the Indian subcontinent, illustrate the puzzling nature of middle-term environmental changes. In both cases the causes of change are complex and are not beginning to be unravelled. But the human reactions and consequences are clear. Food production depends on three factors – the area planted and harvested, the level of husbandry (in terms of selection, fertilizer, and care), and the weather during the planting, growing, and harvesting periods. Climatic fluctuations of the kind examined may disastrously upset the food-producing ecosystems. Both examples also underline the care needed in the interpretation of maps and help to explain why modern geographers are so interested in probability theory. In the kind of environments we have just described, a knowledge of the odds is the first step toward wise resource planning.

El Niño and Pacific Ocean Fluctuations

El Niño is a warm ocean current, which occasionally replaces the normal cold Peru Current that moves north along the western coast of South America. The surge of warm ocean waters recurs every three to five years and lasts from six to eighteen months. The waters normally peak around Christmas time, which is why Peruvian fishermen called the phenomemon 'El Niño' (the boy child). (Between El Niños there are often periods of cooling surface waters in the same area called La Niña.)

The phenomenon begins with a reduction in the trade winds in the tropical South Pacific. This reduces the circulating effect on surface waters in the ocean, allowing warm surface water in the eastern Pacific to accumulate. Where warm water accumulates, the upwelling of colder water from the Peru Current is prevented. Sea temperatures can be raised by as much as 10 °C with heavy rains sometimes ocurring along a normally arid coastline. Research also suggests that salinity and the exchange of heat from the ocean's water to the air in a region to the north of Papua New Guinea, where El Niño is thought to originate, could be an important factor. The pattern of temperature anomalies in the Pacific during a El Niño event is shown in Figure 3.17. This was one of the strongest examples of the effect in recent decades.

The effect of El Niño is twofold. First, there are local effects in the eastern Pacific as the reduction in upwelling reduces nutrients in the surface ocean waters with immediate knock-on effects on the food chain: both fish numbers and seabird numbers are reduced. In consequence the commercial fishing industry is badly affected with severe consequences for the anchovy catch.

Second, there are much wider regional effects within the Pacific and some observers argue that the El Niño effect causes, severe climate disruption elsewhere in the world. The wide variety of disasters linked to the effect, include famine in Indonesia (notably in 1983), bush fires in Australia arising from droughts, exceptional rainstorms in California and even the droughts in the Sahel regon of Africa. Exceptionally dry conditions over Indonesia in 1997–98 during an El Niño peak appear to have contributed to the severity of the huge forest fires that began in September 1997 on the two Indonesian islands of Sumatra and Borneo. Figure 3.17(b) shows the distribution of smoke haze from one of the greatest forest fires in the world's recent history.

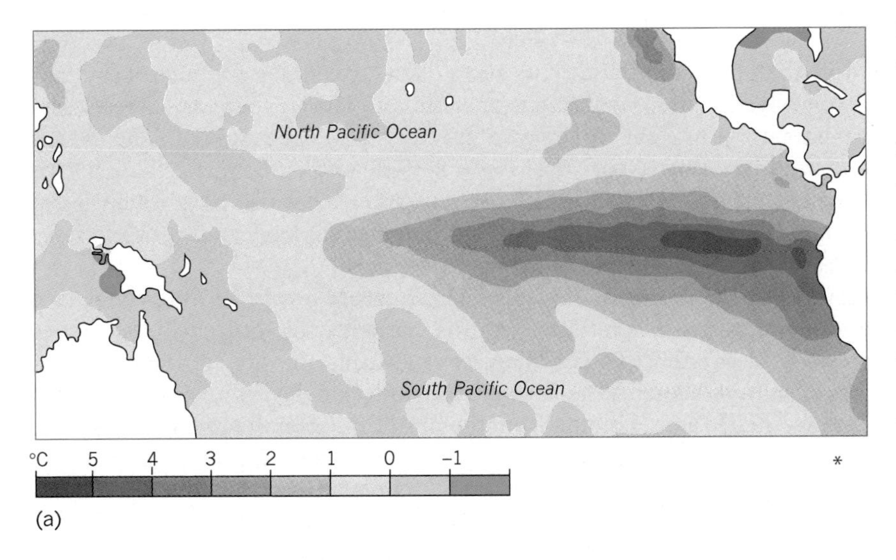

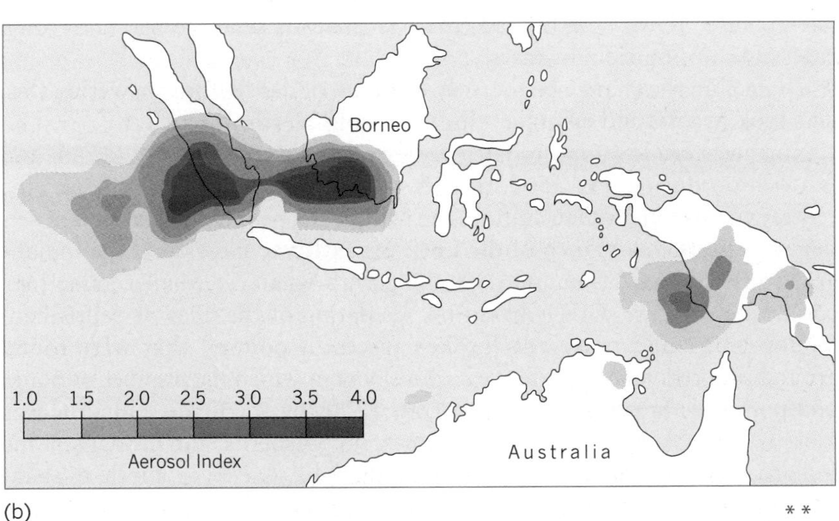

Figure 3.17 El Niño
(a) Identification of El Niño by unusually high sea temperatures in the western Pacific: sea surface temperatures anomalies in the Pacific Ocean in January 1998 at the height of the 1997–98 El Niño. (b) An example of one of the long-distance effects of El Niño in other parts of the earth. Exceptionally wet or dry conditions may trigger environmental hazards. Shown is the smoke haze over Indonesia on 19 October 1997 caused by forest fires. The Aerosol Index is a measure of smoke and particle concentration in the atmosphere.

[Source: United Nations Environment Programme, *Global Environment Outlook 2000* (Earthscan, London, 2000), pp. 33, 90.]

Improved recording of ocean sea temperatures (including the satellite monitoring described in Chapter 22) is allowing better understanding of this short-term climatic oscillation. Gradually the links with other worldwide fluctuations are being established with a view to establishing better forecasts and early-warning systems.

3.4 Mid-term Shifts: Past Evidence

Both the Great Plains and Monsoon India are regions of extreme climatic uncertainty. But it would be wrong to think of this uncertainty as simply variation about a 'normal' climate for each area. The normal may itself show sudden changes, often with severe implications for the peoples whose food supply may be dependent on it, Wisconsin climatologist Reid Bryson argues that such flips in climatic conditions can be found in the past. Let us take two of his examples.

Regional Examples

More than three thousand years ago a distinctive and great civilization, that of the Myceneans, was thriving on a sunny plain in southern Greece. The capital city of Mycenaea, some 95 km (60 miles) southwest of Athens, was the trading centre for the Aegean Sea and much of the eastern Mediterranean. Excavations of the city revealed walls 10 m thick and a kilometre long and the signs of a warlike and sophisticated people whose exploits live on in Greek literature (Figure 3.18).

Abruptly the power of Mycenaea began to decline. In 1230 BC the palace and main granaries were attacked and burned while other tributary cities began to decay. Early archaeologists excavating the sites in the late nineteenth century thought the answer to this decline lay in invasion by other Greeks, the Dorians, coming from the north. But as more research has gone on and more sites have been excavated, this now appears less and less likely. About a decade ago the classical scholar Rhys Carpenter, in a book called *Discontinuity in Greek Civilization*, suggested that an abrupt climatic change leading to loss of crops, famine, and civil disorder was a more likely cause. It was, he argued, the Mycenaeans who burned their own cities, and not outside invaders.

Could climatic change be the answer to the riddle? Before answering this, let's look at a second example, this time from North America.

Northwestern Iowa has today a yearly precipitation of 63 cm (25 in) and is a rich producing area for corn and soybeans. About AD 1200 it was the centre of a thriving Indian culture, the Mill Creek people. Excavation of settlements along this branch of the Little Sioux River shows that the Indians grew corn and ate bison and deer along with whatever smaller game they could catch. Figure 3.19 is based on the dating of the piles of debris containing bones and potsherds (broken pieces of pottery) that were found around the settlements. Note the curious way in which the number of bones and potsherds falls off quickly after AD 1200. By the time Columbus was crossing the Atlantic there were no bones, no potsherds, and no signs of the Indians. All the evidence suggests that the Mill Creek people had abandoned their villages and moved on.

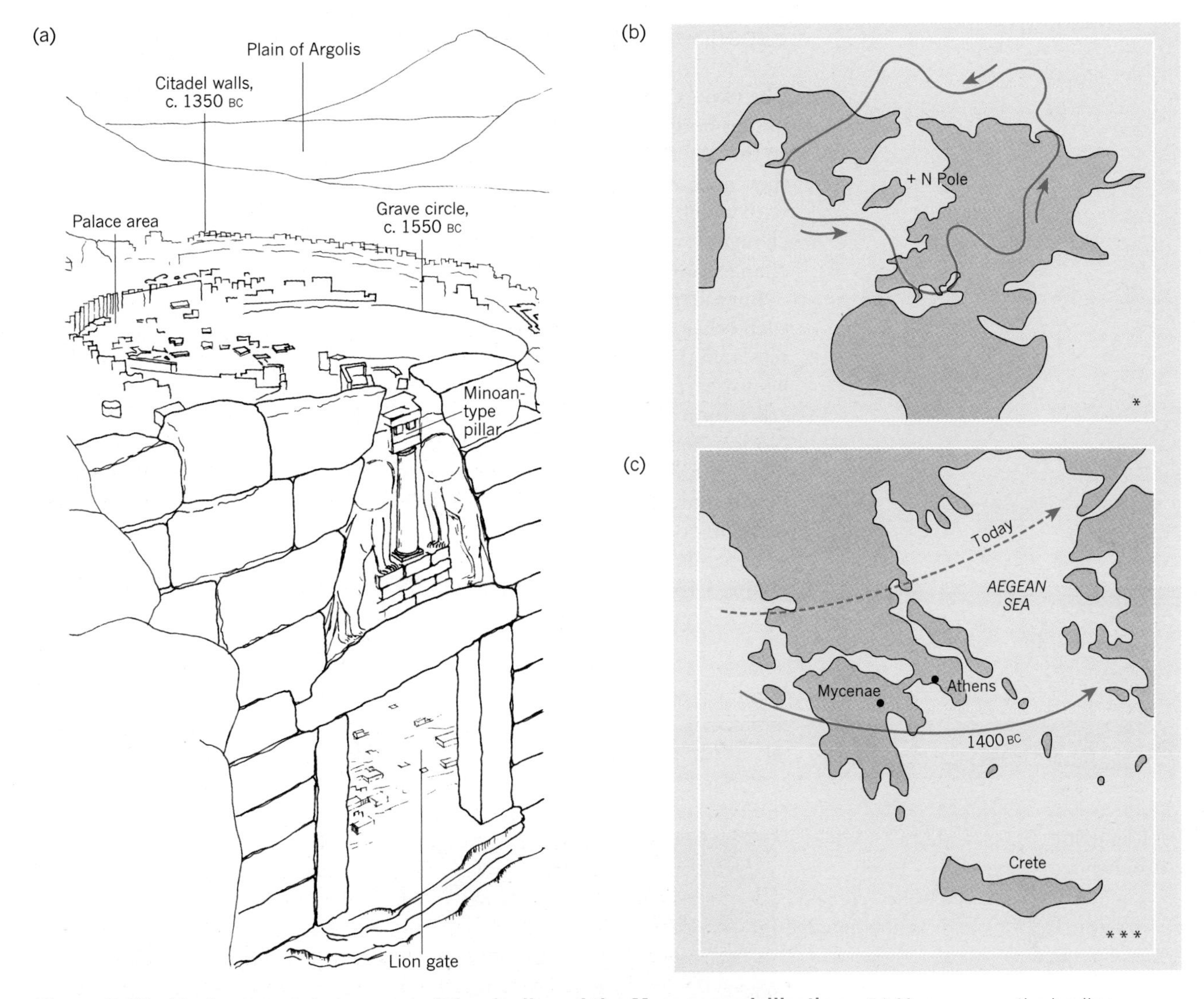

Figure 3.18 Environmental change and the decline of the Mycenean civilization (a) Mycenae was the leading political and cultural centre of mainland Greece from about 1450 to 1200 BC. Thus Homer in the *Iliad* describes Agamemnon, the king, as the most powerful of Greek rulers. The sudden decline of Mycenae has been conventionally ascribed to invasion from outside, but recent research suggests that environmental change may have been a critical factor, with food shortages leading to internal overthrow. (b) One possible hypothesis put forward by Bryson links the decline to changes in westerlies over the northern hemisphere (see also Figure 3.13). Map (c) indicates possible switches in winter storm tracks over the eastern Mediterranean. Such switches are known to have occurred within the last few decades. For a full discussion see the argument put forward by Reid Bryson in *Climates of Hunger*, from which the maps were drawn (details given in 'One step further ... ', at the end of the chapter).

Figure 3.19 Mill Creek ecology Changes in the ecological conditions at Mill Creek in northwest Iowa between AD 900 and 1400 as shown by archaeological excavations. (a) Number of animal bones showing at peak about 1100. (b) Changing percentage share of animal bones showing elk (a woodland species) decreasing and bison (a grassland species) increasing. (c) Number of potsherds – pieces of broken pottery – with an abrupt decline after 1300.

[Source: R.A. Bryson and D.A. Barreis, *Journal of the Iowa Archaeological Society*, **15** (1960), pp. 290–291.]

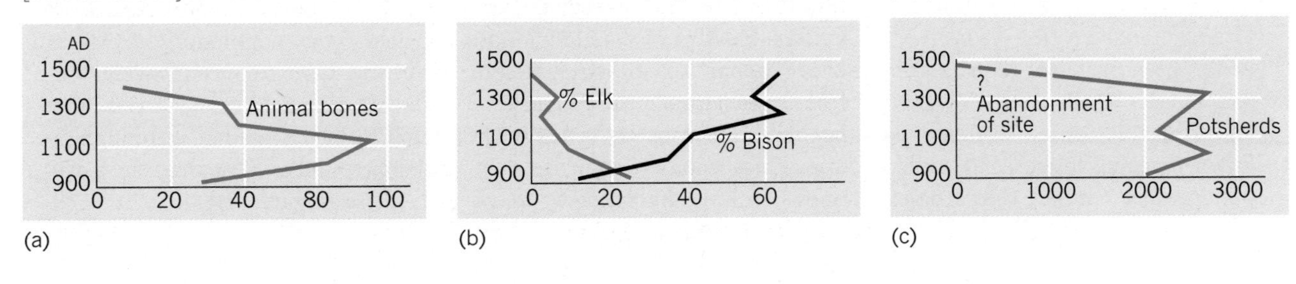

Flips in the Westerlies

The two examples given above are from several cited by Bryson in his fascinating book, *Climates of Hunger*. The title suggests climates that changed so quickly that they were no longer able to support the plants and animals they once did, including those plants and animals (wild or domesticated) with which particular human cultures had built up their numbers. If climates change slowly and progressively, cultures may be able to adjust, but sudden change may be too sharp for the ecological or social system to cope with.

But what links these two peoples, separated in space and time, both with each other and with modern problem regions in the Sahel or India? Bryson sees a vital link in the changing circulation of the westerlies in the northern hemisphere. Look back at Figure 3.13 and note the way in which the westerlies loop and wave their way in a mid-latitude circuit around the pole. Notice that since the waves must 'catch their own tail', they must make a *whole* number of loops – say, three, four, or five – and cannot make any in-between numbers (e.g. 3.12 or 4.76 waves). As slow environmental changes warm or cool the atmosphere, so the area covered by the westerlies contracts or expands and its configuration may change. But the crucial implication of Figure 3.13 is that the pattern can only change by one whole number. (See Box 3.C on 'Kenneth Hare and Canada's restless atmosphere'.)

Box 3.C Kenneth Hare and Canada's Restless Atmosphere

Kenneth Hare (born Wyle, Wiltshire, England, 1919) is Canada's most distinguished climatologist. In 1987 he was appointed a Companion of the Order of Canada for his contributions to Canadian environmental studies for his contributions to the international world study of climatic change. In 1989 he received the International Meteorological Organization Prize from the World Meteorological Organization for contributions to the study of the global cimate and its stability.

After degrees at London and Montreal and war service with the Meteorological Office, Hare held leading positions in English and Canadian universities. From 1979 he has been professor of geography and physics at Toronto University and director of its Institute for Environmental Studies. His earliest publications entitled *The Restless Atmosphere* (1953) integrated the ideas of dynamic climatology being introduced by figures such as Carl-Gustav Rossby (1898–1957) into conventional climatology. Rossby had systematized the upper-level circulation of westerly winds in the upper atmosphere and Hare demonstrated how these ***Rossby waves*** (shown in outline in Figure 3.13) could throw light on the surface-level climatology of the boreal zone in the northern hemisphere. Hare made particular contributions to the climatology of the Canadian sub-Arctic in general and Labrador in particular and showed how better understanding of the circulation of the westerlies could illuminate climatic change in northern Canada. His work on Canada was brought together in *Climate Canada* (1979), which provided a model for the kind of geographic climatology that builds bridges between the geophysics of atmospheric science and the surface reality of climate as experienced by Canada's population. More recently, Hare has been working on global climatic change and the implications of global warming for changes in the Canadian environment (see Figure 19.15).

Bryson argues that it is precisely these jumps from one pattern to another that may abruptly change the climate of some mid-latitude areas. Since we do not have good climatic records for these early time periods, it is impossible to be absolutely sure whether this was what happened to the Myceneans or the Mill Creek peoples. But if we look at the modern record and argue that what *has* happened *can* happen, then it is possible to show that just the kind of flips required have occurred over the period for which records are available. Sudden shifts in the westerlies can move around the storm tracks along which the rain clouds move and sharply alter the pattern of drought or flood.

The patterns, once established, may last for long periods, and there is evidence that there has been a 200-year drought in the United States corn and spring wheat belt within the last thousand years. The only certainty is that climate will change again one day. Ironically the climate in the Iowa area went back to a moister pattern soon after the Mill Creek people pulled out.

But what is the significance to these studies of past climatic change to our modern world? Bryson argues that the present pattern of crops has been developed very efficiently for what we consider to be 'normal' climate today. Given increasing world population (see Chapter 6), the effect on an 'abnormal' year such as 1972 on world food production is enormous. But Bryson goes on to show that our present climate is *not* normal. If we take the period from 1931 to 1960, then since 1880 three out of four decades in the northern hemisphere were colder. The chances of more variable weather around the globe are high for the coming century, and as we shall discover in Part III of this book, part of this change may be self-induced by a growing pollution and other atmosphere-affecting human activity.

The Climatic Environment as Hazard 3.5

Each of the changes described in the preceding sections of this chapter brings some degree of hazard to the human population. It is clear, however, that the degree of hazard is not simply a function of the natural event itself. A severe flood in unpopulated Siberia may represent a much less severe hazard than a mild flood in a densely packed city. A severe hail storm that occurs at harvest time may be a disaster; an equivalent storm on the bare fields of winter may pass without comment. The notion of a hazard is one that only makes sense not just in the geophysics of the event, but in terms of its impact on the earth as a home for the human population (Figure 3.20).

Extreme Geophysical Events

A basic distinction can be drawn between changes such as drought which, even if abrupt, come on over a period of weeks and months and those in which the speed of onset is very fast. There are a number of climatic conditions – blizzards, hurricanes, floods – that are felt by the human population within hours. Similarly there may be events within the slow evolution of the lithosphere or the oceans that are of short duration but of high impact (e.g. earthquakes, volcanic eruptions, avalanches, or tidal surges).

(a)

(b)

(c)

Figure 3.20 Uncertainties in the climatic environment
(a) Hurricane Andrew, satellite photo from GES.7 weather satellite on 24 August 1992 just after the centre of the hurricane had passed over Florida. Winds in the storm reached 230 km (140 miles) per hour. (b) Hurricane in Florida, 1953. (c) The Ohio river flood of 1997 near Falmouth. (d) A classic photograph of a tornado taken near Jasper, Minnesota, on 8 July 1927.

[Source: (a) NOAA/Science Photo Library. (b) E.R. Degginger/Science Photo Library. (c) Michael Keating & the *Cincinnati Enquirer*. (d) FLPA/Lucille Handberg.]

(d)

These sudden changes in the environment form a class of *extreme* geophysical events; we can see some typical examples in Table 3.2.

This table shows extremes in the atmosphere (hurricanes), the hydrosphere (floods), and the lithosphere (earthquakes and volcanic eruptions). Note that all these are part of humankind's abiotic environment as defined in Chapter 2. Extreme events may also occur in the biotic environment (see the cholera pandemic shown in Figure 16.3).

Hurricanes Geographers estimate that about one in five of the world's natural disasters is caused by very severe storms in the atmosphere. Their frequency and sudden onset put them near the top of the hazards league in

Table 3.2 Types of extreme environmental events

Abiotic environment			Biotic environment	
Atmosphere	**Hydrosphere**	**Lithosphere**	**Flora**	**Fauna**
Storms (tornado, hurricane, typhoon) Snow blizzard Droughts	Floods, river Floods, coastal Avalanches, snow	Earthquake Volcanic eruptions Avalanches, rock Shifting sand migration	Dutch elm disease in tree populations Wheat stem rust Algal blooms in eutrophic lakes Water hyacinth infestation in waterways	Smallpox pandemic in humans Foot-and-mouth disease in cattle Locust plagues

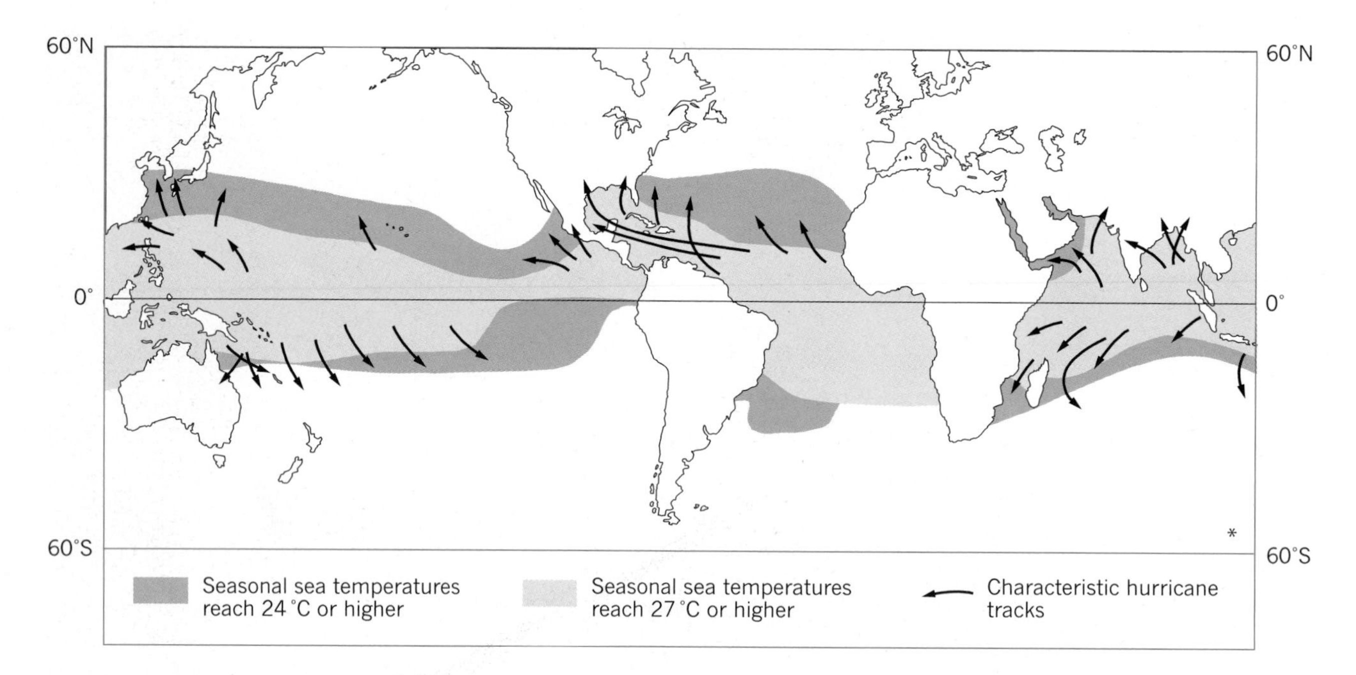

Figure 3.21 Hurricane danger zones Tropical cyclones (hurricanes) form in areas where the sea surface temperatures are high. Most originate in latitudes between 5° and 15° where the sea temperature in the hurricane season is 27 °C (81 °F) or higher. Characteristic tracks of hurricanes are shown on the map by arrows.

terms of loss of life. Violent vortices of this type are termed *tropical cyclones*, or *hurricanes*. They form in moist tropical air between 80 and 240 km (50 and 150 miles) from the equator and move poleward along characteristically sickle-shaped paths (see Figure 3.21).

Tropical storm Agnes killed 118 people along its path up the eastern seaboard of the United States in June 1972.

As far as hurricanes and tornadoes are concerned, the US losses are dwarfed by those of Asian countries. For example, on 13 November 1970 the greatest natural disaster of the century struck the low-lying delta areas of the Ganges and Brahmaputra rivers at the head of the Bay of Bengal. A tropical cyclone with a storm surge and winds of over 160 km (100 miles) per hour destroyed 235,000 houses and 265,000 head of cattle, and led to over 500,000 human deaths.

For example, in the North Atlantic region hurricanes form between Africa and South America and move west and north into the Caribbean, the Gulf of Mexico, and the offshore areas of the eastern United States before curving back toward the northeast. The most commonly affected areas in this region are the Caribbean islands, but Florida may also be hit. Very occasionally, a hurricane will not cross the land until as far north as New England. Considerable research is being conducted on ways of controlling hurricanes by seeding them at an early stage of development to trigger precipitation before they reach critical land areas. At present, however, we can only reduce the damage by more accurate tracking and forecasting of approaching storms. This will allow more time for danger areas to be cleared of people and for at least some vulnerable property to be protected.

Floods Flood hazards occur in two distinct zones: coastal areas and areas bisected by rivers. *Coastal flooding* follows above-average sea levels caused by (a) unusual atmospheric conditions (e.g. the high seas created by onshore hurricane or tornado winds) or (b) earth tremors or volcanic eruptions that set up huge tidal surges. *River flooding*, a more frequent hazard is related to heavy precipitation, rapidly melting snow, and – very rarely – the collapse of natural or artificial dams and the release of impounded waters.

In both coastal and river flooding the hazards are made greater by the attractiveness of such locations as places for human settlement. Some 12 per cent of the US population elects to live in areas where there are periodic floods. The fact that flood losses have topped $1 billion in recent years must be weighed against the advantages – fertility, flatness, etc. – that make areas near the water so attractive. The *floodplain* of a river is created by water spilling over the normal channel limits (often banks or levees) and depositing sediment over the surrounding plain. Under natural conditions, the floodplain, or 'spillplain', will be covered by water for small but rather regular number of days each year. Where there are human settlements, this natural overspill and sedimentation process is interrupted by building artificially high levees and protective dykes. This has the desired effect of keeping a stream within its main channel, but it means that additional debris is deposited on the stream bed, raising the height of the channel and creating a need for still higher artificial levees. Many of the world's major streams now run across their lower floodplains in artificially constrained channels some metres above the level of the surrounding densely settled land. When floods occur under these conditions, the depth of the flooding and its impact on the human population are immense. The effects are particularly severe in densely used floodplains.

Pervasive and Intensive Hazards

How then can we evaluate the climatic environment as a hazard? In a book entitled *The Environment as Hazard* by geographers Ian Burton, Robert Kates, and Gilbert White, seven measures found to be significant in human terms are proposed. The first is the *magnitude* of an event, say, the height of a flood or the intensity of a windstorm. The next four all relate to time: the *frequency* of occurrence, the *duration* of the event, the *speed of onset* from first warning signs to peak, and the *temporal spacing* in terms of randomness or regularity. The last two are more specifically geographical: the *areal extent* over the earth's surface and the degree of *spatial concentration* within that area. Figure 3.22 summarizes the last six characteristics for a

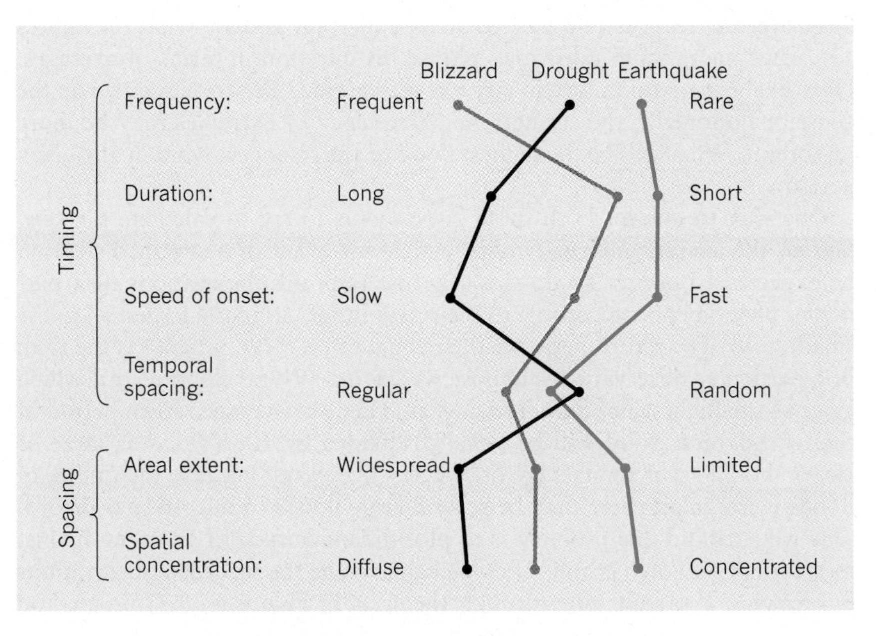

Figure 3.22 Hazard profiles Ways in which it is possible to draw a profile for natural hazard events in terms independent of magnitude. Four aspects of timing and two of spacing are given, each with its range along a representative measure. Sample curves for a drought, a blizzard, and an earthquake are shown.

[Source: I. Burton, R.W. Kates, and G.F. White, *The Environment as Hazard* (Oxford University Press, New York, 1978), Fig. 2.4, p. 29.]

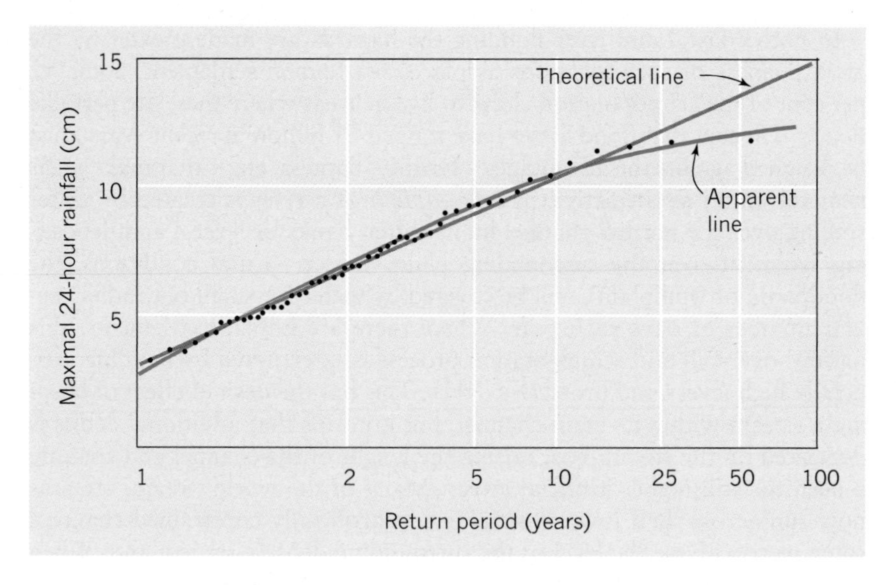

Figure 3.23 Predicting extreme values for natural hazards The graph shows the frequency with which the greatest amount of rainfall in a 24-hour period reached particular levels in any one year during the period 1950–56 in Nantucket, Massachusetts, United States. Each year's record is shown by a point on the graph with the lowest value at lower left and the highest part at upper right. Note that the numbers on the horizontal axis of the graph are not regularly spaced. By drawing the axis in this way the sequence of years is found to fall along a line (labelled the 'apparent line'). This apparent line can be approximated by a straight line (the 'theoretical line'). Extending the straight line allows an estimate to be made of the greatest rainfall to be expected once in a century – about 14 cm (5.51 inches). Further extensions beyond the 100-year mark on the graph are possible but are likely to be of uncertain accuracy.

[Source: D.M. Hershfield and M.A. Kohler, *Journal of Geophysical Research*, **75** (1960), p. 1728, Fig. 1. Published by the American Geophysical Union.]

major drought and a severe blizzard, and contrasts these with the earthquake hazard which we met at the end of Chapter 2. You may like to draw profiles for some other events, say, a major flood, on the same diagram. Note that a basic difference emerges between pervasive hazards, such as drought, and *intensive* hazards, such as tornadoes. Thus tropical storm Agnes, which moved up the east coast of the United States from 19 to 23 June 1972, brought intense hazards along a narrow track for a short duration. But by the time the storm had passed, 118 people had lost their lives and £3.5 billion worth of property damage had occurred – in monetary terms the greatest material disaster from natural causes suffered by the United States to date.

Return Periods

But how frequently is such a disaster to occur? One approach is to look back over the records of a particular type of event and see what the figures say. Thus an agriculturalist may phrase his question in terms of averages. How likely is a crop failure in any ten year period? But to the settler on the coast or floodplain, the strength and frequency of extremes may be more important. What will be the highest flood or the strongest wind in any given period?

One way to answer both these questions is to try to calculate a ***return period***, the average interval within which one event of a specified size can be expected to occur. To do this, we first rank all observations of a particular phenomenon according to their magnitude, from the largest (l) to the smallest (n). The return period is then equal to $(n + l)/r$, where r is the rank of a particular observation. Suppose we have a 49-year flood record which gives us the highest flood level each year. Then the average return period of the tenth-largest flood will be (49 + 1) divided by 10; a flood as large or larger should recur on average once every five years. Because the timing of floods is irregular, there may be several large floods in our 49-year record. One way around this problem is to plot the magnitude of an event against its return period on a graph. Thus we can average the recorded observations by drawing a straight line through them, as in Figure 3.23. This method

Table 3.3 Leading types of natural disaster

Type of disaster	Relative frequency	Loss of life per disaster (average)	Percentage of loss of life from each cause (total = 100%)
Floods	100	828	39
Typhoons, hurricanes	73	1034	36
Earthquakes	41	652	13
Tornadoes	32	51	1
Gales, thunderstorms	15	654	5
Snowstorms	13	130	1
Heat waves	8	292	1
Cold waves	6	259	1
Volcanic eruptions	6	555	2
Landslips	6	221	1

Figures are averages based on tables by Sheetian and Hewitt updated to cover a 25 year period. Note that the relative fequency of floods has been arbitrarily set at 100 to provide a ruler against which the other frequencies can be measured.

allows us to estimate the most likely return ratios on the basis of all the records available. Figure 3.23 shows that we can expect Nantucket to have a heavy rainfall of 10 cm/day (4 in/day) once in a decade and a 14 cm/day (5.5 in/day) rainfall once every century. Of course, these estimates are only averages. The rainfall that comes once every thousand years may still fall next week!

This type of frequency analysis depends on rather simple assumptions. It provides a useful first approximation of the size of the risks a given environment is likely to pose. We assume, for example, that the pattern of floods or heavy rainfalls is not undergoing cyclic changes. If floods of a certain river are getting steadily worse, possibly because of deforestation, then this method will underestimate the size of a 100-year flood. Like rainfall, floods may tend to occur in clusters. The great floods on the Ohio and Mississippi rivers in 1936 and 1937 were paralleled three decades later by the disastrous floods of 1964 and 1965.

The Human Dimension

An approach to hazards in geophysical terms is only part of the story. Some hazards may become more frequent because of human action. A good example is floods in urban areas; storm water now flashes off vast areas of concrete and asphalt straight into streams where once it filtered more slowly through soil and vegetation. For others the geophysics remains the same, but the hazard grows. This is true for coastal floods where an inundation that affected few and was a local agony only a century ago may now affect many and be of international concern. Flooding, whether from rivers or coastal ruination, remains the largest single cause of loss of life from natural disasters (see Table 3.3).

As the human population has grown, so it has tended to move into areas of high attraction but high environmental danger. We saw in Chapter 1 that many cities are located in coastal areas and many more are in vulnerable flood plains. The property values in floodable areas of such cities are very high. The world as a whole has become increasingly dependent on efficient food production from a few regions which, in some cases, notably the mid-latitude wheatlands, may show major swings in their productivity from one harvest to the next.

Hazard research is now a major field in geography, and we shall be returning to it in later chapters (see especially Chapter 11). Here we note simply that it must involve study not only of the range of environmental experience (floods, droughts, cyclones, etc.) but also human reactions.

How do we cope with hazards, both as individuals and as societies? We may simply acccept losses in a resigned way or learn to live with them by sharing risks (say, by insurance or internationalizing disaster relief). Very extreme situations may lead to complete reappraisal of the environment with radical changes in land use or relocation. The stream of 'Okies' (displaced farmers from Oklahoma) moving west to California from the drought-stricken Great Plains in the early 1930s was vividly portrayed in John Steinbeck's *The Grapes of Wrath*. Relocation is the most extreme reaction, and most societies seem to 'stay put and do something different' rather than move on. Study of different reactions may help to evolve new and better ways of coping with hazards.

Reflections

1 Climatic change takes place over a very wide range of timescales, from long-term geological alterations to short-term daily fluctuations in weather. Set up a debate on whether very slow and long-term environmental changes are (a) too remote to concern modern humankind or (b) give us an essential perspective on our tenure on planet Earth. Which side would you support? What arguments would you use?

2 The fact that the earth spins around a tilted axis as it circles the sun has a fundamental effect on our climate. Consider what is happening in Figure 3.3. If the earth's axis were tilted still further from the orbital place (say to, 33½°), what effect would this have on the planet's climate? Where would the Tropics of Cancer and Capricorn and the two polar circles be? What changes would you expect in the climate of your own college town?

3 Rainfall in the tropics has a complex pattern. Use an atlas map to look at the seasonal pattern of precipitation in the tropics between latitudes (10°N and 10°S). To what extent do the maps resemble the idealized model in Figure 3.6? How would you explain the discrepancies?

4 Variations in the loops and curves of the westerlies are a major cause of instability in northern mid-latitude climates. Look carefully at Figure 3.13. Which type of circulation is likely to bring the greatest swings to your own area?

5 Consider the upper map in Figure 3.17 which shows the sea surface temperatures in the Pacific during an El Niño event. Check from newspaper files or Internet sites (see Appendix B) whether a major event is occurring at the time you read this book. What kind of effect is El Niño having outside the Pacific?

6 Climatic hazards come in several different forms. Look at the space–time pattern shown for blizzards and droughts in Figure 3.22. Now take hurricanes and see how their profile fits into that pattern.

7 Review your understanding of the following concepts using the text and the glossary in Appendix A of this book:

El Niño	radio-carbon dating
equinox	Recent epoch
hurricane	Rossby waves
Pleistocene epoch	Sahel
pollen analysis	solstice

One Step Further . . .

A useful starting point is to look at the essays on climate by Andrew Goudie and Kenneth Hare in I. DOUGLAS *et al.* (eds), *Companion Encyclopaedia of Geography* (Routledge, London, 1996), pp. 44–66, 482–507 and then check on any unfamiliar terms in David THOMAS and Andrew GOUDIE (eds), *Dictionary of Physical Geography*, third edition (Blackwell, Oxford, 2000). Two classic accounts of climatology for geographers are provided in R.G. BARRY and R.J. CHORLEY, *Atmosphere, Weather, and Climate,* seventh edition (Routledge, London, 1998) and P. ROBINSON and

A. HENDERSON-SELLERS, *Climatology*, second edition (Longman, Harlow, 1998). Long-term climatic change is considered in A.S. GOUDIE, *Environmental Change* (Oxford University Press, 1992) and its contemporary impacts in H.H. LAMB, *Climate, History and the Modern World* (Methuen, London, 1982).

On climate change and climatic hazards see I.M. MINTZER, *Confronting Climate Change* (Cambridge University Press, Cambridge, 1992) and H. COWARD and T. HURKA, *Ethics and Climate Change* (Wilfrid Laurier Press, Waterloo, 1993).

The climatological 'progress reports' section of *Progress in Physical Geography* (quarterly) will prove helpful in keeping up to date. Look also at *Scientific American* (a monthly) and *New Scientist* (weekly) which carry regular reports on current work by environmental scientists. Journals devoted specifically to climatic topics include *Weather* (quarterly), *Journal of Climatology* (quarterly) and *International Journal of Climatology* (quarterly). For readers with access to the World-Wide Web see also the sites recommended in Appendix B at the end of this book for topics relevant to this chapter.

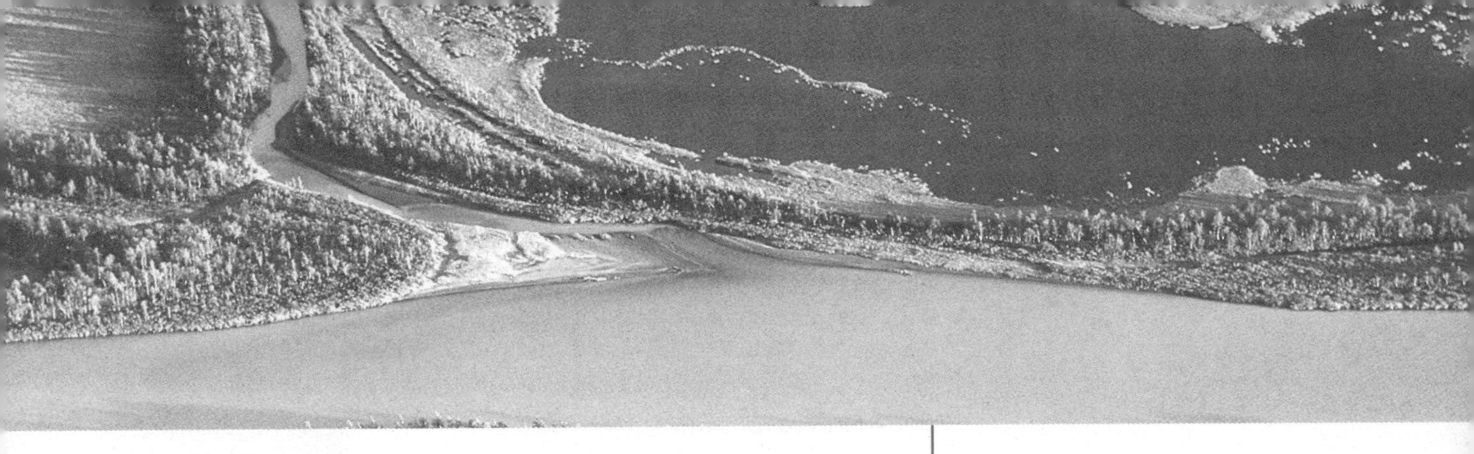

CHAPTER 4

The Biosphere

■ denotes case studies

Figure 4.1 The diversity of the biosphere A summer aerial photograph of a river delta in the Sarak National Park in the Lappland Area of northern Sweden. The drier banks of the delta support coniferous trees giving way to swamps and open water lakes. North is at the top of the photo and the east–west range is around 1.6 km (1 mile). [Source: Tony Stone Images /Hans Strand.]

> We will now discuss in a little more detail the struggle for existence.
>
> CHARLES DARWIN *The Origin of Species* (1859)

The fourth of the earth's great spheres, the *biosphere*, is also its most fragile. In the beginning (4.5 billion years ago), the earth was too hot for life. At the end (roughly the same time into the future) the sun when it comes to the end of its star cycle will engulf life on earth with a fireball. Our biosphere has then a limited duration that is bounded by the existence of water in its liquid state. We live in limited window of opportunity in which the wondrous diversity of life and landscape as typified in the lake vegetation of Figure 4.1 (p. 102) is but a passing fragment.

Exactly when life appeared on the earth is not known but the earliest evidence suggests it was around 3.8 billion years ago. The very earliest life was probably some form of bacteria or blue-green algae. They have simpler structures than other life forms and lack a nucleus (so called prokaryotic life forms). Evolution of the remaining life forms involving a nucleus (eukaryotic forms) may well have had to wait until the new environmental conditions imposed by an oxygen-containing atmosphere were established on the planet. Abundant fossil evidence of life is very much later (0.8 billion years ago).

As we can see from Figure 4.2(a), the earth's life forms evolved in a sequence of increasing complexity. If we now focus on the recent period for which more evidence is available (see (b)) we again see a sequence but it is one punctuated by periods of abundance and interrupted by periods of crisis in which species are reduced or even eradicated. Five such crises are shown. Perhaps the best known is that 65 million years ago (the Cretaceous–Tertiary crisis) which brought an end to dinosaurs on land and ammonites at sea. While the timing of these crises is well recorded, the

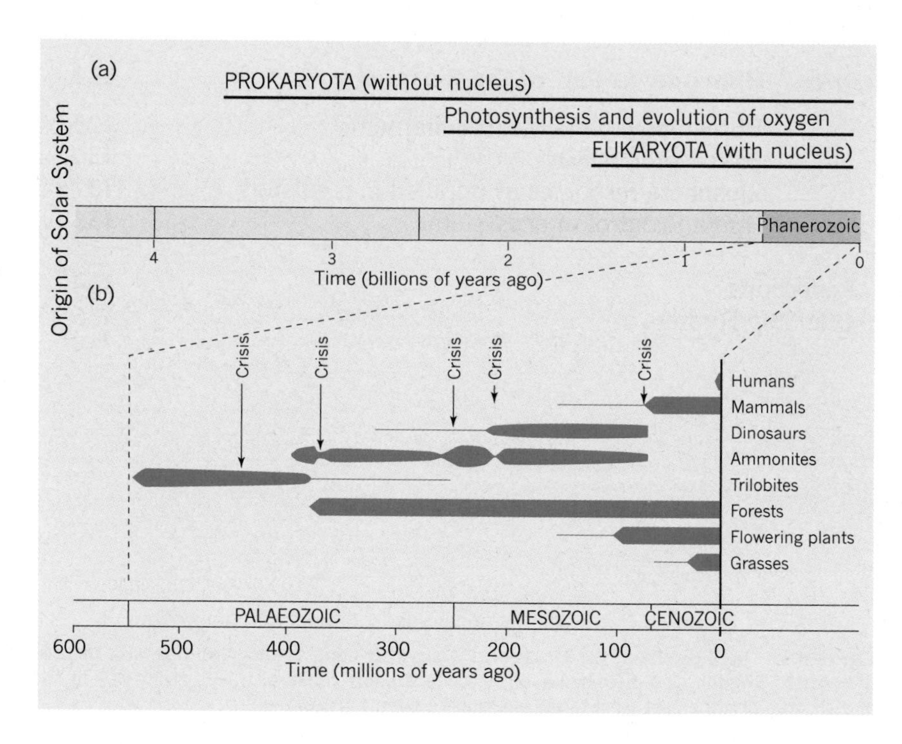

Figure 4.2 Development of life on planet earth (a) Major divisions of time since the origins of the earth. (b) Examples of living organisms within the most recent time period. Note the interruption of evolution by periodic crises with species eradication.

[Source: A. G. Fischer, in I. Douglas *et al.* (eds), *Companion Encyclopaedia of Geography* (Routledge, London, 1996), Fig. 3.1, p. 68.]

causes of such events is a matter of controversy. Hypotheses have ranged from variations in the earth's orbit, through meteorite impacts, through encountering cosmic dust clouds, to the impact of supervolcanoes.

What is clear is that, to judge from recent rates of species extinction, we should now add a sixth crisis on the right-hand edge of Figure 4.2(b). For the advent of humans (see Chapter 5) and the staggering rise of their populations (see Chapter 6) have introduced a wholly new element into the biosphere. *Homo sapiens* has shown an unprecedented capacity for interfering in animal and plant systems, and now poses an unprecedented long-term threat to the stability of the global biosphere. Much of the rest of this volume is concerned with that threat.

In this chapter we look first at the nature of ecosystems (Section 4.1) before going on to consider the great biological realms, the biomes, into which geographers divide the world (Section 4.2). We then take one of those biomes, the tropical rainforest (Section 4.3), and consider it in more detail. Finally, we go back to look at the newcomers on the world stage – the human population – and try to see how they fit into the biosphere (Section 4.4).

Ecosystems 4.1

An *ecosystem* is a system in which plants and animals are linked to their environment through a series of links, some of which form feedback loops. (A glossary of ecological terms used in this section is given in Table 4.1.) The term 'ecological' dates back to 1868, when the German biologist Ernst Haeckel used it in discussing his studies of plants in relation to their

Table 4.1 Key terms in the study of ecosystems

Term	Definition
Biomes	The major environmental zones of the earth marked by a distinctive plant cover (e.g. the subarctic tundra biome).
Carrying capacity	Largest number of a population that the environment of a particular area can carry or support.
Climax	State of equilibrium reached by the vegetation of an area when it is left undisturbed for a long period of time.
Communities	Groups of animals and plants that live in the same environment and depend on each other in some way.
Ecological efficiency	The ability of organisms in a food chain to convert the energy received into living matter.
Ecology	The study of plants and animals in relation to their environment.
Ecosystems	Ecological systems in which plants and animals are linked to their environment through a series of feedback loops.
Food chains	Describe the series of stages that energy goes through in the form of food within an ecosystem.
Food webs	Complex networks of food chains.
Predator–prey relations	Describe the links between the population of one set of animals (the prey) that are hunted for food by another set (the predator).
Seres	The orderly sequence of change in the vegetation of an area over time as it passes through transition stages (seres) toward an equilibrium.
Trophic levels	Main main stages in the food chain, green plants occupying the first level, plant-eaters the second, animal-eaters the third.

environment. It stems from the simple Greek word *Oikos*, meaning 'a house' or 'a place to live in' and serves as a direct link to the geographer's concern with the earth as the home of humanity. You will note close links between geography and ecology at many points throughout this book. In this section we will look at the nature of ecosystems, their structure, and why their understanding is so important in geography.

Figure 4.3 Lakes as environmental features Outdubs Tarn (centre) in the Lake District of northwest England is one example of the lake ecosystem discussed in this chapter. This small lake (about 50 m (165 ft) across) is gradually being filled in by sediment from the stream flowing into it (from the bottom of the picture) and by encroaching vegetation. Mats of reed and mosses are shown as a light ring around the dark waters of the lake itself; scrub timber (darker grey) forms the outer ring, encroaching in turn on the reed–moss layer. The final infilling of this small lake is a slow process in human terms, perhaps a few centuries, but rapid in terms of geological time. Note also the outflowing stream draining to the much larger and deeper lake, Esthwaite Water (left).

[Source: Aerofilms.]

Ecosystems: A Small-scale Example

The easiest way to unravel the structure of an ecosystem is to take a small-scale, familiar example. We could stay on the beach as in Chapter 1 and look at the reactions of plant and animal life to the twice-daily changes of environment in the tidal zone. A still better example can be found by moving inland to a small lake such as that in Figure 4.3.

A lake is a body of standing fresh water. What physical inputs and outputs does it receive? As a participant in the hydrologic cycle, it derives inputs of fresh water from stream inlets and from rainfall and loses water through stream outlets and evaporation. Its most important input is, however, the solar energy it receives from direct sunlight. This will warm the upper layers of the lake very strongly in summer and set up important vertical differences in water temperatures over the seasons.

In addition to the physical processes by which water flows, sediment is deposited, temperatures change, and so on, there are also far more complex biological processes going on in and about the lake. Sunlight furnishes energy used by microscopic green plants in the lake (the ***phytoplankton***) to convert inert chemicals in the water into food. These organisms provide food for small larvae and crustaceans (the *zooplankton*), which are eaten by small fish (see Figure 4.4(a) and (b)).

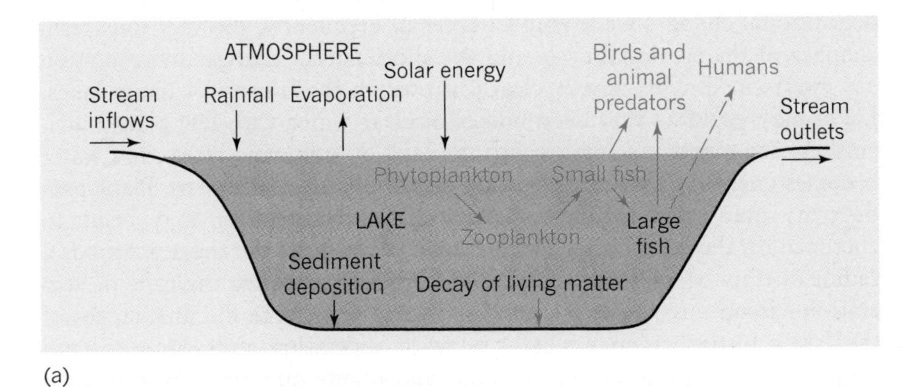

(a)

Water striders
Whirligig beetles
Damsel fly & dragonfly nymphs
Predacious diving beetle
Backswimmers
Frogs
Hydra
Catfish
Mini-crustaceans
Protozoa
Ostracoda
Rotifers
Fly larvae
Water boatmen
Flatworms
Snails
Diatoms
Flagellated green algae
Nonmotile green algae
Chara, Nitella
Duckweed

(b)

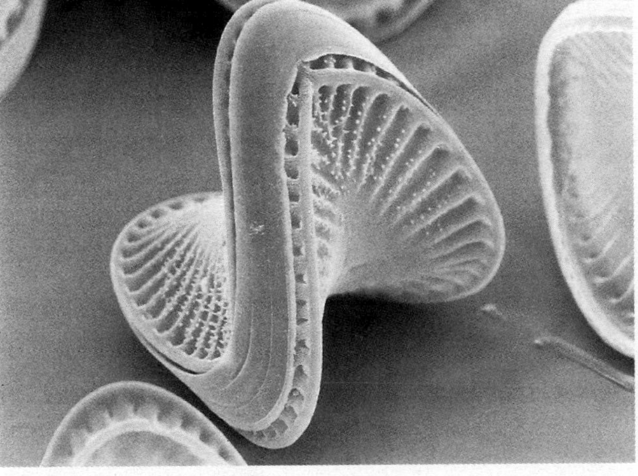
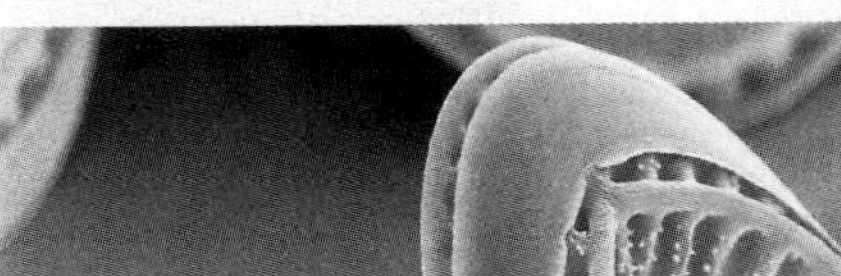

(c)

(d)

Figure 4.4 Lake ecosystems (a) In this simplified flow diagram of the main links in a lake ecosystem, different colours are used to indicate physical and biological processes. (b) A more detailed diagram of a food web shows the many organisms that take part in just one segment of the system. Ecologists find that the more complex the food web is, the more stable the ecosystem is. The critical lower links in aquatic food chains are provided by (c) and (d) microscopic plant life (phytoplankton) and animal life (zooplankton), magnified here to many times their actual size.

[Source: (a) from W.B. Chapham, Jr, *Natural Ecosystems* (Macmillan, New York, 1973), p. 113, Fig. 4.7. (c) and (d) Andrew Syred/Science Photo Library.]

These small fish are eaten by larger fish, which may eventually provide food for animals and for people themselves. Plants and animals die and decay, releasing chemicals back into the lake waters. We show these links in the lake ecosystem in Figure 4.4(a). Of course, this is a highly simplified picture of a process that may involve hundreds of living species and very complex chemical chains. Figure 4.4(b) shows some of the inhabitants of a typical lake.

The long-term history of small lakes is closely linked to the erosional and depositional changes we saw in Chapter 2. Frequently, the very long-term changes of the erosional cycle and the shorter-term changes in vegetation are intertwined. This is very clearly shown in the history of many lakes. Lakes may begin as very deep bodies of clear water with few plant nutrients. As sediments are carried into the lake by inflowing rivers, the water becomes chemically enriched and the depth of the lake lessens. Plant productivity may build up in the lake itself, and vegetation may begin to encroach on the lake margin, with rooted plants near the shore acting as a sediment trap. Mosses and sedges build up, and great floating rafts of vegetation extend into the lake itself (see Figure 4.5(a)). In the final stages of the lake's history, it may slowly fill with vegetation and sediment until eventually a forest is established and land plants take over from aquatic ones. Figure 4.5(b) shows just such a generalized succession of vegetation.

Cycles in the Ecosystem

Not all ecosystems are as clearly defined as lakes. Many have boundaries that are hard to establish and that conceal very important internal vari-

(a)

Figure 4.5 Plant succession as a cause of environmental change
(a) Lakes are often good examples of the intertwining of long-term erosional changes and short-term vegetational changes. (b) Four stages in an idealized sequence of plant succession in a lake environment. These stages are not of equal length: commonly, the later phases of a succession are slower than the earlier ones. In a geological sense, all lakes are ephemeral and have a finite life span.

[Source: FLPA/R. Gutshall/Dembinsky.]

ations. The simple lake ecosystem does, however, illustrate two elements that are critical in all ecosystems – from the smallest to the largest. These are the cycling of energy chemicals (especially carbon) through biological populations, and the linking of these biological populations into food chains. We now look at each element in turn.

Energy Cycles Tracking flows of energy through an ecosystem has proved difficult. Figure 4.6 shows one attempt to measure energy flows in an oak–pine forest on Long Island, in New York state. The energy being stored in this New England type of mixed woodland is estimated to be 2650 g of living matter per m^2 each year, but this is only a rough guess.

If you follow the flows around, you will see that 19 per cent of the energy received is stored by the trees, and a further 2 per cent in the humus of the upper layer of the soil under the trees. Thus the total storage is about one-fifth of the total energy received. The remainder is returned to the atmosphere through respiration. Comparison of different stands of vegetation shows that the stored share falls as the stands become more mature. So, if you abandon a field in New England, it will be shrub-covered within about 15 years, and covered with pine forest within 50 years. If you leave it for more than a century, then oak and hickory will take over from the pines. The bushes and trees simply represent stored energy and, in the early phases of regrowth, the storage is rapid but decreases with increasing maturity. Ecologists believe that very old forests would form a self-perpetuating plant community, called a *climax*. Under climax conditions the input and output of energy flows would be balanced with no accumulated storage. In other words, the forests would show little or no change in mass from decade to decade.

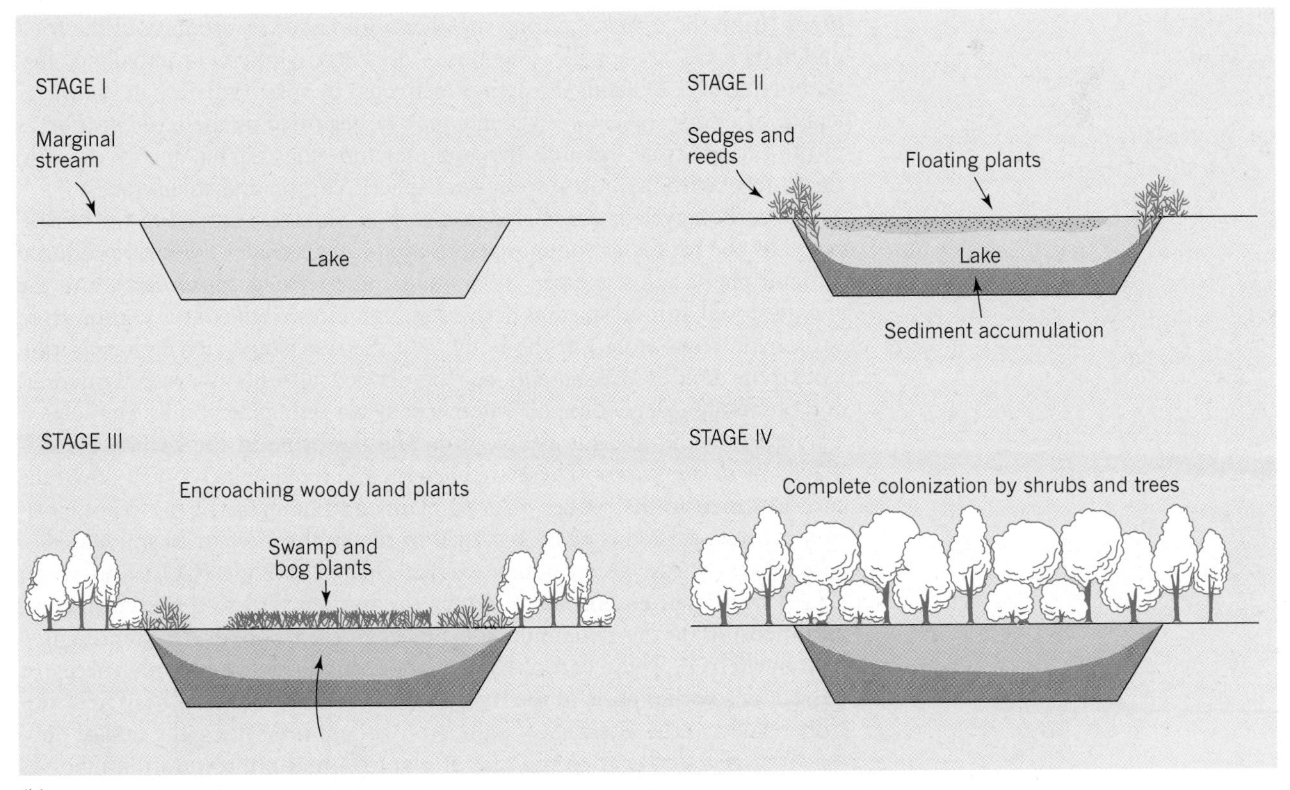

(b)

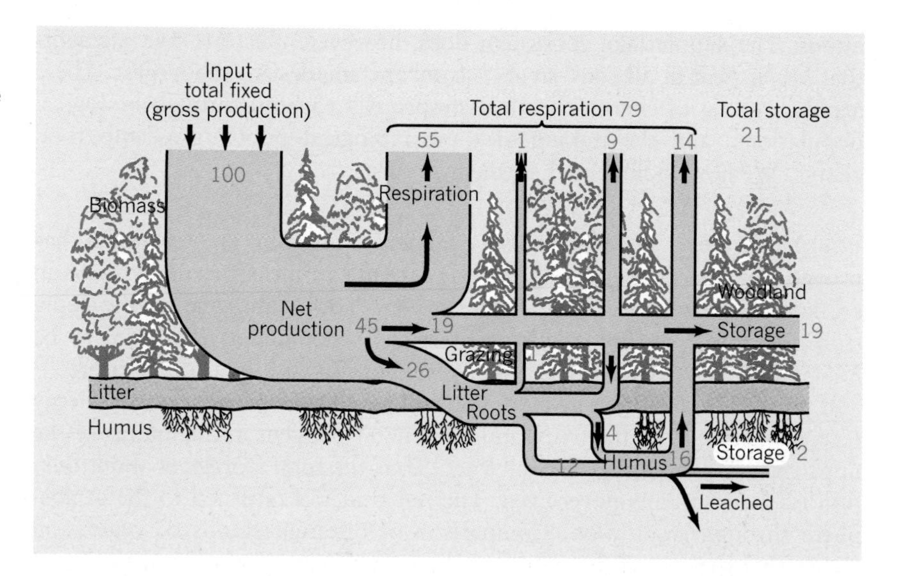

Figure 4.6 Energy flow in a woodland ecosystem The quantities were calculated from a study of oak–pine forest on Long Island, New York. The total energy input of 2650 g living matter/m^2 year is set equal to 100. Of this 79 per cent is returned as respiration, while the remainder is stored in the increased biomass of the forest. Note that storage in the forest (21 per cent) is predominantly in the biomass above the surface.

[Source: G.M. Woodwell, *Scientific American*, No. 3 (Sept., 1970), p. 64.

Nutrient Cycles One essential element in both the lake ecosystem and the woodland ecosystem is the conversion of solar energy into living matter. How is this piece of alchemy achieved? Let us use the *carbon cycle* to illustrate one of the most important aspects of this conversion process. (See Box 4.A on 'The global carbon cycle'.)

We have already seen that carbon is available in the earth's lower atmosphere as carbon dioxide (CO_2). This gas forms a small but vital 0.033 per cent of the total volume of air. It is important climatically as a heat-absorbing blanket, helping to regulate air temperatures near the earth's surface. Biologically, carbon dioxide is essential to plant growth: green plants with the pigment chlorophyll combine carbon dioxide and water through *photosynthesis* (from the Greek, 'putting together with light') to produce all the food materials necessary for life. (See Box 4.A.) Photosynthesis is actually a cluster of interrelated chemical reactions activated by solar radiation at the wavelength of visible light. Green plants may be regarded as the basic *producers* in the carbon cycle, because they manufacture consumable energy (food in the form of carbohydrates) from atmospheric carbon and solar energy.

The carbon cycle is completed and carbon dioxide returned to the atmosphere by the processes summarized in Box 4.A. Consider the food produced by land plants. This is eaten by animals (here termed *consumers*), and the energy stored as food sustains activity at high rates. Some of the carbon from carbohydrates is stored in the body, and the rest is excreted by respiration as carbon dioxide. Consumers can be divided into herbivores, carnivores, and omnivores, depending on whether they eat only plants, only animals, or a mixture of the two (as people do). The final role in the carbon cycle is played by *decomposers*. These are bacteria and fungi, which break down the carbon stored in the tissues of dead plants and animals. In the decomposition process, carbon is again returned to the atmosphere or to soil water.

In other words, green plants extract carbon dioxide (CO_2) and water (H_2O) from their environment, return the oxygen (O_2) to the environment, and incorporate the remaining substances into carbohydrates (represented here by CH_2O). These carbohydrates are decomposed to provide energy or passed on to other parts of the food chain. Rates of photosynthesis are critically related to the intensity of light. At low light intensities, the rate of photosynthesis is slower than the rate of plant respiration; respiration involves the oxidation of the carbohydrates and the release of carbon dioxide and

Box 4.A The Global Carbon Cycle

We can describe the overall process of photosynthesis in the carbon cycle by a simple chemical equation:

$$\text{Light} + nCO_2 + nH_2O \xrightarrow[\text{of chlorophyll}]{\text{in the presence}} (CH_2O)_n + nO_2$$

In other words, green plants extract carbon dioxide (CO_2) and water (H_2O) from their environment. The oxygen (O_2) is returned to the environment, and the remaining substances are incorporated into carbohydrates (represented here by CH_2O). These carbohydrates are decomposed to provide energy or passed on to other parts of the food chain. Rates of photosynthesis are critically related to the intensity of light. At low light intensities, the rate of photosynthesis is slower than the rate of plant respiration; respiration involves the oxidation of the carbohydrates and the release of carbon dioxide and water. At a slightly higher light intensity, the two rates are equal. Above this point, the rate of photosynthesis surpasses the plant respiration rate and carbohydrate products accumulate. The greatest, or saturation, rate of photosynthesis is reached in full sunlight. In addition to light, photosynthesis requires adequate moisture and proceeds most rapidly at temperatures between 10 °C and 50 °C (50 °F and 122 °F).

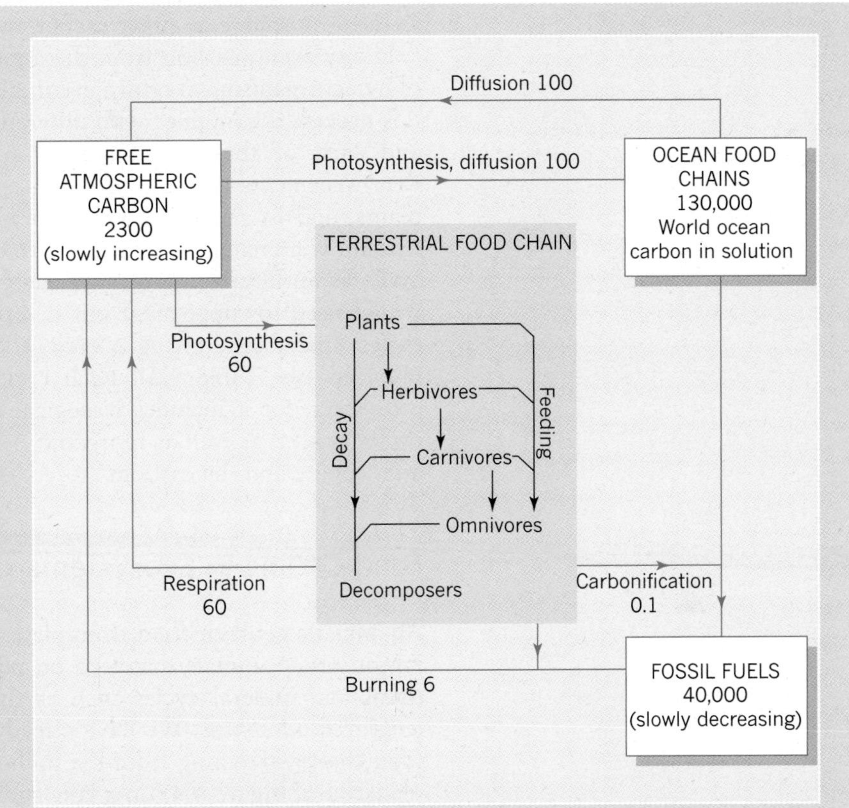

Photosynthesis is a critical element in the wider carbon cycle. The figures indicate the estimated stores (boxes) and annual flows (arrows) of carbon in the world carbon balance in units of 10^9 tons. *Carbonification* is the conversion of dead plant and animal remains into coal, oil, and similar fossil fuels. *Diffusion* refers here to the interchange of carbon dioxide gas between the atmosphere and the oceans by molecular mixing. (Note the different use of the term *diffusion* in Chapter 13.) The carbon cycle shown is only one of the major cycles of important chemical elements through the environment. Similar flows occur in the ***nitrogen cycle*** and the *potassium cycle.*

[Source: After J. McHale, *The Ecological Context* (George Braziller, New York, and Studio Vista, London, 1971), Fig. 21, p. 52. Copyright 1971 by J. McHale. Reprinted with permission.]

water. At a slightly higher light intensity, the two rates are equal. Above this point, the rate of photosynthesis surpasses the plant respiration rate and carbohydrate products accumulate. The greatest, or saturation, rate of photosynthesis is reached in full sunlight. In addition to light, photosynthesis requires adequate moisture and proceeds most rapidly at temperatures between 10 and 50 °C (50 and 122 °F).

Not all producers and consumers are decomposed as soon as they die. Organic matter is stored and concentrated geologically for millions or billions of years as peat, lignite, coal, petroleum, and natural gas. (See Section 10.2 on fossil fuels.) Plants are also burned as fuel by people. Like eating, burning separates the elements in carbohydrates and returns carbon to the atmosphere as either carbon monoxide or carbon dioxide.

In any event, carbon from the atmosphere is circulated through a chain of living organisms to return eventually to the atmosphere. At each stage in this process it combines with different elements in various chemical forms, and each of these combinations is accompanied by energy transfers. Rearrangements of molecules and energy transfers (by photosynthesis in plants, and by metabolic synthesis in animals) are the essential processes that allow human life on the earth to continue. We have selected the carbon cycle, as an illustration of how energy transfers are accomplished, but we would need to supplement our description of it by a description of other cycles, such as the nitrogen cycle, to fully explain the exchange processes involved (see Table 4.2). Each cycle represents an essential link in the ecosystem, for it includes biological elements (producers, consumers, and decomposers) as well as inorganic elements (e.g. carbon dioxide gas in the atmosphere and the carbons stored as fossil fuel).

Food Chains in Ecosystems

All animals get their food from plants, either directly or indirectly by feeding on other animals that feed on plants. Thus, the process of photosynthesis and mineral cycles such as the carbon cycle provide the basis of lengthy *food chains*. We have already encountered a simple example of a food chain, stretching from the millions of microscopic plants (the phytoplankton of Figure 4.4) on a lake surface to human fishermen.

In the world's oceans, fish such as tuna that are caught and consumed by humans are directly dependent on a three- or four-link chain. Phytoplankton are consumed by larvae and shrimps, which are in turn eaten by squids and small fish, which form part of the food consumed by tuna. However, in each case it takes from 5 to 10 food units (calories) of the prey to produce one unit of the predator; this difference is termed a *food-conversion ratio*. One unit of tuna consumed by a human being represents an estimated 5000 units of phytoplankton.

Table 4.2 Major geochemical cycles in the ecosystem

Group	Cycled compound or element	Role in the biosphere
Compound	Water (H_2O)	Major compound of biosphere. Wood is 50% water, many mammals (including humans) are 85% water. Water acts as solvent for other mineral nutrients.
Major elements	Oxygen (O)	Major constituent of living matter (70%). Basic building block. Oxidation important component in growth processes.
	Carbon (C)	Major constituent in living matter (18%).
	Hydrogen (H)	Major constituent in living matter (11%).
Minor elements	Nitrogen (N)	Plentiful in atmosphere but scarce in biosphere.
	Sulphur (S)	Bacteria play major part in releasing sulphur compounds for recycling.
	Phosphorus (P)	Important in photosynthesis. Limiting element in ecosystem growth.

It is useful to represent the levels in a food chain as a series of food pyramids, as in Figure 4.7. Each step of the pyramid is termed a *trophic level* (from the Greek *trophe*, food). The first level (T_1) at the base of the pyramid is composed of green vegetation with energy contained in the plant tissues. The second level (T_2) consists of herbivorous animals that feed on the plants; the third level (T_3), of carnivorous animals that feed on herbivorous animals; the fourth level (T_4), of carnivorous animals such as humans that feed on other carnivorous animals and all the lower tiers. The fifth component is made up of decomposers, which break down the dead tissues of organisms at all the other levels of the food chain.

Biologists have shown us the exact structure of trophic levels for individual communities. For example, they have analyzed the food chains and conversion ratios of over 200 species of fish in coral reefs in the Marshall Islands in the Pacific Ocean. By estimating the dry mass of organisms, from plankton and algae to sharks, they showed that the base of the pyramid (T_1) consisted of producers with a weight of 703 grams (g) per m^2. Above these organisms were herbivores (132 g) and finally carnivores (11 g). Other researchers have tried to estimate the actual energy flows between the different species in a community.

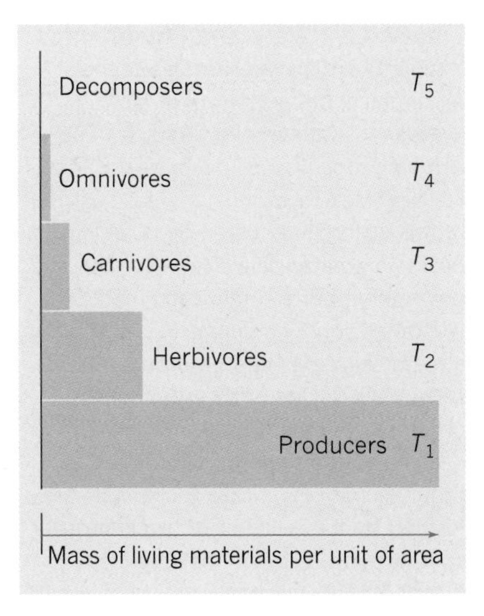

Figure 4.7 Trophic levels The pyramid shows the relative dry weights of living materials typical on the four main trophic levels of an ecosystem.

Selecting Ecosystem Units

Because environmental variation occurs on many geographic scales, geographers have developed systems of regions that can be modified to these scales. One of the most versatile regional units identified so far has been the *watershed*, or *catchment area*, of a stream (see Figure 4.8). Stream watersheds form a convenient unit because they can be simply and unambiguously defined from a topographic map. They are independent of scale, in that large river basins such as the Amazon can be broken into a hierarchic system of smaller basins; like a toy Russian doll within larger dolls, each smaller basin fits exactly within the next larger one. Each subdivided basin can be identified and its streams numbered in a way that provides a measure of relative size.

Watersheds also have other advantages as a basis for regional divisions. Soil-related changes in vegetation reflect location within the watershed. Also, the physical features of a basin directly affect the hydrologic characteristics of the streams draining it. A rainstorm falling on a long, narrow basin is likely to produce a lower peak in the water level of the stream draining it than a similar storm falling on a rather broad, almost circular basin. Watersheds therefore form useful ecosystem units for agencies interested in flood control, navigation, hydroelectric power production, or soil conservation.

The founding of the Tennessee Valley Authority in 1933 (see Box 18.C) saw a trend in using river basins as planning units that has spread around the world. Schemes for the São Francisco River in Brazil, the Snowy River in southeast Australia, or the lower Mekong Delta all involved a combination of planning measures keyed to the use of water resources. Their adoption allowed the competing demands for water – for irrigation, flood control, hydroelectric power production, and navigation – to be dealt with through a single controlling authority. Watershed units are employed in heavily urban areas as water pollution control problems worsen. For all these purposes, the watershed provides a convenient and natural spatial unit.

In Australia, the government's Division of Land Use Research has developed a *land unit* system to aid in the ecological mapping of that country

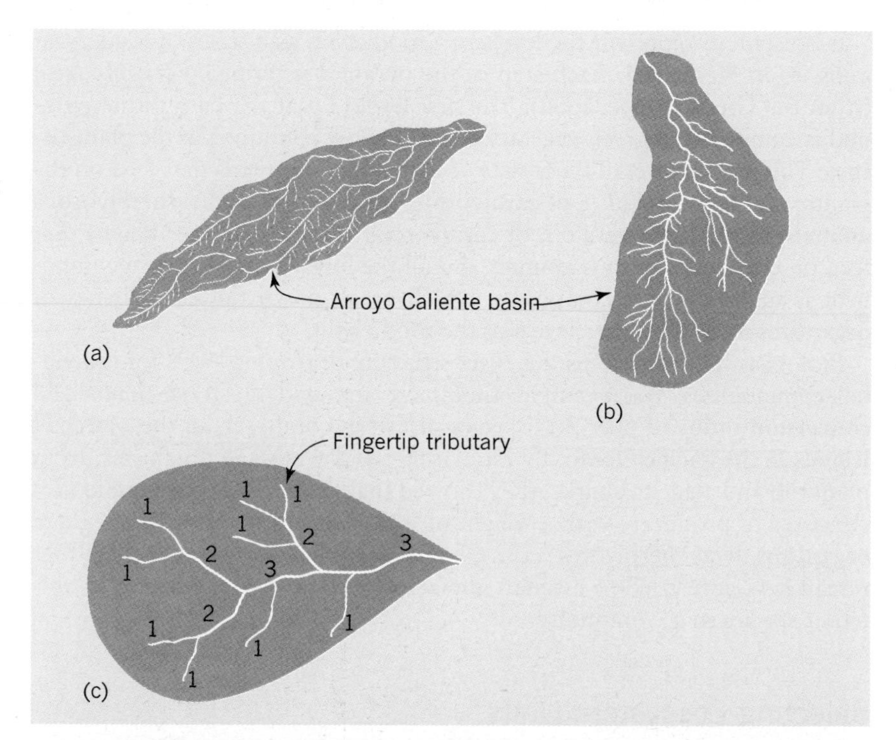

Figure 4.8 Watershed hierarchies Pictures here show examples of the hierarchical breakdown of large watersheds into smaller units. (a) The Arroyo de los Frijoles basin near Santa Fe, New Mexico with (b) a small segment of this watershed, the Arroyo Caliente basin, in greater detail. (c) A further enlargement of a single catchment, illustrates *stream ordering*. A hierarchy of stream segments can be ordered in several ways. One of the commonest ordering systems (Strahler ordering) designates the fingertip tributaries as order 1 channels. Order 2 channels are formed by the junction of two first-order channels; order 3 channels by the junction of two second-order channels; and so on, The watershed in (c) is therefore a third-order unit. (d) Badland gully system at Zabriske Point, Death Valley, California.

[Source: (a), (b) L.B. Leopold *et al*., *Fluvial Processes in Geomorphology* (Freeman, San Francisco, 1964), Fig. 5.4, p. 139. Copyright 1964. (d) Tony Waltham Geophotos, Nottingham.]

(see Figure 4.9). Each land unit describes a local environment that arises from the variation in four elements (climate, geology, soil, and vegetation) and the interactions between them. Land units are mapped from aerial photographs at scales of 1 : 10,000 to 1 : 25,000, and combined into larger units called *land systems* which are shown on maps with a scale of 1 : 1,000,000.

Recognition of land units usually proceeds in the way outlined in Figure 4.9. An initial division is based on the gross geology and land forms. This is broken down into finer terrain types. At a still finer level, the position on a slope is important in determining the type of drainage, the soils, and the vegetation. (See the above discussion of soils.) However, vegetation, and to some extent soils, also reflect the influence of climate conditions. An

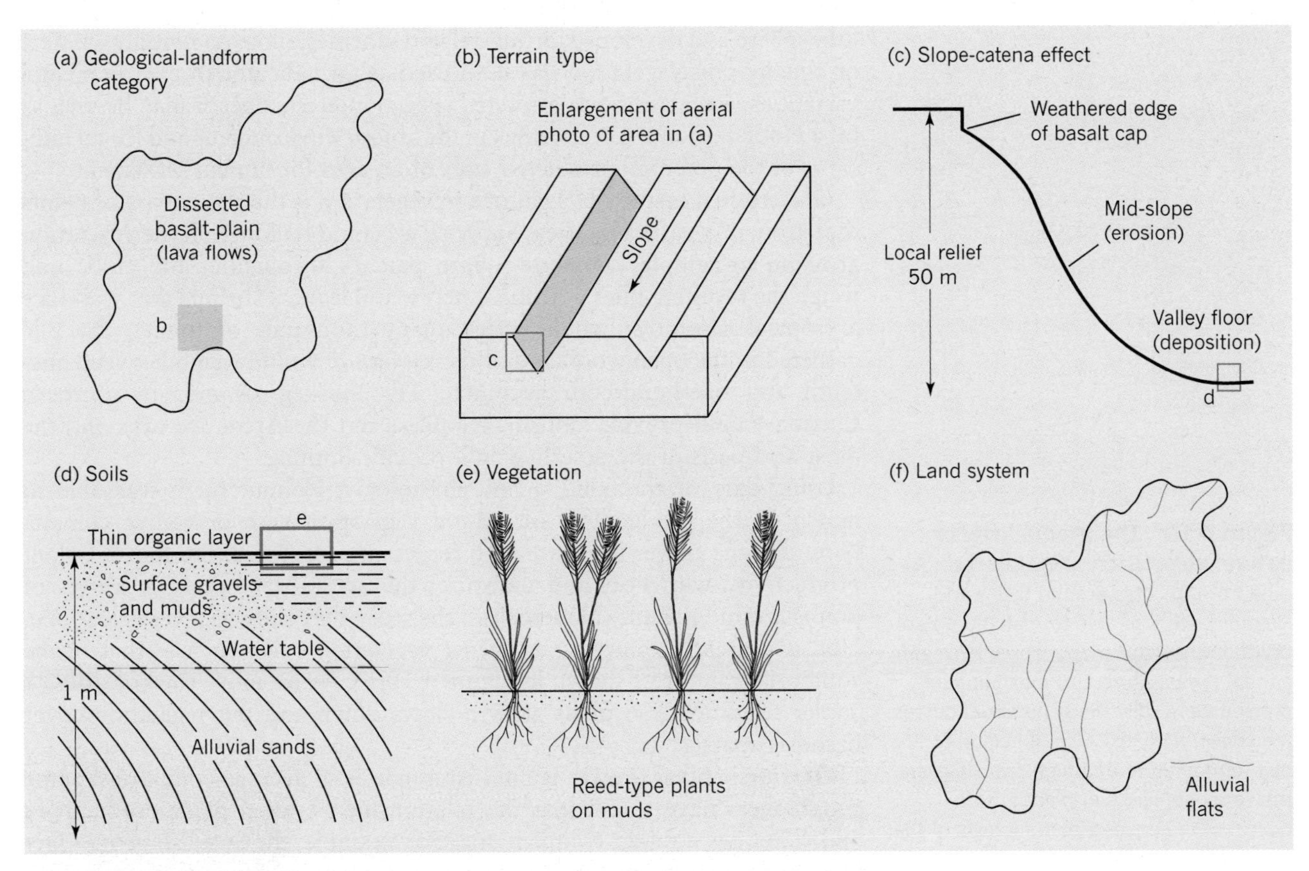

Figure 4.9 Ecological land units Assessment of land potential in Australia is often carried out in terms of land units. Environmental attributes are measured from aerial photographs and at sample sites on the ground to show individual *land units* to be recognized. These combine distinctive vegetation, soil, slope, and terrain conditions within a basic geological-landform category. Land units are combined to give more generalized *land systems* such as the 'alluvial flats land system' shown in (f). Maps of Australia at the scale of 1:1,000,000 showing land systems with their resource appraisal are produced by the Division of Land Use Research of CSIRO, Canberra.

estimate of the capability of each land unit for agricultural production is made, and units are combined into land systems.

Regions Within the Biosphere 4.2

Geographers have long been intrigued with the possibility of reducing the spatial variety of the globe to a single, comprehensive scheme of biosphere regions. We look with some envy at the relative order the labelling of species brought to the previously jumbled world of botany and the periodic table brought to chemistry. Can any similar way be found to classify the mosaic of different environments and ecosystems geographers study?

The Search for a Measure

Many geographers have wrestled with the problem of devising a single measure of environmental differences. Some have concentrated on the lithosphere and developed terrain-related schemes; more commonly, climate or climate-plus-vegetation has been used as an indicator of environmental variations. Vegetation has attracted special interest, since it may be seen as (a) a biotic response to variations in the abiotic environment and (b) an indicator of the potential productive uses of an area for human settlement.

One crude measure of variations in vegetation is the sheer mass of plants that grow in a particular area. Suppose we could bulldoze all the vegetation growing in sample kilometre-square patches throughout the globe and weigh the resulting piles of trunks, stems, and leaves. The pile that was once a tropical *rainforest* would weigh many thousands of tonnes: the pile gathered in the open woodland of the *savannah* would weigh between one-tenth and one-hundredth as much. The muskeg swamps of northern Canada would provide still smaller piles, and the Arctic ice caps and the most arid parts of the deserts would provide nothing.

Using data of this kind, plant physiologist Helmut Lieth was able to match up the productivity of natural vegetation with prevailing climate. Note that the curves in Figure 4.10 represent *average* relationships of plant productivity with heat and moisture; the actual data show a scatter of points around the line. Nonetheless, the story they show is a very clear one. Hot, wet areas produce a very dense vegetation cover such as that of the equatorial rainforest shown in Figure 4.10(b). As we move toward drier or cooler conditions, so plant growth slows down and the vegetation cover becomes sparser.

Of course, plant growth is not just a matter of average conditions. Some geographers have found that the relationship between plant productivity and climate is a rather complex one. For example, the Swedish geographer Sten Paterson has determined that productivity increases with the length of the growing season, the average temperature of the warmest month, the annual precipitation, and the amount of solar radiation, and decreases with the annual range of temperatures in an area. By combining values for these factors into a single index, he was able to give numbers to different places on the surface of the earth, indicating their potential for plant growth. The method and the world map derived from it are described in Box 4.B on 'An index of global productivity'.

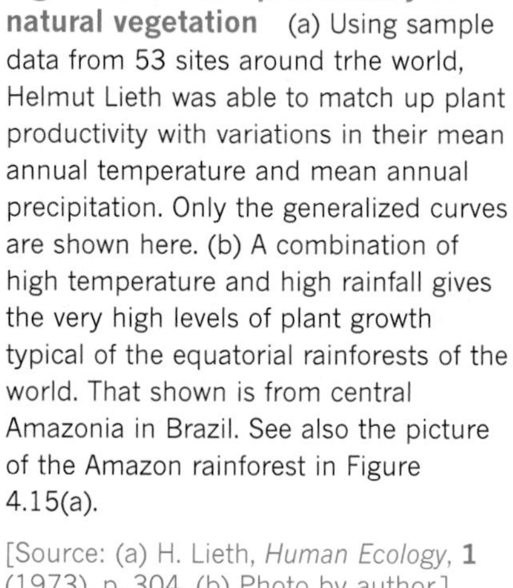

Figure 4.10 The productivity of natural vegetation (a) Using sample data from 53 sites around trhe world, Helmut Lieth was able to match up plant productivity with variations in their mean annual temperature and mean annual precipitation. Only the generalized curves are shown here. (b) A combination of high temperature and high rainfall gives the very high levels of plant growth typical of the equatorial rainforests of the world. That shown is from central Amazonia in Brazil. See also the picture of the Amazon rainforest in Figure 4.15(a).

[Source: (a) H. Lieth, *Human Ecology*, **1** (1973), p. 304. (b) Photo by author.]

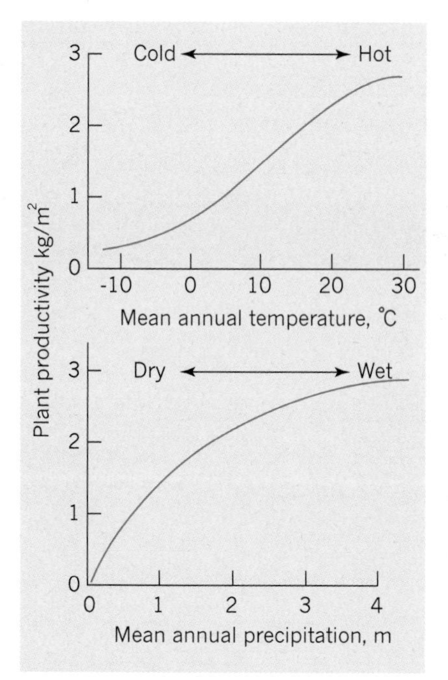

(a)

(b)

Box 4.B An Index of Global Productivity

The earth can be divided into zones of varying productivity, based on the potential for plant growth. (See map.) Here the potential is estimated from climatic elements included in Paterson's index of productivity. Note that the 'A' zone is the most productive and the 'F' zone least productive.

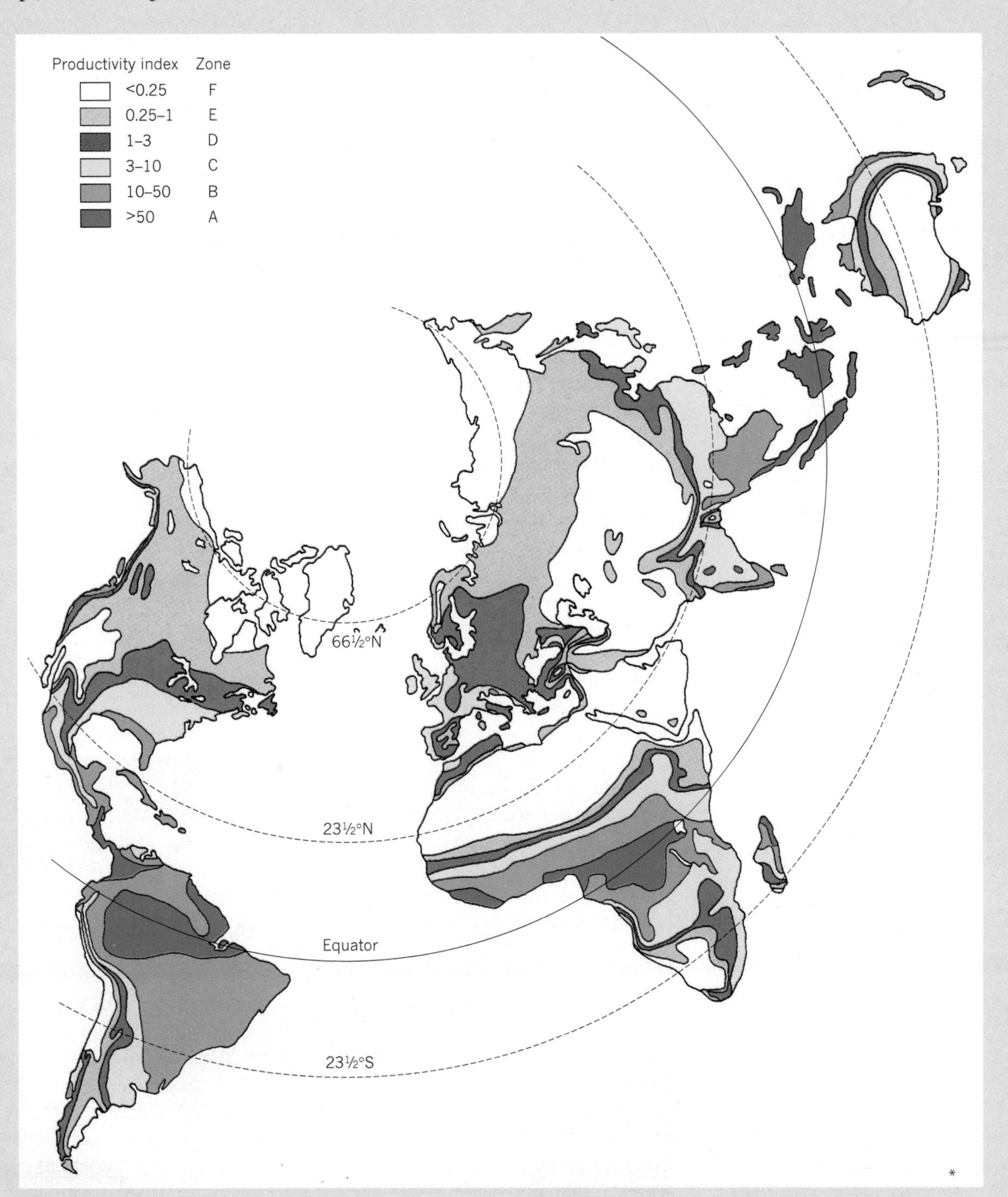

Box continued

The index of plant productivity on which the map is based is calculated using the following equation, the right-hand side of which is a combination of basic climatic elements

$$I = \frac{T_m PGS}{120 T_r}$$

where I = an index of plant productivity;
T_m = the average temperature of the warmest month in °C;
T_r = the annual range in average temperature between the coldest and warmest months in °C;
P = the precipitation in centimetres;
G = the growing season in months; and
S = the amount of solar radiation expressed as a proportion of the radiation at the poles.

The growing season is calculated by counting the number of months in which the average monthly temperature reaches or exceeds the threshold needed for plants to grow (assumed to be at 3 °C). Thus, Portland, Maine, with an average temperature in the warmest month of 19.7 °C, a temperature range of 24.9 °C, a rainfall of 106 cm, a growing season of 8 months, and a radiation value of 0.56, has an index of 3.13. For a full discussion see S.S. Paterson, *The Forest Area of the World and Its Potential Productivity* (Geography Department of the Royal University of Göteborg, Göteborg, Sweden, 1956).

It is important to remember that we are talking about zones of *potential* productivity; Paterson's index measures the plant life that the climate of a particular environment *could* support. The actual plant cover of an area will reflect other factors such as its vegetational history or the degree of human interference with the ecosystem. People may reduce plant growth (e.g. by pollution) or increase it (e.g. by irrigation). Furthermore, broad zonal maps such as that in Box 4.B does not reveal local areas whose productivity is affected by factors other than the ones included in an index.

Nine Basic Zones

In Box 4.B we discussed the varying productivity of the global environment in terms of a six-point scale. This ranged, like school grades, from A to F. Areas in zone A were highly productive; areas in F had almost no potential. (See map in Box 4.B.) We can now take this idea somewhat further by distinguishing three main divisions within the biosphere in terms of types – forested, intermediate, and barren.

The three maps in Figure 4.11 show the land areas of the globe divided into nine basic environmental zones, or *biomes*, ranging from the *polar zone* at high latitudes to the *equatorial zone* at low latitudes. Note that the boundaries shown are only approximate and give a broad-brush picture. The actual boundaries may be very irregular and much modified by human interference. Rather than review each zone's characteristics at length, we have summarized them in Table 4.3. You may find it useful to match up these zones with the productivity ratings given in the extreme right-hand column of the table and already encountered in our discussion of the Paterson Index.

The table indicates each zone's share of the total land area, but this is not necessarily an indication of its importance to humans. For example, the *Mediterranean zone* (only 1 per cent of the earth's land surface) has played a part in human development out of all proportion to its small size; whereas the largest zone, the *savannah zone* (24 per cent of the land surface), has played a minor role.

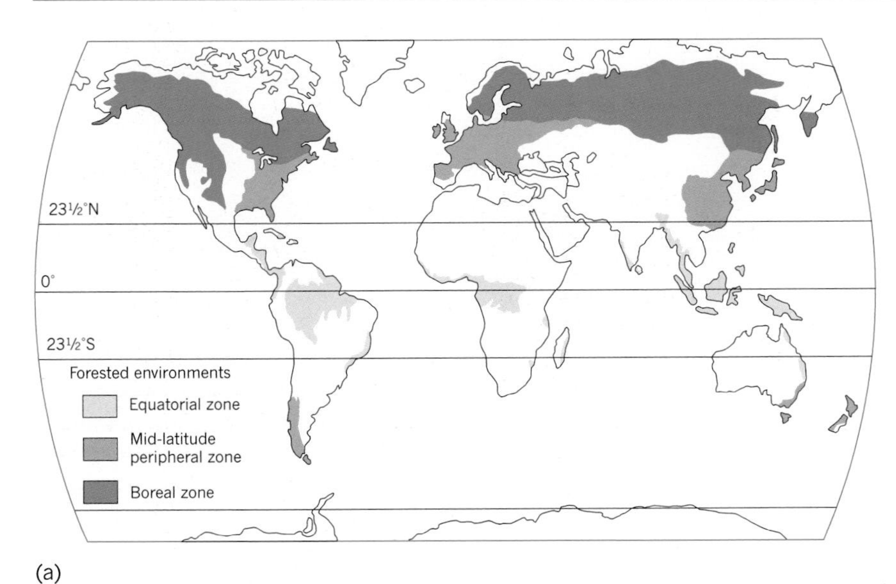

(a)

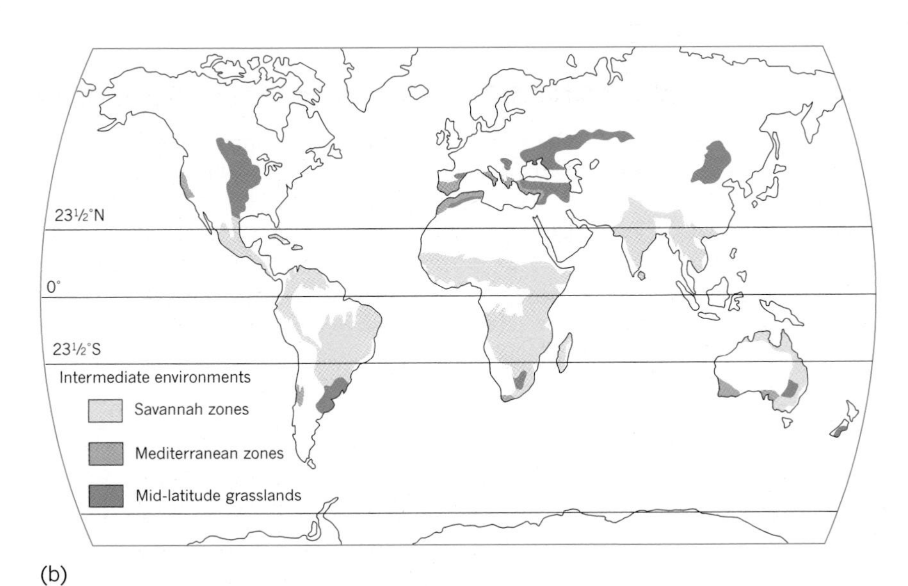

(b)

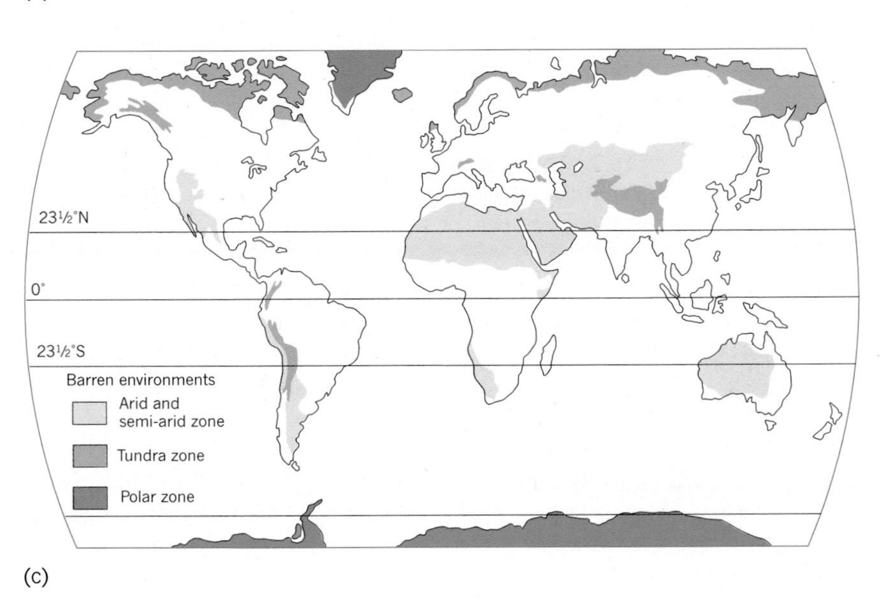

(c)

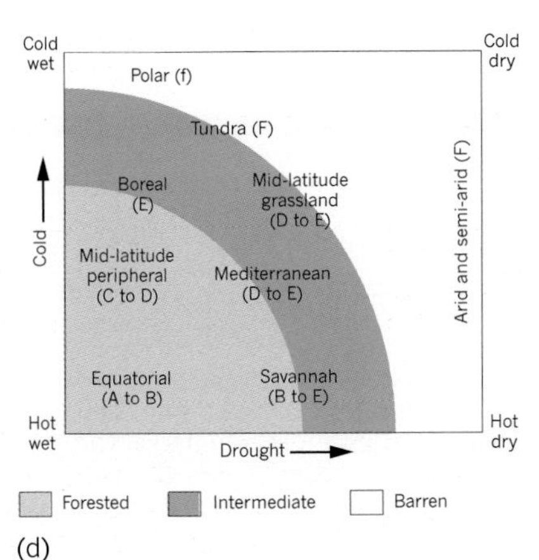

(d)

Figure 4.11 Major land biomes Biomes (a) to (c) are distinguished by their distinctive climates and vegetation. (See Table 4.4 for details.) This scheme is one of many geographers have proposed in an attempt to break down the complex ecological mosaic into simple patterns. Note that it is highly generalized; sharp local differences occur, notably in highland areas. (d) Generalized relationship of the biomes to climatic variation in cold and drought. The letters A to F refer to Paterson's productivity grades as given in Box 4.B.

[Source: M. Vahl and J. Humlum, *Acta Jutlandica*, **21** (2–6), (1949), p. 28, Fig. 1.]

Table 4.3 Major land biomes

Type	Zone name	Percentage of land surfaces[a]	Main areas[b]	Dominant vegetation cover
Forested	Equatorial	8	Amazon Basin, S America (51); Indonesia, S Asian peninsulas (21); Congo Basin, Africa (19).	Natural broadleaved evergreen forest; wide variety of species; swamp forests on floodplains and coasts.
	Mid-latitude peripheral	7	Europe (38); E China (30); E United States (26).	Broadleaved, deciduous and mixed forests, merging with warm, temperate, evergreen forests on eastern periphery.
	Boreal	14	Russia and Scandinavia (62); Canada, Alaska, NW United States (37).	Needle-leaf forests; relatively uniform stands with small number of species (e.g. spruce, fir, pine larch).
Intermediate	Savannah	24	African tropics (48); S America (21); SE Asia (20).	Ranges from open, tall-grass savannah to deciduous monsoon forest; gallery forest along stream systems.
	Mediterranean	1	Lands around Mediterranean Sea (49); S Australia (31).	Evergreen drought-resistant hardwoods and shrubs.
	Mid-latitude grasslands	9	C Asia and E Europe (42); C North America (23); E Australia (15).	Grasslands varying from tall-grass prairie to short-grass steppe with decreasing humidity.
Barren	Arid and semi-arid	21	C Asia (42); Sahara and SW Asia (30); C and W Australia (10).	Widely dispersed, drought-resistant shrubs; salt flats, plant-less sand, and rock deserts.
	Tundra	5	N Canada and Alaska (53); Russia and N Scandinavia (42).	Low herbaceous plants, mosses and lichens.
	Polar	11	Antarctica (87), Arctic (13).	Ice caps; no plant life.

[a]Computed on the basis of boundaries shown in Figure 4.11. They do not correspond exactly with the proportions given in Table 4.4, which were computed from boundaries shown on a slightly different basis. Note that characteristics of all zones are highly general; for examples of internal variations within zones, see text.
[b]The figures in parentheses give the percentage of the total land surface of each zone in each area (e.g. the Amazon Basin makes up 51 per cent of the world's equatorial zone.)

Main man-made changes in vegetation	Main precipitation characteristics	Main thermal characteristics	Productivity rating on Paterson's map (Box 4.B)
Very variable ranges from extensive clearing and cultivation (e.g. in Java) to little impact (e.g. in Amazonia); low to very high population densities	High rainfall (over 100 cm, or 39 in) throughout year; heaviest at equinoxes.	Uniformly high temperatures; little seasonal variations.	A to B
Very extensive clearing and cultivation; medium-to-high population densities throughout.	Moderate precipitation (75–100 cm, or 30–39 in) all year; heaviest in winter or autumn on western periphery; summer maximum in eastern warm temperate area.	Cool to warm temperate; seasonal range increases with continentality.	C to D
Limited clearing on equatorward fringes; very low population densities.	Light precipitation (25–50 cm, or 10–20 in) mainly in summer.	Short, cool summers; very large annual temperature range.	E
Burning, grazing, variable clearing and cultivation; high population densities on flood plains in monsoon Asia, otherwise low.	Variable rainfall (25.200 cm, or 10.79 in) with pronounced spring or summer maximum.	Warm; small seasonal variations.	B to E
Extensive clearing and cultivation, especially in lands around Mediterranean Sea; variable population density.	Low to moderate rainfall (50–75 cm, or 12–24 in); mainly in spring and summer; considerable year-to-year variation.	Warm temperate; moderate annual range.	D to E
Hunting and grazing; main settlements and cultivation in last 150 years; low population density.	Low to moderate rainfall (30–60 cm, or 12–24 in), mainly in spring and summer; considerable year to year variation.	Very strong seasonal variations; cold winters dominated by invasions of polar air.	D to E
Little impact outside small irrigated areas and oases.	Very low rainfall (0–25 cm, or 0–10 in) with considerable year-to-year variation.	Very high summer temperatures; seasonal variations range from moderate in tropics to very high in mid-latitudes.	F
Little impact.	Low annual totals (10–40 cm, or 4–16 in) with late summer or autumn maximum; light winter snowfall.	Severe cold; short, cool summers.	F
No impact.	Low annual totals; little detailed precipitation data.	Extreme cold; no months with above-freezing average temperatures.	F

Each of the nine regions is designed to relate in one simple scheme as many different physical conditions as possible. Thus, the savannah zone has the environment it has because of a mixture of climatic, vegetational, hydrologic, and soil factors. It lies within about 30° of the equator. Its largest single area is a horseshoe-shaped belt in Africa, about one-half of that continent, but it also covers much of southern Asia. While temperatures are high around the year, there is a pronounced summer rainfall and a dry winter season. Annual rainfall totals range from 25 cm (10 in) to 200 cm (79 in) and vary moderately from year to year. In southern Asia, West Africa, and northern Australia, the rainfall is associated with disturbances in the monsoon flow south of the equatorial low-pressure trough. Late summer hurricanes add significantly to rainfall totals in various parts of this zone.

The vegetation of the savannah zone is highly differentiated. It has heavy forests near its boundary with the equatorial zone and sparse shrubs and grasses near its arid border. Savannah vegetation consists of an open patchwork of drought-resistant shrubs and trees with expanses of tall, coarse grasses. Variations on this theme include the thorn forest of East Africa and the dense, semideciduous jungle of Thailand and western Burma. Variations in the length and intensity of the rainy season relate both to the variety of vegetation and to soil and hydrologic conditions. Differing precipitation leads to seasonal variations in soil characteristics and intense differences in the flow levels of rivers draining the savannah and monsoon areas.

Boundaries Between Zones

The zoning system outlined in Table 4.3 and mapped in Box 4.2 is a useful capsule guide to the world environmental mosaic. Terms such as 'equatorial', 'savannah', and 'Mediterranean' serve as a way of summarizing the dominant characteristics of regional climates and soil–vegetation complexes. Unfortunately, there is not complete harmony on how many zones there should be or what criteria should be used to separate one zone from another.

We can see the reasons for this lack of harmony by looking more closely at another of the major zones, the *boreal* zone. Generalizing, we can say that the boreal zone is a climatically determined ecological unit covered (except in settled areas) by forests dominated by coniferous growth. The boundary of this zone is not clearly marked by any abrupt change in the environment. Coniferous forests are widely dispersed in regions outside the boreal zone, in the Mediterranean area and parts of Central America, for example, so a definition of the zone based on vegetation alone is difficult. The poleward limit is conventionally marked by the Arctic tree line. In practice, however, this 'line' is really a belt where trees grow only in the most favourable sites; *muskeg* (waterlogged depressions filled with sphagnum moss) occupies the hollows, while *tundra* (dwarf shrubs, herbs, lichens, and mosses) occupy the more exposed ridges.

In the early part of this century, geographers Alexander Supan and Vladimir Koppen equated this transitional belt with an isotherm of 10 °C (50 °F) for the mean daily temperature for the warmest month. Later geographers confirmed that the northern forest boundary follows essentially a thermal limit, though they found a growth threshold of 6 °C (42 °F). Similar constraints apply to the equatorward limits of the zone, at least in the humid sections. For example, the southern boundary roughly approximates the line along which there is a mean daily temperature of 6 °C in six months of the year. It is here that the broad-leaved forests of the mid-latitude zone begin. However, the boundary between the boreal zone and the mid-latitude grasslands in Siberia and the Canadian prairie areas is controlled by the dryness of the environment rather than by its temperature.

Local Contrasts Within Zones

Even if geographers were in complete agreement on how to draw biome boundaries, another difficulty would still remain. The zones we have described give only a broad-brush picture of an immensely detailed mosaic of environments. Environmental variations occur on many scales, and these variations tend to break up and differentiate the major, subcontinental zones. For instance, within the boreal zone itself we can distinguish sharp contrasts. Canadian geographer Kenneth Hare recognizes three subzones within the boreal zone in the northern hemisphere (see Figure 4.12).

The first subzone is formed by the close main boreal forest, where the crowns of the trees touch. This *closed-crown forest* occupies at least half the area of the zone, except in the driest sites. In the drier areas such as the Mackenzie Basin of Canada, the species of trees found alter, grassy openings occur, and soil tends to be alkaline rather than acid. The second subzone is the *woodland* subzone, where lichen breaks up the closed-crown forest. The open, almost savannah-like woodlands are sometimes called *taiga*, though this word is used by Russian authorities to designate the whole boreal zone. The third subzone is the *forest-tundra* subzone, a mixture of tundra on the drier ridges and woodlands in the valleys. This area is a classic example of an *ecotone*, or transitional belt, where two major environmental zones – the tundra and the boreal – interpenetrate and blend into each other.

Another example of local contrasts within a major land biome is provided by Figure 4.13. This shows how one mid-latitude grassland – the Great Plains – has a major break along the hundredth meridian.

The division of the global land surface into the nine major zones listed in Table 4.3 has three limitations. First, strong internal variations (related to elevation, geology, or groundwater) may occur within zones, and boundaries between zones may be fuzzy. Second, the boundaries of zones are slowly but continuously changing because of long-term alterations in the climate of the present postglacial period. Third, the zonal vegetation described in Table 4.3 describes only the undisturbed plant cover that is either known to exist or assumed to be able to grow should human intervention in the ecosystem cease. In some zones, such as the mid-latitude woodland zone, little or no natural plant cover remains; in others, such as the savannah zone, the exact role of human beings is difficult to assess.

To sum up, in this chapter we have taken a measure of our planet's productivity and used it to look at variations in the geographic environment. We have illustrated how the intricate and complex spatial patterns of our environment can be broken down into a series of patterns of different sizes. Within each fragment a different set of local environmental factors comes into play. The environment described is not, however, a fixed one. In the next section we shall go on to see, with respect to tropical rainforest, how it changes, and what special challenges these changes pose for human occupation and modification of the planet Earth.

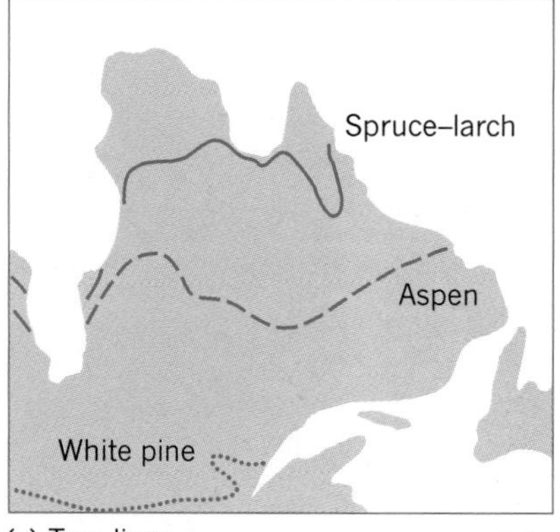

(a) Tree lines **

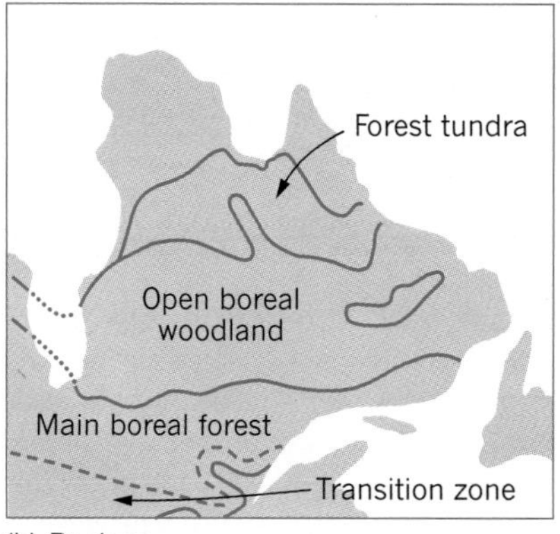

(b) Regions **

Figure 4.12 Variations within major ecological zones ***Treelines*** correspond roughly with subdivisions of the boreal zone in eastern Canada. See the profile of the Canadian climatologist F.K. Hare in Box 3.C.

[Source: F.K. Hare, *Geographical Review*, **40** (1950), Fig. 4, p. 617. Reprinted with permission.]

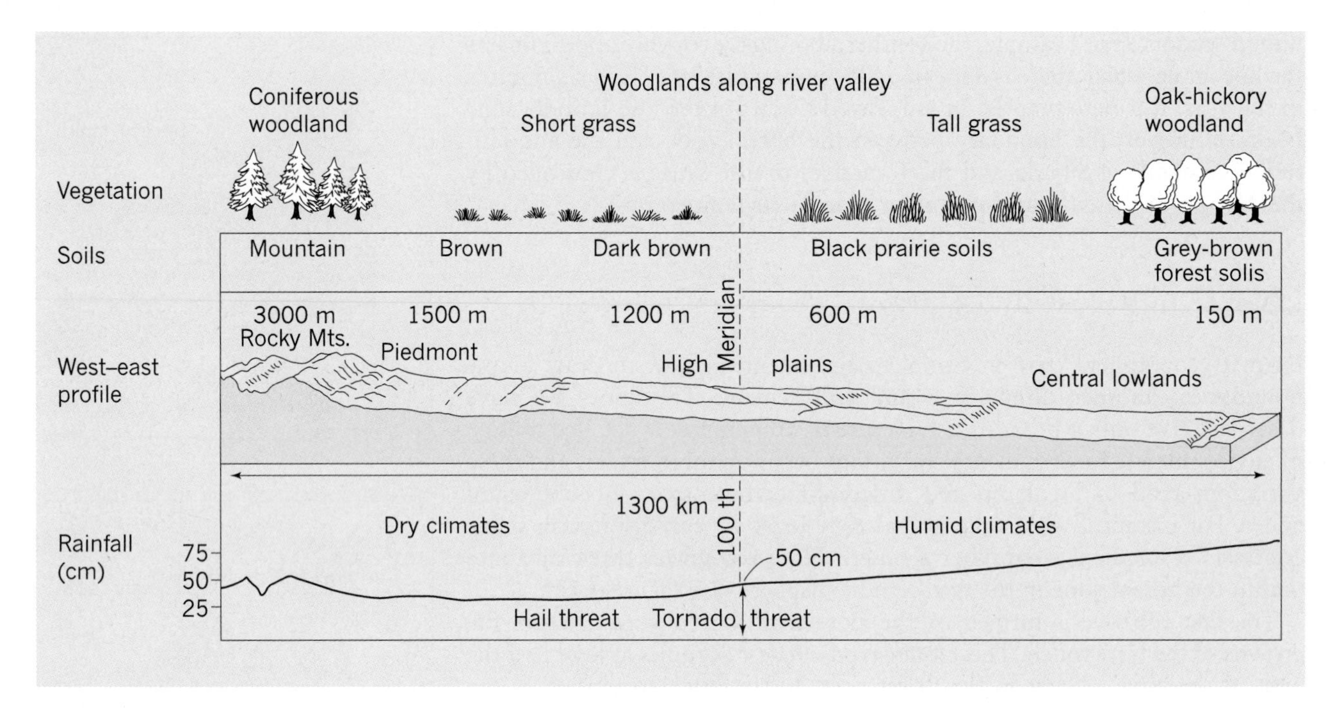

Figure 4.13 Interlocking environmental factors in the Great Plains This idealized cross-profile of the Great Plains of North America shows related changes in climatic, vegetational, and soil conditions. Note that as the moisture situation changes from humid to dry, notable changes occur in soils and vegetation. Soils change from black in the humid east to dark brown and brown on the dry western margins of the Plains. Very critical changes are caused by a layer of salts in the subsoil at the lowest limit to which seasonal rainfall penetrates. This layer is as deep as 1.2 m (4 ft) in the east but rises to only 0.2 m (8 in) in the west. An important dividing line occurs between the 98th and 100th meridians of west longitude, where the layer of salts rises above the 0.8 m (2.6 ft) mark. Along this line black soils give way to brown, and tall prairie grass gives way to short bunch grass. As a result of grazing and wheat farming, much of the original vegetation in the west has disappeared. But the dividing line created by the changing depth of the salt layer remains, separating the productive ex-prairies from the western areas where cultivation is still risky without specialized irrigation or dry-farming techniques.

[Source: W.R. Mead and E.H. Brown, *The United States and Canada* (Hutchinson, London, 1962), Fig. 41, p. 216.]

4.3 The Tropical Rainforest

Nowhere does ecological complexity reach such extremes as in the tropical rainforests among land biomes and tropical coral reefs (see Chapter 2) amongst marine ecosystems. We look here at some geographic aspects of the former.

Extent and Structure

Tropical rainforests are defined literally as the area lying between the lines of tropics with trees as their dominant life form. In practice, as Figure 4.14 shows, they extend outside the tropics and consist of a diverse mix of vegetation. Before recent inroads, the tropical rainforest existed as a broad equatorial belt almost continuous across the major land masses excepting high mountains and the dry areas of East Africa. Today it exists in three major blocks in Amazonia, in west and central Africa, and the Malaysian

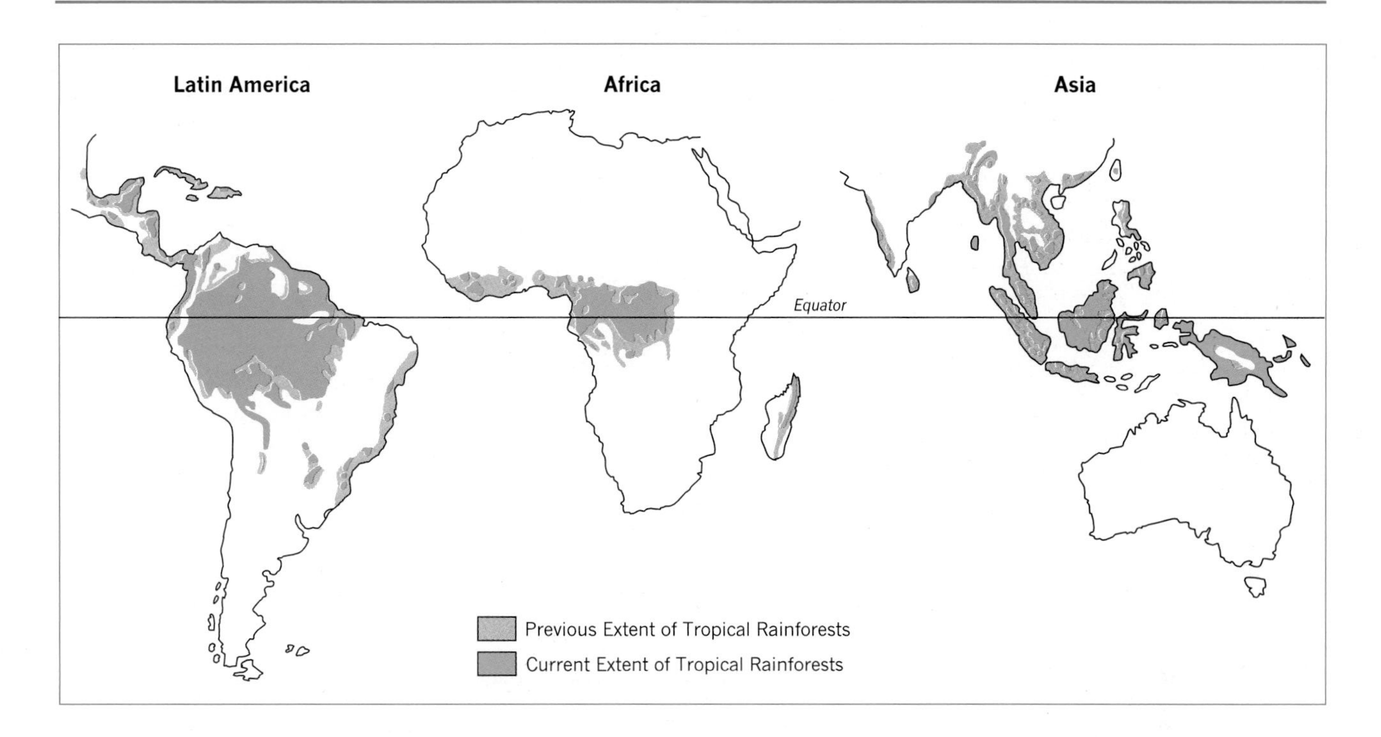

Figure 4.14 Global distribution of tropical rainforest The present distribution of tropical rainforest indicating the main areas lost since about 1900. Note the contrasts in the degree of loss in the three main zones. Compare also with more detailed maps of the Brazilian rainforests in Figures 11.3 and 20.22 later in the book.

[Source: based on various sources as given in T.C. Whitmore, *An Introduction to Tropical Rainforests* (Clarendon Press, Oxford, 1998).]

peninsula and archipelago in southeast Asia. Outside these blocks lie areas of middle America, southeast Brazil, Madagascar, India, and coastal Queensland.

There are three broad groups of tropical forest.

1. *Evergreen* tropical forests are the most widespread, mainly occurring in lowlands but extending above 1000 m where cloud formations and high levels of water saturation allow.
2. *Deciduous* tropical forests extend into areas with greater seasonal variations in rainfall and are characterized by an increasing proportion of deciduous trees as the climate becomes drier and more seasonal.
3. *Mangrove* tropical forests fringe the rivers, estuaries and coastlines with a wide range of adaptation to waterlogging and to saline conditions. Fine distinctions can be drawn within the forest zones based on the probable origin of both the trees and their dependent plant and animal populations. See Box 4.C on 'Wallace's line'.

We shall concentrate here on the classic tropical rainforests which fall into the first category. This plant formation has the chief characteristic that it is an evergreen, broadleaved forest, with a huge diversity of plant life forms and of plant and animal species. The structure of the forest is probably the most complex of all ecosystems. The taller trees form a canopy at about 30 m (90 ft) height with emergent trees extending above the canopy to 50 m (150 ft). The trees often have unusual characteristics such as buttress forms at their base (probably related to the need for stability in shallow-rooting species) and large oval leaves with a characteristic 'drip tip' profile (well adapted to rapidly shedding water). The species diversity of the forests (see Figure 4.15) is legendary with as many as 350 tree species per hectare. Even this number is exceeded if small trees, climbers, and epiphytes are added. The rainforests of the Malaysian archipelago have around 30,000 species (many still undescribed). The orchid flora of New Guinea alone is estimated at nearly three thousand species.

Box 4.C Wallace's Line

Alfred Russel Wallace (born Usk, Wales, 1823; died Dorset, England, 1913) was a British explorer and naturalist. He became famous by independently reaching the same explanation as Charles Darwin for species evolution. He also laid the basis for study of plant and animal geography. He spent much of his adult life collecting in the tropics: five years in the Amazon Valley of South America and nine years in the East Indies (present-day Malaysia and Indonesia).

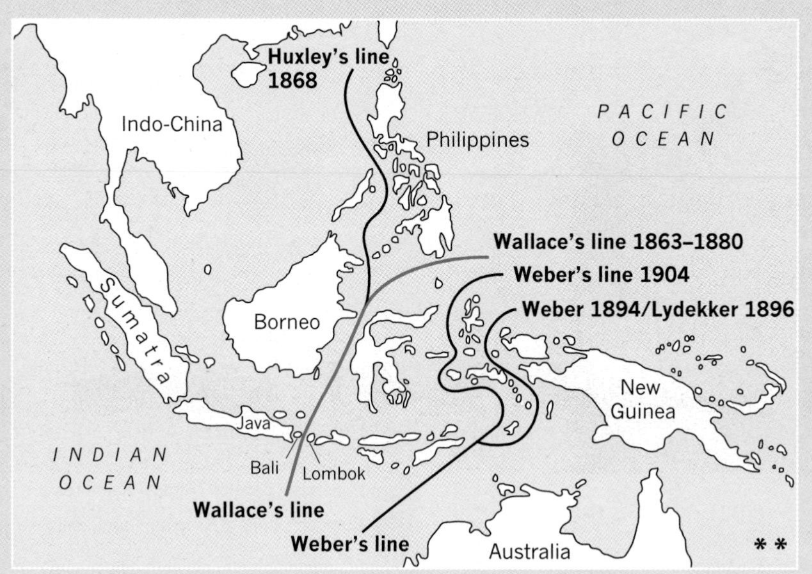

Wallace's researches convinced him that no two species are identical if they develop under different climatic and geographical conditions, even though they may have originated from a common ancestor. Wallace put forward his ideas in a series of volumes of which the most significant were *The Malay Archipelago* (1869) and the *Geographical Distribution of Animals* (1876). Wallace was much admired by Charles Darwin whose ideas he paralleled but he has, surprisingly, been given much less prominence by historians of science.

The regional system proposed by Wallace in his *Geographical Distribution of Animals* book has survived to today. He argued for a division of the world into six zoogeographic regions, each defined by their distinctive faunas. The names of the six regions were adapted from the continents: they are termed the *Palearctic* (mainly Eurasia and Africa north of the Sahara), the *Nearctic* (North America), the *Neotropical* (South and Middle America), the *Ethiopian* (Africa south of the Sahara), the *Oriental* (southeast Asia), and the *Australian.*

Wallace's line is the name still given to the zoogeographic boundary put forward by Wallace in 1858 which runs through the middle of the Malay archipelago and which Wallace argued marked the meeting of the Oriental and Australian realms. It separates the plant and animal life of the Oriental Region from that of the Australian Region. It runs from the Philippines westwards, separating Celebes from Borneo and Bali from Lombok. The location of the line is shown in the map.

The original line was based heavily on Wallace's field observations of bird distributions (notably species of parrots and cockatoos) but with later research Wallace moved his line to the east. As the map shows, alternative lines have been proposed by other biogeographers. The richness and uniqueness of the Malay Archipelago led biologists to propose that it should be deemed a separate realm (termed Wallacea), named in honour of Wallace. Plate tectonic studies have suggested that the unique assemblage of organisms in this area, notably on the island of Celebes, could be explained in terms of its complex geological history. See the discussion in T.C. Whitmore, *Wallace's Line and Plate Tectonics* (Clarendon Press, Oxford, 1981).

Figure 4.15 Structure of the tropical rainforest (a) Amazon rainforest with rain clouds hanging in the upper canopy. (b) Profile diagram of primary mixed forest, Shasha Forest Reserve, Nigeria. The diagram represents a strip of forest 61 m (200 ft) long and 8 m (26 ft) wide. Fifteen species of tropical hardwoods at three height levels are shown.

[Source: (a) South American Pictures/Tony Morrison. (b) Redrawn from P.W. Richards, *The Tropical Rain Forest* (Cambridge University Press, Cambridge, 1964), Fig. 15, p. 126.).]

(a) (b)

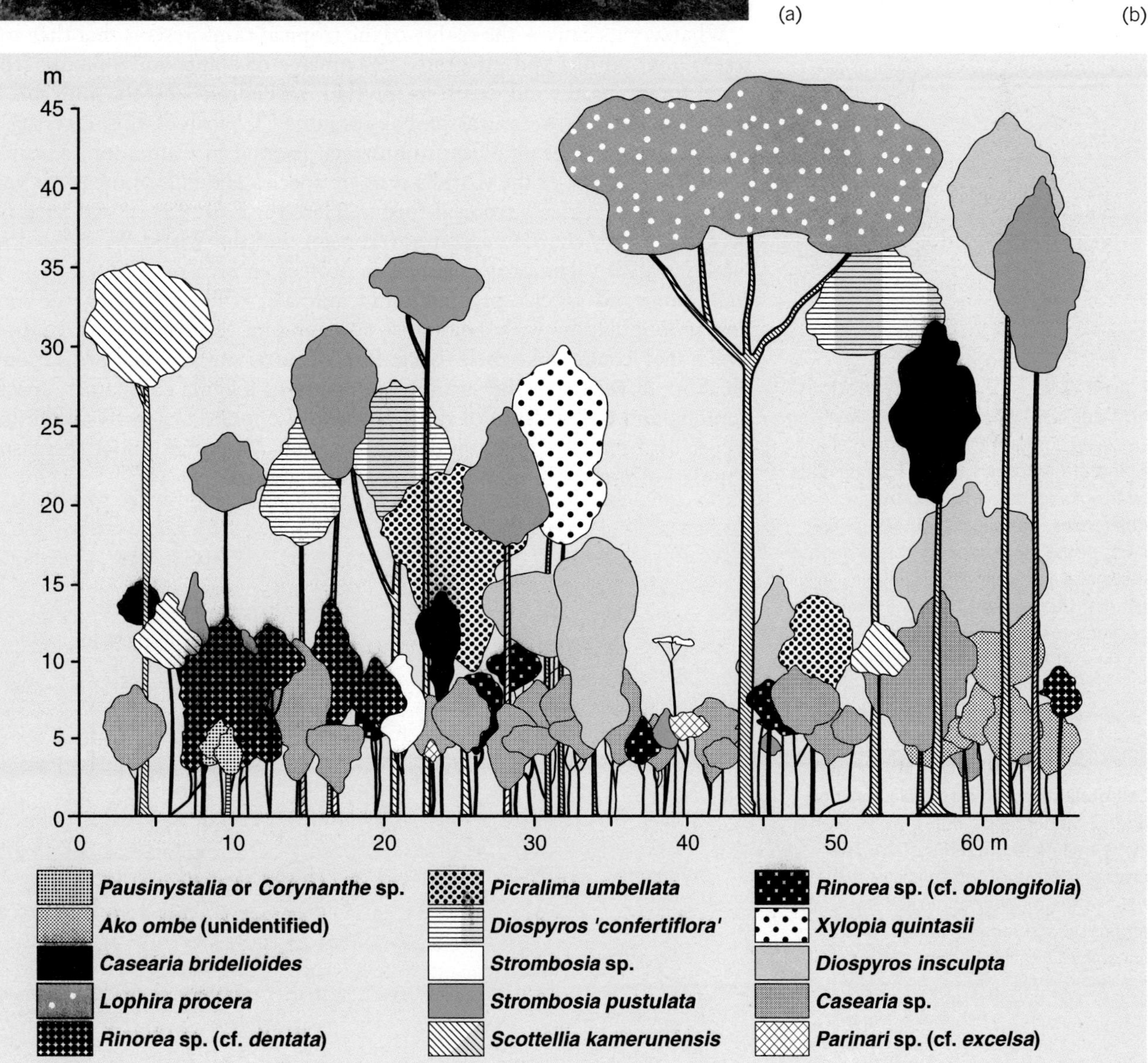

Origin of Diversity

The cause of this biological diversity is disputed. Some scholars favour a historical (non-equilibrium) explanation, others see the cause in the present (equilibrium) conditions. Certainly one of the major factors is the huge age of the forests. Geological evidence suggests that tropical rainforests have existed as a plant formation since at least the Tertiary period. Their extent has varied over time from as far north as present Greenland to New Zealand, but the geographic extent was severely curtailed by the climatic changes of the Quaternary period when glacial phases pinned back existing biomes towards their tropical baselines.

Other causes of diversity relate to present rather than past causes. These equilibrium hypotheses stress that the rate of species formation is far higher in the warm, wet conditions of the tropics and that extinction rates are lower. Evolution proceeds faster given the larger number of generations per year and the greater turnover of populations, allowing increased selection. Extinction rates are thought to be lower given the diversity of environments and the constancy of the physical environment.

Whatever its causes, the reality of the tropical rainforests is that they represent a wondrous and integrated assemblage of plants and animals, from huge forest hardwood trees to myriad specialized insects and micro-organisms that is a critical global resource. Typical of this diversity is Korup, one of Africa's oldest rainforests, located in Cameroon. Korup is home to a quarter of the world's primate species and half of the plant varieties found in Africa's tropical forests. Here the Korup National Park has been created by the Cameroon government and the World-Wide Fund for Nature (WWF). The project aimed to protect an area of rainforest and its eight thousand species of plants and animals, while providing the local human population with land in a surrounding 'buffer zone' area from which they could make their living. Reports now suggest such protection is breaking down under the sustained pressure of logging companies, species hunters, and the pressure of rural poverty. We look later in this book (see especially Chapter 11) at the attacks on that ecosystem under increasing population pressures from human uses and abuses.

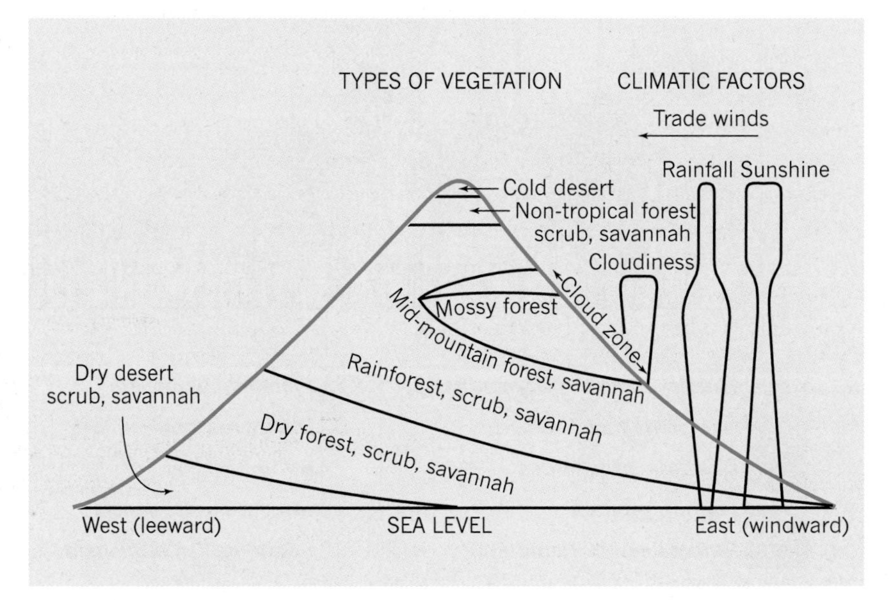

Figure 4.16 Timberline variations and ecological zoning This idealized cross-section of a typical high oceanic island in the tropics shows the changing zones of vegetation caused by differences in the elevation of various parts of the island and by the location of the island with respect to the rain-bearing trade winds. Note the contrast between the height of vegetation zones on the wet eastern (windward) side and the drier western (leeward) side. *Savannah* is open forest with scattered trees mixed with scrub and grassland to give a parklike pattern. Mossy forest is a belt of damp forest in the cloudy zone on the windard side with mosses on the living trees and ground surface. The changing width of the columns at the right of the diagram indicates the relative amount of cloudiness, rainfall, and sunshine at each altitude.

[Source: S. Haden-Guest *et al.* (eds), *World Geography of Forest Resources* (American Geographical Society, New York, 1956), Fig. 58, p. 620.]

Local Variations

We can spot the effect of altitude on rainforest variation within small areas as well as on the world scale. Figure 4.16 shows the variations in zones of vegetation on a small tropical island. Note that the effect of elevation is complicated by the direction in which moisture-bearing air is travelling and by variations in cloud cover with elevation. The simple relationship between height and variations in vegetation is disturbed on the windward side of the island by a characteristic 'cloud forest' belt. The leeward side of the island is much drier, and the forest belts start at different elevations. Similar rain-shadow effects relate to forests outside the tropics. The prevailing moisture-laden winds from the Pacific are visible in the varying forest levels in the southwestern United States (e.g. in the Sierra Nevada or White Mountains).

Humanity as Part of the Biosphere 4.4

In opening this chapter, we referred to the human population as a very recent, but increasingly dominant part of the biosphere. We now try to place our own species within that broader biological framework.

Human Environmental Requirements

The long evolution of *Homo sapiens* as an animal species has given us some highly specific environmental requirements. Like an atmospheric fish we swim through an oxygen-rich gas found only near the surface of one of the minor planets. In view of the multiplicity of physical and chemical conditions in the known universe, humans are a remarkably specialized creation, with only a tenuous toehold on survival. In a reduced oxygen level of our surroundings, we start to pant; in an increased proportion of hydrocarbons, we start to cough; immersed in water, we drown within a few seconds; deprived of water, we atrophy and die within a few days. But as denizens of the planet Earth we appear to be much more robust. Given an unpolluted atmosphere, our tolerance of climatic conditions (precipitation, wind, and solar radiation) actually found on earth is reasonably good.

The climatic element to which we show the greatest sensitivity is probably temperature. We are warm-blooded mammals with an average body temperature around 37 °C (98.6 °F). Prolonged exposure to conditions that raise or lower this normal body temperature more than a few degrees leads to permanent tissue damage and death. The extreme ranges of body temperature ever recorded for a living person are 44 °C (111 °F) and 16 °C (61 °F). Let us compare these with the actual variations in the surface air temperatures on the earth given in Table 2.3. This shows that our extreme tolerance is only one-fifth of the earth's temperature range. Note also that we are much better suited to the hotter ranges of the planet's surface than to the colder. Extreme cold is a much more significant limit to human existence on the earth than extreme heat.

How we can measure whether an environment is tolerable for human life? Many attempts have been made to assess climatic environments using simple combinations of temperature, humidity, radiation, and wind-speed measurements. One simple index used by the US National Weather Service is the temperature–humidity index (THI) (see Figure 4.17).

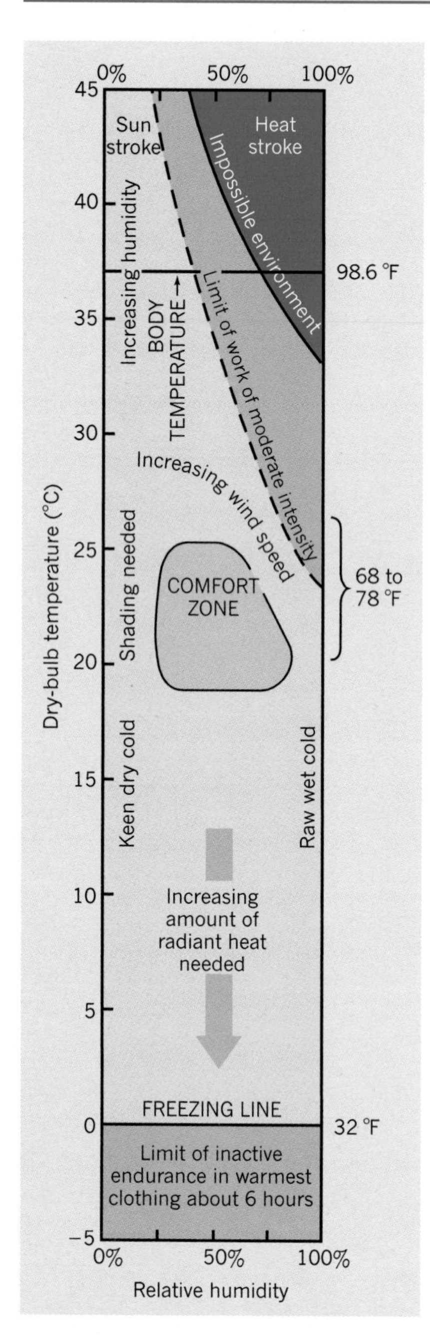

Figure 4.17 Human tolerance of climatic ranges The graph shows comfort, discomfort, and danger zones for inhabitants of temperate climatic zones. Dry-bulb temperatures in °F are shown to the right of the diagram. Note how limited is the comfort zone in the middle of the diagram.

[Source: after V. Olgay, from R.G. Barry and R.J. Chorley, *Atmosphere, Weather, and Climate* (Methuen, London, 1968), Fig. 7.1, p. 251.]

This index is based on temperature readings on two Fahrenheit thermometers, the bulb of one of which is kept permanently damp. Evaporation will make the temperature recorded on the wet-bulb thermometer lower than that recorded by the normal dry-bulb thermometer. When the air is humid, there is little evaporation and the readings on the two thermometers are similar. The THI is the sum of the two readings multiplied by a constant (0.40) and added to another constant (15). If the two thermometers record temperatures of 70 °F and 65 °F, the THI will be $0.40 \times (70 + 65) + 15$, or 69. At a THI of 75 in still air, about half the people in an office feel discomfort; at 80 few remain comfortable; and at 86 regulations suggest that all workers (at least in US federal buildings) be sent home.

These *comfort zones* pertain to still air. As Figure 4.17 shows, winds reduce the effect of high temperatures and humidity and make conditions more bearable. On the other hand, when temperatures are low, strong winds considerably increase the level of discomfort. Thus, any index of human comfort must clearly include a chilling factor related to air speeds.

However sophisticated an index is, it can only describe average reactions. Individuals vary considerably in their ability to withstand stress. Our sex, body characteristics, genetic heritage, degree of acclimatization, and cultural background all affect our environmental tolerance. Most indexes have been tested on urbanized North Americans, and we would expect the reactions of Nepalese, Kikuyu, or Eskimo groups to be somewhat different.

Biome Productivity

The earth we have studied in the last two sections is an ecosystem. That is, it contains an intricate and delicate network of cycles and feedbacks that have both non-living elements (the atmosphere, hydrosphere, and lithosphere of Figure 2.2) and living elements. In this section, we turn to the question of the productivity of those cycles. How far do they provide the basic foodstuffs on which the human population is dependent?

Forecasters at Resources For the Future (RFF) predict that, given the present levels of solar energy, at the present the most food that could be produced by photosynthesis is about 10^{11} tonnes per year.

Let us see how the RFF estimate was produced. The forecasters reasoned as follows. The prime source of energy on our planet is the sun, which radiates electromagnetic energy waves and high-speed particles into space. Since this constant emission represents almost all the energy available to the earth (except for a small proportion from the decay of radioactive minerals), it can be used to estimate the total amount of energy available to human beings. Green plants store solar energy through photosynthesis, and thus we can estimate the theoretical totals of dry organic matter (i.e. plants minus moisture) that the earth could produce. But dry organic matter is not always edible. Even in the case of croplands, well below half of the gross product may be edible. As for grazing lands, an energy-conversion factor of 12 to 1 must be used to convert the energy consumed by stock (i.e. the grain) to the food value of the stock to humans (i.e. animal products). Thus, the original figures must be revised downward to a maximum of around 10^9 tonnes per year. This gives a rough order of magnitude to overall global productivity.

Table 4.4 provides a breakdown of this total productivity between different global environments. It separates out the production of organic matter (at the trophic level T_1 in terms of Figure 4.7) from the matter edible to humans (at level T_4). Note that more organic matter is created by photo-

Table 4.4 Productivity of global environments

Environment	Area	Dry organic matter	
		Maximal production per year	Maximal edible production per year
Forest			
Tropical rainforest	4	30	6
Temperate deciduous forest	1	2	1
Temperate coniferous forest	3	13	3
Taiga	1	2	–
Total	9	47	10
Grassland			
Humid grasslands	3	11	10
Arid grasslands	4	8	5
Total	7	19	15
Cultivated land	2	8	71
Other land			
Wetlands and swamp	1	3	–
Tundra	2	1	–
Hot desert	4	–	–
Cold desert	3	–	–
Total	10	4	–
Oceans and lakes			
Deep sea	65	20	–
Shelf, lagoon	5	3	3
Fresh water	1	–	–
Total	71	23	3

Theoretical estimates of maximal production by phosynthesis in different types of environments. All figures in the table are percentages of the world total. Because of rounding, the total for each column may not equal 100 per cent.
Source: data from R.U. Ayres, *Science Journal*, **3**, No. 10 (1967), p. 102.

synthesis in the tropical rainforests than in any other environment on earth. Forest regions as a whole account for 40 per cent of the total plant production of the globe, and the oceans for another 20 per cent.

However, the figures on the table are only estimates, and they refer only to the production of organic matter on the first trophic level. Most of the organic products of forests are forms of wood, and, with our present technology, very little of these products can be converted to an edible form. As for the organic largesse of the oceans, it is of little benefit to humans because we lack the technology to harvest it properly. The long food chains in the sea, the poor food-conversion ratios at each point, and the present wasteful fishing methods mean that we actually derive little food from the oceans. In practice, about 70 per cent of the food we eat comes from cultivated land, and the world outlook for food production in the immediate future is going to depend essentially on improving the yield of land already under cultivation. The probable contribution of new virgin lands is actually quite marginal. In the longer term, the vital areas for increasing the food supply are the tropical forests and ocean shelves. We return to the issue of food production in the discussion of population in Chapter 6.

Biosphere Resources as Population Regulators

Human life is maintained by an inward flow of nutrients and an outward flow of wastes. Figure 4.18 shows the estimated daily inputs and outputs

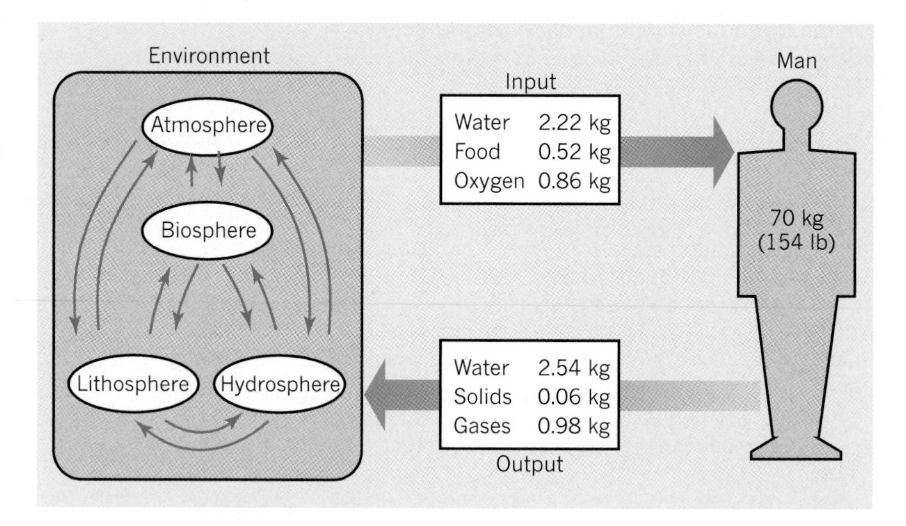

Figure 4.18 Human dependence on the environmental support system Human life depends on a continuous exchange of materials with the earth's environment. Figures given in the chart are the estimated daily inputs and outputs for an adult male.

for a medium-sized male of 70 kg (154 lb). Inputs in the form of water, food, and oxygen allow the renewal and growth of body tissue, and provide the energy for breathing, blood circulation, and movement. The energy consumed is transformed to heat and cycles back to the atmosphere. Excretory matter from the metabolic process forms the outputs, partly solids, partly liquids, and partly gases which complete the input–output cycle.

The energy needed to maintain human life is measured in terms of calories. Such daily needs are related to age, sex, body weight, and the amount of work to be done. Whereas a 2-year-old child can subsist on 1000 calories a day, a pregnant woman would need twice that amount. A man carrying out very heavy labour will require between 3000 and 4000 calories a day. Energy needs are simply a quantitative measure, whereas the food inputs must have certain qualitative requirements. For example, our diet needs to contain around 10 per cent of its content as protein.

Food supply is one of the critical input mechanisms by which the environment controls all animal numbers (including human numbers too, in the long run). Consider Figure 4.19, which shows two ways in which a population may grow. In (a), changes in numbers are constrained by the food supply. Increased numbers born in one season mean less food for each member of the population, give a reduced chance of survival, and return the population to its original position. If you follow the sequence of arrows, you will see how population tends to return to an equilibrium level. Loops of this kind are described as *negative* (or self-regulating) feedbacks and are typical of stable populations.

Figure 4.19(b) shows by way of contrast an uncontrolled situation.

Figure 4.19 Feedbacks and population control The diagrams show how feedback mechanisms regulate population size. In (a) negative feedbacks lead to self-regulation. In (b) positive feedbacks lead to changes that are self-sustaining.

[Source: W.B. Clapham, Jr, *Natural Ecosystems* (Macmillan, New York, 1973), Fig. 104, p. 12.]

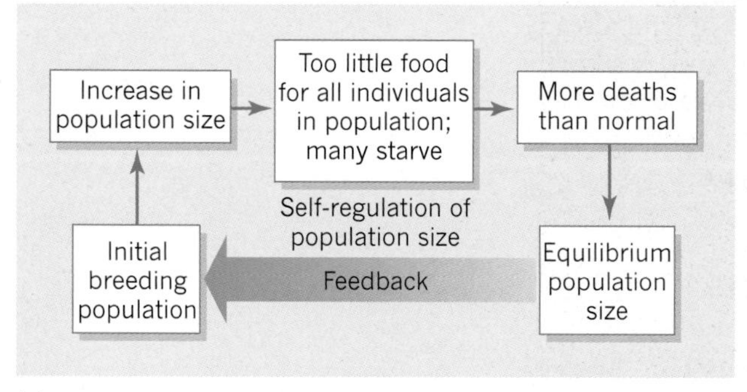

(a)

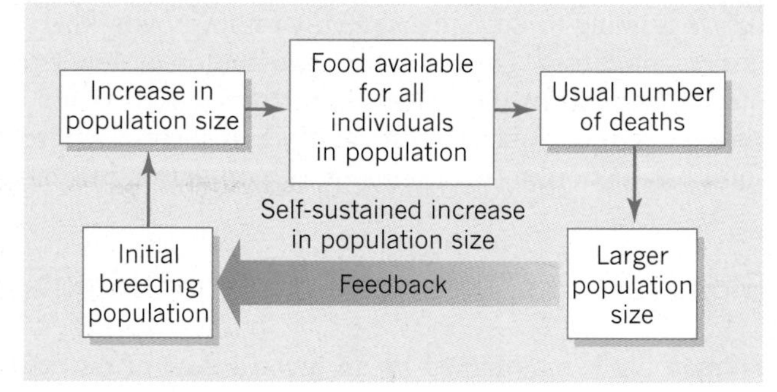

(b)

In this case, food supply is abundant, there are no checks on growth enforced by starvation, and so the population grows rapidly. These unstable situations are characterized by *positive* (or self-reinforcing) feedbacks.

Human Control of Ecosystems

A final example of the interactions between humanity and the rest of the biosphere is in ecosystem control. Bush fires provide a simple example system, which may be controlled, at least in part. Consider Figure 4.20(a) which shows a fire as a cascade-type system. Each season litter accumulates from the trees to build up combustible material. In dry summer conditions a fire trigger (say, a lightning strike) may start up a fire. This spreads to pro-

Figure 4.20 Bushfires as uncontrolled and controlled systems A bushfire model (a) with outbreak patterns under natural conditions (b). Human interference by controlled burning (c) may reduce the severity of fires. Attempts at complete protection (d) may increase fire severity because of litter accumulation. This may be compensated by the relative infrequency of fires. Note that the model shown is greatly simplified. (e) Uncontrollable bushfires as a result of 'slash and burn' techniques in Tanzania.

[Source: (e) PANOS/Netocny]

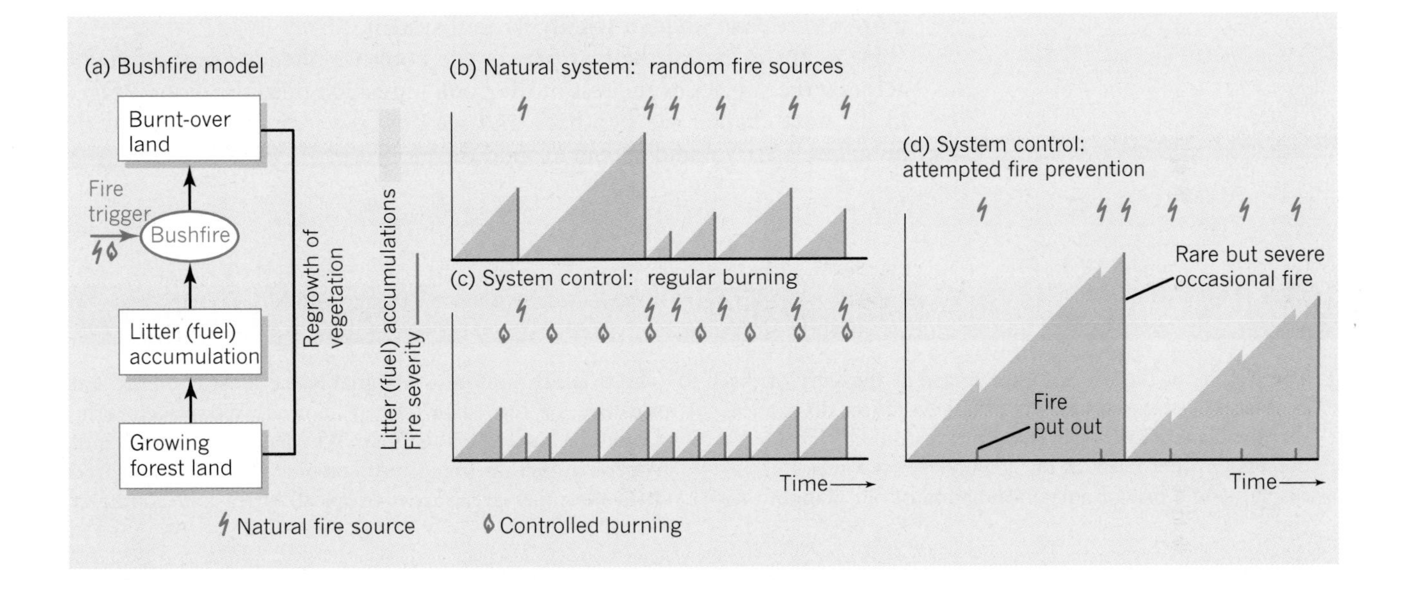

(e)

vide burnt-over areas. Since the fire burns up the litter, this reduces the chance of another fire until the vegetation has grown and litter again accumulates. The pattern of litter accumulation under natural conditions is given in Figure 4.20(b). Six natural sources of fire are assumed over a time period of about 50 years. Fires vary in intensity, depending on the length of interval since the last fire.

Two strategies for controlling this natural system are considered. In Figure 4.20(c), there is regular preventive burning every five years: this keeps the fires at low intensities and under some measure of control. In Figure 4.20(d), a complete prevention system is attempted. The natural fires that start are extinguished wherever possible. Note, however, that since there are so few fires, litter goes on accumulating, so the chance of a large uncontrolled fire goes up. This is shown for the third fire in Figure 4.20(d). In this case, the fire could not be put out and a massive burnout resulted. Clearly natural systems can be only partly controlled and, without full understanding, attempts at control may lead to making the problem even more severe than under a wholly natural system.

Here, at the end of the first part of the book we already see the human actors – the subject of the rest of the book – pushing onto the global stage. In the next chapter we step back and see how the main play begins with humanity's arrival and spread around the globe.

Reflections

1. The origins of life on earth are traced at the start of this chapter. Given the complexity of this evolution, do you feel that the planet Earth story is unique, or could life evolve on planets in other parts of our galaxy? What kind of catastrophes could bring human occupation of our planet to an end?

2. What do you understand by the term *ecosystem*? Using the diagrams in Section 4.2 as a guide, trace the main links in one other typical ecosystem. Do you think that the term ecosystem should be used only for natural plant and animal communities? List the (a) advantages and (b) dangers of viewing human communities as ecosystems. What would be the closest parallel to plankton in a collegiate ecosystem?

3. Consider Paterson's map of potential productivity (Box 4.B). List (a) the advantages and (b) the disadvantages of measuring potential fertility in this way. What additional information would you like to have before deciding on the productivity of a particular environment?

4. The main global divisions of the biosphere in terms of land biomes is given in Table 4.3. In what region does your own home fall? What subdivisions within your home biome would you want to suggest to make it more realistic?

5. The tropical rainforest is arguably the most complex and sensitive of the earth's great biomes. What factors lie behind its current loss in area? Why has more of the South American section (mainly Amazonia) survived, compared with the other great areas (tropical Africa and Southeast Asia)?

6. Humanity is an integral and dependent part of the biosphere. Consider the reactions of human beings to variations in temperature indicated in Figure 4.17. What do you think would be the limits of human settlement on your own continent if people were unable to construct artificially heated shelters? For how many months of the year would your own college town be habitable?

7. Review your understanding of the following concepts using the text (including Table 4.3) and the glossary in Appendix A of this book:

biome	ecosystem
boreal	food chains
carbon cycle	Mediterranean
climax	trophic levels
comfort zones	Wallace's line

One Step Further ...

A useful starting point is to look at the essay on the biosphere by Alfred Fischer in I. DOUGLAS *et al.* (eds), *Companion Encyclopaedia of Geography* (Routledge, London, 1996), pp. 67–85. The recent widespread interest in ecology has led to the publication of a large number of excellent introductions to the field. A comprehensive review of biogeography is given in I.G. SIMMONS, *Biogeography: Natural and Cultural* (Arnold, London, 1979), R.J. HUGGETT, *Fundamentals of Biogeography* (Routledge, London, 1998), and C.B. COX and P.D. MOORE, *Biogeography*, sixth edition (Blackwell Science, Oxford, 2000).

For specialist topics see M.A. HUSTON, *Biological Diversity* (Cambridge University Press, Cambridge, 1994), E.O. WILSON, *The Diversity of Life* (Harvard University Press, Cambridge, Mass., 1992), and T.C. WHITMORE, *An Introduction to Tropical Rain Forests*, second edition (Oxford University Press, Oxford, 1998).

The biogeographical 'progress reports' section of *Progress in Physical Geography* (quarterly) will prove helpful in keeping up to date. Look also at *Scientific American* (a monthly) and *New Scientist* (weekly) which carry regular reports on current biogeographical work by environmental scientists. Specialist journals include the *Journal of Biogeography* (quarterly) and *Ambio: A Journal of the Human Environment* (quarterly). For readers with access to the World-Wide Web see also the sites recommended in Appendix B at the end of this book for topics relevant to this chapter.

PART II

The Human Population

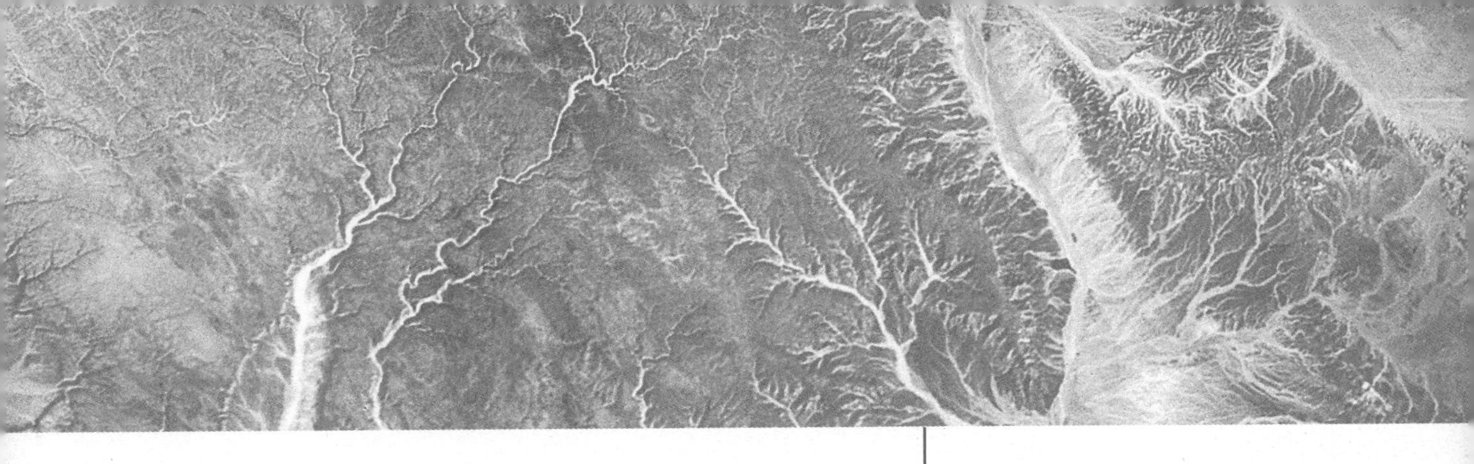

CHAPTER 5

Human Origins and Dispersals

■ denotes case studies

Figure 5.1 Locations of the agricultural revolution The Nile River was one of the locations in the Fertile Crescent of the Middle East which saw the origins of irrigation agriculture. LANDSAT photograph of the river Nile today as it snakes past Luxor (grey area on the right bank, one-third of the way up the image). The major features are the dense bands of cultivation which manage to cling to the flanks of the river through a highly developed irrigation system, assisted by the upstream Aswan Dam which curbs the spring floods on the Nile. The loop in the centre of the river contains the Valley of the Kings. North is at the top of the photo and the east–west distance is around 160 km (100 miles). [Source: Earth Satellite Corporation/Science Photo Library.]

> 'I will tell the story as I go along of small cities no less than of great. Most of those which were great once are small today; and those which in my own lifetime have grown to greatness, were small enough in the old days.'
>
> HERODOTUS (*ca* 440 BC)

The occupation of the planet Earth is somewhat like the scenes from a stage play. The last three chapters have described the stage on which that play is set: the world of continents and oceans, of forests and deserts, of biological life in all its varied forms. But all this is a prologue to the play itself. In this chapter, the curtains are drawn back and the human actors – you and I, our families, our neighbours, our countrymen, our ancestors – come onto the planetary stage. Act I begins.

In this chapter, we look at some of the historical questions that surround the human occupation of the earth. When did the human population arrive on the scene (Section 5.1)? Where did humans come from? Just where in the world did they originate? Was it in just one location, or did we spring up in many places? If the former, then how did we spread from one continent to another?

In the second part of the chapter (Section 5.2) we look at the stages of human settlement. How did the two great revolutions – in agriculture and in urban living – occur and where? Were the great river valleys of the Middle East the cradles of farming and cities (see Figure 5.1 p. 138) as long held? Or were other areas more likely as some geographers have argued?

The later sections of the book bring the human story closer to the present. Section 5.3 considers the great regional hearths of population in the river valleys of China and India, in the high plateaux of Central and South America, and in the peninsulas of Western Europe. We ask why the populations there grew and why some were centres of expansion while others turned in on themselves. In Section 5.4 we look at the vast overseas expansion of one of those centres, Europe, over the last 500 years. What form did that expansion take? What was its impact? Finally, in Section 5.5 we look at the same process, but from the other end of the telescope. We take North America and consider the Euopean impact, contrasting the pre-Columbian and post-Columbian worlds. That spread process of the human population brings us up to the modern period.

5.1 Origins of the Human Population

We can split the question of origins into three parts. Where did the first clusters of the human population form on the earth? Where were the first centres of agricultural innovation? And, finally, where, and at what stage, did urban cultures come into the picture? The order of these questions is a historical one.

Questions of Timing

The present state of knowledge of human evolution suggests that our species is very much a newcomer on the world scene. Let us take the earth's age as 4.5 billion years (see Chapter 2). The first living forms, algae and bacteria, originated about 2.2 billion years ago. Just which date is assigned to the origin of the human species depends on which skeletal remains archaeologists are prepared to call human. The ancestors of modern humans did not begin to break away from the rest of the animal kingdom until 5 to 8 million years ago and several humanlike species emerged during the last 3.5 million years of the most recent geological period (the Pleistocene).

To give a simple yardstick, let us equate the 4.5 billion years to a Biblical seven days. Then the hominids (the group within the primates to which we relate) do not appear on the planetary scene until a quarter to midnight on the seventh day. For *Homo sapiens*, our own recognizable species, the tenure is much shorter: only of the order of 100,000 years. Using our week-long yardstick, that would place our arrival on the planetary scene at just thirteen seconds before midnight!

The complex process by which that evolution occurred is only partly understood. Studies of DNA show the closest links are with chimpanzees and gorillas. With living species, the degree of molecular difference can be used (with certain assumptions) as a 'molecular clock' to estimate the time that elapsed since any two lineages separated. It is this type of research that points to the zone of 5 to 8 million years ago. But that separation marks only the beginning of the evolutionary story. Fossil record of human remains show a complicated story in which a score of different 'human' species can be recognized: more are likely to be added as archaeological excavations proceed in areas of the world that so far have been little studied. Figure 5.2 shows a phylogram or family tree of that evolution. Each human species is prefaced by the term '*Homo*' with the second term often describing the location where the skeletal remains were discovered. Thus *Homo rudolfensis* refers to material excavated by Lake Rudolph in East Africa. Earlier references in the chart are to Australopithecenes (e.g. *Australopithecus ramidus* excavated in Ethiopia in 1992), creatures that appear to show rudimentary human specialization.

For many decades, human evolution was seen as a ladder with primitive species being 'replaced' by a later, more advanced one. Modern humans

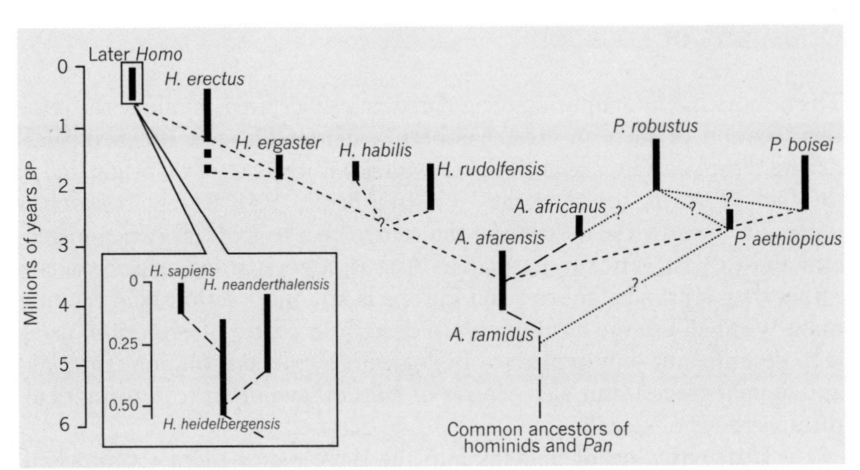

Figure 5.2 Family tree of early hominids Branching tree showing the relationship of hominids (humanlike species) for which their is fossil evidence. Note that *H.* is an abbreviation for *Homo*, *A.* for *Australopithecus* and *P.* for *Pan* (members of the chimpanzee family). Only one hominid species, *Homo sapiens*, has survived to the present day. Years BP indicate years 'before present'.

[Source: B. Wood, in I. Douglas *et al.* (eds), *Companion Encyclopaedia of Geography* (Routledge, London, 1996), Fig. 4.2, p. 88.]

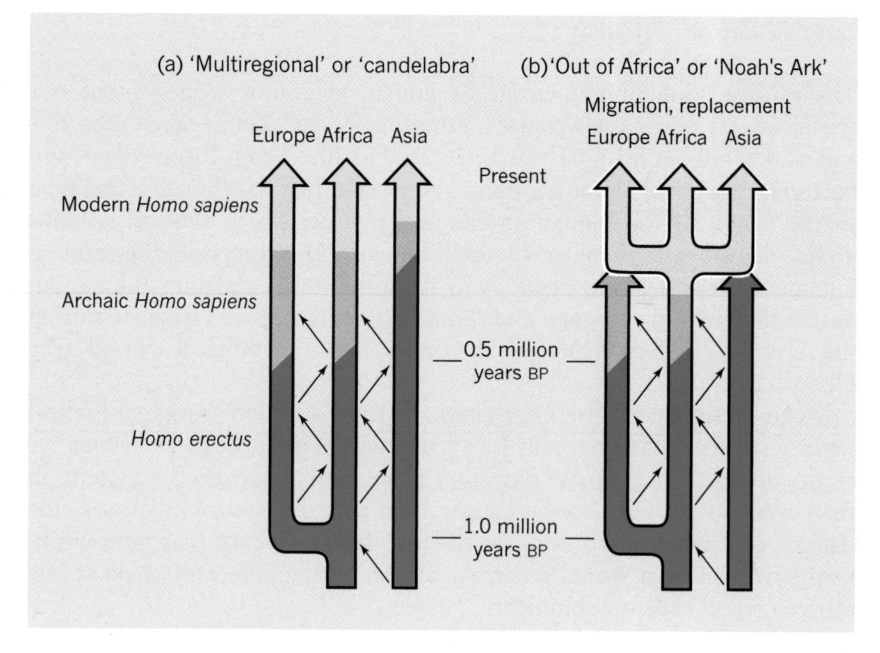

Figure 5.3 Competing ideas on the origin of human species (a) 'Candelabra' hypothesis with parallel developments in the three Old World continents over the last million years. (b) 'Out of Africa' hypothesis with late branching of the African stem into the other continents. This was thought to occur about a quarter of a million years ago. Some links between the continents occur under both hypotheses.

[Source: modified from B. Wood, in I. Douglas *et al.* (eds), *Companion Encyclopaedia of Geography* (Routledge, London, 1996), Fig. 4.7, p. 103.]

were of course placed at the top of the ladder of ascent. But as Figure 5.2 shows, the recent evidence from the fossil record shows great overlap in time with different human species existing in parallel. The diagram looks more like a bush than a ladder.

Still later stages of evolution remain matters of speculation. Figure 5.3 shows two competing hypotheses for the geographical origin of modern humans. The 'multiregional' or 'candelabra' hypotheses show modern *Homo sapiens* emerging in parallel streams in the three old-world continents of Africa, Europe and Asia. By contrast the 'Out of Africa' or 'Noah's Ark' hypothesis shows modern humans emerging solely from Africa and replacing, by migration, the other human species.

Whatever the details of their evolutionary history, there is no doubt that modern humans had dispersed across the Old World by 25–35,000 years ago. By that time they had also managed to cross the barrier of water between the Asian mainland and Australasia. The date of entry into Australia is being steadily revised by modern archaeological exploration of that continent. It may have been as long as 60,000 years ago.

Questions of Location

The notion that human beings originated in a single area, let alone the possible location of such an area, has been a matter of acute archaeological debate. From current archaeological evidence it looks as if we originated in the Old World rather than the New (see Figure 5.4). Recent research is tending to narrow the location of the source area to tropical Africa in general and to East Africa in particular. Asia is now regarded as a secondary rather than a primary source, and Europe is no longer seriously in the running. We shall use the term *hearth* to describe a centre of evolution, using it to describe not only centres of biological or genetic evolution (for plant and animal species) but also centres of cultural evolution (e.g. for agricultural methods or city living).

The differentiation of humans into the three major races – Caucasoid,

Box 5.A Island Colonization Models

Since Charles Darwin's *Origin of Species* (1859), islands have held a special fascination for biogeographers. Darwin saw the possibilities that island chains offer for research during his visit to the Galapagos in 1835. Since then, an increasing amount of research is being done on the spread of human populations through island chains. (See Section 16.4, which reports research on the colonization of Polynesia.)

The diagram throws some light on the question of island colonization by using a model first developed to explain the varying number of plant and animal species found on different islands. Typical relations between a continental land mass and some offshore islands are shown in (a), where the figures indicate the number of different species of plants or animals found on each. The number of species on each island is determined by the *immigration rate* (the number of new species that arrive from the continent over a given time) and the *extinction rate* (the number of existing species on each island that fail to establish themselves and die out in a given time). Immigration rate is inversely related to the distance of the island from the continent (b), so that nearby islands have more species arriving (whether by wind, wave, or animal dispersal) than do remote islands. The upper curve shows that this simple relationship may be modified by 'stepping-stone' islands which allow species to migrate by a series of shorter steps. Extinction rate is directly related to the size of the islands (c) so that more species die out on small than on large islands. This fact is largely due to the narrower range of ecological conditions normal on small islands. If we now combine these two factors – island accessibility and island size – we have a reasonable explanation for the kind of variation in species numbers so generally observed. Island A is both near the continental source and large in area and so has many species, while island B is both small and remote and thus has a low number of species. Island C is also small and remote but is better linked to the continental source via the intermediate chain of islands, so it has a higher number of species than B.

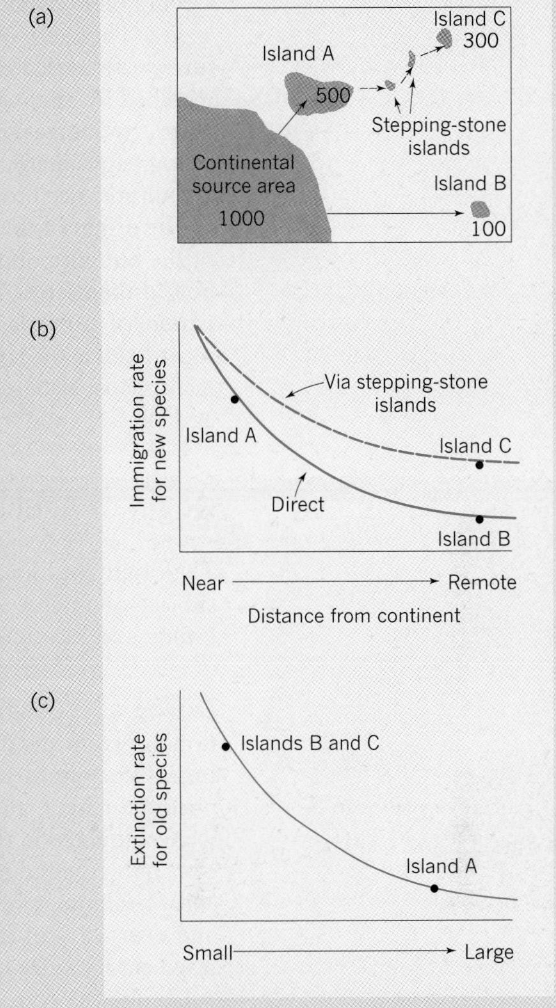

Generally, large islands that are close to the continental land masses have a much richer range of fauna and flora than islands that are small and remote. So far as human beings are concerned, the same kind of rules appear to apply. If we measure human variety by means of genetic terms, it is the small, remote islands that display the smallest range in blood types or biochemistry. Whether the model would also help us to understand cultural variability is more debatable. The original model was put forward by R.H. MacArthur and E.O. Wilson, *The Theory of Island Biogeography* (Princeton University Press, Princeton, N.J., 1967). The wider role of island colonization models is discussed at length in A.D. Cliff, P. Haggett and M. Smallman-Raynor, *Island Epidemics* (Oxford University Press, Oxford, 2000), Chapter 2.

5.2 Agricultural and Urban Origins

Geographers commonly divide human culture into four distinct technical stages. These are (1) food-gathering and hunting cultures, (2) herding cultures, (3) agricultural cultures, and (4) urban cultures. Each stage is matched by an increasing complexity of material goods and social organization, by increasing ability to support high population densities, and by ever greater interference with the natural environment (see Chapter 9). Not all cultures need to go through all stages.

The origin of the food-gathering and hunting cultures is the same as that of the human population itself; the first human groups in East Africa supported themselves in this way. Little is known about the earliest *domestication* of animals and the origin of herding cultures; some authorities regard this as a late, rather than early, stage in human cultural development. Most debate has centred on the origins of the third cultural stage – agriculture – and it is to this controversy that we first turn.

Origin of Agricultural Hearths

The origin and location of the world's *agricultural hearths* have been the subject of intense academic debate. We consider here questions of both timing and location and of their impact.

Timing and Location Archaeological evidence indicates the domestication of plants and animals by 8000 BC in the hills of what are now Iraq and Iran. Other finds reveal some similar activity in scattered spots in India, northern China, and central Mexico. It seems likely that wheat and barley were cultivated in the Middle East at an early date, and that the cultivation of corn by the Indians of Central America came later. Little is known of the early beginnings of rice cultivation in Asia, but new archaeological finds and new ways of dating finds may yet enable us to revise and rewrite the fragmentary story of the development of agriculture.

Despite the scarcity of firm evidence, there has been plenty of conjecture on the location of the first agricultural communities. In the sweeping survey *Agricultural Origins and Dispersals*, Berkeley geographer Carl Sauer argued for separate hearths of domestication in both the Old and New Worlds, outside the conventional hearth areas. (See Box 5.B on 'Sauer and agricultural origins'.)

As Figure 5.7 illustrates, the *Sauer hypothesis* places the Old World hearth in South Asia and the later New World hearth in the valleys and lowlands of the northern Andes. Sauer chose these areas on the basis of five criteria. First, the domestication of plants could not occur in areas of chronic food shortages; the domestication of crops and animals implies experimentation, and a sufficient abundance of food so that the experimenters can wait a while for results. Second, hearths must be in areas where there is a great variety of plants and animals and thus a large enough gene pool for experiment and hybridization to occur. Third, large river valleys are unlikely hearth areas because their settlement and cultivation require rather advanced techniques of water control. Fourth, hearths must be restricted to woodland areas where spaces can readily be cleared by killing and burning trees; grassland sod was probably too tough for primitive

Box 5.B Sauer and Agricultural Origins

Carl Ortwin Sauer (born Warrenton, Missouri, USA, 1889; died Berkeley, California, USA, 1975) was an American geographer, widely regarded as the leading cultural geographer of his generation. A native of the Missouri Ozarks and a University of Chicago graduate, Sauer moved to Berkeley in 1923 and was for over 50 years on the Berkeley faculty. He built one of the most distinctive and distinguished graduate schools in the United States (see Box 7.A on 'The Berkeley School').

Sauer's earliest regional writing was on the Middle West and his reports on the Upper Illinois Valley (1916), the Ozark Highlands (1920), and the Kentucky Pennyroyal (1927) are classics. After moving to California he became fascinated with Baja California and the American Southwest, and began a longstanding research interest in Spanish America, particularly Mexico and the Caribbean, which led to major reports published in the *Ibero-Americana* series which he founded in 1932 with the anthropologist Kroeber. Sauer made occasional methodological statements on the nature of geography (e.g. the influential *Morphology of Landscape*, 1925) but he would seem to have regarded them as interruptions to his substantive field studies. Sauer's main research centred on the early relations of humans and plants. Sauer's very extensive writings all show a sensitivity to the diversity of human cultures and the way in which they have adapted to and changed their landscapes over time. His research ranged widely in space and time and was marked by a deep concern with the use of the environment in a humane way, and a belief that rural peoples over the centuries had learnt much that contemporary society must not forget. In *Agricultural Origins and Dispersals* (1952), Sauer presented a synthesis of archaeological knowledge and speculation to argue for new locations where the domestication of plants and animals had already begun. Many of his more speculative and controversial views on the location of agricultural hearths and on the antiquity of human colonization of the New World have been supported by recent archaeological evidence.

In his productive retirement years, Sauer published three major books, *The Early Spanish Main* (1966), *Northern Mists* (1968), and *Sixteenth Century North America: the Land and the People as Seen by the Europeans* (1971), each exploring transatlantic contacts and images. He died with a fourth volume on seventeenth-century views of the North American environment almost complete. A review of Sauer's work is given by John Leighly in his preface to *Land and Life: Selection from the Writings of C.O. Sauer* (University of California Press, Berkeley, 1963).

[Source: Photograph courtesy of the late Mrs C.O. Sauer.]

AGRICULTURAL ORIGINS AND DISPERSALS

BY

CARL O. SAUER

CHAIRMAN, DEPARTMENT OF GEOGRAPHY
UNIVERSITY OF CALIFORNIA
BERKELEY

SERIES TWO

THE AMERICAN GEOGRAPHICAL SOCIETY
NEW YORK · 1952

cultivators. Finally, the original group of cultivators had to be sedentary to stop crops from being consumed by animals. The main nomadic groups probably did not meet this requirement; nor, probably, did the areas they inhabited (see Figure 5.7) meet the first four requirements.

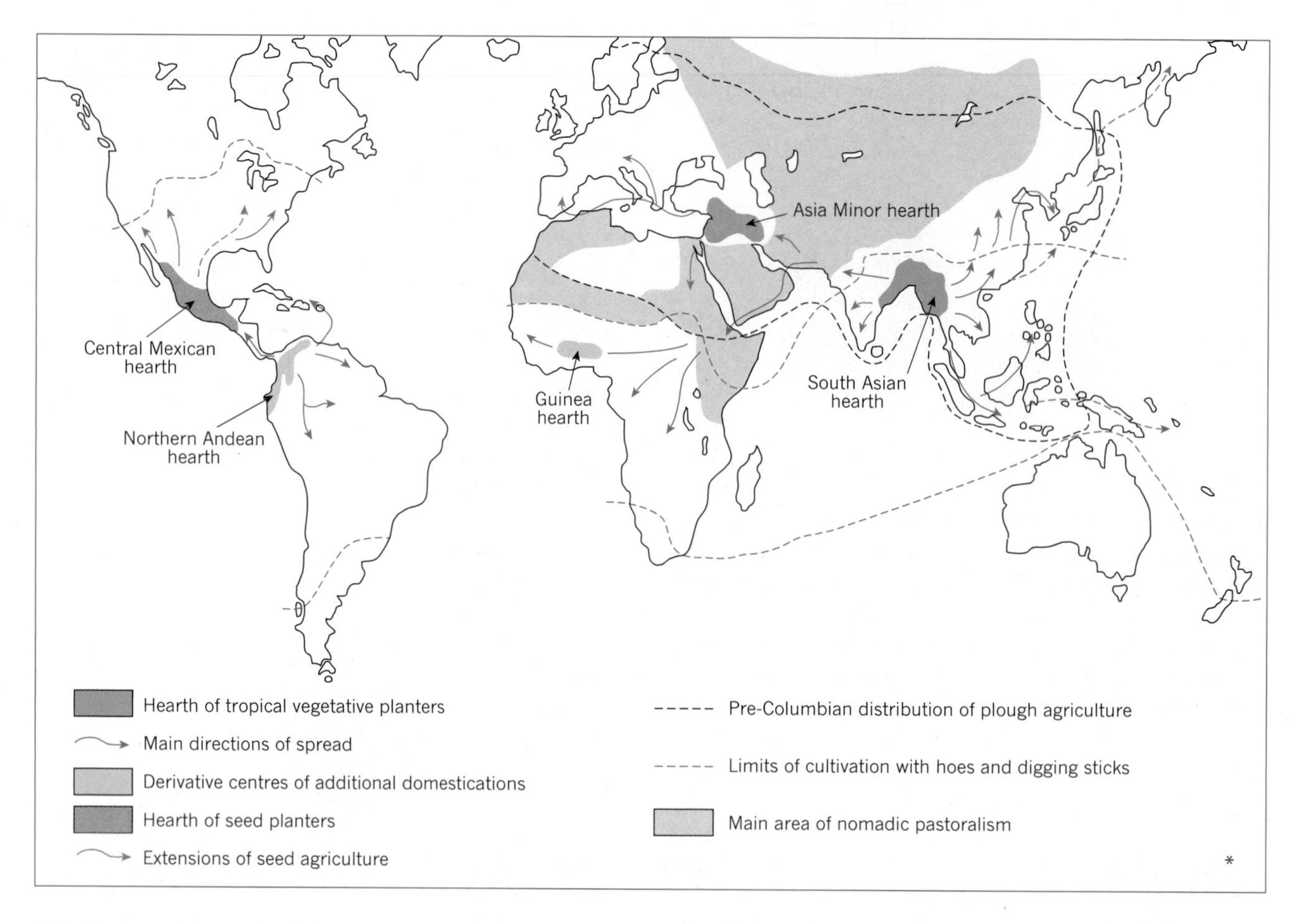

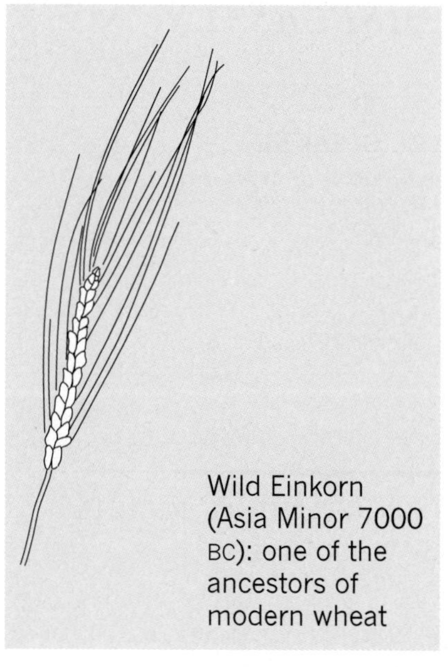

Figure 5.7 Agricultural origin and dispersal This world map shows in a highly generalized form the supposed main features of agricultural diffusion in the pre-Columbian world. It is based on the views of geographers Carl Sauer and Eduard Hahn and geneticists such as N.I. Vavilov. Note that there are two main hearths shown in the Old World: (1) tropical southern Asia as a centre for agricultural systems based on reproduction by vegetative planting (e.g. subdividing an existing plant into several parts, each of which grows into a new plant); and (2) subtropical Asia Minor as a centre for agricultural systems based on reproduction by planting seeds. This fundamental distinction between the two types of plant propagation is repeated in the two centres (again, tropical and subtropical) suggested for the New World. The probable main lines of spread around each hearth are shown together with a secondary hearth area in West Africa. The map also shows the important technological distinction between methods of cultivation based on hoes and digging sticks (found in both Old and New Worlds) and that based on the plough (found only in the Old World in pre-Columbian times). Areas of nomadic pastoralism are those where herding of animals by migratory peoples had been established in the Old World. Compare this map with the detailed table of plant origins given later in this chapter (Table 5.3).

[Source: E. Isaac, *The Geography of Domestication* (Prentice Hall, Englewood Cliffs, N.J., 1970), Fig. 3, p. 41.]

By combining these criteria, Sauer chose as his hearth areas the most probable environments for agricultural innovation. They had the climatic range to induce diversity and the rivers to provide a regular supply of fish for a sedentary settlement. Here wild plants were developed by centuries of selecting, propagating, and dividing. In Sauer's view the seed agriculture of the Middle East, China, and Central America is a much later, and more sophisticated, outgrowth of the activity at the two earlier centres of this type of agriculture in central Mexico and Asia Minor. The lively and sometimes hostile response of archaeologists to this view suggests that the debate is still wide open. Sauer's special contribution was to argue that, since evidence is so thin from this very distant period, we need to think hard about the geographical conditions under which domestication might occur.

The Spatial Impact of the Agricultural Revolution Whatever the precise location of the first agricultural communities, the impact of permanent agriculture on the spatial organization and density of the human population is clear. It increased the reliability of the food supply as well as the volume of food, so that more people could be supported by a given area. They no longer needed to concentrate totally on food production and could branch out to non-agricultural crafts. As food surpluses became available, goods began to be exchanged. Pottery, weaving, jewellery, and weapons were bartered and traded over long distances.

The impact of these changes on human spatial organization was twofold. First, the centrifugal forces that scattered small numbers of people over large areas were weakened; isolation gave way to contact, and some degree of agglomeration into settled agricultural villages became possible. Second, the population densities rose in some areas to levels several hundred times greater than those of the pre-agricultural communities. For example, we know from archeological evidence that the hill-farm communities of northern Mesopotamia had densities of approximately 70 people/km^2 (180 people/square mile) around 8000 BC.

By 4000 BC the total population of the globe had probably reached around 90 million. The greater part of this population was probably concentrated in areas where village agriculture was intermingling with and slowly replacing food-gathering and hunting. Such areas certainly included a belt stretching from Western Europe and the Mediterranean through the Middle East to western India, northern China, Indonesia, and Central America. Outside this area population changed little from its pre-agricultural pattern. The extreme zones of the Arctic and Antarctic, together with the more remote ocean islands, remained wholly unoccupied.

Origin of Urban Hearths

Although the evidence on the origin and early growth of cities is more plentiful, its interpretation has led to academic controversy hardly less acute than that over agricultural hearths. Specific evidence of urban forms is available for several sites in the *fertile crescent* for the period 3000 to 2500 BC. Calculations based on the size of these built-up areas yield probable populations of around 50,000 for Uruk and 80,000 for Baghdad in that period. Less controversy surrounds the time of early urban centres (although new excavations in Asia Minor indicate that they may be older than they were first thought to be) than how they fit into a development sequence.

The Development Sequence Figure 5.8(a) shows a highly generalized version of the traditional view of human developing resource-organizing technology. There are four main stages (primitive food-gathering and hunting, herding, agriculture, and urbanization), linked by three processes (the domestication of animals, the permanent cultivation of crop plants, and the trading of goods). The position of these stages and processes with respect to the time line at the left reflects the pattern of archaeological evidence.

Archaeologists are increasingly divided over the ways in which urbanization fits into this sequence. Figure 5.8(b) shows a conventional 'linear' view in which urbanization is a late stage in the developmental sequence, dependent on the buildup of a food surplus as a result of the increasing production by agricultural communities. More generally, herding is regarded as an incidental side-shoot, contributing little to the sequence (see Figure 5.8(c)). But is this the correct order? Town planner Jane Jacobs has entered the fray with a controversial book called *The Economy of Cities*. She emphasizes (1) the increasing evidence of highly specialized and long-range trade (e.g. in obsidian axes) among human populations in the basic food-gathering and hunting stage and (2) the increasingly early dates assigned to cities.

Utilizing this evidence, Jacobs questions both the assumptions of the conventional 'linear' view. In the *Jacobs hypothesis* (Figure 5.8(d)) urbanization is an early response to trade and exchange, and permanent agriculture a by-product of the food needs and hybridizing environment of the city. There are other scholars who question the assumption that cities originated primarily for *economic* reasons. Leading urban historian Lewis Mumford cites documentary evidence from the cities of ancient Egypt which suggests that they were founded as centres of royal or priestly power. The view of cities as *Zwingburg* (control centres) rather than marketing or manufacturing centres may be a correct interpretation of their role in the pre-industrial Near East. However, by the time cities appeared in the eastern Mediterranean area, during the third millennium BC, their role as control centres was becoming inextricably entwined with their role as centres of interregional trade.

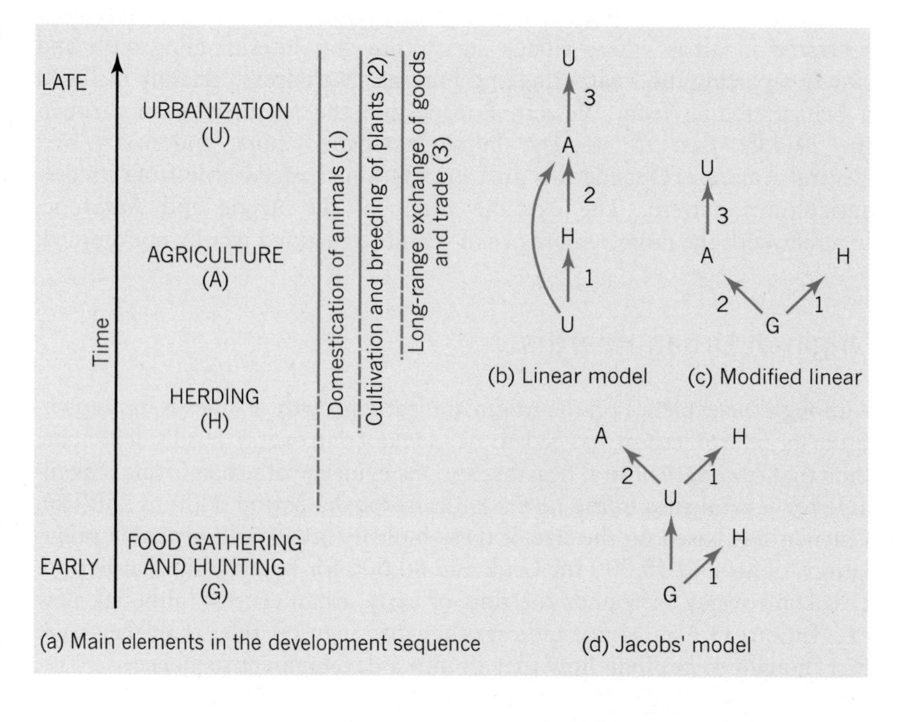

Figure 5.8 The position of cities in the developmental sequence
Diagram (a) shows the traditional main stages and processes in the developmental sequence. This development starts with food gathering and hunting and ends with urbanization. Diagrams (b), (c), and (d) provide alternative models of the place of the cities in human development. Of course, the models shown here are highly simplified. Herding, for example, has a very complex origin and probably developed in different ways in various areas.

The Location of Urban Hearths If we leave alone the muddy waters of when and how cities began, we are left with the question of where they began. Unfortunately, our notions of the spatial distribution of early cities necessarily reflect the concentration of archaeological activity. The patterns observed depend in part on where archaeologists have chosen to look for evidence of urban centres and on fortuitous factors such as the durability of building materials and the preservation of foundations. Most successful cities have experienced so many cycles of building and rebuilding on the same sites that traces of their early outlines are difficult to establish.

We know from the available evidence that urban development appears to have begun in four major river valleys: (1) the land lying between the Tigris and Euphrates rivers in the Mesopotamian area of the Middle East, (2) the valley of the Nile in Egypt, (3) the area near the Indus system in western India, and (4) the Hwang Ho (Yellow River) Valley in northern China. Table 5.1 gives the location and dates of the main areas of early urban cultures.

Prehistorians Childe and Wittfogel have seen a special link between the emergence of an urban civilization and the practice of large-scale irrigation. According to their *hydraulic hypothesis*, the environmental problems posed by the agricultural development of large, seasonally flooded river valleys could only be solved by integrating the collective efforts of many small communities. The construction and maintenance of large-scale irrigation works – dams, reservoirs, feeder canals – demanded the mobilization of large quantities of labour. Fair allocation of water between competing communities also called for the presence of a supervising authority which could plan the distribution of water use over both space and time.

Although the early emergence of Mesopotamia and Egypt as urban centres lends support to the hydraulic hypothesis, more recent work suggests the story may be a much more complex one. In a detailed study of another

Table 5.1 Main urban nuclear areas

Type	Zone	Location	Early urban culture	Representative city
Primary nuclear areas	Fertile crescent (Middle East)	Mesopotamia (Tigris–Euphrates valley)	Sumerian (2700 BC)	Ur, Uruk
		Nile Valley	Egyptian (3000 BC)	
		Indus valley	Indus (2500 BC)	Memphis, Thebes
				Mohenjo-daro, Harappa
	East Asia	Hwang Valley	Shang (1300 BC)	Anyang
	Americas	Yucatan Peninsula	Mayan (AD 500)	Palenque, Tikal
		Central Mexico	Aztec (AD 1400)	Tenochtitlán
		Peru	Inca (AD 1500)	Cuzco
	West Africa	Niger Valley	Yoruba (AD 1300)	Ife
Secondary nuclear areas	Southern Europe	Aegean islands and Peninsula	Aegean (2000 BC)	Knossos Mycenae
		Italian Peninsula	Etruscan (400 BC)	Felsina, Rome
	South and East Asia	Mekong Valley	Khmer (AD 1100)	Angkor
		Japan	Yanato (AD 600)	Naniwa
		Central Burma	Pyu (AD 700)	Sri Ksetra
		Ceylon	Sinhalese (AD 1000)	Polonnaruva

The dates are indicative of the middle period of the culture. The division between primary and secondary nuclear areas is based on Paul Wheatley, *The Pivot of the Four Quarters* (Edinburgh University Press, Edinburgh, 1971).

area of irrigation agriculture, the North China plain, geographer Paul Wheatley finds the emergence of urban form and spatial organization directly planned to reflect prevailing ideas about the universe. (Look ahead to Figure 7.10(c).) Since this symbolic role, found in all the other nuclear urban areas, appears to predate the growth of large-scale irrigation, it may well be that urban forms helped to encourage large-scale irrigation rather than being dependent on them.

Urbanized hearth populations of over 125 people/km^2 (325 people/square mile) existed in limited sections of agricultural districts as long ago as 4500 BC. By the beginning of the Christian Era, the earth's total human population had grown to around 300 million, roughly doubling since the beginning of urbanization. At this time the main lines of the earth's present pattern of population were beginning to be blocked out.

The overwhelming majority of the world's population was located in three huge clusters. Probably the largest concentration of people was in the Indian subcontinent, which had over 40 per cent of the estimated world population. The second largest concentration was that part of China within the Han Empire; perhaps 25 per cent of the world's population was concentrated primarily in the delta plains of the Hwang Ho. Outside these two largest clusters lay the ancient Roman Empire, extending from Western Europe and the Mediterranean to the Middle East, and including the long-established poulation concentrations of the Nile Valley and Syria. These three areas contained well over four-fifths of the world's population. In the fertile alluvial parts of these areas, the population density reached levels in excess of 1000 people/km^2. Outside these areas, population continued to be thinly spread over the land surface, and only a few high-density pockets (e.g. in central Mexico) broke up the pattern of primitive agricultural settlements and food-collecting groups.

Epidemiological Implications The crowding of human populations brought many unexpected side effects. For example, many of the so-called 'crowd diseases' (the infectious diseases of childhood such as measles and chickenpox) depended on high population densities for their survival. As we shall see in Chapter 16, measles is caused by a virus that depends on surviving in a human host (for about two weeks) and then being passed on to another susceptible member of the population.

The statistician Maurice Bartlett has calculated that it takes a population of around 250,000 to provide a continuous reservoir to maintain the disease. Below that threshold there might be occasional outbreak but no long-term endemic state. Figure 5.9 shows a reconstruction of the probable path of the disease spread. By AD 100 cities such as Alexandria, Rome, and London (to the west) and Loyang (in China) were large enough to support the disease.

But where did the disease come from in the first place? The most likely source is from the cross-over of the virus from animals to humans. Two animals that were domesticated by early settlers and lived in close association with human families were the dog and cattle. Although neither species supports measles, they do support very similar infectious viruses: distemper (in dogs) and rinderpest (in cattle). Either could have led to a disease that has been an inseparable part of human history for the last 5000 years. Such diseases illustrate the sensitive links between human changes in the environment and the rest of the biosphere, including the micro-organisms in that biosphere.

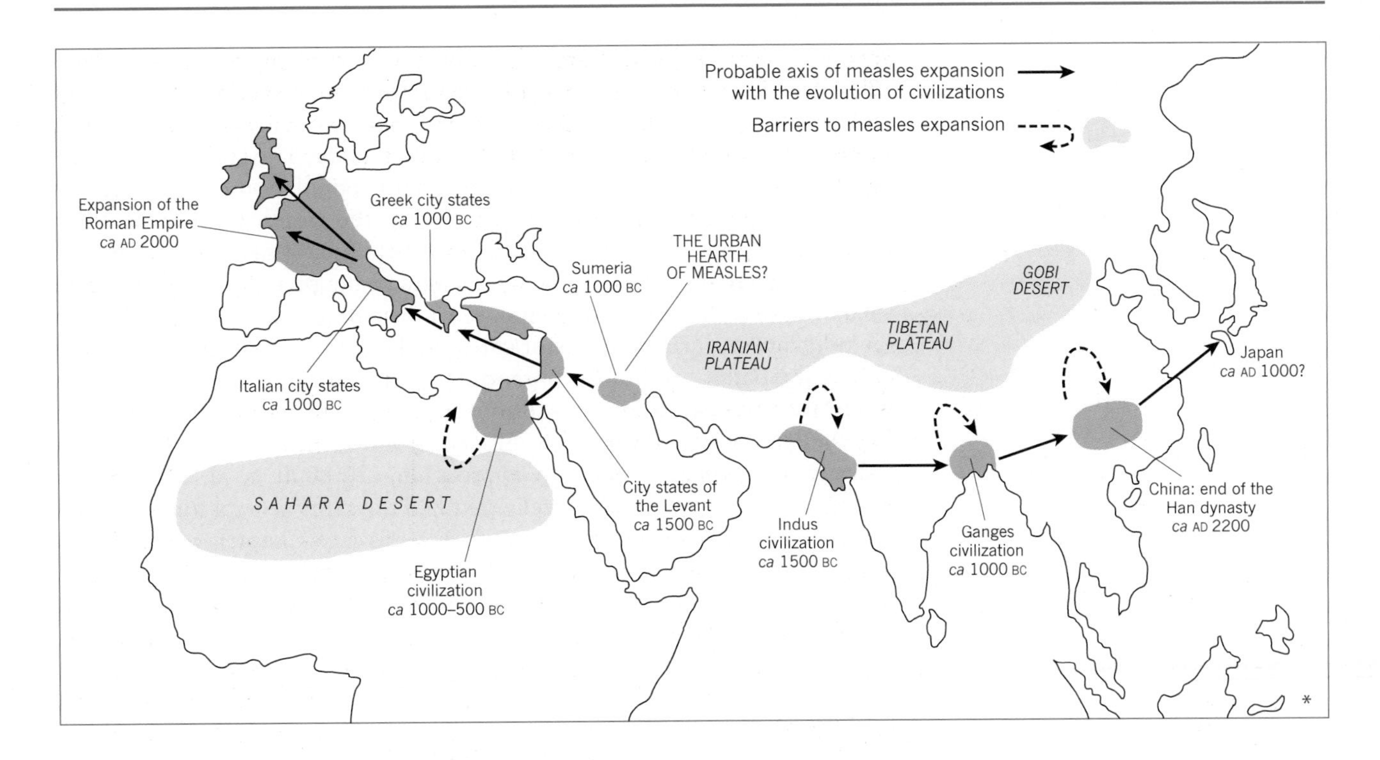

Figure 5.9 Early spread of infectious diseases Hypothetical reconstruction of the original hearth of a human viral disease (measles) and its probable spread from the Middle East. Probable axis of measles diffusion and physical barrier to measles expansion from its likely hearth in Sumeria.

[Source: A.D. Cliff, P. Haggett, and M. Smallman-Raynor *Measles: An Historical Geography of a Major Human Viral Disease, 1840–1990* (Blackwell, Oxford, 1993), Fig. 3.1(a), p. 51.]

The Major Regional Hearths 5.3

In settling questions of regional origins the main problem geographers face is lack of evidence. But as the evidence on the recent historical period builds up, the challenge shifts to making a comprehensive and convincing story of known but perplexing events. Each of the centres of urban culture we have recognized grew in population during the Christian Era, and each merits special study. Here, however, we choose to follow the history of only one of them, western Europe.

The European Hearth

The reason for selecting western Europe rather than China, for example, is less chauvinistic than it may appear. The growth and spread of European culture is well documented; it provided, in fact, a pattern for much of the non-European world. In some areas such as the Americas and Australia the pattern has been followed closely; in others such as China, it has had scarcely any effect; in yet others such as Africa and South Asia, we still await the final results of contact with the west. In looking at the geographical spread of western European culture, we begin with its internal crystalization. Then we move on (in Section 5.4) to its overseas transfer.

Between the beginning of the Christian Era and AD 1500, the world's population probably doubled to around 500 million. This increase was particularly noticeable in the world's third major population concentration – the area of the former Roman Empire. Here the largest gains were registered in the more recently settled areas of western and east-central Europe, areas that were to emerge in the later stages of the period as modern states such as France, Britain, and Poland. For Europe, the improved information on population during the historical period allows us to see much more clearly the pattern of human organization. We can, for example, trace the westward spread of urban institutions in the Roman Empire (Figure 5.10). For the largest city, Rome, housing densities within the known city limits indicate a maximal population approaching 200,000 in AD 200. Below Rome extended an urban hierarchy that had the same general form we shall see later in characteristic modern city systems (see Chapter 14). The collapse of the Roman Empire and the reduction in the scale of political and commercial organization from a subcontinental to a local level was followed by a breakup of urban and regional links.

With the slow revival of trade in early medieval Europe came the emergence of a fine mesh of small inland cities whose sites were often chosen because of their good defensive positions (e.g. the *bastides* of southwest France). Populations remained surprisingly small. For example, by AD 1450, Nuremburg, Germany, an important inland town, still had a population of only 20,000. Even London, which by AD 1350 had regained the level of population it had reached in Roman times, contained only around 40,000 inhabitants. Below these principal cities came a regular hierarchy of smaller cities, not unlike that which exists in the modern world.

At the end of the medieval period, western and central Europe was firmly structured into an organized regional system of cities. At the top of the hierarchy stood the trading cities of emergent industrial areas such as

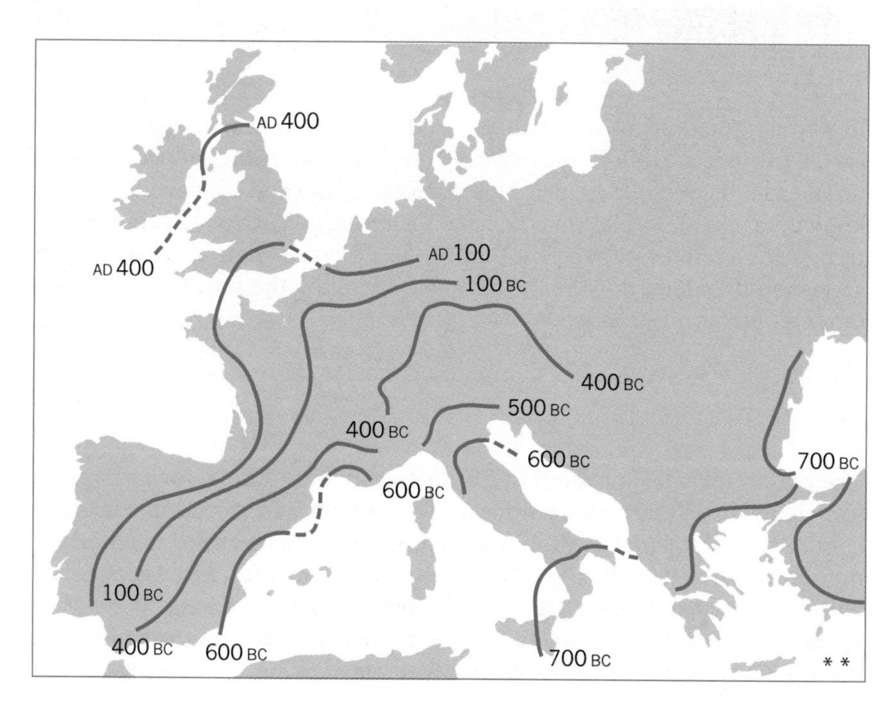

Figure 5.10 The spread of urban organization across Europe The earliest European cities occur in the extreme southeast corner of the continent where cities such as Knossos in Crete or Mycenae in southern Greece were established by around 2000 BC. Major extension of the city-building in Europe accompanied the spread of the Roman Empire. The generalized contours show cities spreading north and west from the Aegean over an 1100-year period of history.

[Source: N.J.G. Pounds, *Annals of the Association of American Geographers*, **59** (1969), Fig. 6, p. 148.]

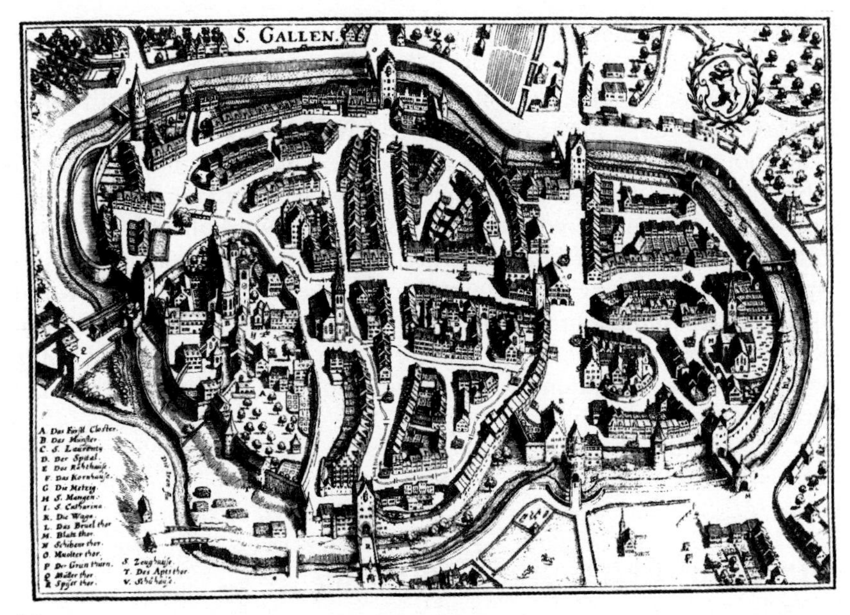

(a)

(b)

Figure 5.11 The growth of the West European city St. Gallen is a city in northeast Switzerland. Although the Romans colonized this part of Europe, the founding of St. Gallen is ecclesiastical in origin and later in date. It developed around a Benedictine abbey founded in the early seventh century. For the next four centuries this was to be the most famous educational institution north of the Alps, and it remained an important seat of learning throughout the medieval period. Along with the growth of the abbey came increased political and economic importance for the city. It was walled in the tenth century, became a free city in 1304, and joined the Swiss federation in 1454. An early seventeenth-century print (a) still shows the medieval core of the city with its walls and gateways. But in 1600 its population was probably still below 5000 people. Growth in the period since then to its present 80,000 has been associated with its role as a commercial centre for the surrounding area (the canton of St. Gallen), its long-established textile industry, and new industries such as glass and metalworking. With this growth the city has sprawled well outside its original walls and now extends east–west along the Steinach Valley. The photo (b) shows the medieval core within the modern city today; forested valleyside slopes north and south of the city show as very dark areas. In comparing (a) and (b) note that north is to the right of the print but to the top of the photo.

[Source: (a) H. Boesch *et al.*, *Villes Suisses a Vol d'Oiseau* (Kummerly & Frey, Berne, 1963), pp. 212, 216. (b) Courtesy of Swiss National Tourist Office.]

London, Flanders, Lombardy, and Catalonia. Below was a network of smaller inland cities like that in Figure 5.11, often playing an important role in trade and administration. The network of cities was growing in two ways. First, by spatial expansion through colonization and the establishment of new cities in eastern Europe. Second, by the growth of smaller cities around fast-growing centres such as Venice and Genoa, both booming as trade with the Levant increased.

Asian and American Hearths

Outside Europe lay other city hierarchies that were also highly differentiated and expanding. In the rest of the Old World the main areas of urban civilization were in eastern China and northern India. In the New World only central Mexico and the Peruvian valleys were urbanized, and these areas had much smaller populations than the Old World hearths of urban culture. Of the five hearths in existence in 1500, the European one was to experience the most significant expansion in the next 400 years of world history. Three of the four remaining hearths came directly under European influence in that period; only the Chinese hearth remained untouched by the major spatial reorientation of world trade produced by the growth of the European centre.

5.4 European Expansion Overseas

Over the period from 1450 through into the present century, western Europe was involved in a sequence of overseas expansions. We look here at the spatial character of that expansion: from early settlements on the oceanic margins to later penetration of the continental interiors.

Rim Settlements

The first phase of European overseas expansions, that of trans-oceanic ***rim settlement*** (or coastal settlement) lasted from the original Age of Discovery in the fifteenth century to the early part of the nineteenth century. Different European peoples took the lead at different times – the Spanish and Portuguese earlier, the French, English, and Dutch later – in establishing settlements along the coastal rims of the Americas, Africa, and southern Asia. The settlements were of three main kinds: trading stations, plantations, and colonies of farm families. All clung, limpet-like, to the edge of the continents.

Coastal Trading Stations Small trading posts were established widely on the coasts of India and southern China. Ports such as Goa, Madras, and Canton served at times as points of exchange through which the products of the two great Asian urban civilizations of India and China could be brought to the growing urban markets of Western Europe. Trade was mainly in luxury articles such as spices, tea, and hand-crafted products such as silks. The number of European settlers at any one time was quite small compared with the size of the indigenous Asian population, and only in India was it possible to exercise any real political control over the large hinterland areas that served the ports. Counterparts to the Indian ports, trading stations, were established on a minor scale in Malaysia, Indonesia, and East and West Africa.

Tropical and Subtropical Plantations Plantations were originally established to grow spices and sugars, but the range of crops was later expanded to include the production of various foodstuffs (coffee, cacao, bananas, etc.). The earliest *plantation settlements* were generally on ocean islands such as Madeira, offshore islands such as Zanzibar off the east coast of Africa, or coastal strips such as the Baixada Fluminense around Rio de Janeiro in Brazil. Inland extensions of these settlements appeared largely in the nineteenth century. Such plantations demanded intensive labour, and European settlers primarily occupied only organizational roles, while the non-European population provided the field labour. When the indigenous population was unable to supply workers, slave labourers were brought in from other areas (see Figure 5.12). After the abolition of slavery, indentured labour on long-term contracts became a substitute. For example, India provided the major source of workers in the sugar-cane plantations of British colonies from Trinidad in the Caribbean, to Natal and Mauritius in the Indian Ocean, to Fiji in the Pacific Ocean. The current population mix in tropical America, East Africa, and parts of Malaysia and Australasia is largely a legacy of tropical plantation settlements.

Mid-latitude Farm-family Settlements The third type of settlement associated with the period of European expansion is the farm-family colony of European migrants in the middle latitudes. This type included the settlements, mainly of English- and French-speaking groups, on the northeastern seaboard of North America, as well as later settlements in

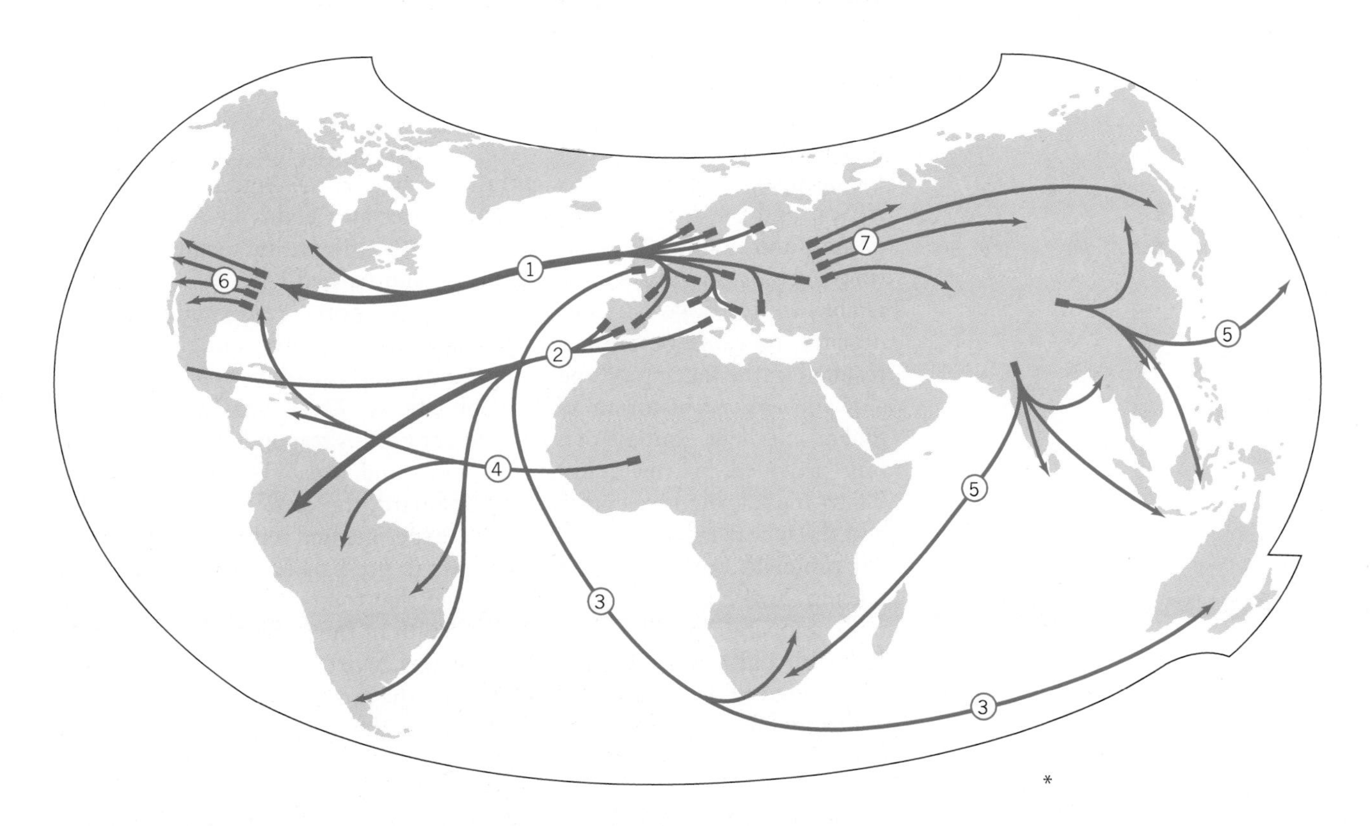

Figure 5.12 Intercontinental migration The map shows the main currents of intercontinental migration since the beginning of the sixteenth century. Only the main flows are shown, and in a highly generalized form.

[Source: W.S. and E.S. Woytinsky, *World Population and Production* (Twentieth Century Fund, New York, 1953), Fig. 27, p. 68. Copyright © 1953 by the Twentieth Century Fund.]

① Migration from all parts of Europe to North America
② Migration from southern Europe to Latin America
③ Migration from Britain to Africa and Australasia
④ Shipment of African slaves to the Americas
⑤ Indian and Chinese movements
⑥ Westward colonization in Anglo-America
⑦ Eastward colonization in Russia

Australia and New Zealand. These settlements were in marked contrast to the tropical plantations because of their dependence on an influx of Europeans. Moreover, their agricultural products were destined for the local market rather than for export to Europe. Different groups from various parts of Europe (Swedes, Germans, Irish, etc.) brought to their overseas settlement some of the distinctive characteristics of their home areas. The layout of farms, villages and towns, the cropping patterns, and the farming technology often reflected traditions and practices in the original homelands. Even today the different ethnic backgrounds of farmers can be traced in the design of their farm buildings; the Pennsylvania barn shows just such a rich variability and has been carefully mapped by cultural geographers.

Continental Penetration

The second phase of European expansion, that of *continental penetration*, began early in the nineteenth century and lasted until about World War I. This phase was accelerated by rapid industrialization in the European hearth, the development of transport innovations such as the railroad, the growing overseas migration of Europeans, and a rapid increase in the rate of exploitation and trade in non-European resources. Its chief impacts on the distribution of population were the springing up of industrial cities in the mid-latitude colonies and the inland penetration of the agricultural frontier as the rich, mid-continental grassland zones were exploited for grain or stock production.

Mid-latitude Grassland Settlements

The nineteenth century witnessed the occupation of the prairies and the pampas in the Americas, the veld in Africa, and the Murray-Darling and Canterbury plains in Australia. The pattern and timing of settlement was greatly affected by technical innovations such as the railroad, refrigeration, and barbed wire. Railroads made it cheaper to move agricultural products to ports, refrigeration made it possible to preserve meat for long-distance shipping, and barbed wire resulted in the fencing of open rangelands

If you look at historian W.P. Webb's book, *The Great Plains*, you will find an excellent example of how different groups responded to the resource mix provided by this mid-latitude grassland area. The pre-Columbian Plains Indian, the sixteenth-century Spaniard, the cattleman in the 1840s, and the wheat farmer in the 1880s all experimented with the Great Plains environment. In the withered hedgerows (a pre-barbed-wire experiment at fencing, British style), in the bankruptcies that followed long droughts, and in the blowing topsoil, we see the aftermath of the experiments that failed. Each cultural group saw in the Great Plains environment different possibilities, linked to the current technology and the group's cultural background.

Meanwhile, on Europe's eastern continental border the Russian state was expanding its settlement of the steppe grasslands at a comparable pace. In the tropics the demand for plantation products accelerated, and the movement of non-European peoples, first African slaves and later indentured Indians and Chinese, into plantation areas such as the Caribbean continued. Trading contacts with the Orient increased and intensified as western countries extended their political control through both treaties and military occupation. More European ports were established on the coasts of China.

Mining and Mineral 'Rushes' The mid- and late-nineteenth-century gold discoveries brought a rapid rise in both the white and the non-white population of mining areas. In some areas with an environment favourable to agriculture, mineral strikes provided the trigger that set off continued migration and long-term settlement. California provides the most well-known example. The discovery of gold at Sutter's Mill on the American River in 1848 led to a rush to establishing claims. 'Forty-niners' poured into California from the east and from many distant parts of the world. In the next twelve years California's population leapt from around 26,000 to approaching 400,000.

The American experience was paralleled by Australia. The 1851 gold strikes at Ballarat and Bendigo, in Victoria, brought in 250,000 prospectors in the next five years (see Figure 5.13). By 1855 more people lived in Victoria than had lived in the whole of Australia before the discoveries. But in marginal environments such as the subarctic zone the population and settlement that followed mining did not last. In 1898 some 30,000 prospectors moved down the Yukon River to the new gold strikes; today the total population of the whole Yukon territory is only half that number. Mining was associated with the movement of Chinese populations. Tin mining in Malaysia in the late nineteenth century brought Chinese settlers into that area. (See Box 5.C 'The Chinese diaspora'.)

The search for gold in the nineteenth century was followed by a search for oil in the Middle East in the early decades of this century. Mining strikes and oil exploration continue to be part of a phase of settlement expansion but the emphasis is now on shifts in capital investment (e.g. drilling rigs) rather than major population movements.

Figure 5.13 European overseas migration Gold rushes in the middle of the nineteenth century brought major increments of population to Australia. This contemporary print shows diggings at Sheep Station Point, River Turon, during the 1853 gold rush.

[Source: Mary Evans Picture Library.]

Box 5.C The Chinese Diaspora

In the text discussion, we have concentrated on overseas expansion as viewed from a European hearth. But Europe was not the only centre. China also was a focus from which population moved to new lands in what is called the Chinese diaspora. (Diaspora is a term originally applied to the dispersion of Jewish populations from Israel.) The earliest references to Chinese communities overseas derives from thirteenth-century Cambodia as coastal traders. Another community was recorded at Tumasik in 1349. The longest continual settlement is probably that of Malacca on the western coast of Malaysia. Descendants known as 'Babar' Chinese maintained both culture and language despite intermarriage with the Malay population and successive changes in colonial government through Portuguese, Dutch, and British administrations. Migrants to Nanyang (the 'Southern Ocean', now southeast Asia) were almost entirely from southeast China, the provinces of Kwantung and Fukien. Their outward migration reflected (i) increasing population pressure, (ii) increasing alienation between the southeast and successive northern Chinese regimes, (iii) the inability of China to extend its southern boundaries overland, and (iv) the availability of seagoing craft and geographical knowledge of southeast Asia. The two Chinese source provinces had a wide range of dialects and migrants tended to go to areas where speakers of their own language had settled. So Hakka peoples went in large numbers to Borneo and Hokkien speakers to Java and Malaya. A major exception were the Cantonese-speaking traders who were found widely wherever commercial opportunities presented themselves. As with European migrants, the expansion took many forms which shifted over time. Many of the wealthiest Chinese served as traders, entrepreneurs, and middlemen, but from the seventeenth century, Chinese migrants were increasingly prepared to work in the mining industry and on plantations. The map shows the main areas of Chinese population concentration in the countries of southeast Asia.

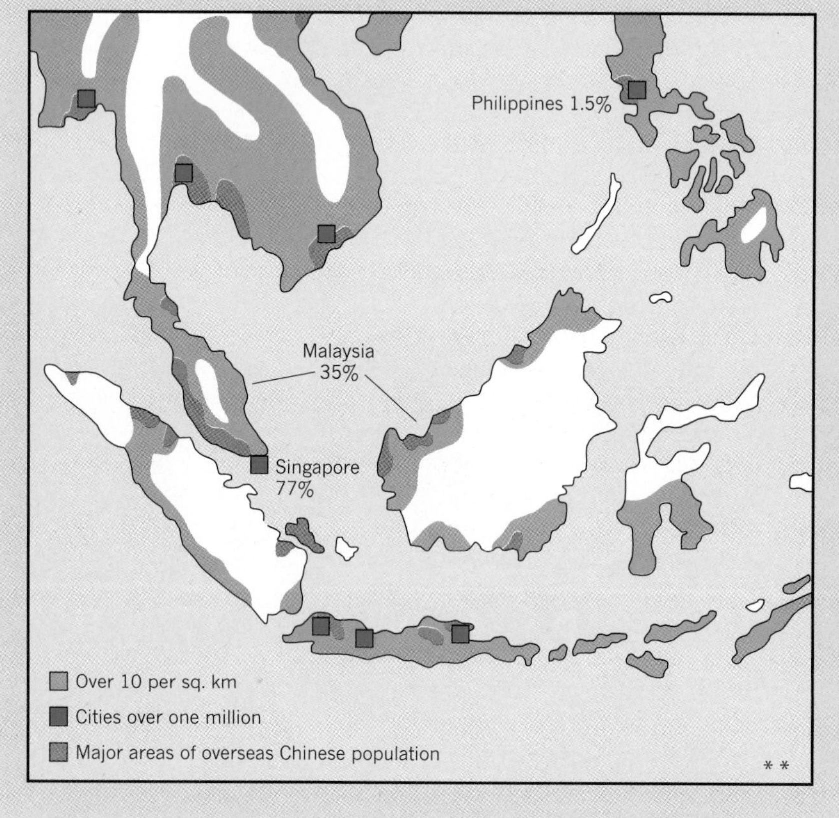

The map indicates the main areas of overseas Chinese population and the proportion they make up of the present population. Over the five countries for which comparable demographic data are available, the 12 million identified as Chinese make up 6 per cent of the total. But this proportion varies from three-quarters in Singapore to 1.5 per cent in the Philippines. In the latter country, the definitions of 'Chinese' is difficult since many ethnic Chinese have interbred, adopted Catholicism, and speak Filipino languages.

The success of Chinese overseas is seen in estimates of gross national product (GNP). Although the population of Chinese overseas worldwide is probably not more than 50 million, it has an estimated GNP which is two-thirds of the size of that of the 1.2 billion Chinese in mainland China. A useful discussion of the Chinese diaspora is given by Rupert Hodder, *The West Pacific Rim* (Bellhaven Press, London, 1992), pp. 43–56 and a fuller account in J. Cushman and Wang Gungwu, *Changing Identities of the Southeast Asian Chinese* (Hong Kong University Press, Hong Kong, 1988).

Table 5.2 Decolonization of areas formerly under west European control

Continent	Spain	Portugal	France	England	Netherlands
North America	Mexico (1821)	–	Haiti (1804)	United States (1776)	–
South America	Chile (1810)	Brazil (1822)	–	Guyana (1966)	Suriname (1975)
Africa	Western Sahara (1975)	Angola (1975)	Mali (1960)	Kenya (1961)	
Asia	Philippines (1898)	Timor (1976)	Vietnam (1954)	India (1947)	Indonesia (1949)
Australasia	–	–	New Caledonia (1980)	Fiji (1970)	–

Each cell contains one example of a country together with the date of decolonization. Some European powers (e.g. Denmark, Italy, Germany) are not represented.

Political Withdrawal

A third phase of political withdrawal began as early as the mid-nineteenth century with the independence movements in Latin America. The former Spanish colonies of middle and south America broke away between 1810 and 1825. Brazil became independent from Portugal in 1822. As Table 5.2 shows, the second half of the twentieth century showed almost complete withdrawal by the other European powers (Britain, France, and the Netherlands) from colonies established in Africa and Asia.

We might say that a third phase of economic consolidation but political withdrawal began around World War I and appears to be continuing today. It was marked by a shift of economic power from the original European hearth to the United States and Soviet Russia. In addition there was a political withdrawal of formal European control of much of Africa and Asia, signalling a virtual end to the British, French, and Dutch overseas empires. Population movements from Europe to mid-latitude countries such as the United States, Australia, and Argentina have continued, and they have not been balanced by the counter flow of black population to western Europe. Despite lessening political control, the presence of European capital, culture, and means of communication in much of Africa and Southwest and South Asia remains a fact of life. In Latin America Europe's economic role has been largely replaced by the United States. The rise of Japan as a prime industrial and trading power and the increasing role of China are bringing new waves of regional expansion based on East Asian hearths.

With the breakup of the European overseas empires came a succession of political problems. Let us take a single example from Southeast Asia. Indo-China has been part of the French area of influence since the missionary activity of the Catholic church after 1650. In the mid-nineteenth century the southern and then northern parts of the region became part of France's overseas colonial empire. At the end of World War II a communist republic was set up, declaring itself independent from France in 1945. Conflict with France was finally ended in 1954, when South Vietnam became a separate state (south of the 17th parallel). The conflict between the North and the South drew the United States into the vacuum left by the French withdrawal. Its progressive involvement ended after years of warfare with the withdrawal of American forces early in 1975 and the North's occupation of the whole of South Vietnam.

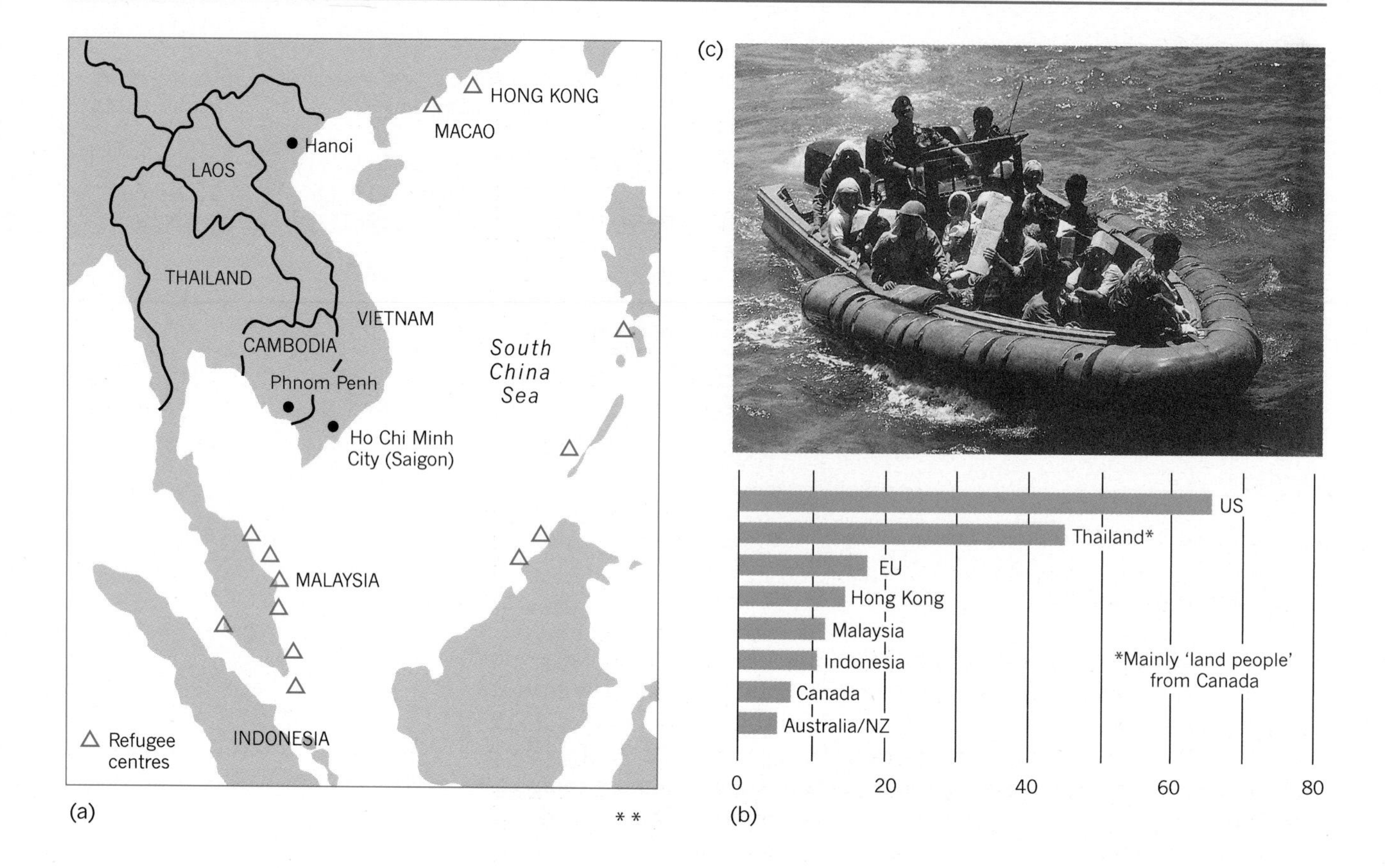

Figure 5.14 Enforced migrations: the Vietnamese 'boat people' Starting in 1975 and continuing into the early 1980s, the flow of more than a third of a million refugees from Vietnam by sea illustrates one of the most spectacular examples of the aftermath of empire. (a) Main refugee centres in Southeast Asia. (b) Leading destination of all Indo-Chinese refugees permanently resettled (total = 100). (c) Vietnamese boat people in 1989 being intercepted by Marine Police at sea off Hong Kong.

[Sources: (a), (b) Data from UN High Commission for Refugees.
(c) © MAGNUM/P.J. Griffiths.]

The fall of Saigon, the capital of South Vietnam, in April 1975 initiated one of the most spectacular and tragic of enforced migrations. Frightened by the prospect of their future under a Communist government, nearly a third of a million people attempted to leave Vietnam by sea. How many of the sad human cargo survived is not known with accuracy, but French estimates range from 30 to 60 per cent. Small, unsound, and overcrowded boats stood a low chance of survival under storm conditions. The tragedy was confounded by the high proportion of children among the passengers and the inability of neighbouring countries, especially Thailand and Malaysia, to cope with the vast number of refugees that wanted to land.

Movement of refugees continued through the late 1970s with the peak surge coming in the first half of 1979. The United States has been in the front rank of countries offering permanent homes to Indochinese refugees (see Figure 5.14), but perhaps half of the surviving 'boat people' still remain in temporary camps in Asian countries. Because of the problems of incorporating them into the local economy, the population remain in crowded, unsanitary conditions subject to malaria, tuberculosis, and parasitic infections.

The boat people of Vietnam provide a recent example of the demographic adjustments that have accompanied the European withdrawal from former imperial holdings. Estimates for the early 1980s suggest there are more than 14 million refugees displaced from their homes around the world, and of these perhaps half can be directly attributed to the withdrawal process. The conflict in Lebanon (another former French colony) in 1982 may be directly traced back to British withdrawal from Palestine after World War II and the set of problems surrounding the establishment of Israel and the following Israeli–Arab conflict in that country.

The Geographic Legacy

In this section we have argued that European expansion provided a spectacular, but by no means unique, example of cultural spread. But what was the summed effect of 500 years of European expansion? On the spatial organization of the world community, the effect was enormous. For one thing, the expansion involved a transcontinental movement of around 95 million people. Over two-thirds of these were Europeans moving to temperate latitudes (notably to the United States). Another 20 per cent were Africans forcibly transported from their homelands, especially to the American tropics and subtropics. The remaining fraction of over 10 per cent were Asiatics. The pattern of Asian population movements is more diffused because there are considerable Asiatic populations growing in parts of Africa, in the Caribbean, and in parts of the Pacific.

Along with the changes in population went a massive exchange and mixing of the world's crops. Table 5.3 gives a list of just some of the leading crop plants now used by human beings and their probable areas of origin. As we saw earlier in this chapter (see especially Figure 5.7) the precise origins of agricuture occurred so far back in human history that we can only make reasoned guesses on their actual location. What is clear is that 500 years of European expansion turned the pre-Columbian pattern of crop distribution upside down. American crops such as the potato and tomato were to become regular farm crops in Europe while Old World crops such as coffee and wheat were to become major crops in the Americas. Indeed, your own backyard is now likely to contain a variety of plants unquestionably richer and more diverse than seemed possible to our more continent-bound forebears of the pre-Columbian period.

Along with the interchange of population and of crops went a fundamental reorganization of wealth. Figure 5.15 shows in a very generalized form the map of world income today in comparison with that of population. The twin-peaked North Atlantic centre symbolizes the extreme degree of financial control exercised by institutions such as Wall Street, London's 'City', Paris's 'Bourse', or the Zurich banks in the organization of much of the world's resource development. Certainly this dominance may now be past its peak with the emergence of a separate Russian centre and the promise of new centres of financial power in the Middle East, Japan – and later China. Nonetheless, much of that control remains and the inequity between world population and world income that persists today is partly a reflection of the superimposition of a world urban system – centred

Figure 5.15 The legacy of European spatial organization
These generalized maps show the pattern of world income and world population in the middle of the twentieth century. Note the similarities and differences between the two.

[Source: W. Warntz, *Macrogeography and Income Fronts* (Regional Science Research Institute, Philadelphia, 1965), Figs 19, 24, pp. 92, 111.]

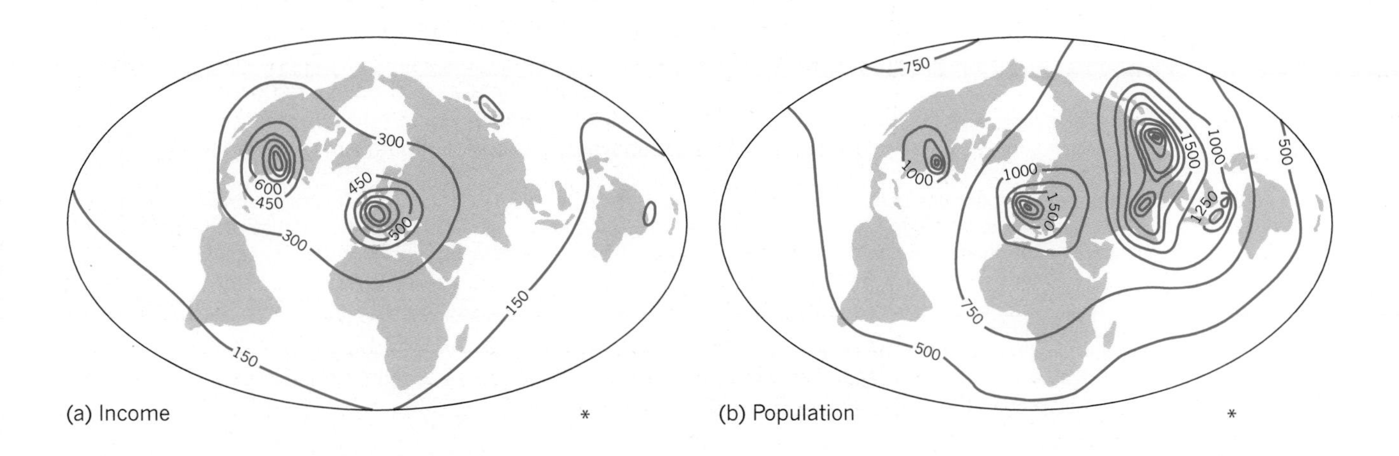

Table 5.3 Sixty leading crop plants and their probable areas of origin[a]

Hemisphere	Continent	Subcontinent	Crop plants (Classes I to IX[b]) with region of origin where known
Old World	Asia	East Asia	VIII Vegetables: rhubarb (China), soybean (China)
			IX Fruits: peach (china)
		S Asia	VIII Vegetables: cucumber (India)
			IX Fruits: date palm (W India), melon
		SE Asia	I Beverages and drugs: teas
			III Root crops: taro (New Guinea)
			IV Grains: rice
			V Sugars: sugar cane (New Guinea)
			VI: Fibres and oil plants: broad bean
			VIII Vegetables: broad bean
			IX Fruits: banana (Malaysia), citrus fruits, coconut, mango, watermelon
		SW Asia	I Beverages and drugs: opium poppy
			IV Grains: barleys, oats, rye, wheats
			VII Forage plants: alfalfa
			VIII Vegetables: carrot, pea
			IX Fruits: fig, grapes (Turkestan), quince
	Europe	Mediterranean basin	VI Fibres and oil plants: flax
			VIII Vegetables: beets, cabbage
		SE Europe	VI Fibres and oil plants: bluegrass
			IX Fruits: apple (Caucasus), pear (Caucasus), plum
	Africa	E Africa	I Beverages and drugs: coffees
			IV Grains: sorghums
			VI Fibres and oil plants: cowpeas
			VIII Vegetables: broad bean
			IX Fruits: melon
New World	Americas	N America	VI Fibres and oil plants: sunflower
			VIII Vegetables: squash
		Meso-America	II Ornamentals: dahlia, marigold
			III Root crops: cassava, sweet potatoes
			IV Grains: amaranths, maize
			VIII Vegetables: kidney bean, lima bean, red peppers, scarlet runner, squashes
			IX Fruits: avocado, papaya, strawberry
		S America	I Beverages and drugs: cacao (Orinoco basin), quinoa (Andes), tobacco (Plate basin)
			II Ornamentals: bougainvillea (E Brazil)
			III Root crops: cassava (tropics), potatatoes (Andes)
			VI Fibres and oil plants: cotton (Ecuador), peanut (E South America)
			VIII Vegetables: gourds, lima bean (tropics), red peppers (tropics), tomato (Andean America)
			IX Fruits: papayas, pineapple (E South America), strawberry

[a]There are many hundreds of plant species used by humans. This highly selective list is designed to illustrate some of the types of plants used and their probable areas of origin. Cultivated plants are very difficult to classify botanically because of the amount of hybridization and the ancestry of only a few has been fully worked out. The source areas should therefore be regarded as reasoned guesses from incomplete evidence in most cases.

[b]The nine crop groups are: I Beverages and drugs, II Ornamentals, III Root crops, IV Grains, V Sugars, VI Fibres and oil plants, VII Forage crops, VIII Vegetables, and IX Fruits.

I am grateful to Professor Jonathan Sauer of the University of California in Los Angeles for commenting on this list, which was originally derived from Edgar Anderson, *Plants, Man and Life* (Melrose, London, 1954), Chap. X and C.D. Darlington, *Chromosone Botany and the Origins of Cultivated Plants* (George Allen & Unwin, London, 1963).

first on Western Europe and then on a combined and widened North Atlantic core – upon much of the remainder of the world.

Geographers have sought to build a spatial model of the diverse pattern of European overseas expansion. We pick up their attempts when we look more closely at models of migration in Chapter 13 and at the subject of economic growth in Chapter 18.

Post-Columbian America 5.5

In the first part of this chapter we looked at the origins of cultural regions on the global scale. In the fourth part we took one culture – that of western Europe – and traced its spread. In this fifth and final section we reduce the scale of our investigations again and look at the question of the persistence of cultural elements in one of the areas of European overseas settlement, the United States.

Pre-Columbian Patterns

European settlement of North America did not start in an empty land. The area north of the Rio Grande contained a population of more than one million American Indians in AD 1500. As we saw in Section 5.1, they had migrated southward from East Asia via the Alaskan bridge in the late glacial period. By the time of the first European contact the native American population had differentiated into a very complex variety of separate subcultures each with its own language and traditions.

Five broad regional groups can be distinguished, those of the eastern forests, the Plains, the northwest coast, the California-intermountain area, and the Southwest. Each region was characterized by differences in farming or hunting systems, in settlement patterns, in languages, and in organizational structures. Examples of three of the five groups are shown in Figure 5.16.

The Sioux were part of the Plains Indian culture group. Originally farmers and hunters in the Mississippi valley forests, they were forced

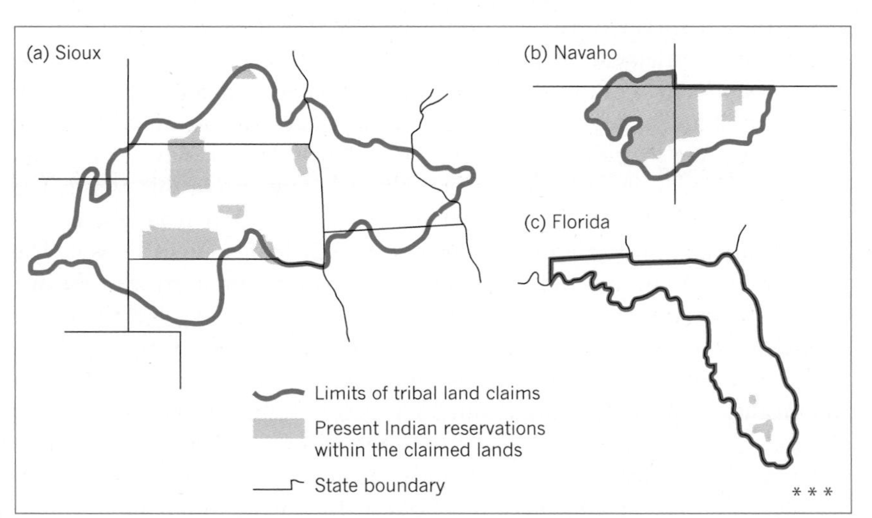

Figure 5.16 American Indian culture areas The detailed culture regions of pre-Columbian America are not known with certainty. Many Indian tribes were mobile and were displaced from their original homelands by European colonists. The maps show three from the several hundred distinct North American tribes: the Sioux in the northern Great Plains, the Navaho of the southwest, and the Florida of the southeast. Only the major reserves can be shown at this scale.

[Source: Tribal land claims are from a map compiled by S.H. Hilliard, *Annals of the Association of American Geographers*, Map Supplement No. 16 (1972).]

further west onto the grasslands as part of the general relocation of tribal areas that followed European occupation of the East Coast. The introduction of horses from the Spaniards moving north from Mexico also allowed a more mobile, hunting economy to be built up. The second example, the Navajo, were part of the southwest culture group. Originally a nomadic hunting people, they settled as northern neighbours of the sedentary Pueblo Indians (*pueblo* is the Spanish word for village) or Arizona and New Mexico and learned to raise corn and weave cotton. Again, the introduction of horses and sheep by the Spaniards modified the economic pattern of the Navajo by emphasizing the grazing and wool-processing side of their culture. Finally, the Florida Indians were part of the Eastern Forest Tribes. Tribal divisions in this area are less well recorded but the groups probably included the Seminole and relatives of the Creek Indians. In addition to hunting and fishing, the people of this area lived by raising such crops as corn, beans, squash, and tobacco. Settlements were both permanent and complex in structure, often with houses built around a central court.

If we look forward to the Slobodkin model in Figure 6.13 we see the range of outcomes for two populations competing for the same resources. Given the advanced technology and organization of the European settlers it is not surprising that competition for the North American land resources led to a displacement of Indians by the newcomers. White men also brought morbid allies in new diseases (measles, smallpox, tuberculosis, and influenza), which killed thousands of Indians. New cultural innovations (e.g. liquor) were scarcely less disruptive. Forest clearance, enclosure of grassland, and decimation of animal populations all changed the ecological balance against which Indian culture had evolved. Even though the Indian populations are now moving back toward the level of four centuries ago and reservations form extensive tracts for some tribes, the pre-Columbian pattern has been erased for major areas of the continent. The Navajo and the Sioux with numbers variously estimated between 50,000 and 100,000 are two of the largest remaining tribes; the Florida Indians number only a few thousand.

Post-Columbian Migration Waves

The population that replaced the Indian came predominantly from Europe and Africa. A survey of the period since 1607 (the date of the first permanent English settlement, at Jamestown, Virginia) shows a series of five distinct *sequent waves* of immigration, each associated with a particular set of migration sources. Between 1607 and 1700 there was an initial wave of English and Welsh, along with a small number of African slaves. Numbers of Dutch, Swedes, and Germans were relatively few during this phase. The period from 1700 to 1775 brought increased migration from the earlier sources and an influx from Germanic and Scotch-Irish sources as well. The years 1820 to 1870 saw increased numbers arriving from northwest Europe (especially Britain, Ireland, the Netherlands, and Germany), but the influx of Africans came to a halt. The 1870s saw the arrival of a vanguard from southern European countries, plus some Asians, Canadians, and Latin Americans. In the half century from 1870 to 1920, the period of the '*Great Deluge*', there was a massive increase in the number of immigrants, and a widening of the source areas to include eastern and southern Europe and Scandinavia. Since 1920, there has been a drop in the number of immi-

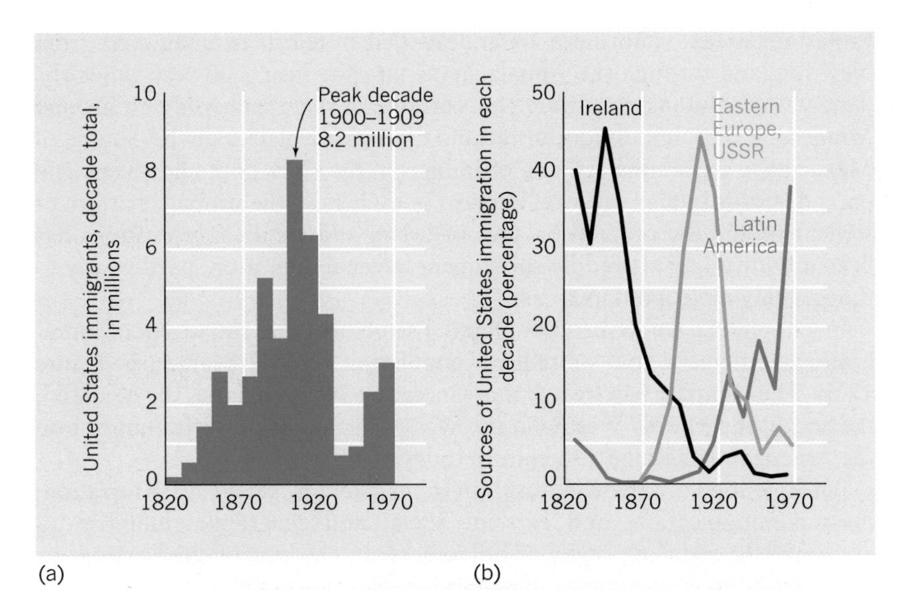

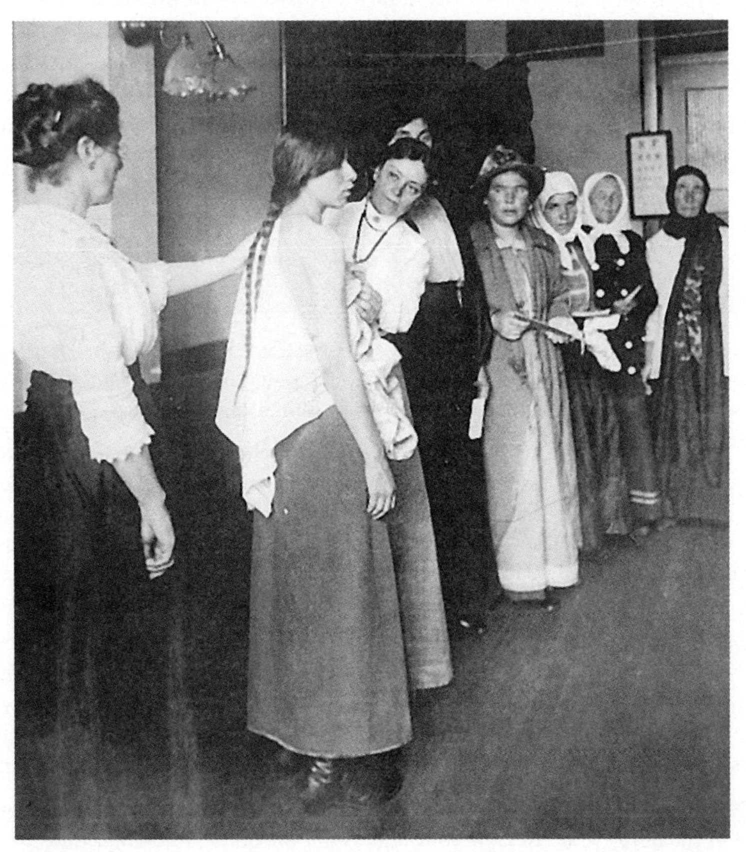

Figure 5.17 Sequences of migration waves (a) The total number of migrants entering the United States over a 150-year period. (b) The changing composition of the migrant stream. The curves indicate the different source areas dominant in the middle of the nineteenth century, the early twentieth century, and at present. Note that the graph describes the percentage of migrants from each source area and not the absolute number. (c) Immigrants being examined by nurses at New York's Ellis Island (*ca* 1910).

[Source: (c) Peter Newark's American Pictures.]

grants, but the pattern of sources has remained wide and there has been a steady rise in the per centage of immigrants coming from Latin America.

While data for the early period are fragmentary, immigration in the period since independence is well documented. Figure 5.17 shows the pattern of numbers and changing origins over the last 150 years. The timing of the various waves of migration is reflected in the original areas of rural concentrations of migrants from different countries. The 'early-wave' Scotch-Irish (the term used to describe both immigrants from Scotland and the

Protestant areas of northern Ireland) settled in a belt running west from New England through the Appalachians into the near Midwest, while the 'late-wave' Scandinavians were concentrated in the upper Midwest around Minnesota. The initial concentrations of African blacks in the South, of Mexicans in the southwest, and of Italians in the cities of the northeast and parts of California are all well known. In each case, the primary pattern of settlement and the original balance of urban and rural concentrations has been muddied and altered by subsequent internal migration, particularly to the growing metropolitan areas.

In Chapter 12 we look at the regional structures that the waves of immigration, settlement, and subsequent population growth created (see Figure 12.5). The complex patchwork that emerged – New England, the Midland, the South, the Middle West, and the West – remain as powerful imprint on the American landscape and culture today.

But how long are these cultural divisions likely to persist? Urbanization, mass communication, and extreme social and geographic mobility all appear to be reducing regional differences. In the rest of this section we shall turn to look at some of these modern forces – population growth and migration (Chapter 6), cultural divesification and mixing (Chapter 7), and urbanization (Chapter 8). In reviewing the changes that operate in the early twenty-first century it is important to set these in the context of the many thousands of years in which the human species managed to survive in balance on this planet in very different circumstances.

Reflections

1. Evidence of the early existence of human groups on the earth is constantly changing with new archaeological discoveries. How secure do you consider the present view that our own species (*Homo sapiens*) probably emerged in tropical Africa?
2. Agricultural hearths are areas in which early plant and animal domestication is believed to have taken place. Where were the main hearths of agricultural thought to have been located? Why there?
3. List the main crops grown on farms and gardens in your own locality. How many of these crops are native to the area? How many are species or varieties introduced by crop exchanges between the Old and New World? (See Table 5.3 for a list of source areas for crops.) Select one crop plant and research its spatial spread.
4. Urbanization is usually viewed as a final stage in the development of resource-organizing technology based on a food surplus from sedentary agriculture. Examine the different developmental sequences shown in Figure 5.8. Which do you think most likely? Can you make a case for any alternative sequences?
5. Find out which early human groups occupied your own area when it was first settled. Have these early settlers left any impact on the present landscape? Did their culture evaluate local resources in a distinctive way?
6. Western geographers studying the transformation of the earth tend to place a very strong emphasis on the worldwide dispersal of European culture. But viewed from the early twenty-first century, that legacy now seems less certain. In what geographical areas did European overseas settlement succeed and persist and what areas was it temporary and transitory?
7. Collect the most recent data you can find to check on in-migration into your own countries. How far are present trends similar to or different from the long-run historical pattern? (See Figure 5.17 for long-term trends for the United States.)
8. Review your understanding of the following concepts using the text and the glossary in Appendix A of this book:

agricultural hearths
continental penetration
fertile crescent
Great Deluge
island colonization models
Jacobs hypothesis
plantation settlements
rim settlements
Sauer hypothesis
sequent waves

One Step Further . . .

The literature describing human evolution and migration and the development of major cultural realms is a very rich one. A useful starting point is to look at the essay on human origins by Bernard Wood in I. DOUGLAS *et al.* (eds), *Companion Encyclopaedia of Geography* (Routledge, London, 1996), pp. 86–106. You can follow this up by dipping into topics of geographical interest in S. JONES *et al.* (eds), *The Cambridge Encyclopaedia of Human Evolution* (Cambridge University Press, Cambridge, 1992) or look at W.W. HOWELLS, *Getting Here: The Story of Human Evolution* (Compass Press, Washington, 1993).

On the origins of agriculture, look at the classic essay by Carl SAUER, *Agricultural Origins and Dispersals* (American Geographical Society, New York, 1952). The origins of the early food-producing centers is also discussed in E. ISAAC, *The Geography of Domestication* (Prentice Hall, Englewood Cliffs, N.J., 1970). The impact of European expansion after the Columbus voyages and the voyages' biological and ecological consequences is engagingly told in Alfred W. CROSBY, *Germs, Seeds and Animals: Studies in Ecological History* (Sharpe, Armonk, N.Y., 1994). Some of the same themes are explored in Ian SIMMONS's two books, *Changing the Face of the Earth* (Blackwell, Oxford, 1989) and *Humanity and Environment*, second edition (Longman, Harlow, 1997). The spread of cities is considered in Lewis MUMFORD, *The City in History: Its Origins, Its Transformation and Its Prospects* (Harcourt Brace Jovanovich, New York, 1961) and G. SJOBERG, *The Preindustrial City* (The Free Press, New York, 1960). Students should also look at the splendid accounts of the evolving geography of the Chinese city and the Western city in P. WHEATLEY, *The Pivot of the Four Quarters* (Aldine, Chicago, and Edinburgh University Press, Edinburgh, 1971) and J.E. VANCE, Jr, *This Scene of Man: The Role and Structure of the City in the Geography of Western Civilization* (Harper & Row, New York 1977). The overseas expansion of Europe and the special case of the United States is described in terms of geographical models of expansion in R.H. BROWN, *Historical Geography of the United States* (Harcourt Brace Jovanovich, New York, 1958) and W.P. WEBB, *The Great Frontier* (University of Texas Press, Austin, Texas, 1964).

The cultural geography 'progress reports' section of *Progress in Human Geography* (quarterly) will prove helpful in keeping up to date. A useful summary of this work is given from time to time in the 'geographical record' section of the *Geographical Review* (a quarterly). Much relevant research on human origins is published in archaeological and anthropological journals. For readers with access to the World-Wide Web see also the sites recommended in Appendix B at the end of this book for topics relevant to this chapter.

CHAPTER 6

Population Dynamics

■ denotes case studies

Figure 6.1 Survivorship in the human population The average life expectancy at birth in many western populations is now in excess of 70 years. Women tend on average to live two or three years longer than men. This Norwegian picture of two older ladies tending their husbands' graves illustrates the reality of survivorship in the 1990s. Norway has one of the most elderly of Europe's populations with one in six of its people aged over 65 years. [Source: © Knudsen/Beyer-Olsen.]

> A finite world can support only a finite population; therefore population growth must eventually equal zero.
>
> GARRETT HARDIN *The Tragedy of the Commons* (1968)

> The optimist proclaims that we live in the best of all possible worlds; and the pessimist fears this is true.
>
> JAMES BRANCH CABELL *The Silver Stallion* (1926)

The elementary facts about the growth of the human population are staggering. We know that some time in the late 1990s the world's total number of people passed the six billion mark. It is on course to add yet another billion within the next twelve years. Every ten seconds the world's population increases by 27 people. This means that in just 70 days we add another city the size of New York City; in every year another country the size of Nigeria.

But geographers need to know that the bare facts need to treated with great care. The numbers are constantly on the change and you need to check a web site, such as that of the Population Reference Bureau (see the list of useful web sites in Appendix B) to catch the latest figures. Then we need to note that different parts of the world are behaving differently: in some long-developed countries population is barely growing at all (see Figure 6.1 p. 170). Even if high growth is occurring, we should note that the world rate of growth is coming down. It peaked in the 1960s and is set to continue to fall. Finally we have to take a view of that growth. Are the new arrivals a burden of extra mouths to feed? Or are they wonderful new increments of energy and talent?

In this chapter we look less at the numbers, fascinating though they are, than at the processes which lie behind them. In Section 6.1 we look at the engines of growth forged by births and in-migration and the countervailing balances of death and out-migration. Why is population changing and how can we best measure it? This leads to a concern with population structure in terms of age and sex (Section 6.2). We then look at the checks on population growth, placing the human species within a wider ecological framework (Section 6.3). Here the questions centre on how other species control their numbers, and whether human beings are in any sense a special or an exceptional case. Third, we look at the facts of world population growth. What happened in the past? In Section 6.4 we look at population change, especially over the last couple of centuries, and try to see why step changes in growth appear to have happened. Finally, in Section 6.5, we look at prevailing global patterns at the start of the twenty-first century. What is happening now, and what are its long-term implications? Is zero population growth possible – or even desirable?

6.1 Dynamics of Population Growth

While the facts of birth and death at an individual level are clear, their effect on the growth and decline of a *population* (that is, a collection of individuals) are less so. Here we look at the processes that shape population growth and the kind of yardsticks we use to measure it. In this chapter we shall be concerned mainly with the human population. Much of the reasoning we shall use could, however, be applied to animal populations as well.

Births, Deaths, and Growth

The total population of any area of the earth's surface represents a balance of different forces (Figure 6.2). If births are more numerous than deaths in any period, the total population will increase. If they are less numerous, it will decrease. This simple relationship is modified by a second force, *migration*. When immigrants are more numerous than emigrants, there will be a population increase. (This assumes, of course, that we are ignoring natural change for the moment.) When emigrants are more numerous, there will be a population decline.

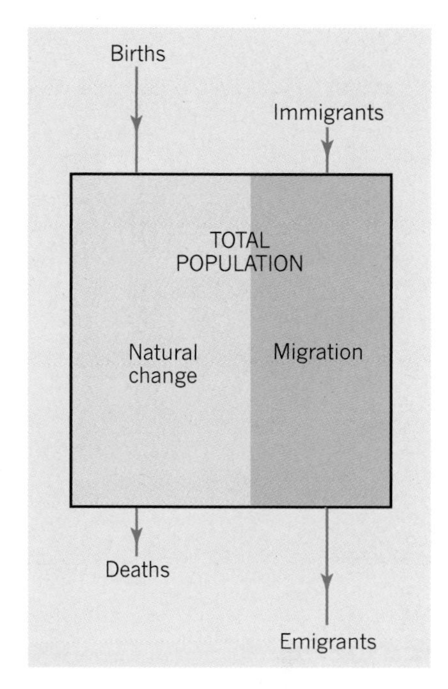

Figure 6.2 Population change
The total size of the population of any part of the earth's surface may be thought of as like the water level in a bath. It is the result of inflows (shown at the top of the diagram) and outflows (shown at the bottom).

As Figure 6.2 shows, net changes in population totals are caused by the interaction of four elements: births and immigrants tend to push the total up; deaths and emigrants tend to bring the total down. Although migration may be the most important factor in small areas (for example, in a small village or a city block), it is less significant on the national level. For the world as a whole, migration is irrelevant because all movements take place within the limits of the recording area. In other words, until interplanetary travel comes along, the planet Earth can be safely treated as a *closed* system for demographic purposes. We shall therefore concern ourselves in this chapter with the natural change component, leaving the more complicated analysis of *open* population systems (in which migration is significant) until later in the book (see especially Chapter 8).

We can illustrate the effects of natural change on a small scale by looking at a single island population. We choose an island that over the period studied, has a very small migration and the population approximates a closed system. Figure 6.3 shows the effects of births and deaths on the total population of the small island of Mauritius in the Indian Ocean. In 1900 the island had a total population of approximately 0.3 million, which increased rather slowly to around 0.4 million by 1950; since then it has increased sharply to over 1.1 million. The diagram shows the changes in births and deaths over the period expressed as birth and death rates per thousand inhabitants per year. Births and deaths were almost in balance until about 1920, when general medical improvements began to lower the death rate. Individual peaks in the two rates are associated with both natural disasters, such as hurricanes and epidemics, and economic fluctuations such as the 1929 depression and the postwar boom of the late 1940s. The period since 1980 has seen a fall in the birth rate, and the island's population of 1.1 million people is now increasing at an annual rate of around 1 per cent.

In describing change on the island we have been talking in terms of 'rates'. How are these rates measured, and how should we interpret them? There are several ways of measuring the vital rates of a population, of which the crude rates of birth, death, and growth are the simplest (see Table 6.1). The *crude birth rate* is defined as the number of births over a unit of time divided by the average population. The *crude death rate* describes the number of deaths per unit of time; *crude growth rate* describes the difference between the number of births and deaths per unit of time, each divided

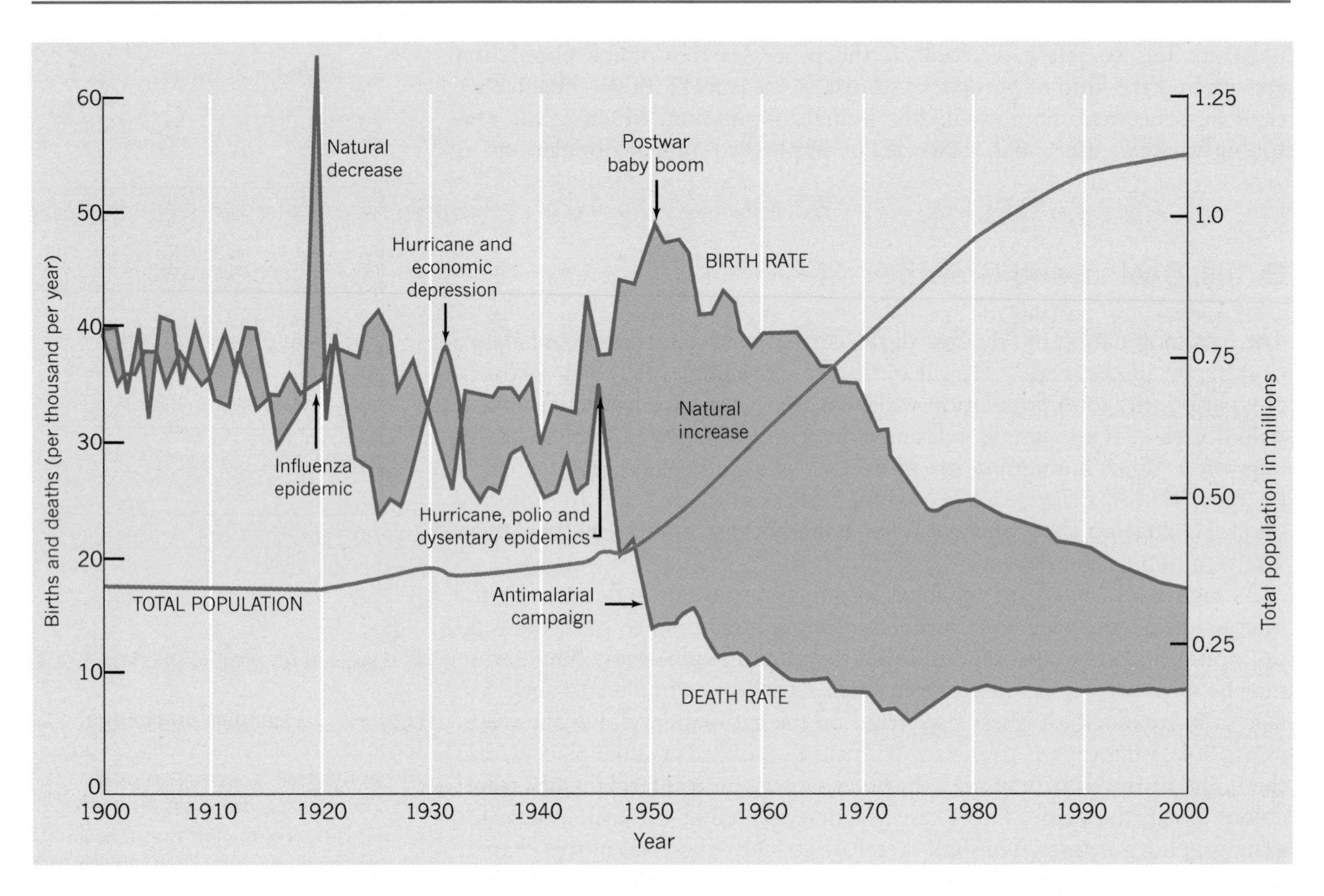

Figure 6.3 Natural components in population change Changing patterns of fertility and mortality are shown for the Indian Ocean island of Mauritius. Related changes in social and economic conditions and natural hazards are indicated. By 2000 the birth rate had fallen below 20 but the death rate remained at around 6.

[Source: Population Reference Bureau, *Population Bulletin*, **18**, 5 (1962), Fig. 1, with recent updating from United Nations sources.]

by the average population in the time interval. Thus if there were 25 births and 18 deaths in a year on an island whose average population during that year was 500, the crude birth rate would be 50 per thousand; the crude death rate, 36 per thousand; and the crude growth rate, 14 per thousand.

These rates are described as *crude* because they fail to take into account such factors as the age and sex of the members of the population, or migration. We should certainly expect an island with a large number of young adults to have a higher birth rate and a lower death rate than an island inhabited only by octogenarians! Hence, demographers have defined and developed much more sophisticated measures of change called *net* rates, which take into account the structure of the population. They are somewhat complex to go into here, but interested students will find references to discussions of these more refined measures in 'One step further . . .' at the end of the chapter. (See also Box 6.A on 'Population growth curves'.)

Growth Rates and Doubling Times

The uncertainties involved in estimating future survival and marriage rates, together with the scarcity of data for much of the world, often force us back to a simpler view of population growth. For example, between 1 October 1967 and 30 September 1968, there were 3,453,000 live births and 1,906,000 deaths in the United States. If we take the estimated mid-year population of the country on 31 March 1968, as 198,400,000, simple arithmetic indicates that for every 1000 Americans there were 17.4 births and 9.6 deaths. The excess of births over deaths was 7.8 per thousand, and the annual rate of natural increase was less than 0.8 per cent.

Table 6.1 Terms used in the study of population

Term	Definition
Birth rates	The proportionate number of births in a population.
Carrying capacity	The largest number of a population that the environment of a particular area can carry or support.
Census	An official counting of the population.
Crude rates	Vital rates which are not adjusted for the age and sex structure of a population.
Death rates (or mortality rates)	The proportionate number of deaths in a population.
Fecundity rates	The biological capacity of females in a population to produce offspring.
Fertility rates	The actual production of offspring by females in a population.
Migration	The movement of people from one area to another.
Migration change	The net change in the total population of an area due to migration.
Morbidity rates	The amount of illness in a population.
Natality rates (see Birth rates)	
Natural change	The net change in the total population of an area due to the balance of births and deaths.
Population pyramids	The age and sex distribution of a population.
Replacement rates	Estimates of the extent to which a given population is producing enough offspring to replace itself.
Reproduction rates	Measure the number of girls born to females in the childbearing age groups (roughly 15 to 45 years) in a population.
Saturation	The level at which the population of an area exactly equals its carrying capacity.
Standardized rates	Vital rates that are adjusted for such factors as the age and sex structure of a population.
Survivorship curves	The proportion of a given population surviving to a particular age.
Vital rates	Changes in the size and structure of a population.

Box 6.A Population Growth Curves

Exponential Growth

Exponential models of population growth describe a simplified situation in which growth (or a decline) is unchecked and the rate of change is constant. We express this simply as

$$\frac{dN}{dt} = rN$$

where N = the number of people;
r = the rate of natural increase (a constant); and
$\frac{d}{dt}$ = the rate of change per unit of time.

The expression states that the amount of growth is related to the size of the population; the larger a population is, the faster it grows.

To simplify the computation, we can rewrite this as

$$N_t = N_0 e^{rt}$$

where N_t = the number of people at time t;
N_0 = the number of people at time 0; and
e = 2.71828.

The constant e is the base of Napierian, or natural, logarithms and is the sum of the infinite series

$$1 + \frac{1}{1} + \frac{1}{2 \times 1} + \frac{1}{3 \times 2 \times 1} + \frac{1}{4 \times 3 \times 2 \times 1} \cdots$$

If we start with 1000 people (N_0) and assume a growth rate of 1 per cent per annum ($r = 0.01$), then we can show, by substituting these values in the equation, that after 70 years ($t = 70$) the original population will

Box continued

have doubled (N_t = 2000). In another 70 years the population will have doubled again. Exponential models show the critical importance of small changes in the rate of natural increase. If we halve this rate (r = 0.005), the population doubles only every 140 years. Despite their simple structure, exponential models are a useful way of describing recent phases of human population growth.

Logistic Population Growth

When a population is allowed to develop in an optimal environment of unlimited size, its growth follows an *exponential* curve. If we now introduce a fixed *carrying capacity*, or *saturation level* (K), the potential for biological growth, or the biotic potential, will be modified by environmental pressures.

We can introduce this environmental pressure into the exponential growth model:

$$\frac{dN}{dt} = rN$$

described above by subtracting

$$\frac{K - N}{K}$$

from rN. We then have:

$$\frac{dN}{dt} = rN\frac{K - N}{K}$$

where N = the number of individuals in the population;
K = the maximum number of individuals allowed by the carrying capacity;
r = the rate of growth per individual; and
$\frac{d}{dt}$ = the rate of change per unit of time.

Both N and K also can be expressed as population densities. This modified growth curve is termed a ***logistic*** growth curve and has a characteristic S-shape. Its calculation is described in Chapter 16 in the discussion of spatial diffusion models. (For a discussion of population models see A.S. Boughey, *Ecology of Populations* (Macmillan, New York, 1968), Chapter 2.)

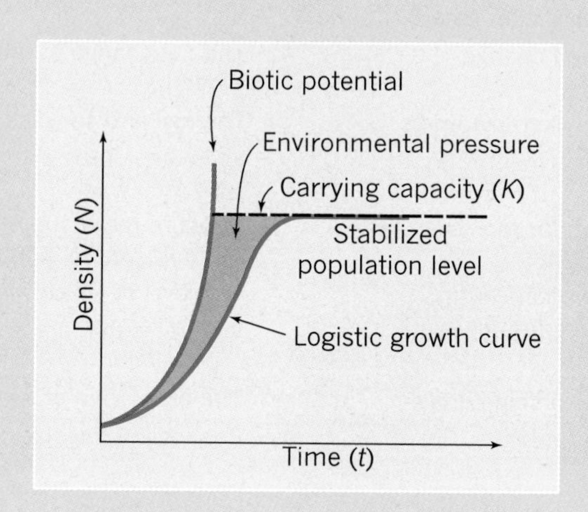

If we were simply to add 8 new individuals each year to the 1000 still living it would take 125 years for the population to double (125 × 8 = 1000). But this is not what happens. The people added to the population also increase at a rate of 8 per thousand. Population grows exponentially, just like money earning compound interest in a bank. (See the discussion of *exponential models* of population growth in Box 6.A.) The doubling time is, therefore, shortened from 125 years to only 87 years. As the rate of natural increase rises, the doubling time decreases sharply. When it is 2 per cent (above the current world rate), the doubling time is 35 years. For some of the populations of tropical Latin American countries with rates over 3.25 per cent, the doubling time is only a little over 20 years!

In an influential book *The Population Bomb* (1969), the Stanford biologist Paul Ehrlich used arguments on exponential growth to show that the then explosion of the human population was bound to be a short-lived phenomenon. He drew attention to the multiplication of the human population over its history with doubling times shortening through 1000 years, 200 years, 80 years to 37 years. Half the recent growth had come in scarcely more than a generation. Ehrlich concluded that if the world's present population continued to grow at the then rate (i.e. doubling every 37 years), by AD 3000 there would be 2000 people piled on every square metre of the earth's surface – land, sea, and ice included. If we extended this nightmare projection far enough, the universe would eventually consist of a ball of closely packed people expanding outward at the speed of light!

Patterns of Survival 6.2

If we turn from Ehrlich's space odysseys to the actual human population, then we need to understand something about its age and sex structure. This is most usefully shown in terms of survivorship curves and population pyramids.

Survivorship Curves

By studying the ages at which the members of a population die, we can establish *survivorship curves*. As Figure 6.4 shows, these tell us the number of survivors of an original group (say all those born in a given year) according to their age at death. If this were a perfect world from which all accidents and infections had been eliminated, so that we all lived into our 80th year, the curve would have an abrupt right angle (as in Figure 6.4(a)).

If all members of a given population have exactly the same capacity for survival, their survivorship curve has this shape. In practice, the curves for real populations have complex forms, but there is a tendency for the populations of advanced countries to have curves closer to the hypothetical right-angled one than primitive populations or those of underdeveloped countries. Figures 6.4(b) and (c) go on to illustrate the estimated survival curves for three populations with low survival rates (those of the Stone and Bronze Ages and of China in 1930) and others with medium or high survival rates (New Zealand and the Netherlands in the 1950s).

We shall see later in this chapter the way in which changes in the causes of mortality have allowed populations on average to live longer. But what are the upper limits to survival? (See Figure 6.5 which shows examples from the ends of the age range.) Study of other animal populations suggests

Figure 6.4 Survivorship curves
Here, idealized survivorship curves (a) are contrasted with actual examples of low survival curves (b) and medium and high survival curves (c) from different cultures and from different time periods. The vertical axis measures the number of surviving members of a population of 1000 births in relation to their age. Age is measured on the horizontal axis in years.

[Source: C. Clark, *Population Growth and Land Use* (Macmillan, London, 1967), Figs 11A, 11B, pp. 39–40.]

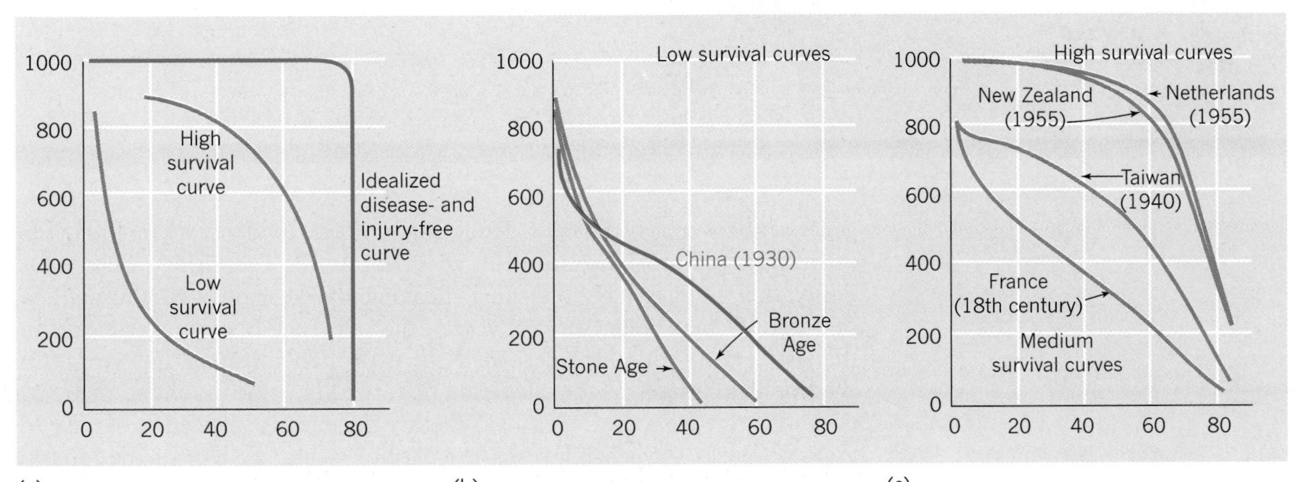

(a)

(b)

Figure 6.5 Ageing The two extremes of the human life cycle: the very young and the very old. (a) Babies at a 1990 reunion of babies born at the London hospital, St Thomas's, earlier in the year. (b) Norwegian seaman born in the first decade of the twentieth century.

[Source: (a) MAGNUM/Steele-Perkins. (b) Knudsens Fotosenter.]

genetic inheritance is the main determinant of age. Evidence of very long life in humans is notoriously difficult to verify as the Cambridge Centre for Population Studies has confirmed. Leaving the legendary Methusalah to one side (Genesis suggests 969 years), the reference books suggest that in the last century the oldest recorded age is probably around 130 years. Of geographical interest is the clustering of very old people in mountain areas, notably in the Russian Caucasus. Whether the exercise needed to survive in areas of lower oxygen improves the cardiovascular system of the inhabitants of these areas is yet to be confirmed.

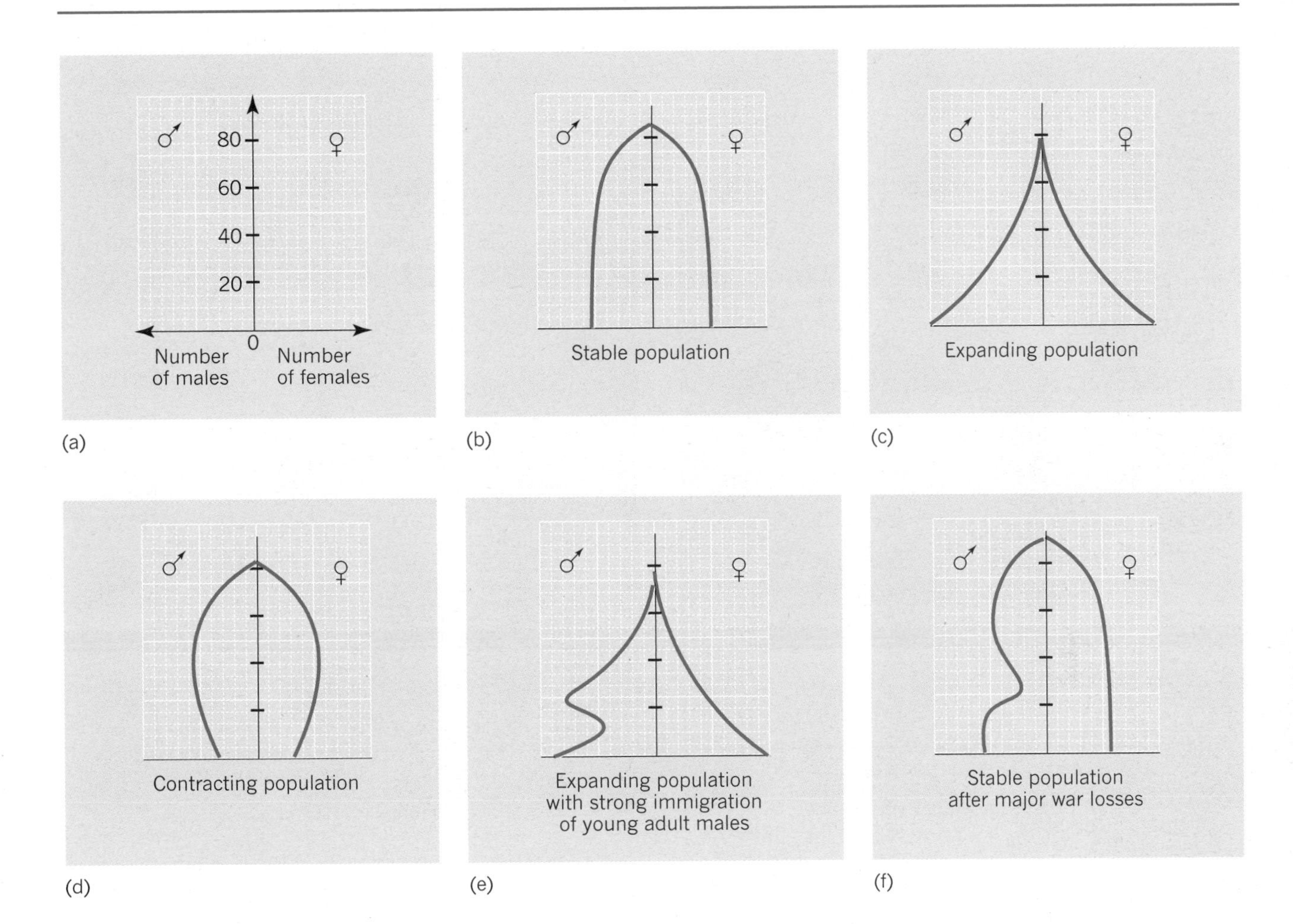

Figure 6.6 Population pyramids Characteristic pyramids for different populations at different stages of growth. These give only the bare bones of the situation and should be compared with population pyramids of actual countries later in the book (see Figure 18.7). Women tend to survive into old age in greater numbers than men and so the curves in the figures shown here are less symmetrical than drawn.

Population Pyramids

Another useful and often-used method of portraying the structure of a population is the ***population pyramid***. This is a vertical bar graph showing the proportion of individuals in various age ranges. Figure 6.6 shows some idealized age pyramids for various human populations. The number of males is measured left to the axis and the number of females right of the axis. We shall come back to population pyramids later in this chapter and also in Chapter 18, where we look at the economic implications of highly skewed population–age distributions for developing countries.

Detailed analysis of the population pyramid for a single country can show a fascinating record of demographic history. Figure 6.7 shows the population pyramid of France covering the hundred years from 1897 to 1997.

Reading from the top of the pyramid (the oldest population) to the bottom (the newborns) six features can be detected. These are numbered on the pyramid.

1 The predominance of females over males in the very oldest age classes.
2 The marked shortage of births due to the Great War during the 1914 through 1918 years.
3 Some recovery in birth rates but still historically low birth rates in the interwar years.

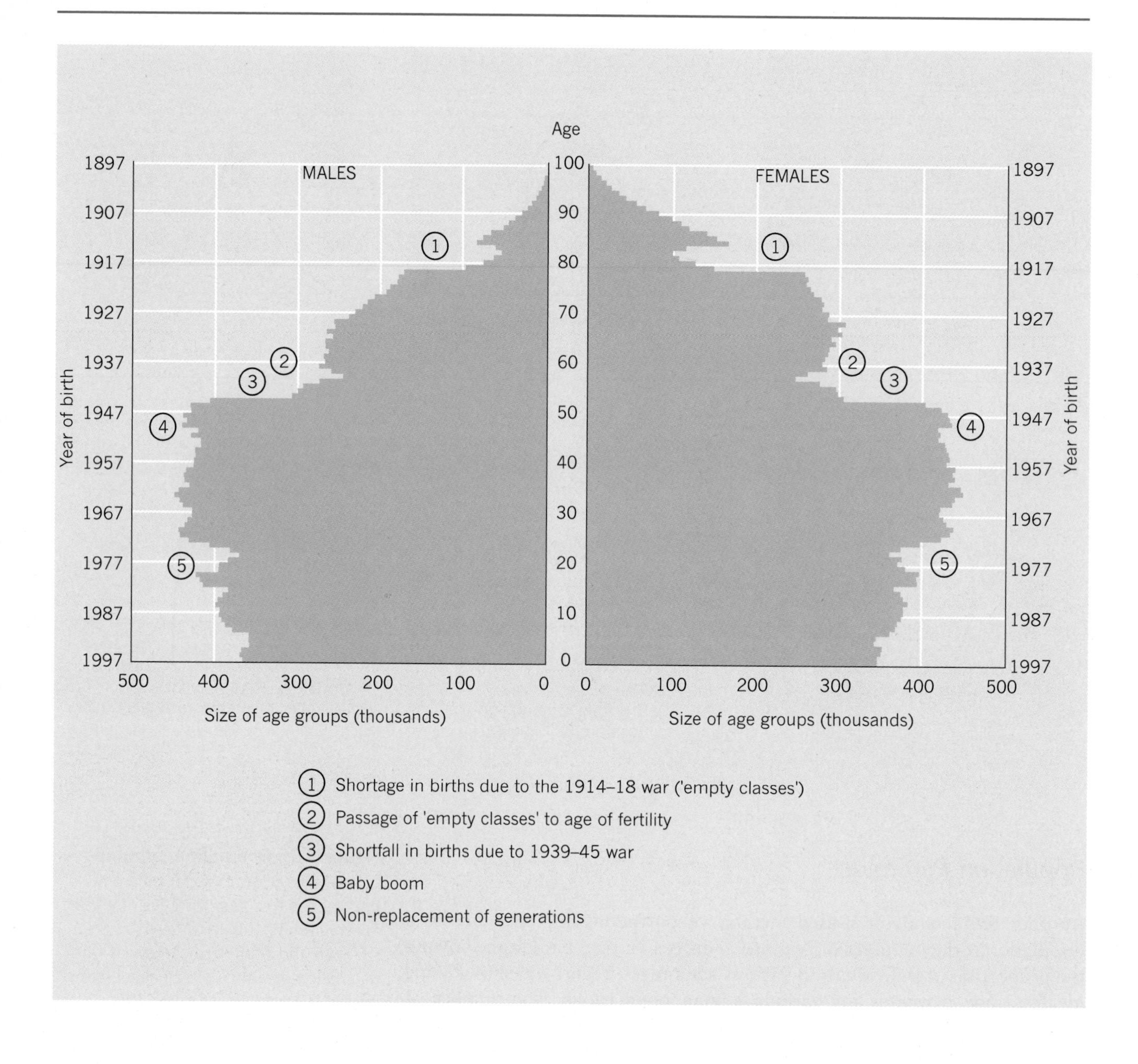

Figure 6.7 The population pyramid of France Reading the population pyramid of a single country (see text). A century of French population, 1897 to 1997. The numbers surviving by year of birth are plotted for males (left) and females (right).

[Source: M.L. Lévy, *Population et Sociétés*, **333**, Fig. 1, p. 2.]

4 A second shortfall of births due to World War II during the 1939 to 1945 years.
5 The postwar 'baby boom'.
6 Low birth rates in the last quarter century with a failure to replace the parental generation.

Still further analysis of pyramids is possible using geographically subdivided populations, e.g. inner-city populations versus rural populations.

To sum up: the population size of any part of the earth's surface is determined by two forces, ***natural change*** (births and deaths) and migration. Generally, the larger the population of an area, the more important natural change becomes, as opposed to migration. At the world level, only natural change is important. Rates of natural change can be measured in several ways and the resulting distribution of population can be shown by survivorship curves and population pyramids.

Ecological Checks on Growth 6.3

The compound course of population growth we have described is representative only of the *biotic potential* of a population, that is, of its theoretical rate of growth when it is allowed to develop in an optimal environment of unlimited size. This is rather like measuring the growth of a favourite indoor plant that we regularly feed, water, and fuss over. Under natural conditions, we would expect this potential for growth to be checked either by some natural limits or, as seems likely in the case of both animal and human populations, by cultural, social, and economic constraints. Here we look at some of the models used to understand these checks and how they operate.

Ecological Feedbacks and the Malthusian Hypothesis

Eruptions of animal populations (from locust swarms to the crown of thorns starfish) abound in the biological record. Figure 6.8 shows one example of a *predator–prey* model; two animals from the Canadian boreal forests (see Figure 4.12) whose numbers follow a cyclic pattern. Look first at the curve (c) for the hares. Note how it rises explosively upward, reaches a peak, and then falls. If you look at the second curve (the lynx) you will

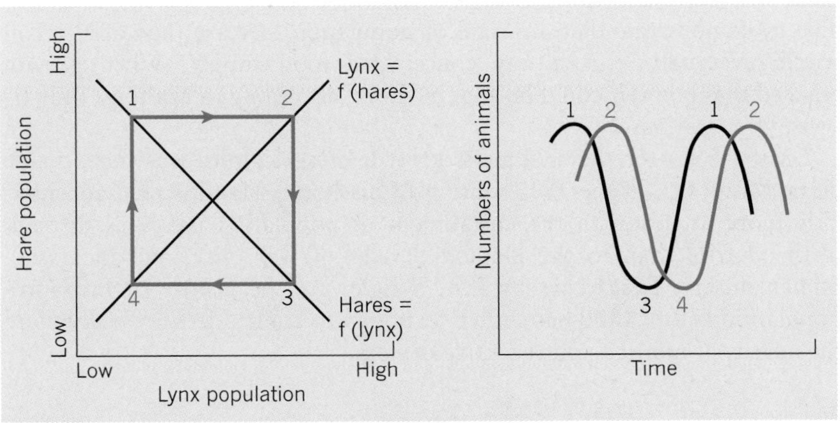

(a) (b)

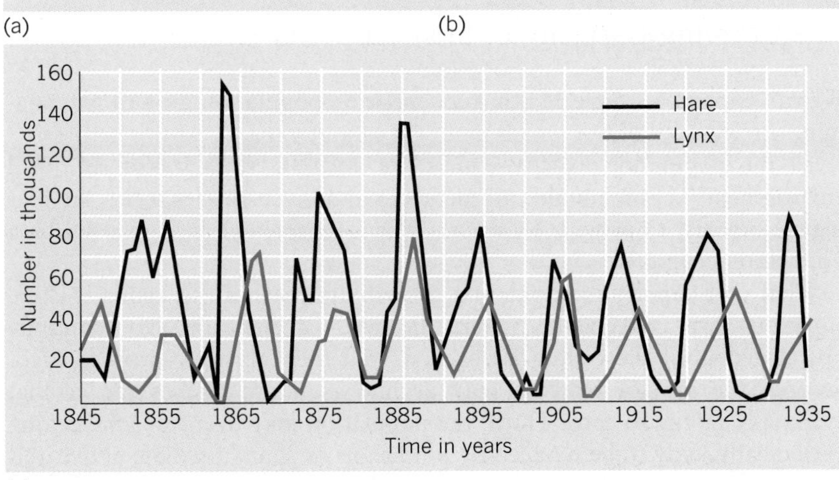

(c)

Figure 6.8 Fluctuations in animal populations (a) Relations between two Canadian wildlife populations, one a predator (the lynx) and another its prey (the snowshoe hare). The hare population is large when the lynx numbers are low (1) and small when lynx numbers are high (3). Conversely, the lynx population is large when there is a plentiful supply of hares (2) but low when hares are scarce (4). (b) The sequence of situations, (1) through (4), gives rise to cycles in the populations of both animals over time. The lynx cycle lags slightly behind the hare cycle. (c) Changes in the abundance of lynx and hares over a 90-year period, determined from the number of pelts received by the Hudson's Bay Company.

[Source: Data from MacLulich; reprinted with permission, from E.P. Odum, *Fundamentals of Ecology*, 3rd edn. Copyright 1971 by the W.B. Saunders Company, Philadelphia. (d) Photographs courtesy of FLPA/Leonard Lee Rue.]

(d)

(e)

see why! Lynx become more abundant as their prey – the hares – grow in numbers. Eventually both populations collapse and a new cycle starts again.

We have already met feedback models that show the kind of forces which bring numbers back into line with the capacity of the local environment to support them. Check back on Figure 4.19(b) which showed that an increase in population size may mean less food per individual, more deaths than normal, and a subsequent reduction in numbers. Conversely, a decrease in population may set in motion a chain of events that raises the number in a group to the original level. Clearly the model is a very rough one, but it broadly fits the observed facts for many animal species.

But what about human beings? The world population has been uniformly increasing over at least the last 500 years. Does this mean simply that the wavelength of the human population is a very, very long one – that it will hit a peak at some point in the next few centuries and then stabilize or decline? Such questions are impossible to answer with confidence without going into questions of human culture, human economics, and human politics – the substance of Parts III, IV, and V of this book. We should, however, at this point, note the analogies between human and other animal populations.

These analogies troubled Thomas Robert Malthus, the English demographer, in writing his now-famous *Principles of Population*, published in 1798 (see Box 6.B on 'The *Malthusian hypothesis*'). Malthus saw dire ecological consequences in the continuing growth of the human population. He claimed that population has a tendency to increase geometrically while the food sources for that population, even with improving agricultural methods, increase arithmetically. Given these assumptions Malthus was able to demonstrate that any rate of population increase (however small) would eventually exceed any conceivable food supply. When growth reached that point, it could be kept in check, according to Malthus, only by 'war, vice, and misery'.

Yet the basis for the arithmetic growth of agriculture was never made clear. Moreover, in the 1817 edition of his book, Malthus paid considerably more attention to the curtailment of population increases through birth contriol than to the gloomy devices of war, vice, and increasing human misery. Researches by E.A. Wrigley on the history of England's population before 1800 has shown an ability to control fertility well before the advent of modern contraceptive devices.

The Carrying Capacity of Environments

We can explain a simple Malthusian check on population growth by imagining a fixed level above which numbers cannot expand. At this saturation level, the population exactly equals the *carrying capacity* of the local environment (i.e. the number of members of a given species it has the biological capacity to provide food for). This is represented in Figure 6.9 by a population ceiling.

What will happen as population growth approaches this ceiling? Three situations are conceivable. First, the rate of increase may continue unchanged until the ceiling is reached, and then abruptly drop to zero. Second, the rate of increase may decline as it approaches the ceiling, eventually falling to zero. Third, the population may overshoot the ceiling periodically, only to be reduced by food shortages, and oscillate above and below the carrying capacity (as in Figure 6.9(c)).

Box 6.B The Malthusian Hypothesis

Modern study of population geography can be traced back to the English demographer, Thomas Robert Malthus (born Dorking, Surrey, England, 1766; died Bath, England, 1834), whose book on the principles of population growth was published in 1798.

Malthus originally decided to be a clergyman and graduated from Cambridge in theology, taking a parish in Surrey in 1796. But in 1805 he moved to a position as professor of history and political economy in the college of the East India Company and held this post until his death. In his *Essay on the Principle of Population,* Malthus proposed that a population would always outrun its food supply in the long run since population grows geometrically while food supplies grow arithmetically. In the diagram, the food supply is originally at a level of 10 units and increases by 3 units in each time period; the population is originally at a level of 0.1 but doubles in each time period. Whatever figures are chosen, in the long run the exponential curve will eventually intersect the arithmetic curve.

Malthus's essay had a great effect on other thinkers, and suggested to Charles Darwin the relationship between progress and the survival of the fittest, a central idea in Darwin's theory of evolution. A review of Malthus's influence is given in D. Coleman and R. Schofield (eds), *The State of Population Theory: Forward from Malthus* (Blackwell, Oxford, 1986).

THE
Life and Writings
OF
THOMAS R. MALTHUS
BY
CHAS. R. DRYSDALE, M.D.
LONDON:
GEO. STANDRING, 7 & 9 FINSBURY STREET, E.C.
120 Pages. SIXPENCE.

[Source: Pamphlet cover from Mary Evans Picture Library.]

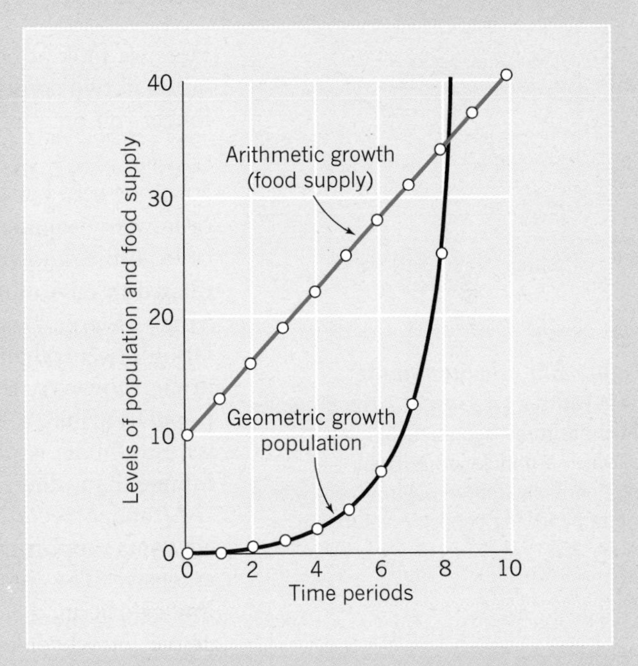

The instantaneous adjustment implied in the first solution seems rather improbable, not least because the mechanism by which such a sudden change might be achieved is unclear. It is unsupported either by empirical evidence on human numbers or by the growing number of studies of other animal populations. The second solution, in which the rate of increase tapers off as numbers approach the critical level, is more plausible. (See the discussion of logistic population growth in Box 6.A.) Such a solution does,

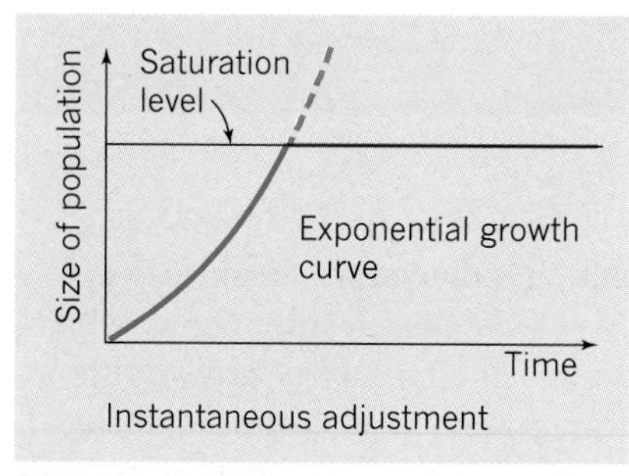

(a)

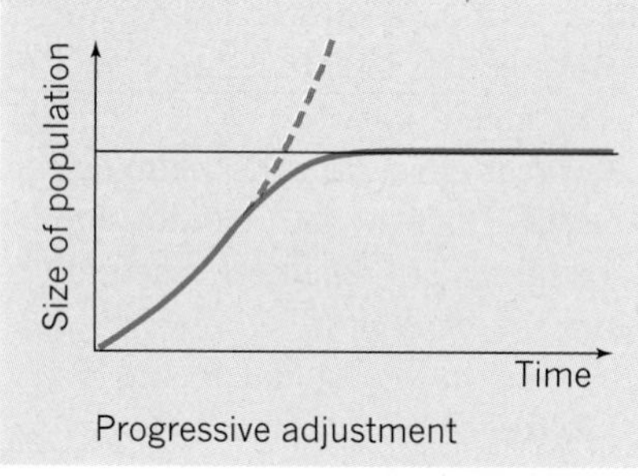

(b)

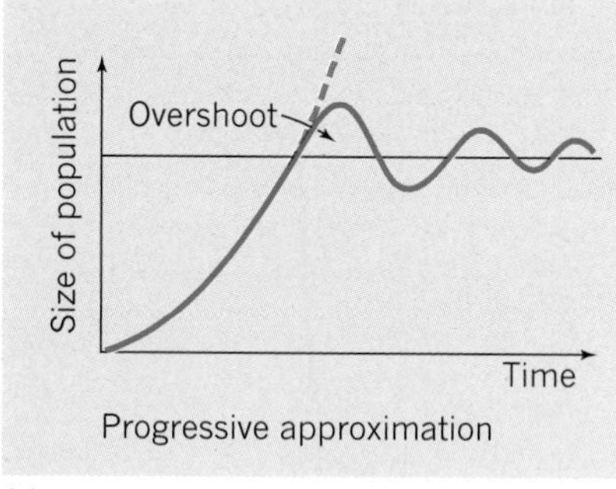

(c)

Figure 6.9 Environmental constraints on growth These graphs illustrate three hypothetical relations between a population growing exponentially and an environment with a limited carrying capacity (saturation level).

however, imply more knowledge about environmental limits and more social control over births than we presently have.

The third possibility as population approaches the critical level is illustrated in Figure 6.9(c). Here, the relationship between the population and the carrying capacity of the environment is reflected in changes in both birth and death rates. Too many people (i.e. a population above the carrying capacity) leads to deaths through starvation and fewer births; this brings the population down. This overshooting and undershooting of the saturation level, as population fluctuates above and below the carrying capacity of the environment, is commonly encountered in animal populations. Periods when a species is abundant follow periods when it is scarce, in a rather regular rhythm.

Because the first accurate census data for human populations became available only in the late eighteenth century, it is difficult to determine which, if any, of these simple models are appropriate to the human situation. Historical trends in the world population reveal that the current exponential pattern of growth is relatively recent. The early period of the human tenure on the earth was one in which the Malthusian constraint of hunger played a key role, and the first two graphs in Figure 6.9 may be more relevant to that time.

Malthusian Checks on Population: Famine

Here we look at how the food supply acts as a check on the human population in two contexts: during specific local famines (Figure 6.10) and in the longer run on a global scale.

Famines and Local Food Shortages

We can find some rough indications of how environmental checks operate by piecing together historical records of local breakdowns of food–population balances. Famines may be closely related to environmental events (as in the case of a drought in the eastern Sahel) or largely unrelated (as was the extensive famine in the refugee population of central and eastern Europe at the close of World War II). We can argue, however, that a poor region with limited food stockpiles whose population has a food intake only slightly above the starvation line and whose climate varies greatly from year to year is more likely to experience famine than other, more fortunate regions.

Although records of famines are difficult to assemble, early historical accounts support the view that famines were once much more pervasive and extensive. The combination of high population densities in rural areas and low caloric intake, in addition to occasional failures of monsoon rains to arrive on schedule, make south and east Asia the world's main disaster areas for famines. The world's worst recorded famine occurred between 1877 and 1879, when an estimated 9 million people died in China (see Table 6.2).

In Europe, the worst disaster of this type was the Irish potato famine of 1845. Large areas with high populations were supported at subsistence levels by a single plant species, the potato. Blight followed by crop failures in 1845 and again in 1846 brought famines of Malthusian dimensions; deaths, massive emigration (800 thousand people moved out of Ireland in the next five years), plus a sharply reduced birth rate lowered the population of eight million (based on the 1841 census) to around half that figure by the end of the century.

(a)

(b)

Figure 6.10 Famine (a) Drought in Bihar, India, in the great famine of 1966–67. (b) Increasing crowding of population would put such pressures on the food supply that, under the Malthusian equation, only hunger, disease, and war could bring numbers back to supportable levels. Famine on the streets of Patna, India, in the 1951 famine.

[Sources: (a) Popperfoto. (b) © MAGNUM/Werner Bischof.]

Table 6.2 Population losses from major famines

Location	Dates	Estimated deaths (millions)[a]
India	1837	0.8
Ireland	1845	0.75
India	1863	1.0
India	1876–78	5.0
East China	1877–79	9.0
China	1902	1.0
China	1928–29	3.0
Former USSR	1932–34	4.0

[a]Famines before the nineteenth century are poorly documented, and estimates of the number of deaths vary widely.

When the forces that trigger famine are environmental, as when the monsoon rains fail, we can regard them as fluctuations in the carrying capacity of the environment itself. Thus, we can discard the idea of a fixed limit on population growth (used in Figure 6.9) and replace it with a variable limit. Figure 6.11 shows a series of changes over time in the carrying capacity of an area.

Environmental changes are of three kinds. First, there are *non-recurrent changes* that may be (1) abrupt (Figure 6.11(a)), such as those following the overrunning of fertile fields by a flow of lava, or (2) more gradual, such as those caused by a deteriorating climate or eroding topsoil. Second, there are *periodic regular changes* (Figure 6.11(b) and (c)), including annual variations in productivity connected with seasonal variations in growing conditions. Changes of this type are caused by, for example, low winter temperatures in the boreal zone or summer droughts in the Mediterranean zone. Third, there are *periodic but irregular changes* (Figure 6.11(d)). That is, environments may have irregular periods of low productivity induced by irregular natural events like river plain flooding. We have already encountered examples of all three types of environmental instability in Chapter 4.

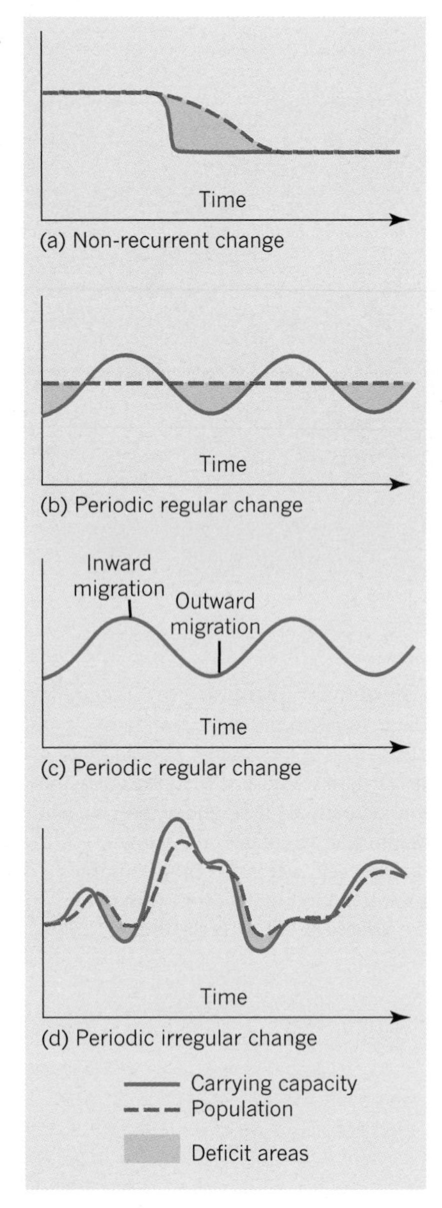

Figure 6.11 Environmental change and population size The graphs show hypothetical responses of population numbers to changes in the carrying capacity of a given region. In (a) the population adjusts slowly to a permanent reduction in carrying capacity over a long time. In (b) the time-span is shorter and the population copes with periodic regular changes by storing food during good years or seasons. In (c) inward and outward migration keep the population in line with the food supply. Where the changes are seasonal regular migration movements of animal stock (termed *transhumance*) may occur. In (d) both strategies are used to cope with irregular fluctuations in the food supply.

Famine and Migration Local human populations respond to changes in the carrying capacity of their environment in different ways. Regular seasonal changes may be coped with either by storing food for use during the season of low productivity or by regular migrations to other areas. (Livestock, for example, are moved from low to high pastures in the European Alps as the seasons change.) Periodic but irregular changes pose more severe problems. If the change is for a relatively short period (as in the case of river flooding), temporary abandonment of the area may solve the problem. More serious climatic changes may be both too lasting and too widespread for evacuation to provide a solution. Here the classic famine symptoms set in. Often, the pattern is reinforced by the consumption of the next season's seed corn for food and the resulting loss of productive capacity in the following season. Longer-term declines usually lead to steady emigration and a cumulative fall in population.

All these responses to environmental change involve some spatial (migratory) movements. There may be outward movements (seasonal, periodic, or permanent) of population from areas where there is a food deficit or inward movements of food from areas of surplus. These strategies obviously apply only in the case of local famines. Such spatial reshuffling of population and resources would not help in the event of a global famine.

Global Food Shortages Setting aside for a moment critical local shortages, how far is the earth as a whole able to support the increasing demands for food that a growing population makes? We can gain some rough idea of the present situation by taking current demands and comparing them with estimates of the earth's ultimate food-producing capacity. The United Nations World Health Organization (WHO) has estimated that the world's people today consume around 10^7 tonnes of food per year. Forecasters at Resources For the Future (RFF) predict that, given the present levels of solar energy and the present distribution of world climate, the maximum amount of organic matter that could be produced by photosynthesis is about 10^{11} tonnes per year. A comparison of these two estimates suggests that only a trivial portion (about one-hundredth of 1 per cent) of the earth's ultimate food-production capabilities is being used.

But how relevant are those estimates really? Organic matter is not always edible. Even in the case of croplands, well below half of the gross product may be edible. As for grazing lands, an energy-conversion factor of 12 to 1 must be used to convert the energy consumed by stock to the food value of the stock to humans. Thus, the original figures must be revised downwards to a maximum of around 10^9 tonnes per year. If we also raise the WHO estimate for food consumption to allow for such things as pre-harvest losses (30 per cent), post-harvest losses (30 per cent), edibility, conversion factors, and non-cropland, then the totals converge rapidly. It would be more realistic to say that the farming industry is operating at less than 15 per cent of its maximal productive potential.

However, the load on the earth's food-producing area is not evenly spread. Most food is produced from a small part of the global surface. Cultivated land constitutes only 2 per cent of the earth's total area (land and sea) but it is capable of producing nearly three-quarters of the world's potential output of edible matter. The grasslands are second in the productivity of edible material. The greatest gap between total productivity from photosynthesis and productivity of edible material is in forest areas. The edible fraction from the oceans and inland waters is quite small; there is little evidence that the world's oceans (despite their enormous area) will make anything but a marginal contribution to world food production in the next generation or two.

Using the most conservative estimates, we can say that the food-producing capacity of this planet is vast, even with our existing technology. Given improved standards of production, its capacity to feed a population much larger than seven billion is not in doubt. Unfortunately, calculations of global demand conceal immense local differences in food consumption (see Figure 6.12). Broadly speaking, the ratio of the well-fed to the undernourished is about 1:6. About 20 per cent of the people in the underdeveloped

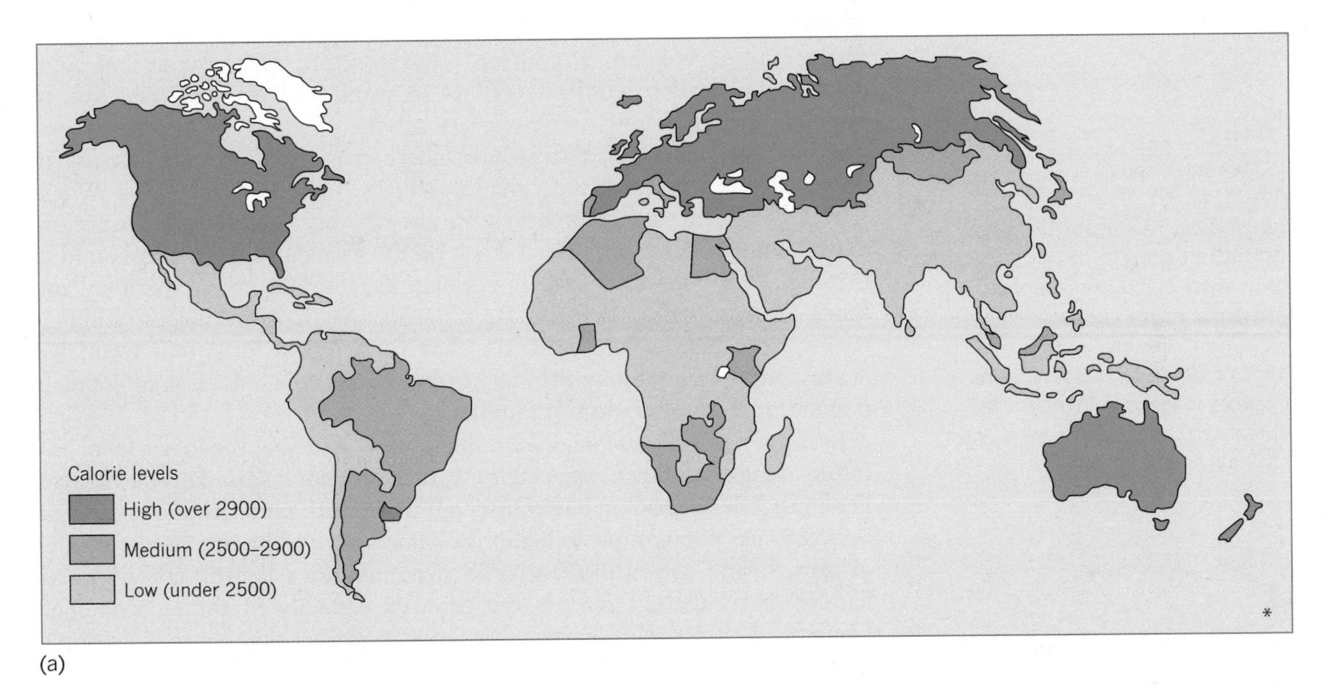

(a)

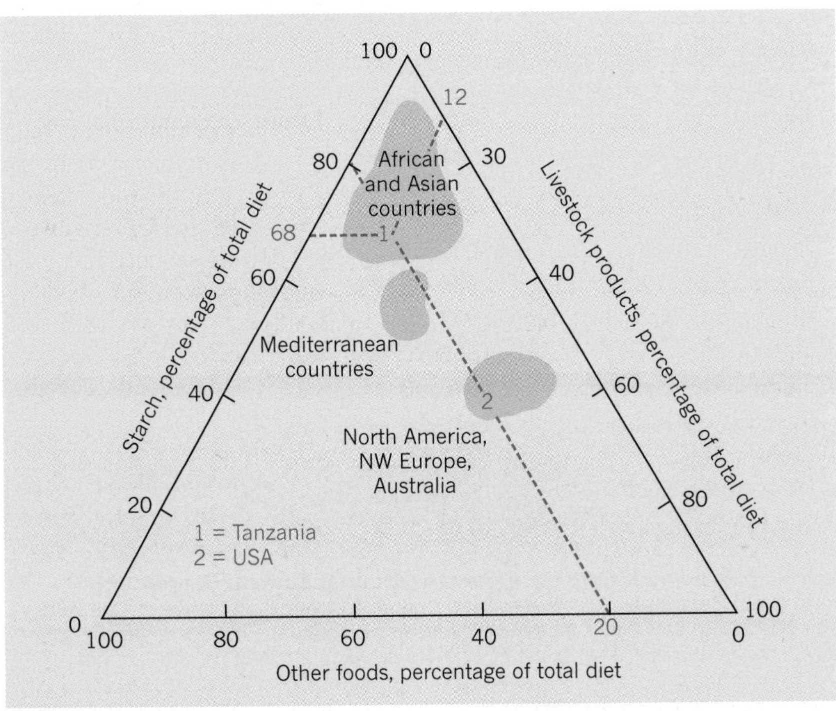

(b)

Figure 6.12 The geography of hunger (a) This map of world calorific intakes shows the tropical areas, despite their fertility, to be considerably worse off than other areas in terms of food consumption. However, the map, based on United Nations estimates for the mid.1960s, uses national data that differ widely in accuracy. (b) Variations within countries are now shown. The triangle shows diet composition for three contrasting groups of countries.

[Source: C.G. Knight and R.P. Wilcox, *Triumph or Triage? The World Food Problem in Geographical Perspective* . (Association of American Geographers, Resource Paper, 75.3, 1976), p. 14.]

countries are undernourished (i.e. receive less than the minimum number of calories they need per day), and some 60 per cent lack one or more of the essential nutrients, commonly protein.

Malthusian Checks on Population: Crowding and Conflict

In our models so far, we have envisaged a simple situation in which a single, homogeneous world population monopolized the exploitation of the available resources. But what happens if we partition the population (into 'haves' and 'have nots', for instance) and allow different groups to compete for resources? We can, of course, follow Malthus and assume that competition will lead to conflict, conflict to wars, and wars to reduction in population. But despite the immensity of war losses in the last 250 years, there is no evidence that they have checked the exponential increase in population. To be sure, individual countries, such as France after World War I, experienced severe checks on growth, but these lasted little more than a single generation. (Check back on the French population pyramid in Figure 6.7.) Wholesale checks on population growth from international and other conflicts would demand far more deaths than the biggest conflicts have yet produced. Note, however, that Table 6.3 includes only estimated deaths and does not allow for low birth rates due to a reduction in the male population and separation of families.

Because the historical record is unhelpful in assessing the losses from any future nuclear conflict, we can only turn to more speculative evidence. Ecologist L.B. Slobodkin has considered theoretical patterns of competition between two populations living in the same area and having different rates of growth and *saturation* levels. Each population's logistic growth curve flattens as its density reaches the carrying capacity of the environment. Figure 6.13 shows representative density curves over time for the two populations.

Figure 6.13 Competition and population growth The graphs show two possible outcomes when exponentially growing populations (N_1 and N_2) with fixed saturation levels (S_1 and S_2) must depend on the same resources to grow. In (a) both populations coexist and survive. In (b) only a single population survives.

Saturation level
S_1
N_1
S_2
N_2
Population size (N)
Time (t)
Coexistence

(a)

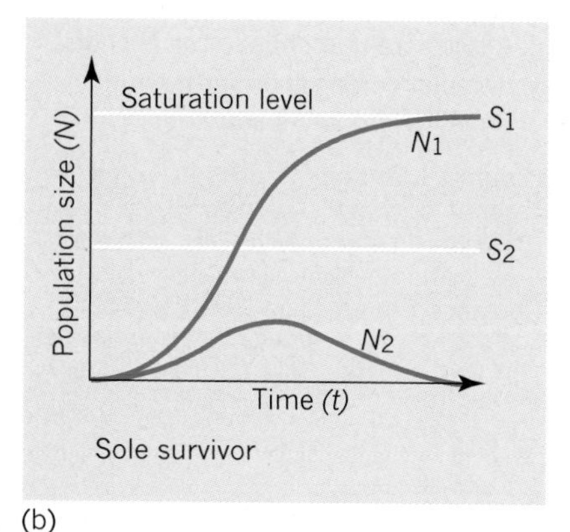

(b)

Table 6.3 Population losses from major conflicts

Conflict	Dates	Estimated deaths (millions)
World War II	1939–45	7.3
World War I	1914–18	7.2
Taiping Rebellion	1851–64	6.3
Spanish Civil War	1936–39	6.3
Ist Chinese Communist War	1927–36	6.1
la Plata War	1865–70	6.0
Indian Communal Riots	1946–48	5.9
Russian Revolution	1918–20	5.7
Crimean War	1853–56	5.4
Franco–Prussian War	1870–71	5.4
Mexican Revolution	1910–20	5.4

Source: Data from L.F. Richardson, *Statistics of Deadly Quarrels* (Boxwood Press, Pittsburgh, Pa., 1960). Richardson gives a figure of 5.8 million deaths for the US Civil War (1861–65), but this is almost certainly an overestimate. Figures generally refer to direct losses of military personnel; civilian losses would inflate those given here.

An equilibrium is established when both cease to grow. But both populations compete for the same resource, so the growth of one is inversely dependent on the growth of the other. Slobodkin's aim was to specify, given these conditions, whether the two populations could coexist in a state of equilibrium or whether one would progressively dominate the available resources to the exclusion of the other. With his theoretical model, two outcomes are possible. Either both populations coexist, but their numbers remain below the saturation levels that would otherwise prevail, or only one population survives and reaches its relevant saturation level. In other, and more complex cases, both populations may fluctuate cyclically. Note the example of the predator (lynx) and prey (snowshoe hare) populations in Figure 6.8.

Although the Slobodkin model refers only to two populations in highly simplified conditions, its ecological implications for the competitive relations between different populations and between subgroups within the same populations is important. It illustrates how we can simulate future conditions of conflict and thus try to avoid them. Furthermore, it underscores the fact that it is not so much the level of resources as our ability to share and distribute them that lies at the heart of the population–resources dilemma.

Non-Malthusian Models

In this chapter we have looked at the most important of the ecological models of human population, the Malthusian. Other views of human population density, notably those of Esther Boserup, will be taken up in the next part of the book (see Section 9.3). We also need to know that the alarmist views of population growth as argued by the neo-Malthusians such as Paul Ehrlich have not gone uncontested.

Economist Julian Simon in his *Ultimate Resource* (1981) challenged conventional views about population growth and resource use. He demonstrated that over the most recent decades and centuries for which relevant data are available, natural resources have become more rather than less abundant and that the real shortage is people to develop them. He uses economic arguments to show that dire predictions are not supportable and that humans are 'the ultimate resource'. We look at these arguments again when we turn to natural resources in Chapter 10.

The History of World Population Growth 6.4

How far do actual patterns of population growth follow the abstract models discussed above? We shall continue to consider the situation at the world level because this is the only level at which we can legitimately ignore migratory transfers of population. In this sense, then, despite its size, the world is one of the simplest population systems.

Box 6.C Living and Dead Populations

Another survivorship puzzle links the relative size of living and dead populations. American artist Laurie Anderson uses the refrain in a lyric 'Now that the living outnumber the dead'. Given the huge growth of modern populations that might seem a plausible hypothesis but one that is untrue, and likely to remain untrue in any forseeable scenario.

If the world population had been increasing at its present rate then the living would indeed outnumber the dead. (Recall the discussion of exponential growth in Box 6.A.) But calculations published by the International Statistical Institute suggest that there were very long periods of human history when growth was not exponential but many deaths accumulated, especially deaths of newborns and children. Using early counts of population in the Chinese and Roman Empires and estimates of the population densities that could be supported by different human habitats, rough counts of world population have been attempted. In 40,000 BC the world's human population was thought to be around half a million. It grew to around 250 million by AD 1000 and to one billion early in the nineteenth century. It is over six billion today.

Using likely death rates, the total number of deaths accumulated since 40,000 BC comes to around 60 billion. The present living population is around one-tenth of that figure. Although the historical estimates are certainly subject to error, these errors are very unlikely to be big enough to change the overall conclusion that in human history the dead substantially outnumber the living. What applies to the human population as a whole may not fit subpopulations. Geography as a professional field grew very rapidly in the second half of the twentieth century and for the population of graduate geographers, the living do indeed outnumber the dead (see Chapter 24). This would be true of most scientific fields today.

Past Patterns of Growth

Conclusions about past population growth might usefully begin with Oxford University demographer Colin Clark's axiom that most of the historical (still more the prehistorical) evidence on population is not very accurate and that, as we go back in time, its accuracy generally diminishes further. It is therefore extremely difficult to estimate the size of early populations. (See Box 6.C on 'Living and dead populations'.)

Archaeologists suggest that, at the beginning of agriculture, the population of the world was not more than ten million. For the beginning of the Christian (or Common) Era the estimate is 250 million, and for AD 1650 it is double that figure, 500 million. As we noted earlier, for periods from the late eighteenth century on, the increasing number of national censuses makes the estimating process somewhat easier. One of the articles of the United States Constitution required that 'enumeration shall be made within three years after the first meeting of the Congress ... and within every subsequent Term of ten Years' (see Figure 6.14).

The actual census conducted in 1790 represents a milestone in demographic history: only Sweden had collected accurate information on such a scale before, and most countries were to lag behind for some further decades. Even today, the United Nations' estimate of world population is very approximate since the most rapid growth is occurring in just those countries of the tropical world where censuses are less frequent and less accurate. Figure 6.15 presents a general picture of the increase from the beginning of the Christian Era to AD 2000.

Viewed in a historical context, the present expansion of the world population by about 2 per cent per year must be extremely rapid. If the expansion is projected backward in time, the population reduces to a single

SCHEDULE *of the whole Number of* PERSONS *within the ſeveral Diſtricts of the* UNITED STATES, *taken according to* " An Act providing for the Enumeration of the Inhabitants of the United States;" *paſſed March the 1ſt*, 1790.

DISTRICTS.	Free white Males of ſixteen years and upwards, including heads of families.	Free white Males under ſixteen years.	Free white Females including heads of families.	All other free perſons.	Slaves.	Total.
* Vermont	22,135	22,328	40,505	255	16	85,539
New-Hampſhire	36,086	34,851	70,160	630	158	141,885
{ Maine	24,384	24,748	46,870	538	NONE	96,540 }
{ Maſſachuſetts	95,453	87,289	190,582	5,463	NONE	378,787 }
Rhode-Iſland	16,019	15,799	32,652	3,407	948	68,825
Connecticut	60,523	54,403	117,448	2,808	2,764	237,946
New-York	83,700	78,122	152,320	4,654	21,324	340,120
New-Jerſey	45,251	41,416	83,287	2,762	11,423	184,139
Pennſylvania	110,788	106,948	206,363	6,537	3,787	434,373
Delaware	11,783	12,143	22,384	3,899	8,887	59,094
Maryland	55,915	51,339	101,395	8,043	103,036	319,728
{ Virginia	110,936	116,135	215,046	12,866	292,627	747,610 }
{ Kentucky	15,154	17,057	28,922	114	12,430	73,677 }
North-Carolina	69,988	77,506	140,710	4,975	100,572	393,751
South-Carolina	-	-	-	-	-	-
Georgia	13,103	14,044	25,739	398	29,264	82,548

	Free white Males of twenty-one years and upwards, including heads of families.	Free Males under twenty-one years of age.	Free white Females, including heads of families.	All other Perſons.	Slaves.	Total.
S. Weſtern Territory	6,271	10,277	15,365	361	3,417	35,691
N. Do.	-	-	-	-	-	-

Truly ſtated from the original Returns depoſited in the Office of the Secretary of State.

TH: JEFFERSON.

October 24, 1791.

* This return was not ſigned by the marſhal, but was encloſed and referred to in a letter written and ſigned by him.

(a)

(b)

Figure 6.14 America's first and recent censuses The United States has conducted a regular ten year census of its population since the late eighteenth century. (a) A page from the modest 56-page pamphlet giving the results of the first United States census, taken in 1790. Thomas Jefferson was required under the Constitution to conduct a census to ensure a fair base for tax apportionment between the states. It has been repeated every subsequent ten years. (b) Two census enumerators, Dorothy Sojourner and Kevin Beatty, check their schedules early on the morning of 21 March 1990 before beginning their count of homeless people in abandoned buildings in Washington, D.C. The most recent census was in March 2000.

[Source: (a) Courtesy of the US Bureau of the Census. (b) Corbis-Bettmann.]

human couple by only 500 BC. In fact, we know that human populations were inhabiting the earth in 500,000 BC. Thus the average rate of population increase in this early period must have been extremely slow. Rates of increase as low as 0.01 per cent per year have been proposed, but any model assuming continuous growth is probably unrealistic and unhelpful. By comparing the human population with other mammalian populations, we can conjecture that primitive populations underwent major fluctuations, and a fluctuating model on the lines of Figure 6.11(d) may be more appropriate.

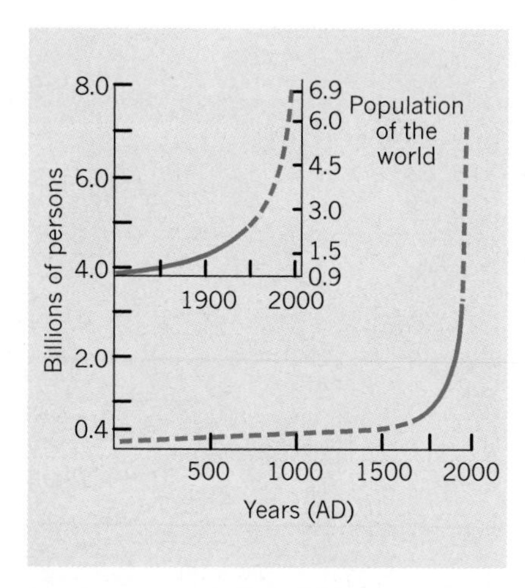

Figure 6.15 World population The graphs show the general pattern of increase in the estimated world population over the last 2000 years and (inset) the last 200 years. The solid line indicates periods for which reasonably good census material is available; the dashed line indicates estimates.

[Sorce: H.F. Dorn, in P.M. Hauser (ed.), *The Population Dilemma* (Prentice Hall, Englewood Cliffs, N.J., 1963), Fig. 1, p. 10. Copyright © 1963 by The American Assembly, Columbia University. Reproduced by permission of Prentice Hall, Inc.]

The Demographic Transition

In the last two hundred years, industrialization and urbanization appear to have caused a transition in the ways in which population grows. This *demographic transition* can be represented as a sequence of changes over time in vital rates (see Figure 6.17). We can recognize four connected phases in this sequence. In the first, or *high-stationary*, phase, both the birth and death rates are high. Although both rates vary, we can assume that the greatest variation is caused by deaths stemming from famines, wars, and diseases. Because the gains in population during a period when death rates are high are cancelled by the losses when death rates are low, the population remains at a low but fluctuating level.

The second, or *early-expanding*, phase is characterized by a continuing high birth rate but a fall in death rates. As a result, the life expectancy increases and the population begins to expand. The fall in the death rate is ushered in by improvements in nutrition, in sanitation, in the stability of the government (which means fewer wars), in medical technology, and so on. The third, or *late-expanding* phase is characterized by a stabilization of the death rate at a low level and a reduction in the birth rate. As a result, the rate of expansion slows down. The fall in the birth rate is associated with the growth of an urban-industrial society in which the economic burden of rearing and educating children tempers the desire for large families, and birth control techniques make family planning easier.

The fourth, or *low-stationary*, phase is a period when birth and death rates have stabilized at a low level; consequently, the population is stationary. This period is unlike the high-stationary phase in that the death rate is more stable than the birth rate.

Because of its unique population records which begin in 1750, Sweden's growth is of special interest. Table 6.4 summarizes the historical record over a 225-year period. Note the way in which demographic rates, age structure, job structure, and geographical distribution are all related.

To what extent do other countries today fit this pattern? Table 6.5 shows the latest UN population indicators for 170 countries giving the top eight on the four measures shown. Note that the highest birth rates and youngest populations are from Africa and Southwest Asia. Conversely, the lowest birth rates and oldest populations are European.

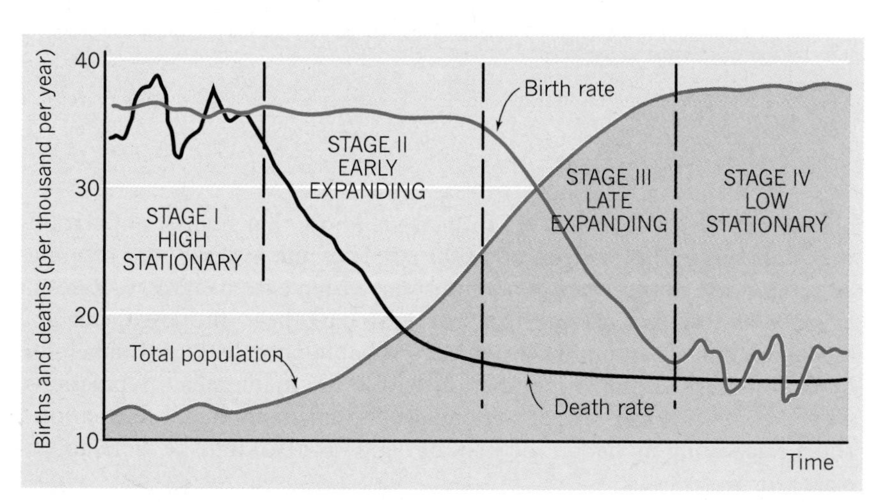

Figure 6.16 The demographic transition The graph shows four stages in a demographic sequence in which industrialization and urbanization are important factors. Note the variability of death rates (due mainly to famines and epidemic diseases) in Stage I and the variability of birth rates (due mainly to cycles in economic prosperity and attitudes to family size) in Stage IV. Compare the general curves shown with the actual pattern of change for the island of Mauritius in Figure 6.3. The sequence for Sweden is given in Table 6.4.

Table 6.4 The demographic transition in Sweden

Stage	Dates	Total population (millions)	Crude rates per 1000: Birth	Crude rates per 1000: Death	Age structure (percentage): Children (0–14)	Age structure (percentage): Old (65–)	Job structure (percentage in agric.)	Urban structure (percentage in Stockholm area)	Stage
	–1750	1.8	36	27	33	6	n.a.	8	
I High stationary									I
	1810	2.5	33	26	32	5	n.a.	7	
II Early expanding									II
	1870	4.4	30	18	34	5	72	6	
III Late expanding									III
	1930	6.3	14	12	25	9	39	13	
IV Low stationary									IV
	1975–	8.2	13	11	20	15	7	19	

Table 6.5 Demographic indication for large countries

Highest crude birth rates (live births per 1000)		Lowest crude birth rates (live births per 1000)		Youngest population (% aged under 15)		Oldest population (% aged over 65)	
1. Somalia (Africa)	52	1. Bulgaria (E Europe)	8.8	1. Liberia (Africa)	55	1. Sweden (W Europe)	18
2. Afghanistan (SW Asia)	51	2. Latvia (E Europe)	8.8	2. West Bank & Gaza (SW Asia)	51	2. Italy (W Europe)	17
3. Uganda (Africa)	51	3. Czech Rep. (E Europe)	8.9	3.Uganda (Africa)	49	3. Norway (W Europe)	16
4. Niger (Africa)	49	4. Estonia (E Europe)	8.9	4. Niger (Africa)	48	4. Belgium (W Europe)	16
5. Angola (Africa)	48	5. Italy (W Europe)	9.0	5. Zambia (Africa)	48	5. Greece (W Europe)	16
6. Yemen (SW Asia)	48	6. Germany (W Europe)	9.2	6. Benin (Africa)	48	6. UK (W Europe)	16
7. Malawi (Africa)	48	7. Romania (E Europe)	9.2.	7. Angola (Africa)	48	7. Germany (W Europe)	15
8. West Bank & Gaza (SW Asia)	47	8. Spain (W Europe)	9.2	8. Burkina Faso (Africa)	48	8. Denmark (W Europe)	15

Values given for most recent United Nations population estimates (*ca* 2000).

But for the world picture, the position is clear. Because of the rapid spread of medical technology, no major countries are found in Stage I. The high-stationary phase characterizes countries with an uncertain and low level of food production; most of the population in these countries engages in agriculture. This phase was universal among the human population during most of its early history, but now is restricted to the more isolated and primitive groups. Most of the populations of the Third World of Latin America, Africa, and South Asia fall into the early expanding phase. Population is expanding rapidly, as environmental and medical technology have brought substantial improvements in life expectancy. Some of the countries in that group have already experienced substantial drops in the birth rate. This decrease is related to socio-economic changes, where people work in their occupations, and is reinforced by family planning techniques; thus the populations of some of these countries appear to have moved into the late-expanding phase. Western Europe, the United States, Canada, and Australia are all examples of countries whose populations appear to be moving into the fourth stage. In Chapter 18 we review the global variation in the stages now reached by different countries and relate these to levels of economic development.

6.5 Current Trends in Population Growth

In this final section we try to look forward into the later decades of the present century. We first consider the problems of making projections, then at the pattern current estimates, and then at the implications of the projected slow-down.

Problems in Making Projections

In looking forward, we have to remember that we are dealing with projections and not promises. It is helpful (even if a little chastening) to look at the projections made in the 1960s of world population totals at the end of that century (see Figure 6.17). All started with the existing population (then around 3.3 billion people) and then built in reasonable assumptions of fertility and mortality. The range of projected outcomes is from 4.5 up to 7.6. But it is reassuring that the United Nations medium projection of 6.1 billion proved correct.

Its different growth curves illustrate that it is extremely difficult to proceed from a general recognition of the type of expansion that is occurring to a precise forecast. Projections are not promises and we cannot know the unknowable. The scourge of AIDS (see discussion in Chapter 20) is already reducing the projected increase in African countries such as Zimbabwe. Other unexpected changes (both positive and negative) may lie in wait as the present century unfolds.

Each projection tends to reflect the trends existing when the forecast was made. There is evidence that improvements in sanitation and medicine, which allowed a decline in death rates, are now levelling off. Birth rates are likely to be more volatile in the future as both fertility drugs and contraceptive devices permit greater control over family size.

Note also that the range of error increases with the time-span of the projection. Most demographic projections for the current century are concerned only with the period up to AD 2030. This range represents only one biological generation for a slow-breeding species like us. So it takes into account the children of the present generation, but not the grandchildren. The estimates are of course recomputed every few years to give a window which is always moving forward.

Global Projections to AD 2100

Given the above warnings, we need to recall that any projection a hundred years (three breeding generations) ahead is bound to be a general scenario. It cannot be a precise projection. Nonetheless, the most likely pattern of

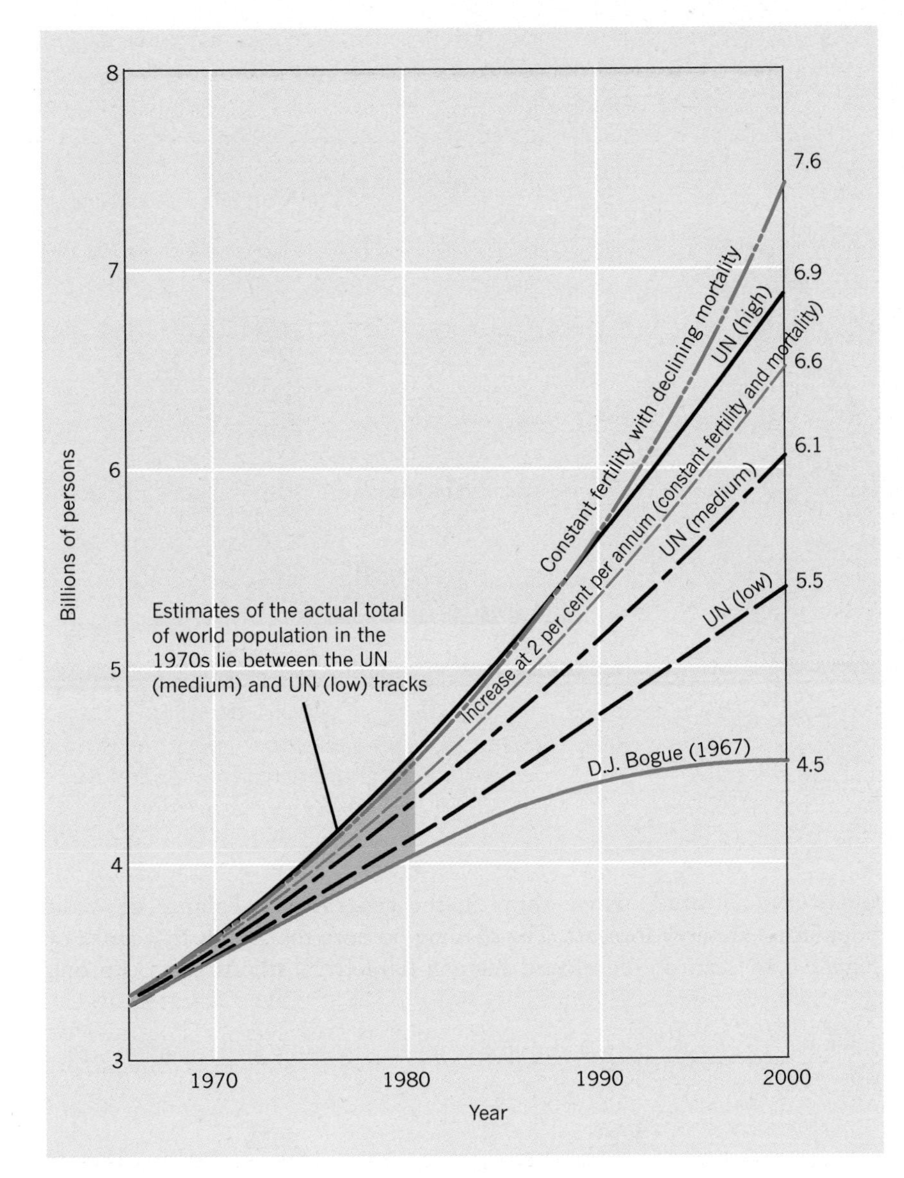

Figure 6.17 World population projections These six estimates of the total number of people likely to be around at the end of the twentieth century were based on evidence available in the 1960s. The highest estimate assumed that fertility would remain constant as mortality declines. The United Nations provided three separate estimates (high, medium, and low), based on varying assumptions. Most recent information available to the United Nations confirms that the UN medium forecast of 6.1 billion for 2000 was the most accurate.

[Source: N. Keyfitz, in *Resources and Man: A Study and Recommendations* by the Committee on Resources and Man of the Division of Earth Sciences, National Academy of Science–National Research Council, with the cooperation of the Division of Biology and Agriculture (Freeman, San Francisco, copyright © 1969).]

world population growth for the rest of the century can be shown (see Figure 6.18). It includes five main features: growth, deceleration, geographical shift, urbanization, and ageing.

Growth World population is expected to continue grow throughout the century. On the middle range of assumptions, the probable total in AD 2100 will be between ten and eleven billion (see Figure 6.18(a)).

Deceleration If we look not at world totals but at the average increase in numbers per decade (Figure 6.18(b)), a very different pattern emerges. This shows the increase reaching a peak in the 1990s. This appears to be a high-water mark in world history with the expected decreases in that augmentation tailing off throughout the rest of the twenty-first century towards zero by the 2090s.

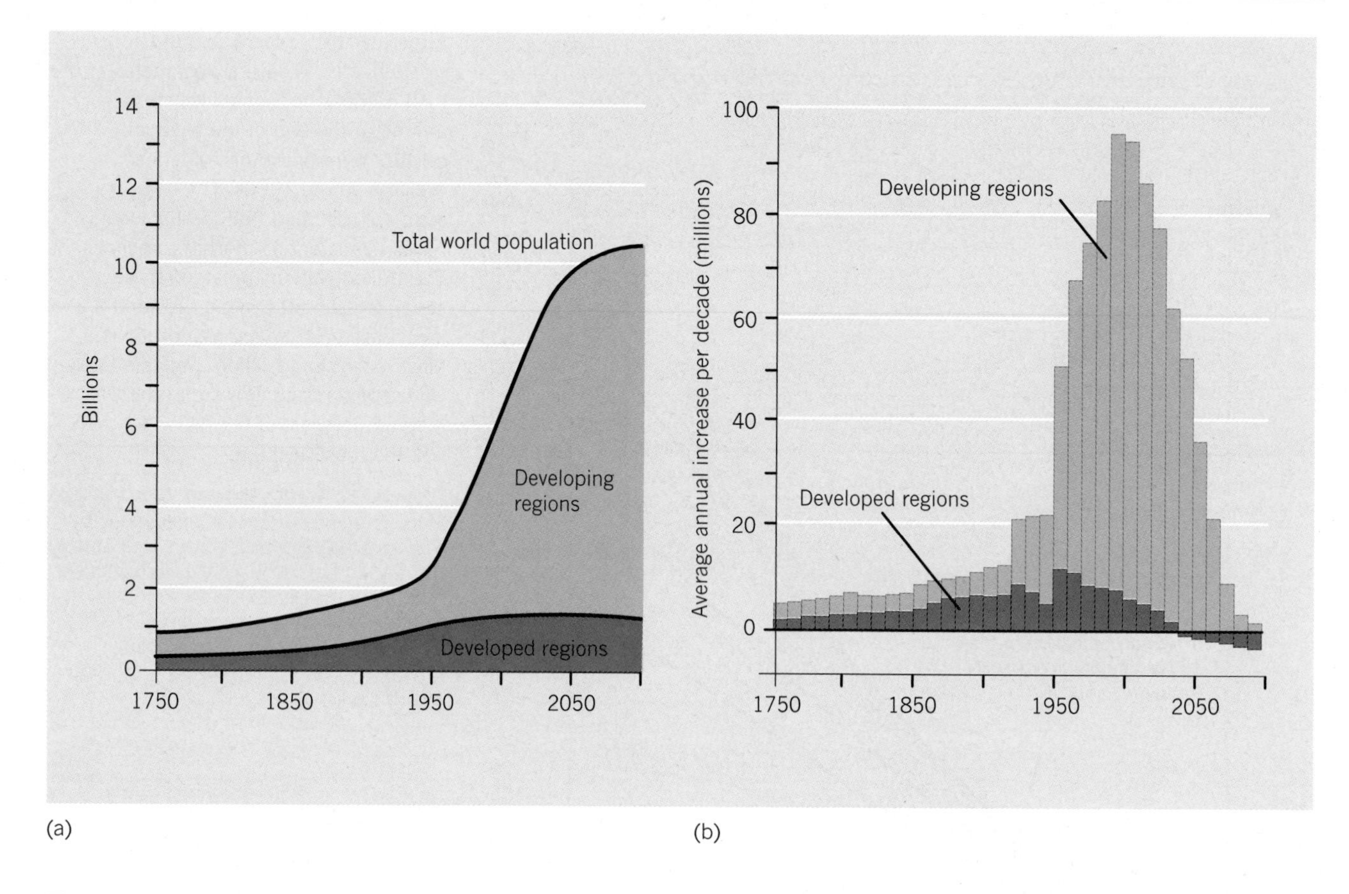

Figure 6.18 World population growth (a) World population growth, 1750–2100. (b) Average annual increase in numbers per decade, 1750–2100. In both graphs the population is split into developing and developed regions. Note that after 2040 the population change of developed regions in (b) is expected to be negative.

[Source: United Nations estimates.]

Geographical Shift Huge shifts in the geographical balance of world population are now forecast. The shading on both the graphs that make up Figure 6.18 separate developed regions (largely North America, Europe,

Table 6.6 The twelve largest countries ranked according to population size with projections for 2050

Rank	1988 Country	1988 Population (million)	2050 Country	2050 Population (million)
1	China	1255	India	1533
2	India	976	China	1517
3	**United States**	274	Pakistan	357
4	Indonesia	207	**United States**	348
5	Brazil	165	Nigeria	339
6	Russia	148	Indonesia	318
7	Pakistan	147	Brazil	243
8	**Japan**	126	Bangladesh	218
9	Bangladesh	124	Ethiopia	213
10	Nigeria	122	Iran	170
11	Mexico	96	The Congo	165
12	**Germany**	82	Mexico	154

Countries shown in **bold** are economically highly developed.
Source: Data from United Nations, *World Population Prospects: The 1996 Revision* (New York: 1996).

Japan, and Australasia) and developing regions (the rest of the world). Only about one-fifth of humanity currently live in the former, four-fifths in the latter. Almost all the new increases in population are projected to take place in the developing world. By the end of this century nearly nine out of ten of the world's population will live in what is now the developing world. Table 6.6 shows the twelve largest countries some 60 years apart. Note that by AD 2050, India will have overtaken China as the world's largest country. By then, the United States will be the only 'developed' country in the world's top twelve.

Urbanization Unevenness will be greatly intensified by urbanization. In the developing regions where most increase is occurring, urban populations are growing at twice the rate of rural populations. The rate of urban population growth is most rapid in just those countries that were least urbanized before. If we keep a constant definition of 'urban' then in 1800 the proportion of world urban dwellers was one in twenty. By 1900 this had increased to one in ten. Over the last century the proportion grew to nearly one half; during the present century it will continue to grow with rural dwellers now in a decreasing minority. The urbanization process is discussed in detail in Chapter 8.

Figure 6.19 Ageing of population Increase of population aged 65 years or over as proportion of total population for selected countries.

[Source: J.I. Clarke, in I. Douglas *et al.* (eds), *Companion Encyclopaedia of Geography* (Routledge, London, 1996). Fig. 12.3, p. 259.]

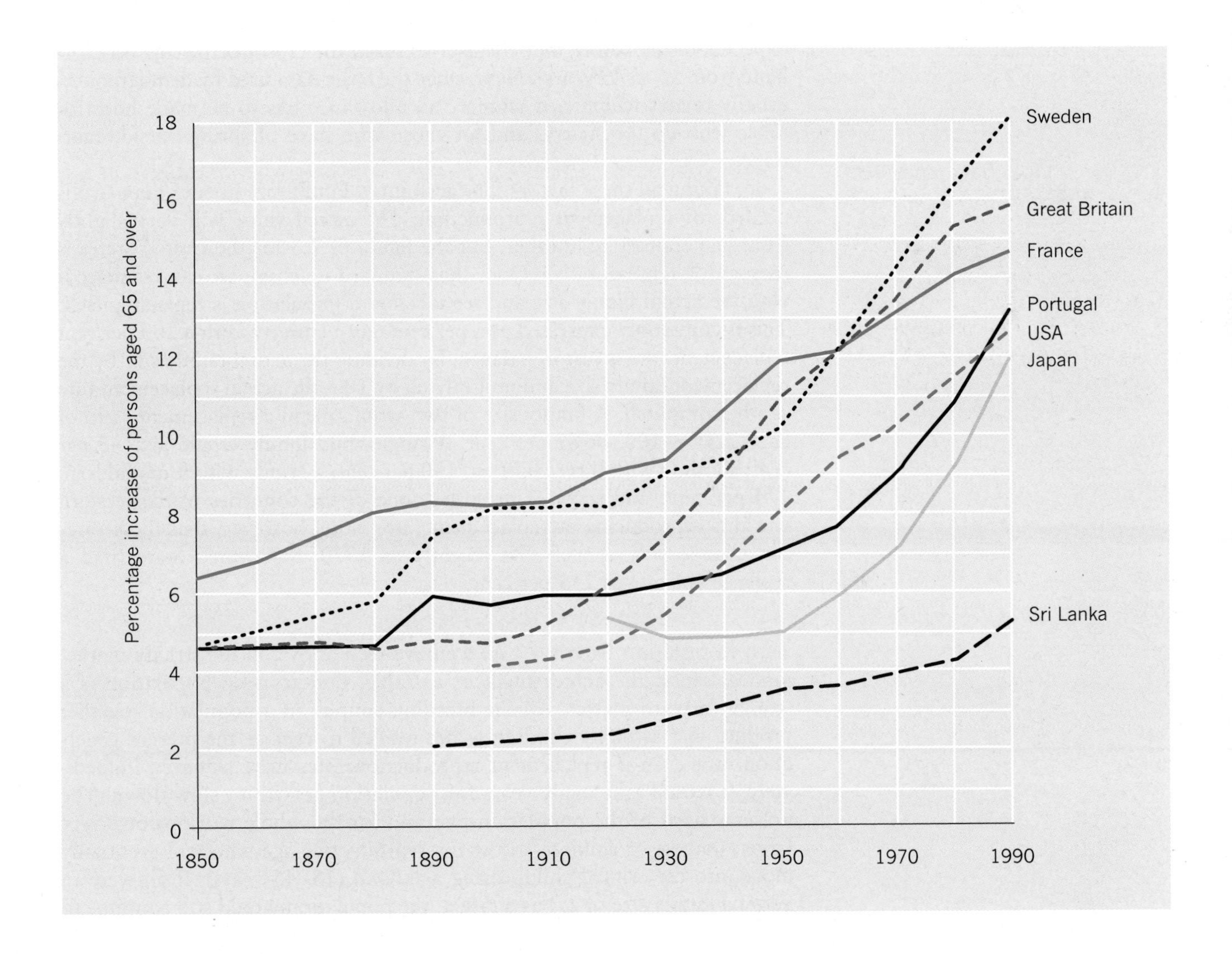

Ageing Projected change in age structures shows the proportion of young people (less than 15 years) reducing and of old people (65 years and above) continuing to increase. In world terms, the cross-over point (where the world's old exceed the world's young) is expected to come around 2030. Examples of this greying process is given for a selection of countries is shown in Figure 6.19. This shows the ageing process is different in level but not in trend in both developed (e.g. Sweden) and developing countries (e.g. Sri Lanka).

Towards Global Slowdown

If the above projections are correct, then the world population during the current century will be faced with major changes. We take up many of these changes in later parts of the book (notably in Chapters 8, and 18–20). Here we concentrate on two demographic aspects: replacement reproduction and zero population growth.

Replacement Reproduction To calculate the ***replacement rate***, we have to take into account several factors. First, we know that the human species produces about 106 male births for every 100 female births. Second, we have to account for the proportion of female offspring who may be expected to die before they themselves reach the reproductive age (say, the band from 15 to 49 years). Next, since the basic data used by demographers usually relates to married women, an allowance has to be made both for those who do not marry, and for those who have offspring outside marriage.

So, taking all these factors into account, a family size greater than two is needed for replacement reproduction. The actual value will vary slightly from one country to another, but the figure of 2.3 for the United States is reasonably representative. Once that standard is set up, we can compare it with the actual family size and see whether a population is replacing itself. This is commonly expressed as a per centage: a country with a 100 per cent replacement rate is just in balance. By dividing the actual family size by the replacement family size and multiplying by 100, the actual replacement rate can be computed. A family size of two would mean a replacement ratio of $(2 \div 2.3) \times 100$, or 87 per cent. A single-child family would give 43 per cent; a family with three children, 130 per cent; a family with four children, 174 per cent; and so on. Currently, none of the countries of the western world is reproducing at replacement rate. Note, however, that these rates can change very sharply. As recently as 1970, the United States had a replacement rate of 115 per cent.

Zero Population Growth? As we have seen above, in the strictly mathematical sense, the achievement of a stable, non-growing population is a straightforward matter. All the breeding couples in a population together produce just as many children as are needed to replace the present generation. But even if replacement reproduction rates were achieved immediately, it would take many years for population growth to slow down. The present shape of the population pyramid for the whole world shows very large numbers of children below the reproductive age who will eventually move into the critical childbearing age-band (15–45 years). If all were to adopt a family size of 2.3 as a target, the population would still continue to

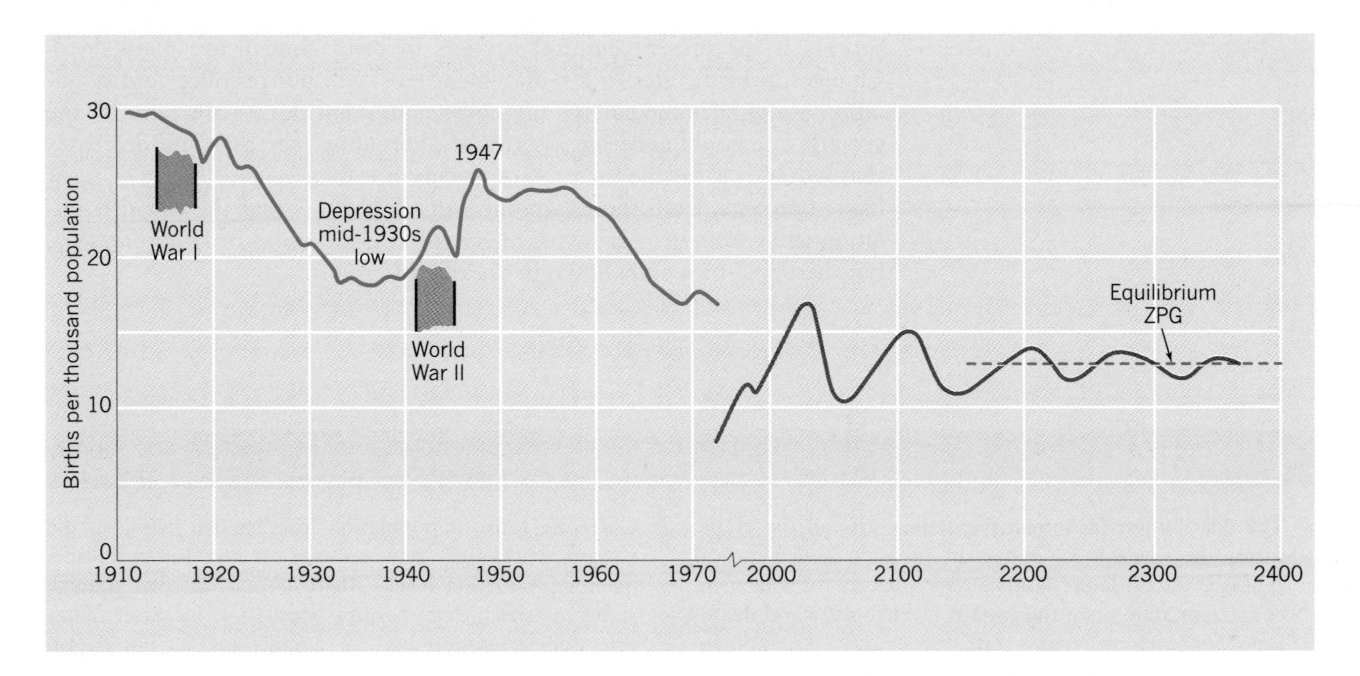

Figure 6.20 Zero population growth (ZPG) On the left are the *actual* birth rates in the United States over the 60 years since 1910. On the right are Frejka's estimates of the fluctuating birth rates needed to maintain the US population at its *present* level (i.e. for ZPG) for the next 400 years. Note that the time scales for the curves on the left and the right are different.

[Source: Data from Population, Reference Bureau and from A. Frejka, *Population Studies* **22** (1968), Fig. 1, p. 383.]

rise to a peak 1.6 times its present level. If the present programmes for birth control through family planning and socioeconomic change were successful in achieving a slow reduction to the replacement rate then the world population would rise to a peak at about 2.5 times its present size sometime in the next century. Thus, there is a momentum about population growth that is very hard to change quickly.

Does this mean that an immediate reduction in population growth is impossible? Research by demographer Tomas Frejka has shown that, for the United States, this is only possible if the target family size is first set well *below* the replacement rate – as low as 1.2 children per family. If this rate were to prevail for a couple of decades, the total population would decline, and the average family size would then need to be raised to above the replacement level. Figure 6.20 shows, on the left, the actual variations in the US birth rate over the last 60 years and, on the right, Frejka's calculation of the see-saw birth rates necessary to keep the population at the 1970 level over a 400-year period. According to Frejka, this long period is necessary for the violent cycles of population increases and declines to be smoothed out.

In the first half of this period, the remedy is as unwelcome as the illness. The effects of attempting to stabilize the population at present levels are scarcely less awful than those of unbridled growth. If Frejka's calculations are correct, the United States would go through alternating phases in which its age pyramids fluctuated violently from a dominance of old folk to a dominance of the young. As Figure 6.20 suggests, these phases would have a period of about 80 years and would become less pronounced as the centuries passed. Enormous economic and social problems would be created in the meantime. This line of alarmist argument is supported by the French historian Chaunu who regards the births lost in the present downturn in western Europe as creating in the next century 'a demographic disaster comparable only with that caused by the Black Death.'

It would appear, then, that despite an increasing awareness that people cannot go on reproducing at present rates for many more years, any substantial reduction in the rate of population growth is likely to be a slow process if the present cultural barriers to birth control are maintained. Changes in birth rates to a replacement level will not produce zero population growth immediately; moreover, an immediate reduction of the growth rate could generate a series of alternating increases and decreases that would last for the next ten or twenty breeding generations. The most likely outcome, from the viewpoint around 2000, is that the world population will continue to grow for the next few generations of *Homo sapiens*, but there will be a slow-down in the rate of growth.

Reflections

1 The distribution of human population around the globe forms the single most important focus in human geography. Using the latest national census figures for your own country or state, map the pattern of birth rates and death rates for each province, state, or county. Attempt to explain the spatial pattern that results.

2 Population pyramids are used to show the distribution of age–sex groups in a population. Construct hypothetical population pyramids for populations that are (a) expanding rapidly, (b) declining sharply, (c) static, or (d) recovering from major wars. Can you match these populations with populations of real countries today?

3 How would you define the term 'overpopulation'? Do you think your own country or state is overpopulated? Why? Suggest and defend an optimum population for your area. How does it compare with the optimum populations suggested by others in your class?

4 In the Malthusian hypothesis, continuing population growth will eventually outstrip the growth of food supply. Set up a debate in the class on the motion that: 'The ideas of Malthus on the balance of population and food supply are no longer relevant.' On which side would you prefer to speak? Why?

5 Currently rates of population increase are falling in the western world and a few countries are now below replacement reproduction level. Where does your own country stand in terms of its natural population increase? What role does migration play in augmenting its population numbers?

6 Look closely at Figure 6.16 and compare it with Table 6.5. Which phase in the demographic sequence best describes the situation in your own country today? Which phase describes its situation a hundred years ago? What factor might account for a change in the pattern of population growth?

7 Review your understanding of the following concepts using the text (note Table 6.1) and the glossary in Appendix A of this book:

carrying capacity
crude birth rate
crude death rate
crude growth rate
demographic transition
Malthusian hypothesis
population pyramid
replacement reproduction
saturation level
survivorship curve

One Step Further ...

A useful starting point is to look at the essay on world population by John Clarke in I. DOUGLAS *et al.* (eds), *Companion Encyclopaedia of Geography* (Routledge, London, 1996), pp. 249–74, and the excellent review of population geography given by Philip Ogden in R.J. JOHNSTON *et al.*, *Dictionary of Human Geography*, fourth edition (Blackwell, Oxford, 2000), pp. 594–602.

Three recent texts which give a concise account of human population and their dynamics as outlined in this chapter are A.G. CHAMPION *et al.*, *Population Geography* (Blackwell, Oxford, 1999), M. LIVI-BACCI, *A Concise History of World Population*, second edition (Blackwell, Oxford, 1997), and H.G. DAUGHERTY and K.C. KAMMEYER, *An Introduction to Population,* second edition (Guilford, New York, 1995).

Malthus's classic 1803 essay has been republished as T.R. MALTHUS, *An Essay on the Principle of Population and a View of Its Past and Present Effects on Human Happiness* (Cambridge University Press, Cambridge, 1990). Malthus's ideas are critically reviewed in D. COLEMAN and R. SCHOFIELD (eds), *The State of Population Theory: Forward from Malthus* (Blackwell, Oxford, 1986).

The population geography 'progress reports' section of *Progress in Human Geography* (quarterly) will prove helpful in keeping up to date. In addition to the regular geographic journals, look at *Demography* (semiannually) and *Population Studies* (quarterly) for substantive reports on current research. The Population Reference Bureau (Washington, D.C.) publishes very useful bulletins and annual data sheets. The United Nations Statistical Office publishes the *Demographic Yearbook* (annual), an indispensable guide to world data on population, and regular reports on world population and world urbanization. A number of sites are recommended in Appendix B at the end of this book for topics relevant to this chapter.

MIVAN

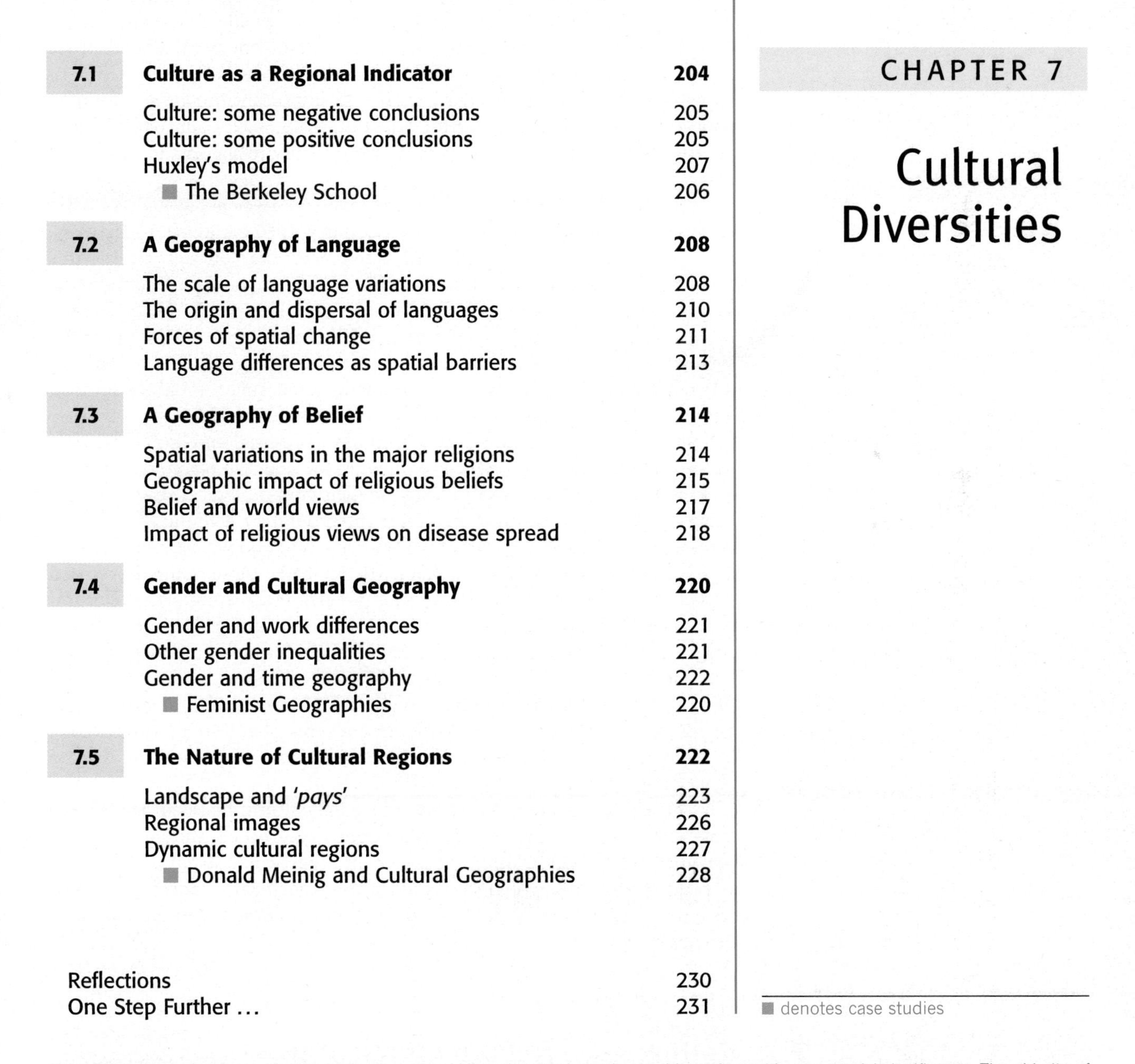

CHAPTER 7

Cultural Diversities

■ denotes case studies

Figure 7.1 Sacred places Human cultures endow certain locations on the earth's surface with very special significance. The old city of Jerusalem is sacred to three world faiths, Jewish, Christian, and Moslem. The Wailing Wall (foreground) is all that remains of the Biblical Temple of the Jews destroyed by Roman soldiers in AD 70. Above it rises the Dome of the Rock, started in AD 691, to mark the spot where the prophet Mohammed is believed to have risen to heaven. The modern secular symbols of radio masts, tower cranes, and barbed wire stand in contrast to the sacred. [Source: © TRIP/A Tovy.]

> Although I am unborn, everlasting, and I am the Lord of all, I come to my realm of nature and through my wondrous power I am born.
>
> BHAGAVADGITA

When the author engaged, usually unsuccessfully, in the schoolyard scuffles beloved of all small boys, the key word to know was 'faines'. Children, happily, are one of the few remaining uncivilized tribes, and – at least in English village schools in the 1930s – they still retained the local dialect truce words. Shouting 'faines' allowed a small boy at the bottom of a pile of struggling bodies a respite or truce. It was important, however, to know your location when you called for time out. In the next county to the west the truce word was 'barsy', and in the next county to the east it was 'cree'! Truce words had their own micro-geography.

Such local differences in schoolyard language over a few tens of kilometres in as relatively homogeneous a country as England are a microcosmic example of the vast cultural differences that shatter the earth's population into a regional mosaic of immense intricacy and complexity. Terms such as Jew and Gentile, Moslem and Hindu, WASP and Afro, Maoist and Mennonite, all serve to recall the countless ways in which the single biological species of humans seeks to separate and divide itself into different stereotypes. Even in places held sacred such as old Jersusalem (Figure 7.1 p. 202), cultural divisions remain undimmed over the centuries.

So in this chapter we move away from the ecological and biological view of the human population which dominated the last two chapters. Here we look at the geographic implications of their non-biological diversity. We begin (Section 7.1) with trying to clear the ground by defining some of the ways in which human culure differs. We ask, 'What do we mean by cultural differences?' and 'How can we define them?' Once these points have been settled, we go on to look at the spatial pattern of three important cultural differences. We look in turn at the geography of language (Section 7.2), the geography of belief (Section 7.3), and the geography of gender (Section 7.4). Here the critical questions are, 'How does culture vary spatially?' and 'Why is it critical for geographers to know about such cultural variations?' Finally, we try in Section 7.5 to use our concepts of culture to build a code by which geographers can separate one part of the earth from another by identifying a system of cultural regions.

In this chapter more than in any other, the cultural biases of the author must show. No one can shake free from deep-rooted and unconscious prejudices that come from being reared in a particular social and cultural setting. Readers are invited to challenge the judgements made here, and to submit for discussion other views of the human cultural mosaic based on their own experience.

7.1 Culture as a Regional Indicator

In our chapter title and throughout this second part of the book, we use the term 'cultural'. Just what do geographers mean by this term? Since *culture* turns out to be a complex word, we shall begin with some negatives before going on to a positive definition.

Culture: Some Negative Conclusions

Let us begin by defining what culture is not. First, culture does not mean simply an interest in artistic pursuits. Though the phrase 'cultural activities' is commonly used to describe musical, literary, and artistic efforts, this is a special use of the term culture in a very limited sense. Although such activities may have a distinct spatial pattern and may define certain cultural boundaries (e.g. the geographic extent of interest in a virulent variety of bagpipe music may provide a clue to the distribution of folk of Scottish ancestry), this view of 'culture' is too narrow for our purposes.

Second, culture does not mean race. *Race* is a biological term used to classify members of the same species who differ in certain secondary characteristics. Humans are a single species (*Homo sapiens*) with a common chromosome number (46), and fertile interbreeding among all the billions of members is possible. Nonetheless, specific biological differences do separate people into recognizably distinct subgroups (see Chapter 5, Section 5.1). Such differences range from (a) variations in external features (e.g. in skin pigmentation, in the shape of the skull, in hair type, and in eyefolds), to (b) differences in internal features (e.g. blood types).

Blood types may be an important indicator of group differences that have been maintained for a long time. Thus the Basque-speaking population of southwest France and northern Spain has a very high proportion of rhesus-negative blood types compared with the population throughout most of Europe. Moreover, rhesus negativism in blood groups is a peculiarly European trait; it is extremely rare among Asians, Africans, and American Indians.

Racial differences of this kind are almost certainly due to long periods of isolation in which genetic variations are accentuated by generations of interbreeding. This hypothesis seems to be supported by the spatial distribution of certain genetically determined diseases. Some of these diseases are highly localized: kuru, a progressive disorder of the nervous system, is so far known only among the Fore peoples of eastern New Guinea. Other diseases have distinct but intercontinental distributions. Sickle-cell anaemia is a disorder of the blood cells found in much of Africa south of the Sahara (and, by transfer, in the black population of the Americas) and in another broad belt running from India to Indonesia.

Whether genetic drift (random changes of gene frequencies) permits long-term adaptation of groups to particular environments is not clear. Some differences have been medically established – for example, differences in the sweating capacity of blacks and whites, the fact that Inuits' skin temperatures remain higher than non-Inuit in cold weather, and the lower metabolic rates of Australian aboriginals at night, which allow them to withstand low night temperatures. The cases of genetic adaptability stand in contrast to the extraordinarily widespread ability of people of all kinds to adapt *technically* to extreme environments by devising life-support systems that range from parasols and the fur wrap to spacesuits for moon-walkers.

Culture: Some Positive Conclusions

Cultural geography has emerged as a distinctive branch of human geography. (See Box 7.A on 'The Berkeley School'.) Considerable efforts have

Box 7.A The Berkeley School

American cultural geography was dominated until the 1980s by Carl Sauer (see Box 5.B), his colleagues at Berkeley, and their students. Sauer moved to Berkeley in 1923 and two years later published his most influential work, *The Morphology of Landscape*. This strongly denounced environmental determinism and suggested a method by which cultural geographers should proceed in their field studies. For the next half century until his death in 1975, Sauer was the focus of a distinctive tradition of work in human geography. Among the better known of Sauer's doctoral students were Andrew Clark and William Denevan (Wisconsin), Fred Kniffen (Louisiana), Marvin Mikesell (Chicago), James Parsons (Berkeley), David Sopher (Syracuse), Philip Wagner (Simon Fraser), and Wilbur Zelinsky (Pennsylvania State). Each went on to encourage a further generation of cultural geographers.

What was the distinctive character of the Berkeley School? Wagner and Mikesell define cultural geography as 'the application of the idea of culture to geographic problems', and go on to suggest four principal themes that define the work of the Berkeley School within cultural geography.

1 *Diffusion of culture traits.* Cultural geographers, like cultural anthropologists of the period, favoured explanation in terms of the diffusion of culture traits rather than independent invention to account for the development of cultures. As we saw in Chapter 5, Sauer was particularly interested in tracing the spread of plants and animals while others have examined such things as house types, names, and ideas as indicators of the diffusion of cultural traits.
2 *Identification of culture regions.* Material and non-material traits were carefully mapped and culture areas were identified by plotting the distribution of such features as building types, ploughs, animals, magazine subscriptions, language, religion, and ethnicity.
3 ***Cultural ecology.*** This was usually studied in historical perspective. Attention was focused on how perception and use of the environment was culturally conditioned.
4 *Regional specialization.* Shortly after arriving at Berkeley, Sauer developed what was to become a life-long interest in Latin America in general and Middle America in particular. While many of his students went on to work in other areas, there remains a strong connection with that region in the work of subsequent generations of his students (e.g. Parsons on Colombia and Denevan on Amazonia).

Many of the ideas that Sauer introduced into the field (e.g. historical reconstruction, culture area, diffusion) were current at the time in German geography and American cultural anthropology. The intellectual debt of the Berkeley School to such scholars as Ratzel (Figure 24.3), Schluter, Hahn, and Kroeber are immense. An excellent summary of the work of the Berkeley School is given by J. Duncan, in R.J. Johnston *et al.* (eds), *Dictionary of Human Geography*, 4th edn (Blackwell, Oxford, 2000), pp. 45–46.

been made by cultural geographers to define culture in a precise and positive way. We can perhaps summarize their view by saying that *culture* describes patterns of learned human behaviour that form a durable template by which ideas and images can be transferred from one generation to another, or from one group to another.

Three aspects of this definition need further accenting. First, the transfer is not through biological means. The *same* newborn child will grow up with quite *different* sets of cultural characteristics if it is reared in different cultural groups. Second, the main imprinting forces in cultural transfers are symbolic, with language playing a particularly important role. (By 'imprinting' is meant the spontaneous acquisition of information, particularly those habits of speech and behaviour acquired in the early years of life.) Third, culture has a complexity and durability which make it of an entirely different order from the learned behaviour of other, non-human animals.

The diversity of human cultures and their innate complexity is staggering. Anthropological research by such scholars as Claude Levi-Strauss has long since done away with the notion that there are any 'simple' cultures. Even in the smallest and most 'primitive' cultural groups (e.g. a small Amazonian hunting tribe or the population of a Micronesian village), there is a massive amount of cultural information to acquire. The child growing up in the simplest society slowly acquires millions of pieces of a cultural pattern, which will be duly passed on (albeit in some modified form) to the next generation. We should note (a) that such transfers are independent of formal education (in the western sense of going to school) and (b) that the transfer process is always incomplete. Thus the culture of the group is always several times larger than the culture of the individual. The most distinguished Harvard professor or the oldest village elder can never hope to acquire, in a lifetime of study, more than some fractional part of the 'genetic code' of the culture of which each is a part.

Huxley's Model

If culture is so complex and all-embracing a thing, is there any hope of disentangling it into simpler, molecular units which we can study and comprehend? Let us look very briefly at the solution of one man. One of the simplest ways of categorizing culture was proposed by English biologist Julian Huxley in a comparison of cultural and biological evolution. The *Huxley model* has three components: mentifacts, sociofacts, and artifacts (see Figure 7.2).

Mentifacts are the most central and durable elements of a culture. They

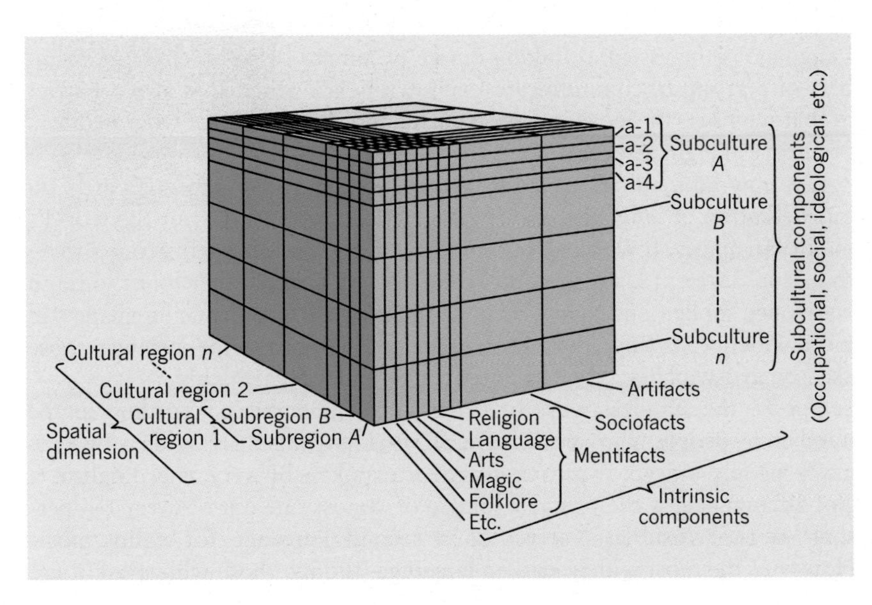

Figure 7.2 The components of culture Wilbur Zelinsky has created a model of culture using a three-dimensional cube which can be analyzed in terms of (a) its intrinsic components, or (b) the cultural characteristics of a given region, or (c) the cultural characteristics of a distinctive group or subculture. Note how the three approaches interlock, so that we can study small cubes within the master cube. For example, the cube of the spatial distribution of terracing in areas of Chinese culture within the cultural 'realm' of Southeast Asia.

[Source: W. Zelinsky, *The Cultural Geography of the United States* (Prentice Hall, Englewood Cliffs, NJ, 1973), Fig. 3.1, p. 73.]

include religion, language, magic and folklore, artistic traditions, and the like. They are basically abstract and mental. They relate to the human ability to think and to forge ideas, and they form the ideals and images against which other aspects of culture are measured.

Sociofacts are those aspects of a culture relating to links between individuals and groups. At the individual level they include family structures, reproductive and sexual behaviour, and child rearing. At the group level they include political and educational systems.

Artifacts are material manifestations of culture. Sometimes termed 'cultural freight', they include those aspects of a group's material technology that allow basic needs for food, shelter, transport, and the like to be filled. Systems of land use and agricultural production are cultural artifacts, as are tools and clothing of a particular design.

Like all such schemes, Huxley's model is only an approximation of reality. In practice, we find aspects of culture in which the three components seem tangled into an intractable knot and others where a strand can be pulled free and studied individually.

7.2 A Geography of Language

Does culture have a specifically geographic pattern? To answer this basic question, we shall take one of the most important of cultural differences, language. In studying language remember that we are using this simply as one example from the multitude of other mental, social, and material examples that might suggest themselves. But the questions we shall ask will be general ones about spatial stability, order, and change. These may provide a basis for your own analysis of other cultural elements that may interest you more.

The Scale of Language Variations

Language is the essential linking device in human cultures, enabling members of a group to communicate freely with each other. It is also a barrier in that members of one language group cannot communicate with members of another language group.

Just how many different languages are spoken today depends partly on our definition of language. Table 7.1 gives the proportions for the world's major languages. If we leave out minor dialects, there are still around three thousand different languages in current use. At least another four thousand were once spoken but have now gone out of use. If we define language still more widely to include the various scientific 'jargons' – for instance, those used by geographers – then the number becomes incalculable.

One of the simplest ways of classifying languages is according to the number of people who speak them. *Global languages* are spoken by very many people indeed; *local languages* are spoken by very few. English, a global language, is the primary tongue of almost one out of every ten persons in the world and serves as a second language for many more. However, the most widely spoken language (though those who speak it are

Table 7.1 Global language groups

Language family	Estimated share of world's language (per cent)	Typical languages	Area
Indo-European	47	German, English, Hindustani	India, Europe, America, Australia, S Africa
Sino-Tibetan	25	Chinese, Thai, Tibetan	East Asia
Afro-Asiatic	5	Arabic, Hebrew, Berber	N Africa, Middle East, Ethiopia
Dravidian	5	Tamil, Telugu	S India, Sri Lanka
African	5	Swahili, Hausa	Africa S of Sahara, but excluding Ethiopia
Malayo-Polynesian	5	Indonesian	Malaysia, Indian Ocean and Pacific Ocean islands
Japanese-Korean	4	Japanese, Korean	E Asia
Uralic-Altaic	3	Mongol, Finnish, Turkish	E Europe, SW and Central Asia
Mon-Khmer	1	Cambodian	SE Asia
American Indian	<1	Over 1200 different languages	North and South America

more concentrated spatially) is Mandarin Chinese, which, with its many dialects, is the language of nearly a billion people in East Asia. Figure 7.3 shows the principal languages spoken on the Indian subcontinent.

Were we to rank the world's languages, putting the global ones such as English and Chinese at the top, these would come about halfway down the list. At the bottom of the list would be the really local languages. Research

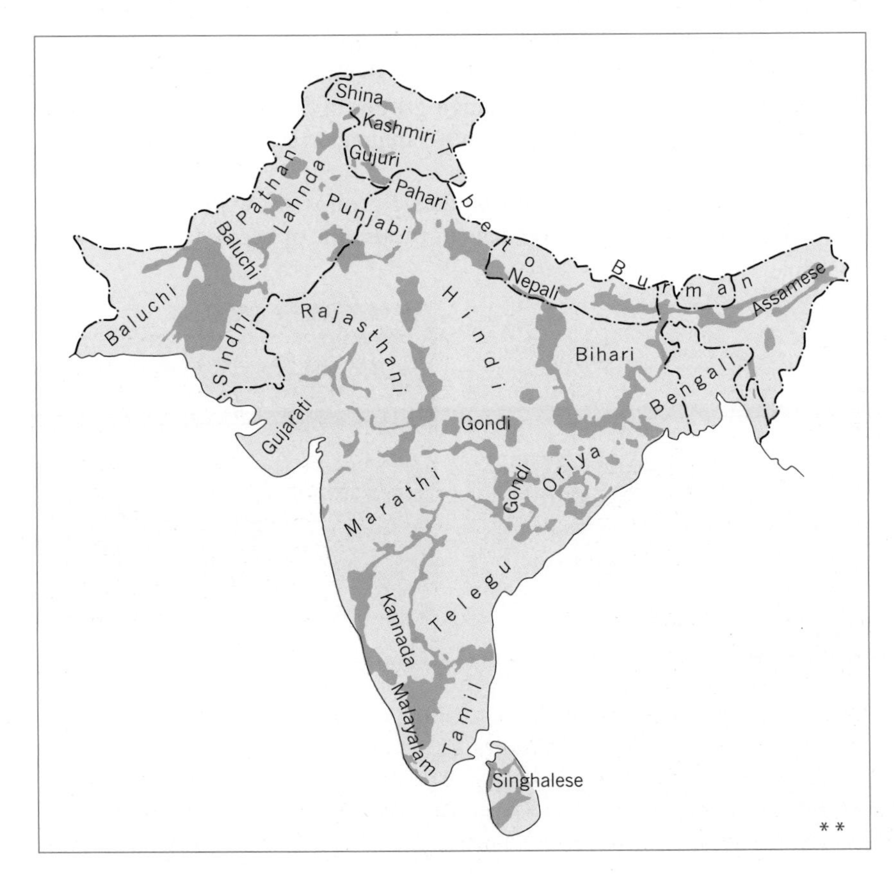

Figure 7.3 Linguistic differentiation of cultural areas
The map shows the principal languages used in the Indian subcontinent. Darker shading is used for areas in which two or more languages are spoken. Note that only the major languages are shown and that English is widely used for government and business. Tamil speakers make up a significant minority in the dominantly Singhalese-speaking island of Sri Lanka.

in New Guinea has revealed some wholly distinct languages (uncomprehended by neighbouring groups) confined to single valleys spoken and understood by only a few hundred people, and having a spatial extent of less than 65 km^2 (25 square miles). Actually, a few languages are spoken by a disproportionately large percentage of the human population. The top 14 languages are spoken by 60 per cent of the world's people. At the other extreme, the bottom 500 are divided among no more than one million people in the remoter parts of Asia, Africa, South America, and Australasia.

The Origin and Dispersal of Languages

Language illustrates clearly a second theme in cultural geography – the origin and dispersal of cultural elements. The questions we must ask to understand this process are about the relation of one language to another.

Extensive linguistic research has revealed that many different languages seem to have emerged from a common stock. For example, it is possible to trace back Indian languages in the northeastern United States such as Cayuga, Seneca, and Tuscarora to a common Iroquoian stock which has some linguistic connections with Sioux language groups further west. However, the language group whose evolution we know most about is the Indo-European family of languages (see Figure 7.4). A wealth of written records in these languages has allowed us to unravel slow linguistic drifts over the centuries. Despite the fact that these languages are spoken by half the world's people they still have many simple, basic words in common. The English word 'mother', for example, is '*Mutter*' in closely related German. It is also recognizable in quite different subgroups – in the Romance group, in the Spanish '*madre*'; in the Balto-Slavonic group, in the Russian '*mat*', and in the Indo-Iranian group, in the Sanskrit '*mata*'. Even in Greek, whose position in the family language tree is still hotly debated, mother is '*meter*'.

The distilling out of the main language groups shown in Figure 7.4(a) was a slow process, taking over tens of thousands of years. Changes over a

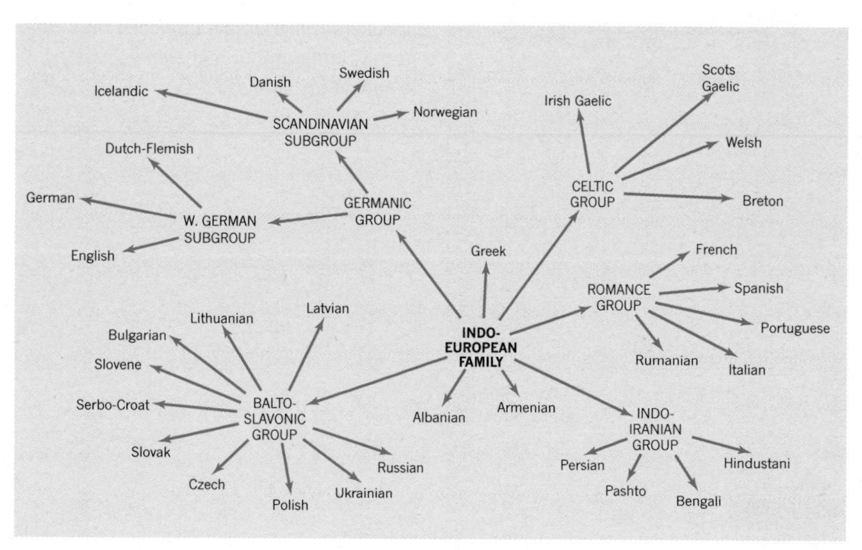

(a)

(b)

Figure 7.4 Linguistic origins and differentiation (a) About one-half of the world's six billion people speak languages in the Indo-European family. The chart shows the main languages in this family and the links between them. Note that some European populations speak languages outside the Indo-European group (e.g. Finnish and Basque). (b) Language overlaps with sign in dual language, Basque and Spanish.

[Source: (b) © The Image Works/Mark Antman.]

much shorter period are noticeable in dialects within a language, but languages themselves are stable enough to provide useful spatial signals of the migrational history of various groups. Thus the language differences between the Gilbert and Ellice Islanders in the Pacific provide evidence of long separation. (See the discussion of *Kon-Tiki* voyages in the Pacific in Section 16.4.)

Forces of Spatial Change

Cultural patterns are clearly not static in either time or space. The proportions of the world's population speaking each of the major languages is changing, and their spatial distributions are waxing or waning. Patterns of language are changed not only by the demographic forces of birth and death – which affect mainly our 'first', or 'native', language – but by the aggressive spread of second languages. Currently, the proportion of English-speaking persons is rapidly increasing in the urbanized and 'westernized' world. Figure 7.5 summarizes the main forces at work hammering out these changing linguistic patterns. Try to work your way through the different branches of this tree and think of examples that would illustrate each type of spatial change in language.

At the same time as global languages are spreading, some small languages are slowly dying out. The Celtic languages of Western Europe, for example, have been losing ground for centuries to the more aggressive English and French tongues. Celtic languages were once spoken in the western parts of the British Isles, the Brittany peninsula in France, and northwest Spain. One of the Celtic languages – Cornish – was confined to the extreme southwest corner of England. Until the fifteenth century, Cornish was spoken over

Figure 7.5 Spatial changes in languages The chart gives a schematic view of some of the main forces causing linguistic changes. Part of eastern Canada is bilingual (speaking French and English). Creole languages (e.g. the French creole in Haiti) have emerged from the mixing of French or Spanish with Caribbean languages. 'Pidgin languages' are very basic forms of English spoken as a trading language in many Pacific Island communities. 'Loanwords' are borrowings due to human communication. (The words jazz and taxi, for example, are loanwords common to very many languages.)

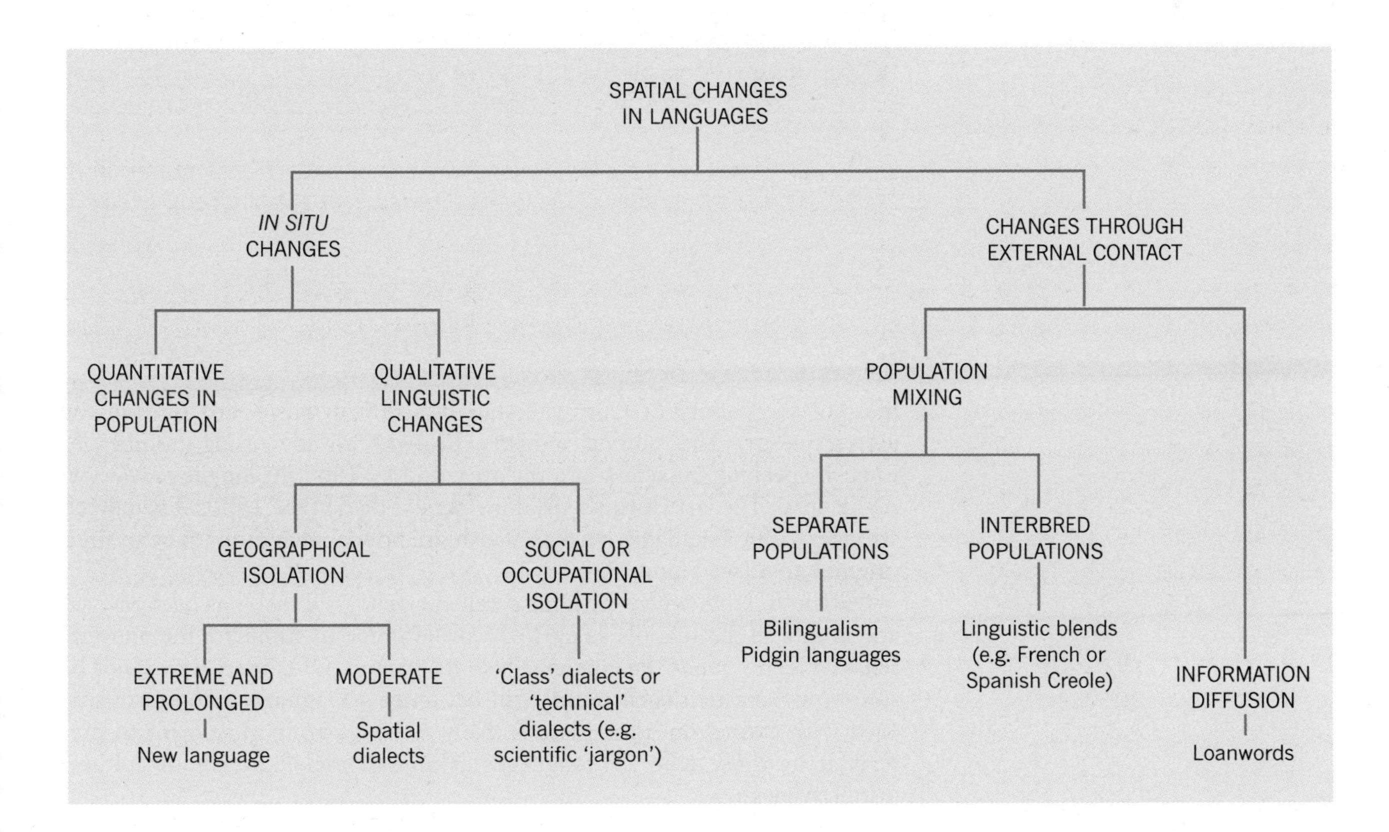

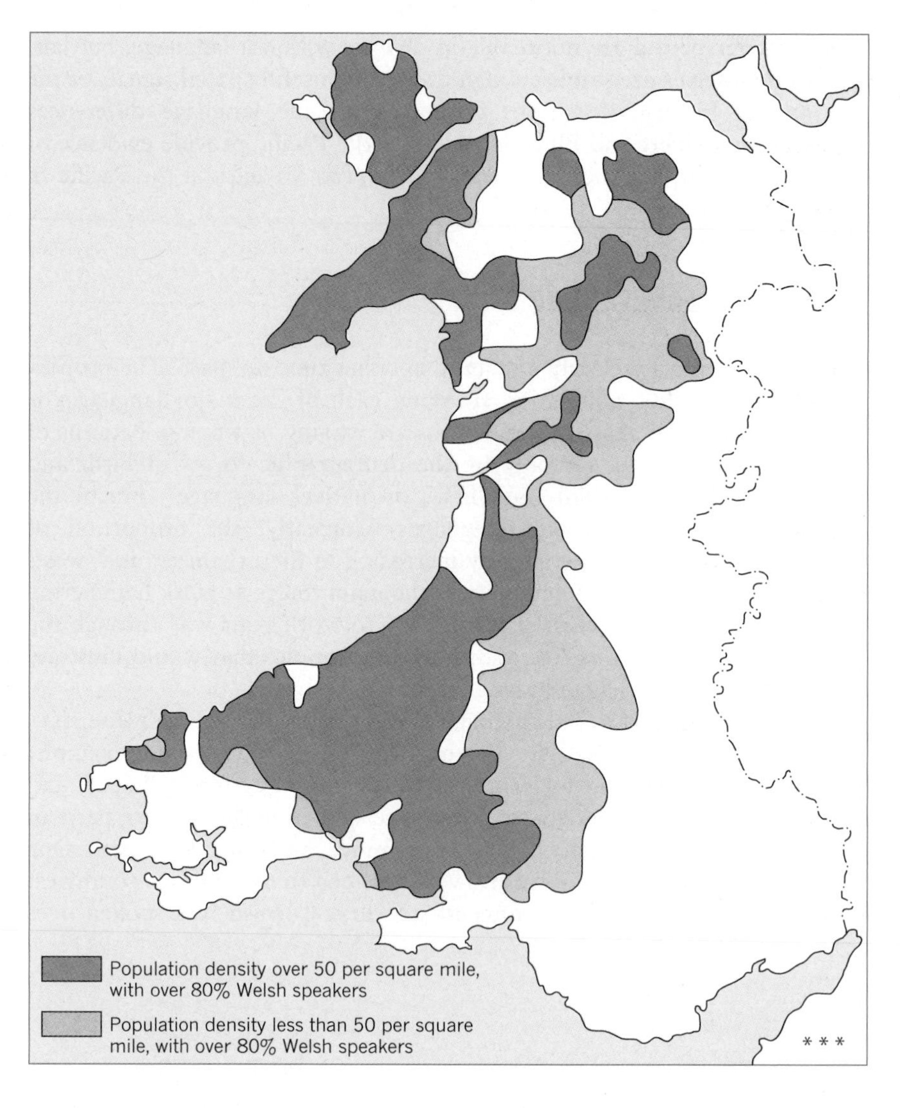

Figure 7.6 Contraction of culture areas By the mid-twentieth century the core of the Welsh language and characteristic Celtic culture had retreated westward and lay well inside its traditional political boundary with England. Vigorous steps are being taken to halt the decline of the language as part of a resurgent Welsh nationalist movement, and a separate Welsh Assembly was established in 1999. But the cultural forces bringing about the decline of local languages are rather pervasive. 'Welsh speakers' are defined as those persons with an ability to speak the language; the use of the phrase does not necessarily imply that Welsh is the primary language in an essential bilingual population. Although the main area of dominant English speakers lies along the England–Wales border and in the industrial south, there is one important exception: southwest Wales (Pembrokeshire) has had a high proportion of English speakers since it was colonized in the eleventh century. (1 square mile ≈ 2.5 square kilometres.)

[Source: E.G. Bowen, *Institute of British Geographers, Publications*, 26 (1959), Fig. 2, p. 4.]

most of the county of Cornwall; but by 1600 it was heard only in the extreme west. The mining industry brought an increasing number of English-speaking outsiders into the area, and by 1800 the language was virtually dead. The last Cornish-speaking person died in the 1930s. Even much stronger Celtic languages such as Welsh are now confined to a part of their original area (see Figure 7.6).

Only in Ireland, where Irish (also called Gaelic, or Erse), has been revived and taught in the schools as part of a programme to stimulate the national sense of identity, has the language held its ground. Of course, we cannot be sure how persistent such trends will be, Once a language becomes threatened with extinction, then there is likely to be a strong movement to preserve it by those who see languages as an essential indicator of cultural distinctiveness.

Language Differences as Spatial Barriers

One important divisive effect of language is to act as a barrier or filter to spatial interaction between regions. This effect can be clearly seen in Canada where there is a strong division between French-speaking Quebec and English-speaking Ontario. In this case the duality is almost complete and the only major zone of language overlap is in the Ottowa valley and in a few areas of 'loyalist' settlement in southwest Quebec.

Some indication of the distorting effect of the Quebec–Ontario border has been uncovered by comparing the observed and expected contact between cities there. ('Expected' contact assumes that city size and distance control the flow of information between cities: see Section 13.2 for a detailed account of how this is calculated.) In one study, contacts between Montreal and surrounding cities were measured in terms of long-distance telephone traffic. The expected interactions given by a population–distance formula are presented in Figure 7.7, along with the actual interactions. The traffic between Montreal and other cities in Quebec was from five to ten times greater than the traffic between Montreal and cities with comparable population–distance values in the neighbouring province of Ontario.

It is difficult to disentangle how much of the Quebec–Ontario difference was due to language differences and how much to the political separation into two different provinces. The strength of the provincial Quebec–Ontario barrier in blocking interaction was overshadowed by the blocking effect of the international boundary to the south. Montreal's traffic with comparable cities in the United States was down to only one-fiftieth that of its traffic with places in Quebec.

Language barriers may have a one-way flap. Many Japanese businessmen read and speak English fluently; very few American businessmen have a comparable facility in Japanese. The one-sidedness of certain interregional flows in trade may reflect, in some small part, the way in which language barriers may be crossed more easily in one direction than in another.

To sum up, language is one of the most important but paradoxical of cultural elements. It binds together certain parts of the human population but separates others. It shows rapid spatial evolution and change and yet is one of the most persistent of cultural features. Certainly, any geographer trying to understand the mosaic of world cultural regions must include language in the analysis.

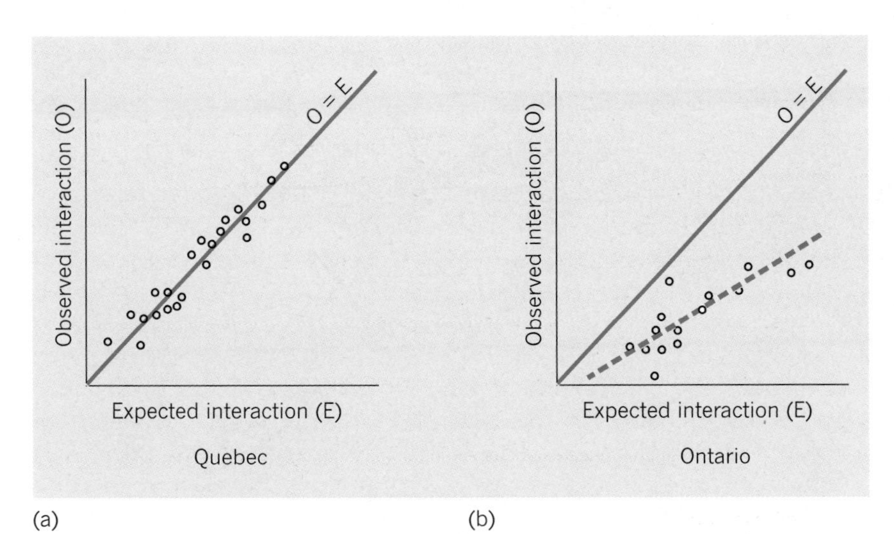

Figure 7.7 Boundaries as filters The impact of the Quebec–Ontario boundary on spatial interaction is measured by the number of telephone calls between the city of Montreal, Canada, and other centres in the two provinces. Actual calls are plotted on the vertical axis against expected calls on the horizontal axis. Expected interaction is given by a gravity model using the size of the other centres and the distance from Montreal. Note how the calls from Montreal to its neighbouring cities *within* the province of Quebec (a) are much higher than those to neighbouring cities in the adjoining province of Ontario (b).

[Source: J.R. Mackay, *Canadian Geographer*, 11 (1958), p. 5.]

7.3 A Geography of Belief

If cultural variations were simply a matter of language differences, the geographer's job would be an easier one. In fact, systems of belief cut clear across language barriers. Albanians and Chinese may hold similar shades of Marxist political belief, congregations in the Democratic Republic of the Congo (Zaire) and Connecticut may share the same Catholic creed.

Spatial Variations in the Major Religions

Each of the world's main religions has a distinctive geography. Christianity's more than one billion adherents are located largely in Europe and the Near East, the Americas, and Australasia. Islam has diffused from its birthplace in western Arabia through the northern half of Africa, central Asia, and India, and into Indonesia. Hinduism and Buddhism are highly localized, the former being largely confined to the Indian peninsula and the latter to East Asia (see Figure 7.8).

We can, of course, divide each major religion into its various subgroups. If we examine the Christian subgroups within the United States, we find a strong zonal pattern. The Roman Catholics are strongly represented in New England and the industrial northeast; the Baptists in the southern states and Texas; the Lutherans in Wisconsin, Minnesota, and the Dakotas; the Mormons in Utah. Even within a metropolitan area geographic differences in religion may occur, with Christian Protestant churches most frequent in the wealthy suburbs.

Why are these variations important in determining the cultural mosaic of the world? Religion's role in group organization, its close relationship to politics and the state, and the attitude of churches toward change and

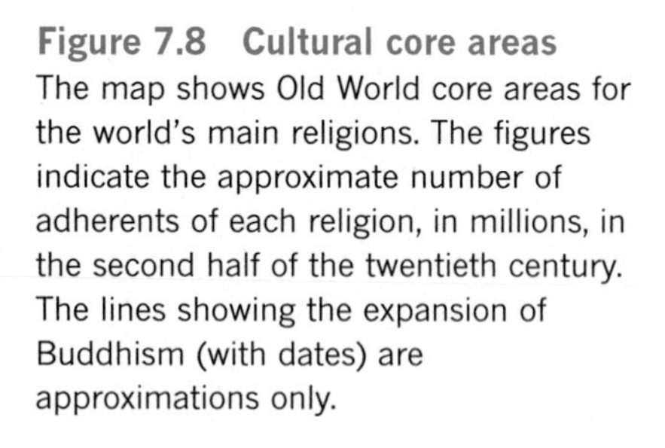

Figure 7.8 Cultural core areas
The map shows Old World core areas for the world's main religions. The figures indicate the approximate number of adherents of each religion, in millions, in the second half of the twentieth century. The lines showing the expansion of Buddhism (with dates) are approximations only.

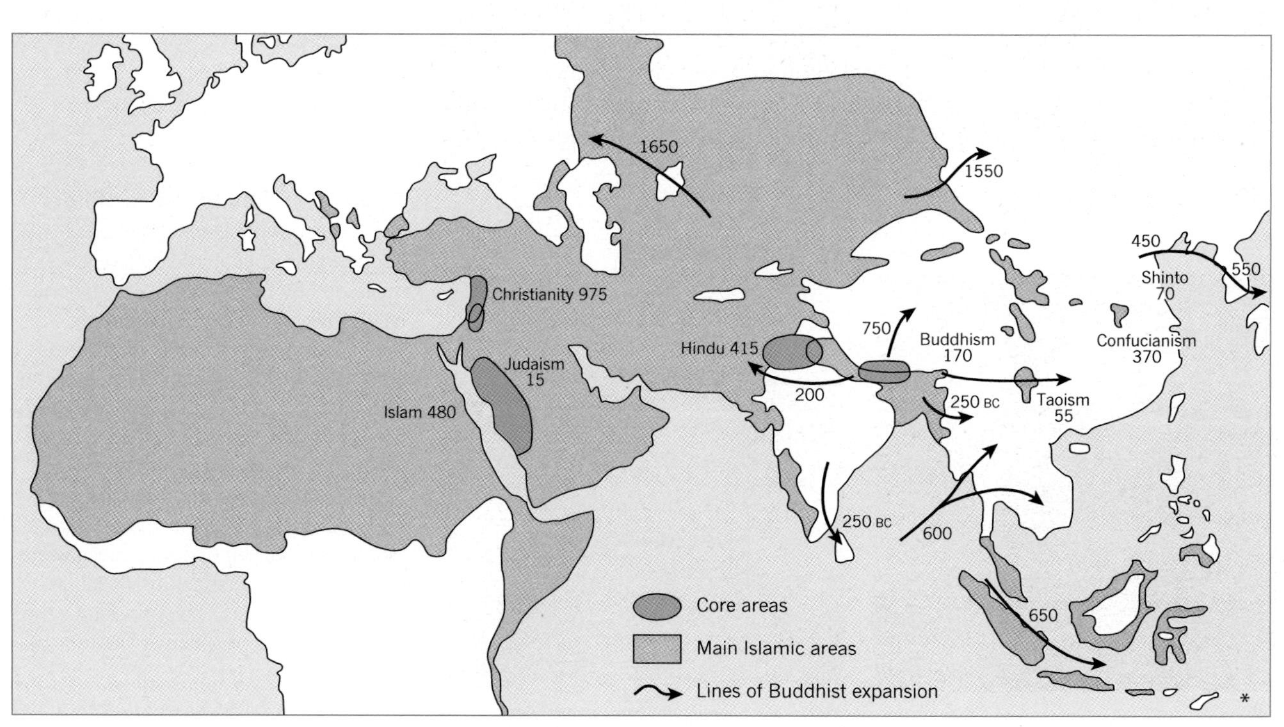

development are part of the answer. Many of the great political conflicts of world history have had a religious basis, and lines of conflict (such as those between Israel and Egypt or within Northern Ireland) may still be drawn up along religious divides. Some aspects of religion's role in determining geographic divisions will be considered in our treatment of boundary conflicts in Chapter 17.

Geographic Impact of Religious Beliefs

Although beliefs may take a bewildering variety of forms, it is religious belief that has played the central part in our cultural differentiation. Like the political beliefs discussed in Part V of this book, religion gives a system of interlinked value that enter into several aspects of human geography. One clear geographical aspect of a belief system is where it gives special value to particular locations. These range from prehistoric monuments (such as Stonehenge) to former hunting lands (the homelands of some Aborigine tribes in Australia). Figure 7.9 gives three examples where valued places have been marked by ceremonial buildings at different scales.

(a)

(b)

(c)

Figure 7.9 Sacred places Some locations on the earth's surface are given special significance in human belief systems (see Jerusalem, Figure 7.1) and marked by specific buildings. Three further examples are given here. (a) In Saudi Arabia, the religious gathering, the Hajj, at Mecca. Pilgrims gather around the Kaaba for evening prayer. Muslims believe it was built by Abraham and the mosque built around it has been enlarged to allow 750,000 people to pray together. (b) In Japan, the golden temple Kinkaku-ji originally built in 1397 (rebuilt 1955) at Kyoto. (c) In England, the small town of Wells in Somerset county has been a centre of Christian faith since the Saxon settlement. The west front of Wells Cathedral built in 1230–1260. The set of medieval statues on the west front is among the finest in western Europe. In an age before print, such decorations played a teaching role, reinforcing the sacred place as an educational centre.

[Sources: (a) © MAGNUM/Abbas. (b) © TRIP/C. Rennie. (c) Michael Holford.]

Other geographic effects may be less clear cut. We take two samples to illustrate the connection.

Belief and the Use of Agricultural Resources Religious beliefs affect agricultural development indirectly through constraints on diet and through the symbolic significance given to animal life. For although humans have been biologically designed as omnivorous animals (to judge from our teeth), a major portion of the world's population restricts its diet in some degree. The world's 300 million Buddhists are generally vegetarians, its 700 million Hindus may not eat beef, and its 20 million Jews may not eat pork. Smaller groups may have still more precise rules; India's Jain communities (with about 3 million people) are forbidden to kill or injure *any* kind of living creature.

As a result of these views, cattle throughout most of India are used only as draft animals and to some extent for milk production. Under the Hindu doctrine of *ahimsa*, the slaughter of cows is prohibited in many of the Indian states. As a result, ageing and unproductive cattle add to the pressure on grazing resources and estimates of the number of 'surplus' cows run to between one-third and one-half of the total.

An extreme view of the importance of cattle is taken by the herding tribes of eastern and southern Africa. Among the Pakot of Kenya, the number of cattle a man has is directly related to his prestige and wealth, and cattle serve as the means of exchange, most notably in the purchase of brides. Numbers rather than quality appear to be important, and this has a bad effect on the standards of livestock and the amount of grazing per animal. Although the attitude toward cattle among the herding tribes has a religious component (cattle having been said to be 'the gods with the wet nose'), the emphasis on numbers appears to have more to do with their convenience as a means of exchange.

More specifically, religious significance attaches to the Moslem view of the pig as an unclean animal. Thus in Malaya pigs are reared for food only in the Chinese enclaves; the native Malay population follows the Islamic code and deprives itself of an important source of food.

Belief and Modernization Religions of most kinds lay heavy emphasis on continuity, tradition, and strict adherence to long-established patterns of behaviour. They have acted, and continue to act, as a vital stabilizing influence, or – depending on one's point of view – an inhibiting drag on change.

Religion is often held to be a major factor inhibiting the spread of family-planning practices. The moral values attached to the human foetus in Roman Catholic doctrine serve as a major barrier to the spread of abortion and certain contraceptive methods. This barrier may operate at the individual and family level for members of the Catholic faith or may become a matter of national policy in countries where there is a strong link between the Catholic Church and the state. Thus contraceptive devices are banned in Ireland, and different attitudes are taken on abortion laws in the various states of the United States.

It is difficult to determine the importance of these attitudes from a strictly demographic viewpoint. Population-control practices are clearly described in the Old Testament and in Egyptian wall paintings dating from 5000 BC. There is ample evidence that human groups throughout history have been able to control family size when this was considered desirable. Attitudes towards what is the most desirable family size are demonstrably more important than which birth-control method is followed. Thus in Europe, a continent with lower birth rates than any area of comparable size in the world (generally about eight per thousand), there is no major difference between Catholic and non-Catholic populations at the national level. Countries in which contraceptives and birth-control information are banned or restricted have birth rates just as low as those where both are freely available.

In other aspects of human behaviour, the influence of religious beliefs on the acceptance of innovations is clearer. Let us take a specific regional example. About 20,000 of the world's 370,000 Mennonites follow the Amish religious code. The Mennonites emerged in the early sixteenth century in Switzerland as a non-conformist branch of the Protestant church, with half their numbers in North America today. The Amish represent an extremist breakaway from that Mennonite movement and are today concentrated in farm communities in certain counties of Pennsylvania and Indiana. Located in the midst of the most highly modernized and swiftly changing regions of the world, Amish communities stand out as islands of tradition. Services are conducted in Pennsylvania Dutch (Palatinate German with some English mixed in), traditional plain clothes continue to be worn, telephones and electric lights are shunned, and the horse and buggy continue to serve as a means of transport instead of the all-pervasive automobile. Here, religious beliefs serve as a cement that continues to hold together a human group with a behaviour pattern more reminiscent of seventeenth-century rural Europe than contemporary North America.

Belief and World Views

At the most generalized level, systems of belief condition our view of the world – and our place in it. We want to order our experiences of the world, and build a picture of the world around ourselves.

Consider Figure 7.10. This shows the world views of two American Indian cultures and of the Chinese. In each case the centre of the world is the human being. Around this is a four-sided framework linked to the four cardinal directions – north, south, east, and west. Note that in all three cases direction is associated with a particular colour. The Pueblo Indians add an animal and the Chinese have a most complex set of associated ideas. Thus the west is associated not only with a colour (white) and an animal (the tiger) but is the direction of metallic autumn, symbolic of both harvest and war. This is also the direction of twilight, of memory and regret, of past mistakes which cannot now be altered.

Ideas of the world have played a significant part in the design of individual houses, settlements, and cities. Many were designed as small-scale replicas of the world picture. Similar ideas permeated Greek civilization while the Christians were later to build their churches with distinct east–west orientation. In China today the world view remains important in building construction. For example, when in the 1960s the new Chinese University of Hong Kong came to lay out its buildings – including a geography department – at Sha Tin, an unusual locational factor had to be considered. Sites were chosen in which *feng shui* ('local currents of the cosmic breath') were harmonious. Chinese landscapes continue to have paths, structures, and woodlots designed to blend with the natural landscape rather than dominate it. This response to the environment is not uncommon. For most of the period of human occupation of the earth, most of its peoples believed that

Figure 7.10 Cultural variations in the world view World views of (a) Pueblo Indians of North America, (b) Maya of Central America (AD 600–900), and (c) traditional Chinese. Note the similarities in that spatial structure is oriented to the four cardinal directions – north, south, east, and west. The associations with each direction vary.

[Source: Yi-Fu Tuan, *Space and Place* (University of Minnesota Press, Minneapolis, 1977), Fig. 9, p. 94, and Joyce Marcus, *Science* **180** (1973), Fig. 2. Reprinted with permission of the American Association for the Advancement of Science. Copyright © 1973 by the American Association for the Advancement of Science.]

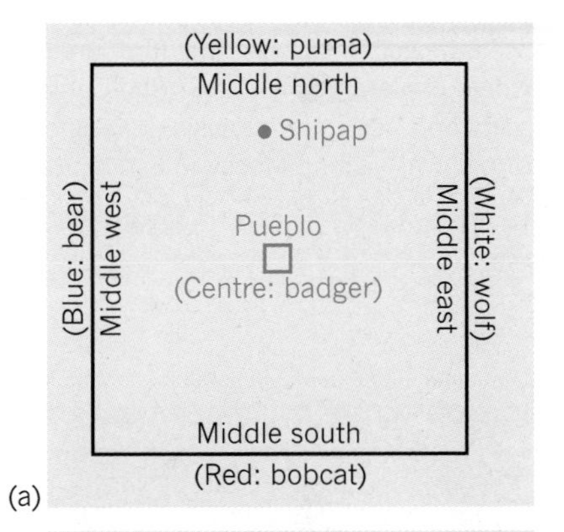

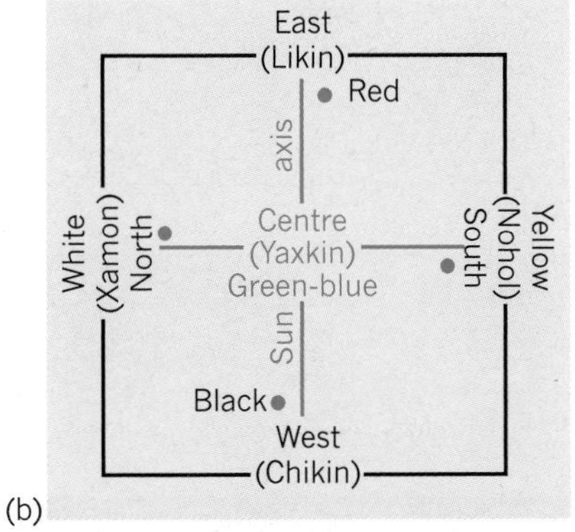

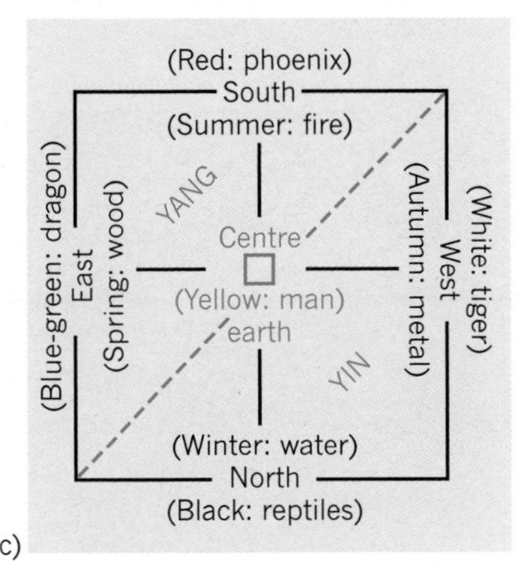

individual natural elements – trees, springs, and hills – had guardian spirits. Before such objects could be used, these spirits had to be raised and mollified.

Impact of Religious Views on Disease Spread

Religious views permeate so many aspects of our life that it may not be surprising to find that this may extend so far as to shape the spatial pattern of disease spread. Two examples of two different diseases at two different geographical scales must suffice.

Figure 7.11 Impact of belief systems on disease spread, I Five of the six cholera pandemics of the nineteenth century were routed via Mecca. The spread reflects the large-scale movement of pilgrims from areas where cholera was endemic (the Ganges delta and parts of Indonesia). See photo of pilgrims visiting Mecca in Figure 7.9.

[Source: A.D. Cliff and P. Haggett, *Atlas of Disease Distributions* (Blackwell, Oxford, 1998), Fig. 1.1(D), p. 5.]

Cholera in Southeast Asia Cholera is one of the classic epidemic diseases of the nineteenth century. It is caused by a bacterium (*Cholera vibrio*) which enters the human gut from infected water and causes acute and life-threatening dehydration. Its traditional hearth was the delta region of the River Ganges and Brahmaputra, located in present-day Bangladesh. For reasons that are still not fully understood, six times during the nineteenth century it erupted from the Indian subcontinent and spread rapidly over both the Old and New Worlds causing great mortality in the cities of Europe and North America. Maps of the five pandemics are shown in Figure 7.11.

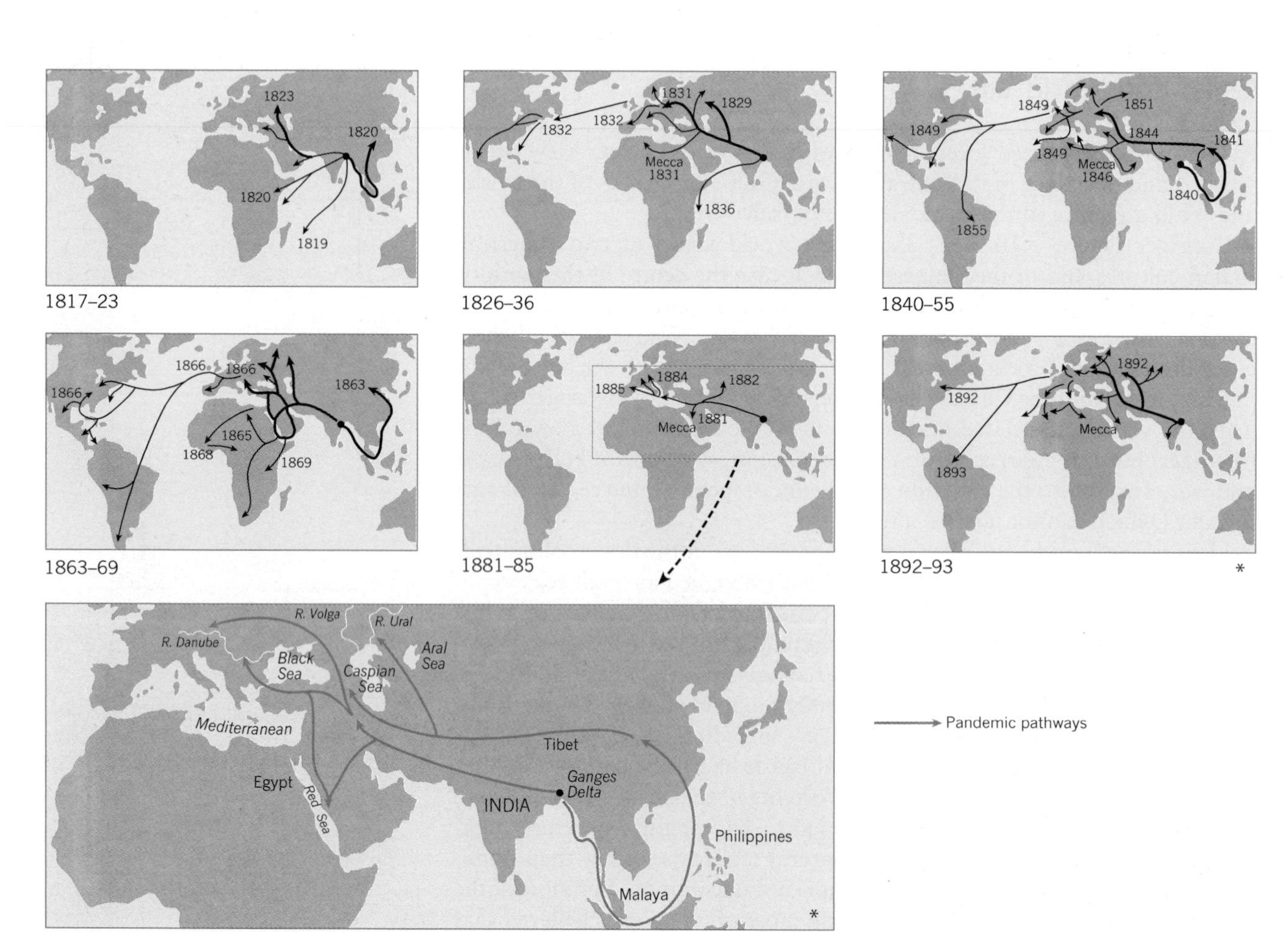

What factors controlled the lines of pandemic spread are not known but improving transport probably played a critical part. Significant in terms of the geography of belief is the travel of Moslem pilgrims, leaving northern India (today, Pakistan and Bangladesh) and the East Indies (today, Indonesia) to attend the Hajj at Mecca in western Arabia. In five of the maps, the vectors of spread from the Indian hearth include that Arabian town.

Measles in the Amish Community As we noted above, the Amish are part of the Mennonite communities. They trace their religious heritage back to sixteenth century Switzerland and they first moved to the United States in 1709 to escape persecution. Today they are mainly a farming community widely scattered over the eastern United States but with a heavy concentration in rural Pennsylvania. Their religious convictions lead to a simple and separate lifestyle using horses and mules rather than tractors (see Figure 7.12). Their distrust of vaccination leaves them with low immunity levels against many of the more common infectious diseases.

As vaccination has rolled back the general tide of childhood infections in the United States (see Section 20.2) so the exposure of the Amish to infec-

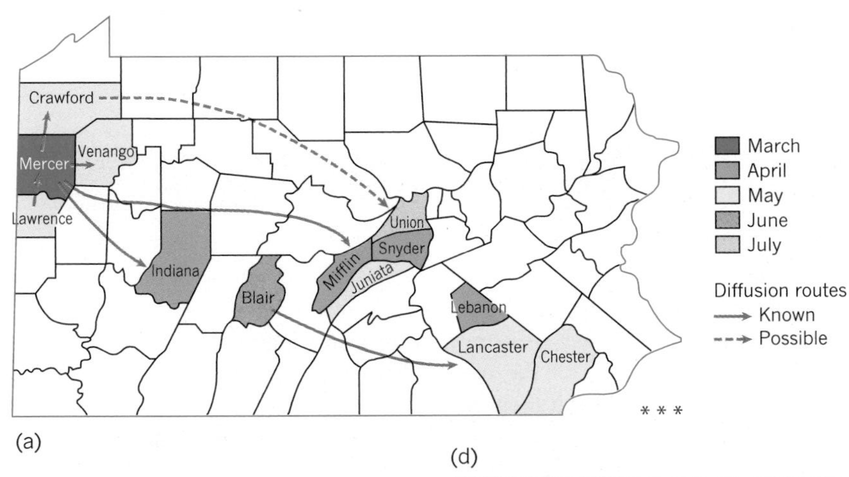

(a)

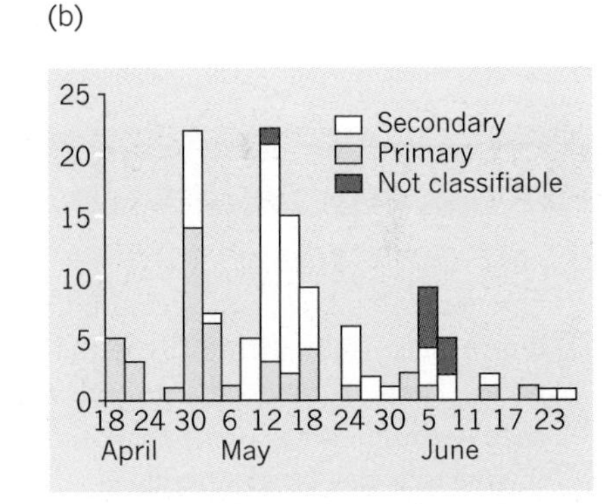

(b)

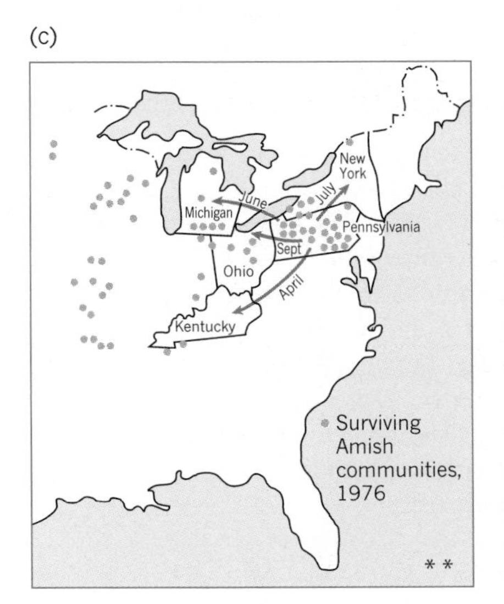

(c)

(d)

Figure 7.12 Impact of belief systems on disease spread, II (a) Outbreak of a highly contagious viral disease (measles) among children of the Pennsylvania Amish community in 1987–88. Dates of rash onset of the first Amish case in each county; vectors identify diffusion routes. (b) By exposure status, date of rash onset among Amish patients in Lebanon County. (c) Spread to Amish communities in other parts of the eastern United States. (d) Amish farmers ploughing with a team of six mules, Gordonsville, Pennsylvania, USA.

[Sources: Data from R.W. Sutter, S.E. Markowitz, J.M. Bennetch *et al.*, 1991, mapped in A.D. Cliff, P. Haggett, and M. Smallman-Raynor, *Measles: An Historical Geography of a Major Human Viral Disease, 1840–1990* (Blackwell, Oxford, 1993), Fig. 9.15, p. 240. (c) TRIP/J. Greenberg.]

tions has become more apparent. Figure 7.12 shows how a major measles outbreak in Pennsylvania schoolchildren in 1987–88 was channelled through Amish communities both within and outside that state. After a few cases in Lancaster county in December the epidemic began in earnest in March in Mercer county in western Pennsylvania. Its subsequent spread though Amish communities in the rest of the state and to other Amish communities in the eastern states is shown by the vectors. The high attack rates and very specific spread vectors show how religious isolation with its social exclusion from nearby non-Amish communities channelled the disease geographically.

7.4 Gender and Cultural Geography

Unlike sex (the biological distinction between males and females), gender refers to culturally learned differences that are attached to masculinity and femininity. Whereas biological sex differences are determined by genetics and shown in anatomical characteristics, *gender* is an acquired identity. Sexual differences have remained broadly constant over time and space. In contrast, gender differences have varied hugely over historical time and continue to vary today from one part of the world to another. (See Box 7.B on '*Feminist geographies*'.)

Box 7.B Feminist Geographies

One branch of cultural geography that has grown in importance since the 1970s is that of *feminist geographies*. These draw on the growth of feminist social science to illuminate geographical questions. Three strands of research can be identified.

First, there is critical work that starts from the facts of the oppression of women in many societies and identifies its specific geographical expressions. One example is the study of the microgeography of women in western society where their traditional roles as carers (of both young children and the elderly) and as 'housewives' have imposed restrictions on their spatial access to paid employment. Another is cross-cultural studies of gender roles in different non-western societies.

A second strand is the study of sexism within geographical institutions. Apart from teaching at the primary-school level, geography remains at the start of the twenty-first century a strongly male-dominated profession. Although there are women explorers of distinction (e.g. Mary Kingsley in West Africa or Freya Stark in Southwest Asia) and leading geographic scholars (e.g. Ellen Churchill Semple or Doreen Massey), they have been numerically much smaller than their male equivalents. This gross under-use of half of the human family is slowly changing. For example, the proportion of women doctorates in human geography in Sweden for most of the twentieth century was 9 per cent; in the 1990s it grew to over 20 per cent.

A third strand is the role of women in changing the focus of human geography. Women geographers have led the movement away from the adoption in human geography of a detached and universal scientific tradition towards one where geographic knowledge is seen as context-bound and partial.

A coming of age of feminist geography is indicated by the founding of specialist publications in the last two decades. The journals *Gender, Place, and Culture* which began in 1994 and the *International Studies of Women and Place* (a monograph series) illustrate this trend. The volumes by S. Hanson and G. Pratt, *Women, Work, and Space* (Routledge, London, 1995) and J. Momsen and V. Kinnaird (eds), *Different Places: Different Voices: Gender and Development in Africa, Asia, and Latin America* (Routledge, London, 1993) show the potential of this approach.

Gender and Work Differences

The United Nation's Report on *The World's Women* estimates that although the world's women compose half of the global population, they perform two-thirds of the world's work hours. This striking calculation takes into account the dual or triple labour that women in many societies typically perform in both the home and the workplace. For example, many women in Third World countries work in three capacities: (1) work within the home related to feeding the family, cleaning the house, tending to dependants (both young and old); (2) working for wages in the formal and informal economy; and (3) growing subsistence crops to feed their families.

Gender differences also come out in the various divisions of the labour force. In the United States today some professions are dominated by women: e.g. registered nurses (94 per cent women), secretaries (98 per cent women), and high-school and elementary school teachers (70 per cent women). In other professions, women are poorly represented: e.g. physicians (18 per cent women) and lawyers (20 per cent women). In higher education, the differences in gender are also marked in colleges and universities where women are much underrepresented especially at the highest levels. In the United Kingdom less than one-tenth of all professorships in geography are held by women.

Patterns of employment show some variations by region. Figure 7.13 compares the proportion of women undertaking different types of work in the United States and sub-Saharan Africa.

In broad terms, the economically developed a country is, the greater the participation of women is in the paid workforce. But that general relationship is also affected by other cultural considerations. Islam has a view of women's place in society which is restrictive outside the home and family so that even very rich countries there may display Third World patterns of female participation.

Other Gender Inequalities

It is difficult to make a reasoned estimate of the global pattern of gender inequalities. But over the last twenty years, a number of United Nations agencies have tried to document the situation and to collate national figures into worldwide projections. On the basis of these elements, a very unequal balance existed at the end of the twentieth century.

Five examples must serve to illustrate the problem.

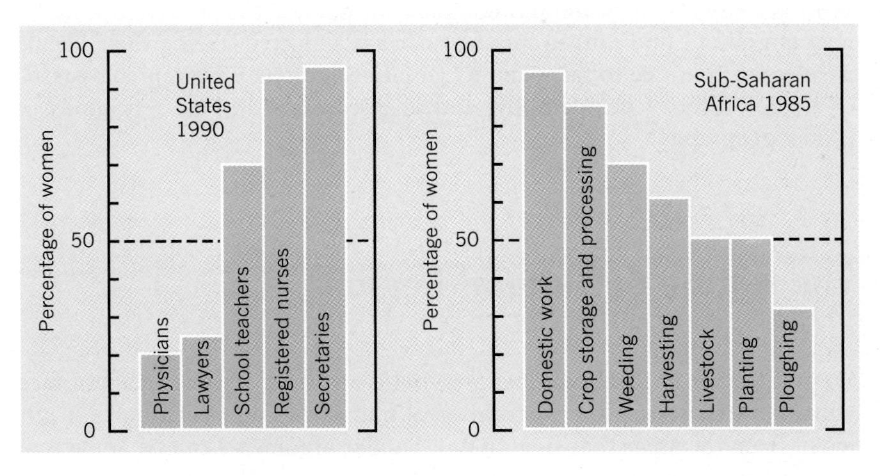

Figure 7.13 Regional variations in female occupations Data on the proportion of women in different types of work in two contrasting regions.

[Source: Data from United Nations surveys. Redrawn from V.S. Peterson and A.S. Runyan, *Global Gender Issues* (Westview Press, Boulder, 1993), Fig. 3.10, p. 74.]

1 Four out of five of the world's refugees are women, along with their dependent children.
2 Two-thirds of the world's illiterates are women.
3 Only 3 per cent of the world's property is owned by women.
4 One-tenth of the world's income is earned by women.
5 Despite the striking contributions of an Indira Gandhi (India) or a Golda Meier (Israel), less than 3 per cent of the world's heads of states over the last half century have been women.

Such differences are deeply rooted in past as well as present societies. Archaeologists can trace such differences in the nature of stone-age wall painting or bronze-age grave goods. Anthropologists record the systems of kinship and marriage (including multiple relationships such as polygamy and polyandry) and inheritance as well as in division of labour. The complex differences in legal systems affect the ways land is owned and divided.

Gender and Time Geography

A third area of geographic concern with gender studies relates to its overlap with time geography. The male dominance in such areas as income distribution or in inheritance is also extended to the control of time as a resource. Figure 7.14 shows the working pattern of a primary school teacher in Sweden over the course of a typical day, coping with the demands of both the classroom and the children at home. A group of geographers at Lund in Sweden led by Tora Friberg and Ann-Cathrine Åquist have been responsible for sharpening up information in this area.

Control of time is reflected in terms of control of space. In western urbanized society, the location of a family home is often determined by the site that is most appropriate to the workplace of the main earner in a partnership. While historically this has tended to be the male partner, this pattern is now changing. The need for joint decisions on location for two partners is one of the factors in sustaining metropolitan growth with migration towards centres where both members may have better job-search opportunities. Deindustrialization and the collapse of male employment in some old industrial areas has undermined the traditional relationship. Questions such as access to flexible transport (normally the family car) has a critical effect on the spatial range within which jobs can be held.

In developing countries, the relationships between gender and employment are equally intricate and complex. In many African societies, males may migrate to find jobs in mining or heavy industry leaving their female partners to continue to farm and to raise the children. Geographic patterns of employment, of demography, and of disease are all here reflections of gender geography.

7.5 The Nature of Cultural Regions

We said earlier in this book that a *region* is any tract of the earth's surface with characteristics, either natural or of human origin, which make it different from the areas that surround it. Similar arguments can be applied to

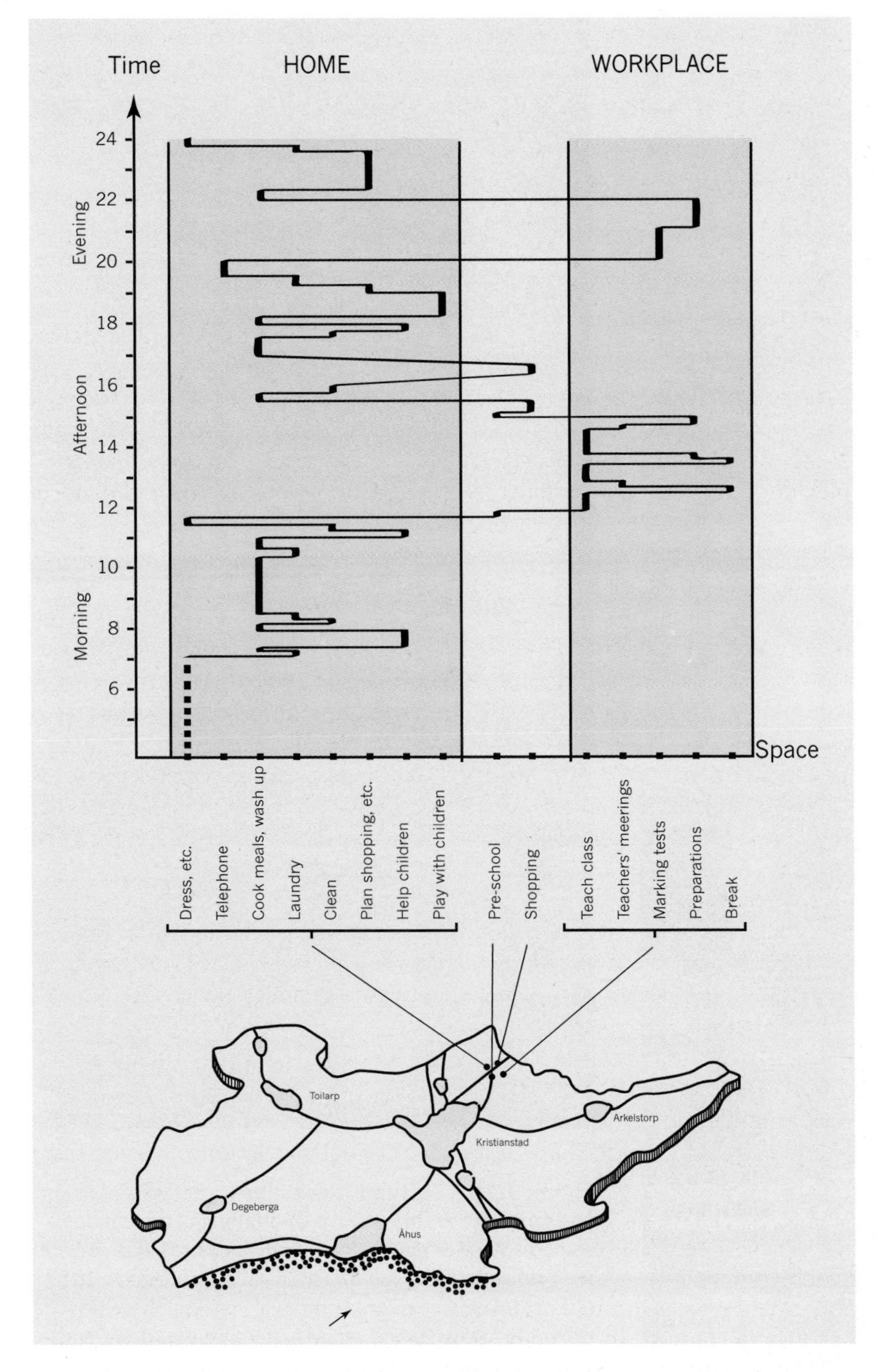

Figure 7.14 Gender in space–time geography Swedish contributions to time–space geography (recall Chapter 1, Box 1.B) have also thrown light on gender differences in travel opportunities. A day in the life of a female primary-school teacher in a village near Kristianstad, Sweden, showing how home duties and job opportunities have to be carefully dovetailed in time and space. Note that the time of day is measured on the vertical axis and location on the horizontal axis. Teaching is confined to the morning, shopping and collecting a child from preschool in the afternoon, with marking and school preparation in the evening.

[Source: T. Friberg, *Everyday Life: Women's Adaptive Strategies in Time and Space* (Lund University Studies in Geography, Lund, 1993), Fig. 9.2, p. 157.]

cultural regions. If we take the individual cultural elements such as language and religious belief and add to them ethnic differences, then we have some of the ingredients for a system of cultural regions. Various proposals for systems of cultural realms have been put forward.

Landscape and '*Pays*'

Geographers show special interest in the visible impact of mentifacts, sociofacts, and artifacts on the earth's surface. As we shall see in Chapter 11,

(a)

(b)

(c)

Figure 7.15 Cultural landscapes Patterns of fields and farms reflect the manner and timing of agricultural settlement. (a) Irregular shapes and sizes of fields in a long-settled area of southwest England. (b) Regular 16 ha (40-acre) fields in an area on the Illinois border settled when the township and range system was used for land surveying (see Figure 21.11). (c) A pioneer Canadian settlement in the Lake St. John lowland with farm boundaries perpendicular to a road.

[Source: Photos courtesy of (a) the United Kingdom Department of the Environment, (b) US Department of Agriculture and (c) the Canadian Department of Energy, Mines, and Resources.]

large areas of the world's landscape have been shaped and patterned by human activity. In rural areas, special attention has been paid to the patterns of fields and farms, roads and boundaries. Different cultural groups have had different ways of settling the land and different ways of fixing boundaries, so that strong contrasts in the *cultural landscape* are observable from the air. Figure 7.15 gives typical examples of such contrasts.

The French word *pays* describes one sort of culture region, with names such as Brie and Medoc familiar to lovers of cheese and wine. It was first used by French geographers in the nineteenth century, notably Paul Vidal de la Blache (Figure 24.4) the Sorbonne professor, to describe small distinctive regions within that country. The distinction was due not just to some striking physical difference or cultural difference, but to a blending of the two in terms of the landscapes of the area.

Other geographers have taken individual elements in the cultural landscape and traced their origin and spread. Evidence on the spread of covered bridges in America has been carefully assembled by Louisiana geographer Fred Kniffen. Originally developed in Switzerland, Scandinavia, and northern Italy, this type of bridge was first used in the United States in an area running from southern New England to eastern Pennsylvania. As the date-lines in Figure 7.16 show, it spread rapidly through the mid-west and into the southern Piedmont, but only very slowly into northern New England. About 1850, when the number of bridges was expanding most rapidly in the East, new centres of growth emerged in the western United States – notably in Oregon's Willamette Valley. Thus do artifacts, such as bridges and barns, fields and fences, house types and street patterns, provide clues to the limits of cultural regions.

(a)

1880 1850 1900 1910 1870 1847 1880 1890 1875 1911-12 1850 1840 1830 1850 1840 1830 1820 1810 1810 1820 1830 1840 1850

---- Lines of diffusion

Areas of exceptional density

(b)

(c)

Figure 7.16 Elements in the cultural landscape (a) A typical example of a covered road bridge, Vermont, New England. (b) Localities with covered road bridges in the United States, with timelines for the bridge's spatial diffusion. (c) Ukrainian church in Alberta, Canada, illustrating the tendency for immigrant groups to bring with them their distinctive architectural styles.

[Sources: (a) [Photo by Corbis-Bettmann. (b) Map from F. Kniffen, *Geographical Review*, 41 (1951). Copyright by the American Geographical Society of New York. (c) Topham Picturepoint.]

In other cases the cultural landscape is littered with artifacts that show the heritage of those who settled it. The distinctive Ukrainian style of church architecture (Figure 7.16(c)) has been transported to areas of Ukrainian settlement in Canada.

Regional Images

Another approach to the cultural region is through its image. How do we view the different parts of our own countries? What is our *cognitive map* (or mental map) of such regions as northern New England or the bush country of Australia? Geographers Peter Gould and Rodney White already have attempted to establish the mental maps of the different parts of Britain. Cross-sections of young people (aged 16 to 18 years) from twenty schools in Britain were used to determine views of the country. All the individuals were asked to rank the counties of England, Wales, and Scotland in terms of their desirability as places to live and work. Statistical analysis of the results for each school yielded surface maps of the areas' perceived desirability (Figure 7.17).

Although the maps for each school differed, they had several common elements. First, each map gave a high rating to areas immediately adjacent to the school. This attachment to the home county was present in all the groups studied. Second, the south of Britain was perceived as being gener-

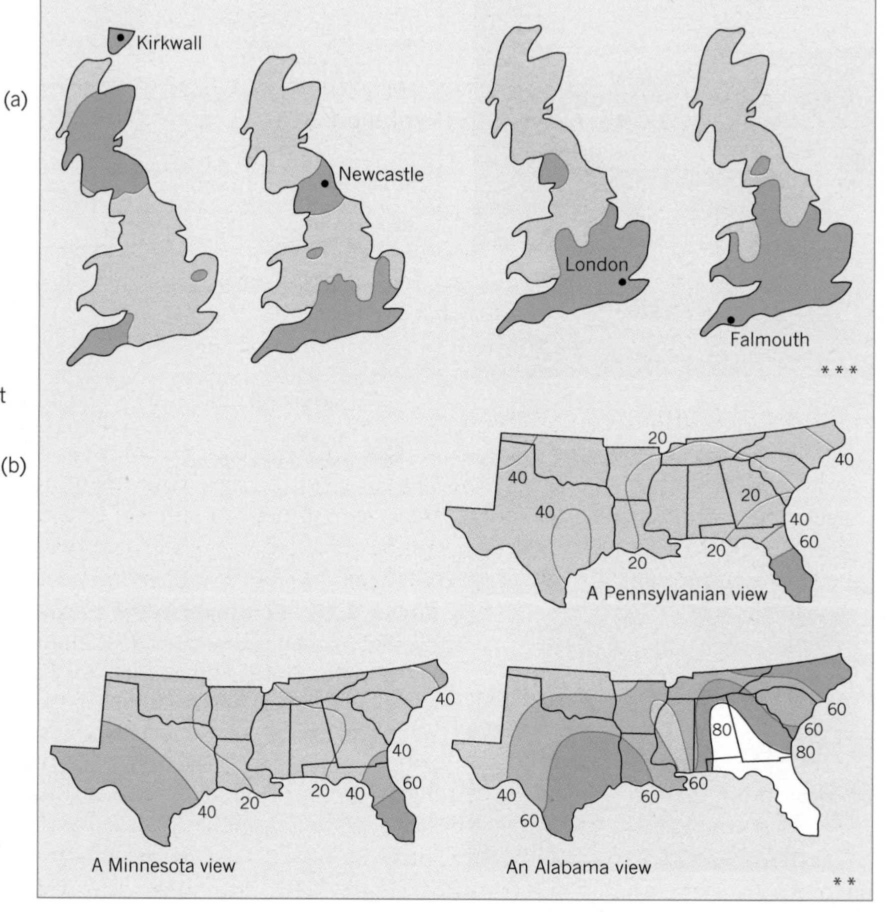

Figure 7.17 Regional images (a) The maps show contrasts in the images of British regions held by senior pupils at four schools in different locations. The 'most desirable' areas of the country according to pupils at each school are shaded. The shading reflects both an affection for the local area surrounding each school and a generally higher ranking of the southern part of the island. (b) Similar maps in which university students on different campuses were asked to mark the states within the United States. Only views of the South and Texas are shown.

[Sources: (a) R.R. White, Pennsylvania State University, unpublished master's thesis, 1967. (b) P. Gould and R.R. White, *Mental Maps* (Penguin, Harmondsworth, 1974), Fig. 4.2.3, pp. 98–99.]

ally more desirable than Scotland and the north, especially by students at southern schools such as Falmouth. Third, London was seen as either very attractive or very unattractive. Other maps indicated a strong preference for well-known vacation areas such as the southwest peninsula.

The notion of regional perception has also been extended to the United States. Let us take Kevin Cox's measure of a group of Ohio students' perceptions of regional differences within the United States. A list of the 48 states of the conterminous United States was given to each student in a class. Each state was taken in turn as an anchor state, and class members were asked to underline the three states considered to be the 'most similar' to it. No maps were available and definitions of similarity were left to each individual to resolve.

From this rapid and impressionistic clustering of states, the major regional differences within the country could be separated out. In order of priority these were: first, a split along the Mississippi separating the 'West' typified by the state of Colorado from the 'East' typified by Connecticut. Second in importance was a distinction between the 'South' (typified by Georgia) and 'New England' (typified by Maine). The distinction was less strong than the east–west split and values merged from north to south rather than breaking sharply at the Mason–Dixon line. Florida was *not* included as part of the South. While the first two distinctions were bipolar, separating two regions of approximately equal size, the third distinction is unipolar. This separates a Great Lakes division (typified by Michigan) from the rest of the United States. A fourth but less significant regional boundary separates a mid-continental area from the rest of the United States.

By repeating the tests of Ohio students on other groups we can see how regional concepts are related to an individual's view of the world. At what age do regional concepts clarify? How stable are they over time? Is there any difference between the male and the female view of the world? We may also compare regional concepts as seen from different locations to test if we each make our distinctions and splits along the same lines. Images and stereotypes, like those of the Old South or the Australian outback, may persist long after the reality has changed. These ideas affect our decisions on migration ('New Zealand is a good country in which to raise children') or investment ('Paraguay is a risky place in which to invest capital'), and so shape the world around us in ways that are persistent and persuasive.

Dynamic Cultural Regions

Although the definition of a culture region is not an easy task, there are cases where the structure is relatively clear-cut. American cultural geographer Donald Meinig has analyzed a number of cases where the area occupied by a culture has grown sequentially from the locality of its origin or hearth. In the *Meinig model* the culture region can be separated into an irregular series of concentric regional shells. (See Box 7.C on 'Donald Meinig and cultural geographies'.)

Core The central area of the Mormon culture region is where the Mormon population is densest, the religion is most dominant, and the history of occupation by the group the longest. This *core* is formed by the Wasatch Oasis, a 210 km (130 mile) strip along the base of the Wasatch Mountains east of Salt Lake City (see Figure 7.18). The area is growing

Box 7.C Donald Meinig and Cultural Geographies

Donald W. Meinig (born Palouse, Washington, United States, 1924) is America's most distinguished historical geographer. A Georgetown graduate, he took his masters and doctorate at the University of Washington. Through the 1950s, Meinig taught at the University of Utah before moving to Syracuse University where he now holds the Maxwell Research Professorship in Geography.

Each of Meinig's early homes eventually led to significant monographs. His time in the northwestern United States was marked by *The Great Columbia Plain* (1968) and his time in the southwest by *Imperial Texas* (1969) and *Southwest: Three Peoples in Geographical Change 1600–1970* (1971). A still earlier research visit to Australia saw a seminal study of the Australian wheat frontier in *On the Margins of the Good Earth* (1962). A characteristic of all Meinig's work is the vivid sense of dynamic historical movements placed in a geographic setting, backed up by the imaginative use of original maps as well as contemporary scenes, maps, and plans.

Meinig's most ambitious scholarly undertaking is a four-volume historical geography entitled *The Shaping of America: A Geographical Perspective on 500 Years of History* being published by Yale University Press. This presents an entirely fresh interpretation of the development of America from its beginnings as a group of precarious footholds on the continent to its emergence in our own time as a global power. The geographer's perspective which Meinig brings to this sweeping topic both complements and challenges conventional histories. Volume I, *Atlantic America* (1986) covers the period 1492 to 1800; Volume II, *Continental America* (1992) covers the period 1800 to 1915; Volume III, *Global America* covers the period 1915 to 1990.

rapidly in population, and for the last half-century has contained about 40 per cent of the total Mormon population of the United States. It is a principal centre of gentile immigration into the essentially Mormon region, but the Mormon–gentile ratio nevertheless has tended to remain rather constant. The links between the core area and the rest of the Mormon region are principally those of information, finance, and organization. The cultural links are symbolized by twice-yearly meetings when, every April and October, representatives from all Mormon communities meet in Salt Lake City to decide church policy.

Domain Areas where the Mormon culture is dominant but with less intensity than in the core, and where local differences in social organization are evident, Meinig terms the Mormon *domain*. The domain extends outside Utah, notably into the river country of southeast Idaho and contains an area more than twenty times the size of the core. It contains a little over a quarter of the total Mormon population, largely in rural settlements, and has a small proportion of gentiles. As Figure 7.18(b) shows, the spatial pattern of colonization was affected by the north–south trend of the mountains and proceeded in a series of 'tiers' from west to east. For the first twenty years after the foundation, settlements were confined to the first tier, but by 1870 (only twelve years later), the second had been occupied and the third was being entered.

(a)

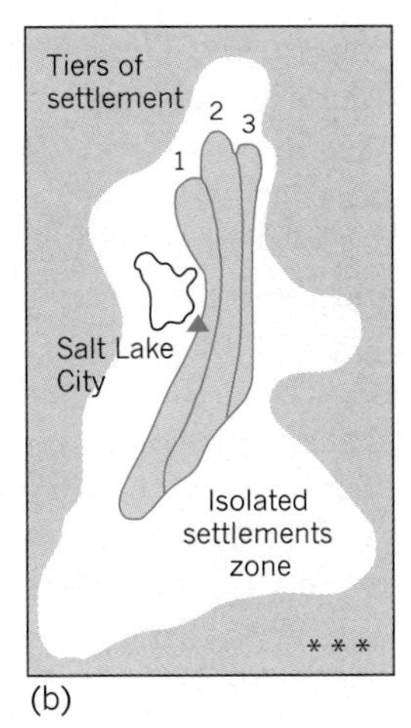

(b)

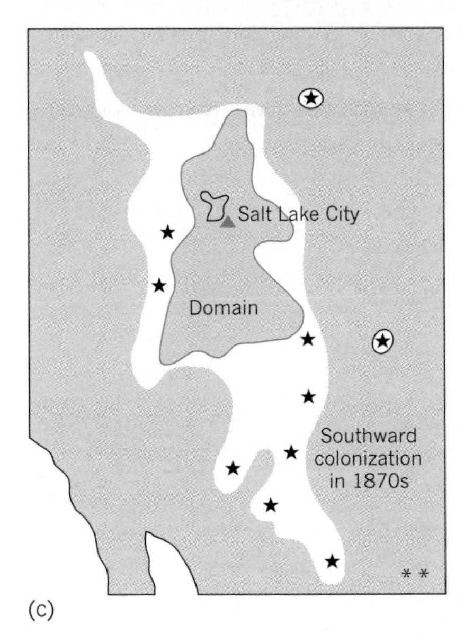

(c)

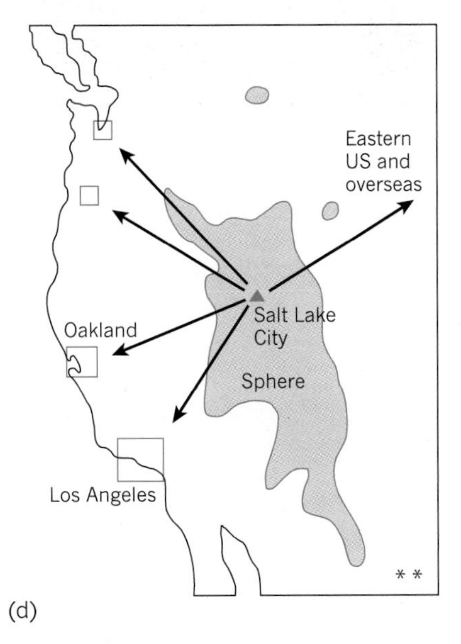

(d)

Figure 7.18 Structure of a cultural region Donald Meinig's study of the Mormon culture region in the western United States forms a fourfold framework. This extends from the *core* area based on Salt Lake City with the Wasatch Mountains behind (a) through the *domain* (b) largely confined to the present state of Utah. Beyond the domain lies the *sphere* (c) of outer colonies, and beyond that the highly dispersed *outliers* (d) of the Mormon church in the rest of the United States and Western Europe.

[Source: D.W. Meinig, *Annals of the Association of American Geographers* **55** (1965), Fig. 7, p. 214. (a) Photo by Corbis-Bettmann-UPI.]

Sphere Outer zones of influence and peripheral contact, where the Mormons form significant local minorities, are termed the ***sphere***. The Mormon sphere forms a fringe all the way from eastern Oregon to northern Mexico, greatly extended in the south and representing the last wave of rural Mormon expansion in the late nineteenth century. About 13 per cent of the Mormon population lives in the sphere, where they form a varying minority proportion of the local population. As Figure 7.18(c) shows, the sphere is made up of discontinuous pockets of settlement, rather than a continuous low-density fringe. Each pocket has its distinctive history. For example, the San Juan colony in southeast Utah was established in 1879 after a famous trek across very inhospitable canyon-intersected country. The 'girdle of wastelands' that surrounded the domain ensured that these pockets would be separate from the major area of colonization.

Outliers Outside the sphere are small outliers (outlying areas) containing the remaining fraction of the Mormon population. The key outliers are in Pacific coast cities, notably Los Angeles. In the last two decades the outliers have been extended outside North America to England, Switzerland, and New Zealand. The outliers are different in several ways from the other three zones. They are urban rather than rural, and the links to the core are long-distance and intercity rather than short-distance and rural-to-city. As Meinig puts it, the outlier is not a further expansion of Zion but a dispersion into Babylon.

The distinction between the first three zones (core, domain, and sphere) and the fourth (outliers) is of general significance to cultural geographers. In terms of spread, the contiguous spread of the nineteenth century has given way to the hierarchic spread of the twentieth. This 'bursting' of the actively growing culture may be seen as something like the process of *metastasis* in biology. A well-known example is that of cancerous cell growth. In metastasis, the cancerous cells begin by local growth but then break away from this original lesion and are carried in the blood or lymph systems to distant parts of the body, where they set up new lesions. Certain culture elements can also get caught up in the powerful circulation movements that go on between large cities and be rapidly transferred to distant parts of the globe. We shall return to this topic in Chapter 16.

Meinig's model clarifies the spatial structure of one of the most distinctive regions to emerge on the subnational level within the United States over the last century and a quarter. In the Mormon cultural area the theological base affects significant aspects of the demography, economic organization, and political viewpoints of the population in the southwest. Although Meinig has used his method only to analyze other areas of Western culture – notably Texas – it has obvious relevance to other non-western cultures as well.

Reflections

1 The human population, although biologically uniform, is broken up by culture into distinct groups at several spatial levels. Consider whether your own county, state, or province can be divided into distinct cultural regions. What bases are there for such divisions? Sketch in some tentative boundaries and compare them with those proposed by other members of the class.

2 The concept of culture as used by geographers describes patterns of learned behaviour that form a durable template by which ideas and images can be transferred from one generation to another. Review Huxley's categorization into mentifacts, sociofacts, and artifacts.

3 Languages vary in spatial extent from the global languages such as English down to small languages understood over only a few square kilometres. List (a) the languages you yourself speak and read, (b) those your parents speak and read, and (c) those your grandparents speak and read. How far does your list reflect birthplaces and family migration? Is the range of languages known decreasing, fairly constant, or increasing over the three generations?

4 Geographic systems of belief also vary spatially in a complex way and have immense importance on attitudes towards resource use and innovations. Consult the Yellow Pages of your local telephone directory and plot the distribution of churches of different denominations in your city or country. Do the different denominations have distinctive spatial patterns? Can you think of any likely causes for these patterns? Check your local findings by looking at maps of church membership for the whole of your country in its National Atlas or a similar source.

5 Use either a place-name dictionary for your own country *or* a local county history to check on the place-names of settlements in your own locality. How far do the names of places reflect (a) the origins and language of their founders and (b) the period when they were founded?

6 Gender differences provide an increasingly important focus in human geography. How are differences in the time geography of males and females (see Figure 7.14) still persisting in your own community?

7 Review the Meinig model of cultural cores, domains, and spheres (Figure 7.18). How far does it provide a useful insight for cultural groups at wider and narrower spatial scales?

8 Review your understanding of the following concepts using the chapter text and the glossary in Appendix A of this book:

cognitive map
core
cultural regions
culture
domain
feminist geography
feng shui
gender
Huxley's model
sphere

One Step Further ...

A useful entry point to cultural geography is to look at the three essays on the geography of language by William Brice, of religion by Geoffrey Parrinder, and of gender by Janice Monk in I. DOUGLAS *et al.* (eds), *Companion Encyclopaedia of Geography* (Routledge, London, 1996), pp. 107–36, 888–905. A brief but useful introduction to cultural geography is given by Denis Cosgrove in his essays in R.J. JOHNSTON *et al.* (eds), *Dictionary of Human Geography*, fourth edition (Blackwell, Oxford, 2000), pp. 134–41. Excellent general introductions to the spatial diversity of human cultural groups are given in Malcolm WAGSTAFF, *Landscape and Culture* (Basil Blackwell, 1987) and by T.G. JORDAN-BYCHOV and M. DOMOSH, *The Human Mosaic: A Thematic Introduction to Cultural Geography*, eighth edition, (Freeman, New York, 1999). Follow these up with a look through some of the papers in an old but still very useful set of readings, P.L. WAGNER and M.W. MIKESELL (eds), *Readings in Cultural Geography* (University of Chicago Press, Chicago, 1962).

For a further discussion of some of the more specialized topics touched on in this chapter, see D.E. SOPHER, *Geography of Religions* (Prentice Hall, Englewood Cliffs, N.J., 1967). For gender geography, the dictionary by L. McDOWELL and J.P. SHARP, *A Feminist Glossary of Human Geography* (Arnold, London, 1999) may prove helpful. Landscape studies are introduced and illustrated in D. COSGROVE and S. DANIELS (eds), *The Iconography of Landscape* (Cambridge University Press, Cambridge, 1988) and mental maps in P. JACKSON, *Maps of Meaning* (Unwin and Hyman, London, 1989).

The cultural geography 'progress reports' section of *Progress in Human Geography* (quarterly) will prove helpful in keeping up to date. Regular research tends to be published in the main geographic journals, but look also at *Landscape* (a quarterly) for articles on the cultural landscape. For readers with access to the World-Wide Web see also the sites recommended in Appendix B at the end of this book for topics relevant to this chapter.

CHAPTER 8

An Urbanizing World

■ denotes case studies

Figure 8.1 Urban sprawl in a world city New York is one of the world's great cities with a metropolitan population at the start of the new century of around 17 million people (still way below Tokyo's 30 million). This view is one seen at night by the crew of Space Shuttle mission STS.59 on 9 April 1994. The large patch of light left of centre is at the west end of Long Island and includes the city districts of Brooklyn and Queens. On the tongue of land just above this is Manhattan (with the dark rectangle of Central Park) and the Bronx. Further left, on the far side of the Hudson River, are Jersey City and Newark. Close to the left edge is Staten Island. North is to the top of the photo and the east–west distance is around 100 km (60 miles). [Source: NASA/Science Photo Library.]

> It isn't size that counts so much as the way things are arranged.
>
> E.M. FORSTER *Howard's End* (1910)

If you had to think of a definition of a city, what would it be? Let us begin with a very simple one. 'A city is a large number of people living together at very high densities in a compact swarm'. Although there are many other significant aspects of cities, let us begin by just concentrating on density. Perhaps the easiest way to visualize what our definition describes is to think of a major rock or pop festival. During the summer weekends these may draw crowds of 100,000 or more to green fields around North America and western Europe. Although newspapers and promoters vary in their estimates of the total attending, it is possible to utilize aerial photographs to measure the numbers per square kilometre of these Woodstock-style events. If we put the densities, conservatively, at about 150,000 per square kilometre, then we need only some routine geometry (and a vivid imagination) to compute that, at these densities, the total world population could be corralled in a ring with a radius of only 96 km (59 miles). If we packed people together at the densities encountered on the New York subway, we could get everyone into a ring with a radius of 12 km (7.5 miles).

The notion of such a seething swarm of humanity might repel you, and the ways in which such a swarm would be fed and watered, let alone sanitized, are wholly unknown. Actually, the world's biggest high-density clusters such as Tokyo and New York (Figure 8.1 p. 232) are several orders of magnitude smaller than the hypothetical world city we have just described. Even so, these *megacities* of above 10 million or so people demand the most complex of life-support and communication systems and pose some of our largest social- and economic-control problems.

In this chapter we shall be looking at the city as a dominating force in the organization of the human population (Section 8.1). Geographers must try to unravel the forces that lead to this dominance. Why do populations cluster into cities (Section 8.2), and what keeps them from clustering even more strongly? What are the trends over time in this regard, and how urbanized is the world becoming? Are cities imploding (Section 8.3) or exploding (Section 8.4)? Why is the city seen as a problem area (Section 8.5)? We can sketch only the most basic elements of the answers to these questions in this book. For those of you who go on to more advanced geography courses, urban geography will form one of the most important and interesting areas of further study.

8.1 World Urbanization

Cities are the world's most crowded places. In New York City, about 55,000 people live on each square kilometre; in Montreal and Moscow, the corresponding population densities are about 52,000 and 49,000. These densities are several orders of magnitude higher than the overall densities for the countries that contain these cities. The United States has, on average, less than 25 people living on each square kilometre. For Canada and Soviet Russia the national averages are lower still – about two and eleven persons per square kilometre, respectively.

While cities are now very much larger than they were some centuries ago (see Figure 8.2), the density contrast between the core (within the city) and the periphery (beyond the city) remains.

Trends in Urbanization

Whatever the precise yardstick we use to measure a city, or urban place, the measurements show that the proportion of the world's urban dwellers is on the increase. If we use a threshold population of 20,000 for our definition of an urban place, then in 1800 only about 1 in 40 (2½ per cent) of the

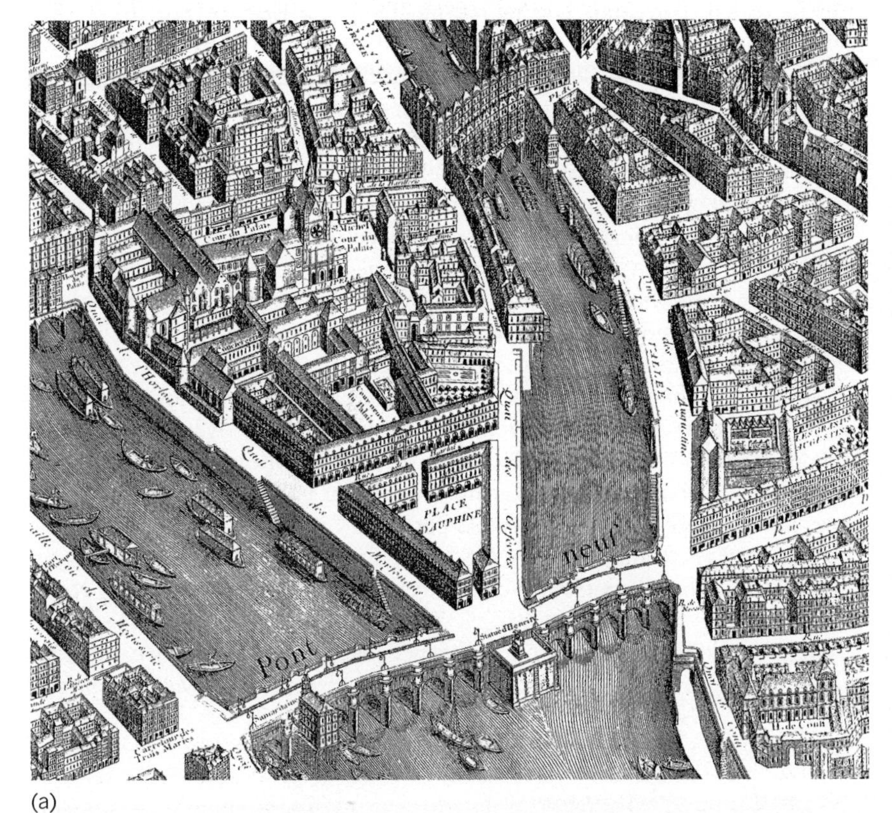

(a)

(b)

(c)

Figure 8.2 Changing form of world cities The closely packed streets of central Paris in the eighteenth century (a), or the walls of medieval Carcassone in southern France (b), or the prospect of compact eighteenth-century Bristol (c), illustrate small and tightly bounded population clusters.

[Sources: (a) © Collection Viollet, Paris. (b) Postcard, *ca*. 1910, Mary Evans Picture Library. (c) Cooke's Universal British Traveller, p. 438, Mary Evans Picture Library.]

world's people were urbanites. By 1980 the proportion had jumped to over 1 in 4 (25 per cent), and it is expected to reach 1 in 2 (50 per cent) by the year 2000.

Trends in the rate of urbanization vary from country to country. For example, the United States is well ahead of the world as a whole. In 1800, about 5 per cent of its population were classified as urban (using the same 20,000 threshold to define an urban place). By 1980 the proportion of urban dwellers in the population had risen to 70 per cent, and by 2000 it was approaching 80 per cent.

Can we detect any general pattern in the urbanization rates for different countries? Figure 8.3 shows a typical curve for a western country. This S-shaped *urbanization curve* has the same kind of logistic form that we will note in other curves later in this book (see Section 16.3). Slow rates of growth in the early part of the nineteenth century are followed by sharp rises in the growth rate in the second half of the century and a progressive slowing down thereafter. During the most rapid period of growth, the key factor in population change was migration from rural to urban areas. Not only were birth rates lower in the cities, but the higher risk of epidemics and degenerative diseases there meant that death rates were higher too. Even when the figures are modified to take differences in the age structure of the population in the two areas into account, London death rates in 1900 were one-third higher than rates in the surrounding rural counties.

It would be tempting to regard the S-shaped western curve as a model for urbanization rates in the developing countries of the world. As Figure 8.3 shows, however, these countries do differ from the already-industrialized countries in two important respects. First, the process of industrialization not only started later, but is proceeding much more rapidly in the developing countries. The figure for the average annual gain in urban population for 34 countries in Africa, Asia, and Latin America over the last two decades is around 4½ per cent. By contrast, nine European countries, during *their* period of *fastest* growth (for most, the latter half of the nineteenth century), had average gains just above 2 per cent. Curiously, the United States, Canada, and Australia, which were hit by huge waves of European immigration, had urbanization rates closer to those of today's developing countries.

There is, however, a second difference between the two urbanization curves in Figure 8.3. Nineteenth-century urbanization was essentially

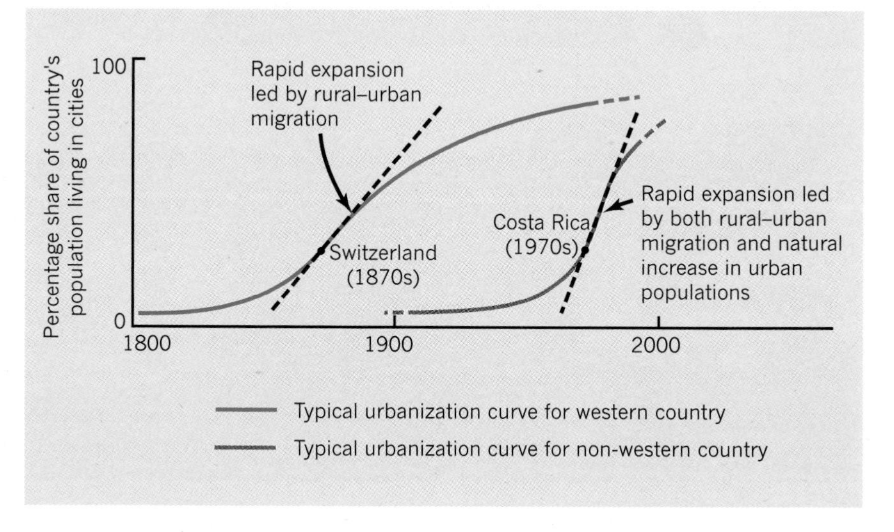

Figure 8.3 Urbanization curves over time On the left in the graph is an idealized S-curve for urbanization in a western country. Geographers are uncertain how reliable a guide this is for projecting the growth of cities in developing countries today, where natural population increase plays a more important role.

'migration-led'; that is, the great proportion of the new urban dwellers were from rural areas. This is less true in today's developing countries. Although the popular images of the rural poor streaming into the shantytowns on the edges of Latin America or African cities are correct, the relative share of migration in the growth of urban populations is smaller in proportion to natural births. In Switzerland in the late nineteenth century about 70 per cent of the urban growth was due to rural–urban migration. In Costa Rica, with a roughly similar proportion of urban dwellers in its population now (as compared to our Swiss example), such migration accounts directly for only about 20 per cent of urban growth. In much of the developing world today the modern city is more healthy than were the cities of Europe and North America a century ago. The birth rates remain higher and the death rates lower. In short, many of the new urbanites are predominantly home-grown.

Changing Spatial Patterns

In the nineteenth century the growth of big cities was a feature of mid-latitude countries (see Table 8.1). In 1900 the four leading cities were London, Paris, New York, and Shanghai. By the early 1920s there were just 24 'millionaire' cities (i.e. cities with 1 million people or more) and only one of these lay within the tropics. In the present decade the number of millionaire cities has increased more than sixfold. One in ten of the world's people now lives in these huge cities, many of which are now located in the tropics.

One useful way of describing the geographic location of the millionaire cities is to look at their latitude. In 1920 the most poleward city was Russia's old capital St. Petersburg (latitude 60°N), and the most equatorward was Bombay (19°N) on the west coast of India. A half-century later, St. Petersburg remains the most poleward city but other millionaire cities extend right down to the equator: Singapore (1°N), with over two million people, is one of Asia's fastest-growing cities.

If we now calculate the *average* latitude (ignoring whether the city lies in the northern or southern hemisphere), the equatorward shift is clear. In the 1920s the average latitude for millionaire cities was between 44° and 45°,

Table 8.1 Population of the world's ten largest metropolitan areas over the last millennium

AD 1000		1800		1900		2000	
Cordova	0.45	Peking	1.10	London	6.5	Tokyo	28.8
Kaifeng	0.40	London	0.86	New York	4.2	Mexico City	17.8
Constantinople	0.30	Canton	0.80	Paris	3.3	São Paulo	17.5
Angkor	0.20	Edo (Tokyo)	0.69	Berlin	2.7	Bombay	17.4
Kyoto	0.18	Constantinople	0.57	Chicago	1.7	New York	16.6
Cairo	0.14	Paris	0.55	Vienna	1.7	Shanghai	14.0
Bagdad	0.13	Naples	0.43	Tokyo	1.5	Lagos	13.5
Nishapur	0.13	Hangchow	0.39	St. Petersburg	1.4	Los Angeles	13.1
Hasa	0.11	Osaka	0.38	Manchester	1.4	Seoul	12.9
Anhilvada	0.10	Kyoto	0.38	Philadelphia	1.4	Beijing	12.4

Population in millions.
Source: 1000–1900 from Tertius Chandler, *Four Thousand Years of Urban Growth: An Historical Census* (Edwin Mellen Press, Lewiston, N.Y., 1987); 2000 from United Nations, *World Urbanization Prospects: The 1996 Revision* (UN, New York, 1998). L. R. Brown and C. Flavin *State of the World 1999* (Earthscan, London, 1999), Table 8–1, p. 135.

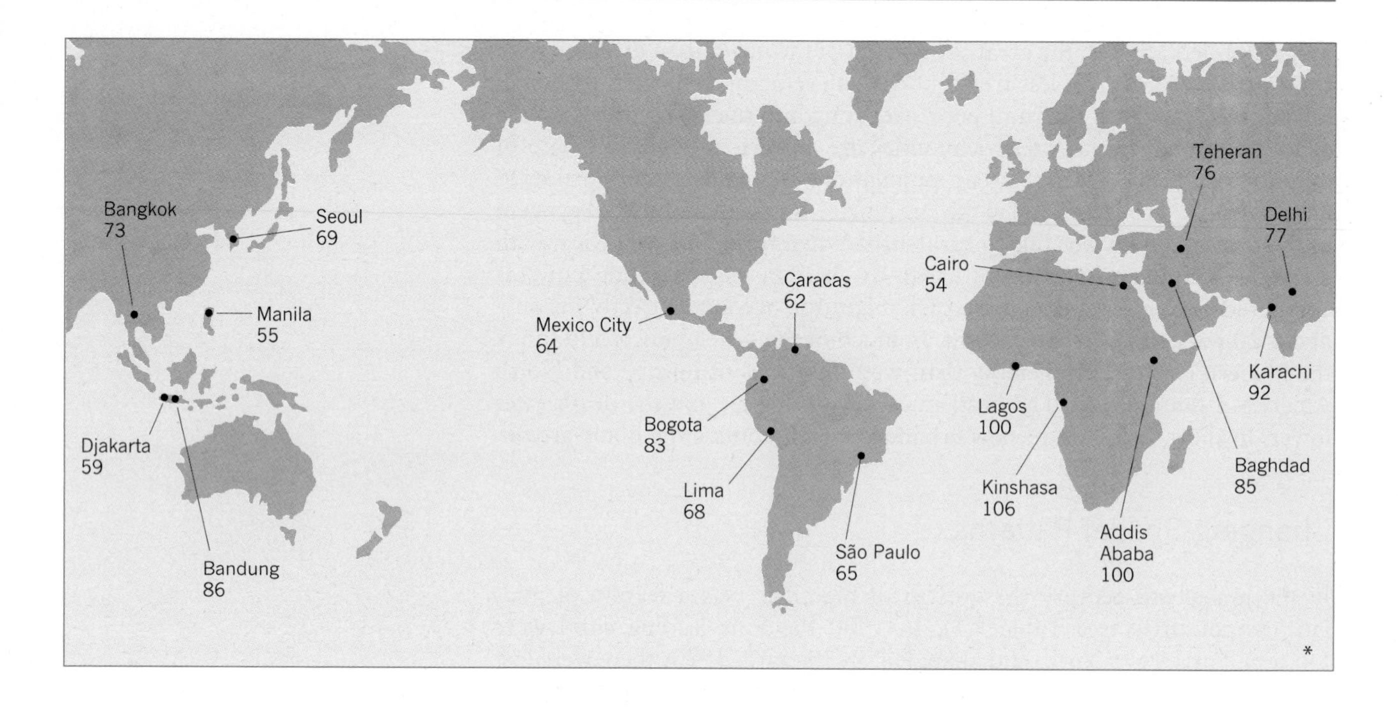

Figure 8.4 The world's fastest growing cities The percentage figures indicate growth in the 1970s. Note that all the cities lie in the less-developed countries.

[Source: Data from *Development Forum*, February (1976), cited in L. Broom and P. Selznick, *Sociology*, 6th edn (Harper & Row, New York, 1977), p. 487.]

about the latitude of Minneapolis in the United States or Toronto in Canada. The average has now moved a further 10° equatorward, to the latitude of Los Angeles or Atlanta. Judging by current rates of growth (see Figure 8.4) the shift is likely to continue.

Table 8.2 shows the distribution of growth rates in five industrial cities at the end of the nineteenth century and compares these with growth rates in five developing country cities in the late twentieth century. All are limited by the exact definition problems we considered earlier in this chapter.

Table 8.2 Rate and scale of population growth in selected industrial cities in the late nineteenth and late twentieth centuries

City	Annual population growth (per cent)	Population added (million)
Industrial cities (1875–1900)		
Chicago	6.0	1.3
New York	3.3	2.3
Tokyo	2.6	0.7
London	1.7	2.2
Paris	1.6	1.1
Developing cities (1975–2000)		
Lagos	5.8	10.2
Bombay	4.0	11.2
São Paulo	2.3	7.7
Mexico City	1.9	6.9
Shanghai	0.9	2.7

Source: Industrial cities from Tertius Chandler, *Four Thousand Years of Urban Growth: An Historical Census* (Edwin Mellen Press, Lewiston, N.Y., 1987); developing cities from United Nations, *World Urbanization Prospects: The 1996 Revision* (UN, New York, 1998). L.R. Brown and C. Flavin *State of the World 1999* (Earthscan, London, 1999), Table 8–2, p. 136.

Dynamics of Urbanization 8.2

Why do cities grow? Can we identify the forces fuelling the present rapid growth of world cities? If we leave aside growth by the natural increase of existing urban populations, we are left with the puzzle of why people increasingly concentrate in clusters so dense that the environment cannot support them harmoniously. Here we try to answer the question in two halves, looking first at why people leave the land ('push' factors) and then at why they cluster in cities ('pull' factors). Together these changes lead to major changes in the major job sectors of the whole economy. Since we shall be dealing with general forces and since each country's urban history is slightly different, Box 8.A 'Urban history of the United States' gives an example.

Box 8.A Urban History of the United States

Urban geographer Brian Berry (see Box 18.A) recognizes a four-stage process in the emergence of cities in the United States: mercantile, industrial, heartland-periphery, and decentralized.

The first stage recognized was a *mercantile phase* beginning with the growth of Atlantic seaboard towns in the eighteenth century. Such towns were generally deepwater ports serving as nuclei of communication and export centres for agricultural hinterlands that produced staples for the world's markets. The hinterlands of Boston, Philadelphia, and Charleston were physically more limited than those of New York. This, plus the relative separation of New Orleans from the main domestic market, allowed New York City – with its middle location along the Atlantic strip and its good internal communications – to move into a dominant position that it retained in succeeding decades. The increase and spread of population inland from these coastal cities followed natural corridors, reinforced by later canal and railroad links, toward the heart of the agricultural-processing regions. With such expansion came the growth of a second generation of inland rail and processing centres such as Cincinnati, Chicago, and St. Louis.

A subsequent *industrial phase*, dating from around 1840 to 1850, took place because of the rapid expansion of manufacturing. The growing demand for iron, and later steel, thrust into locational prominence those areas with (1) appropriate resource combinations (iron ores and coal) and (2) central cities already established during the mercantile phase. Buffalo, Cleveland, Detroit, and Pittsburgh shared these advantages, while other peripheral locations with natural resources (but without ready access to a market) did not. The industrial phase further strengthened the position of New York City and saw the emergence of a major heavy-industry 'ridge' running westward into the heartland of the United States.

This heartland is usually delineated as the area within the Boston–Washington–St. Louis–Chicago rectangle. It had the initial advantages of excellent agricultural resources and a strategic location with respect to mineral resources. We could characterize the period after 1870 as a *heartland–periphery phase*, when contrasts between this core region and other parts of the United States were strengthened. The processes of circular and cumulative causation that increased the wealth of the core area are part of the same processes of regional evolution that brought wealth into the European cultural hearth. We can regard even the spatial changes outside the heartland – that is, the emergence of new peripheral centres in the Far West, the South, and Texas – as direct responses to the needs of the heartland. The resource demands of the peripheral areas fostered regional specialization there, but, at least until World War II, this peripheral growth was essentially dependent on central growth.

Since around 1950 we can detect a *decentralized phase*. In this phase the location of amenity resources (e.g. a sunny climate or unpolluted environment) has

Box continued

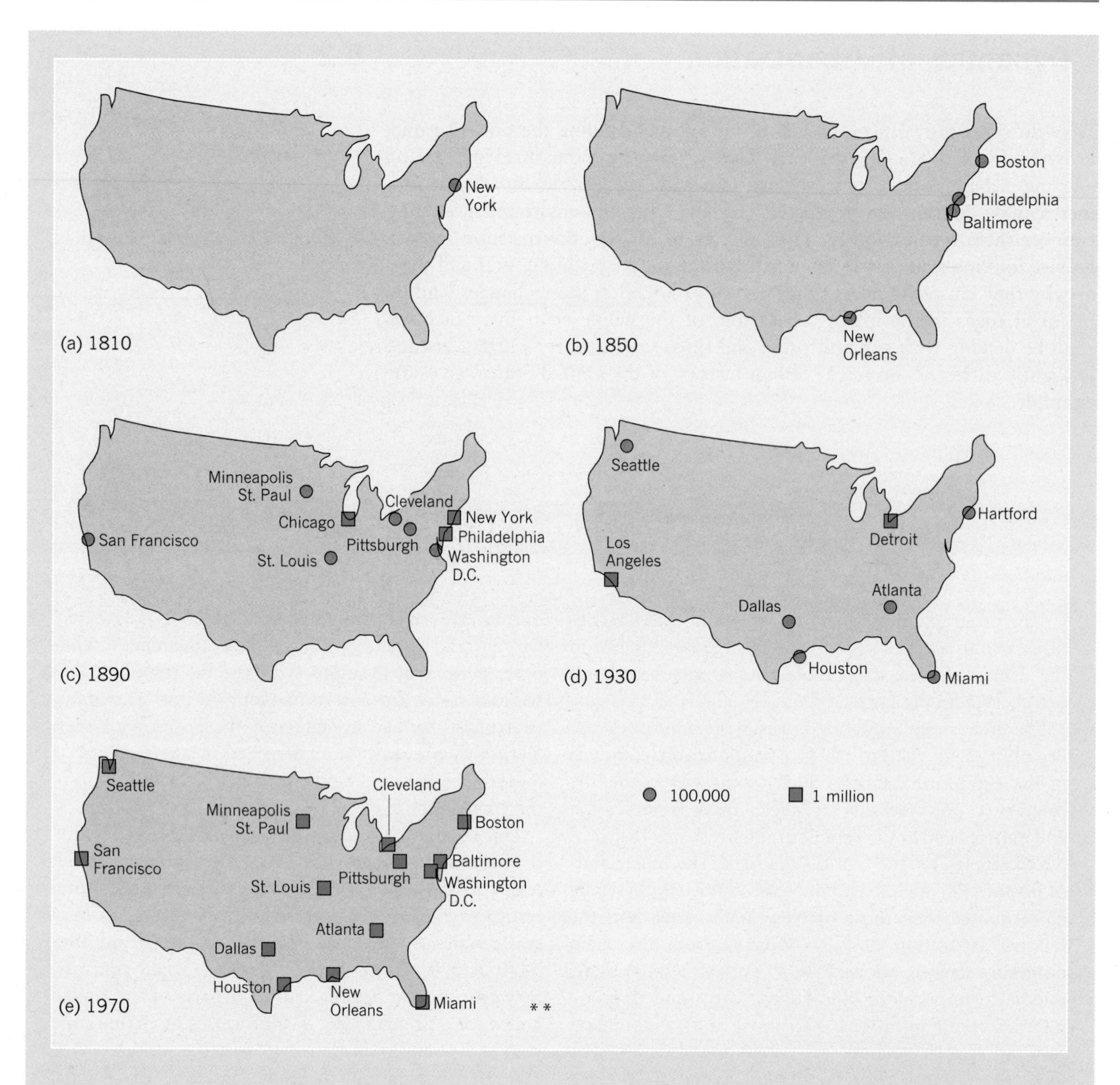

become more important. These resources have stimulated interregional movement of population, bringing rapid urban growth to Arizona and the southwest. They have been responsible for the intraregional rise of small- and medium-sized urban centres with above-average housing, schools, amenities, and so forth. And they have affected local populations by encouraging the suburbanization of manufacturing. These population shifts away from established urban centres are partly related to a rapid overall rise in real incomes. More wealth and leisure have made natural environments (high mountains, persistent sunshine, clear water, extensive forests) a pervasive new influence in the location of population. Of course, in interpreting these signs, we should recall that the total number of people who can respond to this desire is still small and that the area into which the affluent population is moving is largely the western, southwestern, and southern parts of the country.

Among the first group affected was the retirement population. To these retirees have been added workers employed in research and development projects whose location is not greatly dependent on either natural resources or urban markets. Both groups are now more

Box continued

significant elements in the makeup of the US populations. Already research has created important new centres in areas that were previously lagging behind the rest of the United States (e.g. Huntsville in Alabama) or that were sparsely populated (e.g. Los Alamos in New Mexico). Urban growth in states such as Texas and Florida has also vastly increased.

This transference of affluent groups from the outer fringes of suburbia (the 'stockbroker belt') to distant amenity-rich parts of the nation outside the metropolises is one of the latest phenomena in a spatial sorting process. If the US experience is repeated in other areas of the world with high living standards, then some major relocations of population may be in line. The Mediterranean and Alpine parts of Western Europe will probably become increasingly attractive relative to other areas, particularly as barriers to the movements of people and their capital are reduced in an enlarged European Union.

The shifts in growth locations within the United States over the last two centuries may be partly traced to the urban population's changing definitions of natural resources. In the agricultural period the most valued resource was agricultural land, which relied on basic distributions of climate, water supply, and soil. With industrialization the location of mineral resources – particularly coal – became a dominant factor in growth. During the growth of the heartland and peripheral areas, location and good communications were stressed; the current emphasis on leisure activities makes amenity resources an important locational factor. As different amenity resources become important, they provide new directions for the expansion process.

Urbanization: Push Factors ('away from the land')

For most of the world's history the human population was essentially rural. The 'normal' unit of settlement was the small agricultural community, producing (at least in good seasons) the food needed to survive into the next year. As we saw in Chapter 6, over long periods of global history the number of people in such communities was fairly stable, linked to the carrying capacity of the community's land area. Now, one in ten of the world's population lives in cities of a million people or more, and the proportion of people on the land in rural communities is steadily falling.

We can see something of the reasons for leaving the land by considering the recent urban history of the United States. Today fewer than 1 in every 25 Americans lives on a farm; as recently as 1929 the number was 1 in 4! Over the last 50 years, as over the last 200, the rural areas have been net *exporters* of people. Higher birth rates in farm communities have produced a surplus of young people who sought jobs outside agriculture.

We can recognize three important factors in 'pushing' populations off the land, First, if we disallow the period of the frontier expansion, then land is a fixed resource. As the number and size of farm families increased, the extra hands proved valuable in clearing, weeding, and harvesting. Quickly, however, a point was reached where additional inputs of labour brought diminishing extra outputs from crops, i.e. in economic terms, the law of diminishing returns had set in. If all the children had stayed down on the farm, there would have been greater crowding, with further reductions in the productivity of each man–hour. A cycle of crowding, farm subdivision, and rural poverty is a situation common in parts of rural Asia, Africa, or Latin America (see the discussion of shifting cultivation in Chapter 9). But in the United States the surplus of farm population was augmented by a second critical factor, technology. Mechanical progress in terms of the tractor, the combine harvester, or the cotton picker was supplemented by selective crop and livestock breeding and improved cultivation through rotational practices, irrigation, and fertilizers. As a result productivity on the farm has increased at a rate even higher than in manufacturing. The farm population needed to produce a given total of feed or fibre has been greatly reduced.

The long-run decline in agricultural population has been enhanced by a third factor, taste. As the living standards of the American population have risen in real terms, so the proportion of income spent on food products has declined. Investigations show that, internationally, once a certain level of living has been attained, the proportion of additional income in the family budget spent on basic food declines. The parity ratio (the ratio of prices farmers get for their goods compared with prices they pay) has been generally moving against agriculture since at least 1900. Despite great cyclic fluctuations and occasional boom years, agriculture in many developed countries has become a problem for government, not because it produces so little but because it produces so much. Low farm prices have been countered with various forms of subsidies leading to curious anomalies like butter mountains or restrictions in crop area. Low prices are reflected in the fact that 10 of the 25 million people in the United States who are below the poverty line live in rural areas – notably in Appalachia, the Deep South, and along the Mexican border. Canada faces a similar though less pronounced rural poverty problem in parts of the Maritime Provinces and Quebec.

To sum up: the factors pushing population *away* from the rural toward the urban areas appear to be fourfold: (a) a high rural population birth rate, (b) improved agricultural technology reducing the need for farm labour, (c) a swing away from foodstuffs toward other goods as income increases), and (d) relatively low and uncertain prices for farm goods. The relative strength of each factor will vary with different geographic situations: factor (a) is more important in developing countries, (d) more important in highly developed countries.

Urbanization: Pull Factors ('into the city')

Cities that grew just because life on the land was unattractive would resemble gigantic refugee camps. In fact, there are very positive factors pulling people into the cities.

Agglomeration Economies

The major benefits gained from high-density crowding are the so-called *agglomeration economies.* These are the savings that can be made by serving an increasingly large market distributed over a small, compact geographic area. Economies of scale make production costs for each unit low, while the short distance separating buyer and seller in cities cuts back the costs of transporting goods. Clearly production efficiency is directly related to how much is produced. A large output commonly results in lower unit costs because machinery and plants are used more fully and labour can become more specialized. Research and development costs and fixed costs can be spread over more units of production.

Economist Adam Smith's dictum that 'specialization depends on the size of the market' is clearly proven in the modern city. You need to go no further than the Yellow Pages of New York's or London's telephone directories to be aware of the intricate degree of specialization the large metropolitan city permits.

Geographers have dissected the links between extreme specialization and extreme accessibility by examining the flows of information within a city. The persistence of highly specialized office complexes such as New York's Wall Street and London's 'City' district depends on face-to-face contacts with a myriad of specialists as well as electronic flows of information. (The electronic flows are discussed in Chapter 19.) Each individual's activity is dependent on ready access to others, and the actions of the whole group form a complex, or knot, of specialized activities.

Urban Multipliers: Basic and Non-basic The export base of a city, like that of a country, is the activity that lets it sell goods or services or investments beyond its immediate boundaries. Although cities do not have precise boundaries like states, we can nonetheless think of them as having a balance of trade – importing and exporting. Of course, the distinction between the export sector and the rest of the activity of the city is not always clear-cut. Nonetheless, a useful distinction can be drawn between *basic* activities such as manufacturing aircraft engines (to be 'exported' to aircraft assembly plants in other cities) and those *non-basic* activities such as baking bread (to be consumed in the city itself). The *urban base ratio* is used by geographers to describe the ratio of jobs in 'export' sectors of the city economy to the total population. If a city of 60,000 people has 10,000 jobs in its export sector, the urban base ratio is 1:6.

Let us look at the effects of an increase in the export base in a highly simplified situation. Figure 8.5 shows a small Minnesota city, whose export base is iron-ore processing. The ore is not consumed by the local community directly. But the inflow of funds from exporting the processed ore helps balance the 'import' of food, gasoline, and so on, enabling the city to survive as an economic unit even though it is not self-sufficient. Let us suppose that a new market opens up, bringing an increase of 100 jobs in this basic export sector. What will be the effect on the local city?

Figure 8.5 Resource processing and urban growth A 1959 photograph of the new town of Silver Bay (background) that had grown up in this wilderness area of northern Minnesota (north-central United States) in direct response to the location of an iron ore processing plant (foreground). Low-grade iron ore is hauled 70 km (43 miles) by railroad from the Mesabi Range to this coastal site on Lake Superior. After processing, the ore is shipped a further 1000 km (over 600 miles) to steel-making plants around the Great Lakes.

[Source: Minnesota Historical Society/Basgen Photography Duluth.]

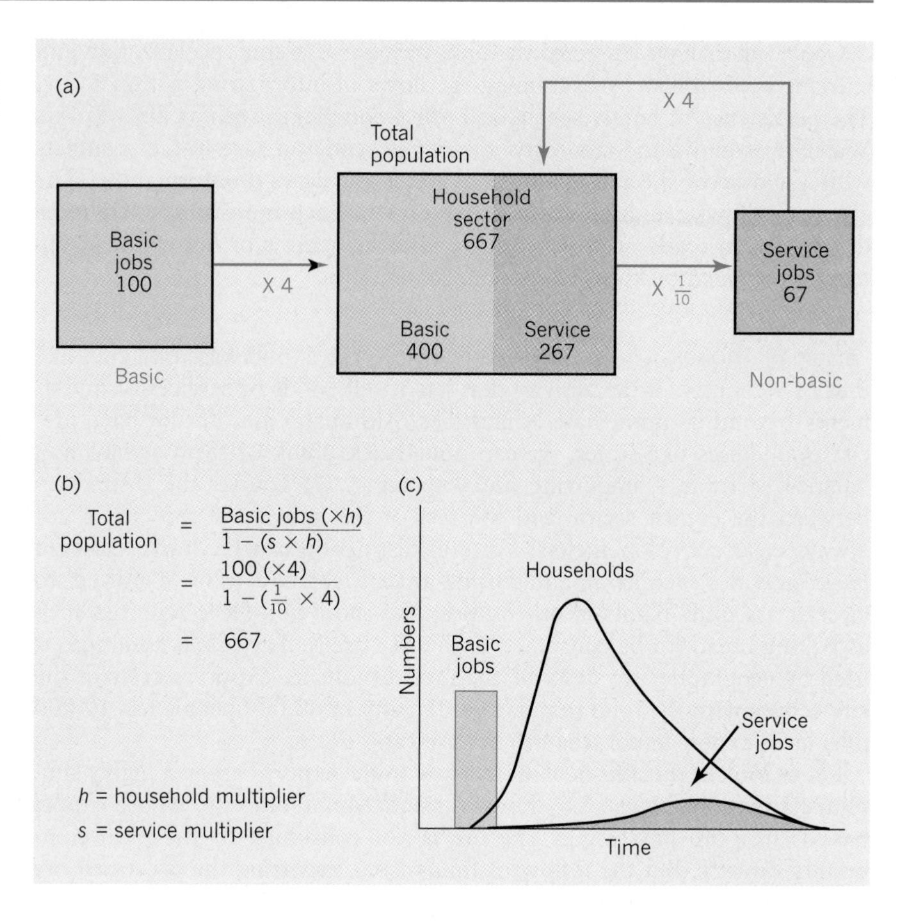

Figure 8.6 The impact of changes in the basic sector on urban growth An increase in jobs in the basic sector of the urban economy has a multiplicative effect on other sectors within the city. In this highly simplified model every basic job creates an increase of four in the population of the household sector (i.e. the worker plus a hypothetical family). So the household multiplier (h) has a value of four. Every increase of ten people in the household sector creates one extra job in the service sector (for teachers, gas-station attendants, doctors, etc.). So the service multiplier (s) has a value of one-tenth. Each person in the service sector is also assumed to have a hypothetical family, which also demands services. And so on! The calculation of the total population is shown in (b). Since there is usually a lag between the establishment of jobs, the arrival of the workers' families and the provision of services, the actual distribution of jobs and households will be as in (c).

A typical cycle of *multiplier* events is shown in Figure 8.6. First, as we have noted, the new market creates a number of entirely new jobs in the local area. If the average family size is four, we can say that every 100 jobs will lead to 400 more people in the household sector. Since these people will demand a set of service facilities – schools, churches, shops, hospitals – the households will, in turn, create a set of service jobs. As a rough estimate we adopt one service job for every ten members of the household sector. Moreover, the workers in the service jobs will have families and create a second, smaller cycle of more households and more service jobs. As Figure 8.6 indicates, these diminishing cycles can be repeated any number of times. If you follow the calculations through, you will find that the 100 new basic, e.g. mining, jobs eventually cause an increase of 667 people in the household sector, plus a further 67 jobs in the service sector.

This model of the domino effect of one activity on another was developed by an economist named I.S. Lowry and is generally termed a *Lowry model*. Two checks may be built into the model to modulate the cycles and thus get a 'truer' picture. First, we can assume that land for residential development cannot be available in the same zone as the mine, and therefore housing must be located elsewhere. Second, we can stipulate a minimum size for service sectors. For instance, we can stipulate that a hospital may be created only when the population in the household sector exceeds an appropriate threshold level. Otherwise, the local population will have to go elsewhere for medical treatment.

Urbanization and Sector Changes

The impact of the new jobs may not be confined to the household and service sectors of the economy. New jobs in mining may lead to associated industrial activities involving the refining, smelting, or processing of mineral ore, or the manufacture of equipment for the mining process. This will cause a proportional drop in jobs in the *primary sector* (a term used to describe jobs in agriculture, mining, fishing, etc.) and a relative increase in jobs in the manufacturing or *secondary sector*. Further growth will lead to an expansion in the wide range of service jobs (in shops, education, hospitals, transport, etc.) in the *tertiary sector*. In the last half-century, the most rapidly growing job sector has been jobs in research and administration, the so-called *quaternary sector*. 'Office' and 'research lab' jobs account for an increasing share of employment opportunities in major metropolitan areas. One way of distinguishing among the four sectors is to take a given natural product and consider which jobs are related to it. For example, timber production gives jobs to a chainsaw gang (in the primary sector), a lathe operator in a furniture-manufacturing plant (in the secondary sector), a furniture-store operator (in the tertiary sector), and a researcher in wood technology (in the quaternary sector).

Figure 8.7 shows the job implications of urbanization at the national level. Sweden's excellent population records allow us to follow the trend from rural to urban settlements since 1750. Since 1850 we can also see the shift in jobs with agriculture declining and industry rising steadily until it

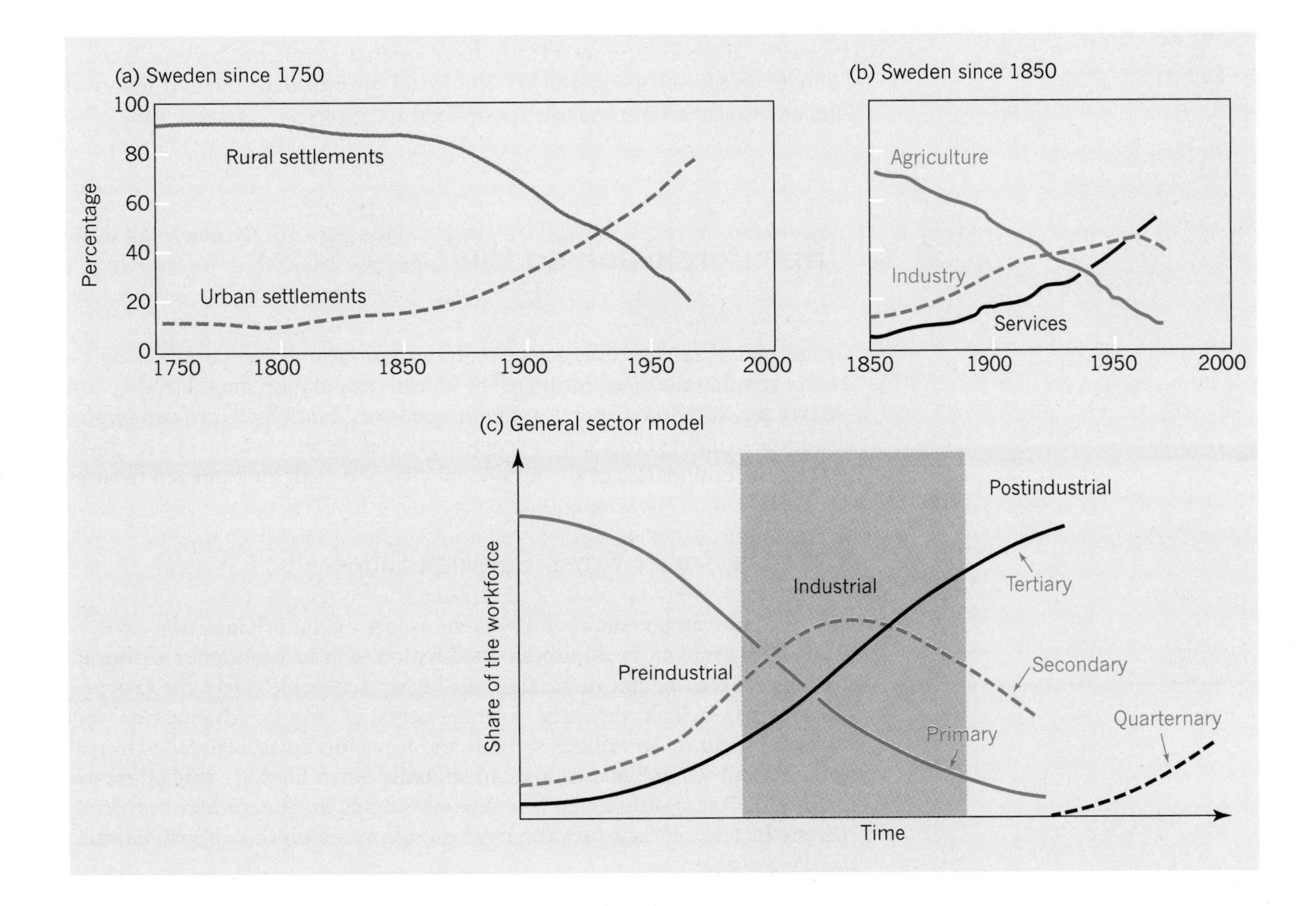

Figure 8.7 Sector model of economic growth The pattern of urban–rural settlement in Sweden (a) is compared with the changing balance of jobs in the three main sectors of economic activity (b) and with a general model of sector change (c).

[Source: Data from *The Biography of a People: Past and Future Population Change in Sweden* (Royal Ministry for Foreign Affairs, Stockholm, 1975), Table A5.]

reaches a peak in the 1960s. Since then it has started to decline and, as the graph shows, service employment has now taken over a the major job sector in Sweden.

Geographers see the Swedish changes as part of a *sector shift* model. Figure 8.7(c) shows the general form of the model as a series of curves, and you should take some minutes out to look carefully at this chart and see what it means. Note in particular the division into three stages termed *preindustrial*, *industrial*, and *postindustrial*. Most western countries would, on the basis of this graph, still be in the industrial stage while most Third World countries would be in the preindustrial stage.

But do countries necessarily have to go through these three stages? Canadian geographer Maurice Yeates thinks not. The evidence for Canada is that it has moved from the first stage straight to the third. Primary activities (agriculture, mining, trapping, forestry, etc.) were very important in the first century of growth but industry has never been a dominant employer. Industrial employment reached its peak (around 23 per cent of the workforce) in the 1960s and is now going down. But service employment has long been important. At the turn of the century it employed around 40 per cent and its share is now approaching 70 per cent. So Canada, unlike the United States, seems to have missed the industrial stage.

At the city level, whether or not the extraction of a given resource will trigger a chain of associated industrial growth depends on a great variety of factors. In many of the coal mining areas of northwest Europe that attracted large populations in the early nineteenth century, the major cities have continued to expand long after the original mining activity has ceased to be significant. The extraction of resources located in central, highly accessible zones (that may already have a dense population) is much more likely to trigger prolonged urban-industrial growth than the extraction of similar resources set in remote, peripheral locations.

8.3 Urban Implosion on the Large Scale

Geographers are especially interested in urbanization as a spatial process – a force shaping the distribution of the world's population map. Here we can see two powerful movements going in opposite directions. One is an *implosive* force bringing cities closer together; the other an *explosive* force pushing out the boundaries of the individual city. We shall look at each in turn.

Shrinking Travel Times Between Cities

The role of changing accessibility is an aspect of the urbanization process of great interest to geographers. For agglomeration economies to occur, markets have to be not only large but highly accessible. Over the last two centuries, in which time the average level of world urbanization has increased tenfold, overall travel times and transport costs have *fallen* in real terms by an even greater amount. New transport technology – whether it is a railroad, Telex, shuttle, jet, or monorail – tends first to connect key cities, thereby increasing their locational advantage in relation to important cities (see Figure 8.8).

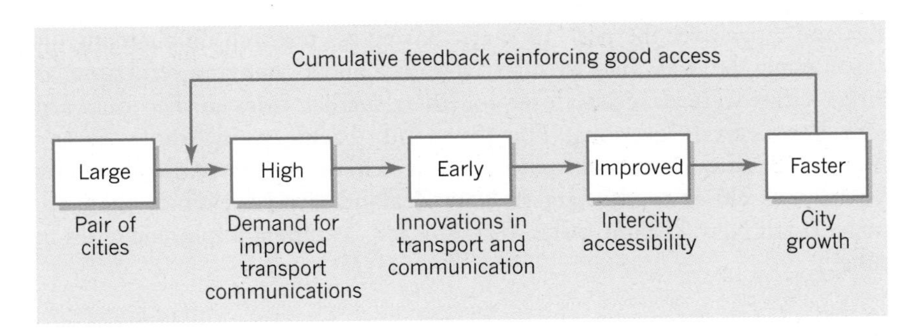

Figure 8.8 Cumulative linkage advantages The set of transport and communication advantages that help important cities retain their importance. This upward cycle is matched by a downward cycle of cumulative deprivation for unimportant pairs. (To find it, substitute the words 'small,' 'low,' 'late,' 'retarded,' and 'slower' in the five boxes. Note that 'retarded' may be relative rather than absolute, but the impact on locational advantage remains a negative one.)

We can illustrate this 'ratchet effect' by measuring changes in the average cost of movement between centres of different sizes. If we use time as a proxy for cost, between the years 1850 and 1900 the distance from New York to California was cut from 24 days (3 days by railroad plus 21 days by overland coach) to 4 days (by direct train). The next half century reduced this distance to 8 hours by DC-6, and to 5 hours by jet (see Figure 8.9). Thus the relative cost of the trip (in hours) has been reduced by over one-half since the 1930s. Further examples of shrinkage are given at the start of Chapter 19.

Yet these changes have not been uniform and the bigger cities gain more from improvements in services over time. As a result, the cost in time of travelling between them is reduced, and they move closer together in relative terms.

We can borrow a term from astronomer Fred Hoyle and describe the effects of this differential spatial shrinkage as an *urban implosion* – that is, the inverse of an explosion. The ways in which large cities converge on each other is shown by the diagram in Figure 8.10. Note that since large cities converge rapidly and the smaller cities converge less rapidly, the larger and smaller cities diverge in relative terms. The speed and self-reinforcing nature of this spatial implosion are principal factors in the current rapid growth of large cities.

At Canterbury University Philip Forer has traced the implosion of New

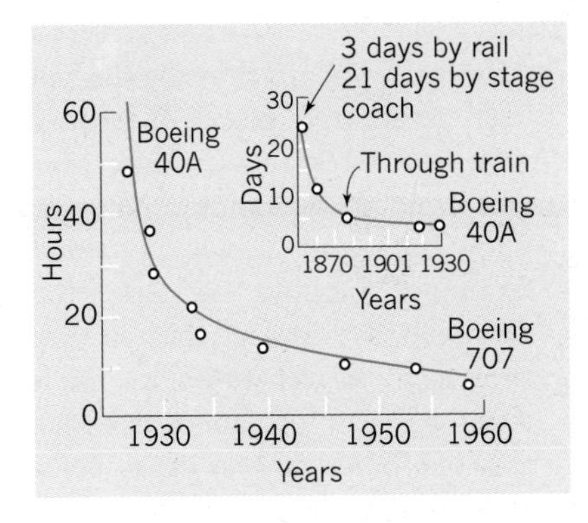

Figure 8.9 Shrinking worlds The graphs show changes in the average amount of time needed to go from New York City to Los Angeles since 1850, using the fastest means of public transport.

[Source: From R.E.G. Davies, *History of the World's Airlines* (Oxford University Press, London, 1964), Fig. 91, p. 509.]

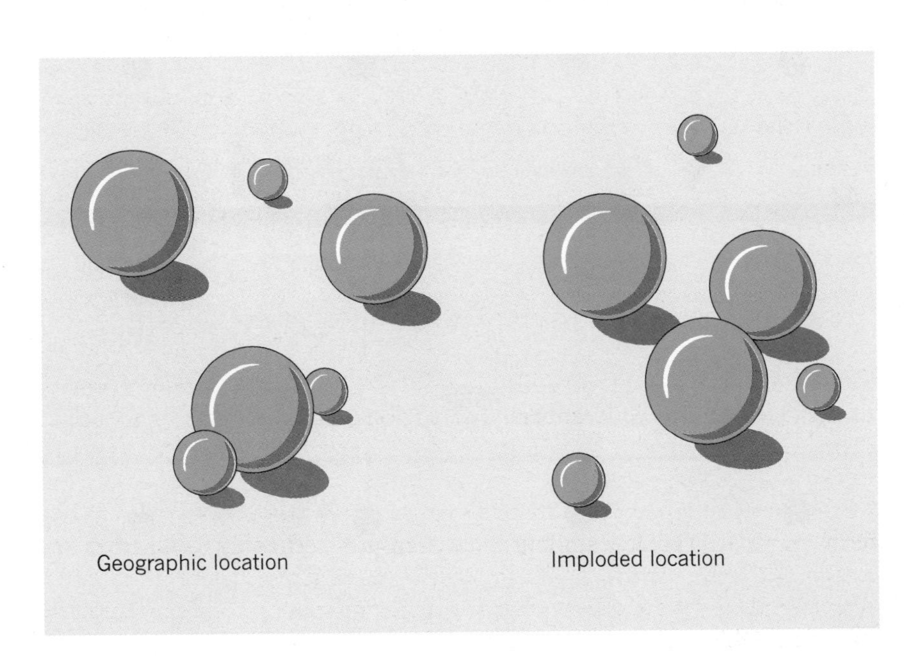

Figure 8.10 Urban implosions Transport innovations are usually first introduced between pairs of large cities between which important flows of people and goods already occur. Such innovations make links between such pairs easier, setting off a cycle of increased contacts and flows. As the second figure in Box 8.B shows, the net effect is to bring larger cities relatively closer to each other. Meanwhile smaller centres become relatively (and sometimes absolutely) more remote. This figure illustrates this process of 'implosion'. Note the convergent movement of the three largest cities (the size of the spheres is proportional to the size of the cities) in comparison with the divergent movement of the three smaller ones.

Zealand cities over the past 30 years. Basing his research on changing air travel times he was able to map New Zealand's changing structure; as larger cities were dragged closer together, smaller cities pushed outward into less central locations. For the South Pacific as a whole Forer's time maps show a curious pattern in which the largest Australian city (Sydney) is closer to the largest New Zealand city (Auckland) than are many small New Zealand cities. (See Box 8.B 'Mapping imploding cities in time'.)

Box 8.B Mapping Imploding Cities in Time

How to plot maps showing the 'correct' position of places in terms of time–space remains an unresolved problem in geography. We can show this by taking a simple example. Suppose we have four towns (p, q, r, and s) which are separated by travel times in hours. From p to q is four hours, from p to r is one hour, and so on as below (assuming travel time is the same in both directions):

p–q four hours q–r two hours
p–r one hour q–s six hours
p–s one hour r–s three hours

If you settle down with a pencil and paper to try to draw this with a scale (say, 1 hour = 1 cm on the map) you'll have a frustrating time! Although we can easily map any pair of towns, the third and fourth simply refuse to fit on the paper in the correct position.

Although there is no precise solution to this mapping problem, there are ways of making a 'best estimate' of the towns' location. One graphical method suggested by geographer Waldo Tobler is outlined in the diagrams and summarized in six steps:

Step 1 Locate the towns arbitrarily on the paper as a set of points labelled p, q, r, and s. In (a) we have chosen a square, but it could be any arrangement.

Step 2 Draw straight lines through each pair of points.

Step 3 On each line, centre a segment with its length proportional to the given separation-time (say 4 cm for the four hour distance between towns p and q). (See diagram (b).)

Step 4 Draw vectors (marked by arrows) from each point to the ends of the segments drawn in Step 3. (See diagram (c).)

Step 5 Average the vectors drawn in Step 4 to give a new set of locations labelled p^1, q^1, r^1, and s^1. The method of averaging is shown in diagram (a) for the first of the towns, and you'll find tracing paper useful for doing this. Since there are three towns the average will be one-third of

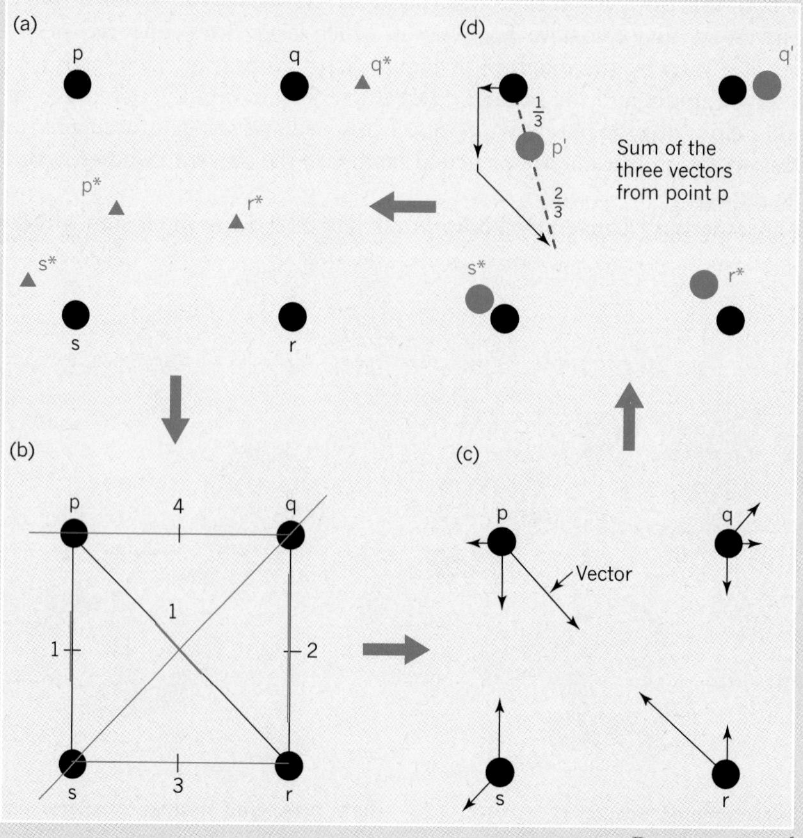

Box continued

The changing spatial relations of cities have given a new interest to ways of mapping change. Thanks to the computer, it is now possible to calculate and represent on maps the location of places in a 'time' or 'cost' space (see the discussion above).

the length of the sum of the displacements.

Step 6 If no points have moved in Step 5, stop. Otherwise use the new locations (p^1, q^1, r^1, and s^1) to start a new cycle starting at Step 2 and going around to Step 6 again.

For interest, the final stopping places are shown by triangles in (a) and marked p*, q*, r*, and s*. If you try the graphical method described above using tracing paper, you'll find it tedious but rather satisfying. Notice how at each cycle the moves in position becomes less and less until no significant shifts are noted. For large numbers of places like those shown in (e)–(g), the graphical method is far too tedious and a computer method must be used instead. The maps that result provide an accurate way of showing, as nearly as possible, how places are located in time-spacing. In principle we can substitute costs in dollars for time if this is more appropriate to the geographic study at hand.

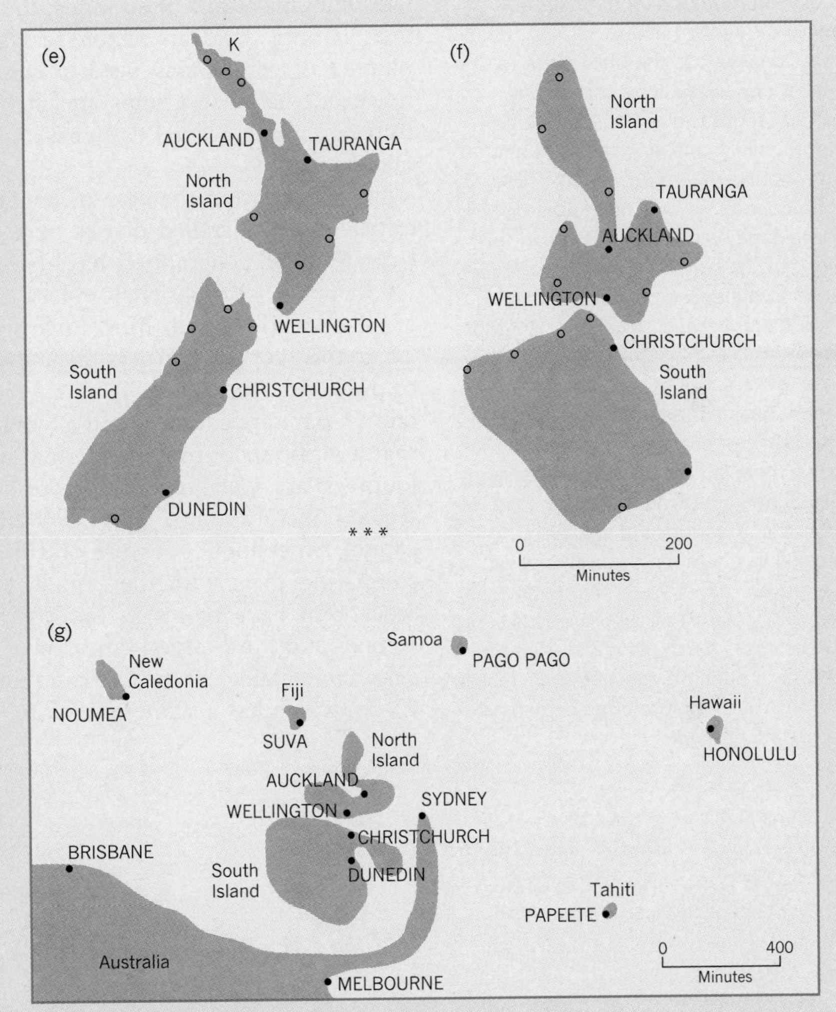

An illustration of the method is shown in Philip Forer's work on New Zealand. A conventional geographic map of New Zealand (e) shows the location of major airports. A time map (f) shows the 'distance' in minutes between the same urban airports. The map is based on travel times and service frequencies in 1970. Note the way in which the three leading cities – Auckland, Wellington, and Christchurch – are clustered near the centre of the map. In contrast, the smaller city of Tauranga is displaced toward the periphery. A larger time map (g) shows the relations between the main New Zealand airports and other major urban airports in the South Pacific. It too is based on air travel times and service frequencies in 1970. Note how large numbers of high-speed flights between Sydney and the main New Zealand cities draw these locations together. These unfamiliar time maps are produced by a ***multidimensional scaling*** of a mass of information on flight times and frequencies between all the pairs of locations shown on the map.

For a fuller description of the method see P. Haggett *et al.*, *Locational Analysis in Human Geography* (Arnold, London, 1977), pp. 326–328.

[Source: The New Zealand example is taken from Philip Forer, *Changes in the Spatial Structure of the New Zealand Internal Airline Network*, doctoral dissertation, Bristol University.]

Contactability: Its Impact on Corporate Organization

Another approach to the urbanization process has been taken by a group of Swedish geographers headed by Torsten Hägerstrand and Gunnar Tornqvist. They note the growing importance of the quaternary sector in the urban growth pattern of the present century. Despite the growth of telecommunications of all kinds, the need for direct person-to-person contacts is critical in this sector. In research planning, for example, a large number of people may need to come together in informal groups. Note, however, that such groups are likely to meet occasionally and irregularly (rather than on a fixed daily basis) and that the membership of the group may be flexible.

In this situation the ease of contact between urban areas may become critical. Using detailed diaries of personal movements of business executives, Swedish geographers have been able (a) to track down the degree of interdependence of different industries and types of activity in different geographic locations and (b) to rank Swedish cities on a *contactability scale*. To do this they use journeys made by private automobile or the fastest form of public transport (see Figure 8.11). Only journeys made between 6 a.m. and 11 p.m. are counted, and a 'contact' is defined as a meeting of not less than four hours during the 'normal' working day (8 a.m. to 6 p.m.). When journeys are weighted to include the cost of contacts, a very detailed regional picture can be built up. If we give the 'most contactable' city – the capital, Stockholm – a score of 100, we can then rate the other cities by comparing them with the capital. Some smaller cities in the vicinity of Stockholm have scores in the 80s and 90s. Göteborg and Malmö, the second- and third-largest cities, have scores of 78 and 70 respectively. The 'least contactable' city is the iron-mining city of Kiruna north of the Arctic Circle, which has a score of only 36.

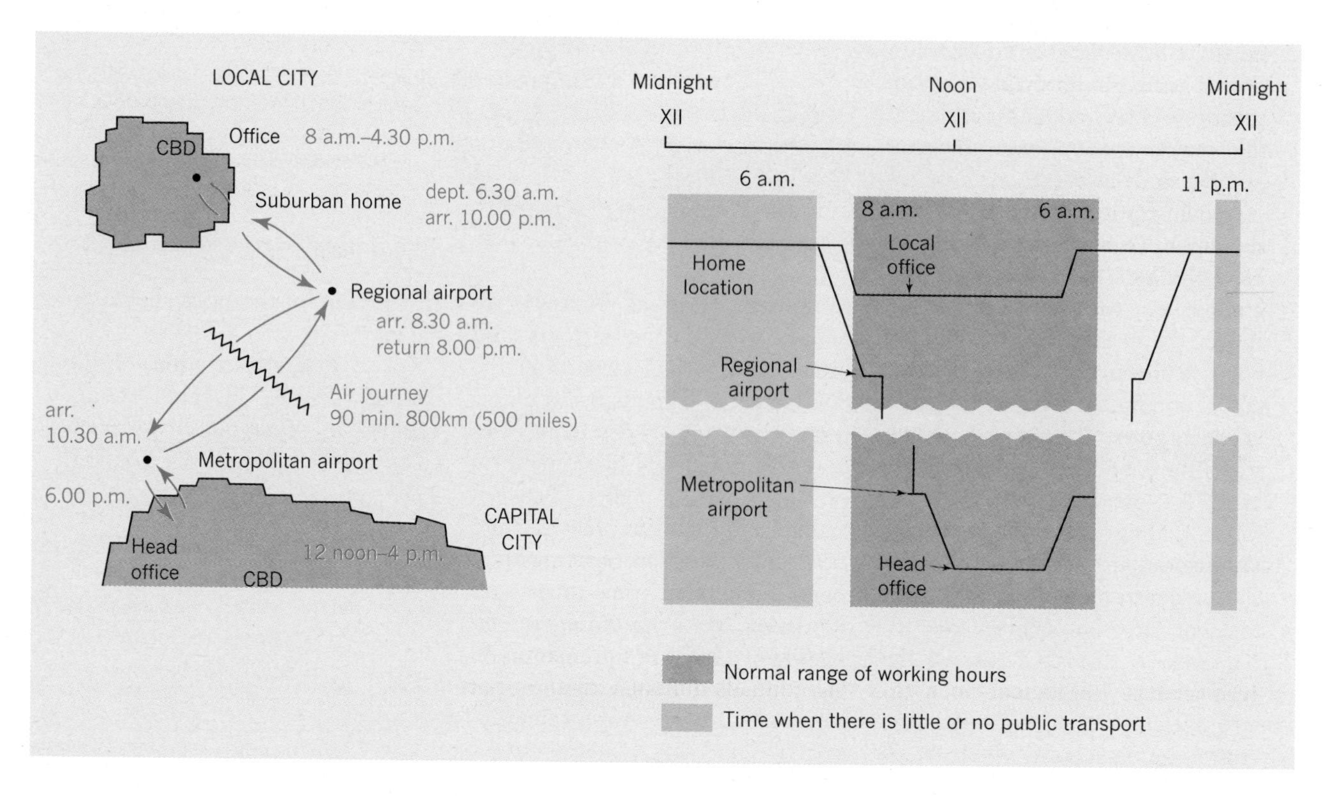

Figure 8.11 Local and regional contact patterns (a) Two typical daily contact cycles for a suburban-living executive. First, the regular daily cycle between home and local office, with a short commuting period (about 40 minutes by automobile). Second, the occasional visit to the head office in the capital city, with a long commuting period (about twelve hours, using an automobile, plane, and taxi) and a short (four hour) conference period. These contact patterns may be plotted on space–time diagrams, as in (b). In this diagram the horizontal lines indicate a static location, and the diagonal lines show travel between locations. The more a line diverges from the horizontal, the greater the speed of travel. (CBD = central business district) Using detailed travel diaries, a group of Swedish researchers led by Torsten Hägerstrand and Gunnar Tornqvist have been able to show the importance of contact possibilities between Swedish cities in the growth of the quaternary sector. Clearly, increased transport speeds allow lower-density living patterns within the daily commuting range and an increased meshing of cities within the occasional daily contact range.

These results are useful in picking out the likely centres of further urban growth in the quaternary sector. Planned changes in transport services (e.g. the introduction of new airline schedules or the opening of new highway links) can be checked for their effect on the contactability rating of different cities. Which cities will benefit? Which will lose out? By repeating studies over time, the changing access of one city to another can be plotted and the course of the urban implosion charted and projected into the future.

One of the most striking impacts of changes in contactability has come through business reorganization. As industrial corporations have grown in size, so the control operations of management have become fewer but larger. (See the discussion in Chapter 19.) Figure 8.12 shows a hypothetical sequence in which the successive takeover of companies leads to centralized head offices. But the ecology of corporate management requires a steady inflow of energy in the form of information and offices tend to cluster in just these central locations where information is richest: in the administrative and financial capitals of a country, or in its major cities.

Empirical evidence on the extent to which corporate headquarters have been separated from the remaining activities has been collected by Berkeley geographer Allan Pred. In the United States the leading 500 industrial corporations show a highly concentrated pattern: in the middle 1970s one out of every three had its headquarters in New York City. Of the remaining corporations, half were concentrated in eight other major regional capitals – Chicago, Los Angeles, Cleveland, Philadelphia, Pittsburgh, Detroit, San Francisco, and Minneapolis-St. Paul. For Britain the situation was even more concentrated. London was the headquarters of two out of three of that country's 500 leading industrial companies. Comparison of the second-ranking centres shows a similar two-to-one contrast between the pattern in the two countries: Chicago had 11 per cent of the headquarters, while Birmingham, England, had only 5 per cent. For Canada the headquarters concentrated in either Toronto or Montreal (roughly in equal proportions but with Toronto growing faster) with Vancouver trailing in third place.

Concentration into high-contact centres is accelerated by the continuing need for the direct exchange of information on a face-to-face basis. The increasing importance of government as a source of major orders or as a body whose legislation affects a corporation's prosperity also points to a capital-city location.

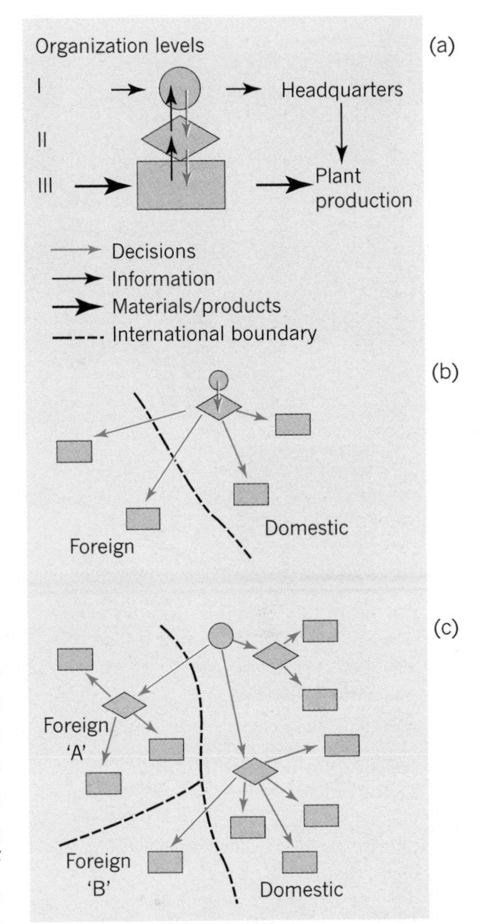

Figure 8.12 Centralization of large offices As industrial corporations have grown so the head offices (circles) have become fewer and more concentrated. Control (arrows) of production plants (blocks) has shifted from local to more distant centres. The critical need of large corporations for information has tended to cause clusters of head offices to form in the major metropolitan centres.

Urban Explosion on the Small Scale 8.4

To understand the second major impact of transport innovations on urbanization, we must recall that people evolved biologically in an environment with a regular cycle of lightness and darkness, and most basic human activities – feeding, sleeping, working, procreating – have a daily rhythm. Not only is the basic biological clockwork that controls human activity important to agricultural communities, but even on the campus, the same daily rhythms can be shown (Figure 8.13).

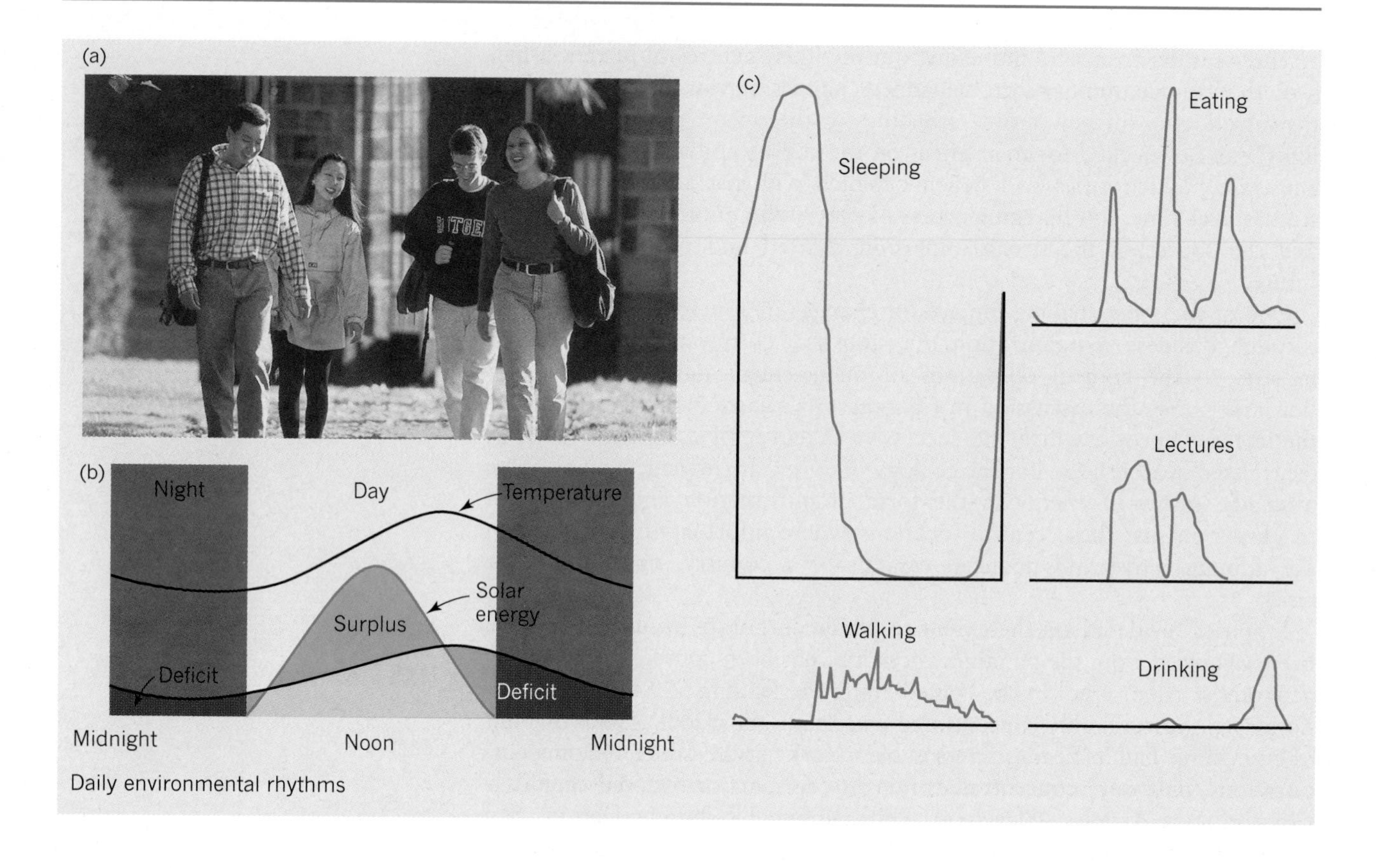

Figure 8.13 Time rhythms in human activity (a) Activity patterns for a sample of university students. (b) Generalized pattern of human activity over a day is related in part to environmental changes in light and temperature. (c) Figures are for Reading University, England, and the concentration of drinking reflects English licensing laws at the time of the survey. The horizontal base lines are time over a 24 hour midnight to midnight period.

[Source: A. Szalai (ed.), *The Use of Time* (Mouton, The Hague, 1970), Fig. 5.1–11A, p. 736; and L. March, *Architectural Design*, **41** (1971), Fig. 14, p. 302. Photo by Bob Krist, Leo de Wys – Stock Photo Agency.]

The Daily Contact Zone of the City

How do these time rhythms affect spatial organization? Basically, they operate through the dominant need of most of the human species for a fixed, or relatively fixed, home base. This home base may vary from an apartment in New York City to a sampan in Hong Kong, but it has the same essential economic, social, and biological role. At the very least the home base provides a place for a stock of household goods. It may also provide a site for the production and rearing of children, and a retreat for the satisfaction of sexual and emotional needs. The need to return regularly to a home base puts a severe constraint on the number of hours that can be devoted to travel to work, and therefore on the distance that separates the elements in any household economy. Even when the adult male breaks away from this pattern and works away from the home base for long periods, the constraint usually remains for females, children, and older folk in the household.

Over what distance does this constraint operate? Before Stephenson's *Rocket* rolled along the rails in October 1829, the only regular ways of moving over land were by manpower or horsepower. Horses and carriages could attain a speed of over 50 km/hour (over 30 miles/hour) over short distances, although in practice the rate of travel was reduced by poor or congested roads to less than a quarter of this figure. How fast or far people moved on their own legs is more difficult to estimate. Modern measurements of our walking speeds on city sidewalks indicate that the average adult pedestrian walks at 5.5 km (3.4 miles) an hour. The values vary with age and sex. Adolescents gallop along at a steady 6.4 km (4 miles) but mothers with young children walk only 2.6 km/hour (1.6 miles/hour). A slope of only 6° cuts speed by over a fifth; a slope of 12° cuts speed in half.

It is not surprising that most communities in the pre-railway period were quite compact, usually with a diameter of not more than a kilometre or two. The average walking time from one end of town to another would thus be about 11 to 22 minutes. This compactness was directly related to the need for communication between different parts of the city and to the limitation imposed by walking or carriage speeds. The communication links between home and work, merchant and scribe, banker and businessman, magistrate and prostitute all had to be spatially short if transactions were to be completed within an acceptably small part of the day.

Urban Sprawl

The technological revolutions in land transport during the past 150 years have successfully reduced the need for compact cities. As steam locomotive speeds increased, and the electric streetcar, the omnibus, the suburban electric railway, and the private automobile succeeded each other, so the links between home and workplace widened (see Figure 8.14). Living above the shop or workshop was replaced by daily commuting over increasingly long distances. Chicago's daily urban system is now reaching out 160 km (100 miles) from the downtown. Today all but 5 per cent of the US population live within the bounds of such daily urban systems.

We can see the dramatic effects of these changes in the exploding size of London (Figure 8.15). In 1800 London, Europe's largest city, could still be

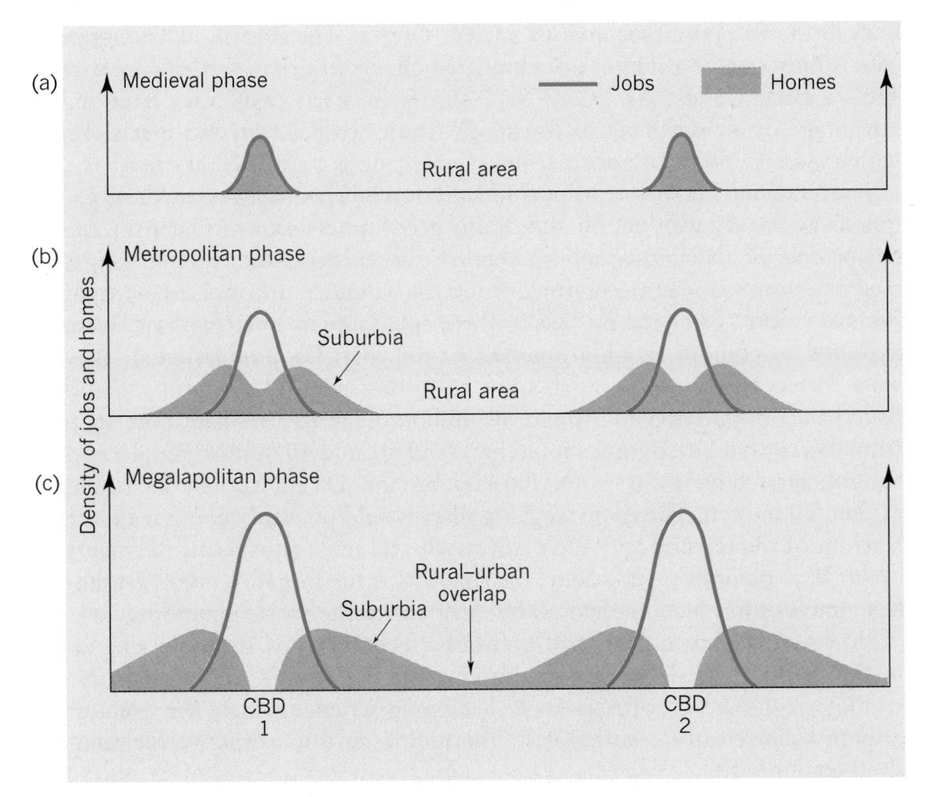

Figure 8.14 A simplified, three-stage model of the sprawling expansion of two cities In the first stage (a) jobs and homes overlap in a small, high-density cluster (like the medieval walled city in Figure 8.2). (b) Transport innovations allow the links between jobs and homes to be loosened. Densities in the central areas of cities appear to fall as less land is available for homes and long-distance commuting creates a daily tidal flow of workers into and out of the city. (c) Continuation of this process leads to further evacuation of the central areas of cities, and suburban sprawl. Corridors of cities develop like those along the German Ruhr and the Boston–Washington axis.

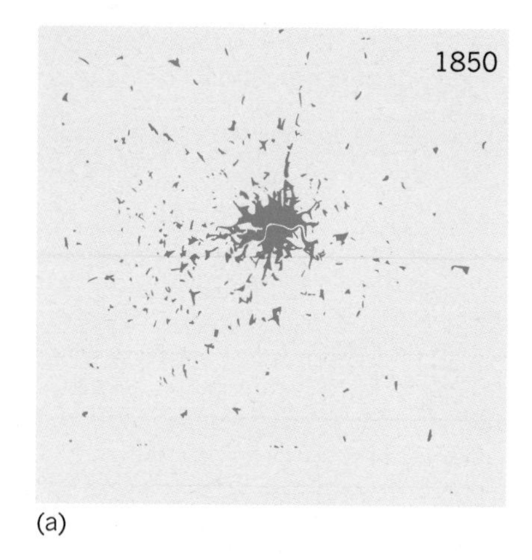

(a)

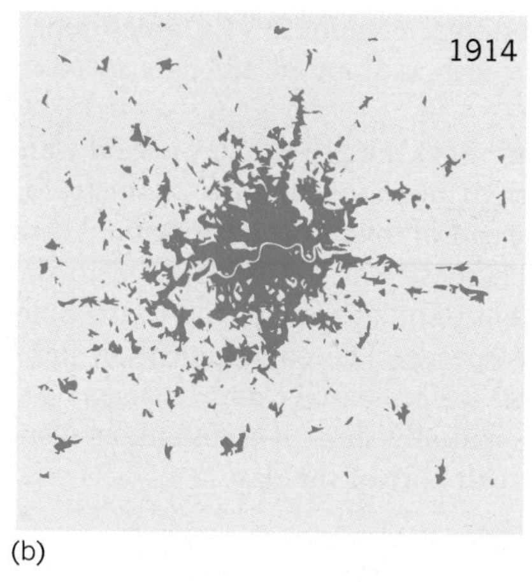

(b)

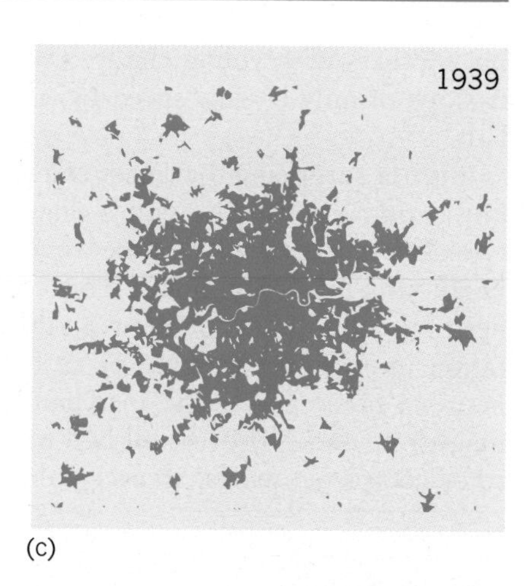

(c)

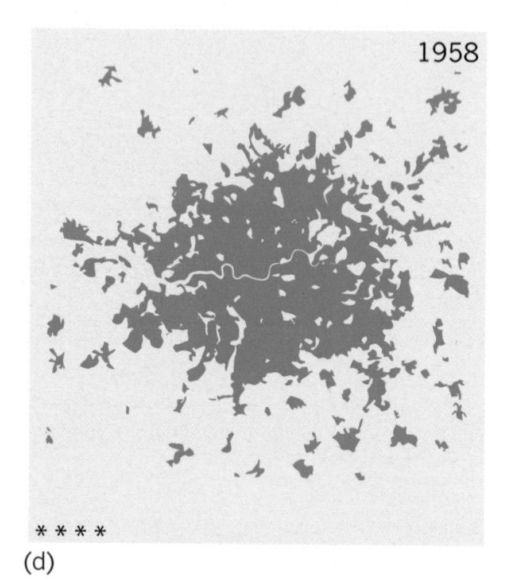

(d)

Figure 8.15 Expanding daily contact zones within a city The maps show the growth of London over a period of 108 years of progressively faster transport.

[Source: D.J. Sinclair, in K.M. Clayton (ed.), *Guide to London Excursions* (Twentieth International Geographical Union Conference, London, 1964), Fig. 9, p. 12.]

crossed on foot in little more than an hour. Even at its widest, the built-up area was only 10 km (about 6 miles) across. As the sequence of maps shows, this area expanded so rapidly that by 1914 its diameter had reached 35 km (about 22 miles), and it approached 70 km (about 43 miles) by 1958. Since then it has shown few changes due to a vigorous policy of protecting further extension into rural areas (the *green belt* policy).

London is modest in size by international standards. Los Angeles (Figure 8.16) already exceeds 80 km (50 miles), and Greater Tokyo 120 km (75 miles) in diameter. If we accept Oxford University geographer Jean Gottmann's idea that the whole network of cities along the east coast between Boston and Washington, D.C., is being fused by factors such as shuttle jets into one vast urban area, then we have a world city of mammoth dimensions, over 600 km (373 miles) in diameter. Gottmann terms the city complex thus created a *megalopolis*. This Greek word, which literally means 'great city', was first used to refer to a city being planned on large lines in the Peloponnesus area of ancient Greece. The Boston–Washington axis is only one of a number of elongated chains of urban settlements that show similar trends. As Figure 8.17 shows, in each case cities have the advantage of a central set of transport arteries (road, rail, and air) along which massive flows of information, people, energy, and freight move.

The tendency for such transport links to forge urban centres into a megalopolis is already well established. Some geographers argued that from the perspective of the mid-twentieth century the existence of three principal megalopolises – sometimes termed *Boswash*, *Chipitts*, and *Sansan* – would become more evident. By 2000 three gargantuan metropolises were expected to contain roughly one-half of the total US population. If they were correct then Boswash, extending from Boston to Washington, should have had a population of around 80 million; Chipitts, the lakeshore strip from Chicago to Pittsburgh, should have had around 40 million people; and Sansan, stretching from Santa Barbara to San Diego, should have had around 20 million. The three megalopolises would probably contain a large fraction of the scientifically most advanced and most prosperous segments of world population; even today, the smallest of the megalopolises (Sansan) has a larger total income than all but a dozen of the world's nations.

However, the last quarter of the twentieth century saw some slowing up in the trend toward big-city coalescence in the United States. In some regions strong counter-forces were beginning to emerge and the greatest growth occurred in a swathe of southern cities and in attractive medium-sized communities.

Figure 8.16 Urban sprawl Los Angeles is often seen as the classic example of urban sprawl and the prototype of the twentieth-century city. It is now the third-largest city in the United States, with an urbanized area that sprawls outside its legal limits to include over eight million inhabitants in the metropolitan area. Although equal in population to Chicago, it has twice its area, and it is nearly ten times as large as San Francisco. This low-density automobile-dependent metropolis has had the fastest growth of any major city in North America. In 1870 it had a population of only 5728 inhabitants; since then, the population has grown to more than one hundred times that figure. This photo shows us the view across the Los Angeles basin to the San Gabriel Mountains, with the civic centre and high-rise buildings around the CBD in the middle distance.

[Source: Photo by William A. Garnett.]

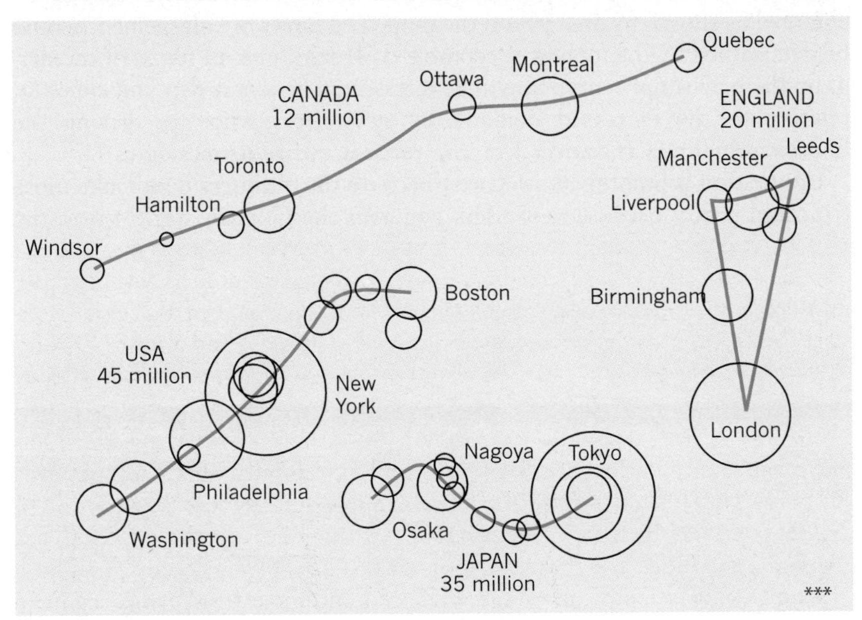

Figure 8.17 Urban agglomerations Four of the world's major elongated urban agglomerations drawn to the same scale. Circles are proportional to city population size.

[Source: M. Yeates, *Main Street* (Macmillan of Canada, Toronto, 1975), Fig. 1.7, p. 34.]

Crossing the Daily Contact Zone

Finding our way across the daily contact zone of an ever-expanding city poses acute problems. The journey to work saps our energy and steals our money and time as we move along a congested route to and from the campus office. Ways of 'solving' the transportation problem of the city are frequently put forward. But such improvements in transportation often bring short-lived benefits. Indeed, some actually increase congestion over the longer term. Why should this paradox occur?

Some insights into the whole process of urban growth may be gained from looking closely at the problem. Consider a typical route into a city centre as shown in the first map in Figure 8.18. The area used by commuters travelling in along the highway (1) forms a 'teardrop'-shaped catchment area, not unlike a river basin. Movements along the highway may be measured in terms of the daily travel time for each trip (the cost of each journey) and the volume of trips each day (the number of journeys). These are plotted as position (1) on the graph. Since in some senses travel is an economic commodity, we can add to the graph the economists' conventional supply and demand curves (S_1 and D_1), indicating that the number of trips and their cost are likely to be related. If the time into the city is short, we may be tempted to go in more often, say, for an evening meal or weekend shopping; if the time is long, we may stay home or shop locally.

The second map in Figure 8.18 shows a major highway improvement. A new multi-lane freeway is built in place of the old highway. Travel times are reduced, and we expect that there will be some small increase in the number of trips from the old users in the existing catchment area (2). In terms of the diagram the supply curve for the old users has shifted downward to position S_2 with the time of trips reduced and the volume of trips increased (as shown by position 2 on the vertical and horizontal axes, respectively).

We might expect this position to be maintained only for a short period. New drivers will learn of the improved highway and shift from their old routes. Others may locate their homes to take advantage of the new highway. So in the longer run we would expect the catchment area to be enlarged as shown by area (3) on the map. The limits of enlargement would be determined by the longest acceptable daily trip time. In terms of the diagram the new supply curve S_2 will be intersected by a new demand curve D_3 relating to the increased demand by new users who are driving the improved highway (position 3 on the vertical and horizontal axes).

Congestion is built up again partly because the old users now make more trips and partly because new users augment the flow. In some senses the

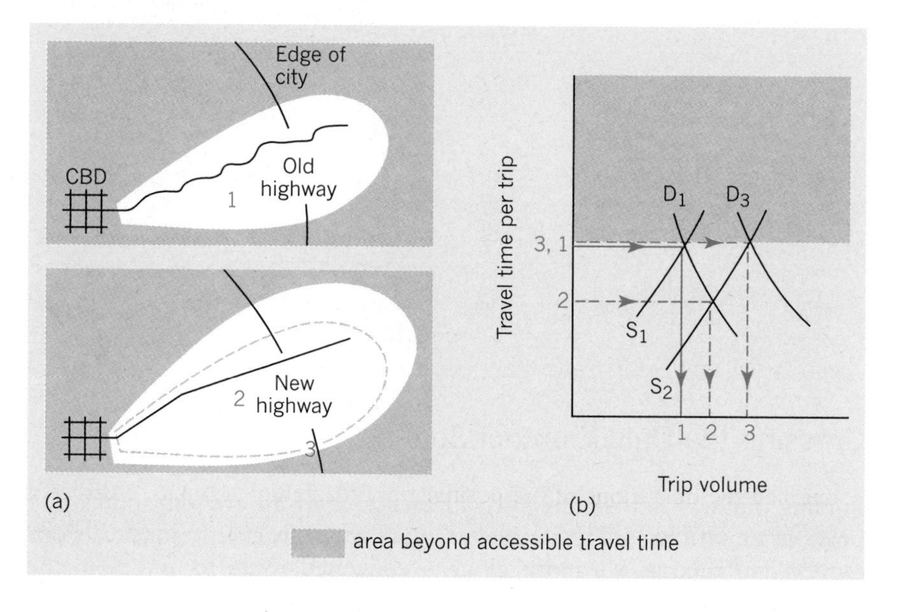

Figure 8.18 Spatial impact of an improved highway link (a) In the diagram the area beyond the acceptable commuting time is shaded. Note that the accessible catchment is expanded by the new highway. (b) S and D describe supply and demand curves. These are shifted to the right by the new highway to enlarge the trip volume from the extended catchment area.

[Source: R.D. MacKinnon, in L.S. Bourne *et al.* (eds), *Urban Futures for Central Canada* (University of Toronto Press, Toronto, 1974), Fig. 13.1, p. 238.]

'solution' has failed, but in another sense it has not because the new highway is now performing more effectively than the old did in terms of the *total* load it is carrying. We must also add the benefits from the other previously used routes, which may now be less congested. Altogether the impact of highway improvements may be complex, and the evaluation may be very different when viewed from an individual rather than from the city in total. In the growing, sprawling city, transport 'improvements' are not likely to bring too many improvements to the individual, yet the city itself gains – and so goes on growing.

The City as a Problem Area 8.5

When representatives from 132 nations assembled in Vancouver in June 1976 to convene HABITAT, the first United Nations Conference on Human Settlements, their focus was on the city as an international problem area. As we note at several places in this book (see Chapters 6, 19, 20), unprecedentedly high rates of population growth and massive rural to urban migration is causing critical urban problems around the world. A key commentator on these trends has been geographer David Harvey (see Box 8.C on 'David Harvey's views of the city'). Here, we divide these problems into those arising from *inward* migration to the city, those from *outward* spread of the urban area, and those from the operation of the city as an *ecological* unit.

Box 8.C David Harvey's Views of the City

David Harvey (born Gillingham, Kent, England, 1935) is the most influential geographical commentator on the evolving western city. A Cambridge graduate who also studied at Uppsala, Sweden, he held the Mackinder Chair at Oxford and is now at Johns Hopkins University in the United States.

Harvey's work has gone through several phases. Originally a historical geographer, Harvey was one of the first British human geographers to embrace the theoretical and quantitative approach being developed in North America. His major contribution was a seminal book *Explanation in Geography* (1969), which outlined the positivist conception of science and argued for its adoption by geographers. But Harvey also concluded that positivism was tied to an ideology of society that precluded major social change. His exploration of other philosophies led to an even more influential book, *Social Justice and the City* (1973), which outlined the limitations of the positivist approach and was the first major presentation, in English, of the case for a Marxist geography. Over the last quarter century, Harvey's prolific writing has widened and deepend our understanding of the western city. A flood of books has included *The Limits to Capital* (1982), *The Urbanization of Capital* (1985), *The Condition of Postmodernity* (1989), *The Urban Experience* (1989), and *Justice, Nature and the Geography of Difference* (1996). In these books Harvey used both theoretical reasoning and empirical evidence to show (a) that the geographic space is continually restructured and (b) that industrial capitalism has been a dominant restructuring force. Currently questions of environmental justice, of alternative modes of urbanization, and how to confront environmental issues in a constructive way lie at the centre of his research. He is working on a book called *Spaces of Hope* in which questions of exisiting versus alternative forms of sociogeographical life will be addressed. An account of Harvey's earlier work is given in J.L. Paterson, *David Harvey's Geography* (Croom Helm, London, 1983).

Problems of Inward Migration: Uncontrolled Settlements

Rapid population growth puts stress on the housing stocks of a city. Much of the very rapid urban growth in Third World countries occurs through *squatter* settlements. These uncontrolled settlements often lie around the periphery of the built-up area, and are made up of temporary buildings (usually built by the squatters themselves) with few public services. Their names vary from country to country: in Latin America they may be called *ranchos* or *favelas*; in Asia, *bustees* or *kampongs*; in Africa, *bidonvilles* or *shantytowns*. Attempts by United Nations agencies to assess the share of such squatter settlements in the total city populations suggest they make up from 10 to 80 per cent. For example, about a quarter of Rio de Janeiro's population is housed in this way (see Figure 8.19).

(a)

(b)

Figure 8.19 Uncontrolled settlements around a Brazilian city *Favelas*, roughly constructed of boards or tin plate, flout most of the city zoning laws, have few or no public facilities (water, sewers, telephones, etc.), and are usually built on land whose ownership is undetermined. Shantytowns of this type are a problem for urban government but also represent a logical, low-cost response to the problems of new migrants to a city. In Rio de Janeiro, over a million people live in shantytowns of the type shown. (a) The Brazilian coastal city of Rio de Janeiro with the statue of Christ the Redeemer overlooking the city. The distinctive Sugarloaf Mountain is located to the right of the statue. (b) View of a *favela* in the Mangeira district of Rio.

[Source: (a) MAGNUM/Abbas. (b) PANOS Pictures/Sean Sprague.]

Attitudes toward the squatter settlements have changed in ways that remind us of the changing attitude toward shifting cultivation in the tropics (discussed in Section 9.2). This was first regarded as harmful, later recognized as a logical reaction to tropical ecology. Much the same shift has occurred over the squatter settlements. As late as 1970, they were described in official reports as a 'spreading fungus' and 'excessively squalid and deprived'. Today a more tolerant attitude prevails as the shantytowns are recognized as performing an important transitional role in urban evolution.

Squatter settlements provide six functions of major importance. They act as reception centres for migrants. They provide housing within the means of the very poor. They provide a variety of small-scale employment. Their social and communal structure provides a cushion for residents during times of unemployment and other periods of difficulty. They encourage self-help in improving the standards of the houses. Finally, they provide a location within range of possible workplaces within the city.

Detailed studies of the squatter settlements show they range enormously in the standards of housing, amenities, and access. For most the standards of facilities are very poor, and the risks of public health problems are correspondingly high. But for some newcomers they provide steps on the urban escalator that probably could not be provided in other ways. Clearance of shantytowns and rehousing the population in more distant areas, as has recently happened in Manila, has generally been expensive for the city government and has pushed the displaced population ever further from job opportunities near the city centre.

Ghetto Formation Uncontrolled squatter settlements are largely a Third World problem. Inward migration of the rural poor in the western world has led to a spatially different though socially similar phenomenon, the ghetto. The term *ghetto* originally described Jewish areas within the medieval cities of eastern and southern Europe. In such cities the Jews lived apart from the rest of the community and in some cases even had a wall separating their area from the rest of the city.

In terms of the modern American city the term ghetto is more generally used to describe a confined part of the city occupied by a distinct ethnic or cultural group. Thus it is typified by the distribution of blacks, Puerto Ricans, and Mexican Americans in many cities. Note that the ghetto is held in place by two sets of forces: the internal cohesion of the ethnic group, and the external 'walls' placed around it by the rest of the urban community. Internal cohesion may be common links of language or culture, and the role of the ghetto community as a transition point for new migrants coming into the city. Although the boundaries around the ghetto are not physical, they may be real enough in terms of the cost of housing or the willingness of families to sell to ghetto occupants. In most ghettos the exclusion process is complex: everybody excludes everybody else.

Typically the ghetto limits are coincident with areas of urban deprivation. Personal income tends to be low, housing crowded, unemployment higher, health poorer, and delinquency rates higher. In St. Louis, no less than 87 per cent of the poverty areas of the city are occupied by non-whites; in Seattle the corresponding figure is 60 per cent. A similar relationship exists in some European cities (see Figure 8.20). But as population density builds, the pressure to break out of the ghetto also rises. Typically the process of spatial extension is a jerky one – a city block or school district switching rather quickly from one group's occupation to another. But at a

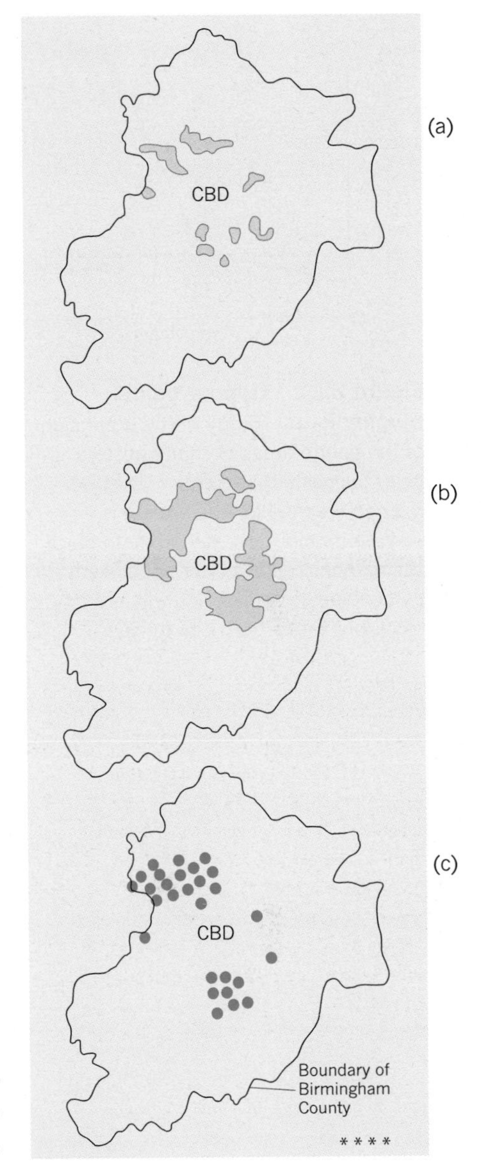

Figure 8.20 Population changes in the inner city Maps show for the city of Birmingham in central England the areas where the percentage share of New Commonwealth population was 10 per cent or higher in (a) the 1961 census and (b) the 1971 census. Map (c) shows the location of primary schools (to age 11 years) where coloured pupils make up half or more of the enrolment. Note the concentration of immigrant population in the ring of old, inner-city housing around the central business district (CBD). The main sources of New Commonwealth migration are the Caribbean and South Asian countries.

[Source: R. I. Woods, *Transactions of the Institute of British Geographers*, New Series 2 (1977), Fig. 1, p. 476.]

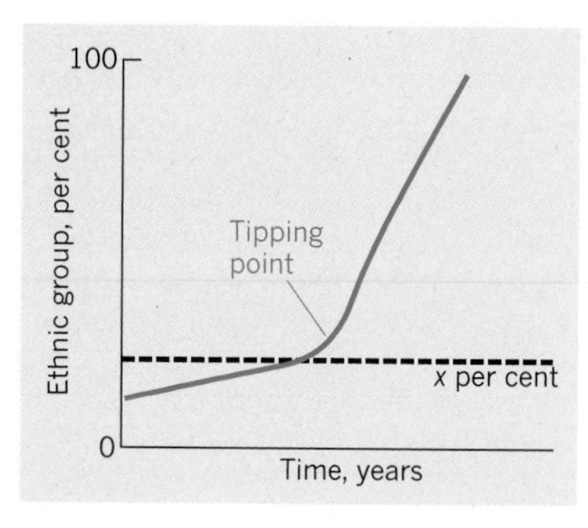

Figure 8.21 Tipping-point mechanisms Study of the expansion of the ghetto areas of many western cities suggests that a critical level (*x*) may be reached at which there is a sudden switch from, say, white to black occupation. The level *x* is not a fixed percentage of households within a city block but varies from city to city.

more general spatial scale we can see how the expanding ghetto follows a definite wedge structure.

In a large American city there may be a variety of ethnic nuclei. Thus Baltimore not only has major black areas but other smaller groups – Lumbee Indians, Appalachian populations, Chinese, Jamaicans, and Cubans. Boston has well-defined Italian and Irish districts. Study of the growth of these ethnic areas usually shows that, even if the pattern is discontinuous, an outward wedge seems to develop. Everyone gradually increases the size of their slice of pie, although the slices may be less well defined the further from the pie's centre.

This process of neighbourhood change may be described in terms of a *tipping mechanism*. When the level of black house ownership or school enrolment reaches a critical level (say, 15 per cent) then there is (a) rapid acceleration in the outward movement of whites and (b) a refusal of whites to buy property in that area. The combination of both factors may together cause very rapid change (see Figure 8.21). Studies of census tracts in Chicago show that up to 75 per cent of dwellings may shift to black ownership over three years once the tipping point is reached. This represents the upper range of 'white flight' behaviour, and turnover will generally be substantially less depending on the age-composition of the population and the condition of the housing stock. No single value for the tipping point exists, and some communities are much readier than others to coexist at particular mixture levels.

Although we have discussed the tipping phenomenon in terms of ethnic groups, the same kind of models may apply to other groups. In Belfast, Northern Ireland, the Catholic–Protestant dispute has also led to separation on a street-by-street basis.

Problems of Outward Migration

We have noted in this chapter that the streams of inward migration were associated with an outward spread of population at lower densities. Two of the problems that this has led to are those of city government and sustaining the central business district (CBD).

City Government

We illustrate the problem of governing the rapidly expanding city by looking at Toronto (see Figure 8.22). As Canada's largest city and commercial capital the city has gone through wave after wave of growth over the last hundred years, building and rebuilding its central core and sprawling outward in a profusion of new factories and homes on the margins of the built-up area. The maps show the rapid expansion of the urbanized area. Growth brought with it a series of problems in terms of such items as highway planning, housing shortages, lack of adequate utilities such as water and sewerage. These conditions led the Ontario legislature to create the Municipality of Metropolitan Toronto ('Metro') in 1954. This was modified in 1966 by the merging of the municipalities into six units – the city of Toronto, and five 'outer' boroughs shown on the map. Metro has brought a series of successes to the government of this prosperous and attractive city: new highways, an extended subway system, housing for the aged, flood-control schemes, and major developments in recreational areas.

But the city has continued to boom. As Figure 8.22 shows, the tidal wave

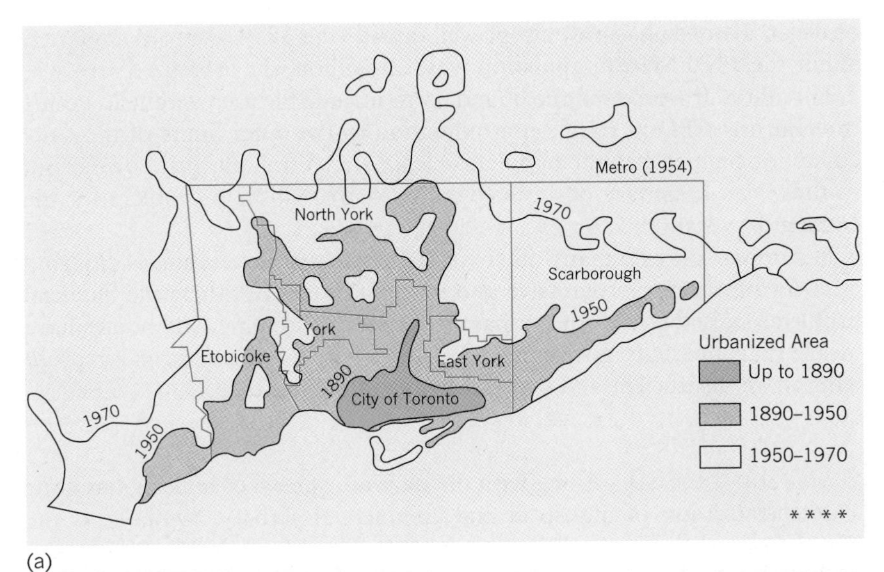

(a)

(b)

Figure 8.22 Governing the expanding metropolis (a) The rapid growth of the Canadian city of Toronto led to the amalgamation of existing communities into a unified metropolitan area (Metro) in 1954. At that time the Metro boundary contained the greater part of the urbanized area. Over the next twenty years the urban tide swept well beyond this boundary. (b) Aerial view of downtown Toronto and shoreline on Lake Ontario.

[Source: (a) After J. W. Simmons and L.S. Bourne, in L. Gentilcore (ed.), (University of Toronto Press, Toronto, 1972), p. 92, Fig. 5.5. (b) Colorific!/Sylvain Grandadam.]

of urban Toronto has now swept well outside the 1954 Metro boundaries. While the 1970 Metro population was 2.0 million, the urbanized area was 2.35 million. If we extend the boundary to include the area within an hour's travel of the CBD (a rough approximation to the outer limits of the daily commuting zone) then the population leaps to 3.5 million. Just how far out to draw this boundary of any expanded Metro will be a problem for the coming decades.

In comparison with many other city areas, Toronto's revision of city government has been a progressive and successful one. In others the financial problems caused by loss of tax-paying industry, shopping, and homes have meant that inner-city government has been left to cope with major problems on an insufficient and sometimes dwindling tax base.

Sustaining the CBD Along with the outward sprawl of housing has gone a decentralization of industrial and commercial activity. Symbolic is the out-of-town shopping mall with its major department stores, its massive parking lot, and its location near one of the superhighways. As suburban development has pushed further from the city, so the ring of shopping plazas has become larger and more diverse. Originally built to cater for convenience and weekend shopping (with an emphasis on food and clothes), the newer centres include more specialized shops.

As this trend has gathered momentum, so the role of shops within the CBD has declined. For example in Chicago the last 30 years have seen the daily number of shoppers in the CBD fall by nearly one-half. Since the late 1960s insurance giants such as Prudential and John Hancock, and major retailers such as Sears Roebuck and Montgomery Ward have been investing heavily in downtown centres, attempting to bring shoppers (as well as jobs and homes) back to the central areas. Private and public investment in the downtown has also led to very substantial schemes aimed at restoring the flagging centres of American cities.

Another approach to sustaining central activity has come through transport innovation. For example, commuters to San Francisco's CBD face special problems. The very broken topography of the San Francisco Bay metropolitan area (3.1 million people and the United States' sixth most populous Standard Metropolitan Statistical Area, SMSA) means bridge crossings between the west and east sides. The year 1972 saw the opening of a new 120 km (75 miles) rail facility for commuters, the Bay Area Rapid Transit District (BART) system (Figure 8.23). This aimed to provide a high-speed commuter service and to divert commuters from cars to trains and to overcome the congestion for cars crossing the bridges into San Francisco from the East Bay Area. Trains are very frequent on the system during the peak rush hours, and are scheduled by computer to run at average speeds of 72 km per hour (45 mph), nearly twice as fast as average commuter speeds by automobile. The system includes a number of interesting innovations in the design of trains, the stations, and control, and set new standards in luxury.

Despite the imaginative scheme, BART has proved only a partial success. Tactical problems relating to fare levels and commuter parking at stations remain to be solved, and much of the original system is still unbuilt. Perhaps the biggest strategic problem relates to CBD employment. As the Bay Area has grown, so satellite centres have emerged outside San Francisco itself, which now has only a third of the region's jobs. So, much of the region's commuting problem lies in cross-movements within the metropolitan area

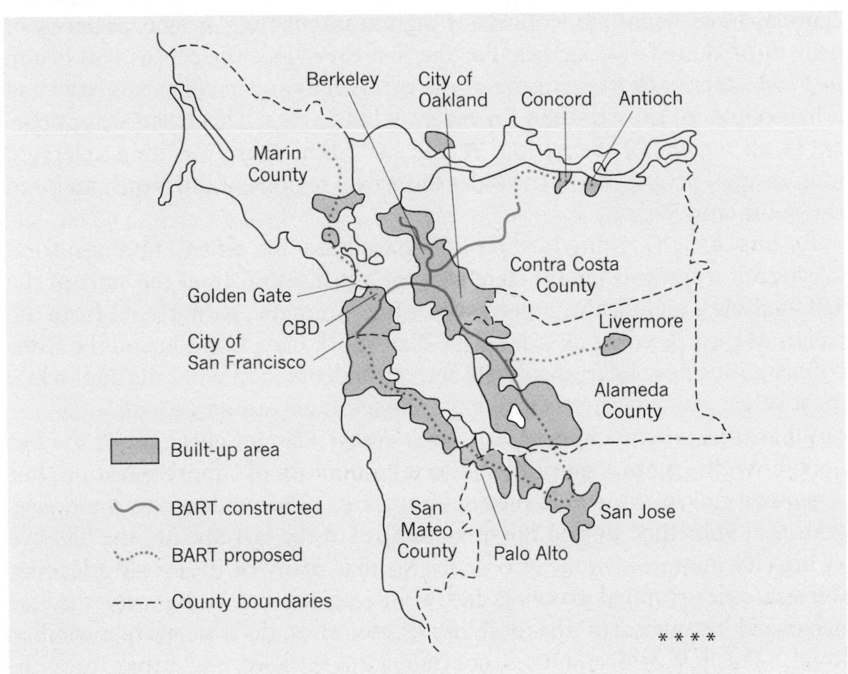

Figure 8.23 Rapid transit solutions to city commuting problems The physical configuration of San Francisco Bay poses special problems for commuters crossing the bridges to get to the CBD area. The Bay Area Rapid Transit (BART) system opened in 1972 represents one attempt at a solution. Earlier studies envisaged most of the inner Bay Area counties being served by the system but San Mateo withdrew from the scheme in the 1960s and Marin County joined to San Francisco by the Golden Gate Bridge, was considered too difficult to serve under existing conditions.

[Source: Quadrant Picture Library/Eric C. Hayman.]

rather than simple suburbs-to-CBD transit. Few of those commuters in the outer Bay Area use public transport, and those who do tend to use the more flexible bus services. Thus the BART system shares, like all fixed-rail systems, the fundamental geographic difficulty that it must focus on the high-intensity traffic moving into downtown. The more jobs decentralize, the harder it is to run fixed-rail systems, except by massive public subsidy.

The City as a Complex System

As with the natural ecosystems discussed in the first part of this book, so the city ecosystem demands flows of energy and nutrients to keep it alive.

Geographer Sherry Olson has described how metropolitan Baltimore – two million people, three-quarters of a million dwellings, and 2000 factories are packed onto 3540 km^2 (2200 square miles) – is sustained by flows. The 150,000 people who move into and out of the city to work each weekday and the two million internal trips are supplemented by the trucks, ships, and planes coming in and out along freeways, harbour lanes, and airways. To this must be added the 1136 million litres (250 million gallons) of water pumped in each day, and draining out along the sewers; the flows of electricity and gas; the 5 television and 26 radio stations and the millions of telephone calls binding the life of the city together. Clearly the city itself can be viewed as a special kind of ecosystem.

The complex worlds within the city are local reflections of the city's role in relation to the outside world. It is to that outside world that we turn later (see Chapters 14 and 15).

The Limits to Urban Growth

Analysis of the results now available from the last censuses of the twentieth century have begun to confirm a significant change in the patterns of growth of United States cities. For the first time since the census was begun in 1790 (refer back to Section 6.4) the rural areas and small towns have had a higher rate of growth than the metropolitan areas. This trend was apparent in all regions of the nation. At the same time there was an accelerated migration of people out of the older industrial regions of the North and into the South and West.

Results for the United States are paralleled by census findings from Western Europe and Japan. Here too, the censuses taken at the start of the 1980s show a significant, indeed sometimes dramatic, turnaround from the trends of the preceding decades. Like New York City, London and the Ruhr conurbations now have absolutely fewer residents than a decade ago, while most other major western cities have slowed their rate of growth.

What then is happening to our large cities? Careful checking of the figures show this is not a quirk of changed boundaries of suburbanization, but a genuine fall in the whole metropolitan area. The evidence to hand suggests that sometime during the final decades of the last century the balance of big city migration went into deficit so that many of the world's leading western cities stopped growing and went into reverse. Results for the less urbanized countries of the developing countries do not show a similar break. Third World countries continued to show strong urban growth, although at somewhat lower rates than forecast. (For example, Mexico City at 2000 has a population of 18 million, well below the fashionable projection of over 32 million made a generation ago.) Whether their growth path will follow the western city pattern is still a major debating point for city watchers. An index of quality of city living made up of some 40 measures of city living (from open space, through job prospects, to crime) put medium-sized cities at the top (see Table 8.3). Vancouver (Canada), Zurich (Switzerland), and Sydney (Australia) occupied the top three slots.

Can we then talk about some limits to the eventual size of cities? We have already seen that a major factor in urban growth is the *centripetal force* of agglomeration economies. But these advantages do not accrue indefinitely as size increases. Lower production costs may be outweighed by increased transport costs as urban areas grow to such a size that raw materials must be imported and finished goods exported over greater distances. Cities may

Table 8.3 Quality of city living

City	Index (New York = 100)
1. Vancouver (Canada)	104.8
2. Zurich (Switzerland)	104.4
3. Sydney (Australia)	104.3
4. Geneva (Switzerland)	104.2
5. Auckland (New Zealand)	104.1
6. Toronto (Canada)	103.9
7. Copenhagen (Denmark)	103.8
8. Melbourne (Australia)	103.7
9. Vienna (Austria)	103.7
10. Amsterdam (Netherlands)	103.7
11. Brussels (Belgium)	103.1
12. Perth (Australia)	103.1

Quality of Life Index is based on a combination of 42 aspects of city living ranging from crime rates and personal security, through housing and transport, to recreational resources and political stability. The index reflects data available in November 1998.

become increasingly congested, internal transport costs rise, and public health dangers from infectious diseases or antisocial behaviour increase. Let us take again the familiar example of the United States. As Table 8.4 indicates, most big cities have severe social problems – high crime rates, drug abuse, poverty, and increasing pollution. These are the *centrifugal forces* that tend to scatter population away from the cities. In general the problems appear to get worse, the larger the city. But, as the table also shows, the rewards (as measured by median income) also increase with city size.

Urbanization continues to increase so long as the benefits to be gained from crowding exceed the costs. As the cost of overcoming spatial separation has fallen, so the relative strength of centripetal forces has grown. Water pipelines, supertankers, and refrigerated ships are symbols of a continuing transport revolution which enables centralized population

Table 8.4 Some implications of city size[a]

Characteristic	10,000 inhabitants	100,000 inhabitants	1,000,000 inhabitants
Median income	90	100	120
Crime rates			
Murder	37	100	310
Rape	38	100	260
Auto-theft	30	100	320
Air pollution	82	100	155

[a]United States data for the 1960s standardized to give characteristics of cities of 100,000 an index value of 100. Note that the median income is probably underestimated due to the 'suburbanization' effect (the fact that those who earn large incomes in the city may reside outside its legal or statistical boundaries).

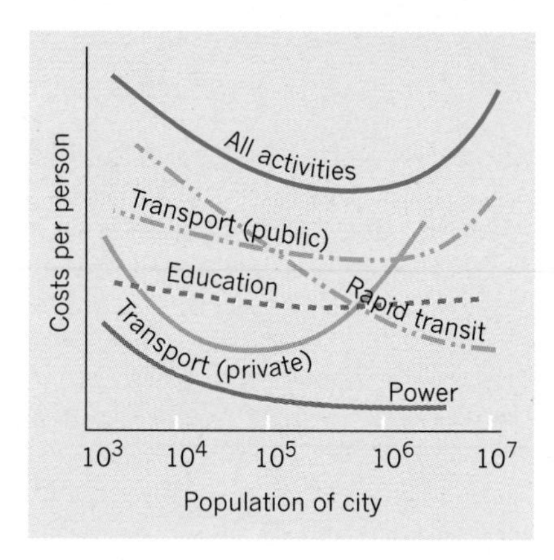

Figure 8.24 Urban economies
The general cost curve for all activities suggests that middle-sized cities may be more efficient than either large cities or small towns. This does not necessarily mean that they are preferable, however, as comparable curves for welfare or satisfaction levels show.

[Source: R.L. Morrill, *The Spatial Organization of Society* (Wadsworth Publishing, Calif., 1970), Fig. 8.01, p. 157. Copyright © 1970 by Wadsworth Publishing Co., Belmont, Calif. Reprinted with permission.]

agglomerations to draw on widely distributed resources; equally important are the sewage lines and disposal services which allow a city to dispose of its massive daily burden of wastes. It is difficult to measure the exact balance of costs of cities, but the probable form of the cost curves as the size of the cities increases is given in Figure 8.24. This figure suggests that there is a threshold over which further increases in size are uneconomic and growth will slow down.

At any one time, one particular bottleneck may be critical. For the eighteenth-century European city, the water supply and infectious diseases were significant barriers to growth; for today's cities, crime, energy, and finance may be higher on the agenda. Geographers view cities as large and complex ecosystems. As in all such systems, life is dependent on the inflow of resources and the outflow of wastes. Over the last few centuries that flow has been at very high levels, and cities have grown accordingly. But what of the future? We shall look at this question again in Chapter 19.

Reflections

1 The proportion of the world's population living in cities has consistently increased over the period since 1700 when reasonably accurate figures are available. But why do cities grow? List (a) the benefits and (b) the costs of high-density crowding of the human population into large cities. Is the balance between the two changing? If so, in what direction is it shifting?

2 Work through the Lowry model as shown in Figure 8.6. Substitute alternative multipliers for those in the diagram (say, 5 for 4, 1/10 for 1/8), and recalculate the effect of 100 new jobs in the basic sector, ignoring the effects of the constraints. How would you improve this model to make it more realistic?

3 Jobs may be divided into four sectors: primary, secondary, tertiary, and quaternary. In urbanizing countries, the proportion of jobs changes systematically over time (see Figure 8.7). Gather information on the proportion of jobs in the primary, secondary, and tertiary sectors for your own country. Do you consider that it is now in the preindustrial, industrial, or postindustrial phase?

4 The organization of cities reflects in part the mobility and biological clock of the human species. How does the need to sleep for around eight hours a day affect the organization of cities? What would happen if we needed sleep once every six hours, or once every six days?

5 The growth of very large cities in western countries has now slowed. List factors you think might be important in explaining this. Compare these with those chosen by the rest of your class.

6 Review your understanding of the following concepts using the chapter text and the glossary in Appendix A of this book:

agglomeration economies	quaternary sector
centrifugal forces	secondary sector
centripetal forces	tertiary sector
Lowry model	urban implosion
primary sector	urbanization curves

One Step Further ...

A useful starting point is to look at the essays on urbanization, western, and Third World cities by Alan Gilbert, David Herbert, and David Drakakis-Smith in I. DOUGLAS *et al.* (eds), *Companion Encyclopaedia of Geography* (Routledge, London, 1996), pp. 391–407, 702–52. A brief but useful introduction to urban geography is also given by Ron Johnston is his essays in R.J. JOHNSTON *et al.* (eds), *Dictionary of Human Geography*, fourth edition (Blackwell, Oxford, 2000), pp. 870–84. Two standard texts in the field are H. CARTER, *The Study of Urban Geography*, fifth edition (Arnold, London, 1995) and T.A. HARTSHORN, *Interpreting the City*, second edition (Wiley, New York, 1991).

The distribution of urban centres worldwide is described in a well-illustrated text, S.D. BRUNN and J.R. WILLIAMS, *Cities of the World* (Longman, Harlow, 1993) while the forces which lie behind urban growth are explored in P.J. BOYLE, K.H. HALFACREE, and V. ROBINSON, *Exploring Contemporary Migration* (Longman, Harlow, 1998). Works on the evolution of urban systems which emphasize the role of contact patterns and the location of corporate headquarters are J.S. BOURNE and J.W. SIMMONS (eds), *Systems of Cities* (Oxford University Press, Oxford, 1978) and A. PRED, *City Systems in Advanced Economies* (Hutchison, London, 1977). The classic geographic study of urban growth in the Boston–Washington corridor is Jean GOTTMAN, *Megalopolis* (MIT Press, Cambridge, Mass., 1964), while a comparable study for the Windsor–Quebec corridor in Canada is M. YEATES, *Main Street* (Macmillan, Toronto, 1975). Look at David HARVEY, *Justice, Nature, and the Geography of Difference* (Blackwell, Oxford, 1996) for a rich source of ideas on the implications of urban change on wider social and environmental issues.

The urban geography 'progress reports' section of *Progress in Human Geography* (quarterly) will prove helpful in keeping up to date. Urban geography now dominates many geographic journals, and you will find something of relevance and interest in most issues. *Urban Geography* (quarterly) is a specialist journal in this area. Those really enthusiastic about urban studies can keep up to date with journals such as *Urban Studies* (quarterly) and the *Journal of the American Institute of Planners* (monthly). For readers with access to the World-Wide Web see also the sites recommended for this chapter in Appendix B at the end of this book.

PART III

Resources and Landscapes

CHAPTER 9

Pressures on the Ecosystem

■ denotes case studies

Figure 9.1 **Intense population pressure on available land** Steep terracing, which is linked to intricate systems of rice irrigation, is used by the Ifugao tribal people in the Bannawol area of Luzon Island in the northern Philippines. It provides a way of using very steep slopes for irrigated rice production. Terraces are known to have been in use for some centuries; their careful maintenance takes a major proportion of the villagers' working days. [Source: Still Pictures/Martha Cooper.]

> Fair-Swooping elbow'd earth –
> rich apple-blossom'd earth!
> Smile, for your lover comes.
>
> WALT WHITMAN *When Lilacs Last in the Dooryard Bloom'd* (1860)

Minimata before 1953 was a tiny, unimportant fishing village off the coast of Japan. In that year, it began to gain worldwide notoriety as a fearful symbol of the consequences of human intervention in natural ecosystems. In that year, many residents of the village went down with a mysterious and deforming disease of the nervous system. Termed the ***Minimata disease***, it was later traced to concentrations of a deadly mercury compound – methyl mercury – in human body tissues. Altogether, 900 people in the neighbourhood of the village were affected by mercury poisoning; of these, 52 died and nearly twice that number were crippled beyond recovery.

Once the cause of the disease had been identified, it was not hard to trace the source of the mercury to the wastes discharged into Minimata Bay by a giant chemical plant. But Minimata was not an isolated case. The same problem of mercury poisoning was later to force a 1967 ban on fishing in 40 Swedish rivers and lakes. In 1970, a scare developed in North America when a graduate student at the University of Western Ontario turned up dangerously high mercury levels in his research on the fish populations of Lake St. Clair. (Lake St. Clair lies on the Canadian–US border northeast of Detroit.) Elevated mercury levels were subsequently found in 30 other states in the United States.

Minimata and mercury provide only one extreme example of the problems that can arise because of human intervention in the tangled web of ecosystems, on which the survival of human populations on the earth ultimately depends. Every time we modify the earth's surface we set off a chain of events, some expected, some accidental. Our opening photograph (Figure 9.1 p. 270) shows such a modification. Here the Ifugao people of the northern Phillipines have built huge flights of terraces to allow irrigated rice cultivation on slopes far too steep to be tilled by any other method.

In this chapter we take a long view over human occupation of the earth and try to place both mercury poisoning and terracing within a single space–time context. We look first in Section 9.1 at the scale and pattern of human intervention and then at the increasing degree of intervention as human population densities were built up. Sections 9.2 to 9.4 trace the impacts of human intervention at low, medium and high population densities. Finally, in Section 9.5 we look at the most recent pollution problems to place Minimata in context and to suggest a geographical framework for studying the increasing number of ***pollution*** incidents.

9.1 Intervention: Benign or Malignant?

Before considering the ways in which humanity intervenes in natural ecosystems, readers may find it useful to have a recap of some of the points made in Section 4.1. There we noted (a) that ecosystems were structured webs connecting the material environment and its biological population, (b) that the main linkages in the ecosystem were food chains running from simple phytoplankton to higher animals, and (c) that the size of biological populations appeared to be controlled by complex feedback mechanisms related to the food supply.

As one of the higher animals, *Homo sapiens* comes at the end of both terrestrial and marine food chains. As both herbivores and carnivores, people are consuming predators of both plant and animal products. Though once a prey to a few other higher animals, humans are now almost entirely removed from this role if we discount the few deaths due to sharks, tigers, and the like. We remain vulnerable, however, to a host of micro-organisms – mostly notably the disease-carrying virus and bacillus populations. (We review their effects in looking at human diseases in Chapter 20.) This position of the human species in the ecosystem has been reinforced by two further critical factors: first, the dramatic exponential increase in numbers of the species, and second, its growing power to modify food chains through technology. To put it simply, our 'natural' position in the ecosystem gave us a potential for dominance; our subsequent technological development and increase in numbers enabled us to capitalize on this potential.

Designs for an Improved Ecosystem

Most human changes in ecological systems have had a benevolent purpose. For in most cases, intervention has been directed at improving the productivity or the habitability of a given environment.

Consider the example of the lake ecosystem discussed in Chapter 4. (Recall Figure 4.4.) How could people modify the lake so as to 'improve' it for their own purposes? If we assume the objective of the improvement to be the basic one of increased food production, then we can easily draw up a list of means. At the beginning of the list will come simple schemes such as that of the selective killing of fish or animals that prey on species with a food value to humans. Thus, we might try to eliminate species such as pike in order to increase the numbers of trout or carp. At the end of the list will come major schemes for environmental reorganization such as draining the lake and using the fertile lake-bed soils for direct crop production. The first type of intervention demands only primitive resources (plus some basic understanding of the ecosystem); the last type demands an advanced technology and a high input of resources, for it entails sweeping a whole ecosystem away to replace it with another.

All such processes result in one or more of three main categories of change. First, there is the impact on other animal populations. Thus, the expansion in the number of *Homo sapiens* has been accompanied by a great expansion of some animal species (usually domesticated ones such as the horse, the cow, and the chicken) that are of direct use to human beings. At the same time some species have been severely reduced in number or wholly eliminated. Figure 9.2 shows the progressive destruction of the North American bison herds on the mid-continental plains during the nineteenth century. At the same time, cattle were being introduced into the same environment and today exist in far greater numbers than the species they replaced.

Second, there is the impact on plant populations. A similar process of selective destruction and expansion of plant populations has led to reorganizations in the balance of plant life. These reorganizations range from rather complete zonal changes, such as the replacement of the mid-latitude

Figure 9.2 Human impact on animal populations The maps show the diminishing range of the North American bison over a 75 year period of European settlement, from (a) before 1800 to (d) 1875. A conservative estimate of the numbers of bison when the first European settlers arrived in North America is 60 million, probably the largest known aggregation of large animals. The bison (or 'plains buffalo') provided the mainstay of the economy of Plains Indian tribes, but was slaughtered at an increasing rate during the nineteenth century as European agricultural settlements spread westward across the continent. By 1900 a low point had been reached, and the species was on the verge of extinction. Since then bison have been protected on government reserves and the number of managed herds now runs to several thousand animals. The dangerous decrease in range and number of the bison has parallels with the current situation for other large animals (notably the whale and the rhinoceros) in other environments today. (e) Bison herd in Missouri, 1856.

[Source: (a)–(d) Data from J.A. Allen, in R.H. Brown (ed.), *Historical Geography of the United States* (Harcourt Brace Jovanovich, New York, 1948), Fig. 9, p. 379. (e) Mary Evans Picture Library.]

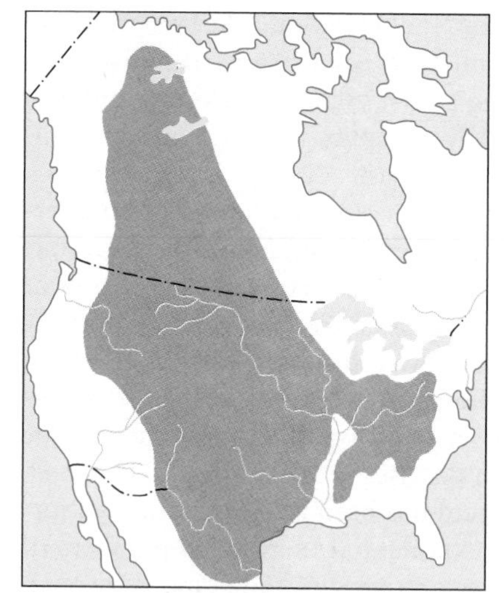
(a) Before 1800

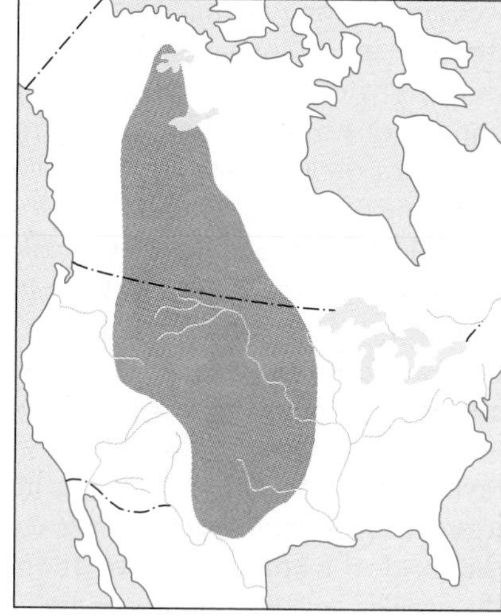
(b) 1825

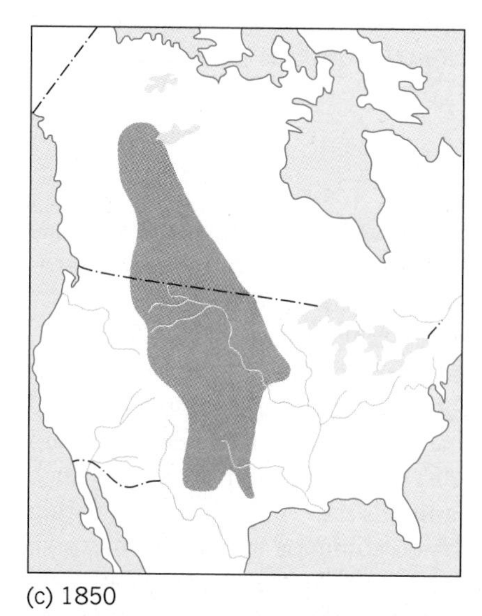
(c) 1850

Northern herd
Southern herd

(d) 1875

(e)

mixed woodlands of western Europe by an intensely cultivated mosaic of cropland, grassland, and woodland, to strictly local changes in species composition. These changes in land use are reviewed at length in Chapter 11.

Third, humans affect ecosystems by directly altering the inorganic environment. This has traditionally been achieved by interventions in the hydrologic cycle. Irrigation schemes ranging all the way from the primitive diversion of streams to massive basinwide projects are one aspect of such intervention (i.e. bringing water to dry areas to artificially boost plant production). The other side of the coin is the removal of surplus water from marshy or waterlogged areas by drainage and reclamation. Perhaps the most dramatic examples of reclamation are in coastal areas. In the Netherlands, the history of land reclamation goes back to early dike-building projects in the eighth and ninth centuries. Reclamation of a large part of the Zuider Zee by the creation of extensive polders (see Figure 9.3) was begun in the early 1920s and is still continuing. In 1957 a plan was adopted for a twenty year programme of reclamation in the Schelde and Rhine estuaries of south Netherlands which is now nearing completion.

As human technological reach increases, the capacity to alter inorganic environments is accelerating. Projects for the major remodelling of terrain

Figure 9.3 Land reclamation One of the most spectacular examples of human intervention in the natural environment is provided by the reclamation of land from the sea in the Netherlands. (a) This extract from a seventeenth-century map of the Zeeland district in southwest Netherlands shows land reclaimed around the town of Terneuse on the southern side of the Schelde estuary. The method of reclamation was the building of protective banks (dikes) to give a series of polders from which water was drained by windmill-driven pumps. The reclamation, which began in the twelfth century and has continued at an accelerated rate in the twentieth century, has added over a half to the original land surface of the country. (b) Oblique aerial photograph of settlement on a Dutch polder.

[Source: (a) Visscher Map of Holland of the Dutch Brabant, 1747; The Hulton Getty Picture Collection Ltd. (b) Aerofilms.]

with massive earth-moving equipment and by the controlled use of explosives and attempts at small-scale climatic modifications are examples of this trend.

Accidental Side Effects

To the changes that follow directed and purposeful human efforts we must add indirect impacts that have occurred without people being aware of them. These accidents range from the spectacular, such as the much-publicized fouling of Lake Erie, to the insidious, such as the slow rise in DDT levels in some living species. (For a review of pollution terms, see Table 9.1.) Books with titles such as *Rape of the Earth*, *The Population Bomb*, and *Silent Spring* have attracted public attention to these growing by-products of human activity. But it seems likely that many second- and third-order effects of humans on the environment remain undetected and surface only as links in complex ecological systems.

Any list of these indirect impacts would probably be incomplete. We can include the following: (1) accelerated erosion and sedimentation following changes in the vegetational cover of watersheds; (2) physical, chemical, and biochemical modifications in soils following cultivation or grazing; (3) changes in the quantity and quality of groundwater, surface water, and inland waters; (4) minor modifications of rural microclimates and major modifications of urban microclimates; (5) alterations in the composition of animal and plant populations, including both the elimination of species and the creation of new hybrids.

The extension and expansion of all five categories are possible. For example, the third category might be expanded to include not only proven

Table 9.1 Important terms used in the study of pollution

Term	Definition
Biodegradable	Adjective applied to pollutants that can be decomposed by biological organisms.
Biological concentration	Process by which organisms concentrate certain chemical substances to levels above those found in their natural environment.
Biological magnification	Chemical substances in food by each organism of a food chain.
DDT	Popular name for *dichlorodiphenyl-trichorethane*, a powerful insecticide developed in Switzerland in the 1930s.
Dioxin	Very powerful poison present in some weed-killers and found to cause certain deformities in foetuses.
Eutrophication	Excessive growth of algae in nutrient-rich waters leading to reduced oxygen levels and the death of many organisms.
Greenhouse effect	Excessive accumulation of heat and water vapour in the earth's atmosphere due to the increased retention of solar energy by polluted air.
Methyl mercury	Highly toxic compound of mercury widely used as a pesticide.
Minimata disease	Name given to symptoms of mercury poisoning first recognized on a large scale in Japan.
Particulates	Tiny particles of solid or liquid matter released into the atmosphere through air pollution.
Photochemical smog	Air pollution caused by the reactions between sunlight and particulates that produce toxic and irritating compounds.
Recycling	Reprocessing of waste products for reuse.
Teratogenic pollution	Pollution causing birth defects.
Thermal pollution	Discharge of heat into waterways causing reduced oxygen levels and disrupting natural biological cycles.

effects on groundwater levels, such as the lowered level of water in the Texan aquifers (water-bearing rocks), but also the uncertain impact of toxic chemicals on lake waters. In a similar manner, human modifications of microclimates are being extended to regional levels by decades of Russian river abstraction in Central Asia. (See the discussion of the Aral Sea in Figure 11.2.) As an extension of the fifth category, the long-term effects of fissionable materials on the genes of animals and plants can only be guessed.

Because a complete roster of impacts, both planned and accidental, is impractical, we shall examine some case studies to highlight the character of the interventions. We shall organize these cases around the idea of population densities; for, in general, the density of human numbers is an approximate indicator of the degree of environmental alteration. Other things being equal, the most crowded parts of the globe are those that have experienced the greatest environmental change.

Human Intervention at Low Densities 9.2

We begin with those areas of the earth in which human populations are low. Today these areas of light intervention are restricted to areas of hunting, grazing, and shifting agriculture. But historically, they cover by far the longest period of human occupation. We look at the role of fire and of grazing and consider the way in which humans helped to create new ecological communities.

Fire and Shifting Cultivation

Typical of the permanent occupation of the earth at low densities are the tropical grasslands and forests. In both of these areas, one of the most important ways of modifying the environment has been by fire. Even before the advent of humans, occasional fires were started by lightning bolts or volcanic action in most vegetated areas except tropical rainforests. The long-term impact of such periodic fires is difficult to judge, but there are some indications that species native to the chaparral of the summer-dry Mediterranean zones and the savannah of the winter-dry subtropical zones evolved in association with fire.

Early peoples undoubtedly caused fires themselves. Apart from accidental campsite fires, there were two basic types of purposeful burnings. The firing of native grassland in the dry season provided fresh growth for grazing herds; forest trees in the humid tropics were burned to permit cultivation. Vegetation in the humid tropics would not burn naturally, but, by girdling or felling, trees could be dried sufficiently in the dry season to burn. The newly opened areas of a forest allowed crops to grow for a few years before production dropped and the plot was abandoned as shown in Figure 9.4.

Shifting cultivation (sometimes termed *swidden*, or *slash-and-burn*) releases large quantities of soil nutrients which can be used by planted crops for a few years. However, these are quickly depleted and the soil and the vegetation can take fifteen to twenty years to recover before cultivation can

(a)

(b)

Figure 9.4 Shifting cultivation in the humid tropics (a) Niger delta of Nigeria showing shifting cultivation. Cultivated plants shown in the foreground are planted among the ashes of the cleared forest. (b) Deforestation in Amazonian rainforest with slash-and-burn cultivation, Rio Alto Nishagua, Peru. To set this in context, turn to Figures 11.3 (eastern Brazil), 15.1 (Rondonia, Brazil), and 20.22 (Rondonia, Brazil) which show at other scales the attack on the tropical rainforests of South America.

[Sources: (a) Sara Leigh Lewis/PANOS Pictures. (b) © Tony Morrison/South American Pictures.]

be repeated. Figure 9.5 shows that if the cultivation cycle becomes too short, as it is likely to do when the population rises and crowding increases, then there may not be enough time for the lands to regain their original levels of fertility. Burning a 40-year-old tropical rainforest provides mass-

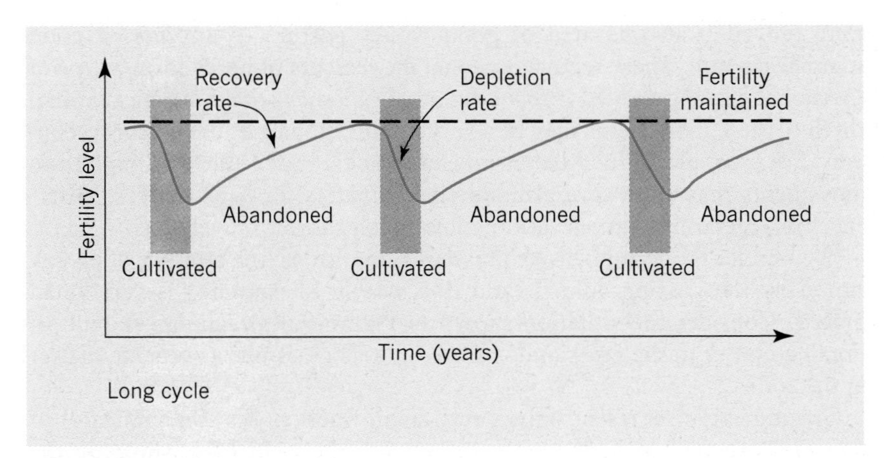

(a)

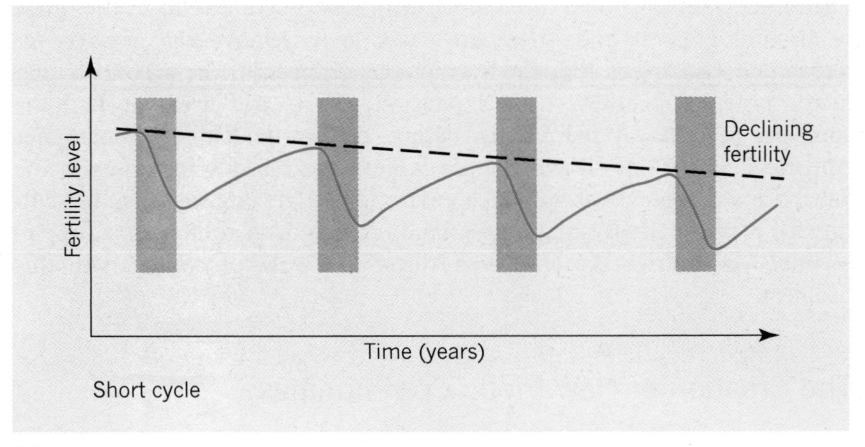

(b)

Figure 9.5 Land rotation and population density The graphs show the relationship of soil fertility levels to cycles of slash-and-burn agriculture. In (a) fertility levels are maintained under the long cycles characteristic of low-density populations. In (b) fertility levels are declining under the shorter cycles characteristic of increasing population density. Notice that in both diagrams the slopes of both depletion and recovery have the same angle, only the time gap is different.

ive doses of nutrients that can be used by planted crops. The main elements, in proportion, are calcium (100 units), potassium (32 units), magnesium (13 units), and phosphate (5 units). Burning savannah woodlands releases less than one-tenth as many nutrients, but a relatively greater share of potassium.

Fire continues to play an important part in crop strategies in many parts of the world. It eliminates unpalatable species and restricts the growth of woody and herbaceous plants. The exact role of fire in the creation and maintenance of grassland areas remains one of the puzzles of modern biogeography. The current trend in research has been toward concentration on the ecological role of humans in the creation of such areas, and we discuss this role in the next section.

The Ecology of Grazing

Attempts to reconstruct the geographic distribution of vegetation at the end of the last Ice Age suggest about 30 per cent of the land surface was then covered with open woodland and grassland. Early people cropped the wild animals of these areas for food by hunting, and for at least the last 5000 years have grazed their own domestic animals over some of these same lands. Typically, the forage for animals varied with the seasons so that herds were moved from one area of good winter pasture to another of good summer pasture. These regular seasonal movements of herds form a type of rotational grazing termed *transhumance*. In biomes with a strong seasonal rhythm, these movements may be a very important part of the pastoral economy. For example, in the Mediterranean biome (recall Table 4.3), migration movements may span several hundred kilometres. In Alpine areas, differences in vegetation with elevation allow much shorter movements.

Even under modern intensive grassland conditions, the amount of energy stored by the grazing animal (and thus usable by humans) is very small indeed. Consider the situation shown in Figure 9.6(a). The great bulk of organic matter in the grassland is stored not as grass but as organic matter in the soil.

Grazing cattle represent only a very small fraction (less than 1/2500) of the total organic stock. If we try to increase the yield by pushing up the numbers of cattle on a grassland area, a point is soon reached where production crashes as shown in Figure 9.6(b). *Overgrazing* reduces the range of pasture grasses, and, in severe cases, may remove the grass cover altogether. Grazing is thought by some geographers to be a contributing factor in desertification – the expansion of the arid areas around the southern margins of the Sahara desert. But as the Danish geographer Annette Reenerg has shown (Figure 9.6(c)), the reasons for crises in the Sahel are a complex web in which environment, grazing systems, and culture all play a part. Resting a grassland, grazing it in rotation, or supplementing the nutrient cycle with fertilizer are ways of coping with this problem.

Figure 9.6 Grassland/cattle ecosystem (a) Flows within an intensively managed grassland/cattle ecosystem. Flows and storages are in calorie units standardized in terms of an original input of 1000 units of radiant energy. Note the very small share represented by grazing cattle. (b) Impact on the yield of livestock of changing the stocking level of cattle. Note the 'crash' in production once a critical density level is passed. (c) Role of overgrazing in the Sahel famine areas of the southern border of the Sahara desert. Note the interaction between environmental factors (left) and cultural factors (right).

[Source: (a) and (b) Redrawn from D.J. Crisp (ed.), *Grazing in Terrestrial and Marine Environments* (Blackwell, Oxford, 1964), Fig. 1, p. 5. (c) Adapted from A. Reenberg, *Det katastrophferamte Sahel*, Department of Geography, University of Copenhagen, 1982.]

The Creation of New Biotic Communities

There are two general effects of low-density human occupation on the composition of biotic communities, whether forest or grassland. First, humans tend to eliminate the more conservative, or aristocratic, elements in the

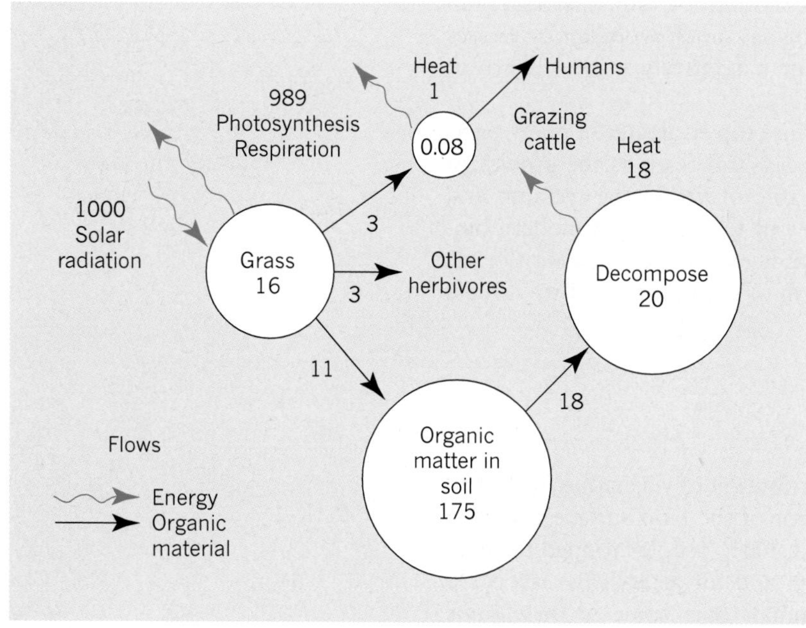

(a)

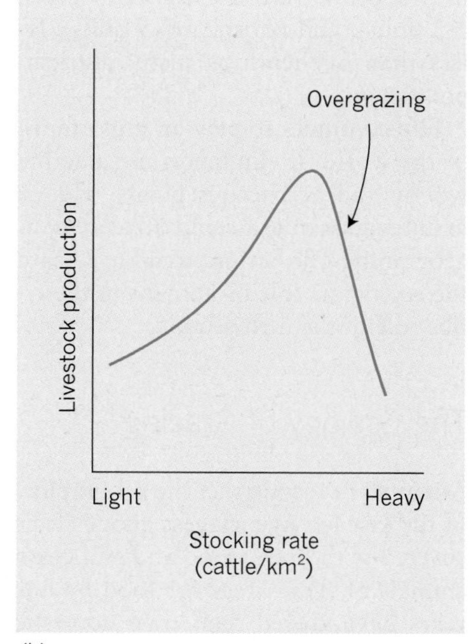

(b)

biotic population – that is, those species with a low tolerance for fluctuations in moisture levels, high nutrient requirements, or little ability to withstand disturbance. Second, humans usually expand the numbers of the less conservative plants that have higher tolerances of drier, lighter, and more variable conditions. Where humans have been active for a long time, plant communities tend to be composed of a small number of extremely vigorous and highly specialized weeds; the secondary forest (jungle) typical of much of the tropics typifies this kind of biotic community. Many of the weeds are widely distributed and originated outside the areas where they now grow. Indeed, their distribution is itself a function of human intervention and the spread of *Homo sapiens*.

The creation of new types of plants, either domesticated or weeds, was probably a slow and continuing process in the human post-Pleistocene occupation of the earth's surface. Figure 9.7 shows the results of St. Louis botanist Edgar Anderson's reconstruction of stages in the spatial evolution and hybridization of one of these weeds, the various species of sunflower (*Helianthus*), in the United States. In this figure we see the mixing of two of the original five species in pre-Columbian times being followed by intercrossings between four of the five species in the present period. The evolution of the common weed sunflowers into what Anderson terms 'superweeds' is continuing today. Mongrelization has increased their ability

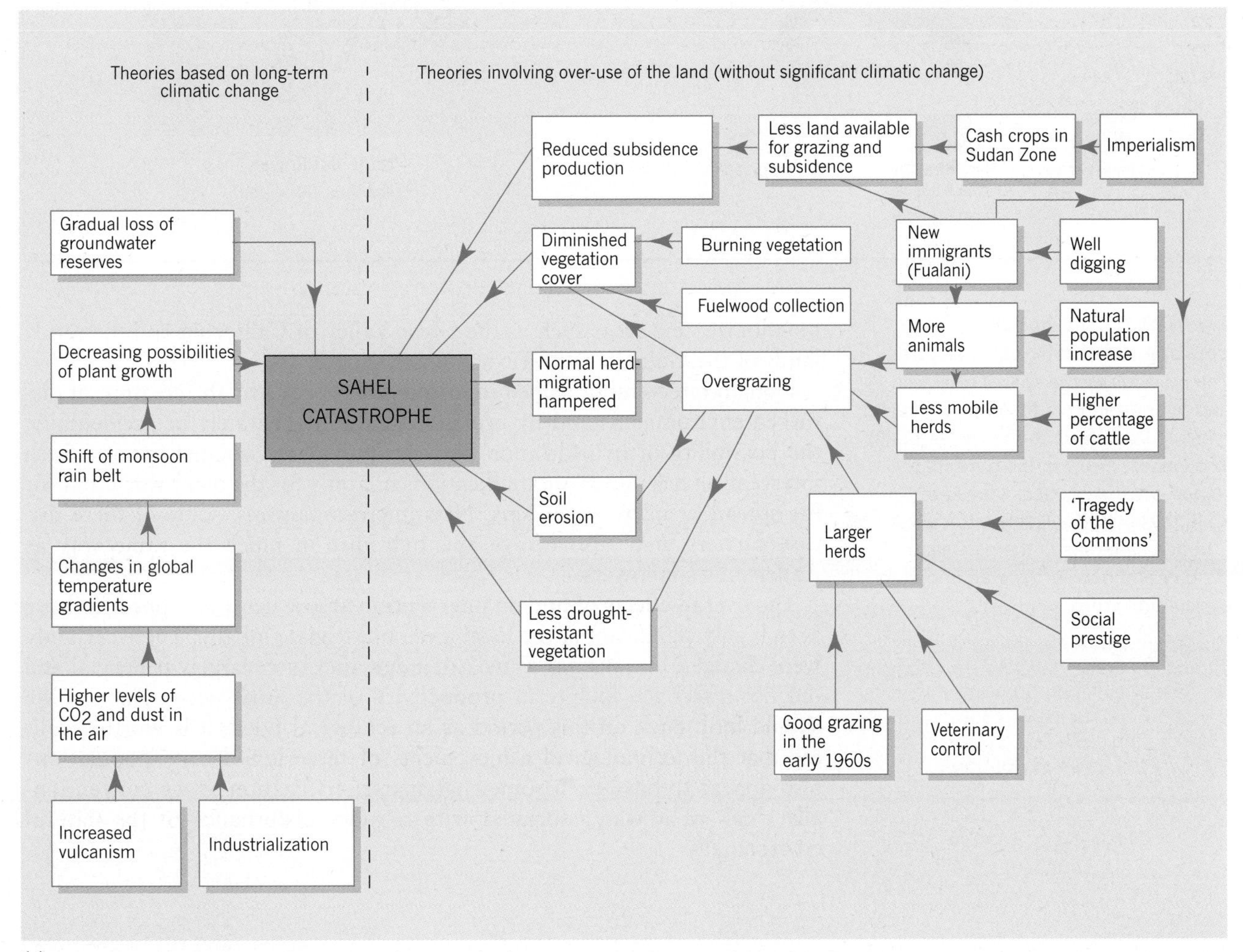

(c)

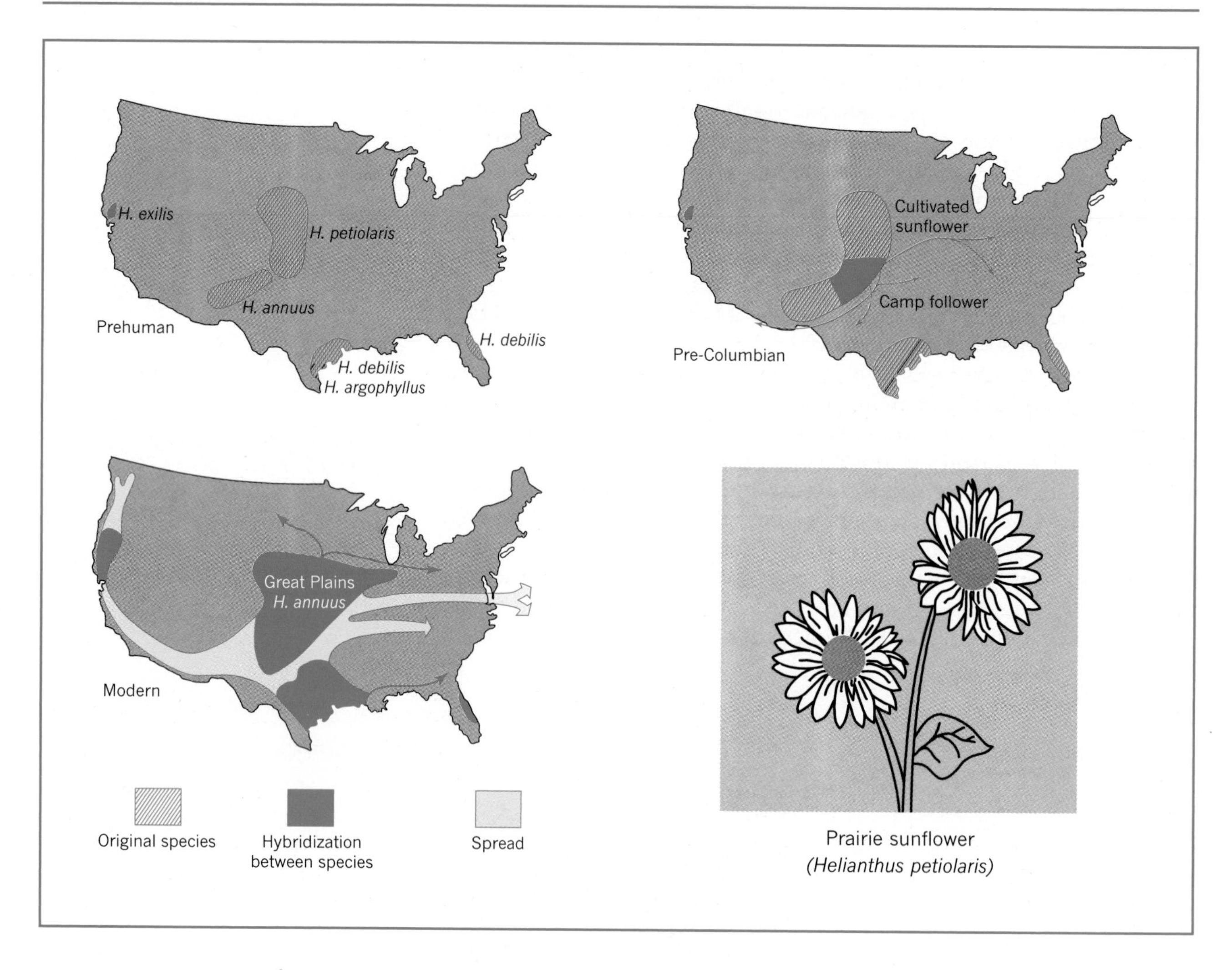

Figure 9.7 The human role in creating new biotic communities The maps show the spatial extension and hybridization of the original distinct types of sunflowers in the United States. The longest arrow in the third map represents the overseas spread of sunflower hybrids, widely introduced into European gardens in the nineteenth century and now forming an important agricultural crop.

[Source: Edgar Anderson, in W.L. Thomas, Jr. (ed.), *Man's Role in Changing the Face of the Earth* (University of Chicago Press, Chicago, 1956), Figs 150–152, pp. 768–769. Copyright © 1956 by the University of Chicago.]

to colonize new areas such as the Great Valley of California and the sandy lands of the Gulf Coast of Texas.

Human intervention helped in forming the species by the creation of disturbed environments and by providing, either deliberately or accidentally, the possibility of hybridization between previously isolated species. Such intervention has important implications not only for the plant world but for the spread of micro-organisms. New micro-organisms, some of them disease carriers, may also evolve and hybridize in much the same way as Anderson's sunflowers.

The overall effects of human intervention at low densities appear to have been highly significant at the local level, but trivial globally. There certainly were changes; but, insofar as we can judge, they were largely beneficial and did not affect the long-term productivity of the areas occupied. Lest we should look back on this period as an ecological Eden, it is worth recalling that the technological achievements of these low-density populations also appear to have been somewhat sparse; civilization as we conventionally think of it was associated with a ruder disturbance of the natural environment.

Human Intervention at Medium Densities 9.3

Between the extremes of the empty lands and the crowded, heavily urbanized areas lie zones with a medium population density. Medium densities encompass all the ranges of population density that support permanent agriculture. These densities may, in fact, vary by a factor of 100. For example, the swidden agriculture of the northern Congo supports a density of about 8 people/km^2; by contrast, the intensive *paddyfields* of the Mekong Delta support about 800 people/km^2.

Agriculture and Rotation Cycles

In Section 6.3 we saw that Malthus thought food supply limited population size. A different view has been put forward by the Danish observer Esther Boserup. (See Box 9.A on 'Ester Boserup and agricultural intensification.) She suggests that a growing population stimulates changes in agricultural techniques so that more food can be produced. So, as population densities increase, the type of agriculture practised tends to change accordingly.

Box 9.A Ester Boserup and Agricultural Intensification

Ester Boserup (born Copenhagen, Denmark, 1910; died Geneva, Switzerland, 1999) was a Danish land economist who has made important contributions to the links between human population and land resources. She worked as a civil servant in Denmark, and later as a consultant and researcher with the United Nations. This gave her extensive field experience in South-east Asia. Classical economists such as Johann von Thünen (see Box 15.C) and David Ricardo had seen two forms of agricultural expansion: *extensive expansion* with the extension of cultivation into new lands which are marginal; *intensive expansion* which enhanced the output of existing lands through the application of better husbandry (e.g. improved weeding, fertilizer, drainage and so on). Both forms were subject to diminishing returns to labour and capital. Boserup proposed a third form of intensification using an increasing labour force to crop farmland more frequently (i.e. to increase the cropping intensity or to reduce the fallow).

Ester Boserup published her ideas in two influential books, *The Conditions of Agricultural Growth* (1965) and *Population and Technological Change* (1981). In both she made fallow reduction a central plank in agrarian intensification, arguing that while fallow reduction is likely to yield diminishing returns, these are more than compensated for by the additions to total output conferred by increased cropping frequency.

Although she did not develop these implications, implicit in reduced fallowing is the changing role of land tenure, the increasing capitalization of the land, and more subtle forms of state–society interaction. Her work has been enthusiastically adopted taken up by some archaeologists and anthropologists who have charted patterns of state formation and social development in terms of her agrarian intensification scheme. Feminist writers have emphasized her work on the integration of women in development. Historical geographers have tended to be more critical of her evidence. See David Grigg, *Population and Agrarian Change* (Cambridge University Press, Cambridge, 1980) for a well-balanced account.

The *Boserup model* proposes a simple five-stage progression in which each step represents a significant increase in both the intensity of the cultivation system and the number of families it can support. Stage 1, *forest-fallow cultivation*, consists of 20–25 years of letting fields lie fallow after 1 or 2 years of cultivation. Stage 2, *bush-fallow cultivation*, involved cultivation for 2 to as many as 8 years followed by 6–10 years of letting lands lie fallow. In stage 3, *short-fallow cultivation*, there are 1–2 years when the land is fallow and only wild grasses invade the recently cultivated fields. In stage 4, *annual cropping*, the land is left fallow for several months between the harvesting of one crop and the planting of the next. This stage includes systems of annual rotation in which one or more of the successive crops sown is a grass or other fodder crop. Stage 5, *multicropping*, is the most intensive system of agriculture. Here the same plot bears several crops a year and there is little or no fallow period.

We can find examples of such systems by taking cross-sections through time or space. Thus, in western Europe, we can trace the change from the forest-fallow system (stage 1) of the Neolithic farmers to the short-fallow cultivation of the medieval three-field system (stage 3), in which one-third of the land area was left uncultivated each year. Present intensive cropping involving multicropping and supplemental irrigation represents a midpoint between stages 4 and 5. In the humid tropics, a cross-section through space reveals all five stages operating today.

The Hollow Frontier

Not all agricultural systems fall neatly into simple rotational patterns. In some parts of the world, human intervention has been abrupt and episodic, with periods of intensive use being followed by periods of abandonment.

One example of such *land-use cycles* is provided by the history of some plantation crops in the humid tropics. For example, the growing of coffee (*Coffea arabica*) was introduced into southeast Brazil in the late eighteenth century. The subsequent process of land-use change is shown in Figure 9.8.

By the beginning of the nineteenth century, coffee-growing was still confined to the coastal lowland around Brazil's capital city, Rio de Janeiro. With escalating world demand, the area under cultivation increased rapidly; and by 1850 the coffee plantations had crossed the coastal mountain belt of the Serra do Mar and had become well established in the foothills flanking the Paraíba River. Geographers have mapped the spread of coffee plantations in the ensuing decades. Forests were felled and burned as the coffee frontier advanced for some 300 km (186 miles) along the Paraíba River to within a short distance of the city of São Paulo itself. Within another generation, the coffee frontier had moved northwest to Campinas and Ribeirão Preto, and with astonishing rapidity the plantation tract along the Paraíba collapsed. The coffee groves were abandoned to weeds and cattle, and the plantation houses and slave quarters to cattle ranches or to decay and the encroaching forest.

The complex environmental changes that were triggered by the introduction and abandonment of coffee plantations are shown in Figure 9.9. The original forest cover that existed in the early 1800s was either cleared and planted with coffee or used as a source of charcoal and construction timber. The diagram identifies five ways in which the environment could be altered, which lead in turn to six main types of land use. Abandonment of areas once planted would, in the short run, bring about secondary forest areas

Figure 9.8 Changes in land use in the humid tropics European settlement of the humid tropics for plantation cropping initiated an intricate cycle of changes in land use. (See also the diagram in Figure 9.9.) For the Paraíba Valley of southeast Brazil, the main plantation crop was coffee (a). With the ageing of the coffee bushes (b) and the progressive abandonment of the hillside plantations (c), the area has become largely grassland. The grasslands are maintained by intensive grazing (d) and regular burning (e).

[Source: Photos by the author.]

dominated by rapidly growing species and, in the long run, some re-establishment of the slow-growing tropical rainforest appropriate to the area's prevailing climate and soil structure.

These land-use cycles have led to a distinctive spatial pattern of

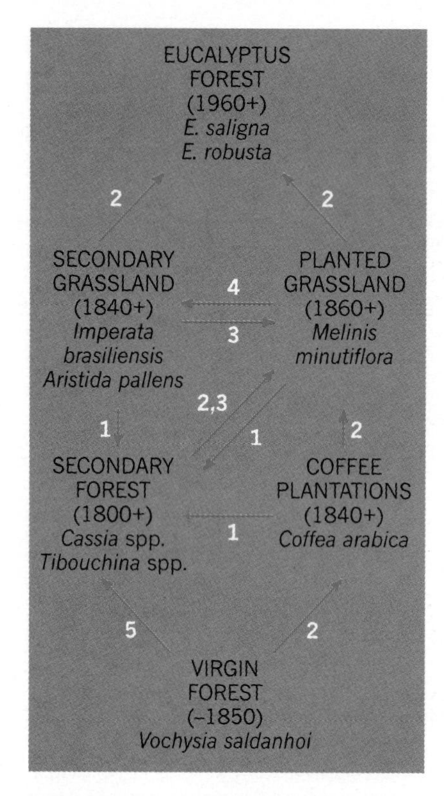

Figure 9.9 Land-use cycles This flow diagram reconstructs the sequence of land use in the Paraíba Valley of southeast Brazil since 1800. Only representative plant species are shown. The numbers indicate how the environment was altered: (1) by abandonment, (2) by clearing and planting, (3) by burning, (4) by heavy grazing, or (5) by cutting trees for charcoal, timber, and so forth.

[Source: P. Haggett, *Geographical Journal*, **127** (1961), Table 1, p. 52.]

settlement. The term *hollow frontier* is a graphic description of a situation in which agricultural colonization proceeds as a wave leaving behind it a trough of worked-over land with a lower density of farm population. The century of movement in Figure 9.10 shows a simplified version of the Brazilian experience. There the centres of colonization were the cities of Rio de Janeiro and Santos. Today the edge of the coffee-growing settlement is some 700 km (450 miles) inland. In the longer term, the hollows may be filled with different forms of cultivation building up population levels.

Although the hollow frontier was first developed in the context of rural land use, it also applies to urban land use. The later discussion on 'surfaces' (see Section 15.1) shows how a similar process is operating today with the spread of cities with areas of former inner-city industry left as derelict areas.

Agriculture in the Ecological Food Chain

Agriculture can be directly incorporated into some of the ecological frameworks we met in Chapter 4. For an edible crop is also the beginning of a natural *food chain*, in which predatory and parasitic organisms compete for energy on every trophic level and divert calories away from human food supplies.

Some idea of the range and complexity of farming systems is given in Table 9.2. The table is a complex one so it is worth looking carefully at how it is constructed. The upper part of the table is concerned with tropical farming systems, the lower part with temperate or middle-latitude systems. Four main types of farming are shown across the table. Different levels of land-use intensity are given down the table. The letters A to D indicate the type of ecosystem each farming system represents. These range from 'wild' types such as reindeer herding or collecting fruit from wild trees at one end, to 'permanent, human-directed' ecosystems such as rice cultivation at the other.

Despite the diversity of agriculture, a very few crops dominate the world's production totals. Grain yields dominate all other farm products in terms of tonnage with the 'big three' – wheat, rice, and maize – making up three-quarters of grain output. Each of these crops shows great genetic diversity and can be fitted into very different natural environments and into various farming practices. Meat and animal products form a relatively small share in the world food production overall, despite their great importance in specific regions.

Table 9.2 also describes each agricultural system in terms of a food chain

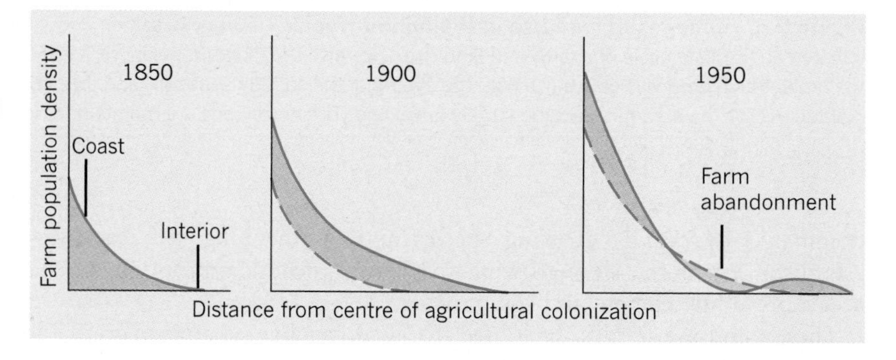

Figure 9.10 The hollow frontier The land-use cycles described in the two preceding figures have left a characteristic settlement pattern in Brazil termed the hollow frontier. New waves of settlers have pushed further inland away from the centres of early agricultural colonization. The abandonment of some of the earlier coffee-growing areas has led to hollows in the density of farm population.

Table 9.2 Farming systems classified as food chains

Intensity scale	Tree crops	Tillage (with or without livestock)	Alternating tillage (with grass, bush or forest)	Grassland or grazing land (consistently in indigenous or man-made pasture)
Very extensive	Cork collection, *southern Portugal* (b)	–	Shifting cultivation, *Zambia* (c)	Reindeer herding, *Lapland* (a)
Extensive	Self-sown or planted blueberries, *north-east USA* (b)	Cereal growing, *Pampas of South America* (d)	Shifting cultivation, *Java* (c)	Wool growing, *Western Australia* (b)
Semi-intensive	Vineyards, *France* (d)	Dry cereal farming, *Israel* (d)	Cotton with livestock, *southeast USA* (d)	Cattle/buffalo mixed farming, *India* (b)
Intensive	Citrus orchards, *California* (d)	Corn belt, *mid-west USA* (d)	Grass beef farms, *England* (d)	Intensive dairy-farming, *Denmark* (d)
Typical food chains (see Figure 9.11)	Tropics **A** Mid-latitude **A**	Tropics **A** Mid-latitude **AB**	Tropics **AC** Mid-latitude **ABCD**	Tropics **C** Mid-latitude **CD**

Landscape types: (a) wild, (b) semi-natural, (c) human-directed (temporary change), (d) human-directed (permanent change)
Source: Simplified version of work by Duckham and Masefield reported in I.G. Simmons, *The Ecology of Natural Resources* (Edward Arnold, London, 1974), Table 8.5, pp. 198–9.

(A, B, C, and D) in which humans are at the end of the chain. The four types of chains are set out in Figure 9.11. In the first, people consume food as herbivores and in the other three as carnivores. The efficiency of chain A, in which persons consume a crop directly is much greater than B, C, and D

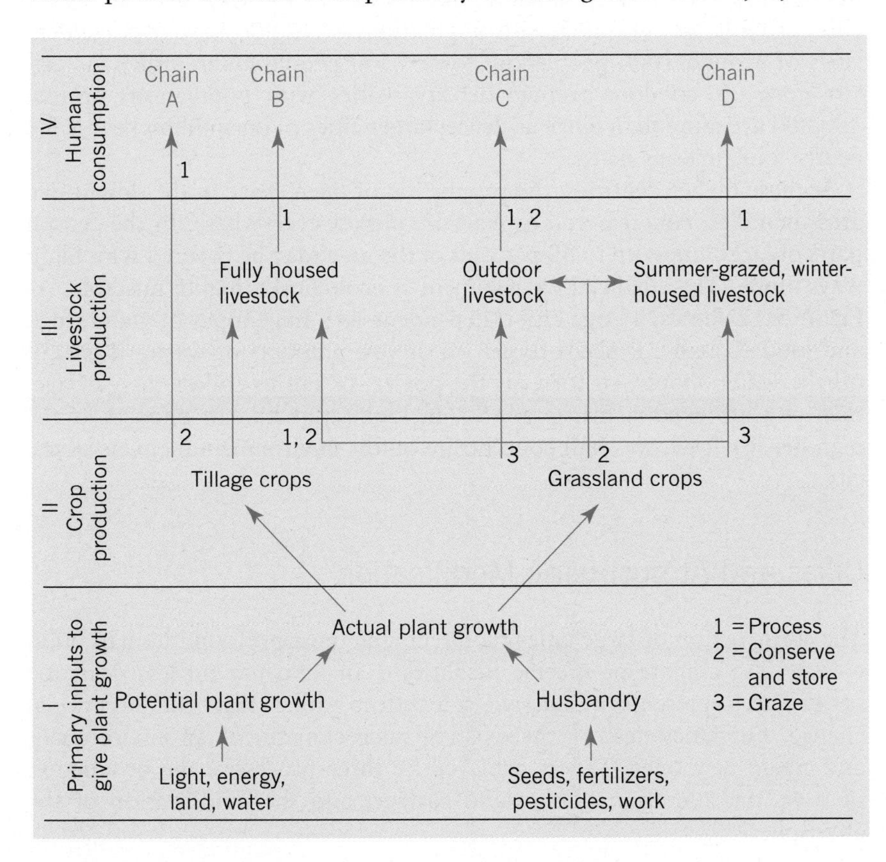

Figure 9.11 Food chains in agricultural production Arrows show the movement of food from the primary inputs (bottom of figure) to final human consumption (top of figure). Regional examples of such chains are given in Table 9.2. Note that whereas in chain A people directly consume tillage crops, in chains B, C, and D they consume animal products. These animal products are in turn formed from either tillage crops or by direct grazing.

[Source: A.N. Duckham and G.B. Masefield, *Farming Systems of the World* (Chatto and Windus, London, 1970.]

where they consume animal products. If you look again at Table 9.2, you will find examples of each of the four chains.

These examples give only a small indication of the rich variety of adjustments between agricultural systems and local environments. Anthropologist Clifford Geertz, in his book *Agricultural Involution*, makes a study of agricultural practices in Indonesia. He draws comparisons between two contrasting farming systems: *swidden* (shifting cultivation) and *paddy* (irrigated rice production). He demonstrates clearly that the systems of swidden cultivation actually simulate the exchanges of elements (among the atmosphere, vegetation, and soils) that occur in the tropical rainforest under natural conditions. Conversely, the terraced rice paddies in the same area represent an artificial system in which elaborate control of water, fertilizers, and weeds is necessary. Thus although the swidden systems seem intuitively to be an unstable method of cultivation (since the land use changes every few years), and rice paddies a stable one, the reverse may actually be the case. Replacement of the natural environment by a different ecology greatly increases agricultural production, but continuous work and the associated dense, rural population are required to maintain it.

9.4 Human Intervention at High Densities

Despite the very small percentage of the earth's land surface occupied by cities, they have a profound effect on the environment. Within and around urban areas, the covering of rural land by city blocks proceeds faster as the cities grow larger. US cities with populations of 10,000 have average densities of around 1000 people/km^2 (over 2500 people/square mile). As city size goes up, so does average density. Cities with populations around 100,000 are more than twice as dense; larger cities of one million people are nearly four times as dense.

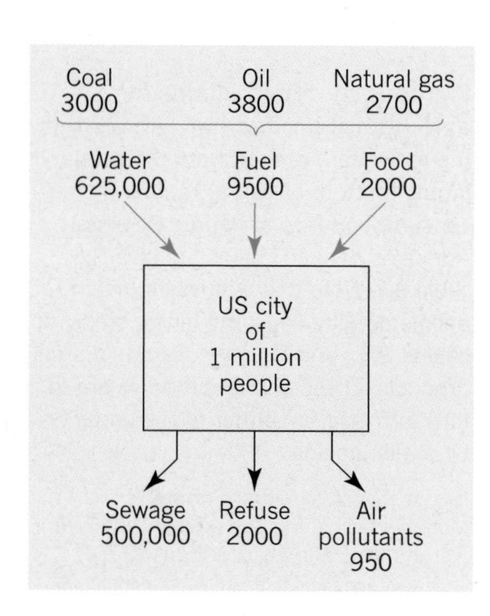

Figure 9.12 The city as an ecosystem The maintenance of city life demands a regular inflow of energy and nutrients and an efficient waste disposal system. The figures given are rough estimates in tonnes for an American city of a million people.

[Source: Based on data by J. McHale in I.G. Simmons, *The Ecology of Natural Resources* (Arnold, London, 1974), Table 13.7, p. 358.]

At these higher densities, the proportion of open space in the downtown area dwindles; concrete and asphalt are almost everywhere. In the central parts of large cities, up to 40 per cent of the area may be covered with highways alone. This artificial environment is ecologically highly unstable. As Figure 9.12 shows, a large city is dependent on a huge input of water, fuel, and food. Equally, it needs to get rid of vast amounts of waste. If we cut off the water supply, or turn off the power, or fail to collect the garbage, then city life is soon disrupted. So, in looking at human impacts at the high-density level, we shall concentrate on the environmental effects of the city.

Cities and Atmospheric Modification

The construction of large cities represents the most profound human influence on the climate of specific localities – an influence no less dramatic because it represents a massive side effect rather than an intentional change. For cities destroy the existing microclimates of an environment and create new ones. This is achieved by three processes: the production of heat, the alteration of the land surface, and the modification of the atmosphere.

Urban Heat Islands The generation of heat within a city generally results directly from the combustion of fuels and indirectly from the gradual release of heat stored during the day in the city's material fabric (brickwork, concrete, etc.). Temperature studies reveal urban ***heat islands***, caused by the fact that city temperatures are generally higher than those of surrounding rural areas. For example, central London has a mean annual temperature of 11 °C (58.8 °F), the surrounding suburbs have a mean annual temperature of 10.3 °C (50.5 °F), and the rural areas have a mean annual temperature of 9.6 °C (49.3 °F) (see Figure 9.13).

These differences are at a maximum with calm conditions or low wind speeds; wind speeds above 25 km or 15 miles per hour tend to blot out the heat island effect. Contrasts for London are at their sharpest during the summer and early autumn, thus indicating that thermal contrasts depend more on heat loss from buildings by radiation rather than on combustion. There are, however, marked variations between cities in different macroclimates and in different topographic situations. In Japan the continued expansion of cities is paralleled by increases in their mean annual temperature (e.g. Osaka's temperature has riden by 2.5 °C (4.5 °F) over the last century), but it is difficult to isolate the effect of suburbanization from other influences on temperatures.

Rugged City Terrain Cities also affect microclimates through their rugged artificial terrain of alternating high and low buildings and streets. Even though we are conscious of gusting winds being channelled along the canyon-like streets between high buildings, the terrain of the city lowers average wind speeds compared with those in surrounding rural areas. Average wind speeds for a site in central London (7.5 km or 4.7 miles per hour) are substantially lower than for a suburban site (Heathrow Airport, London, 10.2 km or 6.3 miles per hour), although there are considerable variations according to the season and time of day. The effect of urbanization on rainfall is uncertain, but there are strong indications that, under certain conditions, cities in middle latitudes can cause sufficient local turbulence to trigger rainstorms. Significant variations among cities in different climatic zones are probable, and more research on comparative urban climatology is required. It would be dangerous to judge all the world's cities on the evidence provided just by London or Los Angeles.

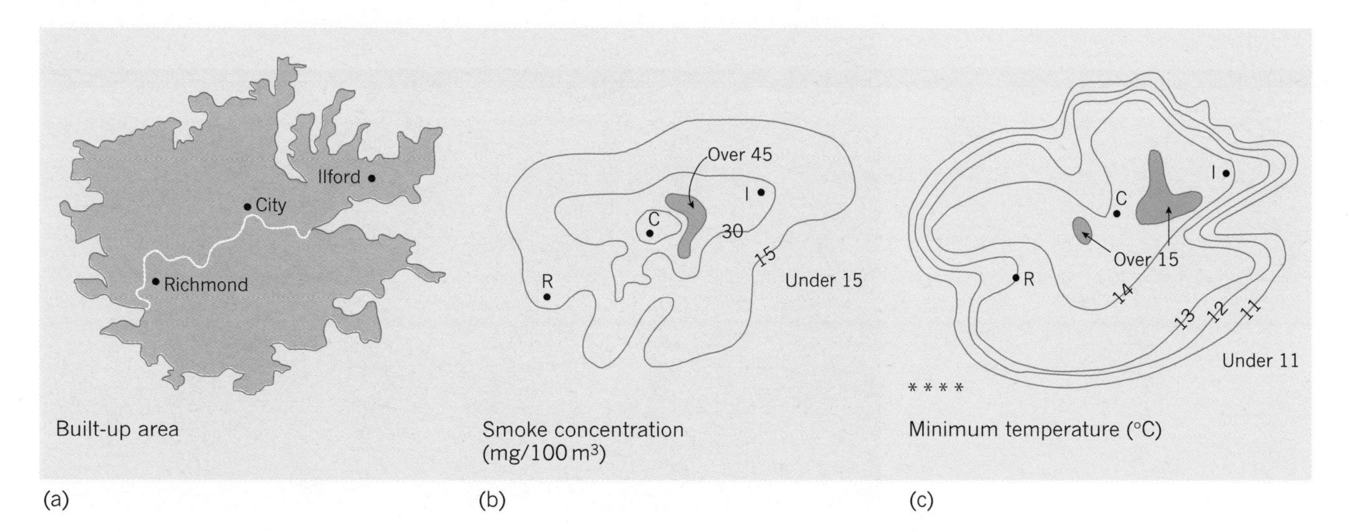

Figure 9.13 Urban climates The three maps show the built-up area of London (a), the average smoke concentration from October 1957 to March 1958 (b), and the minimum temperature on 4 June 1959 (c).

[Source: T.J. Chandler, *Geographical Journal*, **128** (1962), Figs 2–13, pp. 282, 295.]

City Wastes in the Atmosphere The impact of cities on the atmosphere is particularly evident in the context of pollution. City atmospheres are polluted mainly by the emission of smoke, dust, and gases (notably sulphur dioxide). Pollution has three primary effects: it reduces the amount of sunlight that reaches the surface, it adds numerous small particles to the air that serve as nuclei for condensation and hence promote fogs, and it alters the thermal properties of the atmosphere. These three effects often combine to reinforce one another (see Figure 9.14).

Smogs, for example, reinforce the reduction of sunlight. The seriousness of this effect is indicated by the fact that British cities are estimated to lose between 25 and 55 per cent of the incoming solar radiation from November to March. Although certain common features cause a high concentration of

Figure 9.14 Smog Smog at sunset in northern Los Angeles USA. View overlooking North Hollywood, January 1995. See Box 9.B on the mechanisms of smog formation.

[Source: David Frazier/Science Photo Library.]

pollution (low wind speeds, temperature inversions, high relative humidities), there is considerable global variation in the severity and seasonal incidence of pollution conditions. For example, although Los Angeles smog is at its worst in summer and autumn, London smog is a winter phenomenon.

Although *air pollution* is most obvious when it soils buildings and reduces sunlight, its most important effects are on human health. The toxic effects of many pollutants are well known, although they reach dangerous

Box 9.B Smog Formation

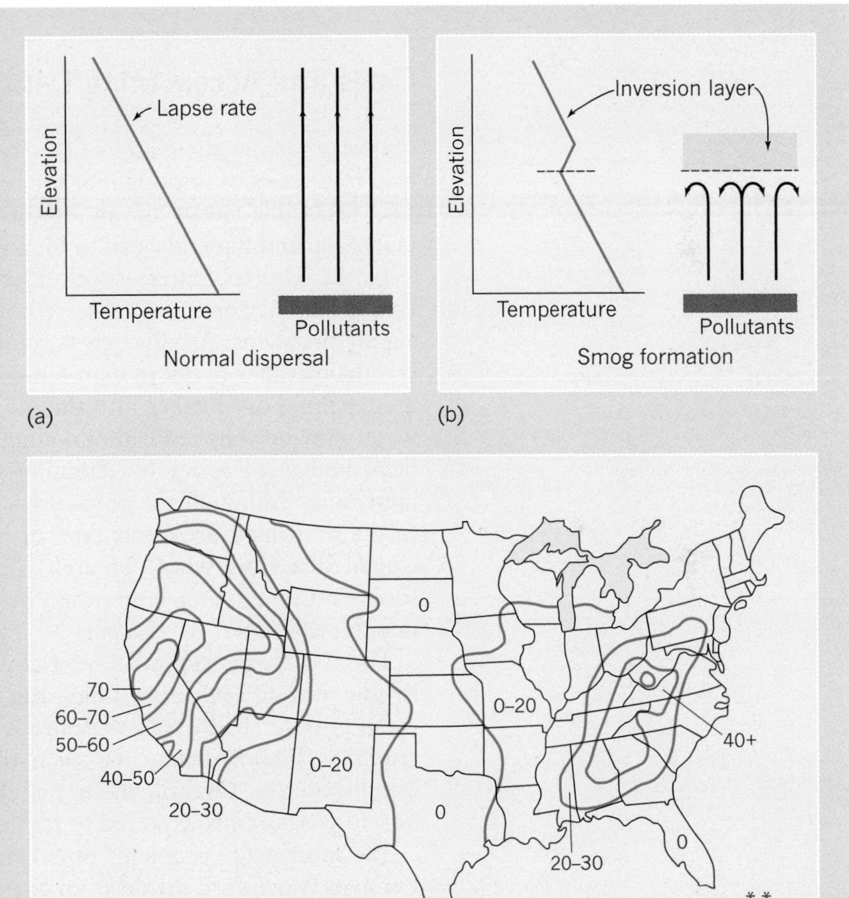

When winds are strong there is seldom noticeable air pollution. Smoke, dust, and gases are rapidly mixed with a large volume of air and dispersed over a large region so that concentrations at any one point remain low. But the calm conditions and extremely light winds typical of high-pressure (*anticyclonic*) conditions favour the build-up of large and dangerous concentrations of pollutants. Normally, air temperature decreases with height (the average lapse rate is about 6.4 °C per km), so that the warm, polluted air over cities tends to rise and mix vertically (diagram (a)).

During anticyclonic conditions two types of *inversion layers* can interrupt this normal vertical dispersal of pollutants. First, a *high-level inversion* at 1000 or more metres (about 3300 ft) can be formed when calm, upper air falls to lower elevations where it is compressed and its temperature rises. Strong and persistent high-level inversions are typical of the eastern end of the Pacific anticyclone that extends over the Los Angeles Basin, particularly in summer. Second, a *low-level inversion* can be formed overnight by the rapid cooling of the ground. These shallow diurnal inversions affect only the lowest 100 m or so (about 330 ft) of the atmosphere.

Inversion layers, from whatever cause, prevent the vertical dispersal of pollutants and raise the concentration levels (diagram (b)). Topographic conditions such as narrow river valleys further constrain any dispersal and accentuate concentrations, as in the Donora Valley disaster in western Pennsylvania (26–31 October 1948) and the Meuse Valley disaster in Belgium (December 1930). The map (diagram (c)) shows the average number of days per year when there are inversions and light winds (i.e. *potential* air pollution conditions) in various parts of the United States. (See R.A. Bryson and J.E. Kutzbach, *Air Pollution* (American Association of Geographers, Commission on College Geography, Resource Paper 2, Washington, D.C., 1968).)

concentrations only under unusual meteorological conditions. The most serious accumulation of pollutants over a large metropolis occurred during the London smog of 5–9 December 1952. During a strong temperature inversion with little air movement, concentrations rose to six times their normal level, and visibility was reduced to a few yards over large areas of the city. (See the discussion of smog formation in Box 9.B.) During the five day period, 4000 deaths were attributed to cardiac and bronchial ailments caused by the smog. The size of this environmental disaster led to legislation (the Clean Air Act of 1956) directly controlling the emission of pollutants through a series of 'smokeless zones' where the burning of certain types of fuel was banned.

Cities and Accelerating Demands for Water

Outside the immediate limits of a city, the demands of its urban population lead to a series of impacts that are no less acute for being spatially distant. The insatiable needs of an urban economy for water, food, building materials, and minerals lead to long-range exploitation of the environment.

In the United States, south Florida and the southwestern states of California, Arizona, New Mexico, and Texas all face severe local water supply problems. As cities grow, *groundwater* water tables are being lowered (sometimes by more than a metre a year) as demand grows and cities are reaching out further into the mountains for fresh water. For example, water may be supplied from flooding a remote valley. The creation of artificial bodies of water by damming for a variety of purposes is rapidly increasing. Although the areas used for this purpose are small in relation to the total land area, this type of land use is important locally. For the United States as a whole, an area of not less than 40,000 km^2 (over 15,000 square miles), an area larger than Belgium) is now covered with artificially impounded water.

The average per capita amount of water used in the world's western cities is now around 600 litres a day. But the demands of the urban public for water are overshadowed by the needs of urban industry. Each tonne of steel requires 100,000 litres, and each tonne of synthetic rubber needs over 2 million litres. Overall, the use of water is growing: it has trebled in the last 30 years and is expected to treble again in the next 30.

The most acute problems now being faced are not so much in the provision of water as in the disposal of polluted water. The average city of half a million people now produces over 1800 tonnes of solid waste each day and a further 200 million litres of sewage. The disposal of organic sewage poses less difficulty than the disposal of the inorganic metallic wastes of industry, and it is to this problem that we turn in Section 9.5.

9.5 Pollution and Ecosystems

In studying the effect of the very dense concentrations of people characteristic of our urban-industrial civilization on the environment, we are returning to the concern with pollution with which we began this chapter. Pollution *does* occur at lower population densities, but at such low rates and in forms so easily broken down that it fails to form the kind of ecological threat posed by the by-products of urban dwellers. Two examples of the range of pollution in space–time terms are given in Figure 9.15.

The Pollution Syndrome

Any historian looking for a word to summarize the last half century is likely to find the term *pollution* (from the Latin *pollutus*, defiled) on the list. But despite the constancy with which it comes up, the term remains difficult to pin down. What do rising mercury levels in the ocean, rising noise around airports, higher temperatures in streams, and higher carbon dioxide levels

(a)

(b)

Figure 9.15 Different geographical scales of pollution (a) Trees in northwest Europe affected by ***acid rain***. Here the effect builds up slowly over several decades. (b) The *Exxon Valdez* oil spill (see Box 9.C) occurring at a specific point in time and spatially more confined. High-pressure hoses with hot water are being used to try and break up onshore pollution.

[Source: (a) © Silvestris/FLPA. (b) Vanessa Vick/Science Photo Library.]

in the atmosphere have in common? Perhaps the simplest answer is that each represents a substance that in terms of human environment, is in the wrong place, at the wrong time, in the wrong amounts, and in the wrong physical or chemical form.

One of the simplest illustrations of this definition is provided by *thermal pollution* in streams. Heat added to water is in no obvious sense a pollutant, and yet it changes the characteristics of that water as an environment just as surely as chemical contaminants or atomic radiation. Where does the heat come from, and just how does it affect an ecosystem?

Nearly all the waste heat entering our streams comes from industrial processes, and over three-quarters of this comes from electric-power generation. A single 1000-megawatt nuclear-generating plant may need around 45,000 litres (a million gallons) of water each minute for cooling. (See the discussion of nuclear-power generation in Chapter 10.) The water emerging from the outlet pipes may be 11 °C (almost 20 °F) warmer than that entering the intake pipes.

The critical problem with warm water is what happens to its chemical structure. Warm water holds less dissolved oxygen than cold water but speeds up the metabolic rates of decay organisms in the water, which *increases* their demand for oxygen. These twin effects cause a marked reduction in the water's oxygen level and a fouling of the stream environment. The warmth of water is also critically related to the metabolic rates of fish populations. Spawning and egg development among salmon and most trout occur at around 13 °C (55 °F). Even small increases in heat may change the timing of a fish hatch and bring young fish prematurely into an environment in which their natural food sources have yet to arrive.

It would be misleading, however, to regard thermal pollution of streams as only harmful. Many fish species (e.g. catfish, shad, and bass) spawn and develop eggs at higher water temperatures (from 24 °C (75 °F) to 26 °C (79 °F)) and grow very rapidly in temperatures up to 35 °C (95 °F). Broad changes in a stream's ecology following thermal pollution may actually increase its overall productivity if we count the increased growth of algae. Future research might well be directed toward making such increases in productivity available to human beings, thereby changing thermal pollution to thermal enrichment.

Chemicals in Food Chains

In the discussion above, we used thermal pollution to illustrate the sometimes equivocal nature of a 'pollutant'. There are other pollutants, however, whose damaging effect is direct and one-sidedly harmful. For example, of the 103 elements in the chemical table, about one in eight plays a prominent part in environmental pollution. Table 9.3 lists the so-called 'sordid sixteen'.

We can divide these elements into three main groups. In the first group are elements such as carbon, oxygen, phosphorus, and nitrogen, which are vital to all forms of biological life but which *can* form harmful compounds. In the second group come elements such as strontium or uranium, which are important in radioactive pollution. In the third group are toxic chemicals such as chlorine and arsenic used in insecticides and toxic heavy metals such as mercury and lead.

It is elements in this last group that have proved the most dangerous to the structure of ecosystems in that their effects are insidious. We noted in

Table 9.3 The sixteen most common elements in pollution[a]

Element	Symbol	Main link to pollution
Hydrogen	H	Constituent in pesticides
Carbon	C	Constituent in atmospheric pollution (carbon monoxide) and pesticides
Nitrogen	N	Constituent in photochemical smog
Oxygen	O	Constituent in atmospheric pollution (carbon monoxide and sulphur dioxide)
Phosphorus	P	Causes water pollution by excessive algal growth
Sulphur	S	Constituent in atmospheric pollution from coal-burning power plants
Chlorine	Cl	Constituent in persistent pesticides
Arsenic	As	Constituent in pesticides
Strontium	Sr	Radioactive isotope
Cadmium	Cd	Heavy metal; water pollutant from zinc-smelter wastes
Iodine	I	Radioactive isotope
Caesium	Cs	Radioactive isotope
Mercury	Hg	Heavy metal; toxic water pollutant from the manufacture of some plastics; pesticide
Lead	Pb	Heavy metal; toxic by-product of burning gasoline
Uranium	U	Radioactive element
Plutonium	Pu	Radioactive element

[a]Arranged in order of increasing weight.

opening this chapter the mystery that surrounded the Minimata disease. Although the chemical plant producing plastics on the edge of Minimata Bay was known to be discharging dangerous mercury compounds into the sea, the concentrations were so low – about 2 to 4 part per billion – as to be harmless if drunk in fresh water. Mercury levels in the affected fishermen were about four thousand times greater than this.

The process by which the concentration of mercury built up from harmless levels in seawater to crippling levels in human tissue is termed *biological concentration*. Each creature in a food chain collects in its tissues the mercury present in its own food, and it eats many times its own body weight in nutrients. Since the element is not excreted or broken down, it is passed up the chain in increasing concentrations. Predators such as humans, at the end of the food chain, are especially vulnerable to mercury poisoning since they consume large amounts of 'mercury-enriched' food in which the levels of the metal are already high. In the case of Minimata, the situation was made worse by the local diet of fish, and particularly shellfish. Oysters have been shown to contain concentrations of insecticides as much as 70,000 times the concentration in seawater.

The widely publicized cases of mercury poisoning could be paralleled by the results of the build-up of persistent pesticides in the food chains. Figure 9.16 shows the build-up of DDT in part of the food chain in the Long Island estuary in the United States. Note the numbers indicating the DDT levels. These range from as low as 0.04 parts per million for plankton, at the bottom of the food web, to levels up to a thousand times greater in bird populations toward the top of the food web.

The effects of insecticides on ecosystems are often complex and indirect. Birds at the top of the predator tree (i.e. eagles and hawks) suffer the most. Yet the diminishing range of such birds is due less to direct poisoning than to the thinning of eggshells, which means eggs fail to hatch successfully. Both the bald eagle and the peregrine falcon have been eliminated from the northeastern United States, although changing habitats are likely to have had as much impact as food-chain pollution in this process.

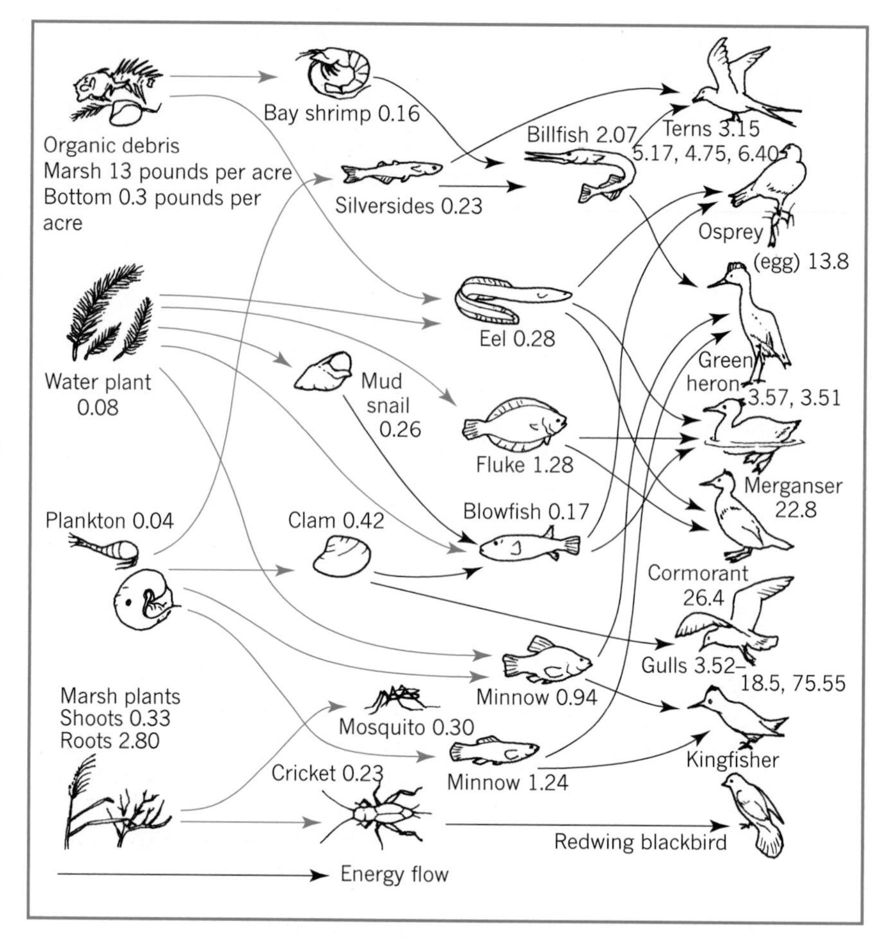

Figure 9.16 Pollution in food chains In this diagram of a food web in the Long Island estuary the coloured arrows show the first links and the black arrows the subsequent links in chains of feeding. The DDT levels found in each organism are given in parts per million. Note the *biological magnification* of these DDT levels by the repeated concentration of this chemical substance by each organism in the food web. Compare the low concentration of DDT in the first column of marsh plants with birds in the last column. In some birds the levels are more than 1000 times higher than in the plankton and water plants. (1 lb/acre ≈ 1.1 kg/ha.)

Dimensions of Pollution in Time and Space

Studies of the lead content in the ice layers in the Greenland ice cap (Figure 9.17) show the sharp upturn in worldwide atmospheric lead associated with the beginning of the Industrial Revolution and with the widespread use of gasoline for automobile fuel. Lead has been one of the most useful ***heavy metals*** for some 5000 years. It has been widely used in pottery, household plumbing, paints, and insecticides. Unfortunately it is also extremely toxic, and the ingestion of excessive amounts can severely affect the human kidneys and liver, as well as the reproductive and nervous systems.

Most lead is heavily concentrated near its source, so that, despite more use of unleaded petrol, downtown city dwellers, in areas where the concentration of automobiles is high, have levels of lead in their blood twice that of their suburban neighbours. It also gets caught up in the general atmospheric circulation of the earth. Fallout from the atmosphere affects all parts of the globe, and the fallout on Greenland shown in Figure 9.17 represents a gross underestimate of lead levels in the urban areas of the globe.

Pollution problems need to be evaluated in terms of four factors as set out in Figure 9.18: (1) the nature and properties of the pollutant, (2) the space–time context of the emission, (3) the specific environments affected by it, and (4) the impacts on ecosystems. It is important in studying the lead pollution problem to know that lead is a highly toxic metal which has been continuously released into the environment for some thousands of years and that, with the advent of leaded automobile fuels, the amount of lead in

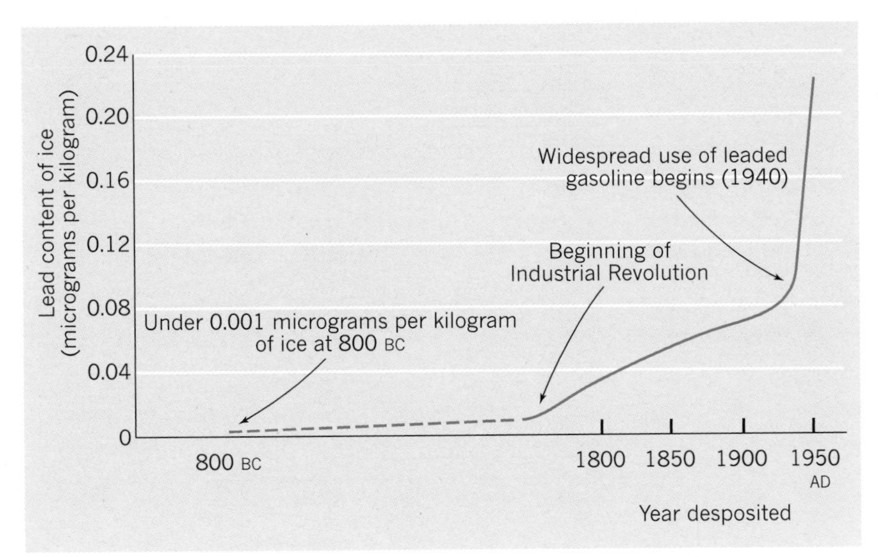

Figure 9.17 Long-term lead pollution The chart shows the lead content of the Greenland ice cap due to atmospheric fallout of the mineral on the snow surface. A dramatic upturn in worldwide atmospheric levels of lead occurred at the beginning of the Industrial Revolution of the nineteenth century and again after the more recent spread of the automobile.

[Source: M. Murozumi *et al.*, *Geochimica Cosmochimica Acta*, **33** (1969), p. 1247.]

the atmosphere has been increasing at an accelerating rate. It is also crucial to our understanding of the lead pollution problem to know that atmospheric lead is absorbed by animals through their lungs, and that its impact is long term and insidious rather than short term and dramatic.

Figure 9.18 summarizes this multidimensional approach to pollution problems. It uses a single example, the oil pollution resulting from the wreck of the supertanker *Torrey Canyon* off the southwest shore of England, to illustrate the dimensional complexity of such problems. A second example of a major oil spill is given in Box 9.C on the *Exxon Valdez* accident. Other instances of pollution, such as noise pollution around a major airport or *DDT* build-ups in marine food chains, can also be analyzed by using the approach in the table.

One difficulty we face in analyzing pollution problems is that the term 'pollution' is now so overworked in the press and TV that it is difficult to establish a balanced view of the degree of environmental threat it represents. The amount of accurate information available on such subjects is less than the strong positions adopted by many people would lead us to suppose. More monitoring of the environment using various sensing techniques is clearly needed. Books such as Rachel Carson's *Silent Spring* (see Figure 9.19) certainly played a critical role in awakening both the scientific community and the political lobbyists to the actual and potential hazards that surround us.

In *Silent Spring* (first published in 1962), Carson describes in graphic terms the effects of exposure resulting from the indiscriminate use of chemicals. She shows how pesticides and insecticides were almost universally applied at all spatial scales from the back garden to widespread aerial spraying of DDT as part of pan-tropical malaria control programmes. The book argues that the world then faced not just an occasional dose of poison that accidentally got into the environment but a persistent and continuous poisoning of the whole human environment. Because of its wide circulation *Silent Spring* caught the public imagination and was credited with tipping President John F. Kennedy's decision to set up a Senate Committee for Environmental Affairs in the United States. It was the first of a series of books – Barry Commoner's *The Closing Circle* (1971) and Denis Meadows' *The Limits to Growth* (1972) were two others – which alerted opinion around the world to the full effects of the human impact on environment.

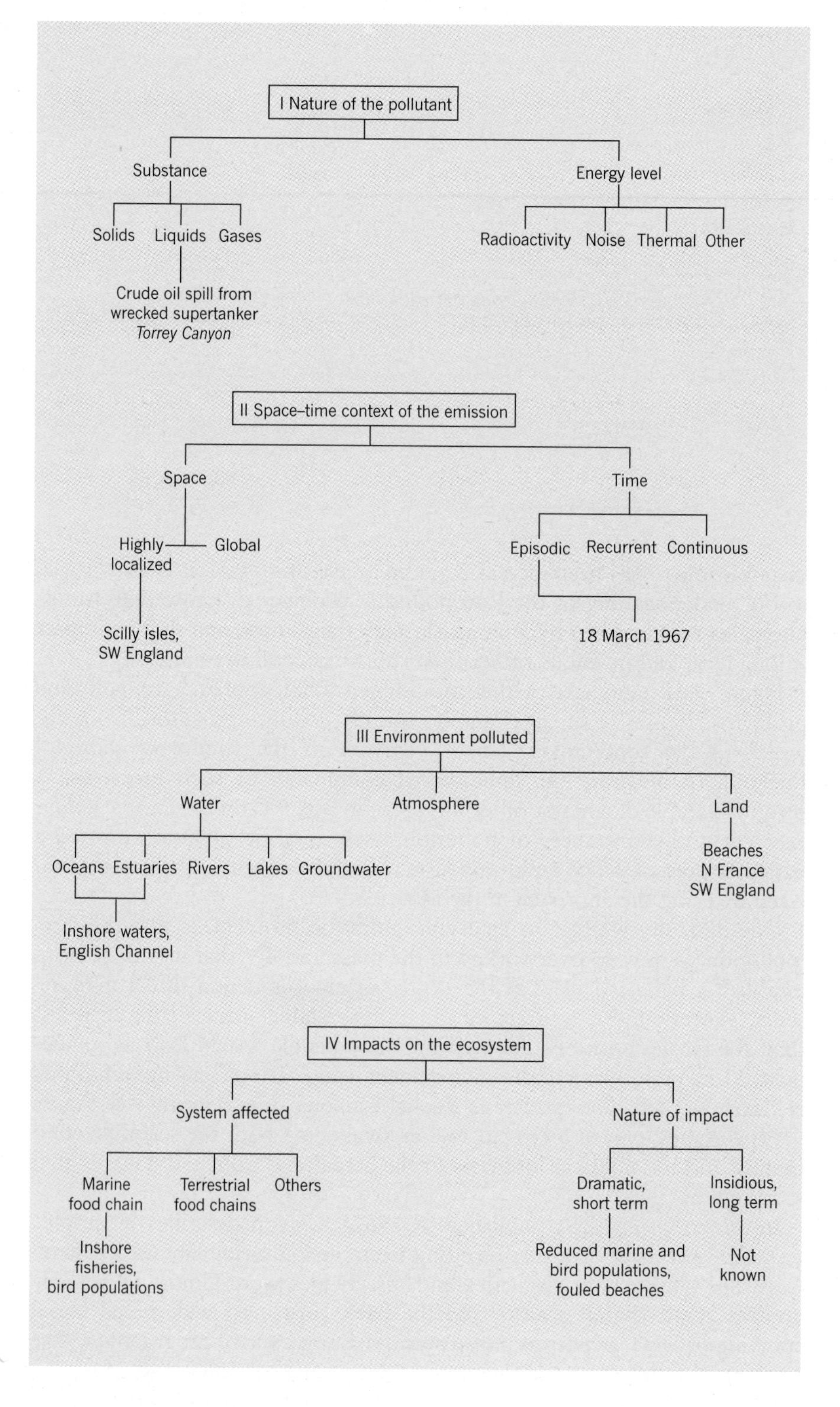

Figure 9.18 Multiple dimensions of a single pollution problem The four dimensions identified are: (I) the nature of the pollutant, a crude oil spill from the *Torrey Canyon* supertanker; (II) the space–time context of the emission, Scilly Isles on 18 March 1967; (III) environment polluted, inshore waters and beaches of the English Channel, SW England and N France; and (IV) impacts on the ecosystem, inshore fisheries, bird populations, and fouled beaches. Compare with the *Exxon Valdez* incident (Box 9.C).

At the same time, we must note that much pollution is very short-lived and that many ecosystems have remarkable powers of recuperation. Pollution control may have to be carefully priced and the undoubted gains from a cleaner environment set against other desirable goals of the human population.

Box 9.C The *Exxon Valdez* Oil Spill

This is one of a series of major oil spills that have caught press attention around the world over the last three decades.

The *Exxon Valdez* ran aground on 24 March 1989, in Prince William Sound, striking Bligh Reef near the Alaskan oil terminal of Valdez. The port of Valdez is the southern terminus of the Alaska oil pipeline which brings the fuel from drilling rigs on the oil-rich northern slope of the state. Five factors contributed to the severity of the spill:

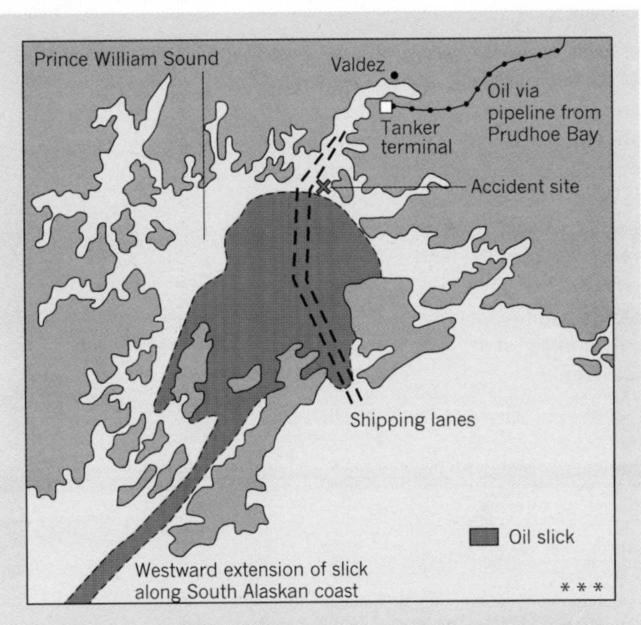

The Exxon Valdez is towed to Naked Island, Alaska, for repairs after the 1989 disaster

£6bn claim over Exxon oil disaster

Joseph Hazelwood: Exxon Valdez captain

by GERAINT SMITH

CLAIMS for an unprecedented £6.7 billion punitive damages have been lodged following the Exxon Valdez oil spill that devastated wildlife and fisheries in Alaska five years ago.

More than 100,000 plaintiffs including fishermen and Eskimos are pressing suits for damages in civil courts in Alaska. These are in addition to the existing claims for compensation of £3.2 billion, and the settlement of £670 million already agreed with the state and federal authorities, according to today's Lloyd's List.

The total cost of the spill, which happened in 1989, is potentially five times the previous largest settlement in legal history — £2 billion in respect of a corporate battle for Getty oil — and dwarfs the £315 million for the Bhopal disaster in India in which 4,000 people died and around 20,000 were injured.

Meanwhile, the Valdez itself, which went aground in Alaska's Prince William Sound while carrying 1.5 million barrels of Brent crude from Sullom Voe, Shetland, spilling 11 million gallons, is still sailing under a different name. It is a target for Greenpeace protesters who claim that the Exxon Corporation, the company running the tanker, is guilty of "environmental crimes".

The huge claim has emerged in legal exchanges before the federal trial scheduled to open in Anchorage on 2 May. These say that the plaintiffs, mainly commercial fishermen and Alaskan natives, are seeking to have an economist, Sam Rhodes, qualified as an expert witness in their case.

Submissions to the court indicate that they will call him to testify that an appropriate range of punitive damages would be in the range of £4 billion to £6.7 billion — the equivalent of a year's earnings worldwide for the Exxon Corporation. But the defence disputes Mr Rhodes's expertise.

There are no guidelines for appropriate levels of punitive damages, in which the Exxon Valdez could prove to be a test case.

Each side is expected to call up to 200 witnesses and, allowing time for appeals, the cases are unlikely to be settled this century.

About 100 plaintiffs, including the City of Weard, Kodiak Island Borough and the native co-operative Chugach Alaska Corporation are seeking about £280 million in compensatory damages.

1 The size of the load: the tanker was fully laden with 1.2 million barrels of crude oil.
2 The proportion of the load lost: one-quarter of the fuel carried by the tanker was lost.
3 The spatial extent of the spill: the oil slick eventually covered 25,000 km^2 (10,000 square miles) of Alaskan coastal and offshore waters.
4 The environmental impact: by mid-1990, 35 thousand dead seabirds from 89 species had been found, but these were believed to represent only a small fraction (estimates range from 10 to 30 per cent) of the total killed. Some 10,000 sea otters, 16 whales and 147 bald eagles were among the larger animals to perish as a result of the spill, with salmon, black cod and valuable herring-spawning grounds also being decimated.
5 The persistence of the effects: oil compounds survive longer in colder waters, harming most fish species and impairing reproduction in marine crustaceans. Wildlife in the region continued to suffer the effects of the spill well into the 1990s.

Controversy also surrounded the circumstances of the accident. Subsequent inquiries showed the tanker was illegally piloted and the 'rapid response' reaction was put into action much too slowly. It was ten hours before the first containment booms and oil-removing equipment reached the scene. Of the seven oil-skimming vessels available, only two were deployed. The massive clean-up operation that took place involved some 11,000 people and cost the Exxon Oil Company, which owned the stricken vessel, well over $2 billion.

(a)

(b)

Figure 9.19 ***Silent Spring*** The best-selling book by Rachel Carson, an American naturalist, was published in 1962. It drew public attention to the growing and potentially devastating use of chemicals (as pesticides, herbicides, and fertilizers) in agriculture. Special attention was paid to DDT, then regarded as a 'miracle' pesticide and contributed to its reduced use. The evocative title of Carson's book refers to the silence which would fall over the land as birds failed to reproduce. Rachel Carson (1907–64) is rightly regarded as one of the outstanding pioneers of the ecological movement.

[Source: (a) cover of Penguin Classics edition of *Silent Spring*, 1999. (b) Erich Hartmann/Magnum Photos.]

Reflections

1 An understanding of the effect of human beings on the natural environment requires knowledge of our position within the structure of ecosystems. Human intervention in the ecosystem is sometimes considered benign, sometimes malignant. List examples of each type.

2 The degree of environmental change is directly related to population density. One response to growing numbers is shortening the time cycle of shifting cultivation. Is shifting cultivation a specifically *tropical* form of agriculture? If so, why? Do you consider it (a) wasteful and destructive or (b) a logical reaction to environmental conditions?

3 Farming systems can be viewed as a chain that begins with the natural environment and ends in human consumption. Take one example from the farming system listed in Table 9.2. Try and represent this as a food chain in the manner shown in Figure 9.11.

4 Pollution may be defined as occurrence of a substance which, in terms of human environment, is in the wrong place, at the wrong time, in the wrong physical or chemical form. Critically examine this definition.

5 Populations at very high densities as in cities provide a focus for pollution problems of varying severity. Los Angeles has a worldwide reputation for atmospheric pollution. List (a) the human causes of this pollution and (b) the natural environmental factors that add to the problem. Do you think that a solution can be found? Why or why not?

6 A single pollution incident such as the *Exxon Valdez* oil spill (Box 9.C) has many dimensions. Using Figure 9.18 as a guide, analyze the multiple dimensions of any single pollution problem you encounter in your own local area.

7 Review your understanding of the following concepts using the text (including Table 9.1) and the glossary in Appendix A of this book:

biological concentration
Boserup model
food chains
heat islands
hollow frontier
land-use cycles
Minimata disease
shifting cultivation
smog
thermal pollution

One Step Further ...

A useful starting point is Ian SIMMONS, *Humanity and Environment* (Longman, Harlow, 1997) which covers many of the themes introduced in this chapter. More formal textbooks are M. McDONNELL and S. PICKETT (eds), *Humans as Components of Ecosystems* (Springer-Verlag, New York, 1993) and Andrew GOUDIE and Heather VILES, *The Earth Transformed: An Introduction to Human Impacts on the Environment* (Blackwell, Oxford, 1997).

The theme of agricultural change in relation to population pressure is reviewed in two books by Ester BOSERUP (see Box 9.A): *The Conditions of Agricultural Growth* (Allen and Unwin, London, 1965) and *Population and Technological Change* (University of Chicago Press, Chicago, 1981). A critical review of these ideas in relation to other approaches is given in David GRIGG, *Population and Agrarian Change* (Cambridge University Press, Cambridge, 1980). The general relations between economic and population growth are reviewed in NATIONAL RESEARCH COUNCIL, *Population Growth and Economic Development: Policy Questions* (National Academy Press, Washington, D.C., 1986)

The agricultural geography 'progress reports' section of *Progress in Human Geography* (quarterly) will prove helpful in keeping up to date. In addition to the regular geographical journals, keep a weather eye on *Scientific American* (a monthly), which gives good coverage to ecological issues. For readers with access to the World-Wide Web see also the sites recommended for this chapter in Appendix B at the end of this book.

CHAPTER 10

Resources and Conservation

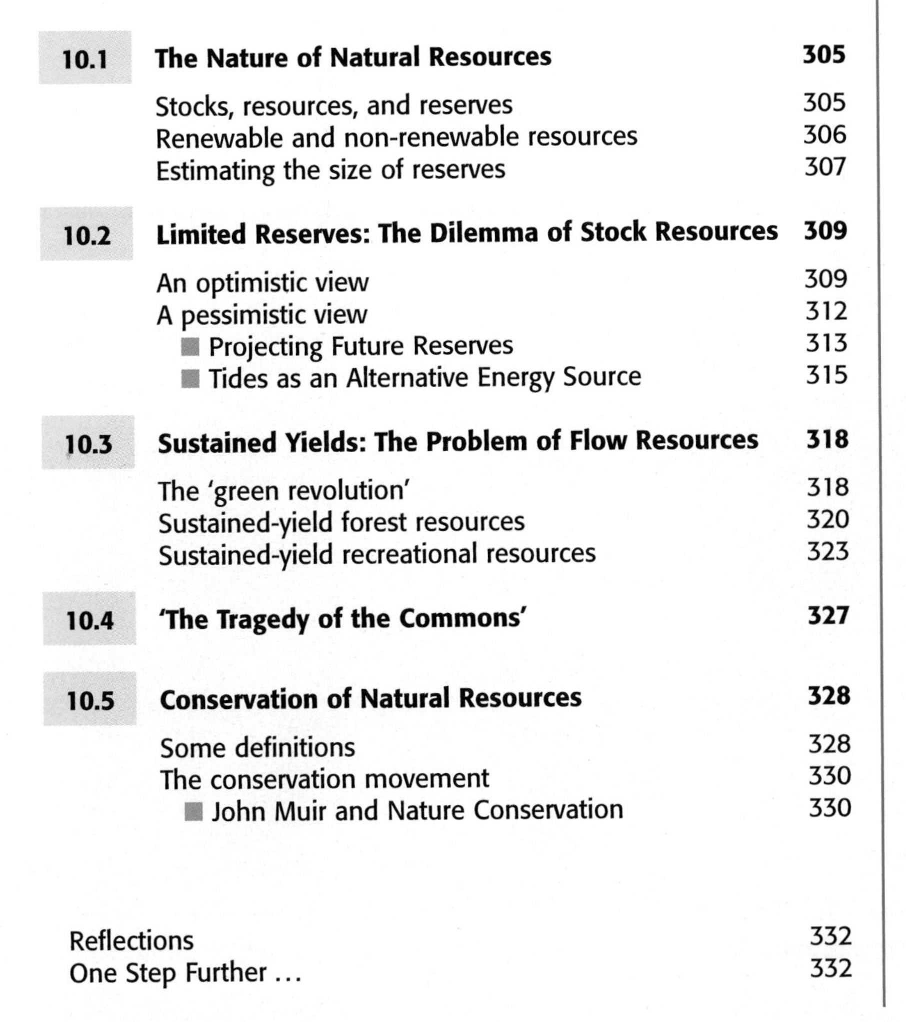

■ denotes case studies

Figure 10.1 Offshore petroleum resources The history of hydrocarbon extraction over the last 140 years illustrates the changing nature of natural resources. This oil platform in the Gulf of Mexico off the Louisiana coast illustrates the shift of the petroleum industry from continental fields towards offshore oil fields on the fringing shelves. Although these account for about a fifth of world production their costs are very high compared with land-based rigs. North America remains an important oil producer but now comes a distant second to the Middle East. [Source: Tony Stone Images/Bob Thomason.]

> Our planet has been aptly called 'Spaceship Earth'. It forms, overwhelmingly, a closed system as far as materials are concerned. Science fiction to the contrary, we have no present basis for believing that this essential isolation will be altered.... This earth is our habitat and probably will be as long as our species survives.
>
> MARSTON BATES *The Human Ecosystem* (1969)

One of the aspects that draws young people to the study of geography is a concern for the future care of the planet. (As we shall see in Chapter 24, the very same interests provides many geography graduates with their subsequent careers.) The nature of the earth's resources and how we best conserve them is central to this chapter.

Our opening photograph (see Figure 10.1 p. 302) is a symbol of resource use: an offshore oil well operating in the Gulf of Mexico. In late 1859 the first production oil well was drilled at Oil Creek near Titusville in northwestern Pennsylvania. The rig struck oil at a depth of 20 m (about 66 ft). Twenty years earlier the oil had been merely a nuisance as a contaminant

(a)

(b)

(c)

Figure 10.2 Some contrasting types of natural resources Three United States examples. (a) Mining resource: an open pit gold mine, Montana. (b) Oil rigs at Oil Creek, Pennsylvania, 1865. (c) Hydroelectric dam (the Hoover Dam, 220 metres (720 ft) high) built on the Colorado River during 1931–36.

[Source: (a) Tony Stone Images/Mark Snyder. (b) Corbis-Bettmann. (c) David Parker/Science Photo Library.]

in salt wells or had been collected from surface seepages and sold in small bottles as 'Rock Oil', a medicine with uncertain curative powers. Twenty years later 30 million barrels were being filled from wells around the world, 80 per cent of it from Pennsylvania (see Figure 10.2(b)). A new era in world fuel resources was under way.

The story of oil is just one of the dramatic examples of the selective use humans have made of the earth's natural resources. We could supplement it with parallel studies – of copper, uranium, even sand – that depict a natural substance rising rapidly in people's estimation to become a valuable resource. Such resources raise various questions that we consider in this chapter. What are natural resources, and how do we measure them? What determines whether or not particular resources will be used? If we do use them, how long will they last? Such issues lead on to the question of conservation, and we look at the issues this in turn raises at the end of the chapter.

The Nature of Natural Resources 10.1

The language in which we talk about the earth's natural resources has become somewhat tangled. In particular, we tend to confuse the concept of potential resources – such as the potential *hydroelectric power (HEP)* of the Amazon River system – with resources that have actually been developed, such as the electric power produced from Niagara Falls. It is useful, therefore, to distinguish at the outset between three apparently similar terms: stocks, resources, and reserves.

Stocks, Resources, and Reserves

The sum total of all the material components of the environment, including both mass and energy, both things biological and things inert, can be described as the *planet's total stock*. In earlier chapters we saw that the prime source of energy of the earth is solar radiation. The fact that the earth receives 17×10^{13} kW per day of solar energy provides some theoretical upper limit to production systems dependent on that source of power. Then too, we argue that all material goods must be derived ultimately from the 6.6×10^{21} tonnes of matter that make up the planet Earth (see Table 10.1).

In spite of its abundance, the vast proportion of the earth's total stock of matter and energy is of very little interest to us. Either it is wholly

Table 10.1 Earth's most abundant materials

Elements in the earth's crust	Percentage by weight	Metals in sea water	Percentage by weight
Oxygen (O)	46.60	Sodium (Na)	10.60
Silicon (Si)	27.72	Magnesium (Mg)	1.27
Aluminium (Al)	8.13	Calcium (Ca)	0.40
Iron (Fe)	5.00	Potassium (K)	0.38
Calcium (Ca)	3.63	Strontium (Sr)	0.01

inaccessible within existing technology (e.g. the manganese nodules abundant in some areas of the deep ocean floor) or it is in the form of substances we have not learned to use.

Resources are a cultural concept. A stock becomes a resource when it can be of some use to people in meeting their needs for food, shelter, warmth, transportation, and so on. The petroleum stocks of Texas were substantially the same in 1790 and 1890, but in the intervening period attitudes toward those stocks changed dramatically. Uranium ores provide a more recent example of the stock-to-resource transformation.

The transformation from a stock to a resource is reversible. Figure 10.3 is an aerial photograph of one of the most valuable resources of Neolithic Britain, the flint-axe mines near Brandon. As iron axes replaced flint, around 500 BC, the resource lost its usefulness and rejoined the unvalued stockpile. Thus, we can define *resources* as that portion of the total stock that could be used under specified technical, economic, and social conditions. Resources as such are determined by human concepts of what is useful, and we can expect *resource estimates* to change with technological and socioeconomic conditions. In this context, *reserves* are the subset of resources available under prevailing technological and socioeconomic conditions. They form the most specific but the smallest of the three categories and are relevant to one period of time only, the present.

Renewable and Non-renewable Resources

Geographers classify natural resources in various ways, as Figure 10.4 shows. The primary distinction made is between *non-renewable resources*, which consist of finite masses of material such as coal deposits, and

Figure 10.3 The changing status of natural resources Flint mines produced a key resource for Stone Age cultures but have been abandoned for over two thousand years. Grimes Graves, in eastern England, was one of the most important English sources of flint. Traces of mining activity are evident in the disturbed ground in the upper-middle section of the unforested area.

[Source: Royal Air Force photo. Crown copyright reserved.]

Flint axe

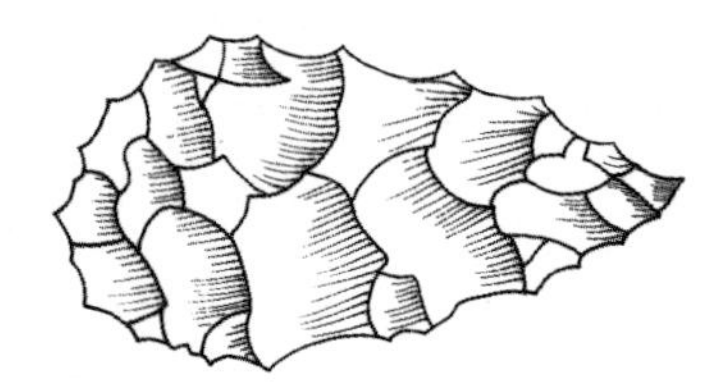

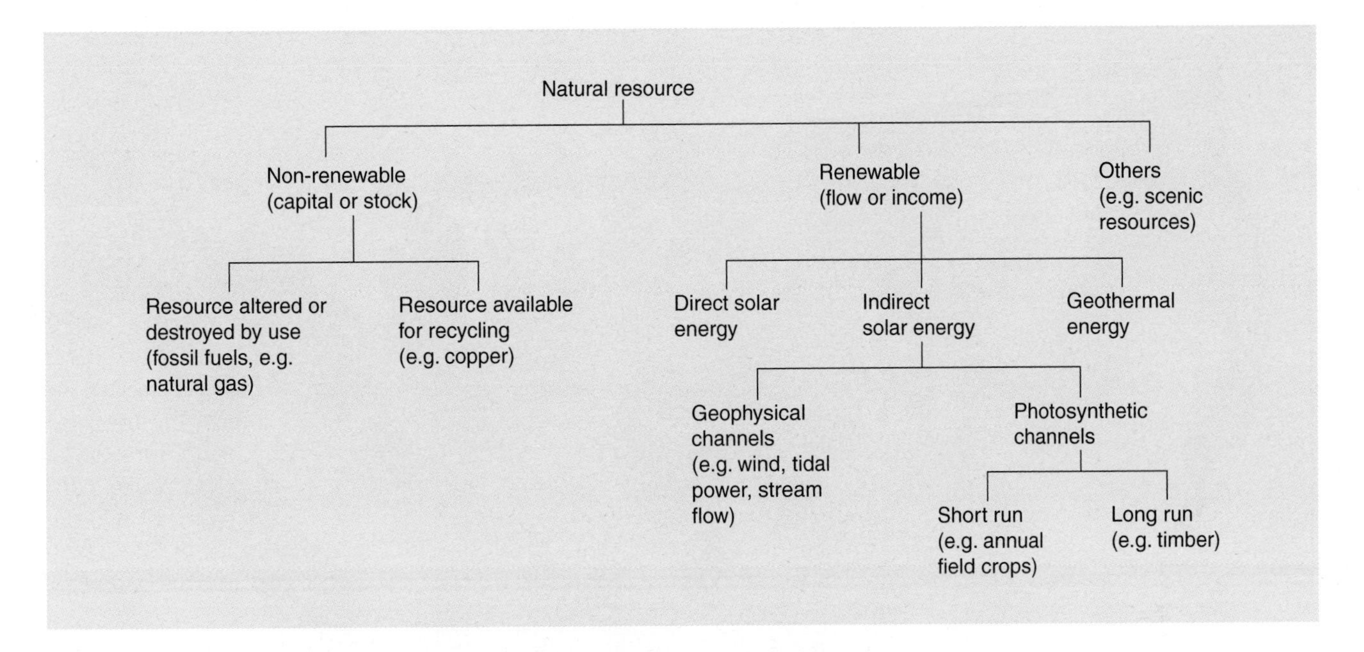

Figure 10.4 Types of natural resources

renewable resources. Non-renewable resources form so slowly that, from a human viewpoint, the limits of supply can be regarded as fixed. Some, such as stocks of coal or metal ores, are unaffected by the passage of time, while others deteriorate. The planet's stocks of refined ore, for example, are reduced by oxidation. The stock of natural gas is reduced by seepages. Renewable, or flow, resources are resources that are recurrent but variable over time; an example would be water power. Flow resources are usually measured in terms of output over a certain time. For instance, the upper limit on world tidal power is about 1.1×10 kW per annum.

Renewable resources can be separated further into those whose levels of flow are generally unaffected by human action and those demonstrably affected. It is difficult to envision human beings ever being able to interfere with the world's potential tidal energy. In contrast, the yield of groundwater resources can be permanently reduced. Overpumping may irreversibly close fissures capable of storing water or, as in the coastal valleys of southern California, allow incursions of saline ocean water. Between these two extremes stand resources such as forests where a reduced flow (e.g. due to over-cutting) can be counterbalanced by subsequent remedial action but only just so long as the soil has not been too badly degraded.

Estimating the Size of Reserves

How do we go about estimating the size of particular reserves? We first need to know the distribution of the resource. Figure 10.5 indicates areas within the conterminous United States in which geological conditions could produce petroleum. These areas are sedimentary basins where the organic products from which oil is derived were once deposited, compressed under other sediments, and preserved. The probable location of specific fields can be narrowed by geophysical surveys and confirmed by trial bores.

Whether a particular oil field will be used depends not just on geological conditions. As Figure 10.6 indicates, we can consider the size of the reserves the field represents as a joint function of four factors.

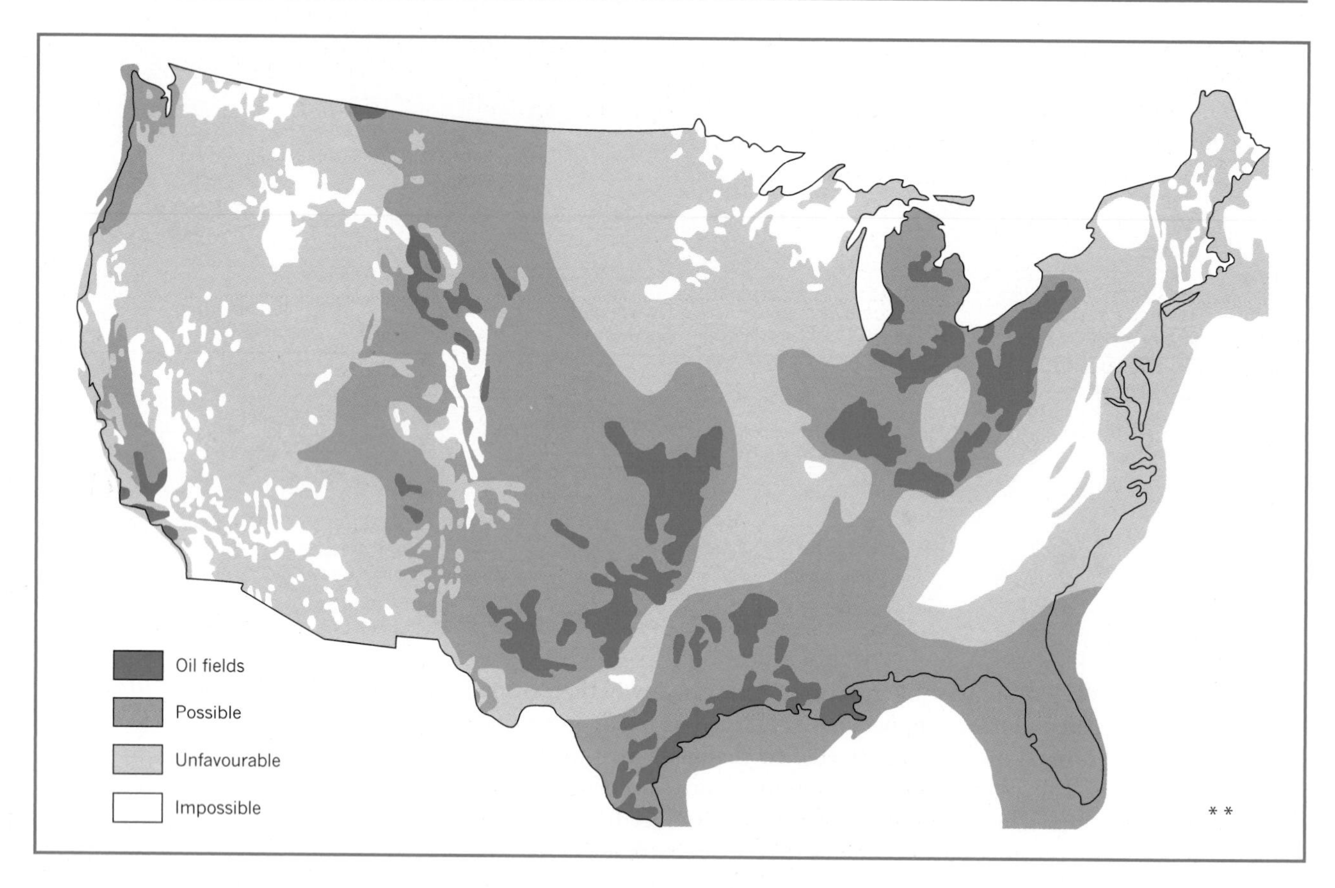

Figure 10.5 Actual and potential resources Potential petroleum-producing areas in the continental United States are classified according to the likelihood of oil finds. 'Possible' areas are those where small commercial quantities of oil can be found.

[Source: US Geological Survey, *Map of the United States Showing Oil Fields and Uproductive Areas* (US Government Printing Office, Washington, D.C., 1960).]

1 The quality of the oil, its chemical characteristics, and its freedom from impurities such as sulphur.
2 The size of the field, and whether it is large enough to justify the capital investment needed to work it.
3 Accessibility of the field, both in a spatial sense (i.e. its distance from refineries or consumers) and in a vertical, geological sense (its depth).
4 The relative demand for oil as indicated by the prevailing price level.

Alteration of any of these four factors can change the size of the reserve. Note the effect of a low price in Figure 10.5 in reducing the estimated size of

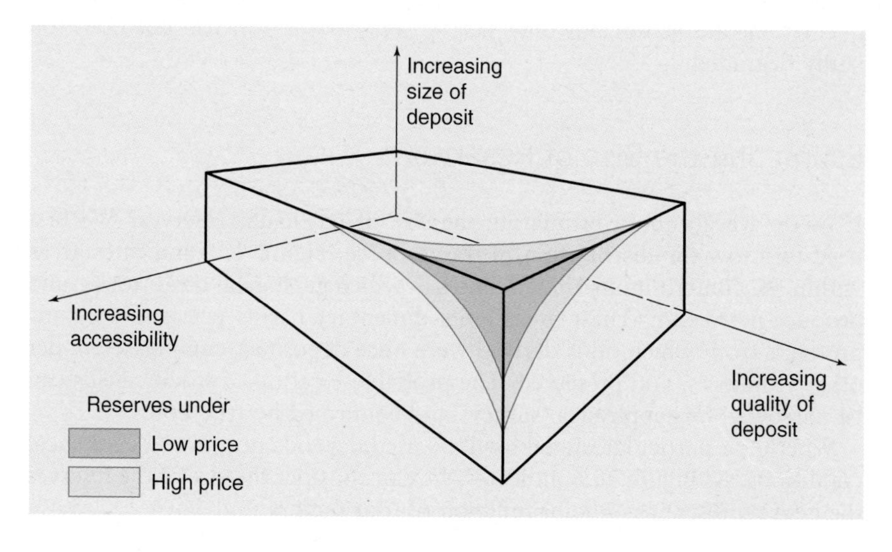

Figure 10.6 Factors affecting size of reserves Hypothetical relationships between the size of a deposit, its quality, its accessibility, and the prevailing price level are indicated. The block indicates the total size of the world stock of a particular resource, and the coloured 'corner' the extent of reserves. High prices *increase* the area of reserves, low prices *decrease* the area of reserves.

the reserve. We could go on to elaborate the relationship various ways. For example, we could define our third element, accessibility, as including strategic accessibility (which depends on who owns the field). Potential oil reserves in the North Sea are much more likely to be exploited because they fall within the ownership of countries (such as Norway and the United Kingdom) that were wholly dependent on imported oil (see Figure 17.23). In cases like this the high cost of exploitation may be counterbalanced by strategic advantages to a country in being able to control its own sources of oil.

Criteria similar to those for estimating petroleum deposits can be used to gain a general idea of the size of current reserves of other resources. In the case of *stock resources* reserves are expressed as a finite total; reserves of *flow resources* are described in terms of the potential output in a particular time period. In both cases estimates of reserves are usually only approximate, refer to a specific time, and must be continuously updated as technical and market conditions change.

Limited Reserves: The Dilemma of Stock Resources 10.2

The extent of human exploitation of terrestrial resources is staggering, particularly in the most recent period of human history. We estimate that the world population doubled between 1800 and 1930 and doubled again between 1930 and 1975. Each human being whose existence is projected in the population models of Chapter 6 is going to require basic necessities such as food, water, shelter, and space (as well as a growing range of non-essential goods). As living standards rise, the pressure on resources created by an exponentially growing population is being increased still further by a rising per capita demand for resources. This demand is being met by a massive consumption of available natural resources.

The combined effect of increased population and resource consumption per capita has been a sextupling in the level of resource extraction between 1880 and 2000. The amount of most metals and ores used since 1950 is in excess of the combined amount used in all previous centuries. A projective study called *Resources in America's Future*, published in the mid-1960s by Resources For the Future, estimated that by AD 2000 the world would need a tripling of aggregate food output, a fivefold increase in energy, a fivefold increase in iron alloys, and a tripling of lumber output. Recent checks show these projections to have been broadly on target. Given the population and changes sketched in Chapter 6, we can expect this massive increase in the use of resources to continue at least into the early decades of the present century.

How long will reserves of the earth's non-renewable resources last? There appear to be two kinds of answer. The first, which concerns the medium term (about 30 years), is primarily based on economics and is generally optimistic; the second involves a much longer historical period, is based on ecological arguments, and is less hopeful.

An Optimistic View

The classic economic test of increasing scarcity is a marked rise in the real cost of a product in comparison with the general price level. How do natural resources meet this kind of test? Long-term trends in fluctuations in price levels for natural-resource products during the nineteenth and twentieth centuries are shown in Figure 10.7(a–c). Because the prices are plotted in ratio terms, the erratic movements represent considerable fluctuations. For example, the real prices of forest products are now over three times what they were in 1870. In comparison, mineral prices have declined, and farm products have increased relatively little. Perhaps the most remarkable fact is that the prices for all resource products varied rather little over the past century. According to this conventional economic index of scarcity, natural resources do not appear to have become significantly scarcer since 1870.

It is important to bear this long-term trend in mind in looking at the occasional sudden leaps in prices so alarmingly reported in the press in the 1970s. Figure 10.7(b) shows price trends over a 30 year period and emphasizes the surge in oil prices following the Arab–Israeli conflict in 1973. To see these changes in perspective, we need to recall that it is the relative price of resources that is at issue in determining scarcity. Figure 10.7(d) also uses a linear rather than a ratio scale on the vertical axis.

To understand the relative stability of natural-resource prices over the long term, we need to think about what happens when a rapid hike occurs in the price of a commodity – as happened in the case of tin in the 1960s or of oil in 1974.

As the price of a resource rises, a chain of compensating movements occurs. First, high prices bring greater care in the way resources are used. Supplies of products with high values per tonne are carefully metered and their use carefully recorded; conversely, wastage results from low-value products. Water is a classic case of the effect of a change in attitudes toward

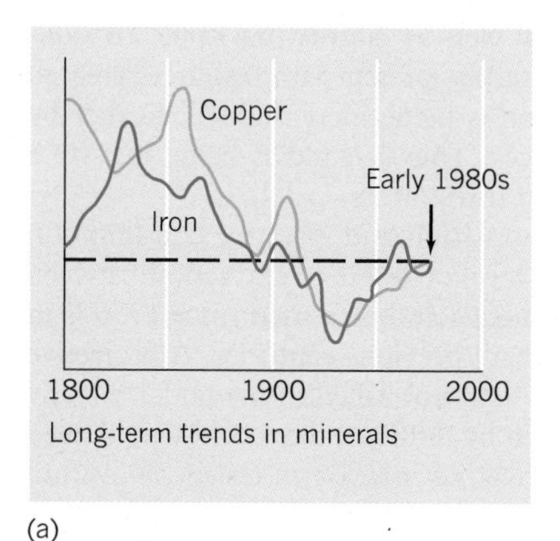

(a)

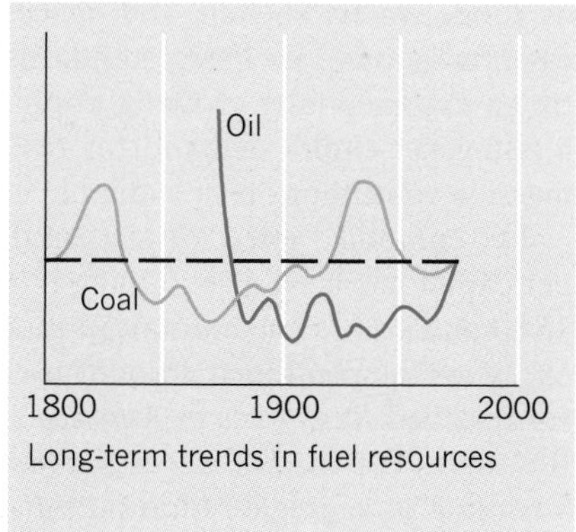

(b)

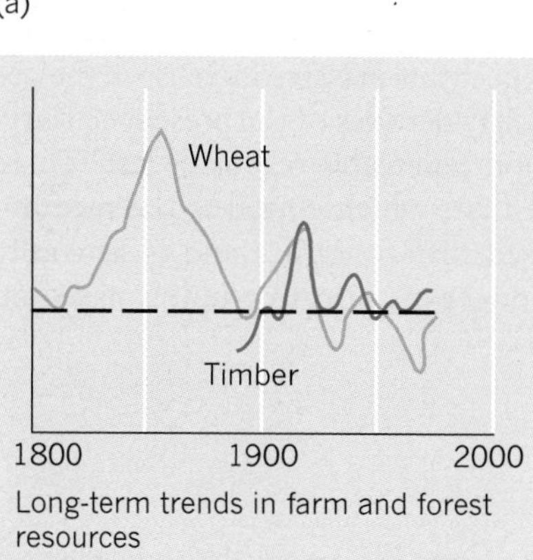

(c)

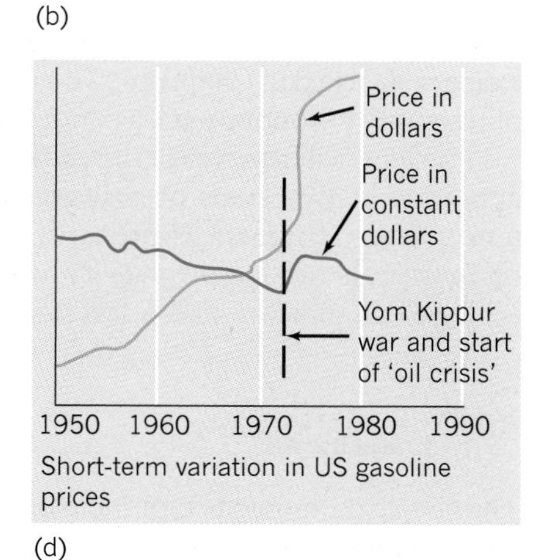

(d)

Figure 10.7 Trends in resource prices Figures (a) to (c) show relative fluctuations in the deflated price of natural-resource products between 1800 and the 1980s. Note that the vertical axes represent ratios. While these show cyclic change, they do not show the expected increase in scarcity. Figure (d) shows both undeflated and deflated price trends in US gasoline for a 30 year period from 1950. Despite the oil crises of the early 1970s, the relative price of oil in real terms has not shown a consistent increase. Despite our intuitive guess that the prices of natural resources have been rising over our lifetime, the deflated prices show cyclic changes about a surprisingly constant level.

a resource regarding the way in which it is used. Once free, its distribution is now metered; and the fact that people must pay for water discourages reckless use of it.

Another reason for stability in the price of resources generally is that various resources may be substituted for one another. An exponential increase in overall demand for one resource may be met by using other resources for the same purpose. The increased demand for textiles is balanced by a switch from natural fibres (such as cotton and wool) to synthetic fibres (such as Dacron, Orlon, and nylon) derived from coal, oil, and even urine. Consequently, each resource has a convex consumption curve. Consumption does not fall because of a physical shortage of the old product but because it becomes more expensive in relation to the new substitutes. Ways in which substitution may occur are illustrated by the substitution that occurred in Germany during World War II, when a variety of ersatz products, many based on chemical derivatives of coal, were produced. The changing pattern of world fuel requirements, another example of substitution, is summarized in Table 10.2.

Still another reason for stable resource prices is a switch in methods of extraction. Even when the same natural resources continue to have the same uses, considerable changes in methods of extraction can occur. In copper mining, since the last century, there has been a radical switch from the selective mining of high-grade deposits to the mass mining of low-grade deposits. In 1900 ore had to have a copper content of at least 3 per cent to be worth mining; today, ore with a copper content of around 0.5 per cent is being used. Oil drilling now involves less wasteful extraction methods than in the 1870s, and so on.

This switch to lower-grade resources significantly affects the total assessment of the reserves of a stock. The amount of the element in the earth may represent the total stock; in practice, however, the recoverable reserves will be a small fraction of the ultimate reserves. Figure 10.8 presents the theoretical distribution of an element as a frequency distribution diagram; generally, the element is so diffusely located that it is economically unrecoverable. Only when geophysical and geochemical processes concentrate the element in certain locations will recovery normally be possible. Other concentrations may be achieved biologically (e.g. as direct organic concentrations lead to coal, lignite, and petroleum deposits) or by mechanical means (e.g. fluvial sorting of gold into placer deposits).

Not all resources follow the simple arithmetic and geometric model in Figure 10.8. Two specific deposits that fail to show the regular (but inverse)

Table 10.2 Changing emphasis on world energy sources

	Percentage contribution to total energy used					
Energy source	**1875**	**1900**	**1925**	**1950**	**1975**	**2000**
Wood, vegetation	60	39	26	21	13	5
Coal	38	58	61	44	27	21
Oil	2	2	10	25	40	39
Natural gas	<1	1	2	8	15	15
Other sources (mainly hydroelectric and nuclear)	<1	<1	1	2	5	20

Source: Data compiled from United Nations sources.

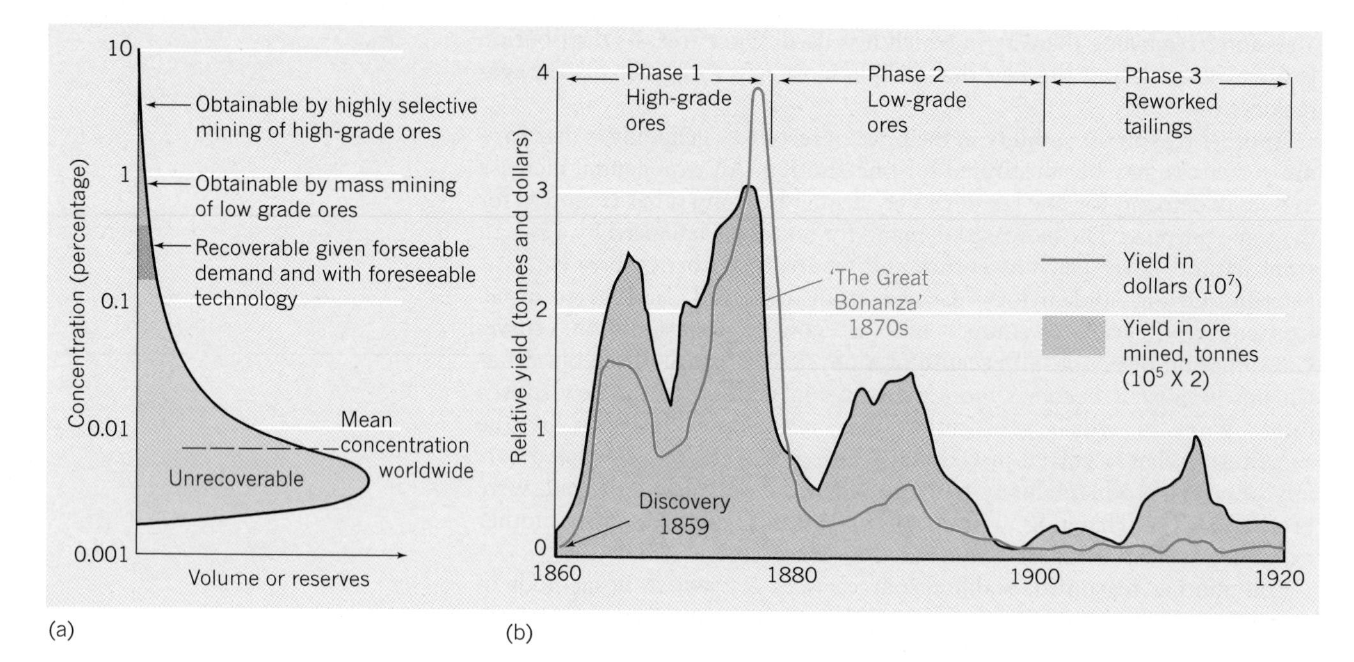

Figure 10.8 Patterns of resource distribution (a) The curve shows the general concentration and volume of mineral resources; the vertical axis is logarithmic. For the upper part the quality of ores (concentration) is inversely related to their quantity (volume of reserves). (b) Silver production from the Comstock Lode in Nevada, western United States, discovered in 1859. Mining has gone through three distinct phases, each using ore of lower quality.

[Source: Data from E. Cook, *Man, Energy and Society* (Freeman, San Francisco, 1976), Fig. 13.3, p. 394.]

relationship of ore quality to ore volume are the lead–zinc ores found in limestone, and mercury. Here there are abrupt changes in concentration ratios. Although the principle that ore resources increase at constant geometric rate as ore quality decreases can be a valuable guide to the abundance of some mineral deposits (notably iron, bauxite, magnesium, and copper) in the earth's crust, it would be dangerous to assume that it applied to all minerals, and more dangerous still to assume that it applied to non-mineral resources. Work remains to be done on developing precise models of the relationship between quantity and quality for the widest possible range of natural resources. Such models would be valuable guides to the probable future availability of some key resources.

A Pessimistic View

If we take a much longer view of resource extraction, then our predictions may be more pessimistic. We noted in Chapter 6 that the present period of rapid population growth and resource exploitation, far from being part of the normal order of events and projectable into the future, is very abnormal. A continuation of the present rate of world population expansion would allow each person a share of only 1 m^2 of the earth's land surface – Antarctica and the Sahara included – in around 525 years.

Population growth is a useful starting point, for we have seen that the consumption of resources is partly a function of the rate of population increase. If we take the rates of energy consumption per capita as an example, the daily minimum needed by a primitive person to keep alive was equivalent to about 100 watts (for food). As other sources of energy, notably firewood, were added, the level rose to around 1000 watts per capita. Here the rate stayed until the continuous mining of coal (beginning about eight centuries ago) and the production of oil (beginning just over a century ago) brought an exponential increase in energy consumption of around 10,000 watts per capita. We can gain some idea of the recentness of the consumption of *fossil fuel* deposits by noting that half the cumulative total (or world) production of coal has occurred since the 1930s and half the oil consumption since the 1950s. Given the rates of consumption typical of the present decade, natural resources that took 100 million years to form by sedimentation will be consumed in about 100 years of industrialization.

The Problem of Dwindling Fossil Fuels Energy resources are the key to the pace of resource extraction. The current form of world society is highly dependent on energy, and the future availability of other resources, organic and inorganic, is indirectly related to energy resources. Models for estimating the likely volume of fuels available have recently been developed. (See Box 10.A on 'Projecting future reserves'.) These models allow us to project complete cycles of production for the major fossil fuels. Despite some variation, the results of the projections indicate that approximately 80 per cent of the petroleum family's resources (crude oil, natural gas, oil from tarsands, and shale oil) will probably be exhausted in about a century. Similar calculations show that roughly 80 per cent of the world's coal reserves will be depleted in about 300 to 400 years. From a historical perspective, the age of fossil fuels will be a limited one, lasting from about AD 1500 to AD 2800 – a very short time even in terms of the brief human occupation of the planet.

Box 10.A Projecting Future Reserves

Assume that the volume of a mineral resource available in any area is finite and that discovery and production follow logistic curves over time. Then the rates of change of discovery, production, and reserves must follow a generally cyclic form (see diagram (a)).

The peak of ***proven reserves*** comes where the curves for the rates of discovery and production intersect, with the production rate still rising but the discovery rate already on the decline. This intersection occurs roughly halfway between the two peaks. With similar curves for actual resources, we can project how far their cycle of exploitation has run. As an example, consider the curves for crude oil in the United States, exclusive of Alaska (see diagram (b)).

This figure indicates that reserves probably reached their peak as early as the 1960s and are now on the decline. But new exploration methods and an ability to use smaller fields has displaced the curve to the right. Problems in calibrating such curves, their applications, and the reservations needed in interpreting them are discussed in M. King Hubbert, *Energy Resources* (National Academy of Sciences–National Research Council, Washington, D.C., 1962).

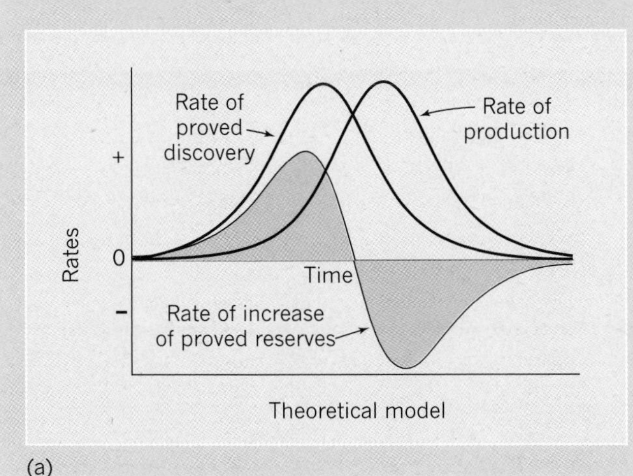

(a)

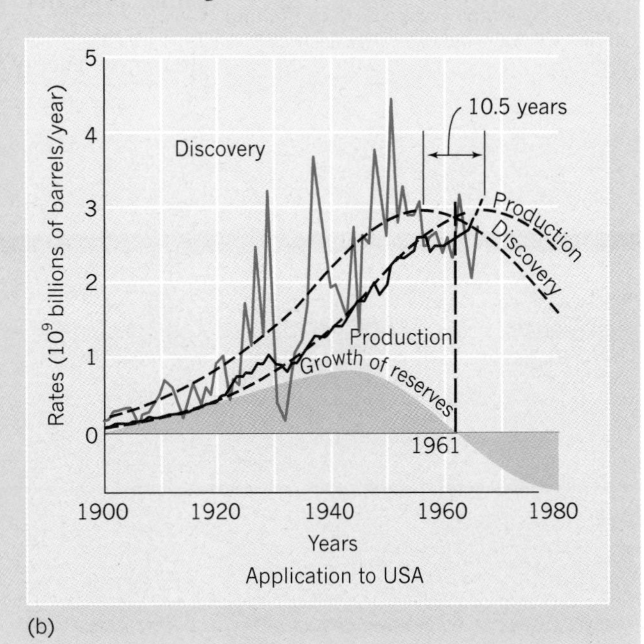

(b)

Oil illustrates well the rapid geographic change in areas of supply that followed the working out of easily accessible fields. We noted earlier that in 1880 world production was 30 million barrels, 80 per cent of it coming from Pennsylvania. The succeeding decades have seen a dynamic shift in both the quantity and the location of supplies. By 1910 the 1880 world total of 30 million barrels had increased tenfold, and by 1950 it had grown by over 100 times. New producing areas had sprung up all over the world: the Caucasus in southern Russia and Dutch Indonesia in the 1880s and 1890s, Texas and Oklahoma in the 1900s, Venezuela and the Middle East in the 1930s. Today the pattern continues to change. To the discoveries in Libya and Algeria in the 1960s, we now must add those in Nigeria, the North Slope of Alaska, and Europe's North Sea in the 1970s. Current additions are being made in the shelf areas of the world off the coast of north Australia and south China. The dominant spatial fact that emerges from the changing historical geography of oil production is the increasingly important role of the Middle East in general and the Persian Gulf in particular. Gulf states now control about three-quarters of the world's petroleum reserves, a share that seems unlikely to change very significantly in the next decade or so.

Alternative Energy Sources According to current projections, within the next two centuries there will be a need for a reliable source of energy to substitute for fossil fuels. Where is that energy coming from? (See Figure 10.9.) We have already noted the use of hydroelectric power (see Figure 10.2) and extension of that source is continuing worldwide and on an ever-increasing scale through such schemes as China's Four Gorges scheme (see Figure 18.19(c)). But hydroelectric power schemes have limitations: the silting of reservoirs behind dams, seasonal variations in output, transfer costs to demand areas, and so on. The closing years of the last century saw an increasing interest in exploring an ever wider range of *alternative energy* sources to fossil fuel: solar radiation, wind power, tidal energy, geothermal energy, as well as atomic fission, and atomic fusion. Since space does not permit a full analysis of all such sources, we take one of them – *tidal energy* – and consider it as a representative example (see Box 10.B on 'Tides as an alternative energy source').

Figure 10.9 Alternative energy sources (a) Geothermal energy: Nesjavellir geothermal power station in south-west Iceland. Surplus water at 83 °C (180 °F) is sent via an 27 km (17 mile) long insulated pipe to the capital city, Reykjavik. (b) Wind energy: wind turbines Palm Springs, California.

[Source: (a) Simon Fraser/Science Photo Library. (b) John Mead/ Science Photo Library.]

(a)

(b)

Box 10.B Tides as an Alternative Energy Source

Tides are the regular rise and fall of ocean waters generated by the gravitational pull of the moon (and to a lesser extent the sun). The pattern and range of tides are very complex because of the resonances set up within different seas and ocean basins. Although small tidal mills were known as early as the medieval period and there have been enthusiastic proponents in both the nineteenth and twentieth centuries, there are relatively few large-scale schemes in operation to study.

One example of a tidal power station in operation is that constructed by the French government at La Rance on the north coast of Brittany. The decision to construct the plant was taken by France's nationalist leader, General de Gaulle, in 1959 and the scheme began producing electric power in 1966. Although it is relatively small in relation to other conventional hydroelectric schemes in France (it just comes into the top twenty), it is a new source of renewable and pollution-free energy and it is currently producing electricity slightly more cheaply than nuclear power plants.

On the northern coast of Brittany two high tides and two lows are experienced every day (actually every 24 hours 50 minutes). On the Rance estuary the range between high and low tides averages 8.5 m (27 ft) over the whole year, with monthly ranges varying by 5 m (17 ft) above or below this average. The effect is that the sea rushes in to fill the estuary – and then out again some six hours later – as a massive cycle of water movement.

La Rance tidal power station captures part of the power in these tidal movements by a series of turbines. The turbines are set in a dam built across the mouth of the estuary and are designed to operate in two directions – when the tide is flowing into the estuary as well as when it is flowing out. The dam is broken by a lock to allow ships to move into the estuary and also has gates to allow control of the water level. In France, as in most western countries, the demand for electricity has a distinct daily pattern. In a typical day it rises to a peak in the morning as lights and domestic appliances are switched on and as industry gets into production. It remains high during working hours, has a strong secondary evening peak, and drops to a night-time low. Demand is higher in winter than in summer.

However, the supply of tidal power is linked to a wholly unrelated cycle. The two tidal peaks sometimes coincide with the morning and evening peaks, but as often are out of phase with them. Various ingenious devices for spreading tidal power over more of the day have been developed by civil engineers. For example, La Rance turbines can

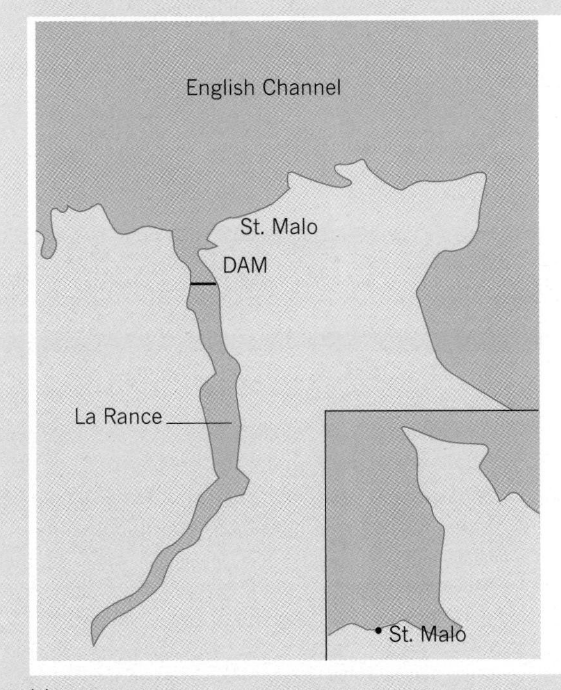

(a)

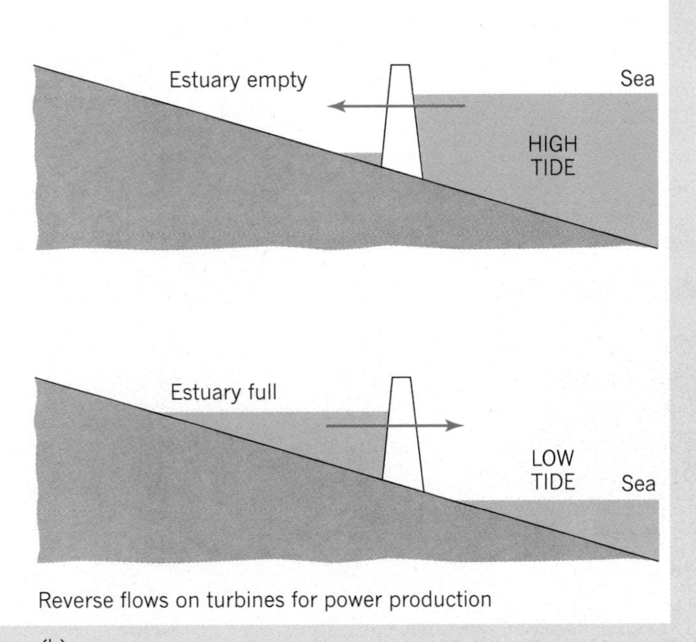

(b)

Box continued

(c)

also be used as pumps and surplus electricity can be used in 'off peak' periods to pump water into the estuary to supplement the outward flow being released during peak periods.

How much power is available from tidal power? Clearly La Rance scheme is a small one and is acting only as one input to the whole French electricity grid. Suitable sites for tidal power schemes exist in many parts of the world. The two prime locations are the Bay of Fundy in Canada's Maritime Provinces and the Severn Estuary on Britain's west coast. In both, schemes have been examined over past decades and found acceptable on technical engineering grounds. However, so far the high capital costs have proved a barrier to power production at competitive price. As the price of conventional fuels increases, so the economic feasibility of tidal schemes seems likely to improve. Their pollution-free character and the recurrent nature of the power source makes them very attractive in principle; they may, however, greatly disturb the ecology of the estuary landward of the dam. If the Fundy and Severn schemes were to go ahead, other sites around the world's coasts which have the appropriate locational characteristics of high tidal range plus nearby power markets may be developed. Sites on the Korean, Chinese, and Indian coasts are particularly promising.

But to put tidal energy in perspective we should recall that it is, even on the most optimistic assumptions, still a minor source of power. It could contribute up to 4 per cent of the world's electricity demands on some estimates, but most forecasters put it at a much lower level. So far it offers no serious bid to reduce our dependence on conventional power sources.

Table 10.3 Potential of alternative energy resources

Type of power source	**Percentage of global total from alternative sources**	**Development level achieved**[a]	**Comment**
Biomass	35	C	Would require massive land use; loss of agricultural land
Hydroelectric	17	C	Already highly developed
Solar (direct)	17	S	Small-scale; energy storage problems
Wind	13	S	Small-scale; energy storage problems
Geothermal	12	C	Many more areas have potential; technology extensive
Ocean thermal	6	T	Still speculative
Tidal	0.2	S	Very localized potential; supply unrelated to demand
Waves and ocean currents	0.05	S	Minor quantities available; not expected to be significant

[a]Development levels: T, theoretical; S, small-scale plants in operation; C, large commercial plants in operation.

Each of the different sources faces a series of development problems. One relates to the scale of production. As Table 10.3 indicates, only hydroelectric power plants are currently capable of producing power on the kind of scale likely to be needed. To meet the needs of a large city, a solar power plant of 10^{10} thermal watts would have to collect solar power over an area of the earth's surface equivalent to 6.5 km^2 (2.5 square miles).

The most important sources are therefore the two nuclear power sources, atomic fusion and fission. Although the earth's resources of nuclear materials (uranium, thorium, and deuterium) are finite, they are large enough so that, given a reasonable rate of innovation in nuclear-reactor technology, the future scale of supply looks promising. The chief limitation to nuclear fuel may lie less in the adequacy of resources than in the safe disposal of radioactive wastes (see Figure 10.10).

The possible environmental threats from nuclear power production are immense and increase in proportion to the increasing number and size of

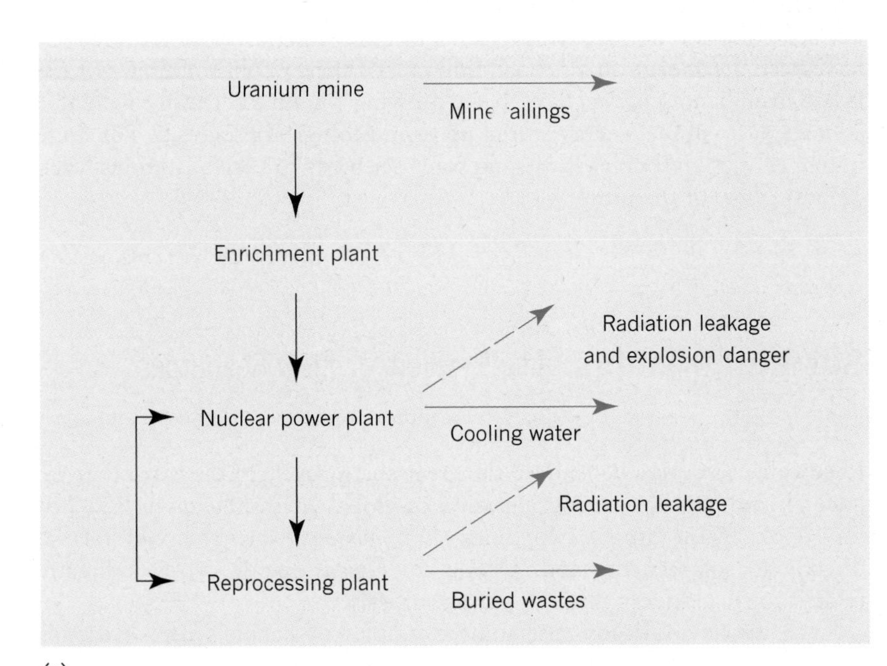

(a)

(b)

Figure 10.10 Nuclear power and environmental hazards (a) Main stages in the mining and production sequence for the use of fissionable fuel for energy production. The main environmental threats at each stage are indicated. (b) Sites of nuclear power plants (stage 3 in the production process) are adjacent to cooling-water supplies and commonly located in rural areas to minimize the population at risk from any potential leakage problems. The Three Mile Island nuclear power plant near Harrisburg, Pennsylvania, is shown.

[Source: (b) Popperfoto.]

the power plants. Radiation release, the amount of buried radioactive waste, and even the possibility of nuclear hijacking are likely to get more rather than less important in the next few decades. Yet the importance of success in nuclear power production is equally vital in the very long run. The world's greatest potential source of energy is not fossil fuels, but the hydrogen locked up in the waters of the world's oceans. Hydrogen as a liquid or gas may well be the element powering our cars and heating our homes by the end of the twenty-first century. The electrolysis of water to produce hydrogen is already being undertaken on a small scale where cheap power is available; if fossil fuel costs continue to rise and nuclear power costs continue to fall, the prospects for extracting hydrogen on a mammoth scale should improve.

Energy looks like the main resource bottleneck over the medium term. We could argue that since energy resources are only one of the sets of natural resources used by humans, a pessimistic view in this area need not prejudice a more optimistic view of other resources. Unfortunately, this is not so. The main human achievements in these other areas of resource use have been dependent on an expanding rate of energy consumption. For the last half-millennium, we have been drawing on rather readily available sources of fossil fuel energy stored up from geological processes. The finite nature of these resources is causing some scientists to take a cautious, even gloomy, view of the future.

10.3 Sustained Yields: The Problem of Flow Resources

Renewable resources depend on the great energy cycle of the earth that we have already met in our consideration of global environments in Part I of this book. These are of two kinds: first, *physical* energy cycles related directly to solar energy and, second, *biological* energy cycles indirectly related to solar energy through photosynthesis.

Since we have already met some examples of people's use of earth's renewable resources related to physical cycles – solar power, water power, and tidal power – in our discussion of energy problems, we shall concentrate here on those related to biological cycles. In this category come plant and animal energy sources as we tap through agriculture, forestry, and fishing. We shall also look briefly at the special problem of recreational resources, which lie on the border between renewable and non-renewable resources. The key theme linking the examples chosen is that of how to maintain and increase the sustained yield from a particular source.

The 'Green Revolution'

Our first example of renewable resources is drawn from crop production in the agricultural sector. As we saw in Section 5.2, the disturbing and mixing of plant species by humans to produce new hybrids has been going on for many thousands of years. Plant breeding, albeit of an accidental kind, has been a camp follower of all human agriculture. Since Mendel (1822–84) laid the foundation of plant genetics with his experiments in heredity, plant breeding has played an increasingly important role in increasing and maintaining agricultural yields; indeed, without it the Malthusian forecast of worldwide famines as a check on human population growth would surely have been fulfilled.

Perhaps the most striking example of the impact of new hybrids in the second half of the last century was the *green revolution*. The green revolution is an evocative term used to describe the development of extremely high-yielding grain crops that allow major increases in food production, particularly in subtropical areas. In 1953 scientists in Mexico began to develop rust-resistant dwarf *wheats* which doubled Mexico's per-acre production in the next decade. After a major drought in India in 1965, Mexican dwarf wheat was widely planted in the northern part of that country with dramatic results in terms of wheat yield in areas such as the Punjab (see Figure 10.11).

Rice development followed a course similar to but more laggardly than that of wheat. The Los Banos research institute in the Philippines was set

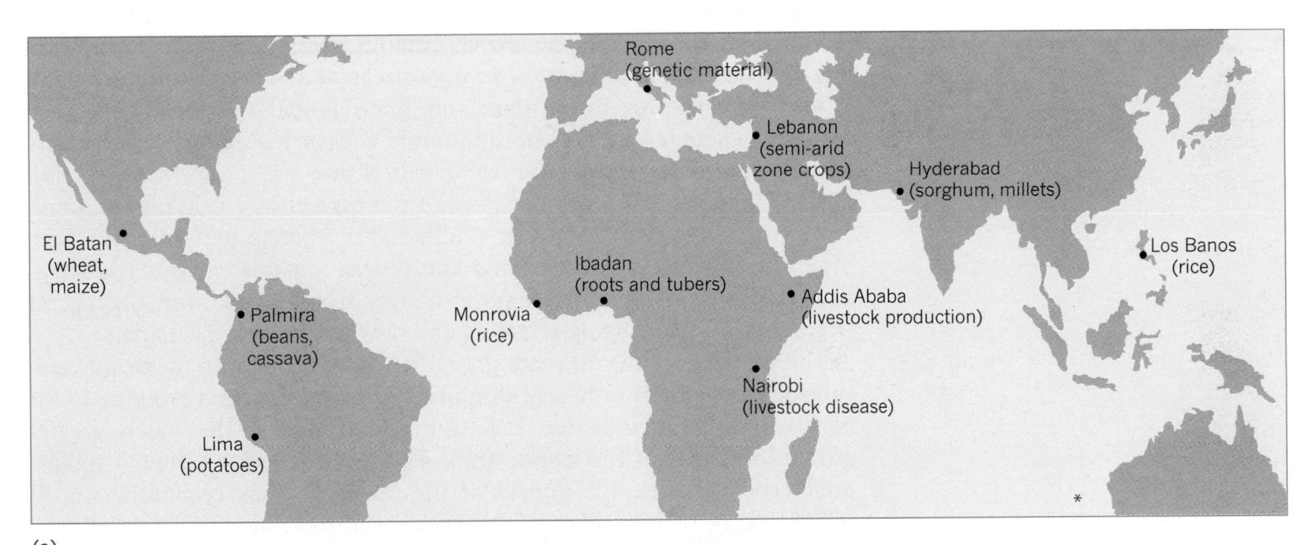

(a)

(b)

Figure 10.11 Network of international agricultural research (a) Major stations established since 1959, many under United Nations (***Food and Agriculture Organization, FAO***) sponsorship, with the leading research interests of each station. The oldest is that of the International Rice Research Institute (IRRI) at Los Banos in the Philippines, home of the new rice strains of the 'green revolution.' (b) Planting rice on peasant's cooperative on Mindinao Island, Philippines.

[Source: (b) Julio Etchart/Reportage/Still Pictures.]

up with Ford and Rockefeller Foundation backing in 1962 to develop varieties of improved rice (IR). The now famous IR-8 variety was spotted in 1965. Its first harvest, from 60 trial tonnes of seeds, produced an astonishing sixfold increase of rice under field conditions. From these beginnings, other varieties of IR with important additional characteristics – better resistance to disease, better taste, and (very important in Asia) better appearance after cooking – were developed. Broadly speaking, the new varieties of rice have produced results wherever they were planted in the humid tropics. About 10 per cent of India's paddy land is now planted with IR varieties, and the Philippines, once a major rice importer, is now nearly self-sufficient in this area.

How far have the 'miracle grains' of the green revolution proved an unmitigated blessing? Any seed that can (a) give two to four times the yield of indigenous grains, (b) has such a shortened growing season that two crops per year are often possible, and (c) has a wider tolerance of climatic variations must be welcome. At the same time, some severe problems have appeared. The high yield is dependent on high applications of fertilizer and insecticides, plus, in the case of rice, copious irrigation. Hence, innovation has been most rapid in the most prosperous areas and among the most prosperous farmers, and, in the short run, interregional and social gaps have widened. There is an urgent need for more widespread adoption of the new varieties in poorer sectors, but the fertilizer and water they need are still beyond the financial reach of many of the agricultural peasants of South Asia. At a quite different scale, traditional marketing patterns have been upset. Countries such as Thailand and Burma – major exporters of rice – have found their traditional markets disappearing. Japan, normally a great rice importer, has bulging elevators and now looks for export areas.

The successes of the 40 years since 1960 have brought to the tropics and subtropics the kinds of benefit that plant breeding has been bringing to the mid-latitude farmlands for a longer period. Given the much greater growth potential of the tropics (Box 4.B) and the lower yields of indigenous crop varieties, the impact of the revolution has been dramatic. It remains to be seen whether demands for fertilizers and pesticides will take the edge off some of the more promising aspects of this important step in the development of the world's agricultural resources. It may well be that since increased agricultural production in this case depends on fertilizer supplies – much of which come from mineral sources – increased yields from a renewable resource may indirectly depend on a non-renewable resource.

The latest phase of the green revolution has been the genetic modification (GM) of major crops. This involves gene splicing to add desirable properties (e.g. high growth) or to lose undesirable properties (e.g. low resistance to rust). Originally applied to soya beans, the GM technique has now been extended to a range of other food crops. The degree of monopoly of leading chemical companies, consumer resistance, and fears about environmental side-effects are still to be worked out.

Sustained-yield Forest Resources

The view of forests as a renewable resource is a relatively recent one. For most of human history on the earth, people's activities have tended to reduce the world's forest cover (see Chapter 11). Heavy inroads were made to clear land, for fuel wood, and for construction materials. Destruction was encouraged by the belief that forest resources were inexhaustible, if not in the immediate neighbourhood, then certainly in the world beyond.

Local devastation of forests and the increasing costs of bringing timber from far afield brought gradual changes in this attitude. As early as 450 BC, official attempts were made to restrict the cutting of the cedars of Lebanon in the eastern Mediterranean. General signs of change did not appear, however, until the twelfth and thirteenth centuries, when drastic restrictions on cutting were imposed in central Europe. Then, gradually, the belief that forests must be protected as a finite and dwindling resource gave way to a belief in new planting and forest-management techniques. By the middle of the eighteenth century, timber was beginning to be regarded as a slow-growing but renewable crop rather than a non-renewable source of fuel and timber.

Modern forest management follows two main ecological principles. The first is that of *sustained-yield forestry* – the continuous production of forest products from an area at some appropriate yield level. This is achieved through planned rotational systems (some with cycles of over 100 years), careful species selection, and protection of the timber crop from both fire and disease. There are presently available a wide range of ways of obtaining a continuous flow of forest products. Fast-growing species in the sub-humid tropics (e.g. some members of the *Eucalyptus* family) may be cropped as frequently as every seven years. Douglas fir in the Pacific Northwest may be cropped by patch cutting (see Figure 10.12), after which timber from areas around each patch naturally re-covers the cleared areas.

The second ecological principle followed in managing forests as a renewable resource is that of *multiple use*. A forest's yield may be measured in more than wood products, and the objective of multiple-use forestry is to maximize the total flow. Thus, timber production may have to be balanced against the role of the forest as a protection against erosion and pollution, as a wilderness or wildlife refuge, or as a recreational area. Not all such uses may be mutually compatible, and we shall look later in the book (in

Figure 10.12 Sustained-yield forestry The early history of lumbering in America shows that forests were treated as a stock resource to be clean-felled. Today conservation programmes ensure that forests are husbanded as a flow resource. Patch cutting is a technique used for Douglas fir lumbering in the Pacific Northwest; it allows natural regrowth of the felled areas from the surrounding untouched ring of forest. Given the high labour costs of replanting, increasing interest is being shown in natural regeneration of trees. This technique shows promising results for the regrowth of hardwoods.

[Source: Corbis-Bettmann.]

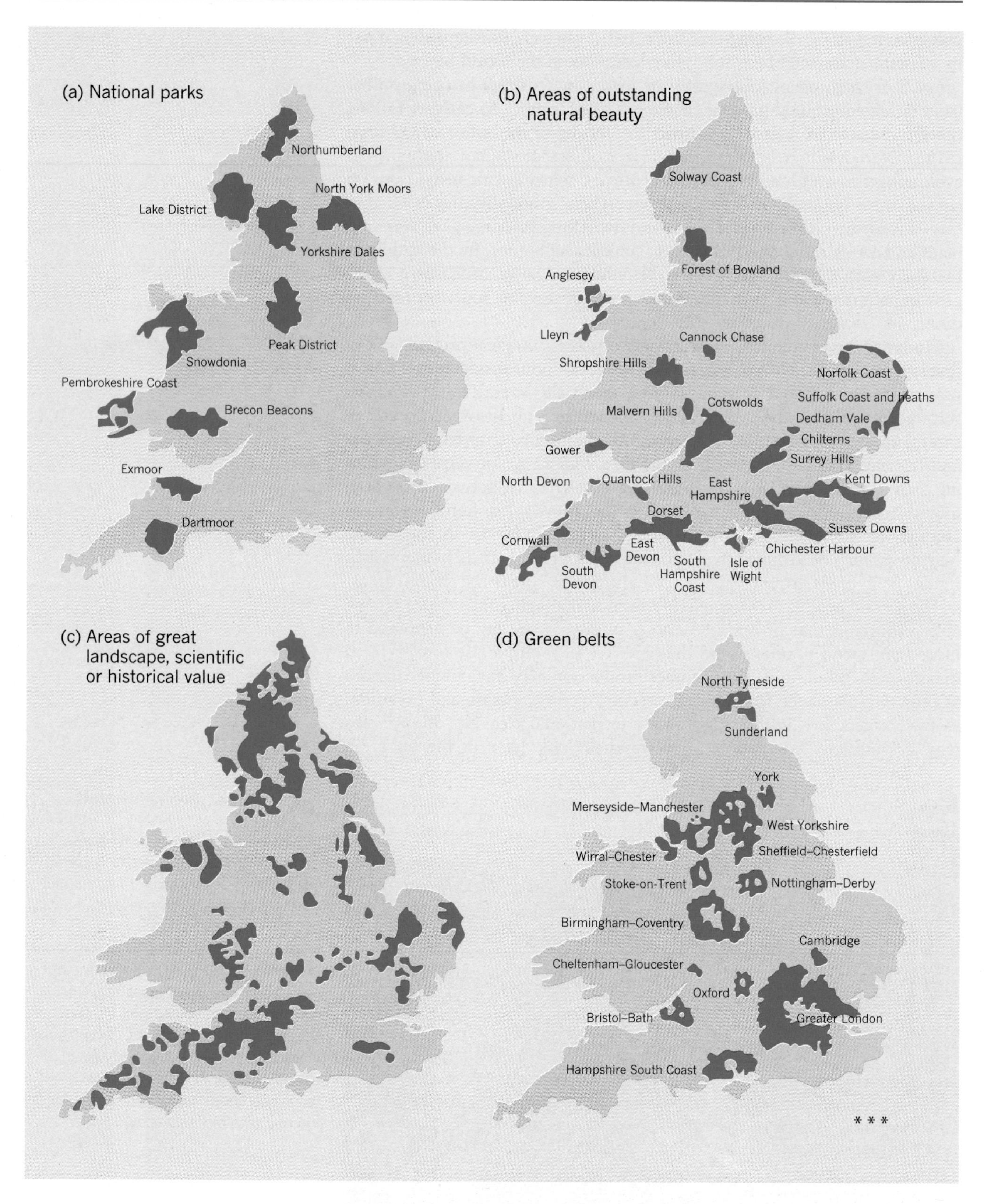

* * *

Figure 10.13 Conservation of recreational resources Most countries have now introduced legislation conserving areas of land for public recreational uses. These maps show the situation in England and Wales around the year 1970. Not all the proposed 'green belt' areas have been formally approved.

[Source: J.A. Patmore, *Land and Leisure* (Penguin, Harmondsworth, 1970), Fig. 62, p. 179.]

Chapter 18) at the planning problems that the use of land for multiple purposes entails. Indeed, the role of a forest area in recreation may be so important that commercial logging may have to be stopped. The precedent set by the United States in 1872 when it established Yellowstone National Park was followed frequently in the last century. The latest United Nations *List of National Parks* now contains the names of some hundreds of parks and refuges in scores of countries around the world.

Sustained-yield Recreational Resources

As the demand for leisure increases, the role of forests as primarily recreational areas is sure to increase. Sustaining the yield of a forest given an increasing number of picnickers, ramblers, and the like may prove even more difficult than maintaining a sustained timber flow. Most countries have now set aside substantial tracts of country in which some aspects of recreational use are safeguarded (see Figure 10.13) but their management poses increasingly severe problems. A shorter working week, increased leisure time and falling travel costs have created a surging demand for increased recreational facilities.

Questions of Resource Demand Table 10.4 summarizes some of the trends in the demand for outdoor activities in the United Kingdom during the last century. The figures for annual growth rates are based on varying periods of time and should be treated with caution. They indicate, however, the nature of the pressures and the special demands for use of water areas.

We can illustrate the impact of demand on a recreational resource by going back to the lake we used to illustrate ecosystem concepts in Chapter 4. Let us suppose the demand in this case is for sailing. On a weekend day with good sailing conditions, a number of folk will bring their sailing dinghys to the lake. Figure 10.14(a) shows this influx of people building up over the day simply as a line that rises until a saturation point is reached. How much sailing enjoyment does each family receive? Clearly, at the beginning (point A in Figure 10.14(b)), the early arrivals have the lake to

Table 10.4 Growth in demand for recreational resources[a]

Growth rate	Urban resources	Countryside resources	Water-area resources
More rapid than population growth	Athletics Golf (8)	Motoring Mountaineering Skiing Camping Horse riding Nature study Gliding (10)	Diving (24) Canoeing (18) Sailing (7) Angling (7)
Similar to population growth	Gardening Major team games Swimming	Walking (3) Hunting	
Slower than population growth	Major spectator sports	Cycling (−2) Youth-hostelling	

[a]General patterns of growth in the United Kingdom. Where figures in brackets are given they indicate the average annual increase as a percentage and are based on different periods.
Source: J.A. Patmore, *Land and Leisure* (Penguin, Harmondsworth, 1972), p. 49.

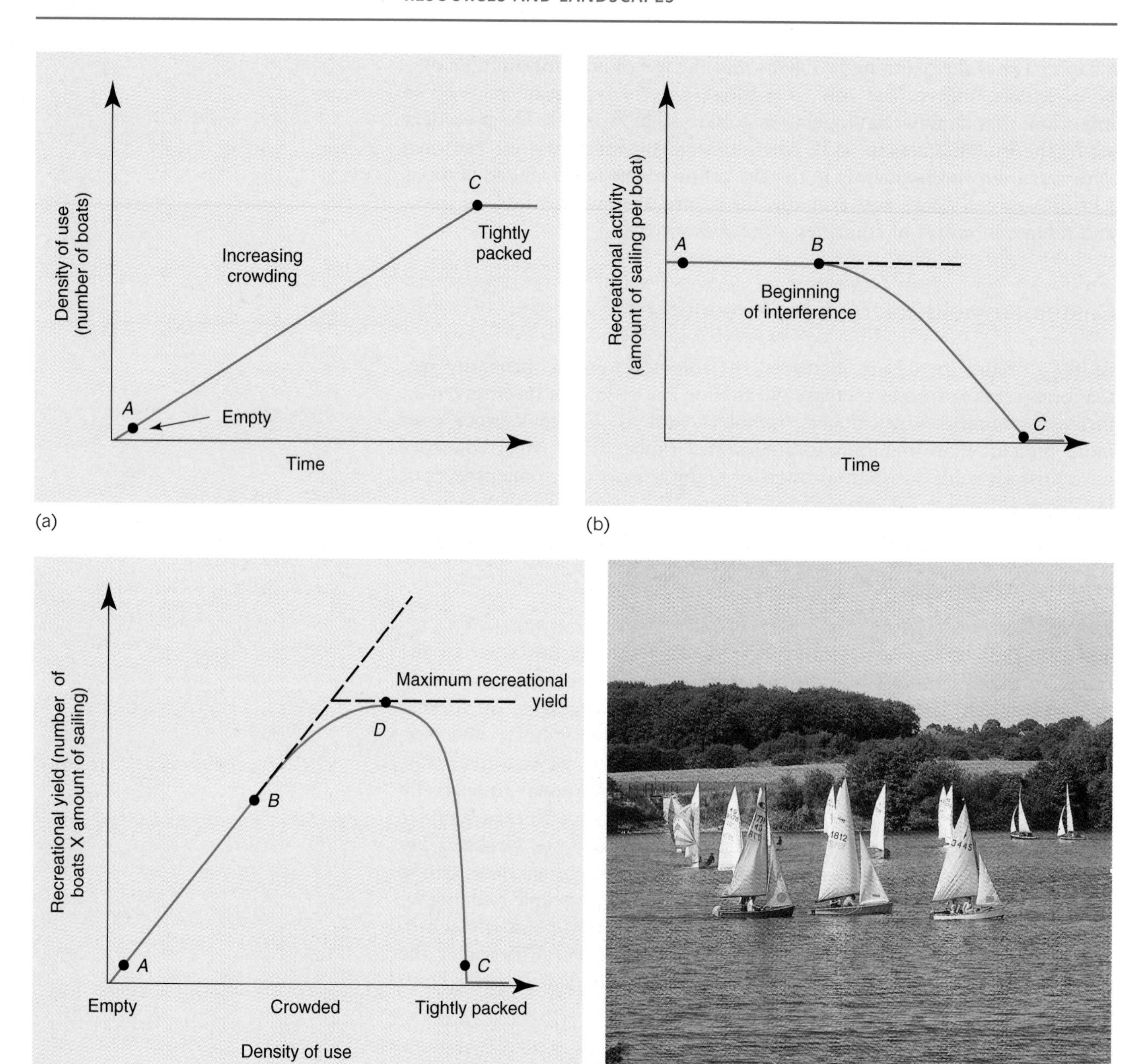

Figure 10.14 Crowding and resource use (a) The graphs show characteristic curves for the build-up of population pressure on a given recreational resource. In this case the increase of sailing boats on a small lake is shown to affect the 'recreational yield' of the lake. Details of the relationships in the three graphs are discussed in the text. (d) Lake with boats sailing on Chipstead Lake, Kent, southeast England.

[Source: (d) J. Allan Cash Ltd.]

themselves. But as numbers build up a point is reached (point B) at which boats begin to impinge on each other's territory and the amount of sailing each family can do is reduced. If numbers were to continue to build up, the lake would finally be so jammed with small boats (point C) that no sailing at all would be possible. At this point the enjoyment of each family would presumably have dropped to zero! (Note that we could apply exactly the same argument to the crowded beach we describe in Chapter 1.)

How does this crowding affect the *recreational yield* of the lake? We can define the *yield* as the 'number of boats × the amount of sailing activity' and plot its curve in Figure 10.14(c). The curve begins at zero (no boats on the lake), rises to a maximum, and then falls back to zero (when the lake is packed with boats and no sailing is possible).

To obtain the maximum yield from the lake, we would like to keep the number of people using it around point D at the top of the yield curve. But how do we do this? A rule of 'first-come, first-served' is one way; charging a fee, rationing, or exclusive club memberships are others, but each has undesirable social implications. Of course, no sensible family would launch a dinghy onto an already crowded lake. Thus, we might expect feedbacks to operate, which would keep the maximum population at around point D. Latecomers would be likely to give up the idea of going sailing or go on to a more distant lake. But if the lake has other uses, if it is used for fishing or as a water supply, then even point D may be too high if it causes too much disturbance for the lake to be used for these other purposes. In these cases, some form of zoning may need to be applied (see Figure 10.15).

Once again, the problems of the lake are representative of other and wider problems. Pressures on beaches, on wilderness areas, and on historical sites all raise broadly similar problems of trying to preserve resources – and yet obtaining the maximum yield from them.

Questions of Supply One difficulty in making assessments of the available supply of recreational resources is that they are cultural assessments. The same qualities may be judged quite differently by different cultures. One of the simplest ways of illustrating this is to show the changing assessment of the same area over time.

Consider Figure 10.16(a). William Brockedon's engraving of the Val d'Isere in the French Alps was made in 1829. It shows a lyrical summer scene with the emphasis on pastoral tranquility in the foreground and the majesty of Mont Blanc beyond. These were the Alps that the painter John Ruskin would write of later as 'alike beautiful in their snow, and their humanity'. This is a view that has persisted until today. It is not, however, the only view of the Alps. Travellers in the eighteenth century, anxious only to cross the Little St. Bernard Pass, would curse 'this awful place' in their journals, and hurry on to the welcoming towns of the Italian plains beyond.

Figure 10.15 Spatial separation of multiple resource uses (a) The artificial Chew Valley Lake created in 1952 to meet increasing demands for water from cities in southwest England. Its location, its pollution, and its multiple uses have made it a source of considerable conflict among different land and water users. (b) The zoning of Chew Valley Lake to allow competing uses.

[Source: Ordnance Survey. Crown copyright reserved. Data from Bristol Waterworks Company. From C. Hartley, *Countryside Community Recreation News Supplement*, No. 3 (1971), Fig. 1, p. 9.]

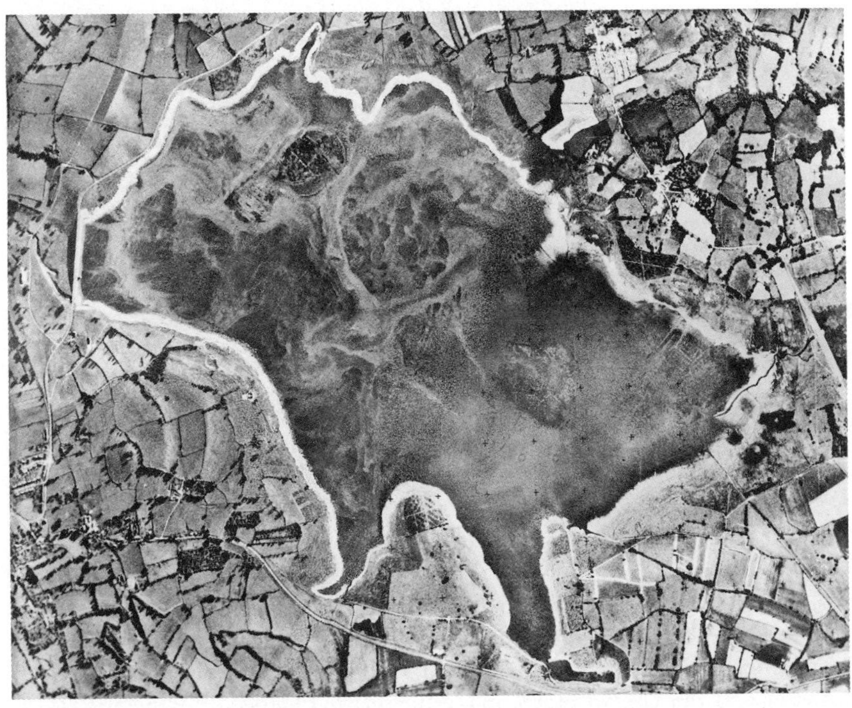

(a)

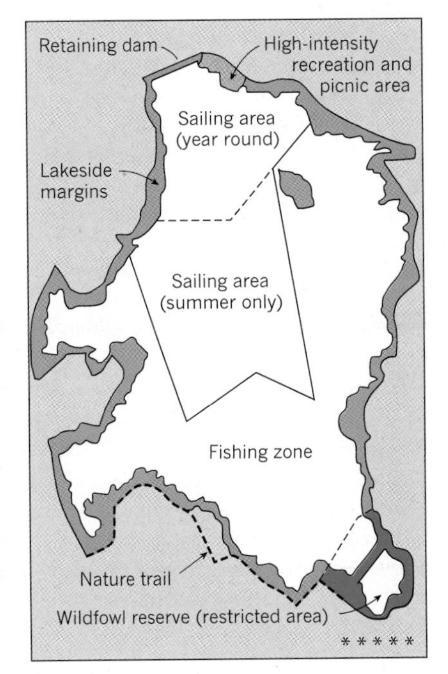

(b)

(a)

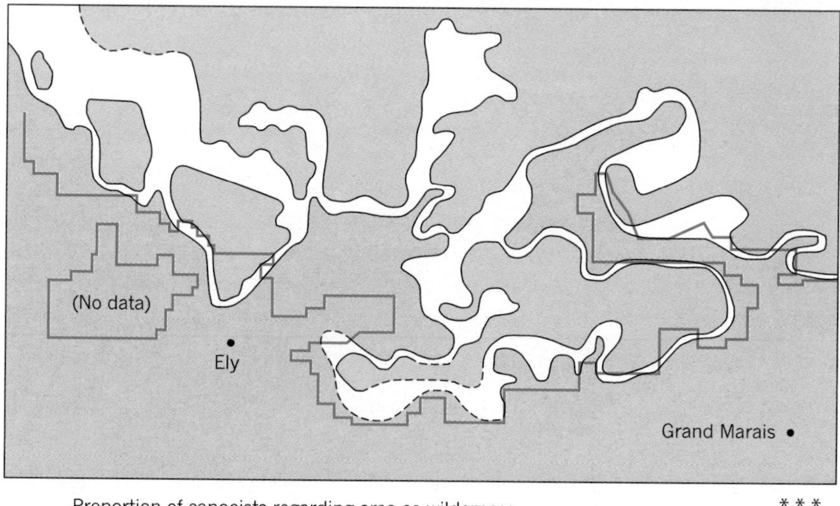

(b)

Figure 10.16 Perception of wilderness areas (a) Once considered a natural hazard, the European Alps are now regarded as a recreational resource. This early nineteenth-century engraving is symbolic of this change in viewpoint. Its sweeping views comprise a number of well-harmonized elements – the sheep and distant church giving an air of pastoral peace and tranquility, the two small human figures underlining the smallness of humanity in relation to the grandeur of nature. (b) Map shows canoeists' perceptions of 'wilderness' areas within the Quetico-Superior Park on the boundaries of Minnesota and Ontario. Shading indicates the percentage of canoeists who recognize the area as a 'wilderness'. This map differs from that constructed by other campers using automobiles or boats.

[Source: (a) Print by William Brockedon, 1829. (b) R.C. Lucas, *Natural Resources Journal*, **3** (1962), Fig. 3, p. 394. Reprinted with permission.]

* * *

So how do geographers pin down what they mean by a recreational resource? Let us begin with the simple case of 'wilderness areas.' Figure 10.16(b) shows an empty forest and lake area on the borders between Canada and the United States made up of the Quetico Provincial Park in Ontario and the Boundary Waters Area in Minnesota. This area has been officially designated as a wilderness by the two state governments, so its boundaries represent the official image of a wilderness area. But how do these boundaries match the view of the people who actually use the area? What areas do the campers, canoeists, and anglers consider to be wilderness, and what essential qualities should a wilderness environment have?

The main research problem is one of converting anecdotal observations and subjective hearsay into some form of quantitative yardstick that allows different assessments of environments to be metered and mapped. The solution adopted was to interview a representative sample of the groups using the area. Respondents were invited to say whether they thought they were in a wilderness area and to indicate on a map where the boundaries of the area lay. The term 'wilderness' was always left to the respondents to define, implicitly or explicitly.

By superimposing the maps of the respondents on one another, we can construct contour maps of the area showing the percentage of parties visiting the area that described it as being 'in the wilderness'. The canoeists' contour maps have an intricate pattern (Figure 10.16(b)) closely related to the waterways and sensitive to distance from the canoe trails. The boundaries of the wilderness for other users (campers, motorists, resort guides) were less sharply defined and were related more to roads, crowding, and noise levels in the less remote areas.

From a planning viewpoint, it is worth recording that all the boundaries drawn by users of the area differed from those drawn by the resource managers of the wilderness area. Such users' responses, when translated into maps, are clearly of considerable potential value in the future zoning of areas for recreational use. Also, these responses imply that some measurement of natural resource potential is possible even for such elusive things as the quality of a landscape.

'The Tragedy of the Commons' 10.4

A critical issue relating to rates of resource use is their ownership. A generation ago, Garrett Hardin wrote a classic paper entitled 'The Tragedy of the Commons' (1968). By 'commons' he meant commonly owned resources and by 'tragedy' he meant a situation in which the depletion of natural resources occurs because (a) individual interests and (b) collective interests do not coincide.

He illustrated his model with a parable that continues to have relevance today. A village has a section of common land (i.e. owned by the whole community) on which all families have rights to graze sheep. As the number of sheep go up, so the productivity (measured by wool or meat) increases to a certain threshold. At that point, the common land begins to be overgrazed and it is in the interest of the whole community to limit the use. But Hardin argued that what is good for the whole community may not be best for the individual. If A grazes 11 sheep (rather than 10), the gain from the extra sheep is greater than the marginal reduction per sheep that comes from edging over the overgrazing limit. The individual gains but the rest of the community loses. Beyond a further point there is a long-term reduction in which the productivity of the whole commonland is reduced through soil erosion. Figure 10.17 shows the general principles involved.

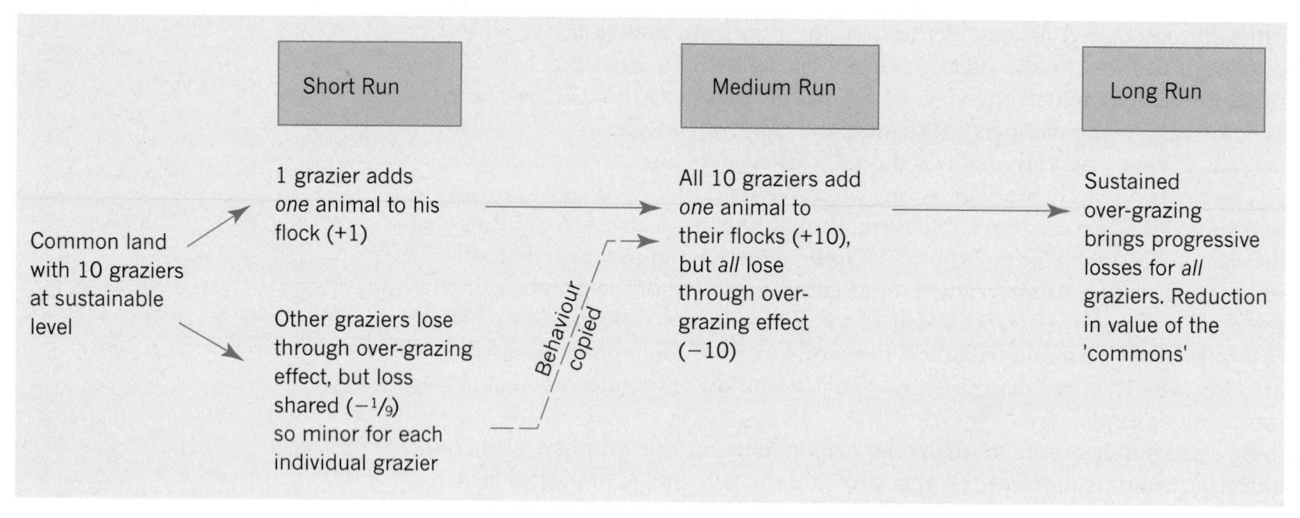

Figure 10.17 Tragedy of the commons Garret Hardin's example of graziers using common land and building up their flocks. So long as the marginal return from the additional animal is positive it is worthwhile for the individual even though this means that the average return per animal is falling. The other graziers will be impelled to follow the example of all others and add to their herds to maintain overall return per animal given the fall in average yield.

Although Hardin gives the example of a common grazing land, the parable has many other examples in overfishing (e.g. reduction of cod fishing on the North Sea, the Iceland waters, the Labrador Banks) or recreational overuse of National Parks. 'Breaking the rules' by one individual or one trawler or one national fishing fleet may bring retaliation from the other users. The overall productivity of the resource is pushed downwards. Efficient use of the resource requires the rationing through limitations on herd size but individuals will not altruistically limit their herd sizes unless they know that all others will also To ensure that they do requires a powerful external organization with the power to enforce optimal use, thereby ensuring the best interests of both the individual and the collective.

10.5 Conservation of Natural Resources

The notion of conserving the earth's natural resources is immediately appealing. No reader would willingly see Great Plains topsoil eroded, Tennessee hillsides filled with gullies, or redwood groves needlessly felled. Indeed, the destruction of animal population, like certain species of whales, has become an emotional public and political issue (see Figure 10.18). But what about a natural gas field in Oklahoma? What are the special merits of leaving the gas in the ground? Who benefits?

Some Definitions

Let us begin by trying to define *resource conservation*. One off-quoted definition runs as follows: '*Resource conservation* is the scheduling of resource use so as to provide the greatest yield for the greatest number over the longest time period.' This definition fits renewable resources nicely and encompasses the targets we met in our discussion of sustained-yield forestry, though it is less easy to apply to non-renewable resources (e.g. our gas field). Restricting the use of a finite resource means that we are saving for future generations some portion of a resource that would otherwise have been used in the present generation.

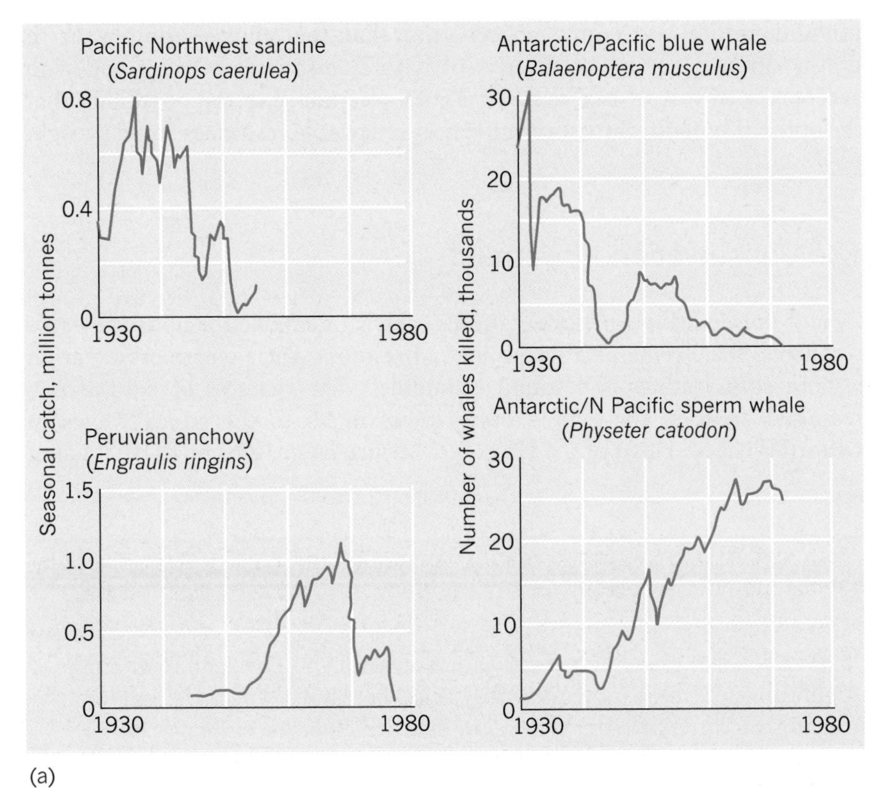

(b)

Figure 10.18 Collapse of overused biological resources (a) Fishing and whaling have been a traditional source of food and raw materials for humans. There have been some striking patterns of rise and fall in some marine species. Note that information is not available for exactly comparable time periods for the four species shown. (b) Whales were common off the coast of Western Europe up to the middle seventeenth century. A Dutch engraving of a stranded sperm whale in 1577.

[Source: (a) I.G. Simmons, *The Ecology of Natural Resources* (Arnold, London, 1974), Figs 9.4, 9.5, pp. 237–8. (b) Engraving by courtesy of the American Museum of Natural History from Kraemer's *Univers et humaniote*. Mary Evans Picture Library.]

Over the short-run, this idea is appealing, since it encourages the use of other resources and stimulates substitution. However, we have already noted that few natural resources of the past were ever fully worked out (some gas fields may stand as an exception); instead they were priced out of production. Such is the rapidity of technical substitutions that resources saved by an over-cautious conservation policy may never be used at all. Indeed, by limiting the investment of the present generation in them, we may actually be making future generations poorer. Clearly, this is a hard view to accept, but if we really wish to make the best use of non-renewable resources, it makes sense to use up the cheapest and best ones first and then to go on to the more inaccessible ones. Timing is of the essence; and a more useful definition of resource conservation than that above is simply that it is 'the optimal timing of the use of natural resources.' And along with acceptance of this revised definition goes acceptance of the possibility that the optimal time for the use of some non-renewable resources could be right now.

The Conservation Movement

Even if preservation for preservation's sake is untenable for non-renewable resources, there remains a wide range of resource areas where preservation is both sound ethics and sound economics. The richness of natural recreational areas in the United States owes much to the ethical views of Gifford Pinchot. Head of the US Forest Service from 1898 to 1910, Pinchot

Box 10.C John Muir and Nature Conservation

John Muir (born, Scotland, 1838; died California, United States, 1914) was an explorer and naturalist who campaigned for forest conservation and whose influence helped to establish National parks in the United States. He moved to the United States as a child and studied at the University of Wisconsin.

Muir tramped the wild places of the United States and also Europe, Asia, Africa, and the Arctic. He was especially attracted to the glaciated landscape of the Sierra Nevada in California and spent six years in the vicinity of the Yosemite Valley in the 1860s. Through his advocacy, the area was given by Congress to California as a public park and recreation area. It was created a National Park in 1890, and has since been extended to cover an area of 3000 km^2 (1200 square miles). Muir kept detailed notebooks of his travels which he subsequently used in his extensive and influential writings: his classic books are *The Mountains of California* (1894), *Our National Parks* (1901), and *The Yosemite* (1912).

The National Parks set of ten stamps issued by the United States government in 1934 was the first of what is now a commonplace theme in postal designs in countries around the world. The National Parks movement, together with other ecological societies (such as the Audubon Society, the Sierra Club, and the World Wildlife Fund), is now a major force in the conservation of the world's flora and fauna. Muir is commemorated today in national parks such as Yosemite, in Muir Woods north of San Francisco Bay, and with a glacier in Alaska which he explored and which bears his name. But perhaps his greatest monument is the ***Sierra Club***, one of the earliest conservation societies, which remains today a major force in American conservation. A map of the distribution of the club within the United States is shown in Figure 16.5(e). [Stamp reproduced courtesy of the US Postal Service.]

caught President Theodore Roosevelt's interest and aroused his enthusiasm. Roosevelt added a greater area of the United States to its protected forest lands than all the presidents before and after him.

Presidential concern with conservation has, however, a long history. George Washington was acutely anxious at the loss of soil from his Mount Vernon estate and had his servants bring up mud from the Potomac River to fill in gullies. Nearly two centuries later, in 1970, President Richard Nixon set up a National Environmental Protection Agency. In the intervening period, and particularly since Pinchot, legislation governing the use of all kinds of natural resources has been passed by the Congress. The National Park Service was established in 1916. (See Box 10.C on 'John Muir and nature conservation'.)

The first international treaty relating to resource protection (with Canada, on protecting migrating birds) was signed in the same year. The 1930s were a decade of special concern with soil erosion and water control and saw the passing of the Taylor Grazing Act (1934), the setting up of the Soil Conservation Service (1935), and the passage of the Flood Control Act (1936). Since 1970, the emphasis has swung to a concern with conserving waste from the use of mineral resources, to protection from pollution (via the 1970 Clean Air Act), and to general improvements in recreational facilities and improved environmental standards.

Federal legislation on conservation in the United States has been important on a global scale (see Figure 10.19). It not only initiated complementary legislation at the state level *within* that country, but was widely copied in other areas. The Tennessee Valley Authority (TVA), set up by New Deal legislation in 1933, served as a basinwide 'demonstration farm' for watershed-control programmes elsewhere (see Box 18.C). The São Francisco Valley scheme in Brazil, the Gal Oya scheme in Sri Lanka, and the Snowy Mountain scheme in Australia are examples of the scores of developments that followed, in some measure, the TVA lead.

Currently, interest in conservation is expanding spatially in two ways. First, at the global level, there is increasing concern with international cooperative action on the wise use of resources and on environmental protection. United Nations agencies are now playing a key role in setting the

Figure 10.19 Wildlife conservation at two different geographic scales
(a) Microscale: conservation of nests of the white stork (*Ciconia ciconia*) in northern Europe. Since the birds are migratory and spend the winter in North Africa, conservation by protection along the flyway linking the two regions is also necessary. (b) Macroscale: Ngorongoro National Park, East Africa. The park is set within an enormous crater of an extinct volcano. The crater is over 20 km (12 miles) across at its widest point; the height of rim is 2500 m (8200 ft) and the floor, 600 m (2000 ft). It is one of East Africa's key reserves for grazing animals and their predators.

[Source: (a) Walther Rohdich/FLPA. (b) R.D. Estes/Science Photo Library.]

(a)

(b)

world on the long road toward acceptance of international standards of air pollution control and joint action on the use and abuse of ocean resources. Second, there is unprecedented conservation activity at the local community level. Locally active groups involved in conservation-minded activities range from the Sierra Club and Friends of the Earth, right through to local Women's Institutes in English villages valiantly protecting their oaks and elms from the chain-saws. This significant increase in public support for conservation measures at both international and local levels represents a most encouraging shift in people's view of their place in the ecosystem. Whether it is enough of a shift in view of the unprecedented inroads now being made on all our natural resources – both stocks and flows – is still in doubt.

Reflections

1 Natural resources are that part of the material environment that has utility in human terms, while all of the material environment constitutes the world total stock, most of it of little interest to human beings. But note that the boundary between resources and rest of the material environment. List examples of 'old' resources that are no longer of interest and 'new' resources that have recently been added to the list.

2 Consider the main characteristics that separate *renewable* from *non-renewable* resources (see Figure 10.4). Can you think of any examples of resources that (a) fall on the borderline between the two types or (b) have changed from one type to the other as human attitudes toward them have altered?

3 Resource estimates change as technology and socioeconomic conditions change. Debate the contention that 'resource shortages are simply the result of prices that are too low.' What happens when the price of a natural resource rises? Does this have geographic implications for the spatial pattern of its production and consumption?

4 Follow up on the problems of nuclear-energy production by reading some of the material suggested in 'One Step Further ...'. What are your own views on the balance between the advantages of this fuel source and its possible environmental hazards? Do you have nuclear power plants in your own area and, if so, where they are located and what is their likely productive lifespan?

5 Evaluate the model of the recreational yield of the lake in Figure 10.14. How would you go about restricting the use of a wilderness area to ensure that it was not degraded by overuse? Who benefits from such restrictions? Who pays for them?

6 A common definition of resource conservation is that resource use should be organized so that the largest number of people receives the greatest yield through the longest period of time. Renewable resources fit well within that definition, but what about non-renewable resources? How do we determine optimal timing of use?

7 Conservation has to be paid for by this generation or those that follow. Think critically about the conservation movement. Would you favour a \$10 or a \$100,000 fine for shooting members of an endangered bird species (e.g. bald eagles)? Is there a price limit on environmental protection in terms of other social goods? How much conservation can we afford?

8 Review your understanding of the following concepts using the text and the glossary in Appendix A of this book:

flow resources
fossil fuels
green revolution
planet's total stock
proven reserves
recreational yield
resource conservation
Sierra Club
stock resources
sustained-yield forestry

One Step Further ...

A useful starting point is to look at the essays on humanity's resources and land production by Ian Simmons and David Grigg in I. DOUGLAS *et al.* (eds), *Companion Encyclopaedia of Geography* (Routledge, London, 1996), pp. 599–619, 651–76. A brief but useful introduction to resources and their management is also given by Phil McManus and Noel Castree in their essays in R.J. JOHNSTON *et al.* (eds), *Dictionary of Human Geography*, fourth edition (Blackwell, Oxford, 2000), pp. 706–10. The broad issues involved in resource appraisal are covered in Canadian geographer Bruce MITCHELL's *Resource and Environmental Management* (Longman, Harlow, 1997).

Important studies of global resources are given in B. GROOMBRIDGE (ed.), *Global Biodiversity: Status of the Earth's Living Resources* (Chapman and Hall, London, 1992) and A.S. MATHER and K. CHAPMAN (eds), *Environmental Resources* (Longman, London, 1995). Two older books that remain useful are I.G. SIMMONS, *The Ecology of Natural Resources*, second edition (Arnold, London, 1980) and the wide range of approaches to the resource–population relationship given in the readings included in I. BURTON and R.W. KATES (eds), *Readings in Resources Management and Conservation* (University of Chicago Press, Chicago, 1965). Some of the classic papers in natural resource study are reproduced in N. NELISSEN *et al.*, *Classics in Environmental Studies* (International Books, The Hague, 1997).

The resource 'progress reports' section of *Progress in Human Geography* (quarterly) will prove helpful in keeping up to date. Geographers have traditionally played a leading role in resource evaluation and in the conservation movement. In addition to the regular geographic serials, you should also browse through *Natural Resources Journal* (quarterly) and one of the more popular resource-oriented journals such as *The Ecologist* (quarterly). For readers with access to the World-Wide Web see also the sites recommended for this chapter in Appendix B at the end of this book.

Small Sea

Syr Darya Delta

Aral Sea

MUYNOQ

Amu Darya Delta

NUKUS

Lake

Sarygamysh

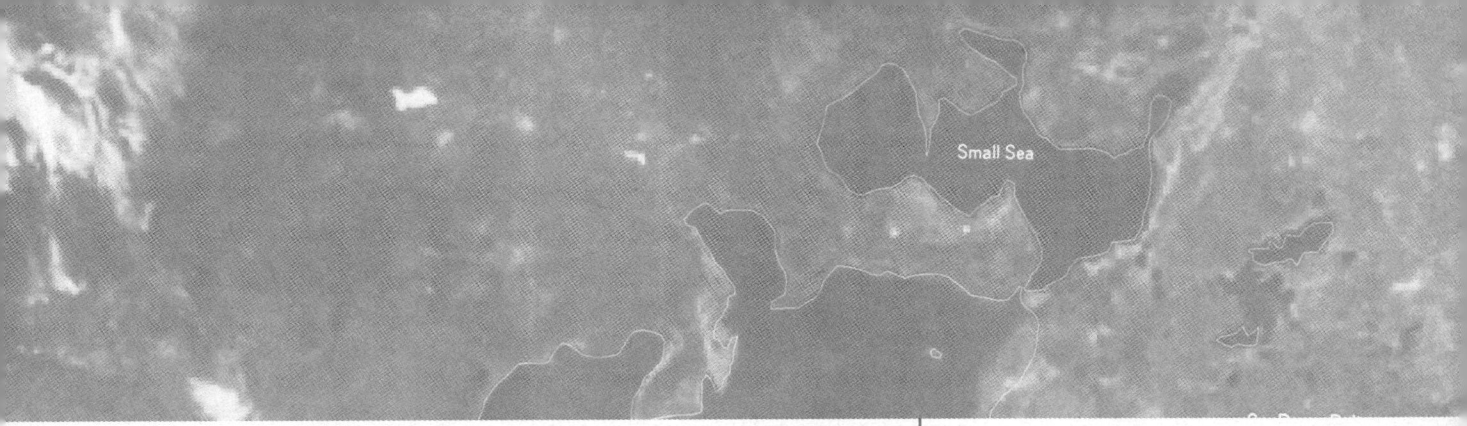

Small Sea

CHAPTER 11

Our Role in Changing the Face of the Earth

■ denotes case studies

Figure 11.1 Human transformation of an inland sea This infrared satellite image of the inland Aral Sea in southern Kazakstan, with a former shoreline overlaid as a white line to show the shrinkage of the sea. In the infrared range water is black, green vegetation is dark grey and bare ground is light grey. The Aral Sea loses water only by evaporation but is shrinking because water from feed rivers, the Amu Darya and Syr Darya, is used for cotton irrigation. Once the world's fourth largest lake, the Aral Sea had lost 60 per cent of its volume by 1996. Its salinity increased threefold and this has killed many types of fish. Sediments were exposed to the air and swept up in great dust storms. Image taken in August 1993; the overlaid shoreline is from 1989. North is at the top of the photo and the east–west distance is around 320 km (200 miles). [Source: Earth Satellite Corporation/Science Photo Library.]

Not here for centuries the winds shall sweep
Freely again, for here my tree shall rise
To print leaf-patterns on the empty skies
And fret the sunlight. Here where grasses creep
Great roots shall thrust and life run slow and deep:
Perhaps strange children, with my children's eyes
Shall love it, listening as the daylight dies
To hear its branches singing them to sleep.

MARGARET ANDERSON (1950)

As we have seen in the preceding chapters of this book, the growing human population has a huge potential for changing the face of the planet Earth. Our opening quotation and our opening photograph (see Figure 11.1 p. 334) in this chapter stand in stark contrast. I was taught at Cambridge by biogeographer Margaret Anderson more than half a century ago. Her poem catches the gentle moments of reflection on the planting of a young tree and picks up the theme of people coaxing environmental change little by little.

Our photograph peers down at an altogether cruder scene. The Aral Sea on the southwestern border of Kazakstan covers 67,000 km^2 (about 26,000 square miles or an area roughly the size of Ireland). Until the last few decades it was the world's fourth largest inland body of water. A huge and longstanding scheme to irrigate large areas of a desert in the former Soviet Central Asia caused the Sea to shrink drastically. From 1960 onwards, the irrigation scheme has increasingly diverted water from two of the lake's major tributaries, the River Amur and the River Syr-Dar'ya (see Figure 11.2). As the map shows, the Aral Sea has lost three-quarters of its water and half its surface area. It shrank more than 100 km (over 60 miles) from its original boundaries and the water level dropped some 13 m (over 40 ft) over a 25 year period. As a result the Aral Sea is now split into two separate basins of bitter, salty water.

The ecology of the surrounding land has been badly affected with violent dust storms along the southern and eastern shores. The remaining sea area is now too saline for fish and too shallow for ships to navigate. In the irrigated desert areas, water in the irrigation channels is now becoming polluted and drinking water being reduced to almost marginal quantity.

Another example at a smaller scale was given in Figure 10.2 where a mine is being gouged and blasted away in successive strips. These massive hollows, exemplified by the copper-rich hillock near Butte, Montana, or the Mesabi ironfields of Minnesota, now form some of the greatest artificial hollows on the planet. Thus the threads of the last chapters – increasing human numbers, human interference in the structure of ecosystem, and the accelerating search for resources – begin to intertwine here, as we see the landscape change.

Landscape change and the human role in changing the face of the earth have been a constant theme in geographic writings across the centuries. In the United States one of the earliest volumes on these topics came from a Vermonter, George P. Marsh. In his *The Earth as Modified by Human Action* (1874) Marsh drew attention to the unsuspected importance of human intervention in shaping what had hitherto been regarded as natural America. In the present century Marsh's ideas have been developed in

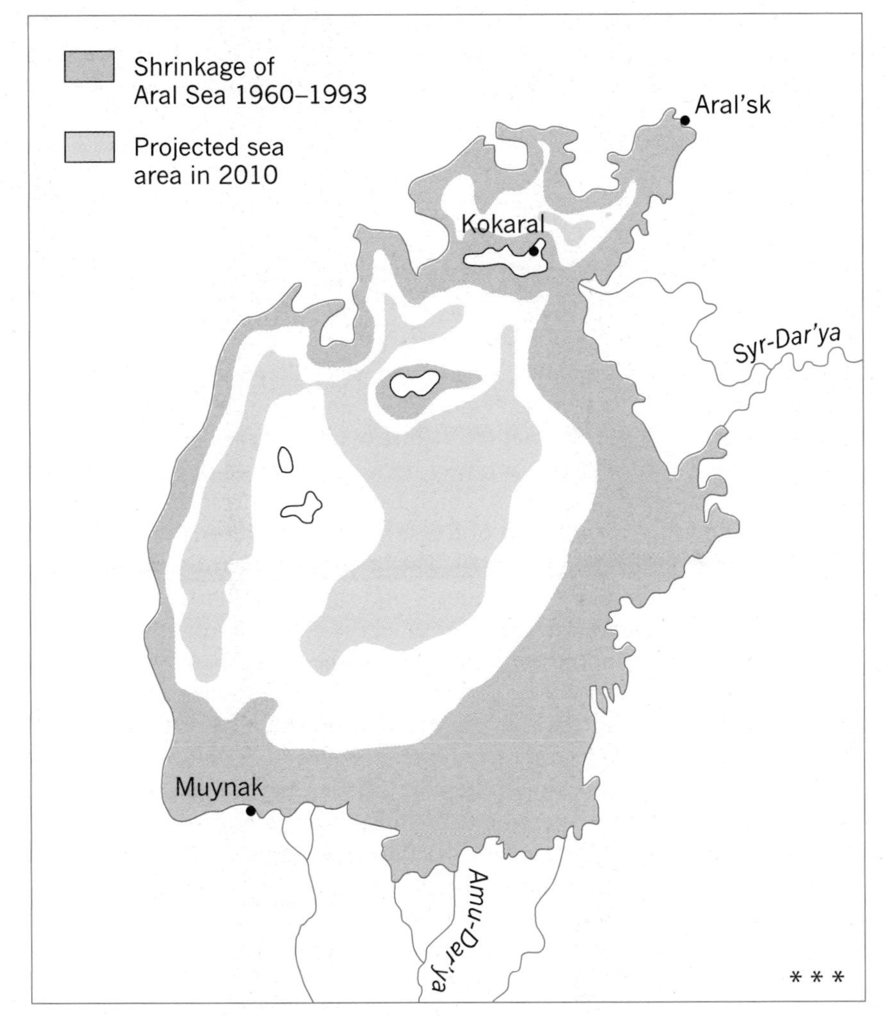

Figure 11.2 Shrinking Aral Sea The Aral Sea (see the satellite view in Figure 11.1) is of great interest and concern to geographers because of the remarkable shrinkage of its area and volume in the second half of the twentieth century. In 1960 the lake covered an area of 68,000 km^2 (26,000 square miles) and had an average depth of 16 m (53 ft). In 1992 it had split into two, the 'Greater Sea' to the north and the 'Lesser Sea' to the south. Historical evidence shows this shallow inland sea has also fluctuated greatly in the past. Past changes are likely to be forced by climatic variations whereas changes in the last few decades are linked to irrigation demands for cotton growing on the inflowing rivers.

two ways: first, by detailed historical reconstruction, geographers have estimated the magnitude of the effects of human intervention. This reconstruction facilitated by modern techniques of measurement and monitoring. Second, academic (and, increasingly, public) concern over the harmful effects of human intervention on environmental quality has created a convergence of disciplines. Contemporary geographic work in this area draws not only on its own long traditions, but on parallel work in such fields as biology and civil engineering.

In this chapter we look at the net effects of all the ways we have influenced our environment in terms of how we have changed the landscape. For it is in the changing face of the land that we can see most clearly the spatial effects of the several processes disentangled in Chapters 9 and 10. However, measuring changes in land use turns out to be a more perplexing problem than it appears at first sight. Environmental change also can have natural causes, and in any situation these two sources of change – human and natural – must be singled out and identified. We shall also, in this chapter, assess the extent of human intervention on different geographic scales, from the global to the local-township level; and we shall pay special attention to the present and future roles of human beings like ourselves in changing the face of the United States.

Finally we look back on the last few chapters and in Section 11.4 consider the 'human–environment controversy'. Just how do geographers see people: dominating the environment, or being dominated by it, or neither?

11.1 Difficulties in Interpretation

We have already encountered, in the last two chapters, some dramatic examples of landscape change. Figure 11.3 shows an aerial view of the Brazilian area discussed in Chapter 9. Just over a century ago, this rolling country in the Paraíba Valley (midway between the cities of São Paulo and Rio de Janeiro) was covered with a high tropical rainforest. In the 1850s it was caught up in the great swath of forest felling and burning that preceded the coffee boom in this part of Brazil. The boom lasted hardly a generation, and by the 1890s the plantation houses and the slave quarters were being turned into cattle ranches. In the meantime, the coffee frontier was moving into virgin country hundreds of kilometres to the west.

The environment that we see in Figure 11.3 has experienced faster and more radical changes in the last century than in the tens of thousands of years humans have previously lived there. Instead of the original forest cover, we now have a mixture of fire-controlled grassland and scrub forest. Replacing the original soils on the steep hillsides and debris-covered valleys is a thinned and denuded skeletal soil. Molasses grass from South Africa and eucalyptus from Australia are helping to stabilize the vegetation community and to make it of some continuing use to people.

The story of the Paraíba Valley could be retold with suitable changes throughout the rest of eastern Brazil (see maps in Figure 11.3). Indeed Brazil's story of widespread deforestation and consequent environmental changes is typical throughout the heavily populated areas of both the tropics and the mid-latitudes. Over much of these areas, the environment is increasingly a human artifact. Some of the effects of our impact on the changing landscape are immediate and apparent. The suburban sprawl of Los Angeles into the San Fernando Valley, the flooding of Lake Nasser above the Aswan dam on the Egyptian Nile, and the reforestation of Scottish moorlands are plainly visible evidence of ongoing processes. It is when we extend our investigation to eras before our own that we run into more difficult problems in interpreting change.

Here we look at two basic questions. First, how do we establish the facts of past change? Second, once we have established the facts of change, how do we know what parts of the environmental change we have measured to credit (or debit) to human activities?

Historical Evidence

As we saw in Section 3.2 various methods have been developed to reconstruct the nature of past environments. Table 11.1 summarizes the methods available for reconstructing changes in land use. Note that as we approach the modern period, the variety and accuracy of the evidence tend to increase. How have geographers used these tools, and what results have

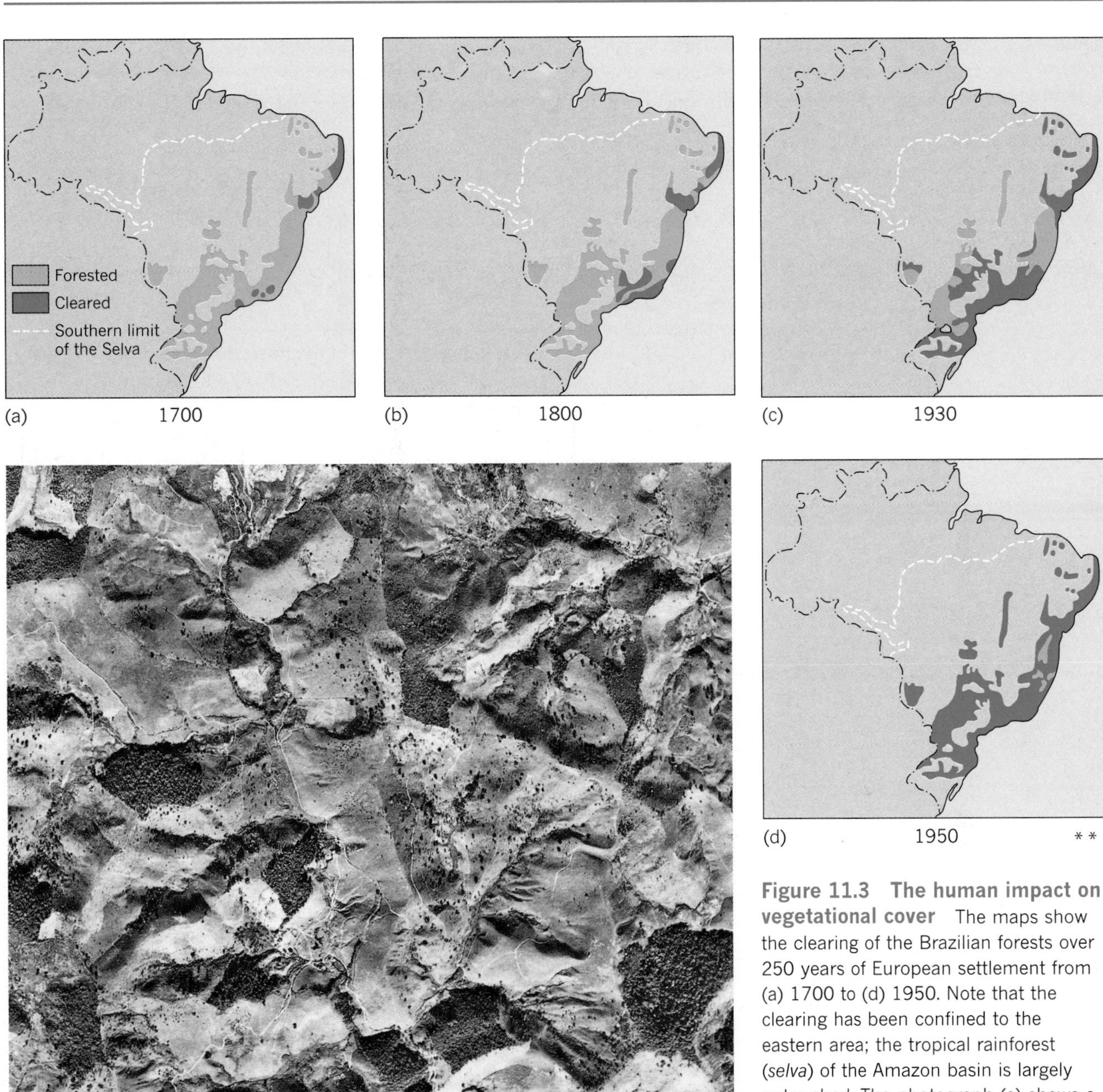

Figure 11.3 The human impact on vegetational cover The maps show the clearing of the Brazilian forests over 250 years of European settlement from (a) 1700 to (d) 1950. Note that the clearing has been confined to the eastern area; the tropical rainforest (*selva*) of the Amazon basin is largely untouched. The photograph (e) shows a sample of heavily eroded terrain in the Rio de Janeiro–São Paulo area. There are still a few remaining woodlots, much degraded remnants of the original forest cover.

[Source: (e) Servicos Aerofotogrammetricos, Cruzeiro do Sul, S.A. After P.E. James, *Geographical Review* **43** (1953), Fig. 316, p.309. Reprinted with permission.]

they found? Let us illustrate the answer by taking a representative case and looking at it in some depth.

European historical geographers, led by H.C. Darby at Cambridge University, have diligently explored manuscript archives from the medieval period in reconstructing the massive changes in the forest cover of Western Europe. The changes can be inferred from an interpretation of such sources as the great survey carried out in Norman England nearly nine centuries ago, during the *Domesday survey* of 1086. For example, one of the

Table 11.1 Methods of reconstructing land use

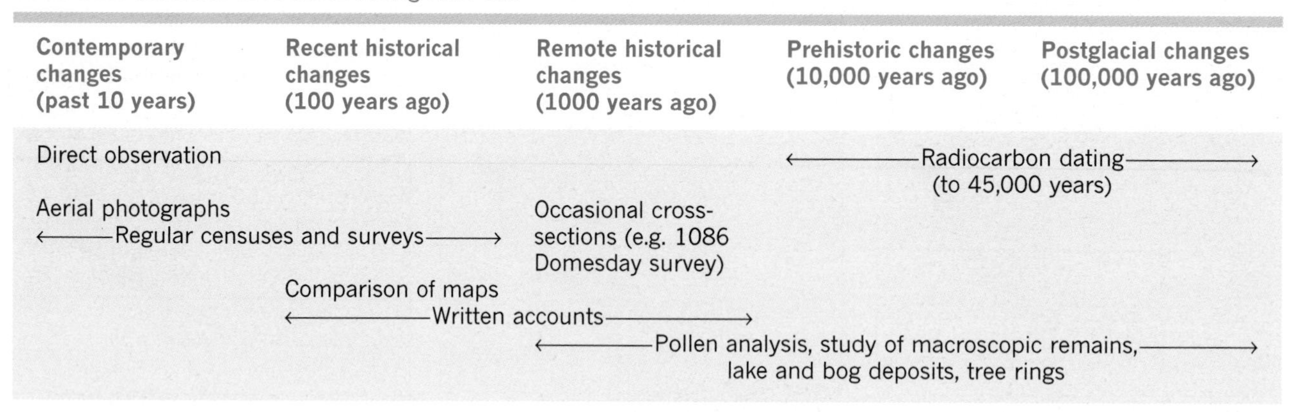

Contemporary changes (past 10 years)	Recent historical changes (100 years ago)	Remote historical changes (1000 years ago)	Prehistoric changes (10,000 years ago)	Postglacial changes (100,000 years ago)
Direct observation			←Radiocarbon dating (to 45,000 years)→	
Aerial photographs		Occasional cross-sections (e.g. 1086 Domesday survey)		
←Regular censuses and surveys→				
	Comparison of maps			
	←Written accounts→			
		←Pollen analysis, study of macroscopic remains, lake and bog deposits, tree rings→		

questions asked by the Royal Commissioners was 'How much wood in this place?' The form of the answer varied among the thousands of replies received from small hamlets, villages, and towns across the country. Some respondents specified the amount of woodland in quantitative terms in one of two ways: by its area or linear dimensions, or by the number of hogs it would support (by feeding on acorns or beech mast). Still other respondents simply stated that there was enough wood for fuel, mending fences, or repairing houses. Darby's patient scrutiny and assembly of detailed and dissimilar information built up a picture of the area of woodlands in early medieval England that shows marked local and regional variations in the pattern of heavily and lightly wooded tracts (see Figure 11.4).

Detailed statistical evidence for such an early period in landscape evolution is highly unusual. Even when documentary evidence is missing, however, some land uses can be determined from place names. Figure 11.5 presents the distribution of names of hamlets and villages in an English county with characteristic woodland place names: *lea*, *feld*, *wudu*, and *holt*. The location of these names is well to the north of a band of names (*tun*, *ingham*, and *ham*) indicating early settlement. There is a contrast between early settlement of areas of light gravel and loamy soils in the south and later settlement of areas of intractable heavy London clay to the north. The distribution of this ***place-name evidence*** acts as a cross-check that supports the evidence of the Domesday records.

The reconstruction of successive patterns of woodland distribution in medieval Europe indicates that the clearing process was not one of continuous advance before the axe and the plough. Working at the Würzburg Geographical Institute in Germany, Helmut Jager has found an oscillating pattern in the upper Weser Basin of Germany. As over much of Western Europe, the destruction of war, great plagues, and population declines were reflected in changes in land use; land returned to woodland at times of reduced economic activity and was cleared during prosperous periods when population increased.

The possibilities and limitations of archives as sources of information exemplify the kinds of characteristic peculiar to a single source of evidence about a past environment. Some of the possibilities and limitations are summarized in Table 11.1 and illustrated in Figure 11.6. In practice, the reconstruction process involves using various types of evidence to verify estimates derived from a single source. It is this cross-checking of evidence from various sources that allows a final sequence of land changes to be confirmed.

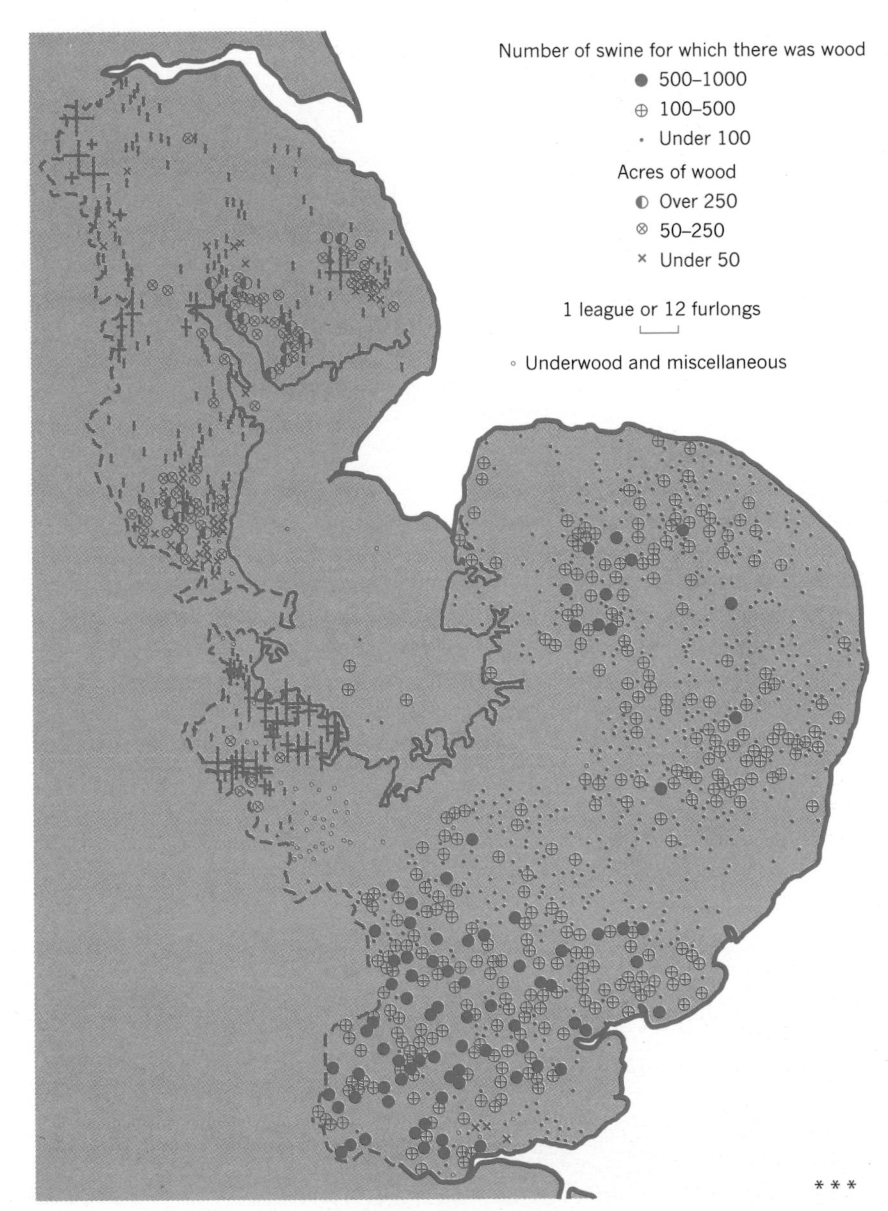

Figure 11.4 Documentary evidence of change This reconstruction of the eleventh-century forest cover of eastern England is based on the Domesday survey of 1086. The different symbols reflect the fact that the extent of woodland was recorded in different terms in different parts of the area. For example, over much of the east, woodland was recorded in terms of the number of hogs (swine) it would support, while in the west it was more common to describe its size in terms of now-obsolete measures of length (leagues (about 5 km) and furlongs (about 0.25 km)). The exact meaning of certain woodland terms such as 'underwood' is uncertain. Despite all these difficulties the main regional contrasts in timber cover stand out. Note that the empty area in the centre of the map, now the English fenland, was largely undrained marshes with some woodland on the few 'islands' within the marsh. Some of these islands had settlements and are shown toward the southern end of the marshland.

[Source: From H.C. Darby, *The Domesday Geography of Eastern England* (Cambridge University Press, Cambridge, 1952), Fig. 106, p. 363.]

* * *

Human or Natural Changes?

Changes brought about by human intervention and those arising from natural causes may be difficult to separate. Consider, for example, the effect on a small watershed of changes in the rhythm and amount of precipitation (Figure 11.7). Suppose there is a poleward displacement of the present climatic zones that brings semi-arid conditions into forested areas. Instead of a year-round pattern of rather light rainfall, we now have a highly irregular pattern of precipitation (occasional large storms separated by droughts). What other changes are likely to follow?

We would expect the droughts and lower precipitation to reduce the moisture available for plant growth and thereby alter the character of the vegetation in the drainage basin. As a result, forest species might be replaced by drought-resistant shrubs. This change would decrease the ability of vegetation to absorb precipitation falling on the soil surface

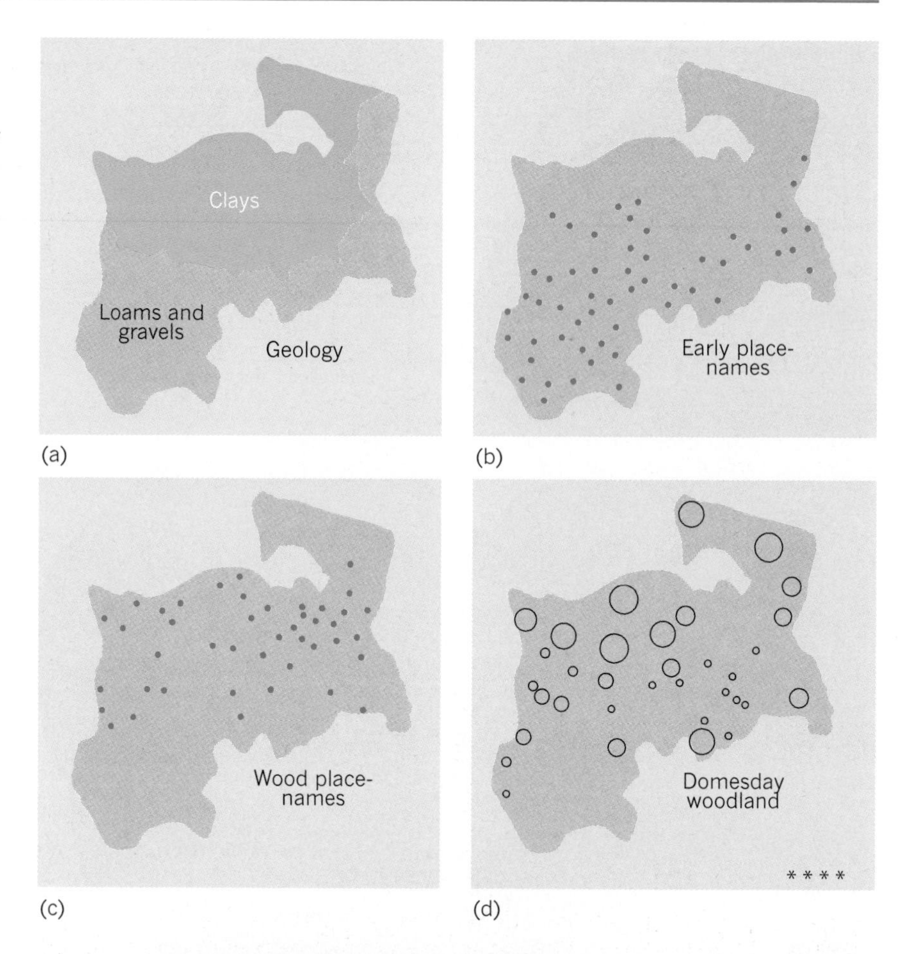

Figure 11.5 Cross-checking of evidence Geological factors (a), as well as place-name and Domesday evidence ((b), (c) and (d)) for the county of Middlesex in southeast England, suggest that there was a heavy forest in the northern half of the country but a more broken cover in the south during the eleventh century.

[Source: H.C. Darby, in W.L. Thomas, Jr. (ed.), *Man's Role in Changing the Face of the Earth* (University of Chicago Press, Chicago, 1956), Fig. 55, p. 192. Copyright © 1956 by The University of Chicago.]

Figure 11.6 Landscape change shown by aerial photographs Aerial views may sometimes show features of geographical interest not apparent from the ground. An old pattern of medieval fields (cultivated in long, parallel strips) shows up beneath the modern field boundaries in an English midland village (Husbands Bosworth, Leicestershire). Note how the low early morning sun casts shadows which emphasize small, otherwise hard-to-see changes in surface relief. Compare with Box 22.B.

[Source: From Cambridge University Collection. Copyright reserved. Aerofilms.]

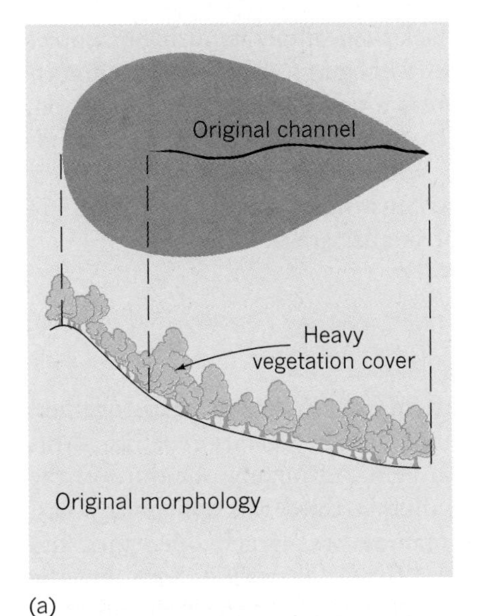

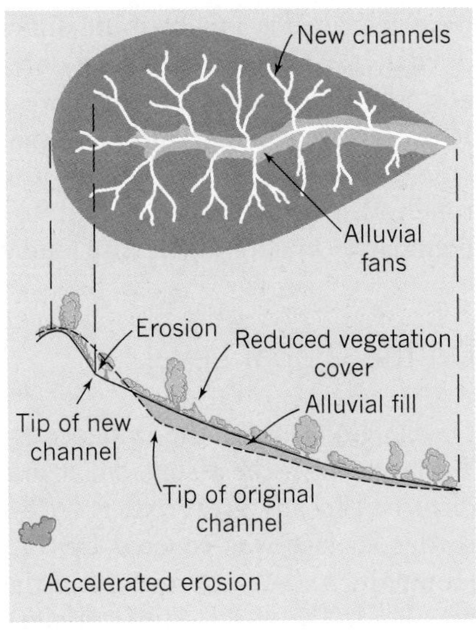

(a) (b)

Figure 11.7 Natural changes in the landscape Environmental changes causing a diminished vegetational cover lead to severe gullying on the slopes and aggradation in the main valley of a small watershed. *Alluvial fans* describe the fan-shaped spreads of eroded materials that form at the lower end of the new channels or gulleys.

[Source: A.N. Strahler, in W.L. Thomas, Jr. (ed.), *Man's Role in Changing the Face of the Earth* (University of Chicago Press, Chicago, 1956), Fig. 124, p. 635. Copyright © 1956 by the University of Chicago.]

during storms, and increase the amount of material washed downslope. Figure 11.7 shows a sequence of morphological changes within the drainage basin. Increased erosion is followed by severe gullying on the slopes, which in turn is accompanied by the dumping of eroded material in the main valleys (aggradation). The number of streams per square kilometre is increased, and the slopes of both the debris-choked channels and the eroded valleysides are steepened.

Such cycles of aggradation and erosion are well documented for many parts of the world. In the semi-arid areas of the southwestern United States, geomorphologists have pieced together a detailed sequence of cycles of gullying (arroyo cycles), although there is considerable dispute over whether these are due to natural variations in rainfall or to alterations in grazing pressures during the post-Columbian period of settlement. More rapid adjustments of the characteristics of channels can be observed over much shorter time periods. Near Ducktown, Tennessee, the denuding of vegetation caused by noxious smelter fumes has transformed a region of few channels and low slopes to one of numerous channels and high slopes. At the same time these gross changes in vegetation and morphology are taking place, more detailed alterations in the structure of the soil and the characteristic pattern of stream flows also occur.

Interpreting environmental change demands, therefore, that we establish *geological norms* – that is, expected long-term rates of environmental change under natural conditions (discussed in Chapter 3). Only then can we begin to assess the ways in which human activities have accelerated or distorted these norms.

The Magnitude of Past Landscape Changes 11.2

How much change has there been in the landscape in the past? To give exact figures on the amount of change induced by the kinds of process we have described is a much more difficult task than simply itemizing examples of change. Many changes have only recently begun to be viewed as important, and so the desire to measure them is a relatively new phenomenon. However, we can make gross estimates by examining the changes in boundaries between major ecological zones. Within these zones we can get a rough idea of the minor variations that must have accompanied shifts in boundaries by looking at case studies of smaller areas.

On the Global Level

On the global level, the evidence is particularly difficult to piece together. Table 11.2 lists the assumed distribution of natural vegetation before it was disturbed to any great extent by human beings. Roughly one-third of the earth's surface was covered by forest; another third was split into polar, mountain, and desert zones; and the remainder was largely open park and grassland. The exact estimates given by different authorities vary, but the general proportions given in the table appear to be of the right order of magnitude.

How have these proportions changed under human impact? Maps of world population density indicate that well over one-third of the earth – the polar areas, the mountains, and the dry zones – is unpopulated or very lightly populated. The massive concentrations of population are in environments that originally supported forests or grasslands. The United Nations *Food and Agriculture Organization (FAO)* estimates that about 10 per cent of the world's land is planted with crops, about 25 per cent is forest, and a further 20 per cent is grassland (covered by grasses, legumes, herbs, and shrubs). It is difficult to compare the estimates of natural vegetation in Table 11.2 with the FAO figures for current land use because slightly different classifications of land were used. What a comparison suggests, however, is that the natural vegetation has been changed mainly in forest areas. Within these zones, the changes have been highly selective and confined mainly to mid-latitude woodlands in eastern North America, Europe, and eastern Asia, and to the monsoon woodlands of South Asia. Large areas of the tropical rainforest and the circumpolar boreal forest remained lightly touched by human intervention until the last few decades.

The reduction in the world's grassland zones has likewise been highly concentrated. The many dramatic changes in mid-latitude grasslands, such as the North American prairies or New Zealand's Canterbury Plains, stand

Table 11.2 Global changes in land use

Original land cover (percentage) (about 10,000 BC)		Disturbed cover (percentage) (land use, in late 20th century)	
Forest	33	Forest, woodland and natural rangeland	36
Open woodland and grassland	31		
Desert	20	Desert	19
Arctic and alpine	16	Arctic and alpine	16
		Cropland	10
		Pasture and meadow	19

Figures are averages based on a series of alternative estimates.

in contrast to the less dramatic alterations of the tropical grasslands and savannah zones. Of course, the estimates here refer only to the changes in cover for very general types of vegetation. As we saw in Section 9.2, significant alterations in species composition can occur in areas where the general appearance of the vegetation apparently has been unchanged.

On the Subcontinental Level

At the subcontinental level the pattern of changes in *land use* becomes clearer. Figure 11.8 graphs more than a century of figures for categories of land use within the United States. The statistics vary in accuracy from one category to another and over the survey period, but they generally improve in reliability during the last half-century. We can see in these figures three types of environmental change.

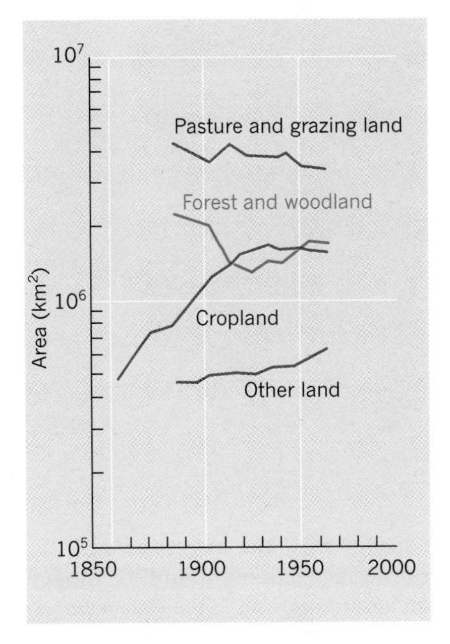

Figure 11.8 Changes in land use in the United States since 1850 Note that the vertical scale is logarithmic.

[Source: Data from M. Clawson *et al.*, *Land for the Future* (Johns Hopkins Press, Baltimore, 1960), Table 4, p. 39.]

First, there was a phase of cropland expansion, which was already under way by 1850 and continued in 1920. The period of this expansion corresponds to the frontier era, when the limits of settlement were extended from the Atlantic seaboard and trans-Appalachian areas to the mid-west, the Great Plains, and the Pacific and Mountain areas. During this period, the total amount of cropland increased fourfold, despite the fact that land was going out of cultivation in the east, particularly in New England.

Second, since 1920 the amount of cropland has remained relatively stable. Additions to existing cropland, such as western irrigated areas, have been counterbalanced by the abandonment of farms and the encroachment of housing on rural areas in the east. The expansion in farmland during this period has been due primarily to changes in the ownership of grazing land; that is, land previously in the public domain (mainly Federal land) or owned by the railroad companies has been incorporated into farms. Much of this change has been concentrated in the Great Plains area. The total area of grazing land as an environmental category has not changed much over this period if we consider both farm pasture and non-farm grazing land.

Third, the composition of forests has changed drastically. Although the amount of forest land outside the farms has diminished by about one-third, the total area of woodland and forest (despite fluctuations) is much the same now as it was in 1850. Since that time the area of virgin timber has shrunk considerably. But the area devoted to forestry in the sense of managed timber resources has grown, and second-growth timber on abandoned farmland is on the increase (see Figure 11.9). The pattern of change, both in the total area of forest land and in the fraction of the original forest land remaining, has been most noticeable in the eastern half of the country. Here forests were cleared both for lumbering and for cropland. Commercial lumbering showed a distinct east-to-west shift across the country, with New England and the northern Appalachians forming the original centre. The first westward movement was into the Great Lakes regions, where white-pine cutting was at its maximum from 1870 to 1890. By 1900 the southern pine region was the chief source of lumber for the national market. The extensive logging of Douglas fir and the shift of emphasis to the Pacific northwest happened largely after World War I.

The key shifts of land use that began with the general European settlements in the 1700s largely came to an end by 1910 or 1920. Since then, most significant changes have centred on the small area termed 'other' land in Figure 11.8. This land includes urban, industrial, and residential areas (outside farms), parks and wildlife areas, military land, land for roads and

(a)

(b)

Figure 11.9 The extension and retreat of settlement and its impact on woodland (a) In the Matanuska Valley, Alaska, modern farm settlement is systematically replacing forest by cropland. (b) In southern Illinois, oak and hickory encroach on abandoned fields. Note the relatively uniform texture on the photos shown by the virgin timber stands in Alaska. In contrast the varied texture of the Illinois woodlands shows trees at various stages of regrowth linked to the different times at which fields were abandoned. In both photos the regular checkerboard pattern of fields and roads indicates that land divisions were laid out on the 'township and range' system discussed in Chapter 21.

[Source: US Department of Agriculture photographs.]

transportation, and so on. The rapid growth of urban areas on the one hand and reserved wildlife areas on the other represents the extreme (and compensating) poles of environmental intervention.

On the Local Level

A small-scale example of changes in land use within the United States is provided by Figure 11.10. This shows the progressive deforestation of a sample area of approximately 10 km^2 (about 4 square miles) in southwest Wisconsin, Cadiz township, over a 120-year period of European colonization. The map for 1831, before agricultural occupation began, was compiled from the original government land survey. Apart from some prairie and oak-savannah in the southwest corner, the area was covered by upland, deciduous, hardwood forest. By 1882 about 70 per cent of the forest area had been cleared for cultivation, and the boundaries of the cleared area reflected the township and range system of land division (see Section 21.2, especially Figure 21.11). By 1902 the forested area had been reduced to less than 10 per cent of the total area; it comprised about 60 small woodlots averaging less than 16 ha (40 acres) each. Continued felling of trees for firewood and occasional saw timber, plus heavy grazing by cattle, caused a decrease to less than 4 per cent by the 1950s. This diminution of forest land was due largely to a reduction in the size of individual woodlots rather than to the elimination of entire woodlots. The number of these remained stable in the twentieth century.

Detailed investigations by ecologists at the University of Wisconsin have highlighted the significant effects of this environmental transformation. Decreases in the amount of water stored in the subsoil as agricultural fields replaced timber caused springs to dry up; the number of permanently flowing streams had decreased to around a third by 1935. The separation of the forest into isolated blocks reduced widespread burning, while fencing diminished the level of grazing within the woods. Both influences led not only to changes in species composition but to a denser forest cover with more mature trees per hectare than under the original, pre-1831 conditions. Animal populations underwent similar changes. Species that were adapted to *edge conditions* (i.e. conditions at the boundary between woodland and open land) benefited from the increased length of the perimeter of the forest caused by its subdivision.

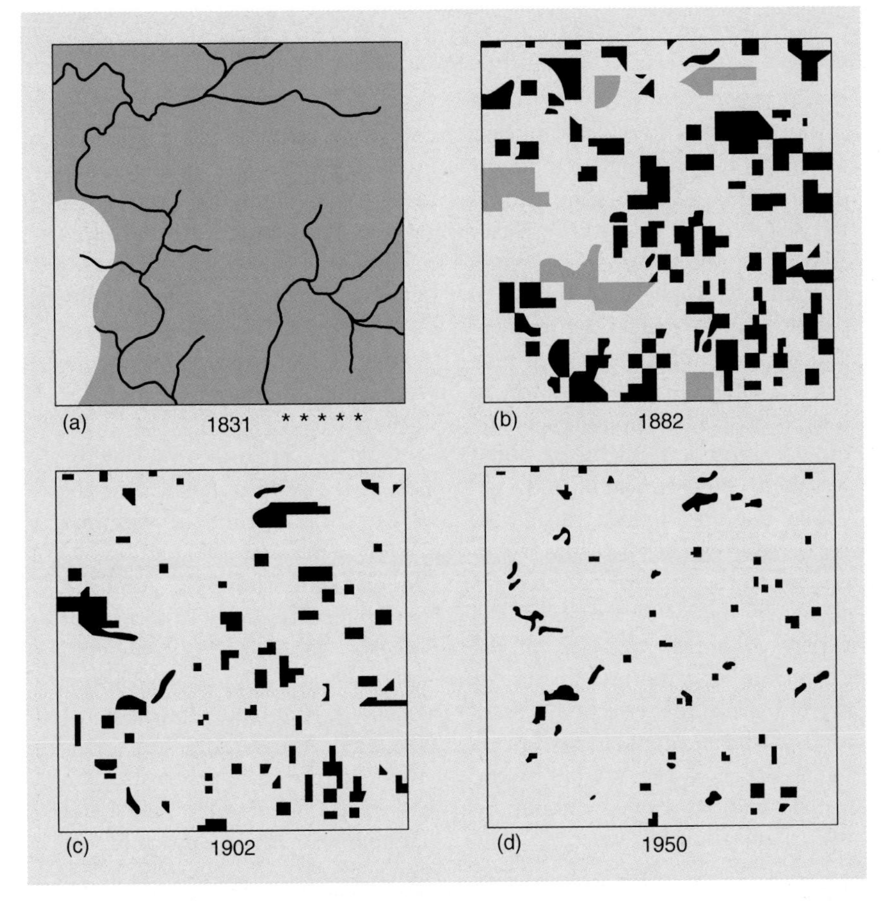

Figure 11.10 Changes in land use on the local scale The maps show changes in the wooded area of Cadiz Township, Wisconsin, since the beginning of European settlement. The coloured area represents land remaining forested or reverting to forest in each year. Compare these maps with one of modern land clearance in the Amazon rainforest (Figure 20.22). Note that in both the fragments of blocks of forest increase the length of the edge conditions.

[Source: J.T. Curtis, in W.L. Thomas, Jr. (ed.), *Man's Role in Changing the Face of the Earth* (University of Chicago Press, Chicago, 1956), Fig. 147, p. 726. Copyright © 1956 by the University of Chicago.]

To sum up: it is clear that gross changes on the global level conceal subtle environmental adjustments on lower spatial levels. This local case study is one of scores of similar investigations of different ecological zones that illustrate the fine level of adjustment within ecological systems, and the key role that human intervention plays in their shaping. Understanding the interplay of human intervention and environmental adjustments gives us a comprehensive view of the nature and magnitude of changes in land use.

Land-use Change as an Ethical Issue 11.3

We have concentrated so far in this chapter on presenting the facts of change. (See Box 11.A on *Man's Role in Changing the Face of the Earth* (1956)'.) But, increasingly, global land use sketched in the previous sections is causing increasing concern. We look at two examples of this: tropical rainforests and wetlands.

Tropical Rainforests

Few changes have caused sharper debate in terms of global land-use changes than the loss of tropical rainforests. The world's rainforests were described in our discussions of the earth's biosphere in Chapter 4 and

Box 11.A *Man's Role in Changing the Face of the Earth* (1956)

It is not often in the history of any science that a book published a half-century ago still resonates with young readers today. *Man's Role in Changing the Face of the Earth* is such a volume.

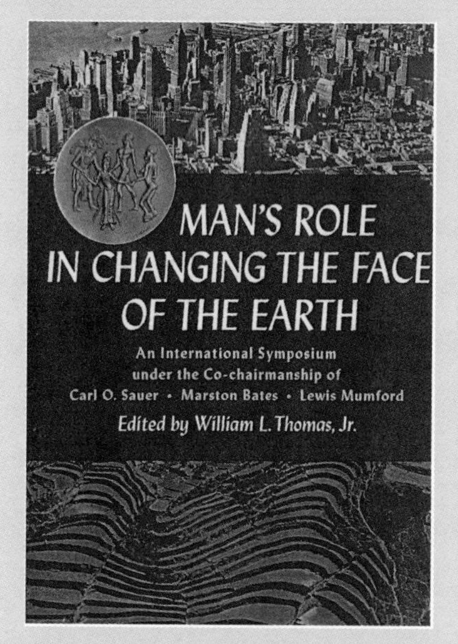

Its historical importance was as the first large-scale evaluation of what has happened and what is happening to the earth under man's impress. (The male bias in the title reflects the period when it was written; 'human impress' would be the more appropriate language today.) It was presented at an international symposium organized by the Wenner-Gren Foundation for Anthropological Research and held at Princeton University, New Jersey, in June 1955. The Princeton papers focused viewpoints from nearly all fields of knowledge upon humankind's capacity to transform the physical–biological environment and upon the cumulative and irreversible alterations of the earth. In these pages 53 contributing scholars provide important insights into what was then a pioneer field of study.

The book is divided into three parts. 'Retrospect' provides background to the modern era of rapid change by tracing the longevity and variety of human changes on the earth from the important first use of fire, through the effects of early food-producing populations in different areas of the world, to the transformations brought about by the urbanization of society.

The authors next considered 'Process' – the methods and agencies of human changes and the means by which humans have altered landforms and the soil, the seas and waters, climate, and biotic communities through human control of physical and biological processes.

'Prospect', the third part of the book, deals with the effects of human actions on the continued habitability of the earth and on the course of humankind's own evolution. It discusses the material and ideological limitations that these actions may produce for future inhabitants of the earth. Each part of the volume also contains a summary report on the ideas discussed by the 70 participants during the symposium meetings.

The co-chairmen of the Symposium were three towering figures of the scholarly world of the 1950s. Carl Ortwin Sauer (see Box 5.B for an account of Sauer's work) who was Professor of Geography at the University of California at Berkeley; Marston Bates, an eminent biologist, who was Professor of Zoology at the University of Michigan, Ann Arbor; and Lewis Mumford, doyen of urbanists, who was Professor of City and Land Planning at the University of Pennsylvania.

Subsequent to the Wenner-Gren Symposium, a number of volumes have been published that take up the same theme in a more modern context. See for example, I.G. Simmons, *Changing the Face of the Earth* (Blackwell, Oxford, 1989) and B.L. Turner *et al.*, *The Earth as Transformed by Human Action* (Cambridge University Press, Cambridge, 1990).

readers may well wish to check back on Section 4.3 to refresh their memory.

There is a general consensus, at least among western observers, that there is an urgent need for conservation. New Zealand geographer John Flenley has summarized the arguments for conservation as tenfold. We may arrange these arguments into four groups.

First, there are powerful ecological arguments.

1 *Loss of diversity*. It has been estimated that one species of animal or plant becomes extinct every minute because of destruction of tropical rainforest. We noted in Chapter 4 the structure of the biome and some biologists have argued that this could lead to the collapse of the

earth's ecosystem. Others stress that we are destroying the forest before we even know its potential to yield useful items. These include drug plants, genetic resources for crop plants, and agents for biological control of pests.

2 *Forest peoples*. Loss is not confined to plant and animal species. The forests are the habitat of forest tribes whose way of life is being destroyed. In this argument, deforestation is a form of genocide.

Second there are a group of more specific environmental arguments based on the physical geography of the forests.

3 *Role as a carbon dioxide sink*. The role of the tropical rainforest in helping to contain the rising atmospheric CO_2 levels was touched on in our consideration of the global carbon cycle (see Box 4.A).

4 *Runoff and flooding*. The tropical rainforests, and especially the montane rainforests, are protection forests. They conserve water and control runoff, thus providing more regular year-round flow in rivers and thus helping to prevent flooding in lowlands.

5 *Soil conservation*. Forest soils do not survive well after deforestation, and some undergo irreversible degradation. Some erosion following deforestation is almost universal, and the Oxford geographer Andrew Goudie considers this the most serious problem of all in the long term. Its negative effects are twofold: destroying the soil at the point of deforestation, also polluting water courses and lakes, damaging fishing, and causing silting up of reservoirs and lagoons.

6 *Climatic reasons*. It has been suspected for some decades that deforestation (through reducing evapotranspiration) might affect precipitation, and thus the entire water cycle. This suspicion has now been confirmed by careful studies in Amazonia.

Thirdly, there are economic reasons.

7 *Timber resources*. Tropical rainforests are a valuable source of high-value hardwood timber. It serves as a renewable resource for producing specialized hardwoods and the world will soon run short if it is destroyed.

8 *Ecotourism revenue*. Tropical rainforests can be a source of income for tropical countries, through the growing importance of ecotourism.

Finally, there are broader and less tangible reasons.

9 *Scientific and educational value*. Because of its size and intricate biodiversity, the tropical rainforest is the most complex ecosystem on earth. It should be retained for scientific research and education.

10 *Philosophical and aesthetic reasons*. The tropical rainforest is unique and it may be argued that the present generation has no right to remove in a few decades something that has evolved over many millions of years. It has a right to exist and its extraordinary aesthetic value should be preserved for posterity.

Of course the ten reasons do not exist just as separate categories. The value of the forest for drugs and as a genetic resource (1) also reinforces the economic arguments of (7) and (8). We should also note that the calls for preservation are coming mainly from western countries. It is also argued, sometimes by countries such as Indonesia that are making major inroads on their forests, that it was Europe that largely cleared its forests in the medieval period as did North America in the nineteenth century.

Increasingly, debates about global land use are seen as part of the international agenda. Here the seminal meeting was at Stockholm in 1972. (See Box 11.B on 'The Stockholm Conference'.) This set up for the first time what was termed a 'Bill of Rights' for planet Earth, a bill that may turn out in the long run to be every bit as important as its United States counterpart for human rights passed by its Congress in 1791.

Box 11.B The Stockholm Conference

The Stockholm Conference is the common name for the United Nations Conference on the Human Environment, held in 1972. This was the first UN Conference on the Human Environment was held in the Swedish capital in 1972. Subsequent meetings have been held at Rio de Janeiro (1992) and Kyoto (1999). In response to mounting global public concern about deteriorating environmental and living conditions, delegates from 113 nations met and produced an action plan of 109 separate recommendations. They also agreed a Declaration of 26 common principles on human rights and responsibilities in respect of the global environment which remains as a guide to influence human actions and policies.

1 The Human Race has the fundamental right to freedom, equality and adequate conditions of life and bears a solemn responsibility to protect and improve the environment for present and future generations. Policies promoting oppression and foreign domination must be eliminated.
2 The natural resources of the earth must be safeguarded for the benefit of present and future generations through careful planning and management.
3 The capacity of the earth to produce vital renewable resources must be retained and, wherever practicable, restored.
4 Man has special responsibility to safeguard and wisely manage the heritage of wildlife and its habitat which are now imperilled. Nature conservation must therefore receive importance in planning for economic development.
5 Non-renewable resources of the earth must be used in such a way as to guard against their future exhaustion and to ensure that any benefits are shared by all mankind.
6 The discharge of toxic substances, or of other forms of pollution, in such quantities or concentrations that exceed the capacity of the environment to render them harmless, must be halted.
7 States shall take all possible steps to prevent pollution of the world's seas.
8 Economic and social development is essential for ensuring a favourable living and working environment.
9 Environmental deficiencies generated by conditions of underdevelopment and natural disasters can best be remedied by accelerated development through the transfer of substantial quantities of financial and technological assistance.
10 Stability of prices and adequate earnings for primary commodities and raw materials are essential to environmental management.
11 The environmental policies of all states should enhance and not adversely affect the present or future development potential of developing countries.
12 Resources should be made available to preserve and improve the environment: additional international technical and financial assistance should be forthcoming in this respect.
13 States should adopt an integrated and coordinated approach to their development to ensure that development is environmentally sound.
14 Rational planning must reconcile any conflict between the differing needs of social development and the natural environment.
15 Human settlements and urbanization must be planned to provide maximum social and economic benefits for all with minimum adverse effects on the environment.
16 Demographic policies, without prejudice to human rights and which are deemed appropriate by governments concerned, should be applied in those regions where excessive population growth rates or concentrations may jeopardize the environment or development process.
17 Appropriate national institutions must undertake

Box continued

the planning, managing and controlling of a country's environmental resources.

18 Science and technology must be applied to the identification, avoidance and control of environmental risks and the solution of environmental problems, for the benefit of all.

19 Education is essential in order to promote enlightened opinion and responsible conduct by individuals, enterprises and communities with regard to protecting and improving the environment.

20 Scientific research and development on all aspects of the environment and development must be encouraged in all nations, and the free flow of data and information must be supported and assisted, with new technologies being made available to developing countries.

21 States have the sovereign right to exploit their own resources but must ensure that their activities do not damage the environment beyond the limits of their jurisdiction or control.

22 All states shall work toward developing viable and enforceable international law regarding questions of liability and compensation arising as a result of environmental damage or pollution.

23 The differing system of values prevailing in countries throughout the world must be taken into account in all respects, policies and judgements.

24 International matters concerning the protection of the environment should be handled in a cooperative spirit by all nations on an equal footing.

25 States shall ensure that international organizations play a coordinated, efficient and dynamic role in protecting the environment.

26 The human race and the environment must be spared the effects of nuclear weapons and other means of mass destruction.

Since the Stockholm Meeting there have been other world meetings to reinforce the Stockholm meeting, notably the UN Conferences on Environment and Development (UNCED) at Rio de Janeiro, Brazil, in 1992 and at Kyoto, Japan, in 1999.

Wetlands

Although less emotive than the rainforests, the loss of the world's wetlands is also a cause for concern. 'Wetland' refers to any land that is intermittently or regularly waterlogged. It includes salt marshes, tidal estuaries, marshes, and bogs.

Wetlands are extremely important ecosystems. They tend to be highly productive sources, both of food and fuel (peat in mid-latitudes, mangrove forests in tropical). They are crucial environments to over 60 per cent of the world's fish catch, and serve as major breeding and migrating grounds for the world's wildfowl. Many wetlands also play a role in water purification and supply. They are threatened by a variety of human activities, including drainage for agriculture, dam and canal construction, peat cutting, forestry, and pollution attributable to industry or the overuse of fertilizers and pesticides.

One of the most famous wetlands is the Everglades National Park, Florida (see Figure 11.11), which forms a complex mosaic of vegetation: mainly coastal mangroves and tropical marshes dominated by sawgrass but with some forest on slightly raised areas. A combination of pressures is now threatening the Everglades: flood-control measures, water demands, and ever-swelling visitor numbers. Many other wetlands are increasingly being reclaimed for agriculture, housing, or industry.

Again, there has been international action to try to highlight the problem and to coordinate protective measures. The Ramsar Convention is the common name for the *Convention on Wetlands of International*

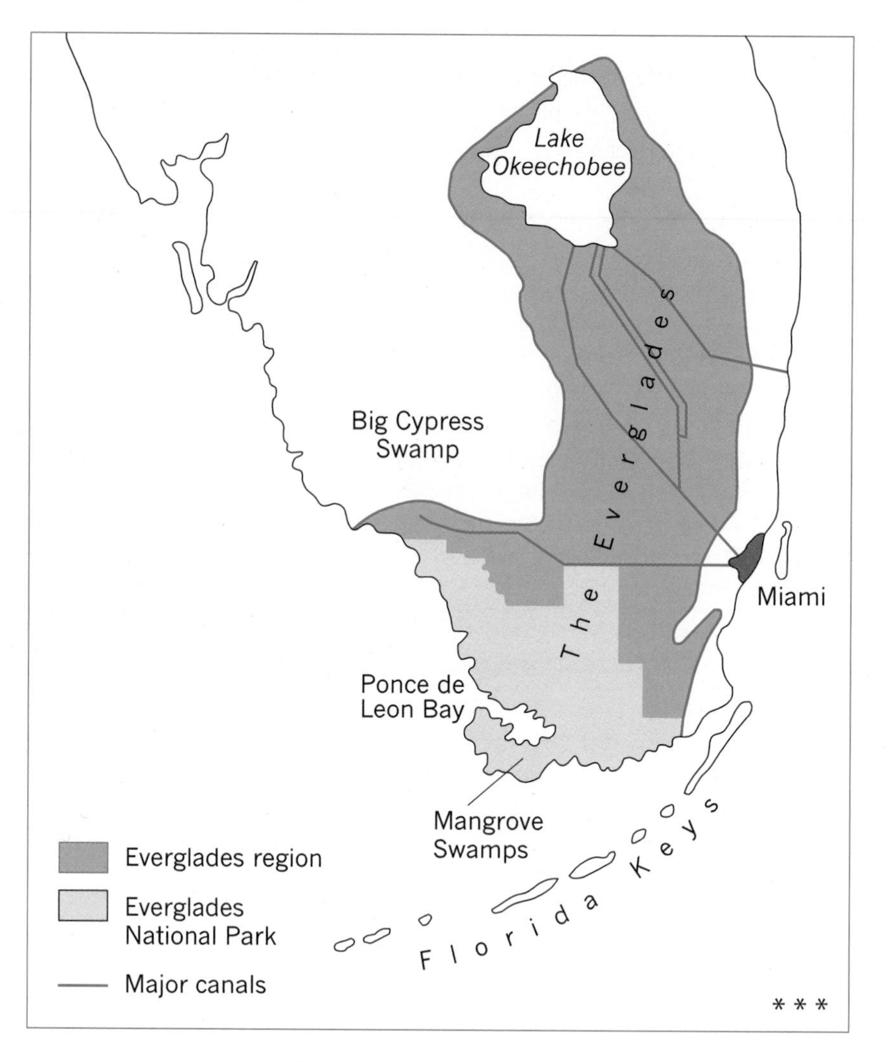

Figure 11.11 Everglades as a wetland area The Everglades occupy a 10,000 km^2 (4000 square miles) area of marshland in southern Florida, in the southeastern United States. Water moves slowly across the limestone-floored basin from Lake Okeechobee in the north to the Gulf of Mexico on the west. The glades are dominated by subtropical saw-grass but extend into mangrove swamps on its seaward edges. The mild climate and abundance of water provide a rich environment for myriad birds (herons, egrets, and ibis) as well as alligators, snakes, and turtles. The area is the largest subtropical wilderness area left in the conterminous United States and as such is now a classic conflict area for different interests in which agriculture, conservation, tourism, and urban growth all place stress on the fragile but rich habitat.

Importance signed in Ramsar, Iran, in 1971. As such it is one of the world's oldest international conservation treaties. What makes it unique is that it protects a specific type of ecosystem – wetlands – on a global basis. By 2000 some 60 countries had signed the treaty. Each country agreed to protect (or replace with one of equal worth) at least one listed site within its national territory. To date over 500 wetland sites, covering in total an area the size of 30 million hectares (75 million acres), have been nominated.

The work of the Ramsar Convention has traditionally been shared by three bodies: the Worldwide Fund for Nature (WWF), the World Conservation Union (IUCN), and the International Wildfowl Research Bureau, but repeated calls have been made for a permanent, funded Secretariat to be established. A number of the Ramsar sites are endangered, despite being on the list. Several sites in Greece are under special threat as they are likely to be drained as a result of development projects funded by the European Union. Other sites in Africa, Germany, Jordan, Pakistan, South Africa, the former Soviet Union and Uruguay were singled out as being under particular threat.

The Human–Environment Controversy 11.4

In the last few chapters we have looked at the ways in which human populations have responded to the challenge set by the global environment. As in Chapter 1, we have found that relation to be a two-way one: people being influenced by environment and changing that environment in turn.

The Historical Debate

Historically, the 'chicken-and-egg' relationship between people and environment has always puzzled geographers (see Table 11.3). Did the environment control us? Or was the environment an opportunity for conquest? Let us go back to the mid-nineteenth century, when English biologist Charles Darwin published his *Origin of Species* (1859). The book, based on worldwide evidence from Darwin's travels, made a strong impact on geographers. Its theme of competition among species for limited resources and the selective survival of the better-adapted species held special attraction for a German geographer, Friedrich Ratzel. In his *Anthropogeographie*, published a quarter of a century after Darwin's book, Ratzel argued that the distribution and grouping of human population on the earth's surface can only be understood in the context of the physical environment. He was particularly concerned with (a) those push-and-pull factors that had led to major migrations and (b) the physical conditions under which major civilizations had been able to develop.

Ratzel's ideas were enthusiastically taken up in the United States by Ellen Churchill Semple (Figure 11.12). She attended his lectures at Leipzig during the 1890s, and on her return home set out to introduce the ideas in American geography. Her *Influences of Geographic Environment* (1911) represents the fullest, best-documented, but perhaps most extreme

Table 11.3 Main schools of geographic environmentalism

School	Brief description
Cognitive behaviouralism	Holds that the impact of environment on people is partly dependent on their perception (cognition) of the resources and barriers it poses. See Figure 11.13.
Human ecology	Envisages reciprocal reactions between human and environment, like those of other plant and animal species. This view is associated with the Chicago geographer *Harlan Barrows* (1877–1960).
Physical determinism	Holds that the environment largely controls human development. It is associated with the German geographer *Friedrich Ratzel* (1844–1904) and his American disciple *Ellen Churchill Semple* (1863–1932). See Figure 11.12.
Possibilism	Argues that the environment offers sets of possibilities but that the choice between them is determined by human beings. The French historian *Lucien Febvre* (1878–1956) was one of the strongest proponents of this view.
Scientific determinism	A variant of physical determinism in which the argument proceeds from the statistical analysis of sets of data rather than from individual case studies. Yale geographer *Ellsworth Huntington* (1876–1947) was the leader of this school of thought.
Stop-and-go determinism	Holds that people determine the rate but not the direction of an area's development. The term was coined by Australian geographer *Griffith Taylor* (1880–1963). See Box 11.C.

The ideas of the various schools are discussed further by R.E. Dickinson in *The Makers of Modern Geography* (Routledge, London, 1969).

Figure 11.12 Ellen Churchill Semple A Vassar graduate of 1882, Semple studied at Leipzig in the 1890s with the leading German anthropogeographer, Friedrich Ratzel. In three influential books published between 1903 and 1931 she expanded Ratzel's ideas on the ways in which geographic conditions influence human behaviour.

arguments for what came to be termed *environmental determinism*. On the opening page of her book she set out her position in no uncertain way: 'Man is a product of the earth's surface. This means not merely that he is a child of the earth . . .; but that the earth has mothered him, fed him, set him tasks, confronted him with difficulties . . . and at the same time whispered hints for their solution.' From this the book goes on to illustrate how the different major environments – oceans and continents, mountains and plains, warm climates and cold – have shaped the history of the human groups that had occupied them.

It was inevitable that there would be a pendulum swing away from the more extreme views of the Ratzel–Semple school. This came in 1924 with Lucien Febvre's *Geographical Introduction to History*. A French historian deeply interested in geographical problems, Febvre argued for an alternative view. He saw the earth's environment as presenting not necessities, but possibilities. By citing examples of quite different human developments in the *same* types of environment, Febvre was able to develop counterarguments to the earlier views.

Environmental determinism was not, however, a dead issue, and the interwar period saw some further major statements. Conscious that Ratzel had been writing at a time of very sparse global information, the Yale geographer Ellsworth Huntington set out to retest some of his hypotheses using better statistical data. In particular he set out to measure more accurately the way in which climate affected the ability of humans to perform work – both physical and mental. He backed up these studies with extensive research, particularly in Central Asia, to try to check out the importance of climatic change in determining major migrations out of that area.

Another challenging variation on the environmentalist theme was introduced by the Australian geographer Griffith Taylor. (See Box 11.C on 'Griffith Taylor and geographic environmentalism'.) His studies in the Antarctic and the Australian outback suggested to him that the environmental conditions indicated certain directions along which a country's development could go. Humans were able to accelerate, slow, or stop the progress of development along a particular path, but not change the path. Taylor suggested we were like a traffic controller in a large city, altering the *rate* but not the *direction* of progress. In consequence of this analogue, this view of human–environment relations is often called *stop-and-go determinism*.

Current Views

The historical debate that lasted from the middle nineteenth century to World War II appears to be based on a fallacy. Like the argument over human intelligence (i.e. whether intelligence is a product of our genes or our early upbringing), it assumes a duality. Environment is in many senses inseparable from people. We can illustrate this idea with an example. One of the classic areas frequently quoted by the environmental determinists was the Australian desert. An area in which most readers of this book (and certainly the writer!) would starve in a few days might be rich in resources to Aborigine tribes with youngsters taught from childhood to recognize potential waterholes and buried food sources such as tree roots or insect grubs. Conversely, the tribal group might be wholly unaware that their hunting grounds are underlain by uranium deposits of great value to the western man.

Box 11.C Griffith Taylor and Geographic Environmentalism

Thomas Griffith Taylor (born London, England, 1880; died Sydney, Australia, 1963) was an Australian geographer who made vigorous contributions to the debate over environmental determinism. Taylor's 'stop–go' hypothesis suggested that while the basic environmental resources of an area and its relative location determined the basic direction of economic development, the other classic factors of labour and capital determined the rate of development.

Taylor emigrated to Australia with his parents while in his early teens. After graduating from Sydney University he specialized in glaciology and worked at Cambridge and in the Antarctic (with the Scott expedition) before returning to Sydney as its first geography professor in 1920. He became increasingly involved in public debates on the potential habitability for human settlement of both tropical Australia and the dry inland region of Australia. Government policy at that time was to attract migrants, but Taylor was critical of the exaggerated claims then being made about the settlement possibilities of these areas. His assessments later turned out to be considerably more accurate than official views, but the controversy played some part in his decision to leave for a Chicago University chair in 1928. The commemoration of Griffith Taylor on an Australian stamp issued in 1976 marks a belated tribute to one of geography's most active and outspoken exponents of the importance of the environmental view.

From Chicago, Taylor moved to Canada to found Canada's first geography department (Toronto). He returned to retirement in Australia in 1951. Taylor's strongly held and vigorously argued ideas permeate his many books. His ideas on the Australian environment were set out in *Climatic Control of Australian Production* (1915) and continued through a number of publications to *Australia: a Study of Warm Environments and their Effect on British Settlement* (1941). His experiences of Canada were reflected in a parallel volume on *Canada: a Study of Cool, Continental Environments and their Effect on British and French Settlement* (1947). On a still broader canvas the links between environment and political and economic organization were set out in *Environment and Nation* (1936) and *Urban Geography* (1949). Taylor's account of his own explorations is given in his lively autobiography, *Journeyman Taylor* (1958).

[Source: Stamp copyright Australian Post Office. Alteration of this image in any way is forbidden. National Philatelic Collection, Australia Post.]

The modern view of environmentalism is that summarized in Figure 11.13, and discussed in an elementary way in Chapter 1. It shows the environment composed of two sections: (a) *natural phenomena* in the sense of the totality of the world, and (b) the *perceived environment* and *hidden environment* as those parts of the natural phenomena known and not known to us. (You might like to think of this as being like an iceberg, with only a fragment of ice visible above the surface.) The two-way relations that go on between the two parts are clearly not restricted to the perceived environment. Minimata disease (discussed in Chapter 9) illustrates a case where there were two-way side effects which only later became a part of the perceived environment.

We show in Figure 11.13(b) a two-group situation. The 'western multinational' person and 'aborigine tribal' person share certain common biological characteristics (shaded) and see part of the harsh environment in the

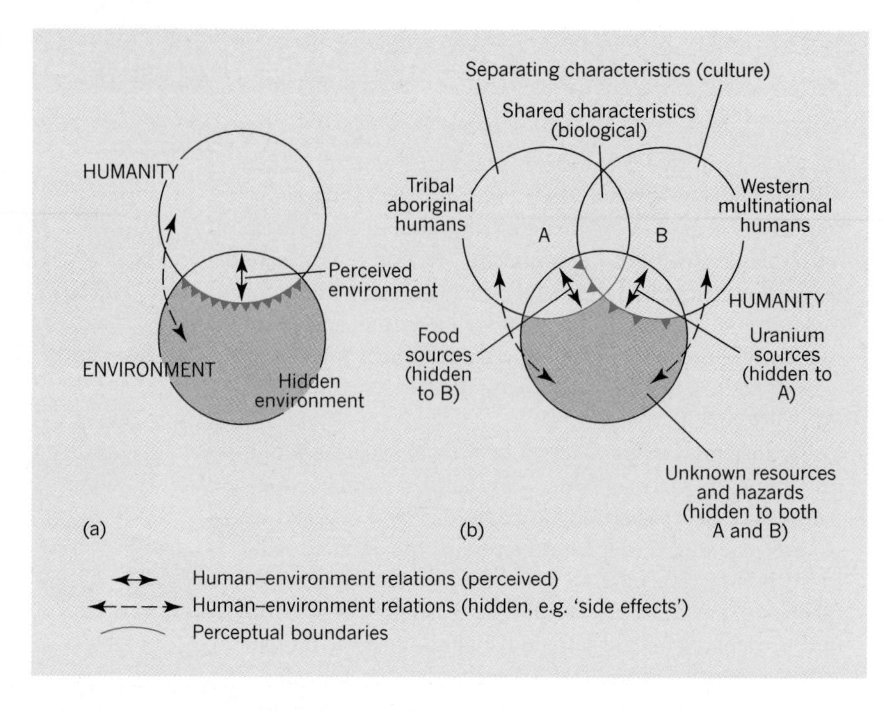

Figure 11.13 Human–environment relations (a) Environments may be thought of in two segments, the perceived and the hidden. (b) Perceptual boundaries may also alter from one cultural group to another.

same way. But the buried eggs and the buried uranium lie in different sets. The critical ingredient separating them is the ***perception*** bundle of attitudes that forms part of their culture.

Even within western society there may be major differences in reactions to a given environment. For example, many human groups occupy environments that are dangerously unpredictable. River valleys may be subject to flood, coastal settlements are at risk of high tides, and semi-arid areas may be hit by drought. Geographers Gilbert White and Ian Burton have investigated the ways in which such dangers are viewed. They began by looking at the likelihood of flooding in different areas and constructed a *risk curve* (Figure 10.14(a)) for the 498 urban communities in the United States that are built on river floodplains (and have flood-frequency data). Some of these urban places were reliable to have floods only once in ten years, while in others waters rose to danger levels dozens of time in a single year. For most cities in the sample, the likely number of floods each year was two or three.

As Figure 11.14 indicates, human responses and adjustment to the known danger of flooding do not increase consistently as the risk becomes

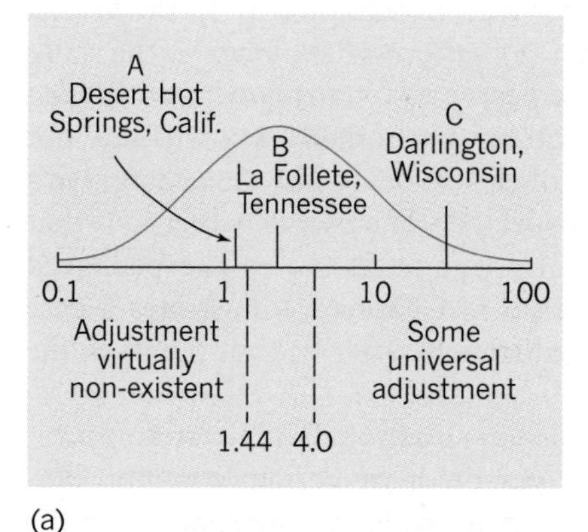

(a)

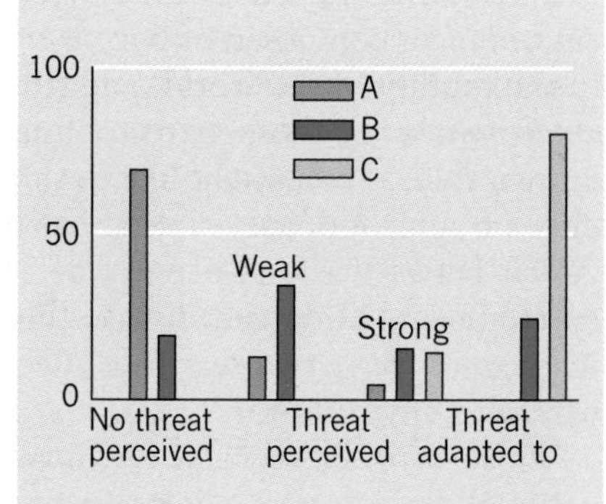

(b)

Figure 11.14 Human perception of future flood hazards (a) The curve shows the frequency of floods in 496 urban places in the United States. Most places for which flood-frequency data were available have two or three floods each year. (b) The degree of adjustment to the hazard in three places with three different experiences of flooding is illustrated. The height of each column reflects the number of respondents in each place who fail to perceive a threat, who perceive a weak or strong threat (two levels of 'perceived'), or who adjust to the hazard. The letters and shading identify the three locations named in (a).

[Source: R.W. Kates, *University of Chicago Department of Geography, Research Paper* **78** (1962), Fig. 9.]

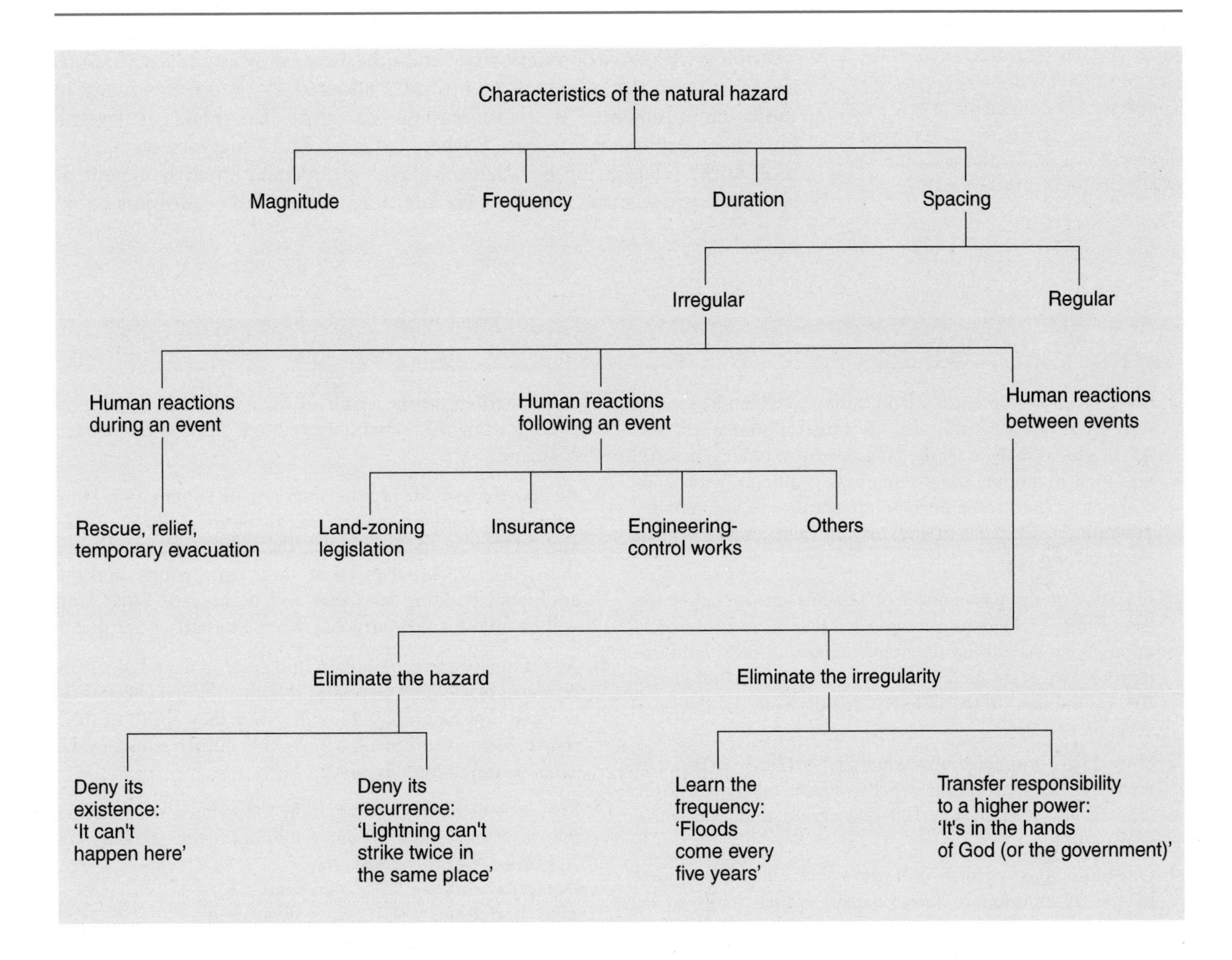

Figure 11.15 Human reactions to irregular natural hazards

greater. Until the environment stress builds up to the point where the likelihood of damage is regular and recurrent, little or no adjustment to the possibility of flooding takes place. Figure 11.14(b) shows the degree to which inhabitants of three United States communities where the likelihood of flooding is different 'perceive' a flood threat. Darlington, Wisconsin, can expect twenty floods in any ten-year period, whereas Desert Springs, California, can expect only one flood in the same period. When the probability of the recurrence of a hazard is high, as it is in Darlington, the danger is widely perceived but evaluated in different ways. Figure 11.15 lists four of the ways in which individuals respond to the possibility of recurring natural disasters. Each response represents an optimistic rationalization for continuing to live in a hazard area. It is interesting that the range of responses is great when the probabilities of a recurrence are moderate; responses tend to be more uniform in high-risk and low-risk areas.

Flooding further stresses the essential interdependence of human being and land. By changing the face of the earth – building levees and dams – we have been able to occupy many of the flood plains of the world's great rivers. In doing so, the potential threat of flooding is *increased* as mud that would have been spread valleywide by the seasonal floods is forced to accumulate in the stream channel itself. To build the levees higher pushes the risk of an occasional big flood still higher, and so on.

Coping with environments, risky and otherwise, shows spatial variations around the world linked to the attitudes adopted by the groups living in those environments. It is misleading therefore to think of overall human–environment relations. Rather, as Figure 11.14 suggests, we have to look at the relationship between particular groups and an environment. It is to these group differences that we later turn in the third part of this book.

Reflections

1 Over much of the earth's land surface, the landscape we now see is not a natural one, but rather a human artifact. At the global scale, only the polar ice caps, tundra, deserts, and high mountain zones remain as genuinely wild landscapes and even some of these are starting to come under pressure. Which parts (if any) of your own country remain relatively wild?

2 Evidence of the past impact of humans on the landscape (see Table 11.1) requires careful research and interpretation. What have been the main changes in rural land use in your own state or district over the last hundred years? Are (a) similar or (b) different trends likely in the next decade?

3 How do geographers know what changes have occurred in land use? Compare the evidence available on changes over (a) a 10-year period, (b) a 100-year period, and (c) a 1000-year period.

4 One of the most sensitive indicators of change are provided by the distribution of forested land (which tends to be reduced by felling) and of marshland (which is often drained and brought into cultivation). Obtain a map that shows the distribution of these two types of land use in your own country. Why are they located where they are? Is their distribution the result of (a) natural environmental conditions or (b) human intervention? How is the pattern changing?

5 Review the growth of your own city or community. How large is its population expected to be in AD 2010, 2020, and 2050? Calculate, roughly, the number of new homes that will be required to house these extra people and the additional building land that will be needed. What land will be used for this purpose? What alternatives are there?

6 Adopt one of the geographers and environmental scientists reviewed in the final section of this chapter and check their writings and biography. How far were they ahead of their time in seeing the difficult choices that human tenure of the earth would eventually pose?

7 Review your understanding of the following concepts using the text (including Table 11.3) and the glossary in Appendix A of this book:

Domesday survey	land use
edge conditions	perception
environmental determinism	place-name evidence
geological norms	possibilism
human ecology	stop-and-go determinism

One Step Further ...

The changing face of the earth under human influence is one of the grand themes in geography. A useful starting point is to look at the two outstanding essays on human modifications of the earth by Ian Simmons and Michael Williams in I. DOUGLAS *et al.* (eds), *Companion Encyclopaedia of Geography* (Routledge, London, 1996), pp. 137–58, 182–205. The standard geographic account of the topics discussed in this chapter is provided in the very large, ageing, but classic volume based on a Princeton symposium by William L. THOMAS *et al.* (eds), *Man's Role in Changing the Face of the Earth* (University of Chicago Press, Chicago, 1956). This is a splendidly conceived book (see Box 11.A) with a wealth of talented authors and is strongly recommended for browsing.

A number of books have followed up the 'man's role..' theme. One of the earliest which remains of value is P.L. WAGNER, *The Human Use of the Earth* (Free Press, New York, 1960). This has been expanded in Ian G. SIMMONS, *Changing the Face of the Earth: Culture, Environment, History*, second edition (Blackwell, Oxford, 1996). A major volume that parallels the Princeton volume in many ways is B.L. TURNER, II *et al.* (eds), *The Earth as Transformed by Human Action* (Cambridge University Press, New York, 1990). For a massive survey of western philosophers' views of human beings as agents of terrestrial change, look through C.J. GLACKEN, *Traces on the Rhodian Shore* (University of California Press, Berkeley, 1967) and for the history of forest clearance see M. WILLIAMS, *Americans and their Forests: A Historical Geography* (Cambridge University Press, New York, 1989).

An engaging book that explores the role of humans and

environment is Yi-Fu TUAN, *Topophilia: A Study of Environmental Perception, Attitudes, and Values* (Prentice Hall, Englewood Cliffs, N.J., 1974). A text that reviews the whole span of human–environment behaviour and its importance in geography is J.R. GOLD, *An Introduction to Behavioural Geography* (Oxford University Press, New York, 1980)

The 'progress reports' section of *Progress in Human Geography* (quarterly) will prove helpful in keeping up to date. Research reports on human impact on environment occur regularly in all main geographic serials. You should also browse through biological journals such as *Ecology* (quarterly), to see something of the research in neighbouring scientific fields, and the *Journal of Historical Geography* (quarterly) to see work on past changes in land use. For readers with access to the World-Wide Web see also the sites recommended for this chapter in Appendix B at the end of this book.

E
D
B
C
A

CHAPTER 12

The Web of Regions

■ denotes case studies

Figure 12.1 **The morphology of landscape regions** The simplest approach to regional division is to use natural variations in terrain. In this LANDSAT image of Los Angeles, in southwest California, the built-up area of the city and its suburbs sprawls over the low-lying Los Angeles sedimentary basin. The dense and regular grid pattern of the road network is clearly visible. The surrounding mountain ranges are vegetated and appear in dark grey, while bodies of water, such as the Pacific Ocean, are coloured black. The smooth coastline of San Pedro Bay, lower right, is broken up by the docks in Los Angeles harbour. Letters indicate main regional divisions: A, Los Angeles basin; B, San Fernando valley; C, Santa Monica range; D, western extension of San Bernadino mountains; E, western edge of Mojave desert. North is at the top of the photo, the east–west distance is around 80 km (50 miles).

[Source: Earth Satellite Corporation/Science Photo Library.]

> The number of different boundaries for any region is equal to the square of the number of geographers consulted.
>
> ANON.

When in Lewis Carroll's *Through the Looking Glass* Alice objects that 'glory' does not mean the same as 'a nice knockdown argument', she elicits Humpty Dumpty's evasive reply: 'When I use a word, it means just what I choose it to mean – neither more nor less.' To judge from their writings, geographers feel more sympathy for Humpty Dumpty than for the disgruntled Alice. And of the many words that geographers use, it is '***region***' that causes more nice knockdown arguments than any others.

We can readily see why such arguments occur. If we look at the photograph that opens this chapter (Figure 12.1 p. 360), we can see at least two ways in which regional boundaries could be drawn. If we concentrate on the enduring features of terrain, then the significant boundaries might be the sharp break of slope that separates the Los Angeles basin and the San Fernando Valley from the hill ranges that enclose them. But if we looked at human occupation, then it would be the edge of the built-up area of Los Angeles as a city that formed the crucial regional boundary. Variations within the city (higher-density cores and lower-density suburbs) might form a second-order boundary. Arguments, Humpty Dumpty-like, might rage between the two approaches. The terrain enthusiasts might rightly argue that the urban scene is only a newcomer on the landscape (the tiny settlement of El Pueblo de Nuestra Señora la Reina de Los Angeles was founded in 1781 and by 1850 had still only 1600 people). Despite the environmental changes noted in Chapter 2, the physical regions of the basin had a validity long before cities appeared and will survive their eventual demise. You can think of your own counter-arguments from the urban geographer's standpoint.

Even when geographers settle on the broad criteria, they do not agree (or not very often) on the precise boundaries of a region. You might like to ponder Figure 12.2, which give some considered geographic opinions on

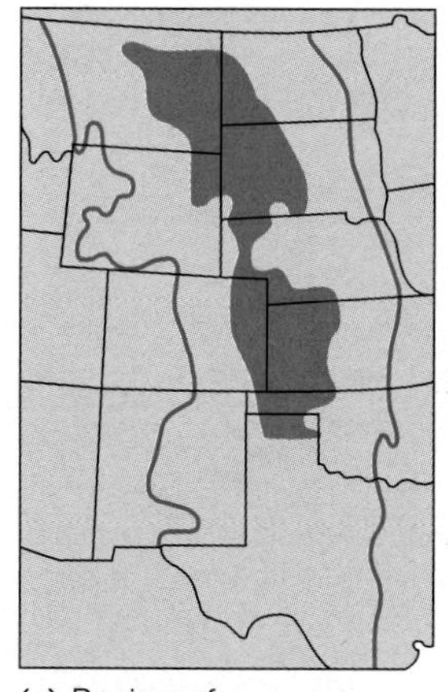

(a) Region of distinctive vegetation

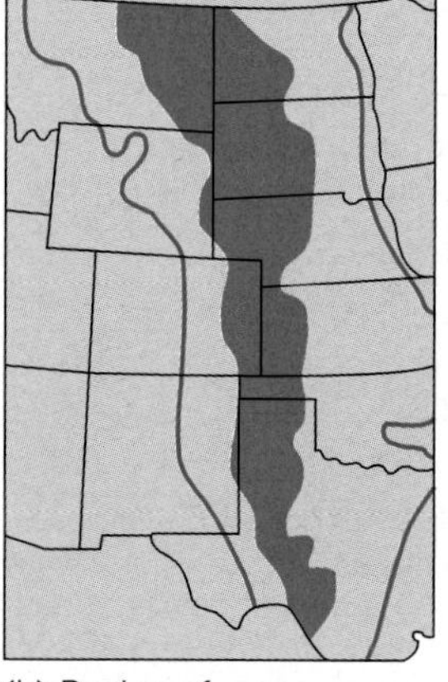

(b) Region of distinctive landforms

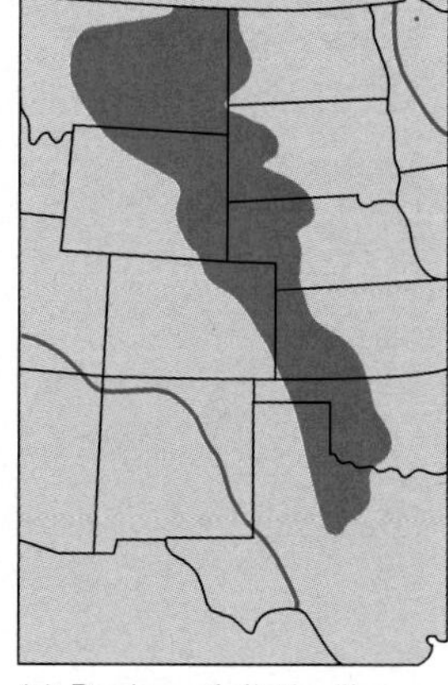

(c) Region of distinctive cultural elements

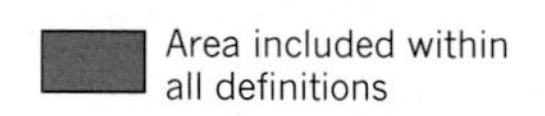

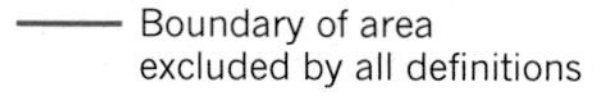

Figure 12.2 Alternative definitions of major geographic regions The Great Plains region in central North America is defined in terms of three of its regional characteristics; its distinctive vegetation, its landforms, and its culture. Each map is based on the definitions given by seven leading authorities. Note that each map shows a core area (shaded) on which all authorities agree and an outer limit (shown by blue line) beyond which none of the authorities considers part of the Great Plains.

[Source: G.M. Lewis, *Transactions of the Institute for British Geographers*, No. 38 (1966), Fig. 11.13, pp. 142–143.]

* * *

the extent of the Great Plains region in the central United States. Three groups of specialists were consulted to draw up vegetation, landform, and cultural regions.

Each specialist drew his or her own map. The shaded areas showed overlap (i.e. agreement between *all* experts) and the outer line the edge of the area that at least one expert considered as part of the Great Plains. In all three maps the disputed area is larger than the agreed area. If you experiment further with tracing paper, however, you will be encouraged to find that there is an area – true, a small one – that all of the geographers and others involved were happy to agree on. There is some area of part of the Dakotas, western Nebraska, and western Kansas that forms the heart of the Great Plains region. Thus relieved, you might like to plunge into this chapter, and the second half of the book.

In this chapter we look first at some of the schemes that have been put forward for regional systems and ask what they have in common (Section 12.1). We then review the uses of regions and consider the different reasons why geographers think they are important (Section 12.2). Thirdly, we look at a set of regions that are based not on hard scientific evidence or other data but exist in the human mind (Section 12.3). Finally we look at some practical issues in region building (Section 12.4).

Types of Region 12.1

We begin our study of regions by taking one part of the world – the conterminous United States (by 'conterminous' we mean the main land area leaving aside unconnected areas such as Alaska and Hawaii). Within this area of 7.7 million square kilometres (2.96 million square miles) we look at two ways of dividing the country. First, we take a division used by physical geographers. Second, we take a division used by human geographers. Then we look at both systems and ask what we can learn about the ways in which geographers approach regional divisions.

Environmental Regions of the United States

The conterminous United States may be divided into seven major environmental regions. Each region is based on the distinctive elements of its physical geography, based largely on the distinctive landforms that arise from its geological structure and erosional and sedimentary history.

From east to west the provinces as shown in Figure 12.3 are as follows. (1) The *Atlantic and Gulf Coast Plains* ranging from New York in the north to the Texas border with Mexico in the south. (2) The northeast–southwest trend of the *Appalachian System* with its low mountain and hill terrain. (3) The highly glaciated *Laurentian Shield* which dominates most of Canada but enters the conterminous United States only through a southern extension into Minnesota and Wisconsin. (4) The great sedimentary basin of the *Central Lowlands* drained by the Mississippi–Missouri–Ohio system and rising west as the Great Plains. (5) The dramatic belt of the *Cordilleran Province* with the north–south mountain lines of the Rockies and the Tetons. (6) The *Intermontane Basins* with the four separate units of the

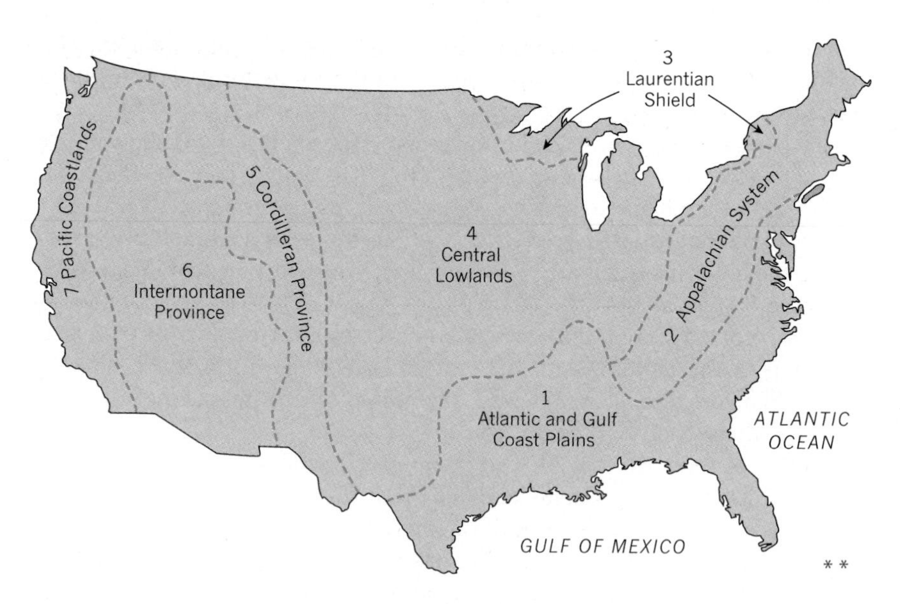

Figure 12.3 Major physical regions of the United States The seven major provinces are based on the physical features (mainly landforms) of the conterminous United States, excepting Alaska and Hawaii.

[Sources: Map modified from more detailed divisions given in the leading regional physiographies, e.g. W.D. Thornbury, *Regional Geomorphology of the United States* (Wiley, New York, 1965) and C.B. Hunt, *Natural Regions of the United States and Canada* (Freeman, San Francisco, 1967).]

Colorado, the Great Basin, the Desert Basin and Range, and the Columbia–Snake basins. Finally, (7) the most westerly province of the *Pacific Coastlands* with its varied landscapes of the Sierra Nevada, the central valley of California, and the coast ranges.

Within each major region other subdivisions can be recognized. Thus the Appalachian System falls into natural parts, a larger southern region and a smaller northern New England section. Figure 12.4 shows a regional division of the southern section with four basic divisions. (I) The gently sloping dissected plateau of the *Piedmont* forms the Atlantic edge to the region. (II) The narrow but high mountains of the *Blue Ridge* rise to around

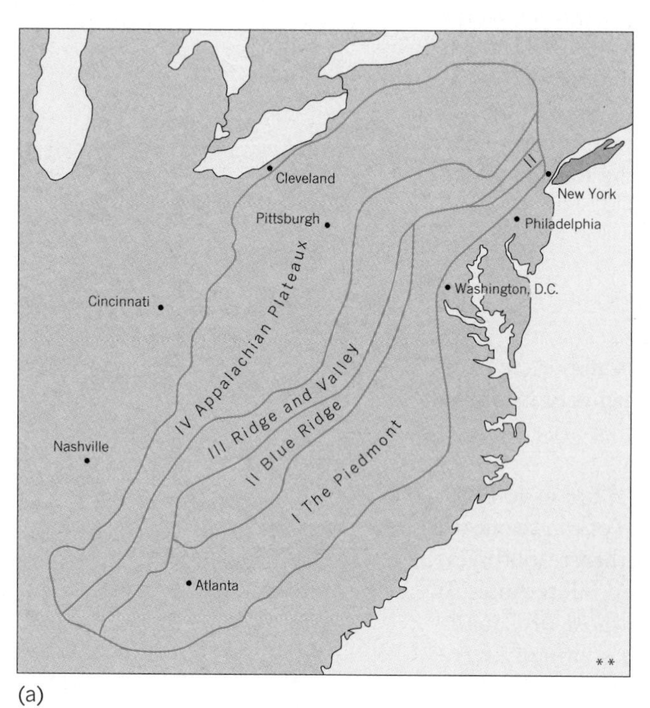

(a)

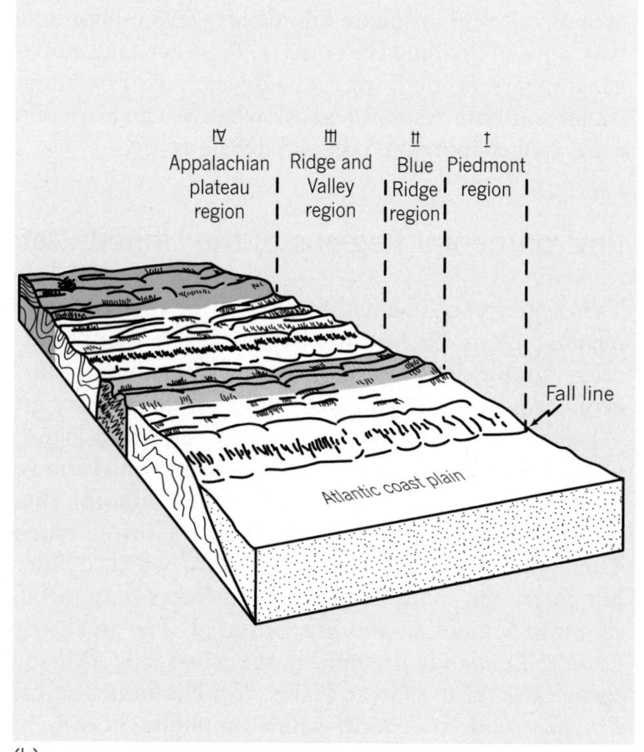

(b)

Figure 12.4 Minor physical regions of the United States Physical divisions of one of the major provinces in Figure 12.3. Here the Appalachian division is subdivided into four subregions based on physical distinctions (a) with a representative cross-section drawn across the line A–B (b).

1800 metres (6000 feet). (III) The distinctive and disturbed folded terrain of the *Ridge and Valley Section* runs parallel with the general trend of the region and with the classic river gaps where streams such as the Potomac cut across the ridges. (IV) Finally the inland province of the *Appalachian Plateau*, with its eastern edge marked by a sharp scarp (the Allegheny front) but its western edge dropping slowly beneath the sediments of the Central Lowlands. The boundaries of these four subdivisions are shown in (a) and a cross-section drawn across them in (b). All show a dominantly northeast to southwest trend from the New York state down to northern Alabama.

Cultural Regions of the United States

How do geographers make sense of the cultural patchwork that has emerged from these waves of immigration, settlement, and subsequent population growth? Many regional systems have been devised to describe the United States mosaic. Pennsylvania geographer Wilbur Zelinsky's system of *cultural regions* is shown in Figure 12.5. Zelinsky uses a fivefold classification. His first division, *New England* (Region I), was largely shaped by English migrations over the period from 1620 to 1830 with the development of settlement in northern New England lagging more than a century behind that of the southern nuclear area. The *Midland* (Region II), south of New England, was settled slightly later (between 1624 and 1850), by a wider variety of migrants. To the English element were added important Rhineland and Scottish–Irish populations in Pennsylvania. In the New York region, Dutch and southern-European migration was more important, as was in-migration from New England.

The most complex and diffuse of the three original hearths of culture on

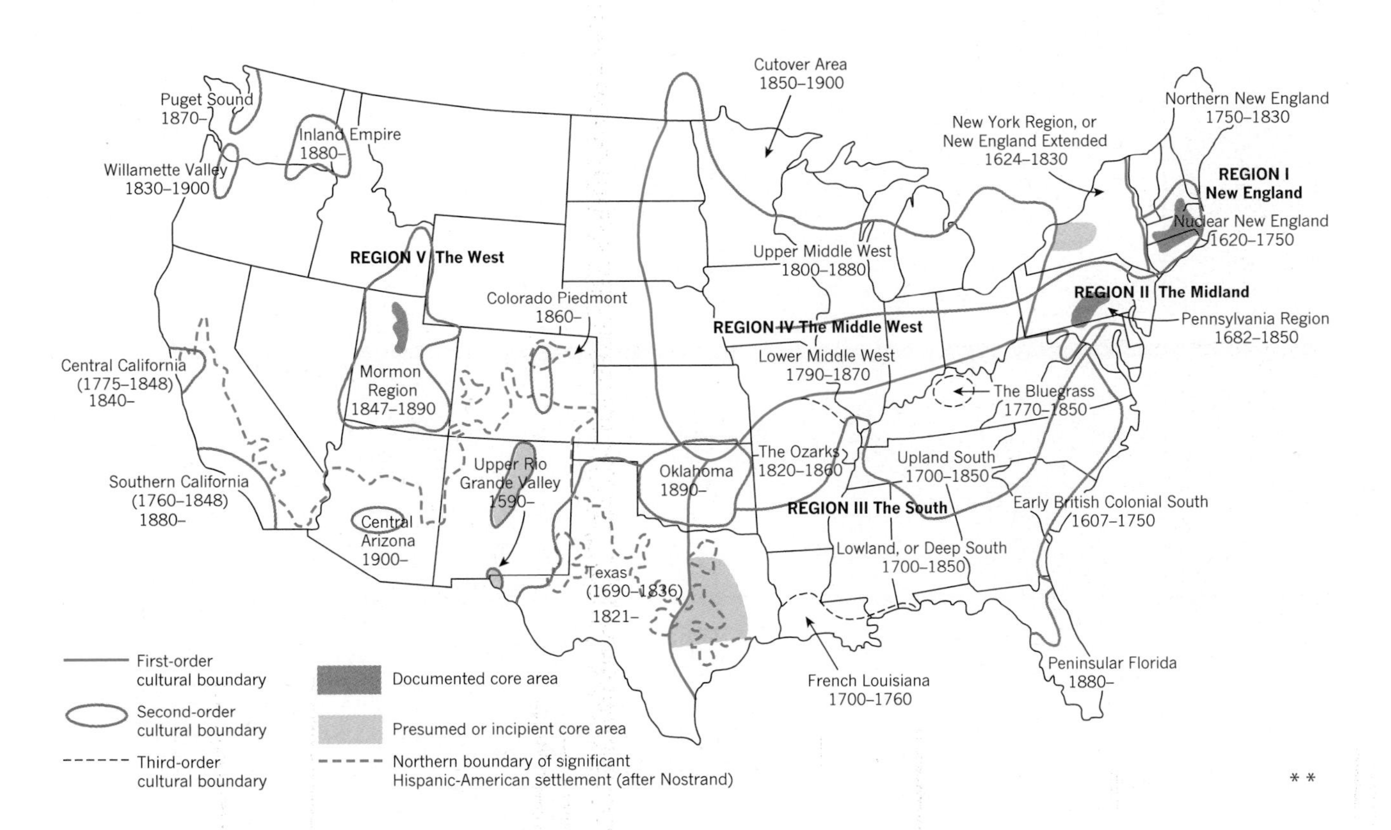

Figure 12.5 Cultural regions of the conterminous United States The map shows Wilbur Zelinsky's division of the United States into five major cultural regions and various subregions. The interregional boundaries vary in importance, and the status of three regions (Texas, Oklahoma, and peninsular Florida) is uncertain. The dates refer to the approximate limits of settlement and the emergence of a distinctive regional character.

[Source: W. Zelinsky, *The Cultural Geography of the United States* (Prentice Hall, Englewood Cliffs, N.J., 1973), Fig. 4.3, p. 118.]

* *

the eastern seaboard is the *South* (Region III). Beyond the narrow coastal strip of English plantations with their African slave population (settled before 1750), the South is divided into two major regions. Each of these has important subregions: Louisiana (in the Deep South) and the Ozarks and Bluegrass country (in the Upper South). The triangular region of the *Middle West* (Region IV) has more definite boundaries. Settled largely in the century after 1790, it was strongly affected by the westward extension of two existing cultural areas – the Midland and New England. Other cultural elements were superimposed on the existing pattern by new waves of European migration (particularly from Germany and Scandinavia).

An attempt to piece together the cultural dependence of the later regions on the earlier ones and to outside sources is presented in Figure 12.6. Only the main lines of influence are shown. Beyond the four main regions that constitute the eastern half of the country stands the enigma of the *West* (Region V). Here Zelinsky chooses to isolate nine subareas with some claim to distinctive cultural identities and to leave the remainder of the West as something of a cultural vacuum. The archipelago of subregions includes those shaped by particular ethnic groups (e.g. the Mexican element in the upper Rio Grande Valley), by religious beliefs (the Mormon region, also described in Figure 7.18), and by resource-exploitation patterns (e.g. the Colorado Piedmont). Outside the five main regions lie three intriguing areas whose status and affiliation are uncertain. Texas and Oklahoma are distinctive subregions that run across the boundaries of major divisions, while peninsular Florida lies beyond the South yet is not part of it. Some of the

Figure 12.6 The origin of cultural region The chart shows the main cultural links between the sixteen cultural regions in Zelinsky's model of the eastern United States (see Figure 12.5). The area's approximate geographic location is indicated on a north-to-south (left-to-right) horizontal scale. The date of the first effective settlement appears on the vertical scale. Texas 'I' refers to Texas during its Spanish-American period (1690–1821) and Texas 'II' to the area incorporated after 1821 into the United States. The arrows indicate major internal migrations and cultural contacts within the United States, and the letters indicate major external migrations from outside the United States.

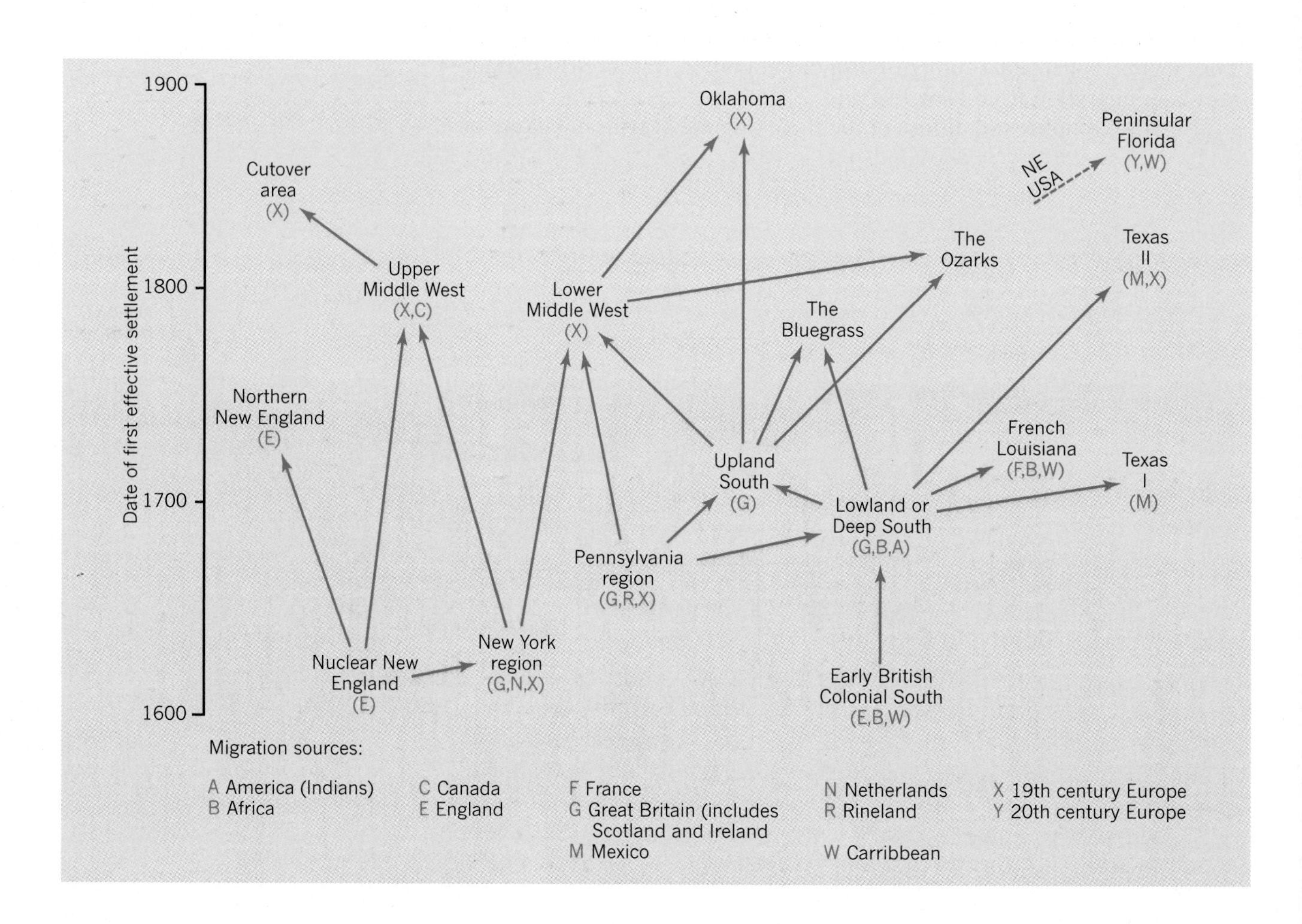

links between these three areas and possible sources of cultural influence are shown in Figure 12.6.

Principles of Regional Division

The word 'area' is generally used to mean a geometric portion of earth-space with no implications of homogeneity or cohesion. In contrast, the region in the sense used in this book is an area in which the character and areal relations produce some form of cohesion. It is defined by specific criteria and is homogeneous or cohesive only in relations to these criteria. Effective regional study is founded on the selection of meaningful criteria.

Single- and Multiple-feature Regions

It is useful to think of regions as arranged in three basic types:

1 Those defined in terms of single features.

2 Those defined in terms of multiple features.

3 Those defined in terms of the totality of human occupation of an area.

Single-feature regions delineate an individual phenomena. For example, we might delineate a region of very steep terrain in which the regional indicator was set to measure slopes over a particular threshold (say a slope of 1 : 4 or steeper).

Multiple-feature regions are defined in terms of a combination of features. For example, we might delineate vulnerable areas within a city in terms of a high rate of unemployment, a high proportion of poor housing, a high rate of street crime.

Total content differentiation assumes a region made up of an interrelated bundle of natural and human features. Whether such regions exist has historically been a matter of critical debate within the field. One group of geographers have argued that there are 'natural regions', the object of geographic analysis being to recognize and delimit the features that are latent on the earth's surface. Such an approach has been likened to searching for a black cat in a darkened room, but knowing the cat was there. Other geographers have argued that no such natural aggregations exist; regional structures are superimposed by the particular geographer who conducts the study. One word proposed by the Harvard geographer Derwent Whittlesey to describe such total regions is ***compage***, drawn from the Latin words for 'fastened' or 'fixed' (-page) and the word for 'together' (com-).

Uniform and Nodal Regions.

No matter what criteria are adopted, geographic regions may also be grouped into two types according to whether they are uniform or nodal. A ***uniform region*** is homogeneous because all parts of the area contain the feature or features by which it is defined. Of course these features may vary internally within the boundaries that are set. Thus our region of steep slopes will include a variety of slopes above our threshold slope of 1 : 4.

A ***nodal region*** contains a focus (sometimes more than one) that serves as the centre of its organization. Such a focus is a centre for communication and is commonly urban in character. Thus geographer John Borchert has delineated a 'Twin Cities nodal region' focused on the combined cities of Minneapolis–St. Paul and extending beyond the home state of Minnesota over a number of northern Great Plains states. The Twin Cities core is

Figure 12.7 Types of region

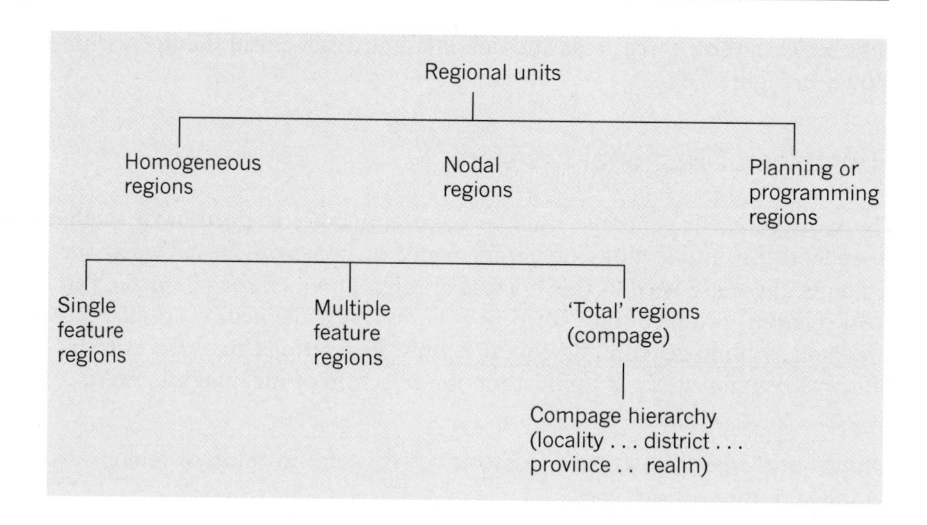

linked to the remainder of the region by ties of different character and intensity (commuting, banking flows, newspaper circulation, and so on).

One critical difference between the uniform and nodal regions lies in their boundaries. Uniform regions tend to be sharp-edged with rather well-defined limits in terms of the categories that generate them. By contrast, nodal regions fade out gradually; this decline is often a function of distance. Nodal regions commonly overlap at their edges with other nodal regions. Figure 12.7 illustrates the main types of regions.

12.2 The Uses of Regions

So far we have used 'region' in a loose way to indicate a focus on one part of the earth's surface. We now need to tighten that up. The central role of the regions has been so widely accepted within the geographical discipline that asking a geographer why we study regions is like asking Christians why they study the Gospels or Muslims why they study the Koran. Geographer Richard Hartshorne has argued for its central importance in geography (see Box 12.A, 'Hartshorne and areal differentiation'). A halting response confirms that often the most fundamental beliefs are intuitive, the hardest to define. But if we stand back and examine the reasons why we do what we do, then several rationales for regional study, rather than a single one, emerge. I identify here four reasons – regions as covering sets, as samples, as analogues, and as modulators – but this is not an exhaustive list. We now take each in turn.

Regions as Covering Sets

A first use of regions is as 'covering sets': this is the term used to describe the jigsaw of areas that completely exhaust a continent or country. The task of the geographer, to study the earth surface as the home of humankind, is confounded by the sheer size of the planet's surface. So we need to be able to reduce the complexity of understanding the whole earth by breaking it

Box 12.A Hartshorne and Areal Differentiation

Richard Hartshorne (born Kittaning, Pennsylvania, USA, 1899; died Madison, Wisconsin, USA, 1992) was the leading exponent of an historical view of the nature of regional geography. A Princeton graduate, Hartshorne spent most of his academic career in the mid-west, first at the University of Minnesota (1924–40) and subsequently at Wisconsin (from 1940). His *Nature of Geography: A critical survey of current thought in the light of the past* (first published in 1939 and revised in 1946) remains today the most complete English-language statement of the basic premises on which geographers work. It has a strong historical emphasis and offers a particularly complete survey of German methodological writing up to that time. In *Perspectives on the Nature of Geography* (1959) Hartshorne presented a briefer restatement and updating of his views. The photograph was taken on his last visit to England in July 1978.

Hartshorne regarded the central purpose of geography to be the study of the areas of the Earth's surface according to their causally related differences, in other words, the ***areal differentiation*** of the earth. He saw geography as closest in philosophy to history. Historians organized their findings with respect to time in *historical periods.* In like manner, geographers organized their findings with respect to space as *geographical regions.*

In a famous diagram (the only one in the *Nature of Geography*)

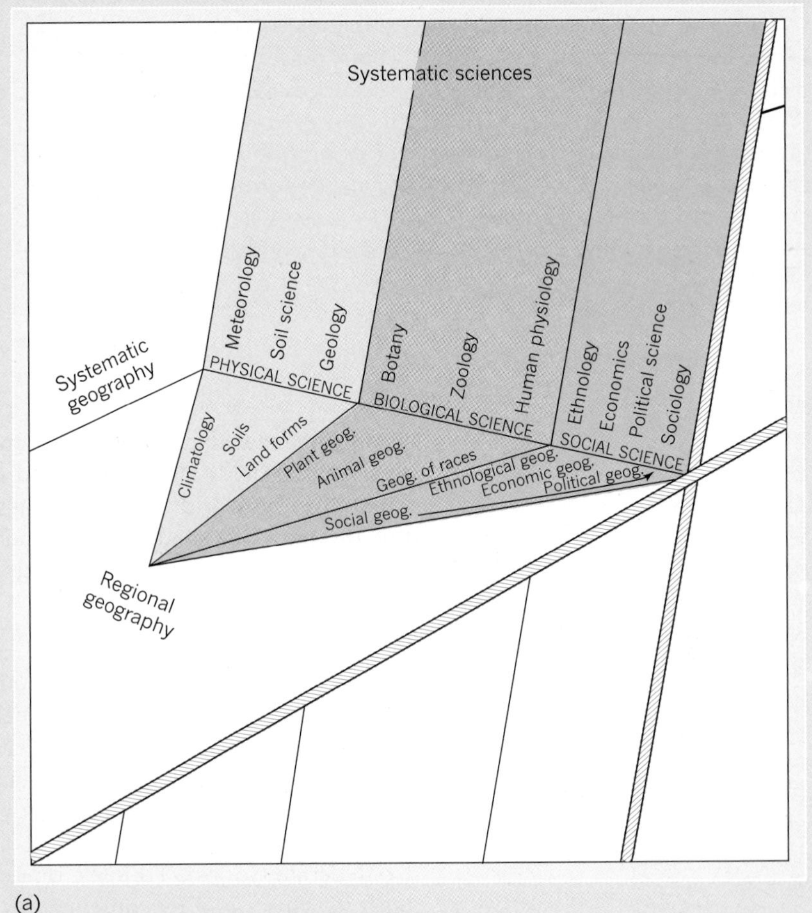

(a)

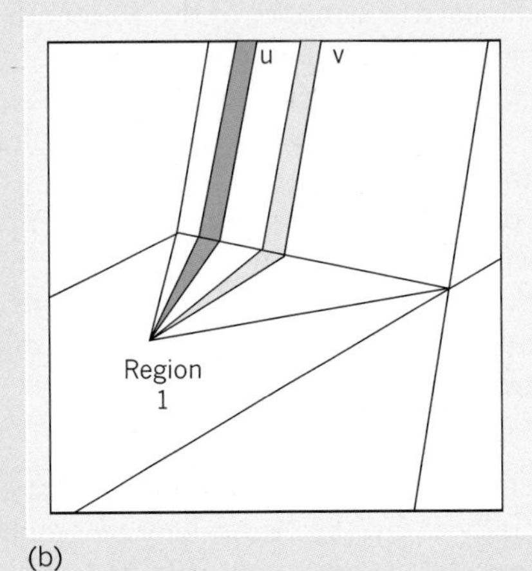

(b)

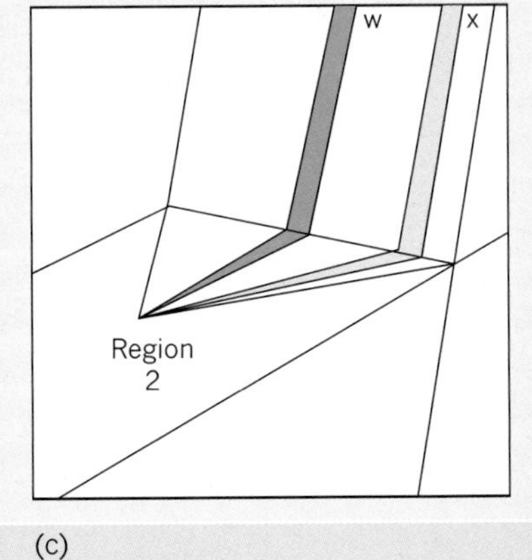

(c)

Box continued

reproduced in (a) he showed the relationship between geography and the systematic sciences as two intersecting planes in solid geometry. As he states: 'Geography does not border on the systematic sciences, overlapping them in common parts on a common plane, but is on a transverse plane cutting through them'. For every systematic science there is a corresponding systematic geography; thus botany has its plant geography, economics its economic geography, and so on. The integration of all the branches of systematic geography, focused on a particular part of the earth surface, is regional geography.

But it is important to note that Hartshorne did not imply that each region had to be studied in the same way. For him, each region does not have to be studied as a litany starting with the physical sciences and going on to the behavioural sciences; ranging from meteorology to sociology as in the diagram. Rather, he saw each region as unique blend of systematic factors. For any particular region at any particular time, its pattern was woven as a specific collection of threads drawn from the systematic threads. This is shown in outline in (b) with a different combination of threads (**u** and **v**) in the first regions as compared to **w** and **x** in the second region (c).

For a review of Hartshorne's work see J. Nicholas Entrikin and Stanley D. Brunn (eds), *Reflections on Richard Hartshorne's 'The Nature of Geography'* (Association of American Geographers, Washington, D.C. 1989).

[Source: Photograph courtesy of Bristol University Department of Geography.]

up into manageable pieces. Geographers need an overall system to divide their spatial fields into smaller and more readily comprehensible sections. These may be world regions at one scale, major parts of the continents to tiny divisions of a small valley at another. We have already seen an example of a world system in the nine biomes (tropical rainforest, savannah, etc.) in Chapter 4.

Comparison with Historical Periods The problem here is directly comparable with problems in other fields: historians use discrete historical periods (the medieval period, the Renaissance, etc.) to give structure to their studies. Again, boundaries are difficult to draw. Even a period as distinct as the American Civil War is not neatly confined between 12 April 1861 (the attack on Fort Sumter) and 26 May 1865 (the final surrender of Confederate troops). To make sense of that war involves consideration of both its causes and its consequences, taking the historian decades outside the narrow confines of the 50 month conflict itself. But despite their drawbacks, both geographical regions (in the spatial domain) and historical periods (in the time domain) are essential to understanding.

In each case, we need a set of rules to establish the protocol for placing an area into a given region, just as a librarian classifies each incoming volume. The rules need to be comprehensive enough to ensure that the whole global region is covered, i.e. that there are no books that we cannot classify. As we would expect, there will be considerable debate over the rules and much anguish over those difficult cases that lie on a boundary and are hard to classify.

Regions of India One of the most complete examples of a regional geography is by the Australian geographer Oskar Spate in his *India*. Spate's regional classification of India (see Figure 12.8) has four levels of increasing detail. There are three 'macro regions' – the mountain rim, the Indo-Gangetic plains and the peninsula. These are divided into in turn into

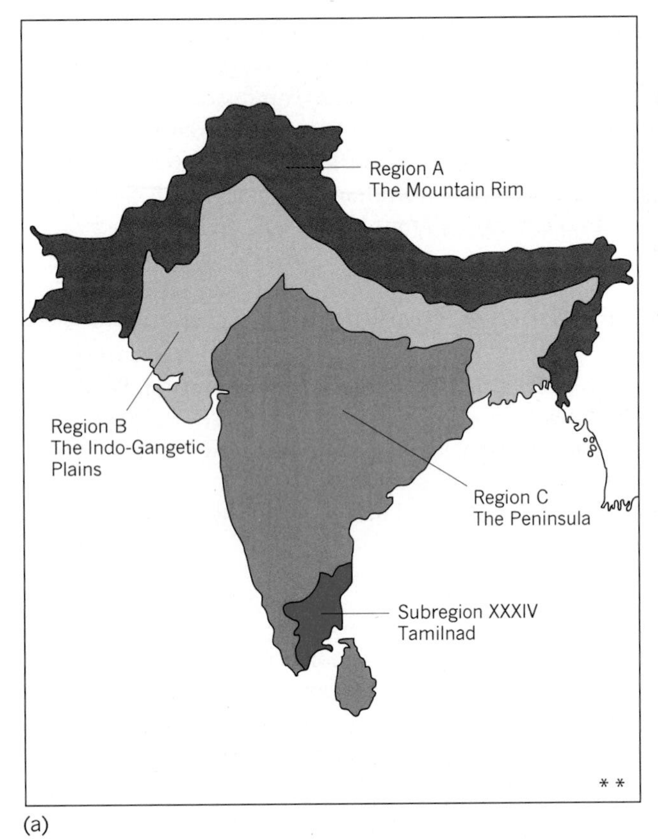

(a)

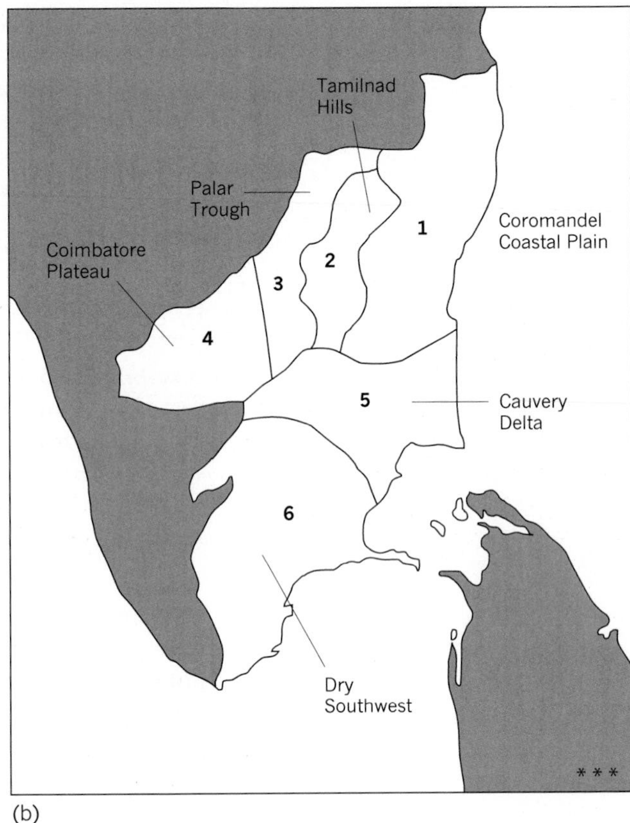

(b)

smaller divisions (34 regions of the first order, 74 of the second order, and about 225 subdivisions of these).

The detailed breakdown shows that one regional master principle is not attempted. Structure is the most obvious guide on the macro-scale. But in an area such as the Indo-Gangetic Plains landforms are of little help, since the divisions are as a rule either too broad or too much a matter of local detail. Here transitional differences in climate or soil are used as dividers in this open plain areas.

On the human side, Spate considers that the best approach in theory might be to relate the regional network to the spheres of influence of towns. But data were lacking and he was dubious of transferring western methods to the Indian scene. Instead, historical identities (e.g. the Tamilnad in south India) are used. These traditional regions are not wholly coincident with physical boundaries, neither are they wholly divorced from them. Sometimes a regional boundary will match a longstanding administrative boundary; more often they cut across them. Some boundaries are transitional zones; some are sharply cut. Some regions are rich juxtapositions of nature and human activity overlapping on the map; others are simply what is left over after more definite regions have been sieved out.

For Spate, the essence is understanding rather than some ideal scheme. We can see this pragmatism by looking at his treatment of one of south India's most distinctive regions, the Tamilnad (Region XXXIV). The Tamil region, roughly equivalent to the state of Madras, is characterized by two dominant themes: the homogeneity of its Tamil culture and the environmental problems set by its rainfall regime (with most falling in the three months of October to December). But within this uniformity, physical environment is very diverse and the fine tracery of its second-order regions

Figure 12.8 Spate's division of India The Australian geographer Oskar Spate wrote a classic regional geography of India in 1954. He divided the subcontinent into three major regions (a) within which a further 37 regions were nested. One of these, Region XXXIV Tamilnad in southeast India, is shown in (b) together with the six subregions into which it is divided.

[Source: O.H.K. Spate and A.T.A. Learmonth, *India and Pakistan*, 3rd edn (Methuen, London, 1967), Fig. 13.1, p. 408.]

and 25 minor subdivisions is mainly based on geomorphology, soil, and climate. But Spate's order is not an imposed or unyielding one; climate enters again to divide the dry southeast from the rest of Tamilnad, and the urban importance of the capital city, Madras, means it must be taken apart from the rest. In its sensitivity to anomaly, transition, and outliers, Spate's scheme is a good example of geographic craftsmanship in which each piece of wood in the inlaid mosaic is carefully fitted and delicately finished.

Regions as Samples from a Population

A second use of regions is in relation to geographic generalization. From the first chapter in this book we have seen how individual cases (whether volcanoes or cities or climatic stations) are used as the raw material from which wider, sometimes worldwide, statements can be made. Thus in Chapter 8 statements about rates of world urban growth were based on the figures for individual cases.

Regions as Exemplars

Small regions are used in a similar way to allow statements to be made about much larger regions or the world itself. In practice regions can be used in two different ways. First, regions serve as *exemplars*. They give local substance to generalization, put flesh on the logical structure, provide a specific example to press home an argument. The broad relationship is shown in Figure 12.9.

Thus Chandler's book *The Climate of London* provides an illustration of the modifying effects of a large urban area on the atmospheric envelope covering it, translating the general physics of boundary-layer climates into the local experience of smogs and gusty streets. The exemplar may be used not only to illustrate principles, but also to illuminate a wider area.

A classic example of using micro-regions to illustrate the characteristics of a much larger geographical area is provided in Richard Platt's *Field Approach to Regions*. In this he illustrates how it is possible to understand the geography of the huge continental area of Latin America by taking well-chosen examples. Thus he uses a single coffee farm (a *fazenda*) in the state of São Paulo to throw light on the historical evolution and current problems of the whole Brazilian coffee belt.

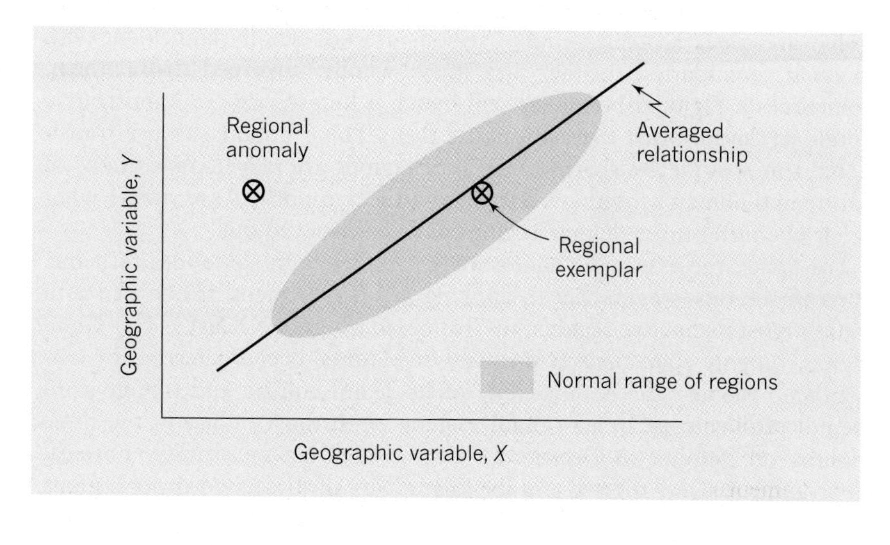

Figure 12.9 Exemplars and anomalies Regions may be used both to confirm general relations (as a classic example) or to illustrate departures from normality. These are termed in the text as regional examples and regional anomalies.

Regions as Anomalies Second, and in contrast to the first argument, regions serve as *anomalies* or residuals. In this case the purpose is to underscore how a local part of the earth's surface departs from a general statement or relationship. In terms of terrain and soils, the distinctiveness of the 'driftless' region of southwest Wisconsin lies primarily in its anomalous character in relation to the glaciated areas of the mid-west that surround it on all sides. Its most striking feature is a negative one, the *absence* of glacial depots, and this makes sense only in respect of broader generalizations about an area in which glacial deposits dominate half a continent.

Anomalies and residuals play an important role in testing and reformulating general models. The failure of a general explanation to make sense in a specific region may be due to limitations in the model itself, or the fact that several influences are coming together in such a way that an expected effect is either heightened or reduced. Wisconsin geographer Glen Trewartha's *Earth's Problem Climates* is a classic example of this approach. He picks out for special consideration those areas that have climatic characteristics not expected on the basis of general models of the earth's atmosphere. Thus we might expect, on most general atmospheric circulation models (see Chapter 2 of this book), that the Amazon Basin would have high rainfall. Why then, asks Trewartha, are parts of it dry? The answer shows how complex are the movements of the intertropical front and leads to an improvement in the general model. This process of generalization followed by exceptions lies at the heart of the learning process and was described by philosopher John Stuart Mill as the 'method of residues'.

Regions as Analogues

Third, regions may be studied as *analogues*. This term describes things that are similar in characteristics to another, so an analogue region is one that has similarities to another. Figure 12.10 shows examples of such analogues. A textbook on the regional geography of the Americas uses five geographical regions in North America to throw light on five comparable regions in South America. In effect the familiar is used to throw light on the unfamiliar.

Climatic Analogues One of the examples in Figure 12.10(b) makes a comparison between terrain types in two areas of Mediterranean climate. This describes the matching of the characteristics of one region to matching regions in one or more other parts of the world. We saw a biogeographic example of this in the biomes of Chapter 4 and a simple research example is provided by a primitive study I once made of the cork oak (*Quercus suber*). This is a type of oak tree native to the western Mediterranean basin. By identifying the trees and the climatic environment in its home range, it was possible to pinpoint, on climatic grounds, areas of western North America where it might also flourish. This illustrates the distinction between the 'benchmark' region from which the characteristics are measured, and the 'analogue' region constructed from the spatial transfer of the characteristics.

Many illustrations of the method are given in the work of Nuttonson's group at the Institute for Crop Ecology in Washington, D.C. Typical is their report on the climate of the Ukraine region of Soviet Russia specified in terms of the critical parameters for growth of grain crops. Climatic stations in the United States are then searched to identify those most closely match-

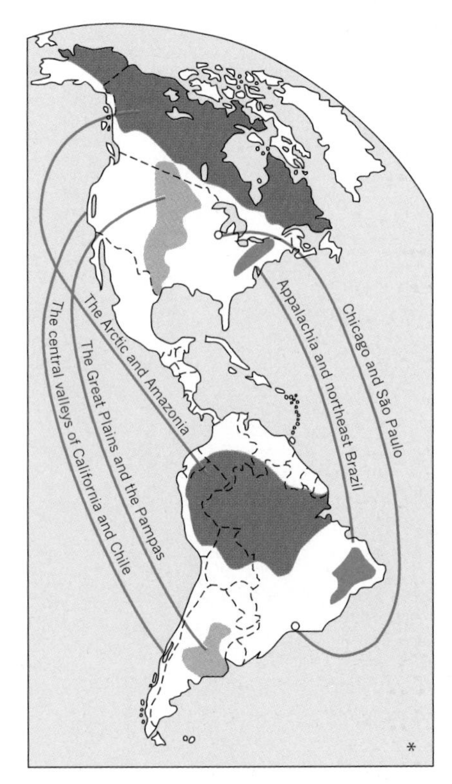

Figure 12.10 Regional comparisons as mirrors Analogues of five geographic regions in the United States and Canada ('familiar' regions) are used to throw light on five comparable regions ('unfamiliar' regions) in Latin America.

[Source: A.H. Siemens, *The Americas: A Comparative Introduction to Geography* (Duxbury Press, North Scituate, Mass., 1977), Fig. 1.11, p. 15.]

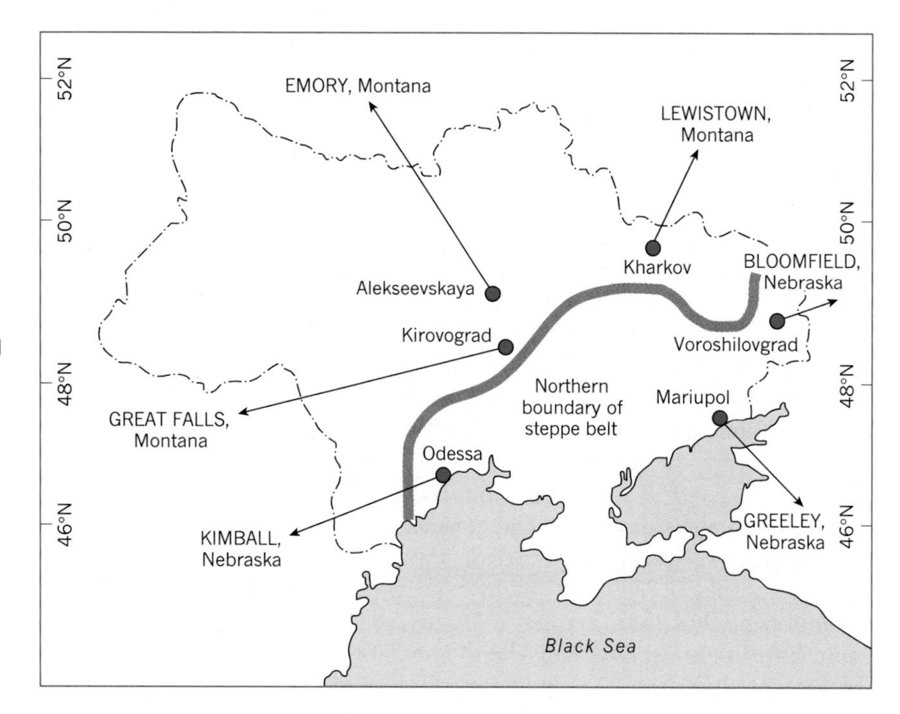

Figure 12.11 Regional analogues for crop exchange Location of climatic stations in the Ukraine. For the six sample stations, the closest climatic analogue in the United States is shown. The technique was used in American experiments with new Ukrainian wheat varieties during World War II to find 'matching' climatic areas within the Great Plains wheat belt.

[Source: Map based on tables in Michael Yakovlevich Nuttonson, *Ecological Crop Geography of the Ukraine and Ukrainian Agro-climatic Analogues in North America* (American Institute of Crop Ecology, Washington, DC, 1947).]

ing the Ukraine. Figure 12.11 shows the resulting regional comparisons between the two areas.

Film Locations A similar search for regional analogues has been followed by film producers. Often the desired location is simply not accessible by reason of cost or political accessibility. Thus Sam Spiegel's *Bridge on the River Kwai* was ostensibly set on the Thai–Burma border, but actually shot on location in the dry zone of southeast Sri Lanka. Carlo Ponti's film *Doctor Zhivago* was not filmed on the Russian steppes but in various 'matching' parts of the world – in Spain, Ontario, and Finland. One extreme case of a low-budget remake of *Beau Geste* substituted the cold dune ridges of Scotland's Culbin Bay for the shimmering sands of the Sahara.

Perhaps I should add a further litmus test for recognizing a latent geographer – the member of the cinema audience who stays on in the emptying theatre to see if the very last line of the credits roll will answer 'Where in the world was it *really* filmed?' I recommend the pastime: you will find some surprises.

Regions as Modulators

Fourth, regions act as modulators: their unique structures modify the ways in which a region behaves over time. Studies of the geographic pattern of business activity show that local regions have activity cycles – in employment, investment, housing starts, inflation, and the like. In some degree the local economy follows the trends in the larger national economy within which it is set. But they do so in a regionally modified way. Those that have a large concentration of industries that are cyclically vulnerable, such as steel and automobiles, may show a marked over-reaction to national

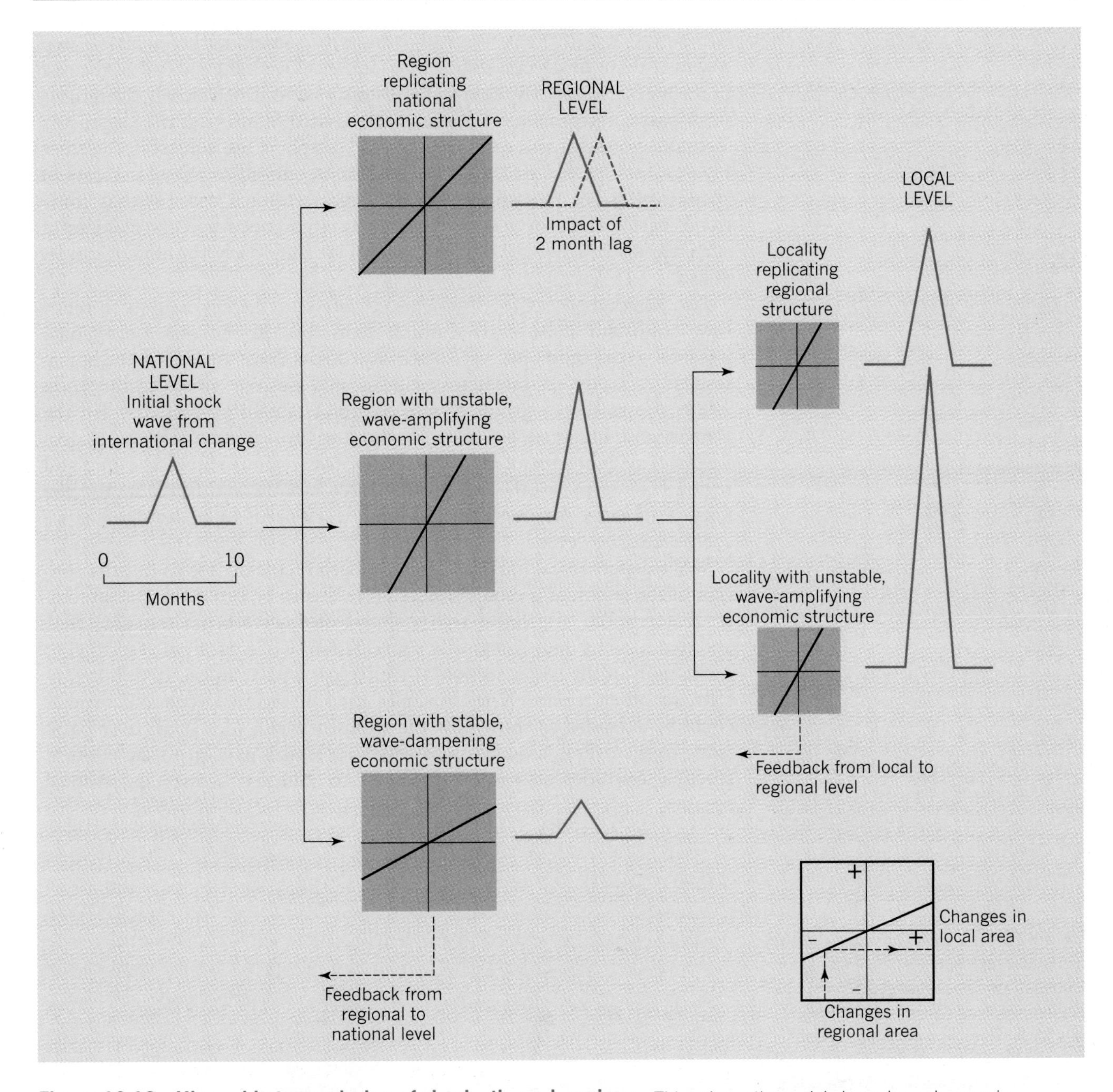

Figure 12.12 Hierarchic transmission of shocks through regions This schematic model shows how changes in economic activity on one level may be transmitted to others. The transformation boxes (shaded) describe the degree of amplification or dampening caused by the economic structure at the regional or local level. The diagram is, of course, greatly simplified. In practice, the waves are highly irregular and may be lagged in time, the shocks may start at levels other than the national one, and the effects of several shocks may intermingle or overlap. The box at the lower right shows how a major fall in economic activity in a regional area is reflected in a dampened form at the local level.

trends. Other areas that have rather stable employment, such as government administration or higher education, may show a muted response to the same tends. As Figure 12.12 shows, we can conceive the local economy of a given region as modulating any pattern, or cycles transmitted from the national or international economy.

Bell-wethers and 'Early Warning' Regions Stock-market analysts keep a watchful eye on certain shares that tend to move a little ahead of the rest of the market. Similarly, electoral pollsters tend to watch closely the returns from particular primary elections in the United States (e.g. the slogan 'As Vermont votes, so votes the nation'). Geographers use similar ***bell-wether regions***, looking for areas that are consistently ahead of others in terms of spatial diffusion. So bell-wethers are areas within a country that show trends earlier than the rest of the country. It is hoped that the changes in such regions will, like the barometer, provide some early warning of storms ahead.

For geographers, as for stock-market analysts, the key issue is to judge how consistent these early-warning signs are. In the case of regional business cycles attempts to isolate lead areas have produced intriguing results. A study of unemployment in mid-western cities in the early 1960s shows that a group of cities around Pittsburgh regularly led the Detroit and Indianapolis areas by three to five months (Figure 12.13). Regional unemployment data for ten British regions shows the midland region leading most other regions by three months and Scotland and the north leading by six months. A lot more work needs to be done before we can be sure, however.

While geographical applications have been relatively few so far, the concept of the region as a modulator could in theory be extended. The prerequisite conditions are that a region shows distinctive behaviour over time and that its behaviour is logically connected in two ways: first, vertically, to oscillations of the larger section of which it is a part; and second, horizontally, to other regions. The principles used in regional economic models could be extended to show how the sediment yields of a small river basin relate to the overall river basin of which it forms a part, or to the political reaction of different electoral regions to national swings in political opinion.

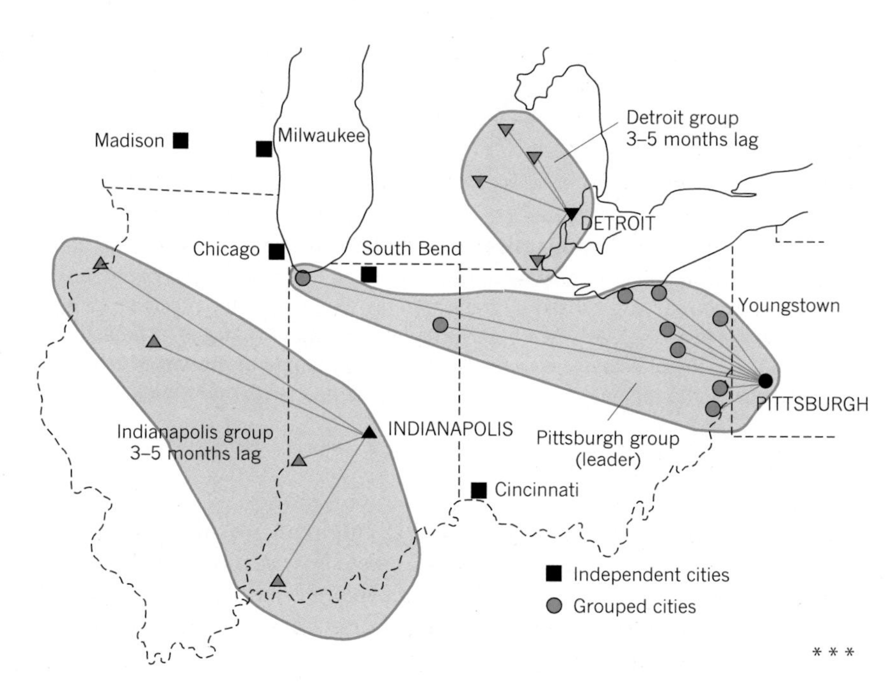

Figure 12.13 Lead and lag regions Leads and lags between 25 cities of the American mid-west based on the study of unemployment time series for each city. Each series was cross-correlated with the others to determine their highest lag correlations. Note the way in which the steel cities of the Pittsburgh group show fluctuations in their employment trends three to five months between the car-making cities of the Detroit group of cities and farm-machinery cities centred on Indianapolis.

[Source: Adapted from data in L.J. King *et al.*, *Regional Studies*, **3** (1969), Table 2.]

* * *

Regional Consciousness 12.3

A third approach to regions is through the human mind. We may recognize regional identity not through textbook map or computer analysis but by the sharp image it leaves in our consciousness. We look here at three examples of this: the image of the city, the literary region and the longed-for homeland.

Regional Images of the Western City

Electronics have shrunk world space to the size of a TV screen. Satellites and television allow each of us city-bound dwellers an unprecedented opportunity to peer into our neighbour's backyard. On an average day's TV programme we find items that range from hotel fires in São Paulo to floods in the Australian outback, to a nature film shot in the Canadian tundra. Never before have our senses been stretched over the whole globe in this way.

But what about our immediate neighbours? How do we view the urban world that is right around us? The evidence at hand from geographers and psychologists suggests that we retain primitive, twisted, and biased pictures of our local 'world' that may be far removed from the 'real' world described in the maps in the preceding section. So we look at some of the early results in an important and rapidly growing area of geographic research on images of the western city.

Mental Maps of Los Angeles Consider the way we look at the city in which we live. How much of it do we know well? How much do we not know at all? Is the geographic view of the city from the central area the same as that from the suburbs? If not, then how does it differ? Figure 12.14 presents three maps of Los Angeles that show the perception of the city by people residing in three district areas within it.

A sample of respondents in each area was asked to sketch maps of the city based on their own travels within the city and the contacts they have made. Their combined views are summarized in the maps. Those living in Boyle Heights, a poor inner section of the city near downtown Los Angeles, have a quite limited view of the city dominated by the nearby city hall, railroad station, and bus terminal (Figure 12.14(a)). Suburban residents of Northridge in the San Fernando Valley had a limited but spatially more extensive view than Boyle Heights residents (Figure 12.14(b)). They had an extensive perception of their own valley and its facilities, but little real familiarity with the main city, beyond the Santa Monica Mountains. The respondents around the University of California campus at Westwood, an upper-class sector just west of Beverly Hills, had a wide-ranging and detailed image of almost the whole Los Angeles metropolitan area (Figure 12.14(c)).

This kind of research indicates something about the warped *mental maps* we have of cities. Groups from higher-income areas have a wider view of the city, and more educated groups have a more accurate view of the city. On the other hand, this research fails to explain why certain regions of the city environment are commonly known, or why some cities produce more positive and memorable mental images than others. One day in Cincinnati may leave a person with a clearer mental map of the city than one day in Kansas City for example. Why?

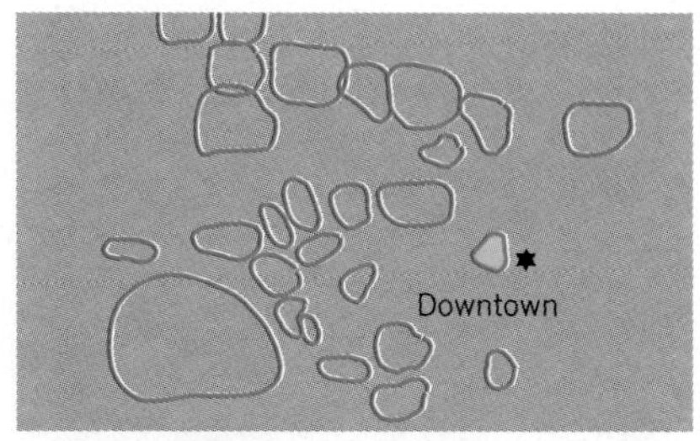

(a) Boyle Heights

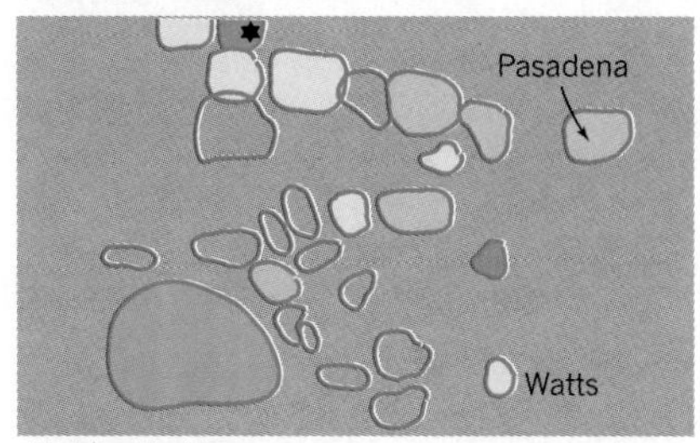

(b) Northridge

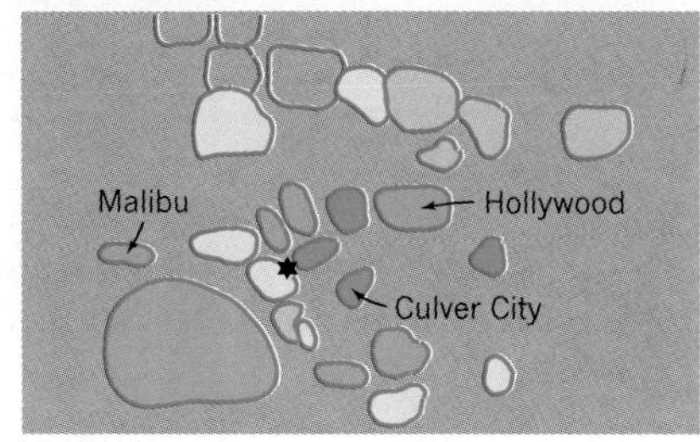

(c) Westwood

Familiarity (per cent)

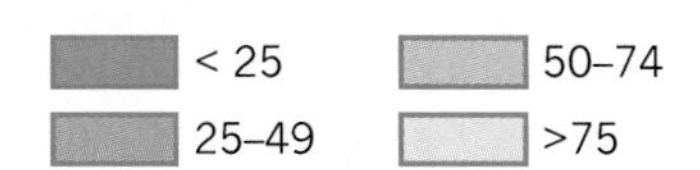

Figure 12.14 Images of a city The maps show Los Angeles as seen by a cross-section of its residents (marked by a star). Boyle Heights (a) is largely black neighbourhood near the downtown; Northridge (b) is a suburban residential community in the San Fernando Valley; and Westwood (c) is a high-class housing area near the UCLA campus. Each cell indicates a different section of the city; tints indicate the proportion of those interviewed who were familiar with that section of the city. Unshaded cells were unfamiliar and lie outside the residents' area of perception.

[Source: Data by R. Dannenbrink, Los Angeles City Planning Commission. From National Academy of Sciences *Publication No. 1498* (1967), Figs 2–4, pp. 107–112.]

Figure 12.15 Images of Boston An aerial view of the city of Boston looking across the Charles River. But how do the residents of the city see it? What landmarks, markers, and districts do they recognize and use to orient themselves? Figure 12.16 suggets some of the answers by research from Kevin Lynch.

[Source: Photograph from Rotkin, PFI.]

Mental Maps of Boston Some clues to the answer to this question have been provided in *The Image of the City* which describes work carried on by Kevin Lynch at the Massachusetts Institute of Technology. Residents of three contrasting North American cities (Boston (see Figure 12.15), Los Angeles, and Jersey City) were interviewed and asked to sketch a map of their city, to provide descriptions of several trips through the city, and to list and comment upon the regions of the city they felt were the most distinctive. With these documents Lynch pieced together the public image of each city held by its inhabitants.

There were considerable, and interesting, variations in the responses of individuals linked to their age, sex, length of residence, area of residence, and so on. But enough common ground was found to allow some citywide generalizations. Lynch organized the common elements (***Lynch elements***) of the mental maps into five types of spatial phenomena. The results for Boston are shown in Figure 12.16.

The five types of elements can be defined as follows. *Paths* (Figure 12.16(a)) are the channels along which we customarily, occasionally, or potentially move within the city. They range from streets to canals and are the reference lines we use to arrange other elements. *Edges* (Figure 12.16(b)) are linear breaks in the continuity of the city. They may be shorelines, railroad tracks, or barriers to movement. The Charles River in Boston and the lakefront in Chicago have all the abrupt barrier qualities of an edge. *Nodes* (Figure 12.16(c)) are focal points within the city. They are commonly road junctions or meeting places. Louisburg Square in Boston or Times Square in Manhattan are typical nodes. *Districts* (Figure 12.16(d))

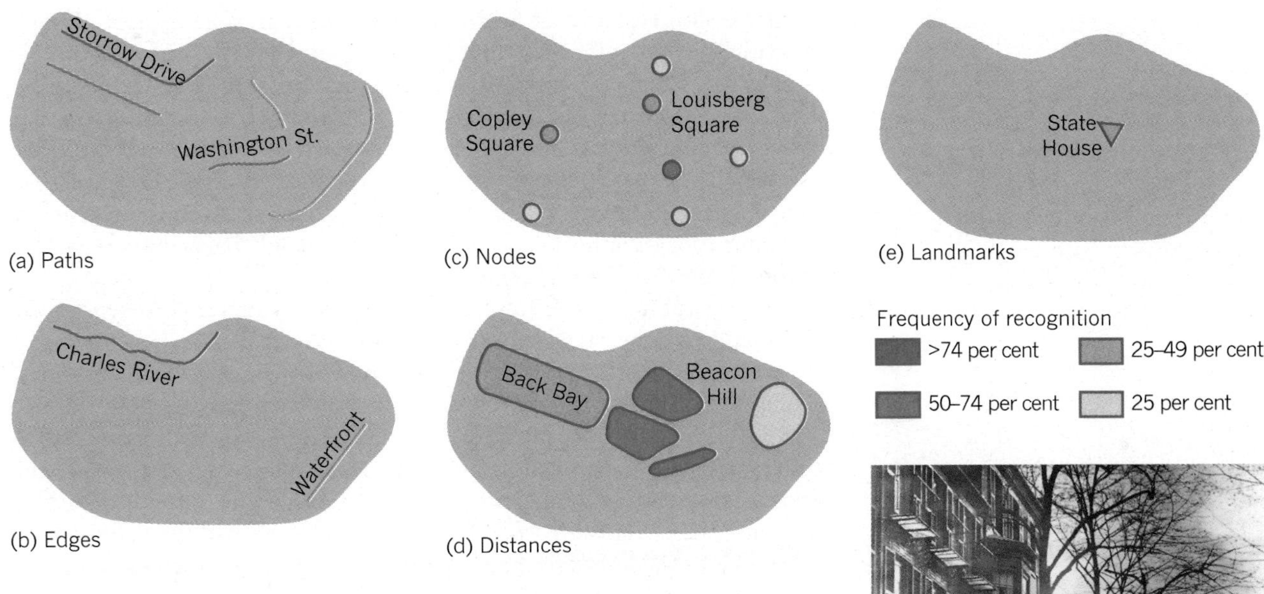

(f)

Figure 12.16 Elements in the mental map of Boston (a)–(e) Five main elements in the image of Boston as seen by the residents. (f) Chestnut Street, Beacon Hill, Boston, before World War I. Typical aristocratic town houses. Boston.

[Sources: (a) From K. Lynch, *The Image of the City* (MIT Press, Cambridge, MA, 1960), Fig. 37, p. 147. Copyright © 1960 by the Massachusetts Institute of Technology. (f) Corbis-Bettmann.]

are medium-to-large sections of the city that we can mentally enter 'inside of' and that have some common identifying character. Beacon Hill or South End in Boston are typical districts. Finally, *landmarks* (Figure 12.16(e)) are also reference points but much smaller in size than nodes. A landmark is usually a simple physical object: a building, a store, a mountain. It may be memorable for its beauty or its ugliness. The gold dome of Boston's state house or Nelson's Column in London illustrates the role of landmarks in giving structure to our image of a city.

The relative strength and richness of these five spatial elements give coherence and character to a city. Cities with strong elements may be interesting environments to live in, despite dilapidation or deterioration. San Francisco, Cincinnati, New Orleans, and Montreal are North American cities that come into this first category. Cities with weak elements may be formless, monotonous, and lacking in character. You may like to provide your own candidates for this type of city.

Lynch's approach has been repeated in many cities around the world on a variety of scales. Dutch geographers have emphasized that although regular street structures help us to find our way easily, difficulties in orienting ourselves arise when the overall structure of a city is clear but the individual elements are too uniform to be individually distinguished. It appears that we construct mental maps more readily in an older European city (a baffling jumble of streets, but with distinctive elements to serve as locational clues to where we are) than in a newer North American city (a regular street pattern, but too many roads looking like each other).

The Persistence of Urban Images But why are these images important? Our answer, as we saw in an earlier discussion of environmental images in Chapter 11, is that images underline action. We make decisions on the basis of what we believe to be true. This is clearly seen in the persistence of elite districts within the city. Now you do not have to live long in any western city to pick out the prestige areas to live. To reside in the Bel Air area of Los Angeles, the South Yarra area of Melbourne, or the Mayfair area of London is to have an address that has been associated with wealth and elegance for several decades.

The Beacon Hill area of Boston is typical of these elite districts. It lies on the western edge of the Shawmut Peninsula on which the city of Boston was established. Originally an area of merchants' housing close to the harbour wharves and centres of commerce and government, it became fashionable in the late eighteenth century. As the city grew over the following 200 years it retained its character and attraction, resistant to downtown encroachment and inner-city decay. For the well-to-do it remains an attractive and convenient location for weekday living, increasingly backed up by weekend 'second homes' in rural New England.

Urban geographer Ron Johnston has tracked a similar persistence for the fashionable southeastern suburbs within the Australian city of Melbourne. He used a resident's inclusion in the annual *Who's Who in Australia* as a definition of high social status. Gridding the city into squares, Johnston found the greatest concentration of prestige addresses was in exactly the same square as a half-century earlier. This square lay in the suburb of Toorak. Like Beacon Hill, Toorak is within a short distance of the CBD. It contains large mansions which have been progressively subdivided as demand for this desirable space increased.

Johnston's study goes on to show that the social pattern in Melbourne is neither a static nor a simple one. The centre of gravity of prestigious residences has moved outward, though at a much slower rate than the outer boundaries of suburban living. Indeed, the convenience of inner-city living has caused a reversal of the suburban-flight trends common in most western cities. As commuting distances lengthened, so higher-income families have been moving back into inner-city areas. This movement is now happening in many western cities. It was in London that the process of upgrading inner-city housing was termed *gentrification*, from the English slang term 'gentry' for the upper-middle class in that country.

Regions of the Imagination

Some of the most powerful regional images stem from the pen of the author or the lens of the film maker. We can think of this as a loop in which a particular artist is inspired by the special character of a region to set their fictional work within its boundaries. The works that result then call the attention of those who read the books or watch the movies. This interest is both a natural reaction to the works but is also seized on by the vacation industry to bring still more people to the region. In these circumstances the boundary between the 'real' region and the 'imagined' region become ever more intertwined. Figure 12.17 gives a schematic view of the process.

The landscape of Britain provides many examples of this process. The poet William Wordsworth drew attention to the English Lake District, the novelist D.H. Lawrence to the coalfield areas of the East Midlands, the modern novelist Cathereine Cookson to the area of Tyneside, in northeast England. Detective novelists Maj Sjöwall and Per Wahlöö use the streets of Sweden's capital city, Stockholm, as their setting, Sara Paretsky uses the streets of Chicago, and Colin Dexter the colleges of Oxford. But of all the examples of *literary regions*, none are stronger than those of Thomas Hardy's and his 'Wessex' novels.

Hardy's Wessex

Wessex appears on the map of England in the sixth century as one of the kingdoms of Anglo-Saxon England: the name literally means 'West Saxon'. Its core area was roughly that of the modern counties

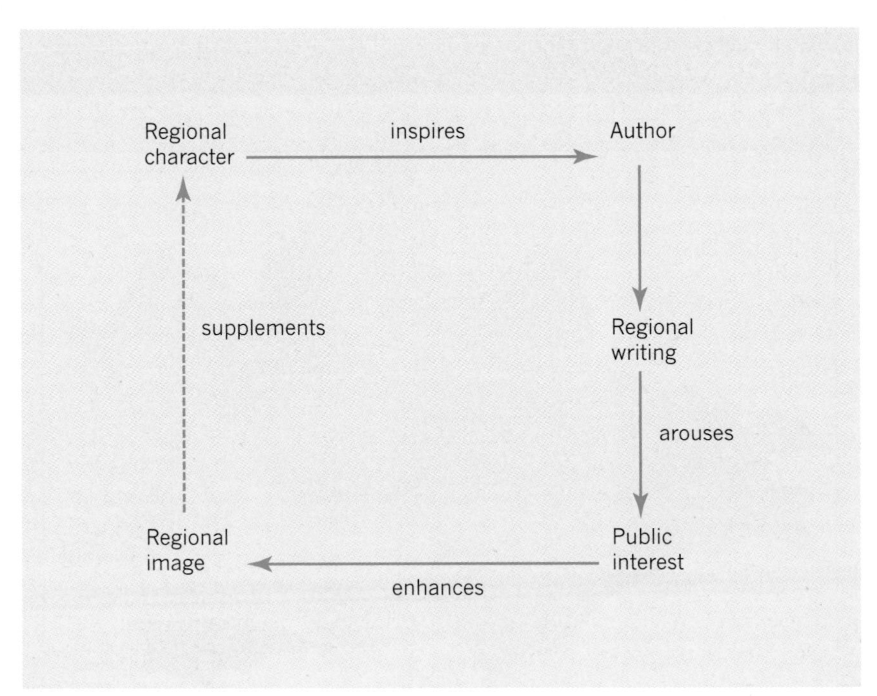

Figure 12.17 Author–region feedbacks Ways in which a region inspires an author who, in turn, draws attention to the region. The cycle described here shows an upward cycle, but negative descriptions could lead to a downward spiral.

of Hampshire, Dorset, Wiltshire, and Somerset, but at times over the next 400 years its political reach extended north of the Thames and west to include Cornwall and Devon. Winchester, its main settlement, was England's original capital.

With the unification of the country the name slipped into the background for the best part of the next millennium. Its revival and its sharpness as a region today are largely due to the writings of a single man, the novelist and poet Thomas Hardy (see Figure 12.18). Born in the Dorset village of Higher Bockhampton in June 1840, Hardy studied as architect but in his 30s abandoned that profession to follow a full-time career as a writer. Although a poet and dramatist of high quality, it is for his novels that he is so well known today. These have two major characteristics. First, a deep gloom in which his characters fight a losing battle against an impersonal force, fate. A sense of the unfairness of life and the unwilling conflict of his heroes and heroines with a hostile and meaningless universe suffuse most of his stories.

A second characteristic is that the stories are set in the landscape so well known and well loved by the author, the region of Wessex in general and the core county of Dorset in particular. (See Box 12.B on 'Hardy's Dorset regions'.) His novel, *Far from the Madding Crowd* (1874), introduced Wessex for the first time and made Hardy famous by setting the beautiful and impulsive Bathsheba Everdene in a closely observed agricultural setting within the rural Dorset landscapes. The books that followed extended his canvas both of characters and of Wessex landscapes. *The Return of the Native* (1878) was admired for its powerfully evoked setting of Egdon Heath, based on the sombre countryside Hardy had known as a child. Three minor novels followed, but in *The Mayor of Casterbridge* (1886) he incorporates recognizable details of Dorchester's history and topography. The busy market-town of Casterbridge becomes the setting for another tragic struggle. The *Woodlanders* (1887) and his two last novels, *Tess of the d'Urbervilles* (1891) and *Jude the Obscure* (1895), which are generally considered his finest but bleakest novels, again show an integral Wessex setting.

Figure 12.18 Literary regions Picture of Thomas Hardy at his Dorchester home, Max Gate, from the early 1920s. Hardy's sequence of novels (listed in Table 12.1) about this part of southern England gave geographical definition to the 'Wessex' region.

[Source: Mary Evans Picture Library.]

Box 12.B Hardy's Dorset Regions

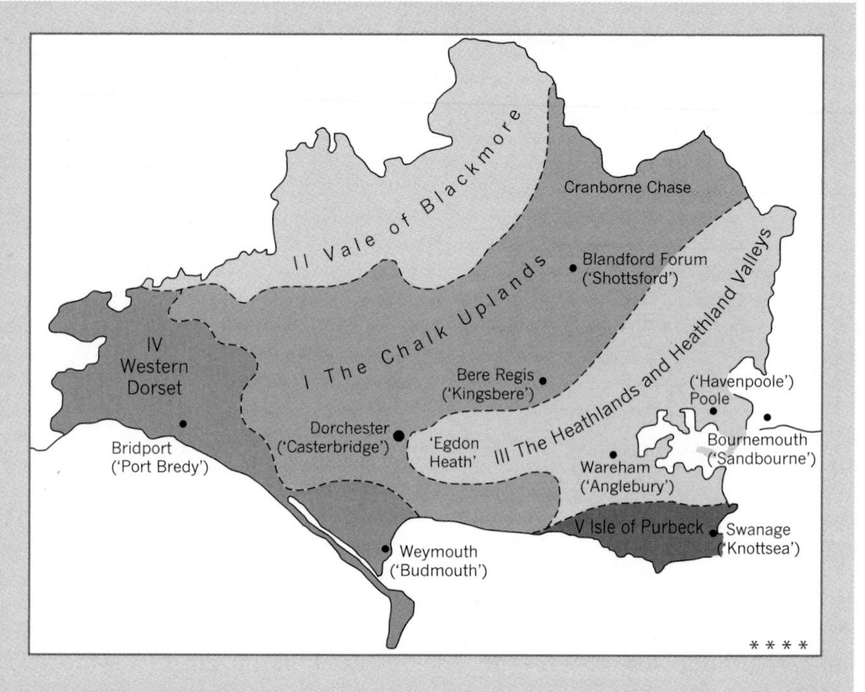

The county of Dorset in southwest England lies at the heart of Thomas Hardy's Wessex. The Cambridge historical geographer Clifford Darby was one of the first geographers to try to unravel the links between a regional novelist and the regions he or she described. Using excerpts from the text of the novels and Hardy's other writing (as in the prefaces to some of the novels) he was able to map the scenes in which the different novels were set. This was consistent with Hardy's own view: 'The series of novels I projected being mainly of the kind called local, they seemed to require a territorial definition of some sort to lend unity to their scene'. The map shows the regions, 'partly real, partly dream country' as Hardy put it, used in the novels.

Five regions are identified within Dorset:

I *The chalk uplands.* Undulating upland rising to 250 m (800 ft). Its scarp slope overlooks the vale of Blackmore and the dip slope descends southwards towards the heathlands. In Hardy's time it was dominated by sheep farming.

II *The Vale of Blackmore.* Damp, well-watered claylands, marshy in places. In Hardy's time the heavy clays supported dense woodland (hence the *Woodlanders* set here) but the areas is mainly dairy farming.

III *The heathlands and heathland valleys.* Underlain by sands and gravels, it was in Hardy's time an area of poor soils and little agriculure. Egdon Heath which figures largely in Hardy's *The Return of the Native* is set in this region.

IV *Western Dorset.* Mainly hilly claylands with varied relief and soil ranging from clay to light sand. A mixture of arable and pasture land.

V *The Isle of Purbeck.* A very distinctive region (hence 'Isle') with a succession of east–west trending ridges. The limestone has given rise to an important quarrying industry. *The Hand of Ethelberta* is largely set in this region.

The regions and their relations to the novels are described in H.C. Darby, 'The regional geography of Thomas Hardy's Wessex', *Geographical Review*, 38 (1948), pp. 426–443.

The continuing popularity of Hardy's novels owes much to their richly varied yet always accessible style and their combination of romantic plots, convincing characters, nostalgic evocation of a vanished rural world through the creation of highly particularized regional settings. The latter has made them especially well suited to film and television adaptation and more people have been introduced to the region through this setting than through the original writing.

In the preface to one of his novels, Hardy described Wessex as 'partly

Table 12.1 Hardy's disguised Wessex place-names

Current place-name (with county)	Hardy's place-name	Number of novels in which the place features[a]
Dorchester, Dorset	*Casterbridge*	13: DR UGT FMC RN TM MC W TD JO WT GND LLI CM TM WB
Weymouth, Dorset	*Budmouth*	13: DR UGT FMC RN TM L TT MC W WT GND LLI CM
Blandford Forum, Dorset	*Shottsford*	9: FMC TM MC W JO WT GND LLI CM
Salisbury, Wiltshire	*Melchester*	9: FMC HE TT Mc TD JO GND LLI CM
Wareham, Dorset	*Anglebury*	6: DR HE RN MC TD WT
Bere Regis, Dorset	*Kingsbere*	5: FMC RN TM TD WT
Bridport, Dorset	*Port Bredy*	5: MC W TD WT LLI
Bournemouth, Dorset	*Sandbourne*	5: HE TD JO WB LLI
Exeter, Devon	*Exonbury*	5: PBE TM W LLI CM
Puddletown Forest, Dorset	*Egdon Heath*	4: UGT FMC RN TD

[a]CM *A Changed Man**, DR *Desperate Remedies*, FMC *Far from the Madding Crowd*, GND *A Group of Noble Dames**, HE *The Hand of the Ethelberts*, JO *Jude the Obscure*, L *A Laodicean*, LLI *Life's Little Ironies**, MC *The Mayor of Casterbridge*, PBE *A Pair of Blue Eyes*, RN *The Return of the Native*, TD *Tess of the D'Urbervilles*, TM *The Trumpet-Major*, TT *Two on a Tower*, UGT *Under the Greenwood Tree*, W *The Woodlanders*, WB *The Well-Beloved*, WT *Wessex Tales**. * = short story collections.

real, partly dream country'. This is shown in his use of place-names, most of which are lightly disguised (see Table 12.1). He stresses that his landscapes are 'drawn from the real, however illusorily treated'.

Geopiety and Remembered Landscapes

One of the factors that gives both force to regional writing, yet sometimes clouds its clarity, is a deep love for the landscape being described. Attachment to a particular part of the earth's surface was a favourite theme of the American geographer John K. Wright. The term Wright invented, ***geopiety***, has been explored further by Yi-fu Tuan who finds it recurring in all ranges of peoples and at all spatial scales from local to global.

This attachment can range in intensity from a general intermittent support for a team in the inter-school sports or Olympic Games, to the passionate intensity of the south Tyrolese for his home valleys expressed in the emotionally charged word *Heimat* (German, literally 'home place'). As one observer recalls:

> When we say the word '*Heimat*' then a warm wave passes over our hearts; in all our loneliness we are not completely alone ... *Heimat* is mother earth. *Heimat* is landscape we have experienced ... our *Heimat* is the land which has become fruitful through the sweat of our ancestors. For this *Heimat* our ancestors have fought and suffered, for this *Heimat* our fathers have died.

The attachment to land may not be confined to a particular quality of the landscape. There are unlikely devotions to places: those born there cannot bear to leave them; but those not born there could never consider living there. The same basic sense of identification with the land is a continuing feature, which remains strong even in a society where the urban proportion of the population is steadily growing. The theologian Paul Tillich recalled that nearly all the great memories and longings of his life were interwoven with landscapes, soil, weather, the fields of grain and the smell of the potato plants in autumn, the shape of clouds, and with wind, flowers, and woods. Tillich goes on in his *Systematic Theology* to develop the notion of the spirituality of space:

> In reality spirit has its place as well as its time. The space of the creative spirit units an element of abstract unlimitedness with an element of concrete limitation. . . . It becomes a space of settlement – a house, a village, a city.

These regional spaces are qualitative, lying within the frame of physical space but incapable of being measured by it. And thus the question arises as how physical space and the space of the spirit are related to each other, i.e. the question of historical space. We should note that love of region, geopiety, *Heimat*, all have their reverse sides. Literature is also strewn with the record of people's hatred and fear of particular places.

For geographers, the landscapes of childhood and youth seem to have a particularly strong impact, and they keep returning to that theme in later life. Thus the landscape of central Sweden has been important in the writings of Torsten Hägerstrand (recall Box 1.B) as have the Ozarks for Carl Sauer (recall Box 5.B).

12.4 Region Building

In this final section (and as a contrast to Section 12.3) we note some of the ways in which geographers go about their business of constructing regions. In practice the techniques are rather complex and increasingly computer-dependent so here we simply note some of the principles involved.

Criteria for Region Building

Regions need to be efficient. The earth's surface is such a large and diverse object to study that geographers need to break it into smaller parts. But the parts must not be too small. A world climatic system that divides the globe into 999 regions may be an athletic geographic feat, but its virtues are lost if all we want to know is whether to issue troops with tropical, normal, or arctic gear. Too much information (too many regions) can be as much of an embarrassment as too little, and geographers are always searching for the right balance. The ideal set of regions is the one that has just enough differentiation to meet our purpose – but not a boundary more.

Let me illustrate the problem by a specific case (see Figure 12.19). Swiss geographer, Dieter Steiner, was faced with the problem of devising a regional system for the conterminous United States. He began with a set of some 70 climatic stations scattered across the country; for each station there were twelve measures of climate (temperature, precipitation, solar radiation). He developed a quantitative technique by which the 70 were reduced to 69, 68, 67, and so on. Each station was placed with the station whose climate it most closely resembled and the 'pair' replaced with a synthetic value representing the shared or combined values.

This method measured precisely the loss of generality as the number of climatic regions was progressively reduced from 70 to 1. As Figure 12.19 illustrates, each successive regional division does not bring proportionate

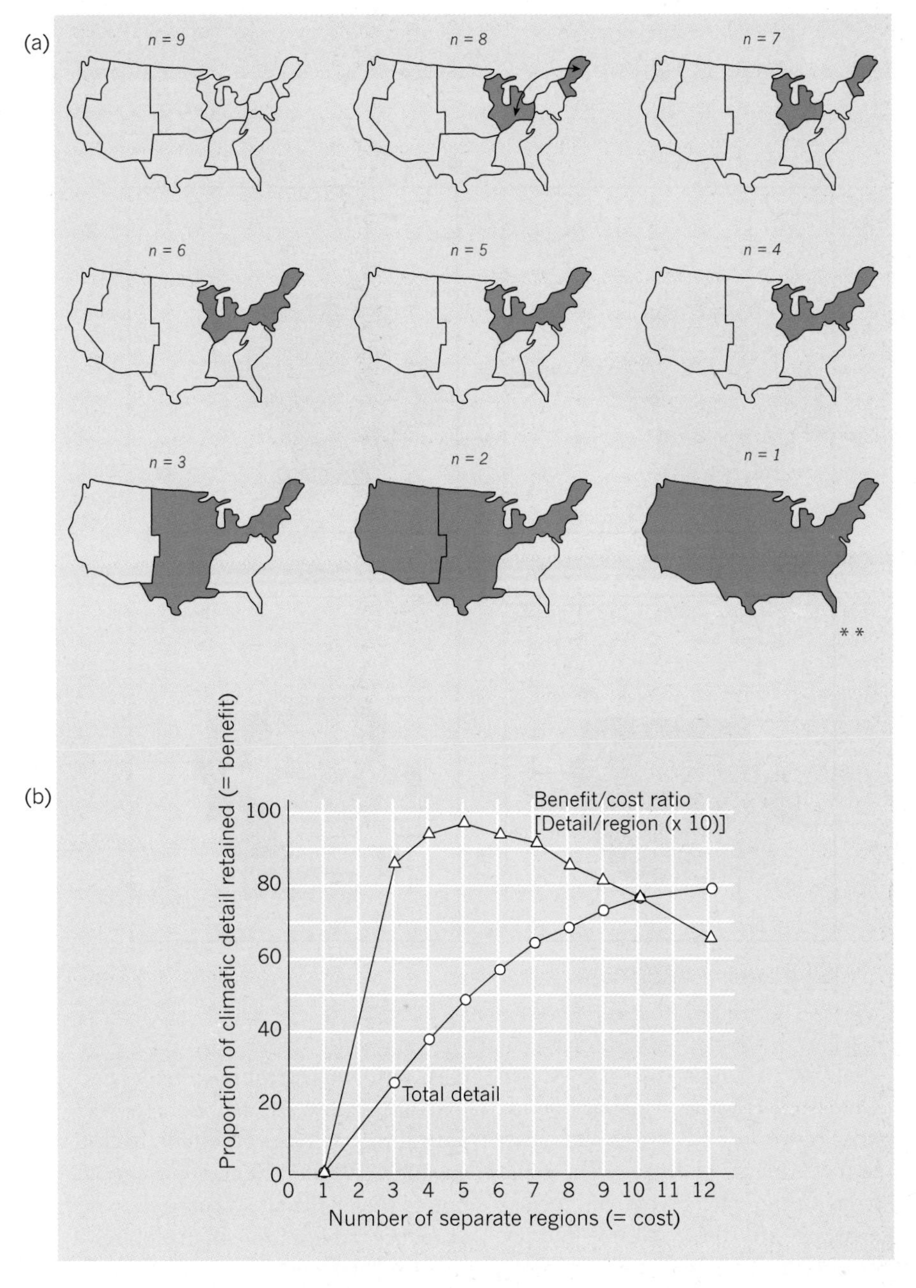

Figure 12.19 Steiner method of climatic regionalization Trade-off between information and complexity in drawing up an efficient system of regions. (a) Steiner's optimum climatic regionalization of the United States based on nine regions ($n = 9$) to one region ($n = 1$). At each stage the regions with the most similar climates are grouped together into larger areas. (b) Impact on the level of climatic detail with changes in the number of regions. The five-region solution has the highest benefit/cost ratio and appears to be the most efficient.

[Source: Data from Dieter Steiner, *Tijdschrift van het Koninklijk Nederlandsch Aartdrijkskrundig Genootschap Tweede Reeks*, **82** (1965), pp. 329–47.]

benefits: some new boundaries bring important insights into the climatic variability of the country, others are more marginal. A 'best buy' regionalization of United States climate on the Steiner scheme appears to be at the third split where five separate regions are recognized.

The Steiner method is only one way of aproaching the problem. It uses a 'bottom up' approach is which data for small areas are progressively combined. A different family of methods starts with the information for the total region and then progressively splits this. Rather like a diamond cutter faced with a large stone to be split, the geographer has to decide (using mathematical methods to help) exactly how best to split the stone. Each part is then split again and again, always trying to optimize the splitting process. A further approach is to redraw boundaries to make them more efficient. This topic is illustrated from American school regions in Box 12.C.

Box 12.C American School Regions

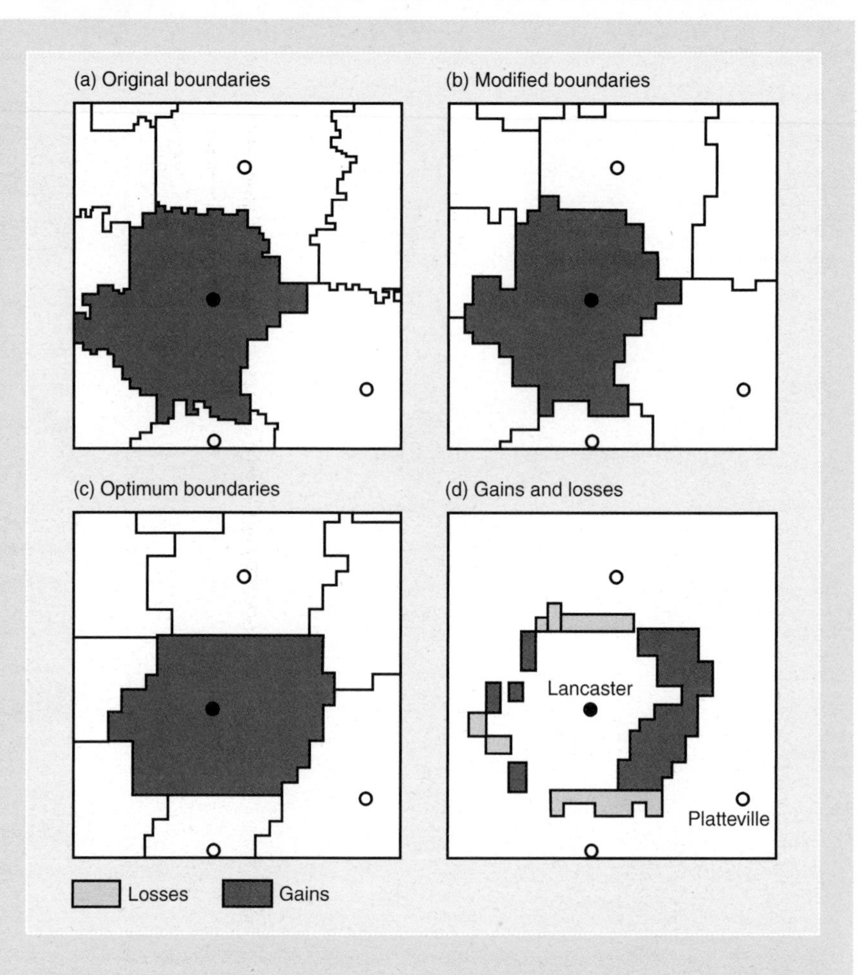

One of the earliest applications of mathematical methods to the region building problem is provided by Canadian geographer Maurice Yeates's study of school districts in southwest Wisconsin. Here the children travel by bus to a central high school from isolated farmsteads (see photo).

Maps (a)–(d) show part of the original school district boundaries studied by Yeates in Grant County, Wisconsin. The township of Lancaster lies in the centre of the map. Yeates's problem was to redraw school disticts so as to minimize the costs of bussing 2900 children to the thirteen high schools, given certain constraints on school capacity (i.e. schools needed a *minimum* numbers of pupils to remain open but also and had a *maximum* capacity for school places that could not be exceeded). To simplify computations, both the schools and the pupils' homes were assumed to lie in the centre of the square-mile (2.5 km^2) section in which they were located: school boundaries have been redrawn on this simplified basis in map (b). Given that there were thirteen schools and 754 square miles of collecting areas, a type of linear programming problem known as the *transportation problem* was set up so as (i) to minimize the total distance bussed to schools by all pupils and (ii) to fill each school to capacity between the two constraints on size set out above. Map (c) shows the optimum boundaries that result from this analysis. As map (d) shows for Lancaster, the new regional boundaries result in both losses and gains in the 'Lancaster' district. The gains from using this type of formal analysis are that school journey times can be significantly reduced and the overall cost of the school bussing programme brought down.

In Wisconsin, the study was only a paper exercise to illustrate the power of the method. But in a number of American cities school districts have been redrawn in reality to maintain 'balance' between school regions with different types of ethnic background. Within England, school boundaries have also been redrawn to reduce the need for small children to cross busy highways. A technical description of the methods used in optimization is given in P. Haggett, A.D. Cliff, and A.E. Frey, *Locational Analysis in Human Geography*, 2nd edn (Arnolds, London, 1977), pp. 477–485.

Box continued

Towards Regional Ecosystems

The real world is made up of an immensely complex mosaic of regions. As we have seen, geographers have attempted to make sense of this by devising a formal system of regions. There is, however, no single or generally

[Source: Topham Picturepoint.]

accepted set of regions that forms a Linnaean system for world geography. (Linnaeus was the eighteenth-century Swedish naturalist who established the modern scientific method of naming plants and animals.) Rather, there are alternatives, each with certain strengths and weaknesses. The most successful types of regional units appear to be those whose spatial boundaries coincide most closely with the ecological or socio-economic features described.

In physical geography the watershed, along with its living community of plants and animals, including humankind, constitute a regional system. Stream watersheds form a convenient unit because they can be simply and unambiguously defined from a topographic map. They are independent of scale, in that large river basins such as the Amazon can be broken into a hierarchic systen of smaller basins; like a toy Russian doll within larger dolls, each smaller basin fits exactly within the next larger one. Each subdivided basin can be identified and its streams numbered in a way that provides a measure of relative size.

Watersheds also have other advantages as a basis for regional divisions. Soil-related changes in vegetation reflect location within the watershed. Also, the physical features of a basin directly affect the hydrologic characteristics of the streams draining it. A rainstorm falling on a long, narrow basin is likely to produce a lower peak in the water level of the stream draining it than a similar storm falling on a rather broad, almost circular basin. Watersheds therefore form useful ecosystem units for agencies interested in flood control, navigation, hydroelectric power production, or soil conservation.

Berkeley geographer David Stoddart finds four advantages to treating such life communities as ecosystems. First, regional systems are *monistic* in that they bring together in a single analytical framework the plant and animal world, the rest of the physical environment, and humankind. Second, they are *structured* in an orderly, and therefore comprehensible, way. Third, ecosystems *function* inasmuch as their structural cohesion depends on continuous cycling of material and energy. Finally, ecosystems have certain *common features* with other systems, and the networks used to construct models of those systems – say, by engineers or physicists – can be extended to ecosystems. Experiments in using electric circuits to construct models of biological systems are an example of how analogies might be developed.

We can also think of the city region in system terms. As with watersheds, city regions need a constant flow of energy to maintain themselves. If we cut off the movement of people, freight, or funds into a city, it will stagnate; if we increase those flows, it will respond by growing in size. The city region is, like the watershed, in a state of balance with the forces that maintain and mould it. We have used the city region as a basic organizational unit in the last four chapters, but it is only one way of ordering the complexities of the real world. A complete integration of such regions with ecological and cultural regions still lies in the future.

Reflections

1 Look at any *two* regional geographies in the library that deal with a part of the world in which you are interested. (There is a huge literature on the regional geography of the major continental areas such as North America or Southeast Asia.) Compare the approaches used by the two geographers. What do they have in common and in what way do they differ? Is there a *right* way to write regional geography? How would you organize your own geographical account of that region?

2 Take your local area (county or state) and propose a regional division based on (a) physical geography and (b) human geography divisions. How and why do they differ? Are there areas of common overlap?

3 One of the purpose of regions is to serve as 'analogues' (see Figures 12.10 and 12.11). What other parts of your own country (or other countries) serve as analogues for the region in which you are now residing? What insights do these throw on human–environment relations in your own region?

4 Individual economic regions are connected to the national and international economy. Look carefully at Figure 12.12 on the regional transmission of economic shocks. Do you consider your own region tends either to accentuate or to dampen changes at the national level?

5 Consider the five main elements in city landscapes proposed by Lynch for Boston (see Figure 12.16). For any other city with which you are familiar, try to identify similar elements that give it structure and regional character.

6 For one English author (Thomas Hardy), the text shows

how literature may tend to accentuate, and even change, the identity of a region (Wessex). *Either* look at any of the Hardy novels (see the list in Table 12.1) to check its regional description. *Or,* for any other 'local' or 'regional' author with which you are familiar in your own country, see how effectively the writing conveys the sense of place.

7 Review your understanding of the following concepts using the text and the glossary in Appendix A of this book:

analogue	literary region
bell-wether region	Lynch elements
compage	nodal region
cultural region	region
exemplars	uniform region

One Step Further ...

The concept of place is explored by Edward Relph in I. DOUGLAS *et al.* (eds), *Companion Encyclopaedia of Geography* (Routledge, London, 1996), pp. 906–23, and a brief but scholarly introduction to regions and regional geography is given by Derek Gregory in his essay in R.J. JOHNSTON *et al.* (eds), *Dictionary of Human Geography*, fourth edition (Blackwell, Oxford, 2000), pp. 687–90. The use of regions for different purposes in geography is explored by Peter HAGGETT, *The Geographer's Art* (Blackwell, Oxford, 1990), pp. 70–94.

The ideas behind regions and the nature of regional geography is described in a scholarly book by the French geographer, Paul CLAVAL, *Introduction to Regional Geography* (Blackwell, Oxford, 1998) and by R.J. JOHNSTON, J. HAUER, and G.A. HOEKVELD (eds), *Regional Geography: Current Development and Future Prospects* (Routledge, London, 1990). Look also at Hilary WINCHESTER, *Landscapes: Ways of Imagining the World* (Longman, Harlow, 1998) which explores the links between regional identity and landscapes. The role of literature in landscape appreciation is given in David DAICHES and John FLOWER, *Literary Landscapes of the British Isles* (Paddington Press, New York, 1979); see the reference to Hardy's Wessex (Box 12.B), pp. 158–71.)

To gain an insight into the way regional geographies are written it is worth browsing in the 'regional geography' section of the library. John PATERSON, *North America*, eighth edition (Oxford University Press, Oxford, 1989) is an excellent example of this genre. Although there are no specific journals of regional geography, *per se*, most of the major geographical journals include papers on regional geography. *Regional Studies* (quarterly) is largely concerned with economic geography and regional development. For readers with access to the World-Wide Web see also the sites recommended for this chapter in Appendix B at the end of this book.

PART IV

Geographic Structures

CHAPTER 13

Flows and Networks

■ denotes case studies

Figure 13.1 The spider's web of transport LANDSAT image of the Zaltan desert region of eastern Libya, showing several oil fields and production centres like spiders in the centre of a web. The radiating lines scratched across the desert floor are car and truck routes and over and underground pipelines linking the fields together. Black smoke can be seen blowing off from the two central oil fields. North is at the top of the photo, the east–west distance is roughly 65 km (40 miles). [Source: Earth Satellite Corporation/Science Photo Library.]

> To arrive at a clear decision on these questions, let us take familiar examples, but set them out in geometrical fashion.
>
> JOHANNES KEPLER *The Six-cornered Snowflake* (1614)

The view of the earth's surface from satellites rarely captures the essential web of linkages that keeps the various parts of the global economy working together. Our opening image (Figure 13.1 p. 392) gives an unusual view of one such network. An oil well in the Libyan desert is pumping crude petroleum to the surface and sending it off by pipeline for coastal refining and export. Around the well is the star-shaped pattern of trails and roads by which other forms of surface transport (motor vehicles and the occasional camel train) link the well with the outside world. Not shown (or showable) is the invisible web of information by which messages via telephone, fax, and e-mail determine the pumping schedules and order spare parts and food supplies. Aerial forms of transport, aircraft, and helicopters, tend to be missed altogether.

This small cell of human activity set in a desert landscape has a distinctive geometry like Kepler's snowflake (in our opening quotation). It has form and structure, somewhat like a biologist's cell. They are essential building blocks from which larger human regions are formed.

These cells have roots that run back well beyond the modern period. Even in the most primitive of early urban settlements, trade and exchange were the distinctive features that allowed population to cluster and urbanization to emerge. What distinguishes primitive from modern systems of exchange are its product range, and its spatial reach, rather than its essential nature.

In this chapter we consider the nature of these cells and the flows that maintain them. Section 13.1 considers the elements that make up a nodal region and the ways in which it may be defined. This leads on to a consideration of flows, their spatial pattern and the complexities of transport space (Section 13.2) and the ways in which these flows can be modelled (Section 13.3). Many flows are channelled into specific networks. Section 13.4 considers the spatial structure of such networks. Finally, in Section 13.5, the impact of changes in networks on regional development are illustrated at very different geographic scales.

13.1 Nodal Regions

We noted in Chapter 12 that a central problem in geographical study is to identify a regional unit appropriate to the analysis being undertaken. In this fourth part of the book we recognize the *nodal* region as our basic unit and we use it as a basis for the arrangement of the chapters in this section of the book.

Elements in a Nodal Region

By a ***nodal region*** we mean the area that surrounds a human settlement and which is tied to it in terms of its spatial organization. A schematic structure of such a region is set out in Figure 13.2.

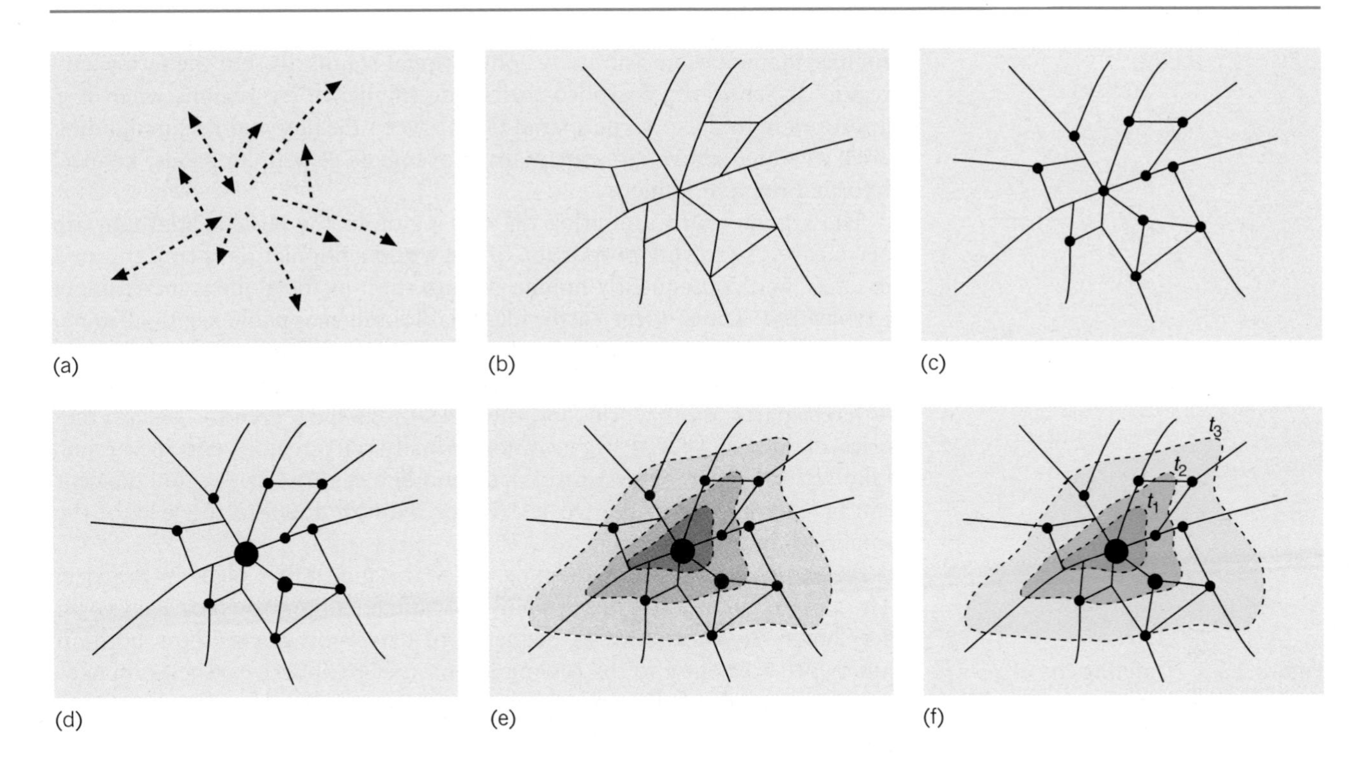

Figure 13.2 Stages in the analysis of nodal regions Basic elements in the geographic structure of regions. (a) Interactions and spatial flows. (b) Networks that entrain flows. (c) Nodes that are established at junctions on the network. (d) Hierarchies that differentiate the role and function of nodes. (e) Surfaces established as gradients around nodes and along networks. (f) Diffusion waves that travel through hierarchies and networks and along surfaces.

[Source: P. Haggett, *Locational Analysis in Human Geography* (Arnold, London, 1965), Fig. 1.5, p. 18.]

One of the difficulties we face in trying to analyze a nodal regional system is that there is no obvious or single point of entry. Indeed, the more integrated the regional system, the harder it is to crack. In terms of Figure 13.2, it is as logical to begin with the study of settlements as of routes. As the regional economist Walter Isard put it: 'The maze of interdependencies is formidable, its tale unending, its circularity unquestionable. Yet its dissection is imperative... at some point we must cut into its circumference'. We choose to make that cut with spatial interaction or movements.

Figure 13.2 shows the six essential elements in the make-up of the nodal region, and relates it to the organization of the chapters in the fourth part of this book. This opening chapter covers elements (a) and (b) in the diagram. It describes the spatial *interactions* between a settlement and its surrounding area in terms of the movements of people, goods, finance, information, and influence. These flows are usually channelled into discrete *networks*, linking together the individual settlements. Chapter 14 addresses elements (c) and (d) in the diagram. The *nodes* are located at critical points on the networks. But since the nodes are not uniform or undifferentiated we also consider in that chapter how they are structured into finely articulated *hierarchies*. Chapter 15 shows how the individual elements – interaction, networks, nodes, and hierarchies – are integrated. It views the nodal regional structure in terms of continuous *surfaces* of indicator values, each describing some aspect of the human geography of the region. Finally, Chapter 16 introduces a time element into the static cultural landscape by considering change over time in terms of *diffusion* processes.

Defining a Nodal Region

Many geographers have chosen as a basic unit a 'city region'. By a *city region* we mean the area that surrounds a human settlement and is tied to it in terms of its spatial organization. (These central settlements may be smaller than cities measured by conventional standards, but the terms 'city region' is arbitrarily extended to include smaller nodal regions when it is appropriate to do so.) The spatial ties between the city and its surrounding area, as stated above, are essentially movements of people, goods, finance, information, and influence.

The arguments for adopting the city region as the basic spatial unit are persuasive. A growing proportion of the world's population is concentrated in cities, and consequently human organization of the globe is increasingly city-centred. Cities form easily identifiable and mappable regional units. Reasonably uniform statistical data have been available for many cities for the last century and a half. Further, city regions stress the comparability of different parts of the world, and thus encourage the search for general theories of human spatial organization. Finally, city regions are hierarchic. Like watersheds, they nest inside one another, and the city-region concept can be enlarged up to the world level or reduced down to the level of the smallest hamlet.

Figure 13.3 shows how defining a nodal region takes place in practice. The various boundaries of the small Australian country town of Tamworth are shown. If you follow the sequence of maps you can see how both an inner core area close to the town and an outer periphery can be estimated.

Can we think of the city region as an ecosystem? Like watersheds, city regions need a constant flow of energy to maintain themselves. If we cut off the movement of people, freight, or funds into a city, it will stagnate; if we increase those flows, it will respond by growing in size. The city region is, like the watershed, in a state of balance with the forces that maintain and mould it. We will use the city region as a basic organizational unit in the next four chapters, but it is only one way of ordering the complexities of the real world. A complete integration of such regions with ecological and cultural regions still lies in the future.

Figure 13.3 Defining the city region If we map the area served by each of a city's professional establishment (doctors, lawyers, etc.) then we get a series of overlapping boundaries (shown hypothetically in (a)) from which we can identify an outer and inner range and a median line (b). Pinning down the limits of the city region is shown for the Australian town of Tamworth (23 thousand people) in northern New South Wales. Tamworth is the marketing centre for a prosperous sheep-farming and wheat-growing area. (c) shows the ranges and median line for Tamworth. Repeating the study for other sectors of city activity and plotting all the medians (d) allows a clear idea of the city region's limits to emerge. Note the important effect of the interstate boundary.

[Source: J. Holmes and R.F. Pullinger, *Australian Geographer* (1973), Fig. 8, p. 221. Photo courtesy of Tamworth City Council.]

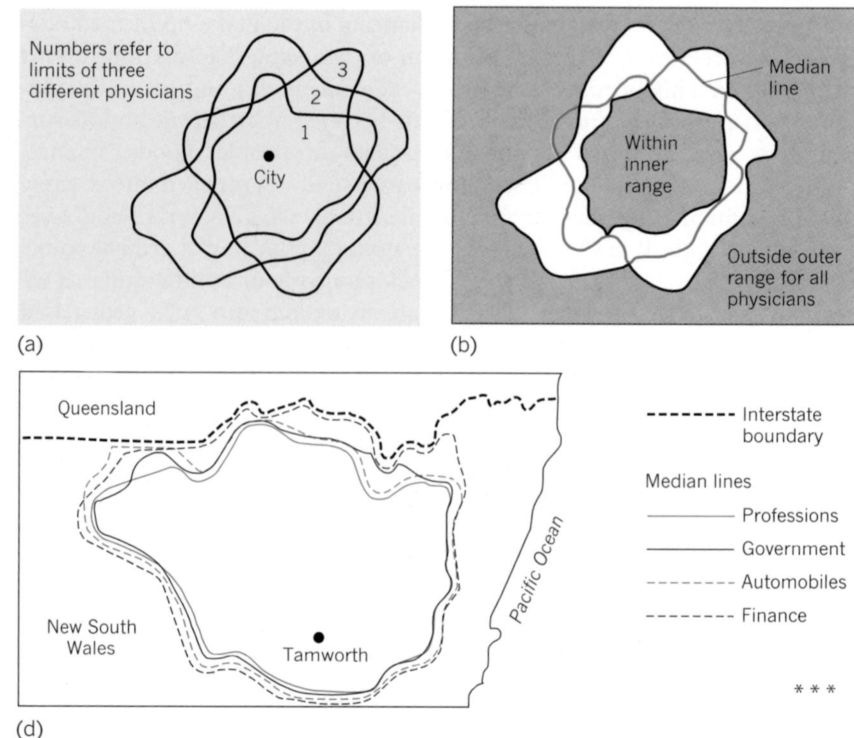

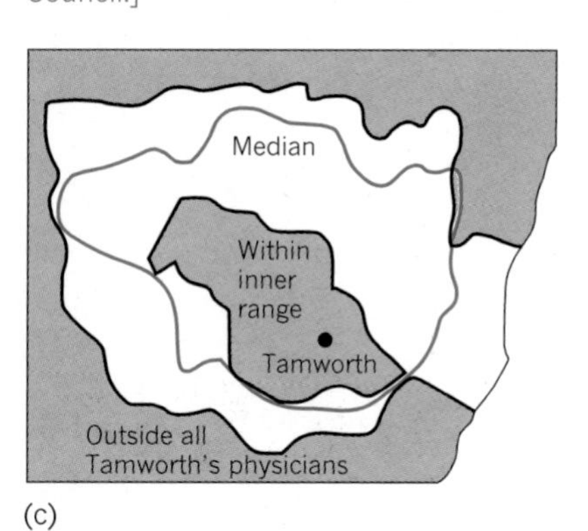

* * *

Transport and Information Flows 13.2

In this section we look at the different types of flows, the spatial patterns they form, and the complexities of the cost surface over which they move.

Types of Flow

We have already mentioned three different kinds of intercity flows: telephone calls, freight cars, and aircraft movements. Just what sort of flows do we have in mind?

We can draw a broad distinction between transport flows and communication flows. *Transport flows* involve the physical movement of something, be it of people or freight, between places. This movement can take place through a series of different transport modes, each with a specific set of advantages and disadvantages. These are summarized in Table 13.1. Thus if we wish to move large cargoes of a bulky commodity, then slow-moving barges are very cheap (on a per tonnage basis) and may be preferable. Conversely, aircraft make up for their very high costs per tonne by their very great speed and freedom from the environmental barriers set by mountain, ocean, or icecap. The relative advantages of the various modes have not remained constant; indeed, only two of the five modes listed in Table 13.1 would have been available for intercity flow 150 years ago. The current trends in both passenger and freight flows are towards more emphasis on highways and airways.

More striking than the changes in individual transport modes has been the much faster increase in communication flows. Unlike transport, *communication flows* do not involve the physical movement of an element between places. Communication is the sharing of information. Although

Table 13.1 The relative advantages of different modes of transport

Mode of transport	Principal technical advantage	Use
Railroads	Minimum resistance to movement, general flexibility, dependability and safety	Bulk-commodity and general-cargo transport, intercity; of minimum value for short-haul traffic
Highways	Flexibility, especially of routes; speed and ease of movement in international and local service	Individual transport; also transport of merchandise and general cargo of medium size and quantity; pickup and delivery service; short-to-medium intercity transport; feeder service
Waterways	High productivity at low horse-power per tonne	Slow-speed movement of bulk, and low-grade freight where waterways are available; general-cargo transport where speed is not a factor or where other means of transport are not available
Airways	High speed	Movement of any traffic where time is a factor – over medium and long distances; traffic with a high value in relation to its weight and bulk
Pipelines	Continuous flow; maximum dependability and safety	Transport of liquids and gases where total and daily volume are high and continuity of delivery is required; potential future use in movement of suspended solids

Source: W.H. Hay, *An Introduction to Transportation Engineering* (Wiley, New York, 1961), Table 8–5, p. 283.

short-range communication is as old as humans themselves, most of the mass communication now flowing between cities is the product of the last two centuries of technology. The nineteenth century was characterized by the invention and rapid spread of 'wire' communication systems. Thus Washington and Boston were the first two cities to be linked by a commercial telegraph line (1844); Europe and America were linked by submarine telegraph cable (1858). After Bell's Boston experiments in 1876, the telephone system spread slowly. By the century's end, Chicago and New York had been linked. The twentieth century saw breakthroughs in 'wireless' communications systems – radio in the 1910s, television in the 1930s, satellite communications in the 1960s, the Internet in the 1990s. Intercity communication using these new modes has been increasing exponentially at rates that make the expansion of world population look slow. For example, the volume of intercity telephone calls is now doubling every decade over most of the world.

We have already commented in this section of the book on the way in which the changing costs of flows have led to two contrary spatial movements – an *implosion* of major cities at the intercity level, but an *explosion* of suburban sprawl at the intracity level. (Compare Figure 8.10 with Figure 8.16.) This drastic change in the geography of the world's leading cities is closely linked to the innovations in transport and communications reviewed above. New airline services or telex links tend to be first established between pairs of cities where the demand is greatest, so reinforcing the already commanding position of the leading cities.

Figure 13.4 The decay in spatial interactions with distance (a) Truck trips around Chicago in the American mid-west. (b) Railway shipments in the United States. (c) World ocean-going freight.

[Source: Data from M. Helvis and G. Zipf, adapted from P. Haggett, *Locational Analysis in Human Geography* (St. Martin's Press, New York, and Arnold, London, 1965), Fig. 2.2, p. 34.]

Spatial Patterns of Flows

If we map the origin and destination of flows, we find that most moves are over a short distance. As an example let us take the hundreds of heavy trucks that roll down the freeways from Chicago bound for other parts of the United States. Most unload their contents a few kilometres from the city; relatively few move a long distance. In Figure 13.4(a) we plot the decrease in traffic with distance from Chicago out to about 650 km (400 miles). Notice that the diagram is drawn so that both the amount and the distance of the flow are plotted not on a linear scale, but on a logarithmic one.

Plots of similar flows on much larger geographic scales have a rather similar pattern. Figure 13.4(b) presents the pattern of rail-freight flows

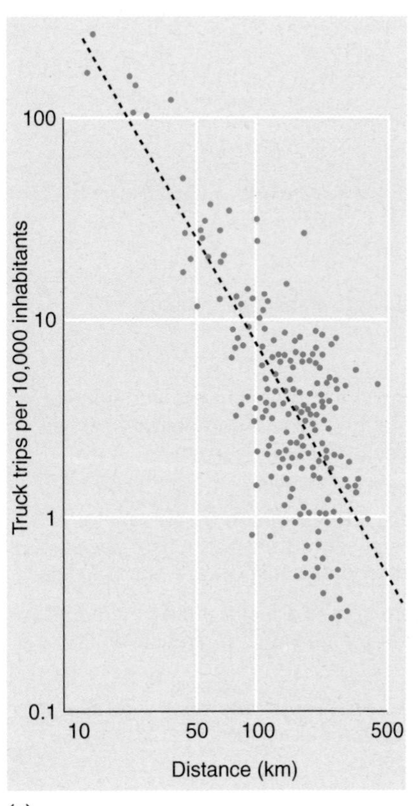

(a)

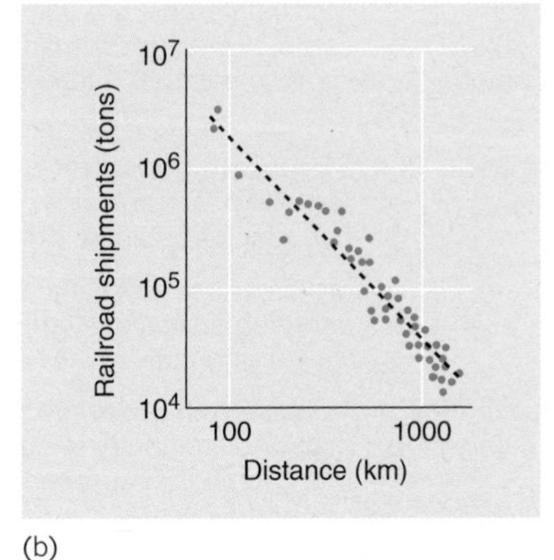

(b)

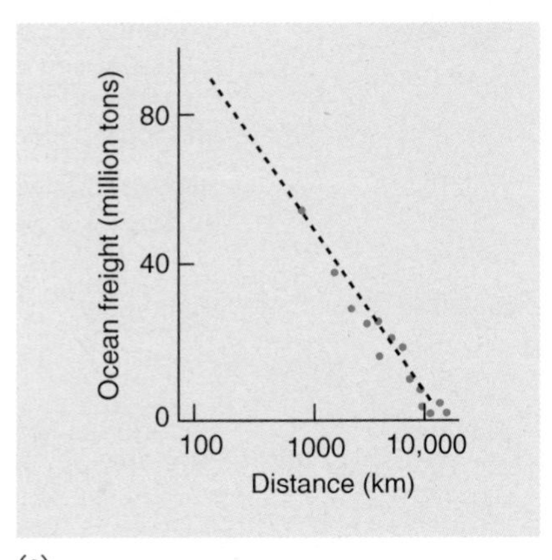

(c)

within the United States, which fall off regularly out to about 2400 km (1500 miles). Figure 13.4(c) shows the pattern of world shipping out to a distance of 20,000 km (12,500 miles). Similar patterns can be found on geographic scales from the world level right down to the level of local kindergarten districts.

The general name for these patterns is *distance–decay rates*. (Some geographers use an alternative term, distance lapse-rates.) As a general rule we can state that that the degree of spatial interaction (flows between regions) is inversely related to distance; that is, near regions interact more intensely than distant regions. The general form of this rule has been firmly established since the 1880s. However, the exact form of the relationship between distance and interaction has been difficult to pin down. On a graph with arithmetic scales, plotting distance against interaction produces a J-shaped curve in which flows decrease rapidly over shorter distances and more slowly over longer distances. Using logarithmic scales on both axes commonly yields an approximately linear relationship, and various alternative mathematical functions can be used to describe such forms. The results of Swedish work on migration indicate that spatial interaction is inversely related to the square of the distance between *settlements*, but this is an approximation of variable empirical findings. (See Box 13.A on 'Distance–decay curves'.)

Box 13.A Distance–Decay Curves

Consider Figure 13.4, in which spatial interaction falls off with distance. One of the simplest ways of describing curves that relate flows and distance is with a *Pareto function*:

$$F = aD^{-b}$$

where F = the flow, D = the distance, and a and b are constants. Geographers are especially interested in the value for the constant b. Low b values indicate a curve with a gentle slope with flows extending over a wide area, whereas high b values indicate a sharp decrease with distance so that flows are confined to a limited area. This formula was used extensively by a group of Swedish geographers in studies of migration between regions on a large variety of geographic scales as far back as the nineteenth century. Their findings showed b values going from as flat as -0.4 to as steep as -3.3. The mean value for all the studies was just below -2 (in fact, -1.94). This figure would suggest that

$$F = aD^{-2}$$

which we can rewrite as

$$F = \frac{1}{aD^2}$$

Apparently, spatial interaction falls off inversely, with the square of distance. That is, the size of a flow at 20 km (12 miles) is likely to be only one-quarter of that at a 10 km (6 miles) distance. This *inverse-square* relationship is analogous to that used by physicists in estimating gravitational attraction. Note in the diagram the contrast between the inverse-square relation when plotted on linear axes (when it forms a straight line) and with logarithmic axes (when the same relationship forms a straight line).

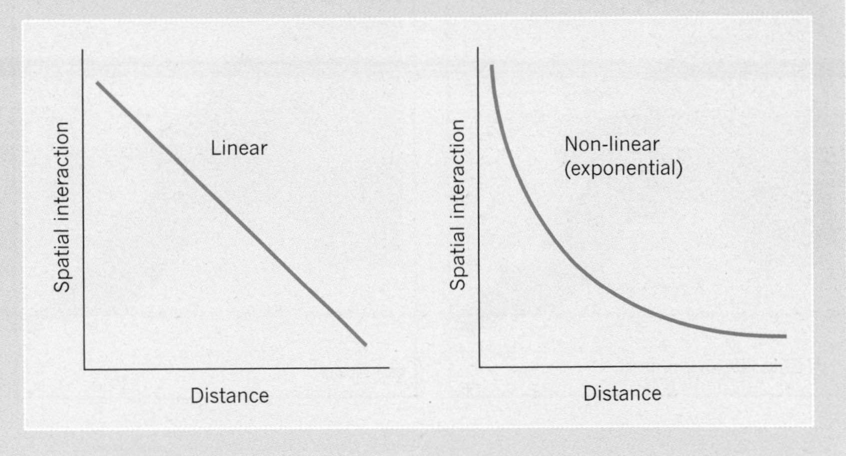

Complexities of Surface Space

In our later discussion of the surfaces (see Chapter 15) we stress geographic distance from the city as a dominating theme. But geographic distance (in the sense of the length of routes in miles or kilometres) is often only a crude measure of the costs of movement. Consider Figure 13.5, which shows actual freight costs per ton from six ports in eastern New Guinea to other points within the territory. The costs form an intricate patchwork related to the volume and type of goods to be carried, the mode of transport (truck, aircraft, or ship), and the degree of competition, as well as the geographic distance to the six ports. The situation in New Guinea is by no means atypical. At any point in time the exact cost of moving resources from one part of the earth's surface to another is a function of a dozen or more different considerations. Think of the different rates quoted for a passenger fare from Boston to London – from scheduled air fares, through reduced air charter rates, to variable shipping rates.

Notwithstanding these complications, we can still make out some rough order in the relationship of costs and distance. First, we find that geographic distance does play a role in determining most rates. Other things being equal, a longer haul costs more than a shorter haul. There are of course exceptions. Within many countries the rate for sending a letter or parcel by internal post varies with its weight but not with the distance it is sent. Similarly, companies may charge a uniform delivery rate throughout their sales area. Such blanket rates do not mean that costs do not increase with distance. The blanket rate subsidizes the longer-distance movements by charging more than is necessary for the shorter-distance movements.

Second, we find that total costs have two elements: a terminal, or handling, element (unrelated to distance) and a delivery, or haulage, element directly related to distance. These two cost elements may vary for different modes of transport, as Figure 13.6(a) shows. A more realistic curve for distance costs is convex and non-linear, indicating that transport costs increase, but at a decreasing rate, with distance (Figure 13.6(b)); that is, the

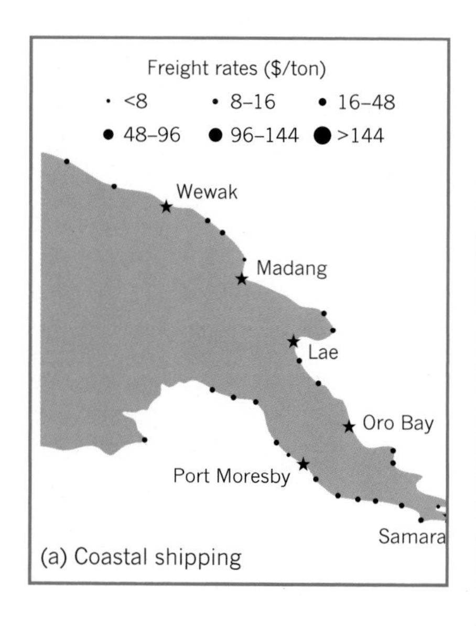

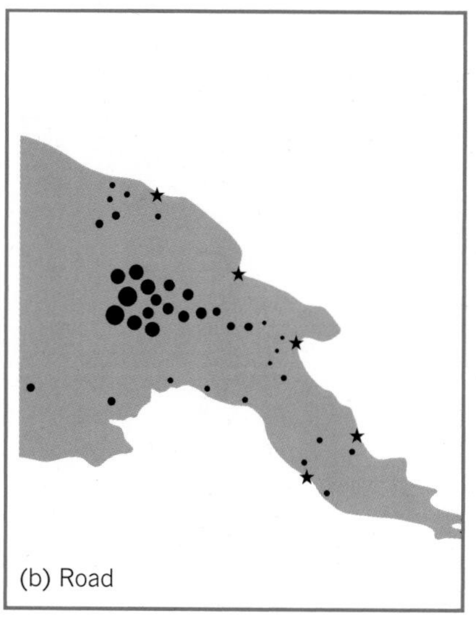

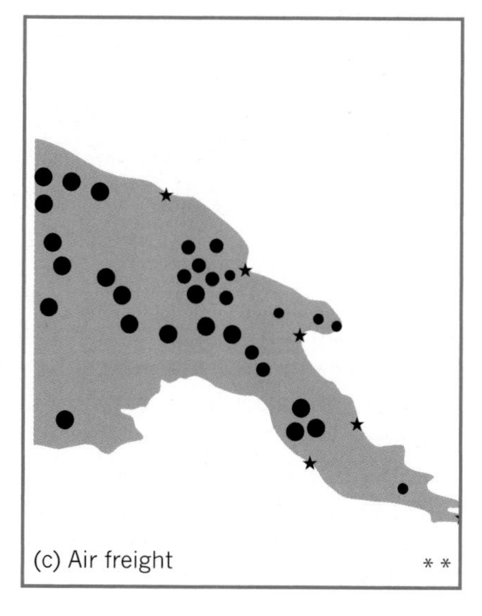

Figure 13.5 Spatial variations in transport costs Freight rates per ton are shown for general cargo from the six main ports (shown by stars) in eastern New Guinea to a selection of inland and coastal locations, using the cheapest mode of transport available. Note the contrasts between the low unit costs but restricted operating area of coastal shipping (a) in contrast to the very high unit costs but flexible operating area of air freight (c). The fact that the road network into the mountainous interior is poorly developed is reflected in the cluster of high costs in (b). The six main ports are marked on the map by stars.

[Source: From H.C. Brookfield and D. Hart, *Melanesia* (Barnes & Noble, New York, and Methuen, London, 1971), Fig. 14.9, p. 357.]

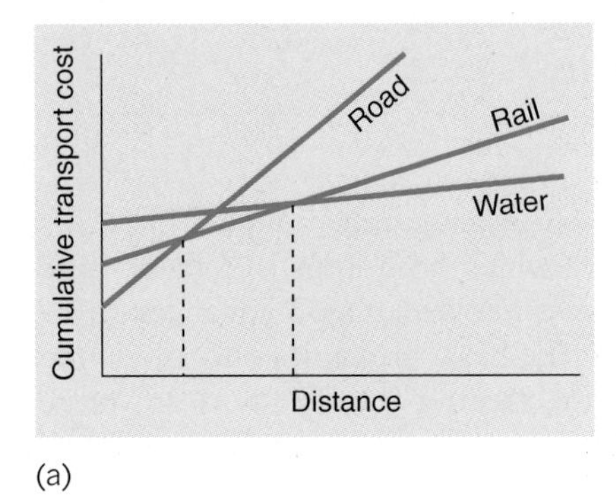

(a)

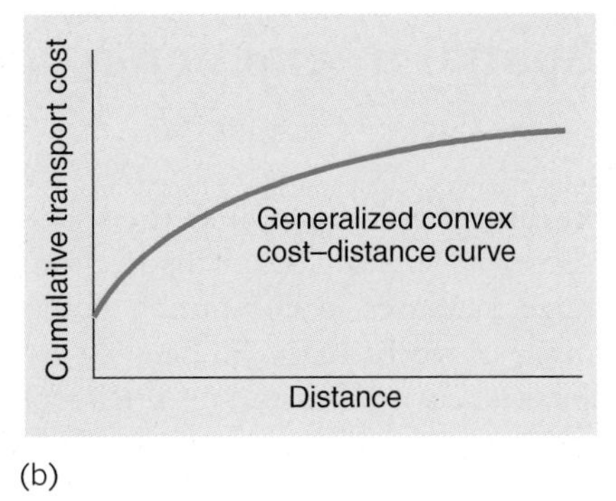

(b)

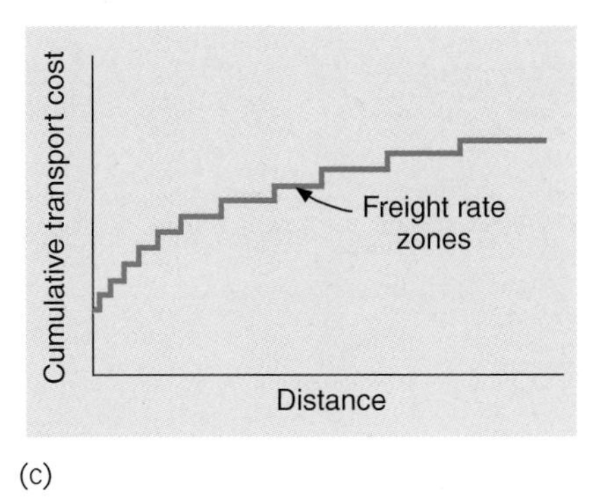

(c)

Figure 13.6 Transport costs and distance Some of the factors that distort a simple linear relationship between costs and distance. (a) Competition between modes of transport. Note the pecked vertical lines which mark a switch in the cheapest mode at that point. (b) Length-of-haul economies. (c) Freight-rate zoning.

cost of moving the first 10 km may be much higher than the cost of moving a similar 10 km stretch between 150 and 160 km away. We can add further realism to our model by breaking up the continuous cost curve in Figure 13.6(b) into a series of steps, each related to a particular level of freight charge, as in Figure 13.6(c).

Three alternative pricing strategies are commonly used to recover the costs involved in transporting a product.

1 *Source pricing*. The price is established at the production point and the customer pays the transfer costs of moving the product. This system is also termed *f.o.b.* (free-on-board) pricing. Customers may be charged freight rates on the basis of the actual distance covered, blanket zones, or a uniform *postage stamp* rate in which the charge levied is unrelated to the distance involved.
2 *Uniform delivered pricing*. The price is the same for all customers regardless of their location. The producers pay all the transport costs involved in shipping their product but recover this by taking the average transport costs into consideration when deciding on the price at which they offer the product. This system is also termed *c.i.f.* (cost–insurance–freight) pricing.
3 *Basing-point pricing*. All production of a given commodity is regarded as originating from a single point, a uniform price is established for all sources regardless of their location, and customers pay the transfer costs from the basing point regardless of the actual location of the producer from which they purchased a product. The most famous case of basing-point pricing was the *Pittsburgh plus* system that operated for some time in the US steel industry. Here customers were charged freight costs from Pittsburgh regardless of the location of the plant from which they actually purchased steel.

Thus it remains broadly true that the economic costs of connecting a city to agricultural zones and industrial-processing centres are a function of geographic distance. The twisting and blurring of freight rates, competition between modes of transport, and the like may be important for individual activities in particular locations. However, the overall picture of a city organizing the space around it – but with its influence gradually fading as we move further away from it – remains. In Chapter 15 we go on to explore the implications of this city-centred organization for other aspects of the world beyond the city.

13.3 Spatial Interaction Models

In this section we look at the ways in which geographers model the flows shown in Figure 13.2. This is an area in which both improved theory and huge increases in computing power have allowed major advances to be made in recent years. So here we set out the basic principles only of *spatial interaction models* and must leave detailed treatment to the more advanced courses in which you can take the ideas further and apply them to actual locational problems.

The Gravity Model

Gravity models are an example of a powerful idea in one field having relevance in another. Sir Isaac Newton, the seventeenth-century English mathematician, was one of many men of genius whose ideas in physics have been later shown to be productive in others. Specifically, Newton's laws of gravitation have been found to throw light on geographers' understanding of the way in which flows occur between cities. This idea was championed by the Princeton scholar, John Stewart, and developed as what he called social physics.

Stewart did not mean that you and I are swept along between cities like molecules in an 'urban gravitational field'. Each individual has a large or small measure of choice. But it does signify that the trillions of telephone messages, billions of freight-car journeys, or millions of aircraft movements that link the world's galaxy of settlements show a tendency, taken as a whole, to move in a way not unlike Newton's physical laws would predict.

More than a century ago observers of social interaction had noted that flows of migrants between cities appeared to be directly related to the size of the cities involved and inversely proportional to the distance separating them. By 1885 the British demographer E.G. Ravenstein had incorporated similar ideas into elementary 'laws' of migration. Although the specific term *gravity model* did not appear until the 1920s, it is clear that nineteenth-century workers were drawing on the relationships formalized by Sir Isaac Newton in his law of universal gravitation (1687), which states that two bodies in the universe attract each other in proportion to the product of their masses and inversely with the square of their distance. Gravitational concepts were specifically introduced by W.J. Reilly in 1929 in discussing the ways in which trade areas are formed. Reilly's ideas were subsequently expanded by researchers concerned with predicting flows in applied fields such as highway design or retail marketing studies.

We can roughly estimate the size of flows between two regions by multiplying the mass of the two regions and dividing the result by the distance separating them. Thus a flow of six units would be produced by two regions with masses of four and three units, respectively, that are two units of distance apart. But what do we mean by 'units' of mass and 'units' of distance?

Mass has been equated with population size in many gravity studies. Information on population to easy to find, and we can readily estimate the size of most population clusters from census figures. However, population data may conceal significant differences between regions that affect the probability of spatial interaction, and the use of some system of weighting has been urged to take these differences into account. Economist Walter

Isard has argued that just as the weights of molecules of different elements are unequal, so too the weights assigned to different groups of people should vary. Weights of 0.8 for population in the deep south, 2.0 for population in the far west, and 1.0 for population in other areas of the United States (reflecting regional differences in travel patterns) were found in the 1960s but have converged since. Multiplication of the population of each area by its mean per capita income is another possible way of refining our measuring stick for mass.

As we have already seen in Section 13.2, *distance* can be measured in several ways. The conventional measure in gravity models is simply the straight-line or cross-country distance between two points. In commuting studies travel time in minutes rather than miles or kilometres may be an appropriate measure. It may take as long to go a short way in urban areas as to go a long way in rural areas. When different forms of transport are available, distance may be measured in terms of the ease and cost of movements also. Fares (for people) and terminal costs and delivery charges (for goods) may be taken into account.

A simple example of the use of the gravity model to estimate flows between four cities is given in Figure 13.7. If you follow the sequence of maps carefully and refer to the caption to check on terms used, then you should see how simple the arithmetic is. You may be puzzled that two separate estimates of flow are given, one using distance (Figure 13.7(c)) and

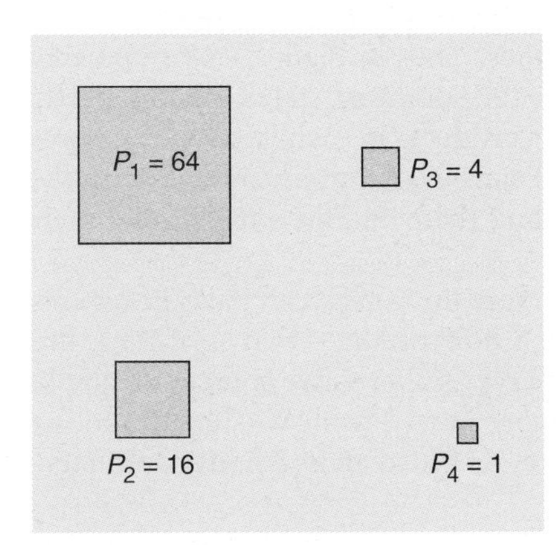

(a) Measure of population (P)

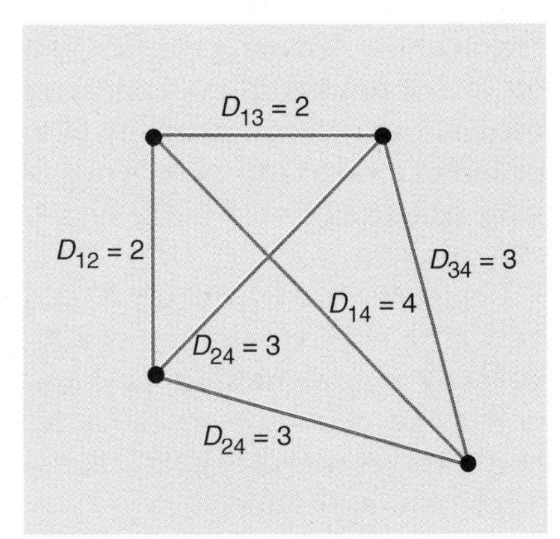

(b) Measure of distance (D)

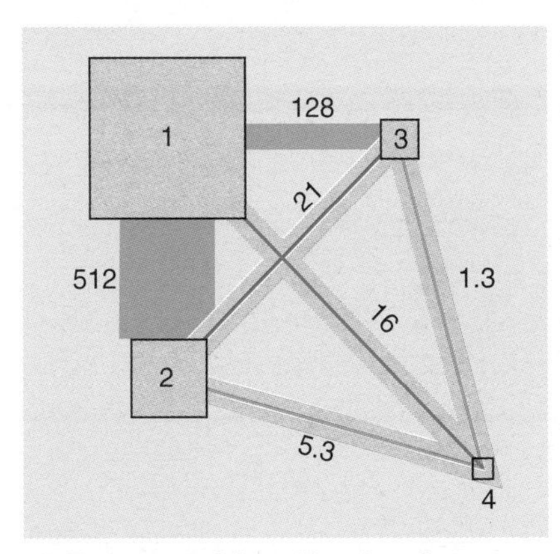

(c) Estimates of flow (F) using distance

$$F_{ij} = \frac{P_i P_j}{D_{ij}}$$

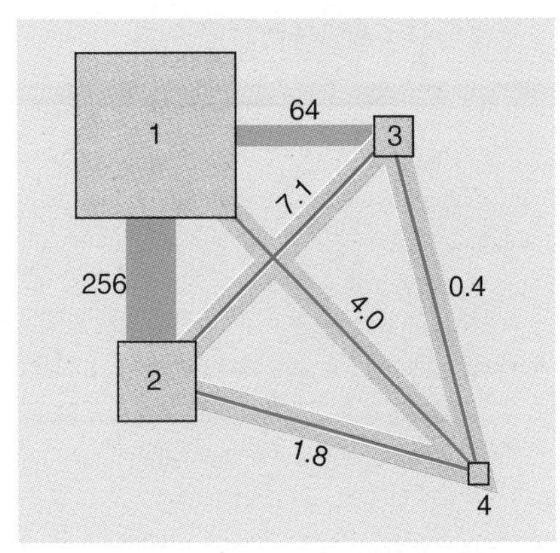

(d) Estimates of flow (F) using distance squared

$$F_{ij} = \frac{P_i P_j}{D_{ij}^2}$$

Figure 13.7 A gravity model of flows between centres Population size (a) and distance (b) are used to estimate the spatial interaction between four centres. In (c) the actual distance between cities and in (d) the square of the distance between cities is used to estimate the flow. Note in (b) that D_{12} refers to the distance between city 1 and city 2, and so on. Since the formula in (c) and (d) is general and refers to flows (F) between *any* two pairs of cities, we use the general term F_{ij} to refer to flows between city i and city j. The same general terms i and j are used in the rest of the formula so that the population of the cities is P_i and P_j and the distance between them is D_{ij}. Check to see that you understand how the figures in (c) and (d) were achieved. For example, flow F_{14} in (d) is 4.0. This is given by multiplying the population of cities 1 and 4 and dividing this product by the square of the distance between them, i.e. $(64 \times 1) \div (4 \times 4) = 4.0$.

one using distance squared (Figure 13.7(d)). Much of the work of geographers has been concerned with estimating just how distance should be measured and how it should be blended into the formulas. It turns out that, although distance squared is the better of the two, we have to use considerably more sophisticated formulas if we are to make estimates of acceptable accuracy. These lie outside the scope of this book but will form an important part of most future courses you select in this field.

The Ullman Model

A different approach to the study of flows between regions is to ask why they occur. We can begin to answer this question by inverting it, or by trying to define the conditions under which flows would *not* occur. For example, if travel between the regions were expensive or if each region were highly self-sufficient, then we would expect rather little to be exchanged. Washington geographer Edward Ullman systematized these notions in a useful model of spatial interaction based on three factors: regional complementarity, intervening opportunity, and spatial transferability. We can illustrate the ***Ullman model*** using his study of interstate flows of lumber in the United States (Figure 13.8).

The first factor on which his model is based, ***regional complementarity***, is a function of the resources available in any particular region. In order for regions to interact, there must be a supply or surplus of resources in one region and a demand or deficit in another. Thus in Figure 13.8 shipments of forest products from Washington to the southeast states are low partly because of the easy availability of forest products in each; conversely, flows of forest products to New York and Pennsylvania are heavy, despite the long distance, because of the high demand there and the small size of their own forest area.

Complementarity will, however, generate flows between pairs of regions only if no ***intervening opportunity*** for a flow occurs – that is, if there are no intervening regions in a position to serve as alternative sources of supply or demand. Seventy years ago little lumber moved from Washington to the northeast because the Great Lakes region provided an alternative and intervening source of supply.

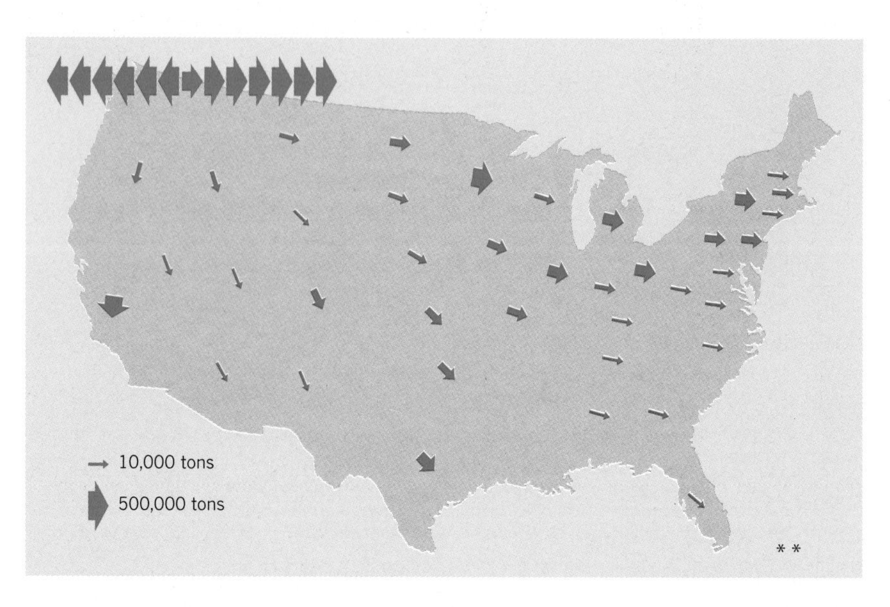

Figure 13.8 Interregional freight flows Width of the arrows is used to show the volume of forest products moved by rail from Washington to other states in the United States in a single year. The row of arrows centred over the state of Washington itself (upper left) shows that most forest products moved very short distances.

[Source: E.L. Ullman, in W.L. Thomas, Jr. (ed.), *Man's Role in Changing the Face of the Earth* (University of Chicago Press, Chicago, 1956), Fig. 162, p. 869. Copyright © 1956 by the University of Chicago.]

Table 13.2 Relative transportability of three lumber products

	High-value veneer logs	Medium-value pulpwood	Low-value mine props
Value (dollars/tonne)	150	20	5
Average length of longest haul by railroad (km)	640	32	8

Source: W.A. Duerr, *Fundamentals of Forestry Economics* (Copyright © 1960 by McGraw Hill, New York), Table 21, p. 167. Reproduced with permission.

The third factor in Ullman's model, ***transferability***, refers to the possibility of moving a product. Transferability is a function of distance measured in real costs or time, as well as of the specific characteristics of the product. Table 13.2 outlines the relationship between the specific value of three types of lumber products (measured in dollars per tonne) and the length of shipment. Local products may be substituted for products that are difficult to transfer in the same way that intervening areas of supply or demand substitute for more distant ones.

Probabilistic Models

Clearly both the gravity model and the Ullman model are gross simplifications. But they laid a groundwork from which geographers have been able to build far more accurate and useful spatial interaction models. One of the outstanding figures in this second wave of probability models is the ***Wilson model*** of Alan Wilson of Leeds University in the UK. (See Box 13.B on 'Alan Wilson and spatial interaction models'.)

Without going into mathematical detail there are three factors that have allowed this improvement. First, the use of advanced *probability theory* in which individual behaviour can be specified and the outcome of human decisions can be incorporated.

Second, the use of realistic *constraints* which confine the answers to the calculations within known boundaries. For example, if we are developing a model to calculate the number of shoppers who will visit a supermarket there is a limit placed on the origins from which they come by the population who live in the origin areas. If we are looking at a journey-to-work pattern at a given factory, the constraint will be placed on the destinations (e.g. the number of jobs in each factory). Modelling for traffic flows within a city or region will demand doubly constrained models.

Third, huge increases in *computing* allow the billions of calculations needed to solve the complex sets of equations that make up the models. Their huge size is due to the fact that the answers cannot be found directly by analytical methods but have to be discovered indirectly by slowly approximating on the solution.

Models incorporating the above changes are now used routinely in a wide variety of applications. Planning new hospital facilities, estimating the impact of new bridges and tunnels, optimizing distribution depots, calculating the impact of higher fuel costs on transcontinental flights, estimating the speed of epidemic spreads give just some of the range.

Box 13.B Wilson and Spatial Interaction Models

Alan Geoffrey Wilson (born Bradford, Yorkshire, England, 1939) is a British geographer who has made fundamental contributions to mathematical models of human spatial interaction. Originally a Cambridge mathematics graduate, Wilson worked in nuclear physics and in the mathematics advisory unit of the UK's Ministry of Transport before moving to the chair of geography at Leeds University. He has since become Vice-Chancellor of that university.

Using models originally developed in statistical mechanics, Wilson built up an alternative to the gravity models widely used after 1945 in urban and regional planning to forecast flows of traffic, freight, and migration. The central idea was to convert the physicist's concept of ***entropy*** (as used in thermodynamics) to situations involving individual members of a population. Entropy is a measure of the amount of uncertainty in a probability distribution or a system subject to constraints. Wilson's entropy-maximizing models allowed (a) more accurate estimates of spatial flows, while (b) removing the need for a formal dependence on physical analogies. Accuracy was partly improved by building constraints into the model to give origin-constrained, destination-constrained, and doubly constrained versions to match different geographical estimation problems. What came to be know as the 'Wilson models' have been extended to a wide number of planning situations, involving both private-sector and public-sector planning decisions. A typical application is the optimization of hospital locations in terms of patient flows and efficient delivery of specialist services. Theoretical links have been established to both linear programming models and catastrophe theory. Many of Wilson's contributions are summarized in *Entropy in Urban and Regional Modelling* (1970), *Urban and Regional Models* (1974), and *Catastrophe Theory and Bifurcation* (1981), and are illustrated in his *Location Decisions and Strategic Planning* (1996).

13.4 Regional Networks

In the previous section we assumed that flows between cities and between a city and its region occurred across a uniform plane. For wave transmissions via radio and TV, this is broadly true. Other flows, however, are generally confined to a series of channels, or routes (Figure 13.9). Geographers see this delicate filigree of transport networks as the arterial and nervous system of regional organization, along which flow signals, freight, people, and all the other essential elements that allow the structure to be maintained. But do these networks have a characteristic spatial structure? And, if so, what controls it?

Networks as Regional Lifelines

Although transport systems form an essential and permanent feature of the

(a)

(b)

Figure 13.9 Junctions for different transport modes Examples of two contrasting types of transport junction. (a) Highway cloverleaf in the United States. This technique was made possible by advances in steel and concrete technology from the 1920s and allowed heavy road transport flows to be merged and mingled without great loss in speed. Note the clusters of warehouses adjacent to the junctions, reflecting their locational attractiveness. (b) Vertical aerial photograph O'Hare Airport, western Chicago. O'Hare's multiple runways and complex terminal buildings allow it to handle more flights than any other of the world's airports. Note the distinctive branching structure of the gates and the cloverleaf junction on the highways to the south of the airport.

[Source: (a) © TRIP/Picturesque. (b) Chicago Department of Aviation.]

economic landscape, the early locational theorists such as Johan von Thünen and Alfred Weber had little to say about them. Yet as early as 1850 a German geographer, J.G. Kohl, created a series of branching networks to serve the settlements in his idealized city region (Figure 13.10(a)). His ideas were taken up by Walter Christaller nearly a century later in his own scheme for a system of cities (Figure 13.10(b)); since then, these ideas have been extended by other workers.

Some features of both the Kohl and Christaller schemes deserve notice. First, the transport networks are *hierarchic* in that they consist of a few heavily used channels and many lightly used feeders, or tributary channels. Like the city systems they serve, the segments of the transport system form an inverse distribution of size with frequency.

(a) Kohl 1850

(b) Christaller 1933

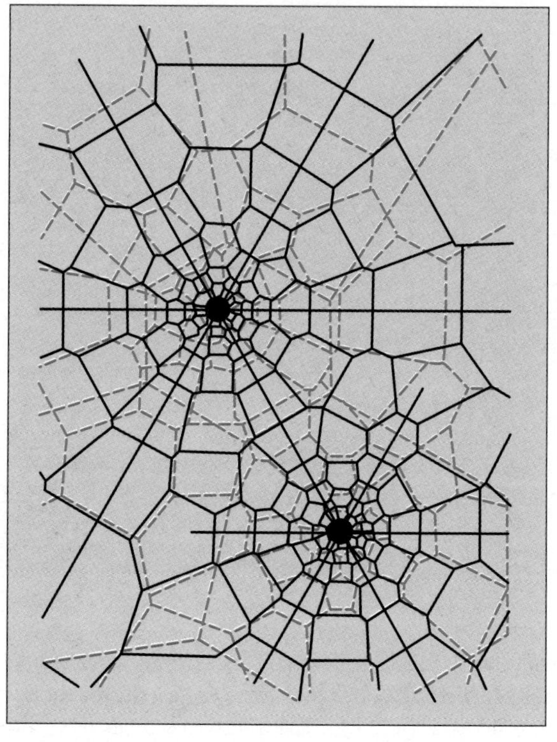

(c) Isard 1956

Figure 13.10 Transport networks for theoretical settlement systems Three alternative schemes proposed between 1850 and 1956. Kohl's network (a) serves a Thünen-like isolated state (see description in Chapter 15). Only the upper half of the circular state is shown and the broken lines subdivided it into identical segments. The Christaller scheme (b) shows traffic routes in a $K = 3$ landscape. (This concept will be discussed in the next chapter; see Figure 14.11(b).) In this case the networks are not symmetrical about the central city: compare the upper right quadrant (a wealthy region in which long-haul traffic prevails) with that of the lower left quadrant (a poor region in which short-haul traffic prevails). Note the contrasts between the straight routes of the former and the zigzag routes of the latter. Minor routes are shown by broken lines. Isard's network (c) has two centres, each of which is surrounded by zones of decreasing population density. Boundaries of the complementary regions are shown as polygons which increase in size with distance from the two centres.

[Source: P. Haggett and R.J. Chorley, *Network Analysis in Geography* (St. Martin's Press, New York, and Arnold, London, 1969), Fig. 3.11, p. 125.]

A second feature also is analogous to river systems, for the transport network has a branching structure in which the angle of branching is intimately related to the flow. The rule governing this phenomenon is familiar. The angle of departure between the branch and the main stem is inversely related to the size of the branch. As the flow on the branch diminishes in relation to the main stem flow, the angle of departure becomes bigger. There is a precise interaction between the shape of the system and the work it has to do.

The number of major routes emerging from each city also has been the subject of research. For inland centres the most frequent number of key routes radiating from a city is six; few cities have less than three or more than eight. These figures are about what we would expect from our earlier discussion of the spacing and location of cities.

Mathematician Martin Beckmann has shown that if a region has a uniform population density and the costs of building a route are everywhere the same, the ideal transport system has a hexagonal honeycomb pattern shown in (Figure 13.11(a). It is assumed in this system that both the origins and the destinations of flows are evenly spread over the region. Beckmann's system is a complicated one and involves some advanced theory that need not concern us here. What is interesting is that the basic honeycomb pattern can be modified to be more realistic.

For example, suppose we retain the idea of a uniform population but assume that we are concerned simply with linking this population with a single source (say, a central place in terms of Christaller's model). In this instance, all the destinations are uniformly spread over the whole region, but the origin is a single point. The best type of transport network is a symmetric honeycomb with holes placed in such a way that the system remains simply connected, as in Figure 13.11(b). By 'simply connected' we mean there are no loops in the system and it still has a basic branching form like a tree. This treelike form is important because it helps us to bridge the gap between Beckmann's honeycombs and Kohl's branching patterns in Figure 13.10(a). The two schemes differ in two ways: Kohl's region is bounded (in fact, it is circular) and has a higher density of settlements in the centre than at the periphery, whereas Beckmann's region is continuous (no boundary is specified) and the population density is uniform throughout. If we modify the Beckmann model by allowing a higher population density (and, thus, smaller honeycombs) near the centre and add a circular boundary, the model has a spatial form very similar to that of Kohl's network. One network is then simply a special case of the other. The missing link between the two is provided by Isard's landscape as shown in Figure 13.10(c) where a honeycomb system of transport links is developed about a pair of centres.

Can geographers find regions in which to test their model of transport networks? Certainly the number of areas of entirely new settlements is fairly small. The pattern of settlement in a Dutch *polder* (i.e. major area of land reclaimed from the sea) may be the nearest approximation we can find to the pattern that would result if a new, empty, and rather uniform landscape were settled all at once rather than slowly over a historical period. The scheme actually adopted there is rectangular, not unlike the pattern of roads laid out in the new territories of the United States during the 1800s. Other small agricultural areas such as plantations also have been designed according to the principle of the Beckmann model.

Network design is important when a new system of roads is superimposed on existing ones. The interstate highway system in the United States and the new highway system of the United Kingdom typify the locational

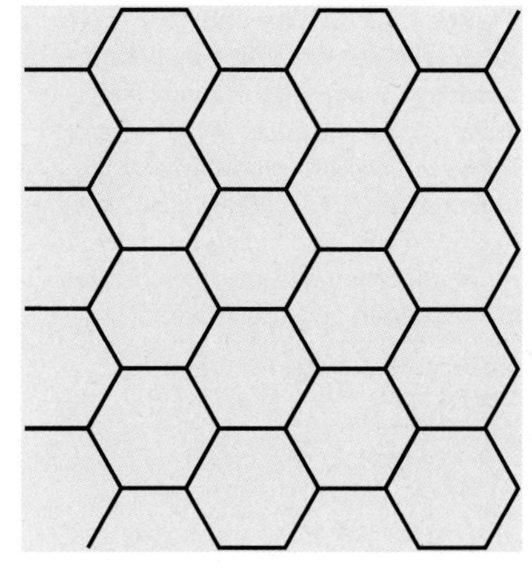

(a) Uniform origins

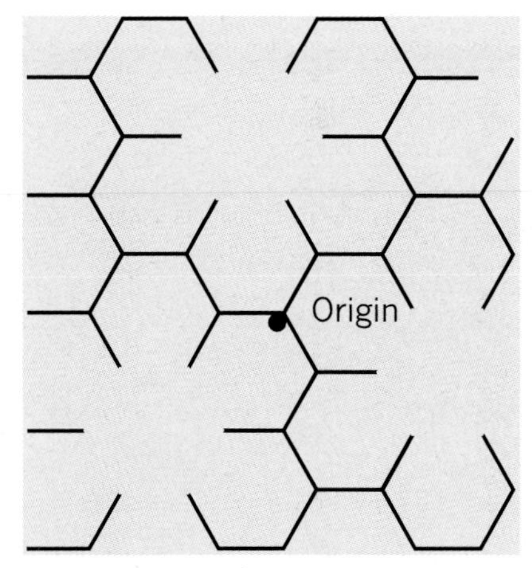

(b) Single origin

Figure 13.11 Optimal transport networks These networks assume a uniform population and homogeneous regions. In (a) there are multiple origins and multiple terminals for flows. In (b) there are a single origin and multiple terminals.

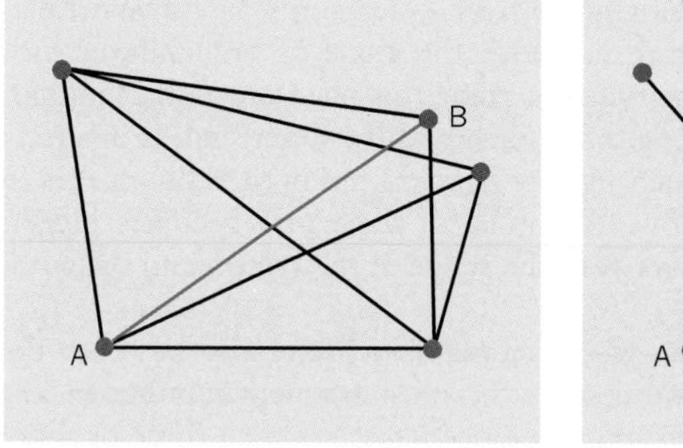

(a) User optimum

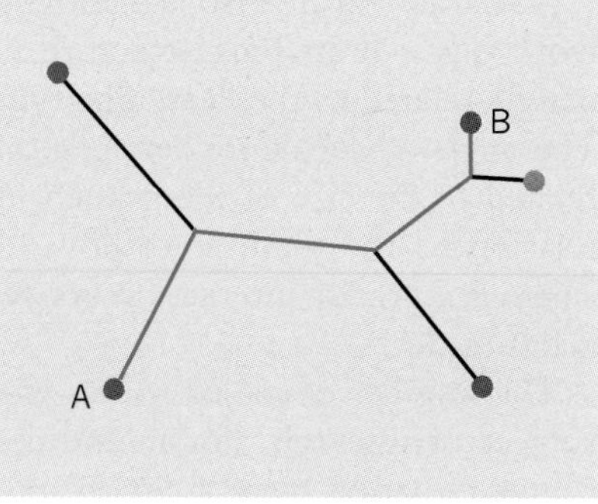

(b)

Figure 13.12 The shortest networks connecting five urban centres Alternative solutions are shown. In (a) the costs of the system to the user are most important and the interchanges are located within the cities. In (b) the cost of building the network is more important and the interchanges are located in non-urban areas.

[Source: W. Bunge, *Lund Studies in Geography C*, No. 1 (1962), Fig. 7.10, p. 183.]

compromise that must be made between the cost of building networks compared with the cost of using them. We can illustrate this need for compromise by considering Figure 13.12, which presents a simple network designed to connect only five towns.

If we design the network to minimize the costs to the user, we will make as many of the links as direct as possible. (See direct link AB in Figure 13.12(a).) But if we design the network to minimize building costs, we will have a different kind of pattern. In this second case the link AB will be much longer, but the *total* length of the network will be much smaller, as shown in Figure 13.12(b). Comparisons with actual transport systems reveal an evolution from the first to the second type as flows increase. Look at a historical atlas, and compare the changing structure of the US railroad network during different periods of its growth. Indeed, there is still a strong contrast between the sparse rail network in the west, where the costs to the builder were very important, and the denser rail network in the east, where greater intercity flows made the costs to the user more significant.

Networks as Graphs

In our discussion of urbanization processes we saw how changes in the relative accessibility of cities had important repercussions on their relative growth. (Take a look back at Chapter 8 to refresh your memory on this point.) So if we add a new transport link to a regional network, we should expect it to affect the relative accessibility of all the cities that are connected to it.

Let us consider a specific example. Figure 13.13 shows the road network of the western part of Ontario, Canada. Each of the 37 nodes represents either a settlement with at least 300 inhabitants or a main highway intersection. The road network, extending from the town of Sudbury to the town of Saulte Sainte Marie, is connected to road systems in the remainder of Canada, but these external links are so few that we can treat the network as a closed system.

Now let us suppose that seven new road links are proposed: links (a) to (g) in Figure 13.13. Which links will have the greatest impact on the total accessibility of points within the area to one another? Also, what local effect will each new link have on the *relative* accessibility of individual nodes within the network?

One way of answering these questions is to turn to a branch of mathematics that treats networks on the most primitive topological level. *Topology* is a branch of geometry that is concerned with the quality of

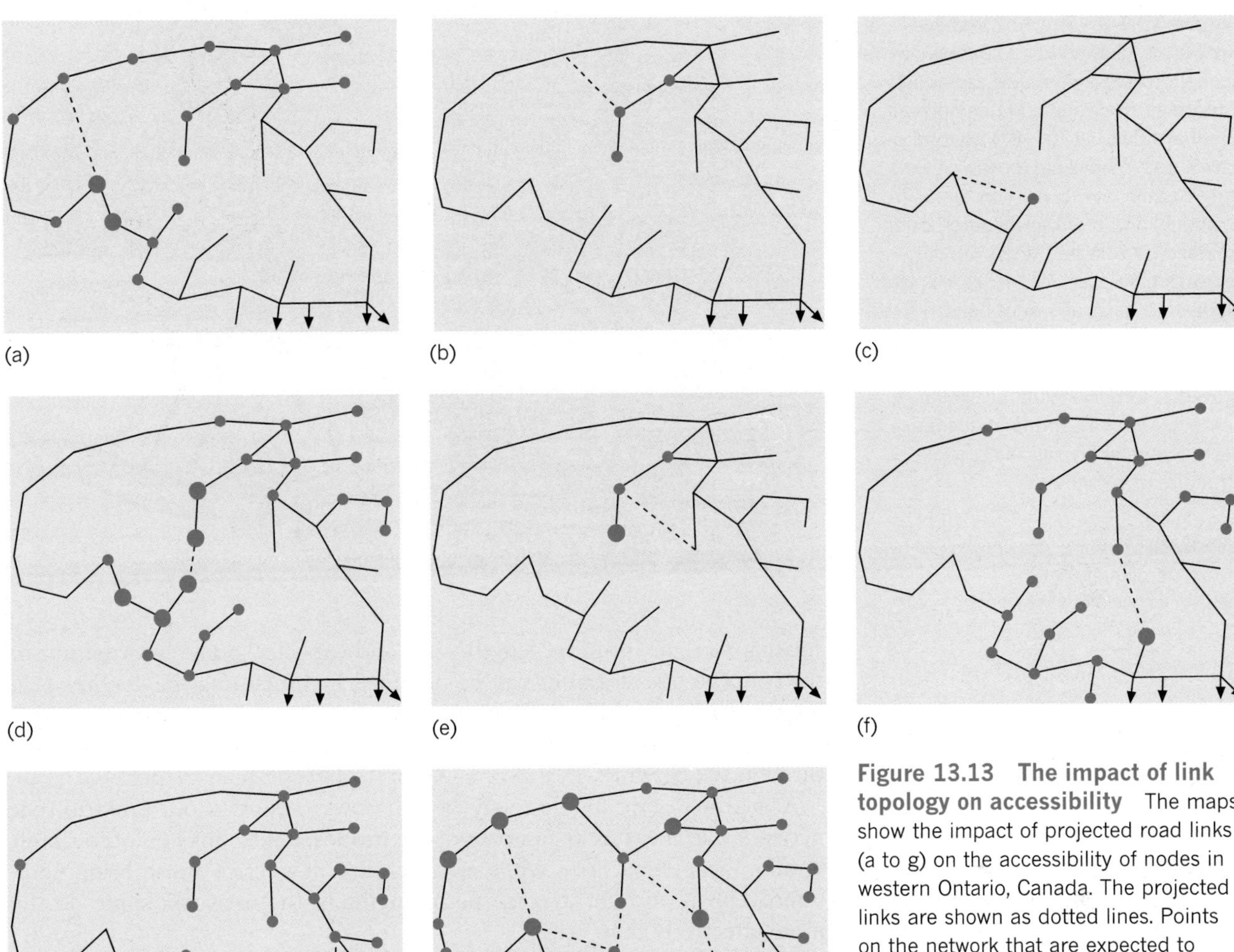

Figure 13.13 The impact of link topology on accessibility The maps show the impact of projected road links (a to g) on the accessibility of nodes in western Ontario, Canada. The projected links are shown as dotted lines. Points on the network that are expected to benefit by the improved connections are shown by dots: large dots indicate major improvements, small dots only minor improvements. Link (d) has the greatest impact on improving access. The final map (h) shows the combined effect of building all new links.

[Source: I. Burton, *Accessibility in Northern Ontario*, unpublished paper, 1963, Fig. 1.]

connectivity, that is, whether objects are or are not connected in some way. Its earliest and most famous geographic application by the Swiss mathematician Euler is shown in Figure 13.14. Note that Euler was not concerned with distance or direction (the usual concern of geometry) but simply with whether a particular path through a network – the seven bridges connecting the different parts of the city of Konigsberg – was possible or not. Euler's puzzling over this problem was to grow into *graph theory*, a branch of mathematics whose concepts are proving of increasing interest to geographers studying regional networks.

To use graph theory, we must reduce the networks to graphs. This reduction involves throwing away a great deal of information about flows and characteristics of routes, but retaining the essential spatial factors of networks, nodes, and links. *Nodes* are the termination or intersection points of a graph. They may be assigned values denoting their location, size, and traffic they can handle, and so on. Depending on the varying scale of the analysis, nodes may be whole cities or street intersections. *Links* are connections or routes within a network. Links can also be assigned values

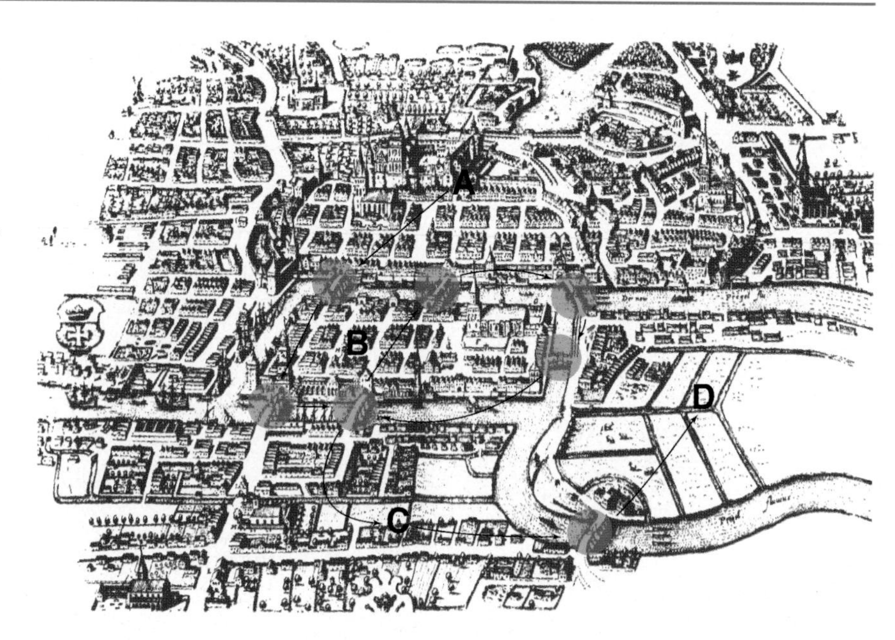

Figure 13.14 The Königsberg problem Sometimes apparently trivial spatial problems can have profoundly important implication. Mathematician Leonhard Euler (1707–83) puzzled over why it was impossible for the citizens of the Prussian city to visit four areas (A, B, C, and D) and cross all its seven bridges (marked by coloured dots) without recrossing at least one bridge. His later studies of the structure of networks laid the groundwork for a major branch of mathematics, graph theory, which has proved of importance in designing computer circuits and is of direct use to geographers analyzing the spatial structure of networks.

[Source: Drawing courtesy of The New York Public Library, The Astor, Lennox, and Tilden Foundations.]

relating to their location, length, size, and capacity. Some information on the connectivity of graphs can be obtained by measuring the *average path length*. Path lengths are determined by the number of steps (or 'hops') between pairs of nodes, moving one link at a time along the shortest path through the network. (See Box 13.C on 'Connectivity in graphs'.)

Armed with this simple ruler, we can now go back to our Ontario road network and measure the impact of the proposed new links on accessibility within the system. For while each of the new links must bring some improvement and cut average path lengths in the network, some do this more effectively than others.

If we build the new links into the network one at a time, we can test just how much each one improves the connectivity of the system. Link (d) (from Folyet to Chapleau) does the most, cutting average path lengths in the network by 9.5 per cent. Next best is link (a). Note that both (a) and (d) short-circuit the network by joining the northern and southern halves. Other links, such as (c), are on the periphery of the network and their building has little overall effect.

How do individual cities benefit from the new links? This is measured by calculating the changes in the average path length for *each* city. Figure 13.13 shows the spatial pattern of improvements in connectivity, city by city, from each link proposed. Note that some links have a very localized benefit (e.g. (c) and (e)). In contrast, (d) brings less striking improvements, but spreads them rather equitably over almost all the eastern nodes. Some projects that are desirable on local grounds may not have the most beneficial effect on the whole network.

Graph theory provides only a first step in the analysis of transport systems. Links must be weighted to reflect the traffic they carry. We also must relax our implicit assumption that each link is just as costly to construct as all of the other ones. The tools needed for a full analysis of transport systems – cost–benefit ratios, spatial allocation models, and the like – are a bit complicated for an introductory course and are generally reserved for courses in transport geography. Graph theory does, however, serve to highlight the delicate balance between the urban system of cities and the regional transport net that links them and binds them together. Each new

Box 13.C Connectivity in Graphs

Graph theory provides an effective way for geographers to measure the degree of 'connectedness' within transport networks. Consider a simple graph consisting of five nodes (A, B, C, D, and E) connected by five links (shown in solid lines).

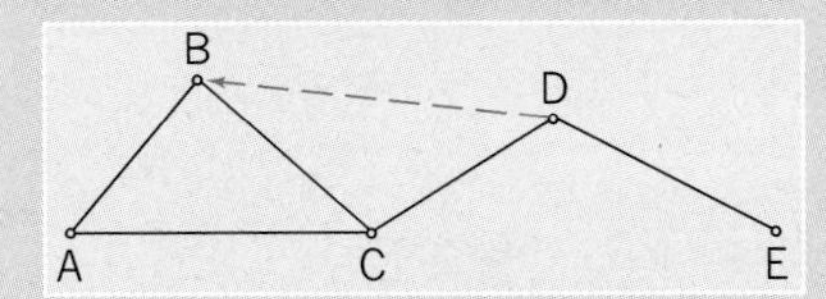

We can summarize the information in the graph in a connectivity matrix. Here the distance between pairs of nodes is expressed in terms of the number of intervening links along the shortest path connecting them, as follows.

From:	To: A	B	C	D	E	Row sum	Average path length
A	0	1	1	2	3	7	1.75
B	1	0	1	2	3	7	1.75
C	1	1	0	1	2	5	1.25
D	2	2	1	0	1	6	1.50
E	3	3	2	1	0	9	2.25
					Total	**34**	**1.70**

The row sum for each node provides a measure of its relative accessibility. Thus node C (with a row sum of five links) is the most accessible and node E (with a row sum of nine links) is the least accessible. The grand total (termed the *dispersion value* of the graph) provides a measure of the graph's size in terms of all the paths within it. Dividing the row sums and the dispersion value by the number of positive values provides a measure of the average path length, which can be used in comparing one network with another.

The connectivity matrix in this simple case is symmetric about the diagonal of zeros because all five links are two-way. If we introduce a one-way link from D to B, it will disturb the symmetry. That is, DB will have a link value of only 1, but BD will have a link value of 2. The effect of the new one-way link is to improve the network's connectivity slightly. (The dispersion value falls to 33.) If we were to reverse the direction of the same link from B to D, the improvement in connectivity would be greater. For a further discussion of graphs including weighted as well as directed links, see K.J. Kansky, *Structure of Transportation Networks* (Department of Geography Research Paper 84, University of Chicago, Chicago, Ill., 1963), Chapter 1, and P. Haggett and R.J. Chorley, *Network Analysis in Geography* (Arnold, London, 1969), Chapter 1.

airline link, pipeline, highway, or shipping schedule tilts the balance of locational advantage this way and that, in a manner geographers must continue to monitor carefully if they are to be able to track – and eventually forecast – changes in the world about them.

Changes in Networks 13.5

In the last section we treated networks as graphs and used some simple examples to show change could be measured. Transport networks are always under change: the completion of the Trans-Siberian railroad, the advent of the wide-bodied jet aircraft, the surge in Internet commerce all

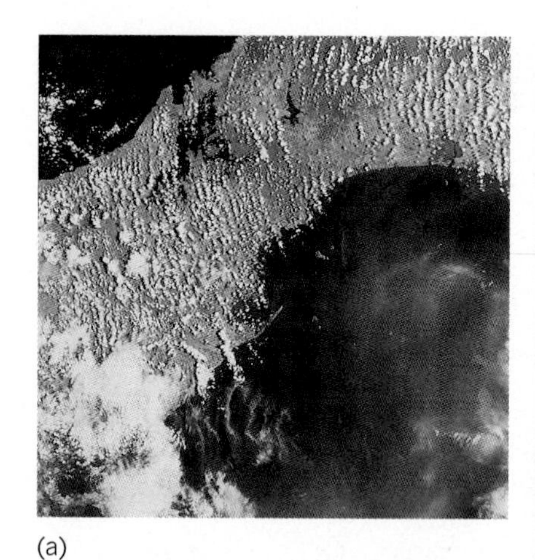

(a)

(b)

(c)

Figure 13.15 The world's principal ocean canals (a) Satellite image of the Panama Canal built in 1906–14 to connect the Caribbean Sea (above) with Pacific Ocean (below). Use was made by the civil engineers of the lakes shown in the northern part of the photograph. (b) Ship navigating locks on the Panama Canal. (c) The Suez canal was built in 1869 to connect the North Atlantic via the Mediterranean to the Indian Ocean via the Red Sea. The lakes that are linked by the canal can be seen in this *Columbia* Space Shuttle view of the Gulf of Suez. (d) The impact of the Suez and Panama canals on distances from the British port of Liverpool.

[Sources: (a) Earth Satellite Corporation/Science Photo Library. (b) © TRIP/R. Belbin. (c) NASA/Science Photo Library. (d) From US Naval Oceanographic Office, Hydrographic Publication No. 117 (1948), Fig. 1.]

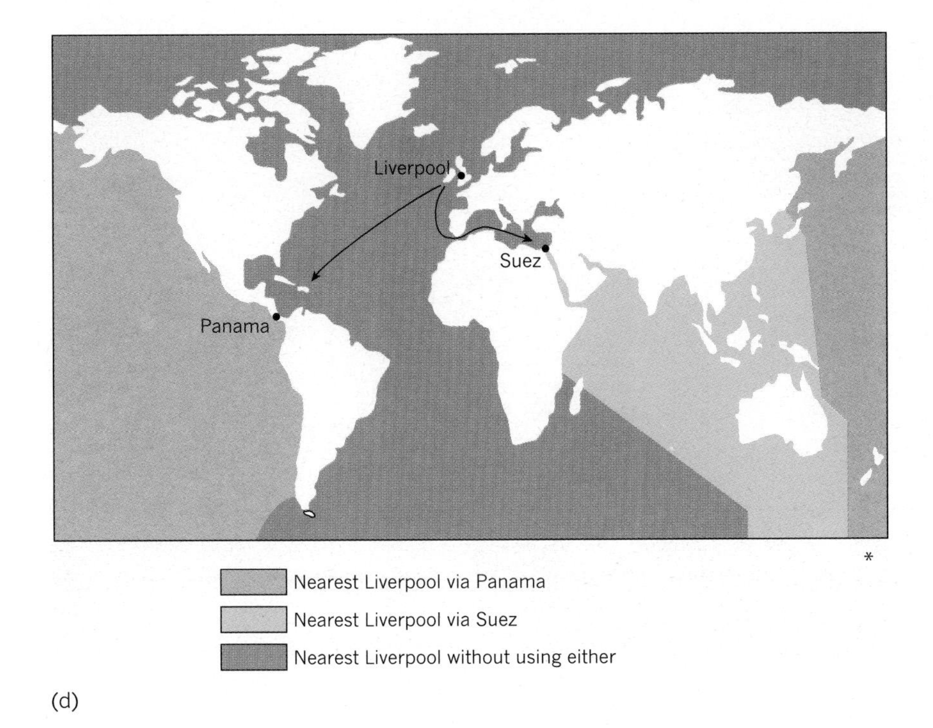

(d)

have important regional consequences. We show here some historical examples of sudden change at scales from the global to the local and consider the abrupt reorientations that follow the establishment of new transport routes. By constructive changes we refer to the building of new routes; by destructive we refer to the breaking of existing routes.

Constructive Changes

Changes that arise from the construction of a new route can occur at a variety of geographical scales. For example, remarkable changes of world distances came with the building of the two great trans-isthmian canals of Suez (1869) and Panama (1914) shown in Figure 13.15.

Prior to the canals' construction, the Eurasian–African landmass had placed a 20,000 km (12,000 mile) north–south barrier between the Atlantic and Indian Oceans. The Americas had an even greater 22,000 km (14,000 mile) barrier between the Atlantic and Pacific Oceans. These barriers were important because they concentrated trade in the northern hemisphere, restricting it mostly to east–west flows. In both instances the barriers pinched down to isthmuses less than 150 km (95 miles) wide. The first breach, Suez, shortened the sea distance between Europe and India by three-quarters. The second canal, Panama, had its greatest effect on the links of the east coast of North America with the west coast and with the western Pacific Ocean. A generalized map (Figure 13.15(d)) shows the shortening of distances to Britain by both canals.

In the same way that the earth's natural resources are intrinsically varied and take on value as humans find new uses for them (or discard them), so the peculiar resource of the earth's space is varied. It is not uniform and continuous like the abstract space of the geometer but torn and twisted. Like conventional resources, particular locations take on value as humans accord them a special use, and lose value as we turn our interest away from them.

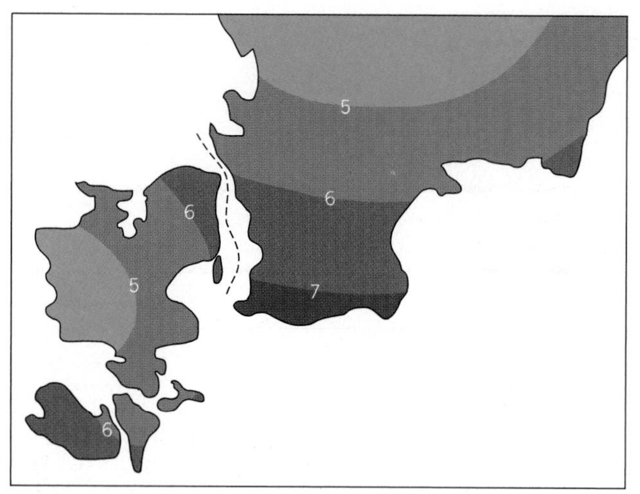

(a) Barrier assumption

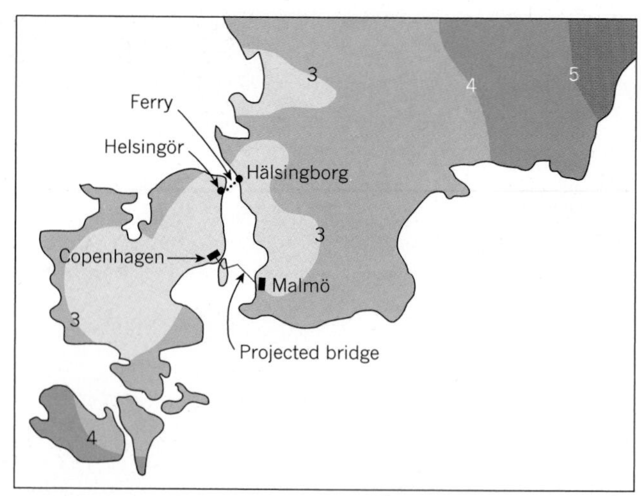

(b) Bridge assumption

Figure 13.16 Projected impact of an international link Hours of travel time by car from any location in eastern Denmark and southern Sweden, in order to reach a market of three million people. Maps show travel times under two assumptions. The Oresund link was completed in 2000. It consists of a tunnel and bridge carrying both road and rail traffic. The rail link now reduces travel time from Copenhagen to Malmö to about half an hour.

[Source: From T. Hägerstrand, *Transactions of the Institute of British Geographers*, No. 42, Fig. 14, p. 15.]

* * *

The completion of the Danish–Swedish road bridge across the narrow strait between the cities of Copenhagen and Malmö is likely to have a dramatic effect on this part of Scandinavia. The Oresund Region with its strong universities and science-related industries is set to become the silicon and biotechnology centre of Scandinavia. Figure 13.16 shows one early calculation of the dramatic reorientation that the bridge building implies.

Disruptive Changes

Locational advantages may of course be reduced as well as increased by network change. Let us take one dramatic example.

On Sunday, 5 January 1975, at 9:30 p.m. the Australian city of Hobart was cut in two. The capital city of Tasmania, Hobart's population at the time was 145,000 spread along the eastern and western sides of the Derwent River (Figure 13.17). On the western shore lay the old city of Hobart with 104,000 people; on the eastern, the fast-growing suburb of Clarence with 40,000 people. Joining them across the kilometre-wide river was the four-lane highway over the Tasman Bridge joining the city to eastern shore suburbs and the airport.

The disaster that struck the bridge was so sudden and unexpected that a few cars ran on into the deep water. The *Lake Illawarra*, a freighter loaded with zinc concentrate, collided with the bridge pillars. The collision brought down a major section of the bridge decking, which fell onto the freighter, sinking it on impact.

Although we have met natural disasters earlier in this book (see Section 3.5), this catastrophic change in transport networks has few parallels. Apart from the immediate disaster of loss of life, the longer-term effects were soon apparent. Approximately 34,000 persons usually crossed the bridge in each direction on a busy weekday. Although over a quarter of the population lived on the eastern shore, this was largely a dormitory area which had mushroomed since an earlier bridge had first opened in 1943. Over 90 per cent of jobs in retailing and manufacturing, and all the entertainments, lay on the western shore.

With the severing of a bridge link, two alternatives were open to eastern shore commuters. One was a 48 km (30 mile) circuitous route to an upstream bridge; another was to queue for a place on the few and now overloaded ferryboats. Both solutions had the effect of suddenly converting a few minutes' drive into a frustrating and expensive 60–90 minute journey. Its effect on both the social and economic geography of the city were monitored by geographers at the University of Tasmania. Findings show how the impacts span a wide range of city life: from changing price levels for houses to shifts in patterns of crime. The two halves of the city were linked together again when the bridge reopened in 1978, and the return to 'precatastrophic' patterns began.

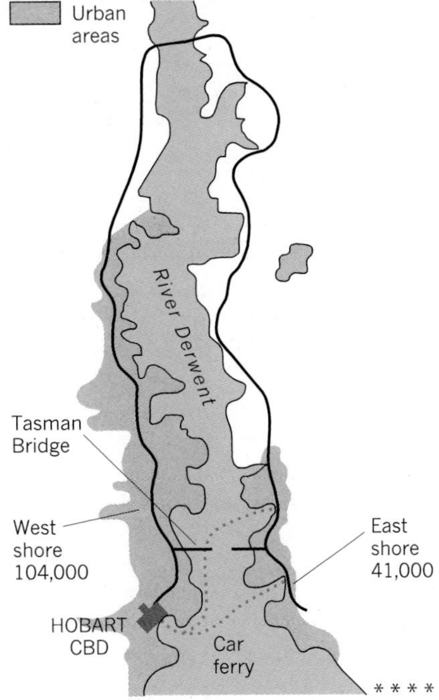

Figure 13.17 Impact of link removal The Tasman bridge linking the two sides of the city of Hobart, Australia, was smashed by an ore-carrying ship in January 1975. Severing this vital link faced east-side commuters with a 48 km (30 mile) additional drive or long queues for ferry boats.

[Source: Photo by author.]

The real world is made up of an immensely complex mosaic of regions. As we have seen, geographers have attempted to make sense of this by devising formal systems of regions. There is, however, no single or generally accepted set of world regions. Rather, there are alternatives, each with certain strengths and weaknesses. The most successful types of regional units appear to be those whose spatial boundaries coincide most closely with ecological or socioeconomic systems.

Reflections

1 Different modes of transport are competitive over different distances and different geographic environments and with different payloads. Take one of the modes summarized in Table 13.1 and consider its comparative advantages and drawbacks.

2 Define the three basic conditions needed for flows to occur in the Ullman model. Illustrate each, using flows into or out of your own state or province as an example.

3 Consider the basic gravity model outlined in Figure 13.7. What changes would you expect to occur in the flows between two cities if (a) both doubled in population or (b) the travel time between them was halved (e.g. by a bridge built to replace a slow ferry link)?

4 Most human movements have a distinctive geographic pattern. Sketch a map of the campus and the surrounding community and plot on it the paths followed by members of the class in getting to the building each day. Compare your results with Figure 13.10. How far does your network of paths resemble Kohl's branching tree? How are the paths of walkers, cyclists, bus riders, etc., different?

5 Spatial interactions tend to follow a distinctive distance–decay function. Gather data on any one pattern of spatial interaction (e.g. the number of guests in a local hotel, the number of visitors to a national park, or the number of students to a state university). Plot the observed number coming from each 'source' area on the vertical axis of a graph and the expected number on the horizontal axis. For the expected number use a gravity model with the relevant population of each source area divided by its distance from the hotel, park, or university campus. Comment on your findings.

6 Consider the two modes of transport in Figure 13.9. Sketch typical networks for each mode and convert to a graph. What differences can you see in the connectivity of the two types of networks?

7 Review your understanding of the following concepts using the text and the glossary in Annex A of this book:

average path length
connectivity
distance–decay curves
entropy
graph theory
gravity models
intervening opportunity model
nodal region
regional complementarity
transferability

One Step Further ...

A useful starting point is to look at the essay by Arild Holt-Jensen on spatial science in I. DOUGLAS *et al.* (eds), *Companion Encyclopaedia of Geography* (Routledge, London, 1996), pp. 818–36. A brief but useful introduction to transport geography is given by Alan Hay in his essays in R.J. JOHNSON *et al.* (eds), *Dictionary of Human Geography*, fourth edition (Blackwell, Oxford, 2000), pp. 855–7. For excellent but brief introductions to the topics covered in this chapter, see R.S. TOLLEY and B.J. TURTON, *Transport Systems, Policy and Planning: A Geographical Approach* (Longman, Harlow, 1995), E.J. TAAFFE and H. GAUTHIER, *Geographjy of Transportation* (Prentice Hall, Englewood Cliffs, N.J., 1973), and A. HAY, *Transport for the Space Economy* (Macmillan, London, 1973). An outstanding historical view of transport changes and their impact on human organization is given in Jay E. VANCE, Jr., *Capturing the Horizon* (Harper & Row, New York, 1986).

A wide range of geographic work is brought together in a set of readings, M.E. ELIOT HURST (ed.), *Transportation Geography* (McGraw-Hill, New York, 1974), while the work of a major scholar in this field is commemorated in E.L. ULLMAN, *Geography as Spatial Interaction* (University of Washington Press, Seattle, 1980). A guide to interaction models is provided by A.G. WILSON, *Complex Spatial Systems; The Modelling Foundations of Urban and Regional Analysis* (Pearson Education, Harlow, 2000). The spatial structure of transport networks and the locational principles that determine their form are discussed in Christian WERNER, *Spatial Transportation Modelling* (Sage Publications, New York, 1985), P. HAGGETT and R.J. CHORLEY, *Network Analysis in Geography* (St. Martin's Press, New York, and Edward Arnold, London, 1969). Transport in the digital age is described in A. KELLERMAN, *Telecommunication and Geography* (Bellhaven, New York, 1993).

The transport and spatial interaction 'progress reports' section of *Progress in Human Geography* (quarterly) will prove helpful in keeping up to date. Research on both spatial interaction and the structure of networks is reported regularly in the major geographic journals, especially in *Economic Geography* (quarterly). The more advanced quantitative work is often presented the *Journal of Transport Geography* (quarterly), in *Geographical Analysis* (quarterly) and in the *Journal of Transport Economics and Policy* (quarterly). For readers with access to the World-Wide Web see also the sites recommended for this chapter in Appendix B at the end of this book.

MetLife

CHAPTER 14

Nodes and Hierarchies

■ denotes case studies

Figure 14.1 Central hubs of the world economic system The central business district of one of the world's great cities. From a population of 200 when bought from the Manhattan Indians by the Dutch West India Company in 1626, Manhattan has grown to be the central node of a metropolis of over sixteen million. The demand for scarce space has forced buildings ever higher with 'skyscrapers' of up to 100 floors and daytime populations of up to 20,000 people per building. Compare Figure 8.1.

[Source: Mike Yamashita/*Telegraph* Colour Library.]

> For some minutes Alice stood without speaking, looking out in all directions over the country... 'I declare it's marked out just like a large chess-board... all over the world – if this is the world at all.'
>
> LEWIS CARROLL *Through the Looking Glass and What Alice Found There* (1872)

A night flight over any densely populated part of the world presents a unique opportunity to see the intricate spatial structure of the human ant's nest. Aboard an intercontinental jet flying at an altitude of 11,000 m (36,000 ft), our concern with cultural complexity slips away. Large cities are visible only as faint clusters of lights spaced many kilometres apart; as the plane loses height, and smaller settlements and isolated farms come into view, sporadic pinpoints of light appear. At night the earth viewed from above looks much like the heavens as seen from the earth. The great galaxies visible to the naked eye dissolve into a host of stars when we look at them through a telescope.

Great cities such as New York (see Figure 14.1 p. 420) form the brightest stars in the sky. But around and between these international giants (see the discussion of 'World Cities' in Chapter 19) there stretches out a vast nebula of settlements of all sorts of size and shape. Figure 14.2 gives just two examples: a small settlement on the Canadian prairie contrasting with a great city on the Mississippi River.

Geographers have long been fascinated by the galactic patterns of human settlement. What forms do they take? Are their forms random and chaotic – or can patterns and formative processes be discerned? If there are regularities, what lies behind them? We look here at some of the answers to these questions, at the models of settlement geographers have built, and how we can use these models to predict changes and to plan more efficient and attractive patterns of settlement.

(a)

(b)

Figure 14.2 Some extremes on the North American settlement spectrum (a) At the lower end lie the small farming communities. Aerial view of the Riceton grain elevators and wheat fields in the Canadian province of Saskatchewan. (b) At the upper end of the spectrum are the great metropolises. Infrared LANDSAT-5 satellite image of St. Louis, Missouri, USA. The Mississippi River separates St. Louis from East St. Louis. Founded in 1764 as a French fur trading post, St. Louis was transferred to Spain, then back to France, finally becoming part of the United States after the 1803 Louisiana Purchase. The city has been a transportation hub for the mid-west since steamboats arrived on the river in the early 1800s. Today, one of the United States top-twenty largest cities, it has a population approaching three million in the wider metropolitan area.

[Source: (a) NFB Collection, Courtesy Canadian Museum of Contemporary Photography, Ottowa/George Hunter. (b) Terranova International/Science Photo Library.]

Defining Urban Settlements 14.1

To answer questions about settlements, we must first puzzle out how to define them. We must try to find ways of describing them and comparing the characteristics of settlements in one region with those of another. Perhaps the simplest solution is to begin by asking how big human settlements are. For if we can define their size, we can go on to compare their magnitudes and relate them to other findings.

Questions of Size

Let us consider the various definitions of urban settlements. The definitions used in legal and administrative documents will tell us precisely what we mean by Topeka, Kansas, or Melbourne, Australia. Unfortunately, the legal and administrative borders of cities are often a historical or constitutional legacy. Typically, the legal city has fixed boundaries that survive long after urban development has exceeded those bounds. Such a mismatch is termed a *bounding problem*. A legal city is often 'underbounded' (Figure 14.3(c)). Parts of the urban area may remain legally outside the city though but share a common boundary with it. Beverly Hills, completely surrounded by the city of Los Angeles, is a case in point. In England, some boroughs still have a municipal status that is a legacy of their former importance and is out of line with the present small size. The number of inhabitants an area must have to be considered urban also varies from country to country. In Iceland places with a few hundred people are termed 'urban', whereas in the Netherlands a population of 20,000 is needed.

A second approach to defining urban settlements is to ignore the legal boundaries and to try to define each settlement in terms of its physical structure. For example, we might define a settlement on the basis of continuous housing, or population above a certain density, or the intensity of traffic. But there are difficulties here too. What do we mean by 'continuous' housing, and what happens when different definitions do not all give the same answer? Figure 14.4 presents some different definitions of New York based on both its legal boundaries and its physical structure. Note that New York City itself (Manhattan, Staten Island, Brooklyn, Queens, and the Bronx) is only a small part of the continuous urban sprawl that is greater New York.

The 'mismatch' between the legal and the physical city becomes vitally important when the legal city, with its static or declining population and limited tax base, has to provide public services such as transport and police for the millions of commuters who cross its legal boundaries to work each day. As the discrepancy between the legal and economic boundaries of the

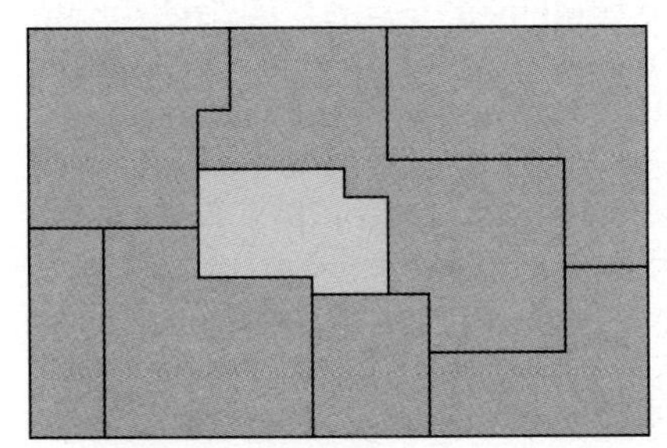
(a) Matched boundaries

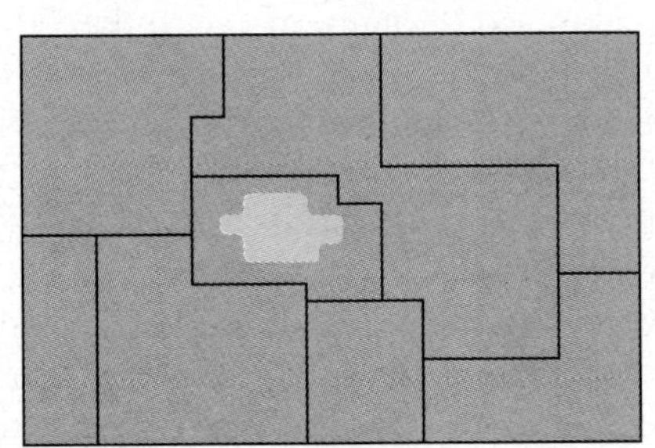
(b) Overbounded city

(c) Underbounded city

Figure 14.3 Difficulties in defining cities The census limits of cities rarely coincide with their actual built-up area (shaded). Matched boundaries are rare; underbounded cities, the most common.

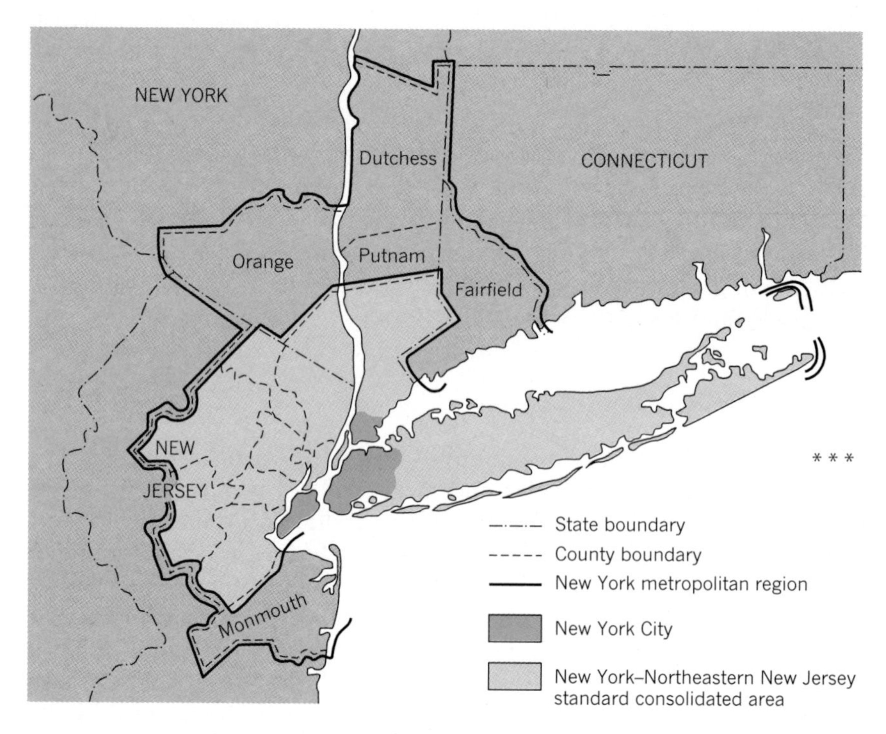

Figure 14.4 Varying definitions of a metropolis The map shows three different boundaries for New York, none of which coincides exactly with the limits of the built-up area. Part of New York City's financial problems stem from the mismatch between its area and the built-up area of the city. As industry and high- and middle-income residents have moved out to the suburbs, so the tax base of the central city has been eroded. See also the photographs of New York in Figures 8.1 and 14.1.

city becomes worse, the pressure for some form of revenue sharing or boundary adjustment grows. This discrepancy also affects our ability to answer even the simplest questions about the size of the city. To take an extreme case, the 'legal city' of Sydney, Australia, in 2000 had a population of only 50,000 while the 'built-up area' of Sydney had a population of over four million. This difference of around 80 times in size is unusual, but important enough to show that definition of settlements is a matter of concern.

Some Possible Solutions

As a result of this problem, international, and indeed intranational, definitions of urban settlements are being standardized. One definition of world metropolitan areas, by demographer Kingsley Davis, runs to twelve pages, including two pages on difficult cases. In the United States the concept of a *Standard Metropolitan Statistical Area (SMSA)* was introduced in 1960 so that metropolitan areas could be defined realistically by using three criteria. It has been modified slightly at each successive census but the principles remain the same. First, a population criterion: each SMSA must include one central city with 50,000 or more inhabitants. Special rules allow contiguous cities (i.e. those directly adjoining each other) and nearby cities (within 32 km, or 20 miles, of each other) to be combined. Second, the metropolitan character of an area is taken into account. At least 75 per cent of the labour force of the county must be employed by non-agricultural industries. Other criteria for SMSAs relate to population density, the contiguity of townships, and ratios between the non-agricultural labour forces of the counties making up the unit. Finally, the integration of the areas that constitute the SMSA is considered. Counties are integrated with the county containing a central city if 15 per cent of the workers living in the county commute to the city, or if 25 per cent of the workers in the county live in the city. This measure of integration can be supplemented by other measures based on the market area for newspaper subscriptions, retail trade, public transport, and the like.

Despite their apparent comprehensives, the SMSA definitions have still not fully solved the urban bounding problem. Improved definitions using county blocks and commuting data continue to be sought by geographers interested in international comparison between cities.

Not only metropolitan areas but small towns and villages as well can be difficult to define. The smaller settlements, however, unlike urban areas, are usually overbounded (Figure 14.3(b)). Similar sets of dull, but necessary, rules must be worked out for these cases too.

Measuring Patterns of Settlement

If we mark the location of cities, towns, and villages on a map, we can see the overall pattern of settlement. Let us look at some examples. Figure 14.5 shows four sample areas from different parts of the United States; each square has an area of 5000 km^2 (2000 square miles), and each dot represents an urban settlement as listed in the United States census. We can distinguish among the four areas in two ways. First, they differ in density (i.e. in the number of urban settlements per square kilometre). For example, the North Dakota area has only 16 towns (3.2 towns/1000 km^2) while the Ohio area has 98 (19.0 towns/1000 km^2). Measurement is no problem, and it is a simple matter to arrange the four areas in a sequence of increasing density, as in Figure 14.5.

The patterns of settlement also differ in a second characteristic which is less easily calibrated. Compare the patterns of the Washington and Minnesota areas in Figure 14.5. Both have similar densities (5.8 and 6.4 towns per 1000 km^2, respectively) but strikingly different arrangements of the towns in space. The Washington towns are clustered, whereas the Minnesota towns are scattered. To measure this second property, geographers have adopted a spacing index (see Box 14.A 'The nearest-neighbour index') that enables them to rank patterns of settlement along a scale from 'highly clustered' to 'highly dispersed'. The values of the index range from a theoretical low of zero, when all the settlements are concentrated at a single point, to a maximal value (2.15) when the pattern of settlement is that of a triangular lattice.

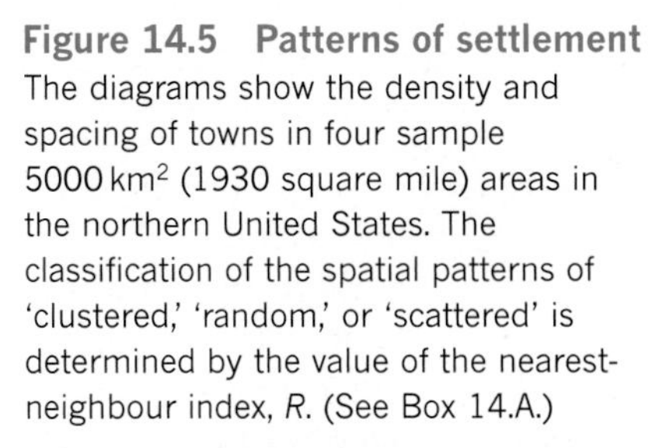

Figure 14.5 Patterns of settlement The diagrams show the density and spacing of towns in four sample 5000 km^2 (1930 square mile) areas in the northern United States. The classification of the spatial patterns of 'clustered,' 'random,' or 'scattered' is determined by the value of the nearest-neighbour index, *R*. (See Box 14.A.)

[Source: L.J. King, *Tijdschrift voor Economische en Sociale Geografie*, **53** (1962), Fig. 3.7, pp. 4–6.]

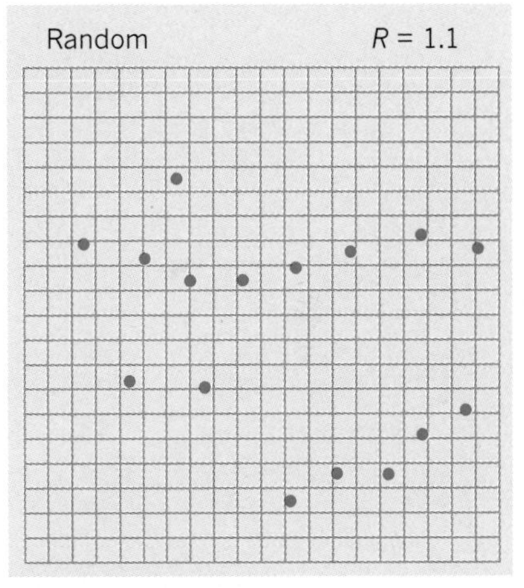

(a) Northwest Dakota
3.2 per 1000 km^2

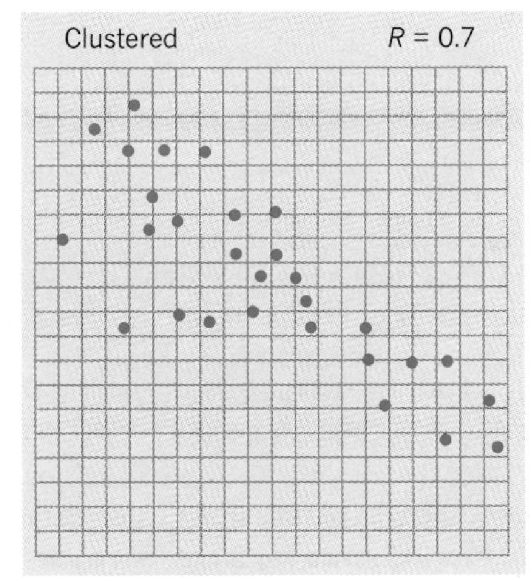

(b) Southern Washington
5.8 per 1000 km^2

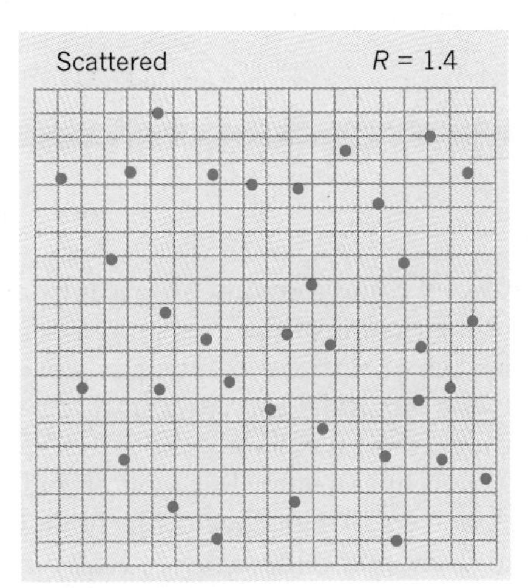

(c) Western Minnesota
6.4 per 1000 km^2

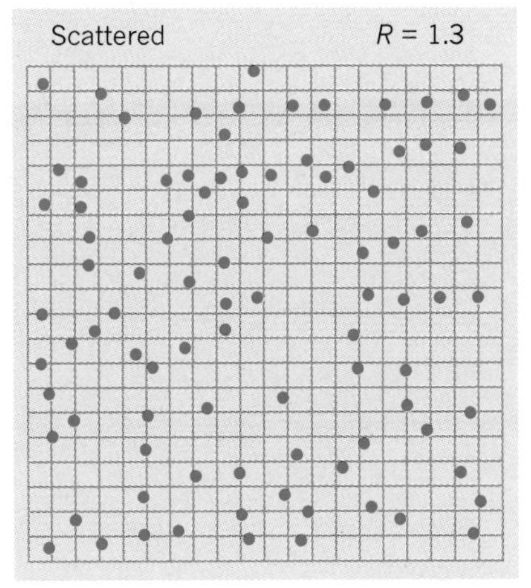

(d) Northwest Ohio
19.0 per 1000 km^2

Box 14.A The Nearest-neighbour Index

Assume a spatial distribution of towns as in Figure 14.5. Using measures first developed by plant ecologists, geographers can define a *spacing index* by comparing the observed pattern of settlements in an area with a theoretical random distribution that is,

$$R = \frac{D_{obs}}{D_{exp}}$$

where R = the nearest neighbour index of spacing;
D_{obs} = the average of the observed distances between each town and its nearest neighbour in kilometres, and
D_{exp} = the expected average difference between each town and its nearest neighbour in kilometres.

The expected average distance is given by

$$D_{exp} = \frac{1}{2\sqrt{A}}$$

where A = the density of towns per km^2. Thus, in an area with an observed nearest-neighbour distance of 3.46 km (2.15 miles) and an observed density of 0.0243 towns/km^2, the ***nearest-neighbour index*** of spacing (R) will be 1.08. Values of 1.00 indicate a random pattern. Dispersed or scattered patterns of settlement have values greater than unity, and clustered patterns of settlement have values less than unity. Nearest-neighbour indexes are discussed at greater length in L.J. King, *Statistical Analysis in Geography* (Prentice Hall, Englewood Cliffs, N.J., 1969), Chapter 5.

Most of the patterns of settlement examined thus far would have values on this scale between around 0.5 and 1.5. These values hover around the value we would assign to a group of randomly generated points (1.0), thereby implying that there are no strong pattern-forming forces deciding how settlements are arranged. In relatively uniform environments the index values for settlements drift towards 2.15 at the uniform end of the spacing scale; conversely, in environments with greater contrasts in population density the values drift toward the clustered end (0). With this index geographers can compare patterns of settlement with different spacing and estimate the probable amount of environmental influence on the location of settlements.

14.2 Settlements as Chains

Once we find a commonly accepted method of defining cities, then analysis of their comparative size and importance can begin. One of the first steps in such analysis is to arrange them in order of population size. Table 14.1 ranks the 20 largest cities in three areas of decreasing size: the world, the United States, and the state of Texas. At first sight, this looks like a dull collection of statistics. But look at Figure 14.6. This plots the size of each of these 'top 20' cities against its rank. Geographers have repeated this process for large and small areas; in each case they have looked for a repetitive pattern in the array of sizes. Have any rules been discovered? What do they tell us about the way such 'chains' of city sizes are linked together?

Table 14.1 Ranks of the twenty largest urban settlements for areas of different geographic magnitude

World metropolitan areas		United States metropolitan areas		Texas metropolitan areas	
Tokyo	28.33	New York	16.71	Houston–Galveston	3.08
Osaka	17.07	Los Angeles	9.07	Dallas–Fort Worth	2.81
New York	16.71	Chicago	7.67	San Antonio	1.06
Mexico City	14.75	Philadelphia	5.60	Austin	0.54
London	12.29	San Francisco	4.80	El Paso	0.47
São Paulo	11.86	Detroit	4.65	Beaumont-Port Arthur	0.38
Beijing	11.40	Boston	3.31	Corpus Christi	0.31
Los Angeles	10.98	Washington, D.C.	3.01	McAllen-Pharr-Edinbur	0.26
Buenos Aires	10.89	Houston–Galveston	2.87	Killeen–Temple	0.22
Ruhr	10.28	Cleveland	2.86	Lubblock	0.21
Shanghai	10.00	Dallas–Fort Worth	2.79	Brownsville–San Benito	0.20
Paris	9.88	Miami	2.41	Amarillo	0.17
Nagoya	9.61	St Louis	2.39	Waco	0.17
Rio de Janeiro	9.7	Pittsburgh	2.27	Longview-Marshall	0.14
Calcutta	8.83	Baltimore	2.15	Wichita Falls	0.13
Cairo	8.54	Minneapolis	2.10	Abilene	0.13
Bombay	8.34	Seattle	1.98	Texarkana	0.12
Seoul	8.11	Atlanta	1.86	Tyler	0.12
Moscow	8.10	San Diego	1.78	Odessa	0.11
Chicago	7.76	Cincinnati	1.65	Laredo	0.09

The figures refer to the estimated population for the early 1980s using United Nations and national census data. Note that the metropolitan areas are made up of clusters of continuous cities: thus New York at 16.71 million includes the New York–Newark–Jersey City consolidated statistical area. Definitions for Texas are slightly different from those for the US.

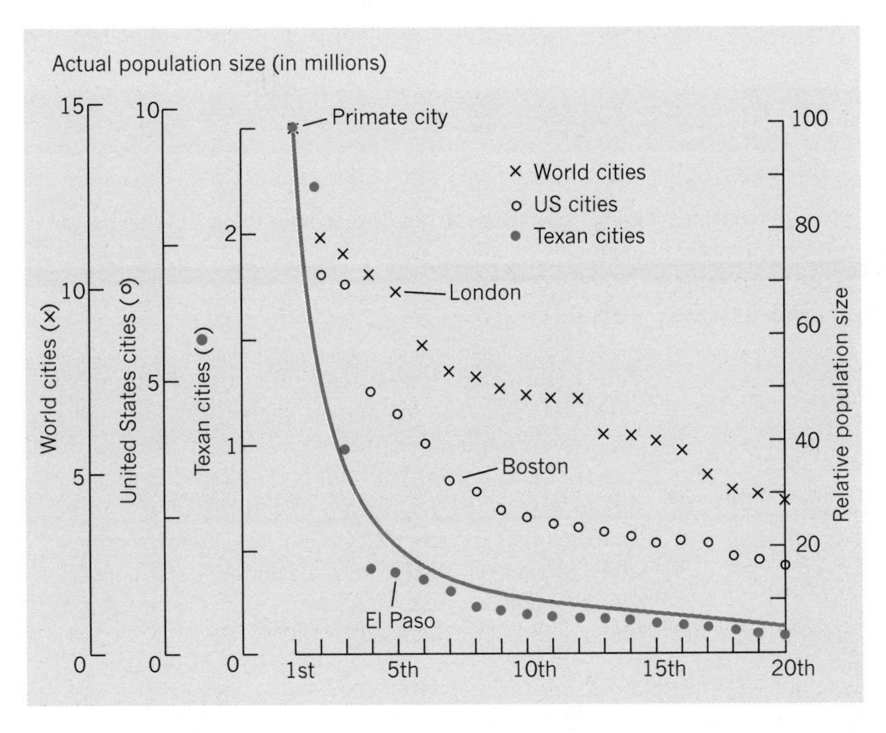

Figure 14.6 City chains The twenty largest cities in three areas (see Table 14.1) are arranged on a population rank–size diagram. The largest city in all three series – the 'primate city' – has a relative population size of 100. Actual population sizes are shown on the vertical scales to the left. The continuous line is an idealized rank–size curve. Note the figures refer to sizes in the early 1980s.

A Rule for the Size Distribution of Settlements

Although some nineteenth-century investigators sought for patterns, one of the first to actually find a significant one was a German geographer, Felix Auerbach, in 1913. He noticed that if we arrange settlements in order of size (1st, 2nd, 3rd, 4th, ..., *n*th), the population sizes for some regions are related. Auerbach found the simplest relationship to be that the population of the *n*th city was $1/n$ the size of the largest city's population. Thus the fourth ranking city was found to have approximately ¼ the population of the largest. This inverse relationship between the population of a city, and its rank within a set of cities is termed the *rank–size rule*. (See Box 14.B on 'The rank–size rule'.) If we apply this rule to the United States and look at Table 14.1, we should expect Philadelphia (ranked fourth) to have one-quarter of the population of New York City (ranked first). Thus, in the early 1980s, Philadelphia should have had a population of 4.2 million; in fact, its population was more than this (5.6 million). In earlier decades of the twentieth century, there is a closer correspondence with Auerbach's ideas but New York City now appears to be falling back in relation to other cities in the US system. The match with the ideal rank–size rule (the curve in Figure 14.6) is better for the Texan cities, but rather poor for the world metropolitan areas.

We can more easily compare the fit between distributions of real cities and the idealized distributions predicted by the rule the axes on which we plot the cities' size and rank are made non-linear. The twenty hypothetical cities in Figure 14.7(a) all conform exactly to the rank–size rule and make an awkward, J-shaped curve. If we transform the values on the axes to logarithmic scale, the curve becomes a straight line, as in Figure 14.7(b). The simple rank–size relationships for the United States and Texas can also be described by a straight line, with a slope of 45° to the horizontal; that is, they have the form predicted by the rank–size rule and shown in Figure 14.7(b). Yet other lines that are equally regular but have *different* slopes have been found. For example, if we look back in time then the curve for the cities of Switzerland in 1960 has a much gentler slope, and that for

Box 14.B The Rank–Size Rule

Assume a set of cities ranked according to size from the largest (l) downward. In its simplest form, the rank–size rule states that the population of a given city tends to be equal to the population of the largest city divided by the rank of the given city; that is

$$P_r = \frac{P_l}{R}$$

where, P_r = the population of the *r*th city,
P_l = the population of the largest city, and
R = the rank of the *r*th city in the set.

This basic formula is often modified by a constant (b) to allow variations from the strict rank–size rule – for example:

$$P_r = \frac{P_l}{R^b}$$

Thus, when the largest city has a population of 1,000,000 and b = 0.5 (a low-angled slope in Figure 14.7), we should expect the population of the fourth-largest city to be 500,000. If b is raised to 2.0, then we should expect the population of the fourth city to be smaller – that is, only 62,500. For an extensive critical study of rank–size rules in a specific regional context, see C.D. Harris, *Cities of the Soviet Union* (Rand-McNally, Skokie, Ill., 1970).

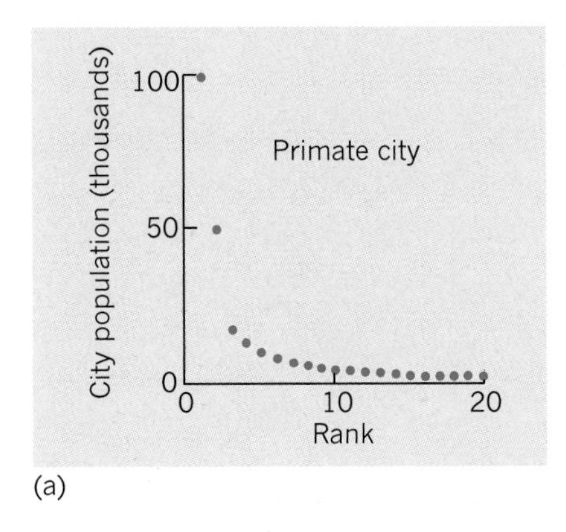

(a)

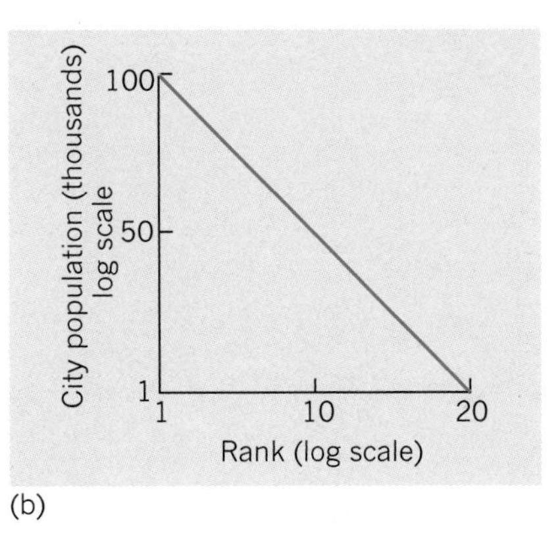

(b)

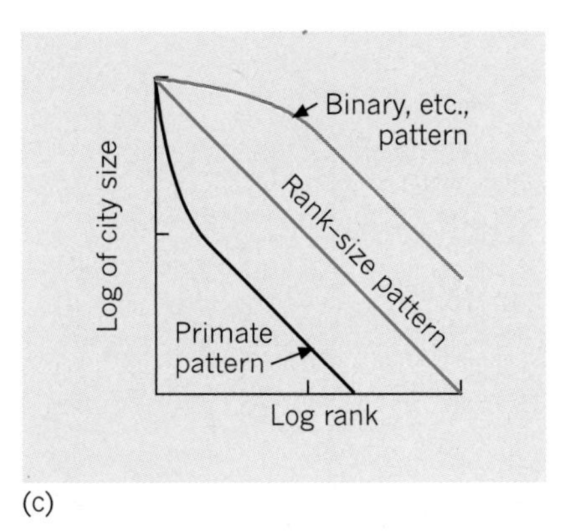

(c)

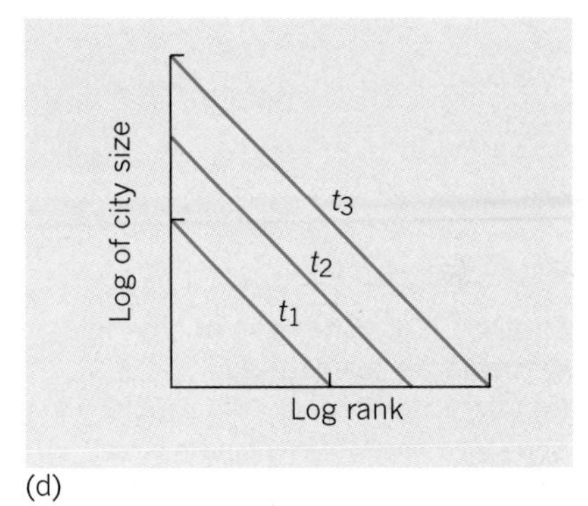

(d)

Figure 14.7 Hypothetical relations between city size and rank Twenty cities arranged in an idealized rank–size pattern are plotted on (a) a graph with arithmetic axes and (b) a graph with logarithmic axes. Figure (c) shows three alternative patterns of city sizes. Figure (d) shows the evolution of idealized city chains as population size increases through three time periods. (See also Figure 14.8.)

urban areas in India in 1921 has a much steeper one. Gentle slopes imply that the decrease in a city's population with rank is extremely slow; steep slopes imply a sharp falloff in size with rank.

Regional Variations in the Rule

Since the size of any one city appears to be linked to the size of all other settlements in a region, we can regard the whole set of cities as forming an interlinked chain running from the largest city to the smallest. These are termed *rank–size chains*. When considering these rank–size chains, some geographers have found it more meaningful to subdivide the distribution of settlements into distinct segments. Australia has a distinct pattern of urban areas, which when plotted on a graph show a flat upper section and a steeper lower section, with the critical break at a population of 75,000. This convex Australian pattern is in contrast to the concave distribution of Russian cities, in which a steep upper section is followed by a flatter lower section. Here the break between the two sections comes much higher than in the Australian pattern; it happens at a population level of around 500,000. In Figure 14.7(c), we distinguish between these two patterns of settlement. The Australian or binary pattern is dominated by a few large cities that are roughly equal in size to a 'tail' of smaller cities conforming to the rank–size rule. Conversely, in the Russian or *primate pattern* the falloff in population among the first few cities below Moscow is sharper than the falloff predicted by the simple rank–size rule. We call this a *primate* pattern to indicate the dominant role of the first, or primate, city. (The term primate is taken from ecclesiastical language where the primate of a church is its superior bishop.)

When historical census figures are available, geographers can trace the changes in rank–size relationships over time. If the total population of a region grew and its cities remained distributed in a simple rank–size sequence, we should expect it to change through time periods t_1 to t_3 in the way described in Figure 14.7(d). The figures for the United States over one and a half centuries are rather stable. The curves for 1790 to the middle of the twentieth century are generally parallel, and there is some evidence of increasing regularity (manifested by the straightening of the curves over time), as we shall see later in Figure 14.19. The statistics for Sweden over the same period reveal a reverse effect; here an S-shaped curve is retained and even accentuated by an overall growth in the population. Figure 14.8 shows the pattern followed by Israel in relation to the expected development of city-size distributions.

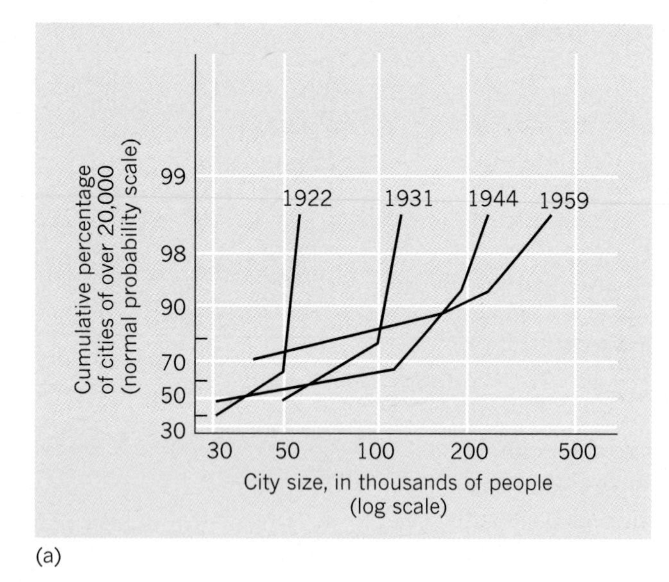

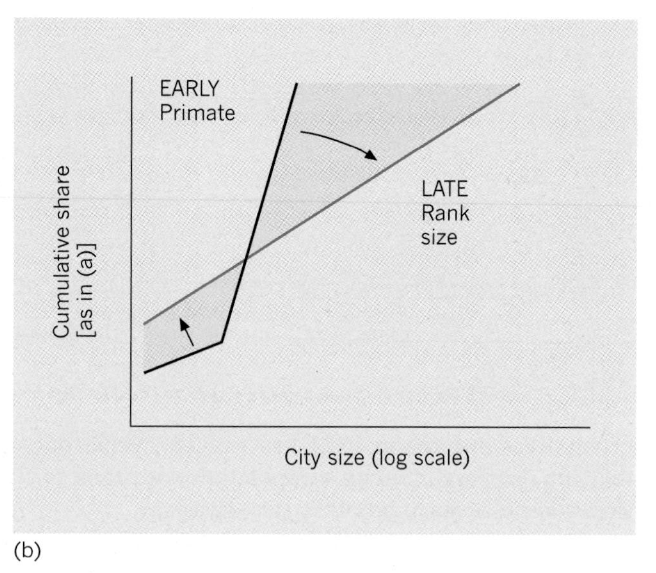

Figure 14.8 Changes in city-size distribution over time
(a) Cumulative share of the population in different classes of city size in Israel (formerly Palestine) over the period from 1922 to 1959. (b) Hypothetical pattern of change from a primate to a rank–size distribution of city size with urbanization.

[Source: Data from G. Bell, *Ekistics*, **13** (1962), Fig. 1, p. 103.]

The Logic of Rank–Size Chains

Enough evidence is now available to prove that regular rank–size chains are recognizable for settlements in many types of regions during different time periods. Why are many towns and cities arranged in this regular fashion?

Geographers are not the only ones to be puzzled by these size distributions. Rank–size rules are not confined to human settlements. Similar distributions have been observed by botanists studying the number of plant species, and by linguists studying the frequency with which different words are used in our speech. The pervasiveness of this type of distribution has led organization theorist Herbert Simon to postulate rank–size rules as the equilibrium slope of a general growth process. We can visualize this process as one in which each unit, such as a city, initially has a random size and thereafter grows in an exponential manner that is proportional to that initial random size. Simon points out that extremely general processes of this kind tend to produce distributions that approximate a regular rank–size form.

Simon's hypothesis has been translated into urban terms by Brian Berry. (See the profile of Berry in Box 18.A.) Berry studied the rank–size distribution of towns with populations of 20,000 or more in 38 countries. Of all the countries, only 13 had rank–size distributions like the ones postulated in the Simon model. These countries were among the largest in the group (e.g. the United States), had a long history of urbanization (e.g. India), and were economically and politically complex (e.g. South Africa). By contrast, 15 countries had primate distributions in which one or more large cities dominated the size distribution and were much larger than we might have expected them to be on the basis of rank–size rules.

In contrast to regular rank–size distributions, primate distributions appear to be products of urbanization processes in countries that are smaller than average, have a short history of urbanization, and have simple economic and political structures. Hence, primate distributions typify the impact of a few rather strong forces. For instance, the impact of imperial status on the large cities in Austria, the Netherlands, or Portugal, each the hub of former empires, is certainly potent. Another powerful force is the superimposition of outside influences on an existing hierarchy. Examples would be the institution of a dual economy (such as a peasant and plantation sector in Sri Lanka) or the influence of a westernized city such as Bangkok on the Thai system of cities.

Simon's model has two principal advantages: it introduces time into the rank–size model by taking into account an urban system's history of development, and it emphasizes the effects of numerous small forces in producing regular structures. Variations from the rank–size model may be caused by the distorting effect of a few powerful forces.

The Christaller Central-place Model 14.3

Two-dimensional patterns of settlement as shown on maps of population distribution invite questions similar to those provoked by one-dimensional patterns. Is any order discernible? If so, what forces lie behind it? Although this problem had been stated by German geographers in the nineteenth century and some tentative hypotheses were made, the main breakthrough did not occur until 1933, when Walter Christaller published his now famous doctoral dissertation on *Central Places of Southern Germany* giving rise to *central-place theory*. (See Box 14.C on 'Christaller and central-place systems'.)

Box 14.C Christaller and Central-place Systems

Walter Christaller (born Berneck, Baden-Württemberg, Germany, 1893; died Königstein, Germany, 1969) was a pioneering German geographer who laid the foundations of a rational theory for human settlements.

His 1933 doctoral thesis at Erlangen University, *Die Zentralen Orte in Süd-deutschland*, was translated into English and published 30 years later as *Central Places of Southern Germany*. It was based on a critical analysis of the size and function of the villages, towns, and cities of Bavaria in terms of his three optimizing principles of market, traffic, and administrative organization. The intricate, crystal-like lattices which Christaller saw holding the settlement fabric together provided new insights into the way the human population organizes itself to exploit the natural resources and locational attributes of a region.

Christaller's ideas owed something to the German settlement geographer Robert Gradmann and to the locational theorists whom we shall meet in other chapters (Johann von Thünen, J.G. Kohl, and – Christaller's old teacher – Alfred Weber). Christaller was, however, something of the 'odd man out' among the German geographers of his generation. His thesis attracted little attention at the time it was published and he did not hold a university post. Much of his working life was spent in the travel business. It was not until the 1950s that his ideas were widely introduced in the English-speaking world. In the later part of his life, Christaller developed a complementary 'theory of the periphery' to explain the apparently anomalous structure of settlements in peripheral locations. For a discussion of Christaller's contributions see B.J.L. Berry and J.B. Parr, *Geography of Market Centres and Retail Distribution* (Prentice Hall, Englewood Cliffs, N.J., 1988).

[Source: Photo from Royal Geographical Society.]

Central Places

The terminology of Christaller's model is straightforward. *Central places* are broadly synonymous with towns that serve as centres for regional communities by providing them with *central goods* such as tractors and *central services* such as hospital treatment. Central places vary in importance. Higher-order centres stock a wide array of goods and services; lower-order centres stock a smaller range of goods and services – that is, some limited part of the range offered by the higher centre. *Complementary regions* are areas served by a central place. Those for the higher-order centres are large and overlap the small complementary regions of the lower-order centres.

Schools provide a good example of a central-place organization. Thus in the United States the local elementary school provides a lower-order centre (to use Christaller's terms) which serves a small part of a city or a single rural community. There is a large number of such schools in any state, and they teach children drawn from only a few square kilometres (i.e. they have small complementary regions). Above the elementary schools come the higher-order services provided by the junior high schools, the high schools, and colleges of various kinds. As we move higher up the educational ladder, the number of centres becomes smaller and their complementary regions become larger. At the top of the ladder stands the state university, sometimes a single institution serving students drawn from the whole state, its complementary region. Other countries will have roughly parallel systems. Education is just one range of central goods and central services that give character to central-place organizations and help to distinguish the central-place functions of one settlement from those of another.

Christaller defined the *centrality* of an urban centre as the ratio between all the services provided there (for both its own residents and for visitors from its complementary region) and the services needed just for its own residents. Towns with high centrality supplied many services per resident, and those with low centrality few services per resident. Christaller found that the number of telephones offers a useful indicator of the range of central goods available in a town. (Today, he might have chosen e-mail connections.) Using telephone data, he defined the centrality of a town as equal to the number of telephones in the town, minus the town's population multiplied by the average number of telephones per population in the town's complementary region. A town of 25,000 with 5,000 telephones in a region with only 1 telephone for every 50 people would have an index of 5,000 – [25,000 × (1/50)], or 4,500. Thus the index basically measures the difference between the expected level of services (i.e. that needed by a town to serve its *own* inhabitants) and the level of services actually measured within the centre.

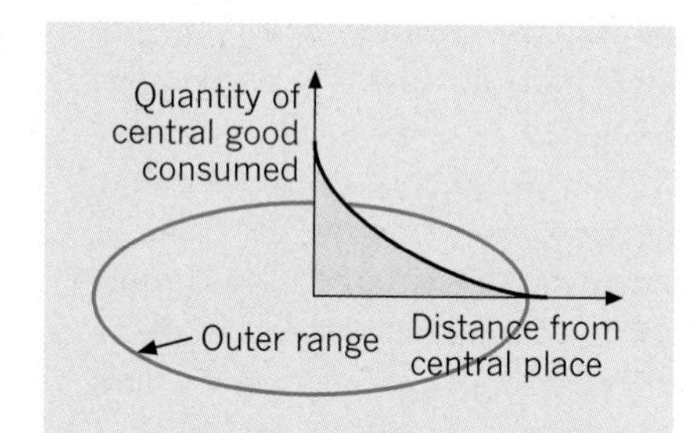

Figure 14.9 Idealized demand zones in the Christaller model With uniform transport costs, the demand for central goods falls with distance from the central place, and the market range (the area within which the goods will be bought) forms a circle.

Later researchers have revised Christaller's terminology to include two simple concepts. The first is *market-size threshold*, below which a place will be unable to supply a market good. That is, below the threshold, sales will be too few for firms to earn acceptable profits. The second concept added was that of the *range of a central good*, the limits of the market area for the good (Figure 14.9). The market area's lower limit is determined by the threshold market size, and its upper limit is defined by the distance beyond which the central place is no longer able to sell the good. If we assume that travel is equally easy in all directions, the range of a good will be a perfect circle. This circle is the outer limit of a *demand cone* in which the quantity of a central good consumed decreases with distance from the central place because of increased transport costs.

Complementary Regions

Given a circular demand cone for central goods, Christaller demonstrated that a group of similar central places will have hexagonal, complementary regions with the central places arranged in a regular triangular lattice. Figure 14.10 depicts the stages by which such a pattern might emerge as population colonized a new area and central places were established (cf. Section 14.5). The final hexagonal patterns follow directly from five simplified assumptions:

1 An unbounded isotropic plain with a homogeneous distribution of purchasing power. (Imagine farms uniformly distributed over a flat plain that has the same fertility everywhere. On this flat plain, travel costs are the same in any direction.)
2 Central goods to be purchased from the nearest central place.
3 All parts of the plain to be served by a central place; that is, the complementary areas must completely fill the plain.
4 Consumer movement to be minimized.
5 No excess profits to be earned by any central place.

The hexagons result from our attempt to pack as many circular demand cones as possible onto the plain. If we require all parts of the plain to be served by a central place (assumption 3), the circles will overlap. But because of our second assumption, that consumers will shop at the closest central place, the areas of overlap will be bisected. A perfect competitive situation will be achieved only when the plain is served by the maximal number of central places, offering identical central goods at identical prices to hexagonal complementary regions of identical size. Only this arrangement ensures that consumers travel the least distance to central places.

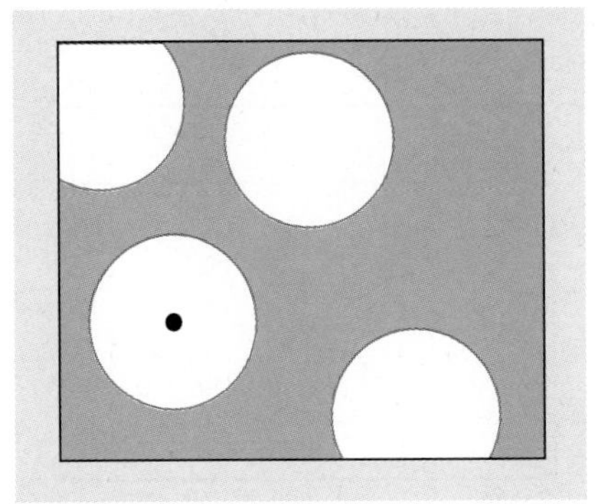
(a)

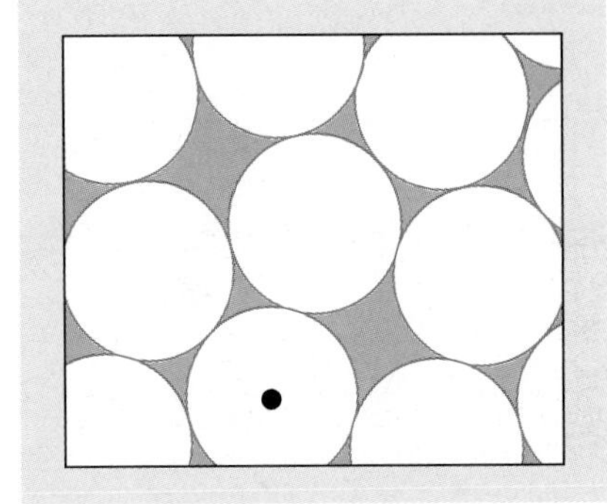
(b)

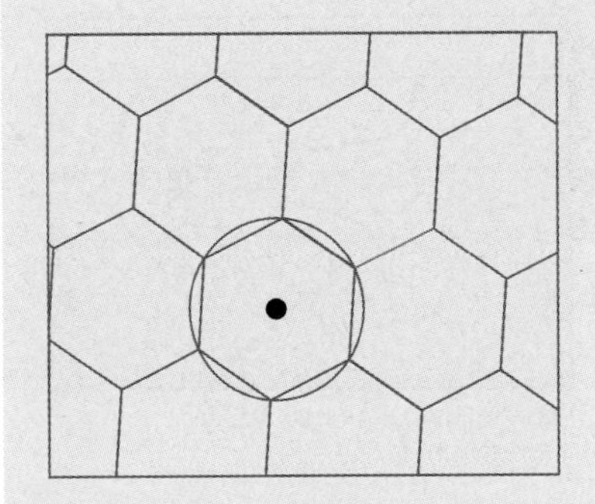
(c)

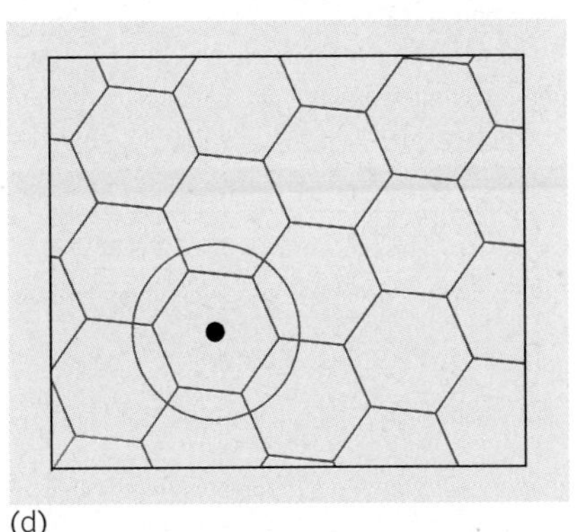
(d)

Figure 14.10 Hexagon formation The overlapping of circular demand cones, along with the close packing of centres, gives a network of hexagonal territories. (Compare with Figure 14.15.)

Central-place Hierarchies

Christaller was able to account for the varying levels of central places within a *settlement hierarchy* by varying the size of the complementary regions, as in Figure 14.11. He discusses three cases.

The first is a *marketing-optimizing* case, in which the supply of goods from central places is as near as possible to the places supplied. A higher-order central place will serve *two* of its lower-order neighbours. It may do this by serving only two of its six equidistant nearest neighbours and thus having an asymmetric complementary region. Alternatively, a higher-order central place may share the same neighbours with two others, for instance, competing neighbours. Note in Figure 14.11(a) how settlement 2 lies on the edge of three complementary regions (those of centres 1, 3, and 4). This arrangement is termed a $K = 3$ system, where K refers to the number of places served, that is, the central place plus two nearest neighbours or the central place plus one-third of each of its six nearest neighbours.

The second case involves a *traffic-optimizing* situation in which the boundaries of the complementary regions are rearranged to allow a more efficient highway pattern than in the first case. As Figure 14.11(b) shows, as many places as possible now lie on traffic routes between the larger towns; for example, the direct route from centre 1 to 5 goes right through centre 2. This situation is represented by the $K = 4$ hierarchy, where a higher-order place serves three adjacent lower-order places. It may do this by dominating three of its six nearest neighbours or by sharing them with another central place of the same order.

The third case Christaller discusses is an *administration-optimizing* situation, in which there is a clear-cut separation of the higher-order place and its neighbouring lower-order centres. That is, each lower-order centre falls clearly within the trade area of a single central place; in Figure 14.11(c), for example, centre 2 falls within the area of centre 1. Such arrangements are likely to be economically and politically more stable than divided settlements. This relationship produces the $K = 7$ hierarchy.

All three cases assume that relationships established for one level (e.g. between villages and small towns) will also apply to other and higher levels (e.g. between small towns and larger cities). They are usually called *fixed-K hierarchies* because the same fixed relationships hold at *all* levels of the settlement hierarchy. This means that we can expand each of Christaller's three central-place variants by building higher and higher levels on top of the basic framework. Consider the situation in Figure 14.11(d), where a second and third $K = 4$ central place is superimposed on the first. As we add each successive upper level, the size of the hexagonal regions increases and the number of places is reduced by a quarter. Thus if a region had 2000 central places on the lowest level, it would have 500 on the next level, and 125 on the next higher level again. If we start at the top, we can put this in a simpler way. In an idealized $K = 4$ school system with three levels or tiers, one junior college would draw students from four high schools, each of which drew children from four elementary schools (i.e. 16 in all). For each of the three cases considered by Christaller, typical sequences would be 1, 3, 9, 27 for the $K = 3$ network, 1, 4, 16, 64 for the $K = 4$ network, and 1, 7, 49, 343 for the $K = 7$ network.

Southern Germany as a Test Area

As we have noted, Christaller developed his basic ideas using southern Germany as his original test area. The theoretical distribution of the status and location of towns in this area, according to his market-optimizing principle, is summarized in Table 14.2. Christaller postulated seven levels of the hierarchy from the level of the hamlet to that of the city (Figure 14.12), each level showing an increase in the area of the complementary region. Approximate populations have been added to the table by extrapolating from Christaller's detailed work on southern Germany. On the upper level of the hierarchy are the *Landstadt* cities with populations of around 500,000 – Munich, Frankfurt, Stuttgart, and Nuremburg, together with the border cities of Zurich in Switzerland and Strasbourg in France. The lowest market centre at the base of the hierarchy has a service radius of a little

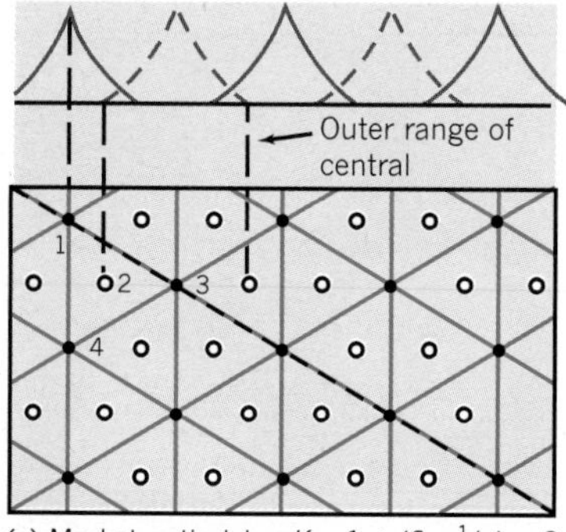

(a) Market optimizing $K = 1 + (6 \times 1/3) = 3$

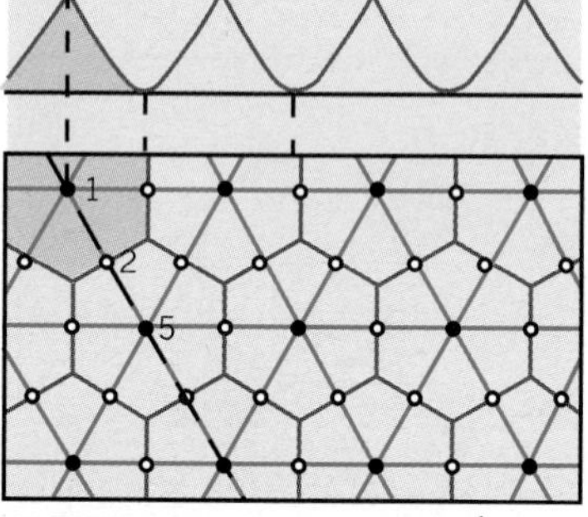

(b) Traffic optimizing $K = 1 + (6 \times 1/2) = 4$

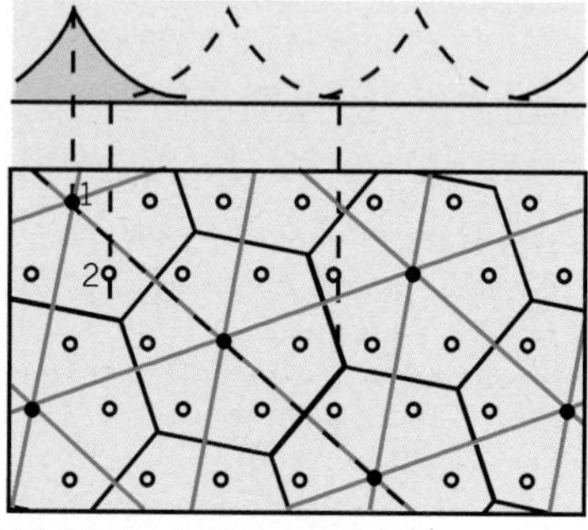

(c) Administration optimizing $K = 1 + (6 \times 1) = 7$

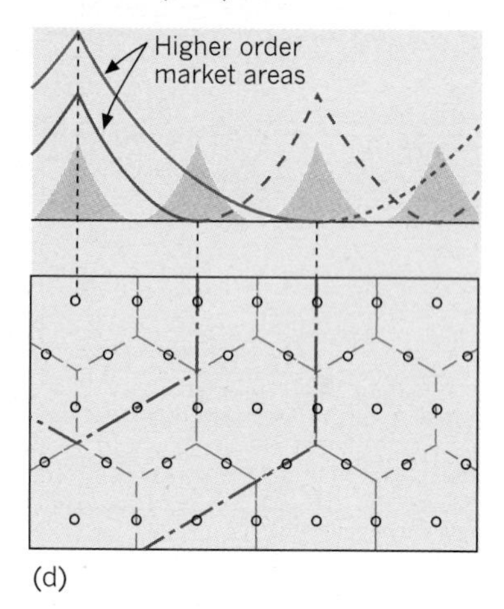

(d)

- ● Central place
- o Dependent place
- —— Boundary of complementary region
- – – Highways between central places

Figure 14.11 Alternative principles of organization in the Christaller model Settlements can be partitioned in one of three basic ways: (a), (b), or (c), by enlarging and rotating the hexagonal cells. Note the way in which the K number depends on whether a dependent place is shared with three other places ($K = 3$), two other places ($K = 4$), or unshared ($K = 7$). The hexagonal cells can be grouped hierarchically to give tiers of higher-order centres. For example (d) shows higher-order centres in a traffic-optimizing ($K = 4$) hierarchy. Note the way in which lower-order centres 'nest' within the market areas of higher-order centres in a manner reminiscent of sets of Russian dolls.

Table 14.2 Status of towns in Christaller's settlement system[a]

			Distance from other towns		Service area	
Type of town	Order	Approximate population	km	mile	km^2	mile^2
Landstadt (L)	Upper	500,000	187	116	35,000	13,514
Privinzstadt (P)		100,000	109	68	11,650	4,498
Gaustadt (G)		30,000	63	29	3,860	1,498
Bezirkstadt (B)		10,000	36	22	1,243	478
Kreisstadt (Kr)		4,000	21	13	414	160
Amtsort (A)		2,000	13	8	140	54
Marktort (M)	Lower	1,000	7	4	47	18

[a]Values are based on a study of southern Germany (see Figure 14.13).
Source: R.E. Dickinson, *City and Region* (Humanities, New York, and Routledge & Kegan Paul, London, 1964), p. 76.

Figure 14.12 South Germany
South Germany was used by Christaller as a test area for his central-place model. Munich, the largest city in the region occupied the top rank in the structure. Munich occupies an L-level position in the hierarchy in terms of the map in Figure 14.11 and Table 14.2.

[Source: Photographs courtesy of Ullstein.]

more than 3 km (2 miles). The application of Christaller's theoretical model to southern Germany is shown in Figure 14.13.

Despite the general agreement between the model and reality, Christaller found several specialized centres – mining towns, border towns, and so on – that deviated from the general pattern. The resources in a particular region or subregion may cause a general increase in the density of settlement, resulting in a closer spacing of centres.

In the pages available here, we have only been able to discuss Christaller's main ideas. From very simple assumptions the model can be further extended and refined to give a more realistic representation of the real-world complexities of human settlement. It is to these we now turn.

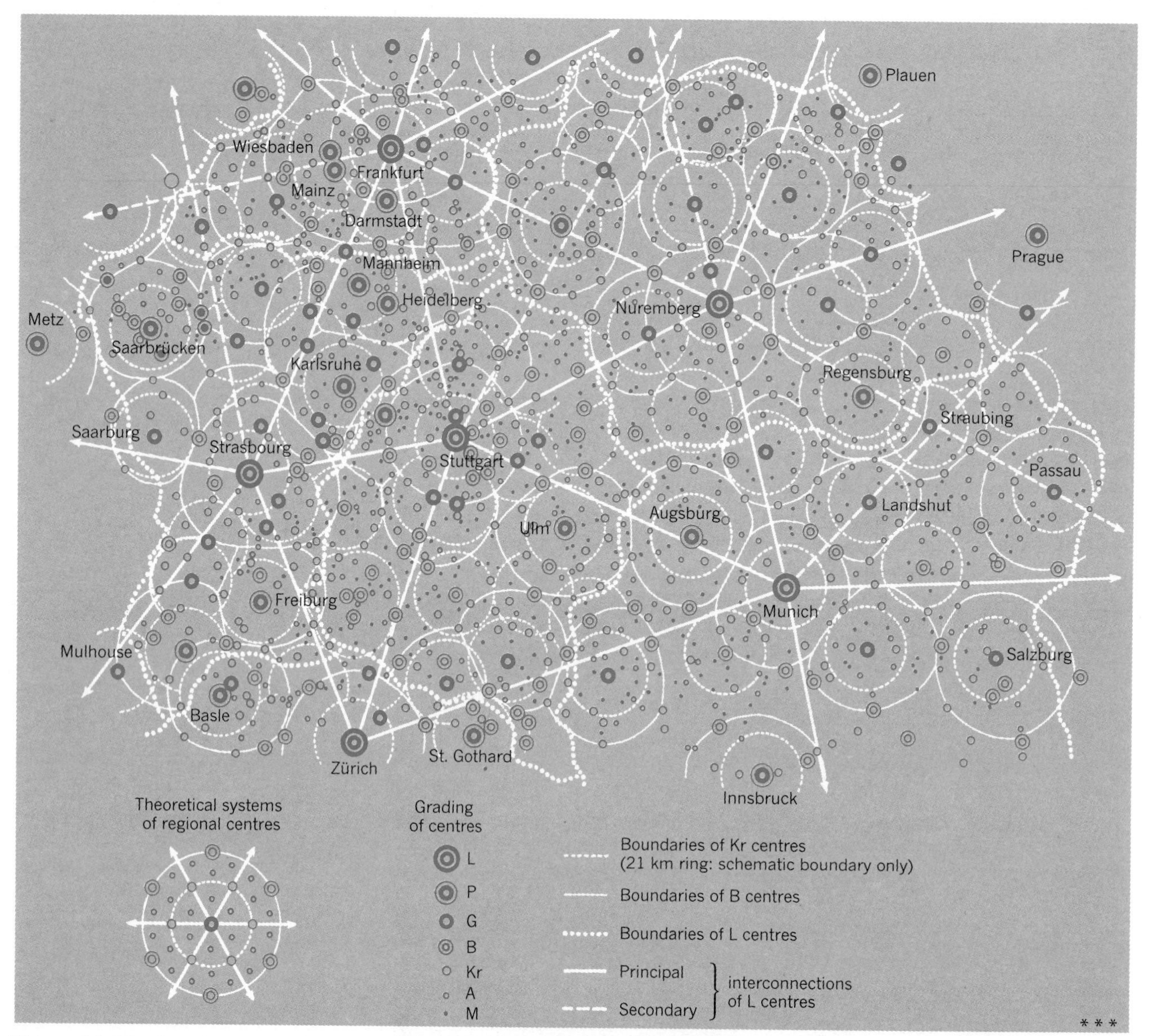

Figure 14.13 South Germany Christaller's map of the distribution of cities, towns, and villages in southern Germany shown by means of seven-level hierarchy. (See Table 14.2.) The map also shows the boundaries of the complementary regions about the four highest levels of centres (drawn as ellipses or circles) and the main routes interconnecting the highest-level centres (drawn as straight lines). The boundary of the area is simply the limit of Christaller's study area; border cities such as Zurich and Plauen lock into a continuing, continent-wide hierarchic system.

14.4 Extensions of the Christaller Model

Since its publication the Christaller model has provoked two main reactions from fellow geographers. First, there are those who have accepted the general argument of his model. Their reaction has been to extend and refine it. Second, there are those who found the Christaller model too rigid and static. They have reacted by trying to build alternative models with a stronger time dimension and a closer correspondence to actual settlement history. In the last two sections of this chapter we look at these two approaches.

Lösch's Modifications

The prime theoretical extension of the Christaller model was created by a fellow German, August Lösch (1906–45), in his *Die raumliche Ordnung der Wirtschaft* (later translated into English as *The Economics of Location*) published shortly before his death. He clarified the ways in which spatial demand cones are derived and verified the optimal hexagonal shapes of complementary regions where the population served was uniformly distributed. However, Losch's main contribution was to extend the notion of fixed-K hierarchies.

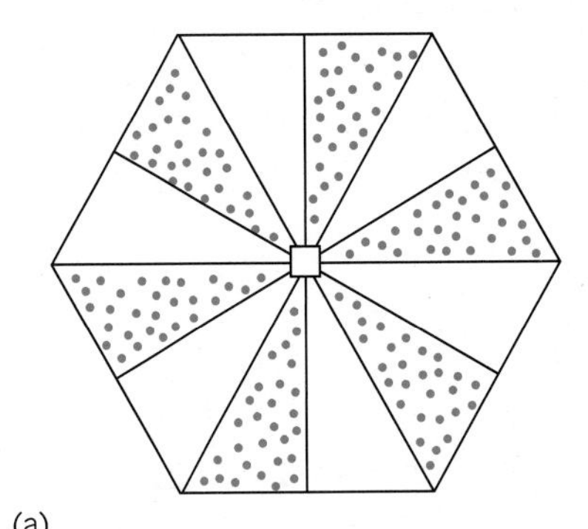

(a)

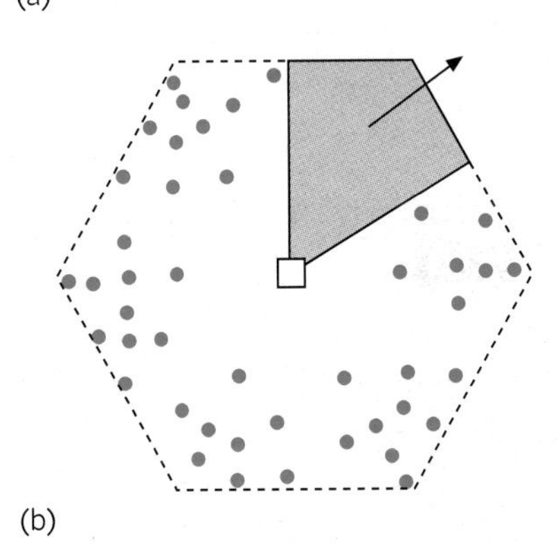

(b)

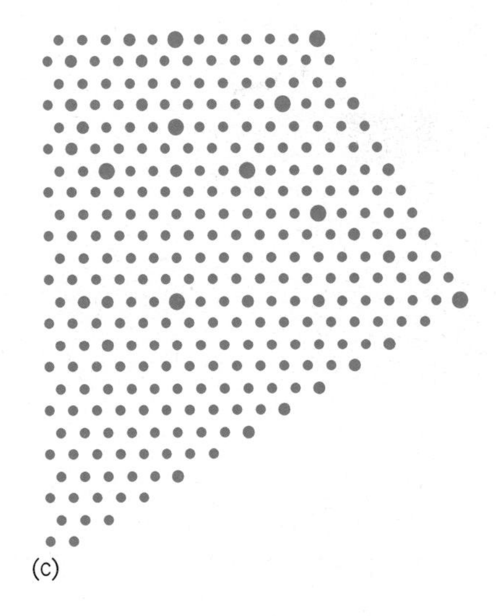

(c)

Figure 14.14 The Löschian landscape City-rich and city-poor sectors in the Löschian landscape. (a) Twelve sectors. (b) Centres with the largest number of functions. (c) An enlargement of a pair of adjacent sectors to show the underlying regular hexagonal pattern; the size of the dot is proportional to the number of functions.

[Source: A. Lösch, *The Economics of Location* (Yale University Press, New Haven, Conn., and Fischer Verlag, Stuttgart, 1954), Fig. 32, p. 127.]

Superimposed Networks Lösch took all the hexagonal networks in Figure 14.11 and extended them to higher orders by superimposing them on a common central place. This common central place is the hub of the settlement system, its single most important city dominating the trade and services in the whole of the surrounding region. Each of the networks was then rotated about this common central city until as many as possible of the higher-order services coincided in the same centres. Such an arrangement ensures that the sum of the minimal distances between settlements is small and that not only shipments, but also transport lines, are reduced to a minimum.

You can envisage this process by imagining that the fixed-K network is drawn on a map. The $K = 4$ network is now drawn on an overlay of transparent tracing paper and pinned to the $K = 3$ map by a single thumbtack through the common central place. By rotating the overlay, many major places on both the $K = 3$ and the $K = 4$ paper are made to coincide. For example, if we have a $K = 3$ school system and a $K = 4$ hospital system, we try to rotate the overlay so that the high school and the doctor's hospital both coincide in the *same* locations rather than being split between two. Losch went on to add the $K = 7$ and still higher K networks to the map, always trying to get as many services as possible to overlap in the same locations.

A simplified version of his final result is shown in Figure 14.14. It shows that the resulting central-place system changed with distance away from the common central hub and was arranged, like a wagon wheel, with alternating sectors. Twelve sectors are produced, six with many production sites and six with few (called by Lösch 'city-rich' and 'city-poor' sectors). In Figure 14.14 the metropolitan centre is the centre of 150 separate fields.

Thus, using the same basic hexagonal unit and the same K concept as Christaller, Lösch evolved a markedly different hierarchy. Christaller's hierarchy consists of several fixed tiers in which all places in a particular tier have the same size and function and all higher-order places perform all the functions of the smaller central places. In contrast, the Löschian hierarchy is far less rigid. It consists of a nearly continuous sequence of centres rather than distinct tiers. So settlements of the same size need not have the same function (e.g. a centre serving seven settlements may be either a $K = 7$ central place or a centre where both a $K = 3$ and $K = 4$ central place coincide), and larger places need not perform all the functions of the smaller central places.

Lösch's model represents a logical extension of the Christaller model. It is based on the same hexagonal unit and hence suffers from the same rigidity, but it yields a relationship between the size and function of central places that is continuous rather than stepped, and therefore more in accord with the observed distributions described in Section 14.2.

Spatial Transformations A number of later workers such as the regional economist Walter Isard have shown how spatial transformations of the Christaller and Löschian frameworks can make them fit much more exactly to observed patterns. If we allow that richer areas are likely to have denser settlement patterns and vice versa then the real world distributions are always going to be much less symmetrical than the models which were assumed to operate in uniform space. One approach used in archaeology is the *XTENT model* which assigns territories to centres based on their scale, assuming that the size of each circle is directly proportional to the area of influence.

Periodic Variations

The permanent provision of central goods implies a high and continuous level of demand. In most peasant societies central goods are provided by markets that are not open every day, but only once every few days on a regular basis. Although periodic markets are now only a small element in the central-place structure of western society and generally sell only agricultural

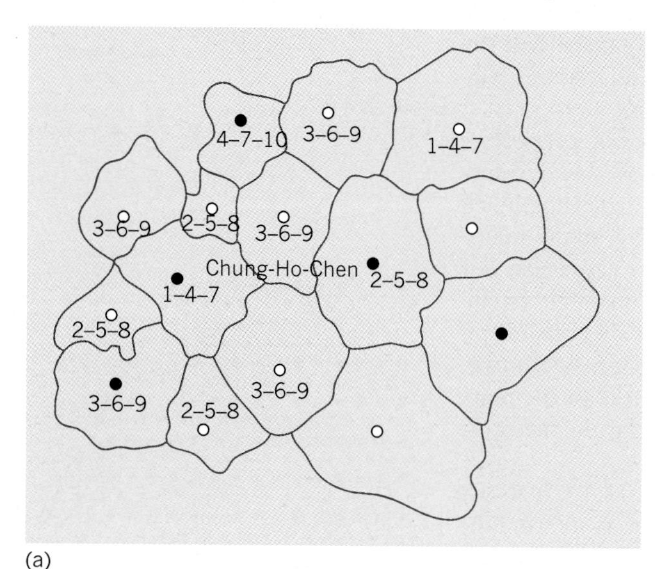

(a)

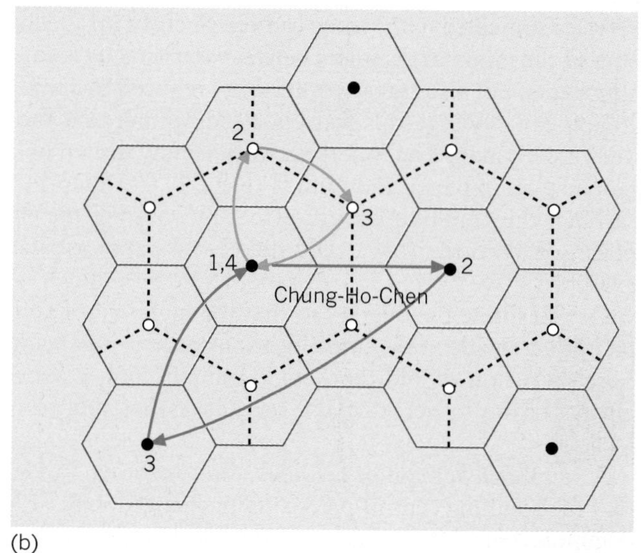

(b)

(c)

Figure 14.15 Periodic central places (a) A map of rural centres in the Chinese provinces of Szechwan, showing the days on which markets are held. (b) The rural centres as part of a $K = 3$ Christaller network. The cyclic movements of traders around a market ring are examples of the space–time meshing of central-place functions. (c) Szechwan marketplace: the town market at Chengdu.

[Source: (a) G.W. Skinner, *Journal of Asian Studies*, **34** (1964), Figs 3, 4, pp. 25, 26. (c) © TRIP/R. Graham.]

products, they continue to be important in most peasant societies and are vital to the exchange structure of two-thirds of the world.

The relevance of Christaller's scheme to periodic markets is demonstrated by the central-place network of rural China. Figure 14.15 shows a portion of the Szechwan Province, southeast of Chengtu, which is simplified to a basic $K = 3$ system of nesting (cf. Figure 14.11(a)). Two levels of the hierarchy are shown: an upper level (*Chung-ho-chen*) and a lower level (*Hsin-tien-tzu*) which has a market area about one-third the size. Periodic markets are superimposed on the system as indicated in Figure 14.15. A ten-day cycle is usually divided into three units of three days. No business is transacted on the tenth day. With synchronized cycles, central goods can be circulated around several markets on a regular schedule and firms can accumulate enough trade to remain profitable. Hence, a central-place system is maintained by *rotating* rather than *fixed* central-place functions.

Periodic market cycles vary widely in length. In tropical Africa, the market week varies from three to seven days. For example, Yorubaland in western Nigeria works on an interlocking system of four-day circuits. Generally the higher the population and per capita income, the greater the total trade and the shorter the length of the cycle. Where demand for goods is high enough, the market opens every day (and thus is a permanent central-place function); where demand is low enough, the cycles become so long that a service ceases to be provided within the region.

Historical research reveals instances of space–time interlocking in medieval Europe, where cattle might be sold or cloth exchanged at large spring and autumn fairs. In the modern international economy the time period may widen to years and the region broaden to involve all the major capitals of the world. The World Fairs and the Olympic Games, with their four year cycle (see Table 14.3), could be viewed as extreme extensions of the periodic case of the Christaller model.

Table 14.3 Summer Olympic Games as a periodic central-place system

Year	Host city	Year	Host city
1896	Athens (Greece, 1)	1952	Helsinki (Finland)
1900	Paris (France, 1)	1956	Melbourne (Australia, 1)
1904	St Louis (USA, 1)	1960	Rome (Italy)
1906	Athens (Greece, 2)	1964	Tokyo (Japan)
1908	London (UK, 1)	1968	Mexico City (Mexico)
1912	Stockholm (Sweden)	1972	Munich (Germany, 2)
1916	*World War I*	1976	Montreal (Canada)
1920	Antwerp (Belgium)	1980	Moscow (Russia)
1924	Paris (France, 2)	1984	Los Angeles (USA, 3)
1928	Amsterdam (Netherlands)	1988	Seoul (Korea)
1932	Los Angeles (USA, 2)	1992	Barcelona (Spain)
1936	Berlin (Germany, 1)	1996	Atlanta (USA, 4)
1940	*World War II*	2000	Sydney (Australia, 2)
1944	*World War II*	2004	Athens (Greece, 3)
1948	London (UK, 2)	2008	?

At the time of writing there were five bid cities for the 2008 Games (Beijing, Istanbul, Osaka, Paris, and Toronto).

14.5 Alternatives and Applications

Not all geographers are happy with the Christaller model, even in its more refined forms. They regard it as essentially a special case in two important ways. First, it is a special case in *conception* in that it describes a closed system where change can occur only from the bottom upward (i.e. with increased rural productivity in the lowest level of settlement leading to an enlarged hierarchy and therefore more higher-order centres). Within that closed system, pure competition for space is allowed. Second, it is a special case in *reality* since it emerged from the study of a particular area (southern Germany, a mid-continental location) with a particular history of settlement in which the 'feudal' organization of agriculture had played a notable part. The Christaller model gives insights to settlement growth only in geographical areas that are rather uniform in character, economically isolated, and historically stratified.

The Vance Mercantile Model

Geographers who accept the above criticisms as valid do not see how the Christaller model can help in studying settlement patterns where the main forces of change are from the outside (i.e. the system is open rather than closed). In these patterns the hierarchy may evolve from the top downward with large seaboard cities, such as those on the east coast of North America in the nineteenth century, acting as centres of innovation for external commercial forces.

Berkeley geographer James Vance's book *The Merchant's World* (1970) represents one of the first major attempts to challenge and augment the Christaller model of settlement structure. Vance approaches the problem through the eyes of the historical geographer, following the twisting path by which actual settlements have evolved – particularly the mercantile cities of America's east coast. There city growth had begun, in a sense, from the top down (the reverse of the Christaller model). Thus Boston was from the start of settlement the point of attachment, the economic hinge, through which the staple goods of New England (timber and fish) were concentrated and shipped back to Europe. If you look at the five charts in Figure 14.16, you can follow the way in which the settlement pattern in the Vance model emerges. Note the establishment of depots for staple collection, the antecedent pattern of land division (refer to Figure 21.11) and the emergence of major wholesaling centres in the interior. Finally, comes central-place infilling in the last phase of the model. Since the ideas stress the importance of trade at each stage, the model is termed the *Vance mercantile model* after the Italian word meaning 'merchant'.

Vance's model applies not just to North America but makes an equally coherent story for lands such as Australia where the 'limpet' cities of Brisbane, Sydney, Melbourne, Adelaide, and Perth were the attachment points from which the central-place development later partially emerged. Thus the attraction of the model is that it introduces a dynamic element into settlement models and supplements, rather than supplants, the earlier work. It also enriches our understanding of those long-settled areas to which the original model best applied, accentuating the role of those coastal cities with the best-developed overseas trade links.

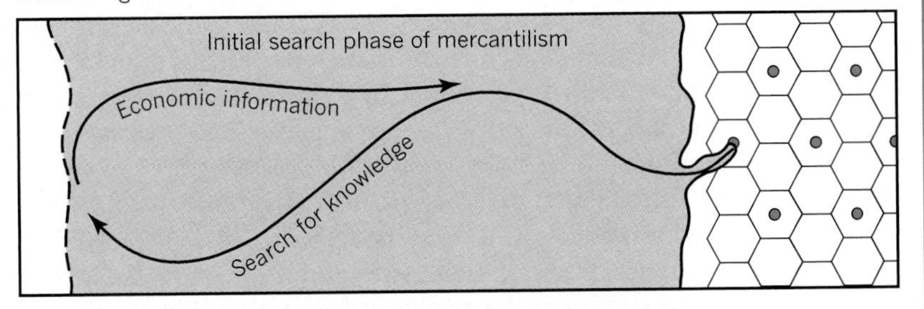

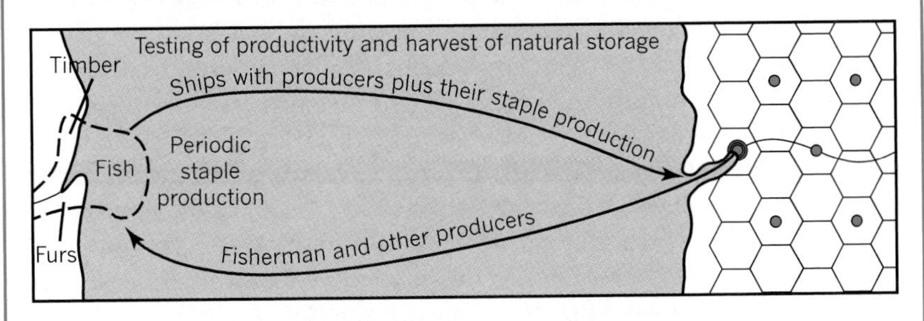

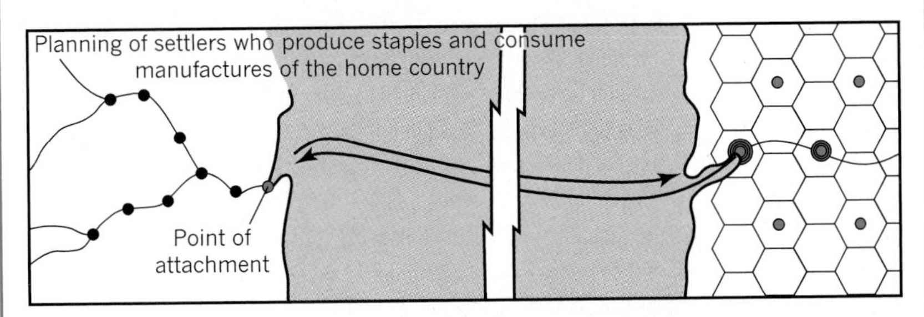

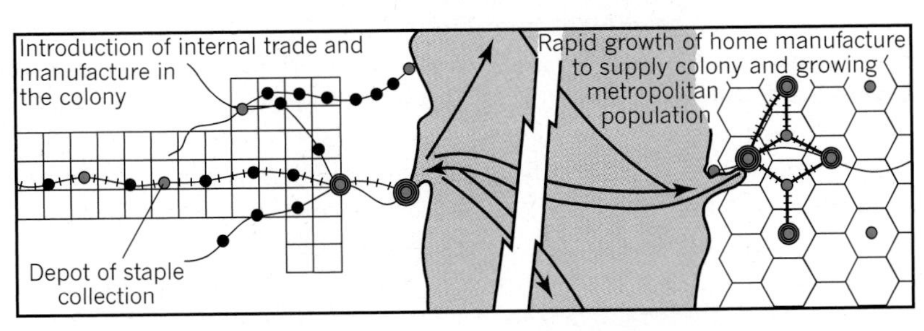

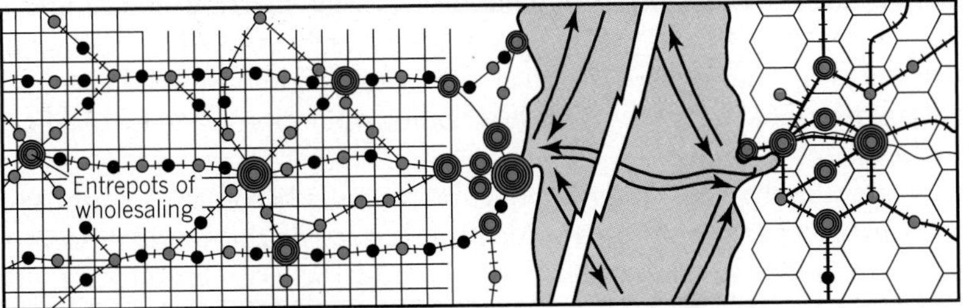

(a)

(b)

Figure 14.16 The Vance mercantile model of settlement evolution (a) Five stages in the evolution of settlement under the mercantile model (left) and the central-place model (right). The term *exogenic* refers to external, and *endogenic* to internal forces. The term *entrepôt* is the French word for storehouse and means a commercial centre for the import and export, collection and distribution of goods. (b) Jay Vance, the Berkeley urban geographer who developed the model.

[Source: J.E. Vance, Jr., *The Merchant's World* (Prentice Hall, Englewood Cliffs, N.J. © 1970), Figure 18, p. 151. Reprinted by permission of Prentice Hall Inc., Englewood Cliffs, N.J.]

South Australia and Catastrophe Models

One innovative approach to central-place systems has been provided by Australian geographer Peter Smailes working in South Australia. He was puzzled by an apparently contradictory observation made in the small, rural townships of that state. When farming was in difficulty, the local central places were not as badly hit as might have been predicted by classical models. Conversely when farming was booming, the local towns did not share fully in the boom.

The area studied by Smailes was a crescent-shaped sweep of semi-arid country in the southern part of the state (see Figure 14.17). Rainfall was low and uncertain, concentrated in the winter months of the year. It was an area of very sparse population: the twelve country townships in the study had an area equivalent to Switzerland but only some 14,000 people. The economy was typified by very large farms ('stations' would be the local term), specializing mainly in sheep and wheat. Farm incomes were at the mercy of two cycles. First, the environmental cycle of rainfall with occasional very bad drought years. Over the years studied, the 'best' years had rainfall three times that of the 'worst' years. Second, the international price cycle for the two staple products: wheat and wool. Australia is very dependent on prices set in Chicago rather than Sydney and such factors as the failure of the Russian harvests or overproduction in North America can cause major swings in price. Again the amplitude of best and worst years was about three to one. Sometimes the two cycles, climatic shocks and price swings, reinforced each other and made matters worse, sometimes they compensated for each other. Cycles in farm income were irregular with a cycle of around four years' duration and income peaks around three or four times the troughs.

Careful study of the income flows showed that in good years, farm income tended to show 'spatial leakage' with profits moving outside the area: e.g. repaying of bank loans, spending sprees in the big city (Adelaide), children sent away to private boarding schools, vacations taken in other parts of Australia. Conversely when conditions were tight then the opposite happened: the shutters on spending went up and more use was made of basic local facilities.

The Australian study shows how the Christaller model can be modified and extended to take into account the effect of both environmental and economic cycles. In effect, use of central-place facilities shifts *up* the hierarchy in good years, *down* the hierarchy in bad years. But in addition to these short-term oscillations there are long-term trends as well. Overall, the rural areas of Australia are losing population as farms increase in size and use labour ever more efficiently. The loss is being partly masked by the fact that the farmers are ageing, selling their farms to neighbours, and retiring to local towns: once that generation leaves the population of many local communities is set to plunge.

Maintaining the network of small rural towns is becoming a major problem in many sparsely peopled agricultural areas of the world – from the Australian outback to the Canadian prairies. Schools are closing, local shops being replaced by catalogue and Internet shopping, and medical practices ever more difficult to maintain. One theoretical approach to this problem is set out in Figure 14.18. This shows how the general relationship between population and central-place facilities postulated by Christaller is in reality complex. The simple relationship shown in (a) and (b) and expanded for a single facility in (c) masks the fact that the threshold is

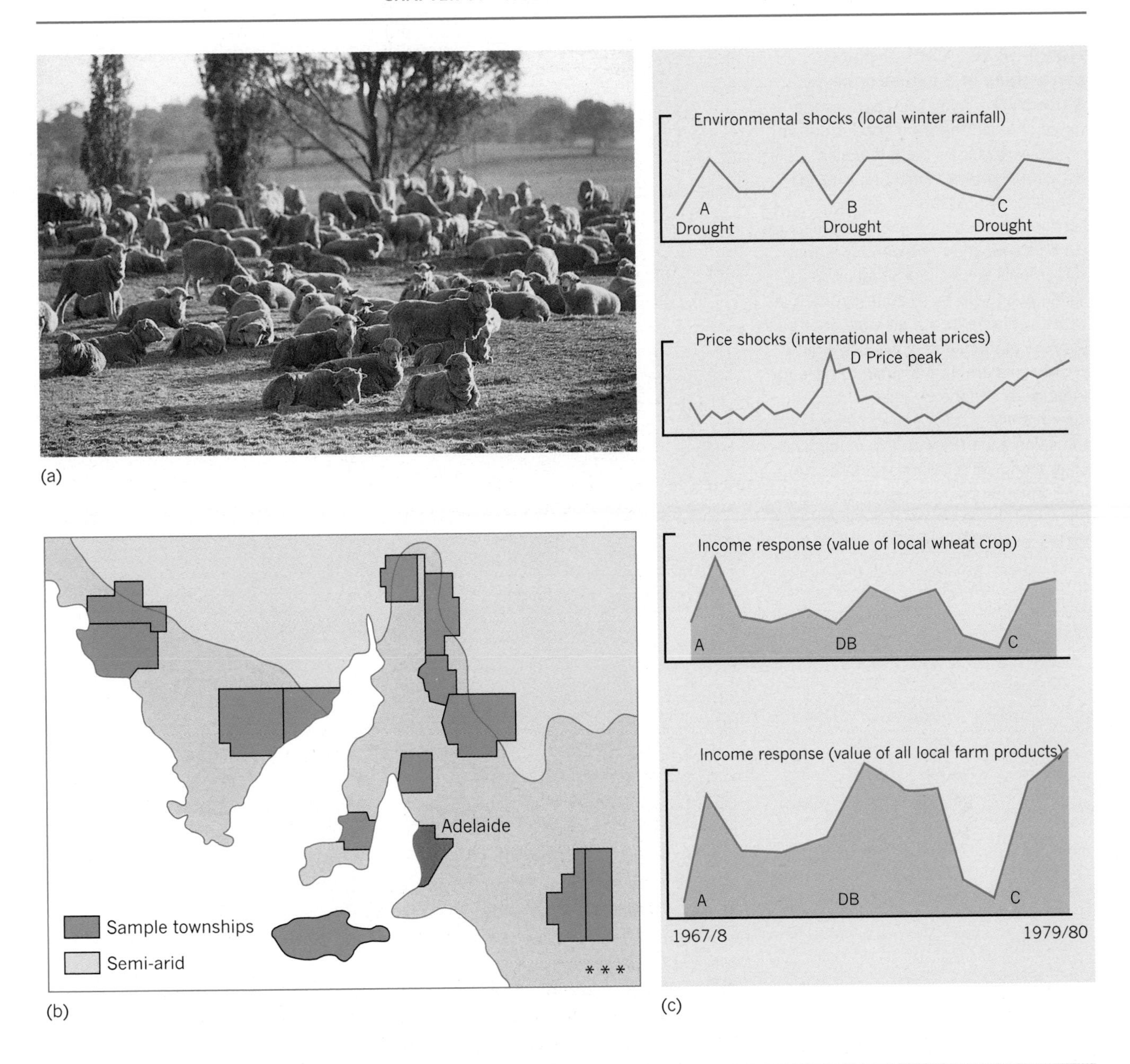

Figure 14.17 Environmental and economic shocks on a rural central-place system (a) Sheep station typifies the semi-arid wheat–sheep zone of South Australia. (b) Location of 14 townships studied by Peter Smailes (see text). (c) Hypothetical environmental, international price, and farm-income cycles. (d) Relationship between farm income and two levels of the central-place system. Δ = symbol for 'change in'.

[Source: (a) © Penny Tweedie/PANOS Pictures. (b), (c), (d) based on research by Peter Smailes, Department of Geography, Adelaide University, on 'Rural crises and community viability in South Australia'.]

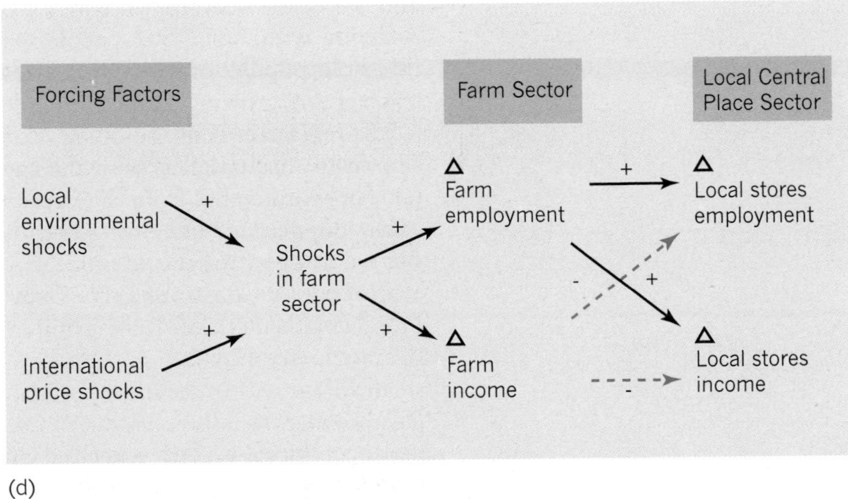

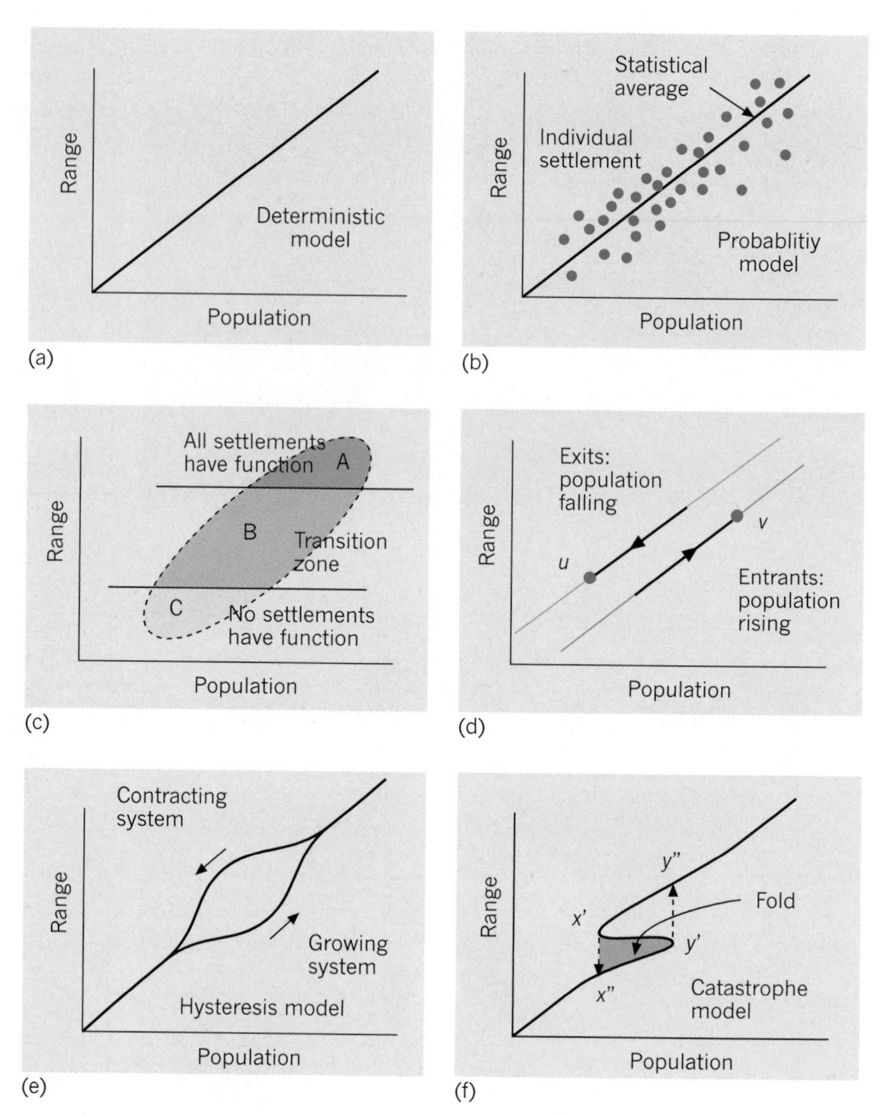

Figure 14.18 Central-place hierarchies in a catastrophe framework In all six diagrams the range of a central-place good is shown on the vertical axis plotted against population on the horizontal axis. (a) Simplified explanation of range of services as a function of community size. (b) Same function but with individual settlements shown. (c) Situation for a single good with three zones: zone A in which settlements are all large enough to support the function, zone C in which settlements are all too small to have the function, and zone B which is intermediate. (d) Expanded picture of the transition zone (B) with expanding and contracting settlements separated. (e) Interpretation as a hysteresis loop. (f) Interpretation in terms of a catastrophe fold. See text explanation.

different for growing and declining settlements. A small village with a dwindling population base may be able to maintain its store for some years after it has passed the uneconomic point; inertia, capital already invested, and age of the operator all play a part. It may just be possible to maintain a facility with, say, 1000 people; but to open a new store would need a threshold population twice that size. So, as (d) shows, the curves for contracting and growing systems are different.

The replacement of the single relationship of (a) by the double relationship in (d) has parallels with the engineering concept of a *hysteresis* loop (e). The word comes from the Greek term for 'being late' and shows how a cause (population) and effect (a central-place facility) can be lagged in time. But we can go further and link the ideas of the hysteresis loop to the concept of a *fold* catastrophe (f). (Catastrophe theory was introduced by the French mathematician René Thom and is also used in the last chapter of this book (see Box 24.C).) On the upper surface of the fold, a shop in a small village with a declining population may close at point x' when the village population falls to level x''. On the lower surface of the fold, a shop may open (point y') only when the village population reaches a much higher level (point y'').

Archaeologists have made use of catastrophe theory in explaining the collapse of old civilizations. One geographer who has explored catastrophe theory to analyze the rise and fall of settlements is Malcolm Wagstaff of Southampton University. He was able to model the collapse of coastal settlements in the Pelleponnes peninsula in southern Greece in the second century AD in terms of a cusp model.

Postscript: Towards Applications

We have left until last the important question of the applications of a settlement model. How can geographers use ideas of city chains and central-place structure? Well, we can use these ideas in two ways. First, the relationship between the size and pattern of settlements summed up by the rank–size rule is stable enough over time for us to use it to project future patterns of settlement sizes. Consider, for example, the regular progression of curves for the United States in Figure 14.19, and try to think what forces we would have to bring in to distort or disturb the pattern. Of course, it is only the whole system of cities that is stable. Individual cities may have widely varying paths of growth. Compare, for example, the constant lead position of New York to the faltering pattern of a city such as Savannah, Georgia. Still greater contrasts separate the rapid rise of Los Angeles with the dropout pattern of Hudson, N.Y.

A second use of the central-place model is in regional planning. The settlement hierarchy in a newly settled area often tends to move from a primate to a rank–size form as population increases and the separate settlements are more closely integrated with each other within the region. For instance, if we were to design such a system for the settlement of the middle-west plateau of Brazil around the new primate city of Brasilia, the later readjustment of the hierarchy could be anticipated and investment in infrastructure (roads, power stations, schools, hospitals, etc.) adjusted accordingly. Central-place theories also have played a role in the designing of hierarchies of shopping and service centres within cities. Increasing use is also being made of hierarchic concepts in the design of key service sectors such as hospital systems.

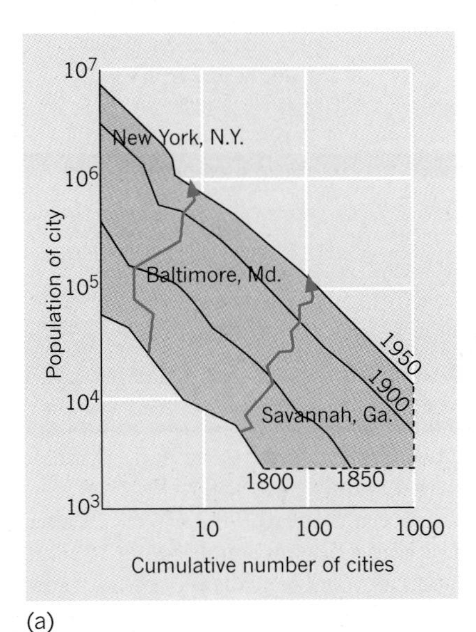

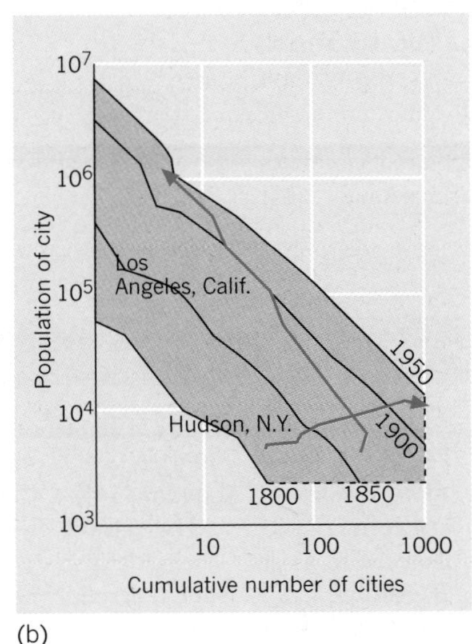

Figure 14.19 Hierarchic evolution
The diagrams show changes in the role of individual cities with a hierarchy. The regular growth of the US urban system over 150 years contrasts with the varying growth trajectories of individual cities.

[Source: C.H. Madden, *Economic Development and Cultural Change* (University of Chicago Press, Chicago, 1956), Vol. 4, Fig. 1. p. 239. Copyright © 1956 by the University of Chicago.]

In beginning this chapter we looked first at the simple chain models of settlements in one dimension and went on to consider hierarchic models in two dimensions. Both types of models are closely linked. Indeed, mathematician Martin Beckmann has shown that the more levels there are in a system, the closer the array of city sizes will come to a continuous distribution. By contrast, regions with only a few units in the hierarchy will have sharply stepped rank–size distributions. Rank–sized distributions must be logical by-products of the central-place system.

Like Alice in our opening quotation the geographer finds the world laid out like a chessboard. In this chapter we have been able to cover only the simplest and most basic moves in the complicated chess game by which cities 'capture' and organize one another's territories into a kind of feudal hierarchy of metropolis, city, and village. Like all models our hexagonal chess set is an oversimplification of reality. To follow the actual 'moves' by which any individual city develops would tax the master skills of a Fischer or a Kasparov.

Reflections

1 Settlements are difficult to define accurately in statistical terms. What is the size of your own community? Look very carefully at your answer to see how it is affected by the bounding problem illustrated in Figure 14.3. Have you over- or underestimated your community's real size? How might you make your estimate more accurate?

2 If settlements are ordered by size within regions, there is usually a regular relationship between a settlement's size and its rank. Gather data for the size of cities in your country or state. Plot their position on a population rank–size diagram like that in Figure 14.6. How closely does the resulting distribution correspond to the one predicted by the rank–size rule? Suggest reasons for any departures from this rule you observe.

3 Central-place functions are not fixed but change over time. Why do so many settlements in western countries appear to be losing their central-place functions (e.g. with bank closures, retail shop closures)? Can you suggest any ways in which their decline might be arrested?

4 In periodic systems, the central-place function circulates around a set of settlements in a regular schedule. What periodic central functions are still found in western countries? List any you can find for your own area, and try to map their tracks over time. Would you consider sporting functions an appropriate example?

5 The kind of formal models presented in this chapter cause debates within geography. Do they (a) fail to reflect the true complexity of city settlements or (b) provide a unique insight into their structure?

6 To what extent does the distribution of schools and colleges in your own city, state, or province have a regular spatial order? Identify some of the factors that you think might be 'disturbing' this order.

7 Review your understanding of the following concepts using the chapter text and the glossary in Appendix A of this book:

bounding problem
central places
complementary regions
fixed-*K* hierarchies
market-size threshold
periodic market cycles
primate patterns
rank–size rule
SMSAs
Vance mercantile model

One Step Further ...

A brief but useful introduction to central-place theory is given by Ron Johnston in his essay in R.J. JOHNSTON *et al.* (eds), *Dictionary of Human Geography*, fourth edition (Blackwell, Oxford, 2000), pp. 72–4. Reviews of the basic theories of settlement hierarchies and their structures are provided in most texts on human geography. See, for example, R. ABLER *et al.*, *Spatial Organization* (Prentice Hall, Englewood Cliffs, N.J., 1971), Chapter 10, and P. HAGGETT *et al.*, *Locational Analysis in Human Geography*, second edition (Arnold, London, 1977), Chapters 4 and 5.

The classic work in central-place theory is Walter Christaller's study of southern Germany, published in 1933. It should certainly be dipped into and is available now in translation. See W. CHRISTALLER, *Central Places in Southern*

Germany (Prentice Hall, Englewood Cliffs, N.J., 1966, transl.). Another classic German work that has more ideas in its footnotes than many books have in their text is August LOSCH, *The Economics of Location* (Yale University Press, New Haven, Conn., 1954).

Some modern theoretical departures are authoritatively presented in B.J.L. BERRY and J.B. PARR, *Geography of Market Centers and Retail Distribution*, second edition (Prentice Hall, Englewood Cliffs, N.J., 1967), Chapter 4, and K.S. BEAVON, *Central Place Theory; a Reinterpretation* (Longmans, London, 1977). For a critical approach to settlement theory with emphasis on the historical evidence and dynamic models, the outstanding book is J.E. VANCE, Jr., *The Merchant's World: The Geography of Wholesaling* (Prentice Hall, Englewood Cliffs, N.J., 1970).

The settlement 'progress reports' section of *Progress in Human Geography* (quarterly) will prove helpful in keeping up to date. Current research is reported in the major geographic journals. Look especially at the *University of Chicago Department of Geography Research Papers* (published occasionally) for applications of central-place concepts. For readers with access to the World-Wide Web see also the sites recommended for this chapter in Appendix B at the end of this book.

CHAPTER 15

Surfaces

■ denotes case studies

Figure 15.1 **Distant tentacles of the world economic system** In contrast to the crowded skyscrapers of Manhattan (see Figure 14.1), the tentacles of economic development reach out to remote areas. LANDSAT satellite image of part of the Amazon Basin within the state of Rondonia in Brazil showing extensive deforestation. The original forest shows up as dark shades, contrasting with the pale tints of the cleared forest. The linear, branching pattern at lower right is typical of the region. North is at the top of the photo and the east–west distance is around 30 km (20 miles). [Source: Geospace/Science Photo Library.]

> The last rural threads of American society are being woven into the national urban fabric.
>
> MELVIN M. WEBBER *Cities and Space* (1963)

Where does a city end? Think about the answer by recalling the last time you drove along a motorway out of a major metropolis. At 1.5 km (1 mile) out from the central business district (CBD) the landscape was probably still urban, dominated by a built-up environment of houses and streets. At 16 km (10 miles) out the landscape was probably semirural; but in this suburban range there would still be massive daily commuting to the city. At 80 km (50 miles) out the landscape may have been truly rural; but the car radio would still carry, faintly, the voice of city radio stations, the farmers would still purchase the city newspapers from their local general store, and occasional family visits will still be paid to the city to carry out their major shopping.

Cities, like old soldiers, do not die – but simply fade away in space. Their spatial dominance lessens as we move further from their centres but is never entirely eliminated. Some minute proportions of a city's mail or telephone calls will reach out even to the far side of the world; its billboard advertising ('Only 1050 miles to Harry's Place') may stretch across a continent. The roads feeling their way into the Amazonian forest (Figure 15.1 p. 448) reflect the demand from urban markets many thousands of miles away.

It is the surfaces within and around this city-centred world that we look at in this chapter. (Recall 'surfaces' within the diagram of Figure 13.2.) We begin by looking at the effect of spatial attenuation within the city itself. We then go on to see how rural land around it is used. Here there is a parallel impact of the city on the processing of natural resources and we study the seesaw of forces that shape the spatial pattern of industry. Finally we look at examples of how these surfaces change over time.

In this chapter shall be concerned with the locational theory that geographers have built up to explain the city-oriented world. Locational theory is one of the three major elements in geographers' research, and we shall be asking questions about why certain activities are located where they are. Since some of the answers turn out to be complicated, you may need to move rather slowly through this and the following chapter.

15.1 Surfaces Within the City

Cities as we know them are not like the dots on the map of the previous chapter. They are places of immense variety, throbbing or quiet, dangerous or secure, impressive or repellent. So here again, as in so many places in the book, we have to put on fresh lenses and change from a focus on cities like small specks on the global surface ('cities in space') to a focus on the worlds within the city itself ('space within the city').

Study of the internal geography of the city is one of the fastest-growing areas within geography so here we are only able to touch on some aspects of geographers' work. We begin with the pattern familiar to most readers and look at the spatial structure of the western city.

The Geometry of Land Values

The basic geometry of the North American city is well known to most readers. At the centre is the downtown shopping district, the banks and offices, the hotels and theatres. Surrounding this is an area of rundown housing, mixed with some industry. Beyond this we run into modest residential areas, typically houses and apartments for office and blue-collar workers. Still further out come the family homes of middle-class suburbanites, thinning out until we come to the golf courses and the rolling acres of the very rich.

In this section we shall probe the familiar pattern and try to see what factors lie behind it. We shall look at the value of land within the city, the use to which it is put, and the population density it supports. Secondly, we shall take Chicago as an illustration of a western city and examine its pattern in relation to existing models of urban structure.

One measure of the value an urban community places on different sites within the city is provided by land values. As an example, Figure 15.2 shows a three-dimensional representation of land values in Topeka, Kansas. Despite minor variations in peripheral areas, the dominating elements in this and in most western cities are the extremely high values attached to land in the central part of the city and the generally steep decline of land values with distance from the centre. The high point in values is found somewhere near the centre of the *central business district (CBD)*. The CBD is usually marked by tall buildings, an extremely high daytime population, and high traffic densities, and geographers have worked out various ways of mapping its precise location. (See Box 15.A on 'Delimiting the central business district (CBD)'.)

If we take a series of such cities, we can draw a general picture of aver-

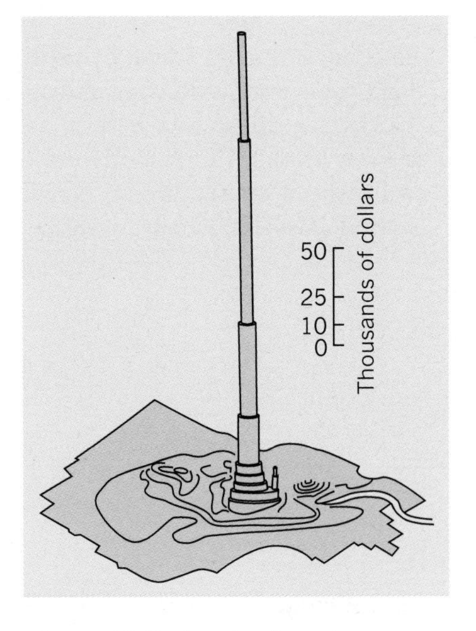

Figure 15.2 Land values in an urban community This three-dimensional map shows land values in the city of Topeka, Kansas. Note the extremely high values in the city centre and the smaller secondary centres.

[Source: D.S. Knos, *Distribution and Land Values in Topeka, Kansas* (University of Kansas, Bureau of Business and Economic Research, Lawrence, Kansas, 1962), Fig. 2.]

Box 15.A Delimiting the Central Business District (CBD)

Despite increasing decentralization of functions to out-of-town locations (e.g. peripheral shopping malls), the central parts of western cities continue to be marked by high buildings, high land values, and distinctive types of commercial activities (e.g. major department stores, corporation headquarters, specialized banks). To map the exact extent of this central business district (CBD), geographers have devised a number of specific measures. One of the most commonly used is that by Vance and Murphy. This needs the following information for the floor area on each building:

1. Ground plan area (e.g. as shown on a large-scale map), A.
2. Total floor area on all levels (i.e. including all the upper floors and basement levels as well as the ground floor), B.
3. Total floor area devoted to specialized central business activities, C. These activities *include* stores, entertainment, banking, and insurance but *exclude* residences, schools, parks, churches, wholesaling, vacant premises, and storage.

From this information two measures can be constructed. First, a *height index* (C/A). Second, an *intensity ratio* ($C/B \times 100$). To take an example: assume a building with a ground plan of 100 m × 200 m (i.e. A = 20,000 m^2) which has total floor area of 78,000 m^2 (=B) on four floors, and of this 45,000 m^2 (=C) is devoted to central business activities. Then its height index will be 45,000/20,000 = 2.25, and its intensity ratio (45,000/78,000) = 58 per cent.

For mapping purposes it is more convenient to sum the areas for all buildings on a given city block, thereby giving a block index. Where the height index has a value of 1.0 or more *and* the intensity ratio is 50 per cent or greater, then a city block is potentially part of the CBD.

Box continued

The line is usually drawn by starting at the city's peak land value intersection and adding in *contiguous* 'CBD blocks' (i.e. those with both measures at or above the critical levels). The exact nature of the area delimited will depend on the precise definition of a specialized central business activity. Lists of such activities with more refined versions of the Vance–Murphy index are given in the standard urban geography texts, e.g. M.H. Yeates and B.J. Garner, *The North American City*, 2nd edn (Harper & Row, New York, 1976), Chapter 12; H. Carter, *The Study of Urban Geography* (Arnold, London, 1995).

age land values (Figure 15.3). The value of land is highest in the city centre and decreases toward the periphery, but the pattern is modified by two additional elements: main traffic arteries and intersections of main arteries with secondary centres at regular distances from the CBD. When we superimpose these three effects on one another in a three-dimensional model, the result is a conic hill whose flanks are disturbed by ridges, depressions, and small peaks. This land-value surface directly reflects the different accessibility of parts of the city and shows where the most intense competition for space occurs (and hence where the higher land values occur).

Land Values and Land Use

What will the effect of these variations in land values be on the distribution of land uses within the city? Let us assume that a city wishes to establish a new university somewhere within its limits. If the university is constructed near the city centre, it will be most accessible by public transport to all students, but it will be taking up valuable land that could be leased at high rents to commercial firms. Conversely, if we choose a green field on the edge of the city as the site of the campus, the land values there will be low and the university grounds can be more spacious. The advantage, however, will be offset by the school's eccentric location. Most students will have to travel farther, some right across the city from the far side.

All types of urban land use entail this type of quandary. As businesses move farther from the city centre, they gain from the cheaper land prices or rents; but they are increasingly divorced from the centre of their potential market and stand to lose money from increased transport costs. We can describe this trade-off between land prices and transport costs by a series of non-intersecting lines, termed by the economist *bid-price curves* (Figure 15.4(a)). Each

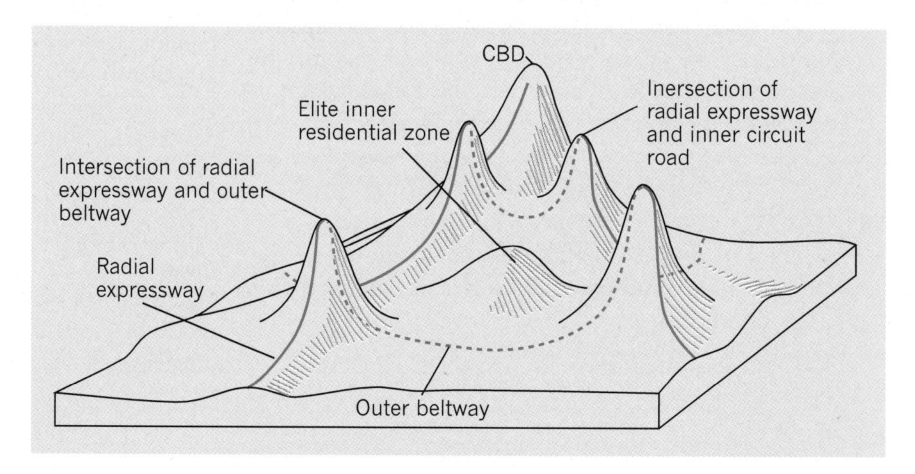

Figure 15.3 A general pattern of city land values This idealized land-value surface emphasizes the high-value ridges that run outward from the CBD along the main arterial roads. Secondary centres on these roads cause local peaks. The CBD itself is the highest peak at the rear of the diagram.

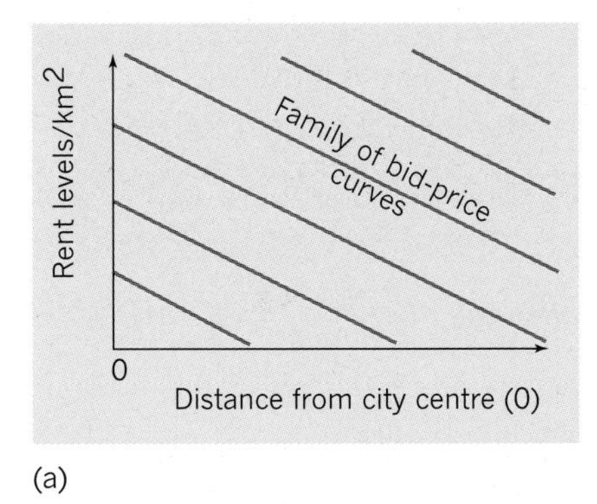

(a)

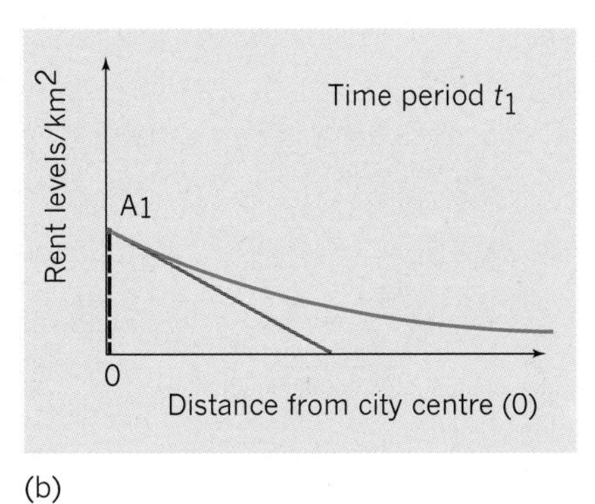

(b)

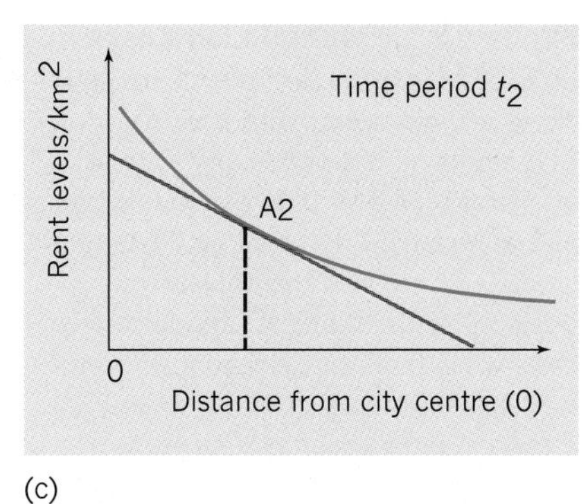

(c)

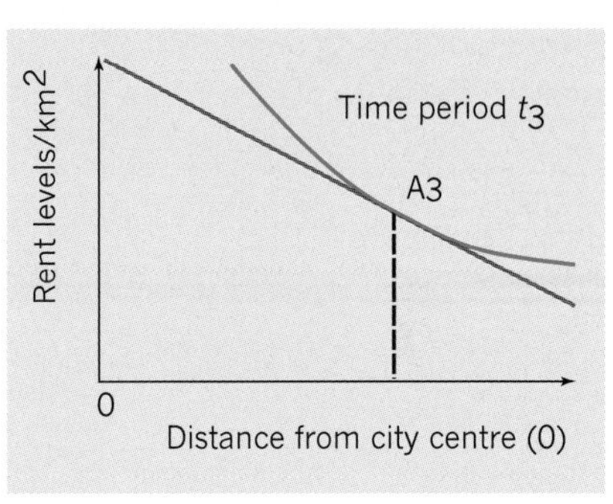

(d)

Figure 15.4 City land-use models For any urban land use – be it a cemetery, a motel, or a college – a location represents a trade-off between the convenience of being near the accessible city centre and the high rent levels we must pay to locate there. If we describe our assessment of these two competing goals by a family of bid-price curves (a), we can compare these with the actual rent levels (blue line). Note how the equilibrium position (A_1) indicating the best-trade (and thus the best location) may shift (to A_2 or A_3) as the city grows and land values rise (in time periods t_1 to t_3). Thus certain types of urban land use (e.g. single-family homes) may be squeezed out towards the margin of the city as it grows.

line represents the rent values that exactly balance the increased transport costs due to locating away from the city centre. If we assume that transport costs increase regularly with distance, then the bid-price curves will be parallel straight lines sloping downwards with distance from the city centre. Note that the lower lines within the family of parallel lines represent lower rent levels and therefore are always preferred over higher lines.

By superimposing bid-price curves on the actual rent curve for a given city, as in Figure 15.4(b), we can determine both the point where the best trade-off is reached and the corresponding rent level. The best location will be where the actual rent curve just grazes the lowest possible price-bid curve. At this point (A_1) the values of both the actual rent curve and the bid-price curve are equal. If we think of this diagram as three dimensional, the bid-price curves become a series of cones centred on the CBD, and we see that A_1 is not a point but a ring of possible locations around the city centre.

As a city grows larger, land values increase, especially at the centre, so the land-value curve becomes both higher and more concave, as in Figure 15.4(c). Hence, the ring of optimal locations for land-using activities is forced outwards from near the city centre (A_1 in time period t_1) to successively more peripheral locations at A_2 and A_3. We can trace this effect in the displacement of activities like manufacturing from inner city to suburban locations.

Each type of land-use activity in a city has a distinct pattern of bid-price curves. Those activities that gain greatly from locating 'where the action is' near the city centre will have steep curves. Theatres, insurance brokers, publishers, are all examples of activities that depend on a high degree of contactability and need accessible locations. Conversely, other activities may not be greatly affected by their location but may be very anxious to avoid high rentals: these will have bid-price curves with a gentle slope. Activities with steep bid-price curves will be able to cling to the steep slopes in land values around the CBD; others are gently angled and find a foothold only in the remoter parts of the city. Figure 15.5(a) illustrates hypothetical curves for banking and insurance offices (which typically cling tenaciously to downtown locations) and golf courses, which are usually located on the fringes of cities and sensitive to rises in land values. Note that the cross-section shown here is a very simple one; if we add outlying secondary centres to the figure we can see that banking may be carried on there too, albeit on a smaller scale (see Figure 15.5(b)).

Land-use Mosaics, Chicago-style

Land use in urban areas is a major research area for the urban geographer. Probably the most thoroughly dissected city is Chicago, where social

Figure 15.5 Multiple land uses
An extension of the arguments in Figure 15.4 to two types of land use. (a) Characteristic bid-price curves for an intensive high-rent land use (banking) and an extensive low-rent land use (golf courses). Note how the equilibrium position for the first is in the downtown area, while that for the second is on the outskirts of the city. (b) This shows how a varied land-values gradient with secondary peaks may allow banking to find a series of equilibrium positions at varying distances from the CBD.

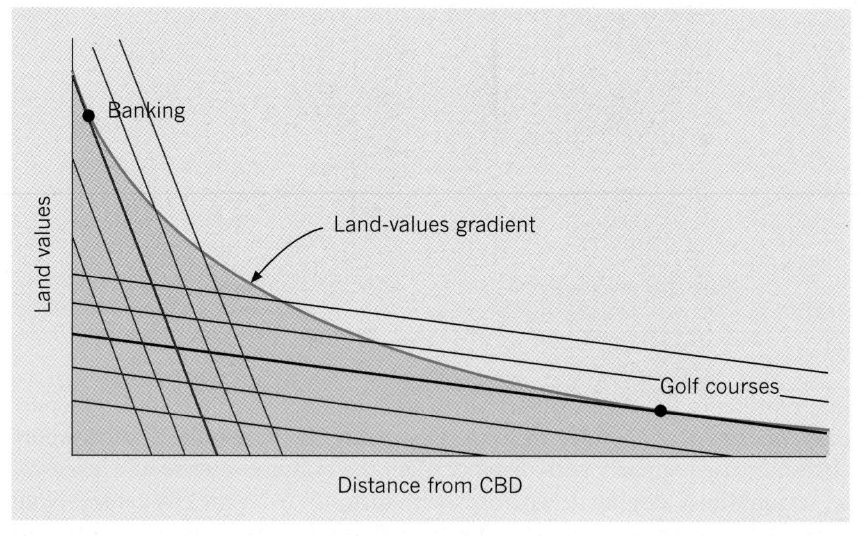

(a) Multiple land uses

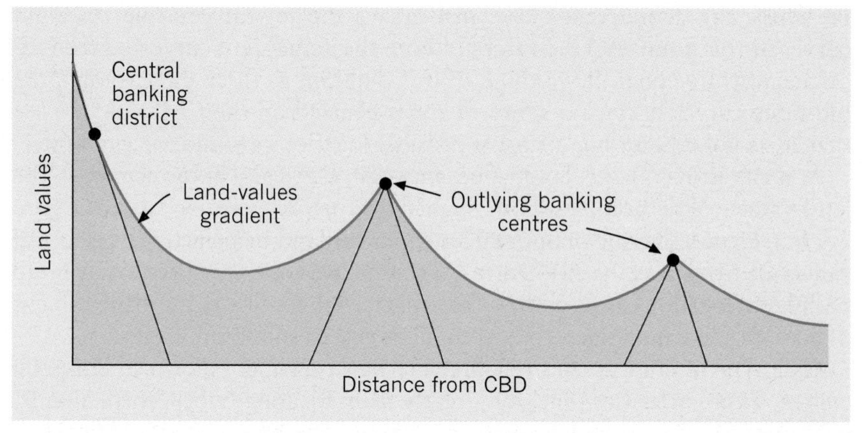

(b) Multiple centres for single land use

scientists such as Robert Park and E.W. Burgess began a trail of research in the 1920s that was followed up a half-century later by geographers such as Harold Mayer and Brian Berry. If we take an aerial photograph of Chicago from the south, looking toward the CBD in the Loop (Figure 15.6), we are viewing a city whose structure has formed a touchstone for studies of large metropolises elsewhere.

Can we make any sense of the complex mosaic of land uses shown by Chicago? Figure 15.7 shows one guide to the pattern provided by geographers. Look first at the top row of the diagram. This shows an idealized Chicago as a circle centred on the CBD with Lake Michigan to the right. The CBD is the key feature, marked out by a score of factors (its high land values, its soaring daytime population, its high buildings, its age, and so on). Around this CBD the city's housing has developed as a series of wedge-shaped *income sectors* (upper left), which fan out as we move outward. The drive north or south from the CBD along the lakeshore will take you

Figure 15.6 Chicago Poet Carl Sandburg's 'City of the Big Shoulders' has a special fascination for geographers. From the 1920s on, its spatial structure has been studied and dissected by a series of distinguished researchers. The key to its semicircular structure is the CBD, whose high-rise buildings form the famous 'Loop' area (named from the loop of elevated railroad tracks that encircle it). The zones and sectors that flare out from the CBD are summarized in Figure 15.7. The transportation lines that are so prominent in the photograph have an important effect on the city. Chicago grew up in the 1830s at the strategic southwest corner of Lake Michigan, and the canal connecting the Chicago River to the Illinois River (constructed in 1848) brought a rapid rise in the city's role as a freight-hauling centre. In the same year, the first railroad link was completed and the stage was set for a period or remarkable economic growth linked to the Chicago's dominant position on the expanding railroad network of the mid-west. Its population grew from around 30,000 in 1850 to over the one million mark by 1890. Today, with over seven million people in the metropolitan area, Chicago is America's third largest city. To meet the vast transit needs of the metropolis, and intricate system of urban highways like the Dan Ryan Expressway has been constructed.

[Source: Tony Stone Images.]

through much higher-income housing than a drive to, say, the northwest. Once housing of a particular type had been established in a sector, it tended to persist as new housing of similar quality was built. Thus the city expanded outward in this wedge-shaped fashion.

A second factor is a series of ring-shaped *age and family-sized zones* centred on the CBD. The three zones shown mark a regular progression from the apartment-dwelling population of Zone I (typically, people in their 20s or an older generation in their 50s or beyond) to the suburban family-home population of Zone III (typically, families with young children). On the right of the top row is the third factor, *ethnic segregation*. This shows the presence of two distinctive black wedges to the south and west, in the inner part of an otherwise white city.

If you follow the arrows from the first three factors, they lead to a fourth diagram of Chicago. This is made by overlapping the first three factors to give a mixture of wedges and zones. Thus we get a white, high-income Zone III in the north of Chicago (the Evanston area). Note how the two black zones distort the simple model: each black zone is itself subdivided into three zones related to age and family size, and each is an area of relatively low income despite its occurring in middle-income sectors in terms of our first diagram (upper left). Indeed, an area such as the black south side of Chicago is in some ways a detailed replica of the whole city, finely divided into a complex mosaic of income, age, and family-sized units.

Of course, Chicago is not the simple ring shown in the upper half of Figure 15.7. So we introduce a more accurate map of the city showing its *differential growth* between 1840 and 1960. This fourth factor is also blended into the model. Note how the zones within the various sectors are now displaced in our map showing differential growth of sectors. Finally, a fifth factor, the *main workplaces*, can be added to the model: the main commercial centre plus three types of outlying industrial area are shown. Since to add all these to the model would be too complicated for our final diagram of the city (lower right), we include just some examples of the way in which industry helps to shape the land-use mosaic of the city. Two industrial satellites are shown, the Skokie area in the north and the area near O'Hare airport in the west. South of the city the heavy industrial area of Gary is also shown.

The patterns shown in Figure 15.7 show how various types of spatial segregation interact in shaping the mosaic of land use within a city such as Chicago. Spatial segregation affects not only the clustering of social groups

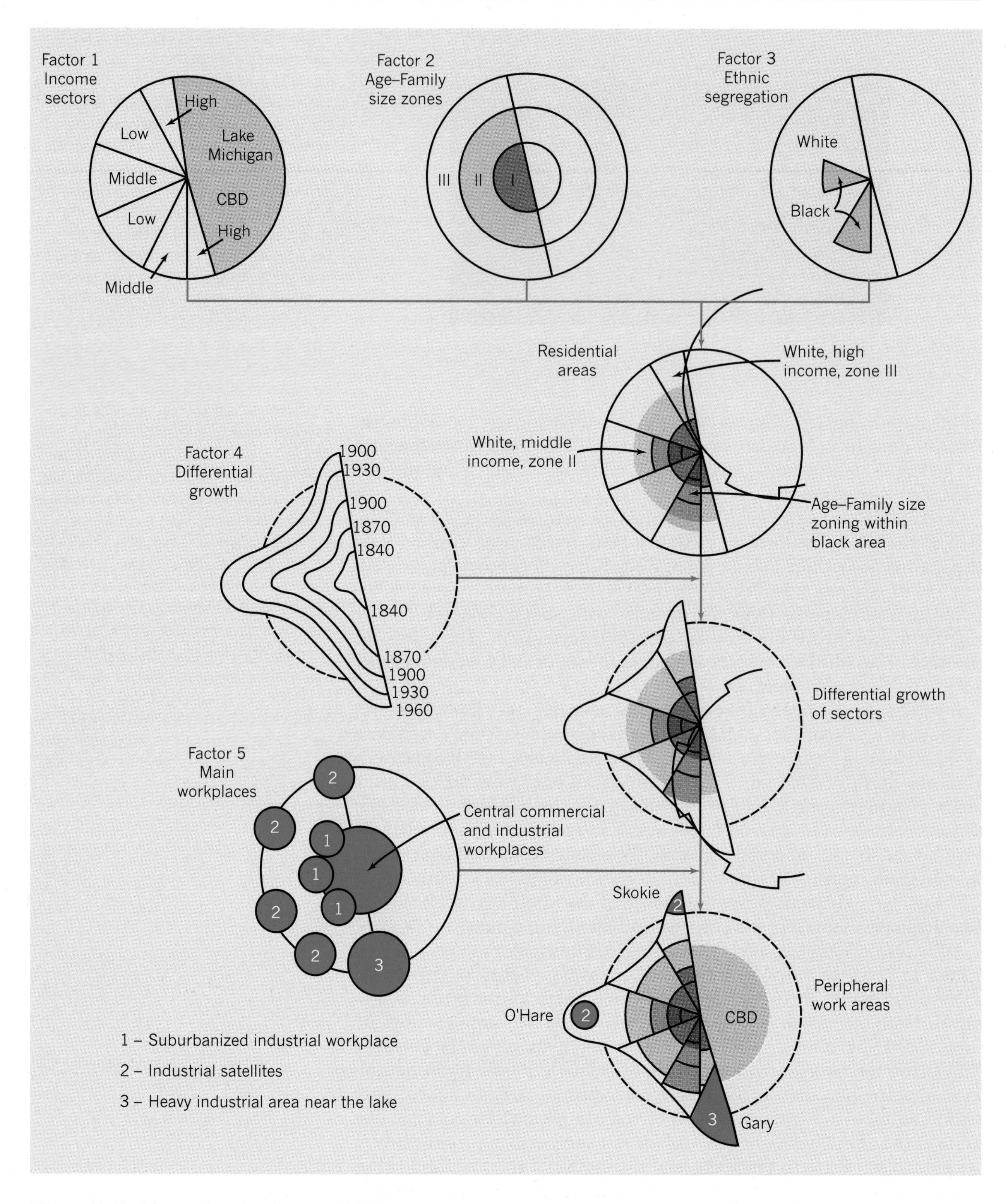

Figure 15.7 The regional structure of Chicago Much of the variety within Chicago's semicircular urban structure (itself the product of a lakeshore location) can be explained by the combined effect of five factors. The arrows show the ways in which factors are cross-bred together in this simplified model of the city. For a description of each factor and its impact on the city, see the text discussion.

[Source: P.H. Rees, University of Chicago, unpublished master's thesis, 1968.]

as in the black ghetto on the south side of Chicago or the Italian quarter to the northwest; it also plays a role in decisions to group certain land uses, such as light industry, in particular sections of the city.

Cross-city Comparisons

If we take a single western city such as Chicago and map its changing population density over time, we find that the population tends to spread out like a slowly melting ice-cream cone, covering ever wider areas but at ever lower densities. Figure 15.8 shows the pattern of population change for Chicago during a 100 year period by a series of population-density gradients that steadily decline in steepness from around 1860 to 1920. The change over the last half-century has been less spectacular but has included a marked reduction in density at the centre of the city.

How typical of the spatial structure of cities is Chicago? Cross-cultural checks on population declines with distance from the city centre have been intensively studied, and investigators have reached some general conclusions about the shape of these slopes. For example, economist Colin Clark studied population-density gradients for a group of 36 cities from Los Angeles to Budapest for a 150 year period from 1807. He found that all the curves could be described as *negatively exponential* – that is, as decreasing sharply at first with distance from the centre and then getting progressively flatter. (See Box 15.B on 'Urban density functions'.) In western cities the decline in population density is reflected in a familiar sequence of housing types, with high-rise apartment blocks near the CBD and low-density housing on semirural tracts of land on the suburban periphery.

Extension of these cross-cultural studies to non-western cities shows that we still have a lot to learn about the ways cities evolve. For example, one striking feature of an Indian city such as Calcutta is the continued increase in densities near the centre and the stability of the density gradient with urban growth. The falling degree of compactness and crowding that characterize the growth of western cities is not seen in Calcutta, where both tend to remain constant. Given the same increase in population, the periphery of the non-western city expands less than the periphery of the western city.

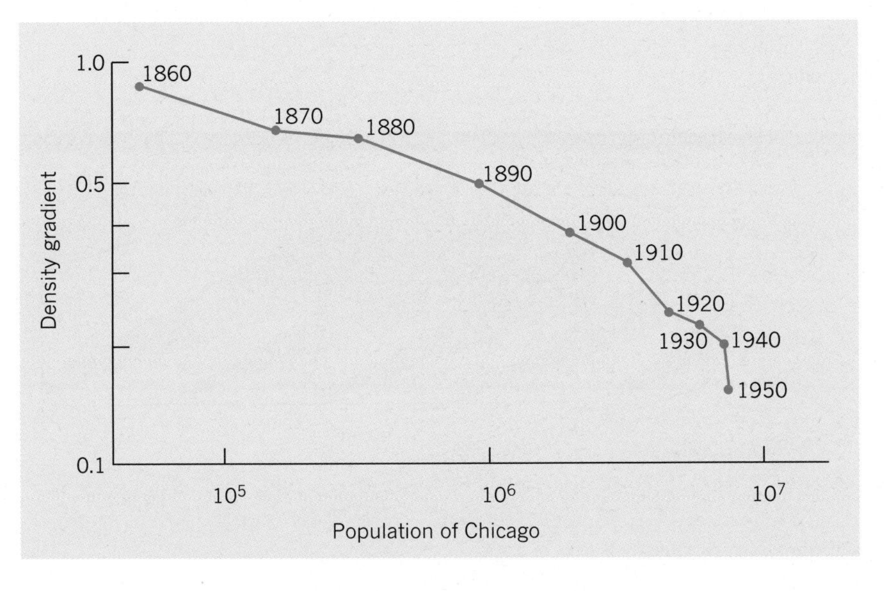

Figure 15.8 Changing urban density gradients over time The average slope of Chicago's population density away from the city centre was steep in 1860 but has declined steadily as the city expanded. Compare this graph with the map of Chicago's outward spread in Figure 15.7.

[Source: P.H. Rees, University of Chicago, unpublished master's thesis, 1968.]

Box 15.B Urban Density Functions

Study of scores of urban areas throughout the world led demographer Colin Clark to suggest a general model for the decline of urban population densities away from the CBD. He proposed a *negative exponential* form in which population decreases at a decreasing rate with distance; that is,

$$Z_{ij} = Z_0 e^{-bd}$$

where Z_{ij} = the population density at distance d from the CBD;
Z_0 = a constant indicating the extrapolated population density at distance zero, that is the centre of the city;
e = the base of the natural logarithms (2.718);
b = a constant indicating the rate of decrease of population density with distance; and
d = the variable distance.

Thus with a central density (Z_0) of 1000 people/km^2 at the centre and $b = -1.0$, we should expect a density of 368 people/km^2 at 1 km from the CBD, 135 people/km^2 at 2 km, 50 people/km^2 at 3 km, and so on. Plotting the curve on paper will give you a clear idea of the concave shape of the curve. A comparison of Z_{ij} and b values allows us to compare different urban structures easily. Clark's work has been extended by others to allow more complex density functions to be matched and compared. (See M.H. Yeates and B.J. Garner, *The North American City*, 2nd edn (Harper & Row, New York, 1976), Chapter 10.)

Figure 15.9 summarizes the variations in both time and space that characterize the two types of city. As the transport revolution that shaped the western city makes its impact elsewhere, more cities may conform to this pattern.

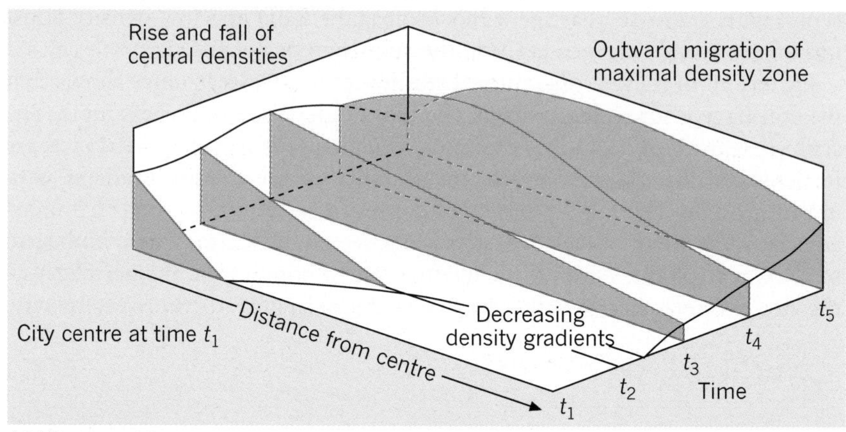

(a) Western cities

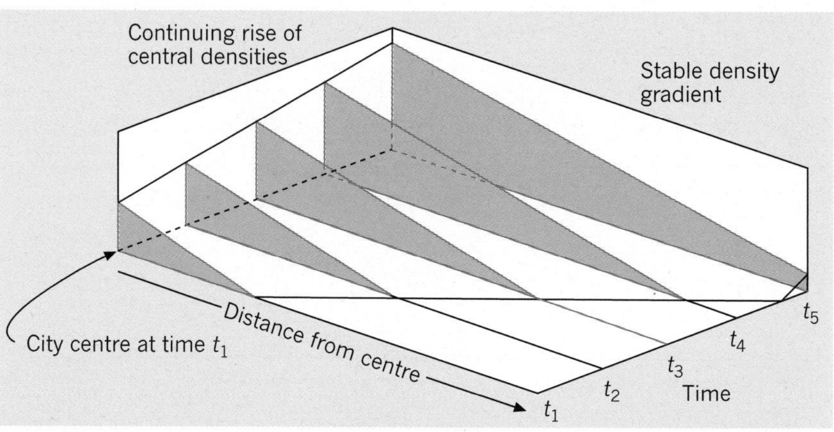

(b) Non-western cities

Figure 15.9 Western and non-western cities Cross-cultural contrasts are evident in these urban-density gradients for a sample of western and non-western cities. In both diagrams population density (on the vertical axis) is plotted against distance from the city centre and time (t_1 to t_5).

[Source: From B.J.L. Berry *et al.*, *Geographical Review*, **53**, (1963), p. 403. Reprinted with permission.]

What kinds of forms will replace the present urban-industrial centre? If we follow the argument of Figure 8.7, then we should expect jobs to be heavily concentrated in the tertiary and quaternary sectors. This means that most jobs in the city would be in offices and research laboratories; such jobs would be highly dependent on massive inputs of information and on the availability of skilled people.

One indicator of the postindustrial city is the distribution of those highly qualified people who form the workforce of the quaternary sector. For example, one of the highest concentrations of PhDs in the United States is not in the large industrial cities, or academic New England or Southern California, but in North Carolina. There an area known as the Research Triangle has grown up based on three nearby university towns: Raleigh, the state capital; Chapel Hill, the attractive state university town; and Durham, home of Duke University. Since 1950 active promotion has brought to the area some 50 research laboratories for industrial corporations and federal organizations with 25,000 new jobs. Newcomers include IBM, Burroughs Wellcome, Monsanto, and the Environmental Protection Agency. The attractive environment, the lack of big-city problems, and the scale economies of common servicing arrangements suggest that the Triangle is likely to retain its distinctive role as a quaternary complex.

New capital cities, such as Australia's Canberra, form a major example of postindustrial cities. They are characterized by low population densities, small differences in living standards and social differentiation, a reliance on personal transport (the automobile) rather than public transport, and a high level of recreational amenities. It is too early to know, however, if such cities are merely the favoured by-products of the industrial city. A move towards a more general pattern of postindustrial urbanization could pose enormous problems in terms of land use and employment opportunties. Mechanization and automation in manufacturing industry have drastically reduced the need for labour. We might expect similar changes in the tertiary sector in the decades ahead as the impact of new IT technologies is felt.

Surfaces Beyond the City: The Agricultural Landscape 15.2

The earliest attempt to correlate land-use patterns with the spatial relationship of a city to its surrounding region was made by a German locational theorist, Johann Heinrich von Thünen. (See Box 15.C on 'Thünen and the isolated state'.) In his classic work on the location of agricultural land-use zones first published in 1826, he not only laid the foundation for refined analysis of the location of agriculture, but also stimulated interest in a much broader area of locational analysis.

Principles of the Thünen Model

The *Thünen model* of the isolated state is modelled on the agricultural patterns of nineteenth-century Mecklenburg. The basic form of the land-use patterns he envisaged is shown in Figure 15.10, and the characteristics of each of the zones are presented in Table 15.1. The patterns are a series of

Box 15.C Thünen and the Isolated State

John Heinrich von Thünen (born Oldenburg, Germany, 1783; died Mecklenburg, Germany, 1850) was both a pioneer locational theorist and a practical farmer. In 1810 Thünen, at the age of 27, acquired his own agricultural estate, Tellow, near the town of Rostock in Mecklenburg on the Baltic coast of Germany. For the next 40 years, until his death in 1850, he supervised the cultivation of the Tellow estate and amassed a remarkable set of records and accounts that provided the empirical basis for his published theories.

In his classic work on the location of agricultural land-use zones, *Der Isolierte Staat in Beziehung auf Landwirtschaft*, first published in 1826 (and later translated as *The Isolated State*), he not only laid the foundation for refined analysis of the location of agriculture, but stimulated interest in a much broader area of locational analysis.

In his work, he imagined an isolated city, set in the middle of a level and uniformly fertile plain without navigable waterways and bounded by a wilderness. In this model he showed how concentric zones of agricultural production would be formed around the central town. Perishables and heavy products (in relation to value) would be produced near to the town; durable and lighter goods on the periphery. Although agriculture has changed dramatically in the last 150 years, his models still give useful insights into the geographic patterns observable today.

Thünen regarded locational rent (*Bodenrente*) as the key factor sorting the uniform area of his isolated state into distinct land-use zones. The location is given by

$$L = Y(P - C) - YD(F)$$

where L = the locational rent (in \$/km^2),

Y = the crop yield (in tons/km^2),
P = the market price of the crop (in \$/ton),
C = the production cost of the crop (in \$/ton),
D = the distance to the central market (in km), and
F = the transport rate (in \$/ton/km).

Thus, for a crop yielding 1000 tons/km^2, fetching \$100/ton at the central market, and costing \$50/ton to produce and \$1/ton/km to haul, the locational rent at the city centre would be \$50,000/km^2. At 10 km distant it would be only \$40,000 km^2. At 20 km distant it would be down to \$30,000/km^2. Beyond 50 km production would be at a loss. The competition of two crops (i and j) for the same area depends on their yield (Y) and relative profitability ($P - C$). When the condition

$$1 < \frac{Y(P - C)_i}{Y(P - C)_j} < \frac{Y_i}{Y_j}$$

obtains for crops i and j, they form two distinct spatial zones; crop i dominates a circular area adjacent to the city, and crop j occupies a ring-shaped zone immediately outside it. The symbol $<$ means 'is less than'. Any other relation of the terms in the equation above results in the two crops being reversed so that j occupies the inner ring, *or* one crop dominating all available land to the complete exclusion of the other, *or* both crops being grown side by side with no spatially differentiating zoning. For further discussion, see E.S. Dunn, *The Location of Agricultural Production* (University of Florida Press, Gainesville, 1974), Chapter 1.

[Source: Photo courtesy of Katz Pictures Ltd – The Mansell Collection.]

concentric shells ranging from narrow bands of intensive farming and forest to a broad band of extensive agriculture and ranching to an outer 'waste'.

To understand the form of these spatial patterns, we need to review the six assumed conditions that dictated it:

1 The existence of an isolated state cut off from the rest of the world.
2 The domination of this state by a large city that served as the sole urban market.
3 The setting of the city in a broad, featureless plain that was equal

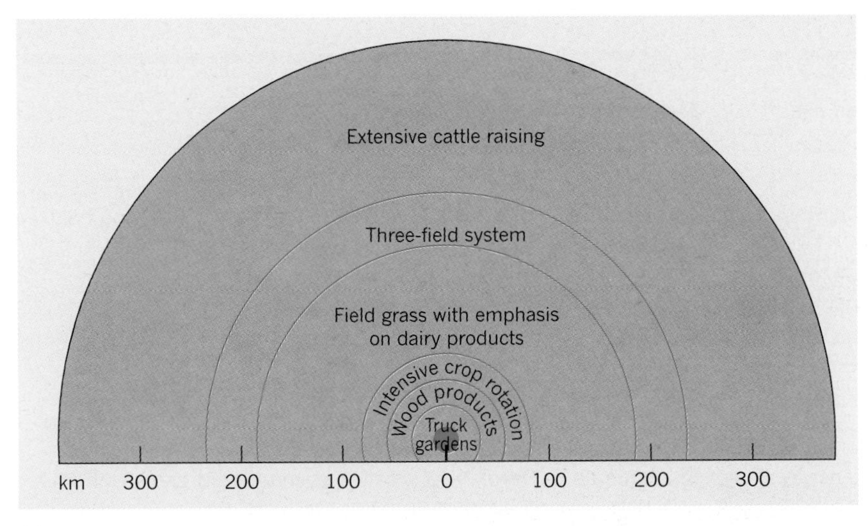

(a)

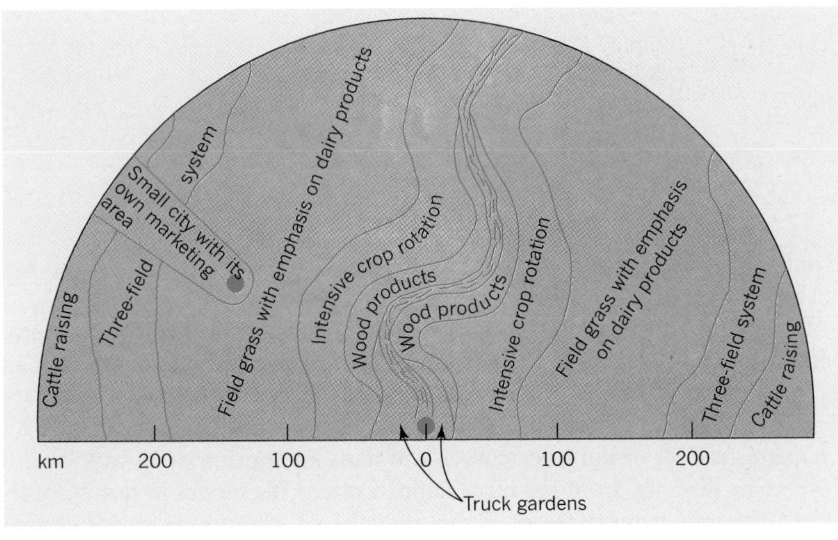

(b)

Figure 15.10 Land-use zones in Thünen's *Isolated State* (1826)
(a) Thünen's original land-use rings. (b) Thünen's rings modified by a navigable river bringing cheaper transport costs. The difference between the two diagrams reflects the contrast between ***isotropic*** space (a) with transport costs equal in all directions and ***anisotropic*** space (b) with transport costs unequal along the river. The distance figures were added to Thünen's original model by Leo Waibel in 1933, in his *Probleme der Landwirtschafts-geographie*.

everywhere in fertility and in which the ease of movement was the same everywhere so that production and transport costs were the same everywhere.

4 The supplying of the city by farmers who shipped agricultural goods there in return for industrial produce.
5 The transport of farm produce by the farmer himself, who hauled it to the central market along a close, dense trail of converging roads of equal quality at a cost directly proportional to the distance covered.
6 The maximizing of profits by all farmers, who automatically adjust the output of crops to meet the needs of the central market.

These assumptions are, of course, unrepresentative of actual conditions either in the early 1800s or, indeed, now. Why, then, did Thünen make them? To understand this, we have to go back to our initial discussion of the role of models in science (Section 1.5). The objective of models is to simplify the real world in order to understand some of its characteristics.

Table 15.1 Thünen's land-use rings

Zone	Percentage of state area	Relative distance from central city	Land use	Major product marketed	Production system
0	<0.1	−0.1	Urban-industrial	Manufactured goods	Urban trade centre of state; near iron and coal mines
1	1	0.1–0.6	Intensive agriculture	Milk, vegetables	Intensive dairying and trucking; heavy manuring; no fallow period
2	3	0.6–3.5	Forest	Firewood	Sustained-yield forestry
3a	3	3.5–4.6		Rye, potatoes	Six year intensive crop rotation: rye (2), potatoes (1), clover (1), barley (1), vetch (1); no fallow period; cattle stall-fed in winter
3b	30	4.7–34	Extensive agriculture	Rye, animal products	Seven year rotation system: field grass with an emphasis on dairy products; pasture (3), rye (1), barley (1), oats (1), fallow period (1)
3c	25	34–44		Rye, animal products	Three-field system: rye, etc. (1), pasture (1), fallow period (1)
4	38	45–100	Ranching	Animal products	Mainly extensive stock-raising; some rye for on-farm consumption
5	–	Beyond 100	Waste	None	None

Source: P. Haggett *et al.*, *Locational Analysis in Human Geography* (Arnold, London, 1972), Table 6.4, p. 205.

Thünen's model permits us to do just this by sorting out some of the key factors (albeit in a simplified form) that caused land-use rings to form.

Given these assumptions, Thünen was able to demonstrate that rural land values would decline away from the central city in the same way urban land values do, though at much lower rates and with gentler slopes (see Figure 15.11). Like each urban land use, each agricultural land use has a characteristic set of bid-price curves and finds an appropriate location with respect to distance from the city. Thünen stated his model in terms somewhat different from those of urban models (see the discussion of zoning mechanisms in Box 15.C), but the processes by which eventual land uses are determined are identical. A product that is bulky (i.e. has a large tonnage per unit of area) or is difficult to transport will have steep bid-price curves and will be quite sensitive to displacement from the market. Conversely, one that is lighter (i.e. has a low tonnage per unit of area) or easy to transport will be less sensitive to displacement. As Figure 15.11 indicates, land-use boundaries occur at the intersections of bid-price curves. Land use in the remoter areas is adapted to take into account the poor accessibility. A smaller tonnage of a given crop may be grown on each unit of area by using extensive farming methods; for instance, long fallow periods may be used to restore the land's fertility rather than fertilizers. Alternatively, a product may be shipped in a more compressed form (e.g. as cheese rather than liquid milk), or animals may be used to 'concentrate' a crop (as when hogs raised on corn are marketed in place of corn.

We can extend the basic Thünen model to explain situations quite unlike the ones it was first proposed to explain. Although the original study was of ring formation around a single node, this is merely one geometric case. If we substitute a linear market for a central one, the model still explains the formation of zones, but in straight parallel bands. Land-use zones along

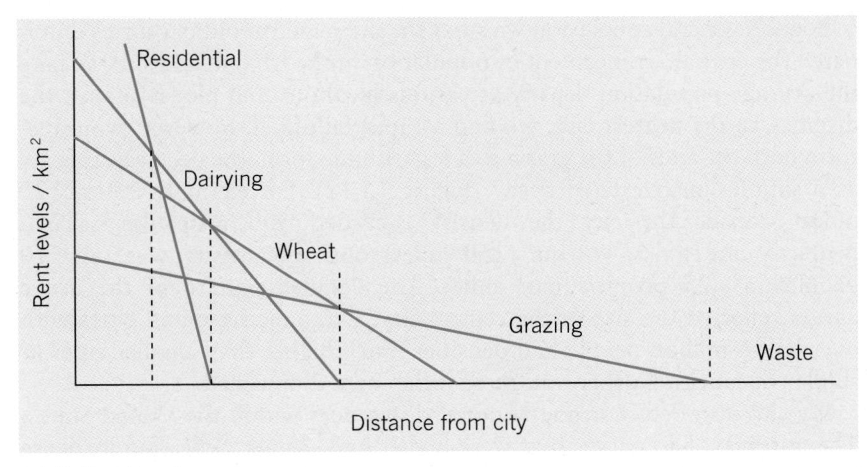

(a) Optimal land uses

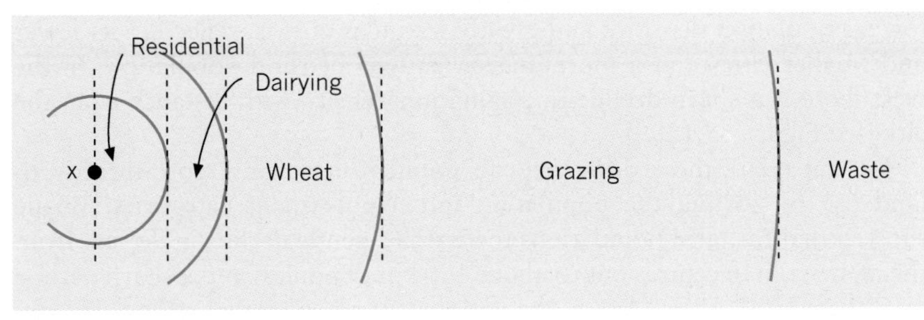

(b) Land-use rings

Figure 15.11 Ring formation in the Thünen model Diagram (a) shows how the bid-price curves for four types of land use have slopes of different steepness. (To understand this figure more easily, you may find it useful to check back to Figure 15.4 where we first looked at the idea of bid-price curves.) Use of land for dairying has a steeper curve than that for wheat farming, indicating that dairying is a more intensive way of using the land and one that stands more to gain from locating near the city. But dairying is itself displaced by land use for residential housing on the fringe of the city. The broken lines in (a) mark the breaks where one type of land use 'outbids' the other in terms of the rent levels it can afford to pay for a convenient location near the city centre. If you follow these broken lines down to diagram (b), you will see that four distinct land zones are formed. ('Waste' indicates land not in use because it lies too remote from the city.) If you complete the arcs of the circles shown in (b), a series of land use rings are formed around the city.

a coastal strip or transport axis are also common variants of the conventional Thünen rings. Several alternatives of the ringed model were discussed by Thünen himself. By introducing a navigable river on which transport is speedier and costs are only one-tenth that of land transport, a minor market centre with its own trading area, and spatial variations in the productivity of the plain, he was able to explain considerable variations in land-use patterns (see Figure 15.10(b)). Once we allow these kinds of variations and add to them the wider rings that the technological advances in transport make possible then the Thunen model can provide insights into contemporary patterns of land use on larger spatial scales.

City-centred Zoning in the United States

Attempts to compare the Thünen model with the real world are hindered by the difficulties geographers have had in drawing up unambiguous definitions of land use. Different crops are grown in different environmental zones, so the Thünen rings are disturbed by other zones related to the ecology of crops rather than the accessibility of cities. Geographers have tended, therefore, to use population density as a substitute for land use and to look for gradual changes in density instead of sharp discontinuities between distinct land-use zones.

Demographer Donald Bogue has investigated how population distributed itself around 67 major cities in the United States at mid-twentieth century. Using the census figures for counties, he analyzed the changes in population density with respect to distance as far as 800 km (500 miles) from the city.

Bogue's general conclusion was that the main metropolitan centres dominated the spatial arrangement of population in the United States. If we take the average population density at various locations and plot it against the distance to the nearest city, we find a rapid falloff. If, however, we transform both the axes of the graph to a logarithmic form, the decrease appears as a simple linear rate of decay (Figure 15.12). For example, 40 km (25 miles) outside the city the density exceeded 500 people/km^2 (1250 people/square mile); 400 km (250 miles) out, the density was only 10 people/km^2 (25 people/square mile). The detailed pattern of the decay curves reflected the size of the central city. Large metropolitan cities with over half a million people had densities much higher than smaller cities at similar distances. Farther out, these differences diminished.

We can also detect strong regional differences within the United States. The northeast had curves that were higher and steeper, reflecting its dense network of cities separated by rural pockets of relatively low population density such as northern New England and Appalachia. The south had lower population densities and greatly irregular curves, reflecting its fewer and smaller cities and a more uneven pattern of rural population. In the west there is a sharp decline in population density with distance from the city (see Figure 15.12(c)).

We can relate these differences in population density more directly to land use by sorting the population into employment categories. Bogue showed that for farm populations there was a gentle decline in density from the metropolitan centres out to about 150 km (93 miles) and a sharp decline at 500 km (300 miles). Again, the national trend concealed strong regional contrasts. The density in the south, for example, changed little with distance. Industrial land use as approximated by industrial employment was much more dependent on distance. Employment declined very sharply with distance from the major metropolises, but the decline is arrested around 50–100 km (30–60 miles) out before dropping sharply again. The brief halt in the decline probably represented a concentration of specialized manufacturing towns at this distance.

Figure 15.12 Density gradients around cities Variations in population density with distance from the nearest city (metropolis) are shown for three regions of the United States in the middle twentieth century. Both axes of the graphs are logarithmic to allow details of the middle parts of the density curves to be shown more clearly. This also allows easier comparison among the three regions. If plotted on arithmetic scales, all the curves would appear to be L-shaped, emphasizing only the dramatic drop in population density within the first few miles of distance from the metropolis.

[Source: D.J. Bogue, *Structure of the Metropolitan Community* (Scripps Population Institute, Ann Arbor, Mich., 1949), p. 58.]

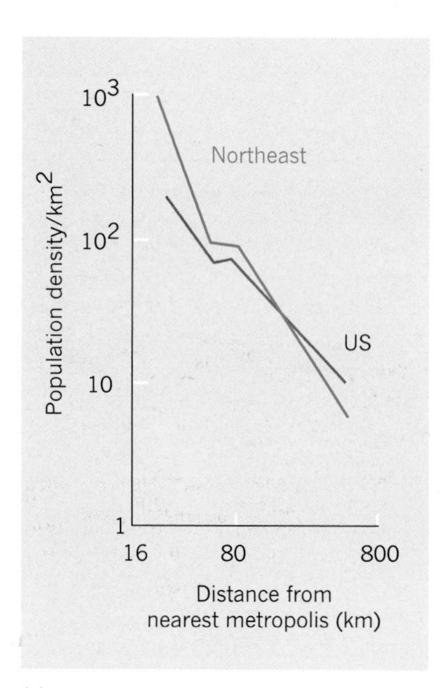

(a)

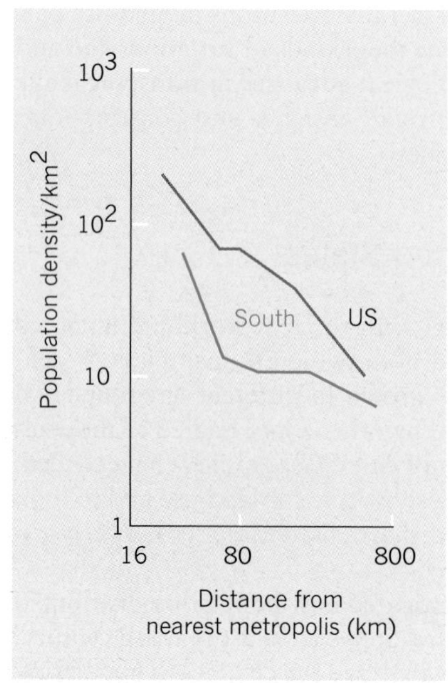

(b)

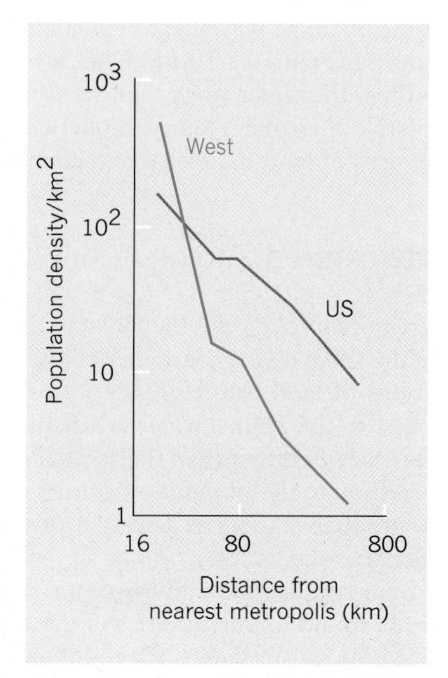

(c)

One important feature of Bogue's study, which has been picked up in later work, is that population density depends not only on distance from the metropolis but also on the direction in which other cities lie. If we divide the area around the city into wedge-shaped sectors, then sectors with routes running to other cities represent ridges of high population density compared with those not containing such routes.

Similar research in different countries but on comparable geographic scales indicates that the findings for the United States are reasonably typical. Population density the world round is sensitive to both the distance from and the direction to the centres of cities.

Zoning in Farm Communities

Geographic studies of land-use zones at the micro level of the village and farm (Figure 15.13) have been done as well as studies of subcontinental zoning. Here, too, the effort required to use an area of land is going to increase with the distance from the centre of the community. If we take a small community, the individual family farm, the time taken to reach the most distant fields will be greater than the time needed to visit the home paddock adjacent to the farmstead. By tracking individual farmers from dawn to dusk, comprehensive diaries of their movements in space over time can be compiled. What do these show? In the Netherlands arable plots only 0.5 km (0.3 miles) from farmsteads receive about 400 hours of care per hectare annually. At a distance of 2 km (1.2 miles), the care level drops to 300 hours; at 5 km (3.1 miles), it dwindles to only 150 hours. We can convert such figures into costs by adding information on the jobs carried out in these movements between farmstead and field. In Punjab villages in Pakistan the cost of ploughing increases by around 5 per cent, the cost of hauling manure rises 10–25 per cent, and the cost of transporting crops

Figure 15.13 Agricultural land use The detailed pattern of crops at the field and farm level provides evidence from which the models of land-use zoning described in this chapter are tested. Distinctive zoning at the micro-level, such as these fields in the Quezaltennago district of Guatemala, is repeated at larger geographic scales all the way up to the world scale. Here high population growth has made Maya Indian landholdings too small to feed a family, forcing the owners to see work as migrant coffee pickers.

[Source: Paul Harrison/Still Pictures.]

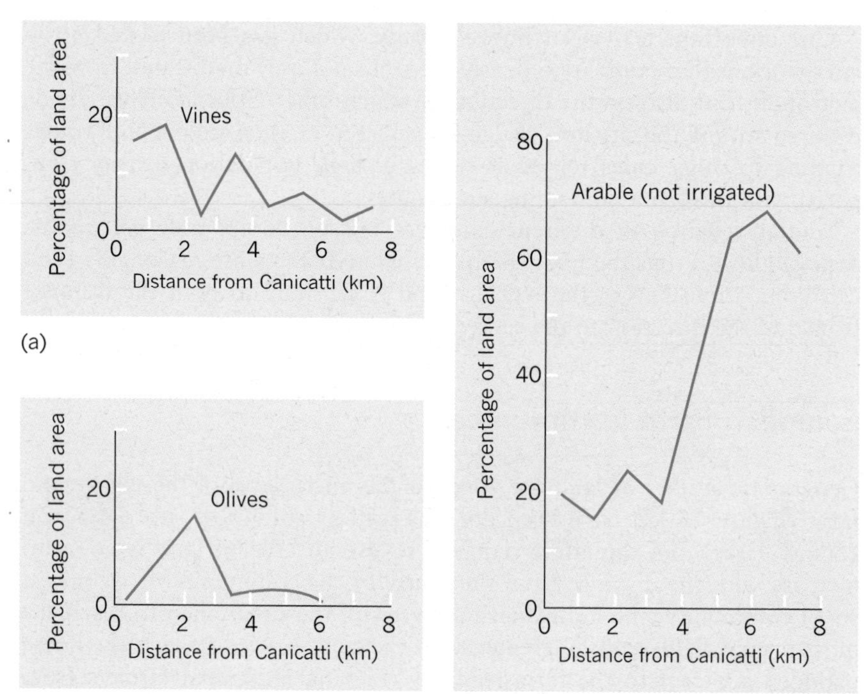

Figure 15.14 Distance and rural land use The diagrams show changing crop patterns with increasing distance from the Sicilian village of Canicatti. Vines and olives tend to be grown close to the village while arable crops are more remote.

[Source: M.D.I. Chisholm, *Rural Settlement and Land Use* (Hutchinson, London, 1966), Table 6, p. 57.]

rises 15–32 per cent every time we move an additional half-kilometre from the village.

How rapidly costs increase with distance is a matter for debate; different rates have been reported in different studies. What does seem clear is that locational adjustments begin to occur at distances as low as 1 km (0.6 miles) from the community centre and that the costs of operation rise sharply beyond about 3 or 4 km (2 or 2.5 miles). We can see the form of these locational adjustments in the distinctive zones of land use around villages. Figure 15.14 shows the sequence of crops extending a distance of 8 km (5 miles) from the Sicilian village of Canicatti. Note that the growing of olives and vines falls off rapidly with distance; beyond 4 km (2.5 miles) the open, arable land is cultivated mostly for wheat and barley. Some clue to the 'sorting' process by which the decision is made to grow certain crops on certain locations is provided by the average number of man-days in fields near the village but declines to less than 40 in remote fields at distances of 8 km (5 miles) or more.

Geographer Mansell Prothero has described a similar zoning around villages in northern Nigeria. He distinguishes four zones. The first is an inner garden zone with close interplanting, a continuous sequence of crops, and intensive care. A second zone at 0.8–1.2 km (0.5–0.7 miles) out is continuously used (mainly planted with Guinea corn, cotton, tobacco, and groundnuts) and fertilized. A third zone with an outer boundary at 1.6 km (1 mile) is used for rotation farming; that is, the land is cultivated for three to four years and then allowed to return to bush for at least five years to get back its fertility. Finally comes the fourth zone of heavy bush. Within this zone are isolated clearings, in which the three-zone sequence of the main villages is reproduced.

Studies similar to those in Sicily and Nigeria reveal a variety of responses by farmers to distance. Sometimes the response leads to sharp land-use zoning, as in Figure 15.15; in other instances farmers may grow the same crops over a wide range of distances but plant them in different combina-

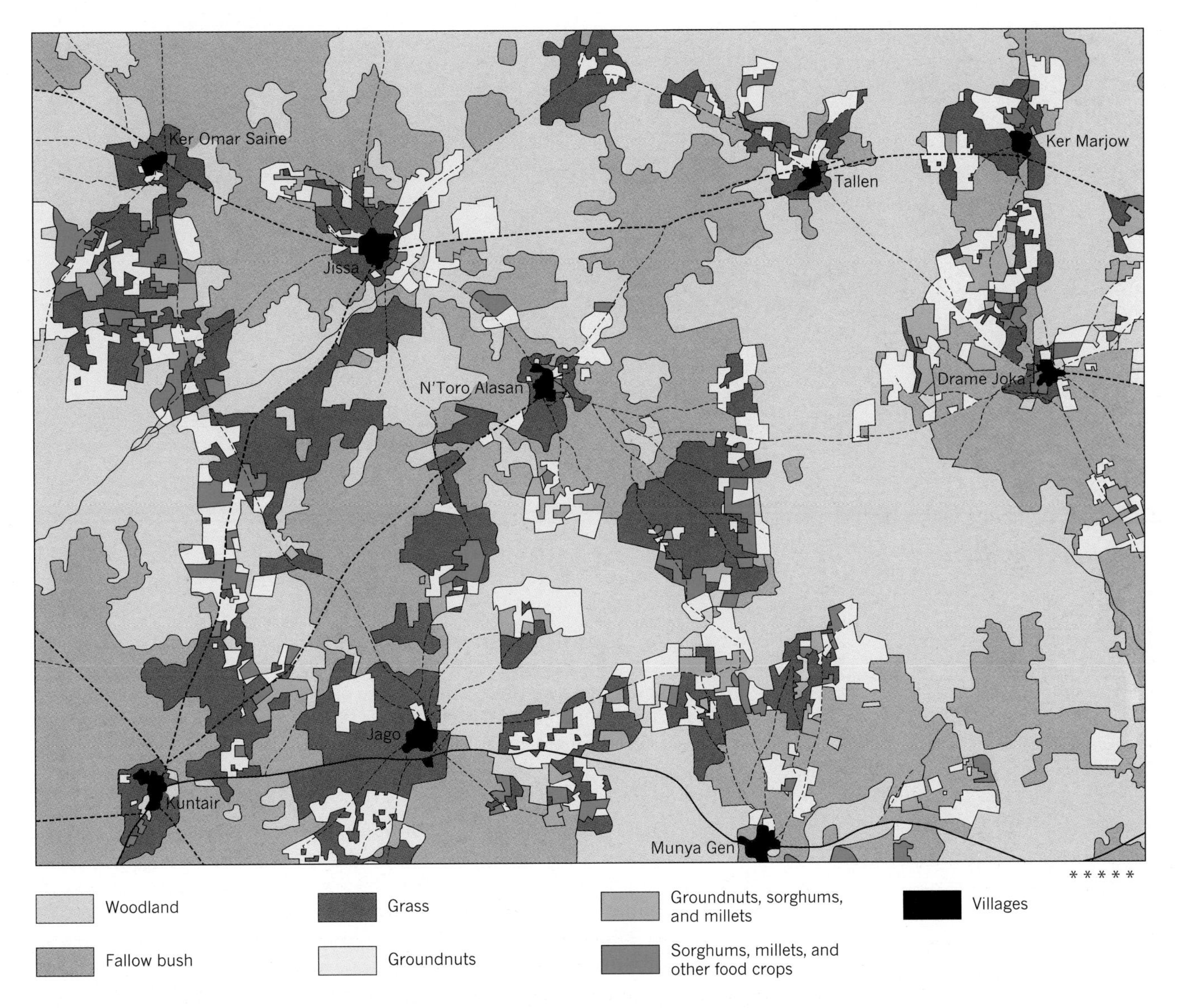

Figure 15.15 Rural land-use banding in the African tropics The map shows areas of permanent and shifting cultivation in the Kuntair area, Gambia, West Africa. Note the relations of the different zones to the location of the nine villages named. The total distance across the map is about 5 km (3 miles). Woodland occurs as a residual zone at the margins of the cultivated land and along boundary areas between villages. Fallow bush is woodland that has previously been used for cultivation, and where the land is now lying idle while recovering its fertility. Groundnuts (peanuts) are an important cash crop, while sorghum and millet (both grain crops) are consumed locally. These crop areas (see key) tend to be located on the woodland margins in newly cleared patches of land with relatively high fertility.

[Source: Directorate of Overseas Surveys, Gambia, Land Use Sheet 3/111, 1:25,000, 1958.]

tions and give them varying degrees of care. Moreover, the simple symmetry of the zones is usually interrupted by local variations in terrain, soils, patterns of ownership, and the like.

Zoning at the International Level

During the nineteenth century, a major urban-industrial nucleus emerged in Western Europe and eastern North America. This nucleus may be thought

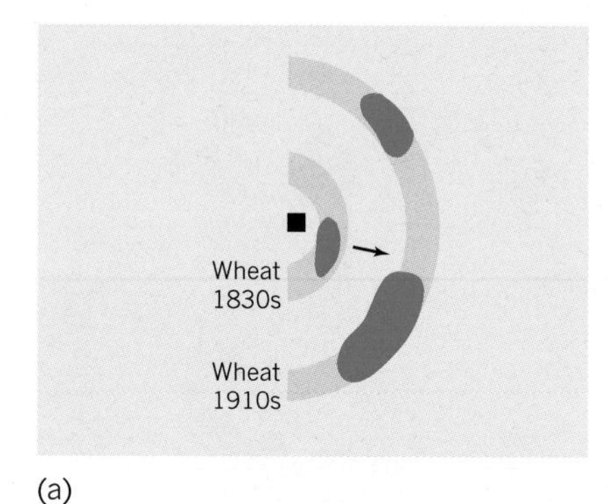

(a)

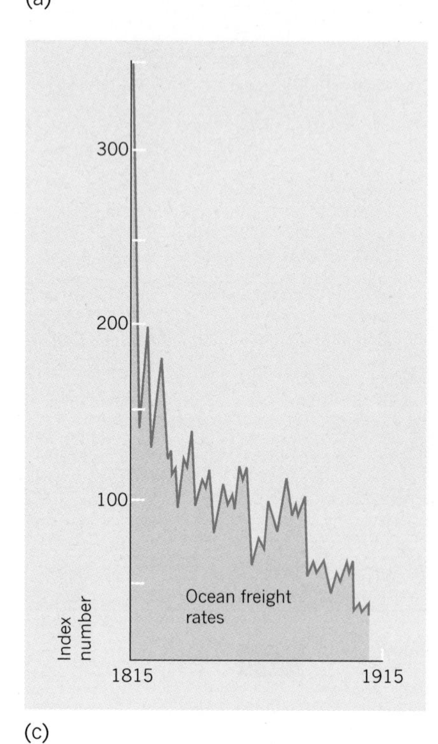

(c)

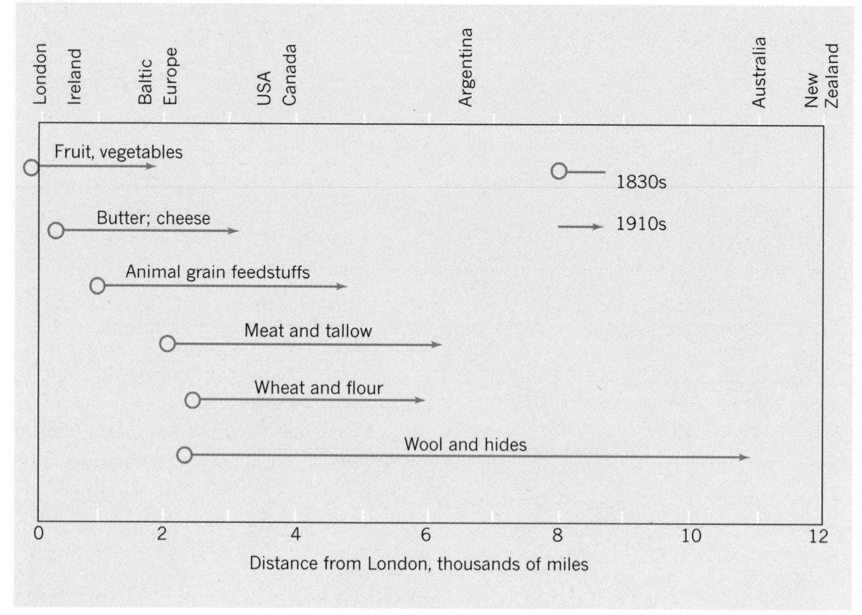

(b)

Figure 15.16 Spatial expansion of land-use zones (a) Outward movement of source of wheat for a central city to serve distant areas. (b) Change in average distances from London over which various types of British agricultural imports moved between the 1830s and the 1910s. (c) Decline in freight rates on American exports as measured by a standard index (1830 = 100) in real cost terms.

[Source: Data from J.R. Peet, *Economic Geography*, **45** (1969), Table 1, p. 295.]

to play a similar role in the international economy to that played by the single city in Thünen's isolated state. Geographer Richard Peet has looked at this notion for one country in the 'nuclear' area, the United Kingdom (see Figure 15.16). Over the period from 1800 to the outbreak of World War I in 1914, the British population increased fourfold. Over the same period living standards increased and food consumption per capita rose. Where did this food come from?

Figure 15.16 shows part of the answer. The broad pattern followed is an outward shift of the source areas from home production to imports from overseas areas as demand increased. If we take the case of wheat (Figure 15.16(b)), the average distance from which imports were drawn rose from 3850 km (2400 miles) in the 1830s to 9500 km (5950 miles) by the outbreak of World War I in 1914. In the mid-nineteenth century wheat for England came from Prussia and the Black Sea ports. Russia rose to importance as a granary in the 1860s until supplanted by the wheatlands of North America in the 1870s. The wheat frontier moved westward into the Great Plains and had reached western Canada by the 1890s. Australia comes on the scene as a major producer at the turn of the century.

The outward movement of the zones was partly related to increased demand at the centre, partly to the overall downward trend in shipping costs as shown in Figure 15.16(c). Although the 'rings' are greatly fragmented by continents and oceans, the dynamics of the expanding Thünen model remain visible.

Surfaces Beyond the City: The Industrial Landscape 15.3

We have seen how cities, as centres of agricultural consumption, play a dominant role in shaping rural land uses. But cities are also the great consuming centres for all natural resources. What influence does the city as a marketplace have on the location of resource processing and thus on industrial patterns in the world beyond the city?

This question of the location of resource processing, with all its population-multiplying implications (Figure 8.6), has long attracted investigators. One of the simplest but most penetrating analyses of this subject came from the German spatial economist Alfred Weber, who published his original study of industrial location in 1909. The *Weber model* paid particular attention to the loss of weight or bulk involved in resource processing. He demonstrated that this weight loss played a significant part in the location of certain industries. Industries with manufacturing processes that involved a large weight loss were found to have a *resource orientation* (i.e. located near the natural resource that was to be processed). To convert wood to pulp and paper involves a 60 per cent weight loss, and so pulp and paper mills tend to be located near major forested areas rather than in the heart of the large cities in which the bulk of their products is eventually consumed. Conversely, an industry such as brewing has a *market orientation*, since its product is very weighty in relation to the malt, hops, and other materials that go into it. (Water is assumed to be generally available; hence a need for it is assumed not to affect location decisions.) This analysis is clearly greatly oversimplified, but much current theory, although different in terminology, sophistication, and method, follows Weber's basic line of thinking and is therefore termed *Weberian*. We first look at the Weber model in a single spatial dimension and then go on to more realistic two-dimensional analyses.

Location on a Line

Let us assume that a single resource is supplied at location R and a single urban market for that resource is located at city M as in Figure 15.17. We further assume that transport costs increase regularly with distance from the supply point and that all other costs (labour, power, taxes, etc.) are fixed and equal everywhere. Figure 15.17(a) shows a simplified situation with a single resource supply point (R) and a single city (M) located on the horizontal axis; costs are marked on the vertical axis. If the loading costs are equal, and delivery costs from R to M are identical functions of distance, then the total transportation costs have the form shown by the heavy line. Transportation costs may be at a minimum at either the supply site or the market site. At all other possible locations on the line of haul between the two (RM), the transportation costs are equal, but higher, because they involve an additional loading or unloading charge.

The processing of resources commonly involves changes in their mass, weight, or value that affect unit transportation costs. The inequality in transit costs for raw and processed goods is illustrated in Figure 15.17(b) by the steeper slope of the delivery-cost curve from R. Transportation costs are lowest from the supply point, R. When transportation costs are higher for

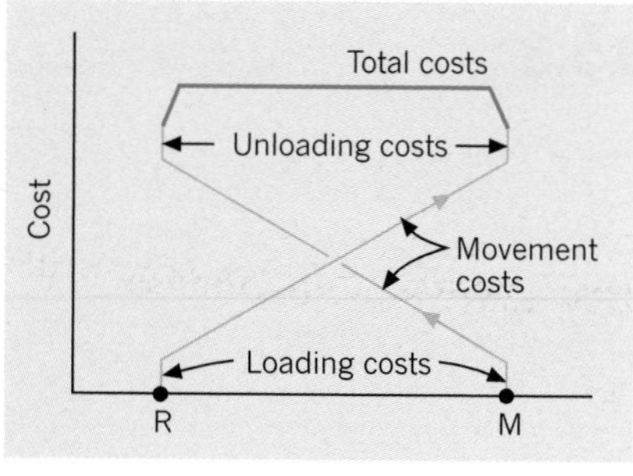

(a)

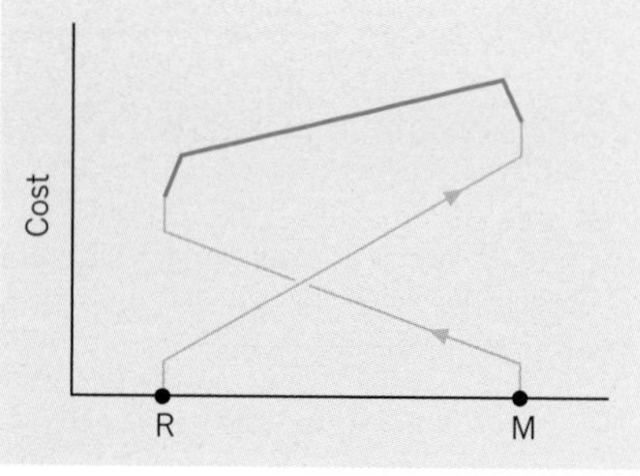

(b)

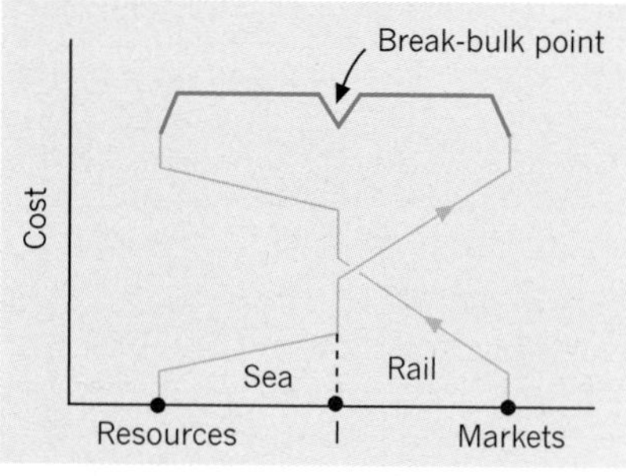

(c)

Figure 15.17 Least-cost locations Simplified examples of the least costly location for a resource-processing facility are shown as a 'tug-of-war' between the resources located at R and the markets located at M. Cost on the vertical axis is plotted against distance along a line between R and M on the horizontal axis. The three different cases shown are discussed in the text.

the finished product, the reverse situation obtains and the lowest-cost point moves to location M. To make the situation a little more realistic, we can introduce two zones (land and sea) with different transport costs. Because the commodity being moved must go by two modes of transportation, the costs of loading, unloading, and reloading at the intermediate point (I in Figure 15.17(c)) may be substantial. Thus all three points (R, I, and M) share the status of *least-cost transport points*. By modifying some of the assumptions in Figure 15.17 we could go on to introduce further and more complex least-cost patterns.

We can see something of the tug of war between these supply and demand points by considering the location of oil refineries. In the early period of oil production (to 1920, approximately) there were advantages to be gained from locating refineries on or near the oil fields themselves. Transport costs were high and the demand was mainly for kerosene, with about half the crude oil being dumped or burned as waste. As the advent of larger pipelines and supertankers reduced crude-oil transport rates, and as the range of petroleum products and the capacity of refineries to use more of the oil grew, it became more advantageous to locate refineries near the point of consumption. The vastly increased refinery capacity on the seaboards of western European countries reflects both the changing economics of location in the industry and the changing political situation. Market-based refineries can deal with a wider selection of crude-oil suppliers, thus reducing the risk of supplies being cut off from a single source; and they can be massive enough for the economies of scale since they can draw on various small oil fields for supplies. The jump from the simple diagram in Figure 15.17 to the complexities of refinery location shows how large is the gap that more advanced locational theory must try to bridge. Some of the terms used in industrial location theory are summarized in Table 15.2.

Location in a Plane

We can move a small way forward if we change the focus of our analysis from a one-dimensional line to a two-dimensional plane. Figure 15.18(a) presents a simple situation in which we have two resource supply sites (R_1 and R_2) and a single market city M. If we begin by assuming equal transport costs per unit of weight, the costs from each of the three points can be represented by a series of equally spaced, concentric, and circular contours called *isotims*. Each isotim describes the locus of points about each source where delivery or procurement costs are equal. Here the isotims are circular about each of the three points because transport costs across the *isotropic space* are everywhere the same. Total transportation costs can be computed by summing the values of intersecting isotims. Lines connecting points with equal total transportation costs are termed *isodapanes* and, like isotims, may be regarded as contours showing the cost terrain of a particular region. In Figure 15.18(a) the lowest point on the isodapane is equidistant from each of the three resource supply points.

If we relax the assumption of equal transport costs per unit of weight, then the isotims around each point may vary. Figure 15.18(b) shows a situation in which the movement costs from R_2 are twice those from the other two sites. This situation distorts the shape of the isodapane surface and displaces the point with the higher transport costs (i.e. toward R_2). Weber compared this displacement process to a set of scales that are weighted too

Table 15.2 Terms used in the study of industrial location

Term	Definition
Agglomeration economies	Savings to the individual manufacturing plant that come from operating in the same location. These may come from common use of specialist servicing industries, financial services, or public utilities.
Heavy industries	Industries that have (a) finished products that have low values per tonne, (b) a high material index (see below) and (c) a high tonnage of materials used per worker.
Isodapanes	Contours of total transport costs.
Isotims	Contours of transport cost for a single element in the manufacturing process.
Light industries	Industries that have (a) finished products that have high values per tonne, (b) a low material index (see below), and (c) a low tonnage of materials used per worker.
Market orientation	The tendency for certain industries to locate near their market. Typical is the brewing industry, which has to add a bulky material (i.e. water) to the finished product.
Material index	Given by the total weight of localized materials used per product divided by the weight of the product. Most industries have an index greater than 1 and are described as 'weight-losing'.
Resource orientation	The tendency for certain industries to locate near their source of localized raw materials. Typical are the mineral-processing industries where a great deal of waste material in the ore can be removed before the refined material is shipped.
Ubiquitous materials	Materials which can be found in any location and therefore play a minor role in locational decision making.

heavily in one scale pan. He suggested that this process might be illustrated by a simple physical model in which the pull of alternative sites is represented by physical weights attached to a circular disc by a system of cords and pulleys. Although clumsy in a practical sense, the idea of locational weights helps us to visualize the many competing forces involved in locational decisions.

Isodapane maps prove useful in practice, and a Swedish example of their use is given in Figure 15.19. This map shows the total costs of transporting pulp wood from various assumed manufacturing centres in southern Sweden. The map emphasizes the relatively advantageous location of the central lake lowland and the coastal strip at its two ends. Costs increase

Figure 15.18 Isodapanes
Contours of the total transportation costs on a plane. In these simplified situation there are two resources (R_1 and R_2), a single market (M), and a uniform plane. Transport costs around each of the three points (R_1, R_2, and M) are shown as concentric circles (*isotims*). Equal isotims about each point gives a midway compromise point of lowest transport cost in (a). Doubled costs from R_2 draw this compromise point toward it in (b).

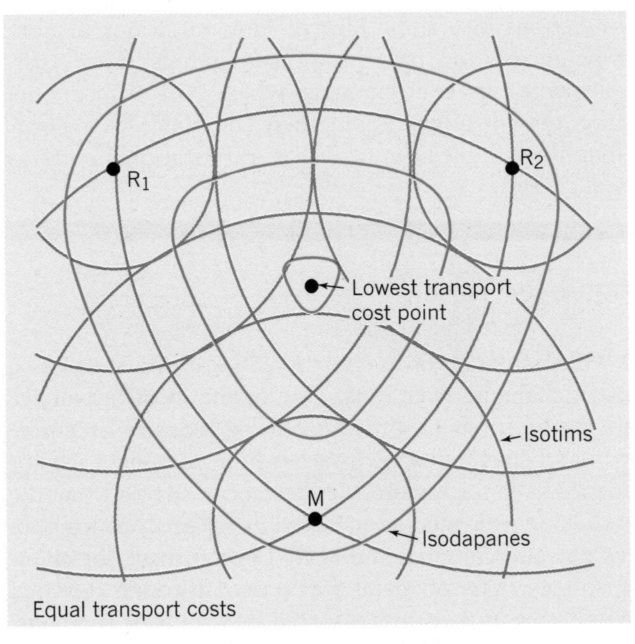

(a)

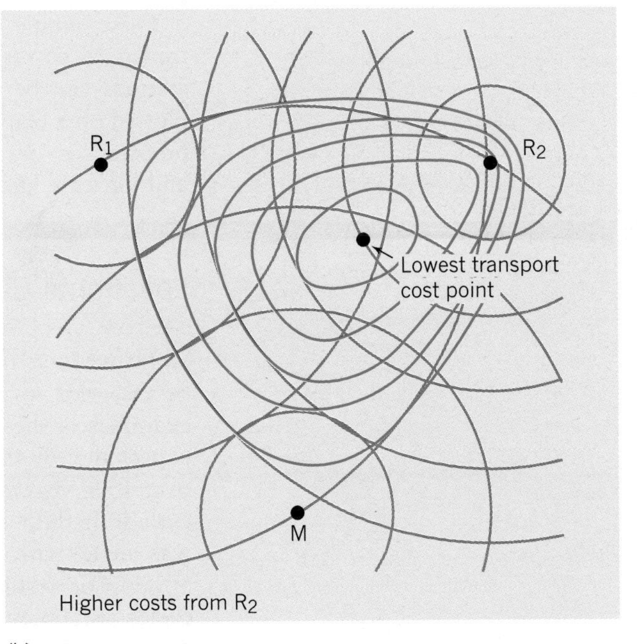

(b)

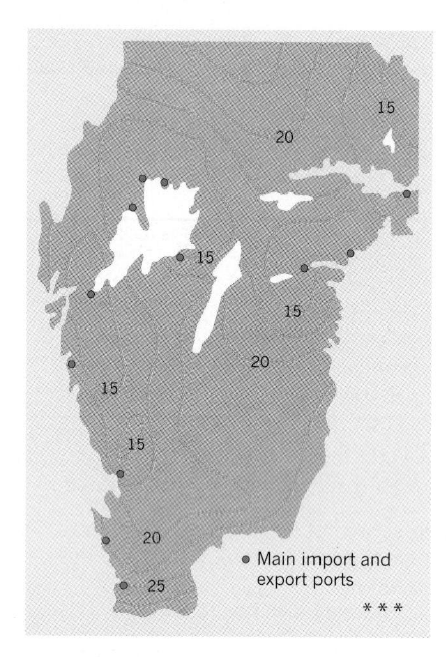

Figure 15.19 Isodapanes in Sweden Contours of total transportation costs for paper-making in southern Sweden. The figures include the total cost of transporting pulpwood, coal, sulphur, limestone, and the paper itself in knoner per tonne of output. The lowest-cost zone runs across the middle of the map and to the northwest.

[Source: O. Lindberg, *Geografiska Annaler*, **35** (1952), Fig. 20, p. 39.]

sharply toward the interior and to the north. Of course, isodapane maps indicate only transport costs; to get a more complete picture we need to consider other costs not related to the transport of goods.

Space–Cost Curves

So far we have analyzed the location of resource processing sites simply in terms of transport costs. This is an oversimplification, as Weber recognized; for location is affected by three other kinds of costs. First, there are labour and power costs that vary over space and are dependent on location. Second, we have costs that are largely independent of location. That is, there may be advantages to be gained from producers joining together to share the overhead costs of marketing and research or to encourage local suppliers to specialize in certain goods. Third, government legislation may cause spatial variations in costs through subsidies or taxation.

We can combine these non-transport costs with transport costs in locational analyses of resource processing. In Figure 15.20 *space–cost curves* are used to create a two-dimensional profile of the distribution of both types of costs. In the first case (Figure 15.20(a)) non-transport costs are considered uniform irrespective of location, whereas transport costs vary systematically about a central location (A). If we assume a horizontal demand curve at a common market price, then the area of profitable production forms a shallow, saucer-like depression where we can locate the point of least-cost production (A) and the spatial margins of profitability. When we allow factors such as labour or power costs to vary, the area of profitable production can be altered. Thus, we can have a second production outlier with low non-transport costs forming a geographically isolated pocket, as in Figure 15.20(b). Changes in the third group of non-transport factors, government intervention, are introduced in Figure 15.20(c). By placing heavy taxes on locations near A and subsidizing peripheral areas, we ensure considerable alterations in the spatial distribution of profitable locations.

These simple cross-sections give some idea of how variations in non-transport costs can be introduced into locational analysis. The cross-sections can be supplemented by contour maps whose cost elements are plotted on a plane rather than on a line. Again, these combinations are only first steps in unravelling the complexities of real-world locational patterns and intricate locational criteria.

Locational Change over Time

So far we have used Weberian analysis to study fixed locations. It can also be extended to industry changing over time. One of the most instructive examples of the changing locations of a manufacturing industry over time is provided by the iron and steel industry. Iron has been in human use for over 4000 years. Evidence from the eastern Mediterranean suggests that the earliest method of producing iron was the so-called direct process. Iron ore was mixed with limestone and charcoal and heated in a furnace for about a day. The resulting spongy mass of metal was passed through repeated cycles of hammering and reheating as it was forced into tools or weapons. This iron is what we would today call wrought iron.

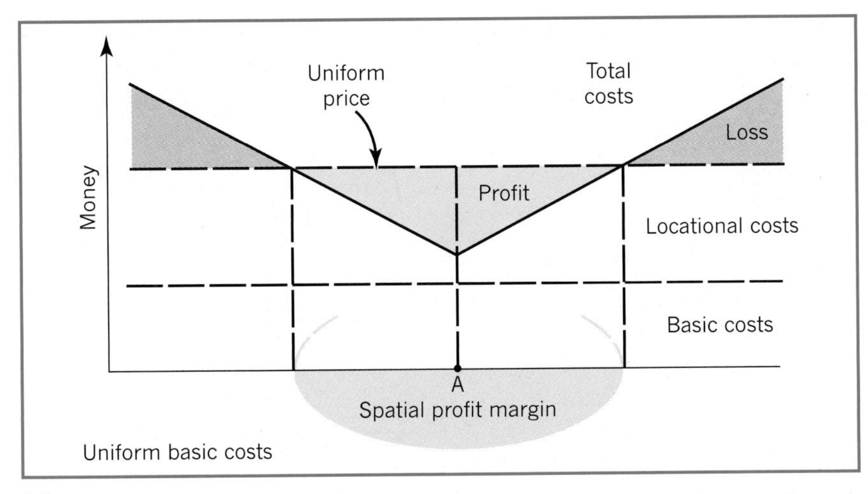

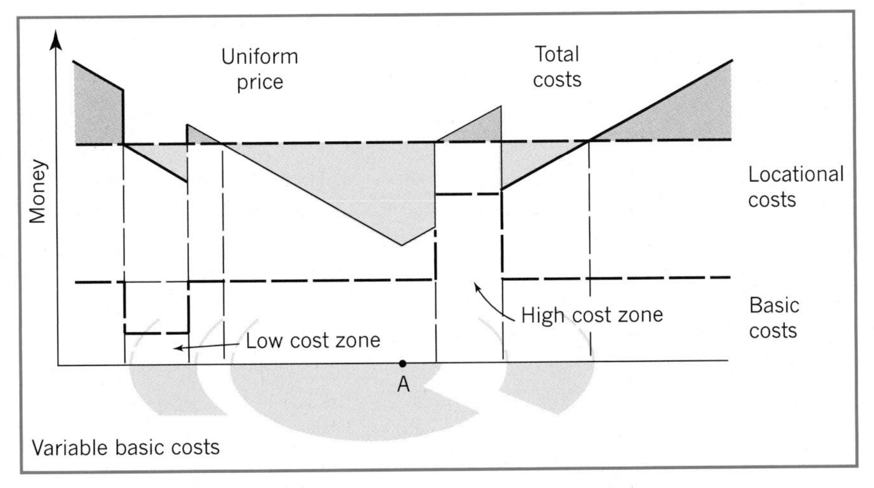

(b)

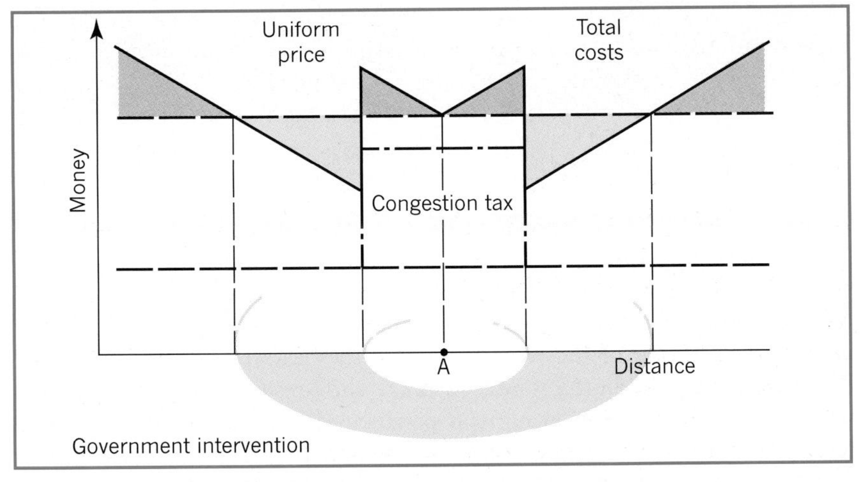

(c)

Figure 15.20 The effect of non-transport costs on location All three diagrams show a cross-section through an economic landscape in which transport costs vary systematically with distance from a central location (A). In (a) the basic (or non-transport) costs are uniform. In (b) the basic costs vary because of local production conditions; note the effect of low-cost and high-cost areas on both the cross-section and the plane (below). In (c) basic costs are modified by government-imposed congestion costs. In practice basic costs are also likely to vary over space, being lower (because of greater economies of scale) near the market centre (A).

Phase One The locational pattern associated with iron making by the direct process can be seen from medieval Europe. This first locational phase was essentially one of small-scale and widely dispersed production (see

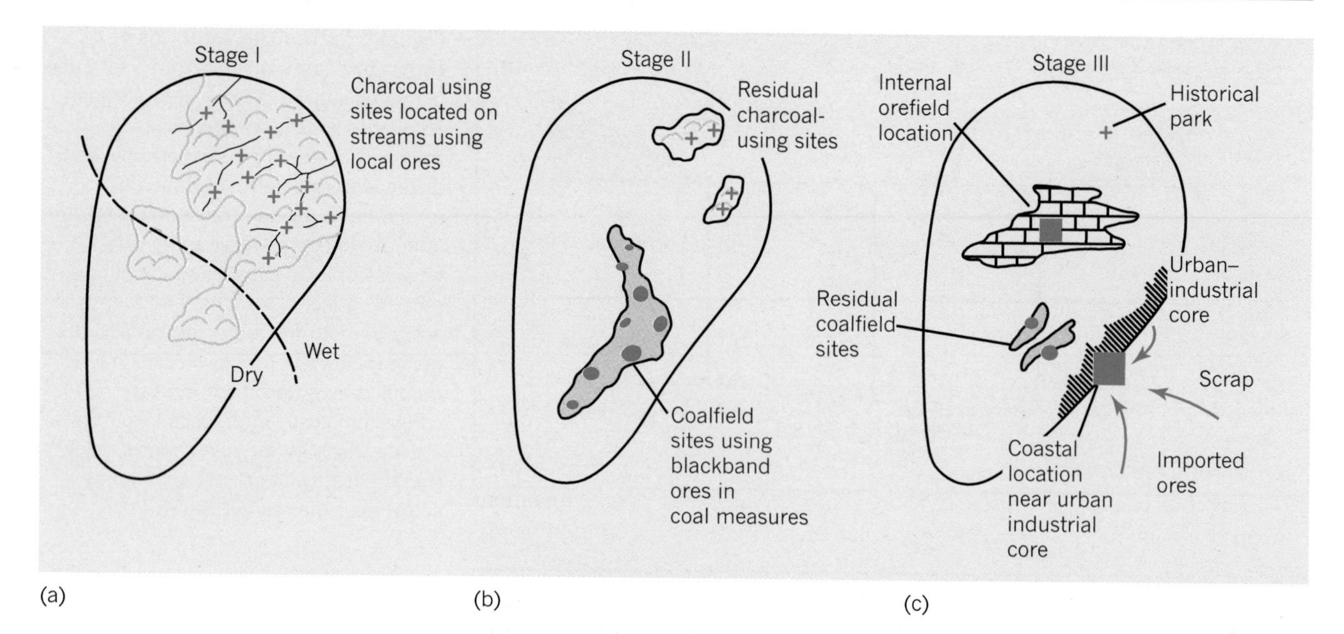

Figure 15.21 Evolution of industrial location patterns (a)–(c) An idealized sequence of three stages in the growth of the iron and steel industry. (d) An example of a Stage III coastal location for a modern iron and steel plant. Sparrows Point near Baltimore, site of a Bethlehem Steel plant using imported Venezualen and Labrador iron ores.

[Source: Corbis-Bettmann/Paul A Souders.]

Figure 15.21(a)). Small deposits of iron ore occur at or near the surface in a wide range of geological formations. Iron making was critically dependent on a large supply of wood, since every ten tonnes of iron smelted required about a hectare (two acres) of timber to be felled for charcoal. The progressive clearing of the woodlands (see the description in Chapter 11) throughout the Middle Ages was to make the presence of forests an ever more critical factor in locating iron production. For example, by the sixteenth century the only two areas of England that could support major charcoal-using industries were the Forest of Dean in the southwest and Wealden Forest in the southeast. A secondary locational factor was a regular supply of water to supply energy for the furnace 'blast', using bellows, and for hammers for forging the iron. High transport costs ensured that iron was made largely to satisfy local demand, and so the scale of operation remained small. This first phase of iron making was replicated on the North

American seaboard in the eighteenth century as New England and the middle Atlantic states also developed their own charcoal-based iron industries dependent on the eastern forest.

Phase Two The second locational phase in iron making is associated with larger-scale productions concentrated on the coalfields (Figure 15.21(b)). By 1500 attempts were being made to replace charcoal as a fuel by coke, and to smelt the iron ore in larger blast furnaces. Successive improvements in the smelting technique using coal meant that by the early nineteenth century coal had largely replaced wood as the essential fuel. The location of iron-making on a coalfield site now held many attractions. First, under the prevailing technology eight times as much coal as iron ore was needed to produce a given quantity of smelted iron so that, in Weberian terms, coal had a very high *weight-loss location index*. It was clearly much more advantageous to take iron ore to the coal, rather than vice versa. Movement of iron ore was not always necessary for a second advantage of many coalfields was the presence of iron ores interbedded with the coal seams (the 'black-band' ores). Older iron ore sources were often abandoned in favour of those located in or adjacent to the coalfields. Third, the coalfield regions in parts of Europe had long traditions of iron making associated with forging and reprocessing iron using either charcoal (e.g. the Sheffield region of England) or coal itself (e.g. Middle England). These forces of locational change in favour of the coalfield areas were probably running at their height in the late eighteenth and early nineteenth centuries. The rapid growth in demand as the Industrial Revolution accelerated was marked by the emergence of booming steel-making areas such as those of Pittsburgh in the United States, around Sheffield in England, and around Liege in Belgium. Here production was increasingly in large-scale units with finished products being transported over longer and longer distances.

Phase Three The third locational phase in iron making is the most complex (see Figure 15.21(c)). It runs from the late nineteenth century through to the present and is closely bound to the continuously changing technology of steel manufacturing. It is marked by a redispersal of the industry away from the coalfield areas, and in particular toward (a) orefield sites and (b) coastal sites. Coalfields became less significant as the amount of coke needed in blast furnaces was steadily reduced: today only about half as much coke is needed as iron ore. Exhaustion of the blackband iron ores in the coalfield areas added to this trend. As the attraction of coalfields declined, so other areas rose in importance. From 1878 a technical breakthrough in steel manufacturing (the Gilchrist–Thomas process) meant that the rich but hitherto unused limestone orefields could be used. Where orefields were located near major market areas (e.g. the Jurassic orefields of eastern France and Luxembourg), major iron and steel industries emerged. Where orefields were located in more remote areas without a local market (e.g. Schefferville on the Canadian shield, or Mauritania in Saharan Africa), the ore was exported. Growing international trade in iron ore has matched the decline in production from home orefields. Other than Russia, all the major steel producers – the United States, Japan, Germany, and the United Kingdom – are now all heavily dependent on imported ores. In consequence, coastal sites where materials (including imported ores) could be easily assembled have become important. In North America the Great Lakes complex of Cleveland, Detroit, Chicago, and Gary (in the United

States) and Hamilton (in Canada) using first Lake Superior and later Labrador ores, overtook the Pittsburg area in steel production in the 1920s. On the middle Atlantic coast the Fairless and Sparrows Point sites draw on South American ores. Similar coastal concentration has occurred in Britain with the growth of the South Wales coastal plants.

The locational pattern of iron and steel production continues to be in a transitory state today. Plants are increasingly specialized, drawing on more distant and richer ores and producing materials for specialized world markets. Technical change continues to reduce the dependence on raw material location and increase the importance of the market, although with an increasing volume of scrap metal in circulation the major urban-industrial centres now serve a double role as both market and source area. Ore enrichment has allowed the same earlier raw material sources to be re-evaluated, leading to the revival of iron ore production (e.g. Minnesota).

Thus the locations of the steel industry today represent a partly inherited pattern in which past locational forces as well as those of the present may be seen. With increasing size the huge investment in a single steel-making plant may give it an inertia sufficient to maintain its location long after the forces that created it have weakened. Both the company (in terms of capital investment) and the government (in terms of the workforce) may have a strong vested interest in preserving such inertia.

In our discussion of urban, rural and industrial land use and the Thünen and Weber families of models we have stressed geographic distance in a static framework. In the next chapter we go on to look at the dynamics of changing surfaces in terms of spatial diffusion waves.

Reflections

1 The central business district (CBD) of a city is the core with highest land values (see Box 15.A). Is your local city's CBD growing, stabilizing, or declining in importance? Give reasons for your opinion. What evidence would you need to assemble to test your views?

2 Chicago is often used as a template for urban structure based on a subtle combination of distance and sector direction from its CBD (see Figure 15.7). How far does the city in which you reside conform to or contrast with the classic Chicago pattern?

3 Early in the nineteenth century, the German locational theorist Thünen developed a theoretical model of geographical variations in land use. Examine the sequence of land-use zones in Thunen's isolated state (Figure 15.10). List the reasons why (a) truck farming is close to the city and (b) extensive cattle raising is carried out on the far periphery of the region. Do any of these factors affect location decisions in agriculture today?

4 Transport costs for the bulk movement of agricultural products continue to decline in real terms. What effect do you think this has on the changing importance of environmental resources (e.g. climate and soils) as a factor in the location of crop areas?

5 The spatial economist Weber showed that industrial location is highly dependent on loss of weight or bulk in resource processing. Give examples from your local area of any industrial concerns which appear to be (a) resource-oriented and (b) market-oriented. Is there an 'in-between' category? How can the firms in this in-between category be fitted into a locational theory?

6 Problems faced by cities tend to be of three main kinds: those arising from inward migration, from outward migration, and from the operation of the city as an ecosystem. Review for any single city with which you are familiar the role of these three factors.

7 Review your understanding of the following concepts using the chapter text and the glossary in Appendix A of this book:

bid-price curves
CBD
isodapanes
isotims
market orientation
resource orientation
space–cost curves
Thünen model
Weber model
weight-loss index

One Step Further ...

A brief but useful introduction to location theory is given by Meric Gertler and Ron Johnston in their essays in R.J. JOHNSTON *et al.* (eds), *Dictionary of Human Geography*, fourth edition (Blackwell, Oxford, 2000), pp. 460–7. The classic theory of ring formation by J.H. von Thünen has been translated in P.G. HALL (ed.), *Von Thünen's Isolated State* (Pergamon, London, 1966) Good accounts of its role in the structuring of land use in rural areas is given in M.D.I. CHISHOLM, *Rural Settlement and Land Use*, third edition (Hutchinson, London, 1979), Chapters 4 and 7. An economist's view of location theory is given in Paul KRUGMAN, *Development, Geography, and Economic Theory* (MIT Press, Cambridge, Mass., 1995).

Changing patterns of agricultural land use in intercity areas are described in H.F. GREGOR, *Geography of Agriculture* (Prentice Hall, Englewood Cliffs, N.J., 1970) and W.B. MORGAN and R.C. MUNTON, *Agricultural Geography* (Methuen, London, 1971). A modern view of industrial location is given in D.M. SMITH, *Industrial Location: An Economic Geographical Analysis*, second edition (Wiley, New York, 1981) and R.C. ESTALL and R.O. BUCHANAN, *Industrial Activity and Economic Geography*, second edition (Hutchinson, London, 1970), and a useful selection of background readings is provided in G.J. KARASKA and D.F. BRAMHALL (eds), *Locational Analysis for Manufacturing: A Selection of Readings* (M.I.T. Press, Cambridge, Mass., 1969). For an excellent discussion of locational decision-making in the context of resource use and economic theory, see P.E. LLOYD and P. DICKEN, *Location in Space*, second edition (Harper & Row, New York, 1977).

The locational theory 'progress reports' section of *Progress in Human Geography* (quarterly) will prove helpful in keeping up to date. Research on agricultural and industrial geography is generally reported in the regular geographic journals, especially the leading journal, *Economic Geography* (quarterly) and the Dutch economic geography journal *Tijdschrift voor Economische en Social Geografie* (bimonthly). Understanding developments in locational theory now demands considerable mathematical competence; browse through the *Journal of Regional Science* (quarterly) to get some idea of what lies beyond the highly simplified models presented in this chapter. For readers with access to the World-Wide Web see also the sites recommended for this chapter in Appendix B at the end of this book.

CHAPTER 16

Spatial Diffusion

■ denotes case studies

Figure 16.1 Innovation waves Adoption by farmers of new technology and practices has been tracked by geographers as a series of innovation waves which have a distinctive structure in both time and space. The adoption of contour stripping as an anti-soil-erosion measure was pressed on farmers by the United States Soil Conservation Service from the 1920s. It remains today as an important conservation practice. This view is of contour farming in a holding by the Mississippi river in southwestern Minnesota.
[Source: Paul Chesley/Tony Stone Images.]

Because I know that time is always time
And place is always and only place
And what is actual is actual only for one time
And only for one place

T.S. ELIOT *Ash Wednesday* (1930)

During the Great Depression of the late 1920s, American agriculture began to run into many problems associated with soil erosion. The problem was particularly acute in areas with vulnerable soils, steep slopes, and intensive storm activity. Trying to persuade farmers to adopt the kind of conservation practices needed to preserve soils and prevent gulleying forced the US Soil Conservation service to study how conservative-minded farmers came to adopt good practices (typified by the contour planting shown in Figure 16.1). It was found that adoption tended to spread out around demonstration farms, like ripples spreading around a pebble thrown in a pond. Once sceptical farmers could see on the ground what could be achieved, then by word of mouth conservation spread.

In another part of middle America waves of a different sort were spreading. Figure 16.2 shows a map based on some remarkable data assembled by the German spatial economist August Lösch, whom we met in Chapter 14. This shows the growth of the business depression of 1929–31 spreading out from Chicago across the American mid-western state of Iowa.

A third example is the spread of a disease. In 1905 the El Tor strain of

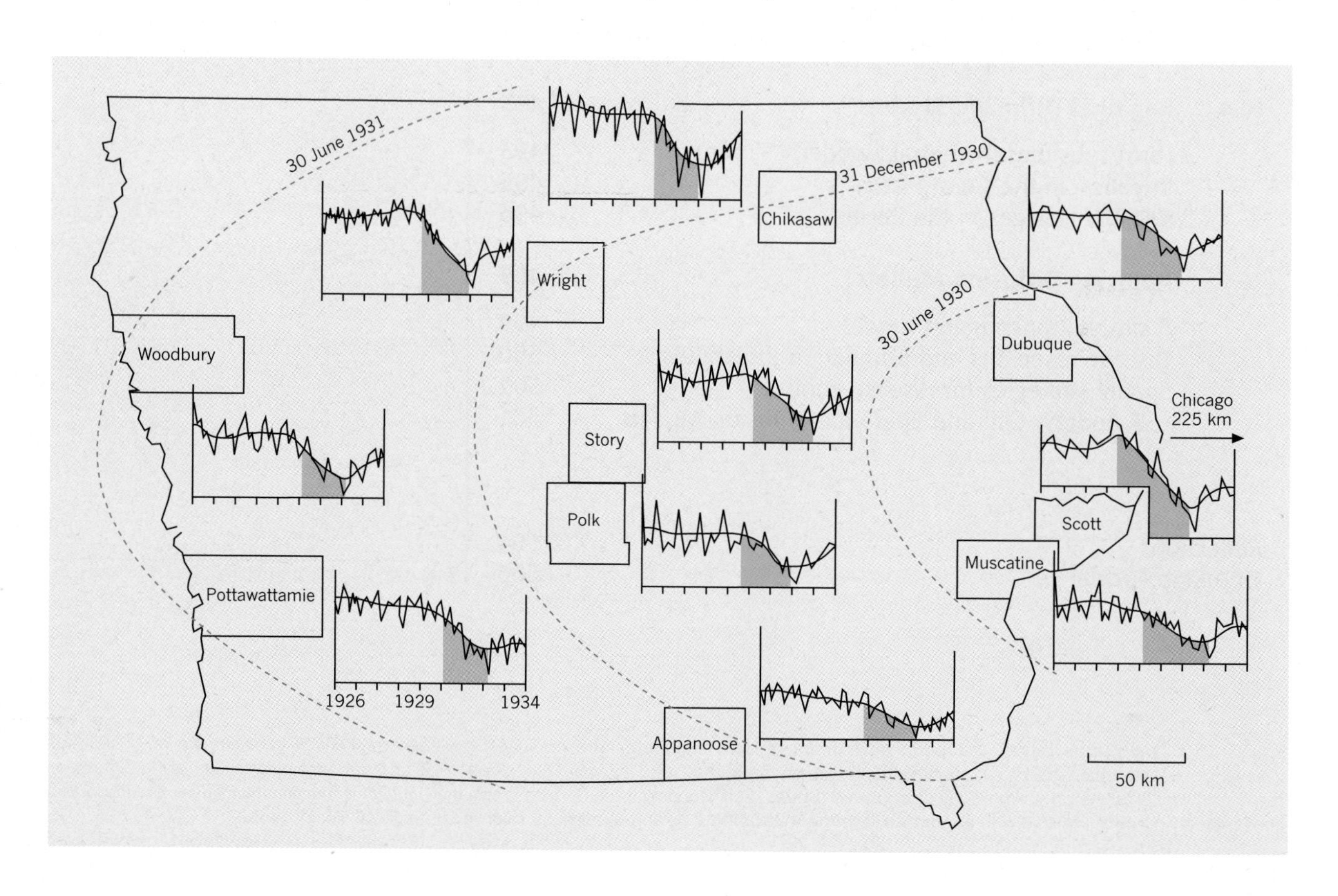

Figure 16.2 Economic impulses as diffusion waves Spread of the business depression of 1929–31 through the state of Iowa in the central United States. Graphs show indices of business activity with their running means for ten counties. Time contours show time of arrival of the depression 'front'.

[Source: Drawn from data from August Lösch, in A.D. Cliff, P. Haggett, J. K. Ord, and R. Versey, *Spatial Diffusion* (Cambridge University Press, Cambridge, 1981), Fig. 2.6, p. 13.]

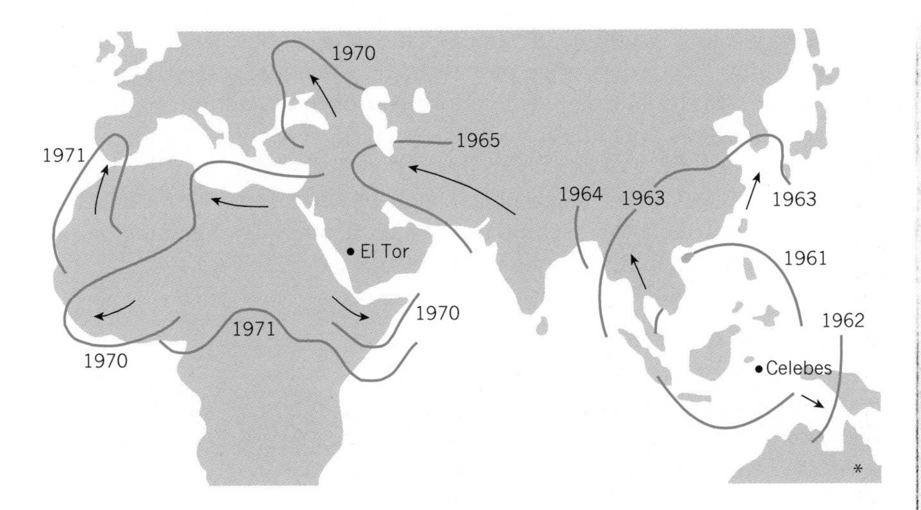

cholera was first identified in the bodies of six Muslim pilgrims at a quarantine station outside Mecca. (El Tor was the name of the quarantine station.) In the 1930s the same strain was recognized as an endemic in the Celebes, which has a largely Muslim population. For another 30 years there was little news of El Tor until, in 1961, it began to spread with devastating speed outward from the Celebes. By 1964 it had reached India (replacing the normal cholera strain endemic in the Ganges delta for centuries), and by the early 1970s it was pushing south into central Africa and west into Russia and Europe (see Figure 16.3). The seventh of the world's great cholera outbreaks was getting into its stride.

Waves as varied as outbreaks of cholera or fashion items have swept around the world in record time. Among the inconsequential waves were the brief western passion for all things Japanese in the 1880s, and the late 1970s craze for skateboards among children on both sides of the Atlantic. Things as different as influenza epidemics and oral contraceptives, bank rate charges and computer data banks, Dutch elm disease and fire ants, have one thing in common. They originate in a few places and later spread over a much wider part of the world.

Why are geographers interested in such diverse things? Principally because their spread provides valuable clues to how information is exchanged between regions. Where are the centres of diffusion – and why? At what rates do *diffusion* waves travel, and along what channels? Why do some waves die out rapidly and others persist? Some innovations may move slowly and quietly, like a tide lapping over mud flats. But rapid innovations are studied most frequently, not because of their intrinsic importance (indeed, they sometimes tend to be trivial), but because we can see the whole cycle of diffusion in a relatively brief time period.

The speed with which global changes now occur is clearly linked to rapid communication channels. In Chapter 5 we looked at the slow readjustments of the worldwide regional mosaic to the ponderous movements and migrations of human population. In this chapter we turn to the much more rapid changes that can occur through the diffusion of cultural elements or epidemic diseases, and we see how the ripples from information or infection explosions in one part of the world find their way into another. In this chapter we try to present some of geographers' more recent research on these topics and their use of computer models. (If you prefer to avoid these subjects, read only Sections 16.1 and 16.4.) Such models have far more than academic interest. If we wish to speed up the diffusion of certain cultural

CHOLERA CASES SPREAD TO PERSIA

By Our Diplomatic Staff

The cholera outbreak in the Middle East is continuing unabated. More than 60 new cases were reported yesterday in Jordan and Syria and the Teheran authorities said for the first time that the outbreak had spread to Persia.

But there is a comparatively low death rate. Jordan has reported no death so far, and in Syria, where there have been more than 2,000 cases, the toll is 68 dead.

Figure 16.3 Spatial diffusion The map shows the spread of the El Tor strain of cholera from the Celebes during the decade 1961–71. This strain of cholera appears to be *endemic* (i.e. permanently present) in the population of the Celebes. From this island it erupts from time to time to spread temporary epidemics in surrounding areas. When such epidemics reach major proportions and span several continents in the manner shown on this map, they are termed *pandemics*. In the autumn of 1977 another major outbreak centred on the Middle East. (See press cutting.) Geographers are interested in the paths followed by epidemics and pandemics through the populations of human settlements because of the insights they yield on other spatial diffusion processes.

[Source: Data compiled from press reports and WHO bulletins. Press cutting courtesy of the *Daily Telegraph*.]

elements, like the adoption of family-planning methods, a knowledge of precisely how waves of change pass through a regional system may be of help. If we wish to halt or reduce the spread of other cultural patterns of behaviour, such as drug abuse, or protect certain very fragile cultures from being swamped by western civilization, this knowledge may be helpful there, too. In such cases geographers, as in our case history of a beach in Chapter 1, contribute their spatial viewpoints as one way of gaining insight into a many-sided, multidisciplinary problem.

16.1 The Nature of Spatial Diffusion

In Chapter 1 we saw how population spreads out over a beach. Again, in the last two chapters, we have come across other cases where a spatial distribution that occurs over time has a distinct diffusion pattern. In everyday language the term *diffusion* means simply to spread out, to disperse, or to intermingle; but for geographers and other scientists, it has acquired more precise meanings.

Types of Diffusion

In geographic writing diffusion has two distinct meanings. *Expansion diffusion* is the process by which information, materials, and so on, spread from one place to another. In this expansion process the things being diffused remain, and often intensify, in the originating region; that is, new areas are added between two time periods (time t_1 and time t_2 are both located in a way that alters the spatial pattern as a whole) (see Figure 16.4(a)). A typical example would be the diffusion of an improved crop, such as the IR-8 strain of hybrid rice, described earlier in our discussion of the green revolution in Chapter 10 (Section 10.3).

Relocation diffusion is a similar process of spatial spread, but the things being diffused leave the areas where they originate as they move to new areas. The movement of the black population of the United States to the northern cities from the rural South could be viewed as such a relocation process, where members of a population at time t_1 change their location between time t_1 and time t_2 (see Figure 16.4(b)). In a similar manner, an epidemic may pass from one population to the next. Figure 16.4(c) illustrates the two processes and shows how they can be combined. The El Tor outbreak is an example of diffusion by both processes. The strain diffuses by relocation through some areas (e.g. as it did in Spain, where small outbreaks were recorded in 1971), but it also diffuses by expansion because it remains endemic in the Celebes. In this chapter we are discussing interregional interaction and are therefore mainly concerned with expansion processes. Relocation diffusion is treated more extensively in the discussion of regional growth models in Chapter 18.

Expansion diffusion occurs in two ways. *Contagious diffusion* depends on direct contact. It is in this way that contagious diseases such as measles pass through a population, from person to person. This process is strongly influenced by distance because nearby individuals or regions have a much higher probability of contact than remote individuals or regions. Therefore,

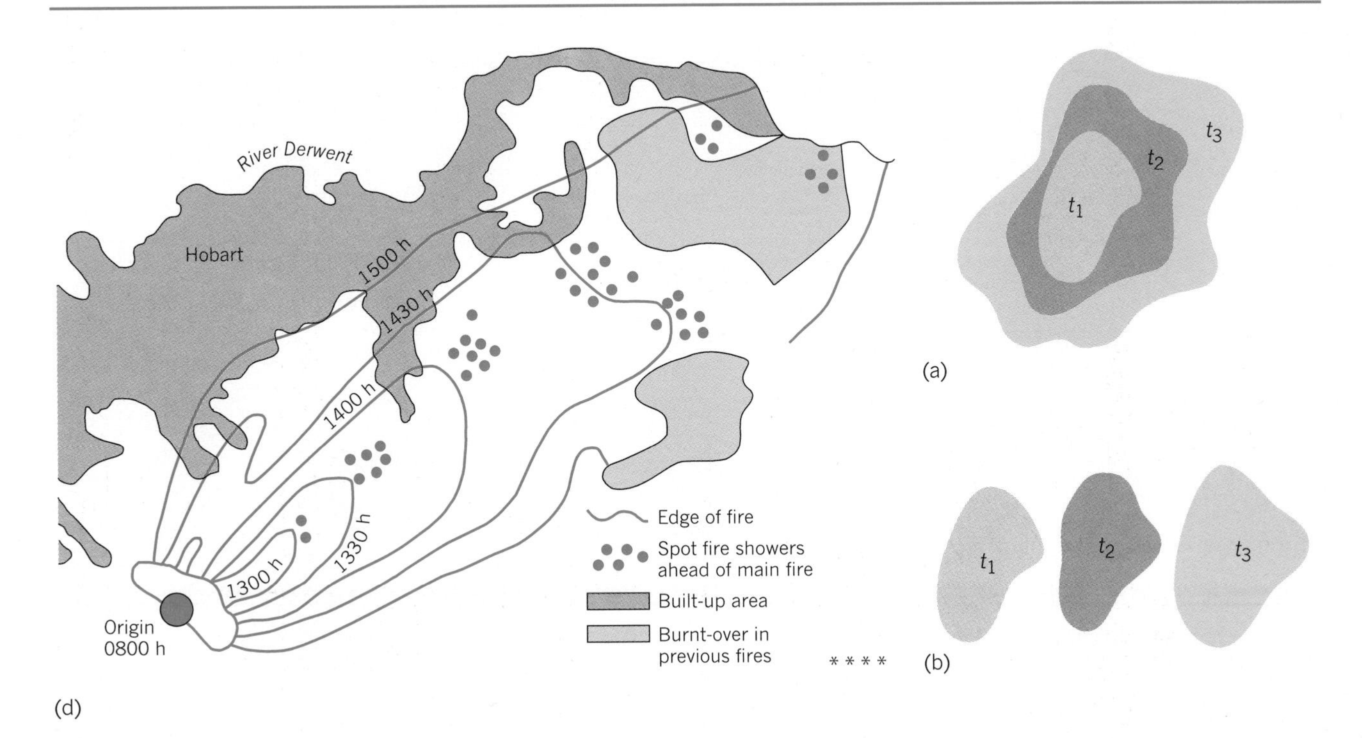

contagious diffusion tends to spread in a rather centrifugal manner from the source region outward. This is clearly shown in Kniffen's study of the spread of the covered bridge over the American cultural landscape, described in Figure 7.16.

Hierarchic diffusion describes transmission through a regular sequence of innovations (such as new fashion styles or new consumer goods such as TV) from large metropolitan centres to remote rural villages. Within socially structured populations, innovations may be adopted first on the upper level of the social hierarchy and trickle down to the lower levels.

Cascade diffusion is reserved for processes always assumed to be downward, from large centres to smaller ones. When specifying a movement that may be either up or down a hierarchy, geographers generally prefer the term hierarchic diffusion. Figure 16.5 demonstrates how diffusion may begin at a lower point in a hierarchy, move slowly upward, and then expand rapidly. We might think of this as a 'Beatles pattern'. A musical style beginning in a provincial city (Liverpool) moves to the national capital (London), then on to other capitals throughout the world. Finally, it reaches the local music store in small towns thousands of miles from its point of origin.

Michigan geographer John Kolars has traced the growth of the Sierra Club as a hierarchic diffusion process. (Note that we came across its founder, John Muir, in Chapter 10.) The club was founded in 1892 in San Francisco and a separate chapter established in 1906 in Los Angeles. For the next quarter of a century growth was confined to California but a New York centre was set up in the 1930s and one in Chicago in the 1950s. With the leap in interest in environmental protection in the last two decades, the Sierra Club has flourished. As we should expect from Figure 16.5(d), this has been accompanied by many new branches being set up in smaller cities. Around Chicago seven new chapters were set up between 1963 and 1973.

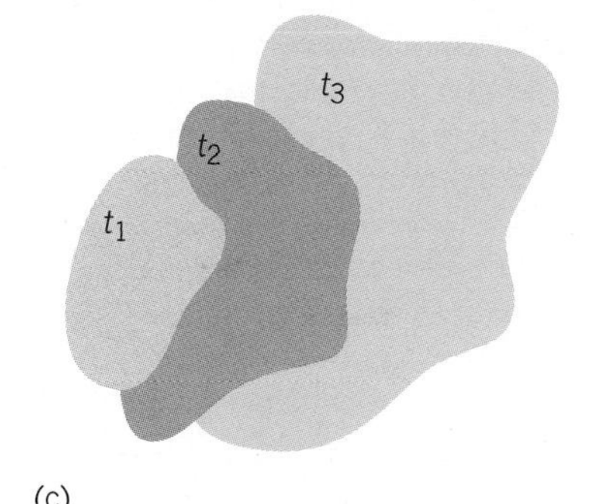

(c)

Figure 16.4 Types of spatial diffusion (a) Expansion diffusion. (b) Relocation diffusion. (c) Combined expansion and relocation processes. (d) Example of a combined expansion and relocation: the spread of the late summer Hobart bush fire in southern Tasmania on the early afternoon of 7 February 1967. The distance between the origin and the coast along the main axis of the fire's advance is about 14 km (9 miles). Spreads of twice the rate shown are common with grass fires.

[Source: Data by A.G. Arthur in M.C.R. Edgell, *Monash Publications in Geography*, No. 5 (1973), Fig. 2, p. 7.]

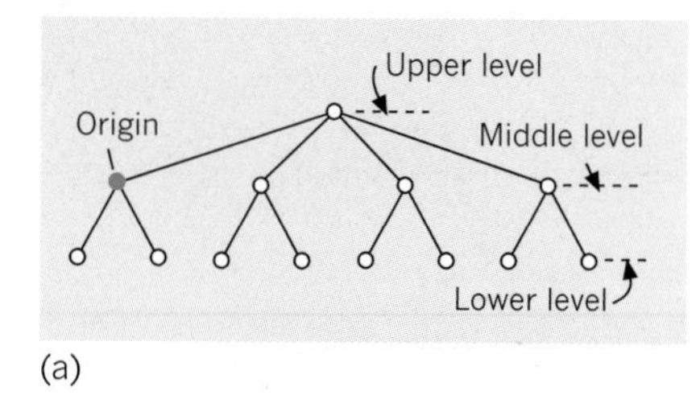

(a)

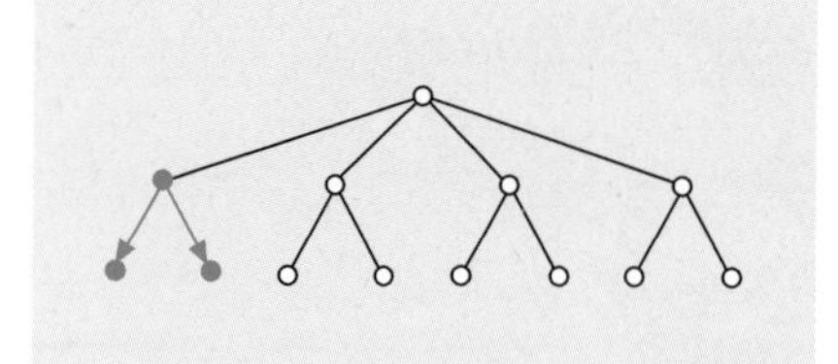

(b) Rapid downward spread from middle level

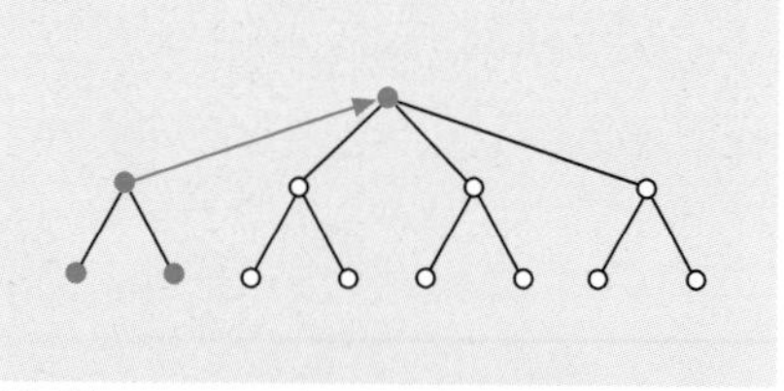

(c) Slow upward spread to upper level

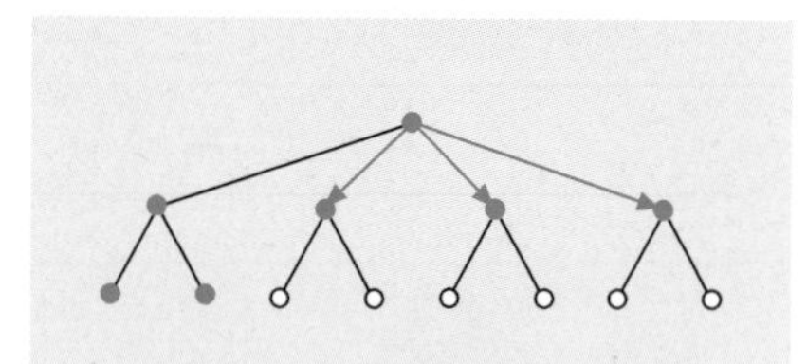

(d) Rapid downward spread from upper level

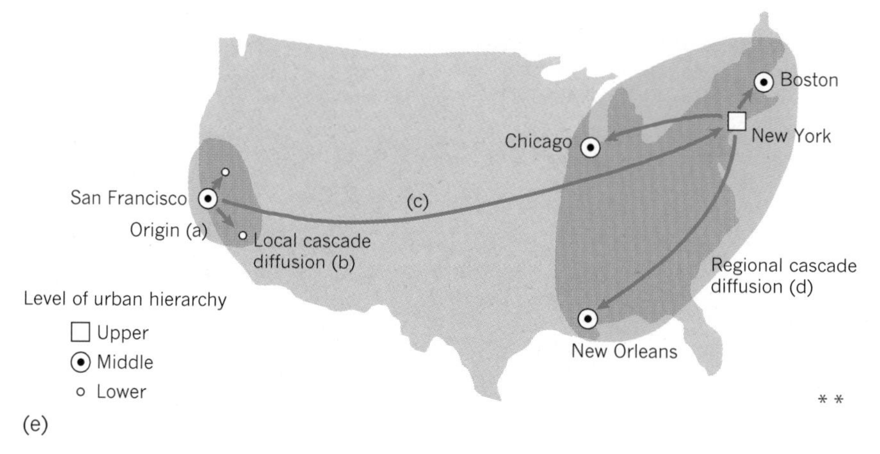

(e)

Figure 16.5 Hierarchic diffusion These diagrams show the spread of an innovation through a hierarchy. The innovation begins on a middle level (e.g. a small county town) and spreads rapidly down to a lower level (e.g. villages in its vicinity) but more slowly to an upper level (e.g. a regional capital). Once there, its downward spread is again rapid. Downward spread through a hierarchy is termed cascade diffusion. The map (e) illustrates the spread process shown in (a) through (d) for a hypothetical diffusion beginning on the west coast.

Diffusion Waves

Much geographic interest in diffusion studies stems from the work of the Swedish geographer Torsten Hägerstrand (see Box 1.B) and his colleagues at the University of Lund. Hägerstrand's *Spatial Diffusion as an Innovation Process* was originally published in Sweden in 1953. It was concerned with the spread of several agricultural innovations, such as bovine tuberculosis controls and subsidies for the improvement of grazing, in an area of central Sweden. This book was the precursor of various practical studies, particularly in the United States.

In one of his early studies of a contagious diffusion process, Hägerstrand suggested a four-stage model for the passage of what he terms *innovation waves* (*innovationsforloppet*), but which are more generally called diffusion waves. From maps of the diffusion of various innovations in Sweden, ranging from bus routes to agricultural methods, Hagerstrand drew a series of cross-sections to show the wave form in profile. Here we discuss the wave in profile and then the wave in time and space.

Figure 16.6 Diffusion wave profile The graph shows four main stages in the spread of an innovation by diffusion. The *innovation ratio* measures the proportion of a population accepting the item.

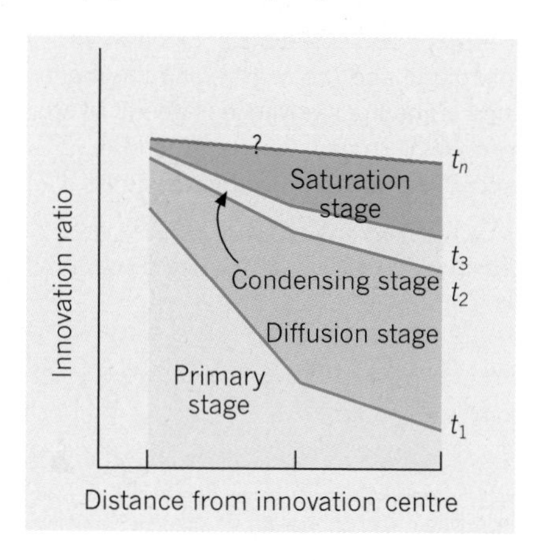

The Wave in Profile

Diffusion profiles can be broken into four types, each describing an *adoption curve* typical of the passage of an innovation through an area. Consider Figure 16.6 which shows the relationship between the rate of acceptance of an innovation and the distance from the original centre of innovation. The first stage, or *primary stage*, marks the beginning of the diffusion process. Centres of adoption are established, and there is a strong contrast between these centres of innovation and remote areas. The *diffusion stage* signals the start of the actual diffusion process; there is a powerful centrifugal effect shown by the creation of new, rapidly growing centres of innovation in distant areas and by a reduction in the strong regional contrasts typical of the primary stage. In the *condensing*

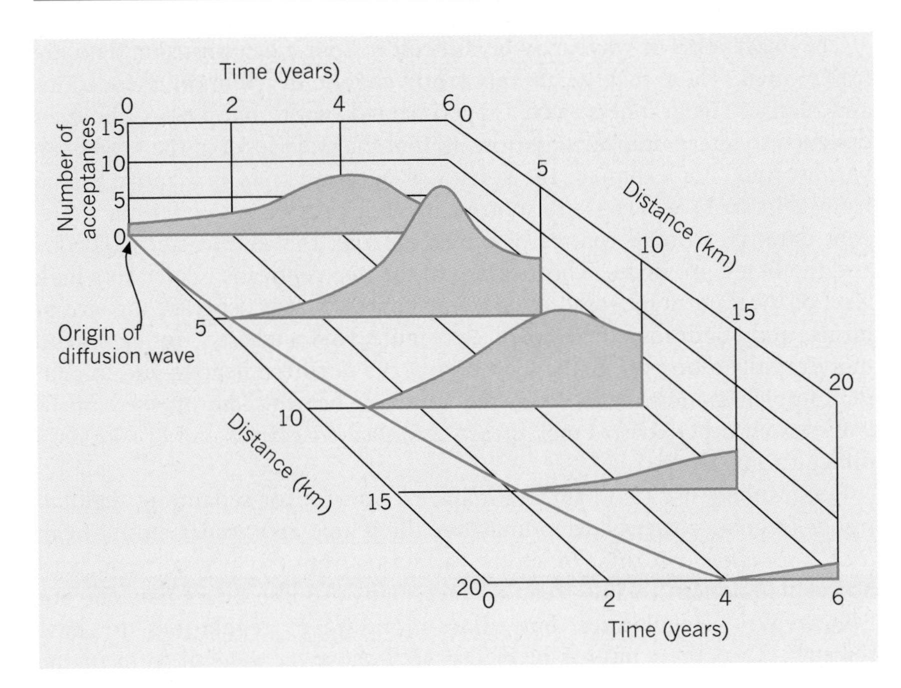

Figure 16.7 Diffusion waves in time and space Waves of innovation change character with distance from the time and point of origin. In the case shown the maximum height of the wave (i.e. the largest number of new acceptances of the item being diffused) occurs at a point 5 km from the origin in space and four years from the origin in time. Although individual waves will vary, some moving very slowly and some very rapidly, a large number appear to follow the general shape shown here.

[Source: R.L. Morrill, *Economic Geography*, **46** (1970), Fig. 12, p. 265.]

stage the relative increase in the number accepting an item is equal in all locations, regardless of their distance from the innovation centre. The final *saturation stage* is marked by a slowing and eventual cessation of the diffusion process. In this final stage the item being diffused has been accepted throughout the entire country so that there is very little regional variation.

Since Hägerstrand's original work, other Swedish geographers have carried out parallel studies to test the validity of this four-stage process. For instance, Gunnar Tornqvist has traced the spread of televisions in Sweden by observing the growth of TV ownership from 1956. Using information obtained from 4000 Swedish post office districts, he demonstrated that television was introduced into Sweden relatively late, yet within nine years about 70 per cent of the country's households had bought their first set. Tornqvist's results broadly confirm Hägerstrand's analysis. The diffusion process slows down, thus indicting the beginning of the saturation phase, at the end of the study period.

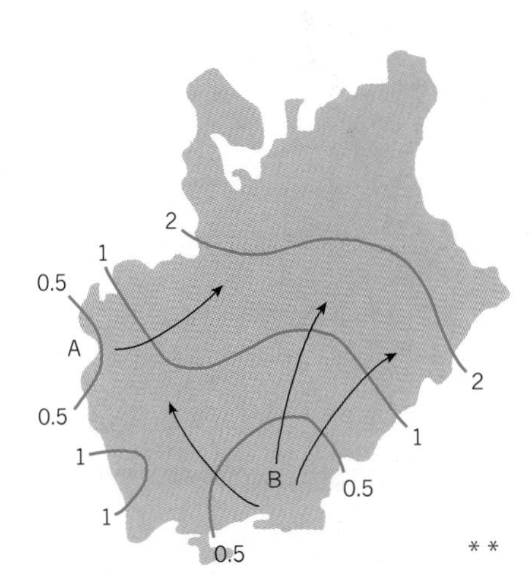

Figure 16.8 Multiple waves The arrows indicate the spread of agrarian riots in Russia from 1905 to 1907, an example of diffusion from multiple centres. This is termed *polynuclear* diffusion in contrast to *mononuclear* diffusion (where diffusion waves originate from a single source).

[Source: K.R. Cox and G. Demko, *East Lakes Geographer*, **3** (1967), Fig. 2, p. 11. Reprinted with permission.]

The Wave in Time and Space More advanced work on the shape of diffusions in space and time has confirmed their essentially wavelike form. Figure 16.7 is based on American geographer Richard Morrill's work. By fitting generalized contour maps (called *trend* surface maps and described in Chapter 23) to the original Swedish data, he showed that a diffusion wave first has a limited height (reflecting a limited rate of acceptance). It increases in both height and extent, and then decreases in height but increases further in total area. The gradual weakening of the wave over time and space is both time-dependent (as the simultaneous slackening of acceptance rates shows) and space-dependent (as the effect when the innovation wave enters inhospitable territory, strikes barriers, or mingles with competing innovation waves shows). The nature of the medium through which the wave is travelling may cause it to speed up or slow down, and a wave travelling from one centre of innovation will lose its identity when it meets a wave coming from another direction.

The exact form of wave may be difficult to spot when diffusion data are first plotted. There may be an apparently chaotic distribution of locations and dates. Geographers have experimented with mapping techniques designed to filter out local variations so that the main form of the waves can be observed. For example, the spread of agrarian riots in Czarist Russia from 1905 to 1910 has a very spotty pattern. Using maps to filter out irrelevant data, two broad centres of unrest emerge: the southeastern Ukraine and the Baltic provinces. The location of the two centres is related to a high level of local tension caused by extreme contrasts in prosperity, the size of farms, and conditions of tenancy. As Figure 16.8 suggests, rioting spread more rapidly along the Baltic Coast from the northern hearth, and in gentler, ripplelike movements from the southern hearth. The intersection of waves from other centres may create complicated patterns and make data difficult to interpret.

Geographers use trend surface maps as a device for separating *regional trends* (regular patterns extending over the whole area under study) from *local anomalies* (irregular or spotty variations from the general trend with no regular pattern). Thus trend surface maps are like filters that cut out 'short-wave' irregularities but allow 'long-wave' regularities to pass through. Thus the contours in Figure 16.8 show the general form of the spread of riots in Russia but cut out confusing local variations.

16.2 The Basic Hägerstrand Model

From his empirical studies Hägerstrand went on to suggest how a general operational model of the process of diffusion could be built. We shall look first at how the first and simplest of the *Hägerstrand models* was constructed.

Contact Fields

If we take any of the examples of spatial diffusion in the past few pages, we see that the probability that an innovation will spread is related to distance. Distance can be measured in simple geographic terms, as when we measure the number of metres between trees affected with Dutch elm disease around the campus. Alternatively, distance can be measured in terms of a hierarchy, e.g. the lower-level centres in Figure 16.5 are two 'steps' away from the upper-level centre. Let us take the first method and use it to measure the spread of information through a human population.

Let us begin by making the simple assumption that the probability of contact between any two people (or groups, or regions) will get weaker the further apart they are. If we call one person the *sender*, we can say that the probability of any other person receiving a message from the sender is inversely proportional to the distance between them. Near the sender the probability of contact will be strong, but it will become progressively weaker as the distance from the source increases. The exact form of this decline with distance is difficult to judge, but the evidence on the telephone calls indicates that it may be *exponential*. That is, it may fall off steeply at first but then ever more slowly. (Check back to Section 6.1 if you wish to refresh your memory on exponential curves.) Thus, we expect the volume of calls to fall off in the ratio 80, 40, 10, 5, and so on with the first, second, third, fourth, fifth kilometres. This is, of course, an idealized decline and actual patterns will be less regular. Geographers term this spatial pattern a *contact field*, drawing their language from the use of gravitational and magnetic 'fields' in physics.

In models of cascade diffusion we can retain the exponential contact field but replace geographic distance with economic distance between cities in an urban hierarchy or with social distance in a social-class hierarchy. Distance may not be symmetric in the hierarchic cases; for example, population migration up the hierarchy (from small to large towns) may be easier than migration down the hierarchy. This implies that the socio-economic distance between levels depends on the direction of movement.

The contact fields in epidemics may be very complex. For instance, studies of measles indicate that the probabilities of contact (and thus infection) *within* a given group like a family or the students at a primary school may be random. However, the probability of contact *between* such groups may be exponentially related to distance in the way we have already described. For example, research on southwest England showed that the probability of measles outbreaks in an area immediately adjacent to an area that had already reported cases was about 1 in 8. With further distance from the infected area the probabilities of infection fell steadily to about 1 in 30.

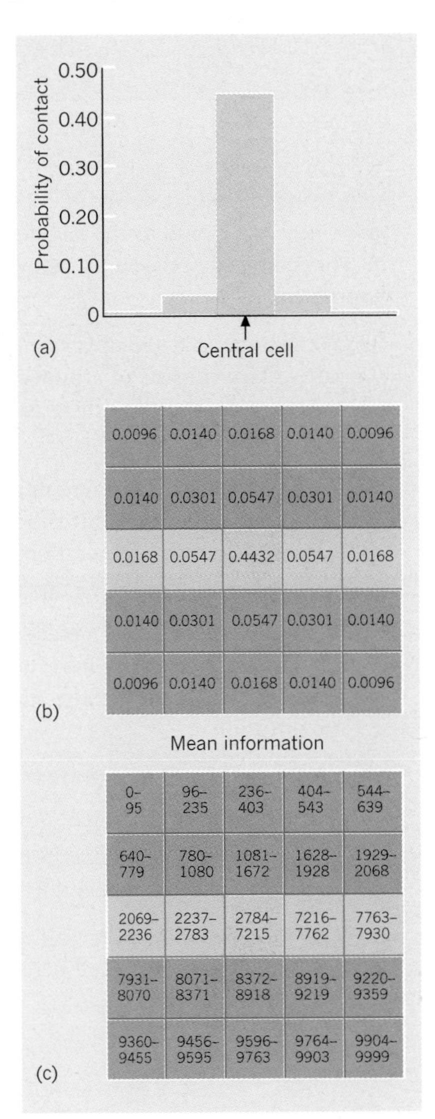

Figure 16.9 Mean information fields in the Hägerstrand model of diffusion The probability of contact with distance (a) is superimposed on a square 25-cell grid (b). The probabilities for all the cells in the grid are summed to give (c) a mean information field.

Mean Information Fields

How can we translate the general idea of a contact field into an operational model that can be used to predict future patterns of diffusion? Hägerstrand considered the problem in his early research, and he formulated various models to simulate diffusion processes. Figure 16.9 illustrates how he used probabilities of contact to determine a *mean information field (MIF)*, that is, an area, or 'field', in which contacts could occur. Superimposing of the circular field shown in cross-section in Figure 16.9(a) on a square grid of 25 cells enabled him to assign each cell a probability of being contacted. As Figure 16.9(b) indicates, the probability (P) of contact for the central cells is very high, in fact, over 40 per cent ($P = 0.4432$). For the corner cells at the greatest distance from the centre, the probability of contact is less than 1 per cent ($P = 0.0096$).

To make the grid operational we use a *Monte Carlo model*. We add together the probabilities assigned to the MIF cells. Thus. the upper left cell is assigned the first 96 digits within the range 0 to 95; the next cell in the top row has a higher probability of contact ($P = 0.0140$) and is assigned the next 140 digits within the range 96 to 235, and so on. Continuing the process gives the last cell the digits 9903–9999, to make a total of 10,000 for the complete MIF (Figure 16.9(c)). As we shall see shortly, these numbers are important in 'steering' messages through our simple distribution of population.

Using the Model

In order to use the model, the basic codes need to be understood. These are set out in Box 16.A 'Rules of the Hägerstrand model'. The key to putting this model into use is in rules 10 and 11. In each time interval the MIF is

Box 16.A Rules of the Hägerstrand Model

We can present the basic structure of the Hägerstrand simulation model in terms of formal rules. The rules given here refer only to the simplest version. They can be systematically relaxed to allow modifications and improvements.

1 We assume that the area over which the diffusion takes place consists of a uniform plain divided into a regular set of cells with an even population distribution of one person per cell.

2 Time intervals are discrete units of equal duration (with the origin of the diffusion set at time t_0). Each interval is termed a generation.

3 Cells with a message (termed 'sources' or 'transmitters') are specified or 'seeded' for time t_0. For instance, a single cell may be given the original message. This provides the starting conditions for the diffusion.

4 Source cells transmit information once in each discrete time period.

5 Transmission is by contact between two cells only; no general or mass media diffusion is considered.

6 The probability of other cells receiving the information from a source cell is related to the distance between them.

7 Adoption takes place after a single message has been received. A cell receives a message in time generation t_x from the source cells and, in line with rule 4, transmits the message from time t_{x+1} onward.

8 Messages received by cells that have already adopted the item are considered redundant and have no effect on the situation.

9 Messages received by cells outside the boundaries of the study area are considered lost and have no effect on the situation.

10 In each time interval a mean information field (MIF) is centred over *each* source cell in turn.

11 The location of a cell within the MIF to which a message will be transmitted by the source cell is determined randomly, or by chance.

12 Diffusion can be terminated at any stage. However, once each cell within the boundaries of the study area has received the message, there will be no further change in the situation and the diffusion process will be complete.

The model was originally set up to simulate the spread of information. It could equally well be used to model the spread of infections in a population; in that case the MIF would be a 'mean infection field'.

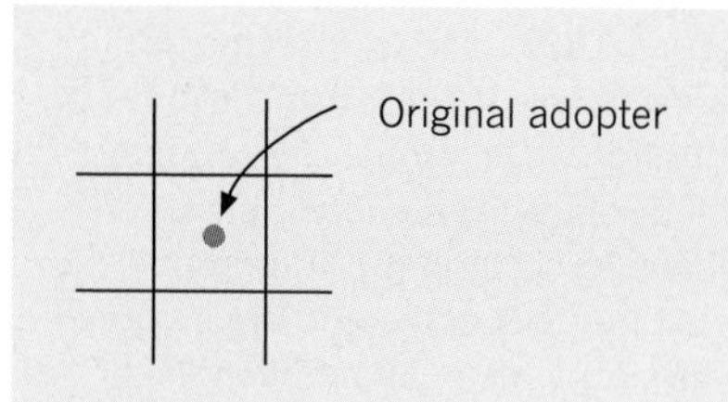

(a) First generation (t_1)

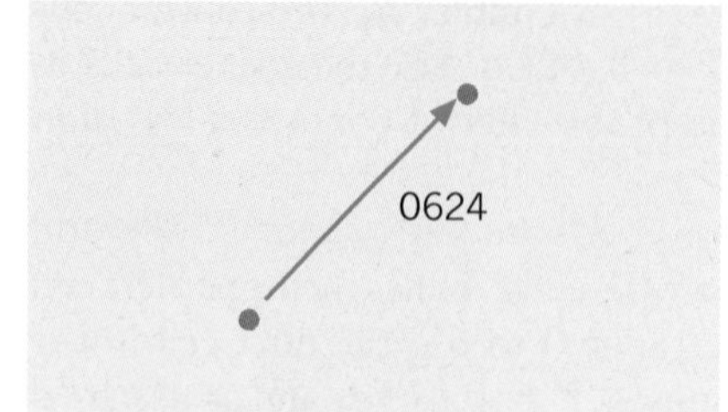

(b) Second generation (t_2)

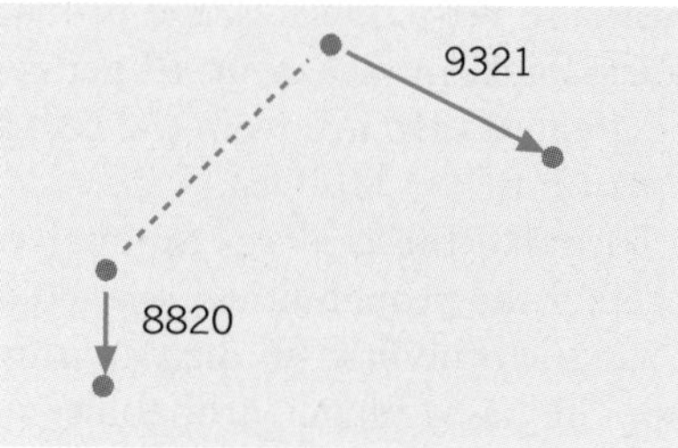

(c) Third generation (t_3)

Figure 16.10 Simulated diffusion The opening stages of the Hägerstrand model are illustrated by a mean information field (MIF). The numbers refer to contacts determined by drawing random numbers. When contacts are *internal* (i.e. with the cell on which the MIF is centred), a circle is added in that cell.

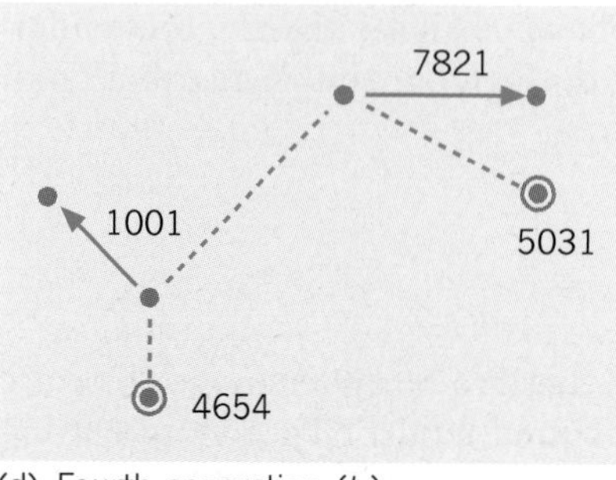

(d) Fourth generation (t_4)

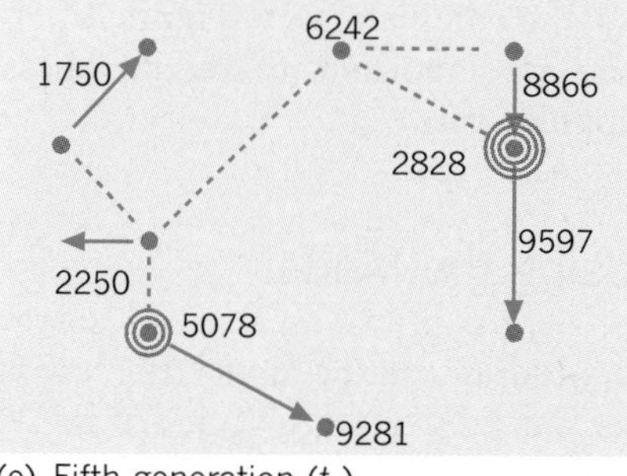

(e) Fifth generation (t_5)

placed over *each* source cell so that the centre cell of the grid corresponds with the source cell. A *random number* between 0000 and 9999 is drawn and used to direct the message, following rules 4 to 6. *Random numbers* are sets of numbers drawn purely by 'chance' (e.g. by rolling a dice). They can be taken from published tables of random numbers, or generated on a computer, or, for small problems, drawn from a hat. We show this process in Figure 16.10. In the first generation the number 0624 is drawn from a table of random numbers and a message is passed to a cell that lies to the north-east of the original adopter, located in the source cell.

Figure 16.10 goes on to present the first few stages in the diffusion process. In each generation the MIF is recentred in turn over each cell that has the message. Because the Hägerstrand model uses a random mechanism, each experiment or trial produces a slightly different geographic pattern. If we ran thousands of such trials (using a computer), we would find that the sum of all the different results matched the probability distribution in the original MIF; that is, we should arrive back at our starting distribution. In order to reap the benefits of the model, it should be applied not to simple, predictable diffusions whose end result is known, but to complicated, unpredictable diffusions whose end result is in doubt.

Modifying the Hägerstrand Model 16.3

If we think about the rules of the basic Hägerstrand model, we can see that they represent a considerable simplification of reality. Areas where diffusions take place are not uniform plains with evenly spread populations; innovations are not adopted the instant a message about them is received; information is not passed solely by contacts between pairs of people; and so on. Hägerstrand was fully aware of these complications, and he used his basic model to provide a logical framework for more realistic versions of the diffusion process. Hägerstrand's later variants of his model contain significant modifications. Others have been added by American researchers.

Some of the modifications introduced into the original model are minor technical improvements in its structure. For instance, regular square cells can be reshaped to fit other regular divisions (hexagonal units have been adopted in some versions), but irregular areas present more of a problem. Adaptation of the contagious diffusion model to cascade processes involves substituting a hierarchy of settlements for an isotropic plain. Probabilities must be assigned to the links between the settlements rather than to cells.

Abandoning the Uniform Plain

Some of the modifications can be simply made. Let us relax rule 1 and assume that the distribution of population is not uniform and that there are a variable number of people within each cell. The probability of contact is then a function both of the distance between the source and destination cells and of the number of people in each cell. Thus, we can multiply the population in each cell by its original contact probability to find a joint product. The ratio between the joint product of any cell and the sum of the joint products for all 25 cells in the MIF gives us a new contact probability based on both population and distance. (See Box 16.B 'Rules for modifying the basic diffusion model'.) We have to buy this added realism at the cost of some tedious arithmetic, particularly because the new probabilities must be recomputed each time we move the MIF grid. On the other hand, such computations can be readily done by a computer.

Although this procedure may seem complicated, we are simply putting *back* into the model the geographic reality that the original assumption of a uniform plain took out. If we were concerned with understanding the spread of cultural artifact (e.g. TV ownership) through a region, one of our first concerns would be the distribution of population and thus of potential purchasers for the product.

Box 16.B Rules for Modifying the Basic Diffusion Model

Weighting Contact Probabilities

If we assume that the probability of contact in a diffusion model is a function of both the distance between the source and destination cells and the number of people in each cell, then we can estimate

$$C_i'' = \frac{C_i' N_i}{\Sigma_{i=1}^{25} C_i' N_i}$$

where C_i'' = the joint probability of contact with the ith cell based on the MIF and population;
C_i' = the original probability of contact with the ith cell based on the 25-cell MIF;
N_i = the number of people in the ith cell; and
$\Sigma_{i=1}^{25}$ = the summation of all $C'N$ values for the 25 cells within the MIF, including the ith cell.

These revised values for the probability of contact (C'') must be recomputed each time the MIF grid is moved to allow for spatial variations in the population.

Innovations and Logistic Curves

The resistance of a population to adopting an innovation usually follows an S-shaped curve (see diagram (a)).

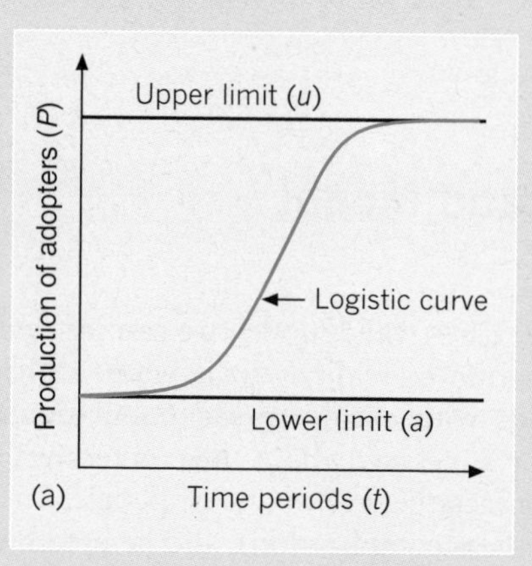

This curve can be approximated by a *logistic distribution* given by the equation

$$P = \frac{u}{1 + e^{a-bt}}$$

where P = the proportion of the population adopting an innovation;
u = the upper limit of the proportion of adopters;
t = time;
a = a constant related to the time when P has reached one half of its maximum value;
b = a constant determining the rate at which P increases with t; and
e = the base (2.718) of the natural system of logarithms.

Thus, with u = 90 per cent, a = 5.0, and b = 1.0, the proportion of adopters will be 4 per cent at t = 2, 28 per cent at t = 4, 66 per cent at t = 6, 85 per cent at t = 8, and so on. As diagram (b) shows, the constant b has a critical effect on the form of the innovation curve. Low b values describe smooth innovation curves (curve 1), whereas higher b values describe rates of acceptance that have a slow initial build-up, explode rapidly in a middle period, and enter a final period of slow consolidation (curve 2). (See P.R. Gould, *Spatial Diffusion* (American Association of Geographers, Commission on College Geography, Resource Paper 4, Washington, D.C., 1969).)

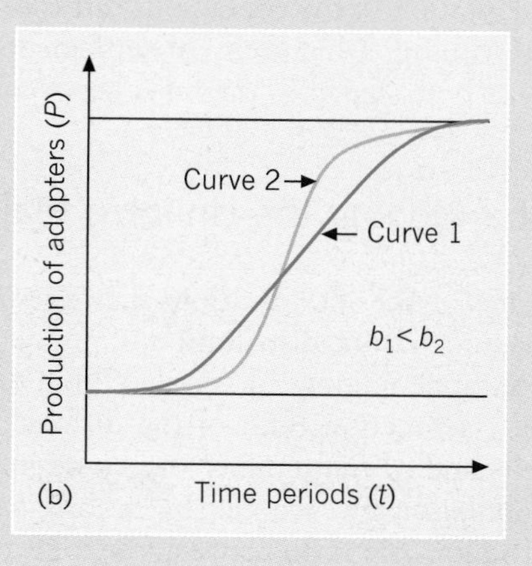

Varying the Resistance to Innovations

In discussing the impact of religious beliefs in Chapter 7 we noted their importance in insulating a group against change. One of the examples we used was the persistence of some seventeenth-century cultural traits in the present-day Amish communities in the United States. Can the model be adapted to incorporate factors of this kind?

Well, it can be if we relax another of the original rules, in this case rule 7. The statement that adoption of an item takes place as soon as a message is received by the destination cell is an oversimplification. From research on agricultural innovations we know that there is generally a small group of people who are 'early innovators' and another small group of 'laggards'; the majority of a population adopts an innovation after the early innovators and before the laggards. In the case of spatial diffusions of population over a territory, this implies that settlements are established sporadically at first. The sporadic phase of settlements is followed by a period when everyone gets on the bandwagon, and eventually by a period of restricted settlements as the number of suitable unsettled locations in the territory diminishes. In the case of spatial diffusions of an innovation throughout a population, there are regional variations in the time of acceptance of the new item or way of doing things. For example, Figure 16.11 uses historical data to show extreme regional variations in the time of adoption of tuberculin-tested (TT) milk by farmers in the countries of England and Wales. In some southern counties, 50 per cent of the milk sold from farms in 1950 was TT milk; the proportion of TT milk in the milk sold in the far southwest did not reach this level for another seven years.

One practical aspect of geographers concern with modelling resistance to spread has been research on ways of speeding up desirable innovations (e.g. understanding of birth control) and slowing down undesirable innovations (e.g. spread of a new disease). Figure 16.12 illustrates how information on AIDS and the use of condoms meets both these criteria.

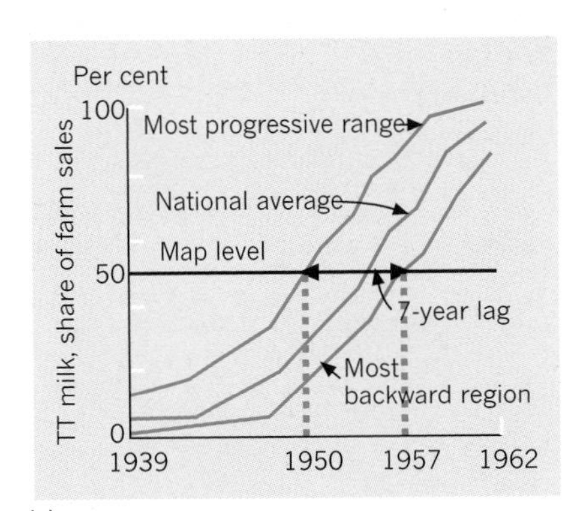

(a)

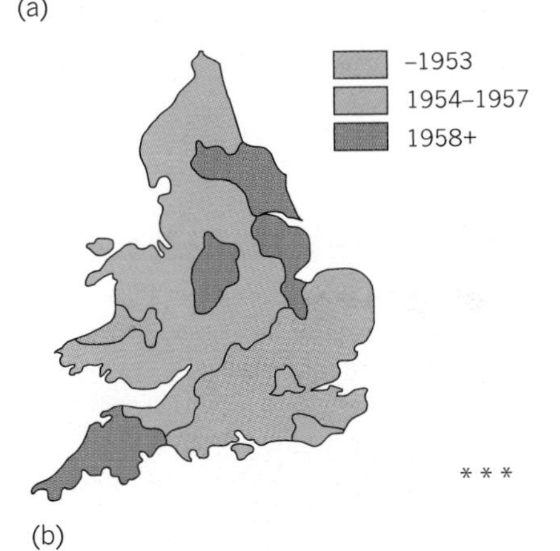

(b)

* * *

Figure 16.11 Resistance to change Regional variations in adoption rates are illustrated by (a) the diffusion of tuberculin-tested (TT) milk on farms in England and Wales. The map (b) shows the year by which each county had achieved 50 per cent TT milk production.

[Source: Data from Milk Marketing Board. After G.E. Jones, *Journal of Agricultural Economics*, **15** (1963), Figs 6, 7A, pp. 489–490.]

Figure 16.12 Social factors in acceptance of or resistance to innovations One practical outcome of geographers' research on spatial diffusion is a concern with ways of speeding up desirable innovations (family-planning clinics) and slowing down or halting undesirable innovations (disease epidemics). The spread of family planning in (a) Rajasthan, India, and (b) Benares, India, illustrates the former concern.

[Source: (a) Michaud and Paulo Kock, (b) Rapho Guillumette.]

(a)

(b)

We can approximate the symmetric course of the diffusion process by S-shaped curves. (See Box 16.B for a discussion of innovations and logistic curves.) Standardized *resistance curves* of this type were used by Hägerstrand to take into account resistance to innovations. After one message, the probability of acceptance was very low (0.0067); after two messages, it rose to nearly one-third (0.300); and after three, to nearly three-quarters (0.700). From then on the rate of acceptance fell again. The probability of acceptance rose slowly after four messages (to 0.933), and still more slowly thereafter. After five messages, even the worst laggards had accepted the item, and the rate of acceptance was 1.000. Like changing probabilities of contact, varying rates of acceptance (or resistance) can be readily incorporated into a computer simulation of a diffusion process. And so, if we have a community highly resistant to change, such as the Amish, we can increase the number of messages sent to any appropriate large number. If a community were wholly resistant to change, the number of messages needed would be infinite!

Adding Boundaries and Barriers

In the original model, messages moving outside the boundaries of the study area were considered lost and had no effect on the situation (rule 9). In later models a boundary zone over half the width of the MIF grid was created so diffusion could proceed by way of these external source cells. More important modifications were involved in the introduction of internal *barriers* that act as a drag on the diffusion process. Like the other modifications we have discussed, such barriers allow observed variations in both the natural environment and the cultural mosaic to be incorporated into the model.

At the University of Michigan, Richard Yuill programmed the Hägerstrand model to simulate the effects of four types of barriers on the diffusion of information through a matrix of 540 cells within a nine-cell MIF. Figure 16.13(a) shows the nine-cell grid with the barrier cells indi-

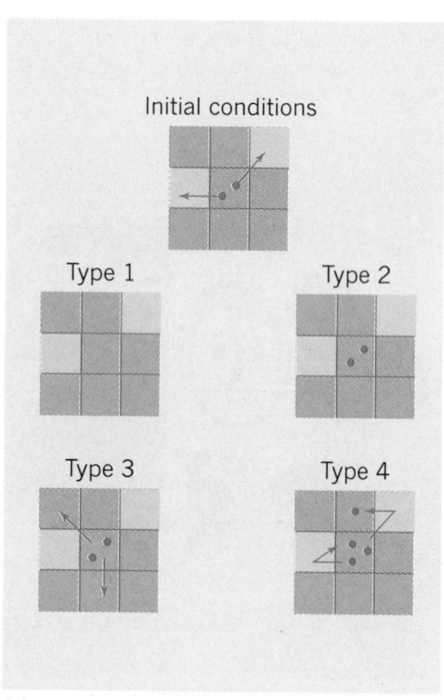

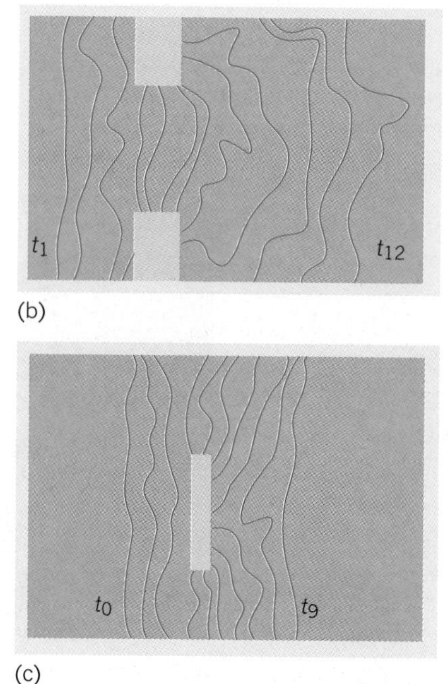

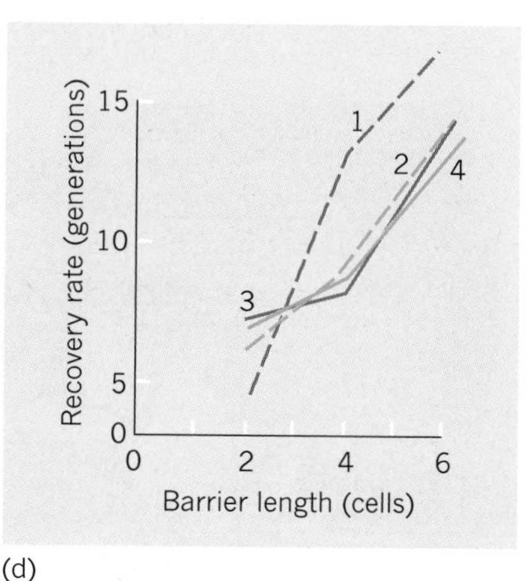

Figure 16.13 Barriers and diffusion waves Four types of barrier cells (a) are used in this simulation model. In (b) diffusion waves pass through an opening in a bar barrier. In (c) diffusion waves pass around a bar barrier. The graph (d) shows the recovery rates around bar barriers constructed from the four different types of barrier cells. 'Recovery rate' is the time taken for the straight line of the diffusion wave front to re-form.

[Source: R.S. Yuill, *Michigan University Community of Mathematical Geographers, Discussion Papers*, No. 5 (1965), pp. 19, 25, 29.]

cated. Four types of barrier cells that provide a decreasing amount of drag are considered: a *superabsorbing barrier* that absorbs the message but destroys the transmitters; an *absorbing barrier* that absorbs the message but does not affect the transmitters; a *reflecting barrier* that does not absorb the message but allows the transmitter to transmit a new message in the same time period (see the arrows in the figure); and a *direct reflecting barrier* that does not absorb the message but deflects it to the available cell nearest to the transmitter.

Each situation was programmed separately and the results plotted. Figure 16.13(b) shows the advance of a linear diffusion wave through an opening in a barrier. The time taken for the original line of the wave to reform determines the *recovery rate*. Varying types of barriers and gaps of varying widths were investigated. In the example shown, the line of the wavefront has recovered by about the eleventh generation (time t_{11}). Another type of barrier is presented in Figure 16.13(c). Here the diffusion wave passes around the barrier and reforms after about nine generations. The recovery rate of a wavefront is directly related to both the type and the length of the barrier it encounters; the curve for a superabsorbing barrier is quite different from the curves for the other three types of barriers (see Figure 16.13(c)).

Yuill's work expands and develops modifications already begun by Hägerstrand. The original model postulated what was, in effect, a row of absorbing cells around the periphery of the study area. The internal barriers in Hägerstrand's model were represented by the lines between the cells. Such barriers could be adjusted to be absolutely effective (i.e. to allow no messages to get through) or 50 per cent effective (i.e. to let one out of two messages cross the barrier). With such *permeable barriers* we can replicate a variety of environments. Thus, the original assumptions of isotropic movement can be brought into line with known patterns. In other words, we can build *low-resistance corridors* into the model to allow faster diffusion in certain directions, and we can also build into the model *high-resistance buffers* to slow down diffusion across barriers.

To sum up, the basic Hägerstrand model can be easily modified to make it fit more closely to the realities of the geographic world. To the changes in population density and barriers discussed here, we can add such further refinements as variations in the 'infectiousness' of the element being diffused.

Regional Diffusion Studies 16.4

Many of the applications of Hägerstrand's model stem from his own pioneering work in Sweden. Here we review two of the applications of the model to regions with contrasting environmental conditions. We look first the spread of cultural attitudes (farmers' attitudes toward farm subsidies), irrigation in the Great Plains, and then at the spread of a cultural group (the Polynesians).

Farm Subsidies in Central Sweden

In the late 1920s the Swedish government introduced a scheme to persuade farmers to forego their traditional practice of allowing cattle to graze the open woodlands in summer. Grazing was proving to be a problem because it restricted the growth of young trees. To encourage fencing and improvements in pastureland, the government offered a subsidy. Figure 16.14 presents computer maps of the central part of Sweden and indicates areas where farmers accepted the subsidy during the years 1930 to 1932.

The maps indicate that in 1930 a few farmers accepted the subsidy in the western part of the region but there were scarcely any takers in the east. The next two years brought a rapid increase in the number of acceptors in the west but little change in the east. The sequence of maps suggests a spatial diffusion process in which distance is an important factor. To simulate this process Hägerstrand built a model using the 1928–29 distribution of adopters as a starting point. The basic model was modified in two ways. First, the potential number of adopters (i.e. farmers) in each cell was added; second, barriers that were 100 per cent and 50 per cent permeable were added to simulate the long north–south lakes that lie across the region. Figure 16.14 compares the simulated diffusion process with the actual one. Because of the random element in the model, we should not expect the simulated pattern to match the actual pattern exactly. But the degree of matching is close, and both the general form of the expansion process and the location of the major clusters of adopters in the western areas are correct.

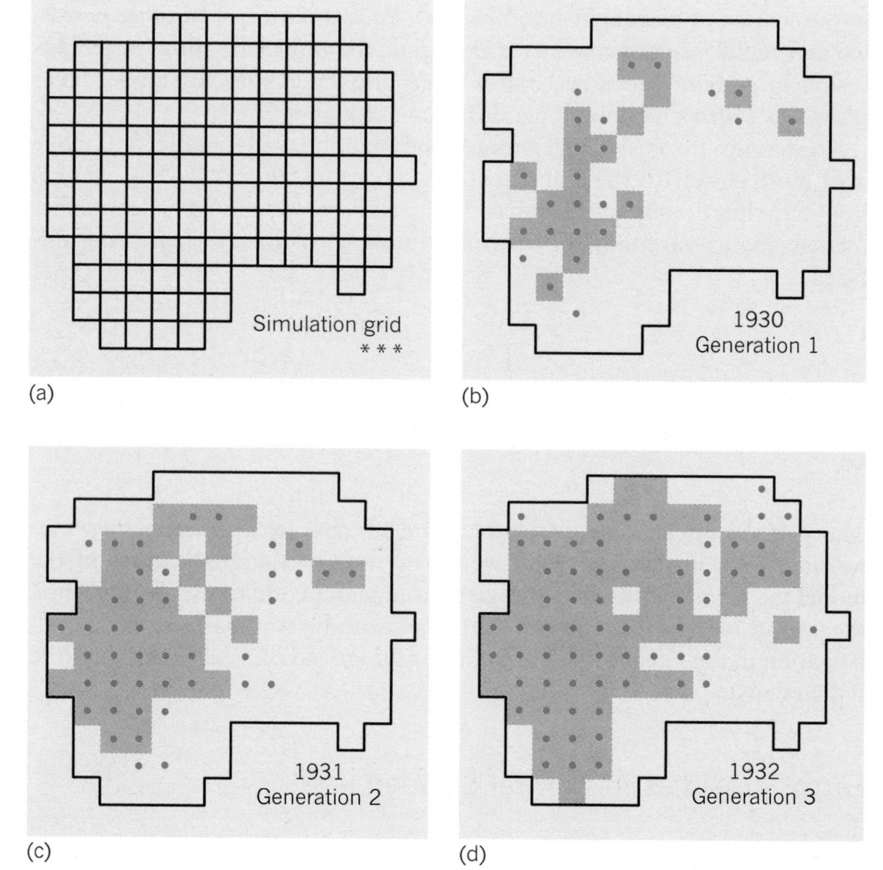

Figure 16.14 Simulating diffusion on a grid Here we see Hägerstrand's simulated diffusion and the actual diffusion of a decision by farmers in central Sweden to accept a farm subsidy. For simulation purposes the test area was approximated by a regular grid (a). The data for the three trial years shown in (b), (c), and (d) are part of a more extended study.

[Source: T. Hägerstrand, *Northwestern University Studies in Geography*, No. 13 (1967), Figs. 6, 9, pp. 17, 23.]

- Actual diffusion
- Simulated diffusion

Irrigation in the Great Plains

An American geographer Leonard Bowden has extended the Hägerstrand model and applied it to agricultural change in the northeastern high plains of Colorado. Drought problems had hampered expansion of cattle farming, and the solution involved irrigation by wells and pumps that tap groundwater. However, the decision to irrigate required large expenses and was not made without consultation with the successful innovators. How did such discussion take place? Bowden proposed that mean discussion fields (MDFs, analogous to the MIFs we have already met in this chapter) could be established by studying telephone calls and social get togethers such as barbecues in the agricultural community.

The actual pattern of irrigation investment at the start of the period included just 41 wells (see Figure 16.15(a)). These were concentrated spottily in the eastern borders of the state between two main rivers, the South

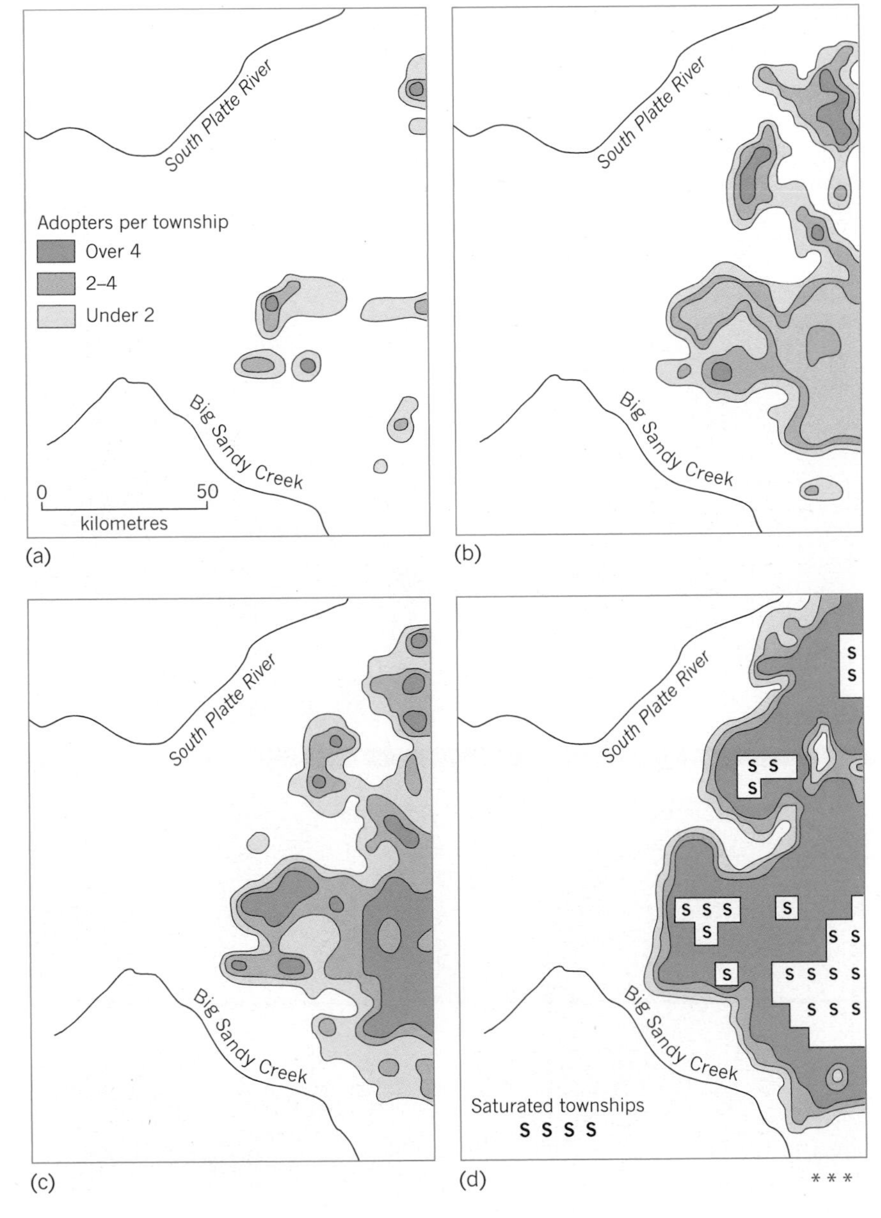

Figure 16.15 Simulation of the decision to irrigate Eastern Colorado on the High Plains of the central United States. (a) Actual pattern of irrigation wells per township in the initial period, 1948. (b) Simulated pattern after fifteen years. (c) Actual pattern after fifteen years. (d) Projected pattern for 40 years on. Note the townships where irrigation wells had reached saturation levels in terms of the groundwater available.

[Source: L.W. Bowden, *University of Chicago, Department of Geography, Research Papers,* No. 97 (1965), Fig. 21, p. 108.]

* * *

Platte and Big Sandy Creek. Bowden devised simulation models similar to Hägerstrand's and began his investigation with the year 1948 as his starting point. An average of ten runs revealed a tenfold increase in the number of wells. As illustrated in Figure 16.15, the resulting distribution was sufficiently close to the 410 wells, the level achieved within fifteen years, that a spatial prediction for the last decade of the century was simulated. For forecasting, however, the researchers permitted only 16 wells per township to avoid excessive removal of irrigation water from the groundwater sources.

Over the last half century, the use of deep tube wells has spread very widely over the semi-arid areas of the world. Figure 16.16 shows a striking satellite photograph of a cluster of tube wells in Saudi Arabia. The distinctive circular dots stand out as irrigated land watered by a centre-pivot system with a long boom sweeping around and delivering water over a circular area. With the rapid growth of demand for water and the depletion of underground resources, the early work by Bowden on irrigation spread assumes new significance.

Kon-Tiki Voyages in the Pacific

When the Norwegian explorer Thor Heyerdahl undertook his now historic voyage on the raft *Kon-Tiki* from the coast of Peru to the Tuamotu Islands, he was carrying out a single experiment: he was testing whether it was possible to cross the Pacific in such a craft. To analyze thoroughly

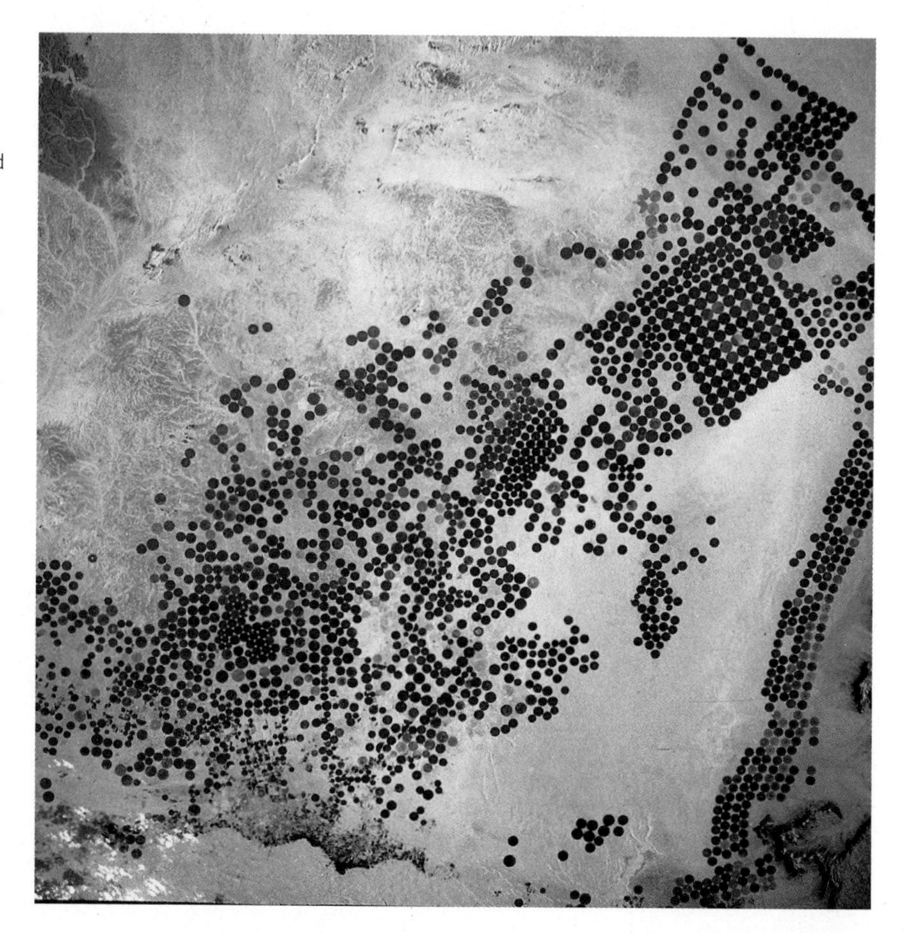

Figure 16.16 Spatial diffusion of new irrigation methods Irrigation scheme at Jabal Tuwayq in Saudi Arabia using centre-pivot techniques. Water is tapped from deep fossil reserves and fed through large rotation booms. This Shuttle photograph shows the characteristic circles of irrigated land standing out clearly against the unirrigated land beyond the reach of the water booms.

[Source: NASA/Science Photo Library.]

the probabilities of contact between South America and different island groups by this means would require too many voyages to be feasible. When direct experiments prove too costly, too risky, or unlikely for other reasons, we may be able to turn to computer simulation for answers to our questions. For example, in World War II, in Project Manhattan (the name for the atomic bomb project), the atomic radiation from bombs was simulated mathematically. Heyerdahl's trans-Pacific migration exemplifies a spatial simulation that can be approached by means of the Hägerstrand model.

The central issue to decide is how the Polynesians came to discover and settle on the islands of the central Pacific. This question has recently attracted considerable attention from anthropologists, navigators, and geographers; but the sparsity of evidence has led to a clash between two schools of thought. The first holds that the colonization process involved intentional two-way voyages and hence a high degree of navigational skill on the part of the Polynesians (see Figure 16.17). The alternative view was that colonization was largely accidental, by travellers drifting off course.

To test the probability of inter-island contact as a result of accidental drifting about, the New Zealand geographer Gerard Ward and a group of investigators at University College London constructed a computer *simulation model* of the drift process. The stages in their computer program are shown in Figure 16.18. There are four main elements in the model:

1 The relative probabilities of wind strength and direction for each month, and of current strength and direction for each 5° square of latitude and longitude in the Pacific Ocean study area.
2 The positions of all the islands and land masses in the study area, together with their sighting radius.
3 The estimated distances that would be covered by ships given various combinations of wind and current strength.
4 The relative probabilities of survival of ships during certain periods at sea.

Figure 16.17 Diffusion mechanisms This eastern Polynesian double canoe was used in inter-island voyaging. The settlement of Polynesia by canoes of this kind is a question hotly debated by anthropologists and geographers. Where did the Polynesians come from and how did they move from island to island? Figure 16.18 shows how computer models of diffusion may throw some light on both questions.

[Source: Drawn by John Webber, artist on Captain James Cook's third voyage. From A. Sharp, *Ancient Voyagers in the Pacific* (Penguin, London, 1957), Fig. 8.]

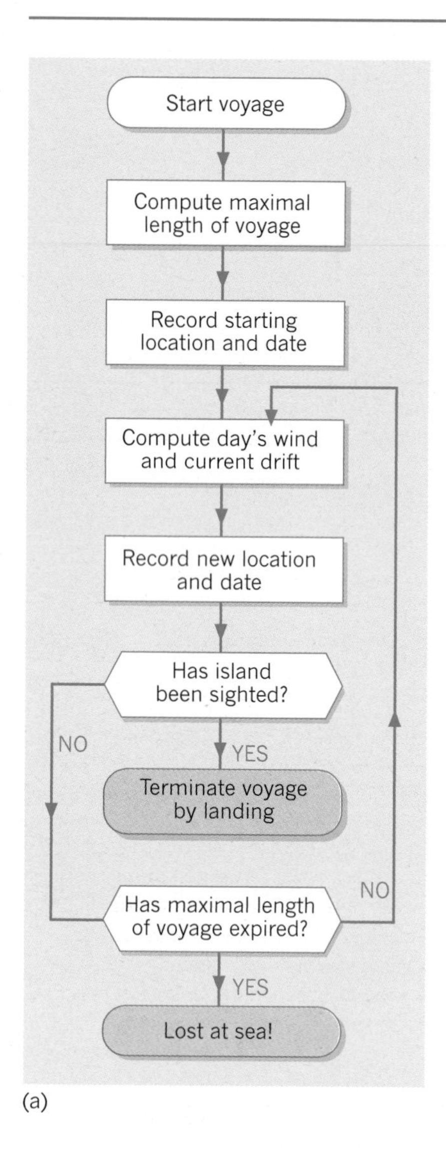

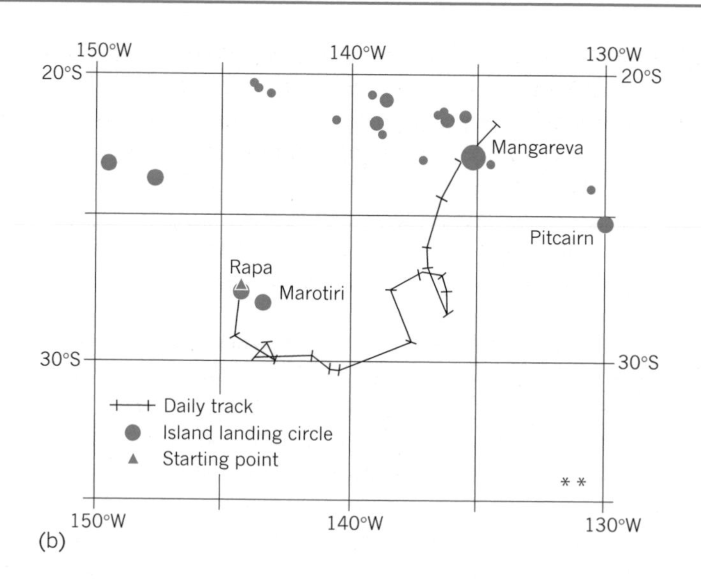

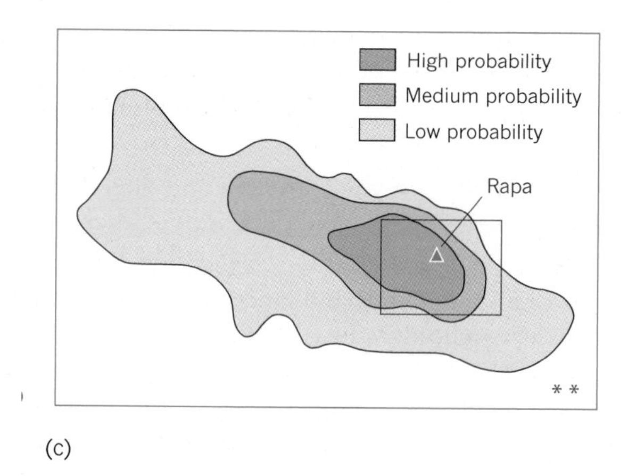

Figure 16.18 Polynesian voyaging
(a) This flow chart shows the main elements in a computer program developed to simulate Pacific 'drift' voyages. (b) Example of a simulated voyage from the island of Rapa in east-central Polynesia over nineteen days of shifting winds and currents to land on Mangareva some 800 km (500 miles) to the northeast. (c) Probability map for all voyages simulated from Rapa. Compare with the Pacific map shown in Figure 5.6.

[Source: M. Levision, R.G. Ward and J.W. Webb, *The Settlement of Polynesia: A Computer Simulation* (University of Minnesota Press, Minneapolis, 1973), Figs 9, 11, pp. 25–26.]

Sighting radius (the distance out to sea from which islands can be seen) was built into the model on the assumption that, once land had been sighted, a landfall could be made. During each daily cycle voyages are started from given hearth areas such as the coast of Peru, and a simulated course is followed until it ends in either a landfall or the death of the voyagers. By simulating hundreds of voyages from each starting point, we can map the relative probability of contact with different island groups as a potential contact field.

The simulation program has already been run for various Pacific Island groups and for locations on the coasts of South America and New Zealand. Preliminary results indicate that the probabilities of inter-island links from the accidental drift of ships differ from one area to another. Wind and current patterns create environmental boundaries that make drifting in certain directions highly unlikely. Some of these boundaries coincide with long-standing anthropological breaks in the geographic pattern of ethnicity and culture such as that separating the Micronesian people of the Gilbert Islands and the Polynesian inhabitants of the Ellice Islands. Other low probabilities of contact coincide with important linguistic boundaries.

The computer model for this research simulates activities that cannot be observed at first hand and are too complex to be simulated by manual calculations. It confirms that certain existing population distributions are possible purely as a result of voyagers drifting off course. However, there remain certain hard cases, notably the Hawaiian Islands, whose settlement remains a mystery.

Applying Diffusion Models 16.5

In this final section we look at one area where diffusion models have been successfully applied to an important area of applied geography: the control of epidemic waves. Epidemics (see the discussion in Chapter 20) are diseases that spread through human populations through processes of contagious and cascade diffusion as illustrated in Figures 16.3 and 16.4.

A Simple Transmission Model

The ways in which diseases are transmitted has attracted mathematical interest from the eighteenth-century Swiss mathematician, Daniel Bernoulli, onwards. To give a flavour of this approach, a very simplified diagram of the spread of one infectious disease of humans through a human population is illustrated in Figure 16.19. Measles is a highly infectious viral disease, which mainly attacks children; the virus persists as an infectious agent in the human body for around two weeks. One attack creates lifelong immunity. It is relatively mild in western countries but is a major killer in the tropics.

This shows that we need to maintain an unbroken chain of *infectives* (humans infected by the disease) if an epidemic is to be maintained. Protection against the spread of infection can be taken at two points in the flow diagram. First, we can use a vaccine. This allows the route from *susceptibles* (those

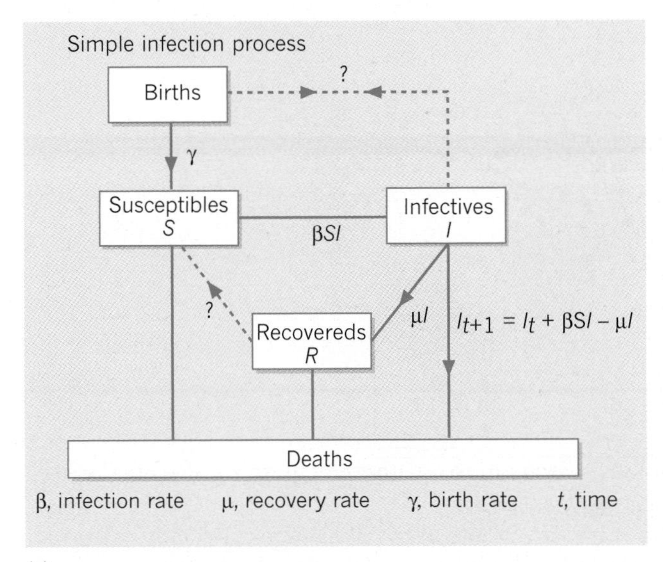

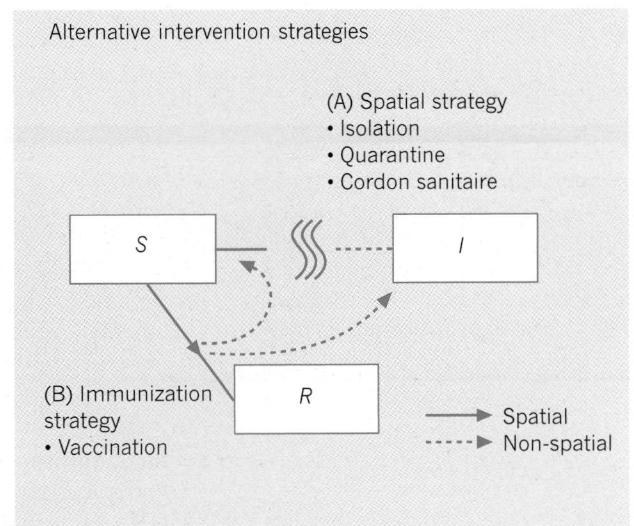

Figure 16.19 Intervention in the spread of an infection Simplified model of control strategies for an infection process. (a) The population is divided into three categories, susceptible (*S*), infective (*I*), and recovered (*R*). Main components in the model based on the Hamer–Soper model of measles spread. (b) Alternative intervention strategies: spatial intervention based on (A) blocking links between infectives and susceptibles and (B) opening of new pathways through immunization which outflanks the infectives box.

Source: A.D. Cliff, P. Haggett, and M. Smallman-Raynor, *Measles: An Historical Geography of a Major Human Viral Disease* (Blackwell, Oxford, 1993), Fig. 16.1, p. 414.]

susceptible to the disease but not yet infected) to *recovereds* (those who have had the disease and successfully recovered) to be short-circuited. This is done by artificially infecting a person with a vaccine of some kind. This provokes a mild reaction that creates antibodies to the disease and allows the establishment of an immune population. Second, we can use spatial protection. This method interrupts the mixing of infectives and susceptibles by erecting protective spatial barriers. This may take the form of (a) isolating an individual or community, or (b) restricting the geographical movements of infected individuals by quarantine requirements; another approach is (c) by locating populations in supposedly safe areas. For animal populations, there exists a further possibility: the creation of a *cordon sanitaire* by the wholesale evacuation of areas or by the destruction of infected herds.

Disease Reservoirs and Population Thresholds

For the infectious disease most commonly used in modelling studies (measles), the critical community size required to sustain endemicity has been studied in detail in a classic paper by the Oxford statistician Maurice Bartlett. Bartlett plotted the mean time period between epidemic peaks (in weeks) for a sample of 19 English towns of different sizes (see Figure

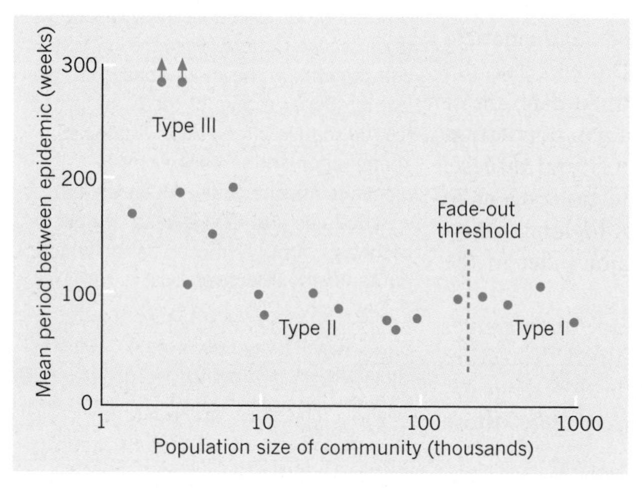

(a)

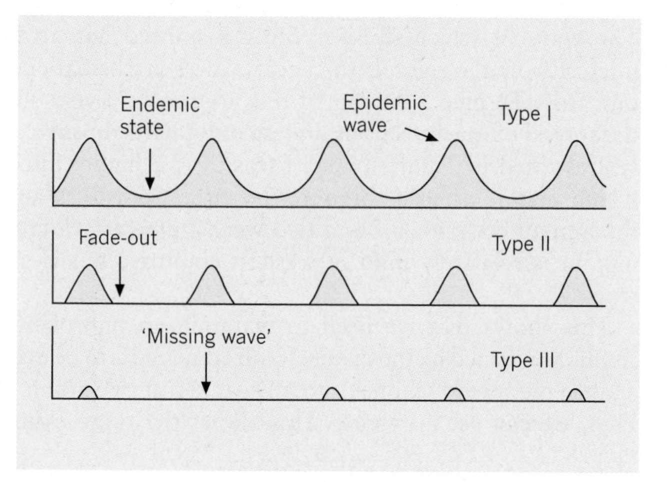

(b)

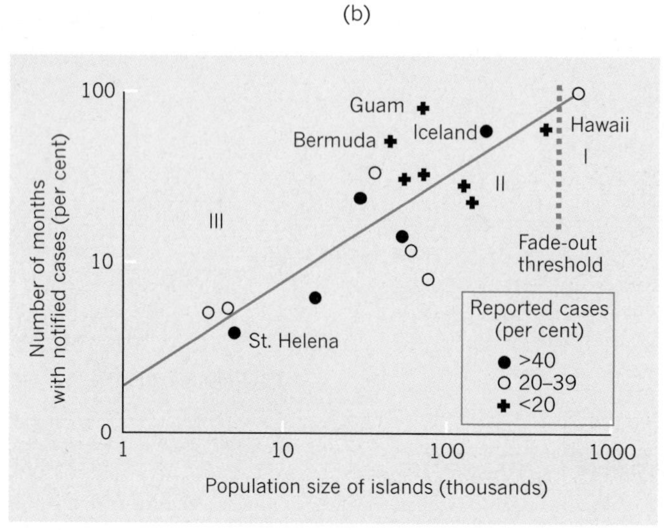

(c)

Figure 16.20 Disease periodicity in towns and islands (a) Bartlett's findings on city size and epidemic recurrence: three groups of patterns (I, II, and III) are identified. The impact of population size on the spacing of measles epidemics for nineteen English towns. (b) Characteristic epidemic profiles for the three types indicated in (a). (c) Black's findings on epidemic recurrence of measles in eighteen island communities.

[Source : A.D. Cliff, P. Haggett, and M. Smallman-Raynor, *Measles: An Historical Geography of a Major Human Viral Disease* (Blackwell, Oxford, 1993), Figs 1-5, 1-9, pp. 7, 12.]

16.20(a)). The time interval between outbreaks was found to be inversely related to the population size of the community. Given the reporting rates for measles at the time, this implied that a population of around 250,000 to 300,000 is required to ensure continuous transmission chains of infection.

Yale epidemiologist Francis Black extended the *Bartlett model* by examining the relationship between measles endemicity and population size in 18 island communities. Black plotted, on logarithmic scales, the percentage of months with notified cases against population size; the population size with 100 per cent reporting denotes the endemicity threshold. Of the islands studied by Black, only Hawaii with a total population (then) of 550,000 displayed clear endemicity. Other islands close to Bartlett's 250,000 value just failed to display endemicity. This may, of course, reflect the difference between the isolation of islands as opposed to the mainland location of Bartlett's cities.

The basic notion of a threshold population, below which an infectious disease becomes naturally self-extinguishing, is critical in devising control strategies. It implies that vaccination may be employed to reduce the susceptible population below some critical mass so that natural breaks in the chain may achieve the rest. As a result, attempts have been made to establish the endemicity thresholds for a variety of transmissible diseases. Once the population size of an area falls below the threshold then, when the disease concerned is eventually extinguished, it can only recur by reintroduction from other reservoir areas. Thus the generalized persistence of disease implies geographical transmission between regions as shown in Box 16.C ('Andrew Cliff and epidemic diffusion models').

Box 16.C Andrew Cliff and Epidemic Diffusion Models

Andrew Cliff (born Grimsby, England, 1943) has pioneered the application of spatial diffusion models to an applied problem, the spread of epidemic diseases. He took his first degree at King's College London, his masters at Northwestern University in the United States, and his doctorates at Bristol. He joined the Cambridge staff in 1972 and is now professor of theoretical geography at Cambridge.

His earliest research was on spatial statistics, notably with his Bristol colleague Keith Ord (their *Spatial Autocorrelation* (1973) was a classic), but later work has focused on applying spatial models to epidemiological problems. Predicting the spatial spread of childhood diseases such as measles with a view to devising control strategies has been a central theme. Communities such as Iceland and the Pacific island groups have been important testbeds for trying out the forecasting models. His research has shown that the Bartlett threshold for measles (or indeed any other infectious disease) is likely to be rather variable, with the level influenced by population densities and vaccination levels. Once the population size of an area falls below the threshold then, when the disease concerned is eventually extinguished, it can only recur by reintroduction from other reservoir areas. Thus the generalized persistence of disease implies geographic transmission between regions as shown.

We can see from this that in large cities above the size threshold, like

Box continued

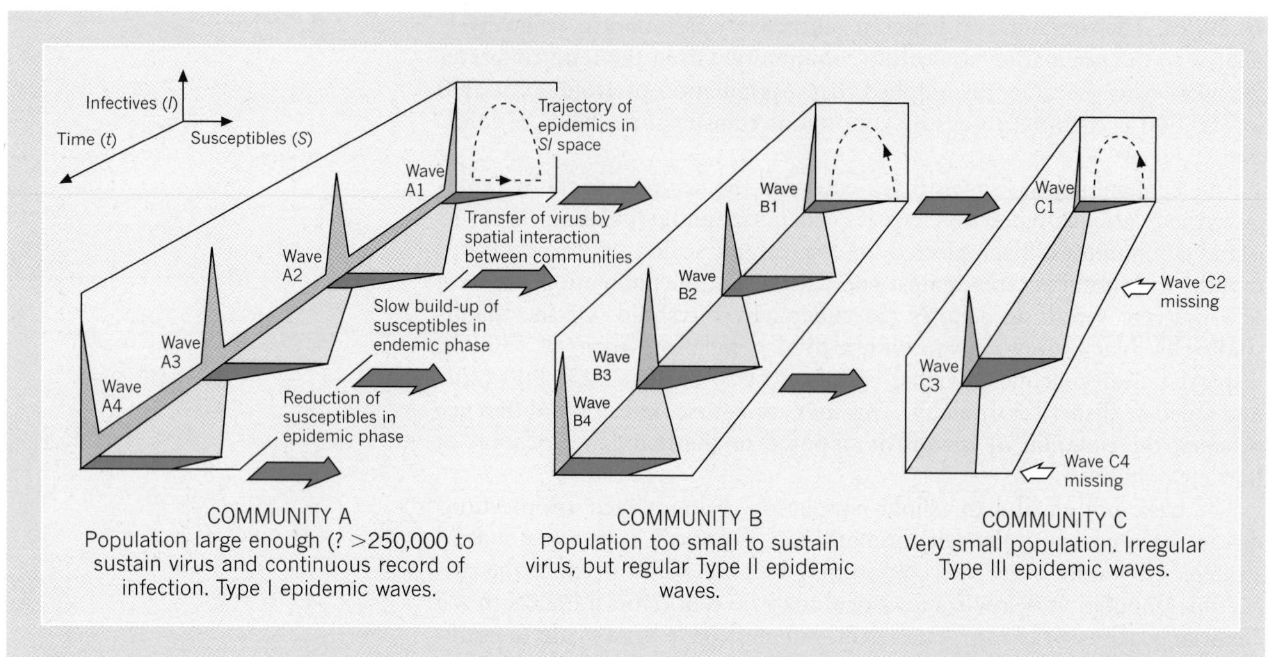

community A, a continuous trickle of cases is reported. These provide the reservoir of infection which sparks a major epidemic when the susceptible population, *S*, builds up to a critical level. This build-up occurs only as children are born, lose their mother-conferred immunity, and escape vaccination or contact with the disease. Eventually the *S* population will increase sufficiently for an epidemic to occur. When this happens, the *S* population is diminished and the stock of infectives, *I*, increases as individuals are transferred by infection from the *S* to the *I* population. This generates the characteristic 'D'-shaped relationship over time between sizes of the *S* and *I* populations shown on the end plane of the block diagram.

With measles, if the total population of a community falls below the ¼-million size threshold, as in settlements B and C of the diagram, epidemics can, as we have noted above, only arise when the virus is reintroduced by the influx of infected individuals (so-called *index cases*) from reservoir areas. These movements are shown by the broad arrows in the diagram. In such smaller communities, the *S* population is insufficient to maintain a continuous record of infection. The disease dies out and the *S* population grows in the absence of infection. Eventually the *S* population will become big enough to sustain an epidemic when an index case arrives. Given that the total population of the community is insufficient to renew by births the *S* population as rapidly as it is diminished by infection, the epidemic will eventually die out.

It is the repetition of this basic process that generates the successive epidemic waves witnessed in most communities. Of special significance is the way in which the continuous infection and characteristically regular Type I epidemic waves of endemic communities break down, as population size diminishes, into, first, discrete but regular Type II waves in community B and then, secondly, into discrete and irregularly spaced Type III waves in community C. Thus disease-free windows will automatically appear in both time and space whenever population totals are small and geographical densities are low.

Isolated communities such as Iceland and the Pacific island communities have been important test-beds for testing out the forecasting models. Cliff's recent work is summarized in A.D. Cliff, P. Haggett, and M. Smallman-Raynor, *Island Epidemics* (Clarendon Press, Oxford, 2000).

Spatial Strategies for Disease Control

If we look at the strategy by which smallpox was finally eradicated (see Chapter 20), we can see the control process as one of the progressive spatial reduction in the areas of the world in which the disease was endemic. Four different spatial control strategies are illustrated in Figure 16.21; in

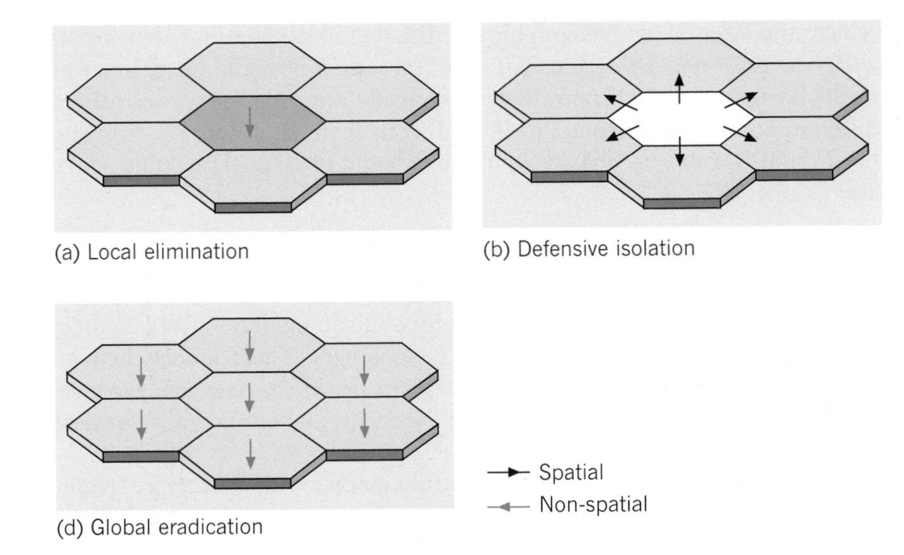

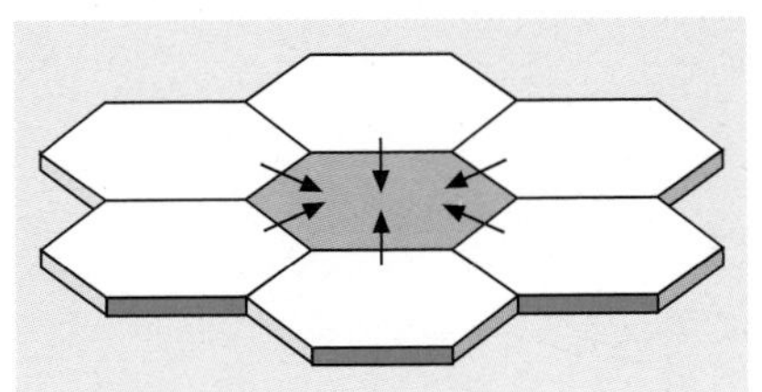

Figure 16.21 Controlling epidemic spread Schematic diagram of four spatial and aspatial control strategies to prevent epidemic spread. Infected areas are stippled; disease-free areas left white. Geographic areas shown arbitrarily as hexagons.

[Source : A.D. Cliff, P. Haggett, and M. Smallman-Raynor, *Measles: An Historical Geography of a Major Human Viral Disease* (Blackwell, Oxford, 1993), Fig. 16-9, p. 423.]

each of the four maps, infected areas have been stippled, while disease-free areas have been left blank.

In the first stage, *local elimination*, the emphasis is on breaking, in some particular location, the disease chain by vaccination. The programmes aimed at eliminating indigenous measles in countries such as the United States and Australia illustrate this phase. Such vaccination programmes may themselves be variable within a country and have a geographic component.

Once an area is cleared of indigenous measles, then there is a need for a second stage, *defensive isolation*, which entails the building of a spatial barrier around a disease-free area. Attempts to erect such barriers were made in the nineteenth century but may be impractical today. We have already noted the difficulties that air travel poses to the use of *quarantine* to prevent infectious cases from gaining access to susceptible populations; the US experience with its measles elimination programme illustrates the point. Attempts to control the spatial spread of such communicable diseases go back to at least the medieval period. By the thirteenth century, most Italian cities were posting gatemen to identify potential sources of infection from visitors to the city. The northern city of Venice, with its extensive trading links with the Levant and the Oriental lands beyond, pioneered the idea of quarantine. It saw the first recorded attempt to place a delay on travel and trade in 1377. Originally a 30 day waiting period (a *trentino*), it was widely adapted by port cities as a defence against the plague and later extended to 40-days (a *quarantino*), today familiar as a 'quarantine' period.

A third stage, *offensive containment*, is a more appropriate approach in these circumstances. This is the reverse of the second case in that the spread of a local outbreak within a larger disease-free area is halted and progressively eliminated by a combination of vaccination and isolation. The Canadian geographer Roly Tinline has explored the use of such *ring-control* strategies for containing outbreaks of foot-and-mouth disease in livestock.

The fourth and final stage of *global eradication* arises from the combination of the previous three methods: infected areas are progressively reduced in size, and the coalescence of such disease-free areas leads eventually to the elimination of a disease on a worldwide basis. Ultimate extinction of a virus from the planet rests on a global vaccination programme to reduce the sizes of the geographically distributed populations that are at risk to levels at which the chains of infection cannot be maintained. In terms of the Bartlett model, this means systematically reducing the wave order of different communities from I to II, and from II to III, eventually bringing the Type III waves into phase so that the fade-out of all the remaining active areas coincides.

We began this chapter with strip-cropping and skateboards and ended with *Kon-Tiki* and pandemic control. We have seen how the general notions of spatial diffusion can be simulated by probabilistic models – most of them developed from the work of Swedish geographers. These models help to throw some light on the process of diffusion by which past changes have occurred – and allow attempts to be made at predicting future spatial trends.

We have seen how very simple simulation models with a structure reminiscent of a family board game can be successively made more complex and more realistic until they allow light to be thrown on some of the most pressing problems of our planet in the new century, the aggressive spread of new infectious diseases. Modern mass communication has made the power and significance of the forces of change we have studied immense and we look at these in Chapters 19 on globalization and 20 on disease control.

Reflections

1 The spread of culture between regions is studied by geographers as a diffusion process. Gather data for your local area on (a) the location and (b) the date of foundation of any one sort of public institution (churches, banks, schools, colleges, and so on). Map the data and try to identify the kind of spatial diffusion processes that appear to be operating. Do contagious or hierarchic processes seem to be more important?

2 The Hägerstrand model operates on the basis of random processes operating within strict rules. Look carefully at the first four phases of Figure 16.10. What would the pattern of diffusion have looked like if the first seven random numbers had been 1920, 8520, 1567, 7803, 3223, 5059, and 2483?

3 Diffusion models help in planning the location of new innovations. Assume you are advising on a locational strategy for a new chain of motels or hamburger restaurants (e.g. Holiday Inns or McDonald's) in your state. Where would you try to locate the first five establishments? Why would you pick these locations? Compare your results with those of others in your class and identify any common locations you all wish (a) to adopt or (b) to avoid.

4 Ways in which the Hägerstrand model can be refined are illustrated by the work of Yuill (see Figure 16.13). List the factors that affect the spread of family-planning information in a country such as India. How many of these factors could you incorporate into Yuill's resistance model?

5 Diffusion models have been applied to the spread of rumours through contact networks. Check how information spreads in a small group by planting a rumour (make it a benign one) with one other member of the class. At the beginning of the next class check (a) who now knows, (b) from whom the information was obtained, and (c) when the 'telling' took place. Try to construct a tree like that in Figure 16.5, showing the way in which the rumour spread through the class.

6 One application of the Hägerstrand model has been to early questions of island voyaging in the Pacific. Trace the loops in the simplified flow chart of Polynesian voyages in Figure 16.18. How accurate is such a model likely to be? How might this type of model be used to trace other cultural diffusion processes?

7 Consider the way in which diffusion models can be applied to understanding the spread of diseases. (See also Chapter 20.) Look at the Bartlett spatial model in Box 16.C and try to link the waves shown there to the descriptions in the text.

8 Review your understanding of the following concepts using the chapter text and the glossary in Appendix A of this book:

adoption curves
barriers
Bartlett model
contagious diffusion
Hägerstrand model
hierarchic diffusion
innovation waves
mean information field (MIF)
random number
simulation model

One Step Further ...

A brief but useful introduction to spatial diffusion is given by Derek Gregory in his essay in R.J. JOHNSTON *et al.* (eds), *Dictionary of Human Geography*, fourth edition (Blackwell, Oxford, 2000), pp. 175–8. An excellent introductory review of theories of spatial diffusion and their use by geographers is given in P. GOULD, *Spatial Diffusion* (Association of American Geographers, Washington, D.C., 1969) and Richard L. MORRILL *et al.*, *Spatial Diffusion* (Sage, Newbury Park, 1988). A more mathematical approach is followed in P. HAGGETT, A.D. CLIFF, and A.R. FREY, *Locational Analysis in Human Geography*, second edition (Arnold, London, 1977), Chapter 7.

The classic Swedish work on diffusion is by Torsten Hägerstrand. This was translated into English by Alan Pred as T. HÄGERSTRAND, *Innovation Diffusion as a Spatial Process* (University of Chicago Press, Chicago, 1968). The 'Polynesian drift' model discussed in the chapter and the application of diffusion models to the reconstruction of past cultural distributions is discussed in M. LEVISON, R.G. WARD, and J.W. WEBB, *The Settlement of Polynesia: A Computer Simulation* (University of Minnesota Press, Minneapolis, 1973), Chapters 1 and 5. A comprehensive review of diffusion theory, including its sociological and biophysical applications, is given in E.M. ROGERS, *Diffusion of Innovations*, third edition (Free Press, New York, 1983) and L.A. BROWN, *Innovation Diffusion* (Methuen, London, 1982). Its application to medical geography in general and the spread of epidemics in particular is reviewed in A.D. CLIFF *et al.*, *Spatial Diffusion* (Cambridge University Press, Cambridge, 1981), Chapter 2, and in their *Island Epidemics* (Oxford University Press, Oxford, 2000). Applications of diffusion modelling to epidemics is reviewed in P. HAGGETT, *The Geographical Structure of Epidemics* (Clarendon Press, Oxford, 2000).

The diffusion 'progress reports' section of *Progress in Human Geography* (quarterly) will prove helpful in keeping up to date. Although research in spatial diffusion is reported in the standard American and British geographic journals, it is worth paying particular attention to Swedish serials. The *Lund Publications in Geography* (an occasional publication) and *Geografiska Annaler, Series B* (quarterly) are of special significance. For readers with access to the World-Wide Web see also the sites recommended in Appendix B at the end of this book for topics relevant to this chapter.

PART V

Geographic Tensions

CHAPTER 17

Territorial Tensions

■ denotes case studies

Figure 17.1 Impact of political boundaries on the landscape Although the long and straight boundary of the 49th parallel forms a wholly artificial boundary between the United States and Canada established by international treaty in 1818 it has a significant impact on present land use. This LANDSAT satellite image of land-use patterns on the border of the USA and Canada. North is at the top. Photographed in the infrared colour range, thickly vegetated mountains appear very dark, as does vegetation on riverbanks. The straight border between Alberta (Canada, at top) and Montana (USA, below) runs diagonally from centre left to upper right. On the US side, patterns of farmland show intense cultivation. Different shaded fields reveal crops at different stages of growth. In Canada, plantations of forest are seen. Satellite image taken in September. [Source: Earth Satellite Corporation/Science Photo Library.]

> My apple trees will never get across
> And eat the cones under his pines, I tell him
> He only says, 'Good fences make good neighbors.'
>
> ROBERT FROST *Mending Wall* (1914)

So far in this book we have been concerned with ways in which the interactions between people and the environment lead to regional differences around the globe. The view taken has been a peaceful one. But the world we live in, the one which geographers have to describe, is not peaceful. Over all of our recorded history the story is one of conflict between the people occupying these regions – whether at the tribal level in earlier periods, or between superpowers today. Much of this conflict is concerned with disputes over territory.

The word 'territory' comes from the Latin word for land, *terra*. So territory means the land owned by or belonging to a town community or state. Figure 17.1 (p. 508) shows land that belongs to the United States and Canada separated by the inter-state boundary running across the northern Great Plains. Geographers use the term territory in a much more general sense, to indicate an area over which rights of ownership are exercised and which can be delimited or bounded in some way. We use the word 'boundaries' to describe the limits of such territories.

Sometimes ownership of territory is formal and may be legally enforced. A householder may legally own the lot on which his house is built, and a nation-state may legally own its lands. At other times ownership may be unstable, and territories may be precariously held by displays of strength at the borders. Ornithologists studying the robin singing on the garden post at the edge of its territory and sociologists noting the gang member painting obscenities on a wall at the boundary of his turf are both recording displays of territorial ownership.

In this chapter we look first in Section 17.1 at the idea of territorial limits and how they have evolved. This leads to considering the most powerful territorial unit (the *state*) in Section 17.2. From the state, we then look at regional partitions that subdivide states and regional coalitions that bind them together (Section 17.3). Boundaries now extend into the 70 per cent of the globe that is water covered and in Section 17.4 we consider these marine boundaries. Finally, Section 17.5 looks at the implications of territorial division for geographic understanding.

17.1 Interpreting Territorial Limits

Territoriality is not a purely human trait. In this opening section of this chapter, we look at the evidence of this trait in other species and compare the behaviour of the species with the ways in which human beings stake out their 'home range' on the earth's surface.

Limits as Median Lines

In Chapter 1 we drew attention to the way in which groups on the beach spaced themselves out, creating, in effect, local 'family territories'. (See the discussion of interpersonal space, Section 1.2.) At many points in this book we have had to recall that, for all our cultural and technical uniqueness, we humans are still an animal species. To understand the intricate and formal way in which we organize our territory, we need to begin by looking over the species fence at the simpler territorial behaviour of other animals. In doing so we shall need to draw on the findings of *ethology*, that branch of the study of animal behaviour in a natural environment pioneered by European zoologists such as Konrad Lorenz and Niko Tinbergen.

Animal Territories: Biological Evidence Many animal species 'stake out' a specific area of space for their activities (e.g. feeding, breeding, or nesting). Territories will be defended very aggressively against other animals of the same species – sometimes by real fighting, but more often through highly ritualized 'displays'. Detailed maps of bird territories show distinctive spatial patterns (Figure 17.2). Note the distinction between discrete and overlapping territories for both isolated and gregarious species. Gregarious species of birds live in colonies, and the territories are those of the whole colony rather than the individual breeding pair. Figure 17.2(d) shows a special case, seabirds whose colonies may be confined to a few traditional nesting sites on islands and whose territories are overlapping areas of the sea.

While the fact that animals have territories is not in doubt, there is considerable debate over its meaning. Just why do territories occur? Two reasons appear the most likely. First, territories help to regulate population density and thus preserve an ecological balance with food supply. Those animals that cannot secure a territory for themselves are forced to migrate

Figure 17.2 Territories of animal populations The map shows four types of 'home ranges' for territorial animals. (a) Solitary animals with sharply bounded and discrete territories. (b) Solitary animals with somewhat overlapping ranges. Core areas around a nesting site for bird populations will be discrete and defended. (c) Gregarious animals in colonies with slightly overlapping group territories. (d) Gregarious animals with highly overlapping ranges – in this case, seabird populations with island nesting sites. (e) Kittiwake colony on a low island in the Kattegat, Denmark.

[Sources: N.C. Wynne-Edwards, *Animal Dispersion in Relation to Social Behavior.* Copyright © by Oliver & Boyd, Edinburgh. Photograph courtesy of Arthur Christiansen/FLPA.]

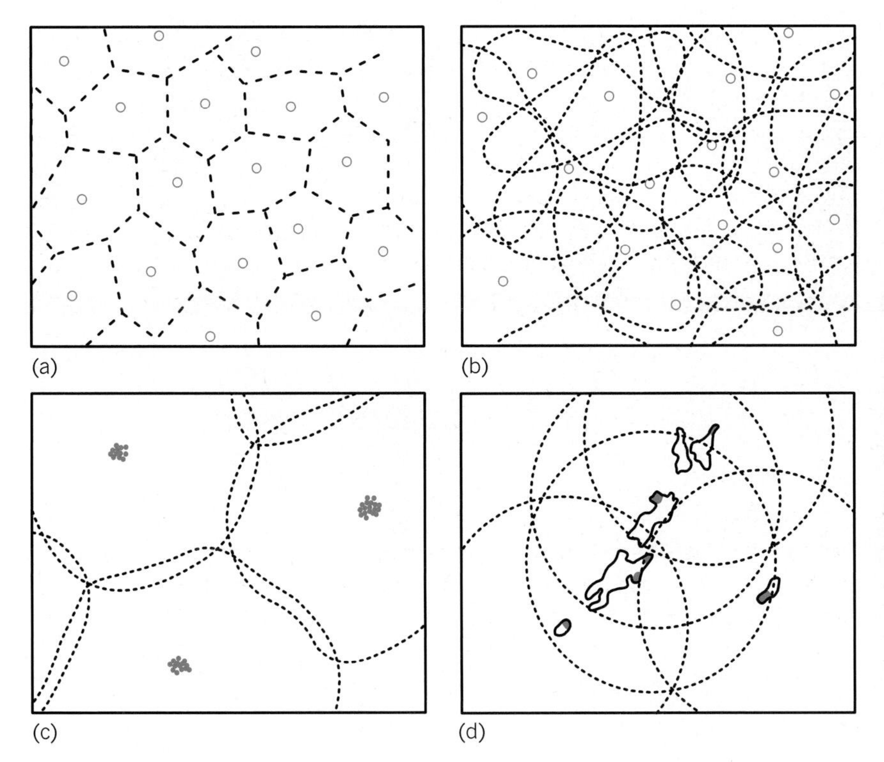

or starve. Second, territories ensure that the strongest members of a population (i.e. those able to obtain and to hold a territory) are the ones that breed and perpetuate the group. Because it forces out the weaker members of a population, territoriality may be an important mechanism in the natural selection process. Note that in the special case of seabird colonies (Figure 17.2(d)), the competition is not for the marine feeding areas but for the few square centimetres of rock ledges in the traditional nesting areas.

Rural Territories: Median-line Models But what have the seabird colonies to do with us? It would be as easy as it would be dangerous to draw facile comparisons between mice in cages and gangs in ghettos, or sororities in suburbia. In approaching the territorial behaviour of the human animal, we choose to begin by looking at some points of similarity with the behaviour of other animals, and then go on to review the differences.

Let us replace our seabirds with a scatter of pioneering farmsteads like those that existed during the colonial period of European overseas expansion. Each farm family clears the virgin forests around the homestead, tilling the most accessible land first and then gradually moving further afield as the family grows and more help is available. If we make simplifying assumptions that ensure that all families are the same size and have the same resources and that the land is everywhere of the same quality, what territorial pattern of farm boundaries will emerge?

Figure 17.3 shows the probable course of events. In the earliest stages each farm can expand in isolation and its territory forms a circle. The untamed forest outside the homestead looks like pastry rolled out on a kitchen table, from which round biscuits have been cut. But as the closer farms expand, their boundaries meet and fence lines are staked halfway between them. In the later stages of farm expansion, only the last vestiges

Figure 17.3 The evolution of farm boundaries The maps show stages in the idealized settlement and colonization of a forest area. The method of drawing boundaries is a follows. First, lines join a farmstead to each adjacent farmstead. Second, each of these interfarm lines is bisected to obtain the median (or midpoint of the line). Third, from this median point a boundary line is drawn at right angles to produce a series of polygons. This type of territorial division is based on two assumptions – that there is a random scatter of farmsteads and that each farmer clears only the land nearest.

[Source: Photo TRIP/G. Hunter.]

Uncleared forest
Farmstead
Farm boundaries

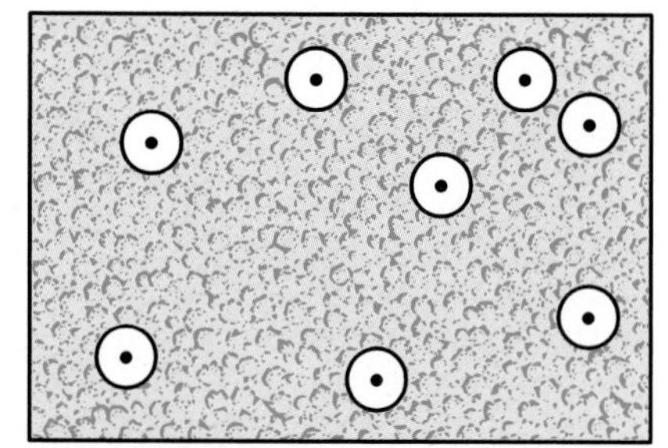

(a) Early phase

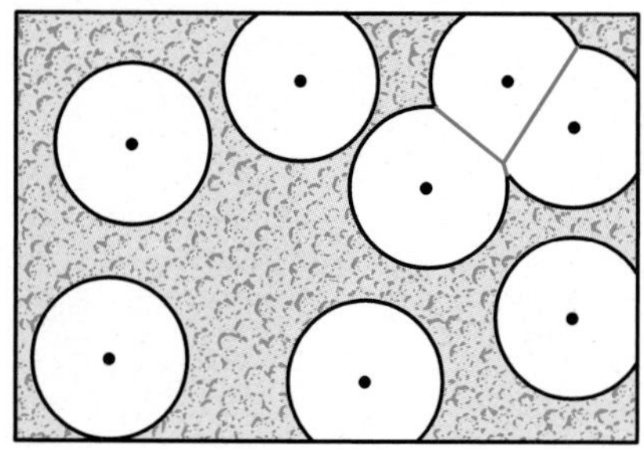

(b) Middle phase

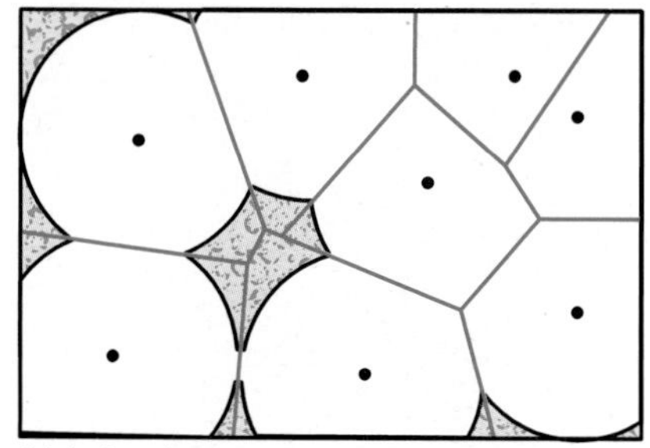

(c) Late phase

(d)

of forest remain and the farm boundaries look like those in Figure 17.3(c). Polygonal territories formed in this way are termed *Dirichlet polygons*, after the German mathematician who studied their geometric properties in the last century. They have the unique quality of containing within them areas that are nearer to the point around which they are constructed (in this case, the farmstead) than to any other points. Essentially, each side of the polygon is a *median line*, drawn at right angles to the line joining two farmsteads at its halfway point.

Clearly our two simple assumptions – that there is a random scatter of farmsteads and each farmer clears the land nearest to him – produce a territorial division of some complexity. It bears some resemblance to the patterns of villages and forests in Gambia, West Africa, that we met in our study of the Thünen model of regional divisions (see Figure 15.15). If the farmsteads had been arranged in a regular triangle lattice, then the farm boundaries would have been hexagonal, just like those of Christaller's complementary regions.

Thus we could regard some human territories simply as median-line divisions of space. Each animal, or gang, or farmer, or nation expands its territory outward until it meets its neighbours. The boundary is fixed at the halfway point between an individual and its neighbours. Unfortunately, this simple geometric view of how territories are formed ignores two significant complications. First, the animals, gangs, and so on may not be uniform but may vary in strength and aggressiveness. Second, the space over which the distances are measured may be not simple and uniform (as in Figure 17.3) but complex and highly differentiated.

Limits as Break-even Lines

Recall that in Figure 17.3(c) the inequalities in farm areas come from the initial irregular location of the farms and the fact that the farmers are assumed to be a homogeneous group. Let us suppose, however, that the farmers were quite different – that some had large families and some had small ones, that some had more resources than others, were more aggressive, and so on. Can geographers incorporate the effect of such differences into a territorial model?

One approach is to use the gravity models we met in Section 13.3. When we have two farmsteads of equal size, we expect the boundary to be exactly halfway between them, that is, at the median point. If they are of different sizes, however, we expect the boundary to be displaced away from the median point in the direction of the smaller farmstead. Just how big the displacement will be can be estimated from a gravity model. (See Box 17.A on 'Estimating boundaries by gravity models'.)

Working with models of competition, regional economists have provided another perspective on how space can be partitioned into territories. Consider the position of the two sellers at centre 1 and centre 2 in Figure 17.4(a). Both produce homogeneous goods and both have to pay the same freight charges, which are proportional to the linear distance the goods are shipped. The costs form an inverted cone about each centre. These are shown as circular contours on the left of the diagram and as V-shaped cross-sections on the right. The boundary (B) is located where the cones intersect; here the cost of goods from both centres is exactly equal.

In the first case the boundary is a straight line, so that a set of centres

Box 17.A Estimating Boundaries by Gravity Models

We can estimate the location of a boundary line between the market area of two centres by using a gravity model like those encountered in Section 13.3. Assume that we have two cities (city 1 and city 2), each with a specific market area (M_1 and M_2) and separated by distance D_{12}. We can estimate a breakpoint (B_2) in units of distance from the second city as:

$$B_2 = \frac{D_{12}}{1 + \sqrt{\frac{M_1}{M_2}}}$$

If we now assume a simple case in which the two cities are 12 km (7.5 miles) apart and both have the same size market ($M_1 = M_2 = 10$), then

$$B_2 = \frac{12}{1 + \sqrt{\frac{10}{10}}} = 6\,\text{km (3.7 miles)}$$

That is, the boundary line occurs halfway between the two equal-sized centres. If we make the two cities unequal in size ($M_1 = 20$; $M_2 = 5$), then

$$B_2 = \frac{12}{1 + \sqrt{\frac{20}{5}}} = 4\,\text{km (2.5 miles)}$$

In this second case the boundary is displaced away from the halfway position in the direction of the smaller centre.

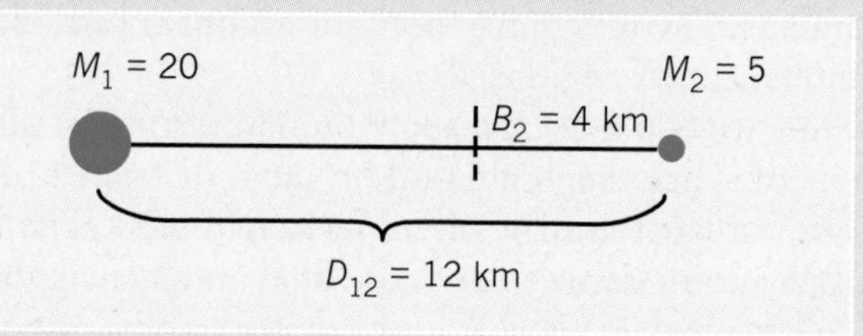

Figure 17.4 Competition for a market territory The diagram shows the hypothetical effect of variations in transport costs and production costs on the location of boundaries between retail centres. There are two centres (1 and 2) producing identical goods and located on a uniform plan where freight charges are proportional to the straight-line distance across the plane. The boundary line between the two market areas forms an *indifference curve* along which prices from both centres are equal and a potential consumer is indifferent to which centre sells him the goods. These curves form hypercircles determined by the equation $P_1 = T_{1x} \times D_{1x} = P_2 + T_{2x} \times D_{2x}$, where P is the market price at the point of production, T_x is the freight rate between a production point and a given consumption point, x, and D_x is the distance. Three different boundaries are shown here. The original market partition model was developed in the 1920s by economist F.A. Fetter and is known as Fetter's model.

[Source: W.H. Richardson, *Regional Economics* (Praeger, New York, 1969), Fig. 2.3, p. 27.]

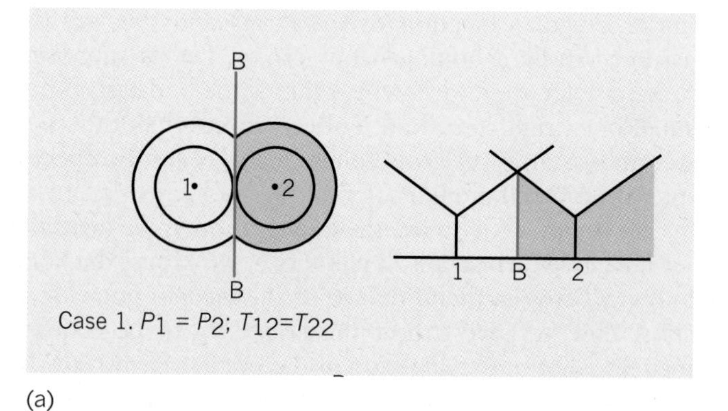

(a)

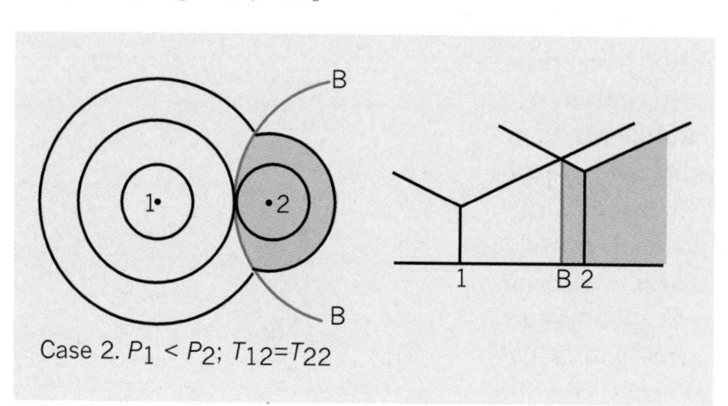

(b)

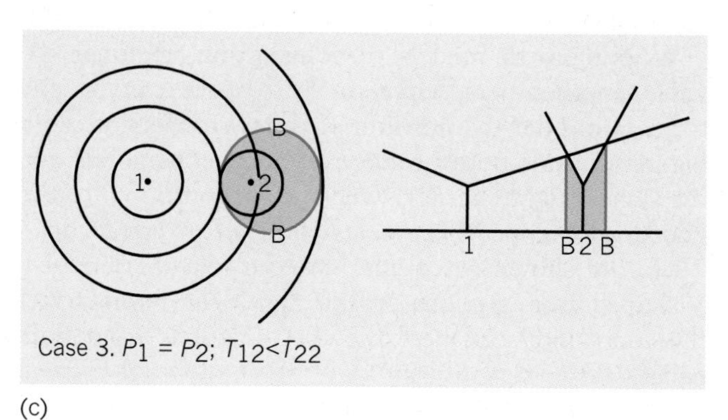

(c)

would form territories in the same way Dirichlet polygons were formed. But we can go beyond the simple polygon model. In the second case, the freight rates remain the same; but the production costs in centre 2 are higher than those in centre 1. The intersection of the two cones is now a curve, and the boundary forms a hyperbola (Figure 17.4(b)). In the third case the position is reversed; that is, there are equal production costs but unequal freight rates. In this last case the boundary forms a circle, and the market area of seller 2 becomes an enclave within the much wider territory of seller 1 (Figure 17.4(c)). The formal structure of the model can incorporate even more complicated variations in both production and transport costs.

So far we have concentrated on certain rough analogies between the spatial form of animal and human territories. What about their purpose? Do human territories also have a role in the control of population density, regulating numbers by limiting the area from which resources are drawn? Certainly in some cultural groups farms cannot be subdivided if the number of potential farmers increases; the legal practice of *primogeniture*, by which the eldest son inherits property and the younger sons leave the estate to make their own fortunes, might, by a long stretch, be seen as ritual ecological behaviour. The second reason for animal territories, survival of the fittest, may also have analogies in the division of market territories by business corporations, with small firms 'going to the wall'.

Limits as Historical Legacies?

What about the ways in which animal and human territories differ? Here we shall emphasize one very major difference – the relative permanence of human boundaries. When we are dealing with unstable or short-lived territories such as the market area of a seller or the domain of a small mammal, we can expect boundaries to be well adjusted to the forces that create them. Thus in Figure 17.4 if seller 1 reduces its production costs more than other sellers, we would expect its territory to increase. But boundaries that are legally defined may persist long after the forces which create them have changed. We have already noted in Section 8.1 the gap between the legal limits of cities and the actual limits of built-up areas or commuting zones. International boundaries also reflect the political balance of forces at the time of their creation. The present boundaries between North Korea and South Korea relate to the military situations that existed in 1953. Different segments of a country's boundaries may date from different periods. In the United States the boundary of Maine with Canada has a 1782 vintage, whereas that of Arizona with Mexico dates from 1853. Figure 17.5 shows the extent to which the boundaries of nation-states in tropical Africa are legacies of European colonial expansion and bear little relation to present cultural and economic realities.

Let us summarize as we conclude the first part of this chapter. We have seen that territoriality is strongly developed in certain animals other than human beings and that it is important in population control and selective breeding. Territories with a similar spatial form arise in human communities, although whether they serve a similar purpose remains a matter of debate. Finally, we have noted the distinctive institutionalized quality of many human territories that sets them firmly aside from their biological counterparts. It is this third type of territory that we shall be concerned with in the remainder of this chapter.

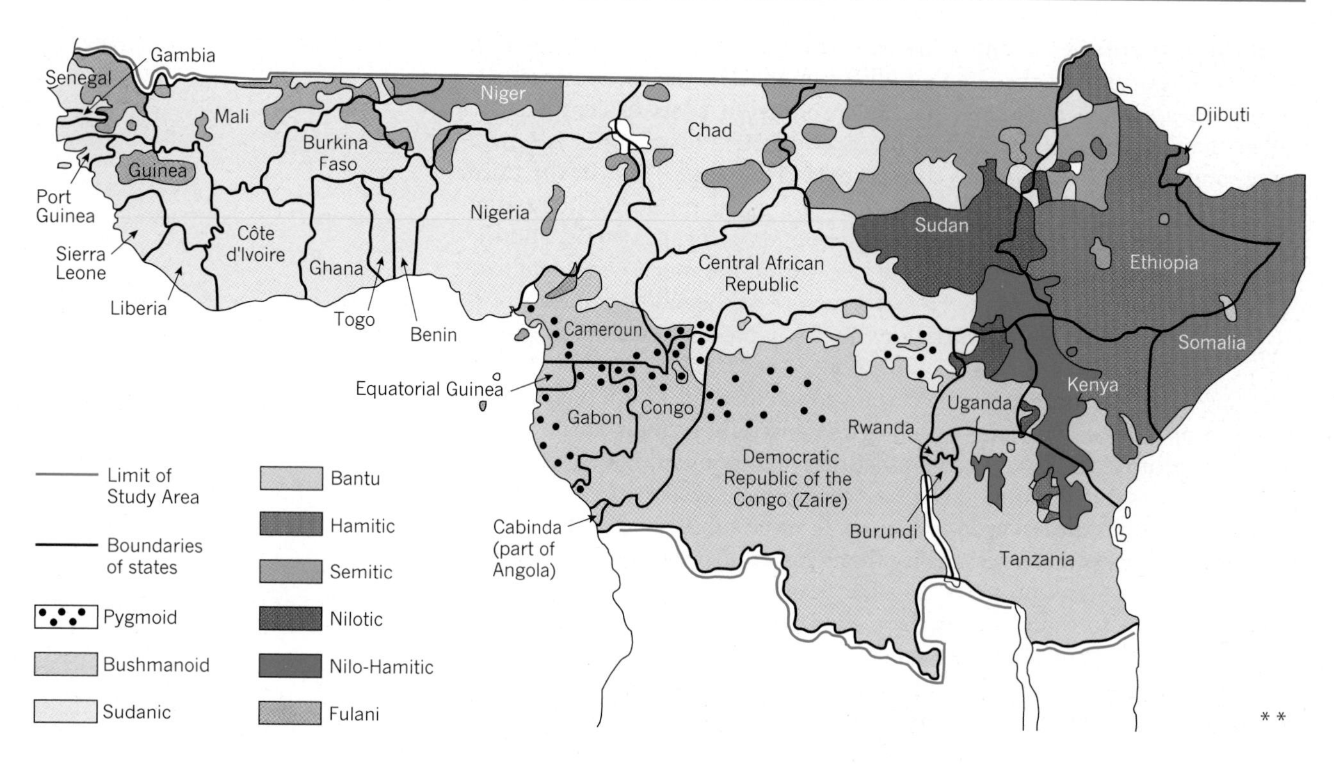

Figure 17.5 State boundaries and ethnic groups The decolonization of tropical Africa in the decade 1955 to 1965 saw the emergence of many new independent states. These states have, however, retained the boundaries of the original British, French, and Belgian, and former German colonies, boundaries which were drawn where opposing forces met during 'the scramble for Africa' in the last two decades of the nineteenth century. As the map shows, the boundaries of the new states run across major ethnic boundaries rather than enclosing homogeneous ethnic groups.

[Source: R.J. Davies, *Tropical Africa: An Atlas for Rural Development* (University of Wales Press, Cardiff, 1970), Plate 10, p. 23.]

17.2 The State as a Region

From the geographer's viewpoint the most evident territorial unit on the world's landscape is the modern nation-state. There are about 200 nation-states in the world today, and they have ranged in size and importance from the former Soviet Union (with one-sixth of the world's land surface) to units of less than a square kilometre. Such *nation-states* cover about 80 per cent of the world's land surface (and the adjacent territorial waters) and divide it into a set of discrete bounded cells (see Figure 17.6). The areas remaining outside national sovereignty are controlled through colonial or joint-trusteeship arrangements.

If we were searching for a single organizational unit in our organization of the world today, there would seem to be simple and persuasive reasons for using the nation-state as this basic unit. Nation-states are the principal 'accounting units' for which comparative statistical data are regularly collected. States are decision-making units in that their central governments can affect the relationships between the population within their borders and the environment. They are clearly defined by boundaries that separate them from their neighbours, and these boundaries form noteworthy discontinuities in the pattern of human organization, sometimes evident in the

(a)

(b)

(c)

Figure 17.6 Boundaries between nation-states While some boundaries are carefully marked and difficult to cross, others are open. (a) The unmarked boundary between Canada and the United States is crossed here by the Alaska highway in the unpopulated northlands of subarctic Canada. (b) The heavily protected 20 km (12 mile) Berlin Wall constructed in 1961 to regulate migration from the German Democratic Republic (East Germany) to the Federal Republic (West Germany). Section between American zone of West Berlin and East Germany on the other side of the wall. 1985. (c) Part of the 2400 km (1500 mile) Great Wall of China built between 200 BC and AD 600 to protect the northern boundaries of the Ch'in and Han empires against nomadic raids. Some idea of the scale of the the Great Wall can be gained from the fact that it is the most readily seen sign of human occupation as viewed by astronauts in outer space.

[Source: Photo (a) from Alaska Stock. (b) Ullstein/Wolfgang Wiese. (c) Mary Evans Picture Library.]

landscape itself. They collect and publish the data we use in building up a picture of the globe.

Despite these advantages there are drawbacks, too. First, there is the difference in size and population (see Table 17.1). How can we usefully compare a country such as Russia with countries such as San Marino or Andorra? Second, large states have immense internal contrasts; in Canada the population forms a thin ribbon of settlement along the southern border – most of the country is virtually unsettled. Third, we cannot always be sure that we are comparing like with like, certainly a basic principle of analysis. In France we find a state where the influence of the central government is rather uniform everywhere. In Indonesia the government has much less control over the outlying, peripheral areas. Fourth, state power is under pressure with the growth of large multinational corporations (see Chapter 19).

The boundaries of states are also a problem. They are often arbitrary geometric lines wholly unrelated to either the natural environment or population characteristics. (Recall Figure 17.5 showing how the boundaries of counties in tropical Africa cut across tribal divisions.) The borders may not enclose contiguous spatial areas. (Consider West and East Pakistan before the creation of Bangladesh in 1971 or the United States and Alaska.) And they may shift violently and abruptly from time to time.

Centrifugal Versus Centripetal Forces

From the geographic point of view we can think of a state as being something like a cell in biology. It has a distinctive shape enclosed by a boundary. It has a nucleus, the capital city. It has internal structure related to differences in population and development. It exports and imports across its boundary. These views of the state in biological terms were current in the last century, but later dropped since they could be used to support the aggressive expansion of a state.

Modern views see the state as a section of the earth's surface whose territorial boundaries are held in place in a state of tension. Figure 17.7 attempts to sketch some of the ideas inherent in the state idea in terms of the kind of human–environment systems discussed throughout this book. It is based on the ideas of Wisconsin geographer Richard Hartshorne who

Table 17.1 The world's largest and smallest countries

Largest countries					**Smallest countries**				
In area	**Million km²**	**Million mile²**	**In population**	**Millions**	**In area**	**km²**	**mile²**	**In population**	**Thousands**
Russia	17.1	6.6	China	1250	Vatican City	0.4	0.15	Vatican City	1
Canada	10.0	3.9	India	966	Monaco	1.5	0.58	Tokelau	2
China	9.6	3.7	United States	272	Nauru	20	8	Niue	2
United States	9.5	3.7	Indonesia	203	San Marino	62	24	Tuvalu	10
Brazil	8.5	3.3	Brazil	164	Lichtenstein	157	61	Nauru	11
Australia	7.7	3.0	Russia	147	Barbados	430	166	San Marino	25
India	3.3	1.3	Pakistan	144	Andorra	453	175	Liechtenstein	31
Argentina	2.8	1.1	Japan	126	Singapore	580	224	Monaco	32

[a]Data refer to the latest estimates for independent countries of the world at the time of going to press.

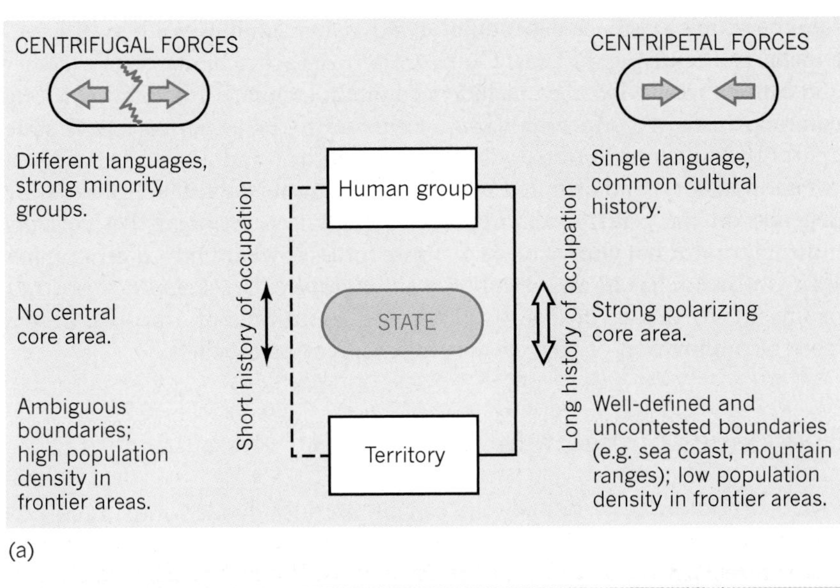

(b)

(c)

Figure 17.7 Centrifugal and centripetal forces in state identity (a) The Hartshorne model of the state held in a dynamic balance between destructive (centrifugal) and constructive (centripetal) forces. Check back on Richard Hartshorne in Box 12.A. (b) Centrigugal forces. The ethnic Kurd population spread across into two Middle Eastern states. Kurdish refugees on the Turkish border with Iraq, 1991. (c) Centripetal forces. Australia's decision in 1901 to replace the individual state capitals (Sydney, Melbourne, Adelaide, Brisbane, Perth and Hobart) by a federal capital located inland in the Australian Capital Territory Canberra is an example of the search for greater state cohesion. The photo shows Australia's old and new parliament buildings at Canberra. The old building in the foreground was opened in 1927; the new one behind it 60 years later.

[Source: (b) Jasper Young/PANOS Pictures. (c) Topham Picturepoint.]

sees the state's existence depending on a dynamic equilibrium between centripetal and centrifugal forces. *Centripetal forces* act to bind a state together and cause it to survive: they include a common language and culture, a long common history, good boundaries. *Centrifugal forces* act to tear a state apart: they include internal divisions in culture and language, a short common history, and disputed boundaries. Continued existence of the state depends on the centripetal forces exceeding the centrifugal. Perhaps the most important but elusive of the binding forces is what Oxford geographer Jean Gottmann has termed a *national iconography* (Greek *eikon* = portrait or image). By this he means the psychological attitude of a people, drawn from a combination of past events and deeply rooted beliefs.

Boundaries of the State

Types of Boundary

Boundaries can be usefully divided into three categories on the basis of when they originated in comparison with settlement (see Figure 17.8).

Subsequent boundaries are those that are drawn after a population has become well established in an area, and the basic map of social and economic differences has been formed. Thus in the case of an old nation-state such as France we may think of the present national boundaries as a rough approximation to the *limites naturalles* of the French nation. The idea of fitting the boundary around an existing ethnic group dominated the thinking of the members of the Versailles Peace Conference after World War I; it played a significant part in drawing the line within the Indian subcontinent between India and Pakistan in 1948; it runs through the demands for a reunited Ireland, yet if we consider the boundaries of most of today's nation-states, we discover that the boundaries represent arbitrary cutoff points. Many of the characteristics that give identity to a nation – language,

Figure 17.8 Three types of boundaries The maps are arranged in a time sequence t_1, t_2, t_3. The three types of boundary are discussed in the text.

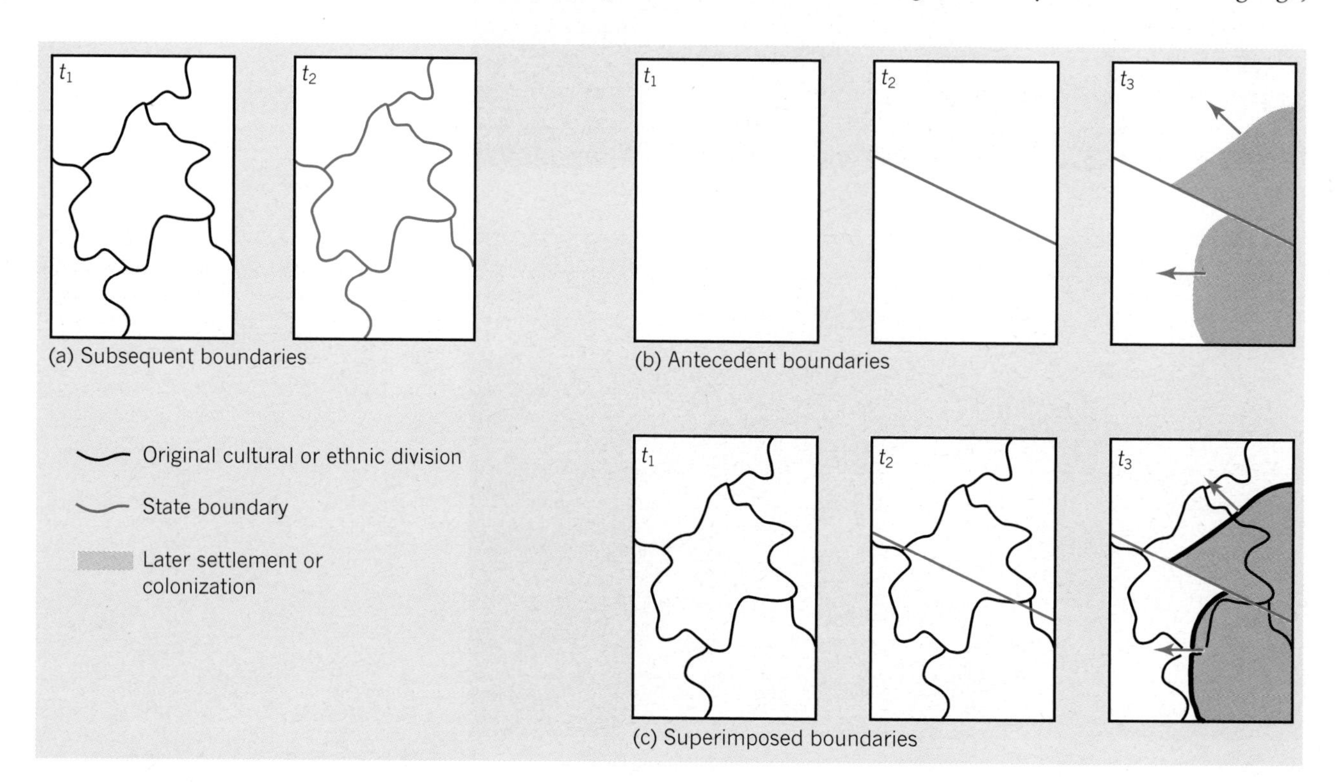

ethnicity, a common history, and cultural traditions – do not end abruptly at its boundaries. Millions of Chinese live outside Asia (let alone in a Chinese state); Israel contains only a small fraction of world Jewry; and so on. Conversely, one very stable national unit (Switzerland) includes four distinct language areas.

By contrast, *antecedent boundaries* precede the close settlement and development of the region they encompass. Groups occupying the area later must acknowledge the existing boundary. The boundary separating the relatively uninhabited land between the United States and Canada, established and modified by treaties between 1782 and 1846, is an example of an antecedent boundary. Antarctica has also been divided up by a cartwheel-like system of meridians running north from the South Pole.

The third type, *superimposed boundaries*, is the converse of antecedent boundaries, in that they are established after an area has been closely settled. This type of boundary normally reflects existing social and economic patterns. The boundary between India and Pakistan, drawn at the division of British India in 1948, is an example of this type of boundary. Many new nation-states in Africa are emerging within a framework of international boundaries laid down by colonial poachers from Europe in an earlier day (Figure 17.5). Nigeria, Tanzania, and Zambia each inherited land areas that make sense in terms of colonial spheres of influence but have little overlap with either environmental or cultural discontinuities.

Drawing Boundary Lines To settle a political dispute with a legal treaty is one thing, to translate that settlement into an identifiable line on the ground is another. The difficulty is most acute on the international level but is also encountered regarding boundaries within the state.

To clarify the problem of dividing land areas, suppose we consider the internal and external boundaries of the conterminous United States. In Figure 17.9 we see that over 80 per cent of the international and interstate boundaries are *geometric*. These geometric boundaries are mostly lines of latitude such as 49°N, which separates the United States from Canada, and north–south meridians such as sections of the Oklahoma–Texas boundary at 100°W. The ease with which these lines can be drawn by standard surveying methods makes them valuable when sparsely settled or empty areas are being divided up.

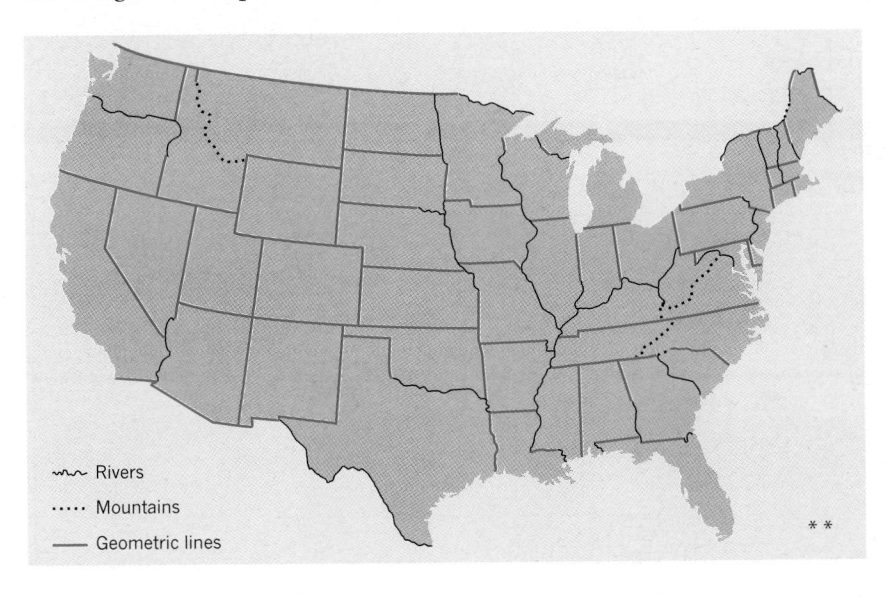

Figure 17.9 Types of boundary lines Different types of lines have been used for the international and interstate boundaries of the conterminous United States. The high proportion of geometric lines typify situations where political subdivision predated close settlement. Note the satellite photo of the Canadian/US border in Figure 17.1.

[Source: N.J.G. Pounds, *Political Geography* (McGraw-Hill, New York, 1963), Fig. 32, p. 89. Copyright © 1963 by McGraw-Hill, Inc. Used with permission of McGraw-Hill Book Company.]

Other geometric boundaries are sometimes used, too. Part of the boundary between Delaware and Pennsylvania is an arc of a circle centred on the town of New Castle. Still other geometric boundaries consist of straight lines between points where locations are known. The 1853 treaty defining the boundary between Arizona and Mexico describes a straight line joining a point 31°21′N and 111°W with another point on the Colorado River. Figure 17.9 also shows some *non-geometric* boundaries; these boundaries generally follow the irregular course of natural surface features. Rivers are the most common natural feature used in drawing boundaries because they are self-evident dividing lines. As we shall see in the next section, they may nonetheless lead to disputes.

Pressure Points Within the State

Why do conflicts arise? An army of historians over the ages has probed the causes of war, and it would be impossible to try to summarize their complex and contradictory findings. We can, however, try to identify those *geographic* considerations that enter the historians' models. Figure 17.10 is an attempt to do so. It shows a hypothetical state (termed here 'Hypothetica') beset by a series of conditions that could give rise to conflicts with its neighbours.

Each condition is indexed by number, and we shall discuss each briefly in turn and give an illustration from the actual record of international conflict. Note that the list we have stays within geographic bounds. The actual source of conflicts goes beyond this to include many non-spatial and non-environmental reasons outside our immediate concern. Figure 17.11 shows examples of conflict drawn from the Middle East. Conflicts that arise in the bordering offshore areas are discussed in the next section.

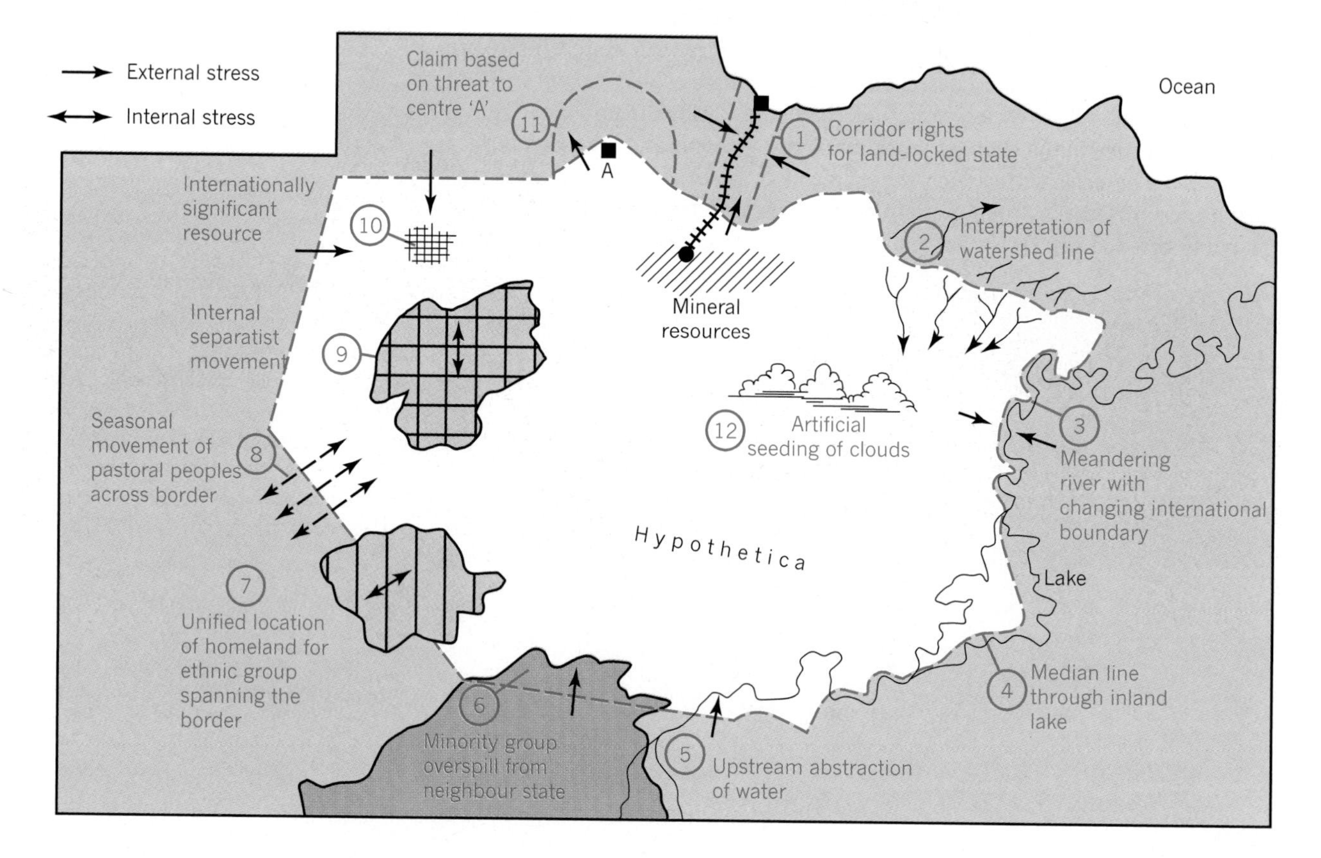

Figure 17.10 Geographic sources of international stress A land-locked country, Hypothetica, with some of the potential trouble spots (1) to (12) identified. Actual examples of the disputes illustrated here are discussed in the text.

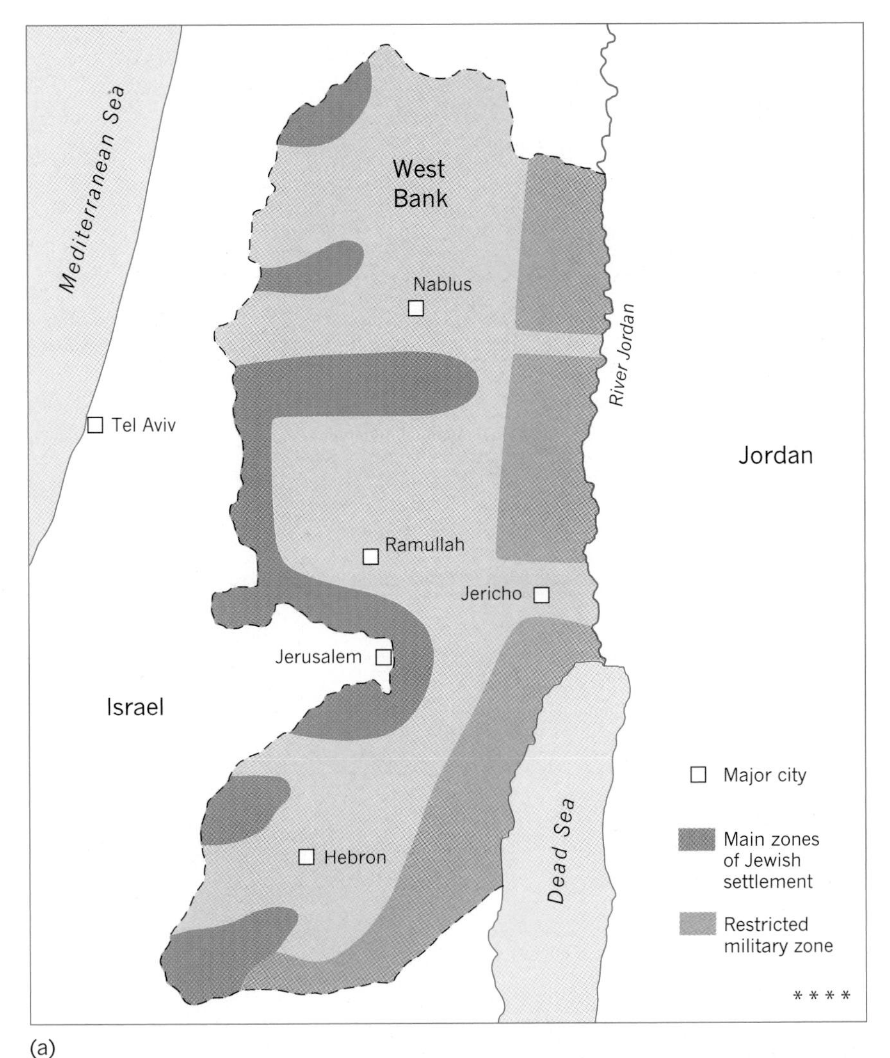

(a)

(b)

(c)

Figure 17.11 Sources of international stress in the Middle East (a) Map of the West Bank within the state of Israel showing the main Jewish settlements within an area largely dominated by Arab occupation. Much of the eastern part of the West Bank is restricted for Israeli military purposes. (b) Historical photograph of Israeli settlers building a new kibbutz in northern Galilee, 1950. (c) Israeli soldiers guarding Jewish settlers on the new settlement of Eshkolot on the West Bank, August 1991.

[Source: (b) MAGNUM/Robert Capa. (c) Popperfoto/Patrick Baz.]

Landlocked States and Corridors The first problem (1) lies in Hypothetica's landlocked position. It has no direct access to the world's oceans and yet needs to ship its raw materials from a tidewater port to its overseas markets. About one-fifth of the world's states are land-locked, and special transit arrangements have to be made for their goods to cross a

(a)

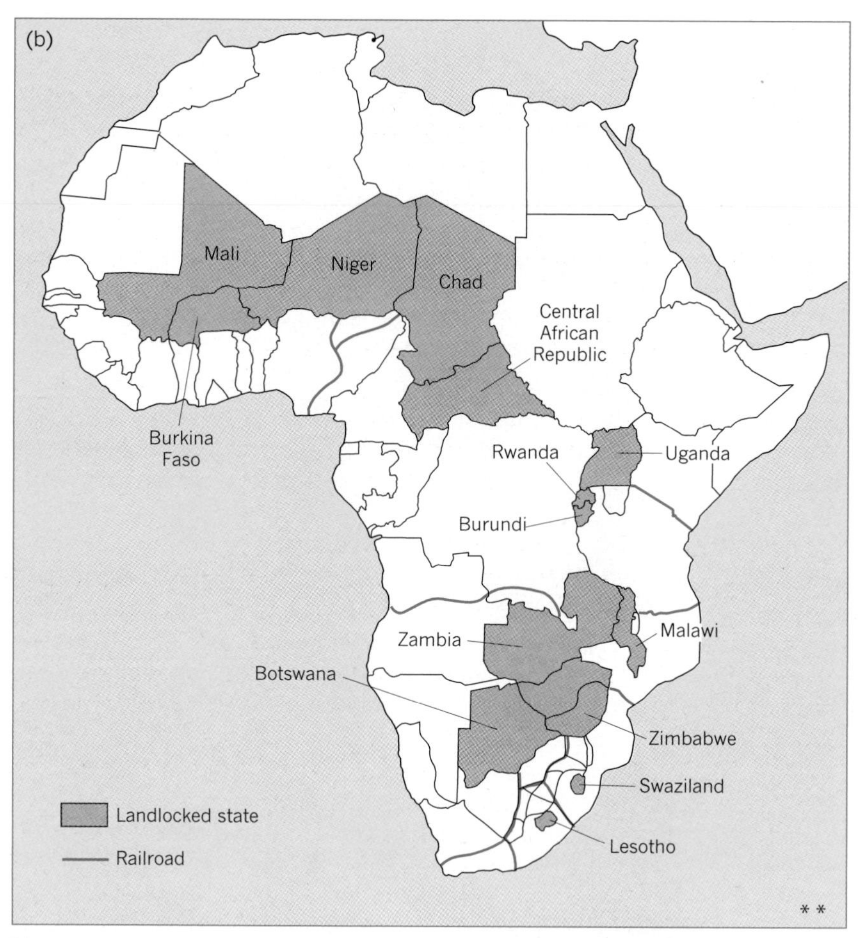

(b)

Figure 17.12 Communication problems for the landlocked state (a) A freight train loaded with tractors makes its way over the Tanzania–Zambia Railway. (b) The 14 landlocked states of Africa and their relation to relevant railroads through neighbouring states.

[Source: R. Muir, *Modern Political Geography* (Macmillan, London, 1975), Fig. 3.4, p. 62. Photo by Colin Garratt.]

foreign country. Figure 17.12 shows the present position of boundaries in Africa. Note the critical importance of the few railroads, and the special transit problems for Zimbabwe and Zambia.

Land corridors have provided one solution. The Polish Danzig Corridor to the Baltic (enshrined in Woodrow Wilson's proposals for European post-war reconstruction in 1918) was described by the French leader Marshall Foch as the 'root of the next war': events in 1939 were to prove him right. It illustrates an attempt by a state in Eastern Europe to secure a narrow strip of land linking its national territory to an adjacent sea. Outside Europe, the Eilat corridor in Israel and the Antofagasta Corridor in Bolivia serve a similar purpose. Note that states may not need to be *completely* land-locked in order to need corridors.

Although most corridors have represented attempts to gain direct access to the sea, some have been aimed at indirect access by way of navigable rivers. Ecuador has a long-standing claim to a strip of northern Peru that gives access to the navigable upper reaches of the Amazon; in 1922 Colombia obtained from Peru the narrow Leticia Corridor, allowing a 120 km (75 mile) frontage on the same river. In general, most landlocked states have secured access to the sea by international conventions that allow the movement of goods across intervening territories without discriminatory tolls or taxes. Switzerland, Austria, and Hungary are European examples of states that use such agreements in their trade.

Conflict over National Boundaries The next four trigger points are all related to rivers and lakes. Problem (2) relates to the use of a watershed boundary in defining an international dividing line. The classic case of this type of dispute occurred between Chile and Argentina. The original 1871 boundary assumed a line joining the mountain peaks would give a watershed, whereas erosion had shifted it well to the east. A mutually acceptable boundary was finally worked out in 1966 and turned on detailed definitions of a master stream. (See the discussion on stream order in Figure 4.8.)

Mountain barriers frequently pose such problems in deciding on the exact location of a divide or watershed. The 1782 treaty defining the northern boundary of Maine in terms of highlands divided rivers running to the Atlantic Ocean from those running north to the St. Lawrence River. But because of the complexity of the hydrology of the area, the boundary line was not finally established until the Webster–Ashburton Treaty was made in 1843.

Hypothetica also conflicts with her neighbours on two other river systems. The detailed demarcation of rivers (3) causes problems for two reasons: the course of the lower portion of a river changes continually, and the river has width and may have several channels. Figure 17.13 illustrates the rapidity of changes in the course of the Rio Grande separating Texas from Mexico. Clearly, the boundaries fixed at one time may produce anomalies a decade later. Boundaries may follow the navigable channel of waterways or, in the case of wide bodies of water (4), some median line between the two shores. A line equidistant from both shores was constructed through Lake Erie by the International Waterways Commission for the Canada–United States border.

The fifth problem relates to river flow coming downstream into Hypothetica from the territory of another state (5). Water may be drawn off or ponded back upstream, with severe consequences for irrigation places downstream. Tension between Israel and its Arab neighbours over the waters of the river Jordan and by India and Pakistan over the Indus waters relate to this hydrologic type of problem, while Welsh nationalists resent the export of water to English cities. So far most disputes have been settled by negotiation. For example, agreement between Canada and the United States has been reached to limit pollution of the Red River before it flows north across the forty-ninth parallel.

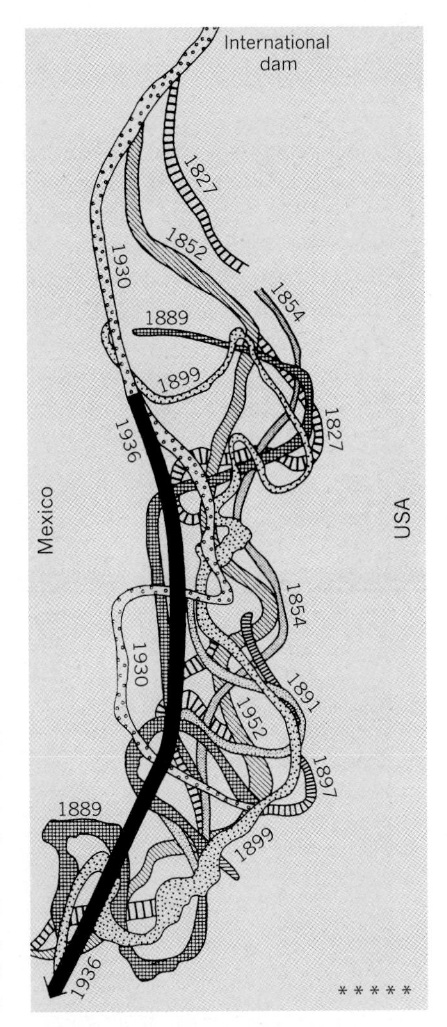

Figure 17.13 River boundaries
Local changes may occur in international boundaries because of geomorphic changes. The map shows shifts in the main channel of the Rio Grande near El Paso from 1827 to 1936. The international boundary between the United States and Mexico runs along the main channel and has therefore changed as the river changed.

[Source: S.W. Boggs, *International Boundaries* (Columbia University Press, New York, 1940).]

Conflict over Minority Groups Perhaps the most explosive and emotive problems in international tension relate to the location of minority groups. Hypothetica exhibits four such problems. Trigger (6) relates to a linguistic minority group lying along its borders. German occupation of the Sudeten parts of Czechoslovakia in 1938 had as its object the union of those German-speaking areas with the rest of Germany. Pressure to redraw the 1926 line dividing Northern Ireland from the rest of Eire to place Roman Catholic populations in the southern state exists today.

Trigger (7) relates to a distinct ethnic group posed halfway between Hypothetica and a neighbouring state. Both the West African states of Ghana and Togo have put forward claims to create a unified homeland for the Ewe people, a distinct African tribe arbitrarily divided by a colonial line dividing British and French spheres of influence in the late-nineteenth century. Where pastoralists cross boundaries in search of seasonal pastures for their flocks (8), small-scale friction can occur. Small changes in the French–Italian boundary in the Alpes Maritimes have eased this problem,

Box 17.B Bowman's New World

Isaiah Bowman (born Waterloo, Canada, 1878; died Baltimore, USA, 1950) was the outstanding American political geographer of the last century. A Harvard graduate, he studied under W.M. Davis (see Box 2.C) and was director of the American Geographical Society (1915–35) and president of Johns Hopkins University (1935–48).

Bowman is probably best known for *The New World* (1921); one of the most influential books on ***political geography***, it described the reshaped world emerging from the boundaries of World War I. Bowman based his analysis of the world's tension areas on experience as leader of the American team accompanying President Woodrow Wilson to the Versailles Peace Conference of 1919. In his *New World* he reviewed, continent by continent, the stress areas of a world just recovering from the horrors of the 1914–18 war. Current problems such as those of the Balkans or the Middle East can be seen in the context of deep-seated difficulties, and some conflicts can be seen to have a spatial persistence that is both intriguing in retrospect and sobering in prospect. Bowman's government service continued with President Franklin Roosevelt and in World War II he was adviser to the US delegation at Dumbarton Oaks (1944) and the conference at San Francisco (1945) that set up the United Nations.

Bowman's role as a government adviser and analyst of ***geopolitics*** represents only a part of his work. Three other significant areas can be distinguished. First, his contribution to the understanding of the regional geography of South America based on extensive field research before World War I and summarized in two monographs, *The Andes of Southern Peru* (1916) and *Desert Trails of Atacama* (1924). Second, as director of the American Geographical Society (AGS) he launched in 1920 the project for mapping the whole of Latin America at the scale of 1:1,000,000. When completed in 1945 the 107-sheet mapping programme had cost over $400,000 but had already resolved boundary disputes in poorly demarcated areas. Third, in the mid-1920s Bowman initiated another AGS programme for a study of the world's pioneer settlement fringes and wrote a key volume on the theme, *The Pioneer Fringe* (1931). As a university president, Bowman found little time for field research but he continued his geographical work on the applied lines set down in *Geography in Relation to Social Sciences* (1934) which gave a measured view of the discipline's problems and potential. His last major task for the US government was to draw up feasibility plans for resettling refugees of World War II in western pioneer fringe areas.

An authoritative biography of Bowman is given in G.J. Martin, *The Life and Thought of Isaiah Bowman* (Archon Books, Hamden, 1980).

THE
NEW WORLD
PROBLEMS IN
POLITICAL GEOGRAPHY
By
Isaiah Bowman, Ph.D.
Director of the American Geographical Society of New York
ILLUSTRATED WITH 215 MAPS AND WITH 65 ENGRAVINGS FROM PHOTOGRAPHS

Yonkers-on-Hudson, New York
WORLD BOOK COMPANY
1921

but it remains unresolved between the states of Somalia and Ethiopia in East Africa.

Distinct minorities within a state (9) form an internal rather than an international problem. The claim for separate identity for the Indian tribal areas within the United States or Canada, the separatist movements of

Quebec within Canada, Scotland within the United Kingdom, or the Basques within Spain all illustrate different levels of internal tension. Such problems become international when they lead to the creation of a breakaway state with its own international status or when external powers are attracted into the arena on behalf of the separatist area.

Other Sources of Conflict The last three conflict triggers in our, by now, trouble-ridden Hypothetica are miscellaneous. Condition (10) relates to where our state contains a resource of international significance. This might be possession of a strategic resource (say, uranium or chrome) in critically short supply or a distinct cultural resource. Thus Jerusalem, currently part of Israel, contains in the Old City sites of central religious significance to the faiths of Jews, Moslems, and Christians (Roman Catholic, Greek Orthodox, and Protestant). From the eleventh-century Crusades onward, states have fought over such holy places. Point (11) is the reverse of (10). In this case, our state believes it has a location so vital to its own identity and survival that it is prepared to go to exceptional lengths to protect it. Thus the Soviet Russian claim in 1939 to Finnish territory in the neighbourhood of Russia's Leningrad, on the grounds that it needed to give its second-ranking city a greater measure of security from possible artillery attack.

A final trigger point (12) is included to remind ourselves that the sources of international conflict are evolving. So far legal disputes over cloud seeding for rain-making have involved only a few Great Plains communities in the United States, but – should technology progress – the consequences could spread upward to the international level.

So far in this section we have tried to identify general classes of trigger points. One of the most outstanding historical surveys of specific problems areas around the world was provided by an American geographer, Isaiah Bowman. (See Box 17.B on Bowman's new world.)

Partition Problems at Other Spatial Levels 17.3

Our third area of concern in this chapter presents a shift of scale: from the international political scene to local, grass-roots politics. For each state is made up of a hierarchy of local areas through which central governments keep in touch with grass-roots needs. Such areas include the state, county, and township in the United States; the province, county, and township in Canada; the county, district, and parish in England; the department, commune, and arrondissement in France; and so on.

Most countries, whether western or non-western, developed or underdeveloped, have intermediate and local authorities with which they share the central authority of the nation-state. The existing sequence is frequently a historical patchwork, and from time to time central governments decide to reform the hierarchy along more relevant lines. The Napoleonic reform of the ancient French system of provinces, the sequence of Soviet reforms since the early 1920s, Salazar's reforms of the Portuguese system in the 1930s, the Swedish reforms of the 1950s, and the British reorganization of the 1970s are each part of a recurring cycle of revision, a cycle in which geographers are increasingly being called upon to help redraw maps.

Local Levels: The Search for Equity

On the lowest spatial level geographers are concerned with how boundaries of local districts are arranged. We noted in Box 12.C the case of creating balanced school regions and Figure 17.14 gives another American example. Here we give another example, electoral regions.

Gerrymanders: The Problem In 1812 Governor Elbridge Gerry of Massachusetts established a curiously bow-shaped electoral district north of Boston in order to favour his own party. In his memory the term *gerrymander* is used to describe any method of arranging electoral districts in such a way that one political party can elect more representatives than they could if the district boundaries were fairly drawn. Since 1812 the number of examples of gerrymandering has grown rapidly, and Figure 17.15 presents two of the most extreme ones.

If we wish to rig boundaries in an unfair way, we have two possible strategies. The first is to *contain* our opponents, to lump all the electors in the

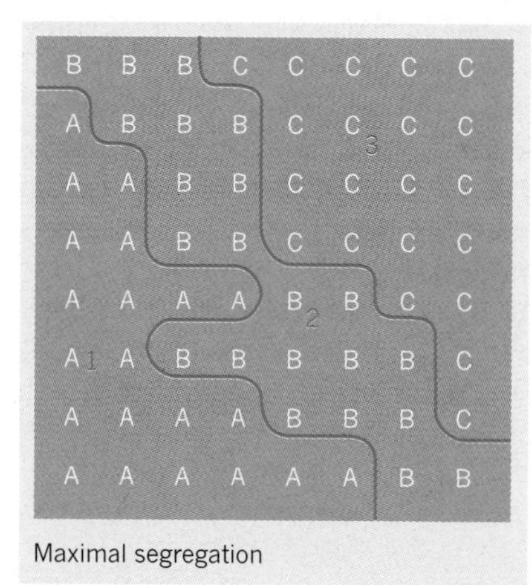

(a)

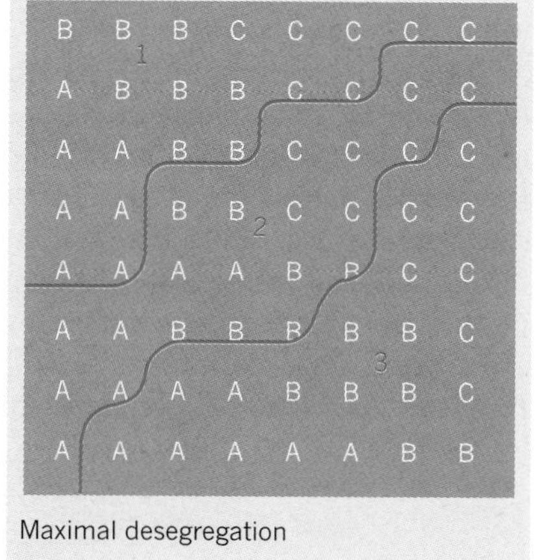

(b)

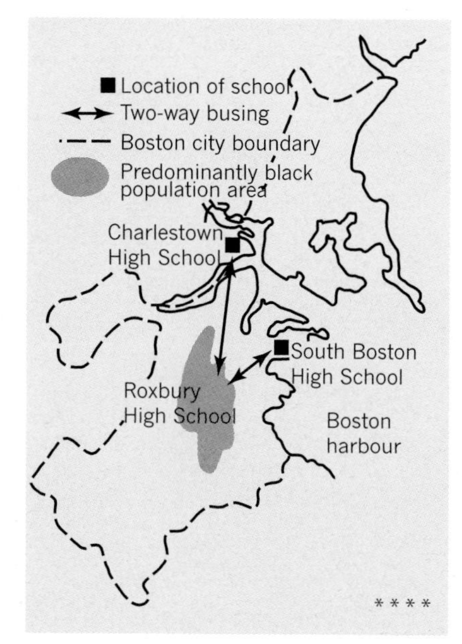

(c)

(d)

Figure 17.14 Alternative region-building strategies In (a) the boundaries create the greatest possible segregation of three hypothetical groups (A, B, and C). In (b) each area has a fair percentage of people from each group. Choices between these two types of zoning often face administrators drawing up school district lines in ethnically diverse areas. In (c) and (d) the problem faced by a school desegregation ruling in Boston is illustrated. (c) Bussing children to school in the United States, June 1963. (d) Protest in St. Louis, Missouri, against the bussing of black children to predominantly white schools. Some 60 buses moved 5000 children each day at this time.

[Source: J.D. Lord, *Spatial Perspectives on School Desegregation and Busing* (Association of American Geographers, Washington, D.C., 1977), Fig. 13, p. 15. Photo by Popperfoto.]

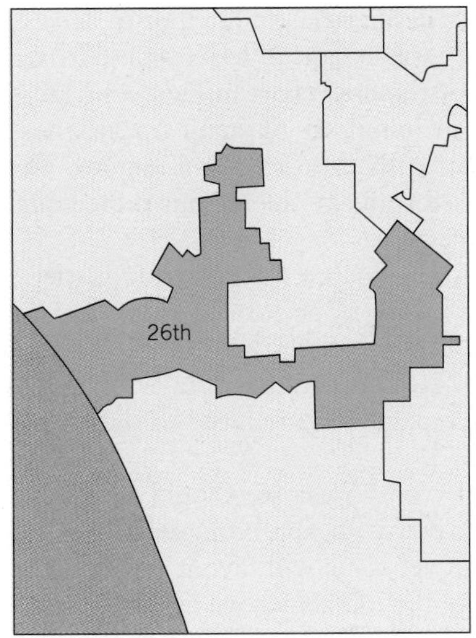

(a) Los Angeles, Calif. * * * *

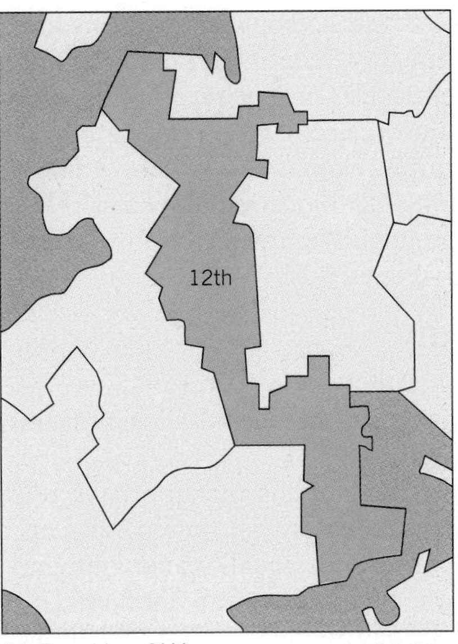

(b) Brooklyn, N.Y. * * * *

Figure 17.15 Gerrymandering in the United States (a) California's 26th electoral district in 1960. (b) Brooklyn's 12th electoral district in 1960. In both cases the boundaries were drawn to give a major electoral advantage to a political party. Following Supreme Court decisions such local gerrymanders have now had to be eliminated.

opposing party within a single boundary so that a representative receives an unnecessarily large majority of votes. The second possible strategy is to disperse our opponents, to split the electors of an opposing party between many districts so that nowhere are they numerous enough to elect one of their own representatives.

Within the United States the location of equitable political boundaries has been a subject of greater interest since the 1962 *Baker vs Carr* decision of the United States Supreme Court. In this case the court ruled that legislative seats must be apportioned and states divided so the number of inhabitants per legislator in one district is approximately equal to the number of inhabitants per legislator in another district. The case was triggered by a group of voters in Tennessee who claimed that their votes were debased by inequities in the state's electoral districts. Here one vote in rural Moore County was equal to nineteen votes in urban Hamilton County (containing the city of Chattanooga). Parallel situations existed in other American states. In Vermont the most populous district had 987 times more voters than the least populous district. Many states had to redraw their district boundaries to comply with the Supreme Court decision.

Gerrymanders: A Solution? It is one thing to identify the problem of unfair electoral boundaries but quite another to discover a solution. Just how can we arrive at a 'fair' spatial arrangement? The boundaries chosen for voting districts are likely to reflect a balance among three considerations. The first is that the voting population in each district should be *equal*. The ideal concept of 'one person – one vote' must be modified to offer approximate equality because uncontrolled factors such as migration and natural change are continuously modifying the voting population within a district. The second consideration is that the voting district should be *contiguous*. Districts are usually in one conterminous unit and ideally should be compact within the sense that communication among the various parts of the district is easy. The third and more debatable consideration is that of *homogeneity*, or balance. Some may think it is desirable for the parts of a district to have a common social, political, or economic characteristic. Others may argue that the district should be a balanced mix representing a wide range of communities rather than a single one.

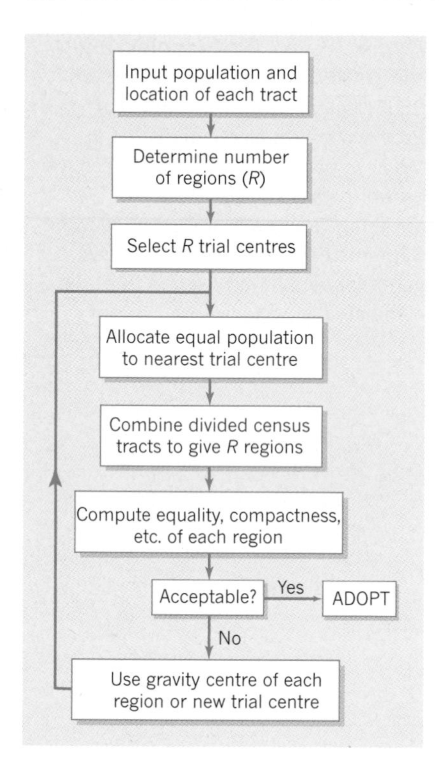

Figure 17.16 Electoral districting This flow chart shows the main elements in a computer program for allocated population in census tracts to a set of compact electoral regions with a similar number of electors in each area. 'Gravity centre' is at the central point within a region which is most accessible for all its population.

Several computer programs have been designed in an attempt to achieve equality, contiguity, and homogeneity (or balance) by a non-partisan method. One such method is outlined by the flow chart in Figure 17.16. It shows electoral ***districting*** by taking an initial set of small tracts whose voting population is known and allocating them to electoral regions. The original allocation is successively adjusted to make the regions more equal in population while keeping them compact.

Figure 17.17 presents the results of using such a method to redistrict a sample part of New Jersey. Six districts were to be formed from over 50 tracts (Figure 17.17(a)). In the first allocation the largest district, with a 12,500-person electorate, was 5 per cent greater than the average for all districts. In the second allocation this discrepancy was reduced to only 1 per cent.

In evaluating this approach to the electoral areas we should recall that numerous district boundaries can be drawn with approximately the same concern for equality and contiguity, and yet some will favour one party at one time rather than another. To resolve the imbalance we need to inspect various boundary maps and adopt the one most in accord with the prevailing concept of political equity.

Regional Levels: The Search for Efficiency

Central governments of states are interested in making the subdivisions within the country as representative and efficient as possible. But what new subdivisions should be substituted for the existing ones? On the basis of what criteria should new maps be drawn? How do we weigh a central government's demand for a few large efficient units against a local outcry for a small unit tailored to the needs of a particular community? The solution to these problems involves not just deciding on new units but arranging them into levels, or tiers. With a *single-tier system*, each local area has direct access to the central government; with a *multiple-tier system*, the local areas must work through an intervening bureaucracy on one or more regional levels.

Figure 17.17 The search for improved electoral districts These maps of Sussex County, New Jersey, in the United States, show the first stages in the allocation of census tracts to electoral regions, using the kind of program shown in Figure 17.16. The electoral districts in maps (b) and (c) are made up by combining the original tracts (a). Figures indicate the population in each electoral district in thousands.

[Source: J.B. Weaver and S.W. Hess, *Yale Law Journal*, **73** (1963), Fig. 1.]

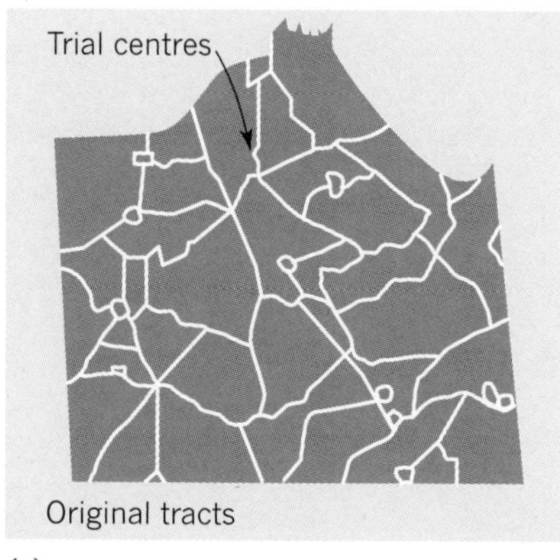

(a)

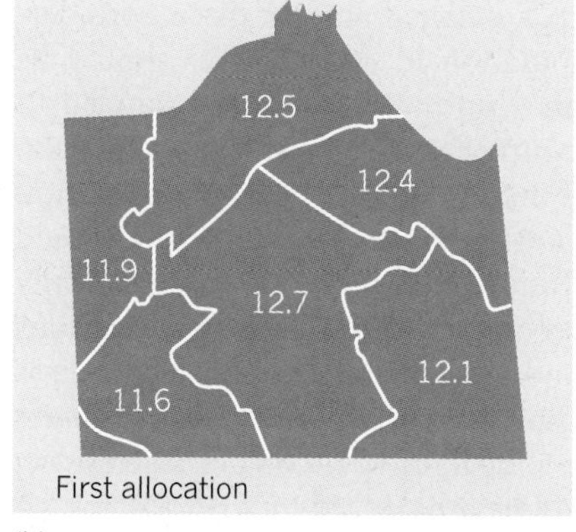

(b)

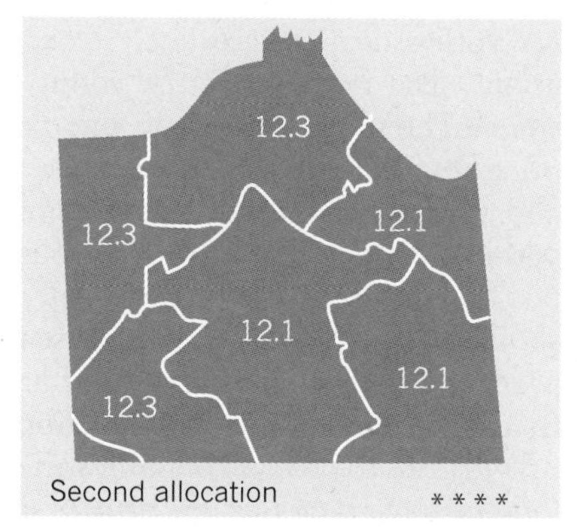

(c)

The British Problem We can examine the situation in a specific context by presenting a historical case of boundary reform in Britain. Until 1974 the country was split into over 1000 local districts, each with a variety of powers and each directly responsible to the central government in the performance of certain administrative functions. The system had been only slightly altered since the nineteenth century, so its lack of relation to the distribution of population and the needs of efficient management was becoming evident.

A Royal Commission was established to prepare recommendations. It proposed three criteria for the new system. First, it wanted the new units to be large enough to provide local government services at a low cost. 'Large enough' depended on the service provided. While 500,000 was the minimum population for an efficient police force, the threshold for education services was 300,000. Lower figures were given by local health and welfare authorities (200,000 people) and by child-care services (250,000 people).

A second criterion for the new distribution was that the new units should be cohesive enough to reflect local community interests and to allow a local voice in political issues. In practice, cohesion and self-containment are measured by journey-to-work patterns, the range of public transport, newspaper circulation, and the organizational pattern of professional, government, and business organizations. This approach yields urban regions connected to rural areas by ties between workplaces and residences.

The third criterion for the delimitation of new units was the present pattern. Whenever possible, existing units and their boundaries, even on the county level, were used as building blocks in order to retain common interests and traditional loyalties, to preserve skill and momentum of existing local authorities, and to minimize changeover problems.

The British Solution Given the three basic criteria described above, we might assume that it would be relatively simple to find a common spatial solution. As geographers would have predicted, this is not the case. The Royal Commission could not reach an agreement and proposed alternative systems. The majority proposal was to divide the country into a few rather large local government areas (61 for England outside London itself) responsible for all local government services. In contrast, the minority proposed a two-tier system with 35 upper-level and 148 lower-level city regions. In this case the upper-level authorities would look after planning and transportation, while the lower-level authorities would control services like education, social welfare, and housing.

We can appreciate the differences between the two proposals by looking at some maps of southwest England (Figure 17.18). The first area shown (Figure 17.18(a)) is similar in size and population to the state of Massachusetts. The existing 12 units vary considerably in size and area. If the majority proposal were adopted, a slightly enlarged southwestern province would be replaced by eight new areas (Figure 17.18(b)), larger and more uniform in size and with less extreme differences in local revenue-raising ability. If the minority proposal were adopted, a two-tier system, set within a redrawn southwestern province (Figure 17.18(c)) would result.

Despite their differences, we can see that certain recurrent geographic features dominate both the traditional map and the two proposed revisions. First, we can identify eight core areas: five of these are largely rural and centre on an existing county town; the remaining three have metropolitan centres in the leading cities of Bristol, Plymouth, and Bournemouth. Second, there are certain persistent boundaries that recur in all three maps and may represent rather fundamental discontinuities in the socio-economic

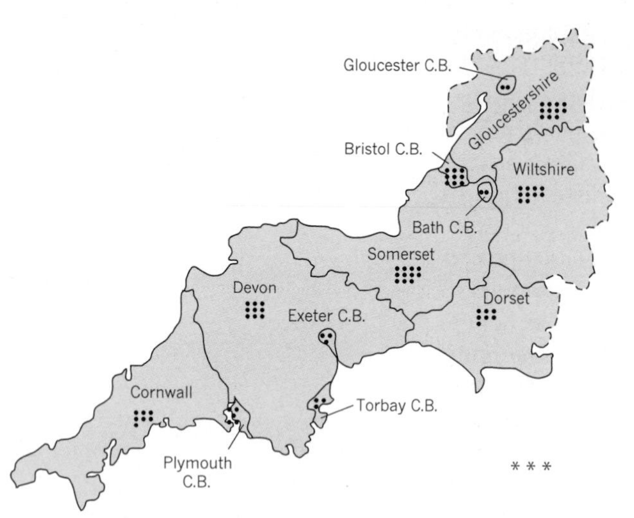

(a) The southwest economic planning region with existing county and county–borough (C.B.) boundaries

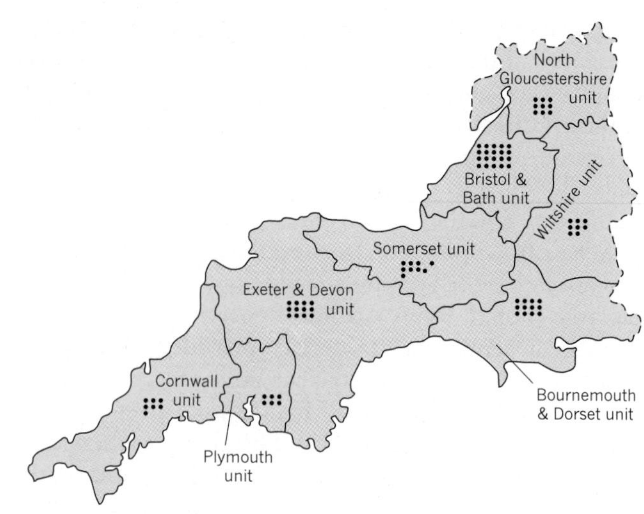

(b) The Redcliffe–Maud proposals for the southwest province, with unitary areas

(c) Senior proposals for the southwest province, with two-tier areas

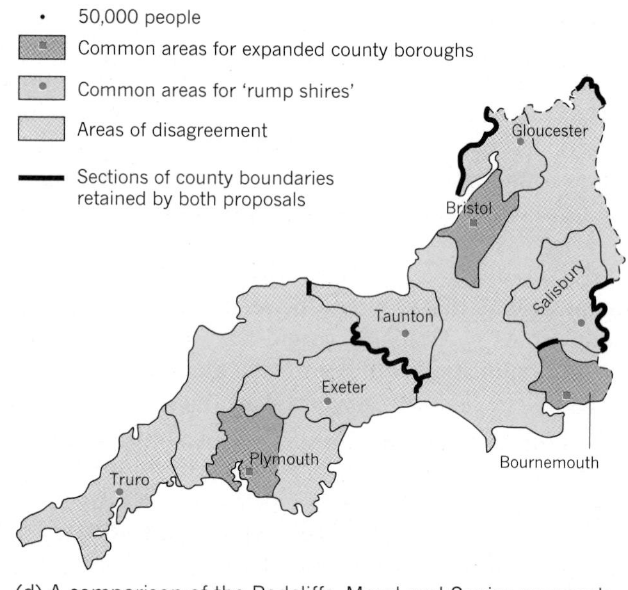

(d) A comparison of the Redcliffe–Maud and Senior proposals

Figure 17.18 Reform of administrative areas The maps show proposed reform of boundaries in southwest England. The first three maps show alternative proposals, and the fourth, the area of agreement and disagreement. 'Rump shires' are the core areas of the old counties or shires (a) left undivided by both sets of proposals (b) and (c).

[Source: P. Haggett, *Geographical Magazine*, **52** (1969), p. 215. Reproduced by courtesy of *The Geographical Magazine*, London.]

structuring of the area. By recognizing recurrent features in proposed regional systems, we can isolate the broad features in any set of solutions and narrow the areas of search. As part of the case of southwest England the main disputes over boundaries occur in three critical zones, as shown in Figure 17.18(d).

The reformed boundaries introduced in the 1970s proved unpopular and in the 1990s further changes were introduced. The old 'shire' boundaries, many of which had existed since Saxon times, proved their resilience.

International Levels: The Search for Common Interest

So far we have looked at the state and its subdivisions. But states may themselves be building blocks in supra-state territories or coalitions. The word coalition comes from two Latin words meaning to grow up together. We look first at the various forms of coalitions. What brings the countries together? In particular what benefits come from removing or reducing boundaries? Last, we consider enforced coalitions in the form of empires.

Types of Coalition Self-interest brings individuals together into groups: the gang in a city, or a trade union in an industrial plant. Likewise a state may join together with other states to pursue some common gain. Figure 17.19 shows some of the groupings at the present time. How can we make sense of this bewildering array?

One approach is to try to identify the purpose of such associations. Let us take the case of the Organization of Petroleum Exporting Countries (OPEC)). This group was formed in 1960 by five leading oil-exporting states. Its purpose essentially was to bring pressure on the major oil companies to raise their prices to the consumer countries and give back a greater share of profits to the producer countries. OPEC's membership now includes the producers of half the world's oil, and it has been strikingly successful in gaining its economic objectives.

A number of the coalitions in Figure 17.19 are, like OPEC, essentially *economic* in function. Thus the European Union (EU), the North American Free Trade Area (NAFTA), and the Association of South East Asian Nations (ASEAN) are all regional trading blocks. The Organization for Economic Cooperation and Development (OECD) is similarly a club of the world's richest countries, formed to foster improved world trade. They are paralleled by groupings for other purposes. For example the countries of the North Atlantic Treaty Organization (NATO) form a coalition for *defence* purposes. Defensive alliances have probably existed at all levels of human organization, from the tribe to super-state as far back as historical records go. It is one of the oldest and most persuasive reasons for union. Third, there are *political* associations of countries wishing to strengthen a common political purpose between member states. The Arab League countries and the Organization of African Unity (OAU) illustrate this purpose.

This threefold division into economic, defensive, and political groups is a useful simplification. Not all organizations fit into the scheme, however. The United Nations is a global association of over 140 states and has economic and policing functions as well as a primary political role. The British Commonwealth is the legacy of a once-powerful economic block, retaining surprising vitality as a cultural grouping with some political aspects. If we plot the states that make up the organizations shown in Figure 17.19, then some overlap in membership is shown up. For example, Nigeria is a member of both OPEC and OAU, as well as of the British Commonwealth. Other countries, such as Switzerland or Finland, go to some lengths to avoid membership of defensive or political blocks.

Figure 17.19 Coalition of states

Geographers are interested in these block arrangements because they modify the dominance of the single state. For example, the growing centralization of function in the countries of the EU means that spatially significant decisions (e.g. over resource use or regional policy) are made in the community headquarters at Brussels, rather than in state capitals such as Berlin, London, or Rome.

Enforced Coalition: Spatial Imperialism Coalitions are not only voluntary. One state may be coerced into political and economic associations with another by conquest or colonization.

In geographic terms, important distinctions can be made between a state's expansion into (a) adjacent or non-adjacent territories and (b) occupied or unoccupied space. Occupied space is where another state already exists; in contrast, unoccupied space is not organized into political units at the state level and may be largely empty of population. Let us take some examples. Peripheral expansion into an already occupied area is illustrated by the absorption by the German state of the Austrian state in 1930. Expansion of Csarist Russia into the adjacent but sparsely populated lands of central Asia is an example of the second case. Non-adjacent expansion by the European Empire included occupation of the Indian peninsula (occupied space) and much of Australasia (largely unoccupied space).

Imperialism is geographically significant in that different colonial powers have left different regional imprints. For example, among the European imperial powers there were consistent differences in the attitudes taken to intermarriage of the colonists with non-European populations of the occupied lands. The racial cohesion among Brazil's 165 million people today stands in contrast to the apartheid that for so long split South Africa's 45 million people. This contrast has part of its roots in differences between Portuguese and Dutch colonial policy three centuries ago.

Imperialism has attracted considerable interest from Marxist writers because of its role in world economic development. Lenin, writing in 1916, saw imperialism as one of the reasons for the persistence of the capitalist economic system in Western Europe – an economic system whose demise Karl Marx had predicted a half-century earlier. Overseas colonies were seen by Lenin to be excellent outlets for capital investment by industrial countries, providing not only sources of cheap raw materials but also captive markets for manufactured goods. Profits from imperialism would allow higher wages to be paid to workers in the imperial homelands (so reducing the likelihood of revolution), but at the expense of lower wages for colonial workers (so increasing the likelihood of revolution). Thus according to Lenin's thesis the impact of imperialism was to spatially transfer the class struggle from a within-state basis to the international level along the *north–south divide*: i.e. developed western countries (the 'haves') versus developing Third World countries (the 'have nots').

Lenin's thesis was not the only model of a global conflict between coalitions. A confrontation between a land-base empire dominating the world's central Eurasian landmass and a sea empire controlling peninsular extremities together with the Americas, Australasia, and Africa was considered by an English geographer, Halford Mackinder, in the early years of this century. Mackinder's recognition of the strategic significance of Eastern Europe resulted in his 1904 dictum: 'Who rules East Europe commands the Heartland, who rules the Heartland commands the World Island, who commands the World Island commands the World' (see Figure 17.20).

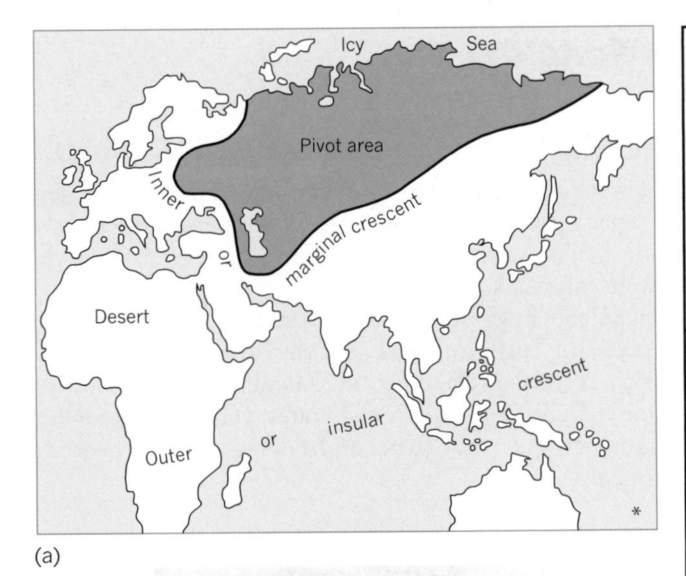

(a)

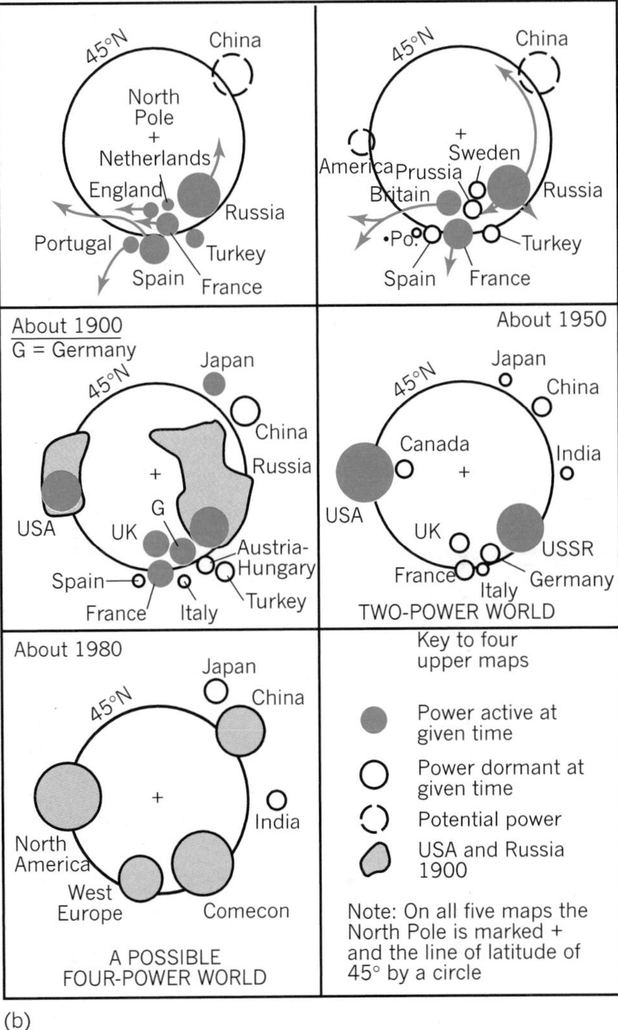

(b)

(c)

Figure 17.20 Mackinder's heartland model (a) English political geographer Sir Halford Mackinder saw control of the 'pivot' area of Russia as being vital to the control of the 'world island' (Asia, Europe, and Africa) and hence of the world. (b) The shifting of political power in the world since the sixteenth century. Comecon is a general alliance of Soviet Russia and the countries of Eastern Europe.

[Source: J.P. Cole *Geography of World Affairs*, 4th edn (Penguin Books, Harmondsworth, 1972), Fig. 4.7, p. 109. © J.P. Cole, 1959, 1963, 1964, 1965, 1972. Photograph courtesy of the Royal Geographical Society.]

Mackinder's heartland model included much of European Russia, Siberia, and Soviet Central Asia, together with what is now northern Iraq and Iran. Mackinder's thinking was adopted by a few German political geographers such as Karl Haushofer; how big a role it played in Hitler's global strategy for World War II is uncertain.

To sum up: coalitions of states have come together for the purposes of economic union, defence, and political association. By reducing the importance of the international boundaries between members of a coalition, considerable economic benefits can be gained, though arguably at the expense of non-member countries. Historically, not all coalitions have been voluntary. Empires represent enforced coalitions in which a single state has dominated other areas. Imperialism has left a geographically significant legacy in terms of population distribution and economic development and is still evolving in different forms. Historian Wallerstein has tried to systematize these developments into a *world system* model.

17.4 Dividing the World's Oceans

Seventy per cent of the world's surface is covered by water. For most of our history the high seas were either a barrier to expansion or, for the last few centuries, corridor space for intercontinental movements. Despite the importance of fishing banks, whaling resources, and strategic islands the oceans themselves were rarely a source of dispute.

Now all this is changing. England and Iceland fought 'cod wars' over fishing grounds; Greece and Turkey quarrel over the ownership of parts of the potentially oil-rich seabed of the Aegean; Canada extends pollution control over areas not recognized by the United States. Suddenly ownership of ocean resources has become a hot issue, and finding ways of peaceful partition becomes urgent.

Territorial Status of Water-covered Areas

Geographers usually divide the ocean into three zones in terms of territorial rights (see Figure 17.21). First there are rights over the immediate *offshore areas* around a country. Second, there are claims for mineral rights in the *continental shelf* around a country. The continental shelf is an area of very smooth, gently sloping ocean floor which fringes all continents. Its limit is conventionally given as a depth of 100 fathoms (183 m, 600 ft) and its width varies from only a few kilometres to 350 km (about 200 miles). Third, there are the *ocean floors* themselves. Ocean floors make up over 90 per cent of the world's ocean-covered areas and vary greatly in depth and smoothness.

Offshore Areas Figure 17.21 shows the conventional legal divisions of water-covered areas in a simplified manner. The starting point for deter-

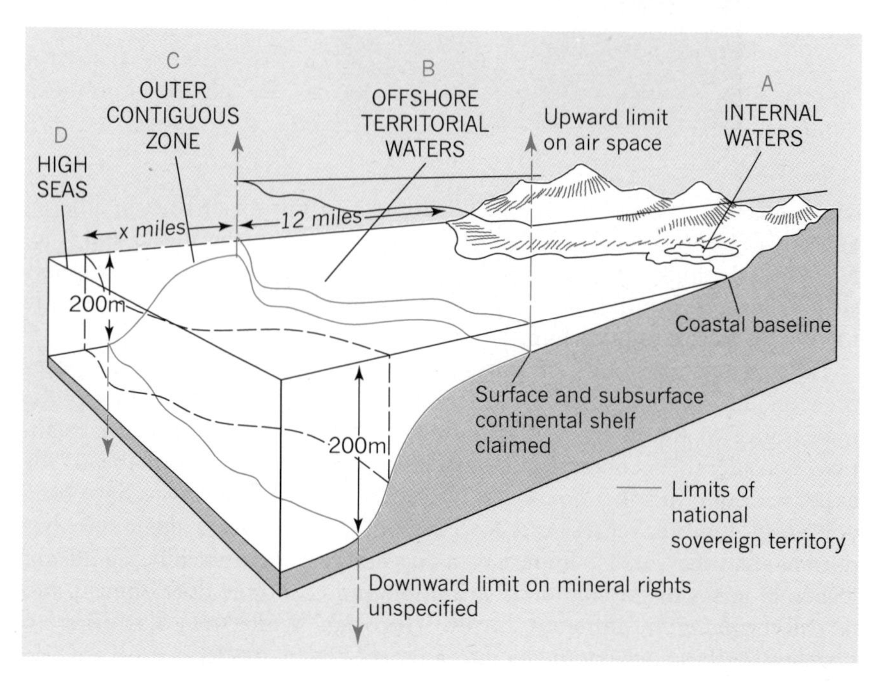

Figure 17.21 Dividing the offshore zone The extension of a state's sovereign territory outward from the coastline. Note the division of the water into four territorial zones – the internal waters, the offshore territorial waters, the outer contiguous zone, and the high seas.

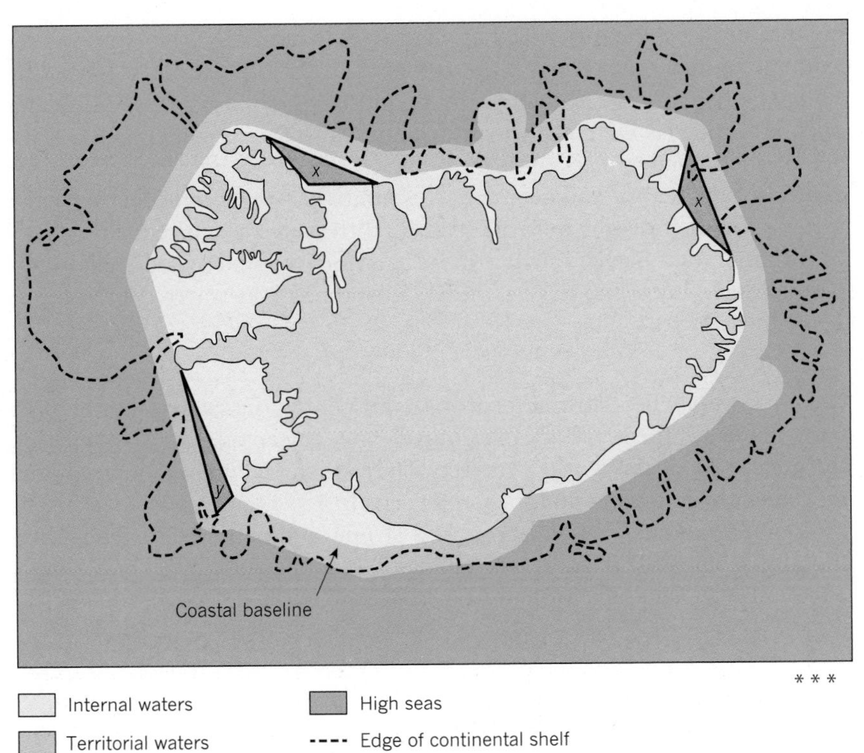

Figure 17.22 Offshore limits This map of Iceland shows the effect of changes in baselines on the limits of internal waters. The 12 mile (19.3 km) limit shown has subsequently been extended to 200 miles (320 km). This unilateral extension by Iceland has caused disputes with Western European countries (notably Great Britain and the former West Germany) who used to fish on the continental shelf outside the 12 mile limit.

[Source: L.M. Alexander, *Offshore Geography of Northwestern Europe* (Rand McNally, Skokie, Ill., and Murray, London, 1963), Map 12, p. 109.]

* * *

mining a state's control over adjacent waters is the *coastal baseline*. All water landward of this baseline is termed *internal waters* and is legally treated as part of the land surface of the state. This generally runs along the coastline itself but may slip across indentations (such as bays and estuaries) and include offshore islands. Iceland (Figure 17.22) provides a good example of this. Note how the coastal baseline sometimes hugs the coast but more frequently runs some miles out from it. In practice, therefore, there is considerable disagreement as to how the baseline shall be drawn. For example, Canada draws a baseline to include the whole of Hudson Bay as internal waters.

Seaward of the coastal baseline lies a country's *offshore territorial waters*. How far offshore such waters should stretch remains unresolved in international law. Early claims of offshore waters were for the limited areas that could be controlled. These ad hoc 'cannon shot' distances have led to a diversity of claims. A survey of the more than 100 maritime states in the late 1970s showed a huge range from Australia, which is bounded by territorial waters only 5.6 km (3.5 miles) in breadth, to El Salvador, which claims 370 km (230 miles). The Philippines claim as territorial waters all the immense sea areas between the islands of the archipelago. A preponderance of states claim a 19 km (12 miles) limit. It was not until 1958 that the United Nations brought together 86 states for the first Law of the Sea Conference. Only slow moves toward the standardization of territorial claims have been made since that date.

Continental Shelves Beyond the limit of territorial waters, two further claims may be pushed. First, special rights may be claimed in a further offshore zone, termed the *outer contiguous zone*. Since the distance limit is nowhere defined in international law, this is marked as width x in Figure 17.21. The main use of the outer zone is in the control of customs, protection of fishing grounds, national defence, and pollution control. An example of such outer zones is the 100 mile (160 km) wide zone claimed by Canada off its Arctic seaboard for pollution protection. A second claim outside territorial waters is to surface and subsurface rights on the continental shelf. Under the Continental Shelf Convention of 1958 claims for control of the seabed and geological resources below (but *not* the overlying waters) for shelf depths out to 200 m (650 ft) were recognized. The physical configuration of these shelf areas is extremely complex, and exploration permits have been granted by the US government in water depths up to 1500 m (5000 ft).

Ocean Floors Beyond the territorial and contiguous waters lie the *high seas*. These are the areas left over outside any current claims and make up the great part of the world's surface. They are areas for free international movement of shipping and for fishing fleets. In 1973 the French government broke international law and provoked strong international condemnation (notably from Australia and New Zealand) by imposing a 116 km (72 mile) ban on foreign shipping around Mururoa Atoll in the Pacific during nuclear tests.

Conflicting Offshore Claims

So far we have assumed each state can push out its claims seaward in isolation. But what happens if it has neighbours also claiming the same sea area? Who then owns it? (See Box 17.C on 'The South China Sea'.)

The question first became important where rivers and lakes run between countries. There the principle adopted was for each state to extend its sovereignty at an equal rate until it met the lake territory of adjacent and opposite states. Thus the method used for Lake Erie was an adaptation of the median line principle used in devising Dirichlet polygons. (See the polygons in Figure 17.3.)

Although simple in theory, the application of this method to seas and oceans is complicated by ambiguous definitions of a country's coastline. Consider again the map of Iceland's territorial waters in Figure 17.22, and note how the irregular shape of the coastline is approximated by a polygon. This polygon, the coastal baseline, is the starting line for all territorial claims. But the baseline can be drawn in different ways (see the two areas marked x), by adopting a small offshore island or rock as one of the bases for drawing the polygon (see area y). Whether or not a bay is regarded as part of a country's internal waters, or an offshore island or sandbar is regarded as national territory, changes the baseline from which the median line with a neighbouring state is determined. It was for this reason that it was not until 1977 that France and England could decide where the median line down the English channel really ran (Figure 17.23).

Shelf Division International boundaries in the continental shelf areas around Western Europe were of little more than academic importance until the middle 1960s. Then natural gas and oil discoveries highlighted the advantages that countries such as the United Kingdom and the Netherlands had gained from what had previously been only paper titles. Figure 17.23(c) shows the effect of median-line territorial boundaries in the North Sea on ownership of natural gas and oil sources. Natural gas is found

Box 17.C The South China Sea

The South China Sea is a major area with political dispute over sea boundaries (see map (a)). Its area of 1.7 million km^2 (0.65 million square miles) is criss-crossed with major navigation routes, bending via Singapore between Japan and the Middle East, and underlain with important oil fields.

Four factors complicate the process of drawing maritime boundaries in this area:

1. This semi-enclosed sea is surrounded by nine peripheral states, which show wide variations in size, political power, and wealth. Those states are Indonesia, Singapore, Malaysia, Thailand, Cambodia, Vietnam, China, the Philippines, and Brunei.
2. Unilateral and overlapping maritime claims have been made by some countries, especially in the Gulf of Thailand and the southern portions of the sea. These disputed areas are shaded on the map.
3. The presence of an extensive group of coral islands (the Spratly Islands; see map (b)) near the centre of the sea which are claimed by up to six countries.
4. The prospect for successful negotiation is reduced by the huge potential of future oil revenues from the hydrocarbon resources and the poor political relations that exist between some of the participant countries (e.g. China and Vietnam).

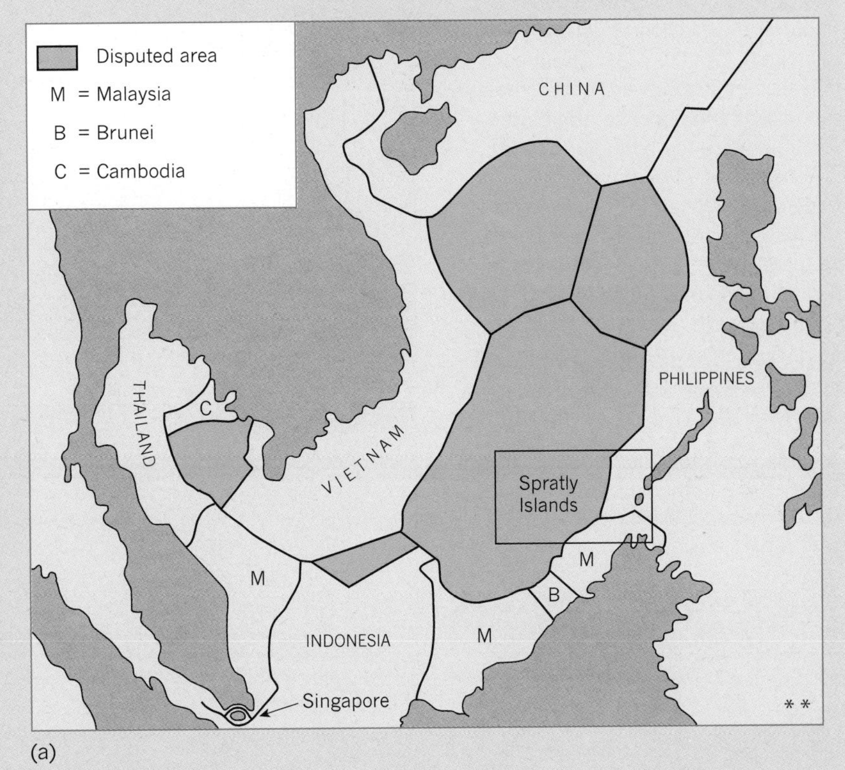

(a)

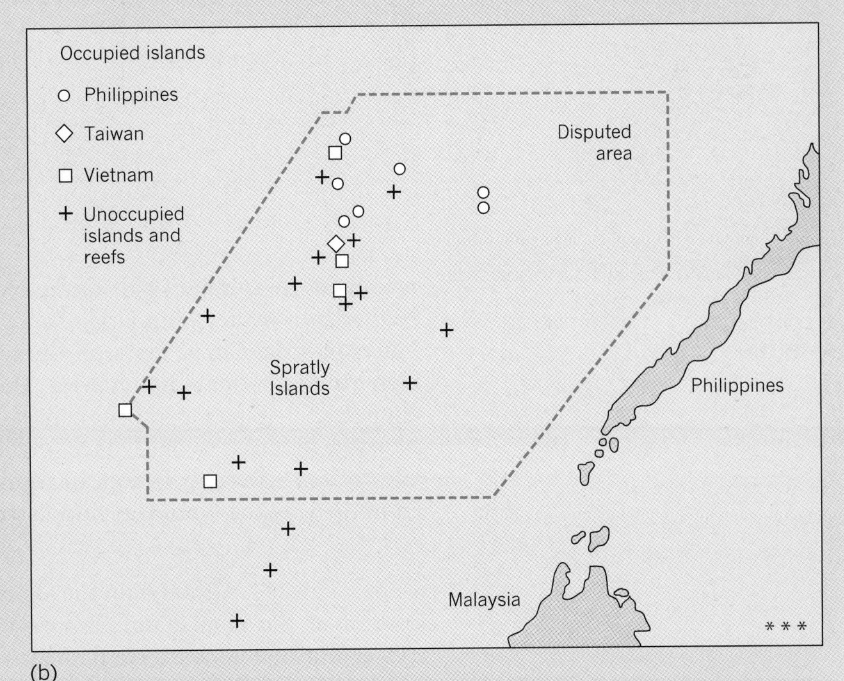

(b)

Despite these tensions, important progress has been made under the ***ASEAN*** umbrella in agreeing navigation routes for the critically important oil tanker routes that cross the South China Sea and Straits of Malacca. See the discussion in J.R.V. Prescott, *The Maritime Boundaries of the World* (Methuen, London, 1985), pp. 209–233.

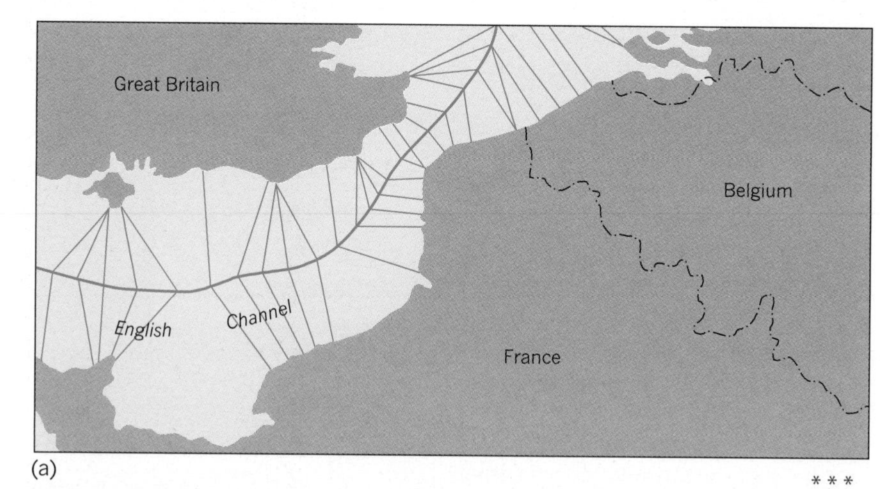

(a)

* * *

(b)

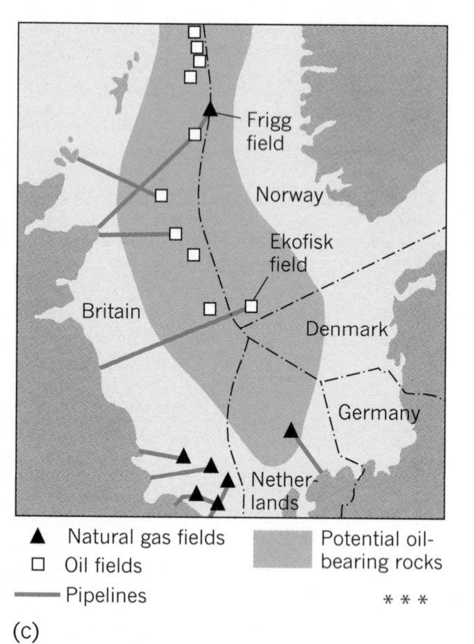

* * *

(c)

Figure 17.23 The principle of median boundaries applied to offshore areas (a) International boundaries in the English Channel. (b) An outward extension of national sovereignty into the oceans, with each nation extending outward until it meets its adjacent and opposite neighbours. Note that while the boundaries in (a) have been ratified by the countries bordering the English Channel and are internationally recognized, those in (b) are merely hypothetical. (c) Impact of median boundaries on share of oil and gas in the North Sea.

[Source: (a) L.M. Alexander, *Offshore Geography of Northwestern Europe* (Rand McNally, Stokie, Ill., and Murray, London, 1966), Fig. 4, p. 58. (b) T.F. Christy and H. Herfindahl, *Hypothetical Division of the Sea Floor*, a map published by the Law of the Sea Institute, Washington, D.C., 1968.]

mainly in the southern half of the North Sea; the earliest strike was in the Netherlands' waters, but larger sources were later found in the British area. Later oil strikes have come in the northern half of the North Sea, many close to international boundaries. The Ekofisk oil field is Norwegian but is connected by pipeline to Britain because a deep trough in the seabed off the Norwegian coast would have made pipeline construction very costly. Clearly some western European countries have benefited more than others from the way the boundary lines have been drawn.

Ocean Floors Although in the short term the greatest interest attaches to control of the immediate offshore waters, the long-term implications of ownership of the continental shelves and ocean floors may eventually be more important. For example, if we allow the present principle of the median line (Figure 17.23(a)) to be extended to the open oceans (Figure 17.23(b)), we find some extraordinary distributions of territory. Portugal, by virtue of its ownership of the Azores and Cape Verde Islands, would stand to gain the largest share of the North Atlantic seabed.

The mineral riches of the ocean floors are only just beginning to be realized. For whereas offshore drilling for oil and gas on the continental shelf had been established from the 1920s, recent decades are seeing an increased interest in the surface deposits of the deep oceans. Minerals such as manganese may accumulate on the ocean flooor in the form of rich nodules. Preliminary estimates for the Pacific suggest that such nodules may contain enough manganese to meet world demand for centuries to come and may be accumulating faster than they could be used. But few hard facts are known and the exploration of the ocean's mineral resources (including dissolved minerals in seawater) promises to be one of the great frontier areas of future research.

If the ocean floors are going to be very important mineral resources in the future, should they be divided in the way shown in Figure 17.23? What about the problem of a landlocked country with no sea boundary and thus no base for a claim? Suggestions at recent United Nations conferences state that the ocean floors out beyond the continental shelves should be divided up into a draughtboard of exploration areas. Then sections of the ocean could be allocated, perhaps on a random basis, between all countries. Although this is only a tentative idea, it will probably be made again as economic and technological problems in using ocean resources are solved. Ownership problems, if not solved by agreement, will be resolved on a power basis.

Mineral resources are not the only reason for heightened international interest in the world's oceans. We have already seen in Chapter 4 the food productivity of the oceans and their sensitivity to pollution. There is therefore increasing concern by United Nations agencies over adequate control of pollution by oil, mineral elements like mercury, or nuclear waste, and the excessive reduction of the population of certain marine animals such as whales.

To sum up: the world's oceans represent the last major frontier area left on this planet. With proper international control their vast potential resources, both renewable and non-renewable, could be harnessed for immense good. But, as the Falkland Islands dispute showed, without international control the oceans may be as great a scene of conflict in the future as were the land areas in the past.

The Impact of Divisions 17.5

In this final section we consider two aspects of territorial division. First, the filtering role of boundaries; second, the calculus of conflict.

Boundaries as Filters

One of the benefits expected to flow from boundary removal and economic linkage between countries is improved trade. This comes from lowering the taxes (tariffs) on goods that cross the international boundary plus the enlarged market. But to understand the importance of removing or reducing a boundary we must first see how it blocks flows in the first place.

Geographers have developed a general graphic model of the blocking action of boundaries (Figure 17.24). When the barrier is a political one,

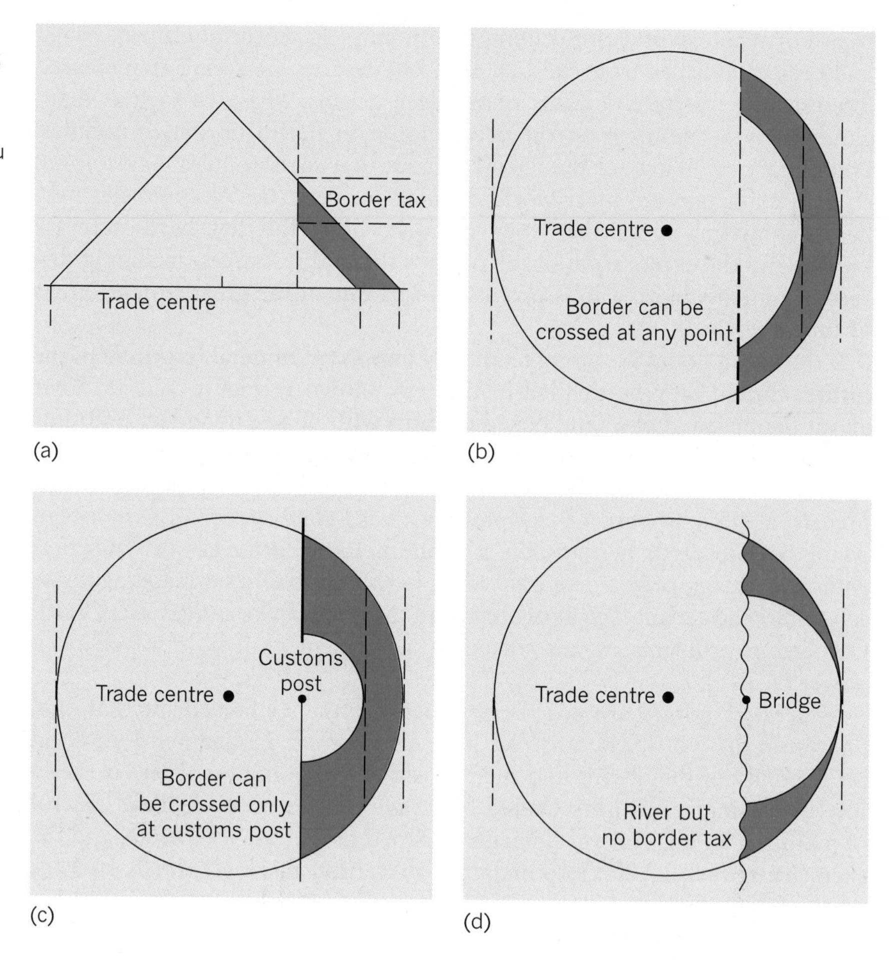

Figure 17.24 Boundaries and flows of trade A general model of the impact of boundaries on spatial interaction. (a) A cross-section of the market area around a trade centre. If you follow the vertical dashed lines, you can see how this cross-section is projected down to (b) to show the same situation in map form. Note how the shaded area indicates that part of the market that is lost by the effect of the border tax. Map (c) shows the still greater loss if the border can only be crossed at a single customs post. The effect of a natural boundary (hence, no border tax) with a constricted crossing point is shown in map (d).

marked by tariff differences, the potential field for trading is restricted in varying ways. Figure 17.24(b) shows the probable form if the political boundary can be crossed at all points along its length; Figure 17.24(c) gives the probable form if the boundary can be crossed only at a customs point. If the boundary is not a political one but a natural feature such as a river, with a single crossing point, the field will probably look like that in (d).

We can throw further light on the effect of tariff walls between countries by adapting some of the conventional economic models of trade. Let us assume that we have two adjacent countries, each of which produces a similar crop (e.g. wheat), separated by a tariff. In each country, the land area devoted to the crop is directly related to its local price. As the price of the crop increases, the amount of land under cultivation increases, and vice versa.

Let us do as economists do and plot the relationship of the crop area to price in terms of supply and demand curves. (Readers who have already taken classes in economics will see that we are greatly simplifying the situation; those who have not need note only the conclusions we draw from the graphs.) Figure 17.25 shows the area–price relationship as an upward sloping supply curve. Price also affects the volume of demand for the crop, but in an inverse way: the demand for a high-priced good is assumed to be less than that for a low-priced good, and vice versa. We can represent this second relationship by a downward-sloping demand curve. The local price for the crop in each country is established when the demand for it and the supply of it are in balance.

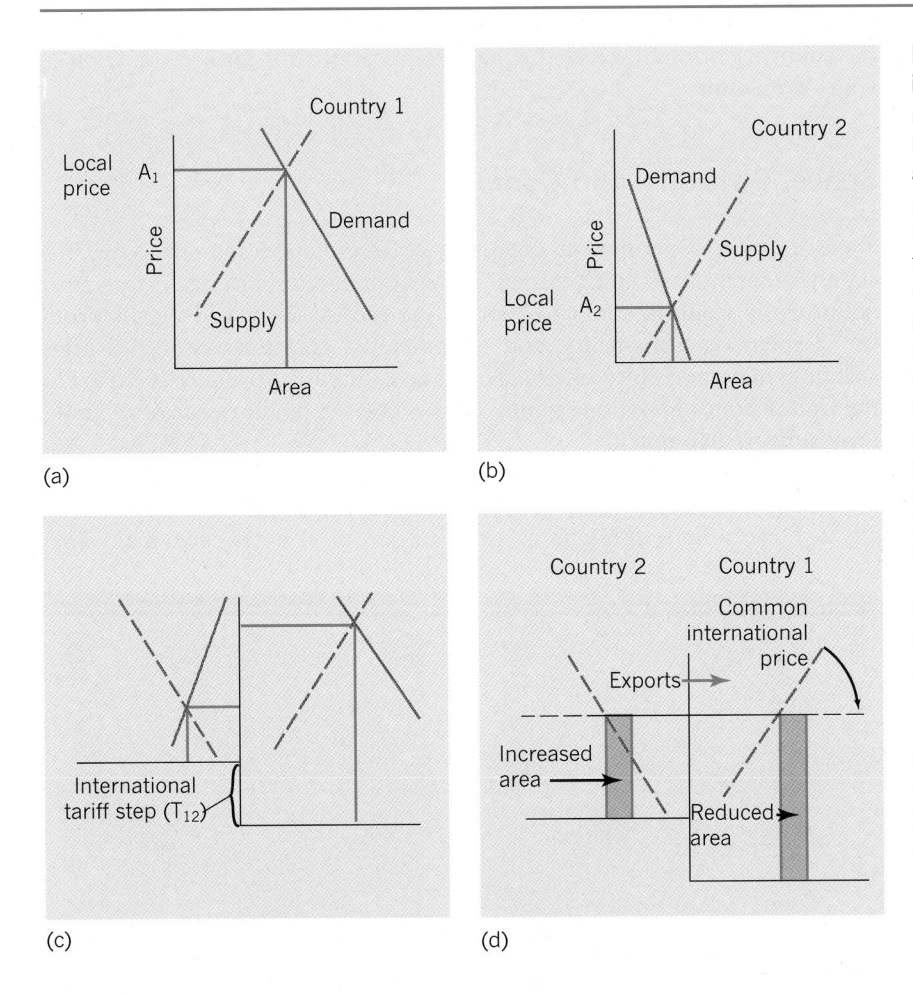

Figure 17.25 Tariffs and international trade A simplified model of the effect of international tariff between two countries on the supply of and demand for a common agricultural product. Charts (a) and (b) show the local conditions in each country in which the local price (vertical axis) and the area of the crop under cultivation (horizontal axis) are related through the conventional economic mechanism of demand and supply. The international tariff between the two countries is shown as a 'step' (T_{12}) in chart (c). In this case the step is not high enough to keep production from being expanded in Country 1 (the low-cost producer) and the extra output exported to Country 2 (the high-cost producer). How high does the step have to be to keep out cheap imports from Country 2? This depends on the relative difference between the local price in Country 1 (A_1 in the diagram) and the local price in Country 2 (A_2). When the tariff (T_{12}) is *less* than the difference in price (i.e. $A_2 - A_1$), as it is in the diagram, then flows will occur as in (a). If, however, Country 2 raises its tariff so that T_{12} is greater than $A_2 - A_1$, no trade will take place.

If we regard the two countries as completely isolated, the demand and supply situation in each will be determined by the internal conditions in each. Let us suppose that in Country 1 farmers get a high price for the crop and a large area is devoted to it (Figure 17.25(a)). In Country 2, conditions are different. The crop sells for a low price and the area devoted to it is smaller (Figure 17.25(b)). What will happen if we drop our assumption of isolation? Logically, we should expect the low-priced crop from Country 2 to find its way into Country 1.

A tariff can prevent this flow of exports, however. We can see this from our model if we place our two diagrams back to back and displace them vertically (Figure 17.25(c)). The amount of the displacement represents the size of the tariff imposed by Country 1. A flow of export occurs only if the difference in the local prices in the two countries is greater than the tariff. If the tariff is low enough for a flow to take place, a general international price will be established. The area devoted to the crop in Country 2 will expand, and the area devoted to it in Country 1 will contract (Figure 17.25(d)).

The analysis here is clearly oversimplified. The model refers to a very basic situation in which there is a single product and there are only two countries. Regional economists such as Bertil Ohlin have developed a complex theory of international trade to explain the flow of many commodities among many nations. Nevertheless, the simple graphical example given here gives us some insight into the impacts of tariffs on trade between countries. You might like to ponder the effects of tariff reductions and the establishment of common international prices (as in the EU) on the crop areas of the countries affected. Does this have implications for farm production in your own country?

Space, Divisions, and Conflict

States exist in a permanent condition of inter-nation tension. Since the supply of territory is finite, pursuit of separate interests by each state must occasionally result in conflict. About 7 per cent of the world's gross product is spent on armaments, and in individual consumer countries arms spending may take up to one-half of its gross national product (GNP). For the United States about one in nine jobs is created by the needs of the military–industrial complex.

It is clear that the world's political map is kept in place at very high cost. But even the stability of that map is perceived as being greater than it actually is. For example, although we think of Europe as having two main wars

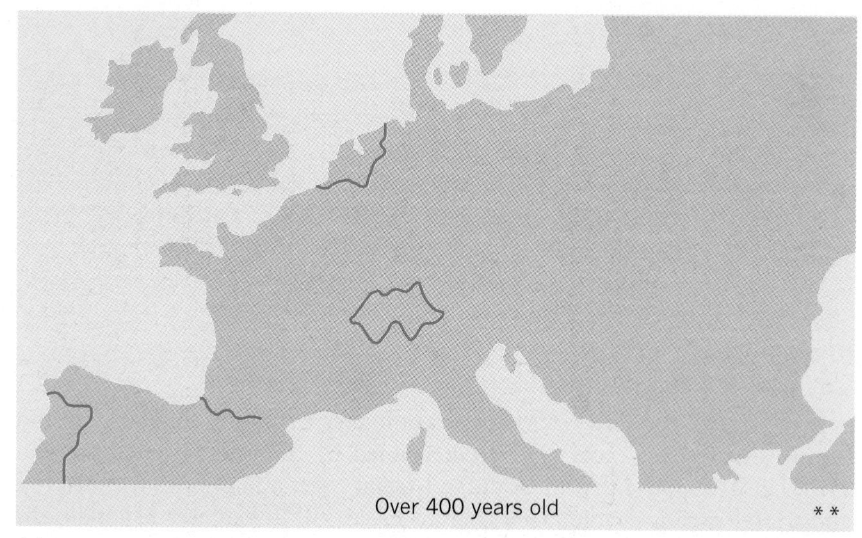

(a)

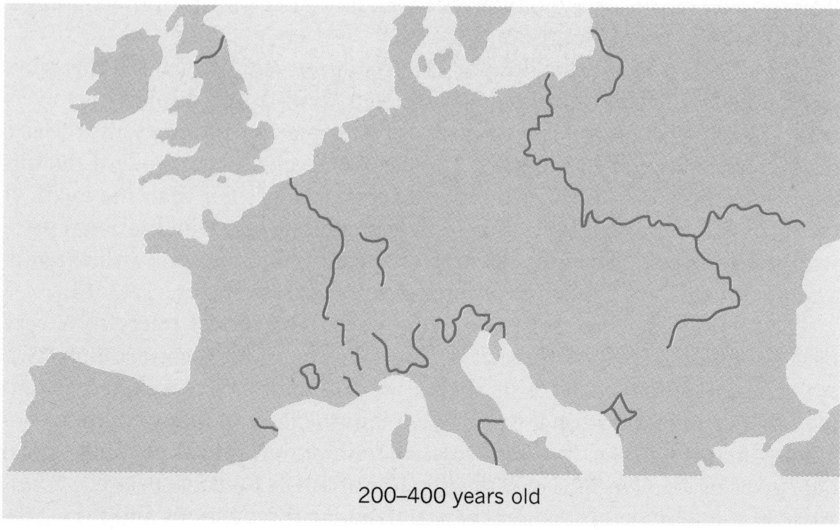

(b)

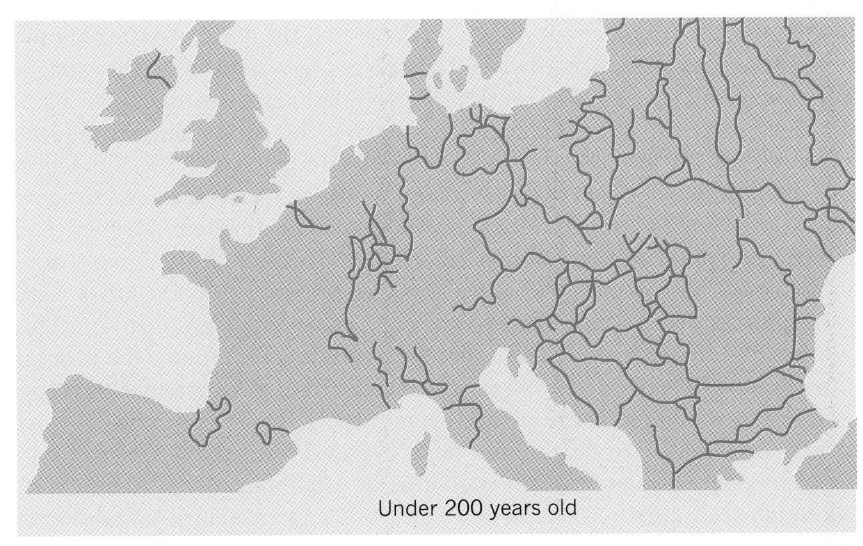

(c)

(d)

Figure 17.26 The stability of international boundaries The maps show contrast in the permanence of land boundaries among European countreis. Some land boundaries have changed little since the fifteenth century. Sea boundaries, such as that between England and France, are not shown. The four oldest land boundaries, shown in (a), are those of Spain and Portugal, of Spain with France in the western Pyrennees, of Switzerland, and of the Low Countries. Note the large number of short-lived boundaries in Eastern Europe. What kinds of factor make for very stable boundaries?

[Source: N.J.G. Pounds, *Political Geography* (McGraw-Hill, New York, 1963), Fig. 7. Copyright © 1963 by McGraw-Hill, Inc. Used with permission of McGraw-Hill Book Company. (d) Ullstein/Jürgen Grabowsky.]

in the twentieth century (1914–18 and 1939–45), an exact count shows a total of seventeen separate conflicts, including five since 1950. Over the years since 1815 Europe has seen 67 boundary changes (Figure 17.26) while a European state has appeared or disappeared on average once every three years. At the world level the start of the twenty-first century sees around 80 of its states in boundary disputes with one or more of their neighbours. Most are likely to be settled by agreement.

One source of conflict not shown on the map is the number of neighbours a state has. Mathematician Lewis Richardson, in a book with the intriguing title *The Statistics of Deadly Quarrels*, thought this might be vital. Written before the intercontinental missile age, the book contends that the potential for any interregional relationship (including conflict) is a function of number of neighbours. Richardson argued that a state such as Germany in 1936, which had nine other nation-states abutting its own territory, had a much higher chance of conflict, according to this line of thinking, than a nation-state such as Portugal with only one neighbour. If territories were closely packed in an unbounded land area, the average number of neighbours could reach as high as six. Such areas might approximate the form of Christaller's hexagonal central-place territories that we met in Chapter 14. The spherical shape of the planet means that if states continue to grow larger, the number of neighbours must decrease until the world is dominated by only two states; then the number of neighbours drops to only one. Although according to Richardson's thesis this arrangement would decrease the potential for conflict, the scale of conflict in a two-state world would be immense.

Although such spatial speculations may play a part in international stress, they are overridden by far more powerful ones that lie outside the scope of this book. You may, however, care to browse through some of the writings in political geography in which these problems are explored in depth. (See 'One step further . . .' at the end of the chapter.)

If spatial factors play an integral part in war strategies, geographers would contend that they should have an equal weight in strategies of peace. Certainly Richardson's conclusions on conflict were aimed at the reduction and resolution of wars. In the early 1960s the founding of the Peace Research Society by a group of behavioural scientists, including geographers, showed a similar sentiment. If good fences do indeed make good neighbours, then geographers need to determine from their research just where fences might best be built.

Reflections

1 The concept of territory is widely observed in many animals. Debate the value of ethological evidence in attempts to understand human territoriality. Do you think that regarding humans as 'naked apes' provides (a) insights into or (b) misleading analogies about our spatial behaviour?

2 Consider the cost boundaries between centres in Figure 17.4. Try introducing a third centre, and sketch the boundaries that might result.

3 Spatial boundaries between states exist in varying states of international tension. Look at the variability in the age of the international boundaries in Figure 17.26. List factors that make for (a) stability and (b) impermanence in boundaries between the countries shown there.

4 Look at Box 17.B on the classic work of Isaiah Bowman on the geographical basis for territorial disputes. Select *three* areas of the world where you might expect international disputes over territory to occur in the next decade. Compare your selection with those of others in your class. Do any common patterns emerge to suggest there are rules that govern such tension areas?

5 Coalitions are groups of states that join together for some common economic, political, or defence purpose. Take one of the regional international organizations listed in Figure 17.19 and find out which countries now belong to it. Look at the distribution of member countries on a world map. Are members contiguous and regional, or discontiguous and global? Does this provide clues to the factors that control membership?

6 Partitions within a state also demand criteria for drawing sets of regional and local boundaries. Gather data for your local area on the boundaries of electoral districts (e.g. senatorial, parliamentary, or city council districts) and on how many votes were cast for various parties at a recent election. Was the result affected by the electoral boundaries? How would you redraw these?

7 International boundaries in offshore and ocean waters are being more tightly defined as their resource potential is recognized. Use the median-line principle in Figure 17.23 and a map to divide a small enclosed sea such as the Mediterranean or Baltic. Would it be fair to use this principle in dividing the world's oceans?

8 Review your understanding of the following concepts using the chapter text and the glossary in Appendix A of this book:

antecedent boundaries	gerrymander
coastal baseline	Mackinder's heartland model
continental shelf	median lines
districting	offshore territorial waters
ethology	subsequent boundaries

One Step Further ...

A useful starting point is to look at the essays by Peter Taylor on global political geography and John O'Loughlin in the geography of conflicts in I. DOUGLAS *et al.* (eds), *Companion Encyclopaedia of Geography* (Routledge, London, 1996), pp. 332–69. A brief but useful introduction to political geography is given by Peter Taylor in his essay in R.J. JOHNSTON *et al.* (eds), *Dictionary of Human Geography*, fourth edition (Blackwell, Oxford, 2000), pp. 594–7. The central idea of territoriality is discussed in Robert SACK, *Human Territoriality* (Cambridge University Press, Cambridge, 1986).

Most of the ideas presented in this chapter are part of the legacy of political geography. Two recent books summarizing geographers' work in this area are Peter TAYLOR, *Political Geography: World-economy, Nation-state and Locality*, third edition (Longman, Harlow, 1993) and K.E. BRADEN and F. SHELLEY, *Geopolitics* (Longman, Harlow, 1998). See also M. MUIR and R. PADDISON, *Politics, Geography and Behaviour* (Methuen, London, 1981) and M.I. GLASSNER and H. DE BLIJ, *Systematic Political Geography*, third edition (Wiley, New York, 1980). Two classical studies by geographers who were intimately involved in the boundary problems that followed World War I are I. BOWMAN, *The New World: Problems in Political Geography*, fourth edition (World Book Company, New York, 1928) and S.W. BOGGS, *International Boundaries: A Study of Boundary Functions and Problems* (Columbia University Press, New York, 1940). Disputes in the world's oceans and seas are thoroughly reviewed on a regional basis in J.R.V. PRESCOTT, *The Maritime Political Boundaries of the World* (Methuen, London, 1985). It is worth browsing through a historical atlas to compare the kaleidoscopic change in some parts of the earth's surface with the relative stability in others. A useful atlas is H.C. DARBY and H. FULLARD, *Cambridge Modern History Atlas* (Cambridge University Press, London, 1971).

The political geography 'progress reports' section of *Progress in Human Geography* (quarterly) will prove helpful in keeping up to date. The main geographical journal specifically concerned with the contents of this chapter is *Political Geography* (bimonthly). Research is also published in the general geographic serials. Journals such as *International Affairs* (a quarterly) and *World Politics* (also a quarterly) often carry interesting papers. The *Journal of Peace Research* (an annual) is devoted to applying academic ideas to the resolution of conflicts. For readers with access to the World-Wide Web see also the sites recommended in Appendix B at the end of this book for topics relevant to this chapter.

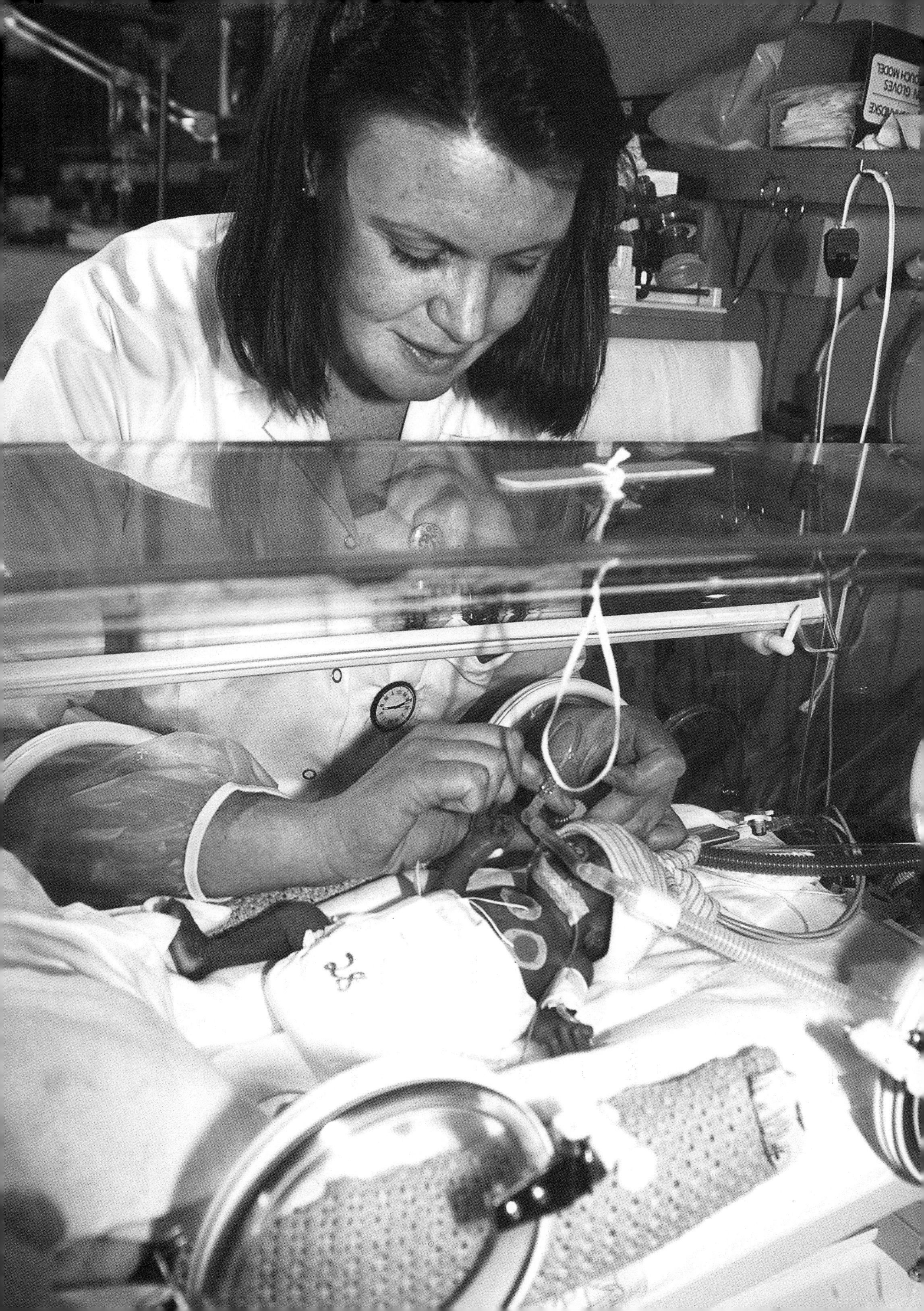

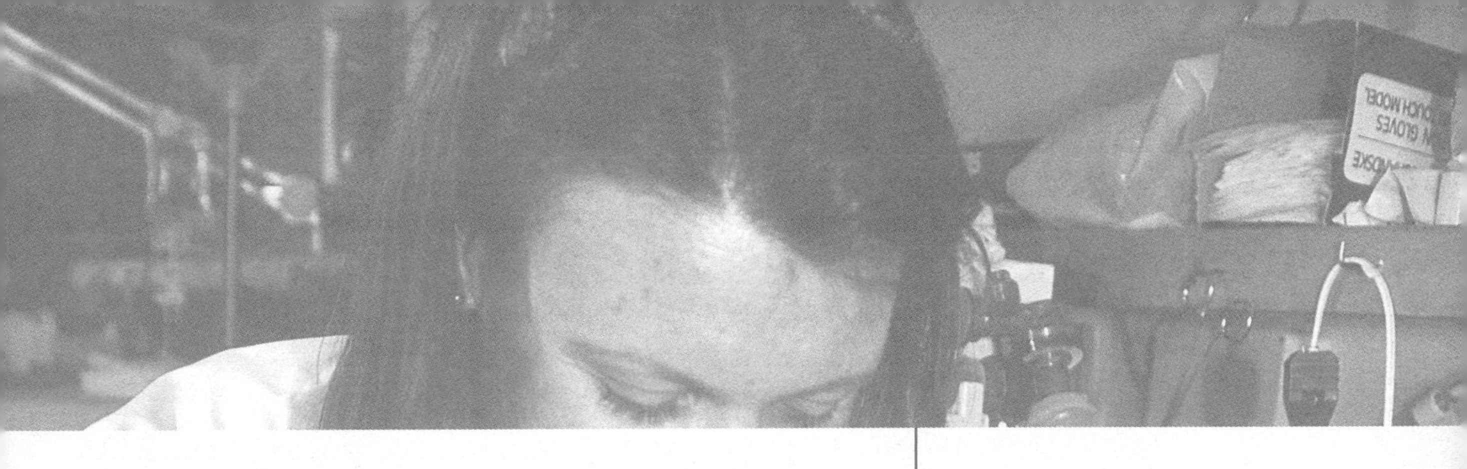

CHAPTER 18

Economic Inequalities

■ denotes case studies

Figure 18.1 Global economic disparities There are few clearer or more poignant indicators of the great economic disparities across the world than in the resources invested in the treatment of newborn babies. This premature baby being carefully tended in an intensive-care incubator in a Norwegian maternity hospital, contrasts with the very high infant mortality rates in developing countries.
[Source: Knudsens Fotosenter.]

> For whosoever hath, to him shall be given, and he shall have more abundance; but whosoever hath not, from him shall be taken away even that which he hath.
>
> The Gospel According to St. Matthew, XIII, 12

Humorists throughout the ages have warned us to choose our parents with care. Geographic humorists might remind us to choose our birthplaces with equal caution! Because it sums up so many economic and cultural considerations, our location in terms of the country within which we live continues to be one of the prime determinants of our life – and indeed has a bearing on whether we will even survive the trauma of birth itself. Figure 18.1 (p. 548) shows the huge resources devoted to care of a newborn child in an economically advanced country (in this case, Norway). Contrast that with the pictures of malnourished children in Chapter 6. Despite great advances through WHO campaigns, the infant mortality rate still varies hugely around the world. Infant deaths per thousand live births is around four (i.e. four babies who will fail to reach their fifth birthday) in the countries of Scandinavia. Compare this with over one hundred infant deaths on the same measure in the poor countries of the African Sahel. Whether we describe this as a development 'gap' or a *north–south divide*, the reality of the geographic contrast in wealth is stark.

In this chapter we review some of the massive differences in *economic geography* between the countries of the world. We try to answer three questions about the geographic importance of country units. First, we look at the main inequalities that exist between states at present (Section 18.1). What is the spatial pattern of rich and poor countries? On what evidence is the pattern based? Second, we ask why these contrasts in wealth and poverty exist (Section 18.2).

We then turn to the contributions of geographers to studies of the economic-development process (Section 18.3). Does development follow a particular geographic pattern and show a specific spatial form? Are the different parts of the world converging? Are countries getting more like one another? Or do the rich get richer and the poor get poorer, as our opening quotation suggests?

When we talk about a country as being developed or undeveloped, we are referring to the *average* conditions of the country taken as a whole. This ignores the strong regional variation within the country. In Section 18.4 we look at regional variations *within* a country, and ask how these internal variations can be measured. Finally in Section 18.5, we look at how variations within countries are tackled. What kinds of regional policy have been attempted and how successful have they proved?

18.1 Rich Countries and Poor

You need hardly take a geography course to learn that some nation-states have populations more prosperous than others. If we make the naive assumption that richness means material wealth, even a layperson will have no difficulty in placing Switzerland in a different box from Senegal. But these are extremes. If we want to distinguish between geographic areas and see how the gaps between the rich and the poor are changing, then we need a reliable ruler that can measure finer differences. What rulers have been developed, and what patterns do they reveal? (See Figure 18.2.)

Gross National Products (GNPs)

One of the first institutions established by the United Nations when it first met in San Francisco in April 1945 was the International Bank for Reconstruction and Development, soon to be known simply as the *World Bank*. In its valuable series of annual reports, the World Bank has charted the economic changes in the global economy over more than half a century. In 2000 it published a major review that surveyed the position at the start of the twenty-first century and we use that to begin our survey.

Measuring the size of a single economy is a complex business and, given great differences in the data available (or more often, unavailable), is particularly hazardous where countries at different stages of development are being compared. The World Bank has teams of economists, accountants, and statisticians who try to make the best estimates and these are used in our discussion here. The key concept is that of a country's ***gross national product (GNP)***. This is the total value of the production of goods and services measured over a year plus the net value of income earned abroad. It is an imperfect measure since definitions vary slightly from country to country and because much subsistence activity lies outside the 'measured' economy. The World Bank tries to get around some of these problems by using a device called ***purchasing power parity (PPP)***. This converts local estimates of GNP to a common yardstick based on the US dollar. Measured by this modified GNP, the global economy at the start of this century measured more than 40 trillion US dollars. (See Appendix B for web sites in which this figure is regularly updated.)

The global total is made up by summing the individual values for each of the 210 economies around the world for which data are available. Of this total, the United States alone (the world's single largest economy) contributes one-fifth of the global total. Together, the ten largest economies

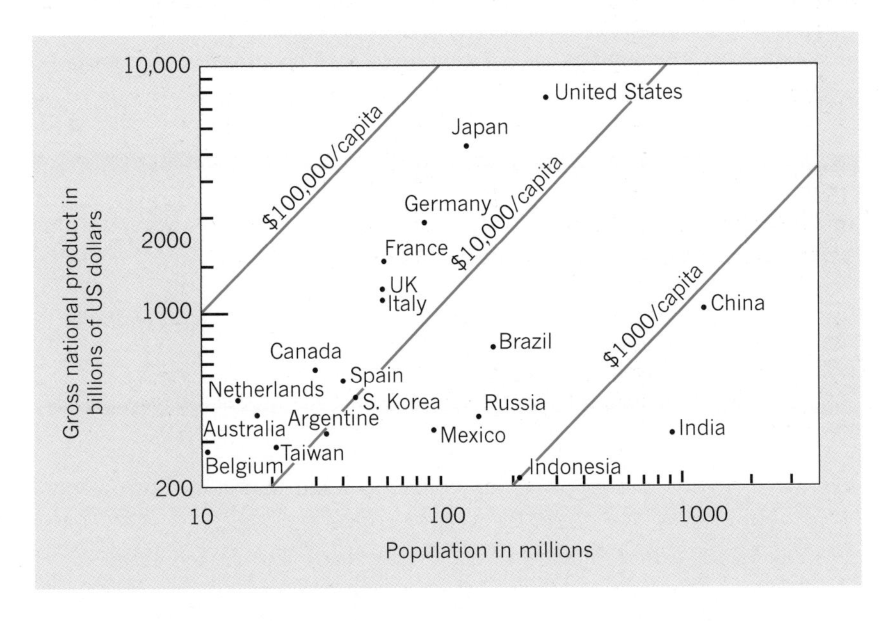

Figure 18.2 The wealth of nations
The populations of the world's 25 largest countries plotted against their gross national products at the start of the twenty-first century. High-, middle-, and low-income countries are on the basis of GNP per capita. Note that both axes are logarithmic so contrasts are much greater in reality than shown here.

[Source: Data from World Bank, *Entering the 21st Century* (Oxford University Press, Oxford, 2000), Tables 1 and 1a, pp. 230–1, 272.]

contribute over 60 per cent of the global total. This highly skewed distribution of wealth partly reflects the distribution of population: half of the 210 countries have small populations of five million (the size of Croatia) or less. But to a still greater extent it reflects huge differences in per capita wealth. The wealthiest *developed countries* (e.g. Switzerland) have GNPs per capita nearly 40 times the poorest (e.g. Mali, in the Sahel zone of Africa), even when allowances are made for PPP. Figure 18.2 summarizes two of the major indicators for the world's 25 most populous nations.

The World Bank divides the 210 countries that make up the global economy into three divisions based on the ratio of GNP to population (see Table 21.1). The table classifies all countries with populations over 30 thousand into three categories. Table 18.1 gives the number of countries in each group by geographical area with an example of each. Note that the average income for the rich countries is $80 and for the poor countries $7 where the US GNP per capita is set at $100 for reference.

High-income countries have with standardized incomes of GNP per capita that are one-third of more of the US level. The 53 countries in this group are divided into two types: (i) the large OECD countries which include all the major economies of Europe, North America, Japan, and Australasia; and (ii) small but wealthy countries that have prosperity because of oil resources (e.g. Brunei), are efficient city states (e.g. Singapore), or are offshore financial centres with strong tourist industry (e.g. Bermuda).

Middle-income countries are defined by GNP per capita of between a tenth and a third of the US level. They are divided into upper and lower groups with a division at around a quarter of the US level.

Lower-income countries have GNPs per capita below one tenth of the US level. Their distribution shows a strong concentration of countries in sub-Saharan Africa and in Asia.

Table 18.1 World Bank classification of 210 countries

Region	**High income**		**Middle income**		**Low income**	**Number of countries**
	OECD	**Others**	**Upper middle**	**Lower middle**		
Sub-Saharan Africa	0 countries	0 countries	5 countries Mauritius ($32)	6 countries South Africa ($23)	38 countries Nigeria ($3)	49 countries
Asia & Australasia	3 countries Japan ($79)	9 countries Singapore ($98)	4 countries Malaysia ($24)	12 countries Thailand ($20)	15 countries China ($11)	43 countries
Europe & Central Asia	18 countries Germany ($71)	8 countries Liechtenstein ($95)	8 countries Poland ($24)	14 countries Russia ($13)	6 countries Armenia ($7)	54 countries
Middle East & North Africa	0 countries	5 countries Israel ($59)	5 countries Saudi Arabia ($30)	9 countries Jordan ($11)	1 country Yemen ($3)	20 countries
Americas	2 countries United States ($100)	8 countries Bahamas ($36)	15 countries Argentina ($35)	16 countries Paraguay ($12)	3 countries Haiti ($4)	44 countries
Total countries	23 countries	30 countries	37 countries	57 countries	63 countries	210 countries
Average income (US = $100)	$80		$27	$14	$7	$21

Classification based on GNP per capita standardized by PPP (see text discussion). All values have been standardized in relation to the United States income per capita (indexed at = $100).

Source: Data taken from World Bank, *Entering the 21st Century* (Oxford University Press, Oxford, 2000), Tables 1 and 1a, pp. 230–231, 272.

Figure 18.3 gives the distribution of the three categories in terms of both population (a) and GNP (b). Note that the high-income countries have only one-seventh of the global population but over half of its wealth. Conversely, the low-income countries have 60 per cent of the global population but only one-fifth of its wealth.

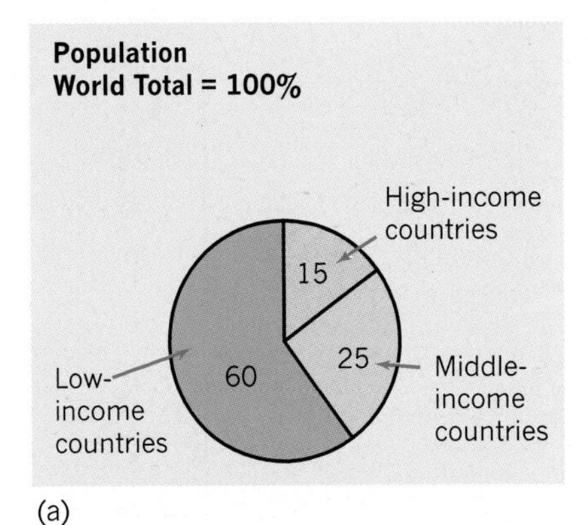

(a)

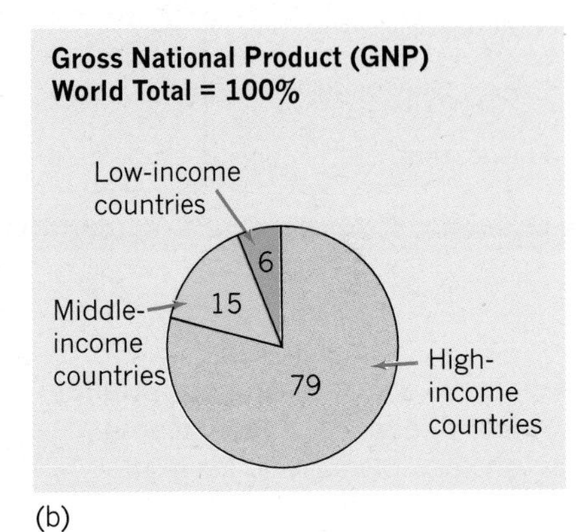

(b)

Figure 18.3 High-, middle- and low-income countries Shares of (a) world population and (b) world GNP made up by 210 countries in 2000 divided into income bands.

[Source: Data from World Bank, *Entering the 21st Century* (Oxford University Press, oxford, 2000), Tables 1 and 1a, pp. 230–1, 272.]

The Search for Other Development Yardsticks

Geographers would ideally like to compute and map some quantitative index that serves as an unambiguous ruler for measuring the economic or social performance of a region. For instance, in terms of Figure 18.2 we might reasonably regard a country as 'poor' in which the low level of health services and nutrition causes many infants to die and a country as 'rich' in which the high level of health services and nutrition prevents infant deaths. If we take available figures for infant deaths per thousand live births in the 1950s, we find Sweden (with 17) and the Netherlands (with 20) at one end of the scale and Tanganyika (with 170) and Burma (with 198) at the other. By the end of the century the figures for all four countries had been reduced. But the two rich countries had reduced their already low figures by three-quarters; the two poor countries had been reduced by about one-half.

But how valuable this single index is in separating poorer from richer countries is open to debate. Difference in single scores may be due partly to differences in the way each country collects population data and partly to how the available wealth is distributed. The World Bank reports a battery of indicators of development. Table 18.2 gives just some of the wide range of ways in which the well-being of a country can be measured and each has its advocates.

If we were to measure all these indicators of wealth, they would not necessarily tell the same story. Figure 18.4 shows how New Zealand and Sri Lanka vary on ten different measures of development. While each profile tells a similar story there are subtle differences.

Ways of combining multiple measurements into a few general indexes are now available. They involve collapsing several measurements into a smaller number of components. The principles involved are set out in Figure 18.5.

Table 18.2 Wealth of countries measured on three different criteria

Criteria I Highest GDP per head[a]		Criteria II Highest purchasing power per head[b]		Criteria III Highest quality of life[c]	
1. Switzerland	43	1. Luxembourg	116	1. Canada	96.0
2. Luxembourg	41	2. Singapore	101	2. France	94.6
3. Japan	38	3. United States	100	3. Norway	94.3
4. Norway	36	4. Switzerland	91	4. United States	94.3
5. Denmark	35	5. Iceland	85	5. Finland	94.2
6. Singapore	33	6. Japan	84	6. Netherlands	94.1
7. United States	29	7. Hong Kong	84	7. Japan	94.0
8. Germany	28	8. Norway	83	8. New Zealand	93.9

[a]Gross domestic product (GDP) per head in thousands of US dollars.
[b]GDP per head standardized to adjust for cost of living differences to give purchasing power parity (PPP). This measure is standardized to set the value for the United States at 100.
[c]Human development index which combines GDP with other measures (e.g. life expectancy, years of schooling, and adult literacy rates) to give an improved though still imperfect measure of development levels.
All indicators refer to the most recent indexes from United Nations tables for the late 1990s.

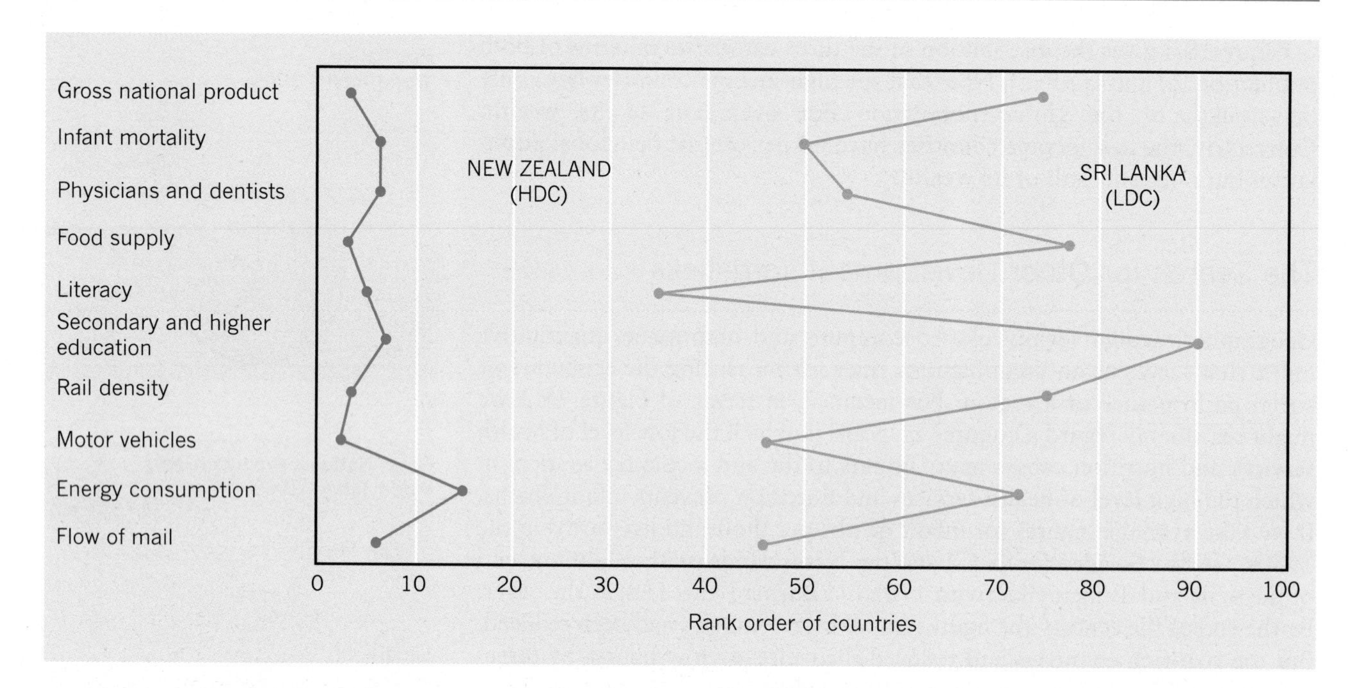

Figure 18.4 Development profiles typical of high- and low-income countries Contrasts in the relative ranking of New Zealand and Sri Lanka measured of socio-economic development. Note the consistently higher performance of New Zealand, which is a highly developed country (HDC), on all measures compared to the more variable, but generally medium to low, ranking of Sri Lanka, which is a less-developed country (LDC). All measures are calculated on a percentage of population-weighted basis to make the scores of different countries comparable.

When multiple measurements are made of the same set of individuals (e.g. countries), we can usually transform the original set of variables into a new set of variables that are independent and account in turn for as much of the original variation as possible.

Using a standard statistical technique called *principal components analysis*, we can identify the long axis of the ellipse (axis I, or the *principal axis*). Note that this principal axis accounts for much more of the original variation within the ellipse than the secondary axis (axis II) drawn at right angles to it. This idea of 'collapsing' a large set of original variables into a

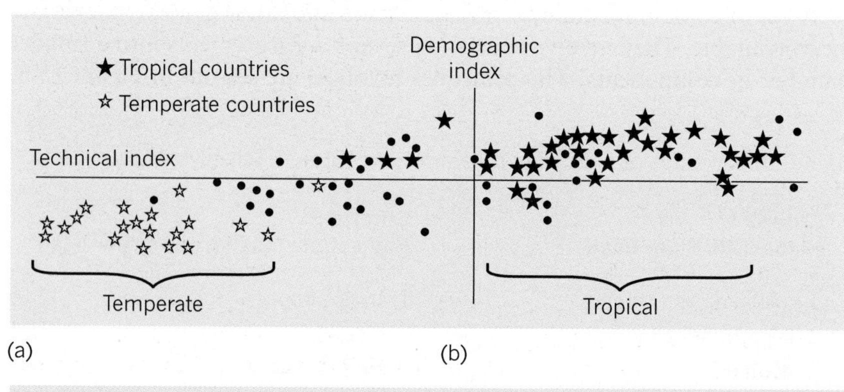

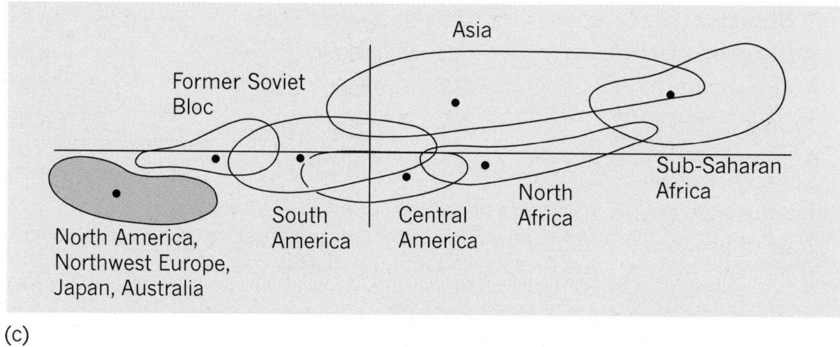

Figure 18.5 Generalized development yardsticks: economic and demographic scales The vertical economic scale accounts for substantially more than the local variation than the shorter demographic scale.

[Source: After N. Ginsburg, *Atlas of Economic Development* (University of Chicago Press, Chicago, 1961), Fig. 3, p. 113. Copyright © 1961 by the University of Chicago.]

small number of basic dimensions or composite variables underlies a large and expanding area of mathematics. Economic geographer Brian Berry at the University of Chicago first used this method to compress 43 different indexes of economic development into a single diagram (Figure 18.5(c)). Note that this figure has two axes. The first and most important is the longer, vertical axis which measures differences in *technical development*. This single index accounts for about 84 per cent of the information in the many original measures. The second and less important axis is the short, horizontal one. This index measures contrasts in the *demographic stage* a country has reached. Together, the two indexes account for 88 per cent of the original contrasts between the 95 countries. (See Box 18.A on 'Brian Berry and urban economic geography'.)

Box 18.A Brian Berry and Urban Economic Geography

Brian Joe Lobley Berry (born Sedgely, England, 1934) is the world's leading urban-economic geographer. After graduating in economic geography from University College London, Berry joined the group working with William Garrison at the University of Washington, Seattle, in the mid-1950s. They made critical contributions to the introduction of economic modelling of spatial organization and the use of quantitative methods in geography. They became the focus of the so-called 'quantitative revolution' in human geography. Berry was subsequently a professor of urban geography at Chicago (1958–76) and of regional planning at Harvard (1976–81), a dean at Carnegie Mellon at Pittsburgh (1981–86). He is now Lloyd Viel Berkner Regental Professor at the University of Texas at Dallas.

Berry's early seminal work was on central-place theory, summarized in his later text on *The Geography of Market Centers and Retail Distribution* (1967). Subsequently at the University of Chicago he ranged widely through human geography, making innovations in multivariate statistical analysis (particularly factor analysis). He made use of such methods in developing an operational field theory of regional geography. Through a prodigious output of research papers and books such as *The Human Consequences of Urbanization* (1973), Berry became one of the most influential human geographers in the world from the 1960s. At Chicago he was also involved in much practical, contract research, being responsible, for example, for the conception and design of a new spatial framework (the Daily Urban System) for the presentation and analysis of United States census data. At Harvard he was involved in early GIS systems (see Chapter 23). His subsequent inquiries extended to urban ecology, growth centre theory, and the concept of counter-urbanization.

At Texas, his interests have turned to the interface between long-run macroeconomic history (the long-wave phenomenon) and such issues as the clustering of innovations that transform economic systems at particular historical phases and in particular locations. He has also examined the way in which pulses of technological change and innovations move outward from centres and regions of high inventive capacity. In 1999 he became the first geographer to serve on the Council of the US National Academy of Sciences.

Author of nearly 500 books and articles, Berry has been concerned with bridging theory and practice, becoming heavily involved in urban and regional development planning in both advanced and developing countries. Recent books include *Long-wave Rhythms in Economic Development and Political Behaviour*, *America's Utopian Experiments*, and *The Global Economy in Transition*.

Berry's original analysis provides a compact description of the differences between states. But how can such differences be explained?

18.2 Explaining Contrasts in Development

An Enquiry into the Nature and Causes of the Wealth of Nations was published by the Scottish economist Adam Smith in 1776. The founder of economics puzzled over why some nations were so well off while others lingered in poverty.

Geographers looking back at Figure 18.5(c) might well think that location had a hand in this. Broadly speaking, all the less-developed countries have tropical locations, and all the highly developed countries have mid-latitude locations. Such a correspondence between variables might lead us to suppose that development is a matter of natural environmental resources – and of climate in particular. Certainly the kind of conditions that we found in the climatic zones E and F in Chapter 4 are major bars to agricultural production. Natural resources do indeed play a key role in development, but we can think of countries (Denmark, Japan, and Israel, for example) that are well developed but that have a rather limited resource base. Equally, Singapore is an example of one of the world's most highly developed economies located squarely within the tropics. So, it is easy to destroy an environmental model of development. But this is not because resources and location play no part in the development equation, but because their effect is not simple. This is equally true of other single-cause explanations: Max Weber's emphasis on the link between capitalism and the Protestant ethic looks somewhat threadbare today. We must look for a more complex explanation.

Interlocking Factors in Development

Nobel-prize winner Paul Samuelson sees as 'four fundamental factors' in understanding development – population, natural resources, capital formation (domestic or imported capital), and technology. These are summarized in diagrammatic form in Figure 18.6. In interpreting the diagram, it is important to note that each factor interlocks with the others. Countries showing sustained economic growth over the last century (e.g. Sweden) have tended to score high on all four counts. However, countries showing slow growth may well have been held back by any one of the critical ingredients.

If we take the first of Samuelson's four factors, population, we note that it also formed the second axis in Berry's development diagram. To understand the links between the population structure of a country and its level of development, you will need to refresh your understanding of the concept of a *demographic transition*, which we first met in Chapter 6. You might like to turn back and look again at Figure 6.16.

Broadly speaking, most of the world's most developed countries are in Stages III or IV of the demographic transition. They have slowly expanding or stationary populations. Conversely, most poor countries are Stage I, and most middle-income countries in Stage II or III. One of the major effects of a country's position in the demographic transition on development is

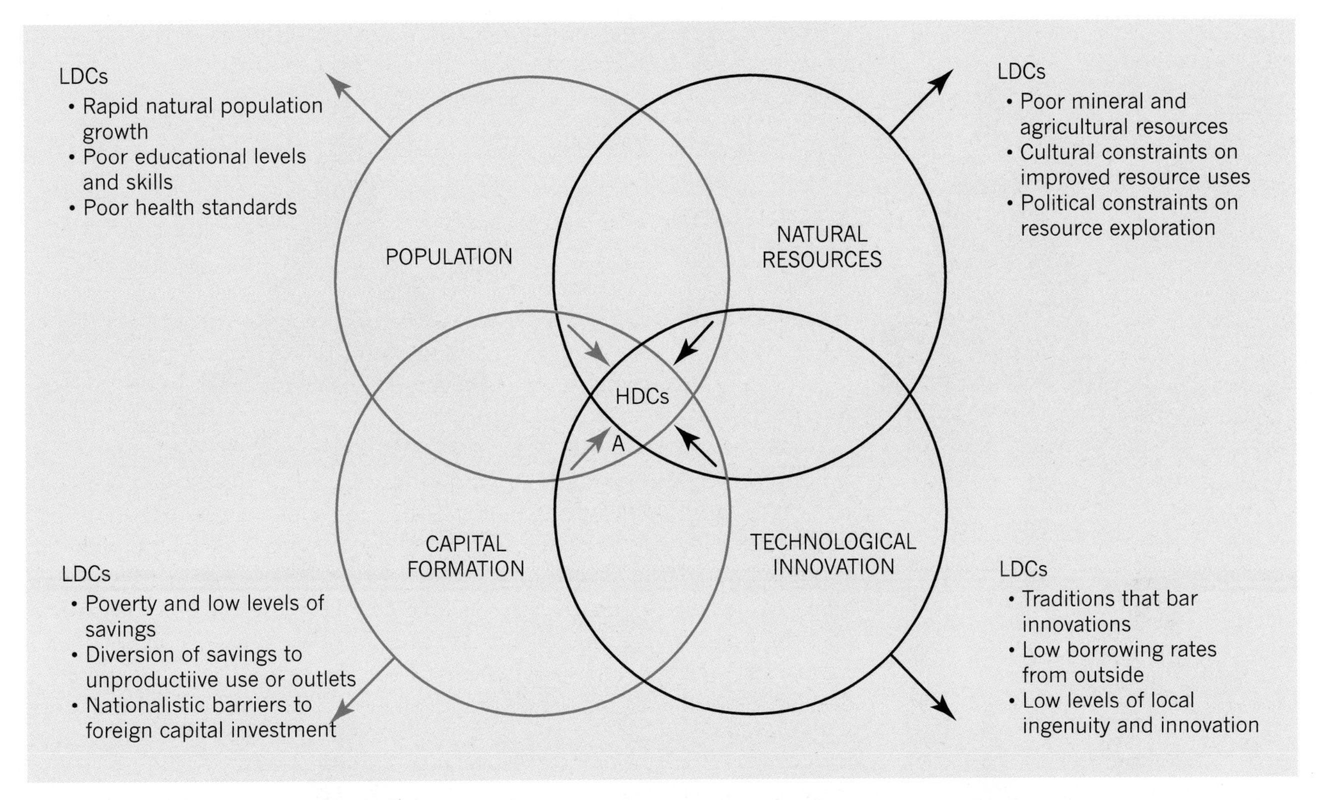

Figure 18.6 Constraints on economic development Four major factors in economic development are shown by a set of interlocking circles. HDCs tend to lie near the central overlap area, although some may attain high levels of development with an overlap of only three of the factors. Japan and Denmark are HDCs with rather limited natural resources and therefore lie in Sector A. What about the LDCs? If you followed the four outside arrows and mentally stretched the circles into ellipses, they could each meet on the reverse side of a sphere. This would place the most developed HDC (e.g. Sweden or the United States) on one pole and the lowliest LDC on the other! Three examples of the ways in which each of four constraints may operate on an LDC are indicated.

related to its age distribution, which affects the size of the active labour force in relation to the total size of the population that must be supported. Consider Figure 18.7, which compares the age distribution of three countries in different demographic phases. Mexico is a country in the second phase, with a population increasing by more than 3 per cent; Japan can be put in the third phase because its population increases by about 1 per cent; Sweden is now increasing very slowly at a rate of less than 1 per cent and is in the fourth phase. Note the contrast in the number of children (aged 0 to 14 years) in the three countries: 44 per cent of the population in Mexico, 30 per cent in Japan, but only 23 per cent in Sweden. Likewise, the number of older folk (aged 65 years and over) is only 3 per cent of the total population in Mexico, but twice this level in Japan, and three times this level in Sweden.

One of the critical questions therefore, is whether a fall in birth rates will accompany the urbanization of the developing countries. United Nations surveys measuring the number of children 0–4 years of age in proportion to the number of women aged 15–44 years (i.e. potentially reproductive females) reveal a general correlation between this index of fertility and the stage of development of a country. Developed countries have generally low index levels, and developing countries generally higher levels. Some demographers have claimed that it is possible to trace the movement of countries with long demographic records, such as Sweden, through the four phases. Others have suggested that birth-rate increases may follow a growing abundance of food before the death rate starts to fall. Will countries presently in the second phase move necessarily and progressively to the third and fourth phases? If so, at what rates?

We could pattern the changes in birth and death rates in Western Europe since 1700 into a general model. These changes reveal a rather consistent S-shaped fall in death rates (from about 3.3 per cent to around 1.5 per cent)

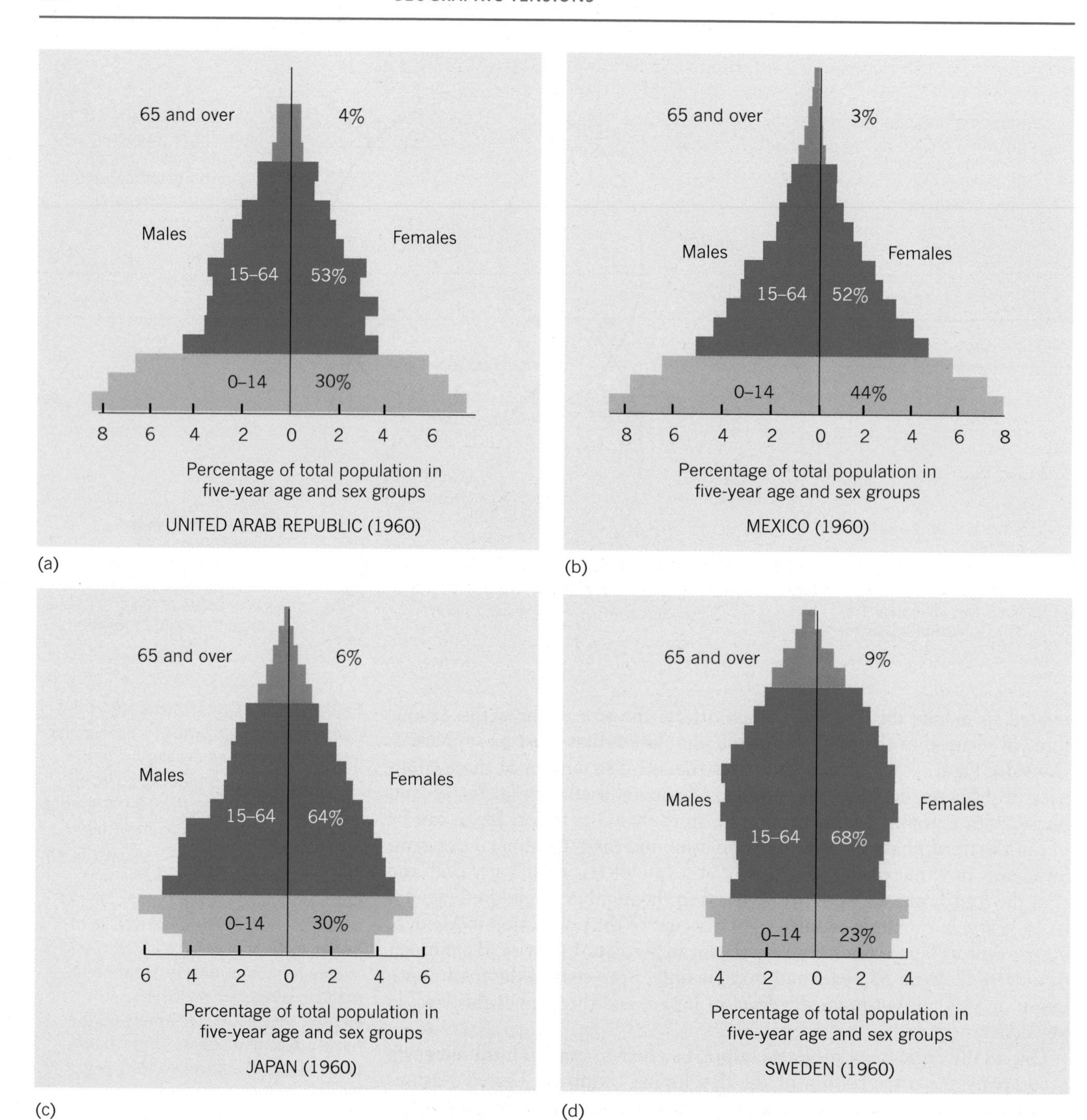

Figure 18.7 Demographic contrasts and development stages Population pyramids show the age and sex distribution of the population of four countries at different stages in the demographic transition. Note that country (a) is an LDC, country (b) is an MDC, country (c) a 'recent ' HDC, and country (d) a 'long-standing' HDC.

[Source: Data from United Nations, *Demographic Yearbooks*, 1957 to 1960. Adapted from J.O.M. Broek and J.W. Webb, *A Geography of Mankind* (McGraw-Hill, New York, 1968), pp. 447–456. Copyright © 1968 by McGraw-Hill, Inc. Used with permission of McGraw-Hill Book Company.]

followed by an S-shaped decline in birth rates (from approximately 3.5 per cent to around 1.7 per cent). The greatest rate of increase in Europe was in the mid-nineteenth century, when the lag between the two rate curves was at a maximum. But post-World War II fluctuations in the birth rate indicate that the demographic transition model is an oversimplification of the demographic situation in advanced countries.

To sum up: the pattern of states around the world shows great contrasts in the levels of material prosperity. Many measures of that contrast are possible, but one useful summary is that of technical development and demographic stage. We now turn to look more closely at that pattern over time and space.

Convergence or Divergence?

Most countries are richer now than they were at the beginning of the century. The real pattern of development is thus a dynamic one, and the question of the direction of change is important. Are the rich countries growing richer – and the poor, poorer? We can use two lines of evidence in trying to decide whether international contrasts in wealth between countries are deepening or lessening: the historical evidence of statistical trends and the arguments of theoretical growth models.

Evidence of Historical Trends

One major block to using historical evidence is its highly variable quality. Estimates for current levels of income and the GNP for most of the less-developed countries are crude, and reconstructions for earlier time periods vary with the assumptions made. Although the present differences between advanced western countries and the *Third World* are clear, the historical trends in those differences are obscure. Even when information is available on income or production, we lack the data on comparative costs needed to translate income or production figures into meaningful comparisons of regional welfare. There is simply not enough quantitative evidence to confirm the impression of an increasing difference between the 'haves' and 'have nots'.

In the case of continental blocks and individual countries the situation is more hopeful. In the United States regional incomes for the nine main census divisions have converged since 1880. The trend was not, however, a steady one: in the 1920s, the regional contrasts in income actually increased rather than decreased. This was, however, an isolated phase linked to the different ways in which parts of the United States withstood the Great Depression. In the case of a smaller spatial economy, Great Britain, during the last 25 years there were rather weak tendencies toward convergence despite a very active regional equalization policy. The gap between the poorest and richest regions narrowed a bit, but the actual magnitude of this gap in Britain, as in most western European countries and particularly Scandinavia, was already narrow. In developing Afro-Asian and Latin American countries, where regional differences in income are greater, the data available are not sufficiently accurate to make firm estimates. The impact of equalization policies on long-term trends within the Soviet Union is not really known either.

Historical and empirical studies provide no strong convergent or divergent trends in regional income. What evidence there is points to rather weak and unsteady changes rather than strong and headlong processes. Regional changes have a complex spatial pattern with different trends operating in various ways on various scales; thus it is probable that divergence and convergence are occurring simultaneously on different spatial levels. What appears to be taking place may be a function of the levels for which data are available.

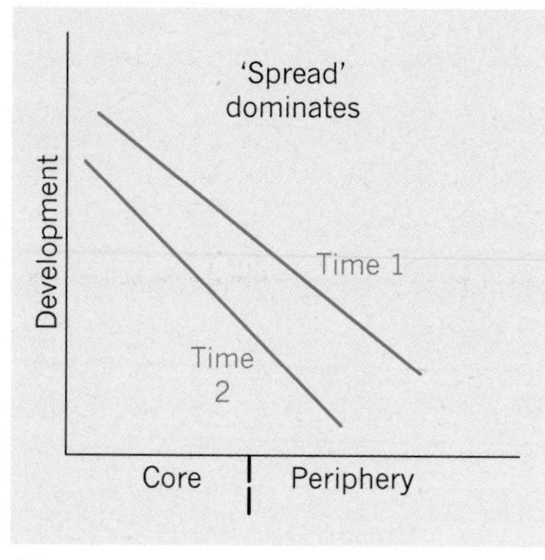

(a)

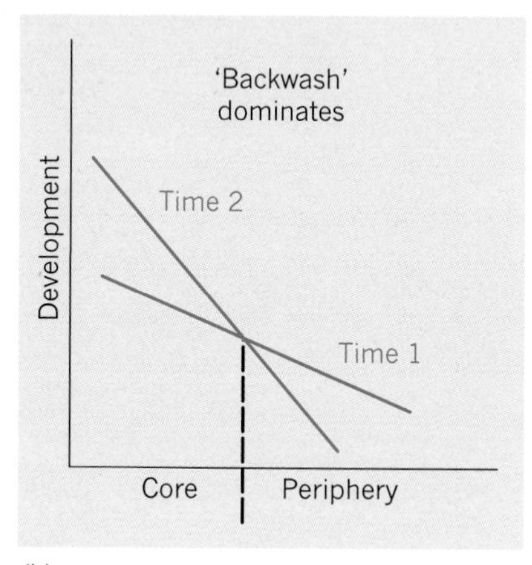

(b)

Figure 18.8 Myrdal's spread–backwash model Changes in the balance between the two sets of forces. In (a) spread dominates so that all areas develop over time with a slight trend toward equalization. In (b) backwash dominates and there is an increasing gap between the core and the periphery.

Evidence from Theoretical Models Theoretical models of regional growth have been largely produced as by-products of general economic theory. They cover only certain parts of the regional growth and have little geographic detail. Here we look at the conditions of a few of the economic models.

Swedish economist Gunnar Myrdal has stressed that economic market forces tend to increase, rather than decrease, regional differentiation. The build-up of activities in prosperous, growing regions influences the less prosperous, lagging regions through two types of induced effects: spread effects and backwash effects (see Figure 18.8).

The positive impacts on all other regions of growth in a thriving region are called, by Myrdal, *spread effects*. This impact comes from the stimulation of increased demand for raw materials and agricultural products and the diffusion of advanced technology. Thus, to give a simple example of a spread effect, the medical services in a poor country may gain from the advances in drug therapy conducted in an advanced country without itself having to meet the high costs of the initial research.

Backwash effects of agglomerated growth are net improvements of population, capital, and goods that favour the development of the growing area. One classic example of the backwash effect is the 'brain drain', typified by movements of medical doctors to the United States from poorer countries. In this and similar selective migration flows the poorer region loses its most highly skilled workers. In a more extreme form, it may also lose its most active population (say, aged 20–40 years), leaving the dependent young and the old behind to be sustained by remittances from the overseas workers.

These two opposing forces do not imply the existence of an equilibrium situation. Indeed, Myrdal maintains that the two effects balance each other only rarely. What is more likely is a cumulative upward or downward movement over a considerable time that leads to long periods of increasing regional contrasts (Figure 18.9).

Although Myrdal's model of economic growth has been criticized for its qualitative nature and lack of econometric substance, the more formal models of regional economic growth *also* fail to demonstrate conclusively the direction of movement. One modern model of the regional system (the Harrod–Domar model) maintains that interregional growth leads to divergence. Rapidly growing regions have high levels of income and net inward movements of labour and capital. In contrast, more traditional models of regional growth indicate that, though fast-growing regions have a net inward movement of capital, income levels in these regions are low and the net movement of labour is outward.

Future Trends

In this chapter we have looked at some of the current patterns of rich and poor countries in economic terms. Most of the world's countries are very poor compared with the United States. Only about ten of the world's countries have gross national products greater than that of the state of California. Given their high standard of living and the concern over population pressures, resources, and pollution that we described in Part II of this book, it is wholly understandable that a *zero economic growth (ZEG)* movement should have arisen in the western world. The idea that we must

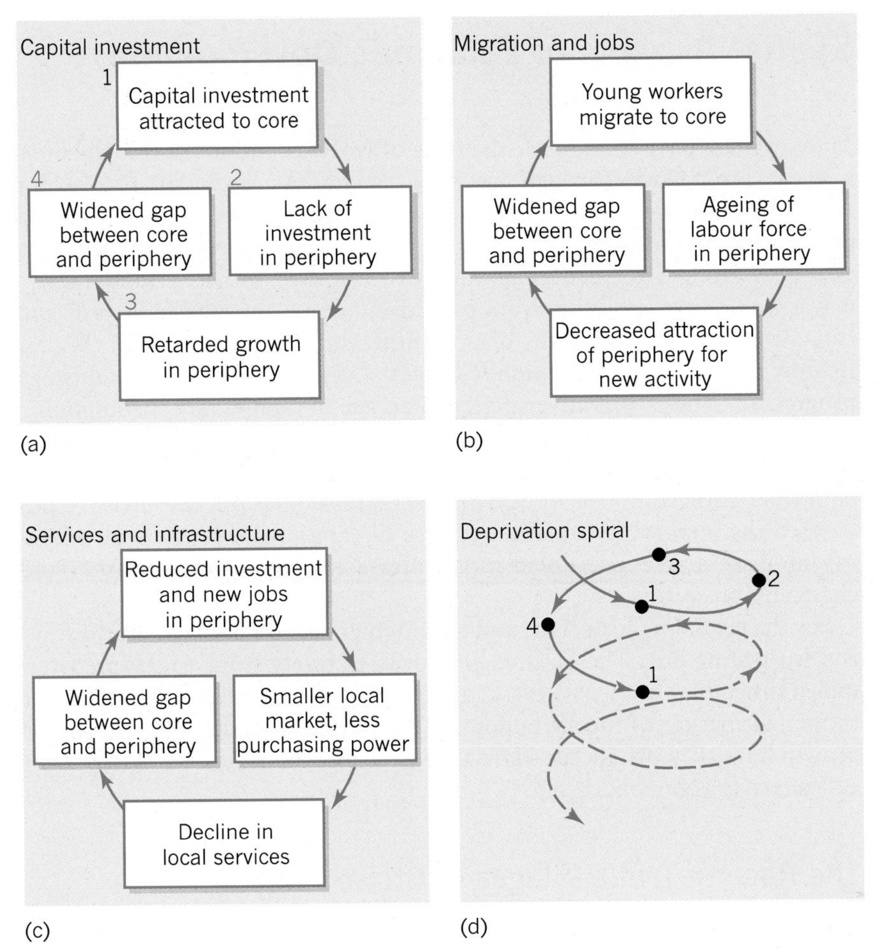

Figure 18.9 Backwash effects in the Myrdal model Three examples of the circular, cumulative, causation process by which the gap between core and periphery areas is widened. The joint impact of (a), (b), and (c) is to give the downward deprivation spiral in (d). Note that the three cycles need not interlink. For example, migrants moving to the core may send back a share of their earnings to their family in the periphery. Also, governments may try to keep up the level of services in peripheral areas to counteract outward movement of capital and labour.

return to a self-sustaining economic system to prevent an ecological disaster has been strongly argued by such figures as Rachel Carson, Paul Ehrlich, and Barry Commoner.

Whatever the appeals of ecological arguments for ZEG a geographer must note that they are not equally convincing all over the globe. Readers of this book from LDCs (and such readers are likely to be very few) will be unimpressed by antipollution arguments when they still want to develop the industry that will pollute and use resources. Concern about pollution is understandably low in countries where famine and disease are the front-line problems. And do the readers from the fortunate lands of North America, western Europe, and Australasia really want their incomes to be frozen for the next few decades – or want half their own country's GNP diverted to international aid?

In the longer run many changes are possible. Once we start measuring growth in terms of net social welfare (NSW) rather than GNP, the developed lands may look considerably less affluent than we now believe them to be. The possibility of regional revenue sharing and equalization appears, currently, to be greater at the more restricted spatial scale of the region than on the global scale. It is to this within-country level of spatial organization that we turn again later in this chapter.

18.3 Spatial Aspects of Economic Development

Many writers have tried to see the facts of world economic development as a linear progression through inevitable stages. As we noted before, for Adam Smith in the *Wealth of Nations* it was the calculus of fixed land and growing population that provided the key to a golden age in which all a country's product accrued to labour. For Karl Marx in *Das Capital* (1867) it was a one-way evolution from primitive culture, through feudalism and capitalism, to the end state of socialism and communism. For Walter Rostow in *The Stages of Economic Growth* (1960) it was again a multistage progression from a primitive society to an age of high mass consumption.

In the historical succession of economic theories, each of the major factors in Figure 18.6 has at one time or another been singled out for the dominant role. Earliest models of development stressed natural resources: Smith stressed the importance of labour, Marx of capital, and Rostow of technical innovation. We also noted in Chapter 8 the role of urbanization and changing job sectors.

For the economic historians and development economists the world story is a frustrating one. The facts of growth have rarely stuck to the predetermined timetables of theory. But geographers have not been wholly immune to the fascination of theory building. What kinds of geographic models of growth have they produced? Have they been any more successful than their colleagues in economics?

The Rostow–Taaffe 'Stages of Growth' Model

We have already seen in this book the kind of development models a geographer builds. In Chapter 16 we reviewed the work of the Swedish geographer Hägerstrand on *spatial diffusion* models, and in Chapter 8, the parallel work on *urbanization* models. Both these sets of ideas underline the approach we shall follow, emphasizing variations in *where* development takes place and the effect it has on the changing spatial organization of the world economy.

The Spatial Growth Model Figure 18.10 shows a four-stage model of the spatial pattern of development of an idealized island country. It is based on the work of a group of geographers led by Edward Taaffe at Northwestern University in the early 1960s and draws heavily on Peter Gould's work on the modernization of West African countries, notably Ghana. It has close parallels with Rostow's division of economic development into four phases: a 'traditional society', a 'take-off' phase, a 'drive to maturity', and a movement 'toward high mass consumption' and is termed the *Rostow–Taaffe model*.

In *Stage I* there is a scatter of small ports and trading posts on the coast. Each small port has a small inland trading field, but most of the interior villages are untouched by the coastal development. Subsistence agriculture dominates the island, apart from the few coastal pockets with trading links to the outside world.

Stage II is the critical stage. It is roughly analogous to Rostow's 'take-off' phase, which is clearly based on the analogy of an aircraft, which can fly only after attaining some critical speed. Other economists have termed this stage 'the spurt' or 'the big push'. It is marked by two geographic charac-

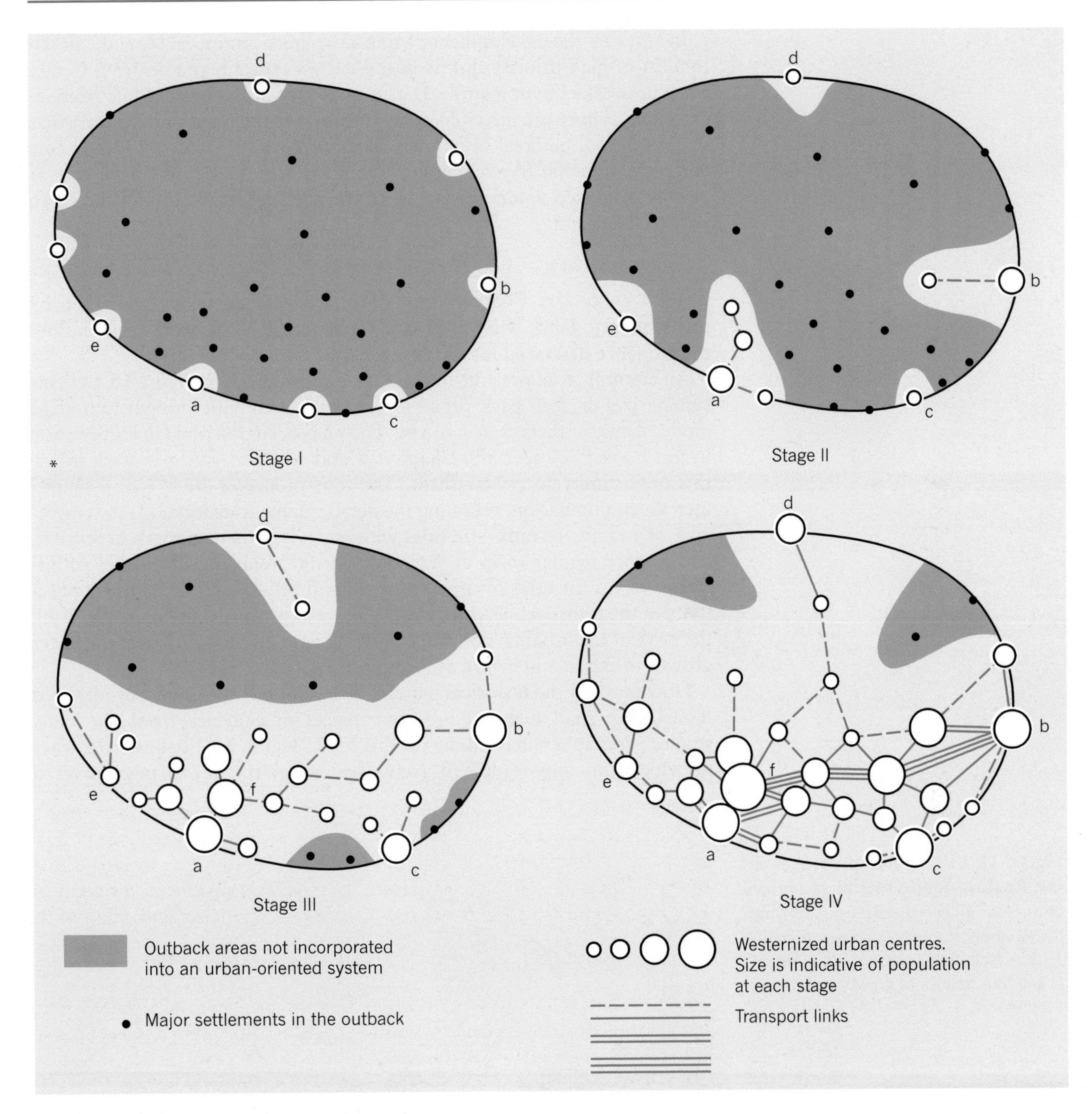

teristics: first, new transport links to the interior to tap new areas of natural resource for export; and second, differential growth of the coastal centres as some expand (centres a and b in Figure 18.10), some maintain their position (c, d, and e), and the remainder are pinched out. Studies of African colonial areas suggest that the tapping of mineral resources and the need for political and military control are key factors in the expansion of transport links to the interior of a developing country.

Stage III is marked by the rapid growth of the transport system about each of the major ports and the emergence of important new inland centres at transport junctions (e.g. centre f). Note also the beginnings of lateral interconnections between a and b. The northern half of the island remains isolated, but the southern half shows rapid growth and urbanization.

Figure 18.10 The Rostow–Taaffe model of spatial structure of economic development An idealized sequence of stages in the economic development of a hypothetical island. Note that Stage II is the critical 'take-off' period in which major transport links are first driven into the interior. The four maps should be viewed as representing the processes of growth discussed in the text rather than as an exact spatial reconstruction of actual events.

In *Stage IV* the development of transport links continues. Note the development of high-priority shuttle lines between centre b and centre f. Centre f has now taken over a's role as a primate city, marking the shift from an external, export-oriented phase to one in which the island country has major internal markets of its own. The north–south transport link is now complete, and the few remaining 'primitive areas' take on a new role as heavily protected wilderness areas for the over-urbanized folk of the south part of the island.

Evaluation of the Rostow–Taaffe Model Two questions are raised by this sequence. First, what processes are shaping the pattern? Second, does the sequence described match the events we actually observe?

An attempt to answer the first question is presented in Figure 18.11. This summarizes the four basic processes that have been built into our four-stage model. None of them is new to you. Each has been described in earlier parts in this book (see especially Chapters 8 and 14), and you may wish to use this opportunity to review them. The four processes are (1) an S-shaped increase in population, reflecting the demographic transition; (2) the emergence of a family of rank–size rules with an early primate pattern in Stage II, and a more regular form in Stage IV; (3) the launching of a series of diffusion waves for rates of urban growth; and (4) an increase in the connectivity of the transport network as development proceeds. In Figure 18.11(c), the peak of the urbanization wave moves inland in sympathy with the faster growth of the inland centre at location f.

How far does the historical pattern of development support our idealized model? We shall mention here two pieces of evidence from the many studies geographers have done on this topic. Figure 18.12 shows the ways in which the importance of New Zealand ports has changed over a

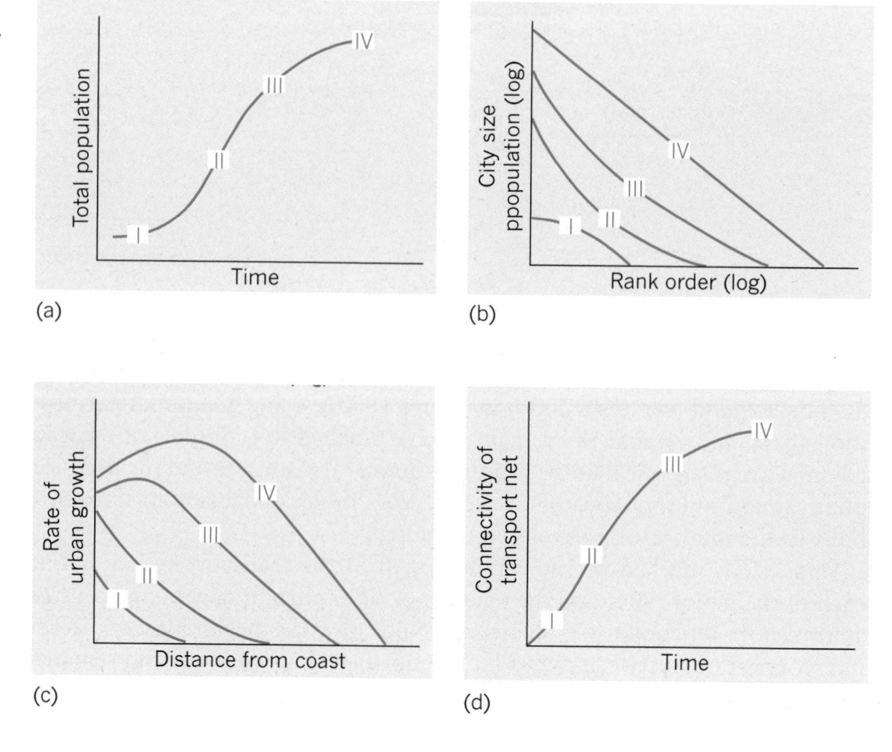

Figure 18.11 Spatial processes in the Rostow–Taaffe model Summary of the four growth processes underlying the geographic changes shown in Figure 18.10 . Note that numbers I to IV relate to the four phases of growth shown in that figure.

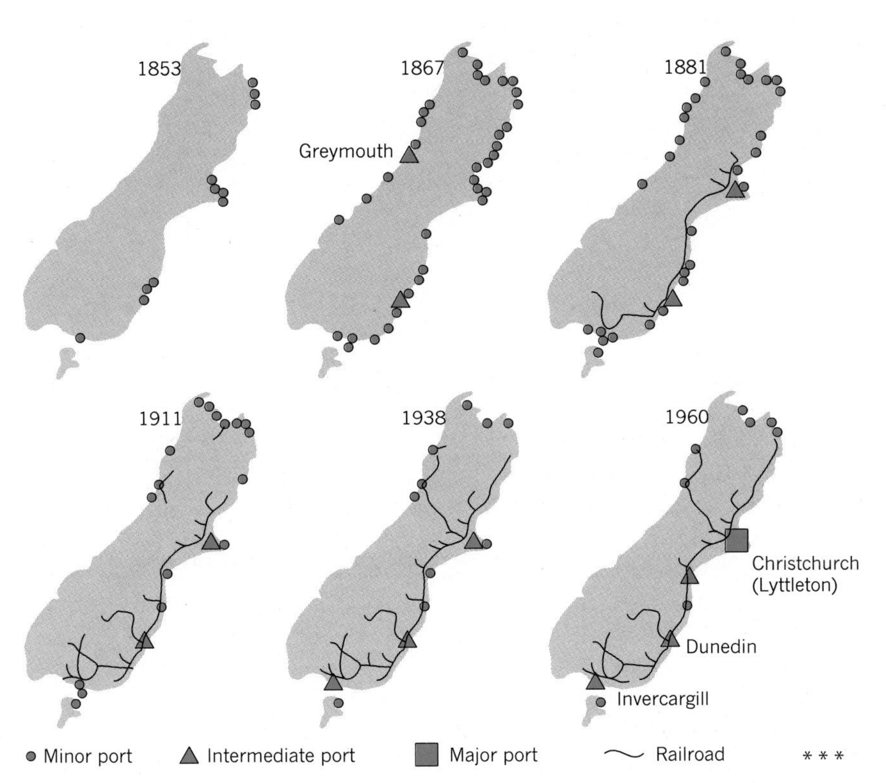

Figure 18.12 New Zealand spatial development A century of development on New Zealand's South Island shows some of the integration of transport links and the concentration of port facilities predicted by the model in Figure 18.10. Note the 'weeding out' process by which the trade of the many small ports in 1867 was progressively captured by the few larger ports. In the case of the South Island there was a strong bias toward overseas export rather than internal exchange of trade.

[Source: P. Rimmer, *Annals of the Association of American Geographers*, **57** (1967), pp. 21–7.]

* * *

100-year period. Note the pinching out of small ports, particularly on the west coast of the South Island, and the increasing dominance of the Christchurch area.

The second piece of evidence is more general. You will recall from the discussion of transport networks in Chapter 13 the ways in which geographers use graph theory. One simple measure of increasing connectivity is the ratio between the number of links in a system and the number of nodes. This is termed the *beta index*. Thus if we have a railroad system with 12 links and 8 nodes, its beta index is 12/8, or 1.50.

Figure 18.13 presents beta indexes for the railway systems of several countries. The values range from around 1.33 to 0.50; when the index is less than 1.00, it indicates that the network is split into several separate subsections as in Stage II (Figure 18.13). Highly developed countries such as France have high beta indexes, whereas poorly developed countries such as Ghana have low beta indexes.

The relation between economic development and the connectivity of transport networks is also shown by the changing position of a single country (Vietnam) over time. In the case of developed countries in the last few decades we should expect the relationship of the railway network to growth to be weaker. The closure of some rail links (and therefore a decreased beta index) would be compensated for, in such cases, by the increased connectivity of other transport links (highways, air routes, etc.).

So the Rostow–Taaffe model forms a useful, qualitative description of the spatial development process. As a pattern it allows geographers to see more clearly when an individual state or region is deviating from the normal course.

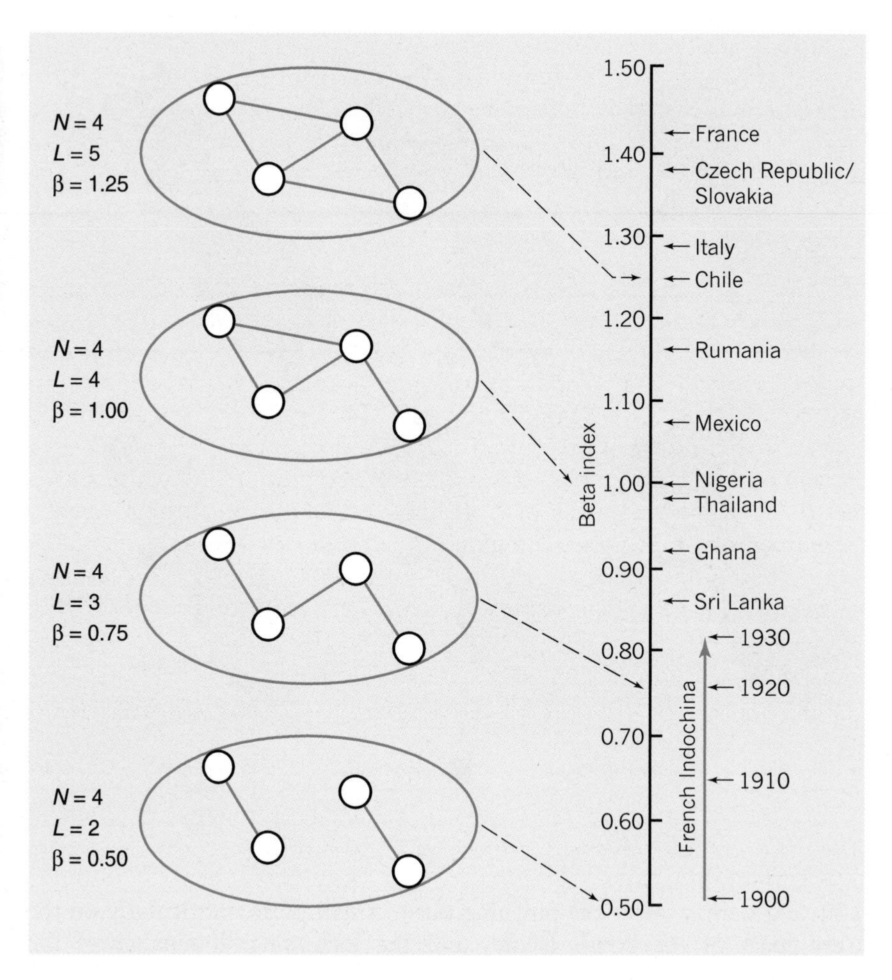

Figure 18.13 Connectivity and economic development Connectivity values (as measured by the beta index) are shown for the railway systems of countries with different levels of economic development. The maps on the right indicate the evolution of a simple transport system and the resulting connectivity values. *N* = the number of modes, *L* = the number of transports links, and β = the ration of links to nodes (i.e. the beta index).

[Source: After K.J. Kansky, *Structure of Transportation* (Department of Geography, Research Paper 84, University of Chicago, Chicago, Ill., 1963), Fig. 25, p. 99. (b) Photos by Colin Garratt and Milepost 92½.]

Friedmann's Core–Periphery Model

An alternative approach to modelling the spatial pattern of economic development has been proposed by planner John Friedmann of UCLA, the *core–periphery model*. He maintains that we can divide the global economy into a dynamic, rapidly growing central region and a slower-growing or stagnating periphery. There are four main regions in Friedmann's scheme.

(1) Core Regions First, Friedmann describes core regions, which are concentrated metropolitan economies with a high potential for innovation and growth. They exist as part of a city hierarchy and can be distinguished on several levels: the national metropolis, the regional core, the subregional centre, and the local service centre. On the international level, the North Atlantic community comprising the metropolitan clusters of both eastern North America and Western Europe may be regarded as a core region for development in the whole western world.

(2) Upward-transition Regions Friedmann's second and third regional elements are growth regions. Upward-transition regions are peripheral areas whose location is relative to core areas, or whose natural resources

lead to a greatly intensified use of resources. They are typically areas of immigration, but this is spread over numerous smaller centres rather than being concentrated at the core itself. *Development corridors* are a special case of upward-transition regions that lie between two core cities. A typical example of an expanding corridor region is the Rio de Janiero–São Paulo corridor in Brazil.

(3) Resource-frontier Regions These are peripheral zones of new settlement where virgin territory is occupied and made productive. In the nineteenth century the mid-continental grasslands of the world provided such a frontier region for grain and livestock production. In the last century agricultural colonization on this massive scale did not occur. New agricultural zones were being opened up but through much effort (e.g. the Soviet occupation of the virgin lands of Siberia, or the colonization of the Oriente, the trans-Andes lowland areas of Colombia, Ecuador, and Peru). Currently, resource-frontier areas are commonly associated with mineral exploitation (the North Slope of Alaska is a good example) and commercial forestry. The continental shelves are likely to be the important frontiers of exploitation through the twenty-first century. In a similar fashion, more intensive development of unused mountain, desert, and island areas for recreational resources is rapidly increasing their status and bringing them into this category.

(4) Downward-transition Regions The fourth element in Friedmann's model is the downward-transition region. These regions are peripheral areas of old, established settlements characterized by stagnant or declining rural economies with low agricultural productivity, by the loss of a primary resource base as minerals are depleted, or by ageing industrial complexes. The common problems of such regions are low rates of innovation, low productivity, and an inability to adapt to new circumstances and to improve their own economies.

Outside the four main types stand a few zones with special characteristics. Regions along national political borders or watershed regions fall into this group. Friedmann suggests that the four main regional types exist on various spatial scales. For instance, downward-transition areas exist on the global level (the rural part of much of the 'underdeveloped world' of Latin America and Afro-Asia) and within the cities themselves (blighted and ghetto areas), as well as on the national level (the Italian south, the Mezzogiorno). They may also vary with respect to the status of the general spatial economy of which they are part. Thus the problems of Appalachia, a depressed area within a core region, must be distinguished from those of depressed areas *within* downward-transition regions.

Friedmann's core–periphery model is linked directly to Thünen's zoning (see Chapter 15). Thünen himself thought of old Western European cities and the growing centres of eastern North America as proving a North Atlantic 'world city' about which global land-use zones developed. By the 1860s the fall in transport costs – both ocean rates and rates for overland travel by rail – had made the sheep and wheat lands of central North America, the Pampas, and Australasia equivalent to the outer rings in the Thünen model. In this historical context Friedmann's scheme fits firmly within the evolving sequence of ideas about the impact of changing accessibility of places on patterns of world development.

Long-wave Models

A third type of spatial approach is to link geographic variations in economic growth to long-wave variations in national economies. As we noted in our discussions in Chapter 16, spatial diffusion occurs not only in time but also in space. The focus or epicentre of economic growth has been changing over the last 250 years and some geographers have seen in this a clue to the striking global shifts in economic growth. (See Box 18.B on '*Kondratieff waves*'.)

Box 18.B Kondratieff Waves

A Kondratieff wave is a cycle in economic activity with a periodicity of roughly 50 years. Shorter-term variations in business activity have long been recognized by economic historians, notably the five year ***business cycle*** (or trade cycle) and the fifteen year building cycle. But it was not until 1920 that the Russian economist N.D. Kondratieff recognized a half-century long upturn and downturn, which he traced back to the late eighteenth century and which has been named after him. Its existence was largely dismissed by other economists until the severe downturn in the late 1920s and early 1930s came in, more or less, on time as Kondratieff had predicted and was followed a half-century later by the milder downturn from the late 1970s. The diagram shows the first four cycles and the start of the fifth Kondratieff.

The most widely accepted explanation of the Kondratieff cycle is that the introduction of new technology causes disruption but, once established, creates an upturn related to new products and jobs employing that technology. Successive Kondratieff waves have been linked to the steam engine (wave I), railways (II), electric power (III), automobiles (IV), and electronics and biotechnology (V). Inevitably, historians have argued about the precise nature of the cause-and-effect chain but the general shape of the waves (whatever their precise explanation) is now clear.

Geographers have been interested in the way in which different Kondratieffs appear to be linked to different regions of the world. Britain was the centre for the first two waves, and joined by Germany and the United States for the third. The two latter states dominated the fourth wave while Japan is taking a leading role in the fifth. Geographer Peter Hall has studied the present wave (the so-called Fifth Kondratieff) and placed it in a long-run context in P. Hall and P. Preston, *The Carrier Wave, 1846–2003* (Unwin, London, 1988). See also M. Marshall, *Long Waves of Regional Development* (Macmillan, London, 1987).

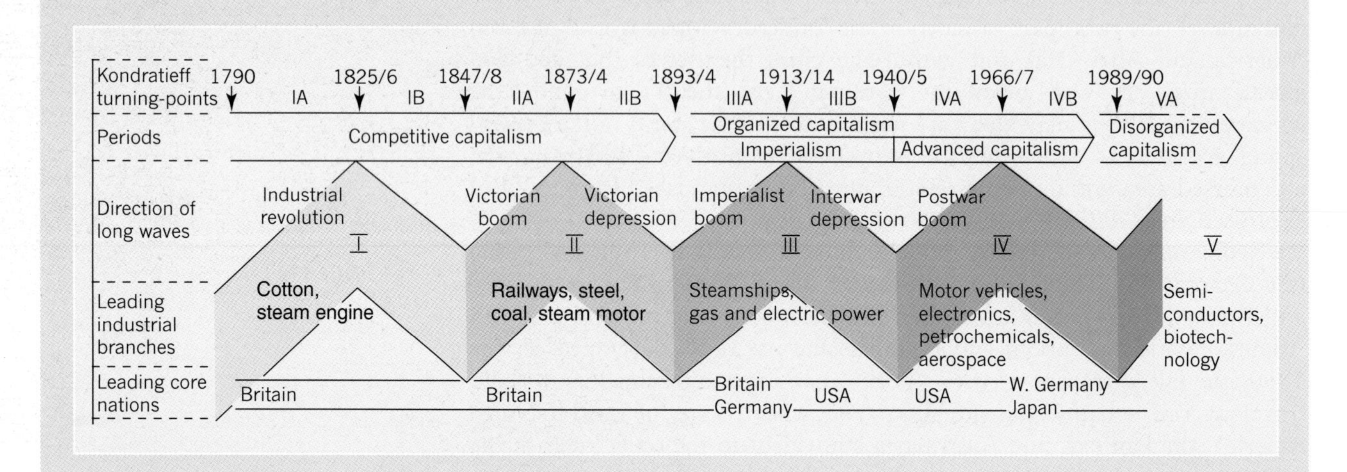

Differentiation within Countries 18.4

Before we look at the details of regional-planning practices, we must investigate three more general issues. What is spatial inequality? What is social welfare? And what is a geographically just distribution? None of these questions can be answered precisely, since the answer to each depends on the reader's view of society itself. This first section may perhaps serve as a framework for a debate that runs far outside the bounds of our particular focus of inquiry.

Lorenz Curves: Measuring Inequalities in Welfare

Our first question is a technical one, and produces much less emotional reactions than the others. Given that differences between regions exist, how should we measure them?

One of the most useful measures of inequality is the ***Lorenz curve*** (Figure 18.14). This is a graphic representation of the distribution of any measure

Figure 18.14 Measures of inequality (a) The distribution of population and income among the four main ethnic groups in South Africa. (b) A Lorenz curve showing the inequalities between the four groups. The shaded area represents an 'inequality gap'. (c) Poor housing conditions in part of the Soweto township, Johannesburg.

[Sources: (a), (b) South African data for 1967 from D.M. Smith, *An Introduction to Welfare Geography*, University of Witwatersrand, Johannesburg, Department of Geography, Occasional Paper No. 11, Table 5, p. 95. (c) Magnum photos.]

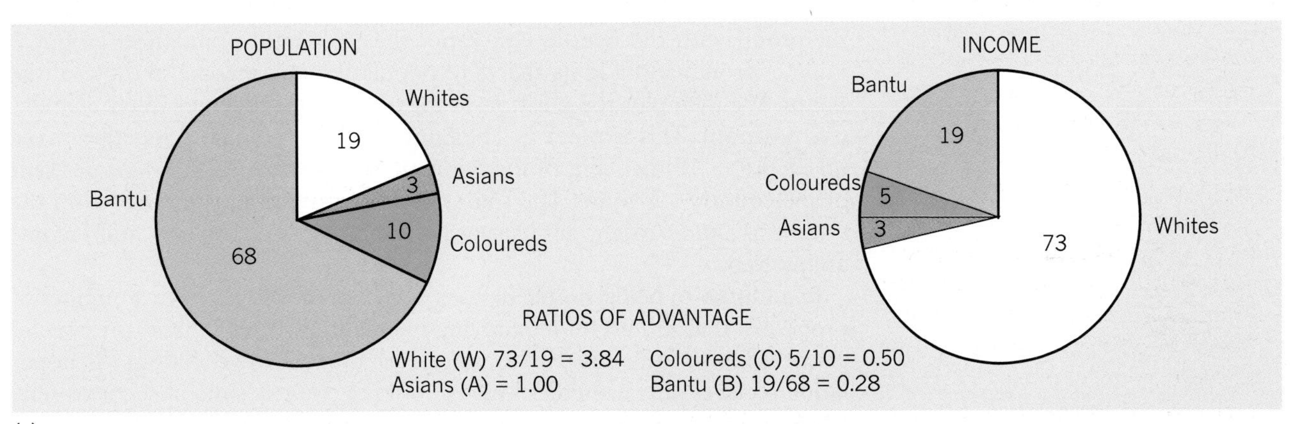

(a)

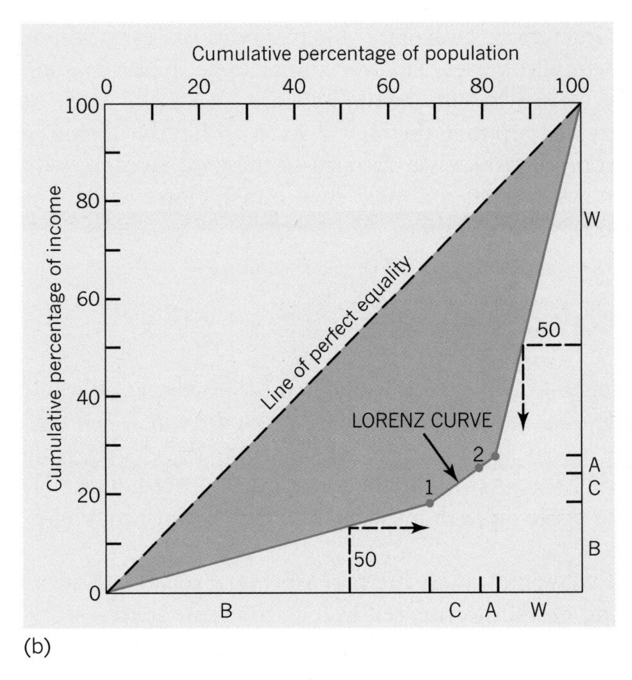

(b)

(c)

of welfare (e.g. income). If it is perfectly straight, the distribution is perfect. The more bowed a Lorenz curve is, the more unequally the proxy for welfare is distributed. The difference between an actual Lorenz curve and a straight line is called an *inequality gap*.

The construction of Lorenz curves is illustrated in Figure 18.14, using the distribution of income in South Africa as a measure of the distribution of welfare there. If we divide up the South African population into major ethnic groups, there are strong contrasts in income per capita. The Bantu people make up about two-thirds of the population and receive about one-fifth of the income. Conversely, the white South Africans make up about one-fifth of the country's population but control nearly three-quarters of its national income.

By dividing their share of the country's income by the per centage of the population they include, we can compute a *ratio of advantage* for each ethnic group. Thus for the Bantu this ratio is 19 (their share of income) divided by 68 (their share of the population), or 0.28. Ratios above 1.0 indicate that a group is better off than the average group in the nation, and ratios below 1.0 that it is worse off. To draw a Lorenz curve, we take the group with the lowest ratio, in this case the Bantu, and plot its position on a graph of population and income (point 1 in Figure 18.14). We then take the group with the next-lowest ratio, the 'coloured' population (about 2 million strong) and add its shares of population and income to those of the Bantu. We then plot the position of the two groups together (their cumulative position). This is point 2. The Bantu and the 'coloured' together make up 78 (68 + 10) per cent of the population and share 24 (19 + 5) per cent of the country's income. The shares of remaining groups are added in the same way, and further cumulating per centages of population and income are plotted.

In addition to being simple to construct, Lorenz curves have a number of properties that are useful in studying inequalities. If we look at the two 50 per cent marks in Figure 18.14, we can see that the lower half of the population receives only around 13 per cent of the country's income; conversely, half the income goes to only around 15 per cent of the population. South Africa is used here only as an example. Lorenz curves for all the world's countries have this characteristic convexity. Inequality exists everywhere, though the degree of inequality varies. This variation can be shown by comparing countries with very different inequality gaps. An LDC such as Thailand has a much greater gap than the HDCs. Even within the advanced countries, there are strong differences in the size of the gap. Sweden, with its progressive taxation system, has a Lorenz curve much closer to the line of perfect equality than the United States.

Alternative Indicators of Welfare

So far we have used income to measure welfare. It is, however, only one index, and a very crude one, of the social welfare of an area. A *social welfare indicator* is simply a term for a statistic which shows the change from 'worse-off' to 'better-off' areas. As you might guess, it has proved extremely difficult to get people to agree on such measures, since they must first agree on what social welfare is.

From the geographic viewpoint, there are two aspects of social indicators we should note. First, no two indicators tell exactly the same spatial story. Consider Figure 18.15, which shows contrasts in the ten provinces of

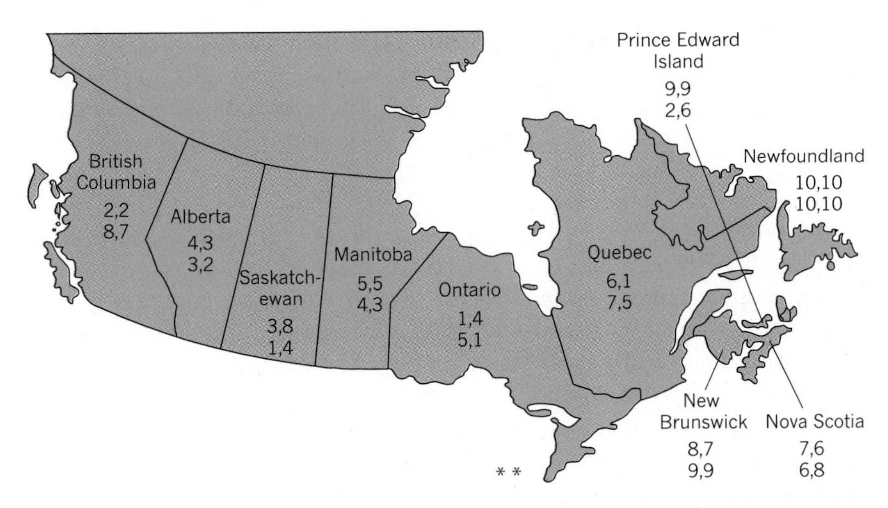

Figure 18.15 Spatial variations in socioeconomic health Each of the ten Canadian provinces is ranked using four alternative measures of income and job opportunities. 'Healthy' provinces have low rank scores, and 'unhealthy' provinces have high ones. Only Newfoundland, which ranks tenth every time, has wholly consistent sources.

[Source: Data from P.E. Lloyd and P. Dicken, *Location in Space* (Harper & Row, New York, 1972), Table 10.5, p. 207.]

Canada in the 1960s. These contrasts have been measured by four different indicators of social welfare: (a) per-capita income; (b) higher education, as measured by the number of college enrolments; (c) average unemployment rates; and (d) average job-participation rates for males.

Each of these indicators tells us something about the 'health' of the provinces in terms of, say, jobs and the potential skills of the population. In Figure 18.15 we have superimposed on each province its position in the marked list of Canadian provinces with respect to each indicator: a 1 indicates a position at the top of the table (the 'best' score), and a 10 indicates a position at the bottom of the table (the 'worst' score). Of the ten provinces, only Newfoundland has a consistent set of scores. The problems of the Maritime provinces show up, as does the relative prosperity of Ontario and Alberta. British Columbia is strikingly well off in some ways but badly off in others.

Ranking is of course a rough measure of contrasts, since small differences may force a province into a low position. 'Roughness' is also a problem in Figure 18.16, which moves south of the border and looks at

Figure 18.16 Interstate differences in the quality of life The graph shows the relative ranking of 3 states (out of the 50) in terms of the nine social welfare indicators. Low rankings indicate a high quality of life; high rankings indicate poor conditions. The nine indicators were derived by statistical procedures from 85 different variables.

[Source: Data from J.O. Wilson, *Quality of Life in the United States* (Midwest Research Institute, Kansas City, Missouri, 1969), p. 13.]

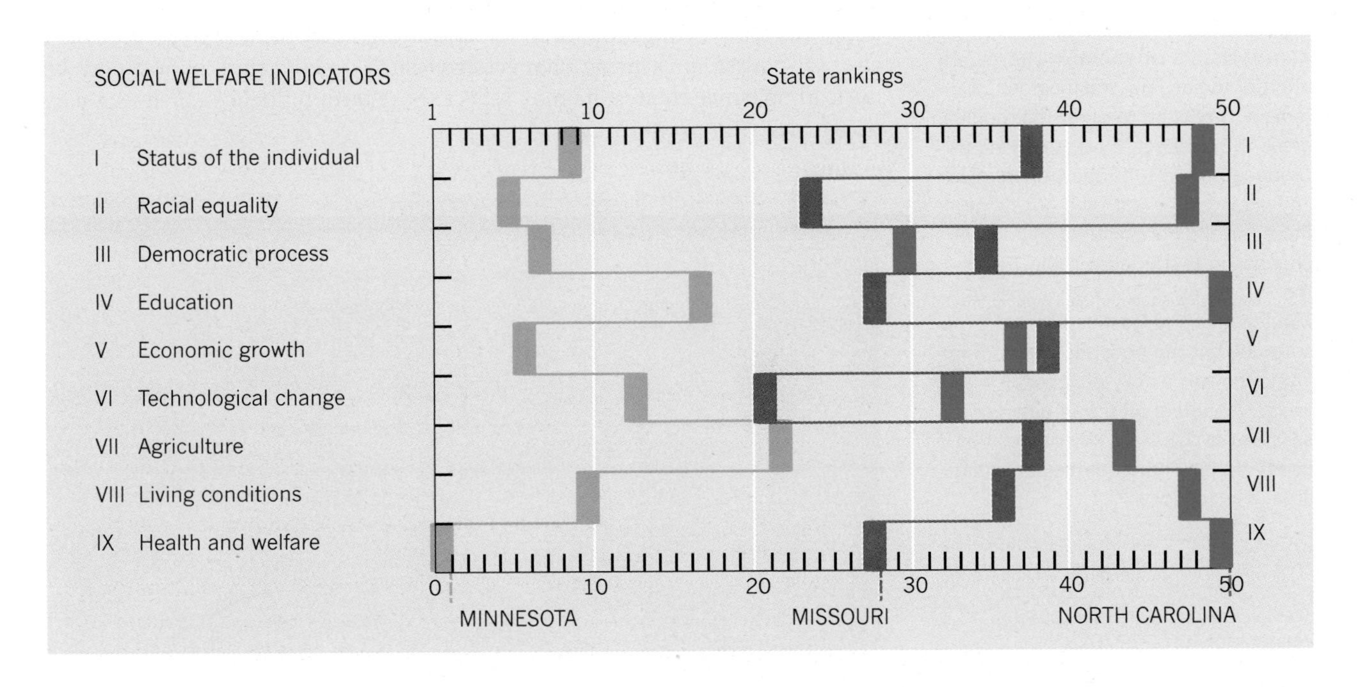

interstate contrasts in the United States. Here the nine measures are much more complex. They were produced by Kansas economist J.O. Wilson from the 'domestic goals' proposed by Eisenhower's Commission on National Goals. Each index was created by collapsing a large number of raw measures of welfare in much the same way measures of wealth were collapsed in Berry's index.

The profile of each state shows its rank position on each index. Profiles are plotted for three states with different average performances. Minnesota is in the upper half of the ranked list of all states, with particularly high rankings on indicators II and IX. North Carolina is in the lower half and Missouri is consistently in the middle of the table. *No* state has an entirely consistent pattern of rankings. For example, California, whose position on the first eight indicators ranges from first to fourth, plunges to fourteenth on the health and welfare index. Hawaii ranks first (ties with Utah) with respect to racial equality but fortieth with respect to technological change.

The second general question raised by social welfare indicators is one of stability. Do we wish to measure inequality in terms of an area's position in any one year or decade? Or should we be more interested in rates of change? Figure 18.17 illustrates the contrast between the answers to the two questions in a study of the quality of life in eighteen metropolitan areas of the United States. Urbanists Jones and Flax, on whose data the chart is based, wanted to see whether living conditions there were getting relatively better or worse. Using seven measures of the quality of life (ranging from housing costs through robbery rates to pollution levels), they came up with the rankings shown on the longer, horizontal axis of the chart. Minneapolis headed the list of high-ranking cities, and Los Angeles trailed the list of low-ranking ones. Washington, D.C., did not do too badly, but San Francisco – a shock to Bay Area fans (including the author) – was next to the last.

How were conditions in these centres changing? The results of a second study on changes during the 1960s are shown on the shorter, vertical axis of the chart. While conditions in a number of low-scoring cities such as New York and Chicago were getting better, the quality of life in Washington, D.C., was getting relatively worse. In San Francisco, according to the study, conditions were deteriorating more rapidly than anywhere else. Of course, measuring changes over such a short time period may be misleading, since the results may reflect short-term influences such as a particular mayoral programme.

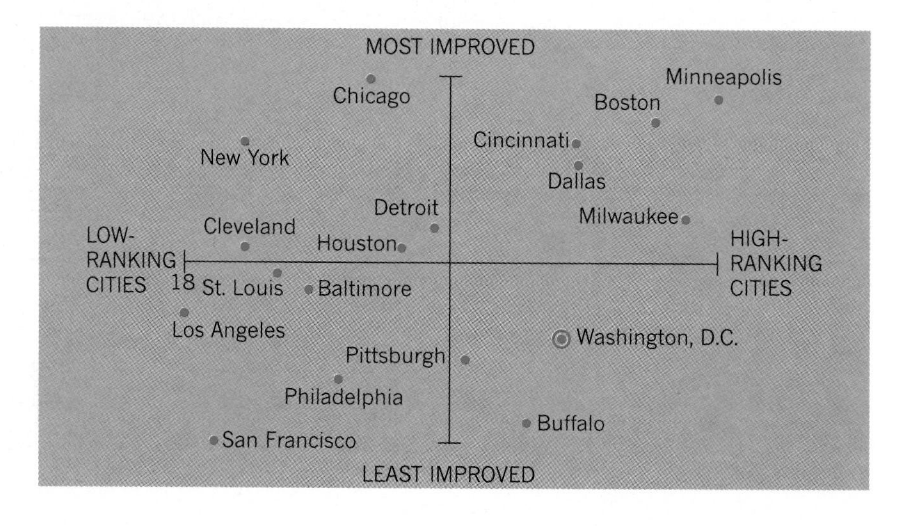

Figure 18.17 Intercity comparisons of social welfare An attempt to see how Washington, D.C., ranked in relation to other metropolitan areas with respect to living conditions produced the results shown here. Each city is ranked on both the quality of life there (on the horizontal axis) and on changes in that quality in the 1960s (on the vertical axis). The high scores of Minneapolis and Boston were no surprise, but the poor showing of San Francisco was wholly at variance with its image. Note that all urban data are sensitive to the exact city boundaries used and how much of the suburban area is included.

[Source: Data from M.V. Jones and M.J. Flax, *The Quality of Life in Metropolitan Washington, D.C.: Some Statistical Benchmarks* (The Urban Institute, Washington, D.C., 1970).]

Spatial Aspects of Social Justice

Let us assume that we have agreed on a proper measure of welfare and that we have found marked inequalities among provinces, states, metropolitan areas, or other regions. What then? A concern with spatial inequalities in welfare must raise, for geographers, ethical issues which have troubled philosophers from Aristotle to Marcuse. As Johns Hopkins geographer David Harvey puts it, how do we achieve 'a just spatial distribution, justly arrived at'?

What kind of claims can the people of any disadvantaged region make on the larger, national community? From the welter of conflicting suggestions three major ideas stand out. First, they can make claims based on *need*. We may argue that all parts of a country should have a basic right to a certain standard of education or medical care, regardless of spatial differences in the cost of providing those services. Postal costs in remote rural areas in most countries are well above the national norm, but postal charges are usually common all across the country. Second, they may make claims based on their *contribution to common good*. Areas that contribute greatly to the good of the whole nation might be expected to receive a greater-than-average payment. A special subsidy to a city faced with the problem of preserving costly old buildings of historical value (e.g. as Amsterdam is in the Netherlands or Venice is in Italy) would be merited because of the city's contribution to the heritage of the whole country.

Third, the people of a region may make claims on the larger community on the basis of *merit*. The environmental challenge to human life varies from region to region. The allocation of extra resources to a region might

Figure 18.18 Enigmas in spatial justice The chart reveals the spatial implications of four alternative approaches to revenue sharing in a simplified five-region country. The arrows indicate the directions of transfers to the subsidized areas (shaded). Rearranging the diagrams (putting the poorer areas at the centre and the richer ones on the periphery) would give us a picture of the problems in big cities today.

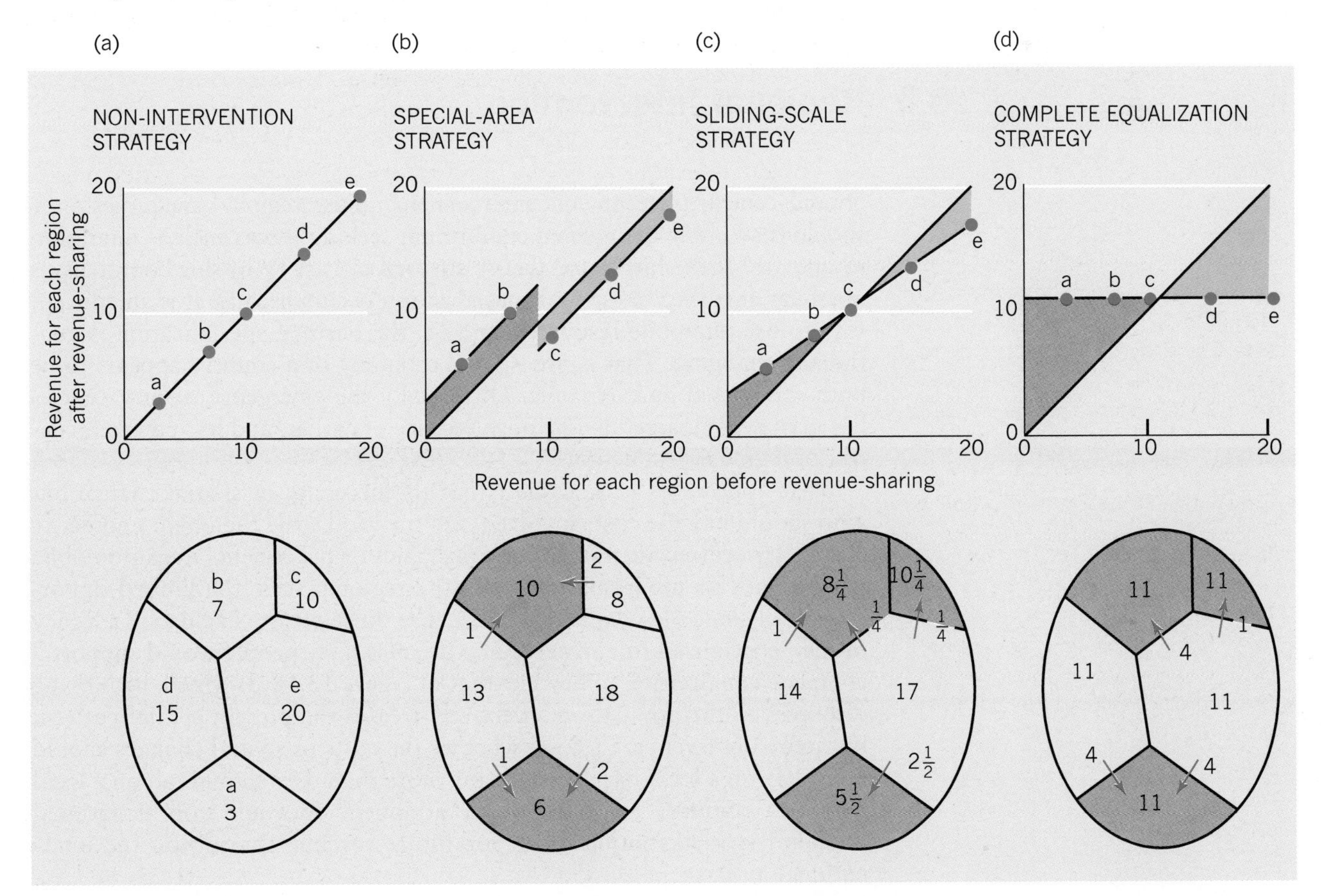

be justified to protect areas of great potential stress from natural hazards (e.g. to reduce crime in high-risk areas in inner cities and ghettos). Of these three bases for the distribution of social resources, need is arguably the primary one, with contributions to the common good and merit running second and third, respectively. Readers interested in following up these lines of thinking should definitely browse through David Harvey's *Social Justice and the City*.

Let us stay with the simple concept of need and look at its spatial implications. Figure 18.18 shows how different philosophies might affect a spatial redistribution of wealth in an idealized five-region country.

Four situations are described. In the first the government's policy is a 'laissez faire' one of non-intervention, with no revenue sharing (Figure 18.18(a)). In the second, problem areas (the two poorest areas) are designated as needing special help (Figure 18.18(b)). This kind of approach – having a special strategy for special areas – poses problems we shall consider later in the chapter. In the third situation, there is a sliding-scale approach to regional needs (Figure 18.18(c)). This is rather like the negative-income-tax approach, in which rich areas subsidize poor areas on the basis of need. In the fourth, a complete-equalization approach is taken and taxes are adjusted so that all areas are brought to the same income-level (Figure 18.18(d)). In each case, a different interpretation of social justice leads to a different spatial pattern of levels of welfare. These cases are, however, highly simplified, and deal with a hypothetical country. We turn now to realistic examples of current regional-intervention policies in western Europe and North America.

18.5 Regional Intervention

Should central governments intervene to adjust regional inequities – or should they allow the normal equilibrium-seeking forces such as migration to operate? If the locational theory stressed in Part IV of this book tells us anything, it is this: (1) spatial specialization occurs because it is an efficient way to use immobile resources; and (2) the basis of specialization is continually changing. That is, the spatial economy of a country appears to be both specialized and dynamic. Historically the emergence of new centres has been accompanied by the obsolescence of earlier, and less efficient, centres of regional production.

What appears to be at issue is not the necessity of spatial change but who should pay the costs involved. On the local scale the blight and decay in our city centres are part of the price paid for the benefits the automobile has brought to suburban areas. On the regional scale the limited opportunities of some old coalfield areas such as those of Appalachia are a legacy of their specialized role in preceding decades. Few people would support a complete equalization policy like that in Figure 18.18(d) which, by moving resources to the population, effectively freezes the present spatial pattern; but more observers are asking whether the costs of spatial changes should fall solely on a local part of a city any more than they should fall on a local part of a country. The logic of the argument does not stop at national boundaries; it has implications for future revenue sharing on the international level.

Tools of Regional Policy

If central governments wish to intervene in the regional growth process, what tools can they use? Three main strategies are currently in use.

The first is investment in the *public sector*. Regional investment of this type spans the construction of entire new cities in underdeveloped regions, such as Brasilia in west-central Brazil (Figure 18.19(a)), to the building of new schools in a city ghetto. It is commonly aimed at improving the basic infrastructure of a region. Transport and power facilities are generally the prime targets for improvement. The dam-building by the Tennessee Valley Authority (Figure 18.19(b)) and the road-building in the programme for developing Appalachia exemplify the emphasis in regional-support legislation. China has also used *special economic zones* located along parts of its coastline to concentrate industrial and economic growth.

Inducements to business in the *private sector* to invest in a region are a second strategy. These inducements may be positive, such as capital grants or tax concessions to industries already operating or willing to operate in undesirable areas, or negative, such as controls and penalties for companies in fast-growing areas. Companies operating in rapidly developing areas may have proportionately higher taxes and face legal restrictions on their expansion.

Figure 18.19 Tools of regional policy
(a) A large-scale national reorientation or resources underlay the building of a new federal capital at Brasilia by the Brazilian government. The new capital lies on the central plateau of Goias, about 900 km (560 miles) inland from the old coastal capital of Rio de Janeiro. Begun in the late 1950s, the new capital city has a population approaching 500,000. (b) Public power and transportation programme sectors are typified by the Norris Dam in the Tennessee Valley Authority programme in the 1930s. (c) Three Gorges Dam: construction of one of the world's largest civil engineering projects, the Wanjiazhai Dam on the Yellow River in Shanxi Province, China. (d) New agricultural settlements in Israel's southern arid zones are financed by the government.

[Source: (a) Erwitt, Magnum. (b) Popperfoto. (c) Popperfoto/Reuters. (d) Courtesy of Israel Information Services © Robert Capa/Magnum.]

(a)

(b)

(c)

(d)

Inducements to *individuals and households* to locate in or leave a region are a third strategy. Migration from a declining area may be hindered by the inability of would-be migrants to sell their houses or land. Compensation to those farmers willing to move (plus aid to enlarge the farms of those who stay) is offered by the governments of both Ireland and Sweden to make migration from agricultural areas with low and declining prospects easier. When the need is to attract population, similar monetary and other inducements are used.

The choice of regional policy tools depends largely on the resources available in the country as a whole and, more importantly, on its sociopolitical system. For instance, Britain tried to resolve the problem of unemployment in its peripheral coalfield areas by loans for new industry (initiated in 1934), building factories and industrial estates (in 1936), providing tax incentives for new industry, including depreciation allowances (in 1937), controlling industrial buildings outside the problem areas (in 1945), and awarding standard grants for new industrial buildings and new machinery (in 1960). The measure of central government control of industry's location in Britain is representative of that in a politically middle-of-the-road state with a mixed economy and a strong commitment to equalizing job opportunities between regions. States following a less socialist policy tend to allow industry to decide its location on commercial grounds; those following a more socialist policy favour a greater degree of direct control by central government.

Defining Areas of Need

To allow regional policies to be introduced, we must draw a clear distinction between problem and non-problem areas. The boundary chosen should clearly reflect the character of the inequality. To take a simple example, in the 1930s the Brazilian government defined the boundaries of the Nordeste (a problem region in the northeast 'bulge' of that country) by rainfall figures. There the key problem was recurrent drought, leading to crop failure and famine, so it made sense to draw a boundary around the aided area in rainfall terms. Regions within the drought boundary received special government help, while regions outside did not.

We encounter much greater difficulties when a problem region is defined by population characteristics and levels of distress. In some countries the basis for preferential regional treatment is unemployment. But unemployment rates are very unstable over time. They tend to underestimate job opportunities; that is, areas of high unemployment also have outmigration and a low proportion of folk in jobs (e.g. there may be no jobs for women in the labour market). This means that the official unemployment figures are too optimistic in problem areas. Thus, before we can establish any realistic index of relative welfare, we need to modify our definition of unemployment to include other variables related to migration or wage levels.

No matter what index of regional need we adopt, we are left with the question of where to draw the line. For example, if we adopt a 4.5 per cent unemployment rate to define the regions that will receive aid, then we invite protests from just below the threshold. An extreme if simplified illustration is shown in Figure 18.18(b) where unaided region c is made worse-off than aided region b. The limit may conceal sharp contrasts within the distressed region itself. The difficulty is made more acute by the facts that the same index may yield different values if different regional subdivisions are chosen. Generally, the smaller the system of subdivisions, the greater is the spread of index values, and vice versa. This gives great scope for gerrymandering the boundaries of a distressed region to make it qualify for government aid.

One reaction to this problem has been the replacement of a twofold division (between regions that receive aid and those that do not) by three or more levels of aid. Such a system has the theoretical advantage of matching the amount of aid to the degree of distress, but is objected to on the grounds of increasing administrative costs. Figure 18.20 shows the situation in Britain (a) at the height of regional intervention in the 1970s and (b) at the end of the twentieth century. There was then virtually a four-level system of regions: (1) areas of rapid growth and high prosperity subject to negative controls on further expansion (e.g. Greater London), (2) 'normal' areas that require neither positive nor negative intervention (e.g. most of southern England), (3) areas with moderate economic difficulties (e.g. Plymouth), and (4) depressed areas (e.g. South Wales) receiving the full range of government assistance. France has a similar five-stage series of zones, with metropolitan Paris and rural Brittany at opposite ends of the scale.

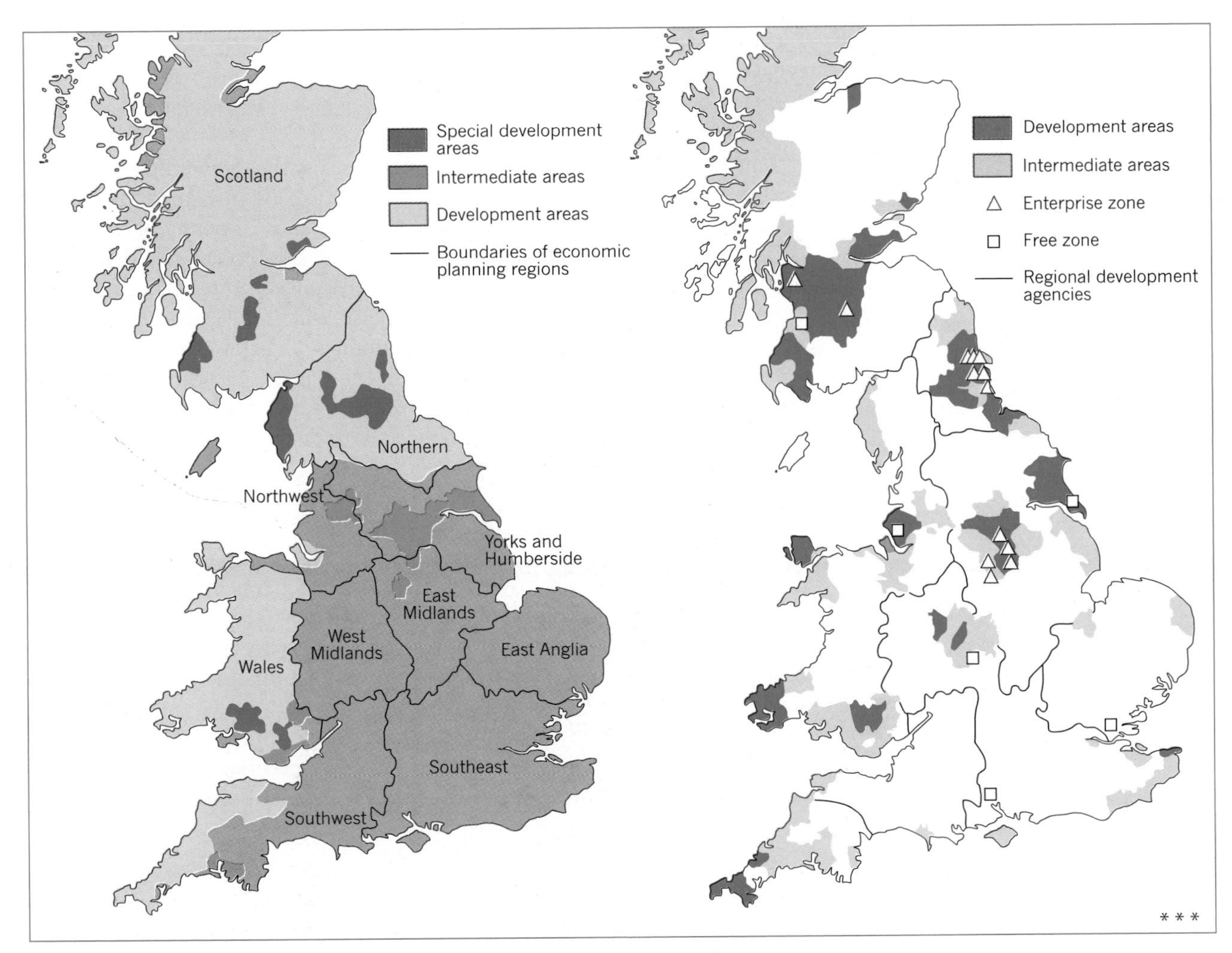

Figure 18.20 Changing zones of government aid (a) In the 1970s Britain was divided into four types of areas, which receive varying degrees of incentives to industrial development. (b) By the end of the century a succession of British governments had juggled the boundaries and played with the particular range of policies followed. Those shown now reflect European Union Policy.

The difficulty of having a fully differentiated scheme of regional assistance is that, at one extreme, it heavily penalizes the very productive regions, while at the other it grants massive help to the least productive ones. If it was successful, it would freeze the existing geographic pattern of production. Such a freezing is, of course, wholly at variance with the change and diffusion already discussed. Indeed, it would be disturbing to think what effect a fully successful regional aid policy would have had on the United States if it had been in operation in 1920, or in 1820! There are clearly some regions (e.g. worked-out mining areas) where levels of unemployment are so high and the prospects so poor that it is hard to justify a policy other than strategic withdrawal. However *within* problem areas discriminatory policies should favour those parts of the problem regions that have the greatest growth potential. It is to this idea that we now turn.

Growth Poles and Withdrawals

Growth Poles The notion of productive points within, or as near as possible to, distressed areas as centres of new investment is an attractive one. It has been strongly urged by French regional economist Francis Perroux in his concept of the ***growth pole*** (*pole de croissance)*. Basically, a growth pole consists of a cluster of expanding industries, which are spatially concentrated (usually within a major city) and which set off a chain reaction of minor expansion throughout a hinterland.

A growth pole in regional development means the deliberate selection of one or a few potential poles in a problem area. New investment is concentrated in these areas rather than being spread thinly 'in penny packets' over the whole area. The arguments in favour of this policy are that public expenditures are more effective when they are concentrated in a few clearly defined areas and that new industries there will stand a better chance of building up enough agglomeration economies to achieve some degree of self-generating growth. Agglomeration economies are the benefits that come from the sharing of common infrastructure (roads, power supplies, water, etc.), the increased size of the labour market, and reduced distribution costs. Although the evidence is conflicting, the necessary population of a growth pole may lie between 150,000 and 250,000 people. Only in cities of this size will there be the basic ingredients of large-scale diversified growth. Whether spatial concentration is an *essential* part of growth pole theory is open to doubt. Some plants – petrochemical or metallurgical smelting plants, for example – may need specific locations away from urban concentrations of population. They bring benefits to the poorer region in terms of increased income (particularly through contributions to local taxation) rather than through more jobs.

In practice a growth-pole policy is likely to run into two kinds of difficulties: the technical ones of selecting the best potential pole, and the political ones of convincing the unsuccessful poles of the wisdom of the policy.

Strategic Withdrawal as a Regional Policy An inverse growth-pole policy may be used when the general economy of an area is experiencing a long-term structural decline. For example, Figure 18.21 presents a local

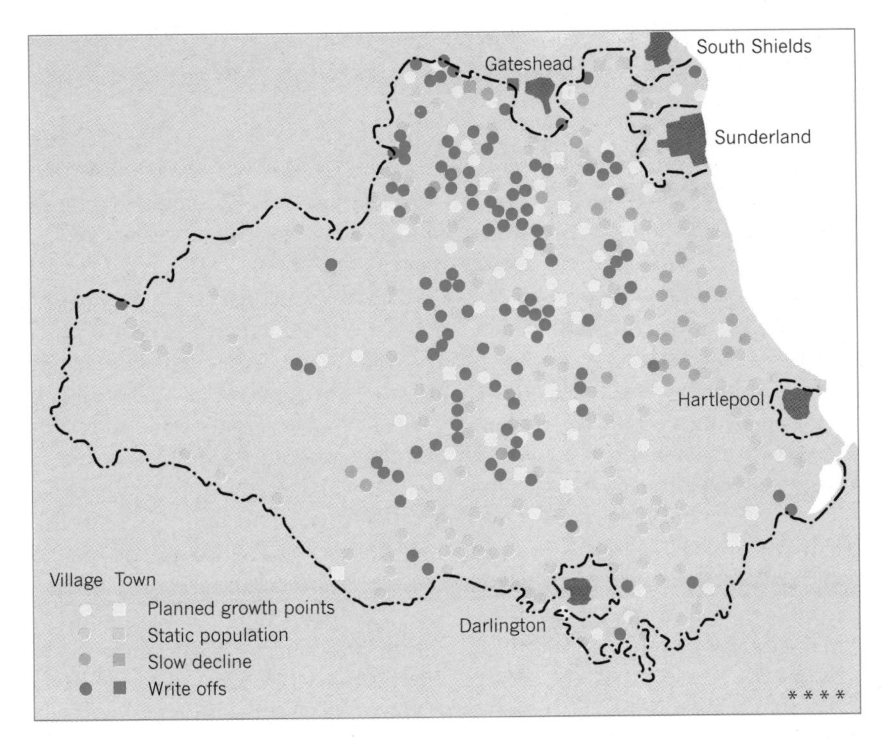

Figure 18.21 A local strategy of spatial concentration and withdrawal This historical map of County Durham in northeast England is based on a development plan adopted in the 1970s. Since then, some settlements have been reclassified. The fourth group of villages, termed 'write-offs' by the plan's critics, comprises mainly small mining villages in which the colliery has closed. In these settlements new capital spending was to be limited to the social and other facilities needed for the life of the existing property. The plan has since been radically revised and is now largely of historical interest.

[Source: Data courtesy of County Planning Department, Durham County Council, England.]

government policy of *strategic withdrawal* toward the mining villages in the Durham coalfield in the northeastern part of England in the 1970s.

This area had its heyday during the nineteenth century when Durham coal was in demand in a country that was a world leader in the Industrial Revolution. Today the mines have all closed, made uneconomic by the competition from more efficient fuels and from open-cast coalfields in other parts of England. Unemployment rates are high, job opportunities low, and there is a steady outmigration from the mining villages. An extreme interventionist might argue that the economic and social waste of unused labour and an underused infrastructure (roads, railways, schools) should be prevented by subsidizing new employment opportunities and by fixing high prices for coal. Conversely, we could argue that migration is an effective solution to the problems of the area, that subsidies can be granted only at the expense of the more productive sections of the country, and that the new generation (and particularly their children) will lead happier and fuller lives once they have established themselves in the more prosperous and job-rich parts of the country.

Government policy, as Figure 18.21 shows, has reflected a mixture of both viewpoints. On the regional level the response has been interventionist. A new town has been built, new industries have been attracted to it, new roads constructed, and so on. On the local level, however, the situation is different. The smaller and more remote villages are being progressively abandoned, and people are being encouraged to move to the larger and better endowed areas. What we are seeing is, in effect, a policy of strategic withdrawal that balances the social costs of non-intervention (borne locally) against the economic costs of intervention (paid largely by the rest of the country).

Box 18.C TVA as a Historical Development Model

The Tennessee Valley Authority (TVA) Act was one of the important laws passed in the first 'hundred days' of the Roosevelt Administration in March 1933. It remains today of interest in creating a template which, paradoxically, has been copied widely in following decades around the world, though not within the United States itself.

The river basin of the Tennesse River covers an area of 100,000 km^2 (40,000 square miles), about four-fifths the size of England. The valley includes parts of eight states: Tennessee itself, plus (clockwise) Kentucky, Virginia, North Carolina, South Carolina, Georgia, Alabama, and Mississippi. It ranges in height from the mile-high peaks of the Great Smoky Mountains to the low, muddy plains where it joins the Ohio River. The impact of the TVA was to spread outside the river catchment itself, affecting an area as large again in adjacent states.

There were four basic elements in the 1933 TVA programme. First, it was to control floods on the river; second, it was required to develop navigation on the river; and third, it was to produce and market electricity 'as far as may be consistent with these purposes'. The key instrument in meeting these objectives was dam building. Nearly 40 were eventually built on the Tennessee and its branches, over half constructed by the TVA itself and the remainder being taken over from existing companies. What marked the new system as it evolved during the next quarter-century was the way the individual dams were integrated into a single system, making it one of the most effectively controlled waterways in the world. Dams were of two general types, high dams in the mountainous upper reaches of the river (largely confined to power production) and low dams in the broad, lower stretches of the river (where power production was combined with locks allowing navigation from one lake to the next).

The TVA progressively moved outside its three main purposes to provide a more integrated approach to the development of the basis. Three auxiliary roles were added: (a) the encouragement of improved forestry and farming practices as part of its water and erosion control mandate; (b) development of outdoor recreational facilities (e.g. fishing, boating, camping); (c) a rural health programme, particularly the eradication of malaria (achieved by 1948). Critics of the TVA accused it of moving too far outside its original brief and usurping the role of other federal, state, and local authorities. The constitutionality of the TVA was challenged by opponents but was upheld by the Supreme Court in *Ashwander vs TVA* (1936) and subsequent decisions. Supporters saw it

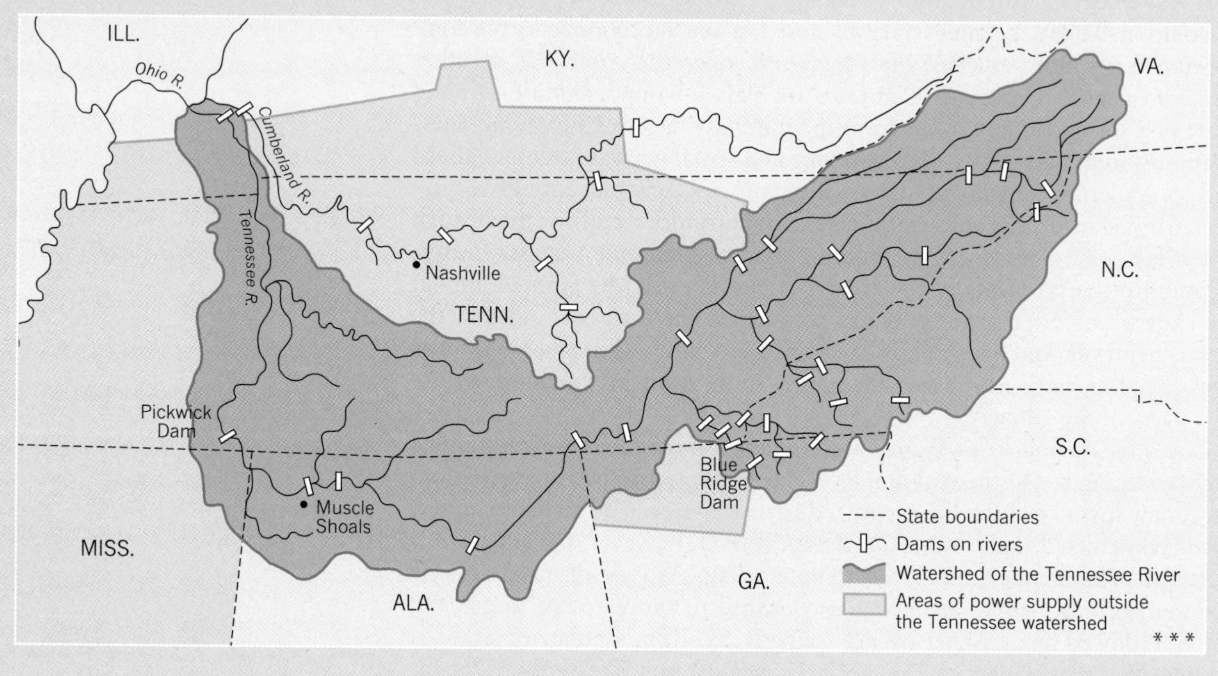

* * *

Box continued

as a new model for tackling the ills of a region previously among the poorest in the United States.

The thousands of visitors to the TVA each year contained many government officials from foreign states who saw there a recipe for regional planning that might fit their own needs. It proved an inspiration for the Damodar valley development in India, the Cauca river project in Colombia, and the Volta scheme in Ghana. At the international level, its influence can be seen in the Mekong Valley project in Vietnam and Cambodia. In Australia, the Snowy Mountains scheme started in 1949 also noted its debt to the TVA although there the major emphasis was on water diversion across watersheds for irrigation use. A good account of the project is given in P.J. Hubbard, *Origins of the TVA* (Harper & Row, New York 1961).

Regional Planning in the United States

The notion of regional planning in the context of the five-year plans of communist states such as the former Soviet Union or the social-democratic traditions of Sweden or Britain is a familiar one. The political institutions of a country such as the United States make it harder for the central government to intervene at the regional level. What kind of regional planning occurs within a more capitalistic state such as the United States?

In the 1930s regional planning in the United States was rather narrowly confined to the development of water resources, with special attention to river-basin planning. President Franklin Roosevelt's establishment of the Tennessee Valley Authority was an outstanding example of this kind of intervention, which was widely copied by other countries. (See Box 18.C on 'TVA as a historical demonstration model'.)

Federal intervention in the economic life of the nation was, however, seen as the thin end of the wedge of a socialist-style intervention in some quarters. Thus it was not until the 1960s and the Kennedy Administration that the Area Redevelopment Act (1961) was passed. Under this act areas with unemployment rates greater than 6 per cent or running at specified levels above the national average were eligible for assistance. Although over one thousand counties were designated to receive aid (see Figure 18.22), the total impact of the act was rather small. Some industrial parks were

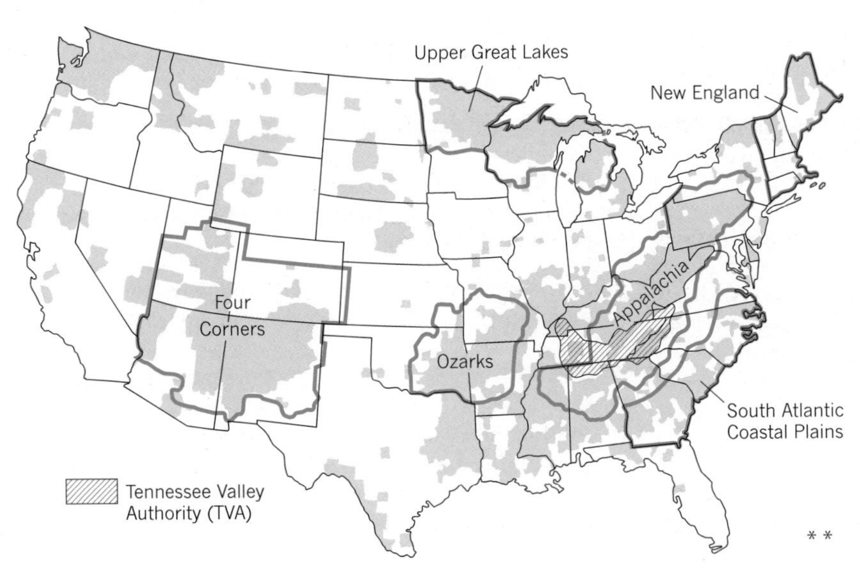

Figure 18.22 US regional planning The map shows two examples of federal programmes in the United States in the 1960s and the 1970s. The shaded areas indicate counties designated to receive aid under the Kennedy Administration's Area Redevelopment Act. Heavy boundaries enclose the six interstate planning regions set up by the Johnson Administration's Economic Development Act. The main federal effort has been in the depressed Appalachian area, where it overlaps with part of the TVA area set up in the early 1930s (see Box 18.C). Since the 1970s federal interest in major interstate development programmes has diminished.

established, but the major share of the act's resources went for recreation and tourist projects.

The major contribution of the Johnson Administration was the passage of the Appalachian Regional Development Act in 1965, which involved federal aid being coordinated on an interstate basis. The main areas of public investment were transportation (particularly through extensions of the interstate highway programme), the development of natural resources, water-control schemes, and social and educational programmes. By 1968 five further interstate areas had been set up under the Economic Development Act.

Aid appears to have been spread somewhat thinly, and the total funds available were, by western European standards, small in relation to the United States' huge GNP. Later administrations (Carter, Nixon, Reagan, and Clinton) gave less emphasis to regional policy. By then public concern had shifted somewhat from the plight of the less prosperous rural regions to the more concentrated problems of the inner cities. Currently it is the plight of the cities and the problems of metropolitan revenue-sharing that dominate the scene. As the suburbs have sprawled outward and industry has decentralized, more American cities have been left to cope with a severe housing blight, high crime rates and other inner city ills, and a dwindling tax base.

Reflections

1 World economic development shows very stark contrasts (see Table 18.1). Why are so many of the world's poorest countries located in the tropics in general and in Africa south of the Sahara in particular? Do you rate the climate of tropical countries (a) a major factor or (b) an irrelevant point in your explanation? Select any one less-developed countries that interests you. Use encyclopaedias and reference books or Internet sites (see Appendix B) to look into its background. Which factors appear to be the most critical in accounting for its low level of development?

2 Measuring the prosperity of a nation-state or region requires more than a single index. Consider the three different listings of the 'richest' countries given in Table 18.2. Contrast the countries that score highly on all three indexes with those that occur on only one or two of the three.

3 How do geographers measure spatial inequality? Assume that a country is divided into five regions, each with 20 per cent of the country's population but with unequal shares of its total income (say, 50, 25, 15, 8, and 4 per cent, respectively). Plot this variation in regional incomes as a Lorenz curve (see Figure 18-14). (Go on to plot actual curves for your own county, province or state, if you have the time.)

4 Demographic and economic development indexes are yoked together in a complex way. For example, what effect does a country's population pyramid have on its level of economic development, and vice versa? Sketch typical population pyramids to illustrate your points.

5 Development is associated with closer spatial integration of the different regions within a country. Consider the changing connectivity of the networks in Figure 18.13. Increase the number of towns (nodes) from four to six, and draw links to illustrate a range of beta index values from 0.5 to 1.5.

6 From the viewpoint of their economic history, each of the world's economies have risen and fallen over the long run. How do you expect your own country's share of the world's wealth to change between now and mid-century? Justify your prediction and compare your views with those of your class colleagues.

7 Consider the ethical implications of uneven economic development within a country. What do we mean by 'a just spatial distribution' (Figure 18.18)? Can a geographer contribute to this debate?

8 Review your understanding of the following concepts using the chapter text and the glossary in Appendix A of this book:

backwash effects
beta index
core–periphery models
gross national product (GNP)
growth poles
Lorenz curves
Rostow–Taaffe model
spread effects
strategic withdrawal
zero economic growth (ZEG)

One Step Further ...

The outstanding and authoritative review of the present state of economic geography is given in Gordon CLARK *et al.* (eds), *The Oxford Handbook of Economic Geography* (Oxford University Press, Oxford, 2000). Another useful starting point is to look at the essay by Peter Dicken on the new geo-economy and by David Smith on quality of life in I. DOUGLAS *et al.* (eds), *Companion Encyclopaedia of Geography* (Routledge, London, 1996), pp. 370–90, 772–91. A brief but useful introduction to economic geography is given by Roger Lee in his essay in R.J. JOHNSTON *et al.* (eds), *Dictionary of Human Geography*, fourth edition (Blackwell, Oxford, 2000), pp. 195–9.

Most of the topics in the chapter come under the general heading of economic geography. Very helpful introductions to that field are provided by Eric SHEPPARD and Trevor BARNES (eds), *The Blackwell Companion to Economic Geography*, third edition (Longman, Harlow, 1991). One of the liveliest and most perceptive analyses of the development process is given by an Australian geographer, H. BROOKFIELD, *Interdependent Development* (Methuen, London, 1975). A world view of regional inequalities and their spatial distribution is given in B.J.L. BERRY *et al.*, *The Global Economy in Transition,* second edition (Prentice Hall, Englewood Cliffs, N.J., 1997) and J.P. COLE, *The Development Gap: A Spatial Analysis of World Povery and Inequality* (Wiley, New York, 1981).

The regional development 'progress reports' section of *Progress in Human Geography* (quarterly) will prove helpful in keeping up to date. World development problems are regularly reviewed each year in the World Bank's *Development Report* (annual) and in the United Nations Environment Programme (UNEP) *Global Environment Outlook* (annual). Regional problems are well covered in the regular geographic periodicals, but *Antipode* (a quarterly) is the forum where the issues of welfare geography and spatial justice are argued in the most challenging manner. You might also like to look at a few of the wide variety of specialized journals: *Regional Studies* (a quarterly), the *Journal of Regional Science* (another quarterly), and *Papers of the Regional Science Association* (a biannual publication) are all interesting. For readers with access to the World-Wide Web see also the sites recommended in Appendix B at the end of this book for topics relevant to this chapter.

Closing
Change
Bid/offer
mid-point
on day
spread

Europe
Austria
Belgium
Denmark
Finland
France
Germany
Greece
Ireland
Italy
Luxembourg
Netherlands
Norway
Portugal
Spain
Sweden
Switzerland

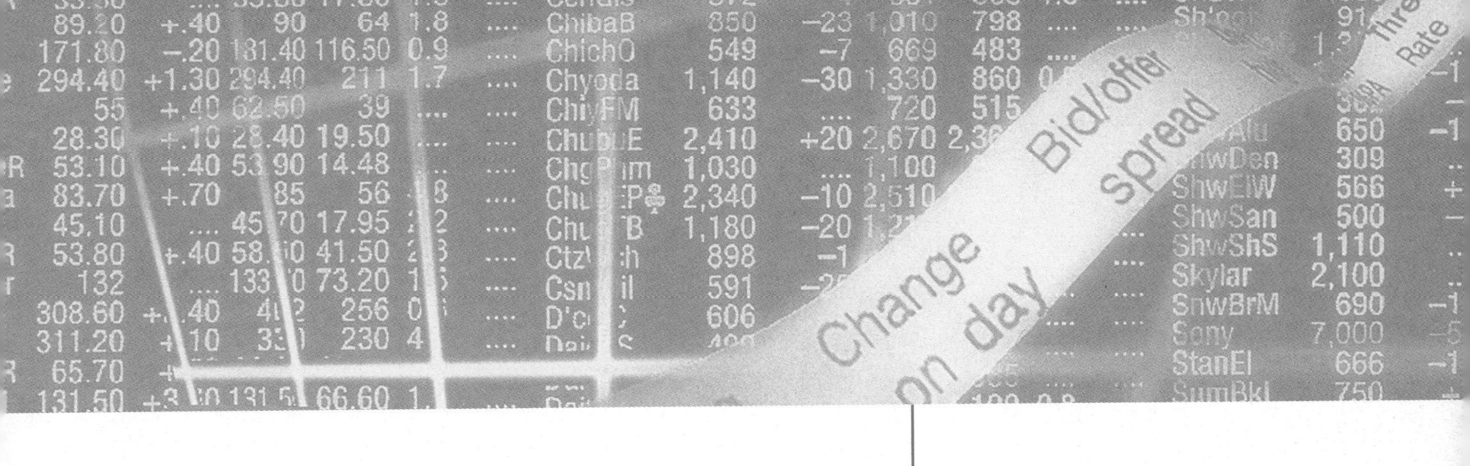

CHAPTER 19

Globalization

■ denotes case studies

Figure 19.1 Integrated electronic worlds Today the stock markets and currency exchanges of the world are linked within microseconds. Huge money flows sweep restlessly from one investment opportunity to the next or suddenly withdraw to avoid risks. This collage of bar charts, stock-exchange prices, and currency exchange figures evokes this world-shaping activity.
[Source: *Telegraph* Colour Library/Peter Sherrard.]

> The world is too much with us.
>
> WILLIAM WORDSWORTH 'The world is too much with us' (1807)

Twentieth-century changes in communication technology still have the capacity to amaze me. I recall vividly as a small boy in the 1930s, my mother's joy at our family home in rural England being linked to electricity (no more Tilly lamps or cooking on a kerosene stove). A telephone was added in the 1950s, TV in the 1960s, fax in the 1980s, and Internet in the 1990s. My five grandchildren are growing up in a world where their spatial separation (two live in Europe, three in Australia) is no bar to regular worldwide connections.

Nowhere has that improved connectivity been more dramatic than in finance. The simple 'hole in the wall' automatic tellers (ATMs) are a symbol. Solid, mechanical and localized they work only by a complex electronic network of communications. Visiting Hong Kong, I place an English bank card in the slot and a few seconds later receive Chinese currency. In those few seconds messages have switched through international checking centres and satellite links, my home account in Cambridge has been checked, the conversion rate calculated, my account debited. What is true of private finance happens even faster on the great stock exchanges that form the marshalling and switching yards of international commerce (see Figure 19.1 p. 584).

In this chapter we look at some aspects of this ever-more connected world at the start of a new millennium. In Section 19.1 we look at the concept that embraces these changes, globalization. We ask what the term means, whether it is new, and how far it is an adequate concept for geographers to use. Section 19.2 reviews the spatial shrinkage that technology has wrought. By how much has the world shrunk? Is that shrinkage general and worldwide or particular and local?

The next three sections look at three other aspects of the globalization process. First, the rise of the large corporation and its challenge to the nation-state as an organizer of geographic space (Section 19.3). Second, the rise of the tropics as the dominant environment for world population, both urban and rural (Section 19.4). Third, the rise of global environmental issues (Section 19.5). Although each is separate, each is linked to the common theme of a world growing more inderdependent and more vulnerable.

19.1 The Globalization Debate

In this section we look at the debate that has grown up around the subject of globalization. We look first at the globalization thesis and then at its critics.

The Globalization Thesis

We can define ***globalization*** as the process by which events, activities, and decisions in one part of the world can have significant consequences for communities in distant parts of the globe. Two parts of the process are usually identified: a spatial part that implies an ever-widening geographic scale or outreach for the process; and a non-spatial part that implies ever more intense linkages over the same geographic scale.

A geographic view of globalization is given in Figure 19.2. This shows a four-stage process in which the local economy is transformed into a global economy. Note the two types of linkages proposed: (a) vertical linkages between the environment and a human population; and (b) horizontal flows between localities and regions.

Globalization as a cultural concept emerged around 1960 when the Canadian scholar Marshall McLuhan coined the term *global village*. By this he meant the impact of new communications technologies on social and cultural life. He argued that time–space compression had so transformed the structure and scale of human relationships that social, cultural, political, and economic processes now operate at a global scale. The whole world had been reduced to the level of a village community.

This implied a linked reduction in the significance of other geographical scales (i.e. national, regional, and local). McLuhan asserted that we live in a world in which nation-states are no longer significant actors. Cultures are homogenized, being served through standardized global products (symbolized by the Coca Cola can or the McDonald's arch) created by global corporations with no specific allegiance to a single place or community.

Later writers extended the concept to cover the political economy. Kenichi Ohmae in his *Borderless World* argued that the nation-state (which we saw in Chapter 17 as the cornerstone of legal control and economic power) has made impotent by new institutions (especially the multinational corporation) which operates without regard to national borders.

An extreme view claims that the global is now the natural order of affairs in today's technologically driven world. Time–space is being compressed, the *end of geography* is arriving, everywhere is becoming the same. Globalization is seen as an inexorable, and virtually unstoppable force. It can be understood and accommodated but not resisted.

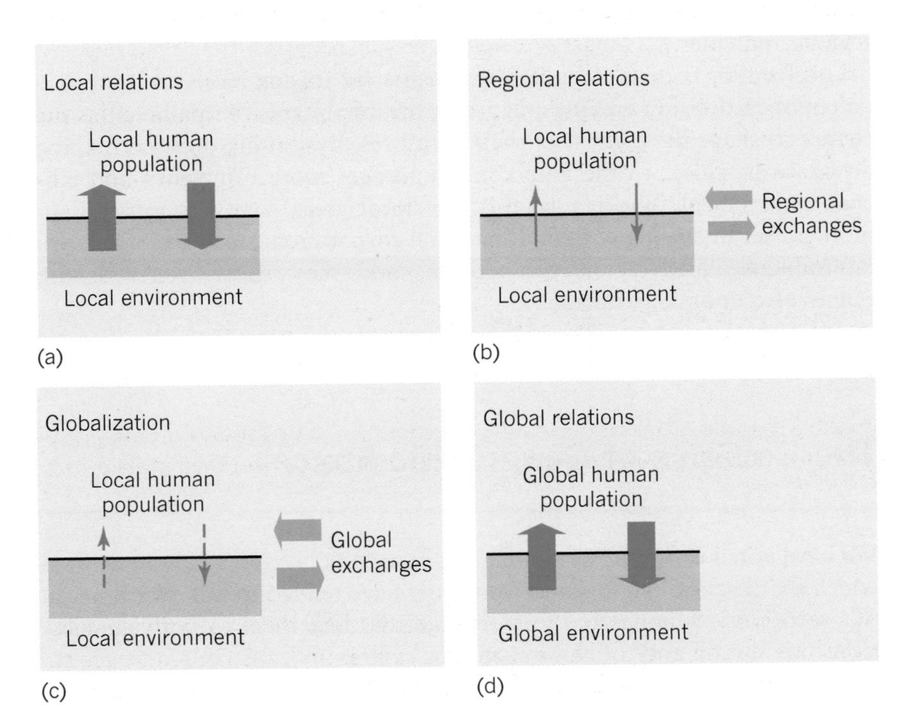

Figure 19.2 Increasing scales of human–environment interaction Evolution from (a) local, through (b) regional to (c) global scales of interaction between the environment and human populations. Note the changing balances of vertical (human/environment) and horizontal (place-to-place) interaction. The effect of globalization (d) is to return to the strong vertical relationships in (a), but at a different geographic scale. The environmental dependance at the local scale has now been replaced by an equivalent degree of dependence – but at the global scale.

Counter-arguments

Although the notion of a globalized world has become fashionable and pervasive in recent years, there are strong opponents. There are two main strands to the counter-arguments.

The first argument is that globalization is nothing new. Critics point out that there was a substantial debate at the end of the nineteenth century – perhaps, the 'first globalization debate' – concerned with the transformation of economic life by the expansionary forces of capitalism. This was best expressed in the economic theory of imperialism developed by Lenin.

In this view, not all that much has changed. We still inhabit an international, rather than a globalized, world economy in which national forces remain highly significant. Internationalization involves the simple extension of economic activities across national boundaries. It leads to a more extensive geographic pattern of economic activity but not to a fundamental change.

A second counter-argument is that global change does not occur everywhere in the same way and at the same rate. The processes of globalization are not geographically uniform. Individual countries, or regions, or localities, interact with the larger-scale general processes of change to produce quite specific spatial outcomes. Under this view, geography is alive and well, and adapting general global forces into distinct patterns. As globalization becomes more intense, so regionalization and localization tend to grow as countervailing forces to offset the effects.

Localization

Most major trends tend to spawn a number of small counter-trends and globalization is no exception. One of these countervailing changes is a growth of *localization*. This term describes the downward shift of power through the spatial hierarchy, from the nation-state down to the level of regions and cities.

Localization is praised by its proponents for raising levels of local participation in decision making and giving the inhabitants of small regions the chance to shape the context of their own lives. By shifting decision making downwards, closer to the voters, it encourages more responsive and efficient government. But against this, the local areas have less rather than more power in dealing with multinational corporations, may be forced into unproductive inter-region competition, and may show ever-widening ranges of economic performance.

19.2 The Collapse of Geographic Space

We have noted earlier in this book (see, especially, Chapter 13) the ways in which the costs of long-distance transport have tended to fall over time. In this section we summarize those changes and link them to both the electronic revolution and to major economic cycles.

Historical Changes in Transport

Two ways in which geographic space has contracted are shown in Figure 19.3. In (a) the way in which the great ocean barrier of the North Atlantic has been reduced is plotted. The Pilgrim Fathers voyage in 1620 took 54 days from England to New England. The nineteenth-century steamship crossings reduced this progressively from three weeks to less than one week. Air crossings from 1920 further reduced this from a day-long series of legs via Iceland and Labrador to the three hour Concorde supersonic flight.

Graph (b) in Figure 19.3 shows similar trends for the continental crossing of the United States from the 21 day stage coach crossing by Wells Fargo in 1850 to five hour crossings by air today. What is critical in both

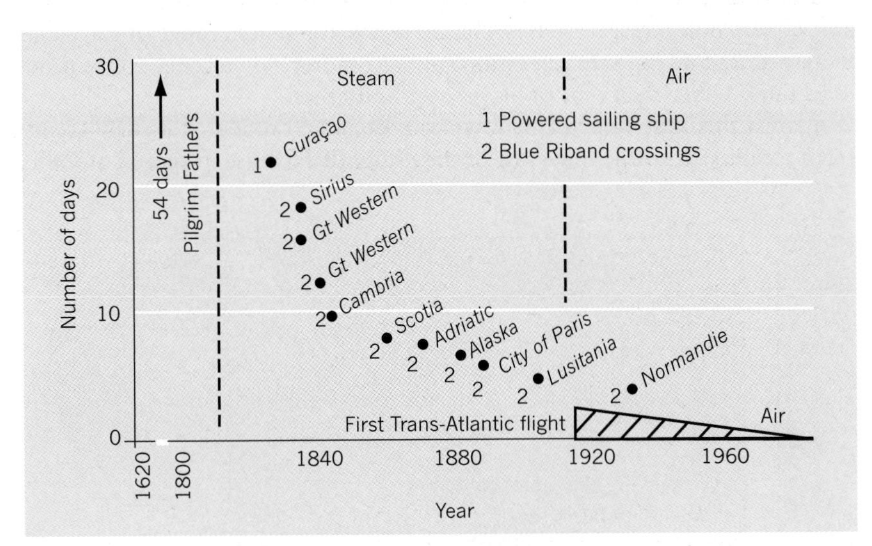

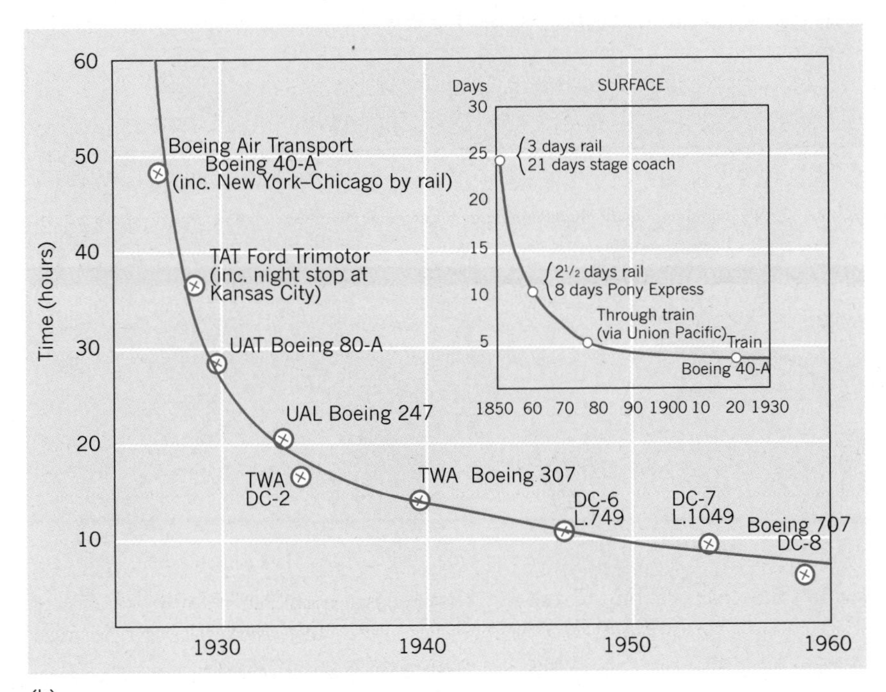

(b)

Figure 19.3 Historical changes in travel Contraction of travel times in (a) crossing the Atlantic Ocean since 1620 and (b) crossing the continental United States since 1850. Number of days taken on the trans-Atlantic (above) and trans-continental (below) journey is plotted against the historical date.

[Source: (a) A.D. Cliff, P. Haggett, and M. Smallman-Raynor, *Deciphering Global Epidemics* (Cambridge University Press, Cambridge, 1999), Fig. 2.3(C), p. 47. (b) R.E.G. Davies *A History of the World's Airlines* (Oxford University Press, Oxford, 1964), Fig. 91, p. 509.]

graphs is the distinctive shape of the curves with rapid falls being followed by shallower gains.

Another way in which travel patterns have changed over recent generations has been shown in an interesting way by the distinguished epidemiologist D.J. Bradley. Using old family records, Bradley compares the travel patterns over four generations: for his great-grandfather, his grandfather, his father, and himself (see Figure 19.4). The lifetime travel track of his great-grandfather around a village in Northamptonshire could be contained within a square of only 40 km (25 miles) side. His grandfather's map was still limited to southern England, but it now ranged as far as London and Cambridge and could easily be contained within a square of 400 km (250 miles) side. If we compare these maps with those of Bradley's father (who travelled widely in Europe) and Bradley's own sphere of travel, which is worldwide, then the enclosing square has to be widened to sides of 4000 km and 40,000 km (2500 and 25,000 miles) respectively. In broad terms, the spatial range of travel has increased tenfold in each generation so that Bradley's own range is one thousand times wider than that of his great-grandfather.

Against this individual cameo, we can set some broader statistical trends from recent years. One indicator of the dramatic increase in spatial mobility

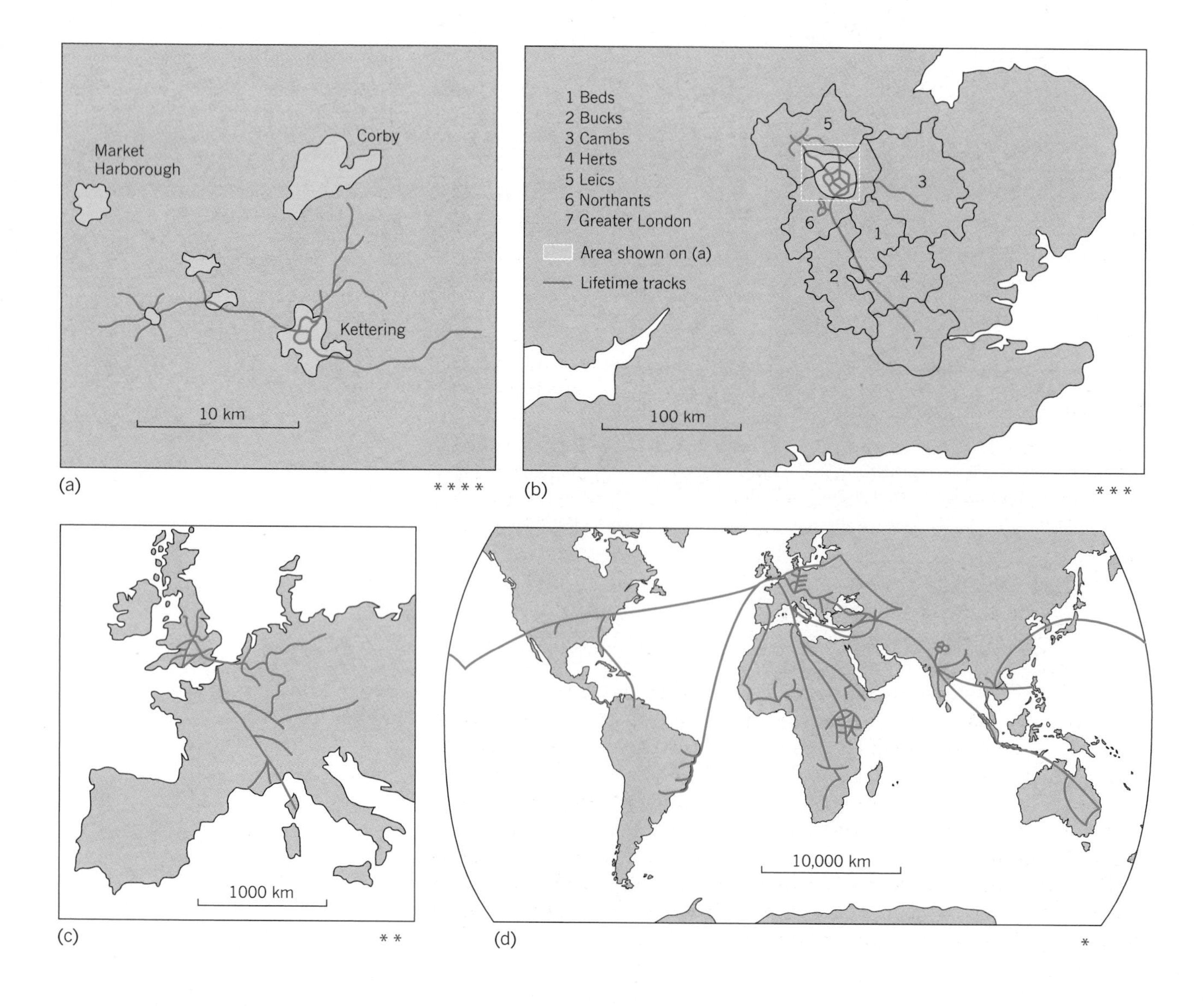

Figure 19.4 Increasing travel over four male generations of the same family (a) Great-grandfather, (b) grandfather, (c) father, (d) son. The 'son' is the distinquished epidemiologist David Bradley. Each map shows in a simplified manner the 'life-time tracks' in a widening spatial context, with the linear scale increasing by a factor of ten between each generation.

[Source: D. Bradley, in R. Steffen *et al.* (eds), *Travel Medicine* (Springer Verlag, Berlin, 1988), Figs 1–4, pp. 2–3).]

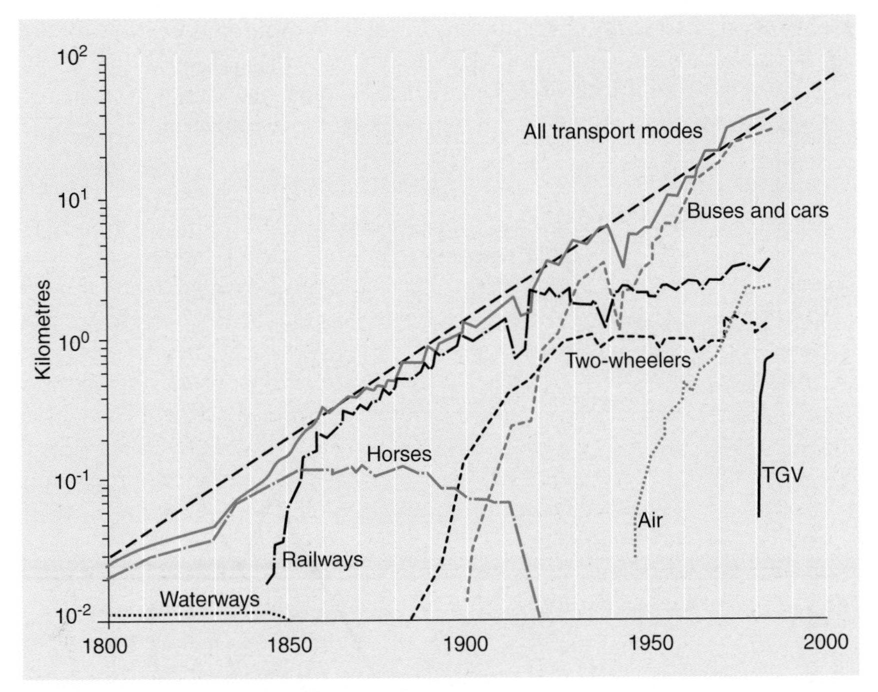

Figure 19.5 Increased spatial mobility of the human population Mobility of the population of France over two centuries. The upper solid line is the sum of all the different transport modes. TGV = high-speed trains. Note that the vertical scale is logarithmic so that increases in average travel distance increases exponentially over time.

[Source: P. Haggett, *Geografiska Annaler*, **76B** (1994), Fig. 4, p. 101.]

is shown in Figure 19.5. This plots for France over a 200 year period the average kilometres travelled daily both by transport mode and by all modes.

Since the vertical scale is logarithmic the graph shows that, despite changes in the mode used, average travel has increased exponentially, broken only by the two world wars. Over the whole period, mobility has increased by more than a thousand-fold. The precise rates of flux or travel of population both within and between countries are difficult to catch in official statistics. But most available evidence suggests that the flux over the last few decades has increased at an accelerating rate. While world population growth rate since the middle of the twentieth century has been running at between 1.5 and 2.5 per cent annum, the growth in international movements of passengers across national boundaries has been between 7.5 and 10 per cent annum.

One striking example is provided by Australia: over the last four decades its resident population has doubled, while the movement of people across its international boundaries (that is, into and out of Australia) has increased nearly 100-fold. If we use as a measure of flux, the ratio between movements and resident population then that flux has increased 200-fold.

One of the factors behind long-haul increases in traffic has been the massive increase in international tourism over the last half century. As we shall note in the final section of this chapter, this is now extending into an ever wider range of environments. The outer limits are being reached in Antarctica. This was first penetrated by European explorers in the eighteenth century. Sealers and whalers were the main visitors in the nineteenth century, while the twentieth century was marked by increasingly intensive scientific exploration of both the coasts and the interior (Amudsen reaching the South Pole in 1911). The last two decades of the century saw the start and acceleration of a wholly new trend – the Antarctic summer tourist. The first cruise ships arrived in the late 1970s and as many as twenty a year now visit the continent, mainly in the Ross Sea area. As Figure 19.6 shows, the number of tourists has increased since 1975 from less than a hundred hardy souls each year to an estimated 20,000 each summer.

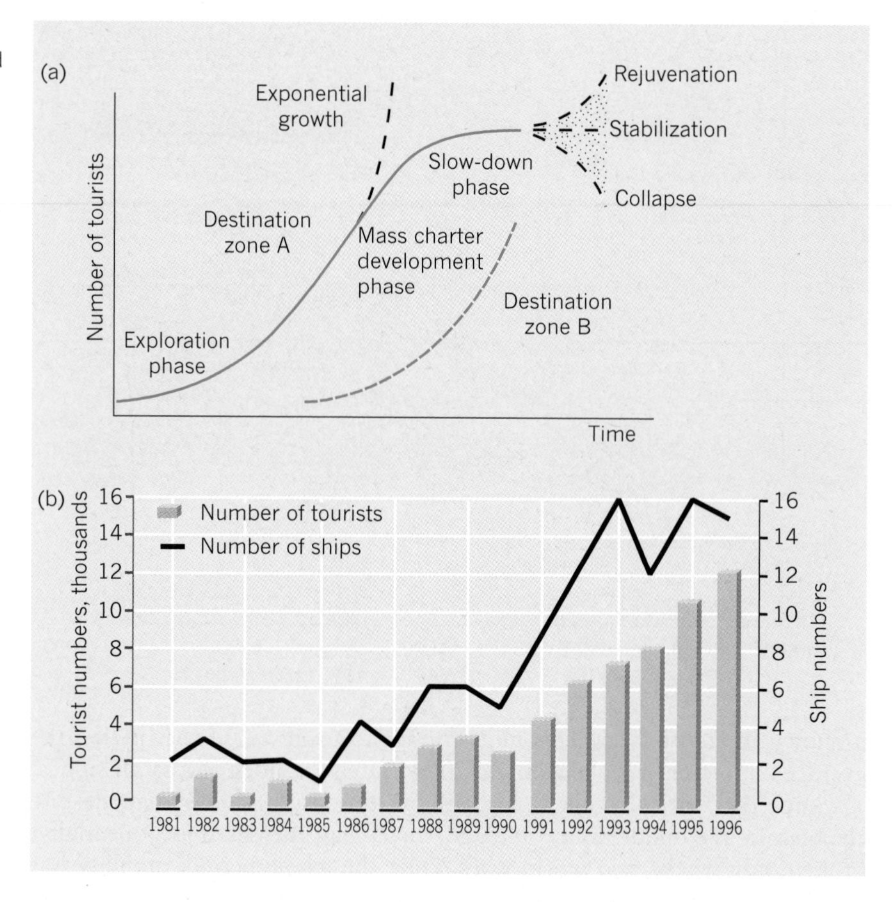

Figure 19.6 Growth in international tourism (a) Generalized model of a cycle of international tourism with a new destination B replacing the old over-exploited destination A. (b) Antarctica represents the outer limits of global tourism as a replacement zone.

[Source: United Nations Environment Programme *GEO.2000 UNEP's Millennium Report on the Environment* (Earthscan, Oxford, 2000), p. 189.]

Revolutions in Electronic Space

Whatever the changes in conventional transport technology, it is in message communication that the greatest revolution has occurred. Technological advances in telecommunication have made it possible to know in an instant what is happening in a stock market or a factory half a world away. An increasing proportion of economic product is *weightless* – that is, can be transmitted via satellite or fibre optic cable rather than transported in a container ship. At the same time, conventional means of transport (shipping goods by water, land, or air) are showing steady cost reductions and the improvements in information technology make it easier to manage.

Figure 19.7 shows in (a) the explosive growth of the *Internet* towards the end of the last century. In (b) the growth of both computers and the Internet can be set aside other technical revolutions. It brings together eight major innovations since 1873 and shows how they have progressively been adopted into consumer use. The curves for PCs and the Internet show a remarkably sharp adoption curves. See Box 19.A 'The Microsoft Corporation', in which the spectacular growth of the world's biggest software corporation is traced.

The Fifth Kondratieff

Some geographers have linked the electronic revolutions of the last few decades to a broad historical model. The theory that in addition to the five

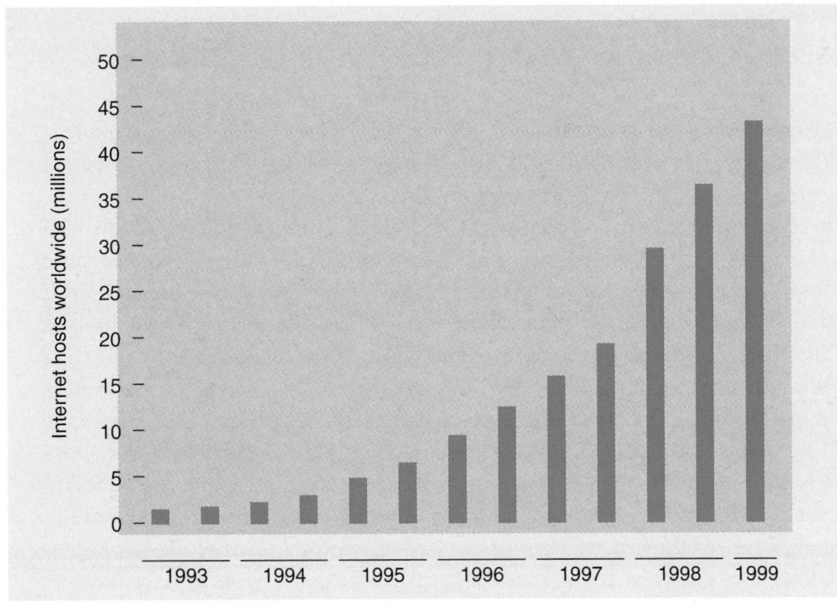

(a)

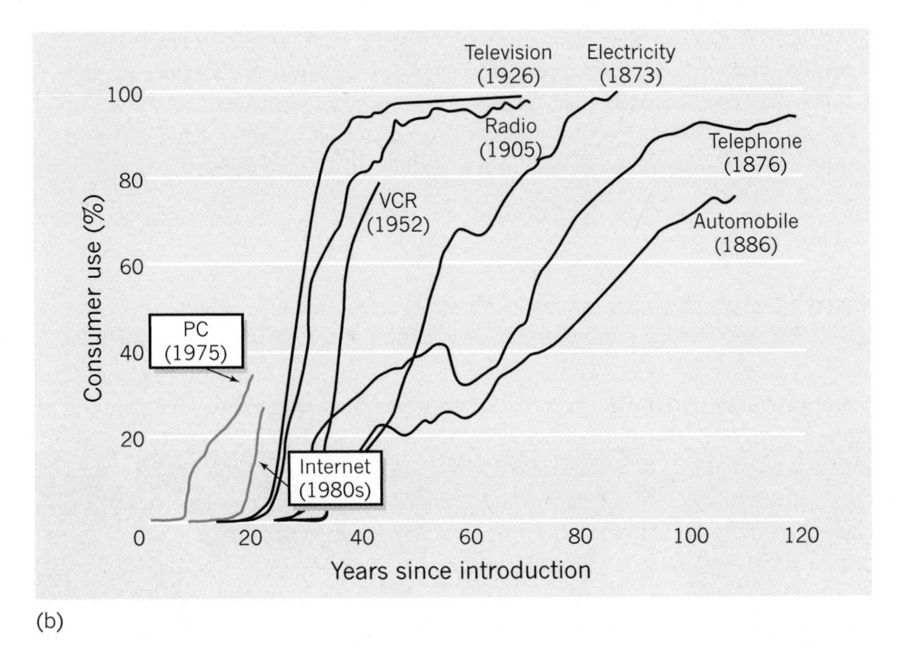

(b)

Figure 19.7 Speed of introduction of new technologies (a) Growth of Internet hosts across the world from 1993 to 2000. Values are given for January and July of each year. (b) Speed of introduction of new technologies. Percentage of consumer use (vertical axis) plotted against years since introduction (horizontal axis). Note the very steep slope shown by Internet use.

[Source: (a) World Bank, *Entering the 21st Century* (Oxford University Press, Oxford, 2000), Fig. 1, p.4. (b) W. Gates *Business at the Speed of Thought* (Penguin, London, 1999), p. 118.]

year *business (or trade)* cycles, there exists a 50 year cycle of economic upturn and downturn. This theory was put forward by the Russian economist Kondratieff in the early 1920s. It was dismissed by many economists until the great depression of the 1880s and 1930s were duly followed (50 years on) by the frequent and severe recessions of the period 1975–92. (For a fuller discussion, see Box 18.B on 'Kondratieff waves'.) The most widely accepted explanation for the *Kondratieff wave* is that the introduction of new technology causes disruption but that once established the technology forms the basis for many new products and jobs. In the 1930s the car was displacing rail while in the 1990s the microchip was replacing mechanical technology. Figure 19.8 shows the role of electronics as a global industry with *Silicon Valley* as its hallmark.

Box 19.A The Microsoft Corporation

Few companies illustrate more clearly the nature of business enterprises in the last quarter of the twentieth century than America's software giant, the Microsoft Corporation. This was started in 1975 when two boyhood friends from Seattle, William H. Gates and Paul G. Allen, converted BASIC, a popular mainframe programming language, for use on an early personal computer (PC). The name Microsoft derived from the words 'microcomputer' and 'software'. In 1980 the American computer company International Business Machines asked Microsoft to produce an operating system for its first personal computer, the IBM PC. They developed MS-DOS (Microsoft Disk Operating System), released in 1981. A decade later, Microsoft released the first of its Windows graphical command programmes which during the 1990s became standard around the world. The success and dominance of the company can be gauged by the fact that by 2000 nearly 90 per cent of all the world's PCs ran on a Microsoft operating system.

In the long history of successful American companies, Microsoft represents one of the most powerful, rapidly growing, and profitable companies of the last two centuries. Unlike the giants of the past it sells not refined raw materials (e.g. Standard Oil) or manufactured products (e.g. U.S. Steel) but information systems. Its main products aid business productivity (e.g. Word, Excel, PowerPoint), provide reference material (e.g. Encarta), or aid Internet searching (e.g. Internet Explorer). Notwithstanding its net income is huge (last checked on the web at well over $20 billion and rising rapidly) and gives it revenues that would place it in the top 50 of the world's countries if measured as GNP. Yet in locational terms it represents a rather conservative organization. Despite a countrywide and worldwide network of sales offices, research and production remains concentrated in Redmond, Washington, near the place where it was started. Its workforce remains at 32,000 (mostly in Redmond), rather small even by comparison with other software companies (Oracle is larger at 44,000). It currently ranks around 80 in the Fortune 500 list of American companies, and around 250 in the Global 500 list.

Figure 19.8 Electronics as a global industry One of the symbols of late-twentieth century economic growth has been the computer industry. (a) Silicon Valley, California. A 1997 view looking north over San Jose airport at centre. (b) Assembly line for electronic equipment.

[Sources: (a) Air Flight Service/Mapping Photography. (b) Selectron.]

(a)

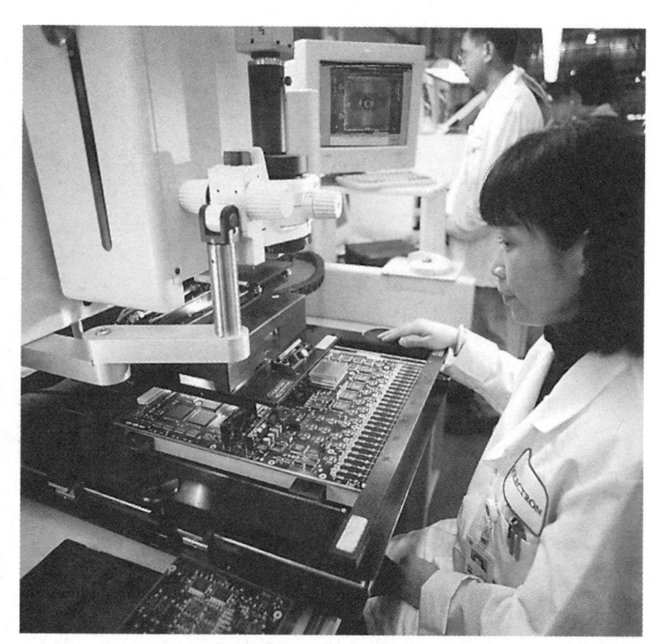

(b)

Changing Geoeconomic Space 19.3

The next three sections look at three other aspects of the globalization process. We begin by looking at the rise of the large industrial corporation and its challenge to the nation-state as an organizer of geographic space. We then consider the spatial ecology of the corporation and the problem this poses for nation-states.

The Rise of the Global Corporation

In Chapter 17 we argued that it is the *nation-state* that, either singly or in coalition, is the strategic political unit of our time. We have seen in the previous two chapters how differences between and within states leave an indelible stamp on the world geography.

But is the state the only decision-making unit that geographers need to know about? Economist John Kenneth Galbraith would say 'no' to that question. In his pioneering *New Industrial State* (1967), Galbraith argues that the large corporation is fast becoming the strategic economic unit of greatest significance. He was able to show that whether we measure at the level of an industry, a country, or the whole world, business activity is becoming more concentrated. For the United States the hundred largest manufacturing corporations now control just over one-half of the total assets of *all* business corporations: 50 years ago it was little more than one-third. In the United Kingdom, two out of five employees work for the hundred largest manufacturing firms – in the early 1930s it was one out of five.

Corporations have grown both by expanding the size of individual plants, but more significantly by expanding the number of their plants and the range of their products. Typically the large business corporation is both multiplant and multiproduct, and so has a highly diverse structure in both product and space.

The Size of Industrial Corporations Measurement of the size of a business corporation may be conducted in different ways. Table 19.1 shows the world's twelve largest corporations based on revenues. *Fortune*, a magazine which publishes regular surveys on the size of industrial corporations both inside and outside the United States, uses three measures. These are the total sales of a corporation in a year, its corporate assets, and the numbers of workers it employs. Figure 19.9 shows the leading 25 corporations. This combined list was made up by taking the 15 top-ranking corporations on each of the three measures. Note that over half the corporations are American with British, Dutch, German, Japanese, Iranian, and Italian making up the remainder. (The state-owned corporations of the former Communist bloc countries are not included.)

In terms of each corporation's products the graph shows a clear distinction between the eight oil companies in the list, which have a high ratio of sales to employees, and the others. These others are companies in electronics, electrical machinery, and automobiles with steel, chemicals, and foodstuffs also represented. Note, however, that the categorization of manufacturing corporations in terms of a dominant product is slightly misleading since corporations are tending to become more diversified.

Let us take as an example General Motors, the corporation that has

Table 19.1 World's twelve largest corporations

Company	Revenues	Industry	Country
1. General Motors	1000	Automobile	United States
2. Ford Motor Company	865	Automobile	United States
3. Mitsui & Co., Ltd	803	Trading	Japan
4. Mitsubishi Corporation	725	Trading (including automobile)	Japan
5. Royal Dutch/Shell Group	719	Energy	UK/Netherlands
6. Itochu Corporation	713	Trading	Japan
7. Exxon Corporation	685	Energy	United States
8. Wal-Mart Stores, Inc.	668	General merchandise	United States
9. Marunbeni Corporation	623	Trading	Japan
10. Sumitomo Corporation	573	Trading	Japan
11. Toyota Motor Corporation	533	Automobile	Japan
12. General Electric Company	511	Electric power	United States

Revenues for 1997 are indexed so that the revenue for the largest corporation is set to a vale of 1000.
Source: *Fortune Magazine,* 'The Global 500 List'. See http://www.pathfinder.com/fortune/global500/index.html for updated versions.

occupied a leading place on all three lists for the last half-century. This corporation has over 120 plants within the United States producing nearly 100 different major products ranging through the familiar automobile and its accessories to domestic washing machines and military hardware. About half the plants concentrated on a single product, but the remainder produced from two to twelve different lines. ITT produces not only a wide range of electronics, but also food, car parts, insurance, and cosmetics.

Spatial Ecology of the Corporation

As the scale and diversity of corporate organization has grown, so too has the spatial diversity. We need here to check back on the work of Swedish geographer Gunnar Tornqvist described in Section 8.3. Recall that he analyzed the spatial structure of the corporation in terms of three levels of organization. At the highest level, *Level I*, the head office is concerned

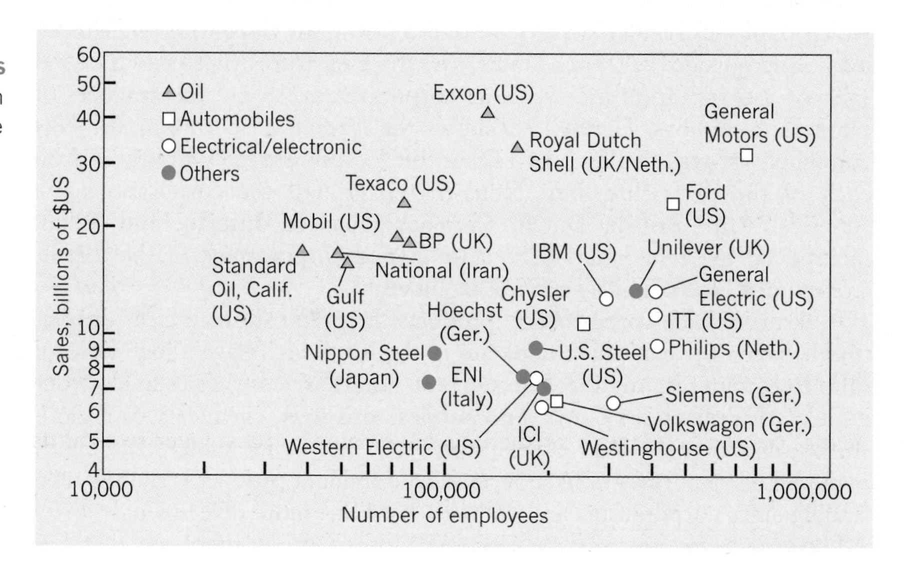

Figure 19.9 Alternative measures of the size of industrial corporations Industrial corporations can be measured in several ways other than their revenues (see Table 19.1). The world's 25 largest industrial corporations were defined here in terms of sales and employment. Note that they do not include banking and insurance combines or institutions within former communist bloc countries. Since data refer to the middle 1970s, the sales position of oil corporations may be slightly inflated. Check from Table 19.1 how many have survived in the top ranks over the ensuing quarter century.

[Source: From data given in *Fortune Magazine*, May 1975. © 1975 Time Inc.]

with the general long-term planning of the corporation's activities. Here the main functions are unprogrammed in that chief executives are involved in highly flexible negotiation with government and public officials and with similar executives in other corporations. The primary locational need is for an information-rich site with easy opportunities for face-to-face contacts and facilities for rapid travel to other major centres. These needs will usually be met only in a national capital city or large metropolitan centre.

At the second level, *Level II*, are the more routinely administrative functions. There is still a need for good location, but contacts can be carried out indirectly via telephone, e-mail, or written communication. *Level III*, the lowest level, is concerned with the routine day-to-day organization of work at the plant level. Locations here are controlled by the manufacturing needs of the product and fall in line with arguments on industrial location presented in Chapter 15. Products dependent on particularly low labour costs may be located in areas where production costs are low, and so on.

In the simplest corporate unit, the single one-product plant, the three levels recognized by Tornqvist may be combined in a single location as shown in Figure 19.10(a). Here the flows of information and decisions take place within the plant, while the input of raw materials and output of finished products reflect the geographic distribution of resources on the one hand and of markets on the other. With corporate growth the situation becomes spatially more complex.

Figure 19.10 shows just two of the possible arrangements of Levels I and II activities as the number of plants and the range of products increases. In the first (Figure 19.10(b)) there is a twofold geographic structure with the head office and administration in one location and the operating units in another. In the second (Figure 19.10(c)) the structure is threefold with complete spatial separation of the three levels. The larger and more diversified the corporation, the stronger the arguments for a threefold structure become.

Figure 19.10 Spatial structure of multinational corporate activity
(a) Tornqvist's concepts of a three-level decision-making hierarchy within a single plant. (b) and (c) Different spatial arrangements for a multi-plant corporation with foreign plants. In (b) only plant production has been located overseas. In (c) middle-level activity has also been located in one of the two foreign countries.

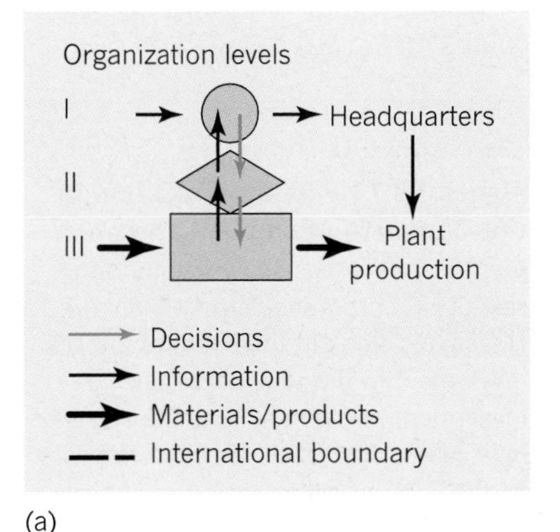

(a)

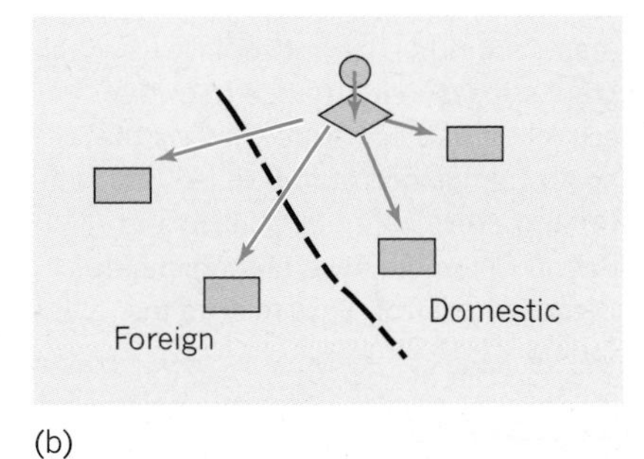

(b)

(c)

Multinationals and the Problem of Foreign Control

The large corporation clearly poses a control problem for central or regional government within a particular state. In communist countries any mismatch between the goals of the corporation and the goals of the state may be reduced by taking over the corporation and converting it into a state enterprise. In capitalist countries, the restrictions may come in the form of legislative action controlling monopoly or prices. The mixed economies of western Europe show various intermediate solutions such as government purchase of stock, full nationalization, and so on.

But these measures apply only to the home-based corporation. How do governments control the large *multinational corporation* whose headquarters lie wholly outside its territory? It was the French politician Jean-Jacques Servan-Schreiber who in *Le Defi americain* (1967) showed the size of the problem for western Europe. He predicted that in the next two decades, the world's third greatest industrial power after the United States and Russia would be not Europe but American industry in Europe. A good deal of Servan-Schreiber's predictions have come to pass. Over two-thirds of all the western European computer industry is American-owned (over half by IBM). In employment terms, the leaders are ITT, Ford, and General Motors, all with over 100,000 employees in Europe.

Although the American-based multinationals are the largest, the problem is a much broader one. For multinational corporations are of increasing importance in shaping development levels all around the world. If we exclude previous communist bloc countries, then multinationals probably account for approaching one-quarter of the world's total industrial output. For the last decade their foreign output has been increasing at something like twice the world growth in GNP. They are much more dominant in technologically advanced industries, those with the highest present and future growth potential (e.g. electronic engineering, chemical engineering, automobiles). About half the multinational assets are American-owned, with European countries (notably Britain and Germany) and Japan controlling most of the remainder.

Foreign multinationals are of concern to state governments in that each has legitimate interests quite different from the other. Plants within a country are, however large, still essentially branches whose ultimate control is in an overseas capital (see Figure 19.11). Such corporations have, at least in theory, the option of cutting their losses in a particular country and investing elsewhere. But there are substantial counter-balancing benefits. Arguably, the foreign multinational represents the most efficient sector in many of the world's countries, and contributes an increasing amount to

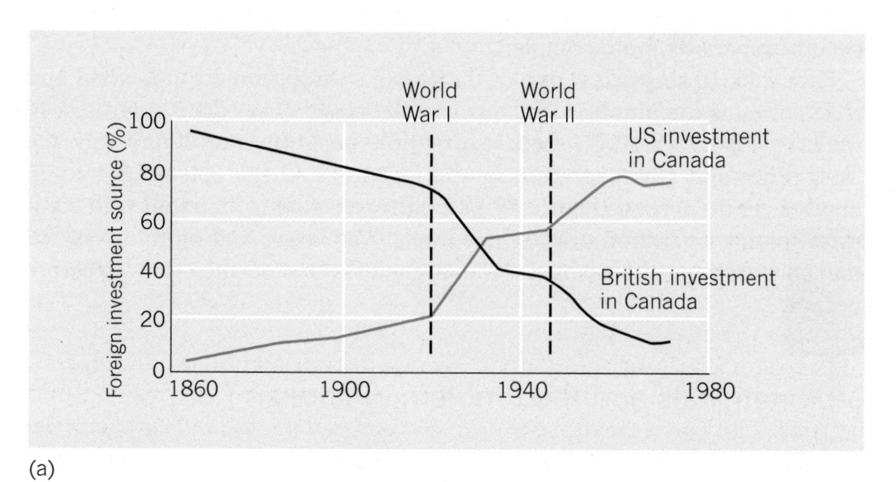

(a)

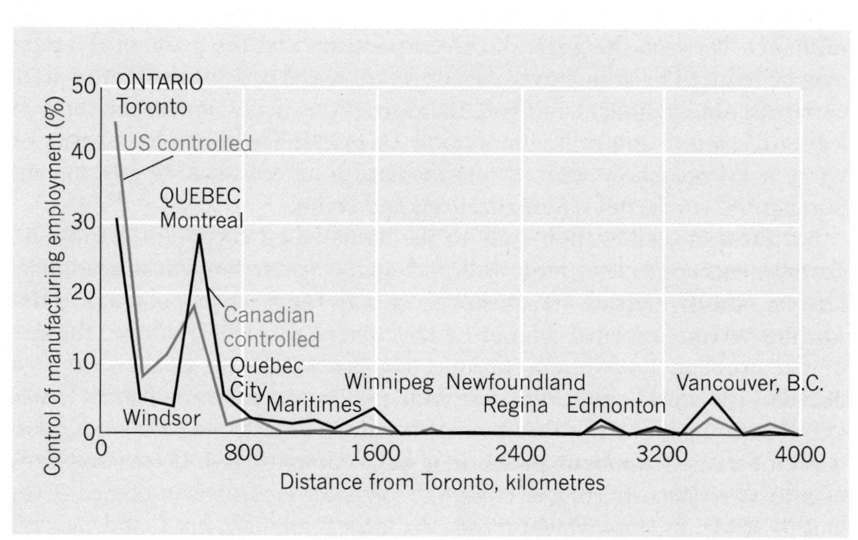

(b)

Figure 19.11 Americanization of the Canadian economy These two graphs show aspects of the important role of US corporations in Canada. (a) Historical trends in the share of the US investment in the total of foreign investment in Canada. Note the surge of investment after each of two world wars as the British share contracted sharply. (b) Spatial shares of Canadian and US-controlled jobs in Canadian manufacturing. The horizontal axis of the graphs measures distance from the economic hub or Toronto. Notice the heavy dominance of the United States in Ontario province; both Quebec and British Columbia have predominantly Canadian control. Data refer to the 1970s.

[Source: Data from B.J.L. Berry, E.C. Conklin, and D.M. Ray, *The Geography of Economic Systems* (Prentice Hall, Englewood Cliffs, N.J., 1976), Fig. 15.6, p. 284.]

their growth. Increasingly, Level II activities (to use Tornqvist's term) are being decentralized to leading cities of those countries in which the multinationals operate.

It would be wrong, therefore, to conclude this section by leaving the impression that multinationals represent some cancerous growth on the world economic scene. Large multiplant, multiproduct, multination corporations arose in the middle of the twentieth century as a logical response to the technologies and economics of the manufacturing process. There remain, however, the inevitable problems of regulating and controlling a twentieth-century economic institution in terms of a unit of human political organization (the state) which has different historical origins and a different geographic scale of operation. Both institutions have legitimately different areas of interest, and geographers will watch with concern which plays the greater role in setting the regional patterns of a changing economic world.

The State as a Corporation

Although the state and the corporation are seen as different institutions, they share some features of geographic interest in common. The role of state intervention in economic affairs is well known in the communist countries of the world, but it is less frequently realized how large is the state's role in western economies. Let us take the United States as an example. From 1902 to 1980 total government expenditure in the United States increased from 7 per cent of the GNP to over 33 per cent. Not only did expenditures increase greatly, so also did government jobs. These now account for about a third of the total workforce in the United States. In one sense, then, the federal government within the United States' economy is acting as the largest single corporation! Trends that are true for this country are also true for a number of European countries and are even more accentuated in the case of the United Kingdom, France, and Italy.

The significance of these broad swings toward additional government employment on the spatial organization of the economy is not yet clearly understood by geographers. Some work has been done on the contentious issue of the impact of defence expenditures. The main defence industries consist of aircraft, steel, electronics, and related services. Studies show that a number of states within the United States have benefited greatly from government expenditure on arms and equipment. California, New Mexico, and Colorado in the west and Washington, D.C., and Maryland in the coast areas of the east have been the main beneficiary regions. A second area of study has been the impact of government expenditures department by department. In Canada study of expenditure by the Ontario Provincial Government Departments showed severe spatial bias in favour of the Toronto region. In this sense, therefore, the government can be seen as an agent within the spatial economy tending to reinforce the advantages of the core region.

There is evidence for a fundamental paradox in government policy. As the state becomes a larger and larger employer of labour, so its own policies play a more important part in determining the spatial structure of the economy. To some extent the growth of government and its centralization may be exacerbating the difficulties of the distressed and marginal areas. In that sense it has contributed to the very problems that it is trying to solve.

19.4 Growth of Global Cities

The second trend in global reorganization is the rise of the global city as a dominant feature in world population. We look here at the changing role of urbanization in reshaping global balance and at *world cities* acting as a focus for money flows.

Redistribution of World Population

To place urban growth in context, we have to go back to the overall increase in world population which we discussed at the end of Chapter 6. Table 19.2 gives a summary picture by world region. We note four important points.

First, the rapid acceleration in growth is very recent. In the past four decades, the world's population has more than doubled. As we saw in Chapter 6, according to the United Nations' 'medium-growth' assumptions, this total is expected to reach 8.5 billion by the year 2025. Although that rate of growth is now decelerating (its peak was at 2.1 per cent per annum in the years from 1965 to 1970) the multiplier of resource use per capita continues to grow with evident environmental implications.

Second, a broadscale geographical redistribution of world population is accompanying this growth. For example, it is expected that some 94 per cent of population growth over the next twenty years will occur in the developing countries. Figure 20.20 shows the present latitudinal distribution of population showing a marked concentration in the northern mid-latitudes. Present and future growth will shift the balance of world population still more towards the tropics and low latitudes. This redistribution will increase the average temperature of the global population by around +1 °C, from 17 to 18 °C (even assuming no increase from global warming). This concentration will place more people than in the world's previous history in areas of high microbiological diversity, potentially exposing a greater share of the world's population to conventional tropical diseases.

Third, the age distribution of the world's population is changing. The proportion of young people is now falling, the proportion of the elderly is rising,

Fourth, the world's growing population is increasingly concentrated in cities. In 1800 less than 2 per cent of the world's population lived in urban

Table 19.2 Percentage of population living in urban areas by world region

World region	1950	1975	1995	2015
Africa	14.6	25.2	34.9	46.4
Asia[a]	15.3	22.2	33.0	45.6
Latin America	41.4	61.2	73.4	79.9
Industrial countries[b]	54.9	69.9	74.9	80.0
World	29.7	37.8	45.3	54.4

[a]Excluding Japan.
[b]Europe, Japan, Australia, New Zealand, and North America excluding Mexico.
Source: United Nations, *World Urbanization Prospects: The 1996 Revision* (UN, New York, 1998); L.R. Brown and C. Flavin, *State of the World 1999* (Earthscan, London, 1999), Table 8–3, p. 136.

communities. By 1970 this had risen to one-third and by 2000 to one-half. Along with the increasing proportion of urban population, the number of large cities and their average density will also have increased. On United Nations estimates, the number of cities with a million or more inhabitants doubled from 200 in 1985 to 425 at the end of the century. Currently there are 25 cities with populations in excess of 11 million.

Urbanization in the Tropics

If we take two of the above trends, the second and fourth, they reinforce each other in a combined trend dominated (a) by developing countries and (b) by the cities located in those countries. Of the urban dwellers of the future, nearly 90 per cent will be living in developing countries. Half a century ago, just 40 of the world's 100 largest cities were in developing countries. By the start of this century that proportion had risen to 65 and that proportion keeps rising (see Figure 19.12).

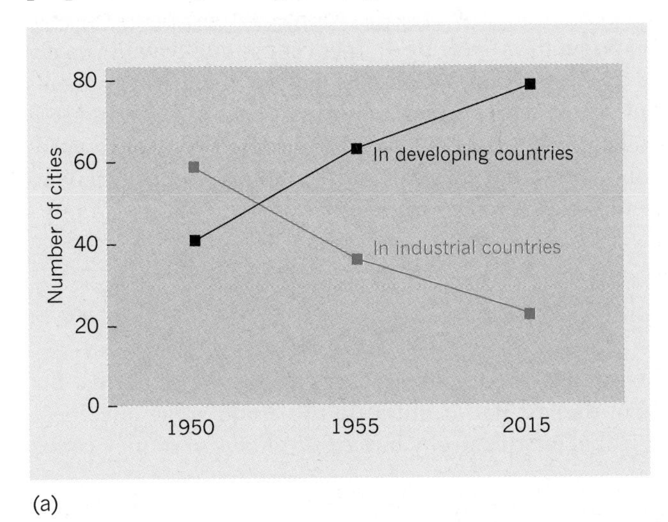

(a)

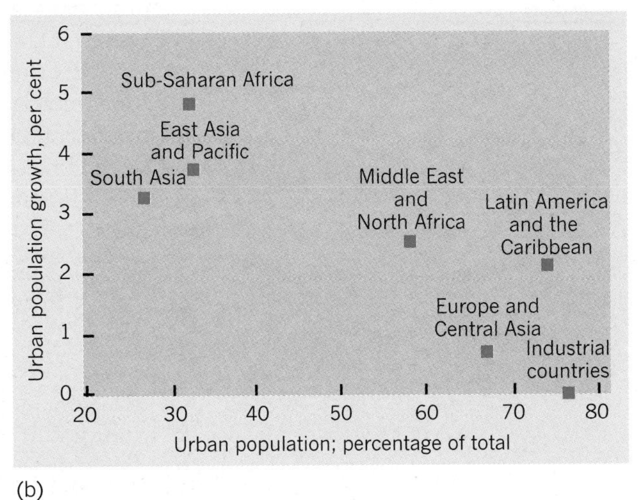

(b)

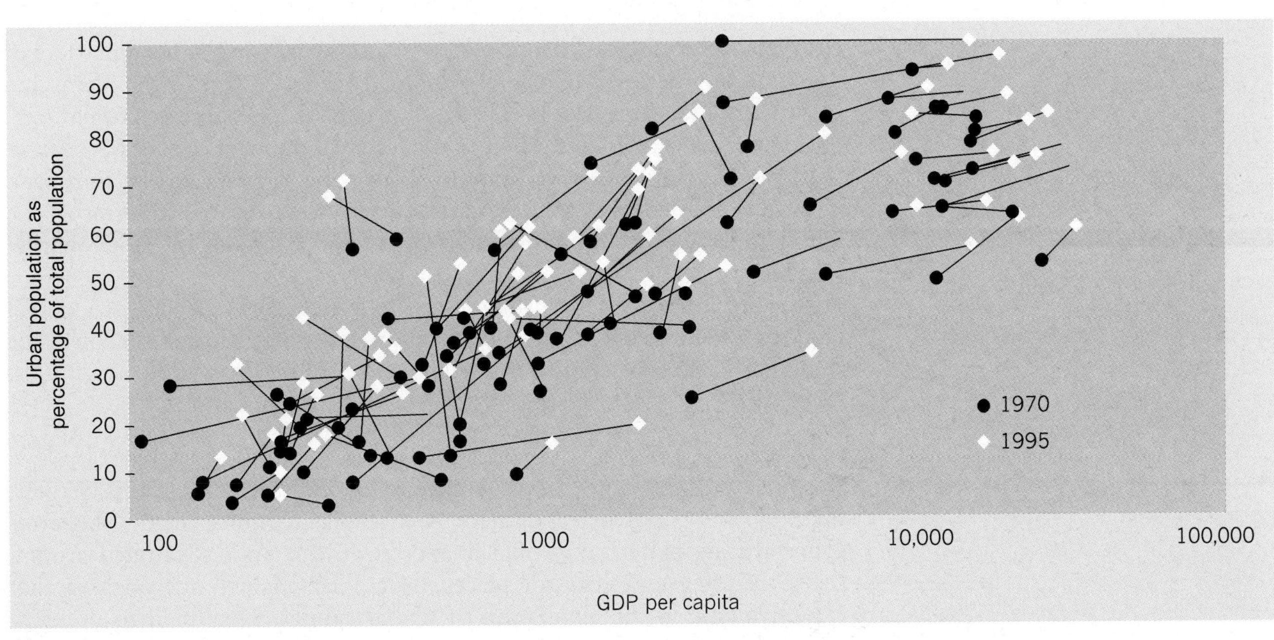

(c)

Figure 19.12 Location of the world's largest cities (a) Changing number of leading cities in developing and developed countries over 65-year period. Figures refer to the world's one hundred largest cities at each of the three dates. (b) Urban population growth by major world region in 1990s. (c) Relationship between urbanization and economic development over a 25-year period between 1970 and 1995. Note that the horizontal scale of gross domestic product (GDP) per capita is logarithmic.

[Source: World Bank, *Entering the 21st Century* (Oxford University Press, Oxford, 2000), (a) Fig. 7, p. 10; (b) Fig. 1.8, p. 47; (c) Fig. 6.1, p. 1260.]

Within the broad divisions urbanization shows a more intricate pattern as shown in Figure 19.12(b). The fastest rates of growth at the end of the twentieth century were in sub-Saharan Africa, South Asia, and East Asia and the Pacific, regions with the lowest levels of urbanization. In the rest of the world cities were continuing to grow but at significantly lower rates.

What make such cities grow? The conventional answer is that cities are an integral part of economic growth and that, as countries develop, so the proportion of population living in cities grow too in a chicken-and-egg fashion. Figure 19.12(c) would appear to support this proposition. It shows change for both developed and developing countries over a 25 year period. The direction of the vectors appear to be uniformly from left to right with both urbanization and GDP per capita rising. But if you look at the lower left of the diagram (where the poorest countries cluster) the vectors are more confusing. In these cases, the two factors are not linked and the driving force in the system in the role of rural poverty in driving population off the land and towards the cities. For many African countries, urbanization has occurred without economic growth. On average, for the period shown in diagram (c), urban population grew by 4.7 per cent annually while its per capita GDP dropped by 0.7 per cent a year. The massive refugee movements and humanitarian emergencies arising from political crises (e.g. Rwanda) or environmental crises (e.g. Ethiopia) have pushed uprooted peoples towards the cities. In general, worsening physical and economic security may be pushing migrants to the relative safety of cities.

Cities as Financial Centres

In the above sections we have talked about cities in population terms. But the greatest changes in the last half-century on the world scene have been the evolution of the global city hierarchy in terms of flows of money rather than flows of people.

Figure 19.13(a) maps the hierarchy of *world financial centres.* It is based on a series of measures which include the volume of international currency clearings, the volume of financial assets and the number of headquarters of the large international banks. Three cities are at the top of the hierarchy: New York–London–Tokyo. The first two have the largest international dealings while Tokyo has a higher domestic component. London, together with the German centre of Frankfurt, play the key role in the European market. The ways in which the largest centres are located in terms of the world time-clock is shown in map (b) with the linking of stock-market trading.

Together, the 25 cities shown in Figure 19.13(b) effectively control almost all the world's financial transactions. As Nigel Thrift has shown, the stock markets are woven in to a complex set of markets and transactions. (See Box 19.B 'Nigel Thrift and the geography of money'.)

Offshore financial centres

We have noted in other parts of this chapter, that for every global trend there is a smaller counter-trend. The ever-concentrating pattern of global finance is no exception. Scattered around the globe are a series of little places, usually islands and micro-states, that exploit niches in the circulation of global capital. They tend to occur in three geographical areas: (i) the Caribbean (e.g. Cayman islands or the

(a)

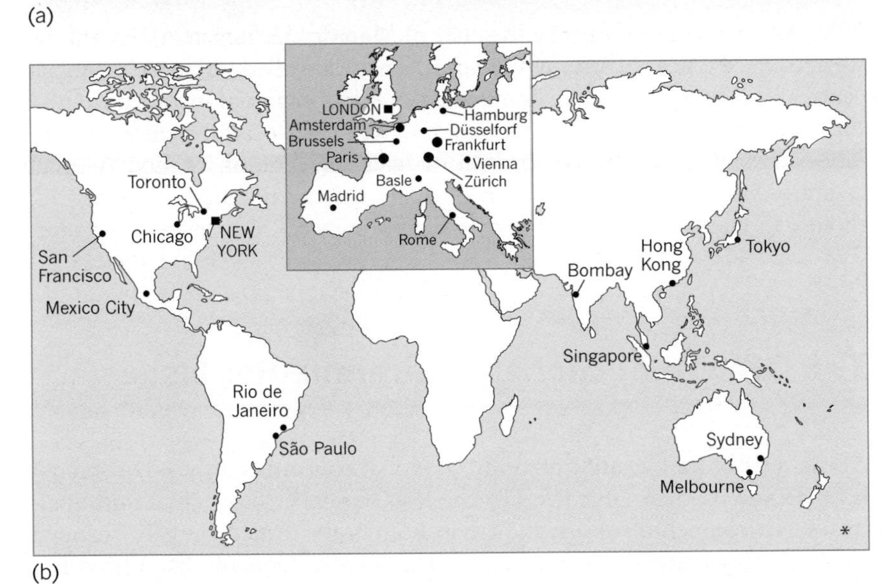

(b)

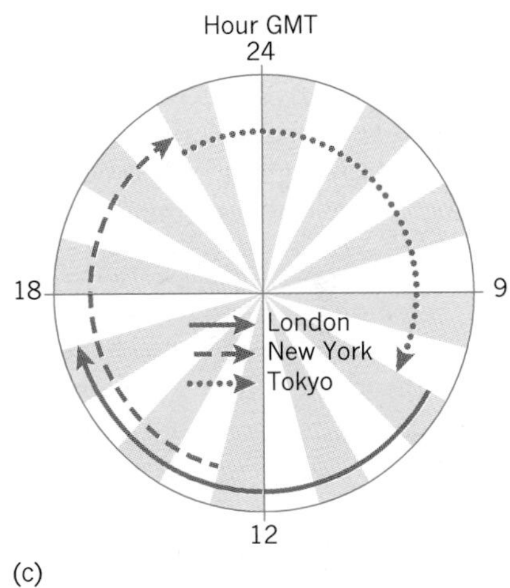

(c)

Figure 19.13 World financial centres (a) New York Stock Exchange Wall Street. (b) Map of the main hierarchy of world financial centres. (c) The timespan of hours during which each of the world's six major stock exchanges is open during a normal working day.

[Sources: (a) Paul Fusco/MAGNUM. (b) H.C. Reed, in Y.S. Park and N. Essayad (eds), *International Banking and Financial Centres* (Kluwer Academic, Boston, 1989), Fig. 16.1). (c) N.J.Thrift, in R.J.Johnston *et al.* (eds), *Geographies of Global Change* (Blackwell, Oxford, 1995), Fig. 2.1, p. 26.]

Box 19.B Nigel Thrift and the Geography of Money

Nigel Thrift (born Somerset, England, 1949) has led a group of geographers concerned with understanding the geography of money. Originally a graduate of the University of Wales, Thrift is Professor of Human Geography at the University of Bristol. His research interests span from the geography of time, through urban growth in Vietnam, to the social and cultural determinants of the international financial system. He has edited *The Socialist Third World* (1987), *Class and Space: The Making of Urban Society* (1987), *New Models in Geography* (1989), and *Money, Power and Space* (1994).

Before the 1980s there were very few attempts to explore the geography of money. But in the intervening years an important new geographic subfield has grown up, concentrating on five main questions.

1 What impact has time–space compression had on the acceleration of money circulation around the world? This work explores broad questions of the role of global circulation of capital in the globalization process.
2 What are the regional impacts of money and finance through the growth of a new financial services industry? Which cities outside the major centres have benefited from this growth and what is required for such dispersion?
3 Why have certain global financial centres been so successful and so persistent? What lies behind the continuing success of the City of London, New York, and Tokyo and what spatial challenge does it face from new centres?
4 In contrast, why and where are groups excluded from financial access? This research focuses on the spatial limits of financial exclusion and the increasing denial of access to poor and disadvantaged groups.
5 How has the growth of financial centres reinforced and broken down gender divisions? Paternalistic structures and 'gender cultures' within the financial industry appear to be slowly changing as the demand for highly skilled labour has intensified.

A good survey of the geography of money is provided by Andrew Leyshon in R.J. Johnston *et al.* (eds), *Dictionary of Human Geography* (Blackwell, Oxford, 2000), pp. 519–521. See also S. Corbridge, N.J. Thrift, and R. Martin, *Money, Power, and Space* (Blackwell, Oxford, 1994).

Bahamas), (ii) western Europe (e.g. Isle of Man or Liechtenstein), or (iii) the western Pacific (e.g. Nauru or Vanuatu). These so-called 'offshore' financial centres tend to provide services outside the risk of national jurisdictions and this give tax advantages. Rightly or wrongly, they are associated with 'hot' money arising from transactions on or over the edge of legality. A small island such as the Caymans has around 550 banks from all over the world, though in many cases there is no physical presence.

19.5 The Growth of Global Environmental Issues

At the start of the twenty-first century, global concerns on the environment have acquired a new urgency. For the last 30 years, the content and quality of environmental observation has been transformed, largely through satellite observation (see Chapter 22). The sheer volume of observation has shifted environmental concern from the minority views of a few far-sighted individuals to a central concern of international agencies. United Nations bodies now serve as a focus for international, governmental, and non-governmental organizations (NGOs) all sharing a concern with the global commons. The UN *Man and the Biosphere Program (MAB)* typifies one such approach, the *World-Wide Fund for Nature (WWF)* another.

Ranking of Environmental Issues

During the preparation of its *GEO-2000 Report,* the *United Nations Environment Programme (UNEP)* carried out a questionnaire global survey. Two hundred environmental scientists in 50 countries around the world were asked to identify those environmental issues which they expected to see emerging as key issues in the new century. The results are set out in Table 19.3.

Half the respondents mentioned climatic change as the leading key issue (see left-hand side of the table). Fresh water (both its scarcity and its pollution) and major land-use changes (deforestation and desertification) also came in the top four. Some issues are closely linked. Thus deforestation (issue 3) and loss of biodiversity (issue 6) show two sides of the same problem.

The right-hand half of the table illustrates eight other concerns selected from the 40 identified. These range ranging from emergent diseases (treated in Chapter 20 of this book) to accumulating debris in outer space. The low position occupied by *sea-level changes* (Issue 36) probably reflects its inclusion within the 'climate change' umbrella rather than its intrinsic importance. For some small Pacific nations (members of the Alliance of Small Island States, AOSIS), a 1 metre rise in sea level would devastate atoll settlements. A similar rise in Bangladesh flood a critical part of its rice-growing delta area, halving that country's production of rice crop and causing an environmental relocation of 70 million people. (See Box 19.C on 'Global sea-level changes'.)

Climate Change

Climate change is now occurring at unprecedented rates and this change is

Table 19.3 Major emerging environmental issues

Major environmental issues	Percentage citing this issue	Minor environmental issues	Percentage citing this issue
1. Climate change	51	15. Ozone depletion	17
2. Freshwater scarcity	29	17. Emerging diseases	14
3. Deforestation/desertification	28	19. Food insecurity	11
4. Freshwater pollution	28	22. Poverty	9
5. Poor governance	27	27. Invasive species	6
6. Loss of biodiversity	23	33. Space debris	4
7. Population growth and movements	22	34. Persistent bio-accumulative toxins	4
8. Waste disposal	20	36. Sea-level rise	3

Source: United Nations Environment Programme, *GEO.2000: UNEP's Millennium Report on the Environment* (Earthscan, London, 2000), p. 339.

Box 19.C Global Sea-level Changes

No environmental change reflects more closely the delicate balance between the human population and the global environment than sea level. We know from our survey in Chapter 2 that the position and level of the world's oceans have varied over geological time. Around 15,000 years ago, the sea was 120 m (400 ft) lower than it is today. The present level is being changed in several ways.

First, the long-run geological changes are continuing. Land continues to rise in areas of glacial rebound (e.g. around the Baltic Sea or Hudson Bay, Canada) where the crust is still adjusting to the removal of glacial ice sheets. Equally, in area of rapid sedimentation (e.g. the southern North Sea) there is a slight crustal sagging as the weight of sediments is accommodated.

Second, human activities are restricting the flow of fresh water back to the world's oceans. It is estimated about half of the world's total flow of water from land to sea is now controlled or influenced by dams, and that proportion is rising. Over the past 30 years, reduction in river discharges to the oceans has led to a drop in sea level of at least 2 cm (¾ in).

Third, sea levels respond to changes in temperature, from whatever cause. As the world warms and cools, so the levels of the seas change slightly in response. Sea water expands when warmed, and polar caps and inland glaciers melt accordingly as temperatures rise. Estimates of changes during the twentieth century, suggest global warming of between 1.5 and 4 °C has occurred. This would have resulted in a sea-level rise of at least 20 cm (8 in). However, overall, sea-level change has been less than this: approximately 15 cm (6 in).

Estimates for the twenty-first century are higher than those for the last. The International Panel on Climate Change (IPCC) estimates that, if no cuts in emissions of greenhouse gases were implemented, sea levels would rise at three to six times the prevailing rate: by 20 cm (8 in) by 2030. This acceleration is likely to be caused by thermal expansion of the oceans and by diminishing alpine and polar glaciers. Models suggest a net gain of ice in Antarctica, leading to a slight drop in sea level, but this will be offset by a small positive contribution from melting ice in Greenland.

It is thought that natural ecosystems, such as ***wetlands*** and coral reefs, could adapt and cope with a sea-level rise rate no more than 2 cm (¾ in) per decade. A rise of 1 m (3 ft) would drastically affect over 300 million people in low-lying coastal areas around the world. Island states such as the Maldives and Kiribati could disappear underwater. A rise of only half this figure would displace 16 per cent of the population of Egypt. In the Netherlands, two-thirds of the country already lies below sea level, and in the southern United States, the Gulf of Mexico would creep 53 km inland. The cost of protecting people and investments from encroaching waters, as calculated by the IPCC, will be over $25 billion per year. To put this figure into context, the GDP for New Zealand is around $60 billion per year.

increasingly being tied into the huge quantities of carbon dioxide, methane, and other greenhouse gases being released into the earth's atmosphere daily. *Global warming* has been occurring slowly since the nineteenth century (see Figure 19.14). The twentieth century was the warmest of the last 600 years, and 14 of the warmest years have occurred in the last two decades.

We are still a long way from being able to predict in any exact fashion the future environmental pattern that will emerge from expected climatic changes. One of the most ambitious predictions is shown in Figure 19.15 for Canada. This uses global circulation models (GCMs) to estimate the effect of doubled CO_2 climates on the existing vegetation belts.

The present-day vegetation (a) is dominated by a broad east–west swathe of Boreal Forest running from the Rockies across to Newfoundland. The various models predict longer and warmer summers with milder winters, probably with an increase in snowcover depth. One predicted map (b) shows that Boreal belt broken into two fragments, each located further towards the north and an invasion of grassland climates extending the Prairies northwards. We should note that climate can change much faster than the biota (the plant and animal communities that are adjusted to prevailing climates).

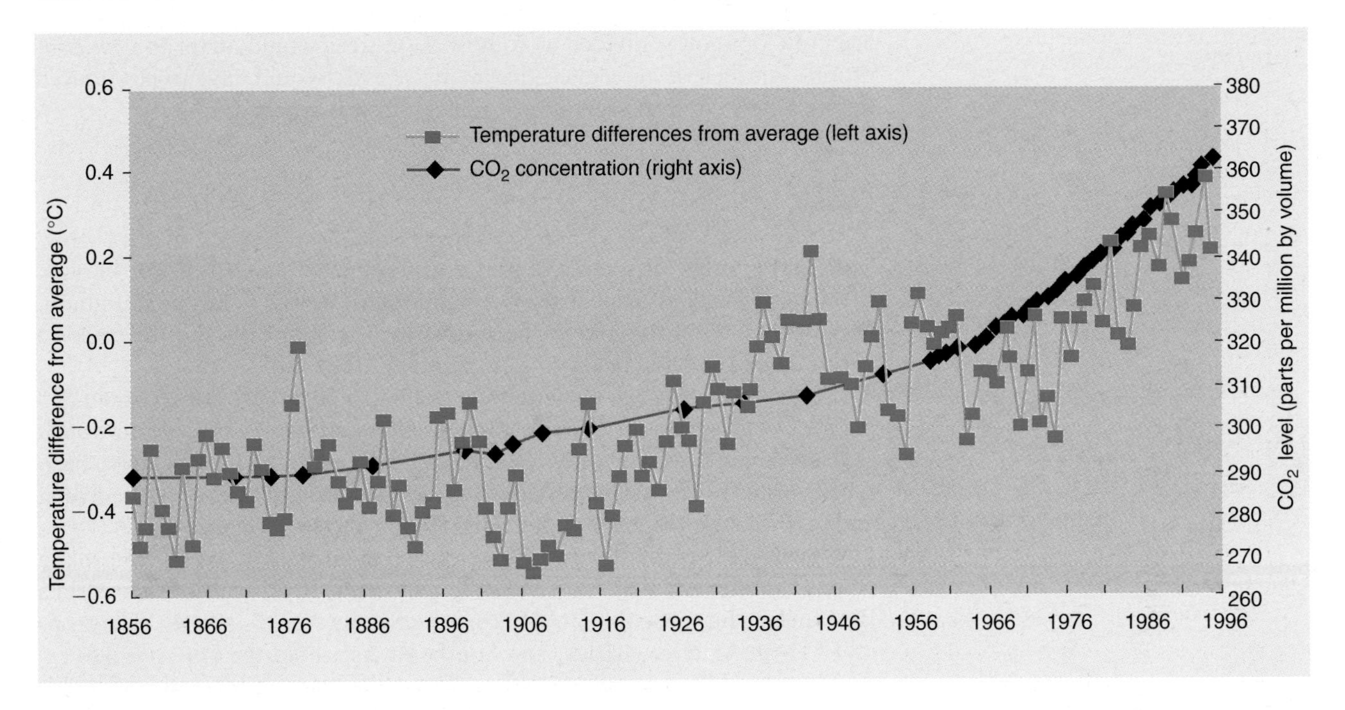

Figure 19.14 Global warming Global changes in average temperature since the mid-nineteenth century. Temperature differences from average (left vertical axis) and carbon dioxide levels (right vertical axis) measured over time (horizontal axis). The broad link between temperature and greenhouse gas concentrations are now generally accepted.

[Source: World Bank, *Entering the 21st Century* (Oxford University Press, Oxford, 2000), Fig. 1.6, p. 41.]

Figure 19.15 Impact of global warming on Canada Predicted northward shift in biomes in Canada which are expected to result from doubling the global greenhouse effect.

[Source: F. Kenneth Hare, *The Canadian Climatic Program* (Environment Canada, Downsville, 1993).]

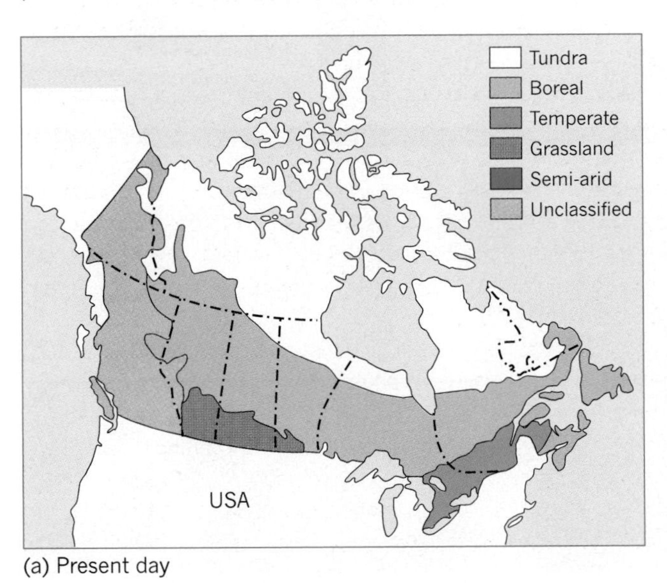

(a) Present day

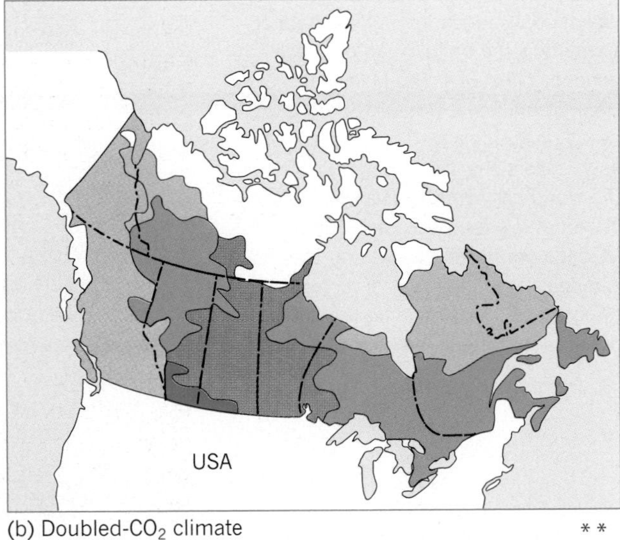

(b) Doubled-CO_2 climate

* *

Scientific opinion is divided as to how soon trees would adapt to new conditions but there is agreement that many stresses would have to be endured before a new climate–vegetation equilibrium was reached.

Loss of Global Biodiversity

The total number of species of plants and animals on earth is not known. Estimates range as widely as five million to nearly a hundred million. Perhaps one-tenth have so far been described scientifically but the fraction varies from one biological group to another.

Larger creatures such as vertebrate animals are rather well recorded (90 per cent of all species are listed), while smaller and less observable ones such as bacteria are rather unknown (perhaps 1 per cent of all bacteria have been typed so far). The role of biological reservoirs and collections in recording and maintaining the gene pool of diversity is shown in Figure 19.16.

What is clear is that the most species-rich environments on Earth are the tropical rainforests which extend only over some 8 per cent of the world's land surface but probably hold 90 per cent of the world's species. The tropics of Latin America, Africa, and Southeast Asia hold the key to ***biodiversity***; the world's driest and coldest areas (the mid-continental and polar deserts) have only very few species by comparison.

The difficulty in defining species and the fact that most are still unrecognized makes calculations of overall biodiversity loss difficult. But we can be clear about limited biological groups. At the start of this century the International Union for the Conservation of Nature (IUCN) estimated that (a) a quarter of the world's mammal species and (b) a tenth of the world's bird species were assessed as globally threatened – that is, at significant risk of total extinction. Table 19.4 gives the threat situation for five different biological groups.

Freshwater habitats are particularly vulnerable. In the United States where so many watercourses are altered by dam building and flow control measures and where industrial pollutant risks abound, many freshwater

Figure 19.16 Protecting loss of biodiversity One of the protections against the reduction in global habitat diversity is the building up of critical genetic collections. The Millennium Seed Bank at the Royal Botanic Gardens at Kew, London, is a leading example of such collections. Here seeds from around the world are carefully classified and stored at low temperatures. Botanical gardens around the world are (like their zoological counterparts) of increasing importance in maintaining the breeding potential of threatened species.

[Source: Kew Enterprises.]

species are under extreme threat: shellfish, 70 per cent of species under threat; crayfish, 50 per cent; and fish, 37 per cent. Paradoxically, part of that threat comes from new biological introductions. For example, the zebra mussel is a small freshwater mollusc native to Russia which was introduced into North America in 1970. It invaded southern Canada and the Great Lakes, and is now found in two-thirds of US waterways. It drives out other mussel species and causes large-scale disruptions in local food webs. It also causes substantial economic damage by clogging the water-intake structures of power plants and water-treatment plants, and by encrusting the bottoms of boats.

Global Fresh Water

Global freshwater use rose sevenfold over the twentieth century, more than twice the rate of global population growth. About one-third of the world's population already lives in countries with moderate to high *water stress*, where stress is defined as water consumption being more than 10 per cent of the renewable freshwater supply. As Figure 19.17 shows, the problem is already most acute in Africa and western Asia. Lack of water is now a major constraint to industrial growth in China, India and Indonesia. In Africa, fourteen countries are already subject to water stress and this number will nearly be doubled in the next quarter-century.

The decline in quantity and quality of the world's freshwater resources threatens to be the dominant cause of tension in the coming century.

Sewage and Pollution

Half the world's population (over three billion people) have poor sanitation which makes them vulnerable to water-borne diseases. In many developing countries, rivers downstream of large cities are open sewers. Counts of faecal *coliforms* (a standard measure of human sewage pollution) in Asian rivers have doubled since the late 1970s and are now 50 times higher that WHO guidelines in streams near major cities. In Latin America only 2 per cent of sewage receives any treatment. As we shall see in the next chapter, diarrhoeal diseases are the second largest source of children's deaths (see Figure 20.19).

Sewage pollution is a side effect of rapid population growth in cities with-

Table 19.4 Threatened animal species

Animals	Critically endangered	Endangered	Vulnerable	World region with highest number of critically endangered species
Mammals	189	375	709	Asia/Pacific (37%)
Birds	176	275	300	Asia/Pacific (34%)
Fishes	172	171	572	North America (31%)
Reptiles	49	83	168	Latin America (43%)
Amphibians	19	33	78	Asia/Pacific (42%)
Total	**605**	**937**	**1827**	Asia/Pacific (194)

Source: United Nations Environment Programme, *GEO-2000: UNEP's Millennium Report on the Environment* (Earthscan, London, 2000), p. 41.

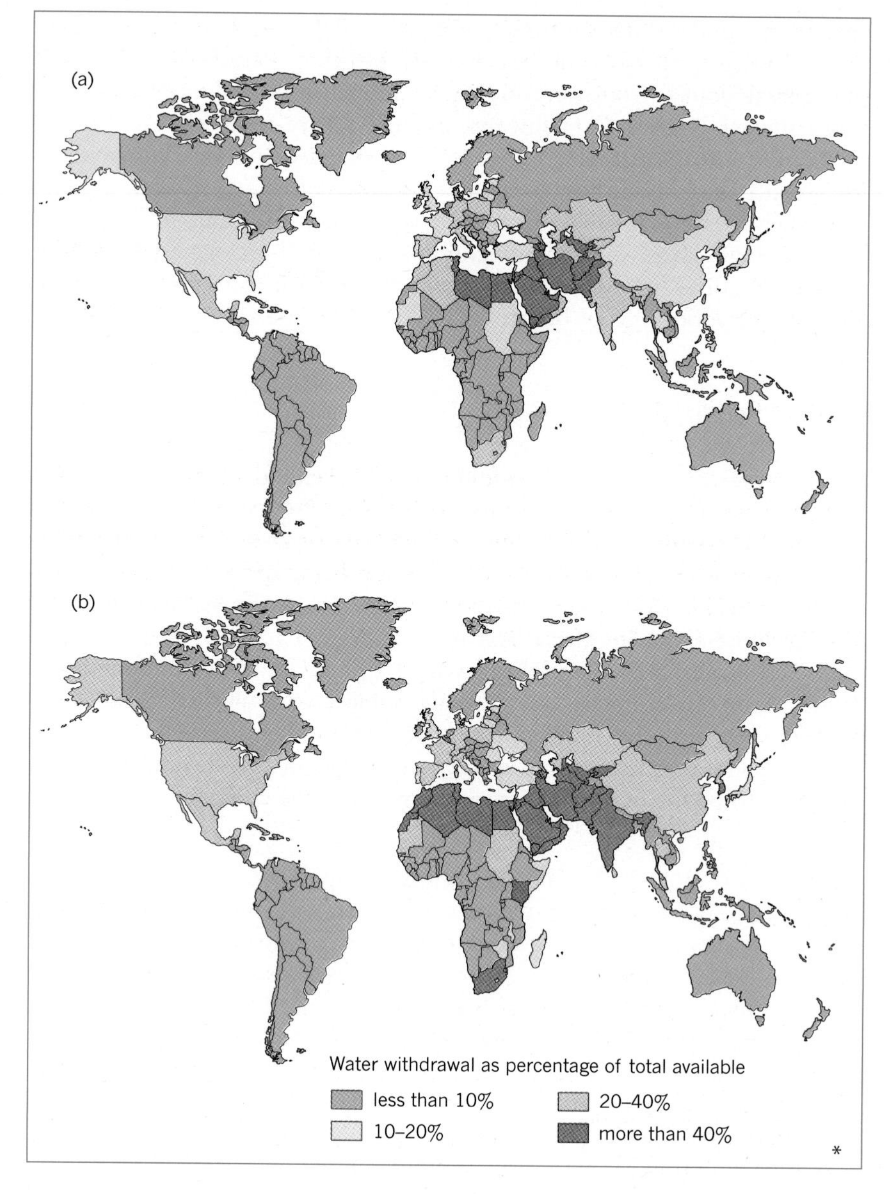

Figure 19.17 Global water stress
Water stress is measured as water withdrawal as a percentage of total water available. (a) World pattern in 1995. (b) Estimated world pattern in 2025 by which time as much as two-thirds of the world's population may be subject to moderate to high water stress.

[Source: United Nations Environment Programme, *GEO.2000: UNEP's Millennium Report on the Environment* (Earthscan, London, 2000), p. 42.]

out drainage infrastructures. The city of Jakarta in Indonesia is estimated to have 900,000 septic tanks, Manila in the Philippines 600,000 tanks.

While sewage pollution is worldwide the largest and most common problem it is not the only one. In industrial countries the intensive use of pesticides and fertilizers has led to chemicals being leached into freshwater supplies. Nitrate pollution exceeds public health standards in many European countries and even in a country as rich as the United States, one in six of its population obtains drinking water from systems that violate health standards (mainly through nitrate pollution). Nitrates are not only dangerous to human health but lead to excessive algal growth in waterways, leading to eutrophication of inland waters.

Industrial wastes are significant sources of water pollution, and often give rise to contamination by heavy metals (lead, mercury, arsenic, and cadmium). A study of fifteen Japanese cities showed a third of all groundwater supplies to be contaminated by chlorinated solvents from industry.

Groundwater Over-abstraction of *groundwater* has also affected both the quantity and quality of water from this scource. This has led to seawater intrusion along shorelines causing salinization of coastal agricultural land. In Bahrain the saline interface between sea water and fresh water is moving inland at rates of over 100 metres a year. In Madras, India, salt water intrusion has moved up to 10 km inland, rendering many irrigation wells along the coastal edge useless.

Use of water is slated to rise in the next few decades. Household, agricultural, and industrial uses are all set to double by the year 2025 and pollutant emissions to watercourses expected to show a fourfold increase.

Scale of Environmental Hazards

One common feature of environmental problems is the importance of geographic scale. John Whittow has shown that there is an inverse relation between the frequency of a hazard and its size. As Figure 19.18(a) shows, human-induced hazards such as vehicle crashes or fires kill small numbers at fairly frequent intervals. Conversely, natural hazards kill greater numbers but occur at less frequent intervals.

However, the relationship is not stable over time. Most environmental records now show that events such as water pollution problems are

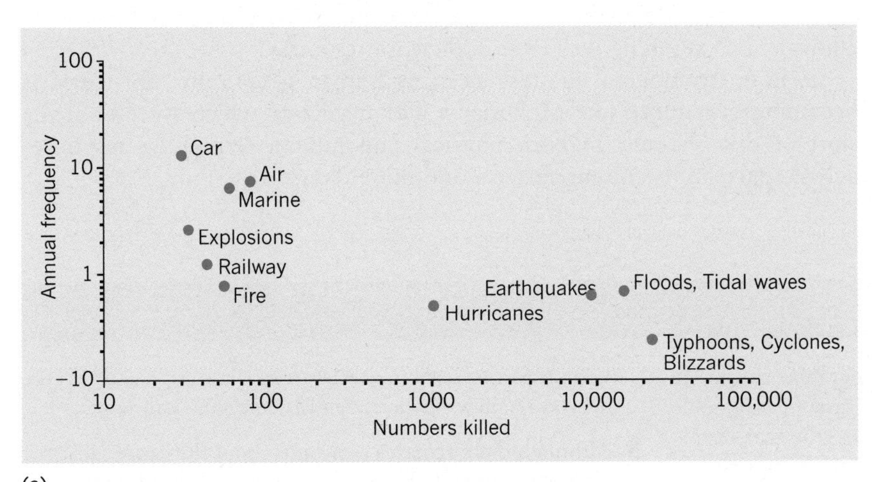

(a)

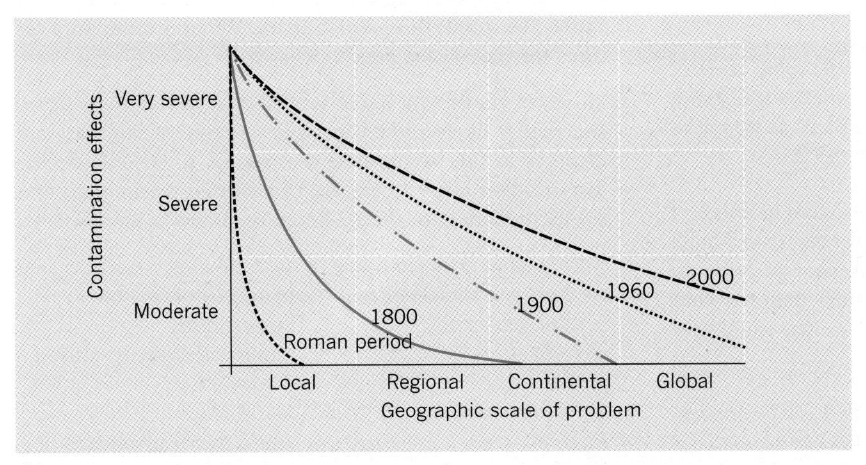

(b)

Figure 19.18 Scale of environmental hazards
(a) Magnitude and frequency of environmental hazards at a world scale. Note that both scales are logarithmic. (b) Historical evolution of the scale of water pollution problems over the last two thousand years.

[Sources: I. Douglas *et al.* (eds), *Companion Encyclopaedia of Geography* (Routledge, London, 1996), (a) Fig. 29.2, p. 623; (b) Fig. 25.2, p.535.]

increasing over time: in number, scale, range, and severity. Figure 19.18(b) show a schematic outline of the historical evolution of such problems.

If we look back to Chapter 3 (see especially Figure 3.17) we can find details of El Niño. There is increasing realization that changes in the Pacific currents have major effects worldwide. The 1997/98 El Niño was associated with several events well outside the western Pacific Basin:

1 North America: unusual jet stream patterns leading to severe storms on the West Coast.
2 South America: torrential rains over southern Brazil and adjacent countries in the Plate basin; severe drought in the Guyanas.
3 Africa: unusually hot weather over southern Africa; exceptional rainfalls in East Africa.
4 Asia: persistent dryness over Indonesia and the Phillipines (prelude to the Indonesian forest fires).

The Indonesian forest fires that began to burn in Kalimantan and Sumatra in September 1997 illustrate the huge scale of environmental pollution that can arise from a major incident. The extent of the smoke haze on a single day is shown in Figure 3.17. Air pollutants from the fire spread east–west along a 3300 km (2999 mile) axis, covering six southeast Asian countries. Smoke haze reached as far north as Thailand and as far south as northern Australia. Air pollution levels in some cities in the region were recorded at eight times the 'unhealthy' standard, with pollution peaks reaching levels equivalent to smoking well over 200 cigarettes a day!

So, in environmental matters as in its human geography, the world is becoming ever more like McLuhan's global village, which we met at the start of this chapter. In both physical and human terms, the responses across space are becoming faster – and more dangerous.

Reflections

1 Globalization is one of the prevailing geographic themes at the start of the twenty-first century. But how far is globalization a new process? In what ways are the transformations brought about by the railways in the nineteenth century different in kind from those of electronic collapse of space at the present time?

2 The chapter argues that although the geographic *scale* of dependence has changed (from local to global), the dependence of humans on their environment remains as strong as ever. Consider the implications of Figure 19.2.

3 Look carefully at the maps of increasing spatial mobility of one family over four generations (the Bradleys) in Figure 19.4. How far is this typical? Check with your parents and grandparents and try to construct a similar map for your own family movements? Does it show a similar trend over time?

4 Why is the control of global finance so heavily concentrated into a few world cities? Is the growing dependence on the Internet and electronic movement of moneys in its various forms likely to accelerate or alter this trend? Where do you see new financial centres emerging and why?

5 Although there remains a dispute about the *exact* causes of global warming, the overall nature of the trend now seems well supported by recent climatic records. Figure 19.15 shows the implications for Canada. What are the implications for your home area?

6 How far would you agree with the statement that 'water shortage is likely to be the biggest source of international disputes in the twenty-first century'. If so, where are the disputes first likely to emerge? Select *two* world areas and compare them with those chosen by others in the class.

7 Review your understanding of the following concepts using the text and the glossary in Appendix A of this book:

'end of geography'
globalization
global village
global warming
Internet
Kondratieff waves
localization
multinational corporation
sea-level rises
world cities
world financial centres

One Step Further ...

A useful starting point is to look at the essay on the new 'geo-economy' by Peter Dicken in I. DOUGLAS *et al.* (eds), *Companion Encyclopaedia of Geography* (Routledge, London, 1996), pp. 370–90, and also his brief but useful introduction to globalization in his contribution to R.J. JOHNSTON *et al.* (eds), *Dictionary of Human Geography*, fourth edition (Blackwell, Oxford, 2000), pp. 315–16. The theme is further expanded by the same author in his major book, P. DICKEN, *Global Shift: The Transformation of the World Economy*, third edition (Sage, London, 1998). The broad theme of globalization is also treated in M. WATERS, *Globalization* (Routledge, London, 1995) and its geographic implications clearly illustrated in R.J. JOHNSTON *et al.* (eds), *Geographies of Global Change: Remapping the World in the Late Twentieth Century* (Blackwell, Oxford, 1995). The dramatic changes in the global economy that underpin many of these changes are covered in P.W. DANIELS and W.F. LEVER (eds), *The Global Economy in Transition* (Longman, Harlow, 1996).

Globalization forms a recurrent theme in the essays both by geographers and by economists in Gordon CLARK *et al.* (eds), *The Oxford Handbook of Economic Geography* (Oxford University Press, Oxford, 2000). On special issues raised in this chapter, see for environmental change P.D. MOORE *et al.*, *Global Environmental Change* (Blackwell Science, Oxford, 1996). The rapid rise of global tourism is treated from a geographical angle in D. PEARCE, *Tourism Today: A Geographical Analysis*, second edition (Longman, Harlow, 1995) The changes in the global balance of the superpowers and the rise of the Asian Pacific region is studied in P. PRESTON, *Pacific Asia in the Global System* (Blackwell, Oxford, 1998). The rise of the world city is studied in Peter HALL, *Cities of Tomorrow* (Blackwell, Oxford, 1996) and the problems of one such city, Los Angeles, in Edward W. SOJA, *Postmetropolis* (Blackwell, Oxford, 1999). See also J.R. SHORT and Y-H. KIM, *Globalization and the City* (Longman, Harlow, 1999).

The global change and globalization 'progress reports' sections of both *Progress in Physical Geography* (quarterly) and *Progress in Human Geography* (quarterly) will prove helpful in keeping up to date. Some specialized journals have been started such as *Global Environmental Change* (quarterly). Regular reports on the changing state of the world are provided by L.R. Brown and colleagues in the *Worldwatch Institute Reports* (annual). The rapidity of changes in the global economy suggest that weekly journals and web sites may be more useful than standard academic journals. Keep an eye on *The Economist* (weekly). For readers with access to the World-Wide Web see also the sites recommended in Appendix B at the end of this book for topics relevant to this chapter.

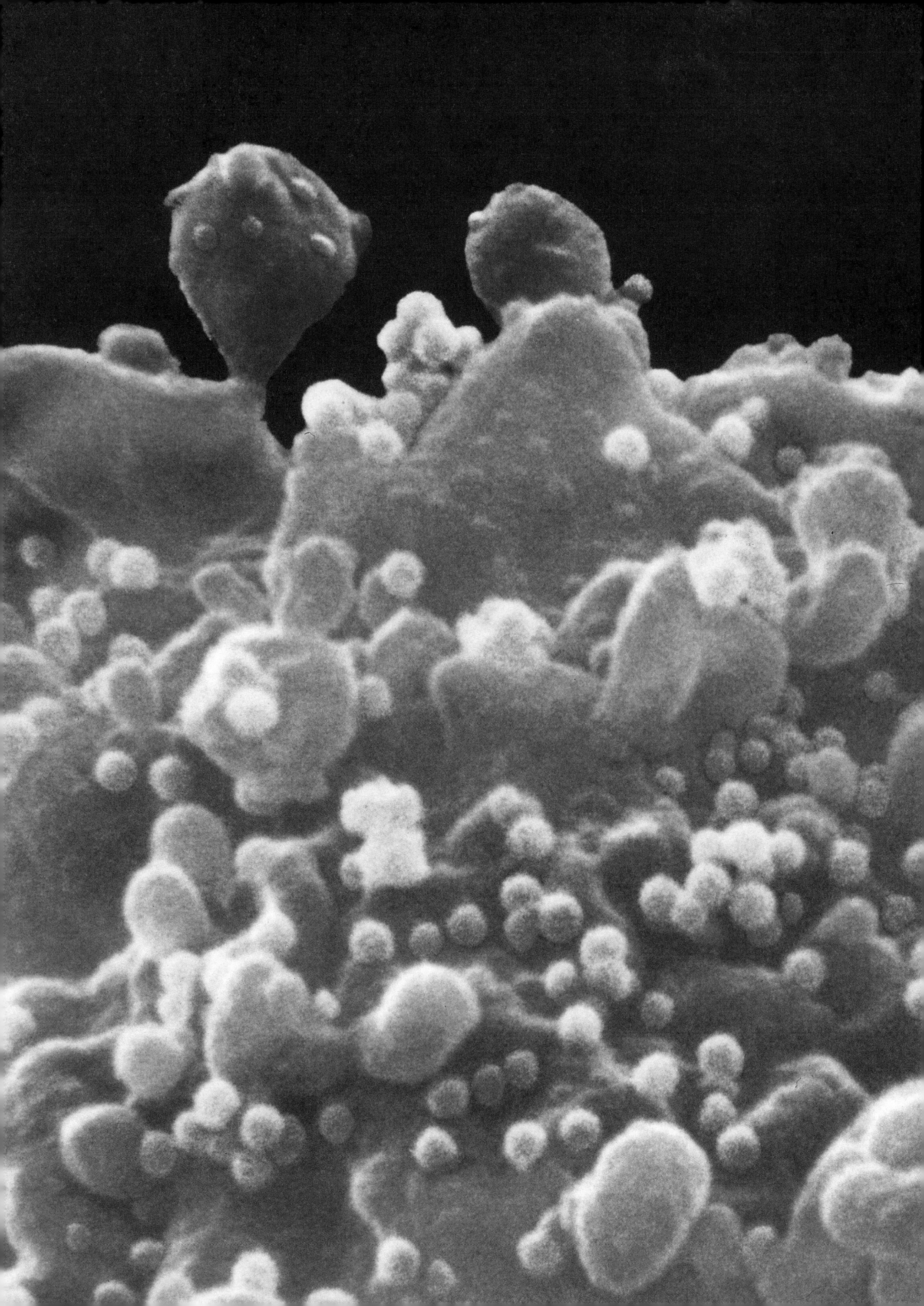

CHAPTER 20

The Global Burden of Disease

■ denotes case studies

Figure 20.1 **Threats to world health** First recognized in the early 1980s, HIV (human immunodeficiency virus) has grown to affect approaching 40 million people worldwide, with 13 million deaths from AIDS already recorded. In some countries in southern Africa, a quarter of the adult population is now infected. This scanning electron micrograph shows the cause of the disease. The surface of a T-lymphocyte white blood cell (purple) is infected with HIV. Small, spherical virus particles seen on the surface are budding away from the cell surface. Magnification ×60,000. [Source: NIBSC/Science Photo Library.]

> Great plagues remain for the ungodly.
>
> *The Book of Common Prayer* (1662)

In the year 2000, some 15,000 babies were born across the world every hour. Most of them survived the trauma of birth and nine out of ten are likely to survive the first five years of life. Half are likely to live to celebrate their 75th birthday in 2075. Just a few will become centenarians and live throughout the whole of the 21st century and see the dawn of the 22nd.

But these are gross averages based on overall global trends. For any individual child of 2000 the future may vary hugely. Some will be well nourished, some born into poverty. Some will be well endowed by their genes, others handicapped from birth. Sadly, an increasing number will be carrying the burden of the human immunodeficiency virus (HIV) (see Figure 20.1 p. 614) whose effects will only become apparent as AIDS in their later childhood.

But these life-chances are not spatially random: there is a distinctive geography of risk which operates at all spatial levels from the most local to the global. Of all the factors that determine the life chances of the child, *where* they were born is likely to play a significant part in their future life. No aspect of geography is a greater source of actual and potential global tensions than the jagged contrasts in life and death chances. No contrast is greater than that bequeathed by good health and poor health, by the absence or presence of disease.

In this last chapter of this section on global tensions, we turn to a consideration of the global burden of disease. We begin in Section 20.1 with a consideration of disease itself. What does this term mean? What causes diseases? How do we measure their impacts on human populations? What are the most important diseases? We lay special emphasis on those diseases that give rise to *epidemics*, outbreaks well defined in terms of time and space.

The next two sections are concerned with the geography of disease itself. We divide this into two parts. First, the changing geography of the 'old' diseases which have plagued the human race for many centuries (Section 20.2). Special emphasis is given to those that are changing most rapidly (e.g. poliomyelitis) rather than those that have proved resistant to change (e.g. tuberculosis). Second, we consider the emerging geography of the 'new' diseases (Section 20.3). Here we pay special attention to the greatest of the new *pandemics*, AIDS.

This leads directly on to the links between disease and those global changes which we considered in the last chapter. So in Section 20.4 we place disease change within the broad context of demographic, economic, and environmental change. Finally in Section 20.5 we look forward into the present century, asking how the pattern of disease is likely to change and how geographers can help in understanding and contributing to such changes.

20.1 Estimating the Disease Burden

In this section we consider the meaning of disease and its causes. What does this term mean? What causes diseases? How do we measure their impacts on human populations? What are the most important diseases?

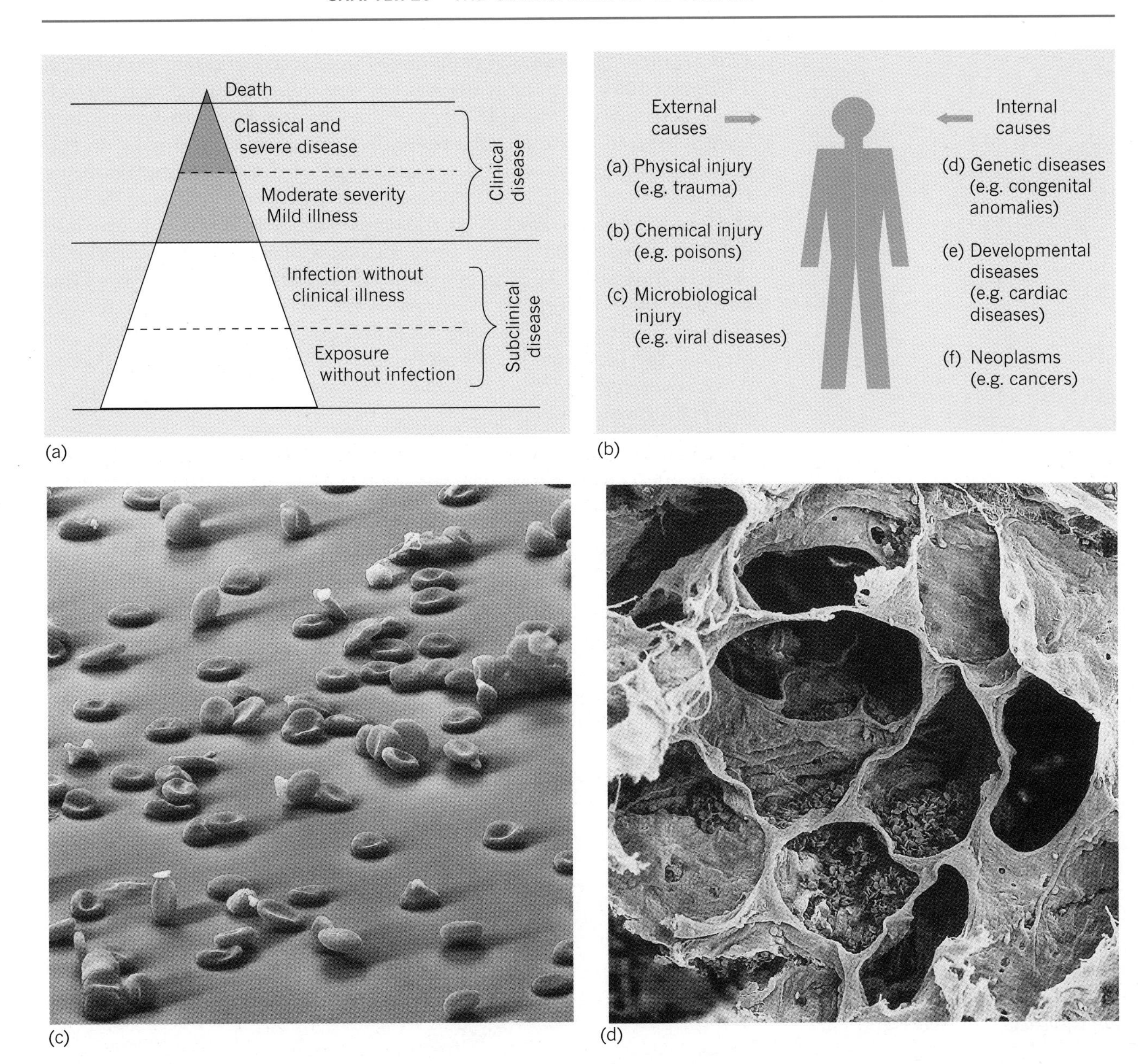

Figure 20.2 Nature of diseases
(a) 'Iceberg' concept of disease. Note that much subclinical illness is hidden. (b) Major pathways of disease entering the human body. (c) Malaria: scanning electron micrograph of red blood cells and *Plasmodium falciparum* protozoa which cause malaria. The disease is spread by bites from infected mosquitoes (*Anopheles* sp.) Magnification ×1200. (d) Lung cancer: haemorrhage in lung cancer. Scanning electron micrograph of alveoli air sacs of the human lung in a smoker with lung cancer. Magnification ×200.

[Sources: (c), (d) SPL Eye of Science/Science Photo Library.]

Disease and its Causes

The term disease literally means 'dis-ease', the absence of 'ease', the opposite of good health. In practice, it is applied to any sickness, ailment, or departure from good health. Most often it is a specific disorder of a specified part of the body (e.g. the liver) or a disorder due to a specific agent (e.g. the measles virus). Figure 20.2 shows in (a) the ***iceberg concept of disease*** in which illness forms the upper part of the triangle and death its vertex. Alongside this (b) shows the ways in which diseases of various types attack the human body. Some are internal, some external, and some a mixture of the two.

Attempts to classify diseases were attempted by the Chinese, Greek, and Arab physicians nearly two millennia ago. Medieval physicians classified diseases in terms of four humours. With the growth of biomedical science in the late nineteenth century, conferences of physicians and statisticians were held to agree on standard definitions of disease. The first *International*

Classification of Diseases was published in Chicago as a slim pamphlet in 1893; a century later the tenth revision now runs to several weighty volumes. Some idea of the wide range of diseases now recognized by medical science is given by turning the pages of a single week's report by an epidemiological agency, we find a wide range of diseases outbreaks being reported. Thus Australia's *Communicable Diseases Intelligence* bulletin chosen for a random week (that ending on 4 April 1994) records over 120 different diseases and agents: these include a major mumps outbreak in Western Australia, 131 cases of hepatitis C, an outbreak of Ross River virus infection in the Northern Territory while its overseas section records Japanese B encephalitis breaking out in Sri Lanka, and a severe malaria outbreak on the Trobriand Islands off Papua New Guinea. Meantime influenza A was sweeping through 21 Russian cities and cholera was continuing to invade northern Mozambique.

Global Estimates

What are the main diseases at the global level? If we begin with the simplest criteria of disease importance, those that result in a human death,

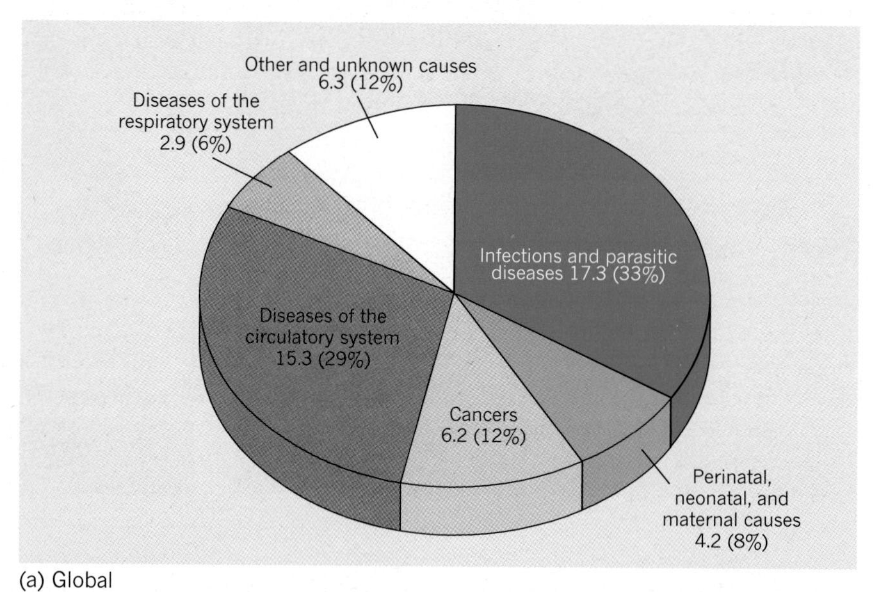

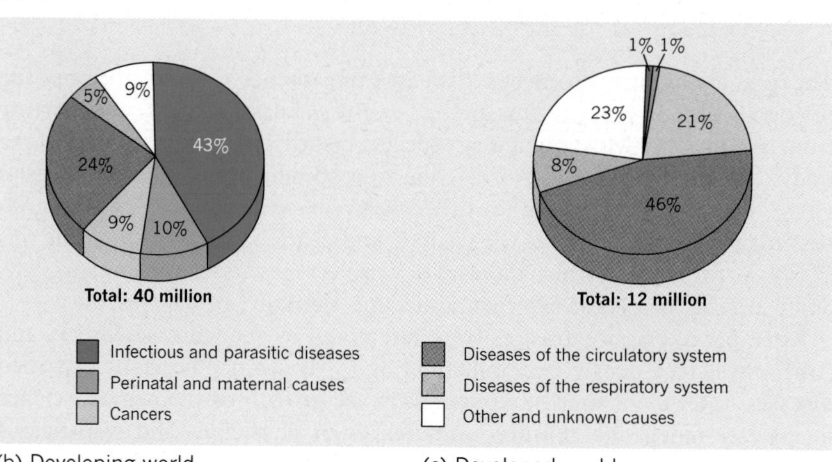

Figure 20.3 Global causes of death Classification of causes of death into six categories. The number of deaths in millions is followed by the percentage of total deaths. (a) Global deaths. (b) Deaths in the developing world. (c) Deaths in the developed world. Note the contrast in the share of deaths from infectious diseases in (b) at 43 per cent and (c) at only 1 per cent. In developing countries diseases of the circulatory system (e.g. heart attacks, strokes) account for 10 per cent of all deaths, but in the developed world this proportion rises to 46 per cent. Perinatal = time immediately before and after birth. Neonatal = newborn child.

[Source: World Health Organization *The World Health Report 1998* (WHO, Geneva, 1999), Figs 5, 6, p. 44.]

then Figure 20.3 shows the situation at the start of the century. This is based on the *World Health Organization's (WHO)* reports which are based in turn on figures supplied by member countries. Accuracy varies from country to country and the totals are rough estimates rather than precise censuses.

The WHO estimates that more than 50 million deaths occurred worldwide in 2000. In (a) the share of six main groups of diseases are shown. The leading cause was infectious and parasitic diseases (such as tuberculosis, diarrhoea, malaria, and HIV) which accounts for one-third of the total. Almost as many (29 per cent) were due to diseases of the circulatory diseases (e.g. coronary heart disease) and a further 12 per cent were due to cancers. Together these three causes account for three-quarters of all deaths. (The contrasts between the developed and developing world are discussed in Section 20.2.)

While death is a stark measure of importance it is not an unambiguous one. Should the death of a child and the death of a very old person be equated? In one life has been cut short, in the other lived (at least, in years) to the full. There are several ways in which an estimate may be made which takes this difference into account. (See Box 20.A on 'Measuring Mortality'.)

Mortality is only one measure of the importance of a disease. Table 20.1 compares measures of mortality in comparison to two other measures based on illness (*morbidity*). This shows that in the year in question infectious diseases occupy half of the top ten places as global killers. Taken together, infectious diseases and parasites take 16.4 million lives a year, ahead of heart disease which kills 9.7 million.

Box 20.A Measuring Mortality

Although counting deaths from a particular cause in a particular year might seem a simple and unambiguous way of measuring the burden of disease, it avoids a central question. Are all deaths equal? Although this will depend on your religious viewpoint, the theological view might be exactly that – all lives and therefore all deaths are indeed equal. Let us think about this a little more. We all have to die at some time, and while the death of a child might be seen as a tragedy (Andrew Marvell's 'cut is the branch that might have grown full straight'), that of someone who is very old and perhaps very sick might sometimes be seen as a blessing. In other words, an argument exists that deaths of the young may cut off more potential years of living than deaths of the old. Clearly this is a huge simplification and leaves many moral questions unanswered.

Agencies such as the World Health Organization and the US Centers for Disease Control have been experimenting with statistics that measure ***years of potential life lost (YPLL)***. This takes into account the year at which a death occurs. For example, if we take the 'life tables' published by most national population agencies we can look up life expectancy. In England in 2000, a 20-year-old male could expect on average to live for another 55 years (60 for a female) whereas an 80-year-old male could expect to live on average for another 7 years. So under YPLL measures the two deaths would be weighted differently, the younger death having a weight much higher than the older.

The effect of measures of this kind is to change the order of some leading causes of death, giving greater weight to the infectious diseases (which especially affect children) and injuries (which tend especially to affect young people) and reducing the relative importance of heart disease and cancers (which tend to affect the old). With illness and disability, similar arguments may be used, giving special weight to diseases that cause a lifetime of disability rather than a brief illness. An introduction to the problems in trying to compute and use these measures is given in Alan D. Lopez *et al.*, *The Global Burden of Disease* (World Health Organization, Geneva, 1996).

Table 20.1 Different definitions of the global disease burden

Criterion I Deaths	Criterion II Morbidity (new cases)	Criterion III Permanent and long-term activity limitation
1 Coronary heart disease (100)	1 Diarrhoea (100)	1 Mood disorders (100)
2 Cerebrovascular disease (64)	2 Malaria (13)	2 Hearing loss (84)
3 ALRI (51)	3 ALRI (10)	3 Schistosomiasis (82)
4 Tuberculosis (40)	4 Occupational injuries (6)	4 Lymphatic filiariasis (82)
5 COPD (39)	5 Occupational disorders (5)	5 Cretinoids (34)
6 Diarrhoea (34)	6 Trichomoniasis (4)	6 Mental retardation (25)
7 HIV/AIDS (32)	7 Mood disorders (3)	7 Schizophrenic disorders (18)
8 Malaria (28)	8 Chlamydial infections (2)	8 Occupational injuries (17)
9 Prematurity (16)	9 Hepatitis B (2)	9 Occupational diseases (14)
10 Measles (13)	10 Gonococcal infection (2)	10 Cataract-related blindness (13)

The leading cause in each of the three categories is set to 100 and other causes expressed in relative terms. Thus in Category I the annual toll of deaths from tuberculosis (2.9 million) are 40 per cent of the deaths from coronary heart disease (7.2 million). ALRI = Acute lower respiratory infections. COPD = Chronic obstructive pulmonary disease. Diarrhoea includes dysentry.
Source: Adapted from the latest edition of the World Health Organization's *World Health Report.*

Table 20.1 also shows another way to measure disease is through disease incidence – the number of new cases of a disease each year. Again, the table shows the dominant position of the infectious diseases. The outstanding cause of illness worldwide is diarrhoea in children under five. This accounts for 1.8 billion episodes a year (and claim the lives of three million children). Acute lower respiratory conditions in children, sexually transmitted diseases, measles, and whooping cough remain major problems.

Yet other ways of measuring the disease burden are in terms of prevalence – the total number of people with a given condition – or the burden of disability that a disease causes. Global figures are hard to find and we know all too little about some major infectious diseases. Such fragments of information as we have at the world scale underscore the role of communicable diseases: e.g. schistosomiasis (a parasitic worm infection of the tropics) has a prevalence of some 200 million people worldwide while 10 million are still permanently disabled by paralytic poliomyelitis (but note the changing situation described in Section 20.3).

National Estimates

An alternative approach to estimating the size of the disease burden is to use figures for a single country. The Carter Center Health Policy Project in Atlanta reworked the available statistics for the United States. It concluded that 740 million symptomatic infections occur annually in the United States, resulting in 200,000 deaths per year. Such infections result in more than $17 billion annually in direct costs, not including costs of deaths, lost wages and productivity and other indirect costs. About one-third of the potential deaths from these causes are currently prevented annually and a further third *could* be prevented by using current interventions.

Figure 20.4 summarizes the Carter Center findings. This plots on a logarithmic proportional scale the number of deaths and number of cases for the main infectious diseases. Note that only diseases causing more than ten deaths or more than 1000 cases per year are shown. Most deaths (32,000) are caused by pneumoccocal bacteria followed by nosocomial (i.e. hospital)

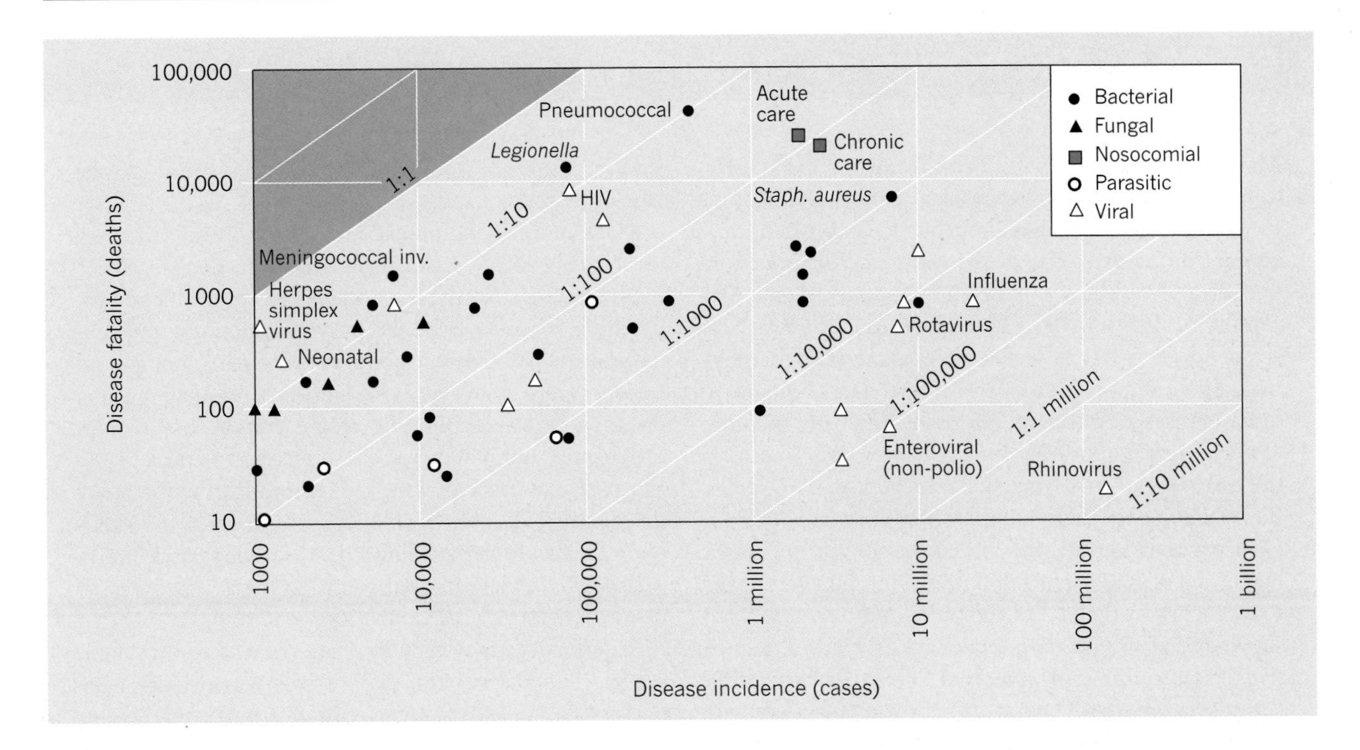

deaths in acute care (26,400) and chronic care (24,700). In terms of morbidity by far the largest number of cases are generated by the the common cold viruses (the rhino viruses) (125 million) followed by another group of viruses, influenza (20 million). The figure shows only the leading 50 of the specific infections used by the Carter Center. Although those not shown have fewer than 10 deaths per year or fewer than 1000 cases, they include many diseases that rank highly on the 'dread' factor. For example one form of meningitis (amoebic meningoencephalitis) was recorded only four times in the United States in the year studied but each resulted in death; rabies killed all ten of those infected; half of the 100 cases infected with the cryptosporidiosis parasite died. The case fatality ratio is shown in Figure 20.4 for the more frequently occurring diseases by the diagonal lines which represent fatality-case ratios. Those shown on or above the 1:10 diagonal include HIV, legionnaires' disease, and meningitis.

Figure 20.4 Deaths and illness for a single country Annual mortality and morbidity from major infectious and parasitic diseases in the United States in the mid-1980s. Note that both disease fatality and disease incidence are plotted on logarithmic scales. The diagonal lines represent the fatality/case ratio with severe conditions in the upper-left quadrant of the graph and minor ailments (e.g. the common cold) in the lower right.

[Source: Drawn from data in R.W. Amler (ed.), *Closing the Gap: The Burden of Unnecessary Diseases* (Oxford University Press, Oxford, 1987), Table 1, pp.104–107.]

The Changing Geography of Old Diseases 20.2

We noted in Section 20.1 that the main global killers remain today the communicable diseases. Most of these diseases are of considerable antiquity and have been part of the human condition since populations first clustered into agricultural settlements. (See the discussion in Chapter 8.) In the nineteenth century, cholera was one of the most dreaded killers (see Figure 20.5). At the start of the present century, many of those diseases remain important today. Two of the greatest killers are tuberculosis and malaria. *Tuberculosis* kills around four million people worldwide, accounting for over 5 per cent of global deaths and making it the fourth ranking cause of death. Tuberculosis largely affects poor people and 95 per cent of deaths

Box 20.B The Resurgence of Malaria

Malaria is a serious, acute, and chronic relapsing infections, marked by periodic attacks of chills and shaking fever, anaemia, spleen enlargement, and often fatal complications. It is one of the most ancient infections known and was described by Hippocrates in the fifth century BC. When malaria first appeared in the Americas is not known, but it seems likely that it was a post-Columbian importation from Europe. The name of the disease arises from the Italian terms for 'bad air' (literally 'mal-aria') indicating the early association between the disease and low-lying, waterlogged, marshy areas. But while the association with swampy or marshy areas has long been known, the role of the mosquito and the malarial parasite were not known until Laveran and Ross's work at the end of the nineteenth century. They showed that the malarial parasite is transmitted by female mosquitoes of several different species of the *Anopheles* family. Malaria arises from infection with one or more of four species of protozoal organisms belonging to the genus *Plasmodium*.

Malaria is today endemic in parts of Africa, Asia, and South and Central America, and was previously more widely spread in marshland areas of Europe and the Mediterranean. Regional studies show malaria is a very complex disease, almost a set of diseases, in which differences in the mosquito vector and parasite result in slightly different ecologies in different areas of the world.

Malaria was a dominant factor in tropical mortality until the 1960s. During the post-war decade, ***DDT*** became widely available and its spraying on mosquito-breeding habitats brought major reductions, particularly on the subtropical margins of North America and Europe. Here, when spraying was discontinued, in some instances malaria did not recur. But in the tropics proper, the campaigns were less successful and a resurgence of malaria has occurred. In the last half century the number of cases has doubled, with today 500 million cases and over 2 million deaths each year. Malaria is ***endemic*** in one hundred countries, but 90 per cent of cases occur in African countries.

An effective treatment for malaria was known long before the cause of the disease was worked out. The bark of the cinchona tree (with the active ingredient, *quinine*) was used from 1700 and from World War II several synthetic drugs (e.g. chloroquine, pyrimethamine) have become available. However, by the late twentieth century, many strains were becoming resistant to drugs, which were thus rendered ineffective. Both the control of mosquito populations and treatment of the disease currently pose critical problems in the tropical world.

Figure 20.5 Nineteenth-century scourges A map of the major cholera pandemics in the nineteenth century was given earlier in the book (cf. Figure 7.11). Cholera is an acute infection of the small intestine spread by faeces-contaminated drinking water. In the cartoon (a), Father Thames is introducing his offspring to the fair city of London. Note the water-borne nature of diphtheria and cholera. (b) Electron micrograph of the bacterium *Vibrio cholerae* which causes cholera in humans. Magnification ×18,200.

[Source: (a) *Punch*. (b) A.D. Cliff and P. Haggett, *Atlas of Disease Distributions* (Blackwell, Oxford, 1988), Fig. 1.1(D), p. 10.]

(a)

(b)

Figure 20.6 World distribution of malaria (a) Geographic distribution of malaria in the mid-twentieth century. (b) Geographic distribution of malaria at the start of the twenty-first century. Note that the staus of malaria in small islands is shown by circles. [Source: World Health Organization, *The World Health Report 1995* (WHO, Geneva, 1995), Map 4, p. 25.]

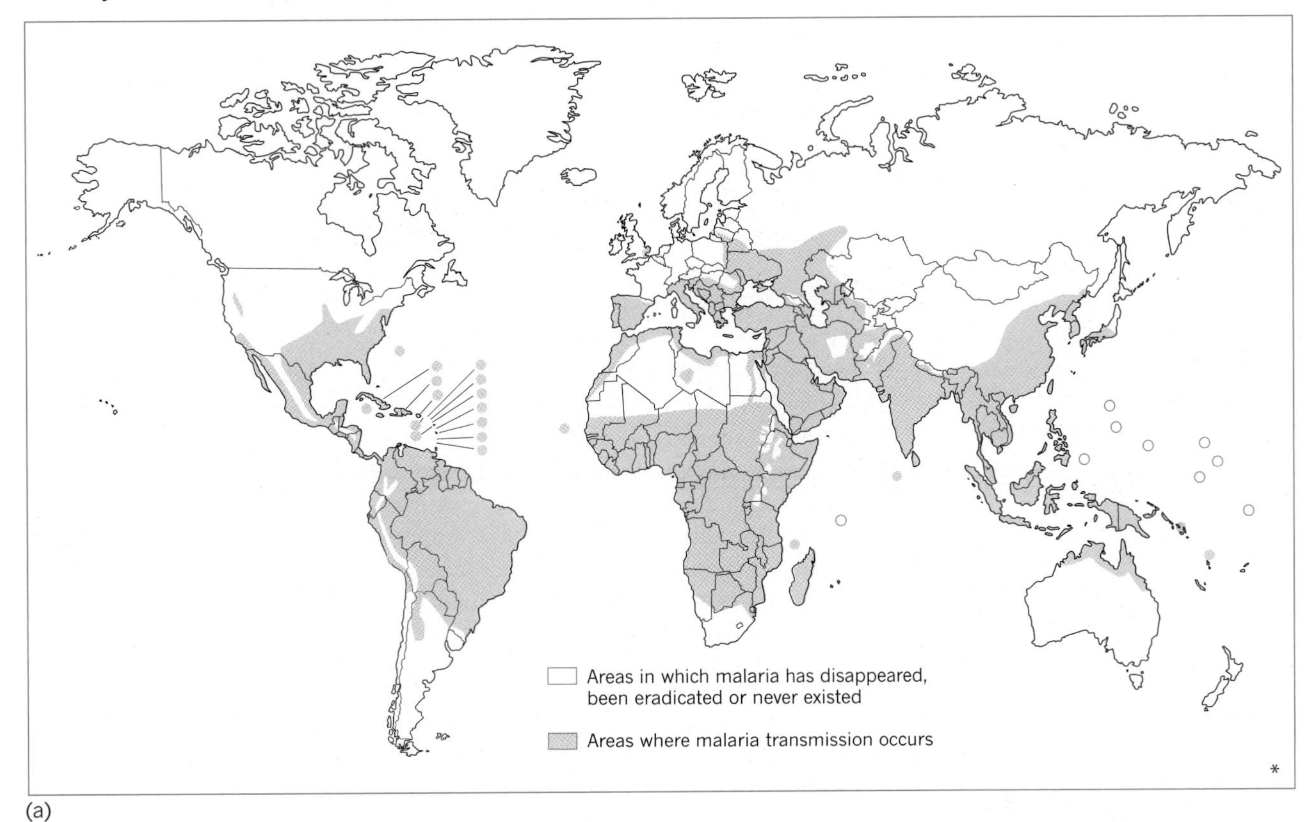

(a)

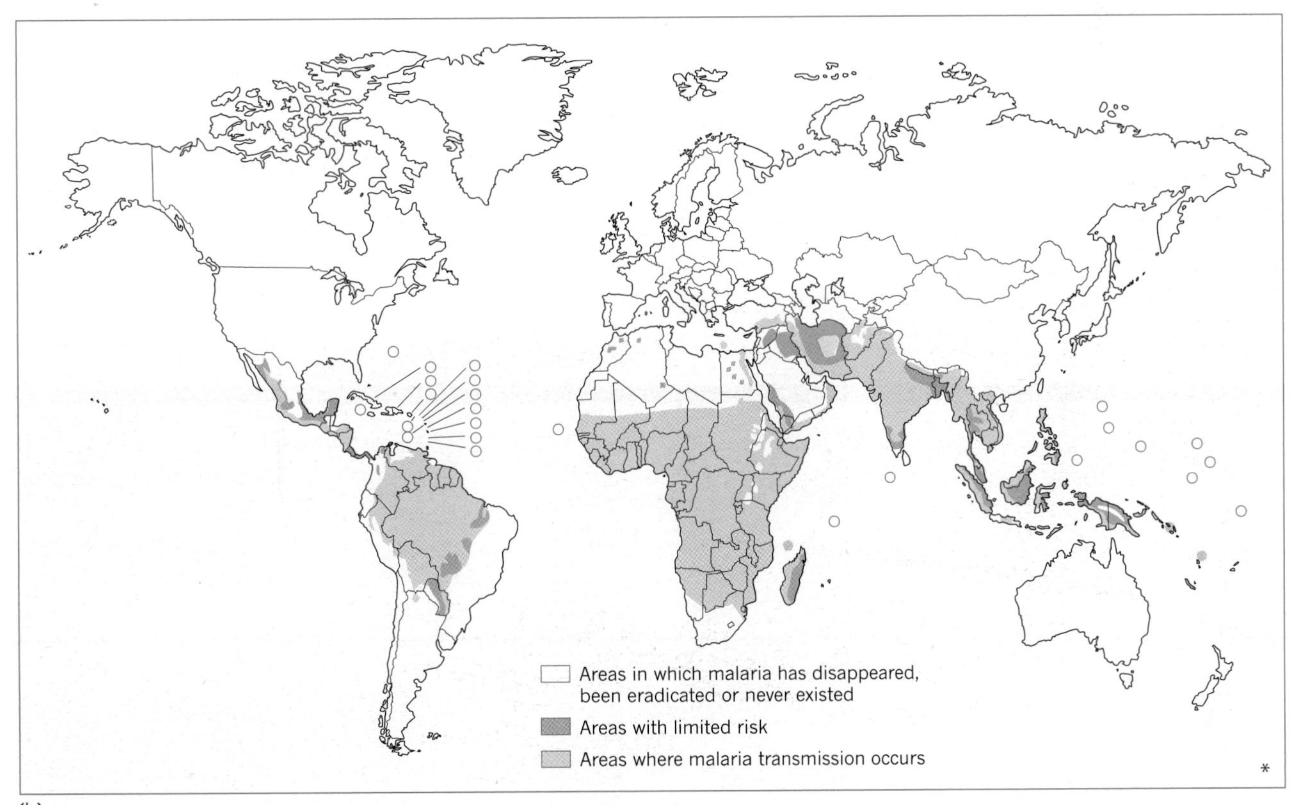

(b)

occur in underdeveloped countries, but over the last decade it has made a spectacular return in the deprived areas of western cities. *Malaria* is by far the most important tropical parasitic disease, causing immense suffering and the loss of life of 2.5 million people a year. Although the geographical area affected by the disease shrunk dramatically since 1940 (see Figure 20.6), gains in control are now being eroded with increasing evidence of drug resistance (see Box 20.B).

But we can set against the huge remaining problems of communicable diseases some examples of great progress. We look here at three examples of success.

Global Eradication of Smallpox

Although the WHO has from time to time conducted major campaigns against infectious disease (notably malaria and yaws) only one disease – smallpox – has so far been globally eradicated. (We confine the term *eradication* to the total elimination of the infectious agent (except, as with smallpox, for preserved laboratory examples); elimination refers to stamping out the disease in a particular country or region but leaves open the possibility of reinfection from another part of the world.) The practical reality of devising, coordinating, and financing a field programme involving more than 30 national governments and some of the world's most complex cultures and demanding environments proved to be of heroic proportions (see Figure 20.7).

Until the mid-1960s, control of smallpox was based primarily upon mass vaccination to break the chain of transmission between infected and susceptible individuals by eliminating susceptible hosts. Although this approach had driven the disease from the developed world, the less-developed world remained a reservoir area. Thus in the early 1960s, half a billion people in India were vaccinated, but the disease continued to spread. Between 5 and 10 per cent of the population always escaped the vaccination drives, concentrated especially in the vulnerable under-15 age group. Nevertheless, the susceptibility of the virus to concerted action had been

Figure 20.7 Centres for global health control Headquarters building of the World Health Organization (WHO) in Geneva, Switzerland. Geneva is the centre for a worldwide network of regional WHO centres. WHO was established after World War II as a specialized United Nations agency following on from the Health Division of the United Nations. The square white building (centre) houses the main meeting room where the General Assembly, with delegates from member countries around the world, meets. The white building at the far right houses AIDS research.

[Source: WHO.]

demonstrated and led to critical decisions at the Nineteenth World Health Assembly in 1966 which agreed on a ten-year global smallpox eradication programme, which was launched in 1967. It started with mass vaccination, but rapidly recognized the importance of selective control. Contacts of smallpox cases were traced and vaccinated, as well as the other individuals in those locations where the cases occurred.

The programme had four main phases in each area targeted. In (1) the *preparatory phase*, before active eradication was started, time was allowed for the epidemiological assessment of the distribution of smallpox and immunity in the local population. For example, epidemiological assessment was organized in India by geographic areas. In each area, the task was to record the location of each village, and whether or not evidence of

Figure 20.8 Global eradication of smallpox WHO Intensified Smallpox Eradication Programme, 1967–77. Countries with smallpox cases in the year in question marked in blue.

[Source: Redrawn from maps and graphs in F. Fenner *et al.*, *Smallpox and its Eradication* (World Health Organization, 1988), Fig. 10.4, Plates 10.42–10.51, pp. 516–37, *passim*.]

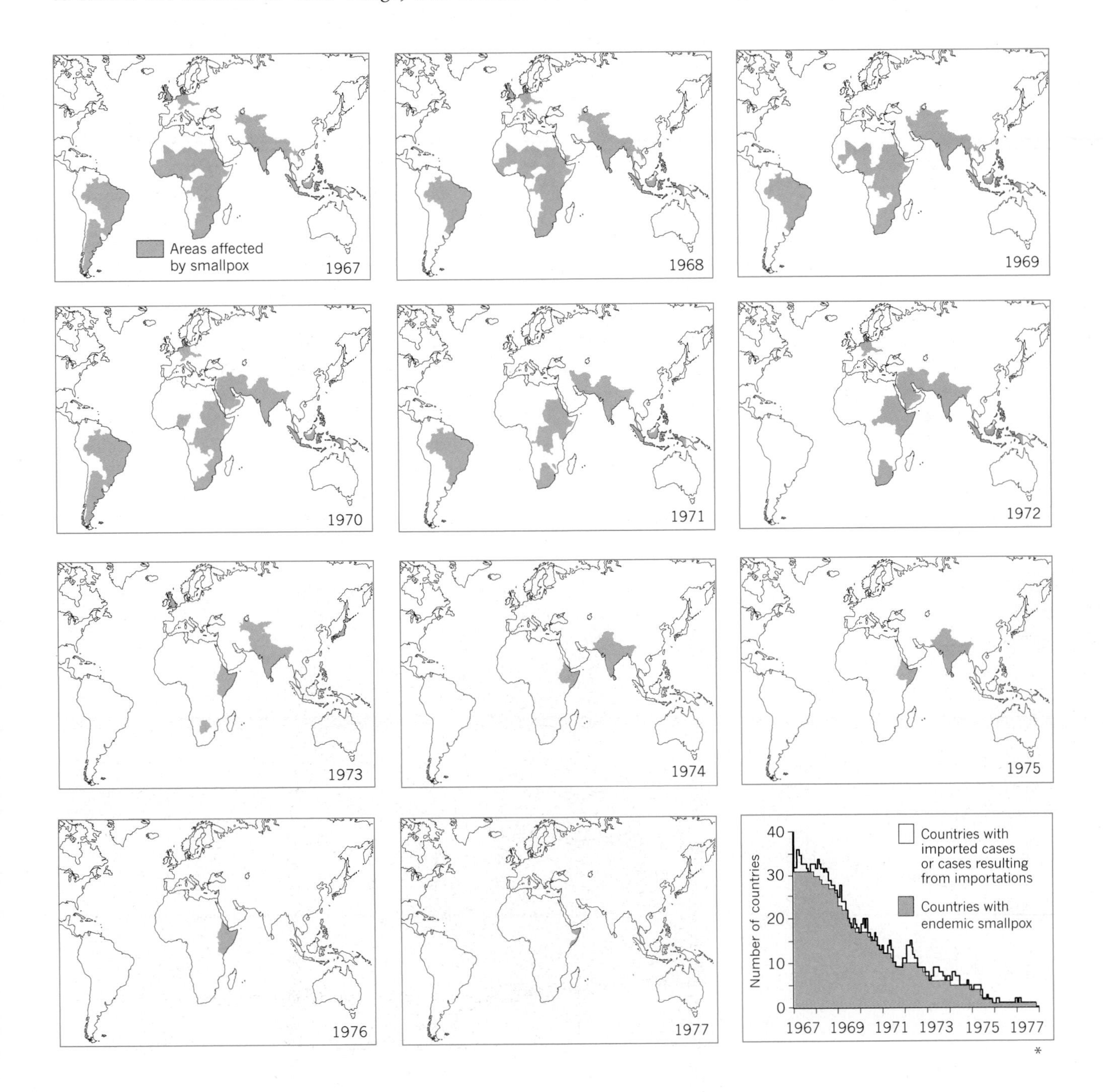

smallpox was to be found in the village population. During the preparatory phase, health-care personnel were also recruited and trained. Education programmes were established to ensure acceptance of vaccination.

In (2), the *attack phase* consisted of systematic mass vaccination and the establishment of a surveillance programme with follow-up vaccination of contacts and individuals in local areas where cases occurred. When smallpox incidence was driven down below five cases per 100,000 people and 4 out of 5 had been vaccinated then, stage (3) the *consolidation phase* was reached. This phase consisted of routine vaccination for newborns and those, such as immigrants, missed in the attack phase. The surveillance network now became critical, with every suspected case followed up by field investigation and action where necessary.

Finally (4) a *maintenance phase* was reached when there was no *endemic* smallpox in the targeted area for more than two years, but while the disease still persisted on the continent concerned. Maintenance vaccination was continued and intense surveillance maintained. Each report of a suspected case was treated as an emergency until the final elimination occurred.

The success of this four-phase programme after 1966, following upon the efforts of individual nations in the postwar period, may be judged from the maps and graphs in Figure 20.8.

By 1970, the disease was retreating in Africa. By 1973, the disease had been eliminated in Latin America and the Philippines; a few strongholds remained in Africa, but most of the Indian subcontinent remained infected. Despite a major flareup of the disease in 1973 and 1974, the hunt by WHO for cases and case contacts continued. By 1976 the disease had been eradicated in Southeast Asia and only a part of East Africa remained to be cleared (see Figure 20.9). The world's last recorded smallpox case was a 23-year-old man of Merka town, Somalia, on 26 October 1977. After a two-year period during which no other cases (other than a laboratory accident) were recorded, WHO formally announced at the end of 1979 that the *global eradication* of smallpox was complete.

Figure 20.9 Smallpox eradication in Africa (a) Surveillance of cases in isolated villages in Ethiopia. (b) The world's last recorded smallpox case (apart from a laboratory accident), 23-year-old Ali Maow Maalin, Merka town, Somalia. (c) WHO Smallpox Zero, 26 October 1979 from cover of *World Health*.

[Source: WHO archive.]

(a)

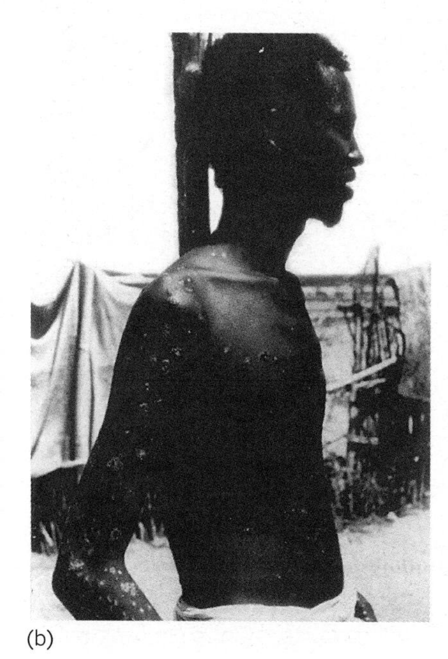

(b)

(c)

Measles Elimination Campaigns

The dramatic success of the WHO smallpox programme inevitably raised the prospect and hope that other virus-borne diseases can be eradicated. In 1974 WHO established its Expanded Programme on Immunization (EPI) with the objective of greatly reducing the incidence of six other crippling diseases (see Figure 20.10): diphtheria, measles, neonatal tetanus, pertussis, poliomyelitis, and tuberculosis. Two other diseases were later added. Table 20.2 summarizes some of the characteristics of the disease and indicates in the final two columns the continental variation in the vaccination levels achieved to date.

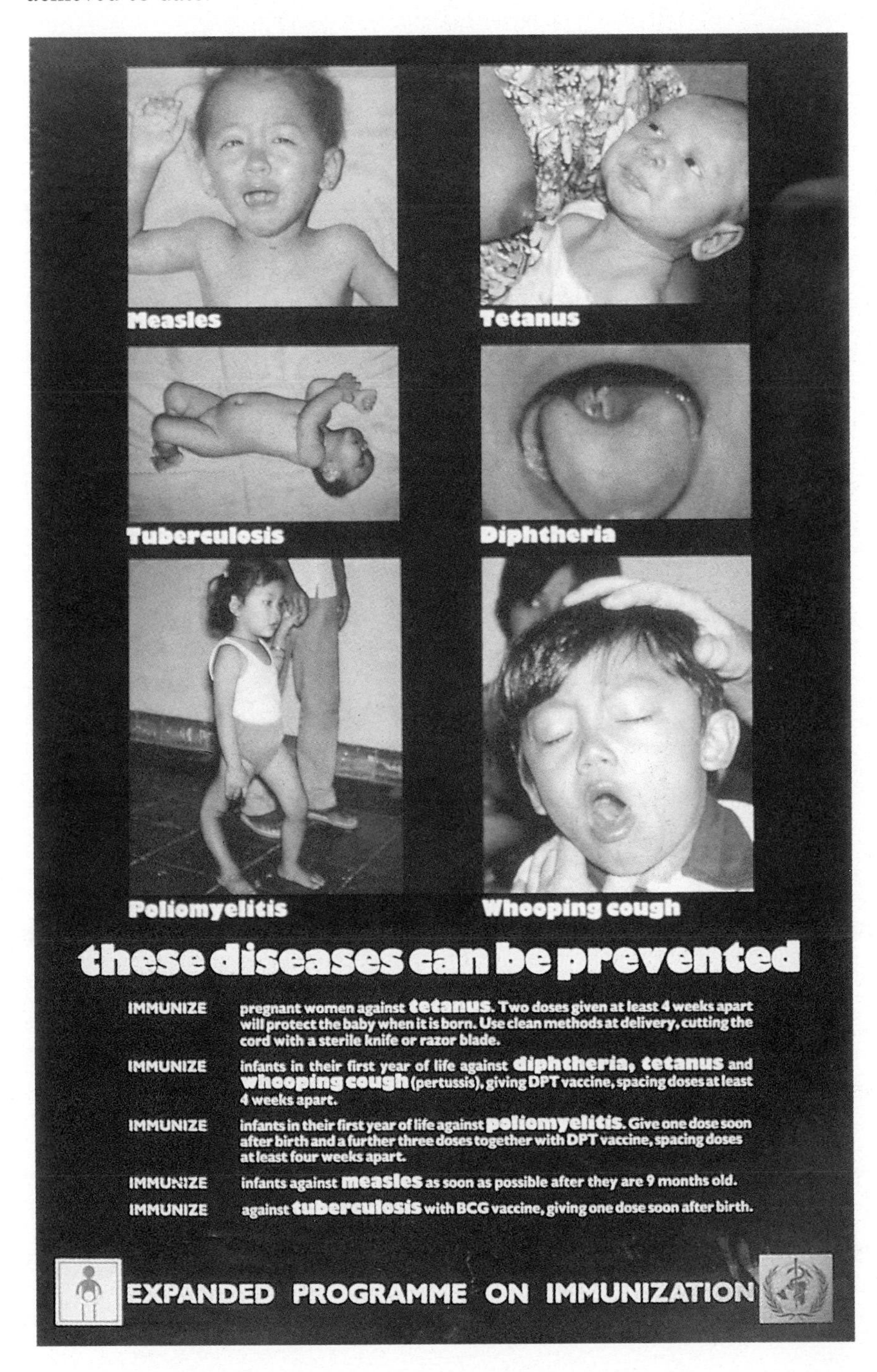

Figure 20.10 Other target diseases for possible eradication Poster of the WHO's Expanded Programme of Immunization (EPI). Examples of children suffering from the original six target diseases are shown. Poliomyelitis and yellow fever have since been added to the list. The character of each disease is shown in Table 20.2.

[Source: WHO.]

Table 20.2 Target diseases in the WHO Expanded Programme on Immunization

Disease	Infectious agent	Reservoir	Spread	Nature of vaccine, its form, and number of doses	Immunization coverage (%) Africa	Immunization coverage (%) Europe
Diphtheria	Toxin-producing bacterium (*C. diptheriae*)	Humans	Close respiratory contact or cutaneous	Toxoid; fluid; 1	50	86
Hepatitis B	Virus	Humans	Perinatal Child–child Blood Sexual spread	HBsAg; fluid; 3	0.15	12
Measles	Virus	Humans	Close respiratory contact and aerosolized droplets	Attenuated live virus; freeze-dried; 1	49	78
Pertussis	Bacterium (*B. pertussis*)	Humans	Close respiratory contact	Killed whole cell pertussis bacterium; fluid; 3	50	86
Poliomyelitis	Virus (serotypes 1, 2, and 3)	Humans	Faecal–oral Close respiratory contact	Attenuated live viruses of 3 types; fluid; 4	50	92
Tetanus	Toxin-producing bacetrium (*Cl. tetani*)	Animal intestines Soil	Spores enter body through wounds, umbilical cord	Toxoid; fluid; 3	35	N/A
Tuberculosis	Mycobacterium tuberculosis	Humans	Airborne droplet nuclei from sputum-positive person	Attenuated *M. bovi*; freeze-dried; 1	68	81
Yellow fever	Virus	Humans Monkeys	Mosquito-borne	Attenuated live virus; freeze-dried; 1	6	N/A

Immunization coverage at 1995. Africa excludes South Africa.
Source: Based on data provided by the Global Programme for Vaccines and Immunization, World Health Organization, 1995, Tables 1–3, pp. 2–5.

Measles was placed high on the agenda for two reasons. First, it shared many of the epidemiological characteristics of smallpox and stood a reasonable chance of eventual eradication. Second, it was a major world killer causing some two million deaths annually, mostly in developing countries.

The United States The roots of the world campaign for measles elimination lie in the United States. In that country, in the early years of the twentieth century, thousands of deaths were caused by measles each year and, at mid-century, an annual average of more than half a million measles cases and nearly 500 deaths were reported in 1950s decade. It was against this background that the ***Centers for Disease Control (CDC)***, Atlanta, Georgia, evolved in the United States a programme for the elimination of indigenous measles once a safe and effective vaccine was licensed for use in 1963.

As we noted in Chapter 16, it is estimated that a population of the order of 250,000–300,000 is required to maintain endemic measles. Work in Africa by McDonald in the early 1960s led him to suggest that one way of reducing the 'at risk' population in large countries below this threshold was by mass vaccination, so breaking the chains of measles infection. In the countries studied by McDonald, he argued that an annual mass vaccination campaign reaching at least 90 per cent of the susceptible children would have the required effect.

In 1966, the CDC announced that the epidemiological basis existed for the eradication of measles from the United States using an *immunization programme* with four tactical elements: (a) routine immunization of infants at one year of age; (b) immunization at school entry of children not previously immunized ('catch-up' immunization); (c) surveillance; and (d) epidemic control. The immunization target aimed for was 90–95 per cent of the childhood population.

Following the announcement of possible measles eradication, considerable effort was put into mass measles immunization programmes throughout the United States. Federal funds were appropriated and, over the next three years, an estimated twenty million doses of vaccine were administered. The discontinuity induced in the time-series of reported cases is shown for the United States as a whole in Figure 20.11(a).

In 1962, the year before measles vaccine was introduced, there were 482,000 cases of measles reported in the United States. Within four years this had been halved and within six years the reported incidence had plummeted to only 22,000. But by the mid-1970s, it was evident that the campaign against measles was running out of steam and that steady increases in incidence were occurring. To remedy this situation, a nationwide childhood immunization initiative was launched in April 1977, followed by the announcement of a programme to eliminate indigenous measles from the United States within five years (the 'Make Measles a Memory' campaign).

Figure 20.11 Measles reduction in the United States (a) Monthly measles cases 1945 to 1986 with annual figures for cases of a neurological disorder (SSPE) for 1968–81 superimposed. Maps show United States counties reporting measles during one or more weeks in (b) 1978 and (c) 1983.

[Source: A.D. Cliff and P. Haggett, *Atlas of Disease Distributions* (Blackwell, Oxford, 1988), Fig. 4.9, pp. 164–5.]

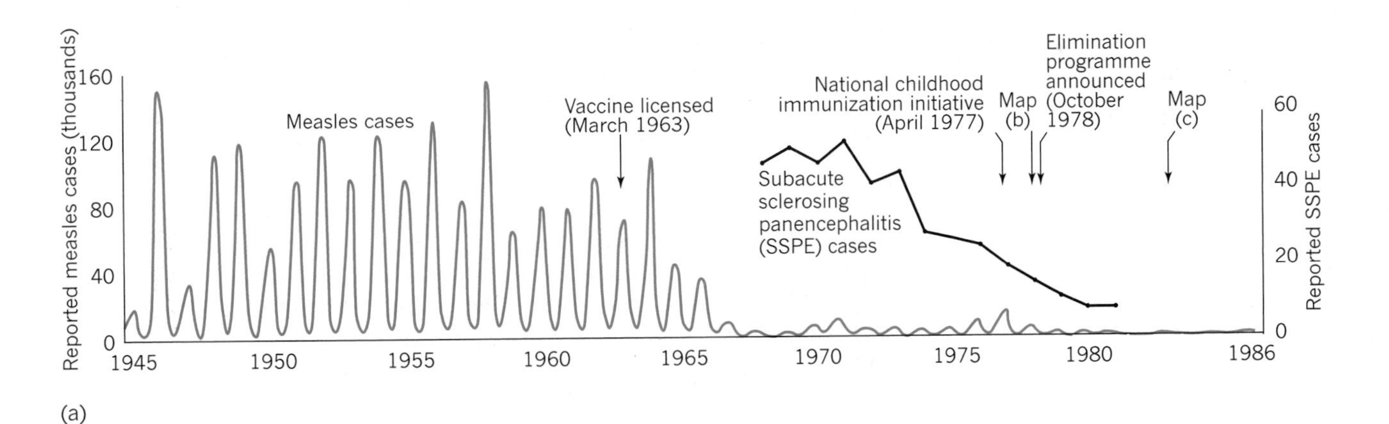

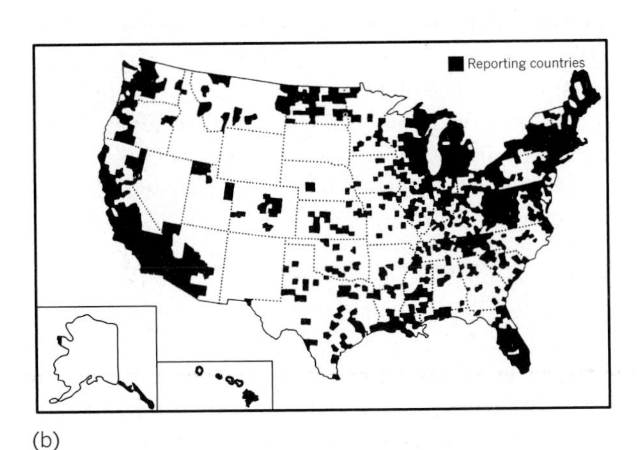

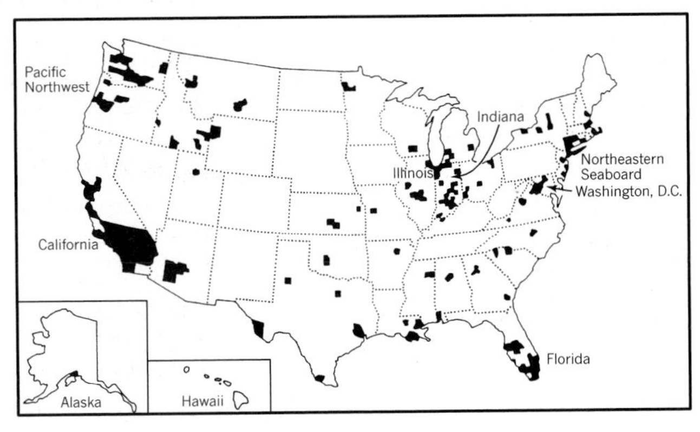

The immunization goal aimed for was again McDonald's 90 per cent of the childhood population.

The geographic impact of this second push against the disease is seen in the two maps in Figure 20.11. The maps show the distribution of counties in the United States reporting measles cases (a) at the start of the campaign (1978) and (b) five years later in 1983. The contraction of infection from most of the settled parts of the United States in 1978 to restricted areas of the Pacific Northwest, California, Florida, the northeastern seaboard and parts of the Midwest is pronounced. The persistence of indigenous measles in many of these regions may be explained by the importation of cases from Mexico and Canada. By 1983, twelve states and the District of Columbia reported no measles cases, and 26 states and the district of Columbia reported no indigenous cases.

Unfortunately, total elimination in the United States still has not been achieved. Vaccination levels have fallen back and the continued importation of measles cases from overseas has resulted in a resurgence of cases. Note the way in which a neurological complication attributed to measles, subacute sclerosing panencephalitis (SSPE), has paralleled the decline in measles occurrence.

Rest of the World The US lead was followed up with varying degrees of vigour in other western countries and pressed in developing countries through the WHO campaigns. But by the end of the century, the position achieved was spatially very patchy. A few countries (e.g. Hungary, Israel, and Sweden) claimed vaccination rates over 95 per cent. In contrast the average coverage for Africa was 50 per cent, Latin America 80 per cent, and South East-east Asia 85 per cent. Worldwide, currently 50 million cases of measles occur each year; up to two million people die, mostly young children in Third World countries. By 2000 the global coverage of measles vaccination remained around 80 per cent, well below the 90 per cent target originally set by WHO.

It has become progressively clear that no matter how high coverage is, the infectivity of measles is likely to make global eradication a very long-term prospect. Even a vaccine coverage of 100 per cent will leave vaccine recipients susceptible unless the vaccine was more effective than the present 80+ per cent level. The US experience shows that vigilant efforts to maintain high vaccination levels, strong surveillance, and an aggressive response against imported cases in measles-free zones are required to hold the ground already gained. Any immediate hopes of the global eradication of measles seems remote but, for measles at least, a feasible scenario is that more developed countries will follow the lead of the United States and try to eliminate measles nationally as an endemic disease. For this to be achieved, however, the great divergence of attitudes to, and of programmes against, measles in the developed world will need to be unified. Whether the coalescence of disease-free zones in developed countries would ever allow a sustained attack on measles reservoirs in developing countries will depend as much on politics and economics as on epidemiology.

Poliomyelitis Elimination Campaigns

Large epidemics of poliomyelitis occurred regularly in the 1950s in all industrialized countries, causing panic among parents and crippling thousands of children every summer. Vaccines introduced in the 1960s virtually eliminated the disease in western countries but it remained a major threat elsewhere. Eleven years after the close of its successful smallpox campaign, the World Health Assembly, meeting in Geneva in 1988, committed WHO to the global eradication of poliomyelitis.

Like smallpox, this target involves not only eliminating the disease, but totally eradicating the causative virus. The goal was made possible by 40 years of research and vaccine development since Enders and his Harvard colleagues succeeded in growing poliovirus in cell culture. The licensing of the Salk inactivated vaccine (1955) and Sabin attenuated live vaccine (1961) was reinforced by the early success of the countries of the Pan American Health Organization which had agreed in 1985 to eradicate the wild poliovirus from the Americas.

The level of global success achieved is shown in Figure 20.12. The map shows that no countries in the Americas reported cases and that Europe, Japan, Australia, and New Zealand were free of cases. Tropical Africa and South and East Asia remained major zones where disease incidence remained high. Overall the level of vaccination worldwide has risen from less than 5 per cent of children in 1974 to over 80 per cent today. Over the same period the number of reported cases worldwide had fallen from a peak of over 70 million to less than 5 million.

The WHO has warned against complacency. Declining polio incidence mostly reflects individual protection from immunization, and not wild virus eradication. For although surveillance is improving, only one in six cases are being officially reported. The global strategy has five components: (a) high immunization coverage with oral polio vaccine, (b) sensitive disease

Figure 20.12 World incidence of indigenous poliomyelitis Situation in 1994, and annual number of cases of poliomyelitis notified in the world, 1974–93.

[Source: World Health Organization, 1995. Map included in pamphlet: 'La poliomelite sera eradiquee', World Health Organization, 7 April 1995.]

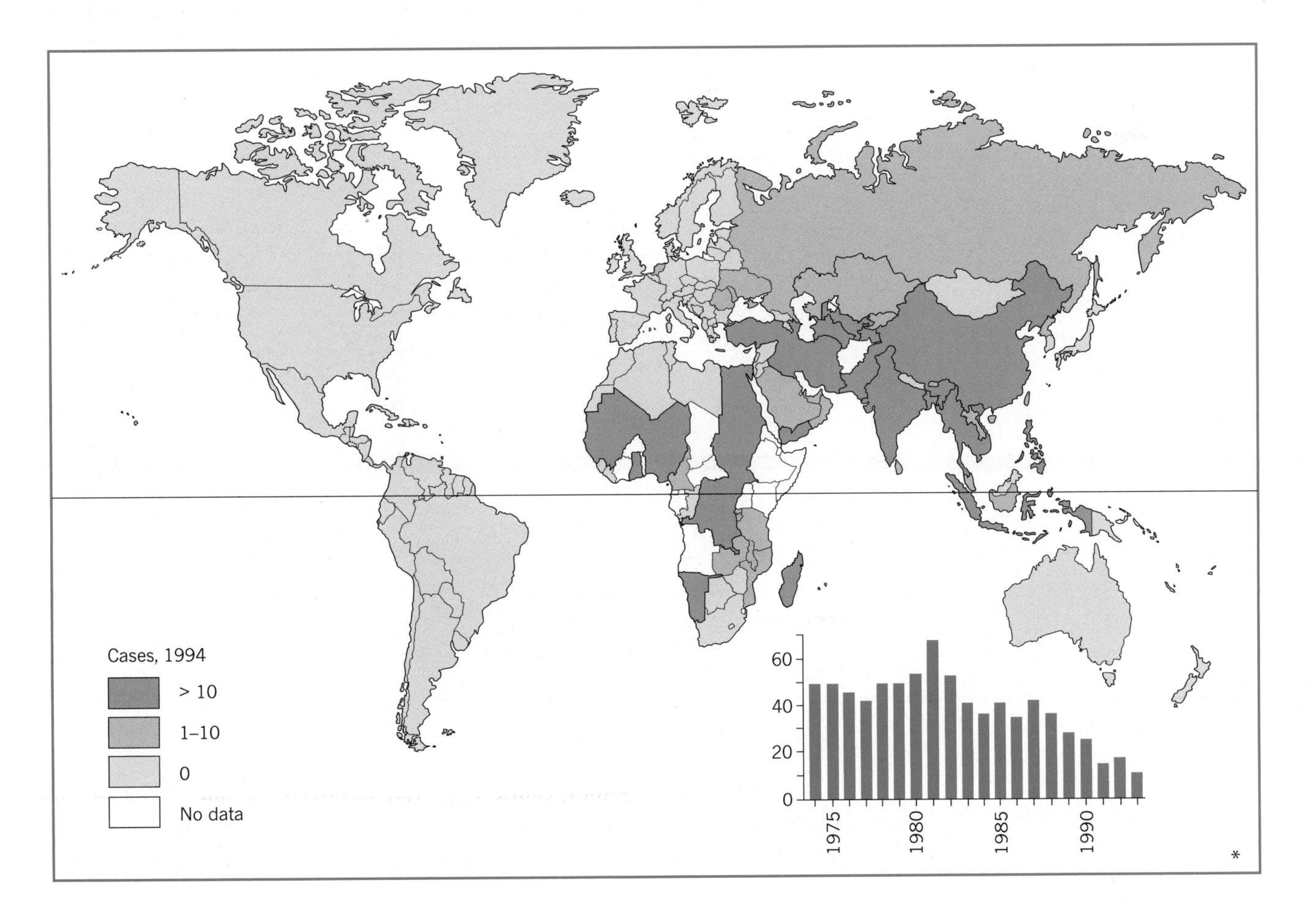

surveillance detecting all suspected cases of poliomyelitis, (c) national or subnational immunization days, (d) rapid, expertly managed outbreak response when suspected cases are detected, and (e) 'mopping-up' immunization in selected high-risk areas where wild virus transmission may persist.

The major cost of eradicating poliomyelitis will be borne by the endemic countries themselves but donor country support will be required for vaccine, laboratories, personnel, and research. Of these, the most urgent need is for vaccine: although each dose of oral vaccine currently costs only 7 US cents, over two billlion doses will be required per year for routine and mass immunization. In the longer run, the economic benefits of disease eradication far exceed the cost. Since the last case in 1977, the United States has saved its total contribution once every 26 days. From 1998, the global initiative has also started to pay for itself. Countries with the lowest polio immunization coverage are nearly always those with internal conflicts, where control programmes have been interrupted and medical infrastructures smashed. Intense efforts by WHO to establish 'days of tranquility' when combatants allowed immunization teams into disputed areas to reach children have been part of the global effort that should allow the disease to be eradicated within a decade.

Other Disease Campaigns

In concentrating on diseases caused by viruses, it is worth recording that WHO campaigns on non-viral diseases has also met with some success. Figure 20.13 illustrates the case of *onchocerciasis* ('river blindness) in West

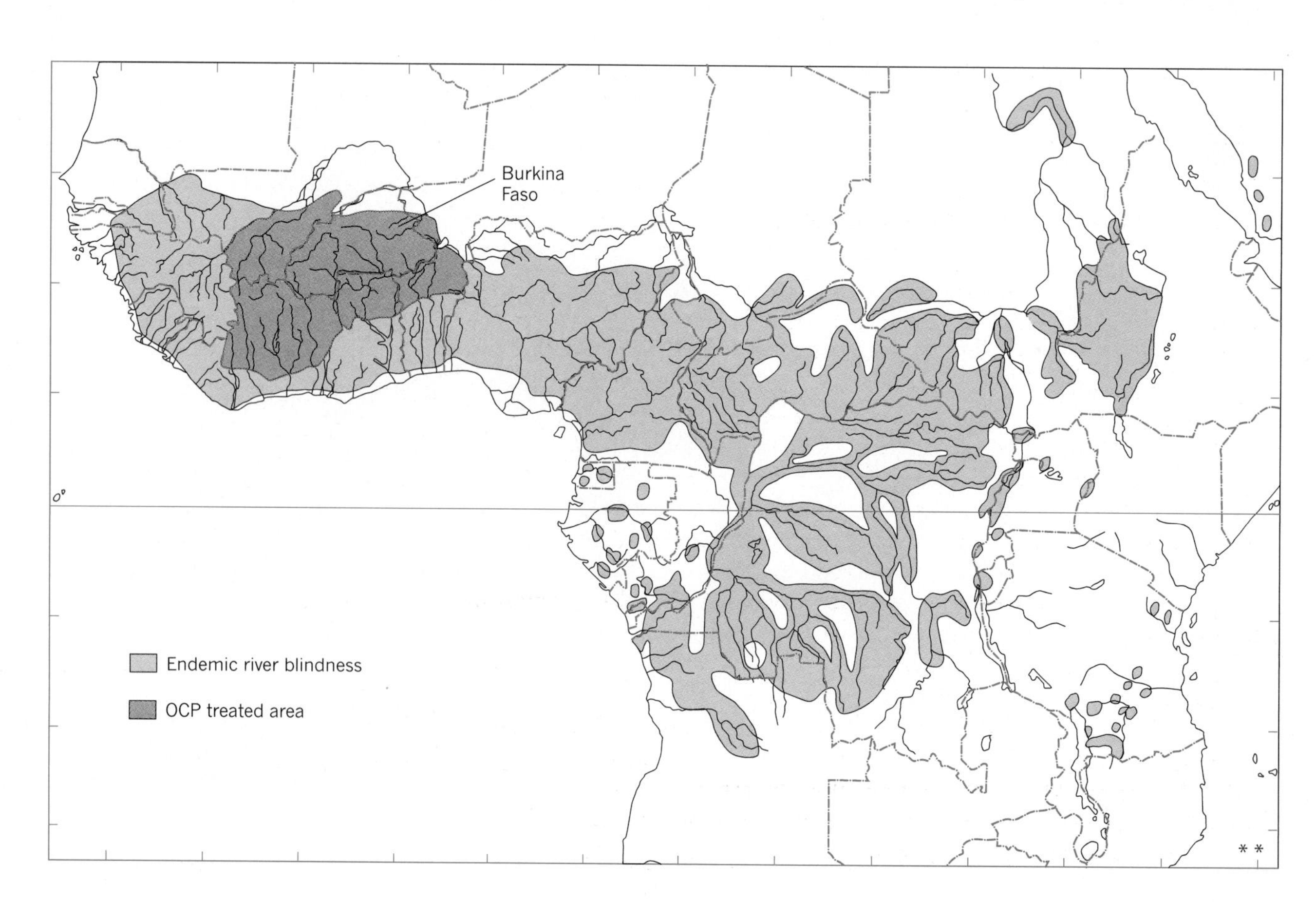

Figure 20.13 River blindness in tropical Africa Onchocerciasis (commonly known as river blindness) occurs widely in tropical Africa in the areas shown. Human infection rates vary greatly over the area with the highest intensity in the Sahel zone on the southern margin of the Sahara desert. Transmission is due to female blackflies of the *Simulium* family which breed in rivers and streams species but are able to infect areas 20 km (12 miles) away from their breeding sites. The area of the WHO Onchocerciasis Control Programme (OCP) involving eleven West African countries is shown.

[Source: World Health Organization, *Geographical Distribution of Arthropod-borne Diseases and their Principal Vectors* (WHO, Geneva, 1989), Fig. 49, p. 119]

Africa by attaching to the optic nerve. It is an infestation with adult worms which may eventually lead to blindness. It is endemic in Africa south of the Sahara, especially in the savanna grasslands from Senegal to the Sudan. Its spread from Africa to Arabia and the New World was likely to have been an unintended product of the slave trade.

In 1975 WHO began a major twenty-year long control programme to eliminate the disease in the Volta Basin in West Africa. The basic strategy was to kill the larvae of the biting fly that carries the worms. These breeding grounds are adjacent to rivers in the Sahel belt and aerial spraying with an insecticide was adopted. This has been supported by treating victims with a worm-killing drug (invermectin). The West African control programme has protected some 36 million people and is now being extended both to Latin America and to other African countries.

The Emerging Geography of New Diseases 20.3

Although the last century saw the global eradication of one major human disease (smallpox) and major reduction in others (measles and poliomyelitis) it has also witnessed the emergence of many new ones. Over these years, numerous diseases caused by micro-organisms have been recognized and recorded in the medical literature for the first time. For example, in 1917, the American Public Health Association published the first edition of its pioneer handbook on the *Control of Communicable Diseases in Man*. It listed control measures for 38 communicable diseases, all those then officially reported in the United States. Since then, the number listed has expanded steadily so that the most recent edition of the handbook, the fifteenth, now lists some 280 different diseases.

The Puzzle of Disease Emergence

Why have new diseases emerged? Most of the 'new' viruses discovered have probably existed for centuries, escaping detection because they existed in remote or medically little-studied populations or because they produced disease symptoms not previously recognized as being due to infectious agents. As with some tropical fevers, they are recognized by western medicine when they impinge on middle-latitude rather than tropical populations.

Improvements in viral detection technology have opened up new frontiers in biology and medicine. It seems likely that the disease list will expand as links between previously unnoticed slow viral infections and chronic conditions (these include neurological problems and cancers) are unravelled.

The discovery of apparently new diseases raises afresh the question of where and when diseases originate, whether they will spread around the world, and whether new diseases will continue to be added to our existing list. Such questions have been given increased significance by the emergence, over the last decade, of AIDS as a major human disease. This has been accompanied by an unprecedented torrent of literature, some of it of geographical interest. Notwithstanding this literature growth, one still unsolved question is when and where the epidemic of AIDS began. Conventional wisdom, based upon a variety of circumstantial evidence, is that the causative agent, the human immunodeficiency virus (HIV), jumped the species barrier from monkeys in Africa – possibly from the chimpanzee in Central Africa for HIV-1 and from the sooty mangabey in West Africa for HIV-2 (see Figure 20.14). Further, while we know that the epidemic spread of the disease in the United States began in 1981, there is increasing evidence that AIDS has been internationally present in a non-epidemic form for many years before the first epidemic cases. Other scholars have argued for a much older origin.

Historical perspective on disease emergence But not all new diseases are caused by improved recognition techniques. Modern examples of 'new' infectious diseases share a number of common features both with each other and with earlier historical manifestations: (a) the onset of the new diseases appears to be sudden and unprecedented; (b) once the disease is recognized, isolated cases that occurred well before the outbreak are retrospectively identified; and (c) previously unknown pathogens or toxins account for many of the new infections.

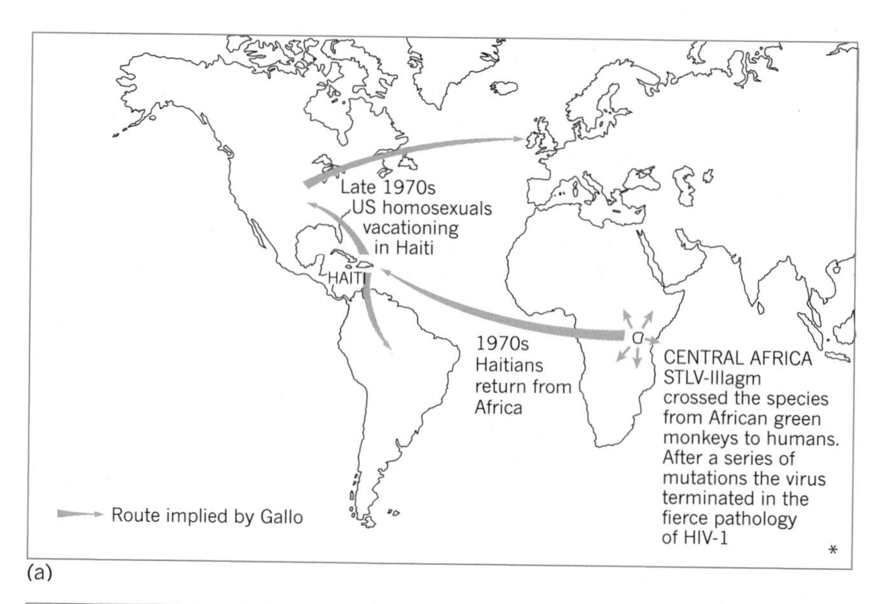

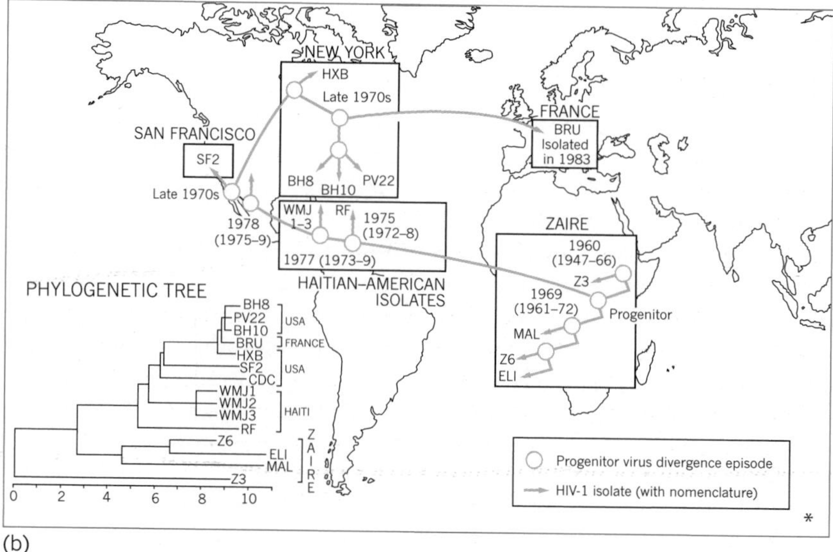

Figure 20.14 The global diffusion of HIV-1 (a) The initial spread of HIV-1 derived by Li and co-workers from phylogenetic histories of virus isolates. This suggests that the earliest event is estimated to have occurred in the Democratic Republic of the Congo (Zaire), central Africa, around 1960 when the isolate Z3 diverged from an African progenitor virus. (b) Shannon and Pyle's model for the global diffusion of HIV-1. While the precise origins and spread of HIV remain controversial, the overwhelming weight of evidence points to an east-central African origin. A variant of the virus (HIV-2) originated in west Africa and spread to the western world via Europe rather than North America.

[Source: M. Smallman-Raynor, A.D. Cliff and P. Haggett, *International Atlas of AIDS* (Blackwell, Oxford, 1992), Figs 4.1(B), 4.1(C), pp. 145, 146.]

Epidemiologist Ampel suggests four factors that may explain these observations:

1 The infection was present all along, but was previously unrecognized and unrecorded. Improvements in viral detection technology have opened up new frontiers in biology and medicine.

2 Pathogens responsible for these new diseases existed in the past but in a less virulent form. Some event, such as genetic mutation, then converted the organism to its virulent form. The classic case of genetic change in an already existing virus was the emergence of the 'Spanish influenza' virus in 1918, which killed more people than the Great War itself.

3 Environmental and behavioural changes provide a new environment in which the disease-causing organisms may flourish. Legionnaires' disease is related to the increased use of cooling towers and evaporative condensers from the 1960s; Lyme disease to the growth of deer population in woodlots that grew up on the abandoned fields of New England.

4 A new epidemic arises from the introduction of a virulent organism into a non-immune population (a so-called ***virgin soil epidemic***) or to the arrival of new settlers in a previously unsettled area. This is a theme with rich historical parallels: one historically important example is the epidemic of smallpox that arrived in Mexico in 1520 with the Spanish *conquistadores*, destroying the native Aztec population.

Thus a historical review suggests that the emergence and persistence of new infectious diseases need very special conditions: there are many cases where diseases probably failed to emerge from contact with a disease-bearing organism. For virus diseases, the historian McKeown has noted that, when an infection comes into contact with a strong human host, three things may occur: the virus may fail to multiply and the encounter passes unnoticed; the virus multiplies rapidly and kills the host without being transmitted to another host; or virus and host populations (after a period of adaptation) settle down into a prolonged relationship that we associate with sustainable diseases. The relative probabilities we can attach to the three outcomes is unknown, but the fragmentary history of puzzling and unsustained disease outbreaks suggests that the third option is the rarest.

The Geography of AIDS

AIDS was recognized as an emerging disease only in the early 1980s. The initials stand for *Acquired Immune Deficiency Syndrome*, the term 'syndrome' indicating a set of conditions accompanying the collapse of the immune system rather than a single disease. The cause of the disease is infection by the *human immunodeficiency virus (HIV)*. Two main strains of the virus have been recognized: HIV-1, probably originating in East Africa (check back on Figure 20.14), and a less aggressive HIV-2 originating in west Africa (see Figure 20.15).

In the last decades of the twentieth century AIDS rapidly established itself throughout the world. It has evolved from a mysterious illness confined to particular groups or distant countries to a global pandemic that has infected tens of millions in less than twenty years. In the WHO world statistics it rose from nil in 1980 to seventh-ranking cause of death by 2000. At the start of the present century, nearly 40 million people around the world are infected with HIV. A third of that number had already lost their lives from AIDS and the current deaths are running at 2.5 million each year. Unless unforeseen changes occur, the 'new' pandemic is likely to endure and persist well into the present century and eventually beyond.

AIDS is dominantly a sexually transmitted disease. The main behavioural characteristics that help the spread of HIV are unprotected sexual activity with different partners. Other channels of spread are injecting drugs with shared needles and through the transfusion of contaminated blood. Women with HIV can also transmit it to their newborn children. The human immunodefiency virus is a 'slow' virus and usually exists without symptoms for many years. But it progressively reduces the T cells that form part of the body's defences against infection. After some years (and it may be a decade

Box 20.C Peter Gould and the Geography of AIDS

Peter Gould (born Surrey, England, 1932; died State College, Pennsylvania, USA, 2000) was one of the outstandingly innovative geographers of his generation. English by birth, he spent childhood years in that country and Germany and was evacuated to the United States during World War II. After military service with a Scottish regiment in Malaysia, he graduated from Colgate and Northwestern universities in the United States. From 1963 he was a faculty member at Pennsylvania State University, much of it as Evan Pugh Professor of Geography. He was a leading figure in the 'quantitative revolution' in geography, illustrating the use of the discipline in many books such as *The Geographer at Work* (1985) and *Fire in the Rain* (1990). In recent years he has turned his attention to modelling the geographic spread of the AIDS pandemic and wrote a major book on its geographic spread: *The Slow Plague: A Geography of the AIDS Pandemic* (1993).

One example of Gould's AIDS work is shown in the diagrams. He took the largest 102 urban centres in the conterminous United States and used air passenger origin–destination data to compute a weighted 102 × 102 transition probability matrix. Probabilities are particularly high among the five largest cities that 'exchanged' by air travel some 13 million people every year in the early 1990s; likewise, probabilities are low among smaller and

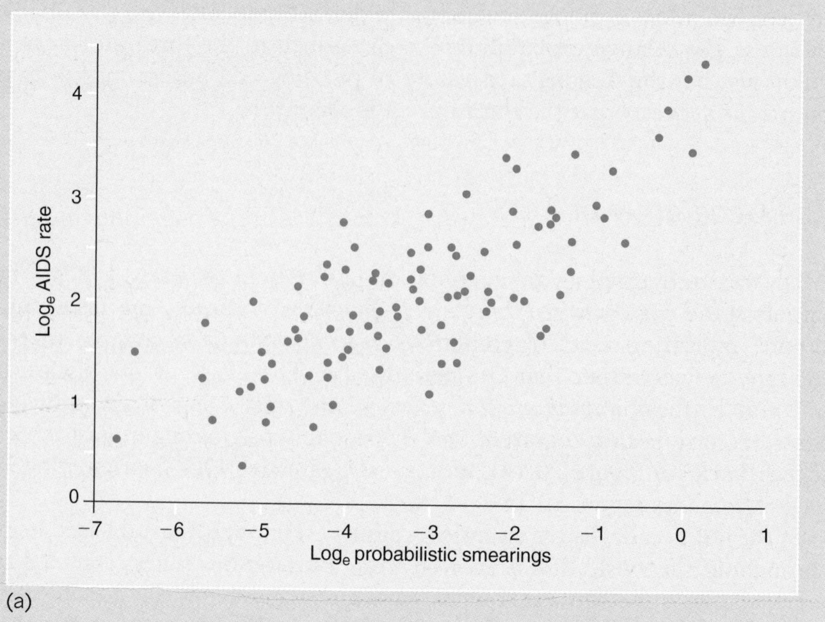

(a)

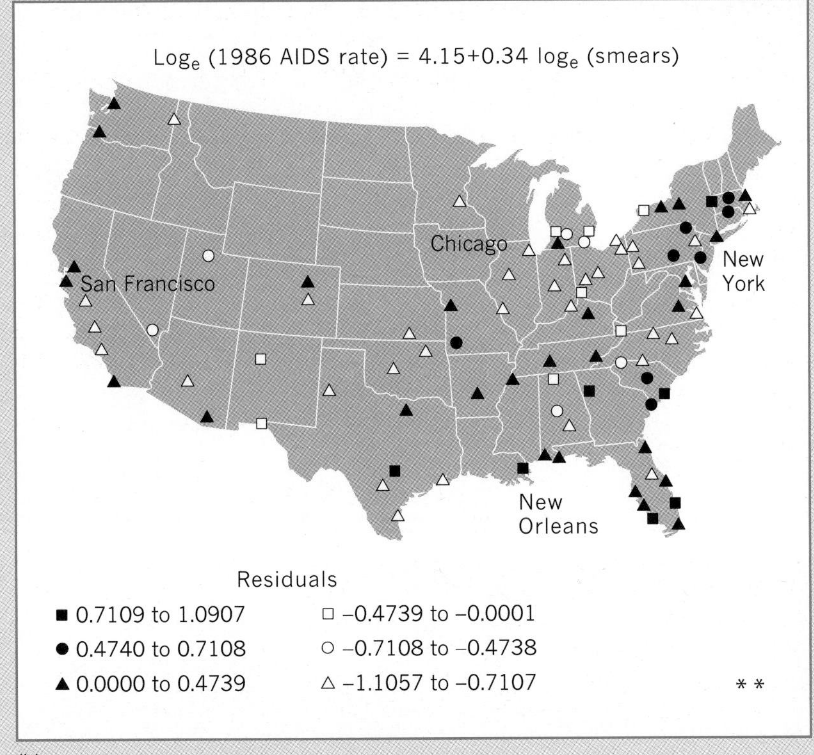

(b)

distant centres with small volumes of population exchange. In mathematical terms, the matrix forms an operator capable of multiplying a state vector. The state vector in Gould's case was the distribution of AIDS cases on a city by city basis at a particular year (1986). After a series of probabilistic multiplications the 'projected' AIDS distribution for 1990 was calculated.

The results of the projections are shown in (a). The projected values from the Gould model (horizontal axis) show a close approximation to the observed AIDS rates (vertical axis) with a correlation of about 80 per cent. Map (b) shows the 'residuals' from the projection, i.e. those cities under- and over-predicted by the model. Many of the negative residuals (towns with AIDS rates overpredicted) lie in the older 'rustbelt' towns of the northeast United States with many blue-collar workers of Catholic and recent immigrant backgrounds. Many southern cities that are tourist destinations come out as positive residuals: they have AIDS rates higher than the model would predict.

Models of this kind, where they can be calibrated on 'old' data from recent epidemic events, can be used to predict the spread of communicable diseases in the future. Russian workers have used such models on a world network of cities to project the spread of a new strain of influenza. Andrew Cliff and Peter Haggett have used similar models to predict the likelihood of measles imports into the several regions of the United States. The construction of global early warning systems for the transmission of communicable diseases are in prospect. For Peter Gould's contribution see his *Becoming a Geographer* (Syracuse University Press, Syracuse, 1999) and for his specific work on AIDS, see his *Slow Plague* (Blackwell, Oxford, 1993).

later before illness is recognized) a distinctive set of conditions occurs, which leads to death.

In western countries, AIDS was first regarded as a 'minority group' disease confined to homosexual or drug-using communities. As the disease has become more prevalent, so AIDS has gradually shifted its pattern to that typical of a heterosexually transmitted infection.

The prospect for AIDS control depends on recognizing the scale of the threat and a political commitment to implementing counter-policies. The key aspect is education on the nature of the condition, on condom promotion, and care for other sexually transmitted diseases. (See Box 20.C on 'Peter Gould and the geography of AIDS'.) Once the disease has been acquired, combination drug therapy has lengthened the life for many patients. Unfortunately, the therapy is very expensive and is inequitably distributed: it is wholly outside the budget of those Third World countries in which incidence is highest.

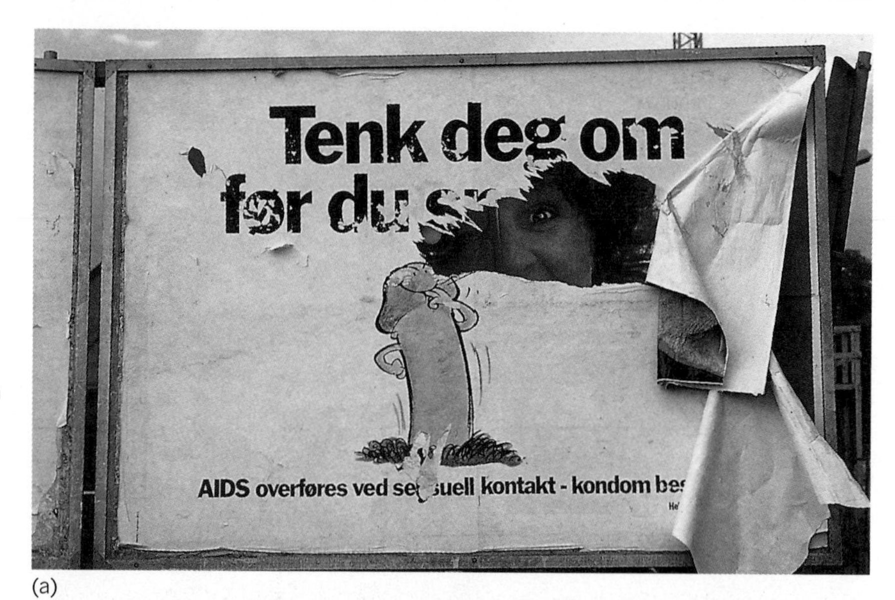

(a)

(b)

(c)

Figure 20.15 Role of education in slowing the spread of HIV
Geographer Peter Gould (see Box 20.C) saw the future spread of HIV as a race between two diffusion waves (see discussion in Chapter 16). The first was the spread of the disease; the second was the spread of information about the disease. Until an effective vaccine is developed, disease diffusion can be slowed only by (i) safer sexual practices and (ii) a reduction in the number of sexual partners. (a) Developed world with Norwegian AIDS poster. (b) Clinic in the Mobarakdi village, Bangladesh, where community health worker shows women how to use condoms. (c) Anti-AIDS sign Gambia, West Africa.

[Sources: (a) Knudsens Fotosenter. (b) Mark Edwards/Still Pictures. (c) Sean Sprague/PANOS Pictures.]

Geographic Distribution AIDS is now prevalent in virtually all parts of the world. But over two-thirds of all the people now living with HIV live in Africa south of the Sahara. Today the most severe HIV epidemics in the world are to be found in the southern countries of Africa. The virus is still spreading rapidly with one in four adults in Zimbabwe thought to be infected. There are signs of improvement in some of the earliest centres of spread in Africa. Uganda has adopted a strong educational programme and its greater use of condoms and the reduction in sexual partners has shown major falls in disease prevalence.

Outside Africa the virus continues to spread. HIV was a latecomer to Asia but its spread in that continent has been swift. Thailand has the best-documented epidemic, especially within the sex industry. Given the huge numbers represented by India and China, even the very low rates of infection there have major implications for world numbers.

In the developed world, both North America and western Europe have seen some stabilization of the disease overall. There remain high urban concentrations in specialized communities, e.g. the gay district of San Francisco or the drug-using district of New York's Bronx. In eastern Europe rapid spread is also associated with specialist drug-using communities. Figure 20.16 shows the sequential spread of HIV in the United States with cities serving as *epicentres* for spread.

Other Emerging Diseases

Although AIDS is the dominating of the newly emergent diseases, it is not alone. Here we illustrate the spatial pattern of two very different diseases, one from the rural tropics and the second from urban America.

African Haemorrhagic Fevers One small group of fevers that have attracted interest in western countries over the last half century are those viral fevers that occasionally break out in Africa. *Marburg fever* erupted in a laboratory in Marburg, Germany, in 1967 after green monkeys from Uganda were imported for cell culture research (see Figure 20.17). Twenty-seven laboratory workers fell ill with a grave illness and seven died. A new rod-shaped virus was identified as the cause. Small outbreaks continue to be reported occasionally amongst both humans and primates in a broad belt across sub-Saharan tropical Africa.

Marburg fever was the first of three African viral fevers all of which share high fatality rates (in the range 20–90 per cent) linked to excessive bleeding (hence the term 'haemorrhagic'), but each has distinctive viral characters and different animal reservoirs. The second, *Lassa fever*, was first identified in 1969 after an outbreak in the Jos region of Nigeria and continues to occur sporadically in west Africa. It appears to be endemic in some wild rat populations. The third, *Ebola fever*, broke out in 1976 simultaneously in southern Sudan in towns along the Ebola River in the northern part of the Democratic Republic of the Congo (Zaire). Small outbreaks have continued from time to time in tropical Africa from Senegal to Kenya. Reservoirs in domestic guinea pig populations have been suspected.

All three fevers have attracted attention from the western press (and even film makers). Outbreaks tend to be localized but very severe (the Yambuka Mission hospital oubreak of Ebola fever in September 1976 affected 318 cases with 288 deaths) and epidemics end as abruptly as they start. November 2000 saw a further severe epidemic in Uganda. Ready links via

Figure 20.16 (overleaf) Transmission of AIDS through the United States urban hierarchy The changing intensity of the AIDS epidemic in 1982, 1984, 1986, and 1988. (a) The effects of hierarchic diffusion are already evident by the first map with major foci on New York and Miami (east coast) and San Francisco and Los Angeles (west coast) serving as ***epicentres*** for the disease. (b) In the second map the disease has moved down to smaller centres and no state is without its AIDS cluster. (c) In the third, spatially contagious diffusion becomes more prominent as urban commuter fields pump the virus into the suburbs of regional epicentres. (d) In the last map clear alignments along interstate highway systems are intensifying, and earlier urban 'beads on a string' are reaching out to coalesce. Gould went on to map AIDS spread at monthly intervals and made a colour film which vividly showed the overwhelming of all US communities with disease. It was used with great effect in HIV awareness programmes in Pennsylvania high schools.

[Source: Redrawn from P.R. Gould, *On Becoming a Geographer* (Syracuse University Press, Syracuse, 1999), Plates 5–8.]

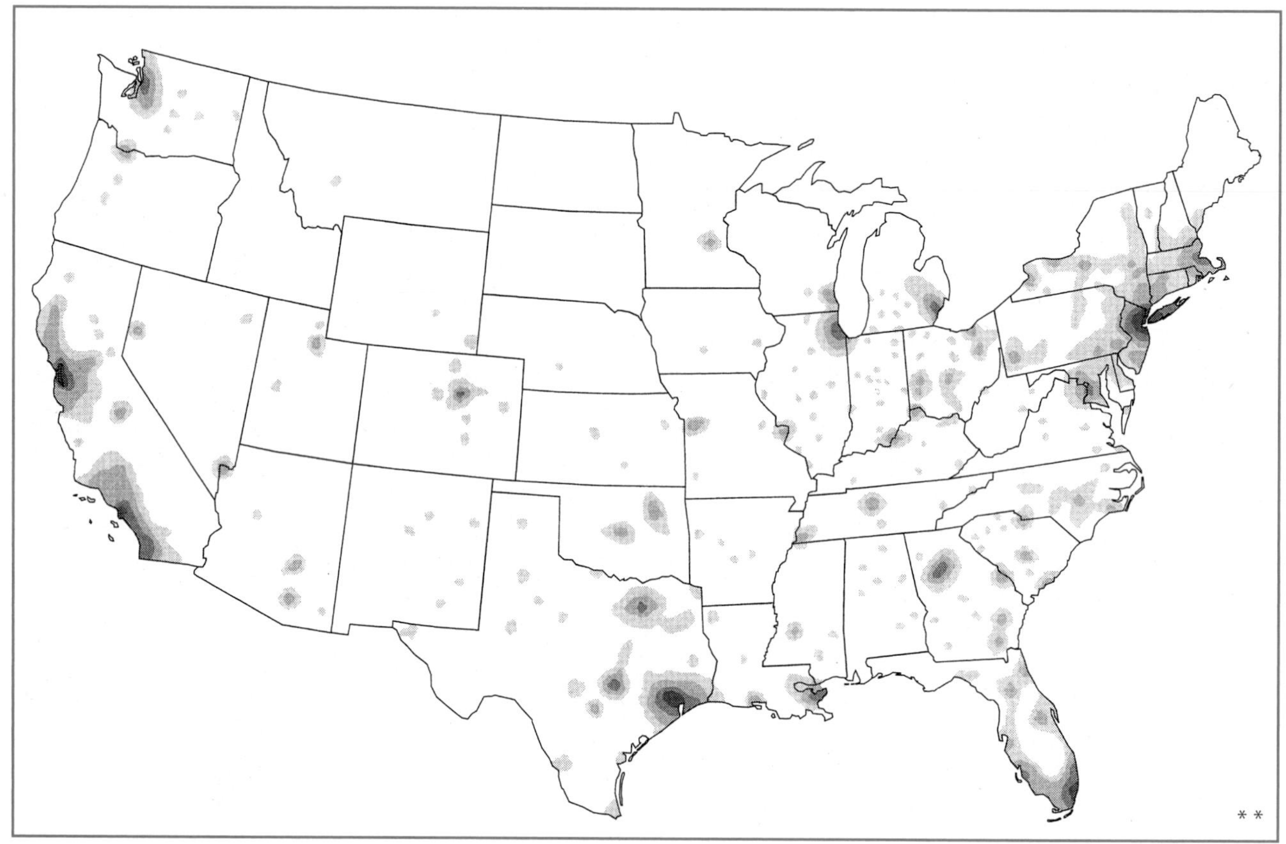

(a) 1984

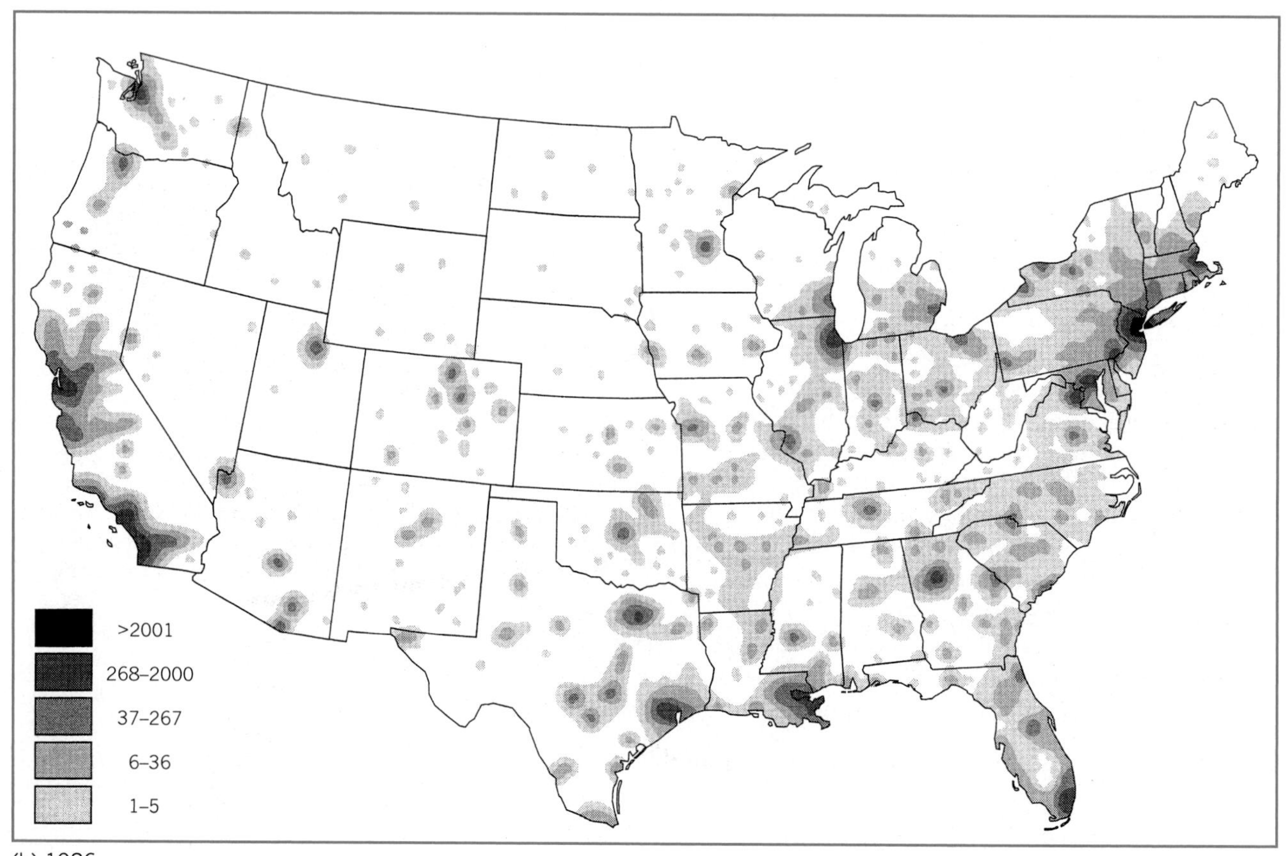

(b) 1986

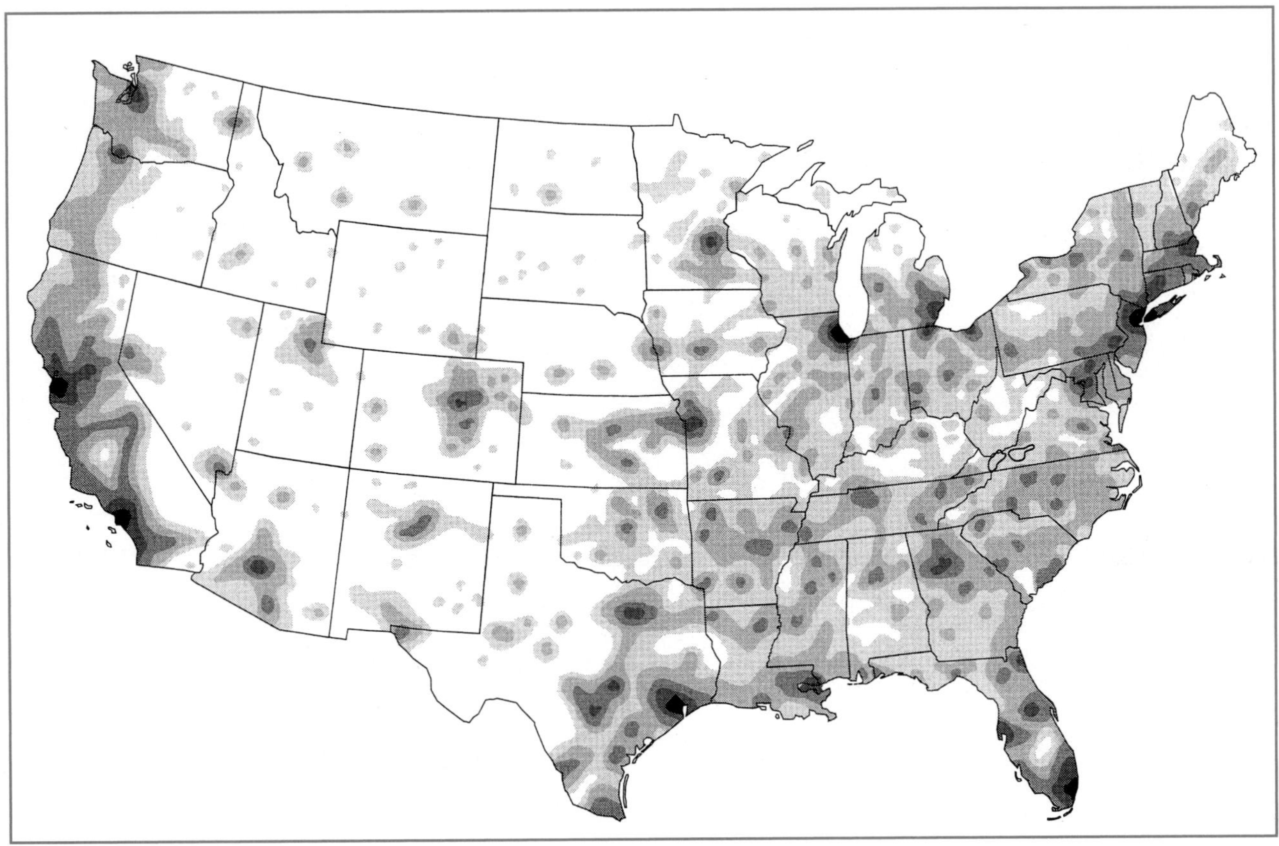

(c) 1988

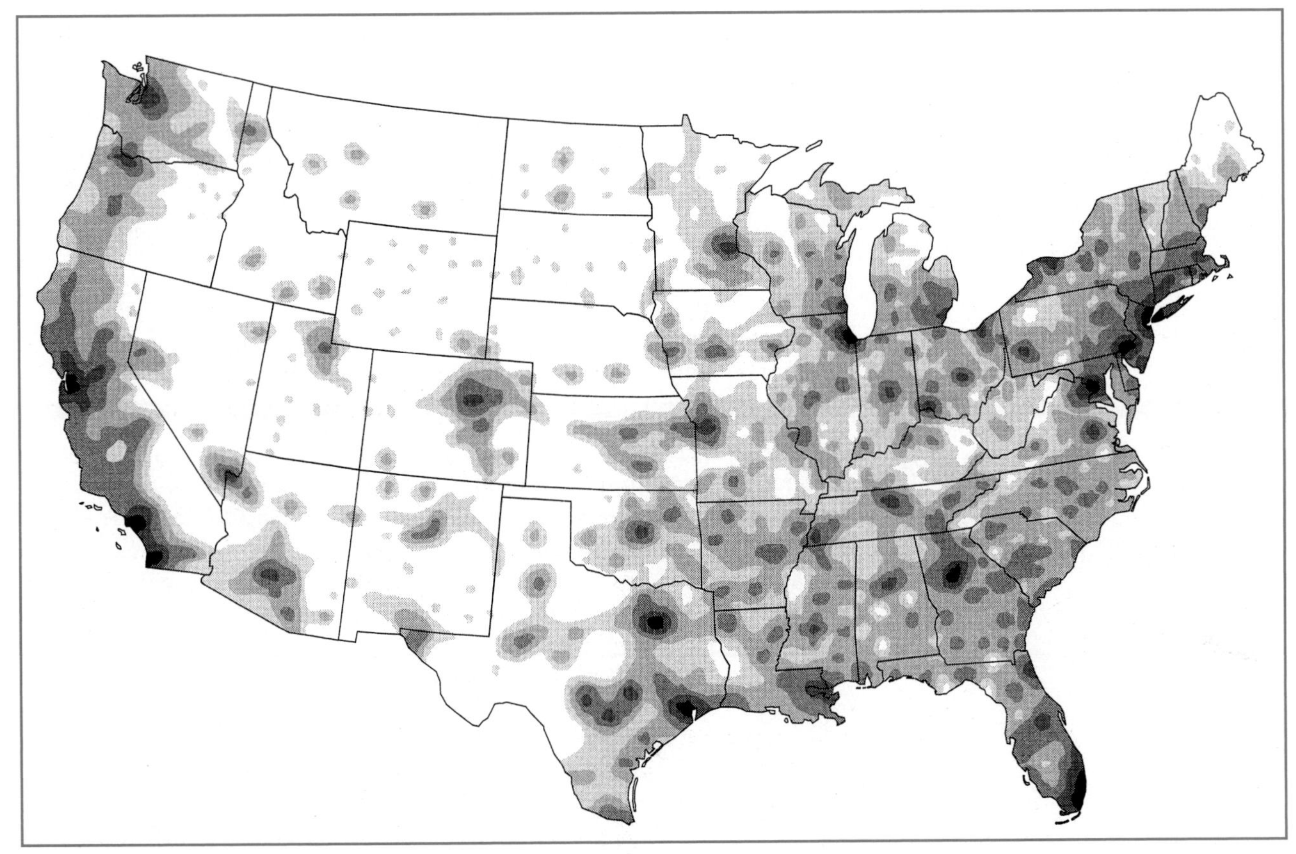

(d) 1990

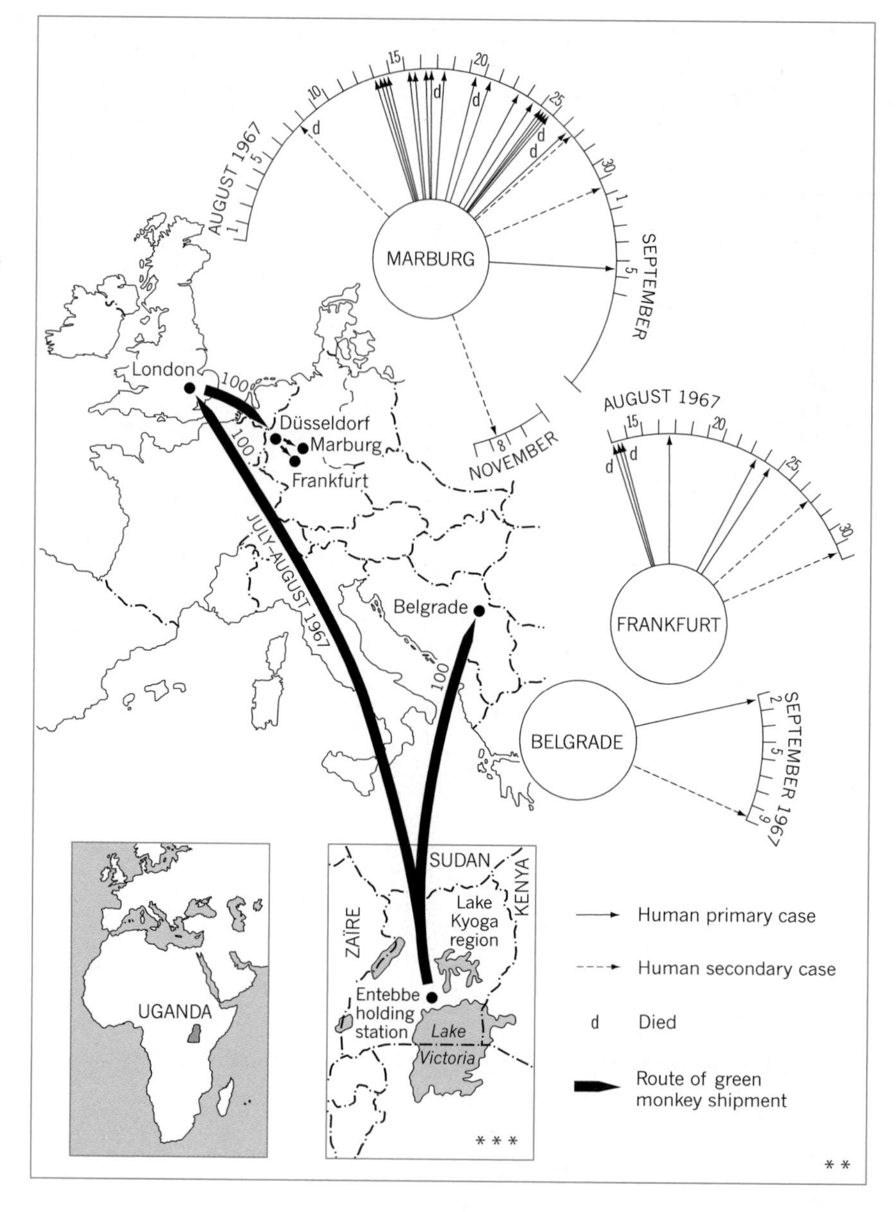

Figure 20.17 Marburg fever outbreak in Europe The vectors show the routes of green monkey shipments from Uganda to Europe. The three city diagrams show the timing of primary and secondary cases with deaths indicated.

[Source: M. Smallman-Raynor, A.D. Cliff and P. Haggett, *International Atlas of AIDS* (Blackwell, Oxford, 1992), Fig. 3.4A, p. 133.]

air transport mean areas well outside tropical Africa are potentially at risk; close surveillance of outbreaks by WHO is being maintained.

Legionnaires' Disease In late July 1976 the Pennsylvania branch of the American Legion (an organization of military veterans) held its annual convention at the Bellevue Stratford Hotel in downtown Philadelphia. Within a few days of the convention ending, reports of illness began (see Figure 20.18). Some 221 cases of illness were identified, almost all with pneumonia and over 30 people died.

Dubbed 'legionnaires' disease', the illness was subsequently identified as been due to a bacterium (*Legionella pneumophila*) which is widely distributed in nature. The bacteria thrive particularly well in hot water systems and in the recirculating water in cooling systems used in many public buildings. It is now a serious pneumonia-causing disease which is responsible for major outbreaks in vulnerable populations (e.g. old people's homes and hospitals). It illustrates how new diseases may emerge from micro-environmental change in urban areas.

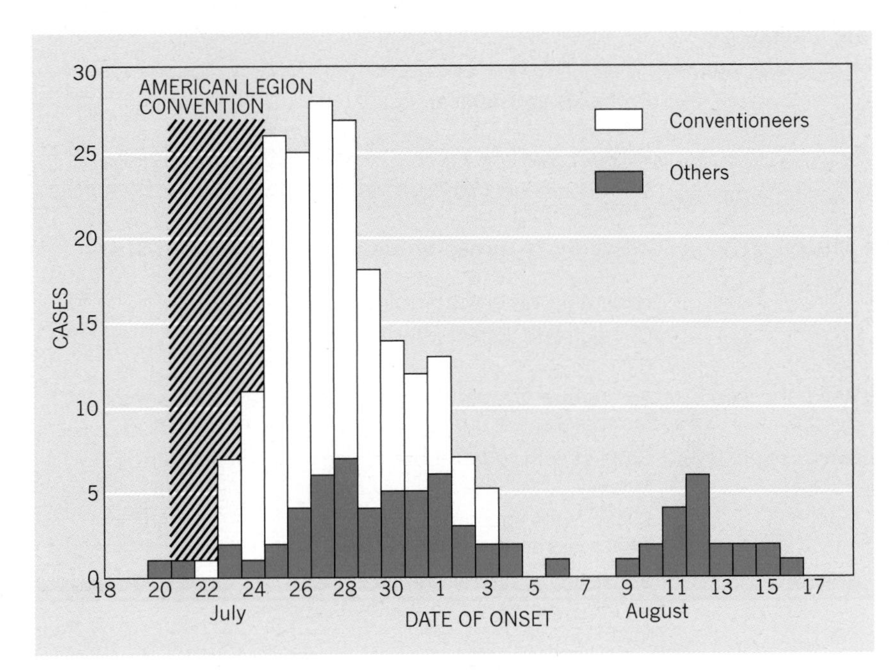

Figure 20.18 First outbreak of legionnaires' disease Distribution of cases in the Pennsylvania epidemic of July–August 1976 arising from a common source at the American Legion convention held at Philadelphia. The cross-hatching shows the duration of the convention with most cases reported after the convention had finished.

[Source: Drawn from Centers from Disease Control (CDC) data.]

Disease in a Changing Geographic Environment 20.4

Diseases spread within a specific historical and geographic context. In the late twentieth and early twenty-first centuries, the environmental context within which disease control is set is changing at a faster rate than any time in human history. Table 20.3 summarizes some of those environmental changes that have disease implications.

We illustrate here the impact of four such changes: (1) economic change, (2) the collapse of geographic space, (3) global land-use changes, and (4) global warming.

Disease and Economic Development

We saw in Figure 20.3(a) the WHO estimates of global causes of death at the start of this century. But this overall picture needs to be modified to take into account the major contrasts in economic development across the world. Forty million of the deaths in 2000 occurred in developing countries (Figure 20.3(b)) and 12 million in developed countries (Figure 20.3(c)). Comparisons of (b) with the global picture of causes of death in (a) show relatively little contrast. Infectious and parasitic diseases rise from a third to 43 per cent of all deaths, and, together with perinantal and maternal causes, account for over a half of all deaths. In age terms, this means that half of all deaths are of children.

In contrast, comparing (c) with the global picture (a) shows major contrasts. In developing countries infectious and parasitic diseases plus perinatal and maternal causes together add up to only 2 per cent. Almost all children survive through to adulthood. In developed countries the great killers are the conditions of maturity and old age: diseases of the circulatory system (46 per cent of all deaths) and the cancers (21 per cent).

Table 20.3 Geographical changes and virus emergence

Geographic change	Disease	Probable mechanism	Location
Increased spatial interaction	Dengue	Disseminated by travel and migration	Worldwide
	Yellow fever	Both virus and major mosquito vector *(Aedes aegypti)*	Africa, Caribbean
	Seoul-like viruses	Infected rats carried on ships	United States
Land-use change: (A) Agricultural	Influenza	Integrated pig-duck farming	China
	Hantaan	Contact with rodents during rice harvest	China
	Argentine haemorrhagic fever	Agriculture favours natural rodent host; human contact during harvest	S America
	Bolivian haemorrhagic fever	Contact with rodent host during harvest	S America
	Oropouche	Cacao hulls encourage breeding of insect vector	S America
	Monkeypox	Subsistence agriculture and forest hunting; increased contact with rodent host	Tropical Africa
(B) Forests and Woodland	Kyasanur forest	Tick vector increased as forest land replaced by sheep grazing	India
(C) Water	Lyme disease	Tick vectors increased as fields replaced by woodland	NE United States
	Dengue, dengue haemorrhagic fever, yellow fever	Water containers encourage breeding of mosquito vector	
	Venezuelan equine encephalitis, Rift Valley fever	Building of dams and irrigation favour increase in vector	Panama, Africa
Military operations	Leishmaniasis	Military camapaigns in new environment (e.g. operation 'Desert Storm')	Gulf, United States
Global warming	Malaria	Extension of thermal range into middle latitudes	Worldwide

The contrasts between developing and developed countries is even more starkly drawn when we look at particular age groups. Figure 20.19 shows the main causes of death among children under the age of five years in the developing world. About one-fifth of the ten million deaths are due to diarrhoea, another fifth to respiratory diseases (ALRI means 'acute lower respiratory infections'), and a further fifth to measles and malaria.

The most striking feature of this figure is the shading. This shows the proportion of the deaths from each cause that is also associated with malnutrition. For example, 65 per cent of measles deaths are linked to malnutrition. Poor feeding stacks the odds against a child's survival. Repeated episodes of infection are associated with loss of appetite and decreased food intake in a downward cycle. The tragedy for developing countries is that mild diseases of the developed world become killers in countries where malnutrition and starvation are the norm.

The cross-section comparisons between disease in rich and poor countries are replicated in comparisons over time. Over the course of the last century, the economic development of such isolated communities as Pacific Islands have seen a disease transition. Childhood fevers have been reduced, but pre-

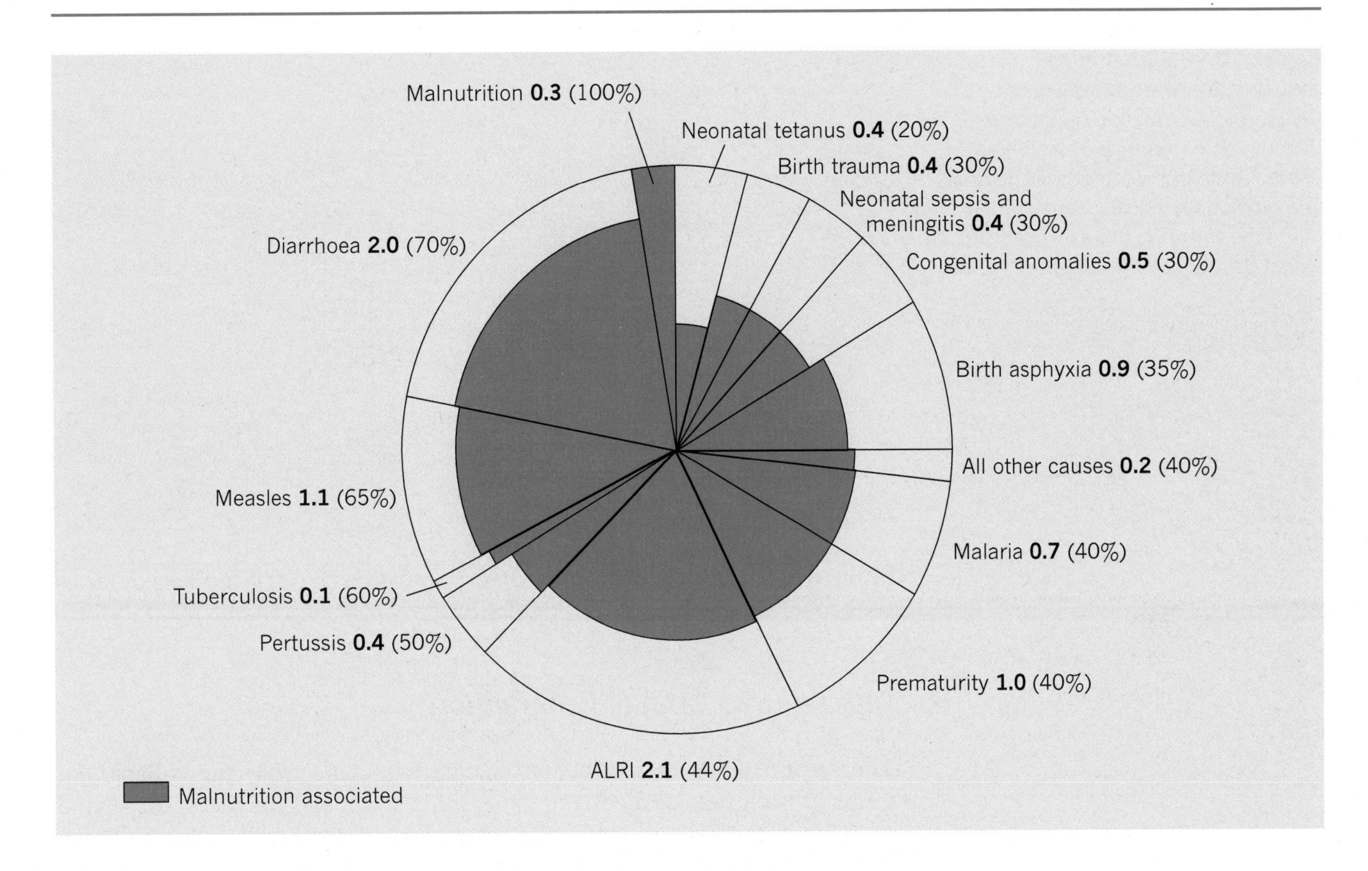

Figure 20.19 Impact of malnutrition on children's deaths Main causes of death among children under age five in the developing world, 2000. The shaded area indicates the proportion of malnutrition-associated deaths. The number of deaths in millions is followed by the percentage of malnutrition-associated deaths in brackets. ALRI = acute lower respiratory infection. Compare with Figure 20.3.

[Source: World Health Organization, *The World Health Report 1998* (WHO, Geneva, 1999), Fig. 8, p. 62.]

viously rare conditions (heart attacks, cancers and diabetes) have increased; problems of malnutrition have been replaced by diseases of obesity. Such changes have suggested to some epidemiologists that the population transition (see Figure 6.16) in demography is replicated by an ***epidemiological transition***. That is, there are systematic changes in disease which mirror the demographic change in age structure.

We noted in the last chapter the major shift in the world's population towards the tropics. It is expected that some 94 per cent of population growth over the next twenty years will occur in the developing countries. Figure 20.20 shows the present latitudinal distribution of population showing a marked concentration in the northern mid-latitudes. Present and future growth will shift the balance of world population still more towards the tropics and low latitudes. This redistribution will increase the average temperature of the global population by around +1 °C, from 17 to 18 °C (even assuming no increase from global warming). This concentration will place more people than in the world's previous history in areas of high microbiological diversity, potentially exposing a greater share of the world's population to conventional tropical diseases.

Urbanization and disease We also noted in the last chapter that the world's growing population is increasingly concentrated in cities. The disease implications of urbanization are complex. Positive effects from improved sanitation or better access to health-care facilities have to be set against the negative effects from increased risk of disease contacts through crowding and pollution. Where rural–urban migration in developing countries results in peri-urban shanty settlements, high rates of intestinal parasitic infections (notably amoebiasis and giardiasis) can result. Each is passed from human to human by the faecal–oral route. They pose an increasing

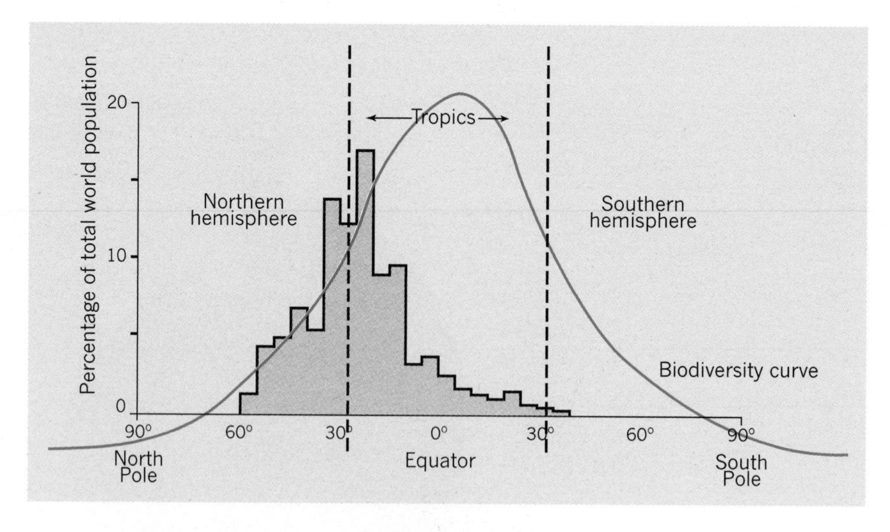

Figure 20.20 Latitudinal distribution of population Geographic distribution of population distribution in terms of 5° latitudinal bands north and south of the equator. The biodiversity curve is approximate and does not make allowances for the global distribution of humidity.

[Source: P. Haggett, WHO Document EMC/IHR/GEN/95.5, 1995.]

health burden as the share of urban populations in developing countries rises towards one-half of all population.

Disease and Spatial Contraction

The second main environmental change has come from the collapse (in terms of both time and cost) of geographic space and the increased spatial mobility in the human population which has accompanied such a collapse. We look first at the evidence for such change and then at its disease implications.

The implications of increased travel are twofold: short term and long term. First, an immediate and important effect is the exposure of the travelling public to a range of diseases not encountered in their home country. The relative risks encountered in tropical areas by travellers coming from western countries (data mainly from North America and western Europe) are given in Figure 20.21.

These suggest a spectrum of risks from unspecified 'traveller's diarrhoea' (a high risk of 20 per cent) to paralytic poliomyelitis (a very low risk of less than 0.001 per cent). Another way in which international aircraft from the tropics can cause the spread of disease to a non-indigenous area is seen in the occasional outbreaks of tropical diseases around mid-latitude airports. Typical are the malaria cases that appeared within 2 km of a Swiss airport, Geneva-Cointrin, in the summer of 1989. Cases occurred in late summer when high temperatures allowed the in-flight survival of infected *Anopheles* mosquitoes that had been inadvertently introduced into the aircraft while at an airport in a malarious area. The infected mosquitoes escaped when the aircraft landed at Geneva to cause malaria cases among several local residents, none of whom had visited a malarious country.

A second short-term factor with modern aircraft is their increasing size. Assume a hypothetical situation in which the chance of one person in the travelling population having a given communicable disease in the infectious stage is 1 in 10,000. With a 200-seat aircraft, the probability of having an infected passenger on board (x) is 0.02 and the number of potential contacts (y) is 199. If we assume homogeneous mixing, this gives a combined risk factor (xy) of 3.98. If we double the aircraft size to 400 passengers, then the corresponding figures are $x = 0.04$, $y = 399$, and $xy = 15.96$. In other words, other things being equal, doubling the aircraft size increases the risk

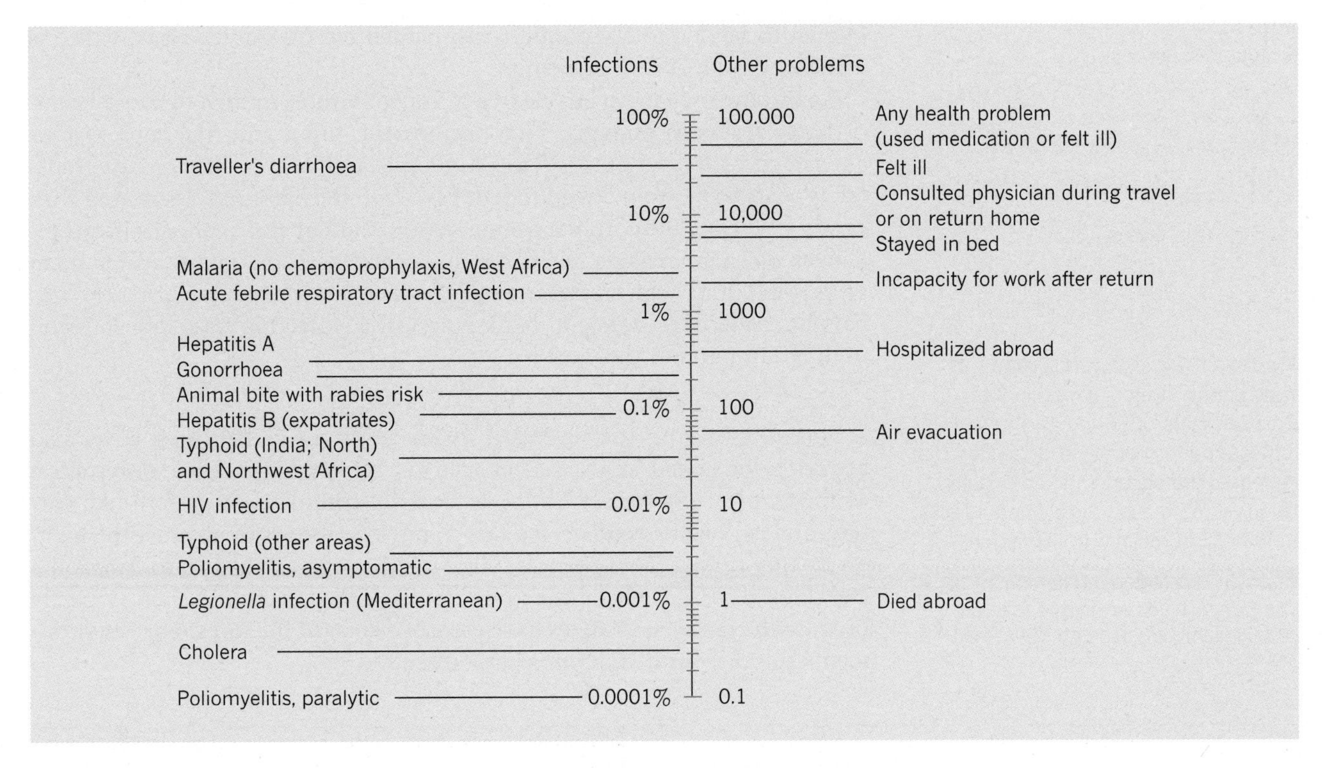

Figure 20.21 Relative disease risk in the tropics Threats posed by communicable diseases to travellers in tropical areas. Note the scale is logarithmic. The left-hand scale is drawn in terms of probable risk based on annual data. The right-hand scale shows relative risks as a ratio with 'died abroad' set equal to 1.0.

[Source: R. Steffen, *International Travel and Health* (Geneva, World Health Organization, 1995), Fig. 1, p. 56.]

from the flight fourfold. Thus the new generation of wide-bodied jets presents fresh possibilities for disease spread, not only through their flying range (linking tropics and middle latitudes) and their speed (giving less chance for symptoms of infection to show), but also from their size.

On a longer time-scale, increased travel brings some possible long-term genetic effects. With more travel and longer range migration, there is an enhanced probability of partnerships being formed and reproduction arising from unions between individuals from formerly distant populations. For example, the probability of occurrence of such a multifactorial condition as cystic fibrosis is reduced; the risk of this condition is somewhat higher in children of consanguineous unions. Conversely, inherited disorders such as sickle cell anaemia might become more widely dispersed.

Disease and Land-use Change

The combination of population growth and huge technological changes has given humankind the capacity to alter environments in ways that are unprecedented in human history. We illustrate the disease implications of three such changes.

Agricultural Colonization

Accelerated world population growth has put pressure on food supplies in tropical areas and has led to the colonization of new environments in the search for expanded food production. *Venezuelan haemorrhagic fever* is a severe and often fatal virus disease only recently identified in the Guanarito savannah area in central Venezuela. Cases were not found in the cities but confined to rural inhabitants of the area who were largely engaged in farming or cattle ranching. Major outbreaks in 1989 and again in 1990–91 had fatality rates of around a quarter. First diagnosed as due to dengue haemorrhagic fever, the disease is now known to be due to a separate virus, named the Guaranito virus, which is associated with rodent reservoirs.

Guaranito appears to be one of a family of viruses known to cause haemorrhagic fevers in humans. They include the Junin and Machupo viruses associated with fever outbreaks in Argentine and Bolivia. In each case transfer appears to be from a wild rodent host with the main risk associated with exposure during the corn harvesting season. Similar seasonal risks from epidemics of haemorrhagic fevers are associated with the family of Hantaan viruses in China which appear to be transferred to humans during the rice harvest. Field mice, rats and bank voles are involved in fever transmission in different parts of the world.

Deforestation and Reforestation Changes in the global forest cover also appear to be linked to disease changes in complex ways. The deforestation of the tropical rainforests has been spatially complex with a fern-like pattern of new logging roads being driven into the forests to abstract the highest-quality timber. New settlers following the logging roads into Amazonia (see Figure 20.22) encountered heavy *malaria* infections. This is partly because the land-use changes have greatly increased the forest-edge environments suitable for certain mosquito species.

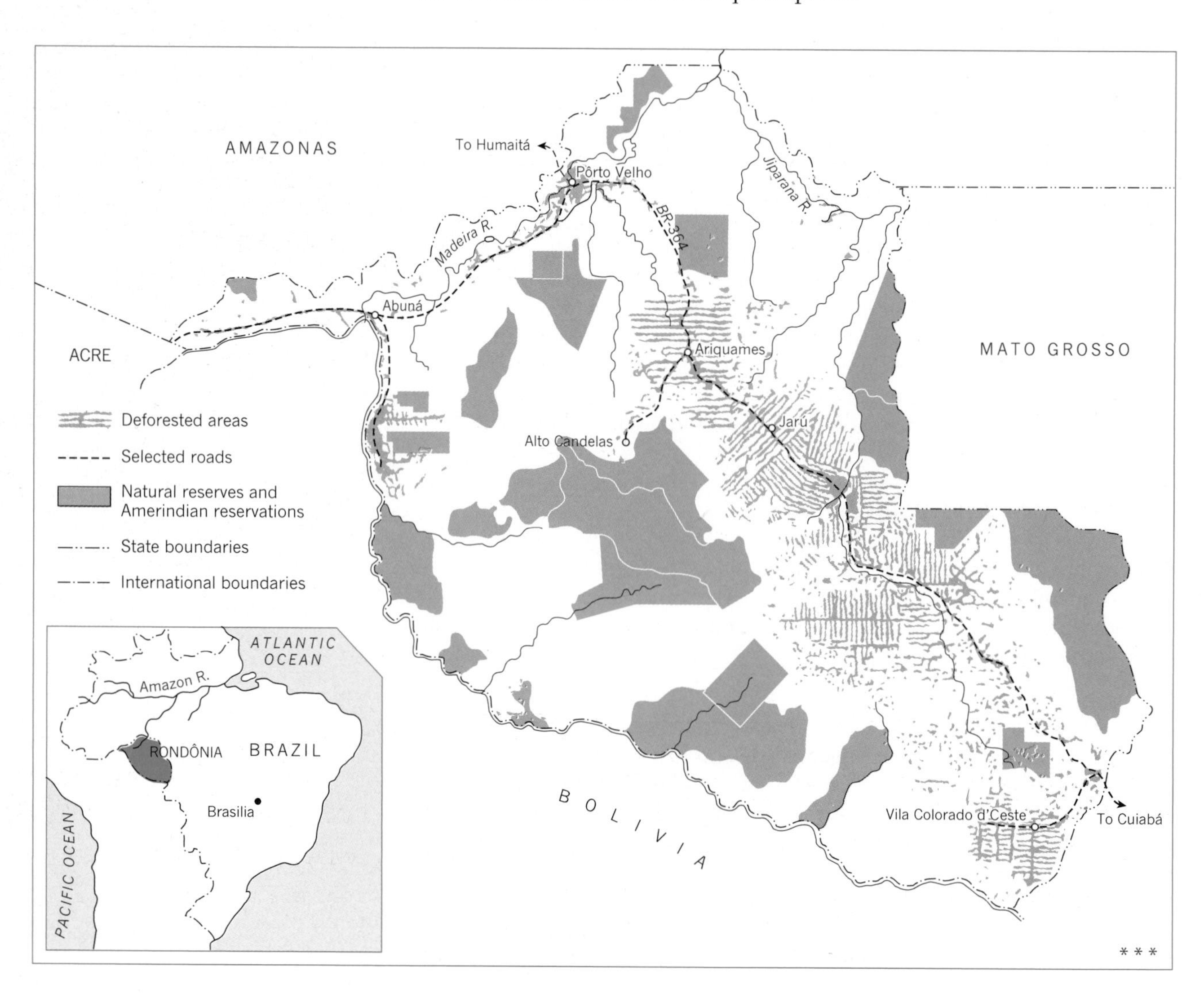

Figure 20.22 Logging roads in Amazonia Deforestation in the Brazilian state of Rondonia showing the intricate pattern of forest dissection that increases the edge conditions which favour some mosquito vectors and add to the risk of malaria transmission.

[Source: D.J. Mahar, *Government Policies and Deforestation in Brazil's Amazon* (World Bank, Washington, D.C., 1989), Fig. 4, p. 32.]

* * *

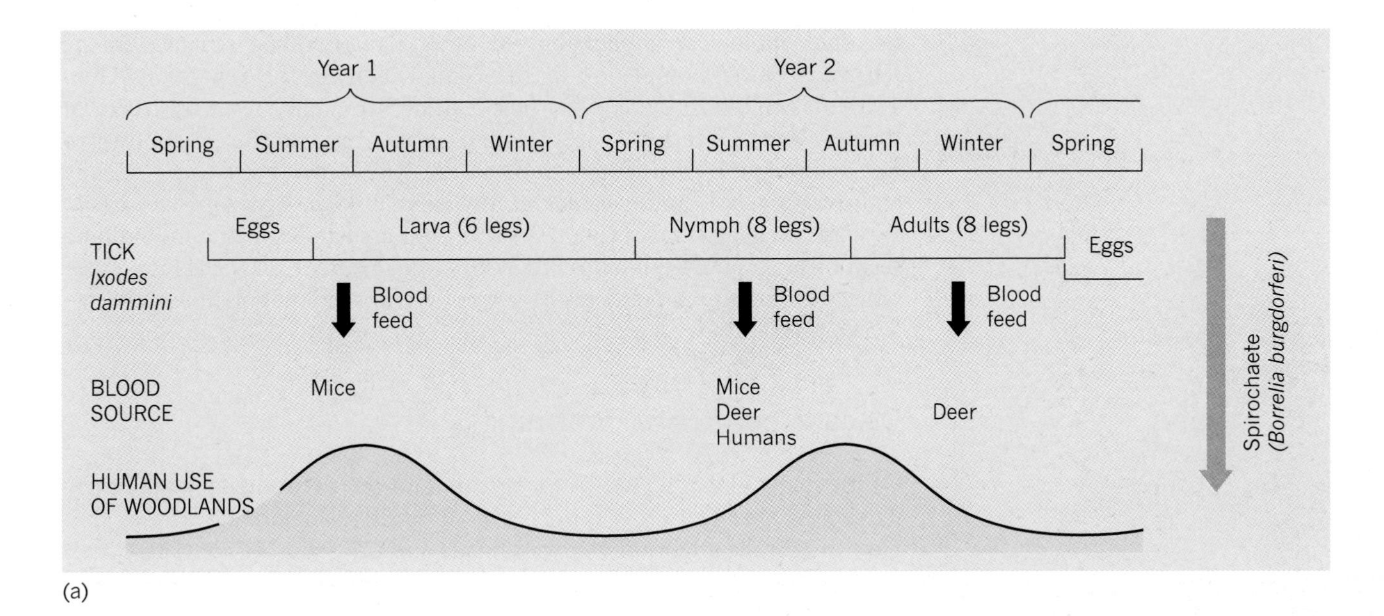

Disease changes can also result from an opposite process in which abandoned farmland reverts to woodland. The classic case is the emergence of Lyme disease, caused by the bacterium *Borrelia burgorferi*. Lyme disease is now the most common vector-borne disease in the United States but retrospective studies suggest it was not reported there until 1962 in the Cape Cod area of New England. The link between Lyme disease and 'Lyme arthritis' was not established until the 1970s with an endemic focus being recognized around the small town of Old Lyme in south-central Connecticut. The critical land-use change that precipitated the emergence or re-emergence of the disease appears to have been the abandonment of farmland fields to woodland growth. The new woodland proved an ideal habitat for deer populations which is the definitive host for certain ticks which spread the bacterium through bites. The complex seasonal cycle of the vectors which involves the ticks, the deer, the white-footed mouse (the reservoir for the pathogen) and human visitors using the forest (see Figure 20.23) illustrates how sensitive is the ecological balance in which disease and environment is held. Epidemic Lyme disease is now an increasing problem in Europe fuelled by reversion of farmland to woodland (partly due to EU set-aside land policies), deer proliferation, and increased recreational use of forested areas. Lyme disease has now been reported from most temperate parts of the world in both northern and southern hemispheres.

(b)

Figure 20.23 Life cycle of Lyme disease (a) Relationship between the two-year life cycle of the disease-carrying tick that causes Lyme disease and human summer-time exposure. (b) Abandoned fields reverting to woodlands around a New England farm provide an environment in which the tick–rodent–deer relationship can flourish.

[Source: (b) John & Lisa Merrill/Tony Stone Images.]

Water Control and Irrigation *Rift Valley fever (RVF)* was until recently primarily a disease of sheep and cattle. It was confined to Africa south of the Sahara with periodic outbreaks in East Africa, South Africa, and, in the mid-1970s, the Sudan. The first major outbreak as a human disease occurred in Egypt in 1977 with 200,000 cases and 600 deaths, the deaths usually associated with acute haemorrhagic fever and hepatitis.

The Egyptian epidemic has been provisionally linked to the construction of the Aswan Dam on the Nile. Completed in 1970, the dam created a 800,000 hectare water body and stabilized water tables so that surface water provided breeding sites for mosquitoes. Whether the mosquito population provided a corridor for allowing the virus to enter Egypt from the southern Sudan has yet to be proved. But the possibility led to concern for the epidemiological implications of other dam building schemes in the African tropics. Completion of the Diama dam on the Senegal River in 1987 was followed by a severe outbreak of Rift Valley fever upstream of the new dam. Over 1200 cases and 250 deaths resulted. But in contrast to Egypt, immunological studies showed the Rift Valley fever to be already endemic in people and livestock in a wide area of the Senegal River basin. Ecological changes favouring the vector and associated with dam building seem to be implicated in allowing both (a) invasion of the virus into a previously virgin population and (b) severe flare-ups in a population with low-level endemicity.

Disease and Global Warming

Of the many global scenarios for disease and the environment in the early part of the next century, it is the health implications of global warming that have caught the attention of governments and press worldwide. There have already been major studies of its potential health implications in at least three countries: the United States, Australia, and the United Kingdom. The WHO also has a committee looking at this issue.

A number of health effects have been postulated as following from a worldwide increase in average temperature from global warming. For infectious diseases, the main effects relate to changes in the geographical range of pathogens, vectors, and reservoirs. So far, few attempts have yet been made to compute the relative burden of morbidity and mortality that would be yielded by these effects. Any such calculation would also need to offset losses against gains that might accrue (for example, reductions in hyperthermia against increases in hypothermia).

The magnitude and spatial manifestations of global warming are still speculative. One of the main conclusions of the report of the *Intergovernmental Panel on Climate Change (IPCC)* is to stress how far research still had to go before reliable estimates of global warming could be identified. But some rough orders of magnitude can be computed from the estimates of the different models that have been used. In global terms, warming appears to range from a predicted rise over the next 40 years of 0.7 to 1.5 °C with a best estimate of 1.1 °C.

We can obtain some idea of the implications of the predicted shift for local mean temperatures with reference to the United Kingdom. Current differences between the coldest (Aberdeen, latitude 57.10°N) and warmest (Portsmouth, latitude 50.48°N) of its major cities is 2.4 °C; this is well beyond the postulated IPCC warming effect by the year 2030. Climate is a much more complex matter than average temperature, but – if the global warming models carry over to the UK – then, by 2030, Edinburgh might have temperatures something like those of the English Midlands and London something like those of the Loire valley in central France. If we accept the much higher estimate of +4.8 °C warming over 80 years, this brings London into the temperature bands of southern France and northern Spain. Provided that these projections are sensible, something might be gained by comparative studies of disease incidence within the UK and adjacent EU countries and disease incidence in warmer climates that match those predicted for the UK.

The biological diversity of viruses and bacteria is partly temperature dependent, and it is much greater in lower than higher latitudes. Conditions of higher temperature would favour the expansion of malarious areas, not just for the more adaptable *Plasmodium vivax* but also for *P. falciparum*. Rising temperatures might also allow the expansion of the endemic areas of other diseases of human importance: these include, for example, leishmaniasis and arboviral infections such as dengue and yellow fever. Higher temperatures also favour the rapid replication of food-poisoning organisms. Warmer climates might also encourage the number of people going barefoot in poorer countries, thereby increasing exposure to hookworms, schistosomes, and Guinea worm infections. But not all effects would be negative. Warmer external air temperatures might reduce the degree of indoor crowding and lower the transmission of influenza, pneumonias, and 'winter' colds.

While modest rises in average temperatures are the central and most probable of any greenhouse effects, they are likely to be accompanied by three other main changes: (a) sea-level rises of up to a metre; (b) increased seasonality in rainfall, thus reducing the level of water available for summer use; and (c) storm frequency increases.

The Future Challenge 20.5

This chapter has only touched on the wide range of issues which surround the global control of diseases. Looking forward into the future we see a series of trends that will influence control measures in the coming decades. While these will contain positive improvements in vaccine power and efficiency, these will be balanced by microbiological resistance.

Diseases in the Twenty-first Century

For young as well as old, the world of 2025 will be very different from that of today. The enhanced life expectancy of children born in the present century reflects the harvest of health improvements introduced in the twentieth century. By then children under five years (around 8 per cent of the world's population) will be outnumbered by the over 65s (10 per cent). By then the world already freed of smallpox should also be free of poliomyelitis, measles, and neonatal tetanus. Most children should be protected from vaccine-preventable diseases.

But the shift in disease in one age cohort will bring changes in others. Along with the decline in children's diseases will come a rise in diseases typical of older age groups: circulatory system and cancers. As circulatory diseases are tackled with drugs and life-style changes so cancers as the 'residual' cause of death will tend to rise. Cancers tend to have four main causes: diet-related (cancers of the stomach and alimentary canal), tobacco-related (lung), infection-related (lymphoma and cervix), and hormone-related (breast). Rapid urbanization associated with sedentary lifestyles, ill-balanced and excessive diets, smoking, and a deteriorating environment are associated with increases in chronic diseases such as diabetes and rheumatoid arthritis. Social alienation in cities may also grow with urbanization and be associated with an increase in mental disorders and, in extreme cases, an increase in suicide. The proportion of population now surviving into extreme old age is likely to bring increases in mental and

Table 20.4 Share of children under five years of age who are underweight in selected countries

Country	Share of underweight (%)
Bangladesh	66
India	64
Vietnam	56
Ethiopia	48
Indonesia	40
Pakistan	40
Nigeria	36
Philippines	33
Tanzania	29
Thailand	26
China	21
Zimbabwe	11
Egypt	10
Brazil	7

Source: World Health Organization, *Global Database on Child Growth*, Geneva, 1997, based on national surveys taken between 1987 and 1995.

physical disabilities associated with stroke, dementia (e.g. Alzheimer's disease) and chronic conditions.

On the unfinished agenda for world health, poverty remains the main item. The priority in reducing the burden of disease must be to reduce it in the poorest countries of the world, and to eliminate the pockets of poverty that exist in more developed countries and in transient groups such as refugees. Improving health and ensuring equity are two sides of the same coin. Wars, conflicts, refugee movements and environmental degradation are not only health hazards in themselves but facilitate the spread of disease. For advanced countries, the search for healthier lifestyles and exploring ways of healthy ageing are likely to move up the agenda. Table 20.4 shows the location of countries where a high proportion of children are underweight and, by implication, malnourished.

Safeguarding the advances made in the last century and exploring the new opportunities stemming from work on the human genome in the twenty-first century are twin objectives in health policy for the coming century.

The Geographer's Challenge to Disease Control

We have noted above that the global burden of disease is inextricably tied in to other aspects of human occupation of this planet. As the environment changes, so the disease pattern alters to reflect those changes.

The challenge to control global disease has a number of geographic aspects:

1 *Disease control is likely to rely and less and less on spatial barriers.* The speed of modern air transport (most of the world's cities are now within 36 hours of each other) and the complexity of air connections (there are now over 4000 airports in the world with regular scheduled services) make the traditional 'drawbridge' strategy increasingly irrelevant. The *quarantine* barriers first set up by the earliest International Sanitary Conferences were modelled to fit a slower mode of travel, notably ships, and fewer connection points.

2 *Rapid reporting and surveillance are likely to be increasingly critical in control.* The use of electronic reporting of disease outbreaks through the Internet allows a more rapid response to outbreaks of infectious diseases than in past decades. Disease is no respecter of national boundaries and international reporting through WHO and its agencies is now critical.

3 *Ever-widening lists of communicable diseases and the high cost of surveillance will make sampling essential.* The legal requirements to notify critical infectious diseases is tending to be replaced by sampling systems in which sentinel practices are used to pick up trends in disease prevalence. This will intensify the legal problems associated with vaccination, identification, restrictions on freedom of movement, and the increasing constraints that these factors pose.

4 *Geographic models will increasingly supplement other epidemiological tools in global control.* In addition to the spatial models discussed in Chapters 13 and 16 (and the GIS methods of Chapter 23), there will be the need to regularly scan the torrent of international and local data for 'aberrant' behaviour. Automatic monitoring of epidemic data allows anomalous events to be highlighted for the epidemiologist to consider. Such anomalies are typified by clinical reports of new resistant malaria strains, to unusual clusters of meningitis cases, to higher-than-average influenza reports.

5 *Disease control and socio-economic development are likely to be more closely tied together*. In a recent *World Health Report,* the Director-General of WHO stated that: 'The world's most ruthless killer and the greatest cause of suffering on earth is listed in the latest edition of WHO's International Classification of Diseases under the code Z59.5. It stands for extreme poverty!' (see Figure 20.24).

(a)

(b)

Figure 20.24 Plenty and scarcity Poverty and associated malnutrition remain the major factors in high infant mortality. Check back to Figure 20.19. (a) Starving children at the International Children's Relief Centre (ICRC) feeding centre, Baidoa, Somalia. (b) Grain mountain in the the United States wheat farming belt. The dilemma of linking areas of malnutrition with areas of oversupply lies at the heart of world health problems.

[Sources: (a) PANOS Pictures/Chris Tordai. (b) United States Department of Agriculture.]

The numbers of the extreme poor have been rising over the last decades at a rate above gross population growth and now estimated at over one-fifth of humanity. For communicable diseases the links between poverty and disease come through many channels: absence of knowledge of protective measures, poor diet, and lack of vaccination, clean water, and sanitation. The correlation runs the gamut of geographic scales from the global North–South contrast between the developed and developing worlds to the local contrast between the affluent suburbs and deprived inner-city ghettos of a western city.

The above five are some of the more important contexts of change against which spatial control is likely to be set. Each century, public health has had to fight disease with the tools available and the constraints imposed at the time. The twenty-first century will be no exception as it prepares to fight both old diseases causing old problems, old diseases causing new problems (e.g. drug resistance) as well as wholly new diseases.

Reflections

1 Why is it so difficult to measure exactly the burden of the world's disease? Consider some of the different ways in which ill-health can be measured (e.g. death, illness, incapacity; see Box 20.A). Why do geographers tend to use the simplest and starkest measure, i.e. death, in mapping world diseases?

2 The main causes of death in the world and its major regions are shown in Figure 20.3. How does the pattern of deaths in your own country fit into this picture? Check the figures in a standard statistical source (see Appendix B of this book for Internet sources) and plot a similar diagram.

3 The world has recently seen major resurgences of diseases (e.g. malaria and tuberculosis) that were once thought to be 'conquered'. What factors lie behind this? How do you view the pessimistic opinion of the great French bacteriologist, Louis Pasteur, that '... in the long-term battle between micro-organisms and humans, I would back the former'?

4 The most striking success of the World Health Organization in the last century was the global eradication on smallpox in the 1970s. How far do you think that success provides (a) a convincing or (b) a misleading template for conquering other infectious diseases?

5 Why has the worldwide spread of AIDS been so rapid? What factors lead to its emergence in Africa and why were some regions (e.g. North America) affected much more quickly than others (e.g. Southeast Asia). How far are the ideas of spatial diffusion, which we met in Chapter 16, relevant to understanding the spread of AIDS?

6 The global environment is now changing at an historically unprecedented rate. How far do these global changes have implications for the emergence of new diseases?

7 Consider the role of malnutrition in affecting disease outcomes (see Figure 20.19). To what extent do you consider that poverty remains the great killer? Look at a disease atlas for your own country (see the Internet sources suggested in Appendix B of this book) and see how far inequalities in health match inequalities in income.

8 Review your understanding of the following concepts using the text and the glossary in Appendix A of this book:

endemic	immunization programmes
epicentres	morbidity
epidemiological transition	pandemics
global eradication	quarantine
iceberg concept of disease	virgin soil epidemic

One Step Further ...

The research reported in this chapter is covered in more depth in Peter HAGGETT, *The Geographical Structure of Epidemics* (Clarendon Press, Oxford, 2000). The spatial distribution of the main human diseases is given in A.D. CLIFF and P. HAGGETT, *Atlas of Disease Distribution* (Basil Blackwell, Oxford, 1988) and the same authors combine with Matthew SMALLMAN-RAYNOR in *World Atlas of Diseases* (Edward Arnold, London, 2001). Useful introductions to medical geography are provided in two standard texts: M.S. MEADE and R.J. EARICKSON, *Medical Geography*, second edition (Guilford, New York, 2000) and K. JONES and G. MOON, *Health, Disease and Society* (Routledge, London, 1987).

Clear explanation and examples of the movement of disease through populations are given in N.F. STANLEY and R.A.

JOSKE (eds), *Changing Disease Patterns and Human Behaviour* (Academic Press, New York, 1980). An outstanding introduction to the geography of AIDS is given in Peter GOULD, *The Slow Plague* (Blackwell, Oxford, 1993) and in G.W. SHANNON *et al.*, *The Geography of AIDS* (Guilford, New York, 1991). Maps of AIDS at geographic scales from the local to the global are given in M. SMALLMAN-RAYNOR *et al.*, *International Atlas of AIDS* (Blackwell, Oxford, 1992). The role of poverty and income inequalities in health differences are treated in R. WILKINSON, *Unhealthy Societies: The Afflictions of Inequality* (Routledge, London, 1996).

The 'progress reports' relating to the geography of health in *Progress in Human Geography* (quarterly) will prove helpful in keeping up to date. The main trends in global health and disease patterns are reflected in the WHO publication *World Health* (quarterly) and the analyses of disease patterns is reported in *Social Science and Medicine* (quarterly). The World Health Organization's *World Health Report* (annual) gives a regular update on changes in disease around the world. For readers with access to the World-Wide Web see also the sites on 'health and disease' in Appendix B at the end of this book.

Underground Tanks
LeakingTanks Area
Solid Waste Landfill
Highways
Streets
Railroads
Rivers

PART VI

The Geographer's Toolbox

60
5
10
15
20
25
30
35
40
45
50
55

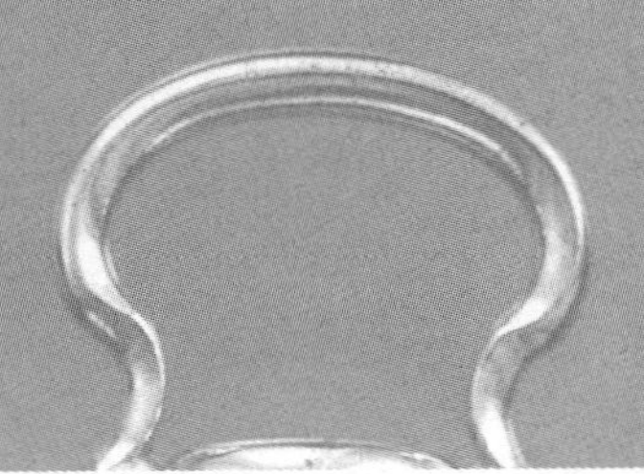

CHAPTER 21

Maps and Mapping

■ denotes case studies

Figure 21.1 Conquering longitude Although the determination of latitude was well established by the Greek period 2000 years ago, the accurate measurement of longitude proved much more difficult. The story of John Harrison's search for an accurate chronometer during the eighteenth century has been told in detail in Dava Sobel's best-selling book, *Longitude*. The photo shows Harrison's prize-winning 4th chronometer held at the National Maritime Museum, London. [Source: National Maritime Museum.]

> Map me no maps, sir, my head is a map,
> a map of the whole world
>
> HENRY FIELDING *Rape upon Rape* (1745)

We began this book by standing on a beach and seeing what we could make of its geography. The world view from the beach is, of course, a very limited one. Even if we scramble to the highest point on the sand dunes backing the beach, we can see only a few kilometres out to sea. If, looking inland, the land is also flat, then our total sweep – from horizon to horizon – still covers only a small disc of the earth's vast surface. We are unlikely, even on the clearest of days, to scan more than 0.0008 of 1 per cent of the total area of the globe. We could fit over one hundred thousand discs this size on the earth's surface without them ever touching one another! Even from a high-flying jet the situation is not vastly improved, though we can then see an area half the size of Texas (or about 0.05 of 1 per cent of the global area).

In other chapters we have used satellite images to show huge areas of the earth's surface and in Figure 2.1, even the whole planet earth itself. But it is worth recalling that these views from outer space are a new privilege, a product of less than a century of high-flying aircraft and space exploration. For most of human history our visual picture of the world was limited by the local horizon. What lay beyond was a matter for, first, speculation, then calculation, and, finally, confirmation. Just where in the world we were located became critical once ocean voyaging was attempted. Dava Sobel, in her fascinating book, *Longitude*, shows how the British navy suffered huge losses by not knowing their exact position at sea and being wrecked on rocks which were thought to be miles away. John Harrison's chronometer (Figure 21.1 p. 658) was a critical tool which allowed the puzzle of location to be solved.

This fascinating trail of conjecture and discovery has gone cold as the pieces of the map have fallen into place and electronic instruments have made the question of finding where you are a trivial one. However, the research methods developed along the trail have proved a very useful legacy for modern geography. To record the characteristics of the earth and to preserve and exchange this information, a comprehensive ***spatial language*** was developed (see Table 21.1). This spatial language described the absolute and relative locations of places either in terms of conventional maps and in other, maplike ways. It allows us to unequivocally assign events to locations. To put it simply, by providing 'a place for everything' it lets us put everything in its place.

In this chapter we present some of the basic grammatical rules of this language and show how geographers use it to describe and record the world beyond the beach. We begin by looking at how human ideas about the world have evolved and how exploration provided solutions (Section 21.1). We then look at how known locations are transferred to the map at very different scales (Section 21.2). The third section (21.3) and fourth section (21.4) look at the mapping cycle. We consider how maps are made. How are maps interpreted? Both parts of the cycle are subject to errors and we try to see how geographers are steadily reducing these. Finally, in Section 21.5, we look at some exciting work on other geographic spaces. It is now possible to make maps using all sorts of bits of information that previously lay outside the map maker's craft.

Table 21.1 Terms used in mapping the earth

Term	Definition
Ellipsoids	Spheroids (figures like a sphere, but not perfectly spherical) with a regular oval form.
Equator	An imaginary line around the earth which is an equal distance from the North and South Poles.
Geodesy	The science that deals with the shape and size of the earth.
Graticules	Networks of parallels and meridians drawn on a map.
Grids	Arbitrary networks of lines drawn on a map. For example, a network of squares may be used to give a simpler reference system than that of graticules.
Latitude	Distance north or south of the equator measured in degrees.
Longitude	Distance east or west from a prime meridian (usually Greenwich in London) measured in degrees.
Meridians	Imaginary half-circles around the earth which pass through a given location and terminate at the North and South Poles.
Parallels	Imaginary lines of latitude on the earth which are parallel to the equator and pass through all places the same distance to the north or south of it.
Poles	The two ends (North Pole and South Pole) of the axis around which the earth spins or rotates.
Prime meridians	Meridians used as a baseline for the measurement of the east–west position of places on the earth's surface in terms of longitude. The most commonly used prime meridian crosses Greenwich in London.

Exploring the World 21.1

Debates abound between cartographers (*cartography* comes from the French *carte*, map or chart) on the earliest world map. Pride of place is usually given to the Babylonians. Several clay tablets with plans of towns or properties survive from around 2000 BC but the first recognizable world map dates from around 500 BC. The small clay tablet in the British Museum shows Babylonia as a disc floating in the sea with Babylon at the centre, other cities shown by circles, the Euphrates river and mountain chains to the north and east.

In this section we look how ideas have evolved since that time as the size and shape of the world and the outlines of continents and oceans was firmly established.

The Size of the World

It is doubtful if our early ancestors were much concerned with the size of the earth. As we have seen, their visual horizon, even from the highest peak in the clearest weather, would have been confined to a few hundred kilometres. Against this background, the intellectual achievements of Greek *cosmology* are remarkable. Their observations of heavenly bodies led them to deduce a spherical rather than a Babylonian-type disc form for the earth. By 200 BC, one of the earliest Greek geographers, Eratosthenes of Alexandria, had made rather accurate estimates of the earth's size. As Figure 21.2 shows, the basic procedures he used in his calculations were simple.

Indeed, Eratosthenes' method was (at least until the advent of the satellite) the same in principle as that used to measure the earth in our own time.

Figure 21.2 The size of the earth Eratosthenes, about 200 BC, estimated the circumference of the earth from the angle at which the rays of the noon sun reached Alexandria and Syene in Egypt.

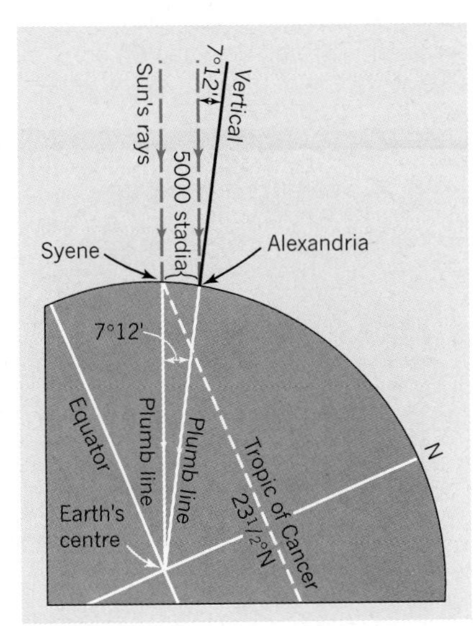

Assuming that light rays from the distant sun were parallel, as for all practical purposes they are, Eratosthenes calculated the differences in the angle they made with the earth's surface at different points and thereby determined its curvature. Specifically, he compared the angle made by the rays of the noon sun at the summer solstice (21 June) at two places in Egypt: Syene, where the sun is directly overhead and the angle is vertical, and Alexandria, where a shadow of 7°12′ is cast. Multiplying the known north–south distance between Syene and Alexandria by the difference in the angles gave him a circumference estimate (based on a Greek measure called the *stadium*, which is about one-sixth of a kilometre) for the earth of 46,250 km (28,740 miles). Since current estimates put the figure at 40,000 km (24,860 miles), Eratosthenes' estimate was extraordinarily accurate.

The Shape of the World

Little improvement was made on Eratosthenes' measurements for another 1800 years. There was an apparent loss of interest in Greek concepts, and their theories about the earth were temporarily replaced, in medieval Europe, by a theological view of the world. Figure 21.3 shows a typical *T-O map* of that period, with the Holy City of Jerusalem occupying the centre of a disc-shaped world.

Typical of the medieval world maps is the Hereford Mappa Mundi which can be seen today in the cathedral library of this quiet English city (see Figure 21.4). It is drawn on a single skin of vellum, a type of leather, in black ink with the gold and red additions now very faded. It measures 112 cm (44 in) from top to bottom and 137 cm (54 in) across. The map was probably drawn in the last fifteen years of the thirteenth century and it reflects the Christian views of that period.

The T-shape is said to represent the crucifixion, the three continents the descent of man from the three sons of Noah. East is at the top with the holy

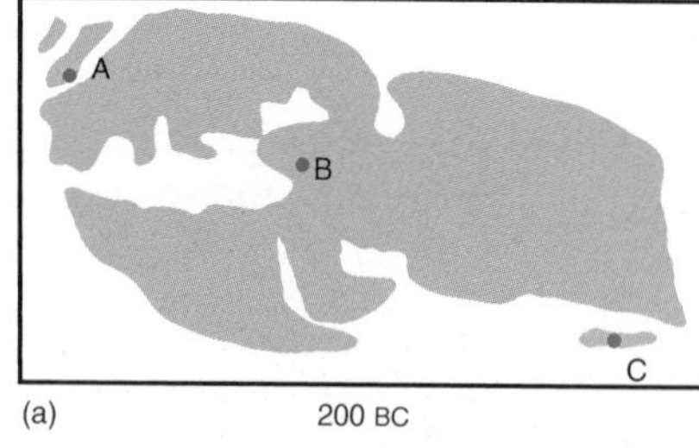

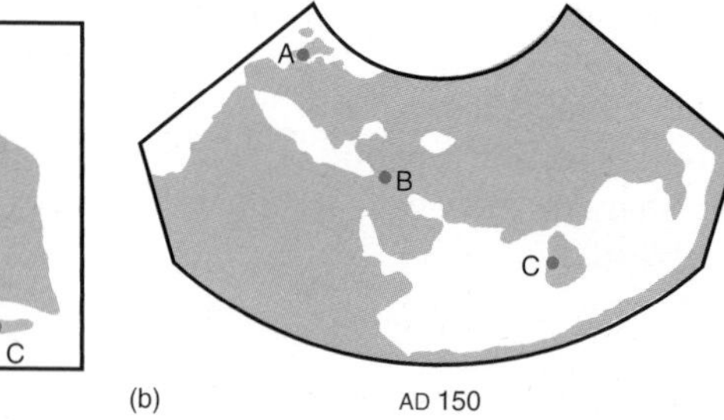

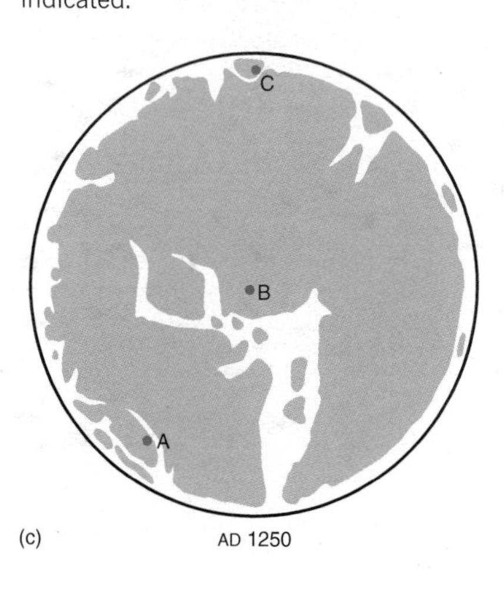

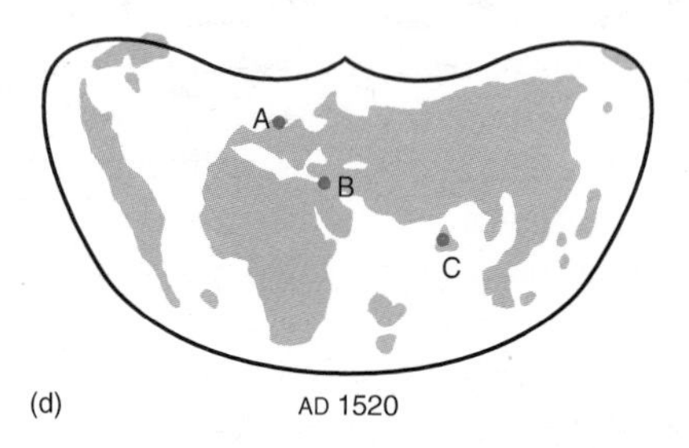

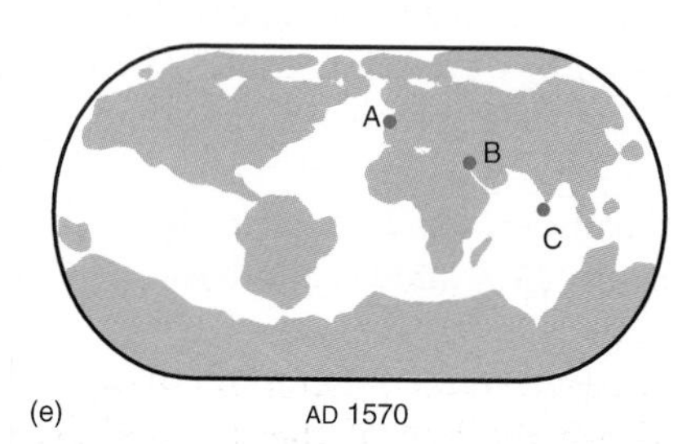

Figure 21.3 Changing views of the world Figures (a) to (e) show representative maps from nearly two thousand years of global exploration. On each map the sites of London (A), Jerusalem (B), and Colombo (C) are indicated.

(a)

Figure 21.4 Medieval world map Mappa Mundi (a) The thirteenth-century world map which served as an altarpiece in Hereford Cathedral, England. It shows the world as an ocean-bounded circular disk with Europe to the lower left, Africa to the lower right, and Asia occupying the upper half. The map is as much a theological as a cartographic statement so the world has the holy city of Jerusalem at its centre and the Garden of Eden at the top. Details of (b) London and Oxford, (c) Rome, and (d) Jerusalem (centre of the map) shown below.

[Source: By permission of Dean and Chapter of Hereford Cathedral Mappa Mundi Trust.].

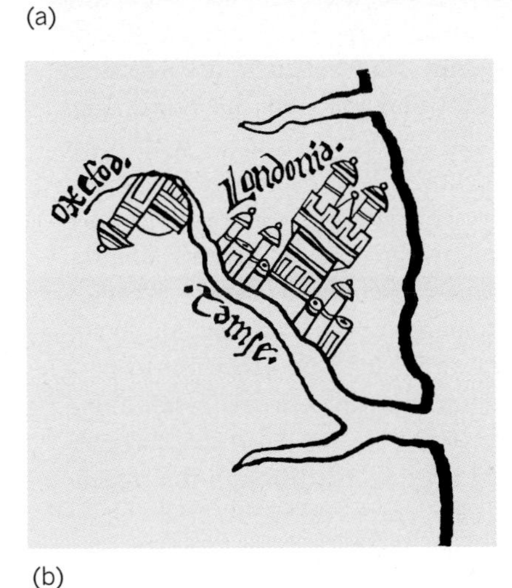

(b)

(c)

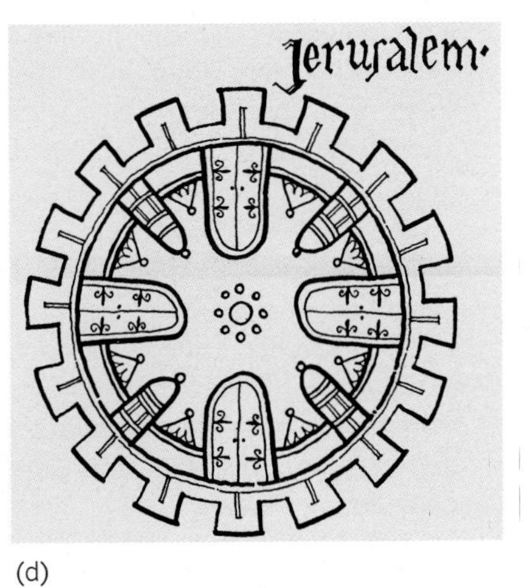

(d)

city of Jerusalem at the centre of the map. The world is enclosed within a circle, divided by the T-shape which separates the three continents (the Americas and Australasia were unknown at this time). Asia is at the top above the bar of the T, Europe to the left and Africa to the right. The

Mediterranean sea forms the vertical bar of the T. As the enlargement of part of the British Isles (b) shows, positions of rivers and towns are clearly shown. Cities such as London, Rome, and Jerusalem (c) are shown by a diagrammatic view. The more distant lands of Africa and Asia are less accurately shown with the assumed position of mountains, rivers, and lakes boldly shown. Cartoons of mythical animals are also detailed.

World Exploration

With the Renaissance and the wave of European overseas voyages by the Portuguese and Spanish (from the fifteenth century) and by the Dutch, French, and English (from seventeenth century), interest in the precise form of the planet was rekindled. Further measurements of the earth were made in the early seventeenth century when astronomer Willebrord Snell's recalculations at Leiden University triggered a succession of increasingly accurate measurements of the distance over ever-longer sections of the earth's curved surface. From the results stemmed a bitter debate; gross variations in the estimated circumference indicated that the earth could not be regarded simply as a regular sphere. Scholarly opinion divided over whether the earth was flattened at the poles or elongated like a football. By the 1730s, evidence from geodetic expeditions was overwhelmingly in favour of the former view. By the end of the century, most of the missing continents including both Australia and Antarctica were being sketched into place. (See Box 21.A on 'James Cook and the mapping of the Pacific'.)

Box 21.A James Cook and the Mapping of the Pacific

James Cook (born Yorkshire, 1728; died Hawaii, 1779) was a British naval captain, navigator, and explorer, said by his biographer to have 'peacefully changed the map of the world more than any other single man in history'. He is best known for his three major expeditions to the Pacific Ocean: these started in 1768, 1772, and 1776 and each lasted for three years. In the course of these he ranged from the Antarctic ice fields in the south to the Bering Strait in the north, and from the coasts of North America in the east to Australia and New Zealand in the west.

Cook's Pacific voyages were preceded by a survey of the St. Lawrence River in eastern Canada and the coast of Newfoundland in 1759 and 1763–67. This work drew him to the attention of the British government and he was given command of HMS *Endeavour* for the first of the Pacific expeditions. The rugged ship, based on coal-carrying vessels at his native Yorkshire port, was only 30 m (98 ft) in length. This left England in August 1768 and rounded Cape Horn into the Pacific Ocean. He reached New Zealand in 1769 and mapped the North and South Islands (see map) with great accuracy. If you compare this with a present-day map, there are only two major errors: in the South Island the Banks Peninsula was wrongly thought to be an island, while Stewart Island was thought to be a peninsula. Cook went to sail along the east coast of Australia, navigating northwards through the treacherous coral reefs on the Great Barrier Reef. Eastern Australia was added to the world map and Cook claimed a large area for Britain under the name New South Wales.

Cook's second voyage in 1772

Box continued

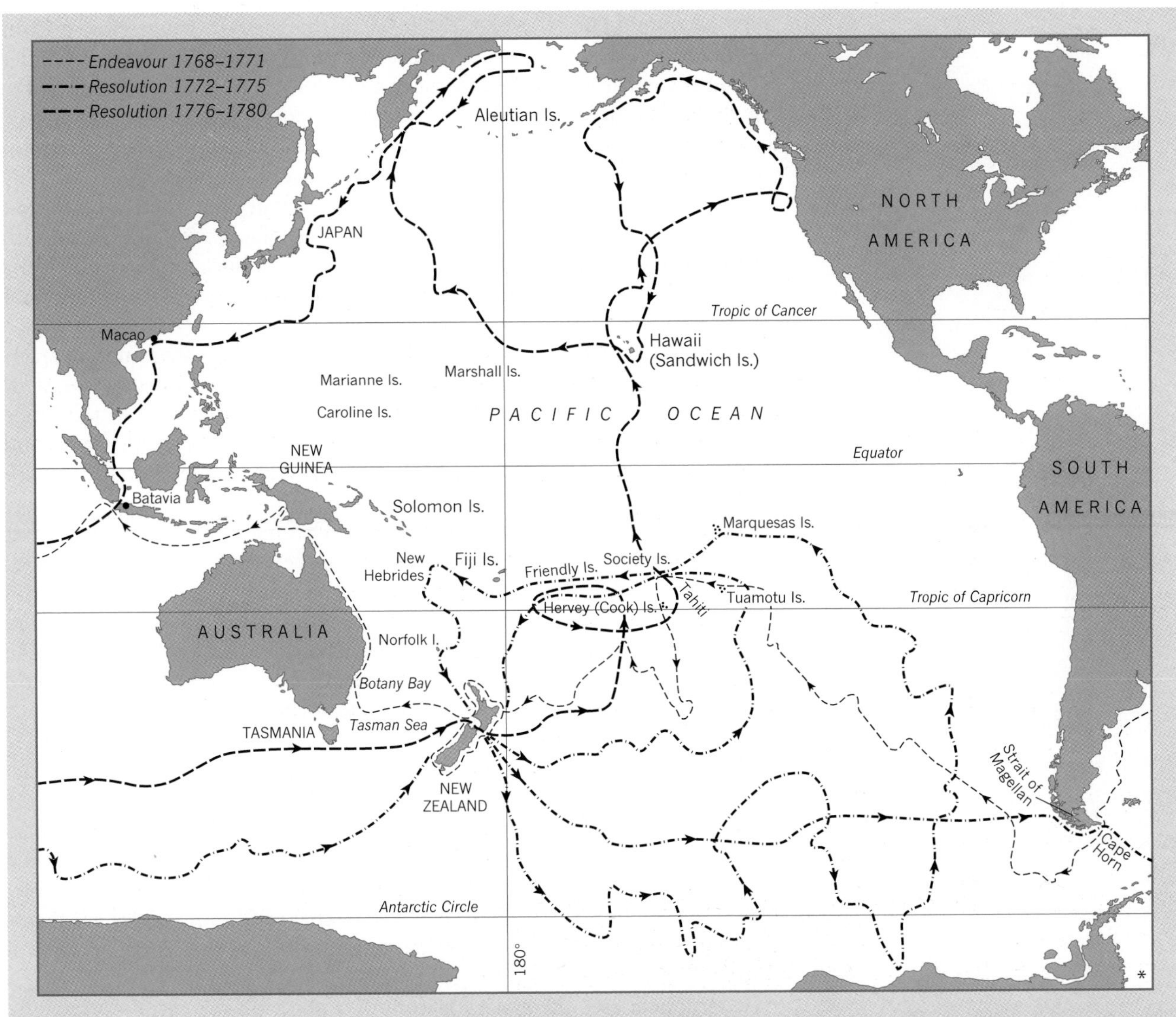

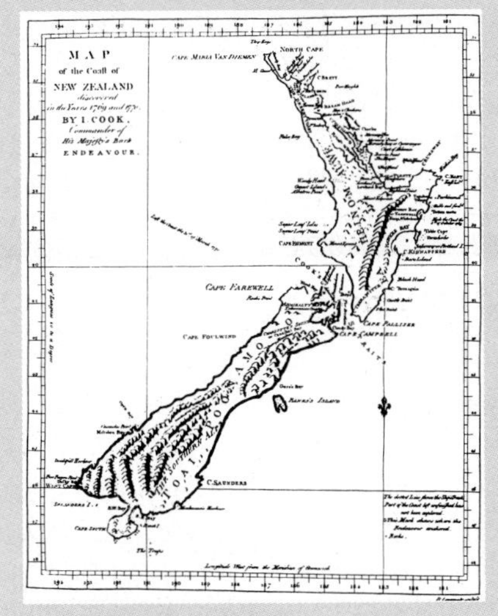

investigated whether or not a continent extended from the South Pacific to the South Pole. Geographers had speculated such a major continent should exist to balance the large land masses known to be in the northern hemisphere. Although he found small islands such as South Georgia, he failed to find a major continent but speculated that a smaller frozen one lay further south in the Antarctic.

His third voyage explored the northern Pacific and the Bering Strait, confirming the absence of a direct water route from the Pacific Ocean to Hudson Bay. He also discovered and mapped the Sandwich Islands (now Hawaii) where he was killed by the islanders in February 1779.

James Cook added precision to the geography of the Pacific Ocean, a huge water body covering half the Earth's surface. Given the improvements in navigation instruments (see Figure 21.1), he was able to set new standards of accuracy in discovery, navigation, and cartography. He also brought innovations in the health-care of ship's crews, and in relations with natives both friendly and hostile.

[Source: The portrait of Cook and maps of his voyages are taken from the definitive account of his life, J.C. Beaglehole, *The Life of Captain James Cook* (Hackluyt Society, London, 1974).]

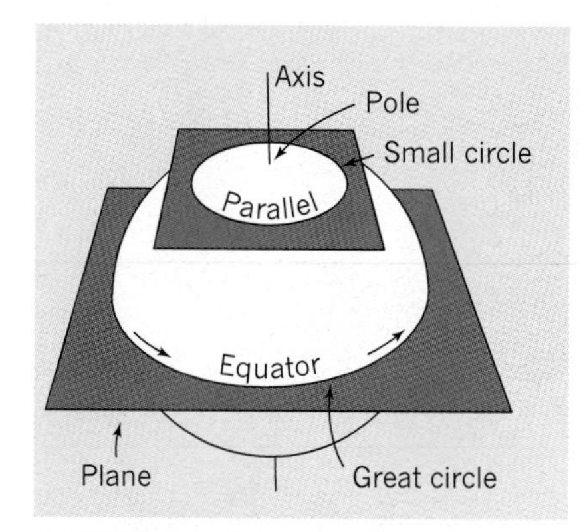

(a) Parallels

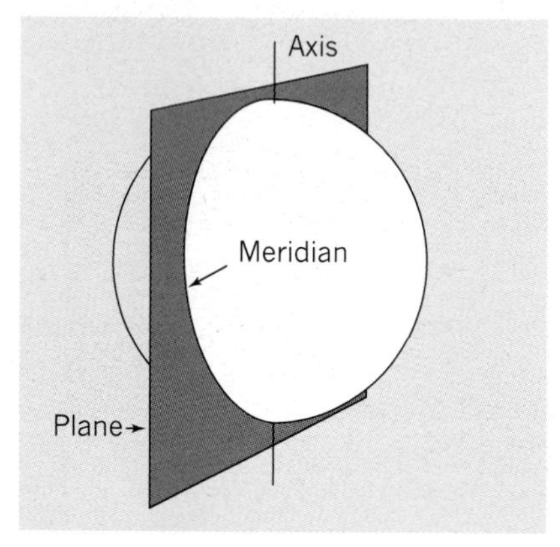

(b) Meridians

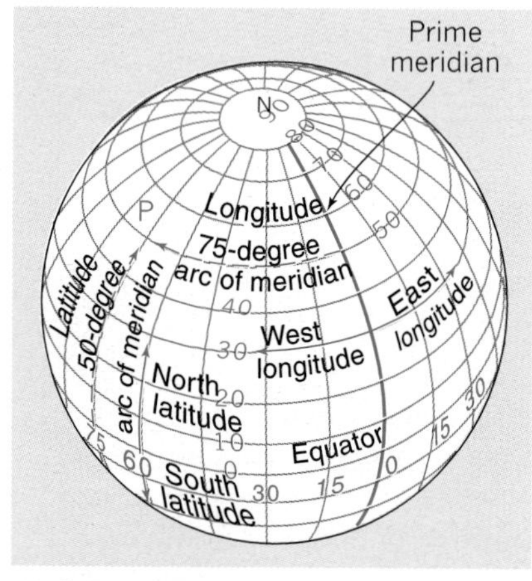

(c) Geographic grid

Figure 21.5 The earth's spherical coordinate system (a) Parallels of latitude lie in planes oriented at right angles to the earth's axis of rotation. (b) Meridians of longitude lie in planes passing through the earth's axis. (c) Combined parallels and meridians form a spherical geographic grid.

To accommodate the spherical form of the earth, a geographic grid of meridians and parallels was adopted. *Meridians* are imaginary half-circles around the earth which pass through a given location and terminate at the North and South Poles. Thus they form true north–south lines joining two fixed points of reference, the geographic *poles* where the earth's axis of rotation intersects the planet's spherical surface. Each meridian is actually a half circle (an arc of 180°). *Parallels* are imaginary lines on the earth that are parallel to the equator and pass through all places the same distance to the north or south of it. They form true east–west lines and are full circles (arcs of 360°). Parallels intersect meridians at right angles (see Figure 21.5 for the earth's *spherical coordinate* system).

Coordinate positions on the meridian–parallel grid are measured in terms of longitude and latitude. The *longitude* of a place is its distance east or west from a prime meridian measured in degrees. It describes the angle between two planes that intersect each other along the earth's axis and intersect the surface of the earth along, respectively, the *prime meridian* and the meridian of the place to be located. Prime meridians have a value of 0°. Angles are measured east and west, reaching a maximum of 180° at the meridian opposite the prime meridian. The prime meridian in worldwide use is the Greenwich meridian (the former site of the Royal Observatory near London, England), but any meridian can be chosen. Italian topographic series use the meridian that passes through Monte Mario near Rome (12°27′E of Greenwich), and some other European countries use the meridian that passes through Ferro in the Canary Islands (17°14′W of Greenwich).

The *latitude* of a place is its distance north or south of the equator measured in degrees. It is given by the angle between the plane of the equator and the surface of a cone that has its apex at the earth's centre and intersects the surface of the sphere along a given parallel. Unlike the prime meridian, the equator is a uniquely determined and natural reference line, rather than an arbitrary one. It also has a value of 0°, and angles are measured north and south, reaching a maximum of 90° at the North and South Poles.

Geographers give the exact location of a point on the earth's surface by specifying its longitude and latitude. In Figure 21.5, the point P has the coordinates 50°N, 75°W.

Degrees can be subdivided into 60 minutes of arc (60′), and minutes into 60 seconds of arc (60″). For simplicity, however, latitude and longitude are generally expressed in decimal parts of a degree. Thus, 77°03′41″ is rewritten in decimal terms as 77.0614°. Modern worldwide reference systems are sophisticated enough to successfully combine the advantages of both the Cartesian and the spherical systems of reference.

Determining Latitude and Longitude

Developing a spherical reference system for the earth is one thing, and determining where a location actually is within that system is another. Finding latitude involves simply measuring the angular elevation of the sun or a star above the horizon, and it was first done with some accuracy more than 1000 years ago. Some of the instruments used in its measurement are shown in Figure 21.6.

The determination of longitude is more difficult. The first requirement is a reliable clock, so that the local noon-time (when the sun is at its zenith)

(a)

(b)

Figure 21.6 Evolution of navigation and surveying instruments (a) Arab-made brass astrolabe from the fourteenth century. Circles marked with angular measurements. By aligning the astrolabe with the horizon, the angular heights of any stars in the sky could be measured. (b) Sextants are used to measure the position of heavenly bodies (sun and stars). Invented around 1730 by the English mathematician John Hadley and the American inventor Thomas Godfrey.

[Source: (a) David Parker/Science Photo Library. (b) Ed Young/Science Photo Library.]

can be compared to noon-time at the prime meridian. The difference between the two gives us our correct east–west position. But the clock must be very accurate. One hour on the clock is recording the time during which the noon-time position of the sun has shifted 1/24th part of its east–west track around the earth. At the *equator* this would mean that an error in time of only 22 seconds on the clock would cause the navigator to be 10 km (6.2 miles) in error in judging his east–west position. It was not until 1761 that this problem was solved when John Harrison's marine chronometer (see Figure 21.1), made in response to a prize offered by Britain's Board of Longitude, provided the breakthrough that allowed position to be determined accurately.

Global Positioning Systems (GPS)

The job of fixing positions with respect to latitude and longitude has become much less tedious in the present electronic era. Figure 21.7 presents one of the electronic systems currently being developed for this purpose. This *Omega system* depends on a pattern of only eight radio stations, the latitude and longitude of which can be given precisely. Each station transmits extremely long radio waves – with wavelengths up to 5 km (3.1 miles) – on exactly the same frequency in a predetermined sequence. Because these waves can be distinguished by the length of their pulse and the sequence in which they are transmitted, ships and aircraft can plot their progress from a known starting position by comparing signals from two stations and using those from others as a check. When it is completed, positional fixes

(a)

(b)

Figure 21.7 Electronics and navigation (a) Global positioning system (GPS). Hand-held satellite navigation receiver uses signals sent by a network of 21 NAVSTAR GPS satellites. Receiver reads these data from up to four satellites and calculates its position in the world with a very high accuracy (within 100 metres). (b) Use of laser beams for the accurate measurement of distances. Geodolite calculates distance to within 1 mm.

[Sources: (a) David Parker/Science Photo Library. (b) National Geographic Society Photographic Laboratory/Bob Sacha.]

based on the Omega system should be accurate to about 1 km (0.6 miles) anywhere on the global surface. Navigational satellites giving more accurate fixes to a few hundred metres may be used in the future.

The determination of positions on the earth's surface has two interrelated phases. The first consists of fixing primary positions with respect to their horizontal location on the geographers' spherical grid system (and their vertical position with respect to mean sea level). Once a network of primary positions is established, a swarm of secondary positions can be fixed from the horizontal and vertical coordinates of the primary position. The trend in current survey work has been to reduce the need for primary survey positions by extending the area of secondary fixes that can be calculated from the primary points. The Omega system represents an extreme reduction of the number of primary positions needed for global coverage.

21.2 Location on the Map

We review here the ways in which locations may be specified on maps. We begin with place-name evidence and then look at two types of graticule: those suitable for local and global maps.

Place-names

The simplest way of specifying a location somewhere on the earth's surface is to give its name. In everyday language we almost always talk about places in terms of the names attached to them: Chicago, Kansas, or Tibet. This way of specifying locations is termed ***nominal specification***. We used it in Chapter 1 to identify particular beaches, and in the preceding section to describe the cities involved in Eratosthenes' calculations. It can describe places that vary in size (from Mount Vernon to Afro-Asia), it can provide distinctive and memorable place-names (such as the Donner Pass in northern California or Xochimilco in central Mexico), and it can be made as complex or as simple as necessary. The typical address on an envelope exemplifies a hierarchical method of specifying location in which areas of decreasing size are nested within each other. Our resistance to the replacement of familiar addresses by postcodes indicates the innate attractions of a nominal system for everyday use.

Another advantage of *place-names* is that they may contain a considerable amount of historical or environmental information. Places founded by particular groups tend to have particular types of names. Witness, for example, the many Spanish place-names in the American southwest and Texas. These names give us clues to the previous extent of a population. When other information is lacking, such clues may be vital. In a similar manner, names may indicate the kind of environment settlers found on first moving into an area. In southern Brazil, names containing the word 'pine' help to depict the original distribution of a type of vegetation now much reduced by deforestation and the clearing of land for agriculture. Work by Zelinsky on the northeastern United States has shown a rich variety in the distribution of stream names. The terms 'creek', 'brook', 'run', and so on each have distinct regional clusters related to the settlement history of the area (Figure 21.8).

If place-names are so attractive and useful, why don't geographers use them all the time? Unfortunately, the disadvantages of nominal specification for scientific purposes outweigh the advantages. First, this kind of language is non-unique because several different locations can have the same name. There are scores of places dedicated to St Paul (i.e. São Paulo) in

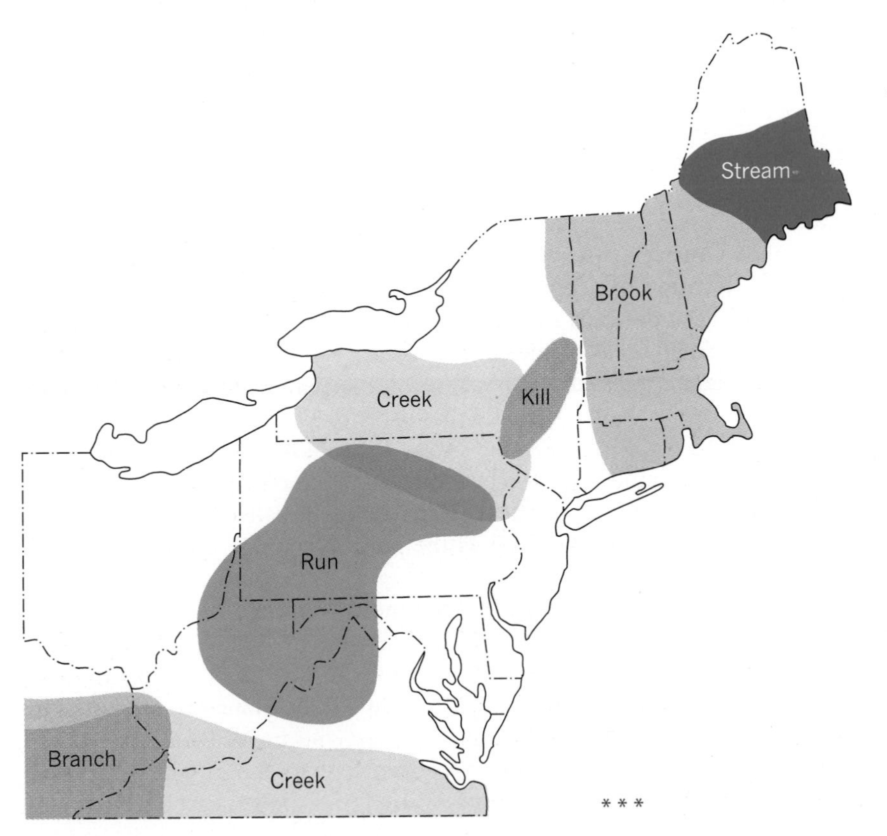

Figure 21.8 Place-names as historical evidence Stream names in the northeastern United States. The approximate areas within which particular terms are dominant are indicated on the map. 'Brook', 'run', and 'branch' are very common terms and strongly associated with particular locations. 'Creek' and 'stream' occur less frequently but are also strongly associated with distinct localities. 'Kill' is a very uncommon term in the northeast as a whole but very prevalent in areas of early Dutch settlement in New York State.

[Source: W. Zelinsky, *Annals of the Association of American Geographers*, **45** (1955), Fig. 20, p. 323.]

* * *

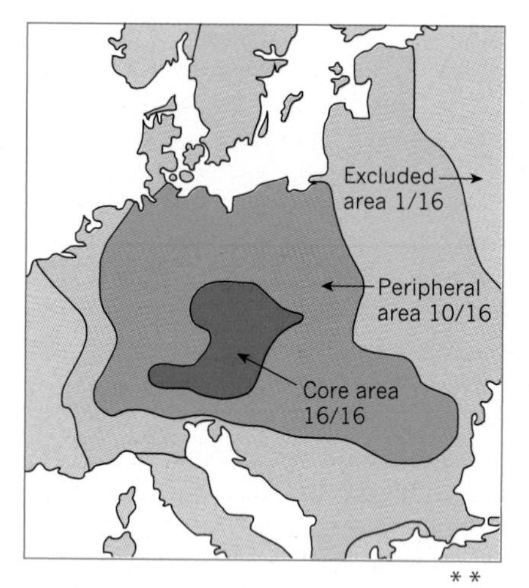

Figure 21.9 Ambiguities in regional names The degree of coincidence among sixteen geographic definitions of the terms 'Central Europe', 'Mitteleuropa', or 'Europe Centrale.' Note the small area of agreement (which all sixteen regarded as the core of central Europe) and the very wide area of disagreement.

[Source: K. Sinnhuber, *Transactions of the Institute for British Geographers No. 20* (1956), Fig. 2, p. 19.]

Brazil, and names such as Newport or New Town occur in literally hundreds of forms in the English-speaking world. Second, nominal terms are unstable. The same location can have different names at different times (such as Breuckelen and Brooklyn), and in different languages. (The German Königsberg is the Russian Kaliningrad.) To correct this problem, many countries have established committees to standardize geographic names throughout the world. In 1890 the United States set up a Board on Geographical Names that published authoritative lists for individual countries. Some idea of the immensity of this task can be gained from the size of standard geographic gazetteers. The Merriam–Webster *Geographic Dictionary* contains approximately 40,000 names, and the *Times Atlas of the World* has 315,000 names. The current London telephone directory, which is actually a list of the 'locations' of subscribers within the city, contains about half a million names.

We can only guess at the total number of world place-names, which probably lies in the trillions. Such a vast array of names creates formidable problems for those concerned with information storage and retrieval, problems that are complicated by the duplication of place-names. Various ways of dealing with these problems are being investigated by experimenters at Oxford University, who are developing systems for the coding, computer storage, and rapid retrieval of information from geographic gazetteers.

Another serious disadvantage of nominal specification is its spatial imprecision. The boundaries of an area may change while the name remains the same. Turn to a historical atlas and compare the Poland of 1930 with the Poland of today. The imprecision on nominal terms for geographic areas is illustrated in Figure 21.9, which shows a series 'Mitteleuropa' (Middle Europe) over a 40 year period. Areas included by at least one geographer extend well outside the limits set by other geographers, and the only part of Europe not included in Mitteleuropa by any of the definitions is the Iberian Peninsula. The only area not in dispute, in fact, is the small core area containing Austria and Bohemia–Moravia.

Local Grids

If we cannot use place-names to specify location, how do we proceed? The geographer's answer is to use *reference grids* – mathematical devices that specify the location of a point in relation to a system of coordinates. The type of grid used depends on whether the area involved is small or large. We can treat small areas of the earth's surface as if they were flat planes; in the case of larger areas the curvature of the planet must be taken into account. Reference grids for small areas use a *Cartesian coordinate system.* The location of a place is specified by its distance from two reference lines that intersect at right angles. The horizontal reference line is called the abscissa or x axis, and the vertical reference line is called the ordinate or y axis. Intersection of the two axes is the point of origin of the system (see Figure 21.10(a)).

Cartesian coordinate systems are commonly adopted by national mapping agencies, and Britain's National Grid system (Figure 21.10(c)) is a typical example. Its origin lies southwest of the Isles of Scilly; its abscissa runs east and its ordinate north. All geographic locations are measured in kilometres, or subdivisions of kilometres, east and north from these reference lines. In practice, references are given first to the 100 km (62 mile) squares into which the grid is divided, second to distances east of the origin,

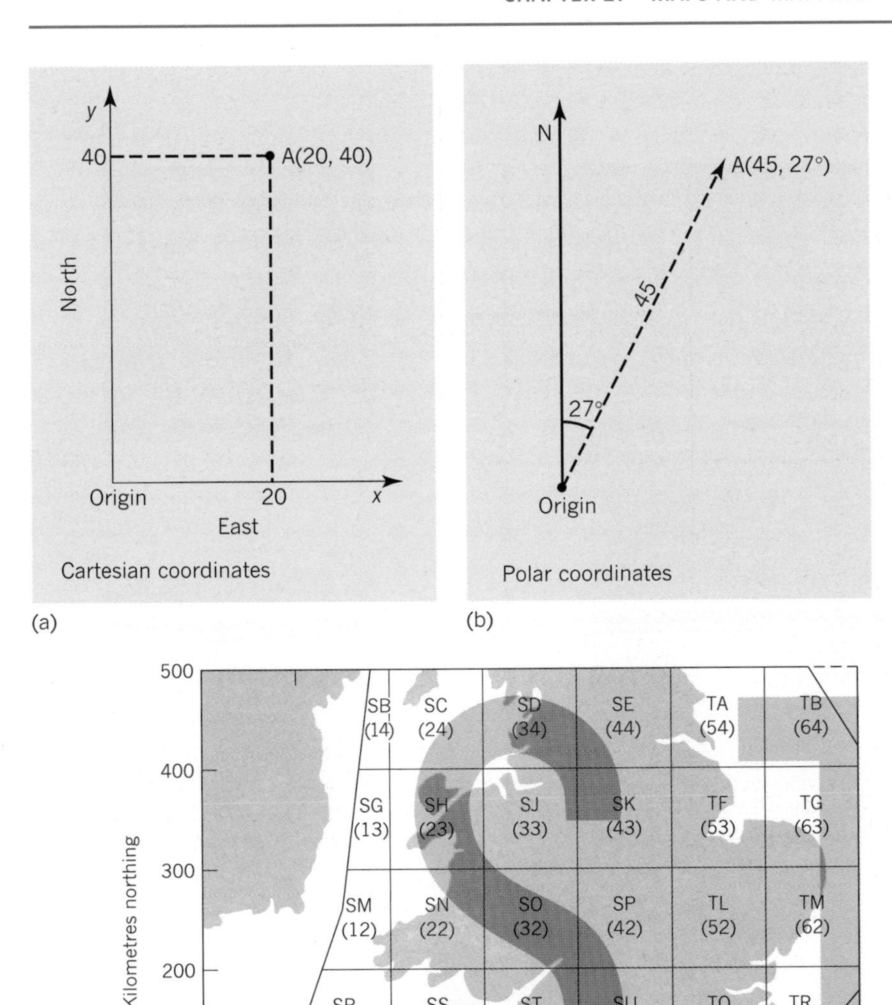

Figure 21.10 Reference grids (a) and (b) The location of A with respect to the origin in alternative spatial reference systems. Note that in the case of Cartesian coordinates the distance east is given first so that the location of A is 20,40 and *not* 40,20. In practice the comma is usually omitted in giving a locational reference so that 20,40 becomes simply 2040. (c) The 100 km squares of the British National Grid. Each grid square has a distinguishing number or pair of letters. Numbers describe the distance east (the 'easting') and distance north (the 'northing') of the southwest corner of each square when measured in hundreds of kilometres from the false origin (e.g. the southwest corner of grid square 31 is 300 km east and 100 km north of this origin). The 'false origin' is an arbitrary point out at sea chosen for the British map users; it stands in contrast to 'true origins' such as the poles or the equator. The letters on the map are part of a wider reference system which ties the British National Grid to other international grids.

[Source: Reproduced from *The Projection for Ordnance Survey Maps and Plans and the National Reference System*, with the sanction of the Controller of Her Majesty's Stationery Office, Crown copyright reserved.]

* * *

and then to distances north of it. Thus the precise reference to the United States Embassy in Grosvenor Square, London, is 51 (TQ) 283808; that is, it lies exactly 528.3 km (330.2 miles) east and 180.8 km (112.3 miles) north of the arbitrary point.

A modified form of the Cartesian system is the method of ***township and range***, which originated in a land act conceived by Thomas Jefferson in 1784. This rectilinear grid is the basis of surveys in most of the United States west of the Appalachians and, in a modified form, much of Canada. It divides land into 6-mile-square (15.5 km^2) townships around a Cartesian axis oriented to the north (see Figure 21.11). Each township is subdivided into 1-mile-square sections (numbered 1 to 36 in zigzag fashion). Each section can be further subdivided into quarter-sections and then into 40-acre (16 ha) fields. Even today the imprint of the township and range system on the agricultural landscape of much of the United States is evident.

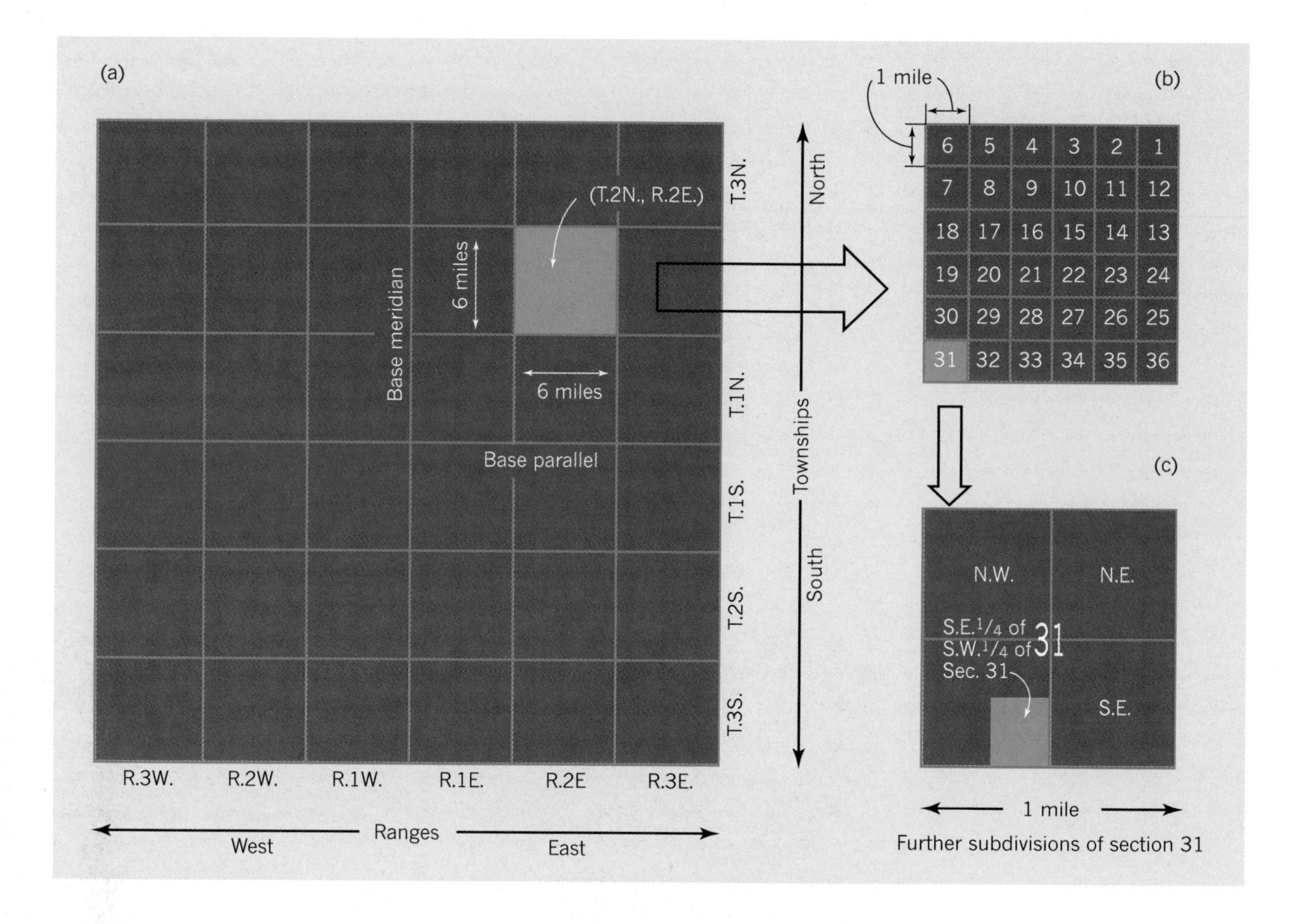

Figure 21.11 A hierarchic reference grid In much of the central and western United States, the location of parcels of land was specified on a rectilinear grid divisible into ranges and townships. Each square of the range and township grid was further divisible into sections and subsections. The 'base meridian' and 'base parallel' shown in (a) tie this grid system into the worldwide graticules of latitude and longitude described in Figure 21.5. The impact of this reference system is still clearly seen in the pattern of roads and fields: see, for example, the Illinois landscape shown in Figure 11.9 earlier in this volume.

[Source: G.C. Dickinson, *Maps and Air Photographs* (Arnold, London, 1969), Fig. 46, p. 124.]

A less common type of locational reference grid is based on a *polar coordinate system* which specifies the location of a place in terms of an angle, or *azimuth*, and its distance from an origin. In mathematical work, angles are specified counterclockwise from a horizontal; in geographic work, angles are measured clockwise from the north, which is designated 0° or 360°. Thus an azimuth of 225° is a line running southwest from the origin. The polar coordinate system is generally used only when the relation of locations to a single origin is important – for instance, in studying distances a population has migrated from a given place of origin.

World Grids

So long as the environment under study is a small one, geographers can use very simple devices for recording location. Once we begin to study very large areas or attempt worldwide studies, then a more sophisticated way of specifying location is needed.

Location on a Sphere

The curvature of any portion of the earth's surface depends, of course, on its size. As the area being mapped increases, so does the deviation from a true north–south direction of parallel reference lines on a flat surface. Thus the extension of the range and township system used in the United States as people moved west brought about an increasing divergence of range lines from their true direction. To minimize this divergence, new grids had to be established for each state.

Once we have established a system of parallels and meridians and have located places in terms of that system, we have the basic ingredients of a world map. To make one we need only to specify a scale and construct a model globe. If we select a scale of 1 : 10,000,000, we can represent the real globe by a sphere 127 cm (50 in) in diameter. Making globes like this is not a difficult task; indeed, people have been constructing them for centuries. Martin Behaim made one of the first terrestrial globes at Nuremberg in 1492, and they are still popular for many purposes.

Despite their attractiveness, however, globes are rarely used by geographers in their research. Most globes are less than 1 m (3.3 ft) in diameter and are too small to be of any real value. There are, of course, a few large globes such as the 39 m (128 ft) Langlois globe in France, but larger globes are expensive and unwieldy to work with. The obvious way around the problem is to substitute flat map sheets for the bulky spheres. How to get a map off the surface of a globe and onto flat paper without tearing or otherwise distorting it is the objective of ***map projections***.

Map Projections: A Round World on Flat Paper

Some projections are very simple to construct. If we place a sheet of flat paper against a transparent globe and shine a light from the opposite side of the globe we can 'project' an image of the earth onto the paper. Records of this type of map projection go back to the second century AD). (See Box 21.B on 'Constructing a polar projection'.)

But the early geographers were to find that there were no easy solutions: such simple maps had very few useful properties. The puzzling problem of mapping the round world on flat paper attracted some of the best mathematical minds of sixteenth- to nineteenth-century Europe, figures such as Gerardus Mercator, Hans Mollweide, and Johann Lambert. Their work is a study in the mathematics of compromise. They showed that, though it is impossible to reproduce, faithfully, in two dimensions all the characteristics of the three-dimensional earth, it is possible to reproduce some of them at the expense of others. We have simply to decide which properties are important and which we are prepared to sacrifice.

In practice, the main decision is whether to select a *conformal projection*, in which the shape of any small area is shown correctly, or an *equal-area projection*, in which a constant areal scale is preserved over the whole map. Although the word 'projection' is used here, in a strict sense it should be applied only to maps that are true geometric projections of a sphere. Through usage the word has come to represent any orderly system of parallels and meridians drawn on a map to represent the earth's geographic graticule.

If we lay out lines of latitude and longitude as a square Cartesian coordinate system, we greatly distort the area away from the equator (Figure 21.12). At 30°N the square unit on the grid is 1.16 times too large in comparison to the true area on the globe, at 60°N the areal exaggeration is twice as large, and at 89°N it is 59 times as large. But it is not only area that changes; the angular relations between points also are greatly altered.

In Figure 21.13 we summarize the characteristics of some of the more commonly encountered projections and comment on their uses. Map projection is a highly specialized branch of geography that is likely to attract

Box 21.B Constructing a Polar Projection

The problems of transforming the spherical earth onto a flat surface can be illustrated by the construction of one of the oldest projection systems, the stereographic. It was known to the Greek geographer, Ptolemy, in the second century AD. In the northern polar case the system consists simply of projecting the parallels and meridians from the South Pole onto a plane tangent to the North Pole.

Note that the meridians and parallels so constructed intersect at right angles, thereby showing the correct shape (*orthomorphism*) over small areas. Scale increases rapidly away from the pole and lower latitude areas are greatly distorted in area. However, any circle on the sphere remains a circle on the map, a property of use in geophysical investigations.

The stereographic projection can be constructed for a plan that may be tangent to the globe at any point on the earth's surface. Thus, in addition to the *polar* case there are *oblique* and *equatorial* cases where the grid of parallels and meridians is projected from a point at one end of a diameter onto plane tangent at the other. Other families of map grids (see Figure 21.13) project onto cones (*conical* projections) or cylinder (*cylindrical* projections) that touch or intersect the globe along circles more important are the non-perspective systems that mathematically transform global coordinates into map coordinations to optimize certain proportions of shape, area, or direction. See J.A. Steers *Introduction to the Study of Map Projections* (University of London Press, London, 1960), Chapter 1.

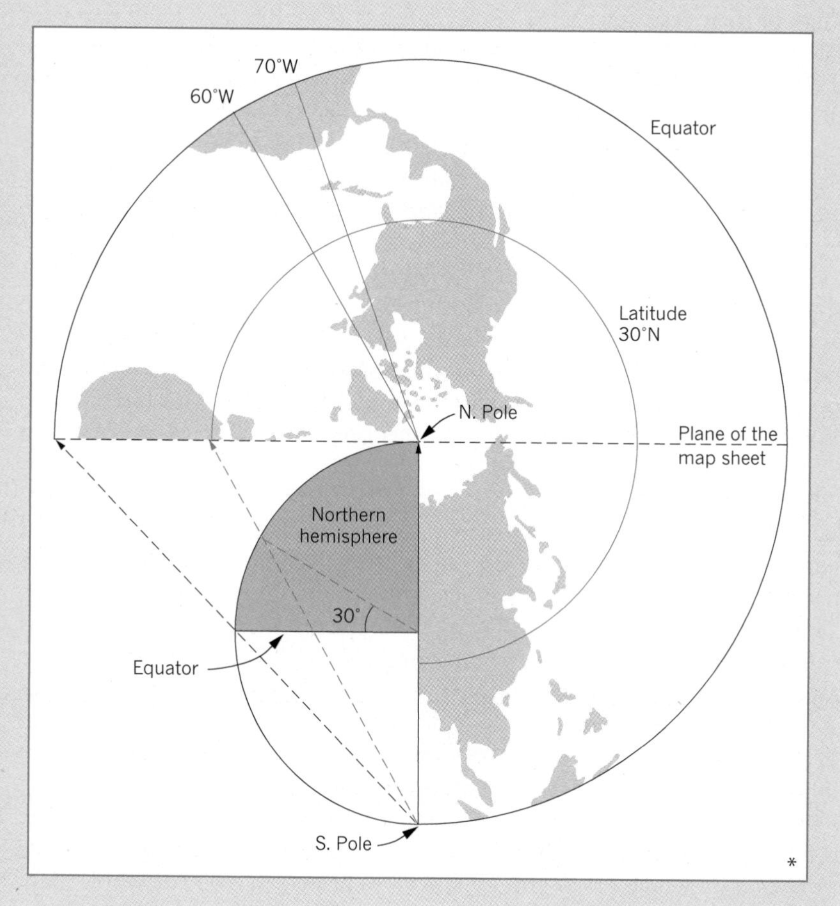

only those with mathematical minds. It is an evolving field, however. The first projections were invented over 2000 years ago; today new projection systems are still being invented to meet special needs such as satellite tracking.

Where maps of the whole world are concerned, the single sheet may be 'interrupted' in order to reduce overall shape distortion. If you turn back to Figure 2.15(a), you will see an example of this. *Interrupted projections* are usually drawn to place the breaks in the world's oceans, thereby enhancing the shape of the continents. Where the world map is to show features of the world's oceans, then the position is reversed and the breaks are placed in the land areas.

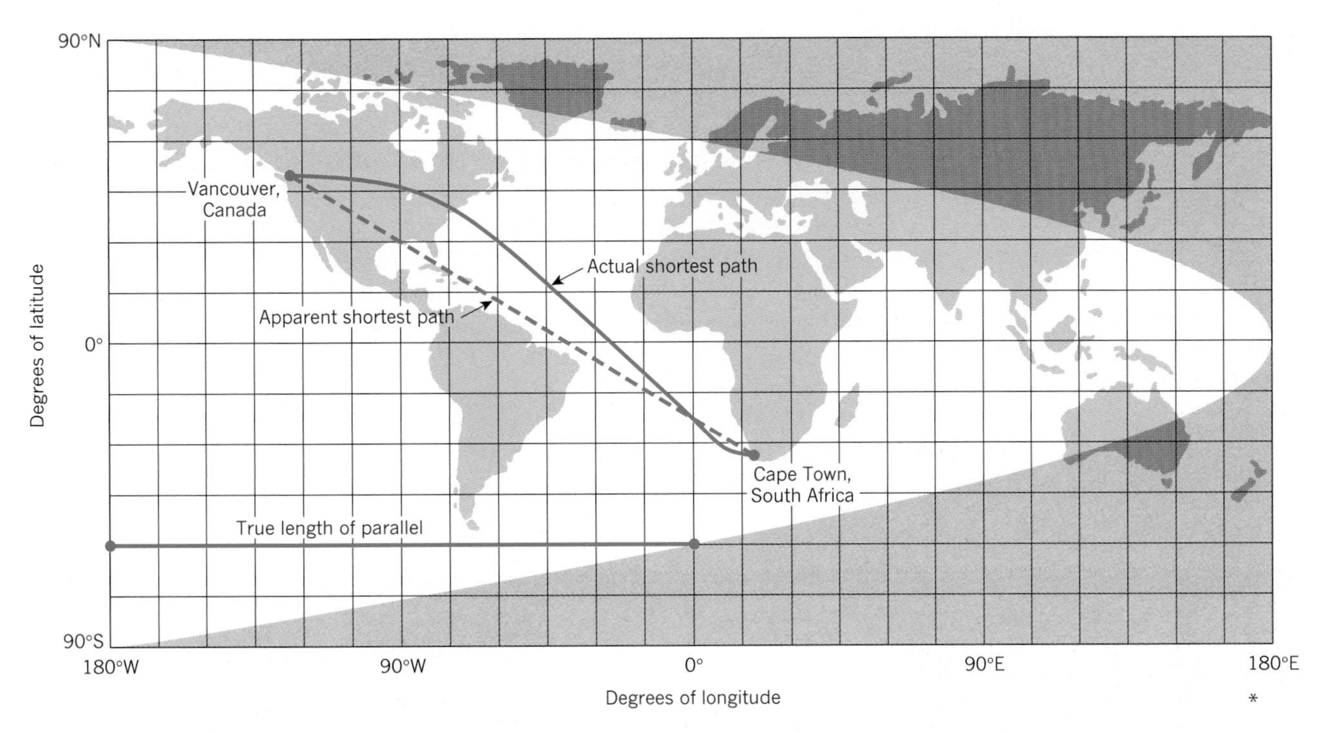

Figure 21.12 The earth's spherical coordinates plotted as a Cartesian grid A simple cylindrical map of the world plotting locations in terms of degrees of latitude and longitude greatly distorts distances along the parallels. Shading outlines the true lengths. Only shortest paths between places on the same meridians or parallels are straight lines.

Figure 21.13 Families of map projections The examples shown are illustrative only. For a full description of projections, their properties, construction, and use see 'One step further ...' at the end of the chapter.

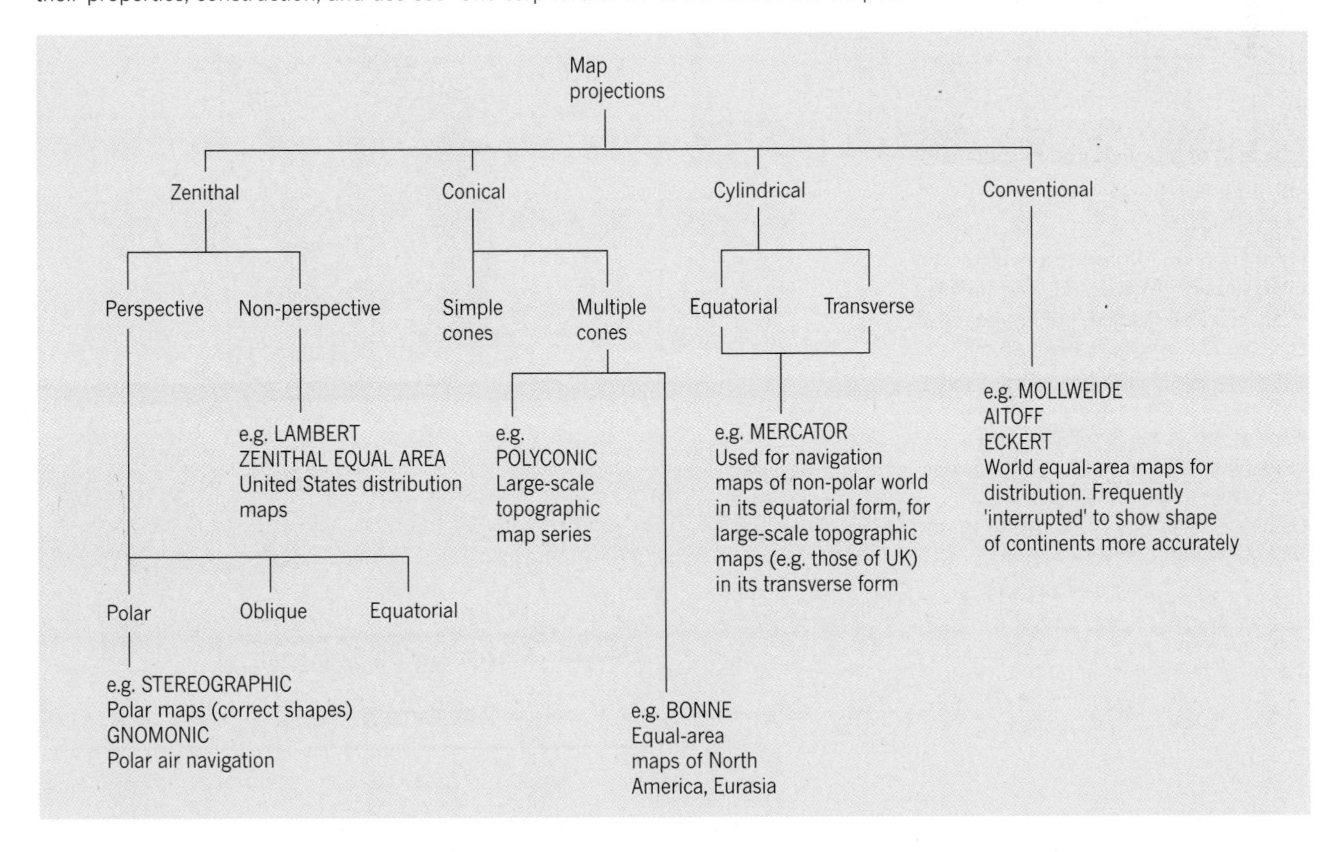

21.3 The Mapping Cycle: Encoding Maps

Making maps is a natural instinct. If we ask a visitor to come home to supper, we sketch a map on a piece of scrap paper to show where we live. With a home in a small village in England, I have to use not only the road numbers (A38, B3130) but also village names (Long Ashton and Chew Magna) and key pub names ('The Bear and Swan' and 'The Pelican'). Authors create imaginary worlds and show them in maps. Robert Louis Stevenson's *Treasure Island* and J.R.R. Tolkien's fictional Wilderland are shown in Figure 21.14.

In this section, we look at the ways that maps are created and the wide range of choices in map making. Different choices can lead the same information to be seen as very different maps.

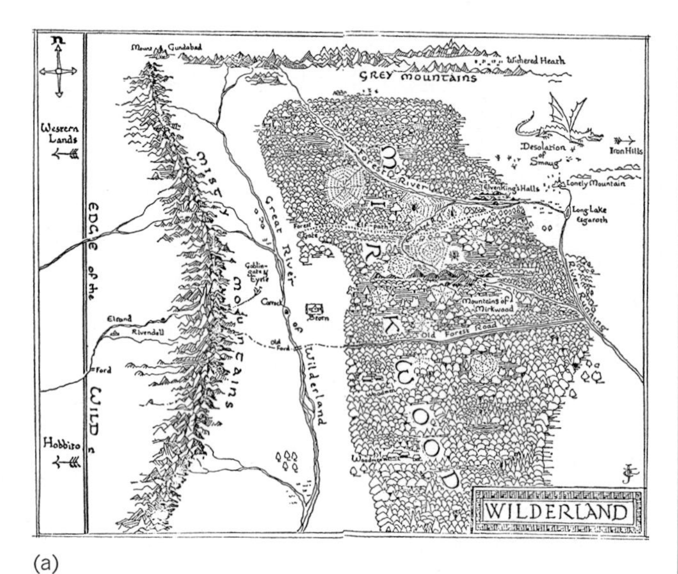

(a)

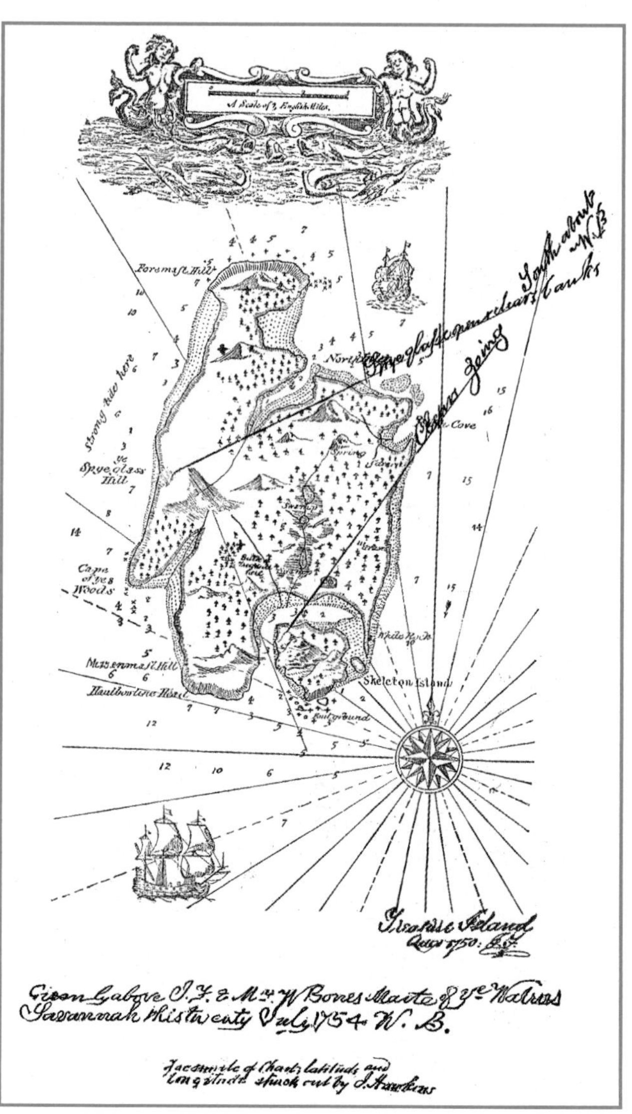

(b)

Figure 21.14 Imagined worlds (a) J.R.R. Tolkein's 'Wilderland', one segment of his fictional Middle Earth, the scene of *The Lord of the Rings*. Early map makers of the real world also used imagination and legend to fill in unexplored areas. Maps of Africa, the 'dark continent', contained mythical rivers and lakes as late as the middle of the nineteenth century. (b) *Treasure Island* map from the book by Robert Louis Stevenson.

[Source: J.R.R. Tolkein, *The Hobbit* (Allen & Unwin, London, 1937), end map. Copyright 1966 by J.R.R. Tolkein. Reprinted by permission of the publisher, Houghton Mifflin Co., Boston. (b) Mary Evans Picture Library.]

Mapping as a Cycle

Maps are one of the ways in which information about the earth's surface is prepared for human use. We are all familiar with the ways in which weather bureaux around the world take weather data and convert it into maps showing lines of equal pressure (isobars) with warm and cold fronts and actual and likely areas of rain. The map we see on our TV screens is the end of one stage in the process; the interpretation of the map provided by the weather forecaster starts the second phase which ends with our decision to delay a proposed journey beacuse of projected snow. We can therefore see the map as the mid-point in an information cycle which has a *map encoding* (map-producing) phase and a *map decoding* (map interpretation) side. The relationship is shown diagrammatically in Figure 21.15.

Note that map encoding involves finding answers to a series of questions.

1 What is the purpose of the map and who are likely to be the end users? A map in a general world atlas may have to serve many uses; a map that shows power lines for repair crews will have very specific requirements.
2 How large an area are we showing on the map? This will, in turn, control the scale of the map, and what – if it is a very large area – map projection we select.
3 What are the conditions under which the map will be used? Field use dictates portability and ruggedness; library and reference use can allow a different format.
4 What are the budget requirements? Issues such as whether colour mapping can be afforded or whether the map has to be in monochrome determine a whole series of design characteristics.
5 What degree of generalization can be introduced? This controls whether data have to be simplified or exaggerated to make a particular point.

Study of such a classic text as Arthur Robinson's *Elements in Cartography* will confirm that these five questions are simply some of the most important of the many scores of design questions which have to be faced in bringing maps into being. Space only allows two of these questions to be followed through to illustrate the encoding process: measurement scales and choice of grids.

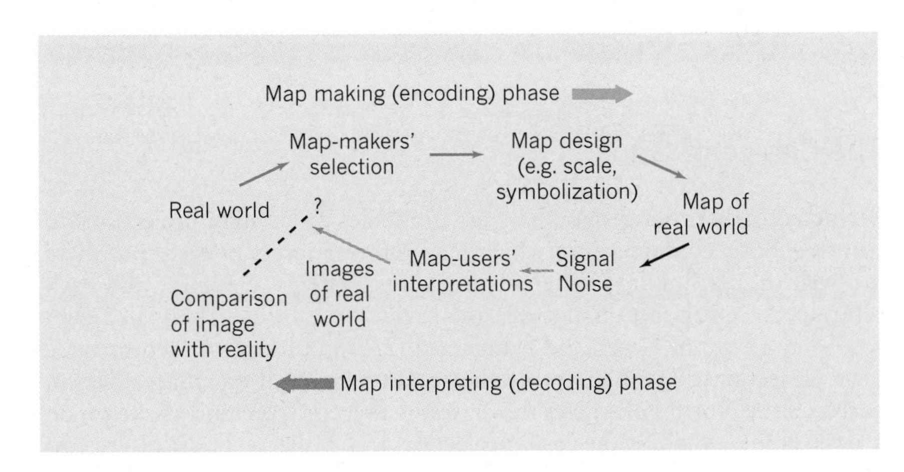

Figure 21.15 Mapping cycle Encoding (map making) and decoding (map interpreting) are two phases of the mapping cycle. Because of errors in both parts of the process the gap between the 'real world' and the map image of the 'real world' may be considerable.

(a) Nominal scale

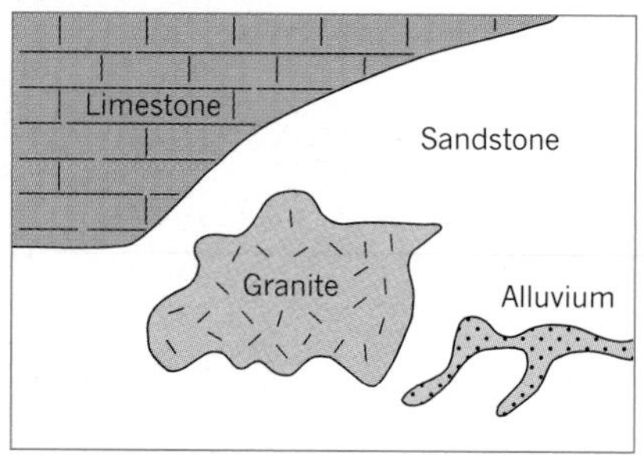

(b) Ordinal scale

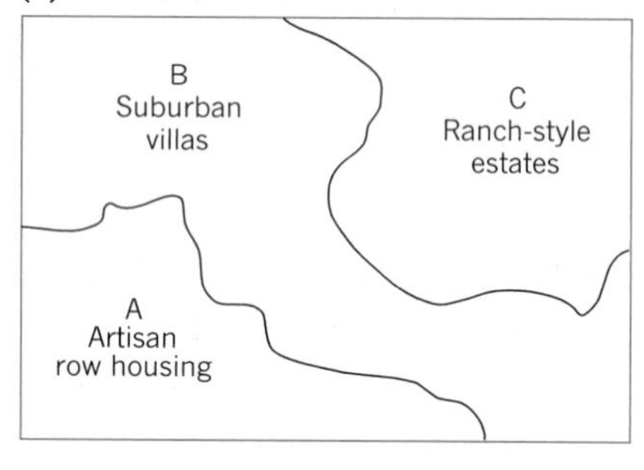

(c) Ratio scale

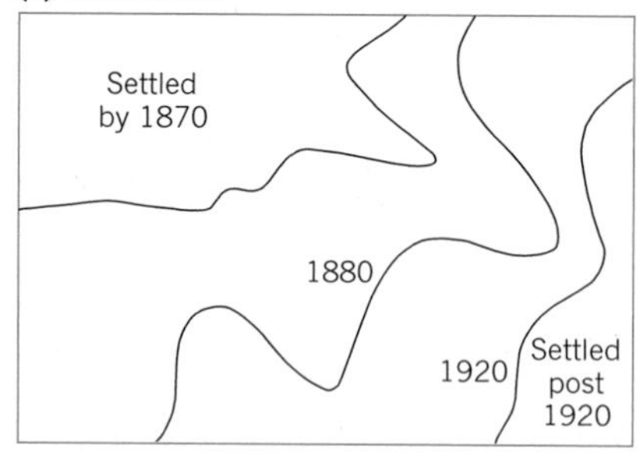

(d) Interval scale

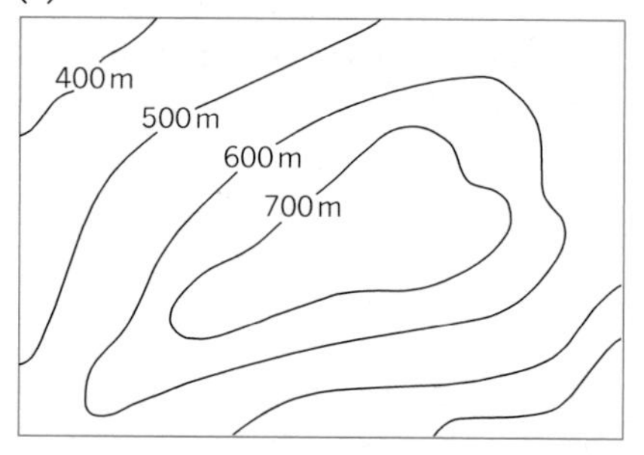

Figure 21.16 Measurement systems Examples of four types of measurement system (nominal, ordinal, interval, and ratio) as applied to map data. Although the latter two cases ((c) and (d)) appear similar, the scales differ in character. The year 1900 cannot be said to be ×1.9 greater than the year 1000 since the origin of the time-scale is arbitrary.

Measurement Scales

Cartographers distinguish four levels of measurement. In increasing order of descriptive richness these are: nominal, ordinal, interval, and ratio.

Nominal scales distinguish between features only on the basis of qualitative considerations. For example, a map of land use may distinguish between swamp, desert, and forest. This scales determines that the three are different but there is no other quantitative relationship.

Ordinal scales involve differentiation on the basis of rank. Thus an English map that shows roads may differentiate between motorways, A-roads, B-roads, unclassified roads, and unsurfaced tracks. Clearly this gives a nominal scale but goes beyond it: the roads are not only different but different in an ordered way (from motorway at the top down to unsurfaced track at the bottom).

Interval scales add still further information than nominal and ordinal scales by specifying a standard unit. Thus a map of temperatures may plot the location of places in terms of degrees Celsius. Assume we have four places with the values A = 19 °C, B = 23 °C, C = 18 °C, and D = 24 °C. We know not only that all four places are different (nominal scale), that they rank from lowest to highest as C–A–B–D (ordinal scale) but we have a quantitative measure of difference. But note that since the zero origin of the Celsius scale is arbitrary (we can have negative values) we cannot say that location D is 1.33 times as hot as location C since the ratio depends on the origin. We can say that a city founded in 1800 is nine centuries younger than one founded in 900; but not that it is half as old.

Ratio scales solve this problem by providing a further refinement on the interval scale, a meaningful origin. Thus, if we have three settlements shown on the map with populations A = 100, B = 200, and C = 400 we can say they are different (nominal scale), that they are ordered A–B–C (ordinal scale), that C has 300 more people than A (interval scale) and that C is ×4.0 the size of A (ratio scale). The difference between ratio and interval measures is that the former has a natural zero point (no population) whereas the latter has an arbitrary zero point. A map of temperatures is thus on an interval scale but a map of snowfall is on a ratio scale.

In mapping, the four scales can be applied to features in each of three dimensions. These are *points* which have zero dimension, *lines* which have one dimension, and *areas* which have two dimensions. Figure 21.16 shows the four levels of measurement cross-classified by the three dimensions. Each cell shows an example of a mapped features which illustrates the range.

The Choice of Grids

We noted earlier that geographers use graticules called map projections to form the basic container into which map information is poured. But there are well over a hundred different projection systems available, and each reproduces a given section of the earth's surface in a different way. In Figure 21.17 the island of Greenland is depicted (2,175,000 km^2) on ten of these map projections. The first (projection a, an azimuthal equal-area system based on the North Pole) gives the closest match to Greenland as shown on a desk globe, whereas the last (projection j, a Miller cylindrical system)

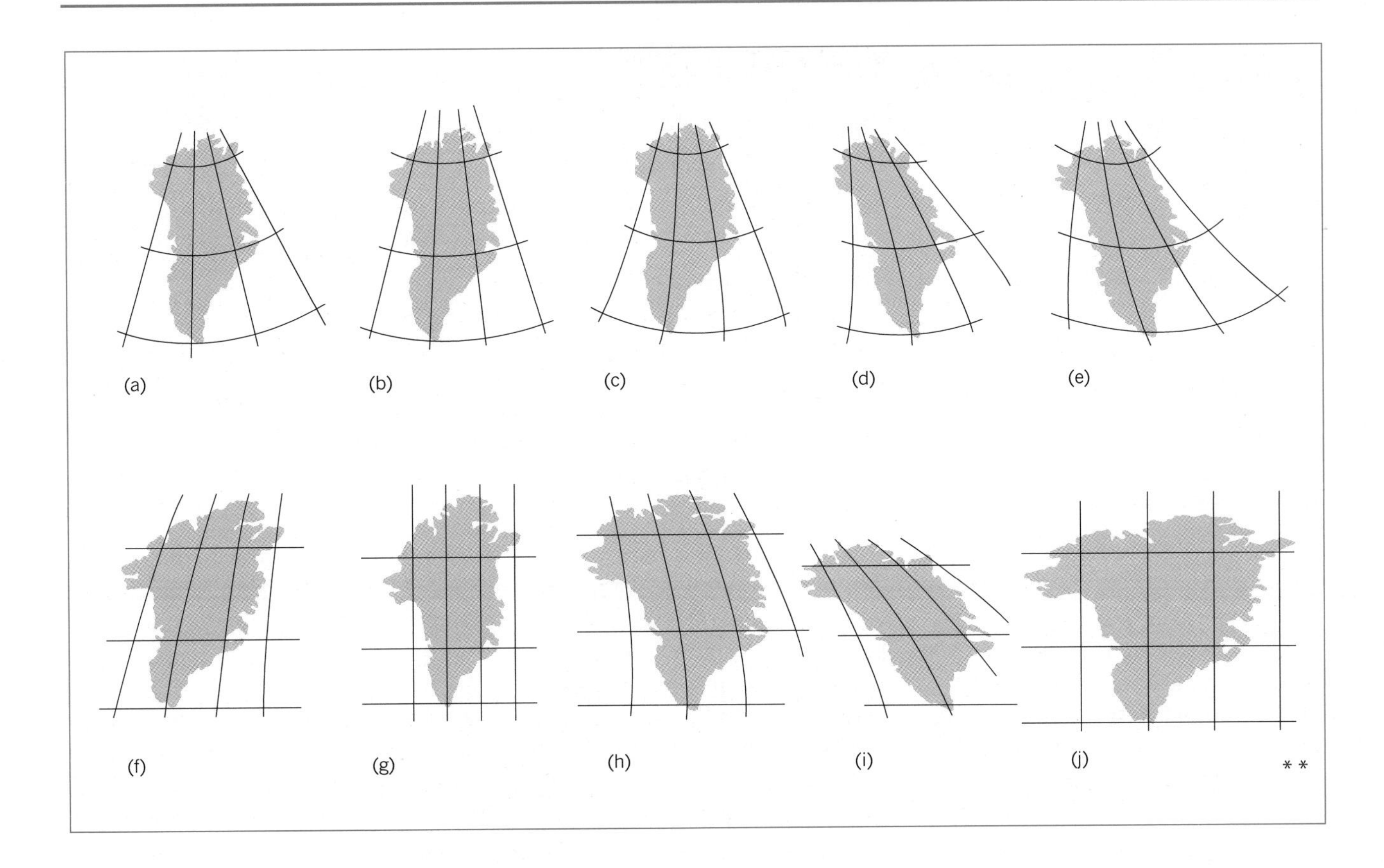

Figure 21.17 Impact of map projections on shape The outline of Greenland shown on ten different projections arranged in rank size from (a) the most accurate representation of its true shape on the globe to (j) the least accurate.

[Source: J.R. Mackay, *Geographical Review*, **59** (1969), Fig. 2, p. 377. Reprinted with permission.]

presents a highly distorted figure. Two noteworthy points on interpreting this diagram are that, first, none of the ten outlines is undistorted (distance, direction, area, or shape cannot *all* be shown accurately at the same time), and, second, each projection is optimal for a selected purpose (i.e. it has some of these properties over part of the globe).

Which projection to choose will depend on the job that the map has to do. The solution to the mapping problem requires a careful balance between the competing advantages and drawback of the different projections. Consider a map of Australian natural resources that is to be published at a scale of 1 : 6,000,000. After exhaustive comparisons of several criteria, a cartographer adopted a modified zenithal equidistant projection. This was centred at a location chosen in relation to the shape of the Australian land mass (at 126°S and 134°E), with true scale along a circle having a radius of 17° of arc from the centre of projection. These features give a maximal linear scale error over the Australia map area of only 1.8 per cent and a corresponding area scale error of only 1.5 per cent. By contrast, the poorest projections considered were yielding a linear scale error of nearly 6 per cent and an area scale error of over 10 per cent.

Mapping agencies for individual countries always adopt projection systems that provide the best advantages for their own users. Maps of the whole of the United States frequently use Albers' projection, while Britain employs a transverse Mercator for its Ordnance Survey series. By a judicious choice of factors such as local prime meridians or minor modifications of scale, a projection system can be tailored to meet the expected uses to which the final map sheets will be put. The choice of projection is always an exercise in the art – or in this case, the science – of compromise.

21.4 The Mapping Cycle: Decoding Maps

So far in the second half of the chapter we have been concerned with building maps (encoding). But map making is only one half of what is termed the 'mapping cycle'. This envisages a circular process in which field data are converted into maps (map making) which are then interpreted by others to give new information or insights about the field (map decoding or map interpretation). We look here at three examples of this decoding process: understanding pattern, detecting trends, and the reliability of maps.

Identifying Pattern

When we look at a map we see a distribution pattern. But is that pattern simply one caused by happenstance? Did it simply happen, by chance, to come out that way? Or is there some significant factor shaping the map in a particular fashion?

Let us look at the map in Figure 21.18(a). This shows a chart of deaths from bronchitis in males some 40 years ago. At that time, London had a smoggy, polluted climate and this respiratory disease was the third most frequent cause of death in males. Each of the 29 circles on the map refers to a London borough, with the size of each circle proportionate to the

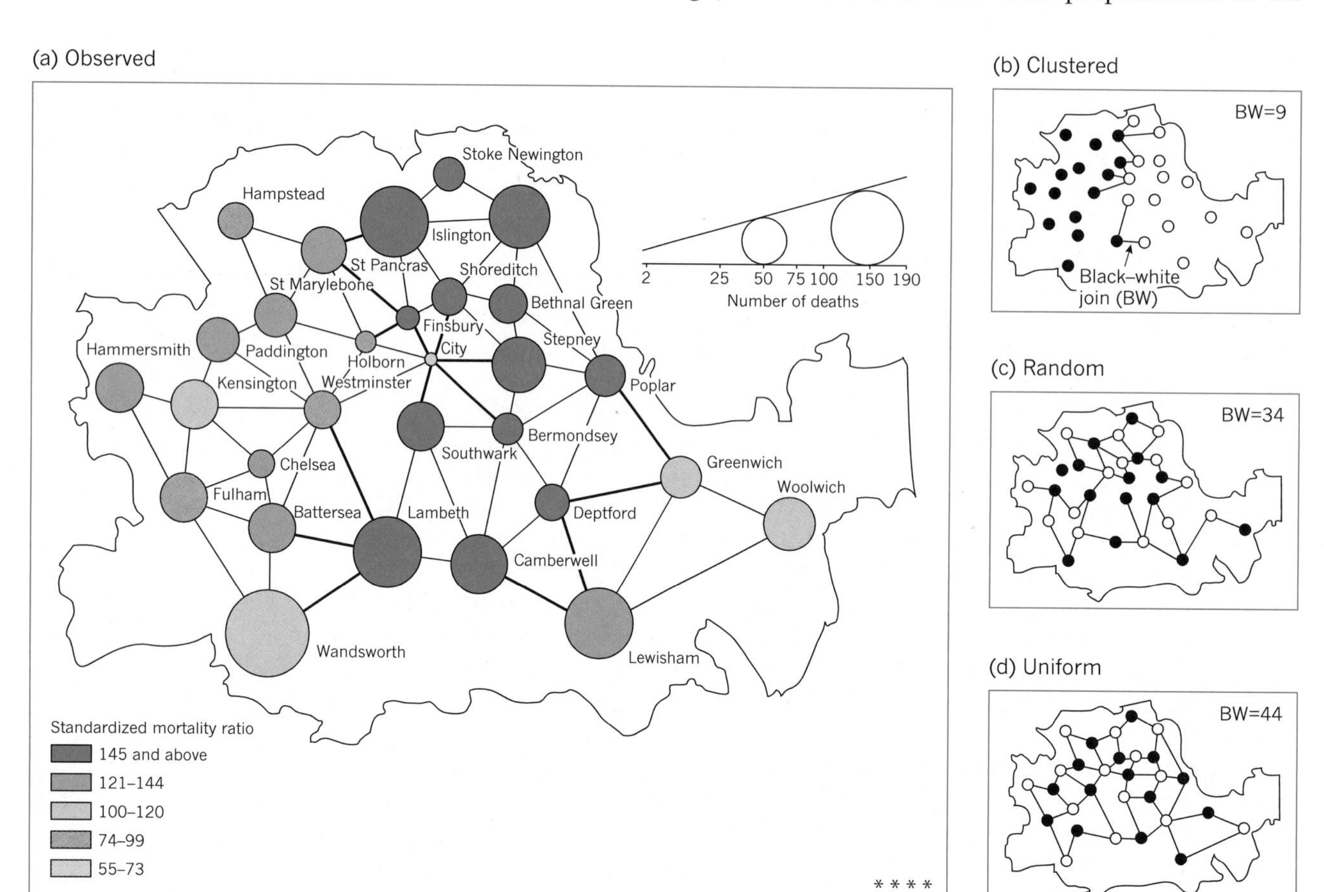

Figure 21.18 Recognition of map patterns (a) Bronchitis in males in the boroughs of London, 1959–63. Circle sizes are drawn proportional to the number of deaths. The actual map is compared with three templates representing clustered (b), random (c), and uniform spatial distributions (d).

[Source: A.D. Cliff and P. Haggett, *Atlas of Disease Distributions* (Blackwell, Oxford, 1988), Fig. 3.12, p. 128.]

number of deaths over a five-year period. Each circle is coded with very high rates (standardized by age) shown as a solid (black) circle. The other boroughs with lower rates are shown by shaded circles. The framework of links between the circles show the London boroughs that are adjacent to each other, defined as sharing a common boundary.

The boroughs with very high bronchitis rates appear to form a distinctive swathe running through central and eastern London. But does this have epidemiological significance? Could it have occurred by chance? One very simple test is to count the pattern of black–shaded (BS) joins, which are shown on Figure 21.18(a) by heavier links. Of the 64 links shown on the map, only 15 are BS. We can now compare this score with other maps which represent known patterns. Map (b) takes the same proportion of 'high' bronchitis but concentrates them in the western side of London. This gives only nine BS links. Map (c) distributes the 'high' values randomly through London and the BS count rises to 34. Finally, map (d) disperses the 'highs' evenly across the London boroughs to give a still larger BS count of 44 links.

It is evident that as the map pattern changes from clustered (b), through random (c), to dispersed (d) so the number of BS links increases. Cambridge geographer Andrew Cliff (see Box 16.C) has used this observation to devise a series of sensitive statistical tests of spatial pattern called *spatial autocorrelation tests*. When these are applied to the original London map (a), they confirm that this map pattern is very significantly clustered from the statistical point of view. There is less than a 1 : 1000 chance that a pattern this distinctive could be due to chance alone.

Figure 21.19 Types of trend surface Shapes generated by linear, quadratic, and cubic equations. The upper sequence refers to two-dimensional curves and the lower to three-dimensional surfaces. *U* and *V* refer to spatial coordinates (i.e. geographic position) and *Z* to the vertical coordinate (i.e. height).

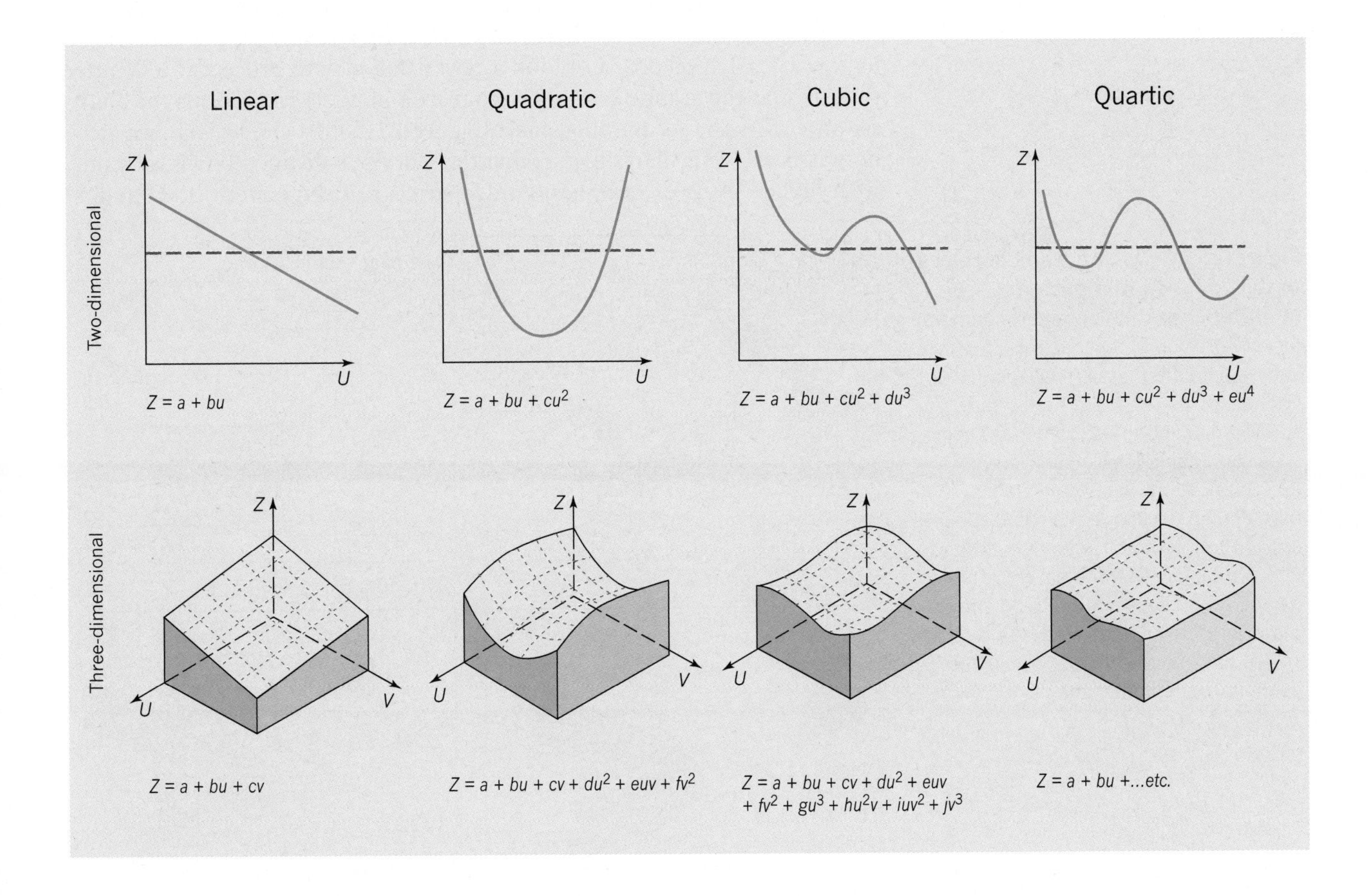

Identifying Trends

A second way of trying to make sense of the map is to try to separate the overall trend across a map (usually called the regional trend) from local effects (called local anomalies or local residuals). There are several mathematical methods that geographers use for constructing these *trend surface maps* but all rely on some form of filtering. The filter sieves out the large, broad-band elements but allows the smaller map effects to pass through. You could think of this like a sieve used by gardeners to separate out the large stones and gravel from the fine material which passes through. Figure 21.19 shows the relationship between the equations and the maps generated in both a two-dimensional and three-dimensional situation.

The best way to understand the method is to look at a specific example. Figure 21.20 shows an area of granite rocks in a glaciated lake-studded area of Quebec, Canada. The contours in (a) show the colour of the granite, a useful clue to the method of formation. But the contours show a very confused pattern. Can we throw any light on how the granite intrusion was formed?

Geologist Tim Whitten used trend surface maps to try to answer the question. The first set of maps in Figure 21.19 shows the result of using three trend-surface maps on the data for the whole granite area. Map (b) shows a linear surface generated with just three terms. That gives a 'shed-roof' type of map sloping towards the north. But this provides a very poor fit to the original data (only just over 2 per cent) so is not relevant. Map (c) shows a cubic surface generated by a more complicated equation with ten terms (see the cubic equations in Figure 21.19).

Notice that the more complicated the granite maps (Figure 21.20), the better the fit. But even the best (map (d)) achieves only 12 per cent explanation. Clearly this does not look a promising way to proceed. So Whitten then divides the granite area into two (east and west) and repeats the analysis on each half. As the final map (Figure 21.20(d)) shows, the impact is now striking. This map suggests that there are two distinct 'bowls' of values with higher levels of explanation: 22 per cent in the eastern bowl and 51

Figure 21.20 Using trend surfaces to decode a spatial pattern
(a) Distribution of colour index contours for an intrusion of igneous rock, the Lacorne granite massif in Quebec, Canada. Maps (b) and (c) are trend-surfaces applied to the data in (a) using linear and cubic forms (see Figure 21.19) applied to the whole granite massif. Map (d) shows the cubic surfaces applied to the massif when it is divided into two regions (east and west). The degree of 'fit' between the trend surface maps and the original map (a) are shown as percentages; the higher the percentage, the better the fit.

[Source: E.H.T. Whitten, *A Surface Fitting Program Suitable for Testing Geological Models which involve Areally-distributed Data*, Office of Naval Research, Technical Report, No. 1228(26).2, Evanston, 1963.]

(a) Original map

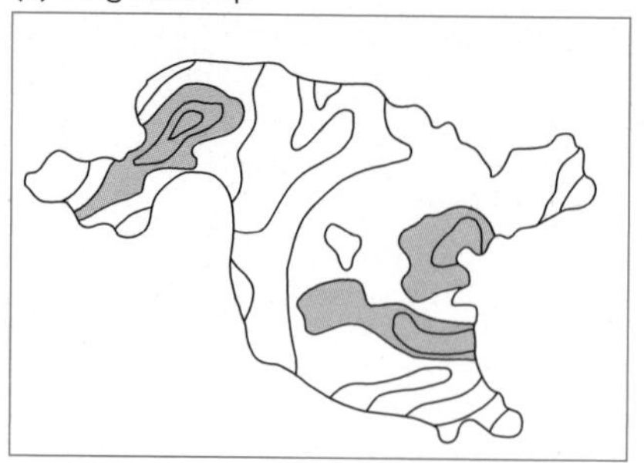

(b) Linear model

(c) Cubic model

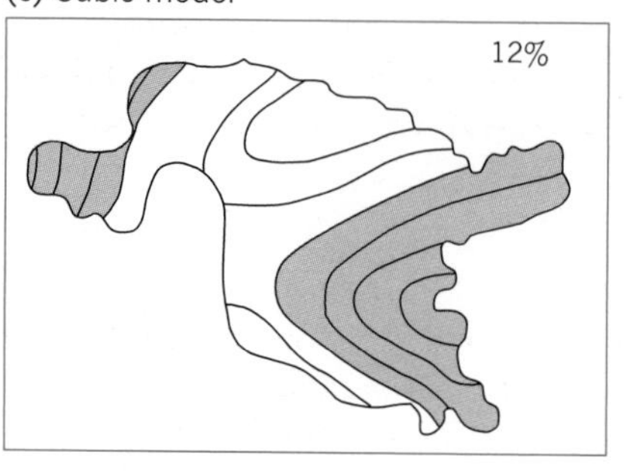

(d) Two-region cubic model

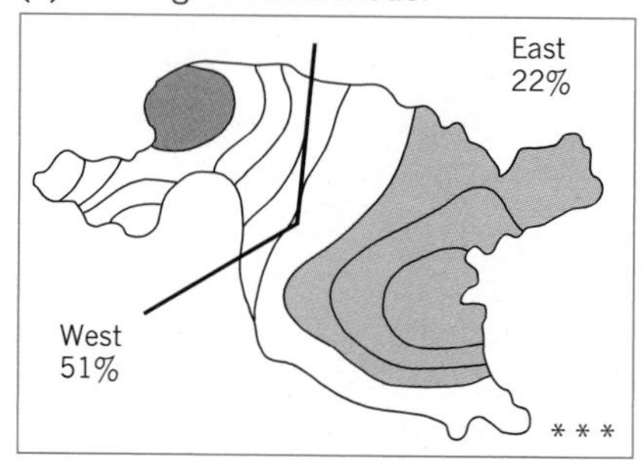

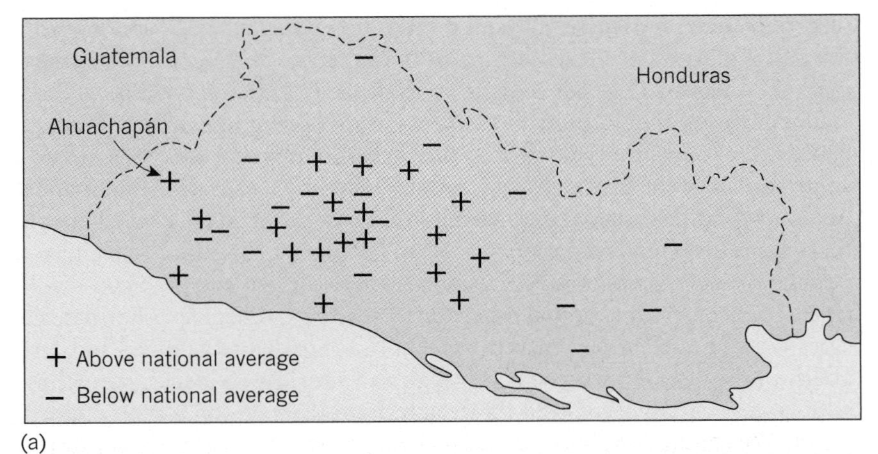

(a)

(b)

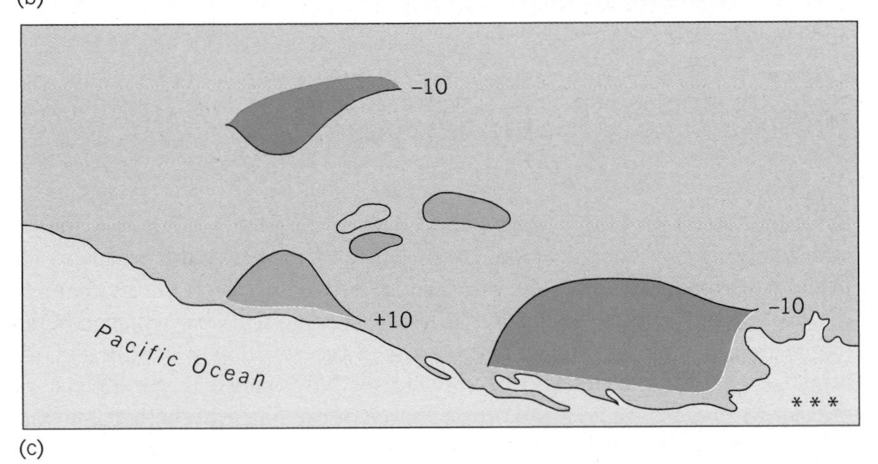

(c)

Figure 21.21 Error and confidence in statistical mapping Use of Bayesian estimators in mapping of (a) the distribution of a blood disease (toxoplasmosis) in 36 locations in El Salvador shown as signs. Map (b) shows contour maps of incidence based on deviations (both positive and negative) from the overall mean value for the country. Map (c) shows the modified contour maps based on James–Stein estimators in which the number of observations and the variability of the observations at each location has also been taken into account. Note that (c) gives a more conservative (but more reliable) view of the variations in disease incidence than (b).

[Source: P. Haggett, 'The edge of space', in R.J. Bennett (ed.), *European Progress in Spatial Analysis* (Pion, London, 1981), Fig. 3.6, p. 62.]

per cent in the western bowl. They imply that the granite was not a single intrusion but consists of two separate intrusions, probably occurring at different geological periods. In short, this example shows how trend-surface maps allow a complicated original map to be decomposed into simpler maps. In turn, these generalized maps allow questions to be answered.

Identifying Significance

We can illustrate what is meant by decoding by considering a specific example. Consider Figure 21.21, which shows the incidence of a unpleasant blood disease (toxoplasmosis) in cities in the Central American country of El Salvador. In map (a) measured rates are expressed as deviations from the national incidence (the average of the rates for all cities). A city showing −0.04 has a rate 4 per cent lower than the country as a whole.

One obvious map to draw from these data is shown in map (b). The incidence value for each city is taken as the best estimator and used to generate a contour map of the disease for the country. But how accurate is that map? If we go back to the original data we find that blood tests were carried out for toxoplasmosis on only a few people: 5000 people in El Salvador (which has a population of 4.8 million). Nor was the sample drawn equally from the 36 cities. Some cities contributed many blood readings, others few. Further, the cities showed extreme contrasts in the number of patients examined and variability of the disease; cities with very high and very low incidences were those with largest variability. Given these facts, map (b) takes on a less confident shape. We can no longer be sure that the highs and lows are a result of real spatial variations in the disease or just a by-product of the sampling process.

Statisticians have provided methods by which these complications can be taken into account. For example, map (c) shows a revised map based on a *Bayesian estimator* (the James–Stein estimator). In some cases the change in map values is considerable; thus the western city of Ahuachapan in western El Salvador is ranked highest with a value of 29 per cent above the national value. But Ahuachapan is demoted by the James–Stein estimator to only twelfth ranking position, only 5 per cent above the national value. The establishment of more accurate map patterns from spatially distributed data has a long history in quantitative geography going back to Choynowski's work on cancer distributions in Poland. Recent work on statistical estimation theory promises to provide a whole new family of estimators (of which the James–Stein statistic is simply an example) on which more accurate maps can be drawn, bridging the gap between cartography and statistics.

21.5 Different Geographic Spaces

Since the Greek geographers first conceived of the planet earth as a sphere and set about measuring its dimensions, one geographic goal has been to fill in the world map as accurately as possible. As we have seen in this chapter, the advent of remote sensing has allowed that goal to be achieved. One might imagine that, as the last blank areas in the world map are filled in, all other geographic puzzles will be solved too. No longer can we debate, as did our forebears, the location of the source of the Nile or whether a northwest passage between the Atlantic and Pacific Oceans really exists.

But even though one set of spatial puzzles has been solved, new ones have been uncovered. As we have just noted in earlier chapters, geographers are becoming increasingly aware of mental maps. We have also seen the ways in which travel times and transport costs can crumple and distort the familiar world map. The conventional world map describes a space that is continuous, isotropic (i.e. movement is equally possible in all directions), and three dimensional. Nevertheless, the real space in which people move is discontinuous, anisotropic (i.e. the costs of movement vary markedly over the map), and change rapidly over time – which means it has four dimensions rather than three. Mapping this real space poses fundamental questions for map makers and requires a reassessment of conventional Euclidean geometry. (See the discussion of *non-Euclidean space* in Box 21.C.)

Box 21.C Non-Euclidean Space

The geometric concepts synthesized in Euclid's *Elements* (written about 300 BC) form the basis of geographic measurement of the globe. Consider, for example, the distance between locations A and B on a plane.

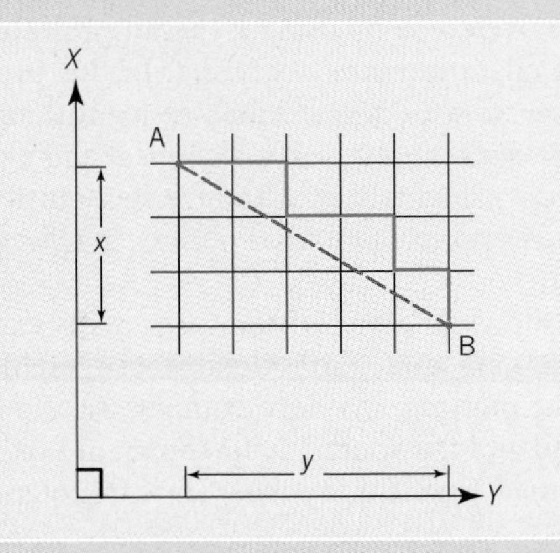

In Euclidean space the distance between these two points (d_{AB}) is given by the Pythagorean theorem as

$$d_{AB} = \sqrt[2]{x^2 + y^2} = \sqrt[2]{3^2 + 5^2} = 5.8 \text{ units}$$

The variables x and y measure the differences between the two locations. If we superimpose on our continuous plan a Manhattan-like grid of streets, the distance from A to B becomes

$$d_{AB} = \sqrt[1]{x^1 + y^1} \text{ or } x + y = 8 \text{ units}$$

as we can no longer walk directly from A to B.

When we compare the formulas for estimating the distance between the same two points in Euclidean space and 'Manhattan' space, we see that the difference lies in the exponents; these have a value of 2 in the first case but 1 in the second. Formal geometries have been developed to handle spaces where distances are both greater and less than those given by the Pythagorean theorem, but so far these non-Euclidean geometries have been little explored by geographers. (See D.W. Harvey, *Explanation in Geography* (Arnold, London, 1969), Chapter 14.)

Spatial Transformations

Some analysts view this break from traditional mapping as paralleling the changes in theoretical physics when concepts of *absolute space* (in which the coordinates reflect the structure of what is being described). The distinction between the two is shown by the maps of Sweden in Figure 21.21. Here we see a transformation from a rectangular coordinate system in

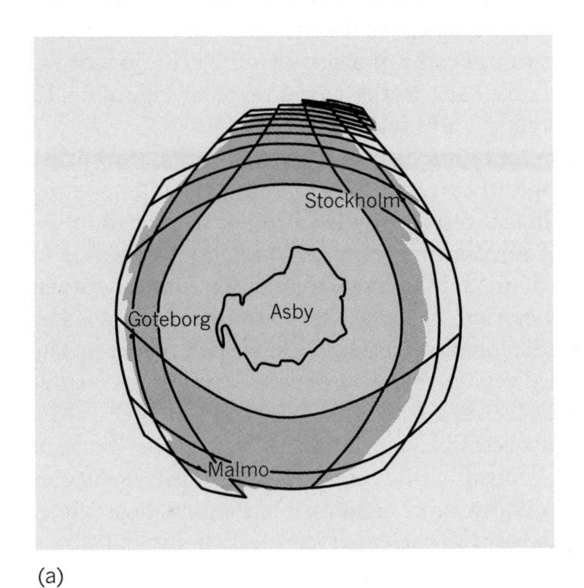

(a)

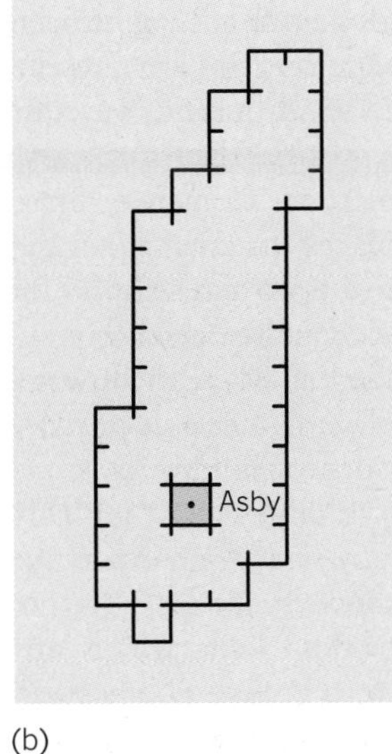

(b)

Figure 21.22 Sweden: a migrant's view (a) A conventional map of Sweden differs greatly from the same space as viewed from a single Swedish parish. (b) The map of Sweden centred on the parish of Asby can show population migration over both short and long distances. Distance from the parish is plotted in proportion to migration, so that nearby areas with strong flows are exaggerated in area. Conversely, distant areas with low migration are shrunk. Note the correspondence between the central cell with curved boundaries and the corresponding square cell cell on the conventional map.

[Source: From T. Hägerstrand, *Lund Studies in Geography, B*, No. 13 (1957), Fig. 38, p. 73.]

absolute space into one in which locations are plotted relative to their direction and distance from a given centre. Swedish geographer Torsten Hägerstrand used this transformation to describe the migration field of a small parish in the south central part of his country. Most of the migration movements were over short distances, but a few migrants travelled large distances, such as 5000 km (3000 miles) to the United States. Local movements could be clearly distinguished only if they were plotted on a large-scale map, but movements to the main Swedish cities and to overseas countries needed to be plotted on small-scale maps. The problem of matching the different scales was overcome by using a special projection centred on the village from which migration was occurring. In this projection all radial distances from the centre were transformed to logarithms, while all directions remain true. The result is a map on which local and international movements can be plotted on a single sheet. Although the transformed map of Sweden is unfamiliar and distorts its familiar outline, it reflects a limited field of space rather accurately.

This simple illustration shows just one of the ways maps can be transformed for different geographic uses. Considerable geographic research effort is going into ways of plotting far more complex relations (e.g. the 'time maps' of New Zealand and the South Pacific shown in Box 8.B). Since this research area is rather mathematical, we must leave its consideration to more advanced courses.

An Imploded Pacific

But why is this holiday puzzle relevant? To translate it into something of a more general concern, let us take an example that spans a hemisphere – the Pacific Basin. This makes up about one-half of the earth's surface and currently the countries around its rim include some of the fastest-growing economies in the world. Conventional maps of areas of this size are difficult to draw since distortion inevitably results in translating a massive settlement of the globe on to a flat, two-dimensional piece of paper; squashing half an orange on to a flat surface will give some idea of the distortion involved. The particular map projection used will determine the nature and extent of distortion introduced. Subject to this proviso, all the projections in common use attempt to ensure that either the locations of points on the globe reflect their relative positions on the globe or that areas of directional relationships are preserved. Look back to the world maps in Figure 21.13 to see different, but equally valid, world map projections.

Multidimensional scaling can be used to convert the Pacific into other metrics. Consider, for example, Figure 21.23(a). This uses lines joining places of equal time, the technical term being isochrones, to plot the relative time accessibility by scheduled airline carriers of twenty-five places in the Pacific Basin; from fast extended 747s criss-crossing the routes between the big cities, to slower local carriers between the small island chains. The place-names associated with the locations used are listed with the map. The isochrones have been standardized so that 100 denotes average accessibility; values less than this indicate superior accessibility and greater values (stippled) demarcate the less accessible parts of the basin. The diagram shows that a large part of the central Pacific centred on the Trust Territories of the Pacific (18) is up to one-fifth more inaccessible than average while, from French Polynesia (4) eastward to the Americas, accessibility falls by a factor of almost two.

The area covered by the maps in (b) is delimited by the box on (a). The maps indicate why the zone of inaccessibility in the central Pacific exists. The left-hand map uses flow lines to show the linkages in seats per week provided by international carriers such as Quantas and Pan Am between various centres in 1975: the right-hand map gives the same information six years later. Not only have route capacities multiplied, but the amount of overflying of the Pacific, by-passing local centres, has increased. For example, the maps suggest that Fiji (3) had less stopover traffic in 1981 than in 1975 and so in that sense had become less accessible.

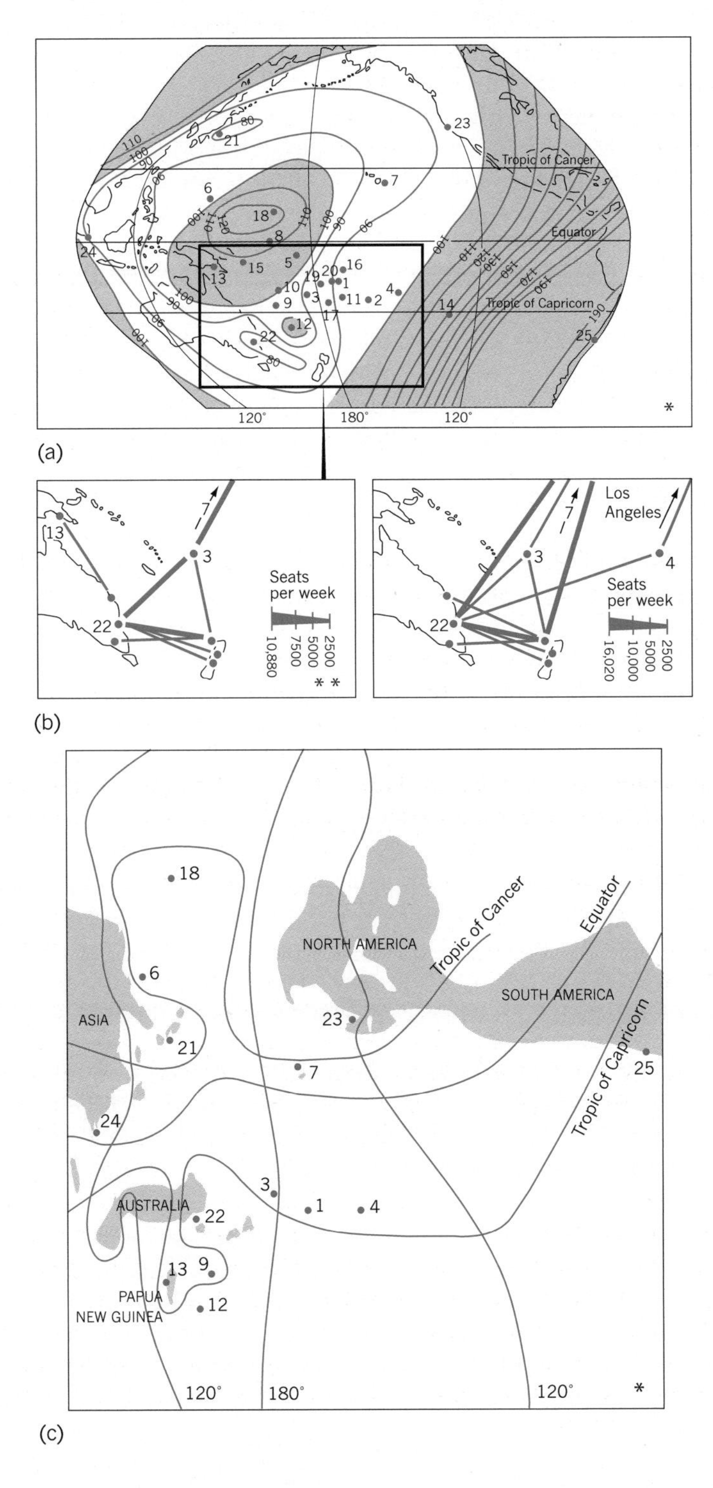

Figure 21.23 Non-linear mapping The hemisphere of the Pacific Basin mapped in time–space. (a) Relative time accessibility in 1975 by scheduled airline carriers between 25 centres in the Pacific Basin. (b) Route capabilities by international carriers in seats per week in 1975 (left) and 1981 (right) within the southwest Pacific. (c) Data on time accessibility in map (a) recomputed by multidimensional scaling to give a time–space map. The key to places mentioned in the text is (3) Fiji, (4) French Polynesia, (13) Papua New Guinea, (18) Trust Territories of the Pacific, (21) Tokyo, (22) Sydney, and (23) San Francisco.

[Source: The map is based on unpublished work by Professor P. Forer of the University of Auckland, New Zealand. The changing air-traffic flows are based on M. Taylor and C. C. Kissling, *Regional Studies*, **17** (1983), pp. 237–50. Combined maps from A.D. Cliff and P. Haggett, *Atlas of Disease Distribution* (Blackwell, Oxford, 1988), Plate 7.2, p. 226.]

To produce the map shown in (c), multidimensional scaling has been used to transform the conventional geographical map (a) into a time metric; the relatively accessible places on map (a) are now plotted closer to each other on map (c), while the relatively inaccessible places have been moved apart. The effect is to push North America and the Far East closer together than they are on a conventional map because of the frequent flights between Japan (Tokyo, 21) and the United States (San Francisco, 23). The inaccessible portion of the central Pacific Basin apparent on map (a) is now mapped at two outposts. The Trust Territories are moved to the north (note the new position of 18). Papua New Guinea (location 13) is no longer located to the north of Australia but, hernia-like, has burst through that continent to be relocated in the south.

Although the new map is unfamiliar, it shows the local forces at work in the Pacific in a dramatic way. In the physical world the earth's crust is reshaped by the massive slow forces of plate tectonics. So technological changes of great speed are grinding and tearing the world map into new shapes. Capturing those spatial shifts is at the heart of modern geography.

To sum up: non-linear mapping provides one example of the ways in which computers allow geographic information to be encoded to exploit the communication properties of a map. It allows a breakaway from the conventional geographic map based on a physical distance separation. It permits distance to be replaced with any other relevant appropriate metric. We have seen very simple examples based on cost, time, and service, and on medieval documents. Such maps are likely to be particularly valuable where they show past or future changes over time. But I hope it will now be clear that the method can incorporate all sorts of information; for example, by measuring attitudes between groups we can produce maps of hate or fear. The literature already includes one map of love; one Shakespeare play (*Romeo and Juliet*) has already been 'mapped in terms of multi-dimensional scaling and Rome and Juliet close together at the centre but the Montagues kept firmly separate from the Capulets' (Figure 21.24).

Afterword: Maps as Touchstones

So central is the map to geographic practice that some observers suggest it can form a diagnostic or touchstone for determining whether a work is 'truly geographical'. For Richard Hartshorne in *The Nature of Geography* the role of the map within geographical writing is universal and its centrality unchallenged:

> So important is the use of maps in geographic work that ... it seems fair to suggest to the geographer a ready rule of thumb to test the geographic quality of any study he is mapping: if the problem cannot be studied fundamentally by maps – usually by a comparison of several maps – then it is questionable whether or not it is within the field of Geography.

The example used by Hartshorne to illustrate his assertion is the contribution made by the American geographer, Isaiah Bowman, to the study of boundary problems and tension zones at the Versailles Peace Conference after World War I. Maps were also considered by George Washington be essential in rebuilding after the War of Independence, and he called for 'Gentlemen of known Character and probity' to be employed in making them.

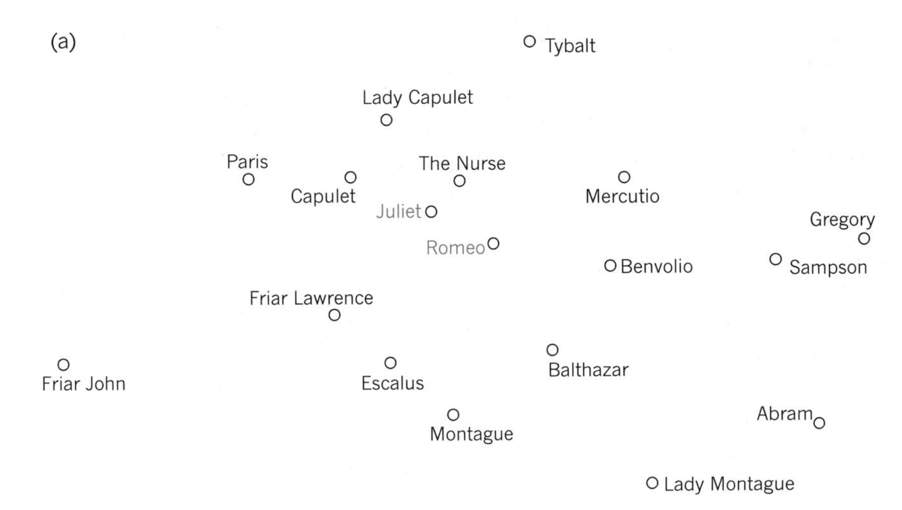

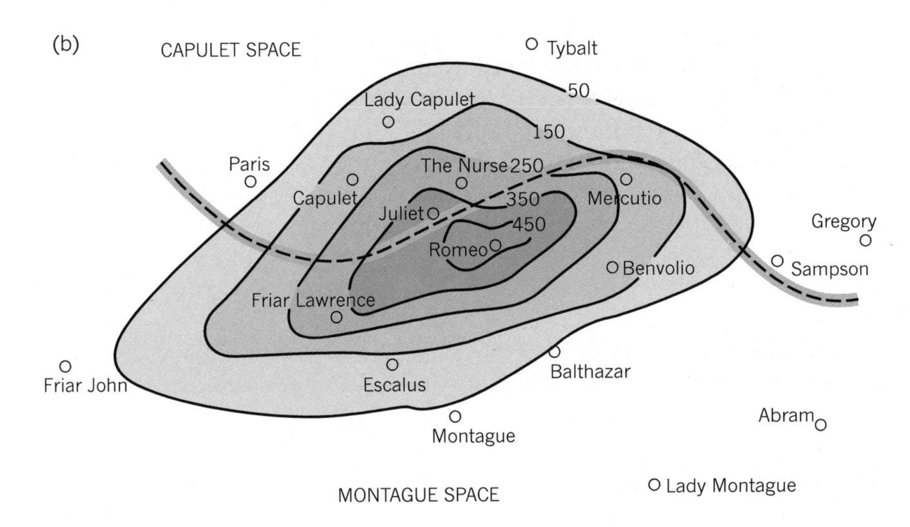

Figure 21.24 Shakespearean space: *Romeo and Juliet* The extreme flexibility of non-linear mapping is shown by Peter Gould's map of the play *Romeo and Juliet* (1610). The input was the number of lines exchanged between the *dramatis personae* and the maps show (a) the location of the main characters and (b) the contour map of number of lines spoken in the play with a 'regional' division into Capulet and Montague space. Readers will not be surprised that my highest role in school productions of Shakespeare was on the outer margins of the 'map'.

[Source: P.R. Gould, 'Concerning a geographic education'. In David A. Lanegran and Risa Palm (eds), *An Invitation to Geography*, 2nd edn (McGraw Hill, New York, 1978), Fig. 16.4, p. 210.]

Although Hartshorne (see Box 12.A) and another leading American geographer, Carl Sauer (see Box 5.B), have been thought to take different views of many issues in geographical philosophy, they were at one on the central role of maps. On pondering why people are attracted towards geography, Sauer writes:

> May a selective bent towards geography be recognized? The first, let me say, most primitive and persistent trait, is liking maps and thinking by means of them. We are empty-handed without them in the lecture room, in the study, in the field. Show me a geographer who does not need them constantly and want them about him, and I shall have my doubts as to whether he has made the right choice of life. The map speaks across the barriers of language.

The use of the term language is interesting, for Sauer goes on to expound, maps may themselves be regarded as a special kind of international language (or set of international languages) for describing the earth and its regions.

If maps are such powerful indication of geographic interest and aptitude, how do we square this with the use of maps by scholar in other fields? Maps may be critically important to the geologist or the civil engineer, and of more than passing interest to biologists or to historians. Clearly we can make maps to show the distribution of phenomena from aardvarks to zygotes, and even the most broad-minded or empire-building of geographers would not over-extend their disciplinary boundaries to include quite so wide a range of phenomena. How far then is the map a distinctively geographic mirror?

If we regard the boundaries of academic disciplines as convenient markers of the limits of focused concern and competence (rather than God-given division of reality), then such questions are secondary. But these still merit a reply and a response must be two-fold. First, maps play a distinctly more prominent and central role in geography than in other disciplines. No other insists that students includes course on map making, map reading, map projections, and the like in their core curriculum. Second, other disciplines turn to geographers for the production of maps, for their collection and care, and for their interpretation and analysis. It is relevant to recall that 'Geographic Information Systems' (GIS) was coined by computer scientists to describe the software they had invented to link and map data for small areas. Thus both internal views (by geographers) and external views (by non-geographers) reinforce a considerable but not exclusive concordance between geography and mapping.

Reflections

1. Determining the size and shape of the earth has interested geographers since the time of the Greek cosmologist Eratosthenes. Measurements since the eighteenth century show that the earth is not exactly spherical. What implications do these small departures from a perfect sphere cause for mapping the world?
2. For any part of the world in which you are interested, check the ways in which it was explored and mapped. At about what time do world maps show the area in a form that is more or less recognizable to the present observer?
3. With Figure 21.8 in mind, consult a topographic map of your own locality, and list the names given to streams and water courses. Then do the same for a different part of your country. Are there any variations between the two lists? What do you think might have caused these variations?
4. Have the members of your class check off on a list of 'possibles' (a) the countries they consider part of the 'Middle East' and (b) the states they consider part of the US 'midwest'. Can you identify a firm core area on which everyone agrees? Are there any 'difficult' borderline cases?
5. What effects has the 'township and range' system used for land surveys had on (a) the rural and (b) the urban landscapes of the central and western parts of the United States?
6. Map projections come in a very large number of forms and none gives a perfect representation of part of the curved earth surface on flat paper. What projections are usually chosen for your own country? Why do countries such as Russia or Chile cause such problems for map makers?
7. Geographers are constantly inventing new ways of mapping human activities onto a spatial background. Consider the non-linear maps shown in Figures 21.23. What do you consider the plus and minus points in presenting data in this way?
8. Review your understanding of the following concepts using the text (including Table 21.1) and the glossary in Appendix A of this book:

latitude
longitude
map projections
meridians
multidimensional scaling
parallels
place-names
spatial autocorrelation tests
spherical coordinates
township and range

One Step Further ...

A brief but useful introduction to cartography and its history is given by Mark Monmonier and David Woodward in their essays in R.J. JOHNSTON *et al.* (eds), *Dictionary of Human Geography*, fourth edition (Blackwell, Oxford, 2000), pp. 59–68. Both authors have made significant contributions to cartography. Monmonier has written a series of innovative books on cartographic method, e.g. M. MONMONIER, *Computer-Assisted Cartography* (Prentice Hall, Englewood Cliffs, N.J., 1982) and *How to Lie with Maps*, second edition (University of Chicago Press, Chicago, 1996). Woodward, with the late Brian Harley, initiated and edited the definitive multiple volume series, *The History of Cartography* (University of Chicago, Chicago, Vol. 1, 1987–) which is now nearing completion.

The standard work that has introduced the principles and practice of map making to generations of students over many editions is A.H. ROBINSON *et al.*, *Elements of Cartography*, sixth edition (Wiley, New York, 1998). For a briefer and basic account of the variety and range of maps used by geographers, see Daniel DORLING and D. FAIRBAIRN, *Mapping: Ways of Representing the World* (Longman, Harlow, 1997) or J.S. KEATES, *Understanding Maps*, second edition (Longman, Harlow, 1996). You might also like to browse through J.J.W. THROWER, *Maps and Man: An Examination of Cartography in Relation to Culture and Civilization* (Prentice Hall, Englewood Cliffs, N.J., 1972). The link between maps and literature and role of places, both real and imaginery, is explored in A. MANGUEL and G. GUADALUPI, *The Dictionary of Imaginary Places* (Macmillan, London, 1980) and J.B. POST, *An Atlas of Fantasy* (Souvenir Press, London, 1980). The history of human efforts to pin down the exact distribution of the world's geographic features is told at length in L. BAGROW and R.A. SKELTON (eds), *History of Cartography* (Harvard University Press, Cambridge, 1964) and summarized more briefly in R. ABLER, J.S. ADAMS, and P. GOULD, *Spatial Organization: The Geographer's View of the World* (Prentice Hall, Englewood Cliffs, N.J., 1971), Chapter 3.

For the most recent developments in mapping, see the *International Yearbook of Cartography* (annual) and the *Journal of Cartography and Surveying and Mapping* (both quarterlies). Other cartographic research is published in *American Cartographer* (semi-annual) and the Canadian Journal *Cartographica* (quarterly). For readers with access to the World-Wide Web a huge range of map sites is now available, see those recommended on 'cartography', 'historical maps', and 'maps' in Appendix B at the end of this book.

A
B
C

CHAPTER 22

Environmental Remote Sensing

■ denotes case studies

Figure 22.1 The Earth's surface from outer space We have used throughout this book many examples of shots taken of our global surface from orbiting satellites. Few illustrate the power of this technique better than this infrared LANDSAT-5 satellite image of San Francisco, California on the western coast of the conterminous United States. In this view of the setting of San Francisco, north is at the top with the east–west distance of around 100 km (60 miles). In the infrared spectrum, vegetation appears dark, water black, urban areas light. San Francisco (A) is the peninsula in the upper left, connected by the Golden Gate Bridge to the land to the north. To the city's east is San Francisco Bay, to its west the Pacific Ocean. Across the bay is the city of Oakland (B). South of San Francisco is the San Andreas fault (C), a major cause of earthquakes in the area. The fault forms a dark line running southeast.
[Source: Terranova International/Science Photo Library.]

> Huck Finn to Tom Sawyer in their flying boat: 'We're right over Illinois yet. And you can see for yourself that Indiana ain't in sight . . . Illinois is green, Indiana is pink. You show me any pink down there, if you can. No sir; it's green.' 'Indiana pink? Why, what a lie!' 'It ain't no lie; I've seen it on the map, and its pink.'
>
> MARK TWAIN *Tom Sawyer Abroad* (1896)

Like Huck Finn, we have hung over the side of the basket and looked down at the puzzling earth surface below. In the main sections of the book we have opened many of our chapters with photographs from the air, looking down at the environmental challenge the earth poses for human beings. We have asked four questions. What is our ecological response? How has it affected the mosaic of world regions? How have these world regions been shaped into a hierarchy of city regions? What conflicts and stresses have been set up between these regions? All four questions underline the critical need for more and more geographically relevant information.

Here we are faced with a paradox: growing information, yet growing uncertainty. Let us take the first point. In the 2000s the level of existing information is, of course, higher by many orders of magnitude than it was a century ago. From wide-ranging surveys of the growth of scientific information we know that information grows in an exponential fashion; that is, the greater the amount of information that exists, the faster it grows. Depending on what we measure, it is possible to estimate, roughly, that the amount of environmental information tends to double within a period of ten to fifteen years or so. If one accepts the general form of this growth curve, the amount of information available to geographers such as Alexander von Humboldt and Carl Ritter in the early part of the nineteenth century was over a thousand times smaller than the amount available to the current generation of geographers. To create the images produced in a few seconds in our opening photograph (Figure 22.1 p. 692) would have required decades of survey and calculation by an earlier generation of mapmakers.

Despite this dramatic increase in data, the demand for geographic information is likely to be far higher in the next quarter century than at any previous time. For, despite its impressiveness, the increase in total geographic information has scarcely kept pace with the growing realization of our ignorance about the earth. Shifts in the geographic focus of human activities – to the jungles of Cambodia, the subarctic steppes of the Alaskan North Slope, or the inner blighted areas of American cities – may show how sparse is the stock of information available. Also, shifts in emphasis on different resources have swung the spotlight to some areas where there is critical lack of information.

In this chapter we focus on a major drawer in the geographers' toolbox and consider ways in which environmental remote sensing contributes to our understanding of the world. We look first at the early development of aerial photography from the 1920s (Section 22.1) and follow this with satellite surveillance from the 1960s (Section 22.2). This leads on to a review of some major areas of application (Section 22.3) and a final review of some of the problems that remain to be solved (Section 22.4). These investigations link back to some of the basic questions we asked in the prologue of the book.

Early Exploration From the Air 22.1

The first manned ascent of a balloon in the United States was from Philadelphia on 9 January 1793. George Washington and a large crowd are said to have watched the event. Today's geographers look back on those early ballooning experiments by the Montgolfier brothers and their friends as a watershed (Figure 22.2(a)).

The first man to try to photograph the earth's surface from a balloon was probably a Parisian photographer, Gaspard Felix Tourachon. His attempts to capture 'nothing less than the tracings of nature herself, reflected on the plate' date from 1858. The mania for balloon photography caught on (Figure 22.2(b)), and by July 1863 even Oliver Wendell Holmes was conceding that Boston 'as the eagle and wild goose see it' was a very different place from that seen by its solid citizens on the ground.

From the late eighteenth century onward, people have been able to view increasing portions of the earth's surface from above and from the mid-nineteenth century to record their view on photographic plates. This was to lead to an ability to construct maps from direct aerial observation rather than laboriously piecing them together from measurements made on the surface. As the airplane succeeded the balloon, and the space rocket the airplane, the horizon continued to recede until, in the late 1960s, the whole hemisphere of the planet came into view. In this section we shall review this latest phase in environmental reconnaissance and the ways in which it has affected spatial language and the world picture developed by maps.

Figure 22.2 The beginnings of air photography (a) Late eighteenth century experiments with ballooning provided the first map-like perspectives of the earth's surface. (b) Jacques Ducorn using a dryplate camera from a balloon in 1885.

[Sources: (a) From William L. Marsh, *Aeronautical Prints and Drawings* (Halton and Truscott Smith, London, 1924). (b) Gaston Tissandier, La Photographie en Ballon (Gauthier-Billars, Paris, 1885).]

(a)

(b)

The Single Aerial Photograph

Photographs of the earth's surface taken from the air present the same problems as those taken from the ground; that is, both give a highly distorted perspective. Each of us, in our first forays with a camera, has come up with prints in which buildings appear to lean drunkenly backwards or portions of the subject's anatomy near the camera are hugely swollen. Identical problems are encountered in photographs taken from the air (Figure 22.3).

(a)

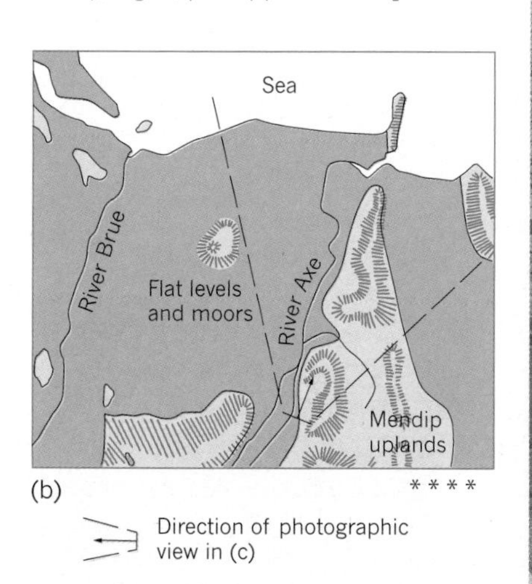

(b)

Direction of photographic view in (c)

Upland areas with slopes indicated by hatched lines

Lowland areas which are flat are subject to flooding

(c)

Figure 22.3 Perspective distortion on aerial photographs
(a) A vertical view of Manhattan from a height of 500 m (1640 ft) shows the extreme distortion caused by difference in height. St. Patrick's Cathedral (in the upper centre of the photograph abutting Fifth Avenue) appears much smaller than it really is in comparison to neighbouring skyscrapers. (b), (c) Aerial views taken obliquely are rarely used for mapping. They have the disadvantage that the scale changes greatly from foreground to background. Foreground features may also block those in the background. In compensation the photographs are excellent for visual interpretation. That shown portrays local variations in environmental quality in the North Somerset Plain, southwest England. Note the sharp contrast between the flat, waterlogged valleys of the River Axe and the steep-sloping hills of the Mendip uplands.

[Sources: (a) Lockwood, Kessler, and Bartlett, Inc. (b) D.C. Finlay, *Soils of the Mendip District of Somerset* (Soil Survey of England and Wales, Harpenden, 1965), Fig. 3, p. 4. (c) Aerofilms.]

Box 22.A Scale on Aerial Photographs

The nominal or average scale of a vertical aerial photograph can be simply represented by

$$S = \frac{f}{H}$$

where S = the scale;
f = the focal length of the camera;
H = the height of the camera above the ground surface.

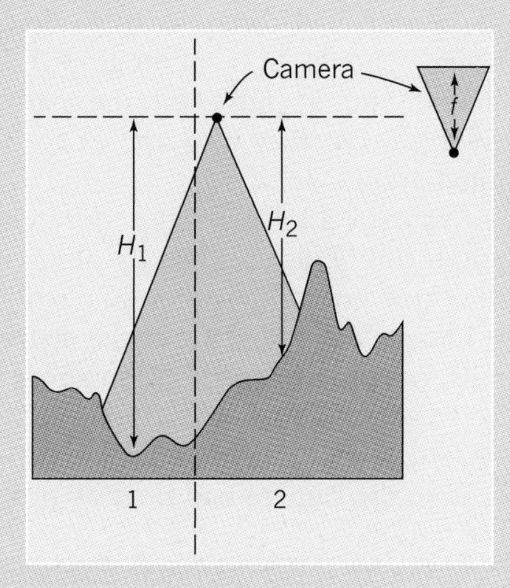

Both f and H are measured in the same units of length. Thus, if the camera has a focal length of 20 cm (about 8 in) and the aircraft is flying at a height of 1000 m (i.e. 100,000 cm, or about 40,000 in), the scale at the ground surface will be 20:100,000, or 1:5000. Areas of higher ground (H_2) are nearer the camera and hence appear larger; conversely, areas of lower ground (H_1) are farther from the camera and appear smaller. Large distortions in scale can be introduced by tilting the camera so the photograph is not truly vertical. This displaces the perspective centre of the photograph from the actual centre and makes the estimation of both scale and height considerably more complex. See D.R. Lueder, *Aerial Photographic Interpretation* (McGraw-Hill, New York, 1959), Chapter 1.)

Note, for example, the way in which the skyscrapers in mid-town Manhattan in Figure 22.3(a) appear to be distorted. The rules that govern such variations in scale in photographs are fairly simple, so long as the camera axis is vertical to the ground surface. (See Box 22.A on 'Scale on aerial photographs'.) As the camera is tilted away from the vertical, more difficult problems occur.

Because the distortion in *aerial photographs* varies with the tilt of the camera, it is useful to place aerial photographs in two main classes: first, *vertical photographs*, in which the camera's axis points directly downward to give a plan of the terrain below, and, second, *oblique photographs*, in which the camera's axis points at a low angle to the ground to produce a perspective view. If the camera is tilted enough to include the horizon, the photograph is termed a *high oblique*; pictures that do not include the horizon are termed *low obliques*.

In practice, it is extremely difficult to guarantee an absolute vertical axis. Thus the first category normally includes photographs whose camera axis is within two or three degrees of the vertical axis. 'Verticals' are the most widely used type of photograph because distortions due to scale, tilt, and height can be readily corrected to produce detailed maps. The vertical aerial photograph is not a map, however. Scale can vary not only from one photograph to the next, but also from one part of the same photograph to another as the terrain below varies in height.

Oblique photographs cover large areas of ground and present few problems in interpretation because their perspective is more familiar to the viewer. 'Obliques' are therefore most popular for illustrative or publicity purposes. They are of limited value for scientific purposes because they contain wide variations in scale, they risk large areas of blocked visibility ('dead ground'), and they possess complex geometric properties from a measurement and mapping (*photogrammetry*) viewpoint (see Figure 22.3(b) and (c)).

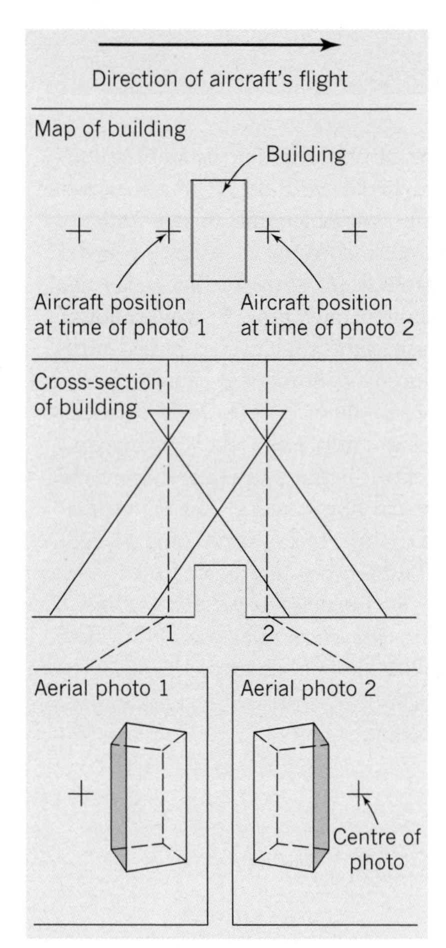

Figure 22.4 Stereoscopic effects Elevation differences in paired aerial photographs. Successive photographs of the same building from two aircraft positions are shown by horizontal displacements of its image on the two photographs.

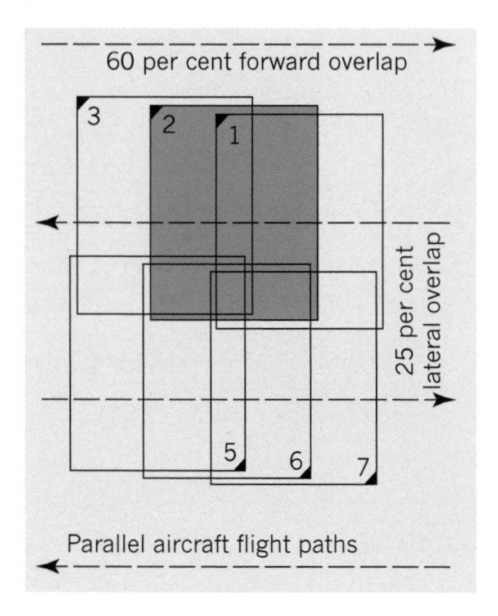

Figure 22.5 Survey flight patterns Aerial photographs used for stereoscopic analysis have both a forward and a lateral overlap. This effect is achieved by a flight pattern similar to the track of a lawnmower.

Stereoscopic Pairs

It is obvious that a flat aerial photograph has only two dimensions, while the terrain it depicts has three dimensions. How is this third dimension, height, shown in a photograph? If we re-examine Figure 22.3(a), we should find a clue. Not only are the tops of the skyscrapers greatly enlarged, but they also appear to be leaning back from the centre of the photograph.

As Figure 22.4 illustrates, differences in the height of objects in a photograph can be determined by observing their horizontal displacement from their true ground position. In a truly vertical photograph this displacement is along lines radial from the centre of the photograph. Objects such as hills, which have an elevation above the mean elevation of the photograph, are displaced outward from the centre, and low points are displaced inward. The amount of displacement is inversely proportional to the altitude of the camera and directly proportional to the variations in height of the terrain.

Although this horizontal displacement is a nuisance because it prevents the direct use of aerial photographs as maps, it permits direct measurements of relief when photographs are taken in overlapping pairs. Figure 22.5 presents a typical series of photographs in a mapping run. Exposures are timed to allow a 60 per cent forward overlap between successive photographs in the same strip. There is also a 25 per cent lateral overlap between adjacent strips. Note that the orientation of the photographs is reversed in each successive strip, because of the flight pattern. The location of photographs in such a run is facilitated by a *titling strip* automatically included at the top of each print, which gives information on the flight number, general location, date and time of exposure, focal length of the camera, flying height, and film type.

Pairs of photographs that overlap as in Figure 22.5 are termed *stereoscopic pairs* (from the Greek *stereos*, solid). *Stereoscopes* are devices that allow us to view two overlapping pictures taken from slightly different points of view at the same time. Viewing the two photographs through a pair of binoculars fuses them into a single image having the appearance of solidity or relief. The horizontal displacements are perceived as vertical displacements in a third (vertical) dimension. Advanced stereoscopic equipment permits an observer to convert vertical images into measurements of height and therefore allows the rapid production of contour maps.

Low cost has been a critical factor in the growing use of aerial surveys for environmental surveillance over the last few decades. In addition to cost advantages aerial photographs may be able to show features that are not visible from the ground and monitor rapid environmental changes. Their early use during World War I for reconnaissance over enemy lines and judging the effect of artillery barrages led to their peacetime use in the 1920s. One of the pioneers in such use was O.G.S. Crawford. (See Box 22.B on 'Crawford and archaeology from air photographs'.)

Box 22.B Crawford and Archaeology from Air Photographs

Osbert Guy Stanhope Crawford (born Bombay, India, 1886; died Hampshire, England, 1957) was a pioneer in British archaeology who saw the potential of air photographs and geographic survey.

During service in France in World War I, he was impressed by the potential value of aerial survey for the discovery of mapping of ancient sites and landscapes. Opportunity for developing the technique in Britain came after 1920 when he was appointed archaeology officer of the Ordnance Survey, a post he held until 1946. Major publications arising from his work in the air and on the ground included *The Long Barrows of the Cotswolds* (1925), *Wessex from the Air* (with A. Keiller, 1928), and *The Topography of Roman Scotland North of the Antonine Wall* (1949).

At Crawford's initiative, the Ordnance Survey also began to issue period maps, beginning in 1924 with the *Map of Roman Britain.* He was also involved from the beginning with the *Tabula Imperii Romani*, a series of maps that would cover the entire Roman Empire. His own fieldwork, including much important work in the Sudan and air photography, continued alongside his official cartography.

Crawford pioneered the process by which features photographed from the air such as soil marks are analyzed in order to work out the types of archaeological features causing them. The diagram shows how such crop marks are formed. Crops grow taller and more thickly where the soil is deeper, as over sunken features such as ditches (right). They may also show stunted growth over shallower soils, such as that over buried walls (left). Sometimes the features show up by differences in colour, sometimes they are picked out by shadows from the low-angled light near sunrise or sunset. Some crops accentuate the differences while others hide them so that features observable at one season or year may be missing on another. (See Colin Renfrew and Paul Bahn, *Archaeology: Theories, Methods and Practice* (Thames and Hudson, London, 2nd edn, 1996) Chapter 3.)

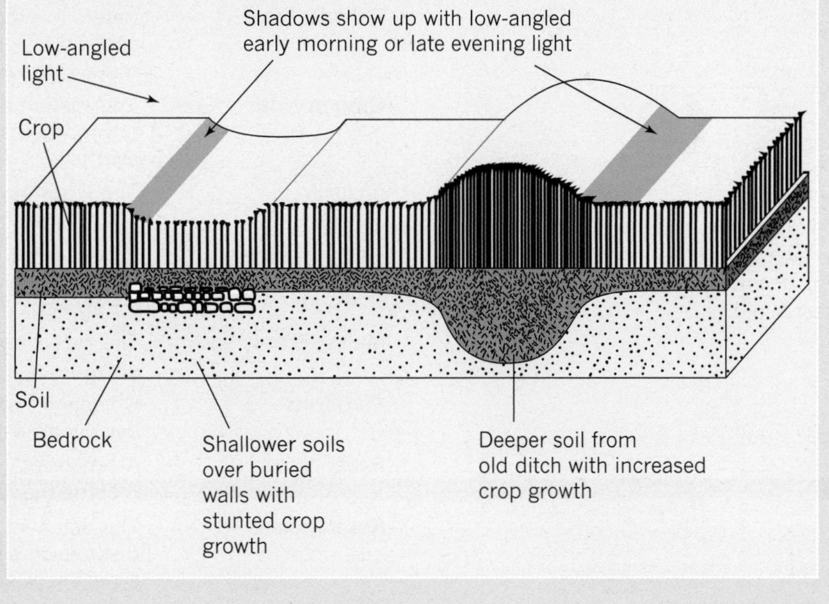

Remote Sensing from Satellites 22.2

It was World War I, with the advent of aircraft and the widespread use of military intelligence, that converted aerial photography from a pastime into a programme. That war led to a growing interest in aerial photography for resource surveys from the 1920s onward. World War II and the cold war that followed it had equally dramatic effects in creating a 'space race'.

This was first seen in the high-flying manned aircraft (such as the U2 spy plane) of the 1950s and 1960s taking photographs from safe elevations as high as 40 km (25 miles). Then over the last half-century the development of satellites and space stations, both manned and unmanned, has provided a new dimension of airborne surveillance. Hundreds of satellites now circle the earth along orbital paths and stay in space for lengths of time dependent on their size and distance from the earth.

The impact of this *remote sensing* on the provision of geographic information has come from two factors: first, the rapidly extending range of sensors, and second, improvements in the spacecraft themselves. With it has come a new language, some key terms from which are summarized in Table 22.1.

Improvements in Sensors

As Figure 22.6 shows, *sensors* for detecting features on the earth's surface fall into two categories: photographic sensors that capture data on film and

Table 22.1 Terms used in remote-sensing studies

Terms	Definitions
Bands	Sections of the electromagnetic spectrum with a common characteristic, such as the visible band.
Enhancement	Refers to processes that increase or decrease contrasts on received images (e.g. photos) so as to make them easier to interpret.
Ground truth	Information about the actual state of any environment at the time of a remote-sensing flight overhead.
Imagery	The visual representation of energy received by remote-sensing instruments.
Line scanning	Produces an image by viewing and recording a picture one line at a time, as on a cathode-ray tube (or TV set).
Multispectral sensing	The recording of different portions of the electromagnetic spectrum by one or more sensors.
Platforms	Objects on which a remote sensor is mounted, usually an aircraft or satellite.
Radar	A sensor that directs energy at an object and records the rebounded energy as radio waves.
Resolution	The ability of a remote sensing system to distinguish signals that are close to each other, in time, space, or wavelength.
Sensors	Instruments used to detect the electromagnetic energy associated with a particular object on the earth's surface.
Signatures	The unique pattern of wavebands emitted by a particular environmental object.
Synoptic images	Those images that give a general view of a part of the earth's surface, usually from high-altitude satellites.
Thermal infrared	Records the thermal energy *emitted* by objects of different temperatures on the earth's surface.

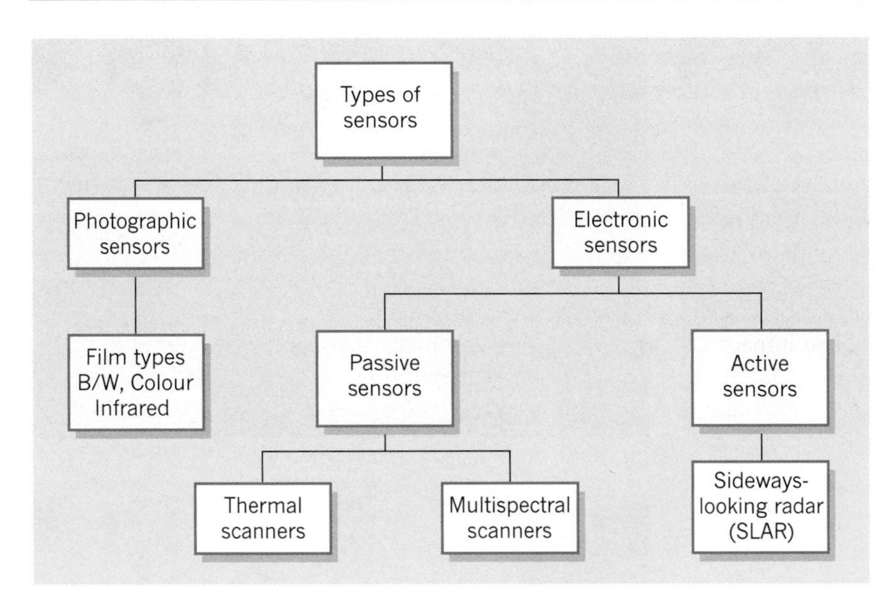

Figure 22.6 Types of sensor

electronic sensors that record data as numerical values. We consider each in turn.

Photographic sensors are instruments used to detect the electromagnetic energy associated with a particular object on the earth's surface. Black and white (panchromatic) aerial photography has for long been the principal sensing technique for the geographic study of the earth's surface. This uses the energy reflected by visible light falling on that surface. However, during the last half-century there have been significant extensions in both the range and the capability of sensors. The array of remote-sensing systems now available has been expanded by the exploitation of different parts of the *electromagnetic spectrum* (see Figure 22.7).

Photographic Sensors Developments in photographic chemistry have vastly extended the amount of information that can be captured on photographs. For example, look at the series of photographs in Figure 22.8. The subject is the same in all three photos, but the film picks up and accentuates different aspects of the scene. Black-and-white panchromatic film (used in the second photo) is sensitive to all colours, whereas black-and-white orthochromatic film (used in the third photo) is sensitive to all colours except red. Thus the panchromatic film represents all colours as shades of grey, and the orthochromatic film records red objects as black. New types of film are sensitive to electromagnetic waves on the borders of visible light.

Geographers' interest in variations in land use make them especially concerned with films that can differentiate between vegetation colours, notably shades of green. *Infrared photography* uses film that picks up varieties of

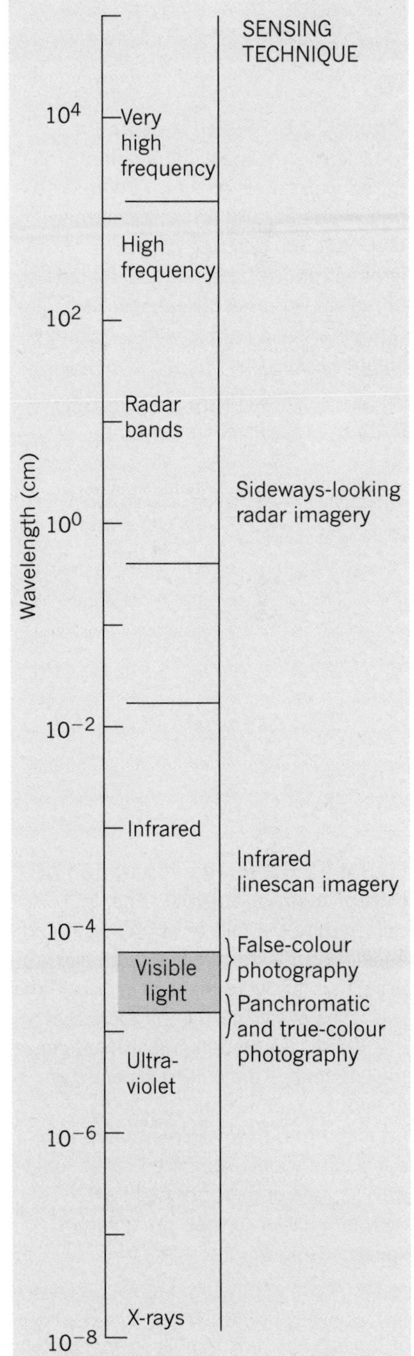

Figure 22.7 Sensing techniques and the electromagnetic spectrum
Remote sensing was at first confined to emissions of invisible light, and conventional photographic techniques were used. Equipment in use today can also detect infrared and radar waves. As more sophisticated sensors become available an ever-wider range of electromagnetic emissions from the earth's surface can be recorded and mapped. *False colour* photographs are used to emphasize and separate important features by giving them distinctive and contrasting colours. In such photos, healthy vegetation, for example, may show up as bright red rather than green.

Figure 22.8 Film types and landscape images Photographs of mixed woodlands using (a) infrared, (b) panchromatic, and (c) orthochromatic film. Note the differing capability of the films to reduce haze and differentiate broadleaf from needleleaf species. Compare these photos with the chart in Figure 22.9.

[Photos by Tony Philpott, Geography Department, Bristol University.]

tone and colour that are not evident in either black-and-white or true-colour photographs. Figure 22.9 shows how the reflective properties of forest-covered terrain vary for emissions of different wavelengths. Differentiation between the two types of forest is clearly much easier than in the visible sections, where the properties of the two types are similar and partly overlap. Most landscape features appear to have similar distinctive

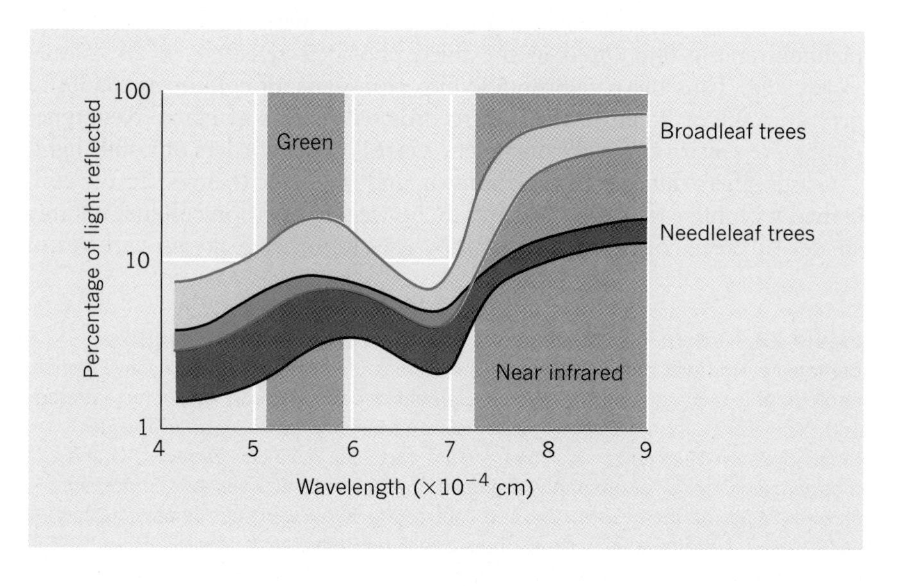

Figure 22.9 Image separation at different wavelengths The light reflected by the foliage of broadleaf and needleleaf trees is shown by the two broad curves. Note that the curves of the two types of foliage slightly overlap at the shorter wavelengths of visible light, making foliage difficult to separate on ordinary panchromatic or orthochromatic film. However, the two curves are widely separated in the longer wavelengths of near infrared light. It is therefore easier to map the different species using infrared film.

[Source: Adapted from R.N. Colwell, in R.U. Cooke and D.E. Harris, *Transactions of the Institute of British Geographers*, No. 50 (1970), Fig. 3, p. 4.]

spectral signatures. Each signature is based on values in each *pixel*, the smallest unit of information on the image.

Electronic Sensors Remote sensing devices detect and record radiant energy emitted by or reflected by ground features and transmitted to the sensing instrument in the form of waves. As Figure 22.6 shows, these sensors may be of two kinds: active sensors (such as radar) and passive sensors such as thermal and multispectral scanners. These newer, non-photographic sensing systems have advantages and liabilities when it comes to the analysis of terrestrial phenomena. Radar sensors direct energy at an object and record the rebounded energy as radio waves. These are usually arranged to cover a wide sweep of landscape by taking oblique images and are known as *sideways-looking airborne radar (SLAR)*. There are two advantages of radar images over conventional photographic images for mapping purposes. First, radar imagery is independent of solar illumination and is unaffected by darkness, cloud cover, or rain. This means that any part of the globe can be scanned on demand, including the zones especially difficult or impossible to photograph by conventional methods – such as the cloud-covered humid tropics and regions of polar night. The time available for radar scanning in the humid mid-latitudes is five or ten times greater than that available for taking aerial photographs of acceptable quality.

Second, radar imagery gives greater detail of the terrain. For example, radar images on a scale of about 1:200,000 provide information on drainage patterns which is roughly equivalent to that derivable from a 1:62,500 topographic map. High-resolution radar can pick up very fine irregularities in the ground surface, and even show differences in subsurface conditions to a depth of a few metres. Figure 22.10 gives an example of a *radar image* of farmland in Kansas in the American mid-west. The various crops appear as different shades of grey; the lightest shades indicate sugar beets.

Passive sensors have now been developed to record signals from a wide range of the electronic spectrum from the visible through to infrared wavelengths. Signals from the ground may be recorded in many ways. Among the earliest were matrix-array instruments with a series of microscopic cells (a 1000 × 1000 array would just cover your thumbprint) which recorded the intensity of radiation received by a number. Numbers were transmitted to a large on-board computer and then transmitted to a receiving station on the earth's surface.

Newer systems are based on linear (rather than matrix) arrays in which the sensors are scanned at very high frequencies. The scanning processes, variously described as 'pushbroom' and 'whiskbroom' modelled on the way you brush leaves from your back yard, are matters for more advanced remote sensing courses.

Figure 22.10 Radar images of landscapes Farmland in Kansas as seen by radar. Different crops are distinguished by various tones. The lightest areas are planted with sugar beets.

[Source: Photo by Westinghouse Electric Corporation.]

Improvements in Satellites

The first American satellite (Explorer I) was launched in January 1958. Remote sensing from manned space flights began four years later when American astronaut John Glenn's took some tourist-like 35 mm photographs during his three orbits of the earth. The Mercury, Gemini, Apollo, and Skylab missions that followed over the next decade also tested various forms of monitoring equipments of increasing power and sophistication.

The first generation of earth observation satellites was formed by the eight members of the TIROS (Television and Infrared Observation Satellites) family launched in the early 1960s. As their names implied, the satellites carried two types of sensing devices. First, there were television cameras that transmitted pictures of the visible part of the spectrum back to earth. The first detailed weather pictures of cloud patterns were received as early as 1960, and TIROS III discovered its first hurricane (hurricane Esther) in July 1961. Second, TIROS carried infrared detectors that measured the non-visible part of the spectrum and provided information on local and regional temperatures on the earth's surface.

From a geographer's viewpoint, the most important development in global satellites was the ERTS. This satellite was launched from California in July 1972. (subsequently renamed as LANDSAT-1). It makes 14 revolutions a day around the earth, its sensors covering a series of 160 km (100 mile) wide strips. The strips overlap, so that the whole surface of the earth is covered once in every 18 days. Thus in the first year of the satellite's life each part of the planet came within range of its sensors 20 times. Of course, many areas were cloud-covered, but it is estimated that in the first year LANDSAT-1 provided cloud-free coverage of about three-quarters of the world's land masses. Pictures reaching the satellite are converted into electronic signals, stored on tape, and then broadcast back to three ground stations, at Fairbanks, Alaska; Goldstone, California; and Greenbelt near Washington, D.C.

Three years later LANDSAT-1 was joined by a second earth resources satellite (see Table 22.2). The further members of the *LANDSAT* family are summarized in Table 22.2. Since each is following an orbit high above the earth's surface the area covered on a single frame is correspondingly large – some 30,000 km^2 (11,600 square miles) or an area roughly the size of Albania or Vermont. Despite this, the image can be 'blown-up' in photographic terms to show an object the size of the Washington Monument in Washington, D.C.

The French *SPOT* system (*Systéme Probatoire d'Observation de la Terre*) began in 1986 with the launch of SPOT-1. This orbits at a height of 830 km (520 miles) on a near polar orbit passing over the same location every 26 days. It has a very high resolving power (down to a 10 m resolution) and can combine a series of wavebands to create accurate images similar in appearance to infrared-colour photography.

Multispectral Bands Unlike the images recorded by earlier satellites, the images received by LANDSAT are passed through a *multispectral scanner*

Table 22.2 Earth resources satellite systems

Name	Launch dates	Altitude	Repeat pattern	Spectral bands	Scanners
Landsat 1–3	1972–84	920 km (570 mile)	18 days	5 band	Multispectral scanner (MSS)
Landsat 4–5	1982–84	700 km (435 mile)	16 days	4 band 7 band	Multispectral scanner (MSS) Thematic mapper (TM)
SPOT 1	1986–	830 km (520 mile)	17 days	4 band	High-resolution visible (HRV)
NOAA	1987–	830 km (520 mile)	17 days	5 band	Advanced very high resolution radiometer (AVHRR)

which is sensitive to different parts of the electromagnetic waveband which we have already met (Figure 22.7).

Each spectral band picks up different information about the earth surface below. For example, water produces a distinctive *spectral signature*. It gives medium values on the green and red bands but low and very low values on the two infrared bands. In musical notation, water would have a sound ♫♫! By contrast, pastureland has the sound ♫♫! Note that both look alike on the first two notes but differ on the third and fourth. While not all features have such clearly defined signature tunes, the principle of using information from all four bands remains. This means that the signals transmitted to earth can be recombined in many ways to bring out unsuspected features of the planet's surface. For example, when the signals are decoded and exposed on photographic film, they may produce false colour images to emphasize and separate important terrain features. For example, vigorously growing vegetation may show up as bright red and diseased crops as pale yellow. Clear water may show up as black, while contaminated water carrying silt and sewage may appear bright blue.

The use of satellites for remote sensing is clearly still in its infancy. At the time of writing work is going ahead in the United States on both a space shuttle and spacelab programme which will provide further opportunity for direct observation by scientists of the earth's surface. The NASA organization is paralleled by a major space-research programme in the former Soviet Union while the Western European countries have combined (from 1975) in the form of ESA, the European Space Agency.

Monitoring Environmental Change 22.3

We have shown earlier how aware geographers must be of environmental change (see Parts I and III). Remote sensing from satellites provides unique opportunities for monitoring this change on a scale unthinkable to an earlier generation of researchers. Here we look at some potential areas of research and then examine more closely a regional example.

Potential Research Frontiers

From the geographic viewpoint, satellites are immensely useful because of their very wide coverage and the fact that difficult or inaccessible terrain presents no bar to data collection. Three areas of research applications seem exceptionally promising at this time.

Global Coverage The first is studies that take advantage of the potential *worldwide coverage* of satellite systems. Typical are proposals for a worldwide study of surface temperatures using infrared scanning systems (Figure 22.11). For despite a growing network of information on atmospheric temperatures, the data on temperatures of the ground surface itself are rather sparse and generally available only for developed areas. Records of daily and seasonal changes in temperature would tell us more about the heat and water balances between the earth and the atmosphere. Such data bear

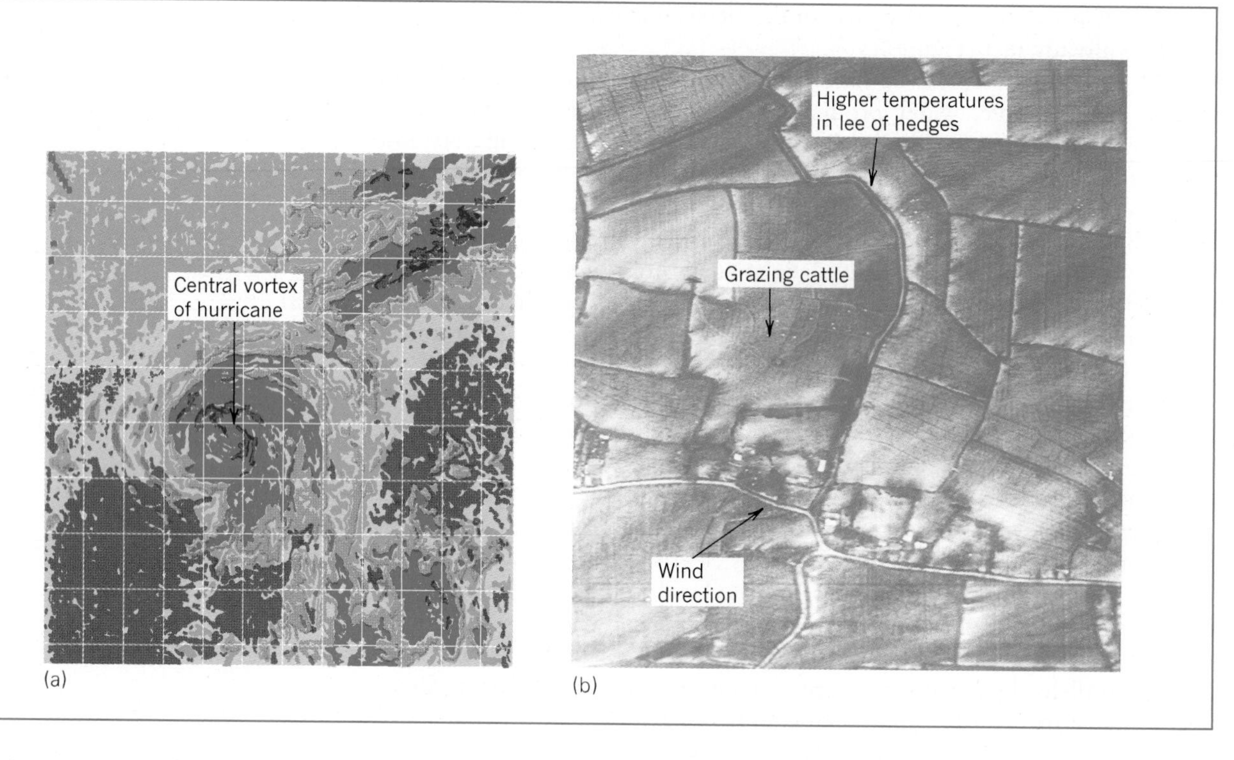

Figure 22.11 Heat landscapes
(a) Infrared data on Hurricane Camille (1969) with differences in temperature showing up as different shades. Note the characteristic vortex shape with violent winds moving around a calm central area. (b) The image shown by an infrared linescan sensor of the River Axe lowlands, southwest England. Note shelter effect of hedges showing in light tones of those ground areas with higher temperature. Grazing animals are seen as light spots by reason of their high body temperature. A conventional air photograph of the same general area is shown in Figure 22.3(c).

[Source: Photographs courtesy of NASA and Dr L.F. Curtis, Bristol University, and Royal Signal and Radar Establishment, Malvern.]

directly on human needs by helping to identify areas where the temperature is suitable for certain crops.

There is a similar lack of worldwide information on precipitation. Rainfall is normally recorded daily by meteorological bureaux' gauges scattered over the earth's surface. Satellites promise the rather exact location of rain and snow as they are actually occurring. Through links with conventional ground stations and surface radar stations, the type and intensity of precipitation can be measured, and improved forecasts can be given.

An ideal system of global watching needs a system something like that shown in a simplified form in Figure 22.12(a). This consists of three elements. First, *geosynchronous satellites* which continuously swing around in time with the earth's rotation and cover its low-latitude areas (effectively latitude 40°N to 40°S). Four are shown, but this number has been increased, with their orbits nearer to the earth and their revolving power improved. Second, non-geosynchronous satellites which cover the middle and upper latitudes. Two are shown following a polar orbit at lower altitudes. Finally, a low-level satellite in equatorial orbit which provides the fine detail not available from the more distant geosynchronous platforms. Completion of such a system from the 1980s has now been made as part of the World Weather Watch Program of global atmospheric research.

A scheme for taking advantage of the global coverage of the satellite system relates to monitoring global land use. The World Land Use Survey was initiated by British geographer Sir Dudley Stamp as long ago as 1950. Its aim was to provide a series of maps, on a scale of 1 : 1,000,000, of the human use of the earth's land surface. Satellites have allowed that static scheme to be expanded to provide regular monitoring of land-use change at a variety of map scales using multispectral scanners (Figure 22.12(b)).

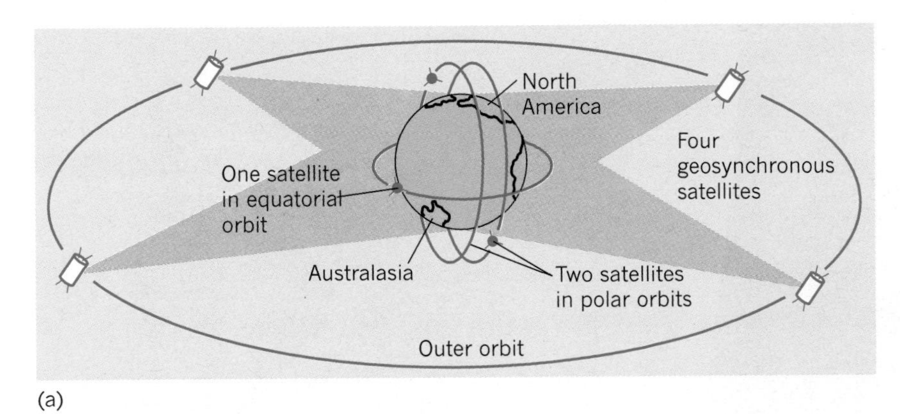

(a)

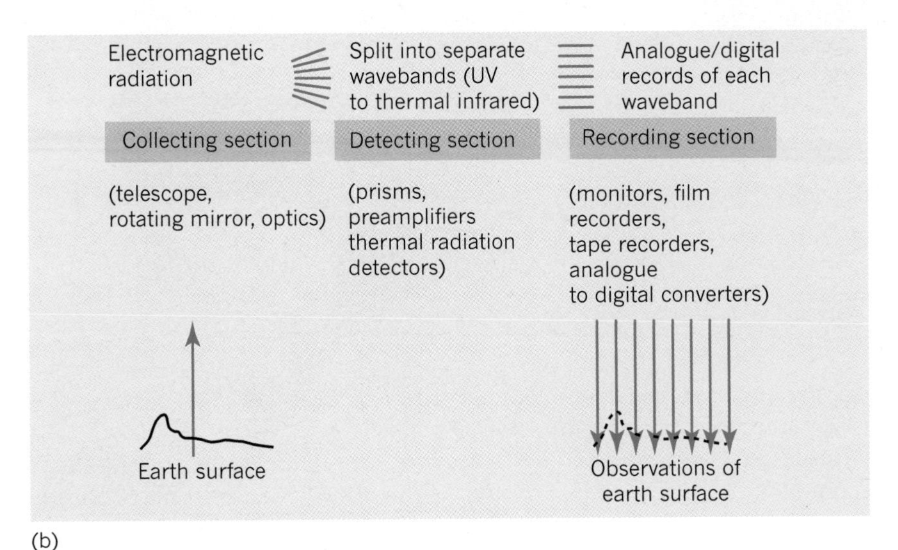

(b)

Figure 22.12 Elements in worldwatch surveillance systems (a) Principles of a proposed global observing system for atmospheric surveillance using seven satellites. (b) Main elements in a multispectral scanner.

[Source: A. Goudie *et al.* (eds), *Encyclopaedic Dictionary of Physical Geography*, second edition (Blackwell, Oxford, 1994), p. 346.]

A second example of global monitoring relates to cloud interpretation studies. An example is presented in Figure 22.13.

Inaccessible Regions The second promising area of research is the study of phenomena that occur mainly in *inaccessible* and therefore *sparsely monitored* parts of the earth's surface. Most of the ice masses of the world, for instance, are located either in polar or high mountain areas where both ground and airborne surveys pose logistical problems.

With satellite surveillance we can regularly record changes in the extent of ice masses, their surface characteristics, or the presence of new snow deposits. Such studies are of value because glaciers contain three-quarters of the world's fresh water and are critically connected to short-term changes in the earth's hydrology and climatology, as well as to long-term changes in sea levels. Studies of glacial budgets (that is, gains or losses in the volume of ice in a glacier or ice sheet), and rates of iceberg measurement at the seaward margins are all feasible with existing satellite and sensor technology. Work is going ahead in this field.

Highly Mobile Features The third promising area of satellite research is the study of *ephemeral* or *highly mobile* phenomena that cannot be recorded with present survey techniques. We can include in this area

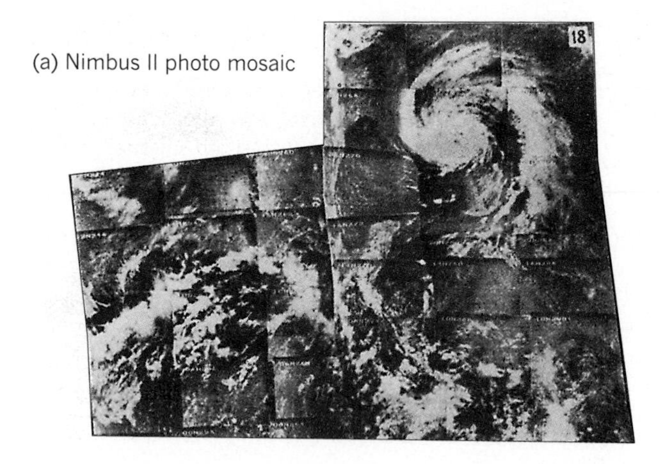

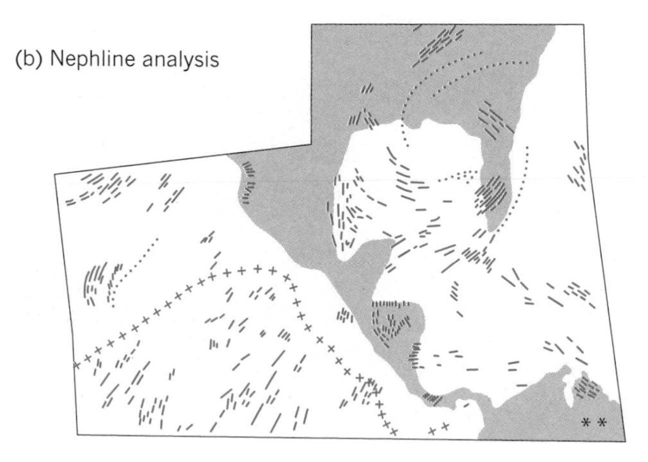

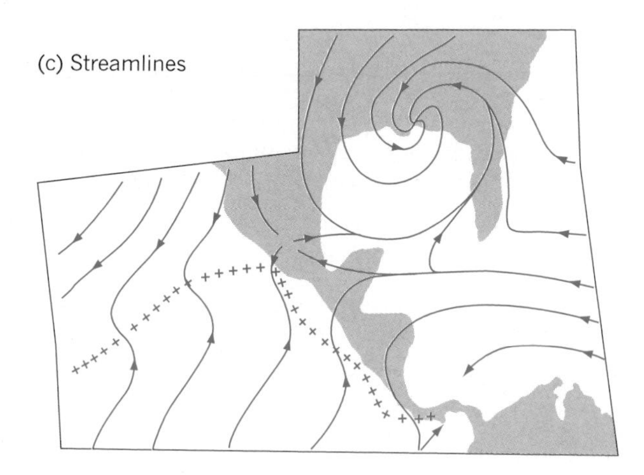

Figure 22.13 Cloud analysis using satellite photographs (a) Nimbus II photographs of the central American region, 11 June 1966, compiled into a ***photomosaic***. (b) The identification of different cloud types, each characteristic of different meteorological conditions. This is termed nephline analysis. The line of crosses (+ + + + +) shows the position of the intertropical convergence on this day. (See Figure 2.11 for the average position of the convergence zone.) (c) The smoothing of nephlines to indicate the main patterns of air circulation.

[Source: E.C. Barrett, *Progress in Geography*, ***2*** (1970), Figs 19, 20, pp. 192–3.]

proposals to monitor the distribution of bush fires and their effect on natural vegetation. High-resolution images with sharp detail also can record the occasional, but ecologically critical, human use of wildlife and recreational areas. Photographs from aircraft are already helping to determine the location and intensity of vehicle parking in national forests and wildlife areas. Satellite photographs also can check on the increasing pressures on these areas and give us early warning of their overuse.

More modest projects use spacecraft for recording changes in the distribution of airborne sediments (e.g. dust clouds and pollution), movements of coastal sediment, movements of traffic in metropolitan areas, and the distribution of sea-going craft. One particularly intriguing suggestion in the realm of historical geography is that sensing devices could help us to reconstruct caravan networks in the western Sahara and Takla Makan deserts. Trails frequented by animals have a surface composition and chemical content unlike the untrodden and unfertilized terrain around them, and it seems possible that such differences are detectable by appropriate sensors.

Early-warning Systems

Keeping the earth under routine surveillance from satellites may allow the development of early-warning systems. Let us look at an example of a US Food and Agriculture Organization (FAO) programme. Since the earliest Egyptian records, locusts (Figure 22.14) have been known to erupt in mass-

(a)

(b)

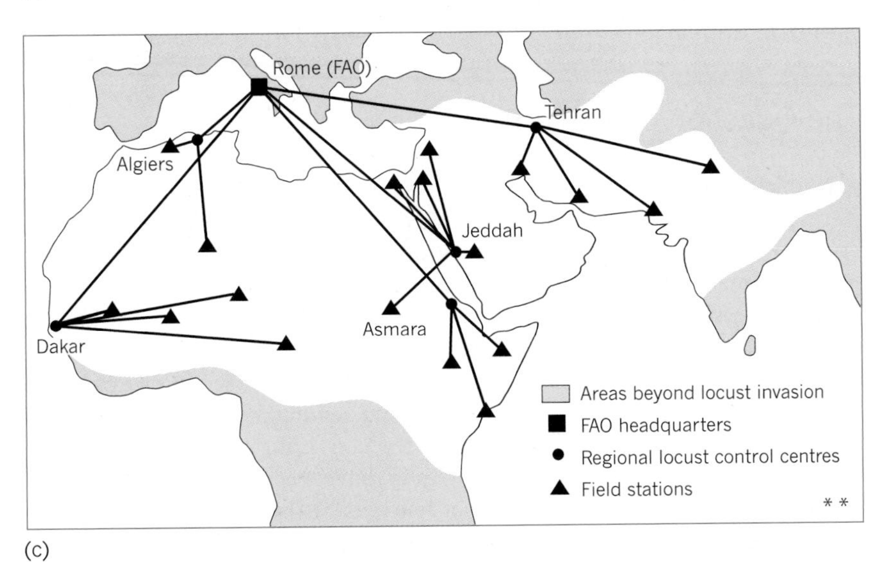

(c)

Figure 22.14 Use of remote sensing in pest control (a) Locust swarm. (b) NASA photo of conditions over the Persian Gulf (black areas) and adjacent land areas (grey). On this day, dry weather dominated with a band of cirrus clouds (white) not leading to rainstorms. (c) Network of locust control centres in North Africa and the Middle East based on the United Nations' Food and Agriculture Organization (FAO) headquarters in Rome.

[Source: (a) FLPA/Pelling. (b) (c) E.C. Barrett, Consultant's Report to the Food & Agriculture Organization of the United Nations on Assessment of Rainfall in North Eastern Oman, Bristol, January 1977.]

ive swarms, devastating crops over hundreds of square kilometres. The Old Testament describes them as one of the plagues that descended on Pharoah's Egypt. Although this member of the grasshopper family lives in the arid and semi-arid lands of the world, its breeding cycle is critically dependent on rain. Rains supply the moist soil conditions which allow the insect to get through the 40 to 100 day gap between egg-laying and the adult stage when locusts then emerge in massive swarms. These are both hungry and highly mobile, having been known to move distances as far as 3500 km (2200 miles) in as little as three weeks.

Information about rain in the potential breeding grounds in the desert margins of the Sahara is of vital interest to the FAO locust control network (Figure 22.14). But information is hard to gain since these areas are very sparsely peopled, there are few meteorological stations, and when rains do come they fall as a random pattern of intense but highly localized storms. So studies have been started to try to use environmental remote sensing. In southern Algeria and central Arabia LANDSAT records have been used to identify areas where *rains are probable* (based on the cloud characteristics), and where *rains have recently fallen* (based on the green 'flush' of plant growth that comes after the rains). Once key areas have been identified from the photos, then the remote-sensing team at Rome can alert the regional headquarters of the locust control programme. Control officers then fly over the areas with light aircraft, or conduct ground surveys in the more accessible areas. If the presence of locusts in their larval stages is confirmed, then control measures can be put under way.

Such early-warning programmes are still in their infancy. Evidence to date of successful applications shows that remote sensing will help to save millions of dollars in crop loss.

22.4 Some Unresolved Problems in Remote Sensing

In this final section we look at two problems in using remote-sensing data that remain only partly solved. The first is the availability of images and the second is the links between the image and reality on the ground.

The Question of Availability

A problem for geographers and other researchers has been how to get hold of remote-sensing images, and the price of the images. There are three general categories of imagery in most countries: (1) imagery secured and held by government agencies which is placed in the public domain, (2) commercially available imagery, and (3) restricted use imagery which is classified as being of strategic or military significance.

Over the last few decades, the trend has been for the second category to grow. Government agencies often make data readily available in the initial development period but then turn over the copyright to commercial agencies. Areas covered by the third category have been generally shrinking, largely by the reduction in US–Soviet tensions at the end of the Cold War but also by the launch of commercial satellites.

Most countries have now established central agencies where the national holdings of aerial and satellite information are catalogued and indexed. For example, in the United States the best place to begin the search is at one of the five National Cartographic Information Centers (NCICs). These are part of the US Geological Survey (USGS) and help individuals to locate photographs, satellite images, maps, and other spatial data. Actual supply of products is usually from another government agency or a commercial company. For example historical aerial photographs are available from the National Archives and Records Library in Washington while LANDSAT images are available from the Earth Observation Satellite Company in Lanham, Maryland. Appendix B gives some useful web sites for those wishing to pursue the question further.

The end of the Cold War in 1992 saw the start of 'open skies' agreements between major world powers. This, plus evolving satellite, recording, and data compression technology, means that images from around the world are now routinely available. An Indian company is now selling images from around the world at a 5 m resolution and Russian company at a 2 m resolution. Pictures down to resolution of 1 m are already available over the Internet from an American satellite company, Space Imaging. Washington, D.C. Its Ikonos satellite launched in 1999 passes over the same point of the earth every three days. If you have a credit card you can order photographs of more or less anything, anywhere.

In 2000 Britain's Millennium Map was launched. This gives a photographic record of the whole of Britain from the air, forming a twenty-first century Domesday Book of the country (see Section 11.1). This was achieved by a team of four pilots flying up and down the country at height of 1400 m (4500 ft) with special computer-controlled cameras taking pictures.

Viewers can pan across the whole of England at a scale of 1 : 1,000,000, or zoom in on a single house at scales less than 1 : 1000. Every outdoor object larger than 25 cm (10 inches) is visible. Customers will be able to type in a post-code or a grid reference into the website (www.getmapping.com) and order a picture of their house, school, or neighbourhood for a price equivalent to that of buying a map sheet. The map will be updated according to demand but images will be revised at least every five years. Other sources of images on the World-Wide Web are listed in Appendix B.

The Question of Ground Truth

In remote-sensing studies, the spectacular gains since 1950 have come from the ability to record all parts of the Earth's surface with ever-increasing accuracy. The fact that images are now routinely available at resolutions that allow individual humans to be plotted gives some idea of the huge gains that have been made.

But, in the end, the image remains as simply an image. It is only when the pattern shown there can be securely matched with the environmental reality at the point on the earth's surface that is being shown that the image can provide useful information. The term *ground truth* has been coined to describe the link between the reality at ground level and the fleeting image caught by a passing aircraft or orbiting satellite. Thus studies of precipitation from cloud studies have to be linked to actual rainfall or snowfall on the ground. As Figure 22.15 shows, although the number of precipitation stations worldwide has increased tenfold over the last century it still remains very 'spotty'. A high density of recording stations tends to reflect both population density and development levels.

In this drive to make valid worldwide statements about the earth as the home of the human species, geographers often assume the need to inspect all corners of the world and to make a complete environmental inventory. When we compare our situation with that of other investigators, we see that this assumption can be relaxed. The geologist makes inferences about rock strata from the records of a few bore holes; the pollster makes predictions about our voting behaviour by interviewing a few thousand citizens, not millions.

In a way the earth's surface is analogous to a population. To be sure, the population is an unusual one because it does not consist of a finite number of individuals (for instance, the six billion individuals who make up the world population) but is continuous. Therefore, when we want to designate an individual within this spatial population, we have to do so in an arbitrary way, by specifying a point location such as 68°30′N 27°07′E, an aerial unit such as 1 km × 1 km, or a census division such as a tract or county. These individuals, however defined, form *sampling units*, from which we can construct a picture of the population as a whole.

Making inferences about a population from a small part of it is a dangerous procedure. We may select samples that are biased in some way or the sample may be too small for a reliable estimate. But how do we decide what

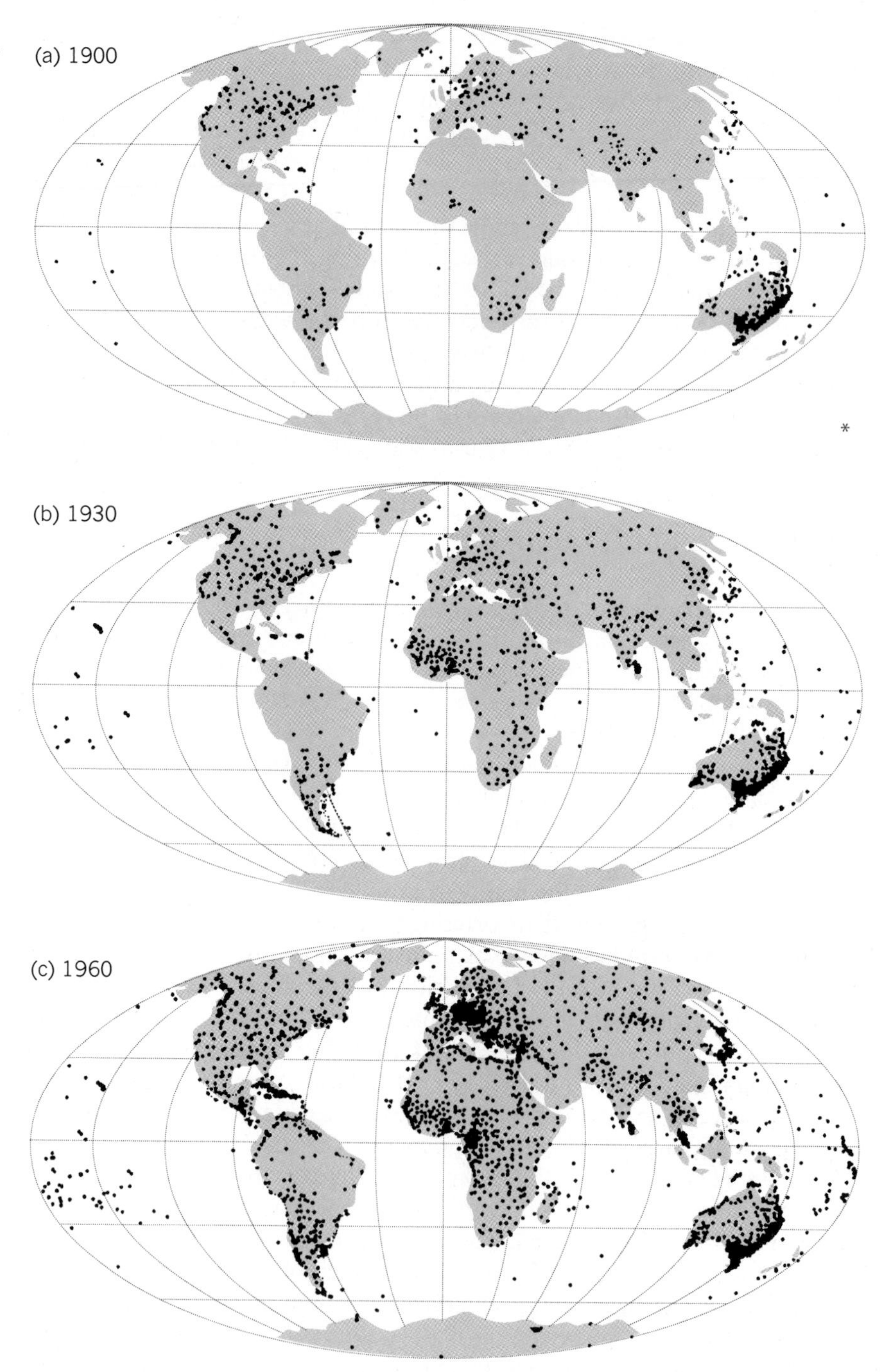

Figure 22.15 Global distribution of ground truth Global distribution of precipitation stations used by the National Center for Atmospheric Research in its worldwide surface climatology project for (a) 1900, (b) 1930, and (c) 1960.

[Source: C.J. Wilmott *et al.*, *International Journal of Climatology*, **14** (1994), Fig. 1, p. 404.]

is 'too small'? Fortunately, many of the rules governing the relationships between a sample and a population were worked out by mathematicians such as Karl Pearson and R.A. Fisher in the first half of the twentieth century. Today there is a large body of well-substantiated sampling theory to which geographers can turn in setting up their investigations. Sampling theory helps us to decide how much error is likely to occur with sample designs of different kinds. For example, in its basic form *random sampling* (Figure 22.16), the accuracy or sampling error is proportional to the square root of the number of observations. This means that if we were to increase the number of sampling points in Figure 22.16 from 25 to 100, we could

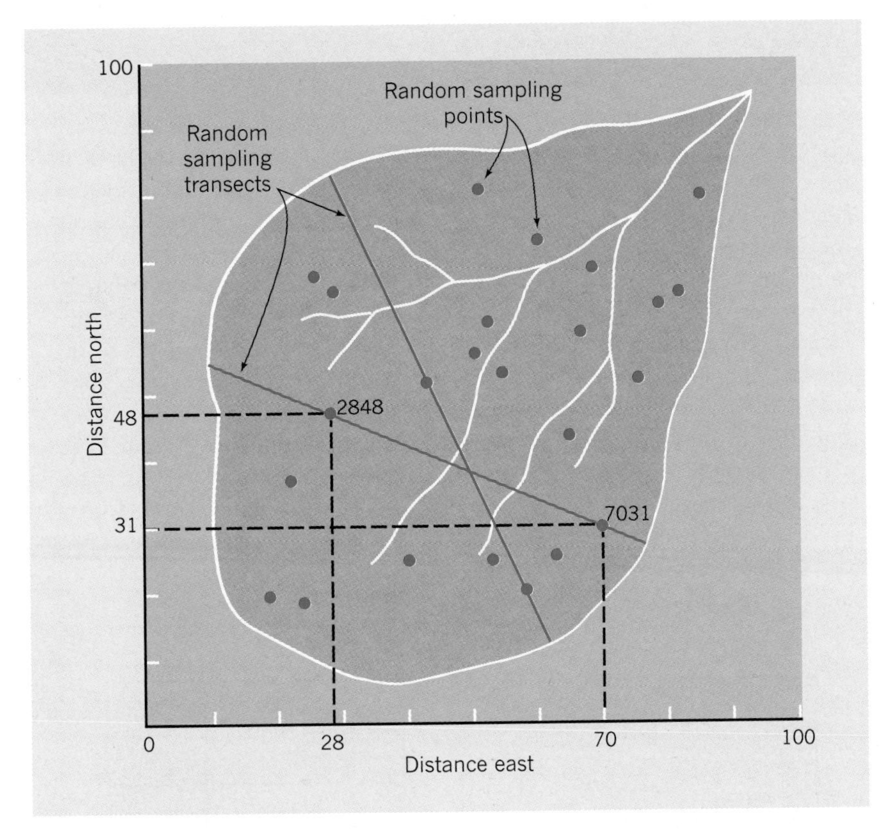

Figure 22.16 Simple random sampling It is frequently too costly to inspect every part of a region before making generalizations about it, so geographers rely increasingly on a series of sample observations which are then matched against remote-sensing images. But how should such observation points be selected? To avoid bias, the location of each sampling points within the watershed shown here is given a pair of random numbers on the two axes of the coordinate system. As an alternative to the point sampling systems random transects across the area may also be used. In these, lines are drawn through pairs of random number coordinates.

expect to improve our accuracy not by four times as much, but by only $\sqrt{4}$, or 2.

Geographers design their sample surveys of environmental characteristics in close cooperation with their colleagues in statistics, and most of the problems they encounter are general statistical ones we need not be concerned with here. Geographic research has been confined mainly to working out the efficiency of different kinds of spatial arrays of sampling points. Each spatial sampling design has particular disadvantages and advantages and is used for specific types of environmental investigations. (See Box 22.C on 'Sampling designs'.)

Conclusion: The Unfinished Map

As we saw in Chapter 21, cartography was one of the earliest areas of research by geographers and played a key part in the geographic training of the Greeks. The last century saw that mapping process transformed as first aircraft (after World War I) and satellites (after World War II) brought new observing platforms for mapping and monitoring the earth's surface.

The huge revolution in global monitoring recorded in this chapter remains incomplete. The world observed by capturing energy pulses from the earth's surface is only part of global reality. They are a necessary but not a sufficient condition for understanding global geography. In the next chapter, we turn to another revolutionary area, Geographical Information Systems (GIS), to see how other types of information can be integrated into mapping and remote sensing. Only when that integration is complete can we move closer to reaching the geographers' goal of understanding planet Earth as the home of the human population.

Box 22.C Sampling Designs

Geographers use different sampling designs to investigate particular spatial distributions. Each involves randomization procedures.

Stratified sampling divides the study area into separate strata, and individual sampling points are drawn randomly from each strata. The number of sampling points is made proportional to the area of each strata. Stratification can be used when major divisions within the area are already known to the investigator.

Systematic sampling employs a grid of equally spaced locations to define the sampling points. The origin of the grid is decided by the random location of the first sampling point. The grid also can be randomly oriented. Systematic location gives an even spatial coverage for mapping purposes.

Hierarchic, *multistage*, or ***nested sampling*** divides the study area into a hierarchy of sampling units that nest within one another. Random processes select the large first-stage units, the smaller second-order units within these, and so on. The final location of points within each small unit can be randomly determined. Nested sampling is useful in reducing field costs and in investigating variations at different spatial levels within the same area but it entails higher sampling errors. For a discussion of the applications and limitations of these and other designs, see B.J.L. Berry and D.F. Marble (eds), *Spatial Analysis* (Prentice Hall, Englewood Cliffs, N.J., 1968), Chapter 3.

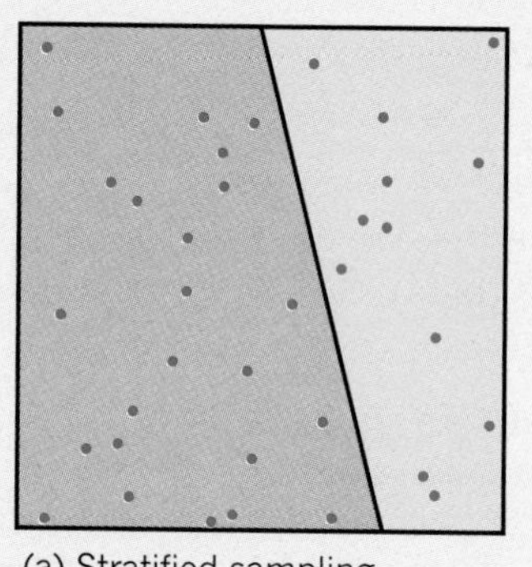

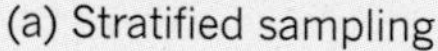

(a) Stratified sampling

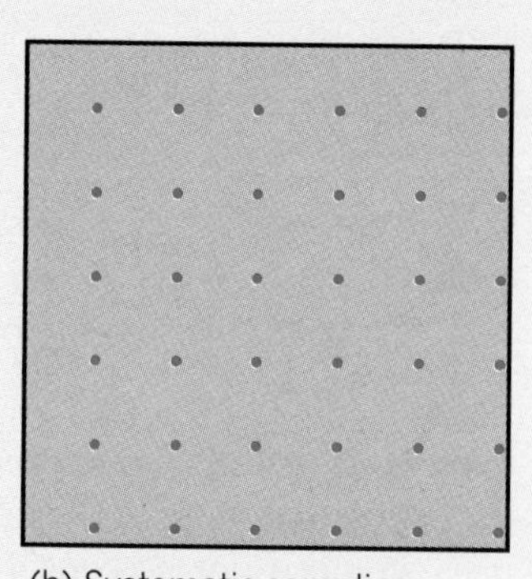

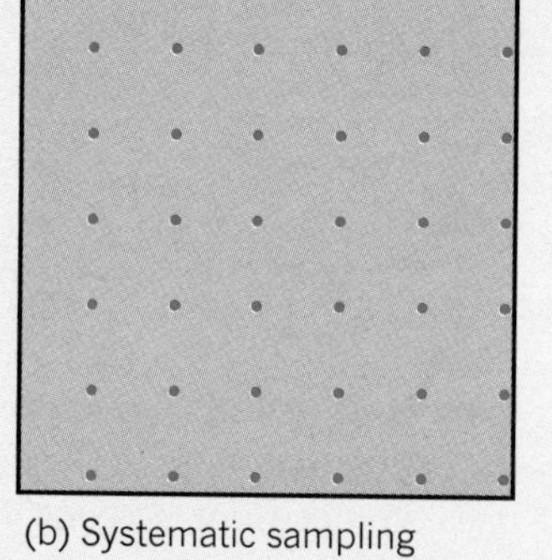

(b) Systematic sampling

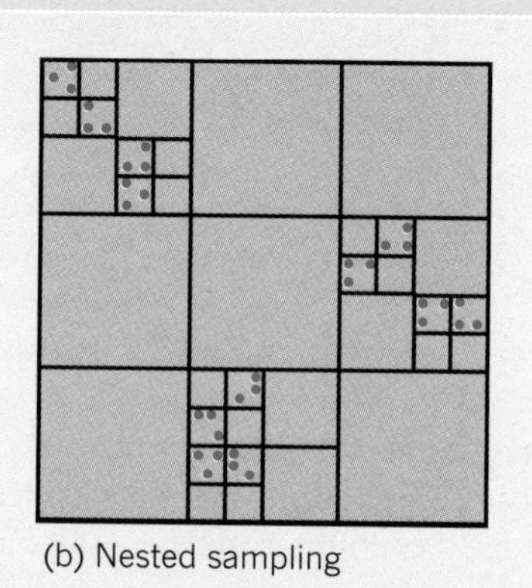

(b) Nested sampling

Reflections

1 Environmental remote sensing has revolutionized our ability to gain useful data from the earth's surface. Which parts of geography have gained the most from this revolution and which the least? Give reasons for your choice.

2 Radiation-sensitive sensors now enable satellites to capture differences in *heat* on the earth's surface on film or other sensor. List ways in which this capacity provides valuable information for the geographer.

3 From time to time there has been much criticism of the amount of federal money invested by the United States in the NASA space programme. Set up a debate in class on the value of this investment. On which side would you wish to speak? Why?

4 Consider the view that, now most of the world is mapped, the geographer can hang up his or her boots. What is the fallacy of this viewpoint?

5 Sampling provide one of the ways in which ground truth can be efficiently acquired to allow remotely sensed images to be checked against the reality on the earth's surface. Review the advantages and disadvantages of the various sampling schemes available.

6 Review your understanding of the following concepts using the text (see Table 22.1) and the glossary in Appendix A of this book:

aerial photographs
electromagnetic spectrum
geosynchronous satellite
ground truth
infrared sensors
multispectral scanner
pixels
random sampling
remote sensing
SLAR

One Step Further ...

A brief but authoritative introduction to remote sensing is given by Michael Goodchild in his essay in R.J. JOHNSTON *et al.* (eds), *Dictionary of Human Geography*, fourth edition (Blackwell, Oxford, 2000), pp. 699–701. Standard accounts of the field are given in E.C. BARRETT and L.F. CURTIS, *Introduction to Environmental Remote Sensing*, fourth edition (Thornes, London, 1999), T.M. LILLESAND and R.W. KIEFER, *Remote Sensing and Image Interpretation*, fourth edition (Wiley, New York, 2000), and G.M. FOODY and P.J. CURRAN (eds), *Environmental Remote Sensing from Regional to Global Scales* (Wiley, Chichester, 1994).

The basic principles of photo interpretation and examples of the use of aerial photographs in research are given in R.A. RYERSON *et al.* (eds), *Manual of Remote Sensing*, third edition (American Society for Photogrammetry and Remote Sensing, Washington, D.C., 1996). Excellent examples of earth photographs from spacecraft are available in a number of NASA publications, for example, N.M. SHORT *et al.*, *Mission to Earth: Landsat Views the World* (Government Printing Office, Washington, D.C., 1977). See also the review in D.J. BAKER, *Planet Earth: The View from Space* (Harvard University Press, Cambridge, Mass., 1990).

The 'progress reports' section of *Progress in Physical Geography* (quarterly) will prove helpful in keeping up to date. Although the results of current research are given in the regular geographic journals, you will need to look at specialized technical journals such as *Remote Sensing* (quarterly), *Photogrammetric Engineering and Remote Sensing* (quarterly), *Remote Sensing of Environment* (monthly), and *International Journal of Remote Sensing* (bimonthly) in order to keep abreast of the most recent developments in remote sensing. For readers with access to the World-Wide Web see also the sites on 'aerial photographs' and 'satellite photographs' in Appendix B at the end of this book, especially those NASA sites that give splendid colour views of the earth's surface from space.

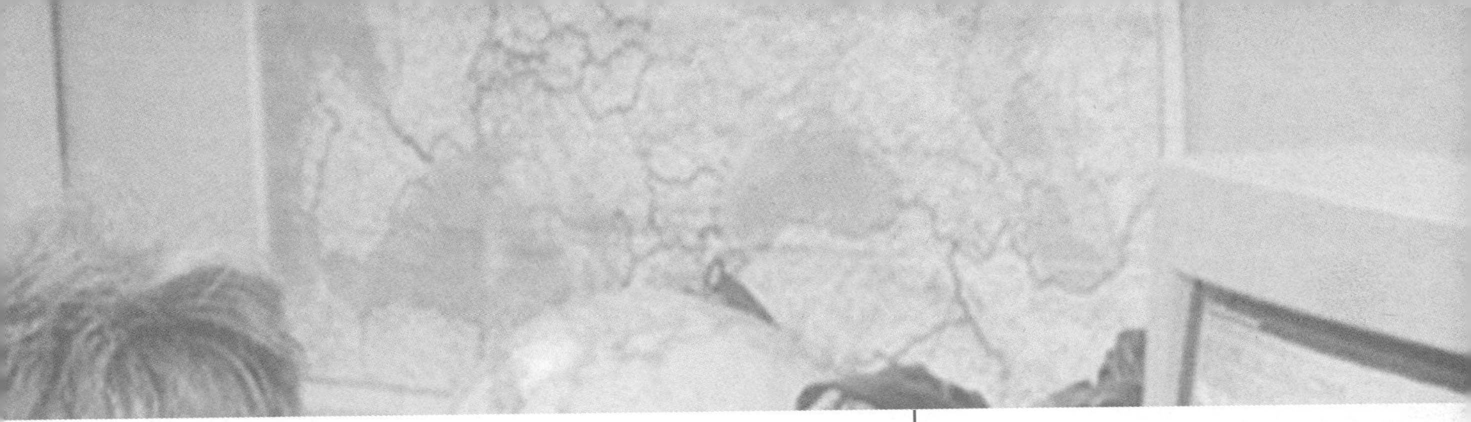

CHAPTER 23

Geographic Information Systems

■ denotes case studies

Figure 23.1 Digital and conventional mapping This photo shows both conventional and emerging mapping media. The wall map and the globe represent the fixed output of traditional cartography, the computer represents the flexibility of GIS. Data are being digitized and entered electronically into a computer database where it can be edited and combined in a myriad of ways. [Source: Integraph.]

> Our confidence in GIS is built on the belief that geography matters.
>
> JACK DANGERMOND President and Founder of ESRI (2000)

One of the impacts of the new and sharpened tools available to geographers since 1950 has been a huge increase in the amount of data about the world. To our forebears the critical limitation to understanding the planet and its geography was the lack of facts. Where was the coastline of Australia? What was the population of Lima? How many people migrated from China into Vietnam? Many of the early maps were full of blanks to be filled in by occasional explorers' reports, a diplomatic office telegram, a tentative census return.

Today the tide of 'data' threatens to overwhelm us. Exact figures depend on how you define data, but if we take just written information then we see the world's great libraries bursting at their seams. Ever more metres of shelving are needed and an increasing amount of the old collection is housed in warehouses, even old aircraft hangers, away from the main library. The rate of new acquisition shows we are dealing with an exponential rate of growth (one in which, as we saw in Chapter 6, the item being counted doubles over some period of time). To conventional data sources, we now have to add the torrent of data from satellite monitoring. This superabundance of information changes its nature from a scarce resource to an oversupply problem. How can we save being drowned in this tide? How do we cull valuable information from the mountains of figures?

As we saw in Chapter 22, geographers have used traditional so-called 'analogue' cartography to collect and store the data about the earth's surface and produce it as maps. Using digital technology, geographers still collect and store spatial data and make map products but in a largely new way. The suite of tools that allows this switch from analogue to digital to occur is GIS (see Figure 23.1 p. 716).

Geographic Information Systems (GIS) have arrived on the scientific scene with huge impact during the past few years. An interest in GIS is now easily the largest speciality interest within the Association of American Geographers. Employment positions in GIS within a widening range of industries provide a large and increasing number of job opportunities for geography graduates.

To understand the key role of GIS in modern geographic analysis, we need to recall the profound effect of the switch from analogue to digital technology on mapping. In its traditional analogue form, the map served two functions: (a) as a convenient storage medium for spatial data about the earth's surface, and (b) as a communication medium to allow geographers to observe and analyze spatial relationships. Both functions were combined in the same map. Digital technology has separated these two functions. First, an unseen *read-only map (ROM)* in which map data are stored. Second, a visual *random access map (RAM)* in which a large number of maps can be created from the ROM store.

In this chapter we look at five aspects of the GIS question. In the first section we seek to define the components in a GIS system (Section 23.1). We then consider the two main forms in which geographic data are held and analyzed: raster and vector forms (Section 23.2). Both forms give rise to problems which are overcome by distinctive GIS functions (Section 23.3). This leads on to a consideration of how GIS is applied. A hierarchy of problems to be tackled is proposed ranging from simple inventories to complex modelling (Section 23.4). Finally, we try to place GIS in its historical context, looking back to its origins and forward to its future impacts (Section 23.5).

Components of a GIS 23.1

Many definitions for *Geographic Information Systems (GIS)* have been written, and no universal agreement has yet been made. We use here the following: 'A GIS is a set of computer tools used to capture, store, transform, analyze, and display geographical data.' There are four elements that are important in a GIS: (1) computer hardware, (2) GIS software, (3) geographic data, and (4) people (see Figure 23.2). We then look at (5) the ways in which the four components are bound into an operational system.

Computer Hardware

By *hardware* we refer to the computer components that form the physical framework on which the GIS is run. This consists of (a) the central processor on which data manipulations and analyses are performed and (b) peripheral devices for storing data, displaying analyses, and creating output (see Figure 23.3). When computer-based GIS first became available, hardware was expensive relative to the other three elements. Over the last few years hardware cost has declined and is today a reducing portion of the total cost of most GIS.

A GIS can be operated in (a) an independent or 'stand-alone' environment or (b) in a distributed environment consisting of a series of personal computers (PCs) or workstations connected by a network. Traditionally, a GIS was often run on a mainframe or in a minicomputer environment, but

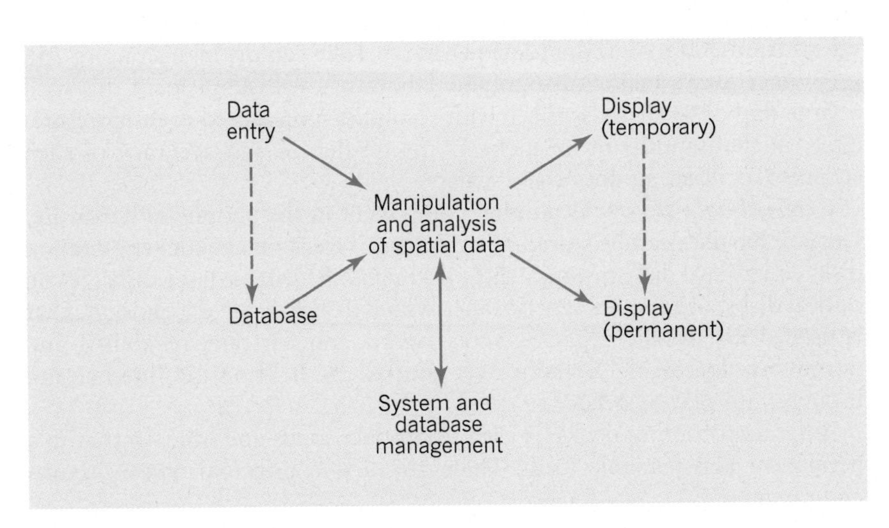

Figure 23.2 Flow through a GIS system A diagram of the flow of information through a generic GIS from data entry to display.

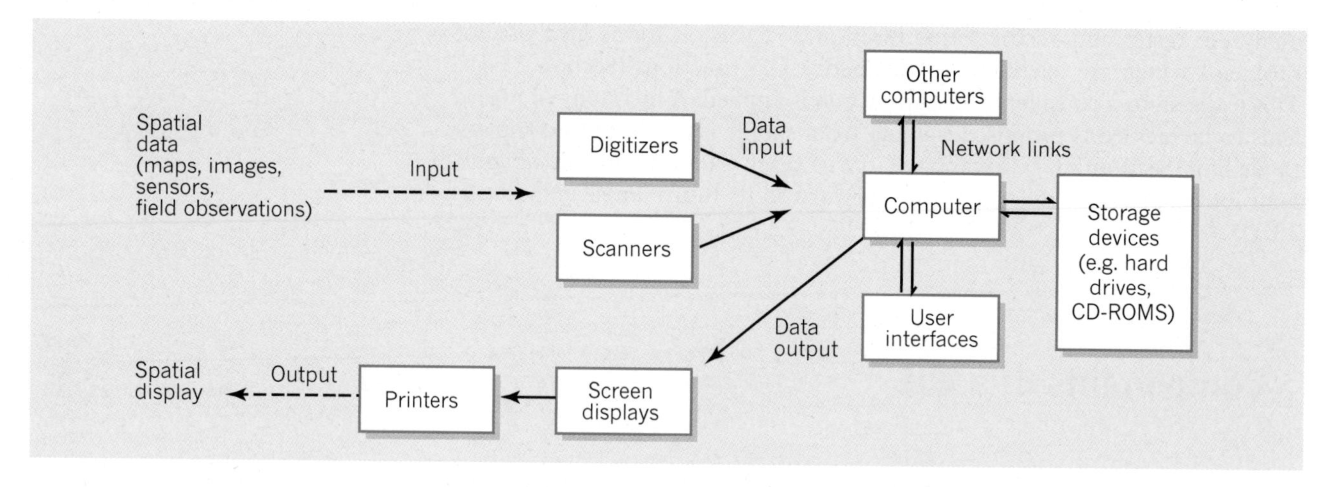

Figure 23.3 Hardware components in a GIS system Elements that make up the hardware components in a GIS system. In practice the complexity of the system will vary from small self-contained systems to larger interconnected systems involving several sets of computers.

the number of these configurations is declining as GIS is made available on an ever-widening number of small, independent PCs. At the lower end of the spectrum, the GIS may simply consists of a PC with its associated monitor and keyboard. In the workstation environment, a large digitizing table and several large monitors may be used.

Whatever system is used, the computer processor needs to have a large internal memory and rapid response and instruction execution times. Both the number of instructions in GIS software and the amount of geographic data stored are usually very large in real-life applications but may be slimmed down for teaching purposes. The size and screen resolution of the graphics output are important factors to consider when selecting the interactive display and editing device.

In all but the smallest systems, the processor and monitor(s) are supplemented by peripheral units. An input or data capture device can be a digitizer or a scanner. *Digitizers* capture data in two or three dimensions. The two-dimensional device is called a digitizing table or tablet. Introduced around 1970, they are today available in many sizes, resolutions, and spatial accuracy. Three-dimensional devices are commonly used for stereo compilation of data from remote-sensing images. These devices are considerably more expensive than the digitizing tablet and require higher operating skills.

Scanners are now available in several forms. Hand-held versions are inexpensive but are not very practical for working with map sheets. Most maps can be handled with page-sized, desk-top scanners with spatial resolutions from 300 to 600 dpi (dots per inch). They can distinguish up to 256 grey tones. But a major cartographic laboratory will supplement this with a large flatbed or drum scanner with resolutions measured even more precisely in thousandths of an inch. As the resolution and accuracy of such scanners increase, so does cost escalate.

External data storage is another component in the peripheral hardware. Geographic data can be stored in a number of electromagnetic ways: (a) on disk, either hard disk or floppy disk, (b) on CD-ROMS or laser disks, (c) on optical disks, or (d) on magnetic tape. Which device to use depends on what is needed for storage capacity, access speed, transfer rate, reliability, and portability. Increasingly, major data sources are held on the Internet and brought into play as needed.

A hardcopy output device is used to produce maps and other output in a permanent and readable form. These are of two principal types: printers and plotters.

1 *Printers* are a familiar part of all computer systems. They can be monochrome or colour and of several formats: dot-matrix, laser, inkjet, xerographic, or light (optical).
2 *Plotters* are used in major cartographic laboratories. They consist of a flat plotting surface or a revolving drum and can be analogue or incremental. Conventionally, they have a plotting head which consists of a pen, a photo-head, or a scribing tool.

A feature of the last decade has been the widening range and falling costs of GIS hardware systems. Simple GIS systems work on PCs (even lap-tops and portables) so long as they have appropriate speeds and storage. But mapping work tends to be slow and the range of data manipulation is limited. At the upper extreme are complex GIS systems as used by the major cartographic laboratories. These have dedicated, high-speed and large-capacity equipment, and a full range of hardware. The cost of GIS hardware thus covers a very wide range. If we start at the bottom with a basic PC for the personal user at a cost of $\$x$, then the 'top of the range' equipment could be as much as $\$1000x$. Most geography departments would have systems at a mid-point between the personal user and the large national laboratory.

GIS Software

Software is the complex sets of millions of instructions that manipulate geographic data with the hardware systems described above. As with hardware, GIS software now comes in a widening variety of forms which are designed for different uses. A summary of four of the more common software packages is given in Box 23.A and discussed further in Section 23.2.

Box 23.A Major GIS Software Packages

Name	Producer	Characteristics
ARC/INFO	Environmental Systems Research Institute, Inc. (ESRI), 380 New York St, Redlands, CA 92373, USA	High-end, full functionality GIS software. Developed over last 20 years to provide the flagship GIS-processing toolbox with over 1000 analytical functions. Versions run on all four classes of computers (PCs, workstations, minicomputers, mainframes). Training necessary for use.
GRASS	Geographic Resources Analysis Support System, GRASS Information Center, CECER-ECA, P.O. Box 9005, Champaign, IL 61826–9005, USA.	Lower-end, inexpensive GIS software. Designed by the US Army to provide computer-based management tools for environmental planners and land managers. Easy analysis, display and modelling of landscape data. Runs on range of workstations (e.g. Sun and Intergraph) and up-range PCs.
IDRISI	Graduate School of Geography, Clark University, 950 Main St., Worcester, MA 01610–1477, USA	Lower-end, inexpensive GIS software. Designed by geographers to stimulate education and research uses of GIS and remote sensing. Handles images up to 32,000 × 32,000 rows and columns. Originally DOS-oriented and can be run on PCs. Rapidly expanding capability.
MGE	Modular GIS Environment, Intergraph Corporation, Huntsville, AL 35894–4001, USA	High-end full functionality software. Allows users to define GIS to meet their specific needs through a set of modular tools (e.g. MGE Dynamic Analyst (Dynamo)). Runs on UNIX workstations and servers.

A distinction needs to be drawn between the individual user or small department and the large commercial or government agency handling massive private or public-sector data sets. In the first, an inexpensive system can be largely used for small projects or for instruction and is readily available at a very modest cost. In the second, a massive software system with a full range of functionality may be required. As with hardware, there is now a wide range of systems on the market with costs ranging a thousand-fold between the bottom and top ends of the marketplace.

Geographic Data

While most attention is often given to the first two elements in a GIS, it is the third, *geographic data*, which is critical. While hardware and software problems have progressively been solved over the last two decades, data problems remain rather resistant. In relative terms, the costs of hardware and software have been falling, while those of data collection and processing have been rising.

What are these data problems and why are they important?

1 Geographic data have been increasing exponentially. Two centuries ago, geographers used to write world regional geographies in multi-volume works that were thought then to capture much of what was known about these lands. Today a flight recording from a multisensor satellites can send back enough data about the earth's surface in a few minutes to fill all the volumes of the *Encyclopaedia Britannica*. In short, information about the earth's surface – from global sweeps to data about a given census tract – is accumulating at rates thousands of times greater than our parents' or grandparents' generation could have imagined. GIS has to cope with a tidal wave of data.

2 Data are having to be brought together with a very wide range and quality of *geo-referencing*. While some data are very specifically tied to an exact location, others relate to much more broadly defined areas. Assembling all these different sources and reducing them to a common format is a major task.

3 Data vary greatly in quality, where quality is measured on a range of scales from accuracy to timeliness. In the sophistication of the GIS system it is easy to overlook the fact that the attractive and colourful output may be resting on unreliable inputs.

4 Data are increasingly being exchanged between one computer system and another. A separate set of problems relate to the 'portability' of data, enabling them to be used for different purposes than those for which they were originally collected.

In its most primitive form, geographical data about the earth's surface comes in three basic forms as shown in Figure 23.4. Simple *points* (also called 'nodes'), *lines* (also called 'arcs' or 'edges'), and *polygons* (also called 'patches').

These are essentially static representations of phenomena given in terms of their spatial (*UV*) coordinates. Points can be represented by a single set of coordinates and exemplified by a borehole or a climatic recording station. Lines need a series of spatial coordinates and a spatial topology connecting them in some prescribed way. Lines are exemplified by transport routes or river drainage lines. Like lines, polygons also need several sets of

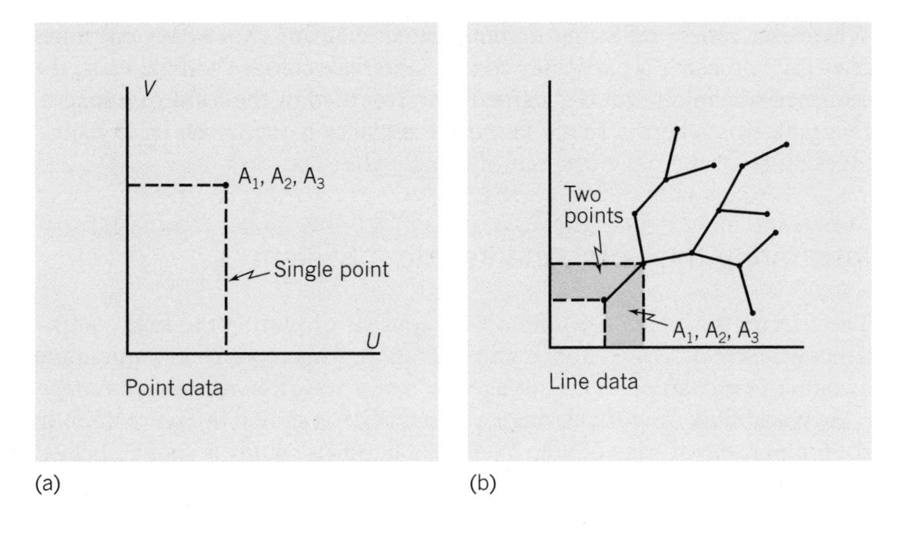

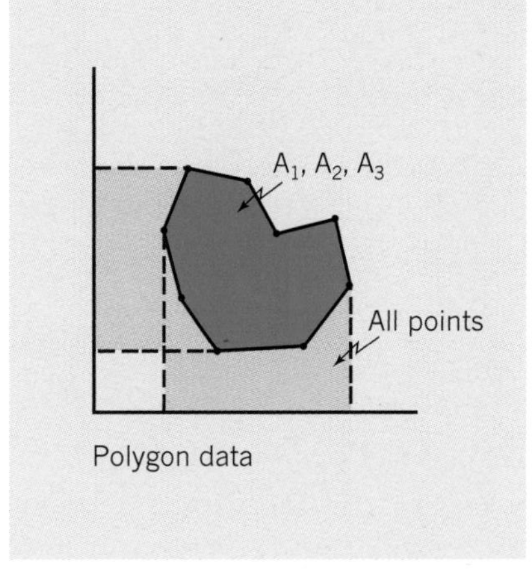

Figure 23.4 Categories of spatial data Examples of three categories of data: (a) points, (b) lines, and (c) polygons. Each category demands successively more data storage in terms of locational information in the *U*, *V* dimension. The attribute data illustrated by the sequence A1, A2, and A3 are non-spatial.

coordinates to identify them accurately. These are typical of a bounded spatial structure such as a lake or a county. Whether a feature is represented as a point or a polygon is partly dependent on scale: on a continental map a city might be represented by a point, at a local scale by a polygon.

People or 'liveware'

We have left until last, the most critical and most expensive of the four GIS elements, people or 'liveware'. These range from highly skilled programmers who create the software and modify the hardware architecture to those who are responsible for routine but critical tasks such as inputting data or digitizing maps.

Given the recent rise of GIS, it was inevitable that those who pioneered the system came from a mixture of different backgrounds: from computer science, from electronic engineering or from user fields such as geography or town-planning. Today there are a growing number of degree courses in GIS and most geography programmes include an essential GIS element in their curriculum. There is now a growing interest in GIS in secondary schools and a number of GIS companies are now marketing simple and inexpensive systems which are capable of being used in the classroom. As the GIS industry continues to grow worldwide, so the range of employment possibilities at a range of levels is likely to rise in parallel.

Ways in which humans interact with the hardware and software have evolved rapidly. Originally, operators could communicate using only specialized and arcane machine codes. This was followed by higher-level languages such as FORTRAN ('Formula translation') or DYNAMO (used for dynamic simulations). Today, operation is much more user-friendly with the development of *graphic user interface (GUI)* functions. These allow the GIS operator to interact with the specific GIS through a series of commands. These are usually of two kinds:

1 Menu-driven systems where the operator has a list of options from which to choose. The menus are usually arranged in a hierarchic form like a branching tree but may also be provided as 'look up' dictionaries.

2 Icon-driven systems where the operator clicks on to options shown by symbols on the screen.

Where the user is repeating a complicated chain of choices several times over (say choosing a particular contour interval, colour shading, etc.), the sequence of choices can be 'learned' and recorded in the form of a macro. Using macros where a single keystroke replaces many scores of decisions takes much of the repetition out of using a GIS.

Integrating the Components into a System

The S-term, the 'system' within a GIS, consists of putting the four components we have described above into operation. This requires as a minimum a computer and an operator, along with access to software and a *database*.

A typical flow of work through a generic GIS is shown in Figure 23.5. In the top part the overlay of map layers in a flood-risk study is shown. Below, the corresponding GIS stages are shown. The flow of map data in a GIS is directed by an operator into a hardware processor. The software is already loaded within the hardware (or can be addressed via the hardware although it may be located in another machine within the network). The operator then selects the functions (in this case, boundary superimposition) that need to be performed by the software. After data retrieval and analytical functions have been performed, the operator chooses the appropriate form in which the output is to be produced. This is usually done interactively on the screen, where final choices of content, scale, colour, etc. are performed. Finally this output is sent to a peripheral hardcopy output device for printing or plotting.

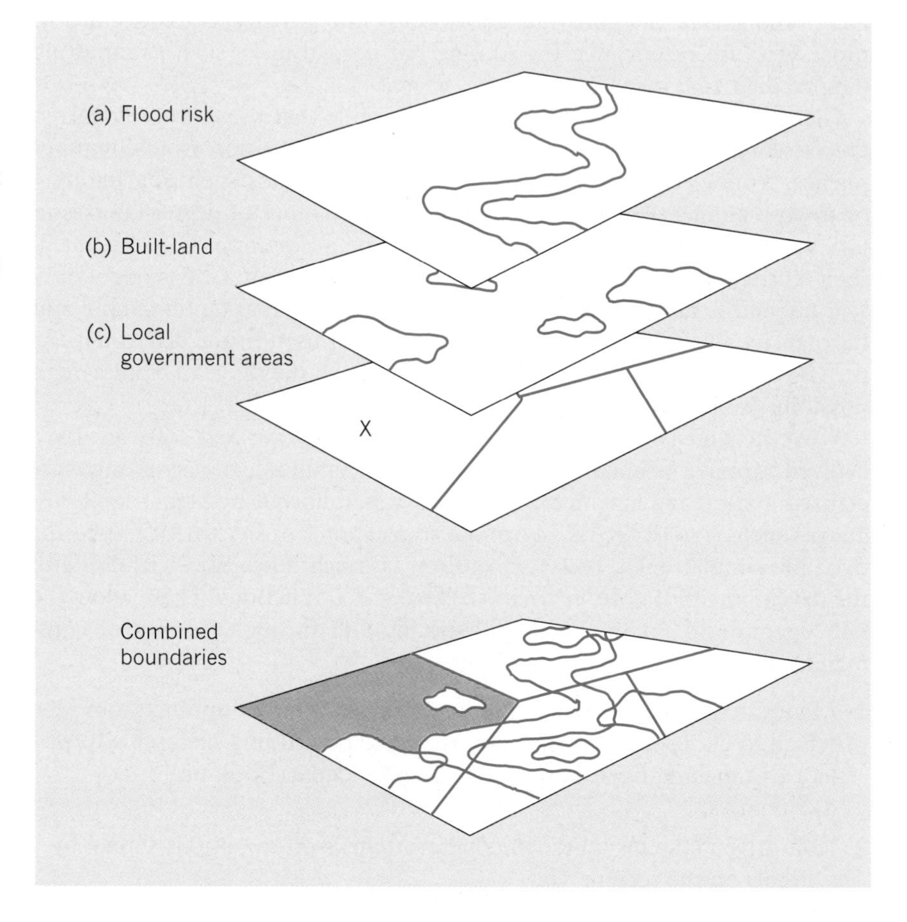

Figure 23.5 Information layers in a GIS system Three layers with flood risk, built land, and local government areas (above) superimposed and combined to give a mosaic of spatial boundaries (below). Each of the resultant polygons in the bottom map contains information from the above layers. Thus the shaded area is above flood risk (a), is not on built-up land (b), and is part of township X (c).

Geographic Data as Raster and Vector Systems 23.2

Two major traditions have developed in GIS for representing geographic distributions. The raster approach divides the study area into an array of rectangular cells, and then describes the content of each cell. The vector approach describes a geographic distribution as a collection of discrete objects (points, lines, or areas), and describes the location of each. Examples of both approaches are shown in Figure 23.6.

Since most currently available GIS software tends to be identified with one or other approach, we discuss each tradition separately. We should note that most software now provides a capability for handling both types of data or converting one into the other.

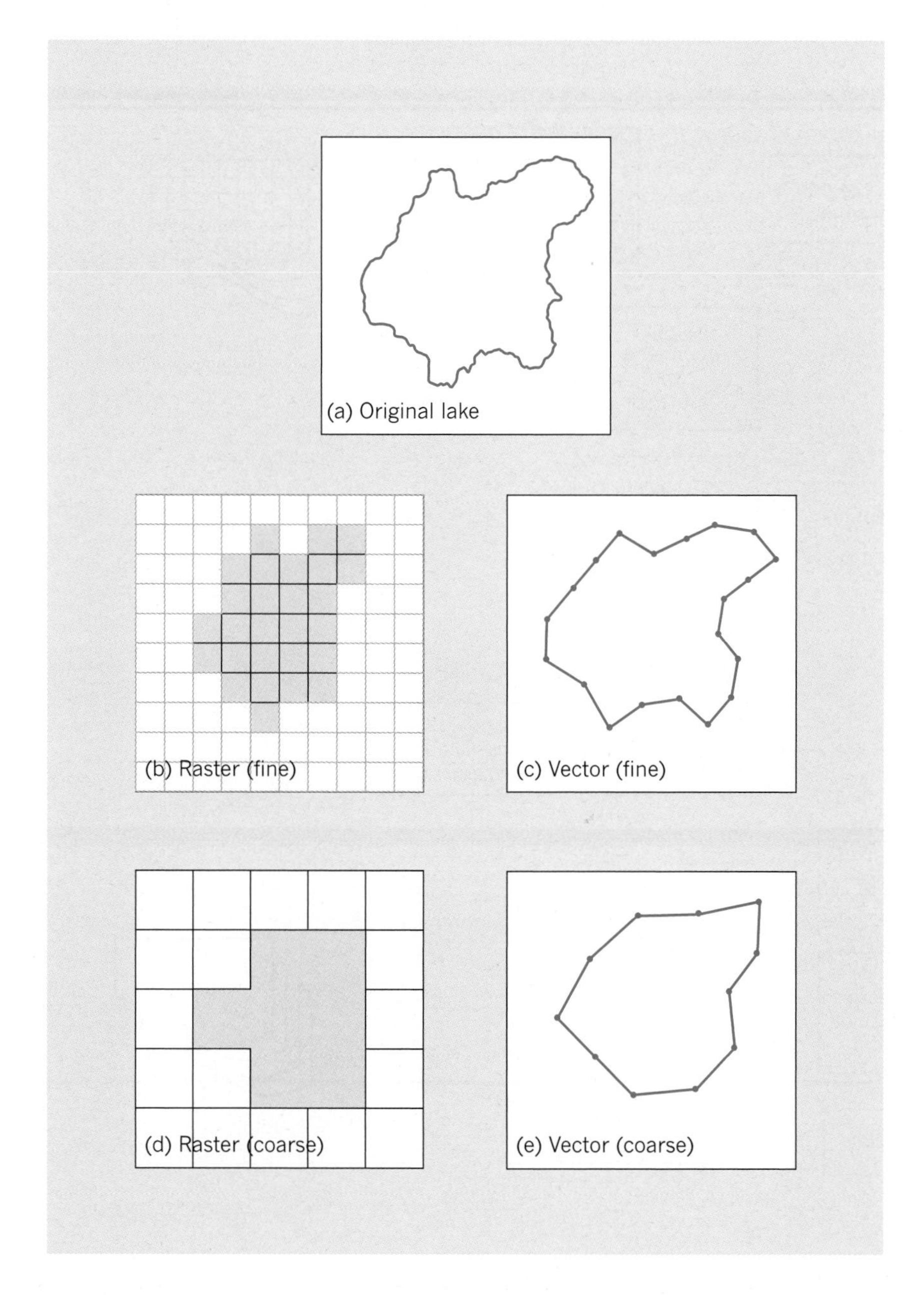

Figure 23.6 Geographic feature in raster and vector structure The figure shows how the same geographic feature is handled by GIS software using raster and vector formats. (a) Lake boundary as original geographic feature. (b) Reproduction of lake in raster format. (c) Reproduction in vector format. Maps (d) and (e) show how the degree of generalization can be made coarser or finer in the two forms of representation.

The Raster Approach

The *raster* approach divides the study area into an *array* of rectangular cells, and then describes the content of each cell. The flow of spatial data to provide raster and vector data is shown in Figure 23.7. Note that the various databases require data to be entered in different ways.

Raster data structures have a number of major advantages.

1 They have a very simple box-like data structure which fits naturally into many concepts of geographic space.

2 This means that attribute data are automatically location-specific in terms of the raster grid and can be easily used.

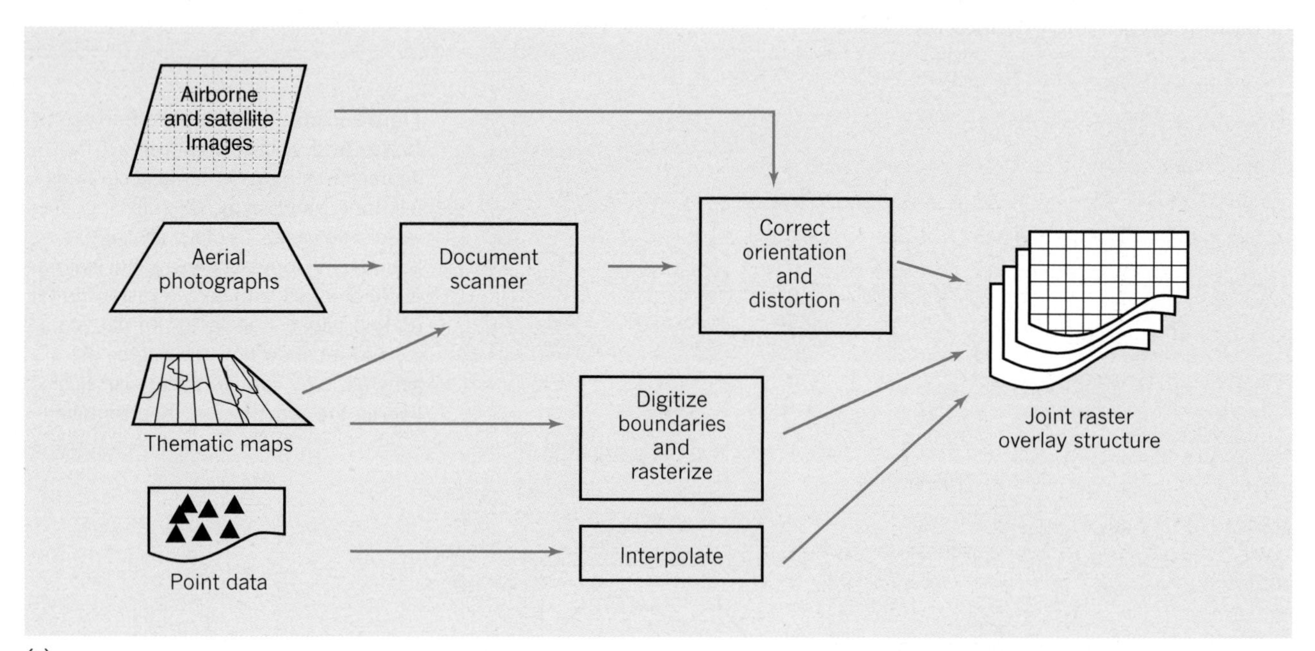

(a)

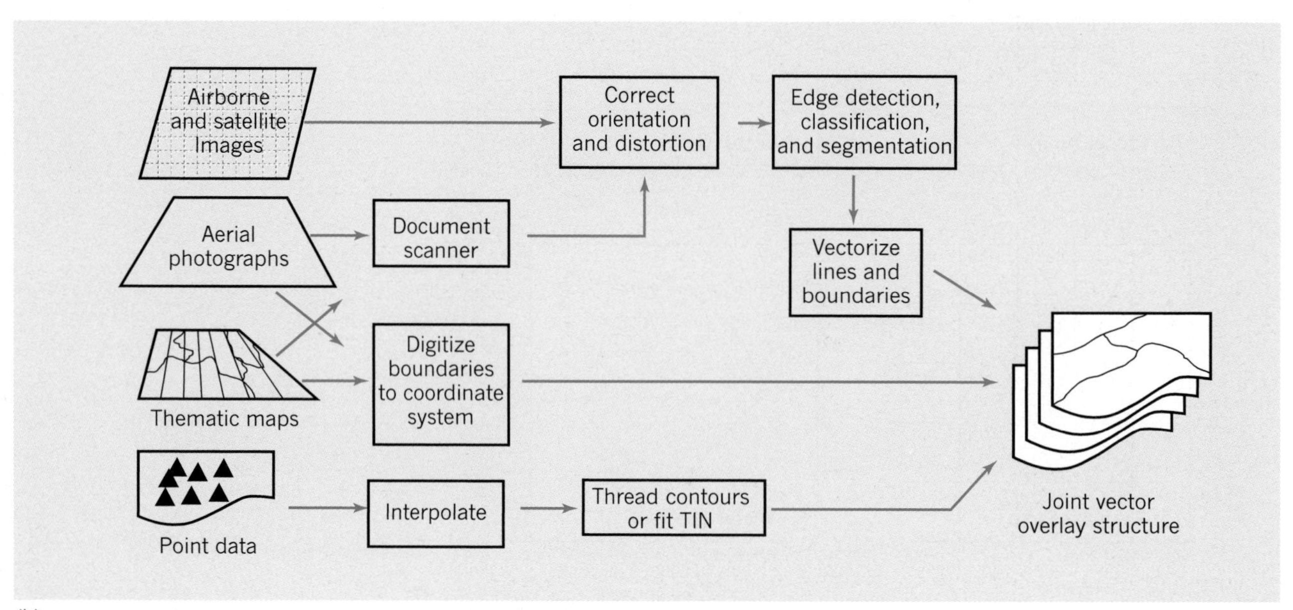

(b)

Figure 23.7 The capture and processing of spatial data to build a GIS database Note the different tracks followed by (a) raster and (b) vector formats. ***TIN = triangular irregular network***.

[Source: P.A. Burrough and R.A. McDonnell, *Principles of Geographical Information Systems* (Oxford University Press, Oxford, 1998), Figs 4.3, 4.4, p. 85.]

3 Many kinds of spatial analysis and filtering may be used.
4 Mathematical modelling is easy because all spatial entities have a simple, regular shape.
5 The software technology for GIS analysis is relatively cheap.
6 Many forms of data are available in raster form, notably the huge amounts on remotely sensed data describing the earth's surface.

Against this is a series of disadvantages:

1 Very large data volumes are generated. A single satellite photograph may consist of many thousands of *pixels*.
2 Using large grid cells to reduce this huge volume of data also reduces the level of spatial resolution and results in loss of accuracy.
3 There is an inability to recognize well-defined structures that occur naturally in a vector format.
4 Crude raster maps with squared outlines appear inelegant.
5 Coordinate transformations are difficult and time consuming unless special algorithms are used and even then may result in loss of information or distortion of the shape of grid cells.

Ways of getting around these problems are discussed in the next section.

The Vector Approach

As we noted above, the *vector* approach describes a geographic distribution as a collection of discrete objects (points, lines, or areas), and describes the location of each. Such an approach develops naturally from traditional cartography, where points, lines, and areas were successively added to the blank map sheet. In addition, a GIS database contains information on the attributes of each cell or object, and on various kinds of relationships between objects. Broadly, the continuous view of space embedded in the raster approach is most commonly associated with environmental and physical science applications of GIS, while the view of space as a collection of described objects that is implicit in the vector approach has found more application in the social and policy sciences, in the mapping industry, and in the management of geographically distributed facilities.

Vector data structures have a number of advantages over the raster format:

1 The data structure is compact and efficient. You need only record a regular line at a few points along its length.
2 The system records natural spatial entities more readily, so gives a more life-like representation of the earth's surface.
3 Topological relations can be readily defined so network structures are easily analyzed.
4 Coordinate transformation (so-called 'rubber-sheeting') is easier with vector than raster data.
5 Accuracy is maintained at all scale levels.
6 Graphics and secondary attributes are readily retrieved and updated.

On the other side of the balance, vector structures have several drawbacks:

1 Data structures are complex to record.
2 High computer power is required to combine several polygon networks.
3 Display and plotting of vector may be time-consuming and expensive, particularly with high-quality drawing, colouring, and shading.
4 Spatial analysis is not possible within polygons (which are considered homogeneous over the whole of their extent) without extra data being drawn upon.
5 Since each spatial entity has a different shape and form (unlike raster where each unit has a similar cell-like form) simulation modelling of spatial processes is more cumbersome.

We turn now to consider some of the operations and functions available within a GIS system to address these problems.

23.3 GIS Problem-solving Functions

At the heart of the success of GIS is a series of critical functions that allows electronic data (whether stored in raster or vector form) to be handled in a geographic form. Many are replicating operations that were done more slowly and cumbersomely by cartographers in the past; others are uniquely available to GIS given the logic and power of the system.

The number of possible functions that are built into a sophisticated GIS is now very large. Some GIS such as *ARC/INFO* have over a thousand functions. It is not possible here to cover all these functions so we have chosen to give some examples of the kinds of operations that illustrate the range and power of the system. Other, more routine, functions are listed in Table 23.1.

Table 23.1 Sample functions from a GIS

Term	Description
Buffer	Creating a buffer zone, of specified width, around a point, line, or area.
Contiguity	Defining contiguity or adjacency measures.
Contour	Contouring a set of attribute values associated with points.
Diffusing	Defining the spread of a phenomenon through permeable barriers.
Direction	Seeking the direction of flow across a surface defined by a set of attribute values.
Distance	Calculating distances between points by space or time.
Interpolate	Interpolating between points, lines, or within polygons.
Intervisibility	Calculating the intervisibility at points on a surface.
Locate	Locating a point or line within a given polygon.
Matching	Matching edges of data that come from different tiles.
Networks	Defining networks and calculating statistics about the created network.
Panning	Panning around an area.
Search	Searching for all features which have a given attribute and which are located within a certain radius.
Segment	Segmenting a surface by Thiessen polygons.
Surfaces	Calculating surface values such as slope, aspect, or gradient.
Windows	Defining a window and extracting the data layers of interest only for that window.
Zooming	Zooming in or out on an area.

Storing Data in Compact Form

Although the memories of GIS systems have been expanding rapidly in recent decades, the amount of data that need to be stored is still massive. Various functions have been developed to get around this problem. For example, the use of *quadtrees* is one way of overcoming the data storage problem for large structures recorded in raster format.

Let us take a simplified example. Consider the area in Figure 23.8(a) which is recorded as a lake in the original map. Reduced to raster format as in (b), it would take 135 separate small units to describe the extent of the lake.

But it is clear that many data items are redundant with many adjacent cells carrying the same information as those which surround them. If we superimpose cells of larger size (i.e. combinations of the original cells), the same information can be recorded as a simpler hierarchic map. The lake is reduced to one 8 × 8 cell, one 4 × 4 cell, ten 2 × 2 cells, and 15 of the original cells. Thus a total of only 27 cells replaces the original 135 cells, a reduction by 80 per cent of the original information load.

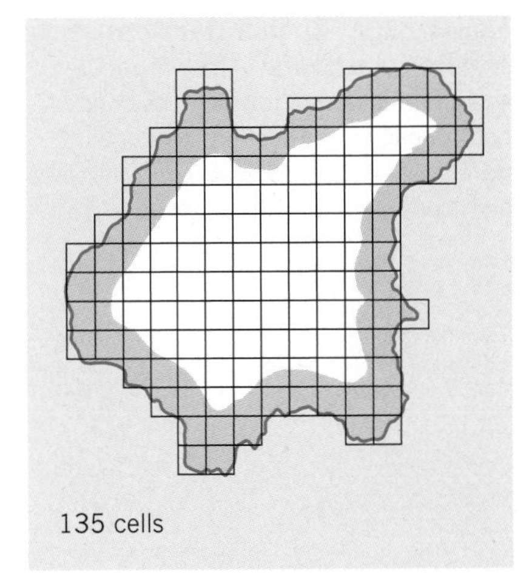

(a)

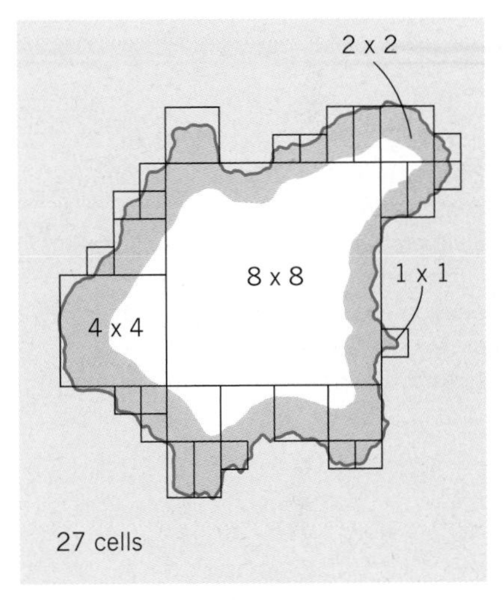

(b)

Figure 23.8 Principles of quadtree structure Use of quadtree methods for compressing raster data in a GIS at different geographical scales. (a) Original cell structure. (b) Structure varied by using larger cells to store uniform information.

Transforming to a Common Spatial Format

Frequently, incoming data are digitized from a printed map. This means that spatial data will be in the coordinates from the given map projection or coordinate system used in the original map. As we saw in Chapter 21, there is a wide choice of such coordinate systems and it will be unusual if all data use a common system. The data in the database must be held in some common locational coordinate system, and you will need to transform new incoming data so that they are compatible with the other data in the database. Many databases accept the geographic coordinate (latitude and longitude) system as the preferable system in which to store data.

Often when we bring data into a database, only a few of the features are referenced to a common coordinate system such as latitude and longitude. Indeed in extreme cases, no feature is referenced to an earth position and features are only referenced to one another in a relative sense.

When this happens, GIS performs a function called *rubber-sheeting* to transform the positions of the data into the desired locational framework. The principles are shown in Figure 23.9. To do so, a few well-defined features are defined. We add their latitude and longitude position as an attribute to their record. Rubber-sheeting then allows a GIS to use a mathematical procedure to obtain an approximated latitude and longitude for the rest of the features.

Overlapping Data Layers

Overlaying one data layer on another, or combining several data layers at once, are now routine operations. Where the data are compatible the operations can be arithmetic *overlays* (in which you perform such operations as addition, subtraction, multiplication, and division). Thus a rural population overlay might be placed over an overlay of cultivated land to produce contours of population density per cultivated hectare. Or a layer showing the distribution of a particular bird species might be placed over a habitat's overlay to study the dependence of the species numbers on particular habitats.

Figure 23.9 Spatial transformation of data framework Use of rubber-sheeting technique to force positions of data points into a desired locational framework. In this case six control points are used to anchor the data.

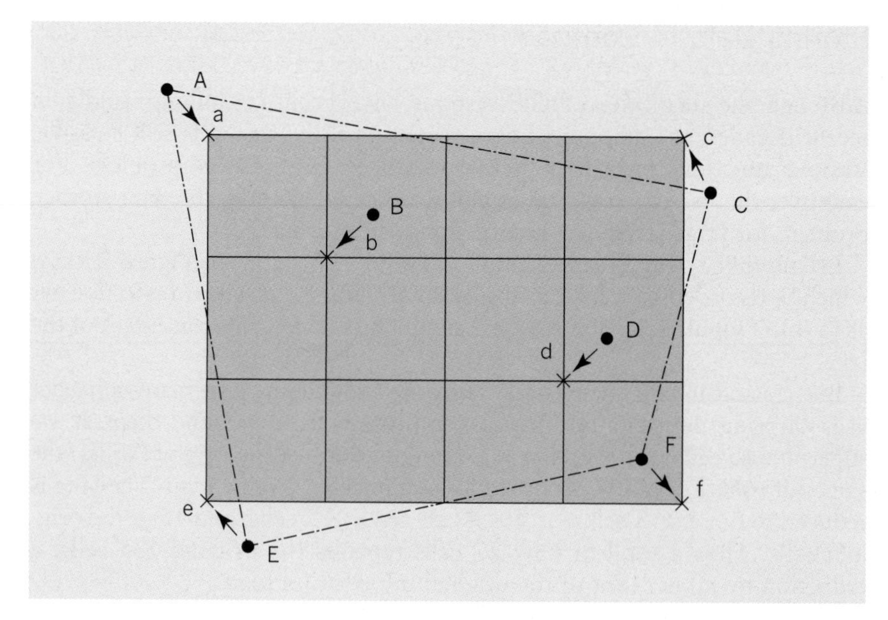

Figure 23.10 Operations in Boolean logic The rules of data combinations and exclusions are illustrated. In (a), (b), and (c) capital letters refer to the presence of phenomena and lower case to their absence. Map (d) shows the superposition of maps (a), (b), and (c).

Equally common applications are logical overlays, in which you find areas that satisfy a set of criteria. This uses *Boolean logic*, a special language of great utility in computer data processing first developed by the mathematician George Boole in the middle of the nineteenth century. It is based 'and/or' operators and 'true/false' propositions (see Figure 23.10). When applied to spatial data it allows land-use classes to be overlapped and combined (e.g. 'take all land that is both liable to flood and has a very high

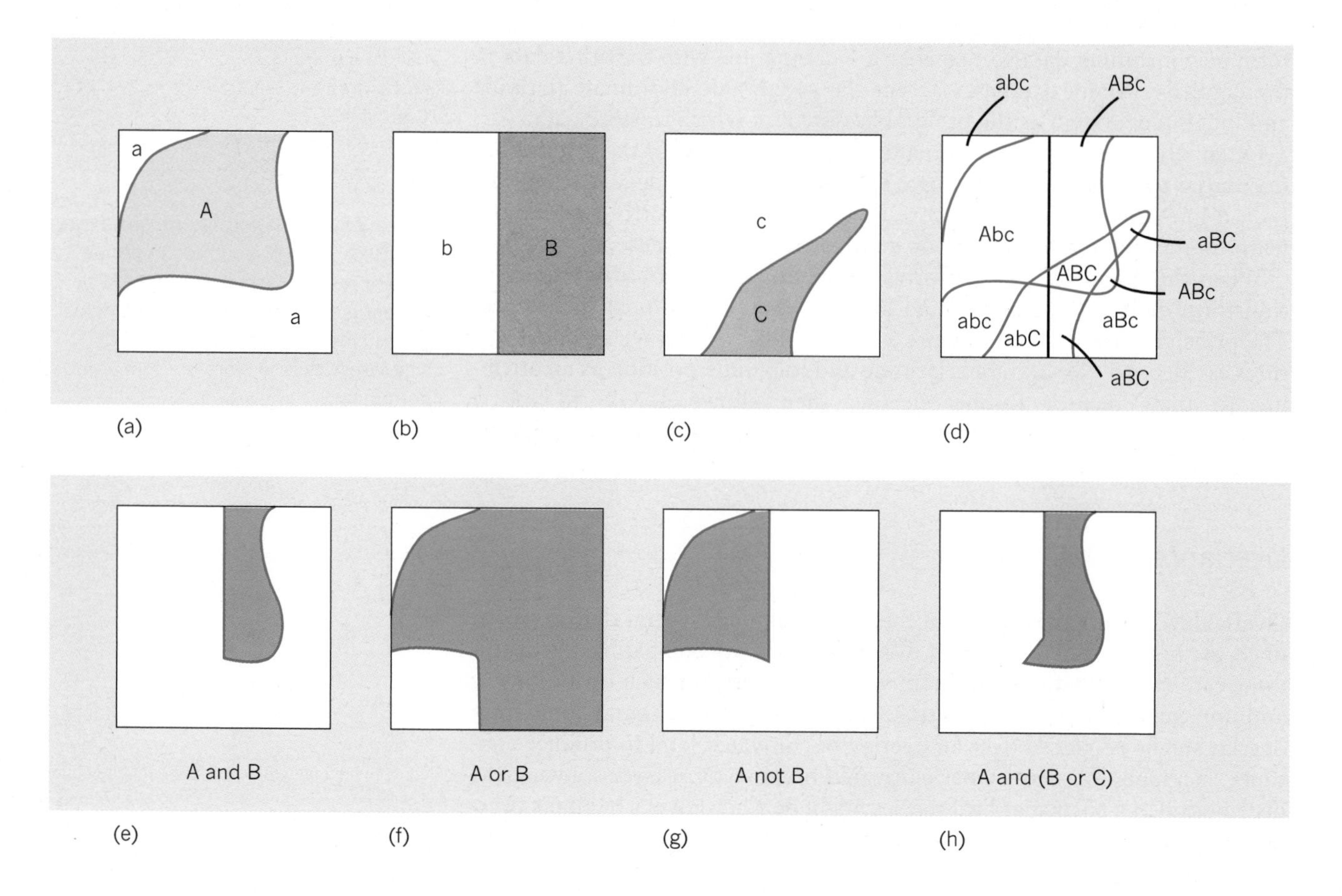

rental value') or exceptions defined (e.g. 'locate the apartment block close to the airport but outside the footprint of high decibel values from the flight path'). It is particularly useful in utility mapping where such sensitive public infrastructures as power lines or gas pipes have to be laid in locations that avoid some localities or environments or have to keep close to others.

Editing Digitized Data

Edit functions include a set of routines that clean up the data in the file. When digitizing is performed manually or rapidly by machine, great care must be taken to avoid small remnants of the digitizing procedure itself remaining in the data file. Figure 23.11 shows five of the errors that can arise. These errors include gaps in the closed shape (where lines fail to connect), overshoot lines, slivers, unconnected nodes, duplicate lines, and spikes.

Most errors arise naturally from the digitizing process and need careful editing. For example sliver errors may arise when two polygons are brought from two different sources and the common boundary is shown in two slightly different forms. The two lines are replaced in editing by a single line.

Once we have digitized data by location into a file, we usually need to add other attributes about the features. This process of adding intelligence

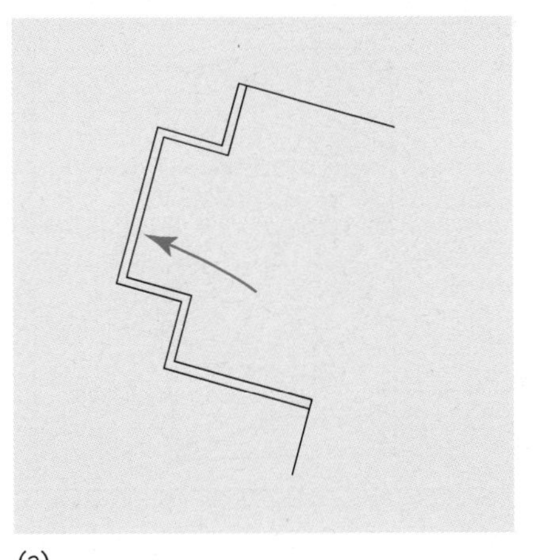

(a)

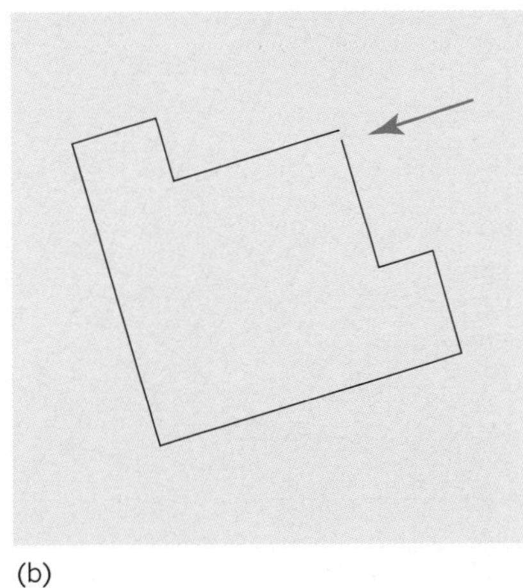

(b)

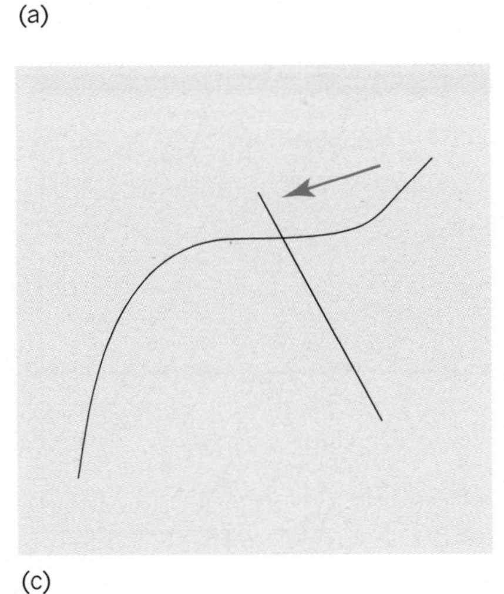

(c)

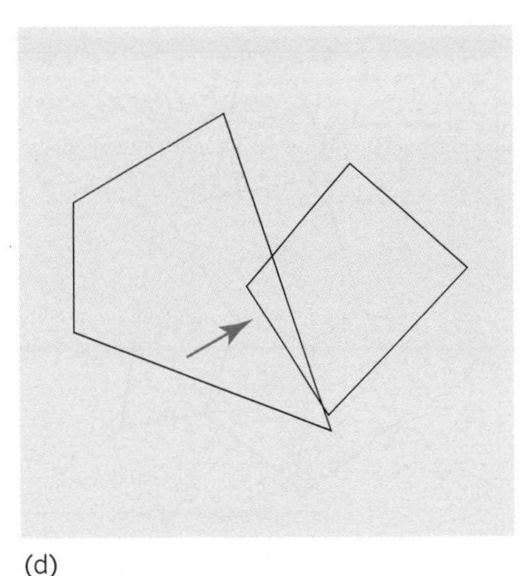

(d)

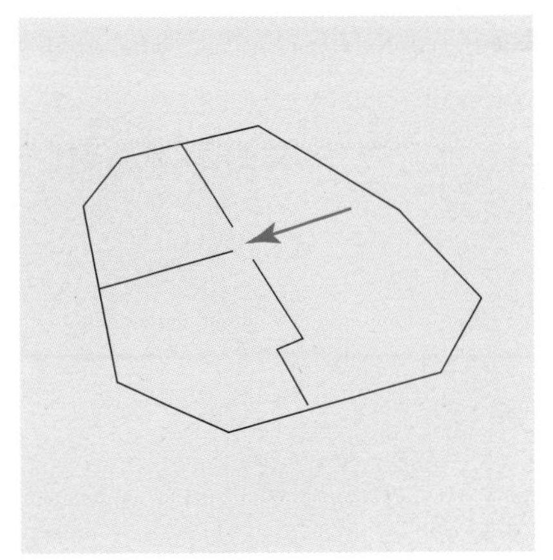

(e)

Figure 23.11 Editing digitized data file errors Digitizing of vector data for entry to a GIS often throws up errors which need correction before analysis can be conducted and maps can be drawn. Five examples are shown: (a) duplicate line, (b) gap in closed space, (c) line overshoot, (d) sliver, and (e) unsnapped node.

Figure 23.12 Thiessen polygons
Application of Thiessen polygons to cholera deaths in the Soho area of London in August and September 1854. (a) Section of John Snow's original map showing coffins piled up outside the house where deaths from cholera occurred. (b) Location of pumps for public water supply and the Soho street pattern. (c) Polygon construction around each pump. (d) Thiessen polygons using direct overland distances. (e) Thiessen polygons adjusted for Soho street pattern. Numbers in (d) and (e) denote the deaths from cholera within each polygon. A micrograph of the cause of cholera, *Vibrio cholerae*, is given in Figure 20.5(b).

[Source: A.D. Cliff and P. Haggett, *Atlas of Disease Distributions* (Blackwell, Oxford, 1988), Fig. 1.16, p. 54.]

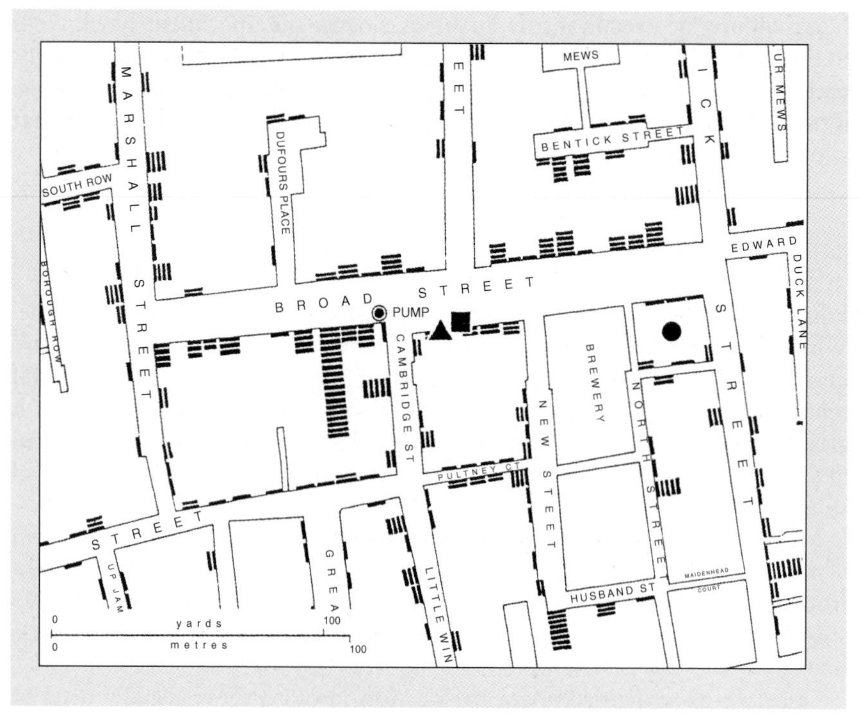

(a)

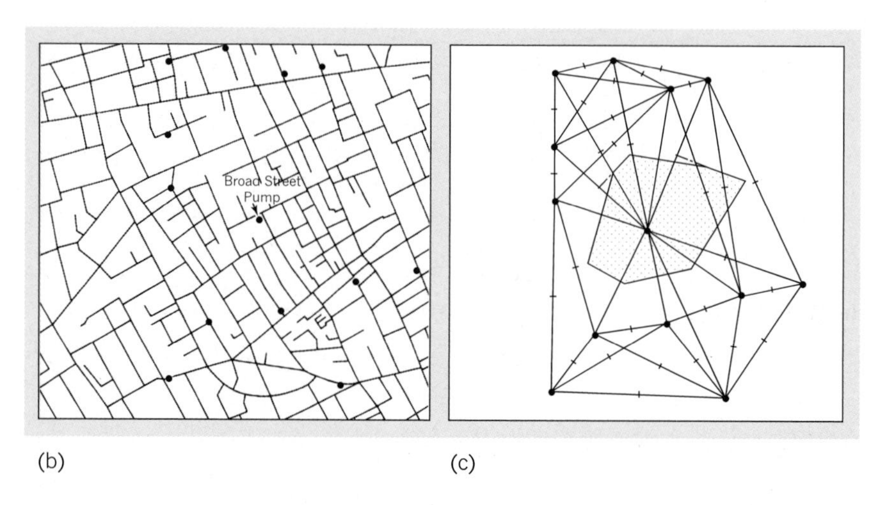

(b) (c)

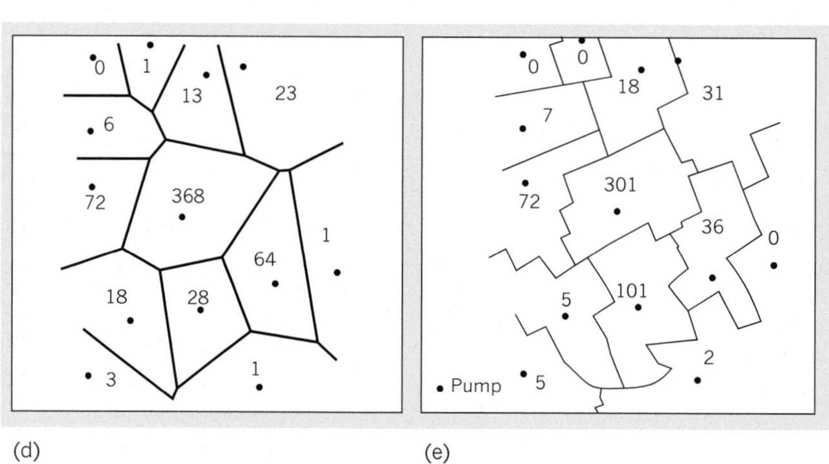

(d) (e)

to the data file is called *tagging*. This tagging may be as simple as adding a place-name to a particular location or a date that indicates how current is the information. Alternatively, we might tag the location with whole blocks of data drawn from a data dictionary already held in the database.

Blending Point and Area Data

Data may occur naturally in point, line, or area forms. But often we need to bring these together in a common format. For example, rainfall may be collected at a series of observing stations (points), but we may need to estimate rainfall over all of the land surface (i.e. the areas between the points). GIS include several ways of solving this problem. We illustrate here one of the simplest, the use of *Thiessen polygons* (also called Dirichlet or Voronoi polygons) and apply this to a classic problem in medical mapping, cholera deaths in the Soho area of London, mapped by Dr John Snow in 1854 (see Figure 23.12). Here some public sector pumps in the streets were contaminated. The procedure for constructing the polygons is shown in diagrams (b) to (d). Three steps are involved: (1) lines are drawn joining a given pump to each adjacent pump; (2) each of the inter-pump lines is bisected to give the midpoints of the line. The location of these bisectors are shown by the ticks; (3) from these midpoints, boundary lines are drawn at right angles to the original inter-pump lines to define a series of polygons.

The polygon around the Broad Street pumps shaded in diagram (c) and has been constructed in this way. Note that all the midpoints on the inter-pump lines radiating from a particular pump are involved in this process.

The basic property of any Thiessen polygon is that it contains all the area of any map that is geometrically nearer to its point centre (here a particular point) than it is to any other point centre on the map. The complete set of polygons around all of the pumps given on Snow's map is shown in diagram (d). The number of deaths within each is also recorded using Snow's data as mapped in Figure 23.12. The excess within the catchment area of the Broad Street pump (348 as compared to the next largest of 64) is apparent.

Changing Data Infrastructures over Time

For geographers interested in tracing historical changes, many data items are collected from a network of legally defined statistical collecting areas. These areas have various labels at different scales: parishes in England, municipalities in Sweden, shires in Australia, townships in the United States, and so on. They form the data infrastructure for which demographic and other social-science information is recorded.

But as populations have grown, so many of the collecting areas have also changed. So the network of collecting areas at one period may not be the same as another. I faced exactly this problem in trying to reconstruct changes in the disease patterns in Iceland. As Figure 23.13 shows, the medical-district boundaries for 1875 are very different from those drawn 80 years later.

One GIS approach to solving this problem is to produce a map that tracks the changes. Look at the hypothetical example in Figure 23.14. This shows in the upper three maps the boundaries prevailing at three different

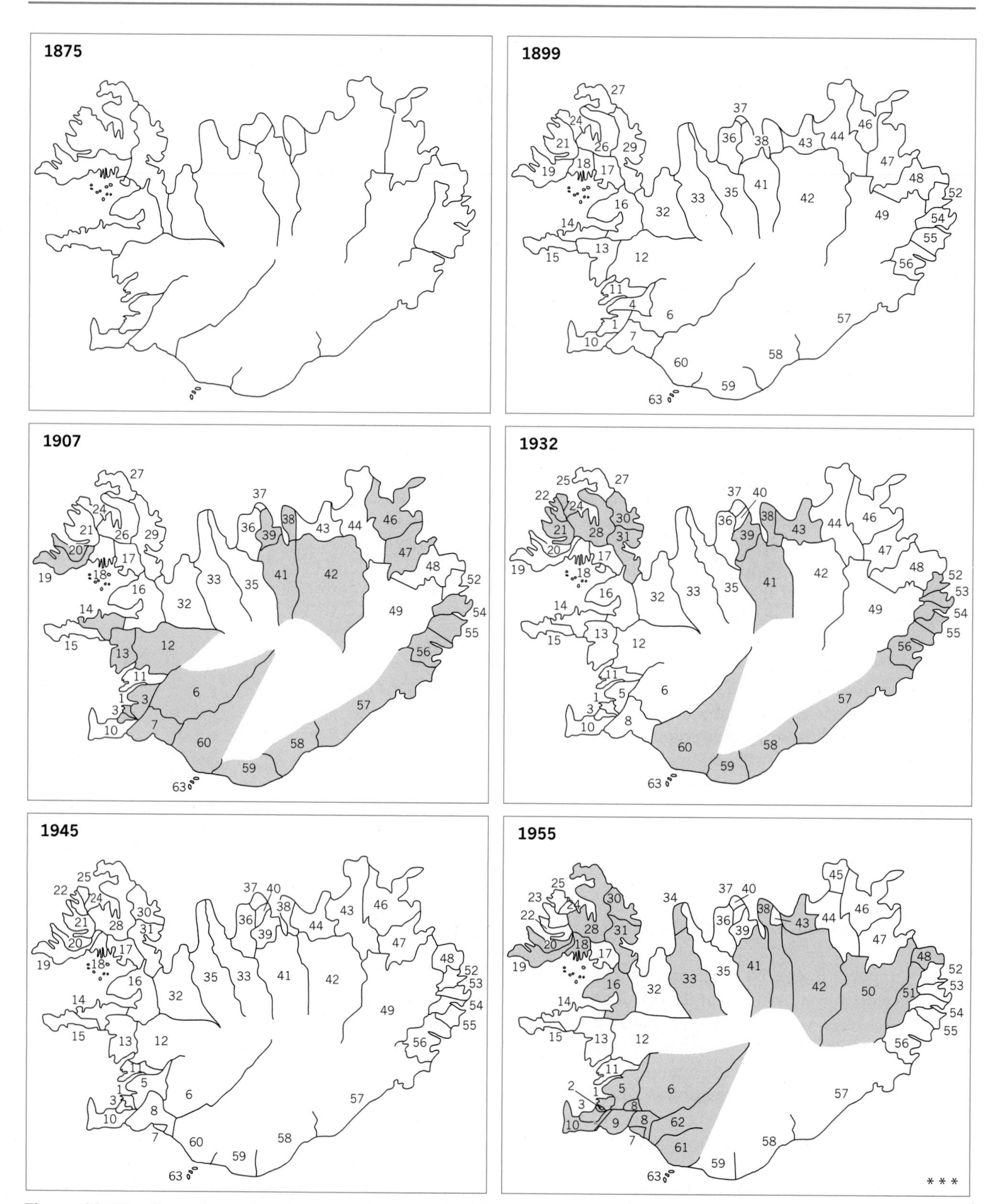

* * *

Figure 23.13 Changing collecting areas for statistical data Changes in the boundaries of recording districts for medical data in Iceland, 1875–1955. Base maps of the medical districts are shown for the dates of major boundary revisions. Blue shading indicates districts affected by boundary changes. Each medical district is made up of the practice of one or more physicians. Boundary changes occur over time by splitting and amalgamating districts in response to population changes.

[Source : A.D. Cliff and P. Haggett, *Atlas of Disease Distributions* (Blackwell, Oxford, 1988), Fig. 2.11(A)–(G), pp. 87–8.]

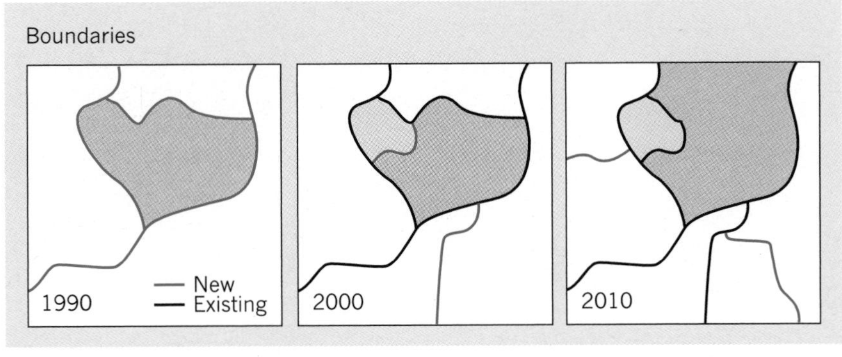

(a)

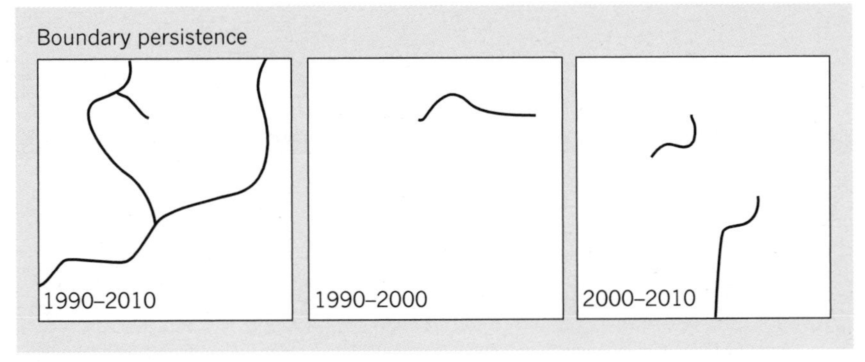

(b)

Figure 23.14 Persistence of boundary segments on a collecting areas map The upper maps (a) show the boundaries at three dates. In (b) the individual segments which make up the boundaries are separated in terms of their persistence over time. A map for any intermediate date (say, 2005) can be drawn by combining the segments which are current for that period.

[Source: A.D. Cliff and P. Haggett, in P. Longley and M. Batty, *Spatial Analysis: Modelling in a GIS Environment* (GeoInformation International, Cambridge, 1996), Fig. 17.6, p. 334.]

dates between 1990 and 2010. If we now combine the three maps we can see how some boundaries are very persistent while others are transitory and come and go over relatively short time periods. By tagging each line segment with its dates we can draw a map for any year (e.g. 2005) by assembling all the segments that were active in that year. Although simple for the elementary case shown in Figure 23.14, the value for very large and complex maps with many thousands of boundaries and many changes will be clear.

Storing the boundary data in this way has two advantages. First, it is now easy to construct a map of statistical areas at any date between the two end-dates simply by assembling the relevant segments. So creating a base map for, say, 1990 or 2000 is now routine. Second, it is easier to create stable areas, which are consistent over the changing map. These persistent areas can be used to give consistent comparisons over time. There is a price to pay since temporal continuity and spatial continuity represent irreconcilable goals. If we are to preserve the maximum account of spatial detail then we can have only very short and broken time series. Once a consistent set of districts has been defined, that is not the end of the story, because the data themselves must be adjusted in a similar manner to tie in to the stabilized districts.

Intervisibiity on a Surface

Topographic data within a GIS can be used to construct three-dimensional block diagrams of the land surface. Ready changes in the viewing angle and orientation of surface allow the topography to be visualized in ways previously restricted to the imagination of the experienced map user.

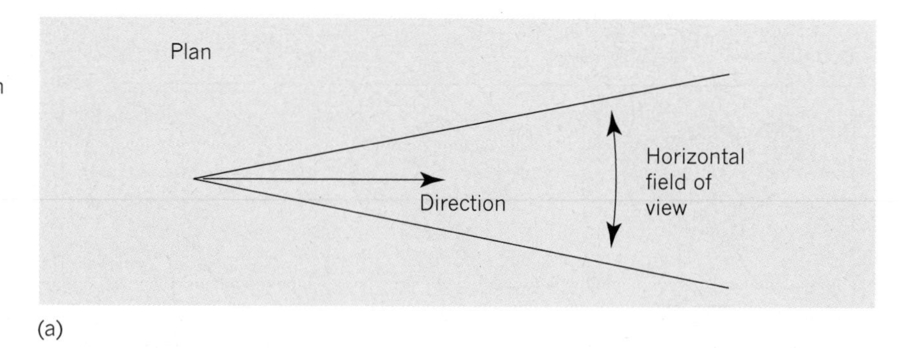

(a)

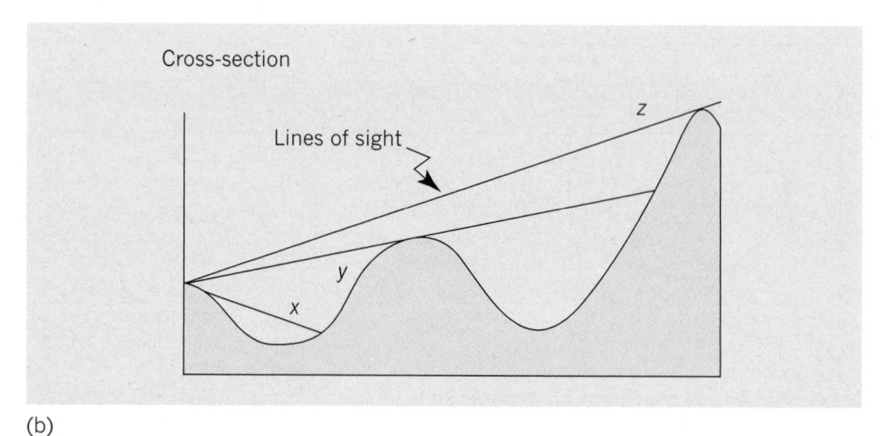

(b)

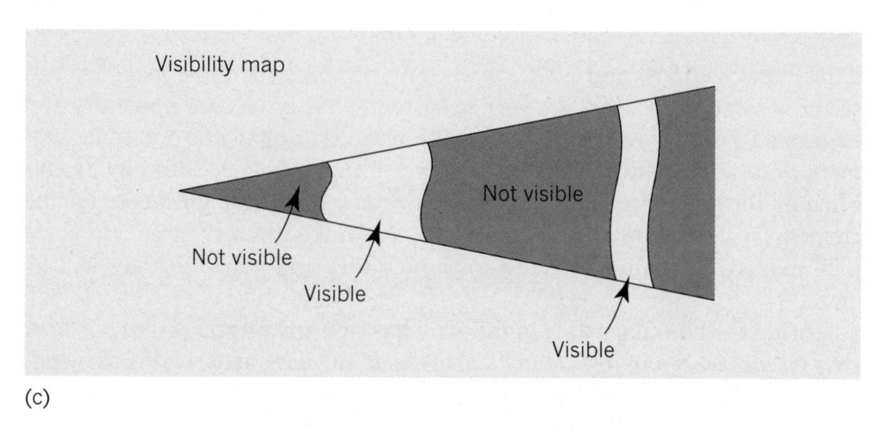

(c)

Figure 23.15 Intervisibility on surfaces Storing elevation data in a GIS allows the visibility of one point from another to be easily computed. (a) Plan view with arc of visibility. (b) Cross-sections. (c) Maps of areas which are visible and not visible from a particular location.

Figure 23.15 shows how calculations on that surface can rapidly identify areas of visibility. Such analysis was originally pioneered for military purposes where the location of surveillance and defence positions within the battle terrain could be optimized. But the method has also proved useful to archaeologists in identifying the choice of settlement sites.

23.4 Hierarchy of GIS Applications

GIS applications can be categorized by several levels of maturity and complexity. Box 23.B shows the domain of GIS in terms of data sources, types of geographic data, and current applications. We describe here five such levels of increasing complexity.

Box 23.B The Domain of GIS

Main producers and sources of data

- Topographic mapping: national mapping agencies, private mapping countries.
- Land registration and cadastral agencies.
- Hydrographic mapping.
- Military organizations.
- Remote-sensing companies and satellite agencies.
- Natural resource surveys: geologists, hydrologists, seismologists, physical geographers, soil scientists, land evaluations, ecologists and biogeographers, meteorologists and climatologists, oceanographers.

Main types of geographic data available

- Topographic maps at a wide range of scales.
- Satellite and airborne scanner images and photographs.
- Administrative boundaries: census tracts and census data; postcode areas.
- Statistical data on people, land cover, land use at a wide range of levels.
- Data from marketing surveys.
- Data on utilities (gas, water, electricity lines, cables) and their locations.
- Data on rocks, water soil, atmosphere, biological activity, natural hazards, and disasters collected from a wide range of spatial data and temporal levels of resolution.

Some current applications

- Agriculture: monitoring and management from farm to national levels.
- Archaeology: site description and scenario evaluation.
- Emergency services: optimizing fire, police, and ambulance routeing; improved understanding of crime and its location.
- Environment: monitoring, modelling, and management for land degradation; land evaluation and rural planning; landslides; desertification; water quality and quantity; plagues; air quality; weather and climate modelling and prediction.
- Epidemiology and health: location of disease in relation to environmental factors.
- Forestry: management, planning, and optimizing extraction and replanting.
- Marketing: site location and target groups; optimizing goods delivery.
- Navigation: air, sea, and land.
- Real estate: legal aspects of cadastral, property values in relation to location, insurance.
- Regional and local planning: development of plans, costing, maintenance, management.
- Road and rail: planning and management.
- Site evaluation and costing: cut and fill, computing volumes of materials.
- Social studies: analysis of demographic movements and developments.
- Tourism: location and management of facilities and attractions.
- ***Utility mapping***: location, management, and planning of water, drains, gas, electricity, telephone, cable services.

[Source: P.A. Burrough and R.A. McDonnell, *Principles of Geographical Information Systems* (Oxford University Press, Oxford, 1998), Box 1.2, p. 9.]

Stage 1: Spatial Inventories

The first level of complexity is one of geographic inventory. Here the intended result is to create a consistent and complete data layer responding only to simple queries, mostly related to location.

One example of a GIS in stage 1 is the 1 : 100,000-scale digital database of US highways and roads created by the US Geological Survey (with input by the US Bureau of Census) for the census enumeration maps to help census enumerators in the field (see Chapter 6). A second example is the capture and archiving of digital records from the LANDSAT satellite system described in the previous chapter. The digital data are received from the satellites, and a number of rectification and enhancement algorithms are performed on the data. The data are then entered into the LANDSAT archive from which they can be used for a variety of research purposes.

Figure 23.16 shows two examples of US data held on the Internet. Map (a) shows an aerial photograph of Boston, Massachusetts, drawn from the *digital orthophoto quad (DOQ)* image. Map (b) shows the USGS standard topographic map (at a 1 : 24,000) scale for part of Yellowstone National Park, Wyoming. Sources of such imagery are given in Appendix B of this book.

(a)

(b)

Figure 23.16 Geographic images on the Internet Two examples of geographic information on the Internet for the landscape of the United States. (a) Digital orthophoto quads (DOQ) of Boston. (b) Digital raster graphics (DRG) scanned from the US Geological Survey (USGS) standard series topographic map of the Yellowstone National Park.

[Source: David E. Davis, *GIS for Everyone* (ESRI Press, Redlands, Calif, 1999), (a) p. 90, (b) p. 91.]

Stage 2: Pattern Description

The second level of maturity adds further analytical operations to those performed in Stage 1. In a level-2 GIS, a geographic pattern is sought within a given data layer, using statistical and spatial analytical tools.

An example would be where we wish to compute statistics about a particular land-use category. For ecological studies we may need to know something about the woodlands that still remain in an area dominated by agricultural activity. How many separate woodland fragments are there? What is their total area? What is the distribution of those fragment sizes: is it made up of a few large forest blocks or many small woodland fragments? What is the length of the perimeter of the woodlands? (This is important for species of small mammals, birds, or insects that have specialized habitats on the woodland 'edge'.) What is the average and the maximum distance from one woodland to another? (This question is critical in establishing corridors for species that need to travel from one safe habitat to another.)

All these measures could be laboriously worked out by hand from an original map using traditional instruments such as a *planimeter*. But this would involve great time and cost. Such answers are readily available in a few microseconds once the data are stored in a GIS.

Combining the technology of Stages 1 and 2 leads to some familiar uses of GIS for route finding. Figure 23.17 illustrates one way in which geographic information on the Internet can be used to determine optimum pathways along streets and roads. The mapped route is reinforced by a series of text prompts for the driver following the route. Different routes

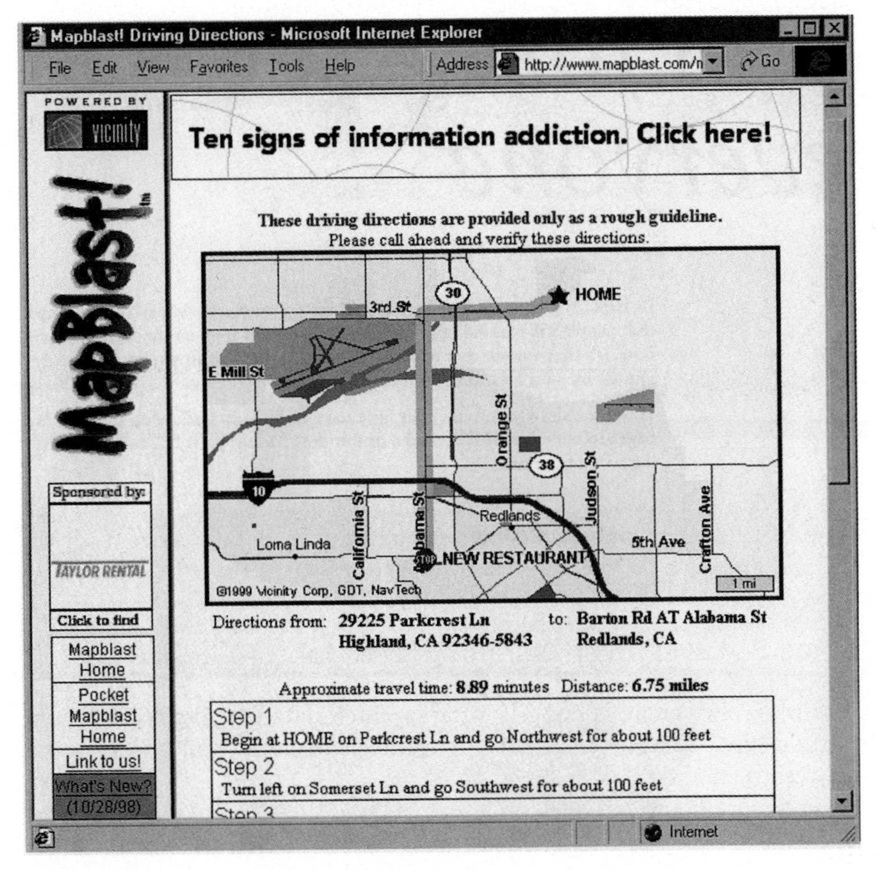

Figure 23.17 Route information from the Internet Millions of people access the Internet every day for geographic information on travel. One of the most common uses are sites such as MapBlast that create maps on demand for journeys. Note the driving instructions below the map. Similar information for the UK is available on web sites maintained by the Automobile Association (AA).

[Source: David E. Davis, *GIS for Everyone* (ESRI Press, Redlands, Calif., 1999), p. 6.]

Figure 23.18 Searching for a well site Maps showing the stages through which sieve maps can be used to identify potential well sites for two villages. (a) Water-bearing rocks (aquifers). (b) Depth of water table within the water-bearing areas. (c) Location of the two villages with (d) overland accessibility from the villages and roads. Maps (b) and (d) are combined in map (e) to suggest potential well sites that minimize both distance to the well and the depth to which it has to be dug in order to reach the water table.

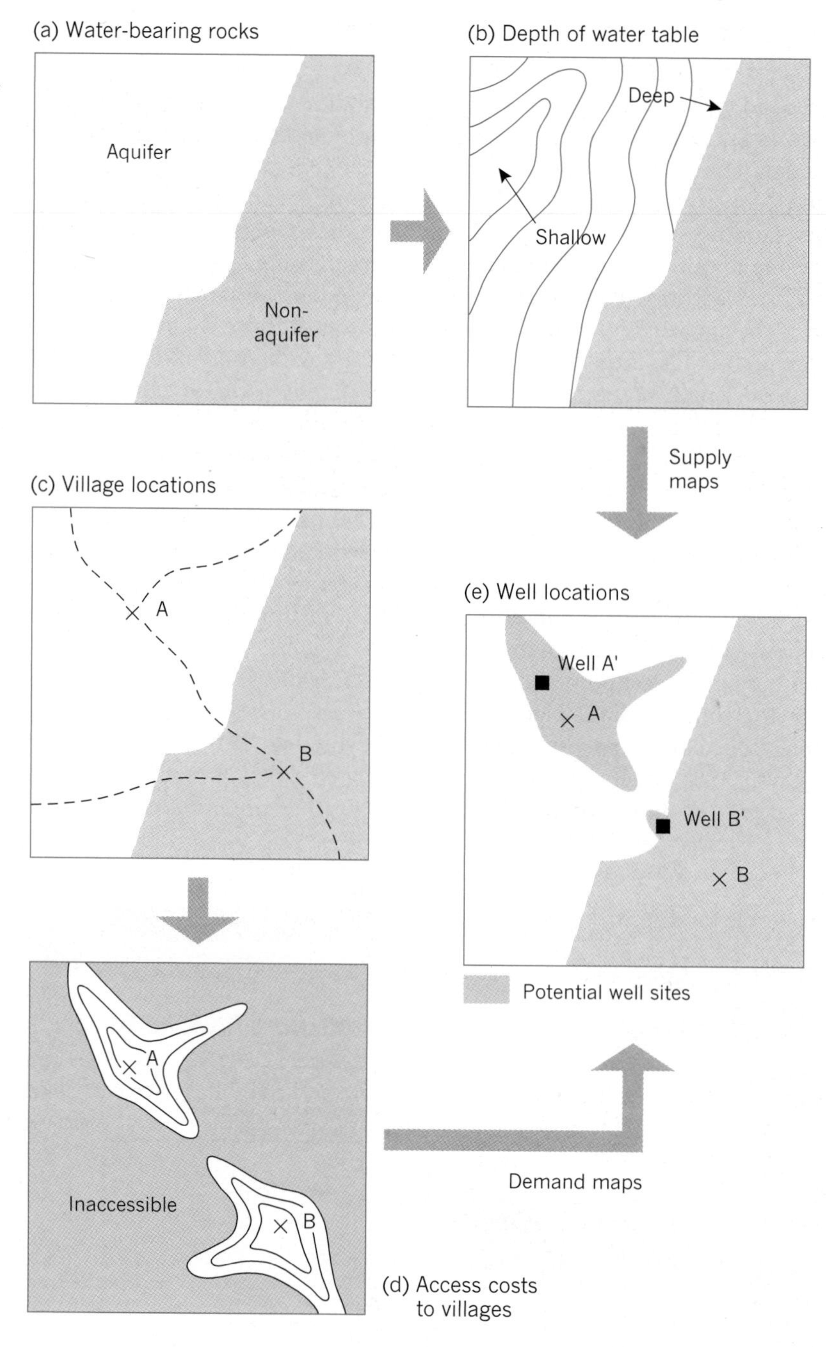

can be followed if a driver wishes to follow the fastest route, to avoid tollways or to dawdle along the most scenic routes. A parallel system for the UK and Europe has been developed by Autoroute.

Stage 3: Layer Overlays and Sieve Maps

A third level of analysis brings several layers of data together to determine relationships. This involves sets of spatial categories and the relations between them.

Let us suppose we are searching for potential sites for sinking a new well

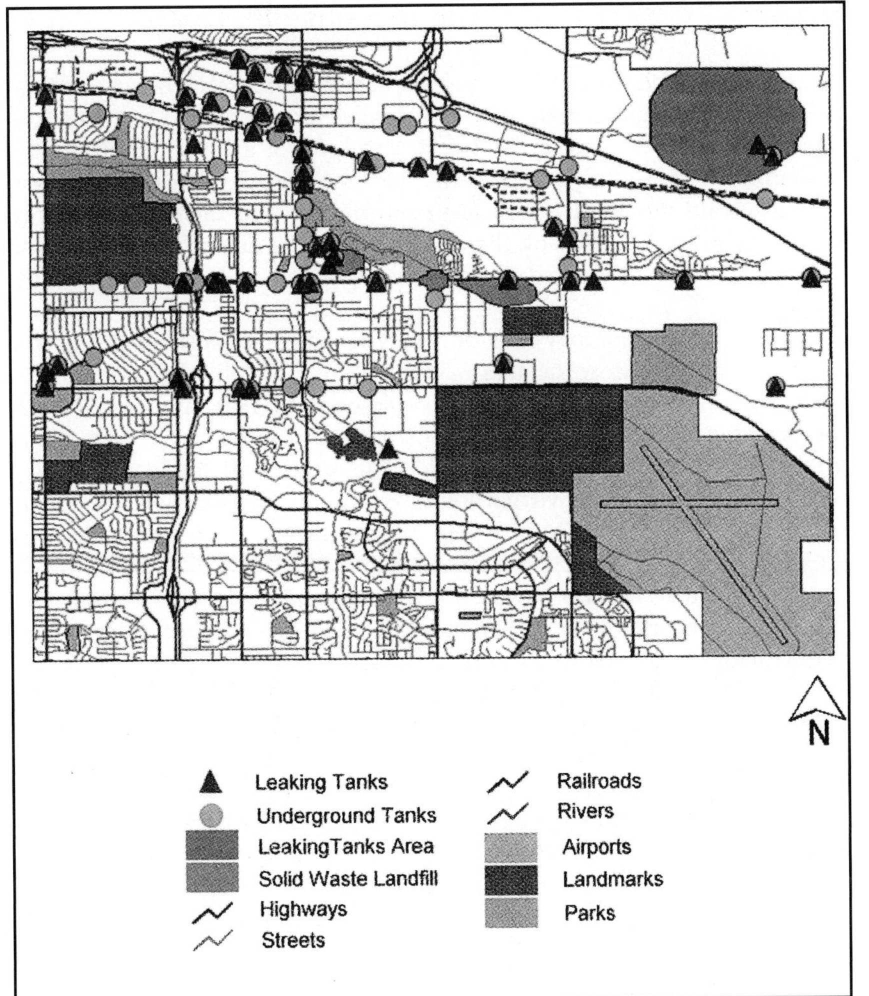

Figure 23.19 Environmental hazards GIS is used to display environmental hazards in relation to at-risk populations. Leaking storage tanks in the eastern side of Aurora, Colorado. Leaking tanks are shown by triangles.

[Source: David E. Davis, *GIS for Everyone* (ESRI Press, Redlands, Calif., 1999), p. 117. Copyright © VISTA Information Solutions, Inc., Geographic Data Technology, Inc.]

to supply a village in India. For the region around the village we can ask a series of questions. Where is the location of water-bearing rocks? What is the depth of the water table (i.e. how deep will the well need to be)? Is the land where the well will be sunk in public or private ownership? How far is the site from a potential pollution source? How long will the water have to be shipped by pipe to get to the centre of the village? As Figure 23.18 shows, the sequence of questions provides thorough *sieve mapping* which allows potential well sites through for final consideration.

Such a search cannot guarantee that an optimal site is available. The search may have to be modified by scoring some criteria as more important than others. Often the site located has to be a compromise, the best available rather than one that fully meets all criteria. A GIS allows many such searches to be made readily and the results of specifying different priorities are clearly displayed.

Another example of multiple layers is given in Figure 23.19, which shows by triangles the location of leaking storage tanks superimposed on a street pattern in the town of Aurora, Colorado. The relationship between such toxic hazards and other key aspects of the local environment is clearly shown. Similar hazard maps relating flood risk to possible insurance claims (homes, factories, etc.) are now a routine part of analysis.

Stage 4: Advanced Display Functions

The display of data from a GIS is now reaching into levels where it is catching up on the images produced by Hollywood screen studios. It is now routine to produce perspective views of a data layer or a combination of data layers. Often a landscape of natural vegetation or city streets is 'draped' over a terrain model (a digital elevation model) and then highlighted to enhance certain features of the display. Names and other information can be added or subtracted at will to simplify or add complexity to the model.

New virtual technology can be used to simulate journeys across the landscape showing how the area would look to a traveller following a particular route. In the years ahead we would expect GIS technology to allow the teaching of world geography by students choosing their own routes around all or some part of the world. The screen and sound system will enable them to see their virtual journeys and, at any location, to call up information to describe the city they can see ahead or the glacier they are crossing.

Nor need such displays be static. Dynamic displays or movies of an area shown in perspective are poignantly visually. Geographer Peter Gould made use of this technique in plotting the spread of AIDS across the United States in the 1980s (look back at Figure 20.16). The ability to show the rapid spread of the disease from several perspective views from differing angles and then to replay those views in sequence created a strong impact on the Pennsylvania high-school classes to which it was shown.

Finally, we cannot overemphasize the growing interactive and dynamic display potentials of GIS. In the analogue cartographic world, it was necessary to specify a map design and then to execute it. Once begun, very little change could be accommodated, and such accommodations were costly. In

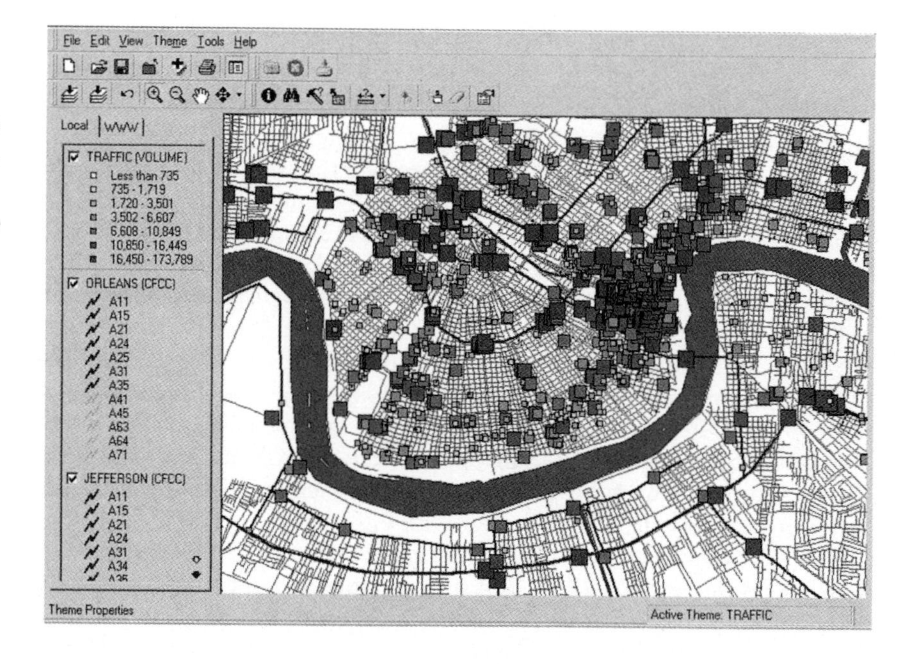

Figure 23.20 Traffic volume on city streets Map of New Orleans, United States, showing average daily volume on selected street locations. The data, from Geographic Data Technology, Inc. are available at ArcData Online.

[Source: David E. Davis, *GIS for Everyone* (ESRI Press, Redlands, Calif., 1999), p. 10. Copyright © Geographic Data Technology, Inc.]

a GIS, you can change the map composition at any time. Also, the chance to interact with information enhances your understanding of it. You can add a data layer and then remove it. You can even flash it on and off while you study the display. The capabilities to zoom in or out on the display, to vary data layers, and to rotate the map in a perspective view add new insights about the map. Users have never before experienced such flexibility, and we are only beginning to learn to take full advantage of it.

Stage 5: Advanced Modelling

The fifth, most advanced, level of GIS use involves a true decision support system. It goes beyond simple locational questions (Stage 1) and sieving questions (Stage 3) to those involving predictions. Models within the GIS system allow 'what if?' questions to be posed and answered.

An example will illustrate the principle. Assume we have data (which may be in real time) of traffic flows on roads within some part of a city. Figure 23.20 gives a typical example. Assume Stage 2 analysis has identified that land is available for three potential supermarket stores within the area of interest. The question then arises as to what impact the building of the supermarkets would have on traffic flows. Routeing and flow allocation models (some elementary ones were discussed in Chapter 13) can be applied and the present and future patterns of flows estimated.

In physical geography, a similar question of environmental impact might arise where alternative sites for small dams are being considered. In this case the flow models would come from hydrographic forecasting models.

In both examples GIS systems are capable of exploring or confirming the relationships of human populations to their regional environments. As GIS systems continue to evolve, so the potential for linking spatial and regional models to basic GIS system grows.

GIS in Perspective 23.5

In this final section we try to place GIS in a historical context, looking both at its past growth and its implications for geography as a whole.

Past Growth of GIS

The first mention of GIS occurred in the geographic and computer literature as early as the mid-1960s. But massive growth began only from 1980 with the introduction of super-mini-computers by manufacturers such as Digital Equipment Corporation (DEC) and Prime Computers. Growth in the software industry followed, led by *Environmental Systems Research Institute (ESRI)* and the *Intergraph Corporation* and who remain the market leaders today. (See Box 23.C on 'ESRI: evolution of a major GIS company'.)

Box 23.C ESRI: Evolution of a Major GIS Company

ESRI was founded in 1969 by Jack and Laura Dangermond as a privately held consulting group. The business began with $1100 from their personal savings and operated out of an historical home located in Redlands, California, the city where Jack grew up. During the 1970s, ESRI focused on the development of fundamental ideas of GIS and their application in real-world projects such as developing a plan for rebuilding the City of Baltimore, Maryland, and helping Mobil Oil select a site for the new town of Reston, Virginia. Jack Dangermond recalls:

> The more projects we completed, the more we learned how GIS methods could contribute to integrating information in new ways. It was during these years that we first observed the potential of GIS technology. We saw how GIS could influence the way decisions were made. We learned that using geography as a framework for integration could benefit the way people approach problem solving in land management and fundamental scientific research and education. We knew GIS could make a difference.

The early years were a struggle because most of the work was being done by the two founders and a small staff. They made the decision early not to take out loans, use venture capital, or go public to raise money.

In its second decade, ESRI moved from a company primarily concerned with carrying out projects to a company building software tools and products. Dangermond realized that to leverage the methods and technologies developed for project work during the 1970s, the company needed to develop commercial software products that others could use and rely on for doing their own projects. The company hired several software engineers and created its first software product.

ESRI's first commercial GIS product, ArcInfo, was launched in 1981. Originally designed to run on minicomputers, ArcInfo offered the first modern GIS efficiently integrated into a single system. As the technology shifted to UNIX and later Windows NT workstations, ESRI evolved software tools to take advantage of these new platforms. This shift has allowed ArcInfo users to switch platforms and take advantage of the principles of distributed processing and data management.

The company held the first ESRI User Conference in 1981, which attracted 18 people and was held at ESRI offices in Redlands. (In contrast, the 1998 ESRI User Conference hosted more than 8000 users and business partners from 90 countries worldwide!) Corporate philosophy continued to focus on developing software and creating useful applications for companies and government organizations. For many years, ESRI was a one-product company, marketing and supporting ArcInfo and a wide variety of related GIS services. Then, in 1986, PC ARC/INFO was developed and released as the microcomputer version of ArcInfo. The 1990s brought more change and evolution. ESRI firmly established itself as the market leader in 1991 with the launch of ArcView GIS, an affordable, easy-to-learn desktop mapping and GIS tool, which shipped 10,000 copies in the first six months of 1992. The company launched the ArcData Program to provide a wide variety of high-quality, ready-to-use data that are compatible with ESRI software. ArcCAD followed in 1992, marking the integration of computer-aided design (CAD) and GIS technologies. Spatial Database Engine (SDE), a client–server product for spatial data management, was acquired in 1994, followed by BusinessMAP, ESRI's first consumer mapping product. 1996 saw ESRI's product line grow again with the release of ArcInfo for Windows NT, MapObjects (mapping and GIS components for software developers), and Data Automation Kit (geographic data creation for Windows), as well as the acquisition of Atlas GIS. This expansion of ESRI's product family gave users a comprehensive set of GIS and mapping software options, and fortified ESRI's position at the top of the GIS market.

GIS is poised for even greater growth during the rest of the decade and beyond. Innovations in computer technology now allow sophisticated GIS operations to be performed on the desktop. Faster and cheaper computers, network processing, electronic data publishing, and improved easier-to-use GIS technology are fuelling rapid growth in the desktop area. Private businesses are adopting GIS technology as a decision support tool. And with the introduction of live mapping applications to the World-Wide Web, anyone with a computer has access to the benefits of GIS technology.

Today, ESRI employs more than 2000 full-time staff, more than 1000 of whom are based in Redlands at the corporate headquarters. ESRI offers employment opportunities to qualified professionals from around the world and has a richly diverse workforce. The Redlands campus has expanded with the addition of a new three-story research and development centre, which opened in early 1996. Further expansion of the R&D centre was completed in the summer of 1998.

[Source: information drawn from the current ESRI web site. See Appendix B.].

The technology of GIS rapidly found practical applications in resource management, particularly forestry, local government, the utility industries, and geodemographics. These uses were largely in the automated measurement and analysis of geographically distributed resources, and the management of spatially distributed facilities. The last two decades of the last century saw widening applications of GIS to a wide range of sciences and social sciences that use geographically distributed data and find value in a spatial perspective. These include archaeology, ecology, epidemiology, geology, geophysics, oceanography, regional science, and of course geography itself.

Like many innovations (see Chapter 16), GIS began slowly in a few centres and then spread more widely and more rapidly. As a country with huge land areas needing inventory and reconnaissance, Canada took an early lead in implementing GIS. A Canada Geographic Information System was developed in the 1960s by Roger Tomlinson as a computer application for the analysis of the data collected by the government's Canada Land Inventory designed to improve land-use policy. A second centre of innovation was the research groups at Harvard (the Laboratory for Computer Graphics and Spatial Analysis), led by geographer William Warntz. One of their earliest and most widely used methods was *Synagraphic Mapping Technique (SYMAP)* pioneering a set of computer programs for creating a diversity of maps by combining alphabetical and numerical keys in fast line printers. By superimposing combinations of numbers and letters on the printer, they developed a series of 'tones' that simulate the grey scales of conventional isarithmic and choropleth maps. The Harvard group conducted basic research into methods for handling geographic data in digital form throughout the 1970s.

The 1980s saw a significant expansion of research once it became clear that GIS had great potential as a tool to support research and decision-making in a wide range of fields. In the United States, the *National Center for Geographic Information and Analysis (NGIA)* was initiated in 1988 with major funding from the National Science Foundation (NSF). This initiative was based largely on the efforts of Ron Abler, then Director of the NSF Geography and Regional Science Program. The US Center exists to conduct basic research in GIS and its applications, particularly in science; funds were awarded to a consortium of the University of California at Santa Barbara; and the State University of Maine. The State University of New York at Buffalo also played a major part in the initiative.

In the UK, the government's Economic and Social Research Council funded the development of a network of regional research laboratories. These were largely based in university geography departments, with the task of promoting GIS applications in a mix of practical and scientific applications in order to build a stronger UK research base. The European Science Foundation launched its GISDATA programme in 1993 aimed at fostering research collaboration among European countries. Programmes similar to these exist in many other countries.

Future Implications for Geography

GIS is both popularizing and transforming geography. GIS allows everyone to access huge amounts of spatial data and to communicate with these data. This ability may well revolutionize our industrialized society and is already affecting the way companies consider their marketplaces. Enterprises as different as oil companies, banks, and car salesrooms are realizing that their marketplaces have a distinctive spatial structure, one that can be readily captured and updated to save on costs of advertising to product delivery. Together with the rest of the 'dot.com revolution' (discussed in Chapter 19), the GIS revolution is starting to change the way corporations think about space.

Such GIS effects are not ethically neutral. Targeted marketing, delivery truck routing, utility repair, and mortgage lending using GIS means that some areas are off target. Both profit and loss regions can be demarcated. The poor financial returns made in some countries, some regions within a country, or some districts within a city may mean a company withdrawing from unprofitable areas. We are already seeing so-called 'red lining' of areas within a city, with banks refusing loans and insurance companies withdrawing cover or insisting on exceptionally heavy premiums.

But there is a positive side. In urban traffic flows, 'smart' vehicles using global positioning systems (GPS) and GIS navigation, can outflank traffic holdups. Paths taken by ambulances or fire engines can be eased. Miniaturization offers the prospect of spatial Braille systems to give improved freedom of travel to blind users. Farmers can scan information from their tractors to adjust seeding and fertilizer application to the spatial pattern of soil depth and site character varying over a large field. Both improved yields and reduced fertilizer runoff into streams may result.

Within geography itself, the switch from analogue to digital cartography through GIS promises a new renaissance not just in cartography but in geography overall. Uses of GIS in geography are limited only by our imagination. The whole range of questions raised in the last 22 chapters of the book can be sharpened, extended, and revolutionized using GIS.

Relfections

1. Use of Geographic Information Systems (GIS) exploded within geography in the 1990s and seems set to continue growing rapidly in the new century. List the factors that lie behind this rapid evolution.
2. Spatial data are usually recorded in a GIS in one of two formats. What do you consider the relative advantages of raster-based versus vector-based systems of recording geographic data?
3. How far does the shifting infrastructure of geographic recording areas (see Figure 23.13) pose problems for GIS analysis? How can such problems be overcome?
4. Explore some of the GIS Internet sites listed in Appendix B of this book. Take any *two* sites and write a critical comparative note on how easy you found them to explore.
5. Show from a hypothetical map how spatial sieving could be used to locate a new road that (a) serves several settlements, (b) avoids traffic noise near housing, and (c) follows a route that avoids the high cost of construction through rugged terrain.
6. How far does GIS raise ethical problems? For example, does the very detailed spatial information on specific parts of a city allow companies to 'red-line' problem areas where some services (e.g. banking or insurance) are very unprofitable? How far do you consider any downside is outweighed by gains in very efficient spatial delivery of other services?
7. Review your understanding of the following concepts using the text and the glossary in Appendix A of this book:

database	raster
digitizers	scanners
overlay	sieve mapping
polygon	Thiessen polygon
quadtree	vector

One Step Further ...

A helpful introduction to GIS including a historical setting is given by Michael Goodchild in his essay in R.J. JOHNSTON *et al.* (eds), *Dictionary of Human Geography*, fourth edition (Blackwell, Oxford, 2000), pp. 301–4. Goodchild is also one of the editors of the great standard reference work in the field, P.A. LONGLEY, M.F. GOODCHILD, D.J. MAGUIRE, and D. RHIND (eds), *Geographical Information Systems: Principles, Techniques, Management, and Applications*, 2 vols, second edition (John Wiley, Chichester, 1999), which is more accessible and usable than its size would suggest.

Valuable shorter texts are N. CHRISMAN, *Exploring Geographic Information Systems* (Wiley, New York, 1997), I. HEYWOOD *et al.*, *Introduction to Geographical Information Systems* (Longman, Harlow, 1998) and C.B. JONES, *Geographical Information Systems and Computer Cartography* (Longman, Harlow, 1997). A more advanced text is Peter A. BURROUGH and Rachael A. McDONNELLl, *Principles of Geographical Information Systems* (Oxford University Press, Oxford, 1998). A very user-friendly text with a companion CD for using GIS methods on your computer is David E. Davis, *GIS for Everyone* (ESRI Press, Redlands, Calif., 1999).

The analysis of spatial data is very well covered in T. BAILEY and A. GATRELL, *Interactive Spatial Data Analysis* (Longman, Harlow, 1995) and P. LONGLEY and M. BATTY (eds), *Spatial Analysis: Modelling in a GIS Environment* (GeoInformation International, Cambridge, 1996). Social implications of GIS are taken up in J. PICKELS (ed.), *Ground Truth* (Guilford Press, New York, 1995).

The 'progress reports' section of *Progress in Physical Geography* (quarterly) has occasional GIS reviews but to keep abreast of the rapidly moving world of GIS it is helpful to scan some of the regular monthly journals such as *GIS Europe* that provide commercial information on GIS products and applications. Keep an eye also on the *International Journal of Geographical Information Systems* (quarterly) and *Computers and Geosciences* (quarterly). For readers with access to the World-Wide Web see also the sites on 'GIS' in Appendix B at the end of this book.

ROYAL GEOGRAPHICAL

Epilogue

CHAPTER 24

On Going Further in Geography

■ denotes case studies

Figure 24.1 On the edge of the undiscovered ocean Isaac Newton's quotation (next page) reminds us that however much geographers have learnt about the earth in the last millennium, we are still on the very edge of understanding. The small girl playing on beach is symbolic. [Source: Elie Bernager/Tony Stone Images.]

> I do not know what I may appear to the world; but to myself I seem to have been only like a boy playing on the seashore, and diverting myself in now and then finding a smoother pebble or a prettier shell than ordinary, whilst the great ocean of truth lay all undiscovered before me.
>
> SIR ISAAC NEWTON (from Brewster, *Memoirs of Newton* 1855, Vol. II, Chapter 27)

One of the odd paradoxes about studying a subject intensively is that you realize how much there is yet to know. Having started on a geography course at Cambridge University a half century ago, I can now see what Sir Isaac Newton meant. For we all stand, like small children, trying to understand a small part of planet Earth in all its beauty and diversity (see Figure 24.1 p. 750).

So, this chapter is addressed to those of you who are now finishing this book. Some of you, perhaps most, took a single geography course and will be going on to your main studies in music or midwifery, in physics or philosophy. I wish you God speed in your studies. But there will be some of you who have found enough in the preceding 23 chapters to interest you. Perhaps you are now considering taking some one or two more courses or perhaps even majoring in geography.

Although many study geography at school level, it is not a large subject at college and university level in most countries. For example, in the United Kingdom, only one in every ten students that studies geography at secondary school level goes on to continue with it at university; in the United States, at the bachelor's level not even one graduate in every 150 in United States universities majors in geography. Even within the social sciences division, in which it is often placed, the numbers of geography majors are dwarfed by those in psychology or economics. On the graduate level, for every doctorate granted in geography there were 10 in physics, nearly 20 in chemistry and engineering, and 25 in education. Within the social sciences division, geographers were only one-seventh as numerous as their colleagues in economics or history.

But this small population of newly hatched geographers should be seen in perspective. We need to see the situation in the United States' position in an international context. Geography as a modern university subject had its origins in the institutes of Germany and France during the early nineteenth century. The German contributions to the field, certainly until World War II, were dominant ones, and a majority of the classic works in geographic literature appeared first in the German language. In the universities of western Europe, and those of Britain and the Commonwealth (to which geography spread, largely, in the first half of this century) the number of geography students is relatively large. Canada and Australia have strong research centres. The situation in Russia is also much more advanced: there geographers outnumber their American colleagues by about three to one.

Also, the situation is always changing. The number of geography students enrolled in North American and West European colleges and universities more than doubled over the 1960s decade. It then stayed on a plateau but has surged again in the 1990s. Membership of leading professional societies such as the Association of American Geographers or the Royal Geographical Society has more than doubled during the same period.

University geography is thus a curious mixture. In size it is perhaps comparable with subjects with small enrolments such an anthropology or archaeology rather than major fields such as mathematics or history. In its recent spurts of growth, however, it behaved more like a modest version of biochemistry or computer science.

This final chapter tries to put the subject in context. We shall look first at the past growth of the field (Section 24.1) to try to understand the path followed in getting to the present position (Section 24.2). We then consider the job situation for geography graduates (Section 24.3). From the present situation we then look forward to the future; we shall project the present drift and contrast this projected future with other targets for the field (Section 24.4). In summary, we shall ask four basic questions: How did geography evolve as a separate field of study? What is its present structure? What do geographers do with their degrees? What is its likely future – if present trends continue? What should its future role be?

The Legacy of the Past 24.1

We can understand the character of geography as an academic field today only if we see it as one scene in a lengthy play. It is useful to divide the play into three acts. The first act is dominated by isolated research by individual scholars, the second by organized research by groups and societies, the third by the incorporation of research into national and international organizations. It is clear that the stages cannot be precisely fixed in time; each is continuing in different subdivisions of the field or in different countries in various stages of growth.

Act I: The Individual Scholar

The first growth period, from the beginning of formal geographic study in ancient Greece to the mid-nineteenth century, can be characterized by geographic studies sporadically distributed in time and space. The number of scholars who would have counted themselves as geographers was always small, and it was only occasionally that clusters of workers formed – in Alexandria in the second century BC, in Portugal in the fifteenth century, or in the Low Countries in the sixteenth century.

The patronage that encouraged these groups to join together usually came from an interest in practical problems; methods of surveying the earth, instruments for marine navigation, map making, and the printing of atlases. In this early period people found the answer to many questions about the general shape of the earth and ways of putting spatial information on maps. The maps of the period include some of the most majestic products of Renaissance Europe (see Figure 24.2). Most of the geographic schools were short-lived, however, and had fluctuating fortunes.

Figure 24.3 shows some of the leading scholars who have contributed to geography since 1750. Remember, in interpreting this diagram, that the number of names that might have been included was less in 1800 than it was in 1900 and much less than in 2000. Indeed, as we shall see later, the general growth of geography over the two centuries spanned by this

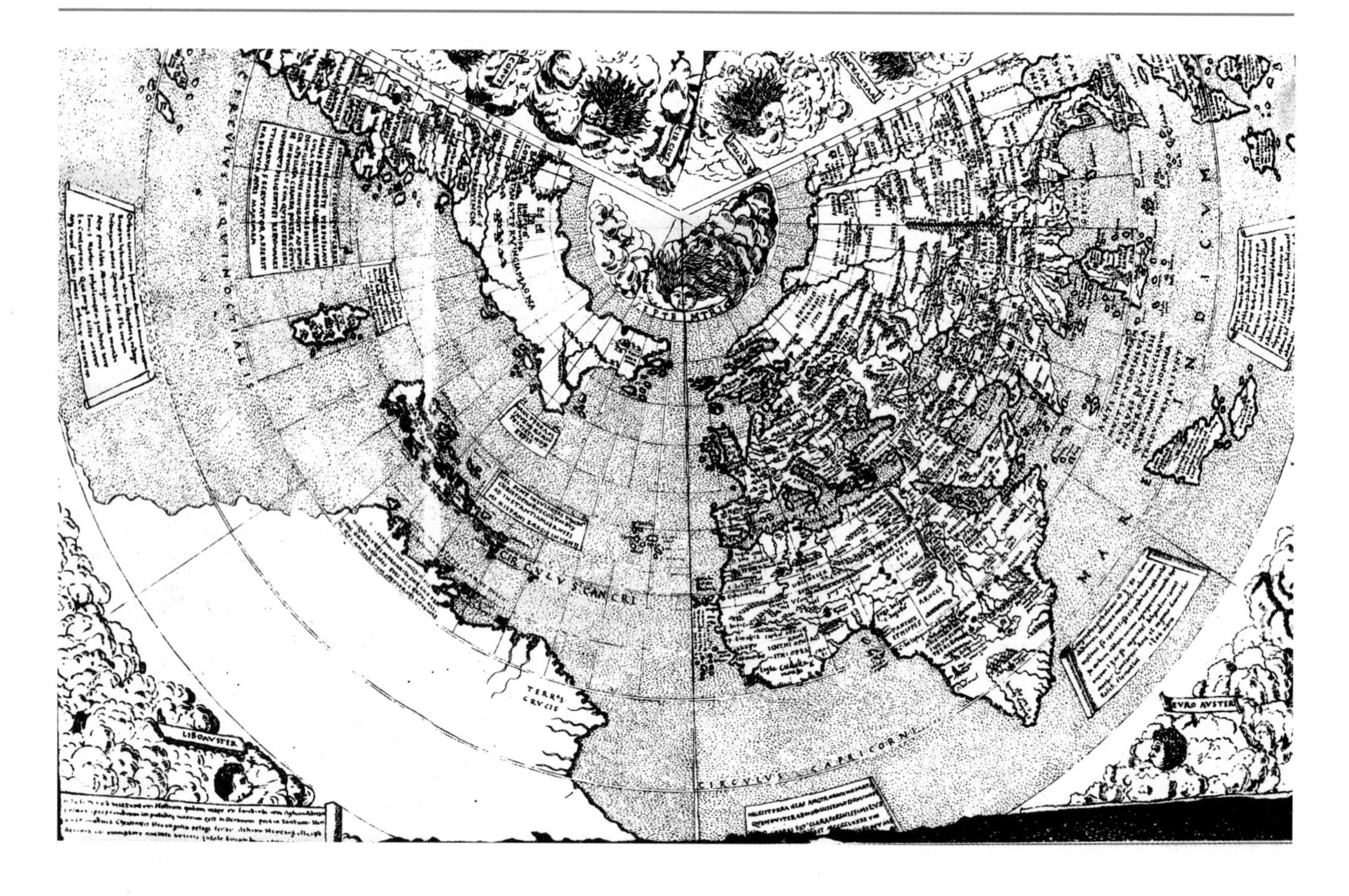

Figure 24.2 Cartographic traditions in geography Map making and exploration of the world played a key role in the early development of geography. This late-Renaissance world map was the first printed map to show America. It was made by an Italian, Contarini, in 1506 and illustrates the compromise between accuracy and adornment typical in maps of this period. For a description of modern developments in cartography, see Chapter 21.

[Source: Map courtesy of the Trustees of The British Museum.]

diagram has been logarithmic. Most of the geographers who have *ever* lived are alive today!

Note too that boundaries between disciplines were loosely drawn. It is not surprising that individuals such as Immanuel Kant, Alexander von Humboldt, and T.R. Malthus played notable roles in the growth of various sciences. The diagram also underlines the significant part played by Germany in the early part of the period (until 1930 nearly half of all geographic publications were in the German language). Inevitably, the diagram reflects the biases of the period. There are very few women geographers in the list, and certain countries isolated by language from the mainstream (notably China) are under-represented.

The exact significance of individuals in an overall growth process is difficult to assess. Science often involves a snowball effect, because of which more than due emphasis is placed on the contributions of a few individuals, for example, on the work of the few physicists or chemists who win Nobel Prizes. This so-called *Matthew effect* (after the Gospel's remark that to him that hath, more shall be given) also occurs in geography, and Figure 24.3 inevitably reflects it. Yet it would be impossible to think of German geography in the mid-nineteenth century without Carl Ritter and Friedrich Ratzel, France without Vidal de la Blache, the United States without W.M. Davis, or Britain without Halford J. Mackinder (Figure 24.4). A number of the leading geographers who have made critical contributions are featured in boxes in each of the preceding chapters.

Box 24.A The Royal Geographical Society

(a)

(b)

The Royal Geographical Society typifies the groups formed in the early nineteenth century to foster the research and exchange of geographic information about a world, much of which remained to be explored (see (a)). It was founded in 1830 for 'the advancement of geographical science', and granted a Royal Charter in 1859. The Society's main aims remain today as:

- to stimulate and support geographic research in the UK and abroad;
- to promote and strengthen the value of geography in formal education and life-long learning;
- to acquire, hold, and disseminate geographic information;
- to encourage a wider public interest, understanding, and enjoyment of geography;
- to advise governments and other agencies on geographic issues.

It recently merged with another geographic society, the Institute of British Geographers (founded in 1933). Together, the enlarged society publishes a number of scholarly journals: *Geographical Journal*, *Area*, and *Transactions of the Institute of British Geographers* together with one popular journal *Geographical Magazine*.

Its historic headquarters is located next to Hyde Park in west London (c) adjacent both to an important university (Imperial College) and to cultural centres (Albert Hall) (d). It houses a major library, map collection, and archives (b). As in the nineteenth century, it remains a centre for global exploration (including the first successful ascent of Mount Everest in 1953) although today these tend to be scientific research in association with the Royal Society in desert environments or in tropical rainforest. The Royal Geographical Society predated the establishment of geography departments in universities, and played a leading role in pressing for geography to be included in the school and college curriculum. It was an important force in establishing geography at Oxford University and Cambridge University in England.

[Source: Photographs by Royal Geographical Society and Aerofilms.]

(c)

(d)

Figure 24.3 Geography, 1775–2000 This chart shows some leading scholars in the most recent period of geographic research. It indicates the emergence of some major schools and the changing national balance of research. (Compare the German names in the nineteenth century with the American ones in the twentieth.) The geographers' names have been mainly taken from a list in Sir Dudley Stamp's *Longmans Dictionary of Geography* (Longmans, London, 1966) but updated for the recent period. Note that the list is largely concerned with western scholarship. No comparative information was available on the growth of, say, Chinese geography. Some names have been included to illustrate the major contributions to geographic thought by scholars from other disciplines, particularly earlier in the period when the boundaries of fields were loosely drawn. Names have been chosen with a view to illustrating the emergence of significant research themes. Inevitably, other geographers would choose other names. Yet, while there might be only a modest overlap, the same general pattern of evolution of research themes over time would probably emerge. Very recent trends (within the last decade or two) are discussed in Section 24.4. Note that the country given after each geographer's name is sometimes nominal since he or she may have worked for long periods in two or more countries.

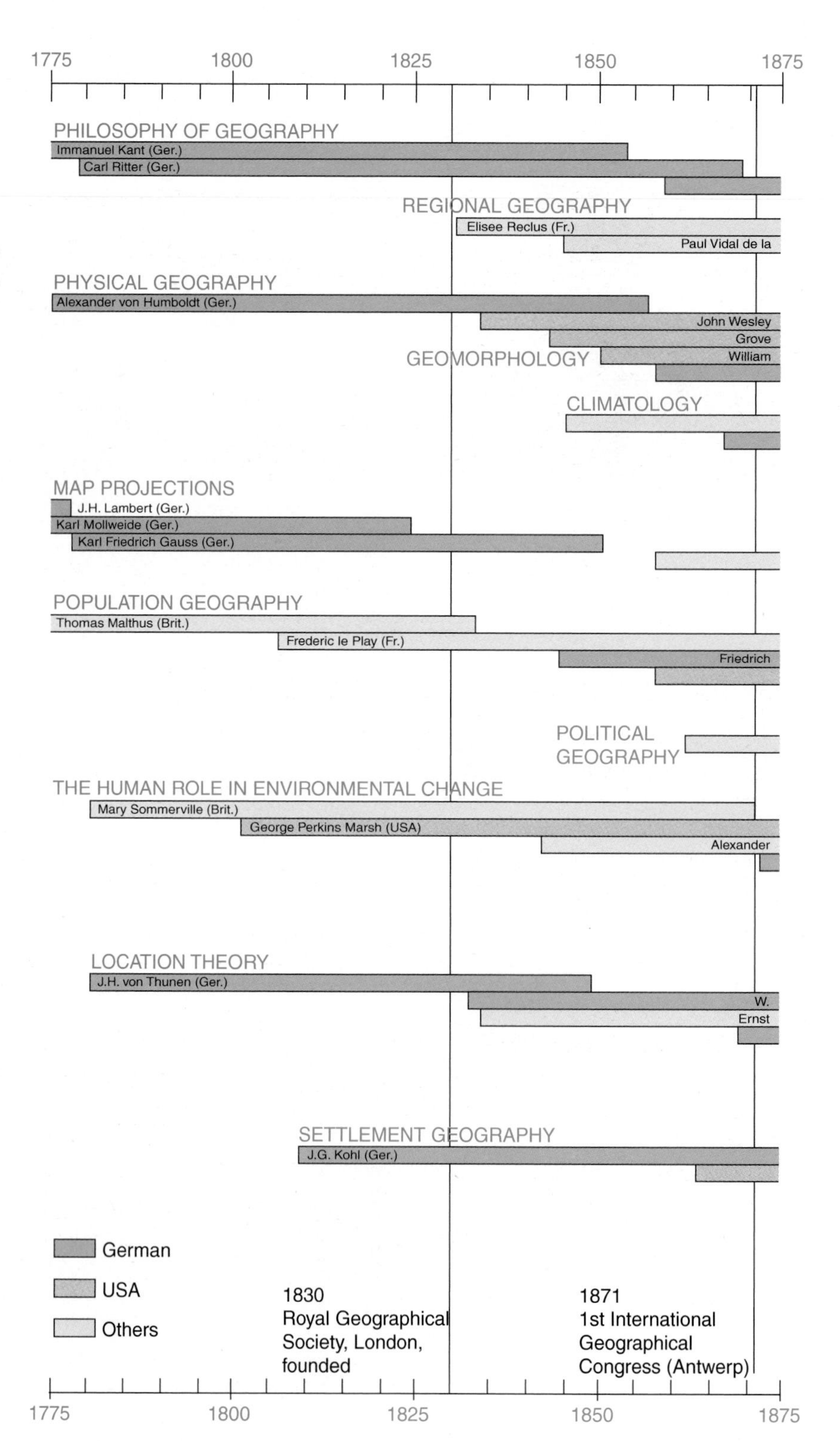

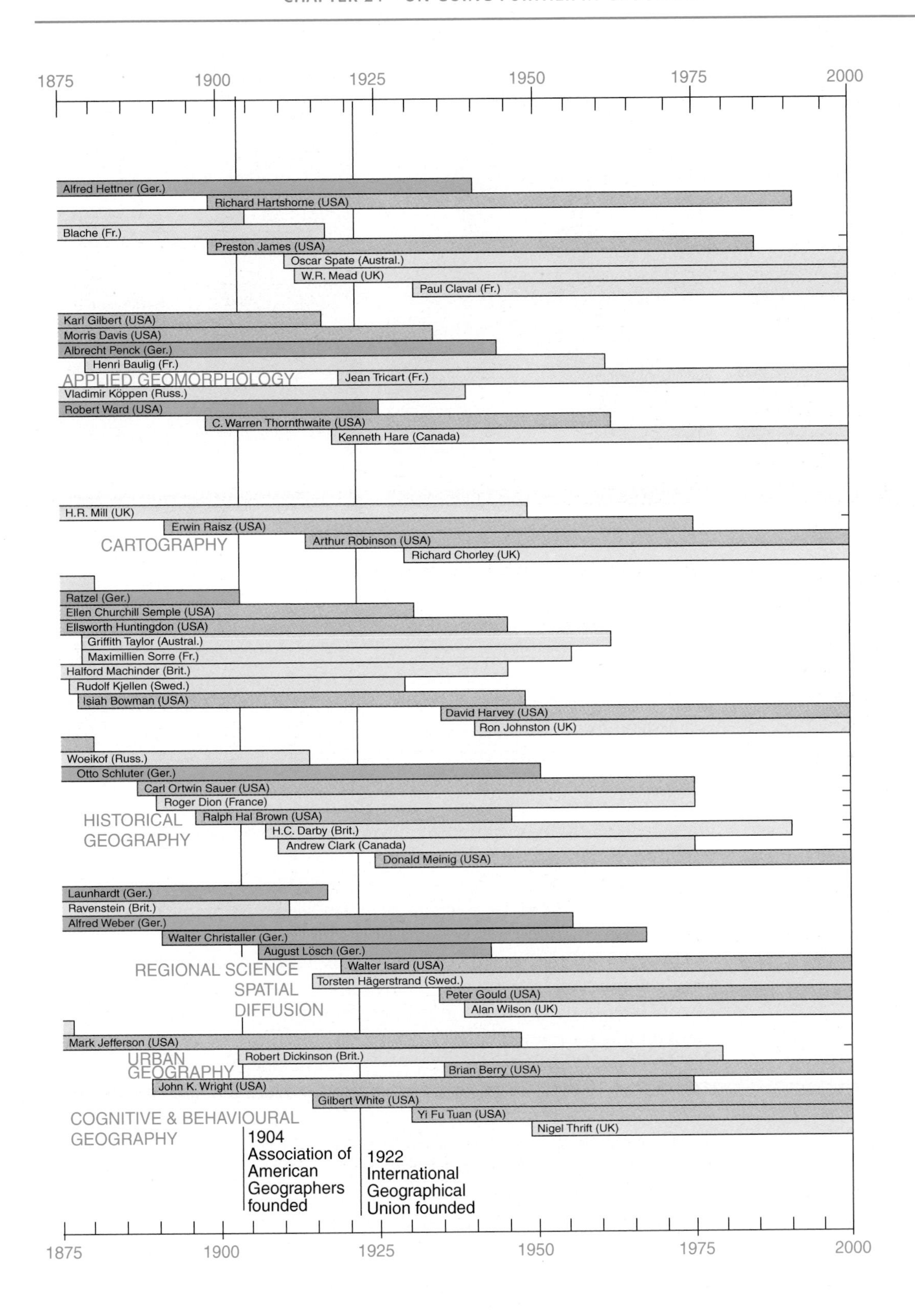
1875
1900
1925
1950
1975
2000
Alfred Hettner (Ger.)
Richard Hartshorne (USA)
Blache (Fr.)
Preston James (USA)
Oscar Spate (Austral.)
W.R. Mead (UK)
Paul Claval (Fr.)
Karl Gilbert (USA)
Morris Davis (USA)
Albrecht Penck (Ger.)
Henri Baulig (Fr.)
APPLIED GEOMORPHOLOGY
Jean Tricart (Fr.)
Vladimir Köppen (Russ.)
Robert Ward (USA)
C. Warren Thornthwaite (USA)
Kenneth Hare (Canada)
H.R. Mill (UK)
Erwin Raisz (USA)
CARTOGRAPHY
Arthur Robinson (USA)
Richard Chorley (UK)
Ratzel (Ger.)
Ellen Churchill Semple (USA)
Ellsworth Huntingdon (USA)
Griffith Taylor (Austral.)
Maximillien Sorre (Fr.)
Halford Machinder (Brit.)
Rudolf Kjellen (Swed.)
Isiah Bowman (USA)
David Harvey (USA)
Ron Johnston (UK)
Woeikof (Russ.)
Otto Schluter (Ger.)
Carl Ortwin Sauer (USA)
Roger Dion (France)
HISTORICAL GEOGRAPHY
Ralph Hal Brown (USA)
H.C. Darby (Brit.)
Andrew Clark (Canada)
Donald Meinig (USA)
Launhardt (Ger.)
Ravenstein (Brit.)
Alfred Weber (Ger.)
Walter Christaller (Ger.)
August Lösch (Ger.)
REGIONAL SCIENCE
Walter Isard (USA)
Torsten Hägerstrand (Swed.)
SPATIAL DIFFUSION
Peter Gould (USA)
Alan Wilson (UK)
Mark Jefferson (USA)
URBAN GEOGRAPHY
Robert Dickinson (Brit.)
Brian Berry (USA)
John K. Wright (USA)
Gilbert White (USA)
COGNITIVE & BEHAVIOURAL GEOGRAPHY
Yi Fu Tuan (USA)
Nigel Thrift (UK)
1904 Association of American Geographers founded
1922 International Geographical Union founded
1875
1900
1925
1950
1975
2000

(a)

(b)

(c)

Figure 24.4 'Modern' geographers Leading scholars from the first three generations of modern geographers. (a) Alexander von Humboldt, 1799–1859, a German polymath who explored South America and introduced biogeographic concepts into geographic writing. (b) George Perkins Marsh, 1801–82, an American conservationist who pioneered our understanding of the role of human action in shaping the earth's surface. (c) Paul Vidal de la Blanche, 1845–1918, a Frenchman who integrated ideas in human geography and wrote model geographies of French regions.

[Source: Photos courtesy of (a) Dr Franz Termer, (b) the Trustees of Dartmouth College, and (c) the American Geographical Society of New York.]

Act II: Groups and Societies

The second period of growth, beginning in the early 1800s, was marked by the organized interlinking of research. One of the earliest methods of linkage was the foundation of societies to foster common interests in geographic research. Such geographic societies generally fall into four groups.

The first group, the national societies, emerged in the early half or middle years of the nineteenth century. These societies have a strong interest in global exploration. For instance, the *Royal Geographical Society (RGS)*, in London, dates from 1830 and marks the merger of several early exploring clubs such as the Association for Promoting the Discovery of the Interior Parts of Africa, founded in 1788. (See Box 24.A on 'The Royal Geographical Society'.) The *American Geographical Society* of New York was founded in 1852 by a group of businessmen to provide a centre for accurate information on every part of the globe.

The second main group of societies is national professional groups, largely dominated by university and research geographers. These groups are usually later in date, smaller in membership, and less catholic in scope than the national societies. The *Association of American Geographers* (1904), the *Institute of British Geographers* (1933), and the *Regional Science Association* (1954) are typical of this group. The third group consists of societies devoted primarily to promoting geographic education in schools; the Geographical Association in Britain and the National Council for Geographic Education in the United States are examples. The fourth and most rapidly expanding set of groups originated during the 1950s. These organizations are subgroups within national professional organizations concerned with a particular aspect of geography (e.g. cartography, geomorphology, or quantitative methods). Geography appears to be following the pattern of other sciences in the rapid growth of this fourth group.

The prime function of the societies remains to foster common research interests through the reading of papers and the publication of journals. The establishment of journals such as the *Annals of the Association of American Geographers* in 1911 represented key breakthroughs in the circulation of research findings. Other journals were published by interested individuals, as was *Petermann's Geographische Mittelungen* in 1855, or by small groups, as was Ohio State University's *Geographical Analysis* in 1969.

The growth of journals provides a useful indicator of the increasing volume of geographic research. As Figure 24.5 indicates, the field has been rapidly expanding since the seventeenth century. The number of all scientific periodicals doubles about every fifteen years, and the number of geographic periodicals increases at about half that rate. The slower rate of increase in geography is typical of all the older, established sciences such as geology, botany, and astronomy because the total increase in scientific publications reflects the high birth rate of new scientific fields such as computer science.

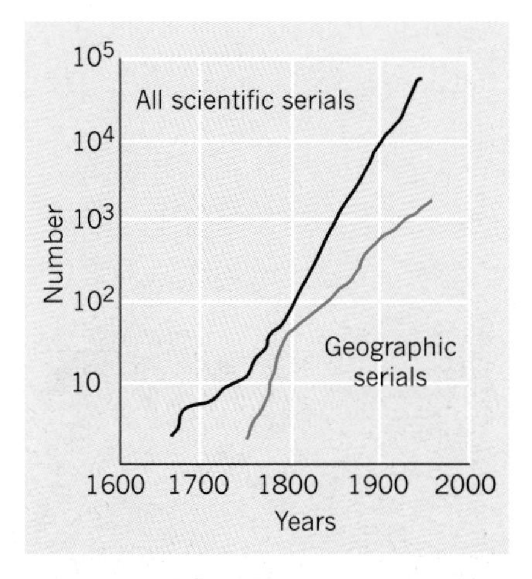

Figure 24.5 The growth of geographic research The chart shows the cumulative total of scientific and geographic periodicals founded since the middle of the seventeenth century. The vertical scale of the graph is logarithmic: the number of both scientific and geographic periodicals has been growing exponentially.

[Source: D.R. Stoddart, *Transactions of the Institute for British Geographers,* No. 4 (1967), p. 3, Fig. 1.3.]

Act III: National and International Organizations

A critical role of geographic societies was to convey the importance of the problems they studied to the rest of the community. Their partial success was marked by the beginning of a third phase of geographic study, overlapping the second, in which geography departments were formally established in major universities and in which some countries set up government-sponsored research centres. Two examples of geographic institutions are shown in Figure 24.6.

In the university sector, Germany again took the lead, with a considerable number of departments established by 1880. Developments in France were only slightly less rapid, but developments in the United States, Britain, and the Commonwealth lagged considerably. New geography departments often showed an irregular spatial diffusion pattern, marked by curious regional concentrations and sparse areas; Figure 24.7 shows the distribution of degree-giving departments in the United States, where a strong mid-western emphasis remains.

Meanwhile, the need for national geographic research centres has led to the emergence of institutes such as Brazil's *Instituto Brasileiro de Geografia e Estatistica* or Russia's *Akademiya Nauk SSSR*, each charged with the investigation and publication of regional data within their vast national territories. The latter has a staff of over 300 geographers working in ten divisions and a massive publishing programme, including the bi-monthly *Izvestiya, Seriya Geograficheskaya*. Even when separate geographic institutes have not been set up, geographic research has been increasingly incorporated into such organizations as Britain's Ministry of Town and Country Planning (now part of the Department of the Environment, Transport and

Figure 24.6 Geographic research centres Two different types of institutions for geographical research. (a) A research and teaching centre within a university. The Department of Geography at the University of Canterbury at Christchurch. (b) A multidisciplinary research centre, the International Institute for Applied Systems Analysis (IIASA) at the Schloss Laxenburg, near Vienna, Austria. IIASA is a major centre for ecological, environmental, resource, and demographic work.

[Source: (a) author. (b) IIASA.]

(a)

(b)

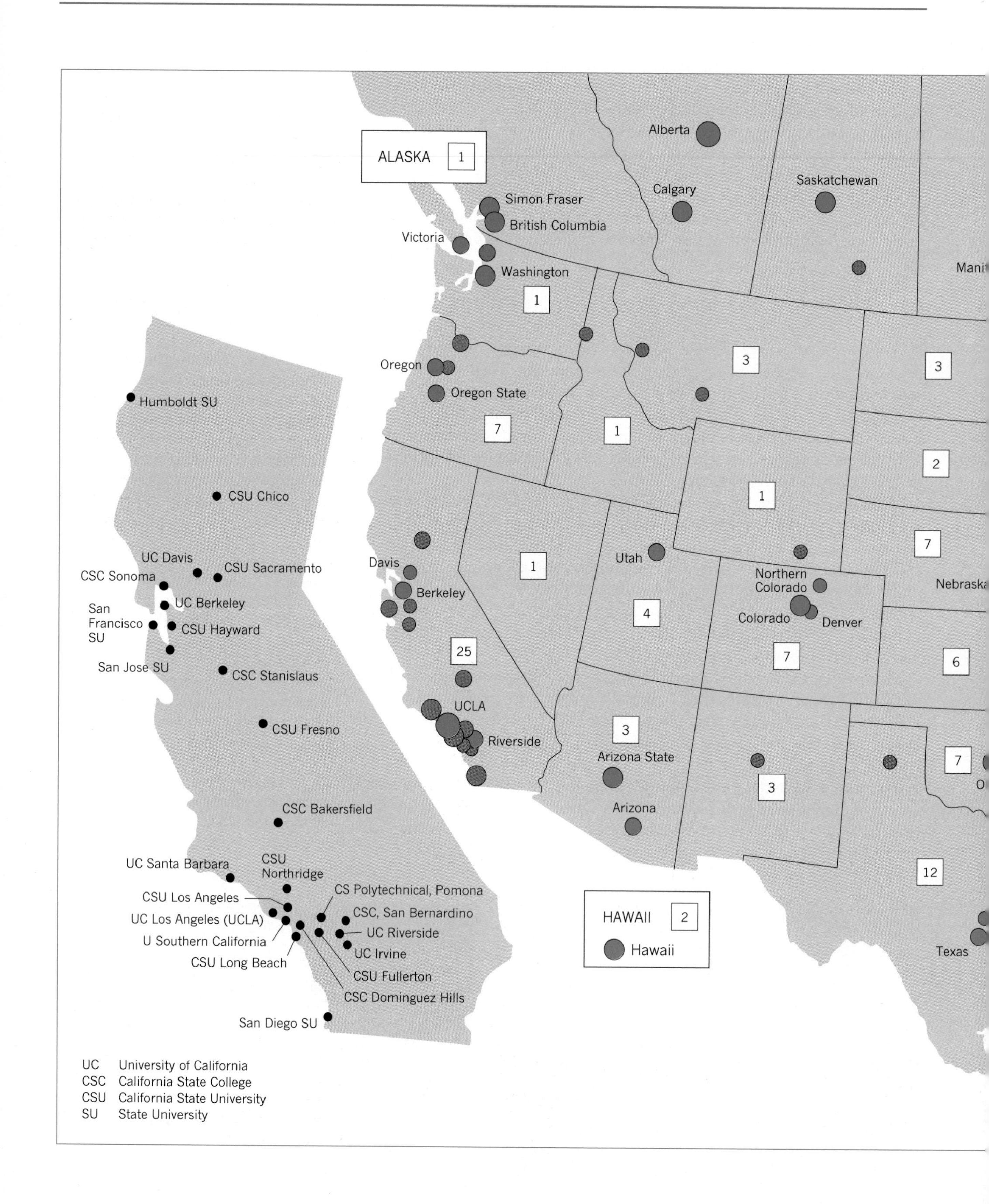
ALASKA
1
Alberta
Calgary
Saskatchewan
Simon Fraser
British Columbia
Victoria
Washington
1
Mani
Oregon
Oregon State
7
3
3
1
2
1
7
Davis
Berkeley
1
Utah
Northern Colorado
Nebraska
4
Colorado
Denver
25
7
6
UCLA
Riverside
3
Arizona State
3
7
Arizona
12
HAWAII
2
Hawaii
Texas
Humboldt SU
CSU Chico
UC Davis
CSC Sonoma
CSU Sacramento
San Francisco SU
UC Berkeley
CSU Hayward
San Jose SU
CSC Stanislaus
CSU Fresno
CSC Bakersfield
UC Santa Barbara
CSU Northridge
CSU Los Angeles
CS Polytechnical, Pomona
CSC, San Bernardino
UC Los Angeles (UCLA)
UC Riverside
U Southern California
UC Irvine
CSU Long Beach
CSU Fullerton
CSC Dominguez Hills
San Diego SU
UC University of California
CSC California State College
CSU California State University
SU State University

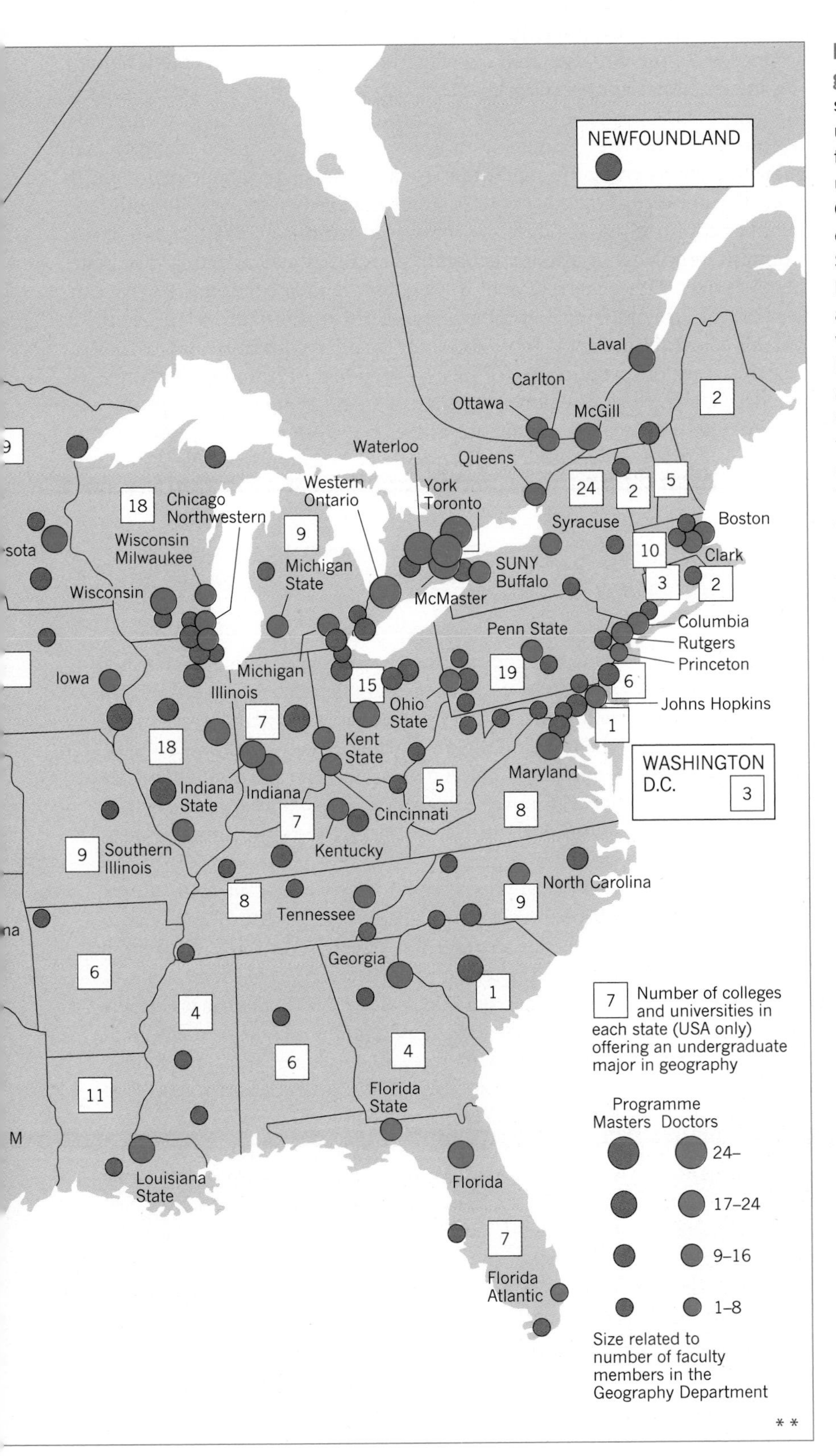

Figure 24.7 A geography of geography departments A cross-section of geography in North American universities in the late 1970s showing the geography departments offering an undergraduate major in the state of California and the location of geography departments in Canada and the United States offering courses at the graduate level. Those with doctoral programmes are named. Where the number of faculty was not available, the location is shown by the smallest of the four circles. Maps are based on data given in 'Guide to Graduate Departments of Geography in the United States and Canada 1976–77' (Association of American Geographers, Washington, D.C., 1976). You will need to consult the latest version of this annual publication to check out more recent changes. Some universities not shown on this map may now offer geography programmes, and the Canadian coverage is less complete than that for the United States. For details of individual departments, their courses, facilities, and staff see the home page of each department on the World-Wide Web (see Appendix B).

[Source: Map from drawing by Andrew Haggett.]

the Regions) and Australia's Commonwealth Scientific and Industrial Research Organization (CSIRO).

Since 1922 the initiation and coordination of geographic research requiring international cooperation has been handled through the *International Geographical Union (IGU)*. This organization holds meetings at intervals of four years. Between congresses it appoints commissions to study special subjects such as arid zones, quantitative methods, or economic regionalization. The number of member countries has now grown to over 70. Different member countries have a different interest in various areas of research; for example, problems in applied geography and regional planning dominate much eastern European research. If we compare member countries, we can see that the size of their geographic research effort is broadly related to their overall scientific budget. However, some smaller countries play a role in research out of proportion to their size. One outstanding example is Sweden, which, with few universities and a small group of geographers, has led research in several important areas. (See Box 24.B on 'Evolution of geography in Sweden'.) The volume and quality of research from New Zealand are also remarkably high in relation to its small number of research centres.

Box 24.B Evolution of Geography in Sweden

Sweden established geography rather later than the colonial powers such as Britain or France which had a long tradition of overseas exploration and military occupation of foreign lands. The ***Swedish Society for Anthropology and Geography (SSAG)*** was established in the 1870s. At its home in Stockholm, reports were given of the great exploring journeys: the voyage of the *Vega* under Nordenskjöld or the overland expeditions of Sven Hedin in Inner Asia in the 1920s were typical. Reports and discussions were printed in the Society's journal *Ymer* (from 1877 onwards). This was supplemented from 1919 by another journal, *Geografiska Annaler,* dedicated to scientific papers. The Society also published the first multivolumed *National Atlas of Sweden* (1953–71) and played a major role in its successor, the new seventeen-volume Atlas completed in 1996.

Although Sweden has universities of great antiquity, notably Uppsala (founded 1477) and Lund (1666), geography did not appear until the very end of the nineteenth century. H.H. von Schwerin began teaching at Lund in 1897 and by 1910 professors had also been appointed to Uppsala, Göteborg, and the Stockholm School of Economics. Growth was slow, with new departments being started at Stockholm University in 1929 and the Göteborg School of Economics in 1964. Like many western countries, Sweden's higher educational system grew very rapidly in the last third of the twentieth century. Existing universities trebled in student size and new universities were widely established across the country. Geography shared in this growth with single-professor departments expanding to multichair institutes within which physical and human geography often became separate from the 1960s. Outside the old universities, a new geography department was established at Umeå (1964) and specialized courses started at Karlstad (1967) and Linköping (1968). To these have been added departments in four of the regional university colleges (e.g. Örbebro and Växjö) and within government research institutes (e.g. within the Institute for Housing Research at Gävle).

Swedish geography holds a high standing within the international geographic community. It has major graduate schools with over 250 doctorates granted in human geography alone since 1900. It also has a tradition of first-rate theoretical and empirical research and pioneered studies in both physical geography (e.g. fluvial studies at Uppsala) and human geography (e.g. time geography at Lund; see Box 1.B). It has distinguished applied work using Sweden's unique demographic data, often made in association with Swedish government agencies. It has also been heavily funded by such agencies as the Bank of Sweden Tercentenary Foundation. See S. Christiansen, P. Haggett, S. Helmfrid, P. Vartiainen, and B. Öhngren, *Swedish Research in Human Geography* (Swedish Science Press, Uppsala, 1999).

The Present Structure 24.2

Geographers have repeatedly tried to define their field at each stage of its growth. For those who like formal definitions, Table 24.1 gives a variety of often-quoted ones. None of them will satisfy all geographers, but all geographers will recognize some common identifiable elements.

Unity and Diversity in Geography

Let us try to summarize what these common elements in definitions of *geography* are.

First, we have seen that geographers share with other members of the earth sciences a concern with a common arena, the earth's surface, rather than abstract space, but that they look at that arena from the viewpoint of the social sciences. They are concerned with the earth as the environment of humanity, an environment that influences how people live and organize themselves and at the same time an environment that people helped to modify and build.

Second, geographers focus on human spatial organization and our ecological relationship to our environment. They seek ways of improving how space and resources are used, and emphasize the role of appropriate regional organization in reaching this end. Their work provides a perspective of our tenure on the earth and various forecasts – both optimistic and pessimistic – of our future on the planet.

Third, we have seen that geographers are sensitive to the richness and variety of the earth. They do not believe in blanket solutions to develop-

Table 24.1 Some definitions of geography

Definition	Source
'Geography is concerned to provide an accurate, orderly, and rational description of the variable character of the earth's surface.'	R. Hartshorne, *Perspectives on the Nature of Geography* (Murray, London, 1959), p. 21.
'Its goal is nothing less than an understanding of the vast, interacting system comprising all humanity and its natural environment on the surface of the earth.'	E.A. Ackerman, *Annals of the Association of American Geographers*, **53** (1963), p. 435.
'Geography seeks to explain how the subsystems of the physical environment are organized on the earth's surface, and how man distributes himself over the earth in relation to physical features and to other men.'	Ad Hoc Committee on Geography, *The Science of Geography* (Academy of Sciences, Washington, D.C., 1965), p. 1.
'Geography ... a science concerned with the rational development and testing of theories that explain and predict the spatial distribution and location of various characteristics on the surface of the earth.'	M. Yeates, *Introduction to Quantitative Analysis in Economc Geography* (Prentice Hall, Englewood Cliffs, N.J., 1968), p. 1.
'Geography is the science of place. Its vision is grand, its view panoramic. It sweeps the surface of the Earth, charting the physical, organic, and cultural terrains....'	*Science*, Review of Harm deBlij's *Geography Book* (John Wiley, New York, 1995).
'Geography is an integrative discipline that brings together the physical and human dimensions of the world in the study of people, places, and environments.'	American Geographical Society *et al.*, *Geography for Life* (National Geographic Society, Washington, D.C., 1994).

ment problems; instead, they feel that policy should be carefully tuned to the spatial variety concealed by terms such as 'tropics', 'Appalachia', and 'ghetto'. On each geographic scale, they seek always to disaggregate and dissect the uniform space of the legislator within the complex space of the real world.

Within this broadly defined area of agreement, different branches of geography have proliferated, each concerned with a limited research topic. Figure 24.8(a) gives a summary of the orthodox division of geography into the study of regions (regional geography) and an analysis of their systematic characteristics (systematic geography). Each may be further subdivided into more specific fields, such as the regional geography of Latin America or urban geography. Some geographers recognize as a separate branch the study of the regional or systematic geography of past periods (historical geography), but others argue that time is an essential component in all geographic studies.

These conventional divisions are important, not least because most university catalogues describe courses in these terms. But perhaps a more helpful way of structuring geography is by how it approaches its problems (Figure 24.8(b)). In this book we have distinguished three different approaches.

1 *Spatial analysis*. The first approach, termed *spatial analysis*, studies the locational variation of a significant property or series of properties. We have already encountered such variations in interpreting the distribution of population density or of rural poverty. Geographers ask what factors control the patterns of distributions and how these patterns can be modified to make distributions more efficient or more equitable.

2 *Ecological analysis*. A second approach to geography is through *ecological analysis*, which interrelates human and environmental

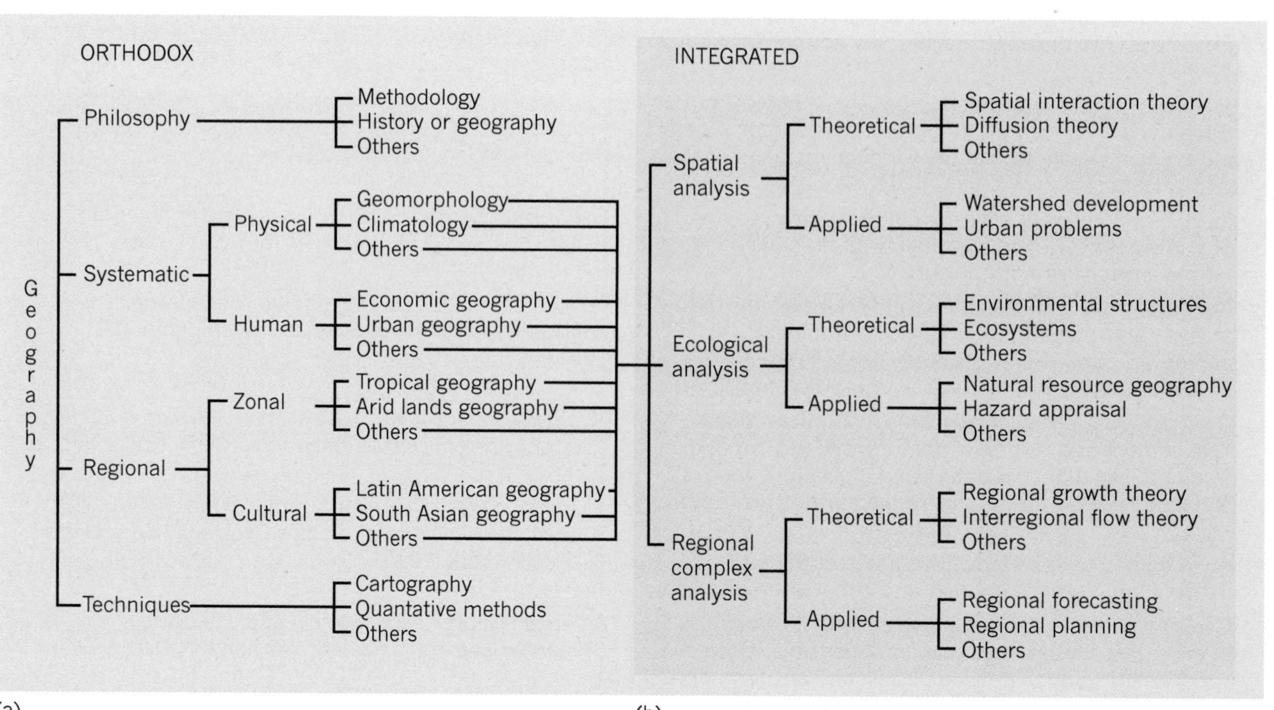

Figure 24.8 The internal structure of geography

variables and interprets their links. We have already studied such linkages in the hydrologic cycle and land-use cycles. In this type of analysis geographers shift their emphasis from spatial variation between areas to the relationships within a single, bounded, geographic area.

3 *Regional complex analysis.* The third approach to geography is by way of *regional complex analysis*, in which the results of spatial and ecological analysis are combined. Appropriate regional units are identified through areal differentiation, and then the flows and links between pairs of regions are established. We have already discussed some of the difficulties in regional complex analysis (Chapter 12) and looked at a few of its possible applications in regional planning (Chapter 18).

The advantage of looking at geographic problems in terms of these three approaches rather than the orthodox divisions is that they stress the unity of physical and non-physical elements rather than their separation. Geographers concerned with water resources or human settlement may find a common ground in the ways in which systems are studied or in their parallel search for efficient regional units.

The work of most geographers falls within the 'triangle' formed by these three approaches to the field. Some may specialize, or, if you like to think of it this way, move toward one of three corners of the triangle. Indeed, the whole subject appears to have zigged and zagged over the decades, sometimes staying in the regional corner (as in the 1930s), sometimes lurching toward spatial analysis (as in the 1950s and 1960s). In the present decade it seems on the move again, now headed for the ecological corner.

Geography and Supporting Fields

Geography is particularly dependent on the flow of concepts and techniques from more specialized sciences. For example, in regional climatology we adapt models originally developed by meteorologists, who in turn draw their concepts from basic physics. Likewise, our models of regional growth borrow from the econometrician.

This dependence underlines the fact that sovereign subjects are about as irrelevant as sovereign states. Good botanists need to be reasonable biochemists, good engineers need to be fair mathematicians, and so on. Familiarity with these supporting fields is normally obtained by taking parallel courses in other departments. Some geographers argue that special importance attaches to mathematics because it provides a common language in which geographers can express spatial, ecological, and regional concepts in a concise and comparable way.

Let us assume that your particular interest is in the humanities and that you are especially attracted to the study of one of the major cultural areas such as China. Then the appropriate courses to support regional courses in East Asian geography might include the history and economic structure of China; clearly, courses in Cantonese or Mandarin would be needed for more serious research. By contrast, if you wanted to specialize in environmental problems, you would need earth science courses such as hydrology or oceanography, perhaps supported by a course in resource economics (Figure 24.9). Which courses you elect to take will depend on a combination of factors: your interest, your ability and previous training, and your long-term expectations. It is to the third of these we now turn.

Figure 24.9 Links between geography and supporting fields The courses indicated are intended to be representative rather than exhaustive. Departmental boundaries are drawn differently in some universities and colleges. For example, geomorphology tends to be taught within geography departments in British universities but within geology departments in America. African studies are used to illustrate area studies; a similar diagram could be constructed, for example for Latin American or South Asian studies.

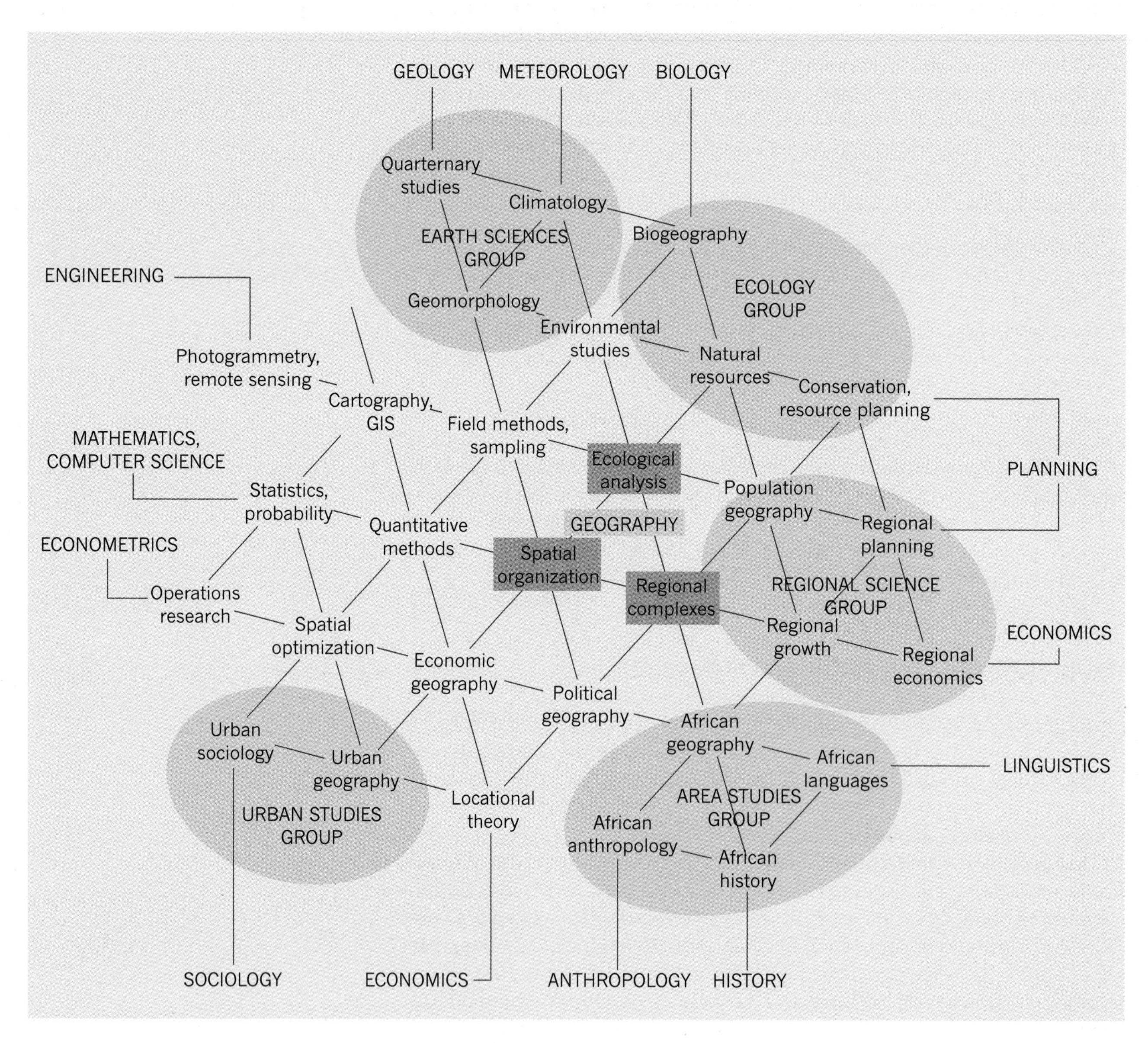

24.3 Jobs in Geography

Geography courses aim to develop a set of marketable skills rather than preparing the graduate for a single job. The job market in the present decade is changing so quickly that if you are looking at only one type of job, that market may undergo significant change before you finish your college programme. It is wise, therefore, to be knowledgeable about other choices and to look at clusters of jobs rather than a single career.

Geographic skills

Some of the skills that geography courses develop have been hinted at in this book and can be taken further by the specialized courses available from most geography departments. Let us illustrate just three. Courses in *computer-assisted cartography* (check back to Chapters 21 to 23) open a cluster of jobs in handling geographically based data and displaying it in directly interpretable map format. The map remains a supremely important medium for passing information, and the microchip revolution is now expanding explosively the possibilities for its use. Similarly, courses in *locational analysis* (see Chapters 13 to 16) develop the skills for finding optimum sites for new facilities in the business world, while courses in *environmental impact analysis* (see Chapters 9 to 11) allow the effects of those decisions to be calculated and considered.

Courses in *regional geography*, particularly when reinforced by the appropriate language, provide another link to the job market. Corporations with overseas facilities, banks with overseas investments, the burgeoning tourist industry, all need graduates with a deep understanding of particular parts of the earth's surface. The potential for graduates with a thorough grounding in the Pacific Rim or the Middle East will be apparent.

Graduate Careers

Although it is now less fashionable to earmark university courses as 'career-oriented' or 'general education', the distinction still lingers. Law, medicine, and geology typify fields where most graduates have a rather well-defined occupational outlet ahead of them. By contrast, fields such as philosophy or political science have a high educational content but less obvious career prospects. Where does a degree in geography take us?

Jobs in geography can be broadly divided into three categories (see Figure 24.10). First, one can find a job that is 'field sustaining' in the sense that it is concerned directly with supporting the continuing study of the field itself. Thus many geographers find jobs in geographic teaching or research. Teaching opportunities exist on all levels from the elementary school to the postdoctoral level. In the countries of western Europe geography is an important secondary school subject, and many graduates go back to teaching on this level. The position of geography in the United States schools is less strong, though an important new curriculum for this level was initiated through the High School Geography Project. The two-year college course formed a significant and expanding job market for geography teachers in

Figure 24.10 Career directions for geography graduates

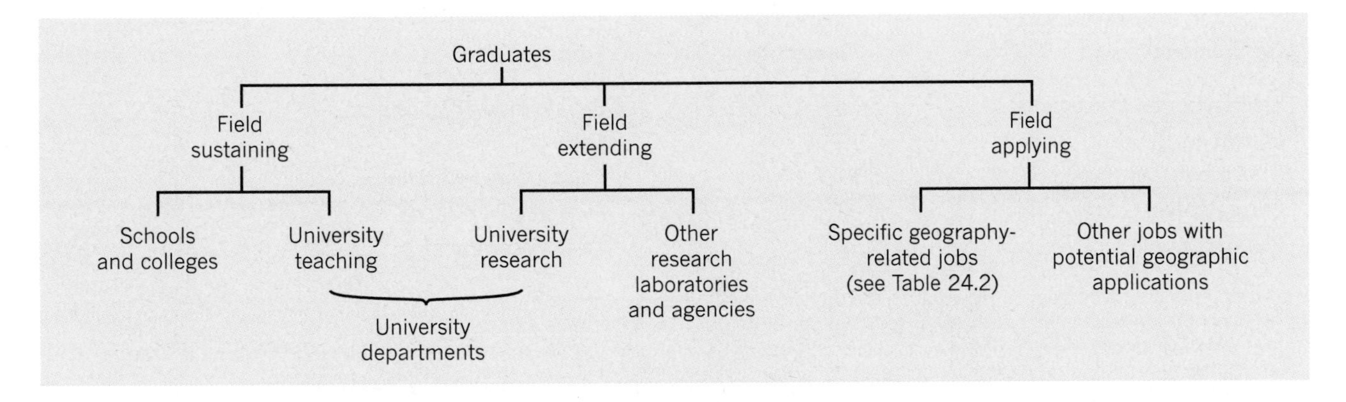

the 1960s and there was a recurrent shortage of PhDs for advanced teaching on the university level at that time.

There is, however, a delicate feedback relationship between the number of graduates produced and job opportunities, and periods of abundant jobs and job scarcity tend to follow a cyclic pattern.

Second, there are jobs outside the academic arena that have direct geographic applications. The main areas (see Figure 24.10) are: (a) cartography and remote sensing, (b) Geographical Information Systems, (c) environmental management, and (d) urban and regional planning. Geographers have a long tradition of public service on all levels, from global agencies to city planning commissions. On the *international* level the tradition goes back at least to World War I, when geographer Isiah Bowman (later president of Johns Hopkins University) was prominent among the geographers advising the American delegation at the Versailles Peace Conference. This tradition continues today, with the British geographic team advising on the settlement of the Argentine–Chile boundary dispute. Geographers are also represented in the specialist United Nations agencies.

On the national level geographers are well represented in the public sector federal departments of the United States; indeed, nearly one-tenth of the American Association of Geographers' membership works in federal establishments (see Table 24.2). This sort of work began in the 1920s, when geographers played an important part in the Soil Conservation Service. In the succeeding years their involvement has widened to include work for several agencies, such as the Geography Division of the Census Bureau and the Water Resources Board. Despite their growing role, however, American geographers are not as well represented in federal agencies as are their colleagues in countries such as Russia, Sweden, and Britain. In Britain many geographers are working in the Department of the Environment, Transport and the Regions, where they are engaged in research that ranges from regional planning to local land use. Below the national level extends a hierarchy of regional commissions and local planning agencies, each with geographers represented on them.

Besides working in the public sector, geographers work in the *private* sector as independent consultants or employees of large corporations. Much of their work is in locational consulting, environmental impact studies, and regional intelligence for investment or marketing. The optimal location of a new regional shopping centre, new airstrips for an expanded flying doctor service, or the environmental impact of a new industrial plant all involve the kinds of problems discussed in this book and analyzed more deeply in more advanced geography courses. More geographers are estab-

Table 24.2 Recent geography graduates survey in the United States

Occupational group	Percentage	Employer	Percentage
Environmental management	13	Private industry/business	40
Education	12	Local government	10
GIS/remote sensing	11	Educational institution	17
Cartographer	8	Federal government	14
Planner	7	State government	11

Based on survey of 564 geography graduates. Education includes both college and school level appointments. Planner includes environmental, transportation, and urban/regional planning. Top five categories shown.
Source: Patricia Gober *et al.*, 'Employment trends in geography', reported in Thomas J. Wilbanks *et al.*, *Rediscovering Geography* (National Academy Press, Washington, D.C., 1997).Tables 3, 4, pp. 203–4.

lishing their own consulting agencies and so multiplying the opportunities for applied research.

Finally, there are those graduates who use their geographic skills indirectly in a very wide range of employment. Here personal qualities and the possession of a degree which shows skills of both analysis and synthesis are more important than geographic skills as such. But this broad category hardly does justice to the range of opportunities that individual geographers – like individuals in any other field – create and develop (see Figure 24.11).

Jobs In Geography New listings are marked with a * **June 2000**

UNITED STATES

INDIANA, GREENCASTLE 46135. The Department of Geology and Geography at DePauw University invites applications for a one-year term position in **Geology and Geography** at the rank of **Assistant Professor** (Instructor for ABD) beginning 15 August 2000. We desire a person who is broadly trained in the geosciences with the ability to teach introductory courses for undergraduate non-science majors. The successful applicant will teach Physical Geography, Oceanography, a seminar for first-year students on Global Environmental Problems, and a Winter Term course/project of their choice.

Applicants should send a letter describing their teaching experience and pedagogy, vita, transcripts of all academic work, and three letters of recommendation. MAY 00-111.

Apply: Dr. Frederick M. Soster, Chair, Department of Geology and Geography, DePauw University, Greencastle, IN 46135. DePauw University is an affirmative action, equal opportunity employer. Women and minorities are especially encouraged to apply.

of academic geoscientists in the Commonwealth of Kentucky. Our faculty, research programs, and facilities provide students with unique curricular opportunities in a variety of subdisciplines such as karst hydrology, climatology, city and regional planning, environmental and sedimentary geology, fossil fuels resources, and international studies. Applied research experiences are offered students through the Department's internationally recognized Center for Cave and Karst Studies, its Kentucky Climate Center, its College Heights Weather Station, and the Hoffman Environmental Research Institute. The Department presently offers degree programs in Cartographic and Mapping Techniques(AS), Geography (BS), Geology(BS), and the Geosciences (MS). Western Kentucky University, with 15,000 undergraduates and graduates, including more than 900 students of diverse ethnic backgrounds and 250 international students from 46 countries, has a strong commitment to achieving diversity among its faculty and staff. It is an affirmative action/equal opportunity employer. We are particularly interested in receiving applications from members of under-represented groups such as women and people of diverse ethnic backgrounds. Detailed information about the Department is available at http://www2.wku.edu/geoweb.

Send letter of application, vita, a separate detailed statement of departmental vision and leadership philosophy, and the names and e-mail addresses of three personal references. JUN 00-128.

Apply: Search Committee, Department of Geography and Geology, Western Kentucky University, 1 Big Red Way, Bowling Green, KY, 42101-3576. Review of applications will begin on 15 August 2000, and will continue until the position is filled.

***MARYLAND, BALTIMORE 21216-3698.** The Department of History, Geography and International Studies at Coppin State College invites applications for tenure-track position of **Assistant Professor** beginning August 2000. Ph.D. preferred, ABD considered. The successful candidate will teach 12 credit hours per semester in undergraduate **Geography and World History.** The geography component of the department supports majors in African

(a)

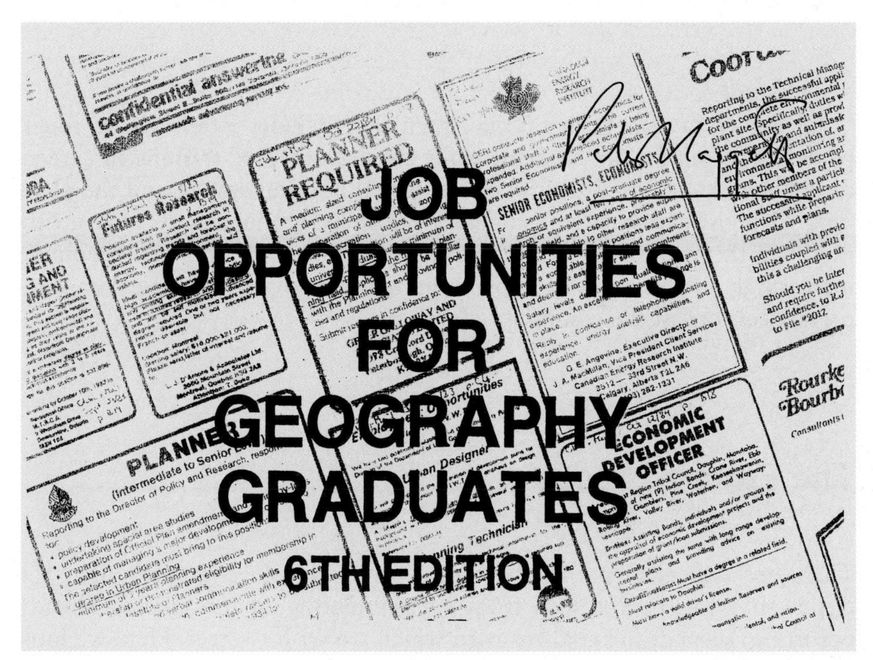

(b)

Figure 24.11 Geographical career opportunities Examples of advertisements for posts for geography students. (a) Field-sustaining opportunities where geography graduates are employed in geography departments. *Jobs In Geography* is a regular publication of the Association of American Geographers, Washington, D.C. It lists mainly university and college posts in North American Institutions. (b) Field-extending opportunities where geography graduates are employed in non-geographic institutions. The Geography Department at Waterloo University, Canada, has pioneered a wide range of internships for its graduates with both public sector and private sector employers.

(a)

(b)

(c)

(d)

Figure 24.12 Four geography graduates Some idea of the range of career opportunities which four geography graduates have followed over the last 30 years. The only link between them is that all are geography graduates who have qualified with first degrees at Bristol University in the UK. (a) Helen Young (graduating class of 1990) now a senior member of the BBC/Met Office forecasting team and a familiar face on British television. (b) Diana Kershaw (class of 1971), head of the planning department for the City of Bristol. (c) Roger Downs (class of 1966) and head of the Geography Department of Pennsylvania State University, USA. (d) David Rhind (class of 1965), director of Britain's major government mapping agency, the Ordnance Survey. The photographs and the positions then held relate to the survey conducted when the Bristol geography department celebrated its 75th anniversary in 1995.

[Source: University of Bristol, Department of Geography, 75th Anniversary Volume: Snapshots from a Family Album, Plate 29.]

All university geographers' mail regularly includes letters (more often, postcards) from former students using their geographic training in careers as unlike as the administration of a regional hospital system and the planning of highways in New Guinea. Geography graduates are also giving more of their spare time to community counselling and are becoming increasingly interested in local planning issues involving ecological or locational decisions at the community level.

24.4 The Future Prospect

There are two elements in the future prospects of geography: first, the trends that can be foreseen from a continuation of movements already occurring; second, the goals toward which we wish to steer. Here we look at each element in turn.

Projection of Existing Trends

Perhaps the only sure forecast we can make about geography is that it will persist. The questions geographers ask are so basic that it is impossible to imagine a world without them, a world where regional differentiation is not significant or where general theories fit local circumstances with sheathlike precision. We certainly expect the causes of spatial differentiation to change, for as we gain uniformity in one sphere we lose it in another. However specialized science becomes, some scholars will still want to integrate and synthesize in the traditional geographic manner.

Beyond this, four more precise trends seem likely. The first projection concerns *quantity*. As we saw in Figure 24.5, the quantity of geographic research as measured by the number of specialized geographic serials has been doubling every 30 years since around 1780. For the increase in more recent years a variety of other measures is available, including membership in geographic societies, enrolments in classes, degrees granted, research published, and so on. These indicators all confirm that though individual decades are likely to have spurts or slowdowns above or below a long-term rate of increase, it would be unreasonable not to expect an increase in geographic research over the next decade.

A second projection relates to the *fission* of geography into specialized subdisciplines. Each individual scholar finds increasing economies of scale by specializing in a limited range of problems; limited time and resources (books, equipment, maps, computing time, or whatever) can be concentrated on in-depth study of a limited topic or region. As a result an increasing number of geographers tend to think of themselves as South Asian geographers, or diffusion specialists, or arid-zone geomorphologists. This trend is not confined to geography. The general botanist or zoologist, not to mention the still more general biologist, has long since been displaced in most universities by endocrinologists, conchologists, or other specialists.

A third projection relates to *quantification*. One of the most striking differences between the research papers in the current geographic journals and those a half century ago is the greatly increased proportion of research using mathematical techniques. In the earlier period, the main mathematical applications were of spherical geometry in cartography and surveying, and of probability and statistics in climatology. Today the range of mathematical models has significantly expanded, and the applications now affect most branches of the field. We find historical geographers fitting polynomial surfaces to the spread of early settlement, or industrial location analysts modelling siting decisions as Markov chains. During the 1960s there was a skirmish between those geographers anxious to innovate with mathematical methods and those sceptical of their usefulness in solving orthodox problems. Today the general acceptance of such techniques, the more complete mathematical training of a new generation, and the widespread availability of standard computers on campuses make the conflict of a decade ago seem unreal. Mathematical methods are now seen as just one of many tools for approaching geographic problems; these methods are appropriate for some tasks, inappropriate for others.

Paradigm Shifts Within Geography

While the first three trends are linear or S-shaped, the fourth represents a more abrupt switch. Human geographers are getting increasingly uneasy about the 'positivist' nature of much geographic research. **Positivism** is a philosophical approach that holds that our sensory experiences are the exclusive source of valid information about the world. This attitude developed in the natural sciences (like physics) but has been borrowed by geographers working in social-science areas. A positivistic approach leads to the discussion of human behaviour in terms of analogies drawn from the natural sciences. Thus in Chapter 13 we discussed human migration in terms of Newton's laws of gravity. Much geographic effort of the 1960s went into trying to explain patterns of human behaviour with neat, lawlike statements. Ultimate causes and the essential nature of phenomena such as migration were put aside as being unknowable or inscrutable.

Given its historical pattern of evolution, it is understandable that positivism should have played a major part in geographic explanations of phenomena. Whatever its virtues in the more physical parts of the field, its effect on human geography was to lead to a stylized, sometimes over-academic kind of research. What then should we replace it with? One group favours renewed interest in *phenomenology*. This approach admits that introspective or intuitive attempts to gain knowledge are valid. It accepts subjective categories as appear to be in the experience of the person behaving. The Great Plains are a 'desert' if the person or group settling them believes them to be one and acts accordingly! Phenomenologists look with scepticism at the attempts at law-like statements so characteristic of the previous decade. Other geographers have developed an interest in *post-modernism* in which previous interpretations are regarded as contingent and partial. They lay stress on open interpretations and plural views. Some of the different schools in human geography are summarized in Table 24.3.

This *paradigm shift* has also occurred in other fields such as social anthropology and social psychology and has reinforced the links between geography and other social sciences. Of course no geographer's work is ever 'purely' positivistic or 'purely' phenomenological. Most geographers adopt a position between these two extremes, with systematic approaches to the field (like that in this book) tending to stress the positivistic side and works with a regional emphasis adopting a more phenomenological viewpoint.

Table 24.3 Philosophical schools in human geography

School	Description
Environmentalism	The view that natural environment plays the major role in determining the behaviour patterns of humans on the earth's surface.
Normative	Describes approaches to geography that establish a norm or standard. Thus, they are largely concerned with establishing some optimum condition (e.g. the 'best' location or the 'best' settlement pattern).
Phenomenology	An existential philosophical school which admits that introspective or intuitive attempts to gain geographic knowledge are valid.
Positivism	A philosophical school which holds that human sensory experiences are the exclusive source of valid geographic information about the world.
Post-modernism	A movement reacting against modern tendencies. It is sceptical of previous theory. Interpretations are regarded as contingent and partial, with a stress on open interpretations and plural views.
Probabilism	A compromise position between environmentalism and possibilism that assigns different probabilities to alternative patterns of geographic behaviour in a particular location or environment.
Methodology, epistemology	Terms used to describe the study of these different schools of thought and their contribution to the philosophy of geography.

Box 24.C Paradigm Shifts in Geography

The years since 1950 have seen two abrupt paradigm shifts in human geography: the move in the middle 1950s towards a logical positivist view and the move in the middle 1970s towards a more phenomenological approach to the field. Such moves appear to be relatively rapid in terms of the timespan of the subject.

Why do such 'jumps' appear more characteristic than gradual, evolutionary change? Let us begin our answer by assuming that a geographer's decision to accept or reject a given paradigm is based on the relative balance of its perceived benefits and costs. (By benefits we mean the insights or analytic advantages that come from a particular philosophical approach.) Acceptance rests on a net balance of benefits over costs, and vice versa. The diagonal line in (a) therefore represents an indifference line where costs and benefits are exactly equal; it marks the boundary between acceptance and rejection. We may go on to represent the same line in a three-dimensional version (as in (b)).

But this simple accept/reject dichotomy as shown in (b) is a gross oversimplification of geographers' actual behaviour. For, where both costs and benefits are very low, we may be indifferent to either outcome, i.e. the paradigm is too weak to be worth either supporting or opposing. Where costs and benefits are very high we may tend to persist with a given decision once we have adopted it. Both these modifications are taken into account in (c) by showing the acceptance/rejection decisions as a surface with a symmetrical fold lying along the axis of the indifference line shown in the two preceding figures.

Adoption of this revised version of the acceptance or rejection of paradigms within geography has some interesting implications. To show these, let us project the surface shown in three dimensions in (c) onto the two-dimensional base delimited by the costs and benefits axes. The fold in the surface now forms a triangle shaded in (d). Note that the shaded triangle may now take on three *different* series of values since it is essentially a triple-sheeted fold in the original acceptance/rejection surface. Two important perspectives on our observed behaviour follows.

Persistence of attitudes once a position has been adopted

Consider the position of a geographer at location 1 in (d). If we assume benefits remain unchanged but evidence on the cost of maintaining an existing paradigm builds up, his or her position shifts horizontally across the benefit/cost space. Since the geographer is on the upper part of the fold shown in (c), he or she changes to rejection not at point 2 when benefits are exactly equal to costs (i.e. the indifference line of (a)) but at point 3 where he or she suddenly falls from the upper 'accept' to the lower 'reject' surface. Contrast this with the position of an initial rejector located at point 4 in (e). In this case the assumptions are reversed and the geographer's position moves over the lower surface of the fold as evidence on the benefits is accrued. Again attitude change occurs not at point 5 but at point 6 when he or she suddenly leaps onto the upper surface of the fold. Of course, the examples chosen are oversimplified in order to make a point; actual patterns might be expected to be much more complex.

Divergence of attitudes from initially similar positions

The position of two geographers located originally at points 7 and 8 is shown in (f). Both the cost and benefits are small, and both people may be expected to be indifferent to the position they hold. However, if the data on costs and benefits accrue over time in the way shown by the parallel paths, we can see how our two geographers diverge. One follows the upper surface of the fold, the other the lower. Note that although given exactly the same evidence, one geographer increasingly accepts the paradigm while the other increasingly rejects it. In other words we dig in our heels to hold our initial positions ever more strongly since there is now so much more intellectual capital at stake.

See P. Haggett, *Mid-term Futures for Geography* (Monash University Publications in Geography, No. 16, Melbourne, 1977).

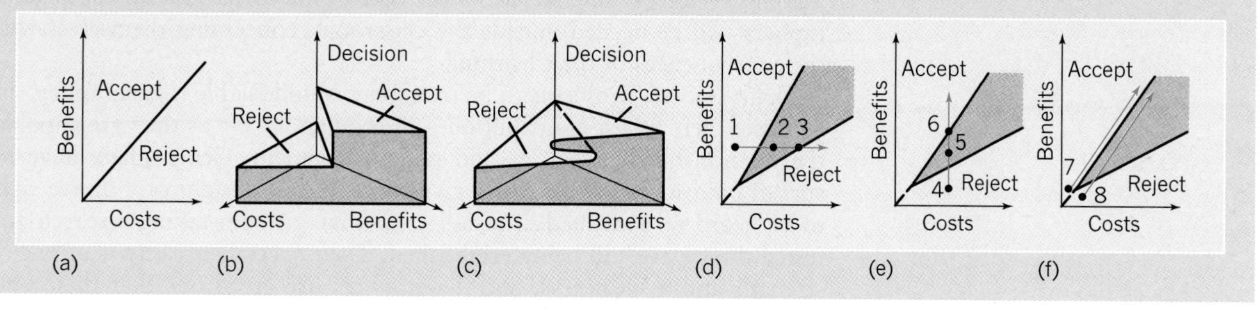

These sudden changes of position or 'flips' in a subject are different in character from the long-term swings. Recently, light has been thrown on our understanding of this subject by the French mathematician René Thom who sees many such changes in the natural world in terms of *catastrophe* theory. By 'catastrophe', Thom does not mean 'disaster', but simply a sudden change. For example, if we progressively lower the temperature of water there is a sudden switch in state from water to ice. If confidence slowly falls on Wall Street, there may be a sudden loss of nerve in which the bottom abruptly falls out of the stock market. Although catastrophe theory is very complex, it is possible to use its simpler models to throw light on such switches within geography. (See Box 24.C on 'Paradigm shifts in geography'.)

New Goals for Geography?

Each new generation of geographers builds on earlier work but reinterprets its goals to match the prevailing scientific and social mores. Judging by the interests of the current generation of graduate students or current issues of radical journals such as *Antipode*, geography is likely to become more strongly oriented towards applied fields and practical problems. The questions debated over coffee cups are outward looking, socially relevant, and action-oriented.

Can we help society to chart a middle way between the mindless exploitation and pollution of the natural world implicit in a short-run materialist economy and the unrealistic pipe dreams of a pastoral, protected, but unproductive world? Can we predict the welfare implications of different types of locational situations or spatial arrangements? Can we help to redraw political boundaries in order to equalize resource opportunities and also minimize the likelihood of future conflicts? What will the world geography or that of the United States look like by AD 2010 or AD 2050? These and many other questions are being asked today.

If these questions are typical of the broad, long-term issues young geographers wish to tackle, there will have to be some significant departures from the presently projected trends in the field. For example, instead of increasing specialization, there will have to be a greater emphasis on extending the ecosystems approach (see Chapter 4), still largely confined to the physical and biological world, to include our own environment-modifying activities. Physical and human geographers will have to spend more time at each other's seminars to exploit their unique opportunities for cooperative research. A rapprochement will be needed between those interested in building quantitative models and those interested in the realities of individual regional complexes. Geographers will also need to stop acting as 'terrographers' and devote more time to at least the continental margins of the ocean-covered 70 per cent of the earth. Finally, more geographers will be needed outside the classroom, concerning themselves with the consequences of their learning.

That there are problems to be overcome is undeniable if geographers are to make their fullest contribution – improving, as far as they are capable, the relationship between us and our environment. Geographers have no special philosopher's stone that gives them instant insight or a divine right to be heard or consulted. The full credentials of painstaking research and tested theory are still being established. Their special curiosity is at once a strength and a weakness, and geographers are conscious that their own studies often lack the rigour of an econometrician's or the temporal perspective of an historian.

We have tried to show areas where these difficulties are being faced and overcome – and others where significant progress remains to be made. The book will have served its purpose if it attracts students from other disciplines into those areas that have, for too long, remained a narrow geographer's monopoly. These students may help to provide new insights and solutions to some of the puzzles that have, for too long, baffled the small group of geographic scholars. We hope that you will accept the invitation – and the implied challenge offered here – and go on from this brief and prefatory book to the rewarding geography courses that lie ahead. As Isaac Newton said in our chapter-opening quotation, an undiscovered ocean remains to be explored.

Reflections

1 Look back at the notes you have made in response to Question 1 at the end of the first chapter. How accurate were your predictions? Compare your own findings with those of the rest of the class.

2 'Adopt' any one of the individuals named in Figure 24.3. Browse through at least two of his or her works, and, if the person is now deceased, read through his or her obituary notice in one of the leading geographic journals. (The *Annals of the Association of American Geographers* is a particularly useful source of information on North American geographers.) List three major influences on the person's work.

3 Figure 24.7 shows North American university geography departments at a particular point in time in the last century. But geography is now taught as a university course at some thousands of colleges around the world. Use some of the Internet sources listed in Appendix B at the end of this book to find the home page of departments in which you are interested in following an undergraduate or graduate course.

4 Geographers tend to make hard work of defining their subject. Read through the definitions of geography in Table 24.1. Which is closest to your own ideas of geography? Why?

5 Geography is often studied at school and college alongside other subjects (see Figure 24.9). Take a poll of such subjects in your own class and debate which subjects in other fields do you consider most useful for geography students to read. Justify your choice.

6 How should the research agenda of geography (or any academic field) be decided? Evaluate the relative importance of (a) experience (i.e. the past history of research endeavours, successful and unsuccessful), (b) the views of the present generation of geographers, and (c) the needs of society in determining this agenda.

7 Geographers go on after graduation into an increasingly wide range of jobs. Consider Table 24.3. What job would you wish to follow after graduation? To what extent does a course in geography provide a valuable background to this choice?

8 Review your understanding of the following concepts using the text and the glossary in Appendix A of this book:

catastrophe
ecological analysis
International Geographical Union (IGU)
Matthew effect
paradigm shifts
phenomenology
positivism
post-modernism
regional complex analysis
spatial analysis

One Step Further ...

A sensitive and insightful introduction to the history of geography is given by David Livingstone in his essay in R.J. JOHNSTON *et al.* (eds), *Dictionary of Human Geography*, fourth edition (Blackwell, Oxford, 2000), pp. 304–8, and at length in his classsic *The Geographical Tradition: Episodes in the History of a Contested Enterprise* (Blackwell, Oxford, 1992). For most readers, the brief list of books on geography given at the end of Chapter 1 will suffice for its philosophy but serious scholars will want to delve just a little into the two classic statements on the field by the leading geographic philosopher of the last century, Richard HARTSHORNE: *The Nature of Geography: A Survey of Current Thought in the Light of the Past* (Association of American Geographers, Lancaster, Pa., 1946) and *Perspectives on the Nature of Geography* (Rand McNally, Skokie, Ill., 1959). The author's brief review, P. HAGGETT, *The Geographer's Art* (Blackwell,

Oxford, 1990) may prove helpful on the methods geographers use.

A historical perspectives on the growth of geography are given in G.J. MARTIN and P.E. JAMES, *All Possible Worlds: A History of Geographical Ideas*, third edition (Wiley, New York, 1993). This shows that geographers' ideas about their subject are constantly changing and evolving. Readers may like to contrast the approaches of two major statements about the field published a generation apart, viz. R.J. CHORLEY and P. HAGGETT (eds), *Models in Geography* (Methuen, London, 1967) and R. PEET and N.J. THRIFT (eds), *New Models in Geography*, 2 volumes (Unwin Hyman, London, 1989). Some of the complex directions taken over the last generation are critically reviewed in R.J. JOHNSTON, *Geography and Geographers: Anglo-American Human Geography Since 1945*, fifth edition (Arnold, London, 1997). The *Annals of the Association of American Geographers* (quarterly, from 1911) and the *Publications of the Institute of British Geographers* (quarterly, from 1935) carry reviews of current changes in the AAG or IBG presidents' 'State of the Union' addresses to the annual conference.

A publication that provides useful information on geography departments in North America is Association of American Geographers, *Guide to Graduate Departments in the United States and Canada* (AAG Washington, D.C., annual). For information on departments in other countries turn to *The World of Learning* (Europa Publications, London, annual). But updated information on geographic societies and departments of geography around the world are best given through addresses and web sites given in Appendix B of this book. The job market in geography and the career options open to graduating geographers are outlined by the Association of American Geographers' Employment Forecasting Committee in T.J. WILBANKS *et al.*, *Rediscovering Geography: New Relevance for Science and Society* (National Academy Press, Washington, D.C., 1997), pp. 187–217. If you would like to see and hear geographers talk about their subject, there is an excellent series entitled *Geographers on Films* (details available from the Association of American Geographers, Washington, D.C. For readers with access to the World-Wide Web see also the sites on 'geographic societies', 'geographers', and 'jobs in geography' in Appendix B at the end of this book.

APPENDIX A

Glossary

The following list gives some five hundred of the most useful terms for a student starting a course in geography using this book. The words have been chosen to give a reasonably broad coverage of the field and have at least ten terms chosen from each of the 24 chapters. Terms in *italics* are defined elsewhere in the glossary. The technical vocabulary of some parts of geography (e.g. the environmental and map-making sides) are far richer than others (e.g. the humanistic and social science sides) so a balance has had to be struck between these different traditions. Each entry has a cross reference to the chapter in which it is used; in some cases such terms may be in figures or tables. For some chapters a few additional terms are added for more advanced topics likely to be met in the literature recommended in 'One step further . . .'.

Any short glossary, however carefully chosen, can be only the tip of the iceberg. Most geographic dictionaries contain not hundreds but many thousands of terms. If you are taking geography further then a useful volume to have to hand is R.J. SMALL and M. WITHERICK, *A Modern Dictionary of Geography*, third edition (Arnold, London, 1995). If your interest lies in physical geography, then an outstanding dictionary is D. THOMAS and A. GOUDIE (eds), *The Dictionary of Physical Geography*, third edition (Blackwell, Oxford, 2000). The best dictionary of human geography is Ron JOHNSTON *et al.* (eds), *The Dictionary of Human Geography*, fourth edition (Blackwell, Oxford, 2000).

A

acid rain Popular term used to describe precipitation that has become acid through reactions between moisture and chemicals in the atmosphere. It normally refers to the excessive acidity of atmospheric moisture caused by polluting emissions of sulphur dioxide and nitrogen oxides from burning fossil fuels. (Chapter 9)

adoption curves Rates at which an innovation is adopted by a population through which a diffusion wave is passing. In the *Hägerstrand model* the rates show distinctive changes over time. (Chapter 16)

aerial photographs Views of the earth's surface usually divided by their angle (vertical or oblique), their level of resolution, and the medium (film or digital) with which they are recorded. (Chapter 22)

agglomeration economies Savings to the individual manufacturing plant that come from operating in the same location. These may come from common use of specialist servicing industries, financial services, or public utilities. (Chapter 8)

agricultural hearths The locations where agriculture was first thought to have evolved. The conventional view was that the river valleys of the fertile crescent in the Middle East (those of the Nile, Tigris–Euphrates, and Indus) might provide the earliest sites. Under the *Sauer hypothesis* a primary location in Southeast Asia and a secondary location in northern South America is proposed. (Chapter 5).

air pollution The presence of gases and particulate matter in the air in high enough concentrations to harm humans, other animals, vegetation or materials. Such pollutants are introduced into the atmosphere principally as a result of human activity. (Chapter 9)

alternative energy Energy obtained from sources other than fossil fuels or nuclear power. The sources generally have low pollution implications and use renewable resources; for example biogas, geothermal energy, hydroelectric power, solar energy, tidal power, and wind power. (Chapter 10)

American Geographical Society (AGS) The oldest major geographic society in the United States. Founded in 1852 by a group of businessmen as the 'American

Geographical and Statistical Society of New York', it played a major part in the exploration and mapping of the Americas. Its massive geographic library is now based in the University of Wisconsin at Milwaukee. (Chapter 24)

analogue Correspondence or partial similarity. In regional geography, analogues refer to regions that because they are similar in known respects are argued to be also similar in other respects. (Chapter 12)

anisotropic An adjective to describe a spatial phenomenon having different physical properties or actions in different directions. (Chapter 15)

anomalies In regional geography, the use of a *region* to illustrate variations from a generalized model. Also termed a residual region. (Chapter 12)

antecedent boundaries Used in *political geography* to describe boundaries that predate settlement and economic development. (Chapter 17)

ARC/INFO A leading GIS software developed by the Environmental Systems Research Institute (ESRI). (Chapter 23)

arctic The northern polar region. In biological terms it also refers to the northern region of the globe where the mean temperature of the warmest month does not exceed 50 °C (0 °F). Its boundary roughly follows the northern treeline. (Chapter 2)

areal differentiation Process by which the spatial covariation of phenomena on the earth's surface can be analyzed to recognize specific regions. Argued by Richard Harsthorne (see Box 12.A) to be one of the central purposes of geography. (Chapter 12)

array In *GIS* a series of addressable data elements in the form of a grid or matrix. (Chapter 23)

Association of American Geographers (AAG) The oldest professional society in the United States for geographers, it was founded in 1904 with W.M. Davis (see Box 2.C) as its first president. It publishes the *Annals* and the *Professional Geographer*. (Chapter 24)

Association of South East Asian Nations (ASEAN) ASEAN was formed in 1967. It is a regional association of non-communist states. (Chapter 17)

atmosphere Major division of the earth's environment consisting of the layer of air surrounding the earth. It is conventionally divided into the troposphere, the stratosphere, and the ionosphere. (Chapter 2)

autocorrelation Statistical concept expressing the degree to which the value of an attribute at spatially adjacent points varies with the distance or time separating the observations. Indices range from positive autocorrelation indicating spatial clustering to negative indices indicating spatial uniformity and dispersion. (Chapter 21)

automated cartography The process of drawing maps with the aid of computer-driven display devices such as plotters and graphics screens. (Chapter 21)

average path length A measure in *graph theory* to indicate the degree of connectivity within a transport network. (Chapter 13)

B

backwash effects In *physical geography* the movement of water back down a beach towards the sea after the swash has reached its highest point. (Chapter 1) In *economic geography* the movement of resources from the periphery area towards the core, particularly marked during the earlier stages of the development process. (Chapter 18)

bands In *remote sensing* the sections of the electromagnetic spectrum with a common characteristic, e.g. the visible light band, the infrared band. (Chapter 22)

barriers A variant of the *Hägerstrand model* developed by Yuill to show the spatial effects on spread of different categories of barrier. (Chapter 16)

Bartlett model An *epidemic* model developed by the British statistician M.S. Bartlett to show the relations between the size of a community measured in population size and the spacing of epidemic waves. Originally based on observations of measles epidemics in British towns, it has since been widened to include many infectious diseases on a worldwide basis. (Chapter 16)

basing-point pricing In this form of *spatial pricing*, all production of a given commodity is regarded as originating from a single point, a uniform price is established for all sources regardless of their location, and customers pay the transfer costs from the basing point regardless of the actual location of the producer from which they purchased a product. The most famous case of basing-point pricing was the 'Pittsburg plus' system that operated for some time in the United States steel industry. (Chapter 13)

Bayesian estimator A type of *map decoding* using prior probabilities based on the mathematics developed in the eighteenth century by Thomas Bayes. (Chapter 21)

bell-wether regions Areas within a country that show trends earlier than the rest of the country. Literally the terms means a sheep (wether) wearing a bell that walks ahead of the flock. (Chapter 12)

beta index In *graph theory* a measure of the connectivity of a transport network. (Chapter 18)

bid-price curves Attempt to explain land use by plotting the rent that buyers are prepared to pay against geographic accessibility to some point. The bid-rent curves slope downwards with decreasing accessibility, usually in a concave way to reflect sharp decreases of rent over short distances. The model has been used for both urban land use and rural land use (as a component in the *Thünen model*). (Chapter 15)

biodegradable Adjective applied to pollutants that can be decomposed by biological organisms. Products made of organic materials such as paper, woollens, leather, and wood are biodegradable; many plastics are not. (Chapter 9)

biodiversity In *biogeography* the number of different species of plants and animals found in a given area. Generally the greater the number of species, the more stable and robust is the *ecosystem* (Chapter 19).

biogeography A branch of *geography* that studies the spatial distribution of plants and animals over the surface of the land. (Chapter 4)

biological concentration The process by which organisms concentrate certain chemical substances to levels above those found in their natural environment. (Chapter 9)

biomass The total mass of all the living organisms in a defined area or ecological system. (Chapter 4)

biome Major environmental zones of the earth marked by a distinctive plant cover (e.g. the subarctic tundra biome). (Chapter 4)

biosphere The thin layer of the earth that contains all living organisms and the environment that supports them. (Chapters 2, 4)

Boolean logic A branch of mathematics named after its nineteenth-century discoverer, which allows operation on spatial sets in *GIS*. Its operations are broadly equivalent to addition, subtraction, and multiplication. (Chapter 23)

boreal The northern cold winter climates lying between the *arctic* and latitude 50°N. The region is dominated by coniferous forest, also known as taiga. (Chapter 4)

Boserup model A model developed by the Danish land economist Ester Boserup (see Box 9.A) to relate increases in population density to the land fallowing systems around the world. (Chapter 9)

bounding problem In urban geography the failure of legal city boundaries to match the actual extent of the city in terms of its built-up area. Since cities tend to grow outside their legal limits, the main problem is one of underbounding. (Chapter 14)

bush fallowing The system of farming in tropical areas whereby the land is allowed to revert to bush to regain its fertility after a period of cultivation. (Chapter 9)

business cycles Recurrent fluctuations in general economic activity. They may be observed in fluctuations in the production or employment within a country or region. (Chapter 18)

C

cadastral map A map showing the precise boundaries and size of land parcels. (Chapter 21)

carbon cycle Sometimes referred to as the earth's 'life cycle' since carbon is one of the three basic elements (with hydrogen and oxygen) making up most living matter. Over 99 per cent of the earth's carbon is locked up in calcium carbonate rocks and organic deposits (coal and oil). The remaining carbon is cycled through plants and animals in a complex process described in Box 4.A). (Chapter 4)

carbon dating Radio-carbon dating is a means of determining the age of some prehistoric remains (e.g. bone and wood) back to about 50,000 years BP. It is based on the fact that carbon-14 decreases at a known and constant rate after the death of the organism (half-life of roughly 5700 years). (Chapter 3)

carrying capacity The largest number of a population that the environment of a particular area can carry or support. The optimum population of a particular species that a given area of land, or habitat, can support under a given set of environmental conditions. (Chapter 6)

Cartesian coordinate system A graticule of lines at right angles used for identifying the location of objects in small areas where the curvature of the earth's surface can be ignored. (Chapter 21)

cartography The practice or science of map-making, based on the French word for map (*carte*). (Chapter 21)

cascade diffusion A variant of *hierarchic diffusion* in which the movement is downwards from upper to lower settlement levels. (Chapter 16)

catastrophes A branch of mathematical topology developed by René Thom which is concerned with the way in which non-linear interactions within systems can produce sudden and dramatic effects. It is argued that there are only a limited number of ways in which such changes can take place, and these are defined as elementary catastrophes. (Chapter 24)

catenas Sequences of soils that vary with relief and drainage though normally derived from the same parent materials. (Chapter 4)

Centers for Disease Control (CDC) The major United States government agency for disease prevention based in Atlanta, Georgia. Originally established as a military malaria-control service, it was reconstituted in the 1940s to cover other diseases and now plays a key role in global disease control. (Chapter 20)

central business district (CBD) The part of a town or city, usually in the centre, where retail stores, offices, and cultural activities are concentrated, and land values and building densities are high. Measurement of CBD using the Murphy–Vance method is described in Box 15.A. (Chapter 15)

central place A settlement or nodal point which, by its functions, serves an area round about it for goods and services. (Chapter 14)

central-place theory Developed by the geographer Christaller in the 1930s to explain the spacing and function of the settlement landscape. Under idealized conditions, he argued, central places of the same size and nature would be equidistant from each other, surrounded by secondary centres with their own small satellites. In spite of its limitations, central-place theory has found useful applications in archaeology as a preliminary heuristic device. (Chapter 14)

centrifugal forces In urban geography, those forces that tend to displace activities towards the periphery of the city. They include high land costs, congestion, travel to work, and higher taxation. (Chapter 8)

centripetal forces In urban geography, those forces that tend to attract activities into the centre of a city. They include accessibility to both suppliers and customers, specialized services, and a wide range of skilled labour. (Chapter 8)

choropleth maps A map consisting of a series of single-valued, uniform areas separated by abrupt boundaries. It is frequently used to show spatial distributions within administrative areas. (Chapter 1)

circum-Pacific belt In *physical geography* the zone around the margin of the Pacific Ocean marked by the edges of tectonic plates with high levels of volcanic and earthquake activity. The plate edges are marked by some of the world's deepest ocean troughs. (Chapter 2)

climax The state of equilibrium reached by the vegetation of an area when it is left undisturbed for a long period of time. It represents a final stage of natural succession in an ecosystem where there is no further net growth in biomass and the organisms are in equilibrium with their environment. (Chapter 4)

coastal baseline A boundary line in *political geography* from which a country's territorial rights to adjacent waters is measured. (Chapter 17)

cognitive behaviouralism Holds that the impact of environment on people is partly dependent on their perception (cognition) of the resources and barriers it poses. (Chapter 11)

cognitive map An interpretative framework of the world which, it is argued, exists in the human mind and affects actions and decisions as well as knowledge structures. (Chapter 7)

coliforms Bacteria (e.g. *E. coli*) found in the human intestinal tract used as an indicator of water quality. Presence indicates that the water supply is polluted, possibly by human sewage. (Chapter 19)

comfort zones In climatic terms, the zone in which human activity is thought to reach a maximum efficiency. The zone is bounded by thermal limits (too cold or too hot), modified by relative humidity (too dry or too humid), and by wind speed (too stuffy or too windy). Although largely studied today in microclimatic terms for building design, the geographer Ellsworth Huntington considered that the earth's macroclimate played a part in the rise of civilizations in optimum climatic zones. (Chapter 4)

communication flows In spatial interaction studies, flows of information (whether by broadcast or by cables) in distinction to *transport flows*. (Chapter 13)

compage A term introduced by the Harvard geographer, Derwent Whittlesey, to denote a region in which many chracteristics overlapped. It was used in preference to the older term 'natural region'. (Chapter 12)

complementarity Term used in the *Ullman model* to describe one of the bases of spatial interaction. Implies that region A produces (or has the potential to produce) goods or services of which region B has a deficit (or potential deficit). (Chapter 13)

complementary regions The dependent area needed to support each level of settlement in a central-place model. Higher-order settlements demand proportionately larger dependent areas than lower-order centres. (Chapter 14)

computer-assisted cartography An early name from the 1960s for the impact of computers and IT on map making; now largely replaced by the term *GIS*. (Chapter 24)

connectivity A term in *graph theory* to measure the degree of connectedness shown by a transport network. A number of indexes have been developed by geographers such as Kansky to measure this quality accurately. (Chapter 13)

conservation The use, management, and protection of natural resources so that they are not degraded, depleted, or wasted and are available on a sustainable basis for use by present and future generations. (Chapter 10)

contagious diffusion A type of *diffusion* process in which spatial contact is consistently maintained. It is also described as neighbourhood diffusion. (Chapter 16)

continental climate The type of climate associated with the interior of continents. It is characterized by a wide daily and seasonal ranges of temperature, especially outside the tropics, and low rainfall. (Chapter 2)

continental drift The complex process by which the continents move their positions relative to each other on the plates of the earth's crust. Also known as plate tectonics. (Chapter 2)

continental penetration A stage in European overseas colonization in which the mid-continental grasslands (e.g. the North American prairies, the South American pampas, the South African veldt) were settled by agriculturalists. In many areas close settlement was related to late nineteenth-century railroad construction. (Chapter 5)

continental shelf The relatively shallow belt of sea-covered land fringing some of the continents. It slopes out to around 200 m before sinking rapidly to the ocean floor. Its exact definition and delimitation are assuming increasing importance in international law as territorial claims extend seawards. (Chapter 17)

coral reef A barrier of coral formed by the accumulation of the skeletons of millions of polyps. Living polyps attach themselves to the skeletons of past generations. Reefs may be an extension of the shore (fringing) or offshore (barrier). (Chapter 2)

cordon sanitaire A protective zone placed around a spreading *epidemic* to prevent its spatial extension into unprotected areas. (Chapter 16)

core Innermost region within Meinig's four-layer culture region. (Chapter 7)

core–periphery models A set of models in *political geography* which specify the power relations between a dominant central core (although this may in turn be dominated from outside) and a dependent periphery. Important contributions to shaping this model have been made by Wallerstein and Friedmann. (Chapter 18)

cosmology The science or theory of the universe (cosmos). (Chapter 21)

crude birth rate The number of live births in a year per 1000 population. The crude rate will be relatively high if a large proportion of the population is of childbearing age and is usually adjusted for age structure to give a net birth rate. (Chapter 6)

crude death rate The number of deaths in a year per 1000 population. It is affected by a population's age structure and is usually adjusted to give a net death rate. (Chapter 6)

crude growth rate In population geography, the vital rates which measure the change in population over time not adjusted for age or sex structure. Typical population growth curves are shown in Box 6.A. (Chapter 6)

cultural ecology A term devised by Julian Steward to account for the dynamic relationship between human society and its environment, in which culture is viewed as the primary adaptive mechanism. (Chapter 7)

cultural regions A *region* whose boundaries are determined by some non-biological characteristic unique to the particular society occupying the area. Cultural regions may be characterized by a particular language or religion or artifact (e.g. type of housing) or some combination of such traits. (Chapters 7, 12)

culture In human geography, the customs, civilizations, and achievements of a particular people. (Chapter 7)

D

database A collection of interrelated information, usually stored on some form of mass-storage system such as magnetic tape or disk. A *GIS* database includes data about the position and the attributes of geographical features that have been coded as points, lines, areas, *pixels*, or grid cells. (Chapter 23)

database management system (DBMS) A set of computer programs in a *GIS* for organizing the information in a database. Typically, a DBMS contains routines for data input, verification, storage, retrieval, and combination. (Chapter 23)

Davisian cycle A model in *geomorphology* developed by William Morris Davis (see Box 2.C) of landscape evolution based on the three principles of geological structure, geomorphic process, and time. The third element lead to a series of stages in the cycle so that landforms went through a sequence of youth, maturity, and old age. Originally applied to humid landscapes, it was later extended to arid, glacial, and coastal landforms. (Chapter 2)

DDT The popular name for dichlorodiphenyl-trichorethane, a powerful insecticide developed in Switzerland in the 1930s. It was successfully used in malaria control in the decades after World War II but withdrawn once its toxic and persistent side effects were understood. (Chapters 3, 20)

decomposers Organisms such as fungi and bacteria, which rely upon the dead tissues of other organisms as an energy source. In order to extract this energy, they break down the organic material, releasing nutrients from those tissues into the environment. (Chapter 4)

deindustrialization The decline of heavy engineering and other manufacturing industries in an area and the corresponding rise of service industries. (Chapter 8)

demographic transition A sequence of demographic changes in which countries progressively move over historic time from high birth and death rates to low birth and death rates. (Chapter 6)

demography The study of the process that contributes to population structure and their temporal and spatial dynamics. (Chapter 6)

dendrochronology The study of tree-ring patterns; annual variations in climatic conditions which produce differential growth can be used both as a measure of environmental change, and as the basis for a chronology. (Chapter 3)

desertification The alteration of arable or pasture land in arid or semi-arid regions to desert-like conditions. It is usually caused by a combination of overgrazing, soil erosion, prolonged drought, and climate change. (Chapter 3)

developed country In *economic geography* any country characterized by having high standards of living and a sophisticated economy, particularly in comparison with developing countries. A number of indicators can be used to measure a country's wealth and material well-being: for example, the gross national product, the per capita consumption of energy, the number of doctors per head of the population, and the average life expectancy. (Chapter 18)

diffusion In *geography*, diffusion is the process of spreading out or scattering over an area of the earth's surface. It should not be confused with the physicist's use of the term to describe the slow mixing of gases or liquids with one another by molecular interpenetration. (Chapters 1, 16)

digital orthophoto quad (DOQ) A map quadrangle made from corrected aerial photographs and available in digital as well as map form. (Chapter 23)

digitizer In *GIS* a device for entering the spatial coordinates of mapped features from a map or document to the computer. A pointer device (e.g. cursor, puck or mouse) is used to locate key points on the map. (Chapter 23)

dioxin A very powerful poison present in some weedkillers and found to cause certain deformities in foetuses. (Chapter 9)

distance–decay curves The attenuation of some activity or process with distance. Enshrined in Tobler's (1970) 'first law of geography: everything [on the earth's surface] is related to everything else, but near things are more related than distant things'. Considerable work has been dome in locational theory in identifying the precise mathematical form of such decay curves: exponential or inverse-square forms appear to fit a wide range of phenomena. (Chapter 13)

districting In *political geography*, the search for optimal electoral areas that combine desirable properties (e.g. compactness) and avoid undesirable properties (e.g. *gerrymandering*). A number of computer districting programs are now available. (Chapter 17)

domain Secondmost region within Meinig's four-layer culture region. (Chapter 7)

Domesday survey A geographically organized survey of England carried out by William I in 1086. For most English villages and towns (but not London and Winchester) the Domesday Book is the starting point for settlement and land-use data and it was extensively used by Sir Clifford Darby, in his multivolume historical geography of Domesday England. (Chapter 11)

domestication The process by which wild animals and plants are brought under human control and are bred to possess special characteristics that enhance their usefulness for human exploitation, particularly food production. (Chapter 5)

dot-matrix plotter A plotter of which the printing head consists of many closely spaced (100–200 per inch) wire points that can write dots on the paper to make a map. Also known as an electrostatic plotter or matrix plotter. (Chapter 23)

dryland farming A method of producing crops in dry areas so as to make maximum use of limited moisture available. It can involve leaving a protective layer of crop residue to reduce evaporation, contour ploughing, and alternating cultivation with a period of fallow. (Chapter 9)

E

ecological analysis One of the three integral characteristics of modern *geography* (along with *spatial analysis* and *regional complex analysis*) which lays stress on human–environment relations. (Chapter 24)

ecological determinism A form of explanation in which it is implicit that changes in the environment determine changes in human society. (Chapter 11)

ecological movement A broad-based political movement and ideology concerned with halting environmental despoilation and creating a society that sustains itself without excessive damage to the natural environment; it includes political parties and pressure groups. (Chapter 11)

ecological niche The position in the ecological community occupied by a particular species defined by the habitat it occupies, what it eats and what it is eaten by. (Chapter 4)

ecology The study of interactions of living organisms with each other and with their environment; the study of the structure and functions of nature. (Chapter 4)

economic geography One of the three major divisions of human geography (along with political and social geography). Economic geography covers the spatial economic studies carried out by geographers: (a) the mapping of economic phenomena (e.g. economic atlases), (b) studying the impact of geography on economic events (e.g. industrial location), and (c) studying the impact of economics on geography (e.g. cost limits on crop distributions). (Chapter 18)

ecosystem A community of plants and animals and the environment in which they live and react with each other. In ecosystems plants and animals are linked to their environment through a series of feedback loops. (Chapter 4)

edge A term used in *graph theory* to describe a link between two vertices or nodes within a network. It is used in determining measures of connectivity. (Chapter 13)

edge conditions A term in *biogeography* for the margins in which two natural plant communities or modified land uses meet. The edge creates special niche opportunities for fauna (including microrganisms) and gives rise to a distinctive linear region. (Chapter 11)

El Niño Appearance of unusually warm water in the eastern Pacific Ocean off the coasts of Peru, Ecuador, and northern Chile. Occurs irregularly about fourteen times a century. It is associated with heavy rains along the normally dry Peruvian coasts and has other long-distance effects around the globe through disturbances in the general circulation of the atmosphere. (Chapter 3)

electromagnetic spectrum Electromagnetic radiation occurs as a continuum of wavelengths and frequencies from short-wave X-rays to long-wave radio waves. Part of this spectrum is of great importance in *remote sensing* with several windows around the visible spectra, the infrared spectra, and the radar spectra, allowing images of the earth's surface to be recorded by a variety of recording systems. (Chapter 22)

endemic A species or a disease that is native to one specific area. (Chapter 20)

'end-of-geography' The view that the reduction in transport and communication costs and associated globalization trends will erode differences in the character of previously diverse regions. (Chapter 19)

enhancement Refers to processes in *remote sensing* that increase or decrease contrasts on received images (e.g. photos) so as to make them easier to interpret. (Chapter 22)

entropy Used in geography as a measure of the degree of disorder or randomness in a system. The *Wilson models* of spatial interaction are based on maximum-entropy estimates. (Chapter 13)

environment The external conditions – climate, geology, other living things – that influence the life of individual organisms or ecosystems; the surroundings in which all plants and animals (including humans) live and interact with each other. (Chapter 1)

environmental degradation The process of depleting or destroying a renewable resource such as soil, forests, pasture, or wildlife by using it at a faster rate than it can be replenished. (Chapter 9)

environmental determinism The view that natural environment plays the major role in determining the behaviour patterns of humans on the earth's surface. (Chapter 11)

Environmental impact analysis (EIA) Studies that assess the probable impact of a new development (e.g. an industrial plant) on its local and regional environment. Such EIAs are increasingly a legal requirement before development can start and have given rise to many consulting companies offering this service. (Chapter 24)

Environmental Systems Research Institute (ESRI) One of the major GIS software companies based in California (see Box 23.C). (Chapter 23)

epicentre In physical geography the point on the earth's surface closest to the subsurface origin of an earthquake shock. (Chapter 2). It is also used in human geography to describe the location of the starting point of an *epidemic* or *pandemic* wave. (Chapter 20)

epidemic A widespread occurrence of a disease in a particular community at a particular time. (Chapter 20)

epidemiological transition A sequence that parallels the *demographic transition*. Countries progressively move over historic time from high infant mortality (mainly associated with infectious diseases) to low infant mortality and death coming at higher ages (mainly through degenerative diseases such as cancers and cardiovascular problems). (Chapter 20)

epistemology The theory of knowledge, especially its methods and ways of validation. In terms of geographic knowledge, it describes the study of the different schools of thought and their contribution to the philosophy of geography. (Chapter 24)

equator An imaginary line around the earth that is an equal distance from the North and South Poles. (Chapter 21)

equinox The date twice each year (in March and September) when the sun crosses the equator, and day and night are of equal length worldwide. (Chapter 3)

ethnicity The existence of ethnic groups, including tribal groups. Though these are difficult to recognize from the archaeological record, the study of language and linguistic boundaries shows that ethnic groups are often correlated with language areas. An ethnic group, defined as a firm aggregate of people, historically established on a given territory, possessing in common relatively stable peculiarities of language and culture, and also recognizing their unity and difference as expressed in a self-appointed name. (Chapter 7)

ethology The science of animal behaviour, the word deriving from the Greek '*ethos*' meaning nature or disposition. (Chapter 17)

European Union (EU) A cooperative economic and political alliance of a number of Western European states, formed by the Treaty of Rome in 1957. Its six founder members had grown to fifteen by 2000. (Chapter 17)

eutrophication The process by which a habitat, usually aquatic, becomes enriched with nutrients such as nitrates and phosphates. This can lead to oxygen reduction and algal bloom. As the nutrient supply (particularly nitrogen and phosphorus) of a lake builds up from the inflow of fertilizer-enriched water from the surrounding fields, some major changes in the lake's biology and chemistry take place. (Chapter 9)

exemplar A *region* used to illustrate a broader relationship or classification. Thus the North American prairies might be used as an exemplar of the world's mid-continental grasslands or Chicago used an exemplar of the spatial structure of western cities. (Chapter 12)

exponential models Of population growth describe a simplified situation in which growth (or a decline) is unchecked and the rate of change is constant. (Chapter 6)

F

feedbacks Links within a *system* that either reinforce (positive feedback) or dampen (negative feedback) an effect. (Chapter 1)

feminist geography A branch of human geography which has grown rapidly in the last quarter of the twentieth century. As Box 7.B shows, it draws on feminist theories and politics to explore how gender relations and geographies are mutually structured and transformed. (Chapter 7).

feng shui A Chinese animist belief that powerful spirits of ancestors, dragons, and tigers occupy natural phenomena and should not be disturbed. Hence special geomancers are required to understand the landscape and choose appropriate sites for such things as burial mounds. (Chapter 7)

fertile crescent Name given to one of the earliest sites of civilization consisting of a crescent-shaped area of fertile land stretching from the lower Nile valley, along the east Mediterranean coast and into Syria and Mesopotamia (present-day Iraq) watered by the Euphrates and Tigris rivers. Agriculture had its origins in the area about 8000 BC. (Chapter 5)

fertility rates Measure the actual production of offspring by females in a population, usually measured as the average number of children born to a women who completes her childbearing years. (Chapter 6)

filter In *raster* graphics a mathematically defined operation for removing long-range (high-pass) or short-range (low-pass) variation. Used for removing unwanted components from a signal or spatial pattern. (Chapter 23)

financial centres See *world financial centres.* (Chapter 19)

fixed-*K* hierarchies Alternative version of Christaller's central-place model which shows settlements arranged in terms of market optimizing (K=3), transport optimizing (K=4), and administrative optimizing (K=7) principles. The integer value of the K-term describes the differences between the tiers of the *settlement hierarchy.* (Chapter 14)

flow resources Natural and renewable sources of good such as crops and fish, natural and renewable materials such as wood and cotton, and natural and renewable sources of energy such as wind, sunshine, and running water. (Chapter 10)

Food and Agriculture Organization (FAO) A specialized agency of the United Nations created in 1945 and based at Rome. It combats worldwide malnutrition and hunger by coordinating food, farming, forestry, and fisheries development programmes. (Chapters 10, 11)

food chain The term that describes the process by which energy in the form of food is passed from one living organism to another. Plants (primary producers) are eaten by plant-eating animals, or herbivores (primary consumers), which in turn are eaten by meat-eaters, or carnivores (secondary consumers), and so on up the chain. (Chapters 4, 9)

fossil fuels Coal, petroleum or natural gas, materials used as fuels that occur naturally, normally as the result of vegetation and organisms that died millions of years ago being compressed into a combustible substance. Peat and lignite are less compressed fuels and less efficient to burn. (Chapter 10)

fractal An object having a fractional dimension; one which has variation that is self-similar at all scales, in which the final level of detail is never reached and never can be reached by increasing the scale at which observations are made. (Chapter 23)

G

gender In human geography refers to the constructed and self-constituted roles assigned to men and women as distinct from their biological differences. A major focus of studies in feminist geography. (Chapter 7)

geodesy The science that deals with the shape and size of the earth. (Chapter 21)

geographic information systems (GIS) A set of computer tools used to capture, store, transform, analyze, and display geographic data. (Chapter 23)

geography Defined by Hartshorne (1959) as 'accurate, orderly, and rational description of the variable character of the earth's surface'. For other definitions, see Table 24.1. It is conventionally divided into three main sectors: *physical geography*, human geography, and *regional geography.* (Chapter 24)

geological norms Attempt to establish a benchmark of change (e.g. soil erosion) over a long-time period to reflect changes which occur in environments undisturbed by human activity. Such norms provide a marker against which human changes can then be measured. (Chapter 11)

geomorphology A subdiscipline of *physical geography*, concerned with the study of the form and development of the landscape, it includes such specializations as sedimentology. (Chapter 2)

geopiety A term in human geography developed by J.K. Wright (1947) to express the thoughtfulness and reverence both for the earth's environment and for the specific character of well-loved regions. (Chapter 12)

geopolitics The study of the influence of the geographical location of countries on their relations with other countries. (Chapter 17)

geo-referencing Tagging of geographic information into a geographic coordinate system. Also described as *geocoding.* (Chapter 23)

geosyncronous satellite A satellite with remote sensors which is placed in an orbit so that it 'hangs' over the same spot as the earth turns. It is capable of viewing nearly a full hemisphere although with much lateral distortion towards the edges. (Chapter 22)

geothermal energy An alternative energy source that is derived from heat sources within the earth's interior. The sources can be tapped to provide power for electricity or domestic heating, either by harnessing steam or hot water rising to the surface through cracks in the rock, or by drilling holes to hot rocks close to the surface and injecting water to create steam. (Chapter 10)

gerrymandering The manipulation of constituency boundaries to favour a particular candidate or party, sometimes used more generally to indicate political bias in the operation of 'first-past-the-post' electoral system. (Chapter 17)

ghetto In urban geography a rundown residential area, usually part of the inner city, in which any underprivileged minority or ethnic group is concentrated. Historically applied to Jewish quarter of a city. Probably from the Italian '*getto*' meaning a foundry, as applied to the first ghetto in Venice in 1516. (Chapter 8)

global eradication In epidemiology, the term is reserved for the worldwide and permanent eradication of a disease. The only example to date of successful global eradication is smallpox. The term 'elimination' is used for disease clearance over a smaller area. (Chapter 20)

global positioning system (GPS) A set of satellites in geostationary earth orbits used to help determine graphic location anywhere on the earth by means of portable electronic devices. (Chapter 23)

global village Term coined by the Canadian Marshall McLuhan to describe the impact of new communication technologies on social and cultural life and the replacement of local by global cultures. (Chapter 19)

global warming Increase in global temperatures resulting from human activities that are enhancing the so-called natural greenhouse effect. (Chapter 19)

globalization The pervasive changes in global economy and global society brought about by new communication technologies and the time–space compression that results from them. (Chapter 19)

Gondwanaland The name given to the ancient southern supercontinent, which was composed of present-day Africa, Australia, Antarctica, India and South America. It began to break up 200 million years ago. (Chapter 3)

graphic user interface (GUI) In *GIS* the ways in which operators communicate with the hardware through mouse and lightpen technology. (Chapter 23)

graph theory The branch of finite mathematics that deals with the structure of networks. It is used by geographers in the analysis of transport networks where the topology of links (or edges) and nodes (or vertices) throws light on their connectivity and inherent spatial structure. (Chapter 13)

graticules Networks of parallels and meridians drawn on a map. (Chapter 21)

gravity models Spatial interaction *models* that have physical analogies with the classic gravity equations. Flows between regions are a complex function of the magnitude of the interacting regions (e.g. their population, or market value) and inversely related to their separation (e.g. distance, time, or travel cost). (Chapter 13)

'Great Deluge' In migration history, the peak period of migration into the United States from 1870 to 1920 with an inflow of over eight million to that country in the 1900–09 decade. (Chapter 5)

green belt An area of open land around a city on which development is prevented or carefully controlled in order to set limits on a city's expansion. (Chapter 8)

green revolution A term coined in the 1960s to refer to the spectacular changes in high-yielding varieties of basic crop plants (especially rice) in the Third World. The term is a contested one and is now used more widely to refer to the set of new production practices (involving fertilizers and pesticides) which have brought fundamental changes to the nature of peasant farming practices in some parts of the tropics. (Chapter 10)

greenhouse effect The process in which radiation from the sun passes through the atmosphere, is reflected off the surface of the Earth, and is then trapped by gases in the atmosphere. The build-up of carbon dioxide and other 'greenhouse gases' is increasing the effect. There are fears that the temperature of the planet may rise as a result; this global warming is expected to have dire consequences. (Chapter 3)

gross national product (GNP) The sum of all output produced by economic activity within a country including net income from abroad. (Chapter 18)

ground truth Information about the actual state of any environment at the time of a remote sensing flight overhead. (Chapter 22)

groundwater Water that has percolated into the ground from the surface, filling pores, cracks, and fissures. An impermeable layer of rock prevents it from moving deeper so that the lower levels become saturated. The upper limit of saturation is known as the water table. (Chapters 9, 19)

growth poles A concept introduced by the French regional economist Francois Perroux in the 1950s. His '*pole de croissance*' envisaged a dynamic and highly integrated set of industries organized around a propulsive leading sector. Its rapid growth was seen as generating wider growth through spillover and multiplier effects in the rest of the economy. (Chapter 18)

gullying A severe form of soil erosion that occurs near the bottom of slopes when water run off removes oil and soft rock to erode a deep channel or gully. (Chapter 5)

H

habitat The external environment to which an animal or plant is adapted and in which it prefers to live, defined in terms of such factors as vegetation, climate, and altitude. (Chapter 4)

Hadley–Ferrel model (Hadley cell) The name given in climatology to the large-scale thermally driven circulation in tropical latitudes and most prominent over the Atlantic and Pacific Oceans. There is one Hadley cell in each hemisphere with heated air rising near the equator, moving poleward aloft, descending at latitude 30° to 40°, and then flowing equatorward near the earth's surface. The cell is named after George Hadley who first described the phenomenon in 1735. (Chapter 2)

Hägerstrand model A series of spatial *diffusion* models developed by the Swedish geographer, Torsten Hägerstrand based on empirical studies of innovations in farm communities in southern Sweden. The models, first developed in the 1950s, were innovative in their pioneering of simulation using Monte Carlo models. (Chapter 16)

half-life The time taken for half the quantity of a radioactive isotope in a sample to decay. Times vary enormously: iodene-3 has an eight-day half-life, but plutonium-239 and uranium-238, the main nuclear fuels, have half-lives reckoned in hundreds of thousands of years. (Chapter 10)

heat islands With the development of human settlements, urban structures cause distinctive urban microclimates to emerge. Heat islands are the relatively warmer temperatures that form over cities (typically 1 °C to 3 °C for a city of ten million inhabitants) as compared to surrounding rural areas. This effect varies diurnally (largest near midnight) and with weather conditions (largest with calm, cloudless weather). (Chapter 9)

heavy industries Have (a) finished products which have low values per tonne, (b) a high material index, and (c) a high tonnage of materials used per worker. (Chapter 15)

heavy metal Any metal such as mercury, cadmium, and lead with a high atomic weight. Widely used in small quantities in industry, they concentrate in soils and water and are highly poisonous to living things. Once they enter the food chain, they accumulate in organs such as the brain, liver, and kidneys of animals, with long-term toxic effect. (Chapter 9)

hierarchic diffusion A type of *diffusion* wave which moves through the urban hierarchy in a cascading manner. Unlike *contagious diffusion*, its geographic pattern of spread may be discontinuous in space. (Chapter 16)

hinterland The area lying inland from a coast that supplies agricultural goods and raw materials to an entrepôt port. By extension, it refers to the area of countryside around a city that lies within its sphere of influence, receiving and supplying goods and services. (Chapter 13)

hollow frontier A land-use sequence in which an agricultural frontier moves forward leaving behind it a tract of worked-out farmland, often with a reduced population. The term has been used at different geographic scales to plantation frontiers in the tropics and middle-latitude farming in New England. (Chapter 9)

homeostasis A term used in *systems thinking* to describe the action of negative *feedback* processes in maintaining the system as a constant equilibrium state. (Chapter 2)

human ecology Envisages reciprocal reactions between human and environment, like those of other plant and animal species. This view is associated with the Chicago geographer Harlan Barrows (1877–1960). (Chapter 11)

hunter-gatherers A collective term for the members of small-scale mobile or semi-sedentary societies, whose subsistence is mainly focused on hunting game and gathering wild plants and fruits; organizational structure is based on *bands* with strong kinship ties. (Chapter 5)

hurricane See tropical cyclone.

Huxley's model A biologist's approach to culture which divides it into three traits: mentifacts (e.g. language), sociofacts (e.g. political system), and artifacts (e.g. housing types). (Chapter 7)

hydraulic hypothesis In cultural geography, the view associated with Karl Wittfogel (1956) that large-scale irrigation stimulated the rise of urban civilizations. (Chapter 5)

hydroelectric power (HEP) Electrical energy generated by the force of falling or flowing water, which is used to spin a turbine. (Chapter 10)

hydrologic cycle The continuous process whereby the earth's fixed supply of water circulates around the oceans, the atmosphere, the soil and the rocks beneath through evaporation, transpiration, precipitation, runoff, and the movement of groundwater. (Chapter 2)

hydrosphere Major division of the environment consisting of water on the earth's surface, mainly as a liquid (97 per cent, the world's oceans and seas) but also in solid (glaciers and ice caps) and gaseous forms (atmospheric moisture). (Chapter 2)

hysteresis From the Greek word for 'coming after', the term is used in *systems* analysis to describe lagged effects where there is a gap between a cause and an effect. (Chapter 14)

I

iceberg concept of disease A pyramid in which subclinical and clinical phases of a disease are separated, with death representing the apex of the pyramid. It implies that for many diseases, the official records are only the tip of an iceberg. (Chapter 20)

IDRISI A widely use *GIS* software developed by the geographers at Clark University. Raster-based, it operates on a wide range of systems from PCs upwards. (Chapter 23)

immunization programmes Large-scale use of vaccines to reduce or eliminate the spread of infectious diseases in human populations. They are usually targeted at particular high-risk groups defined in terms of age, vulnerability, or probability of exposure. (Chapter 20)

infrared photography Produces images that record the infrared section of the *electromagnetc spectrum*. (Chapter 22)

infrastructure The stock of roads, railroads, airports, port facilities, public services, etc. that are necessary to sustain the economic and social activities of a city or a country and are a measure of its stage of development. (Chapter 20)

innovation waves A Swedish term used in diffusion studies to describe the distinctive spatial form of the ways in which new cultural items are adopted. (Chapter 16)

Intergovernmental Panel on Climate Change (IPCC) An international group of scientists brought together by United Nations agencies (notably the World Meteorological Organization (WMO)) to evaluate the evidence on climate change and to formulate response strategies for governments. (Chapter 20)

Intergraph Corporation One of the major *GIS* software and hardware companies in the United States. (Chapter 23)

International Geographical Union (IGU) An international body established in 1922 to coordinate geographical societies worldwide. (Chapter 24)

Internet A system for standardized interconnections between computersof developed during the 1980s by physicists at CERN, Geneva. See Appendix B: Geography on the World-Wide Web. (Chapter 19)

interpolation Estimation of the values of an attribute at unsampled points from measurements made at surrounding sites. (Chapter 21)

intertropical convergence zone (ITCZ) A band of nearly continuous low pressure with high humidity, light and intermittent winds, and rain showers found near the equator. At the meeting point of the two hemispheric *trade winds* it is usually visible on satellite photographs. Its location moves away from the equator with the seasons reaching extreme locations in February and August where it is involved in triggering monsoonal rains. (Chapter 2)

intervening opportunity model A *spatial interaction model* developed by Stouffer (1940) to explain migration movements. It argues that the number of movements from an origin to a destination will be directly proportional to the opportunities at the destination, and inversely proportional to the number of intervening opportunities between the origin and destination. It avoids some of the calibration problems associated with *gravity models*. (Chapter 13)

island colonization model In biogeography a mathematical hypothesis developed by R.H. MacArthur and E.O. Wilson that relates the species distribution on islands to such factors as size and accessibility (see Box 5.A). (Chapter 5)

isodapanes Contours of total transport costs used in industrial location studies. (Chapter 15)

isopleth maps A type of mapping in which lines are drawn to show spatial variations in the values represented. 'Contours' are also used by geographers as a general term for any type of isopleth. Most isopleths are given distinctive names. Isochrone maps show lines of equal time. Isohyet maps show lines of equal rainfall. Isoneph maps show lines of equal cloudiness. Isophene maps show lines of biological events that occur at the same time (e.g. flowering dates of plants). Isotherm maps show lines of equal temperature. Isotim maps show lines of equal transport costs. (Chapter 1)

isostatic uplift Rise in the level of the land relative to the sea caused by the relaxation of Ice Age conditions. It occurs when the weight of ice is removed as temperatures rise, and the landscape is raised up to form raised beaches. (Chapter 3)

isotims Contours of transport cost for a single element in the manufacturing process. (Chapter 15)

isotropic space Space that has the same physical properties or actions in all directions. (Chapter 15)

J

Jacobs hypothesis A controversial view by Jane Jacobs in *The Economy of Cities* that urbanization came as an early rather than late stage in the development of complex societies. (Chapter 5)

K

Kondratieff waves Long-wave economic cycles identified by the Russian economist Kondratieff in the 1920s. The cycles occur at roughly half-century intervals in contrast to the shorter wave business and building cycles. (Chapters 18, 19)

L

landforms Davis meant the physical shape of the earth's surface terrain; structure was the geological composition of the land, and its original elevation above sea level as the result of mountain-building forces. (Chapter 2)

LANDSAT A series of earth resource scanning satellites launched by the United States of America. (Chapter 22)

landscape The appearance of an area, as in landscape painting. The term was introduced into American geography in 1925 by C.O. Sauer (see Box 5.B) in his *Morphology of Landscape*. Sauer differentiated between a natural landscape and a cultural (human-influenced) landscape. (Chapter 11)

land use A classification of land according to its dominant use (Chapter 11)

land-use cycles Sequence of land uses which follow the occupation and abandonment of agricultural land. Although first used in a rural context, the cyclic concept has been extended to the study of changes in urban land use. (Chapter 9)

latifundia Originally large estates formed when the Romans reallocated land confiscated from the conquered communities. Now, used to describe large estates, especially those in Mediterranean countries and their former overseas colonies. (Chapter 5)

latitude Angular distance north or south of the equator measured in degrees. (Chapter 21)

lead and lag regions Areas whose regional economic cycles are either in advance (lead) or behind (lag) the national economic cycle of which they form a part. Areas showing regular leads on a wide range of economic indicators (e.g. employment, production) are termed *bell-wether regions*. (Chapter 12)

light industries Have (a) finished products which have high values per tonne, (b) a low material index, and (c) a low tonnage of materials used per worker. (Chapter 15)

literary region A region adopted by an author and whose writings help to define the location and character of the area described. (Chapter 12)

lithosphere Major division of the earth's surface environment consisting of the crust and upper mantle. (Chapter 2)

localization A countervailing trend to *globalization* describing the downward shift of power from the nation-state to regions and cities. (Chapter 19).

location A particular position on the earth's surface. Locations may be described in terms of mathematical position on a global graticule (e.g. *latitude* and *longitude*) or in terms of relative position in relation to a local reference point. (Chapter 1)

locational analysis The branch of economic geography concerned with the application of spatial models to the understanding of economic phenomena on the earth's surface. (Chapter 24)

logistic Growth curve and has a characteristic S-shape. Its calculation is described in Box 6.A. (Chapter 6)

longitude The angular distance east or west from a prime meridian (usually Greenwich in London but historically other prime meridians were used) measured in degrees. (Chapter 21)

Lorenz curves A measure of the unevenness in a geographical distribution. It is computed by plotting the cumulative values of a feature of interest against the cumulative values of a benchmark such as population or area. (Chapter 18)

Lowry model In urban geography the ideas developed by I.S. Lowry (1964) to generate the distribution of jobs and residences within a city in terms of multipliers arising from changes in basic employment. (Chapter 8)

Lynch elements The view in urban geography that the perceived character of a city rests on a small number of highly visible landmarks. (Chapter 12)

M

Mackinder's heartland model In geopolitics, a model put forward by the English geographer Mackinder to explain the key role of Russia and eastern Europe in world power politics. (Chapter 17)

Malthusian hypothesis The idea put forward by Thomas Malthus (see Box 6.B) in his *Essay on the Principle of Population* (1798) that population, if unchecked, grows at an exponential rate while food supply increases at an arithmetic rate. (Chapter 6)

Man and the Biosphere Program (MAB) A UN programme to protect ecosystems by setting up protected reserves in which the core area of true wilderness is fully protected, but is surrounded by buffer zones allowing some exploitation of natural resources. Local people are involved in the management of these reserves. (Chapter 19)

mangrove A dense forest of shrubs and trees growing on tidal coastal mudflats and estuaries throughout the tropics. Many plants have aerial roots. (Chapter 4)

map decoding Map interpretation or that part of the map cycle which is concerned in translating the map back into real-world features. (Chapter 21)

map encoding Map construction or that part of the map cycle which is concerned with converting the real world into a map representation. (Chapter 21)

map projections The representation on a plane surface of all or any part of the earth's surface. (Chapter 21)

maritime climate An area where the (generally moist) climate is determined mainly by its proximity to the sea. The sea heats up and cools down more slowly than the land, reducing variations in temperature so the climate is equable. (Chapter 3)

market orientation The tendency for certain industries to locate near their market. Typical is the brewing industry which has to add a bulky material (i.e. water) to the finished product. (Chapter 15)

market-size threshold One of the principles in Christaller's schema for central-place settlements. (Chapter 14)

material index Given by the total weight of localized materials used per product divided by the weight of the product. Most industries have an index greater than 1 and are described as 'weight-losing'. (Chapter 15)

Matthew effect A generalization by Ziman of the gospel text that '... to him that hath will be given', to explain the overemphasis on particular scholars in writing the history of science and, by implication, the history of geography. (Chapter 24)

mean information field (MIF) A weighted matrix used in the *Hägerstrand model* to provide a probability surface for simulating the spread of a process of interest. Originally used for the spread of information, it has since been widened to include such processes as the spread of disease through a population. Values within the matrix may be symmetric or asymmetric. (Chapter 16)

median lines Divisions used in defining maritime boundaries so as to draw lines that are equidistant from the coastlines of two or more adjacent countries. (Chapter 17)

Mediterranean An area that has a climate similar to that of the Mediterranean region, namely warm, wet winters and hot, dry summers. The natural type that can withstand drought. (Chapter 4)

megalopolis A giant urban area comprising several large towns or cities. The name was first given to the corridor of urban sprawl along the northeast seaboard of the United States from Boston to Washington, D.C., by the French geographer Jean Gottmann. (Chapter 8)

Meinig model A schema for the evolution of cultural regions developed by the American historical geographer Donald Meinig (see Box 7.C) consisting of distinctive zones of core, domain, and sphere. (Chapter 7)

mental maps The psychological images of places and regions as held by individuals or groups. Their structure is revealed by experiments in ranking of the desirability of different places and the drawing of remembered regions. The techniques of constructing mental maps was pioneered by the American geographer Peter Gould (see Box 20.C). (Chapter 12)

meridians Imaginary half-circles around the earth which pass through a given location and terminate at the North and South Poles. Their location is given in degrees east or west of a prime meridian. (Chapter 21)

microclimate The climate over a small area that is different from the general climate of the surrounding area because it is influenced by local conditions such as a hill or lake. (Chapter 3)

midden The accumulation of debris and domestic waste products resulting from human use. The long-term disposal of refuse can result in stratified deposits, which are useful for the relative dating of human occupance. (Chapter 5).

Minimata disease The name given to symptoms of mercury poisoning first recognized on a large scale in Japan. (Chapter 9)

model An idealized representation of the real world built in order to demonstrate certain of its properties. In scientific work the term 'model' has, to some extent, all three of the meanings. Models are made necessary by the complexity of reality. They are a prop to our understanding and a source of working hypotheses for research. They convey not the whole truth, but a useful and apparently comprehensible part of it. (Chapter 1)

monsoon The wind systems in the tropics that reverse their direction according to the seasons. When they blow onshore they bring heavy rainfall, also known as the monsoon. They are most prevalent in southern Asia, where they blow from the southwest in summer bringing heavy rainfall, and from the northeast in winter. (Chapter 3)

Monte Carlo model A probabilistic mechanism that uses random numbers to drive a simulation model as used in the Hägerstrand model. (Chapter 16)

morbidity rates Measure the amount of illness in a population. (Chapter 20)

mortality The number of deaths in a population measured over a fixed time period. (Chapter 20)

moving averages A simple means of smoothing time series by adding the values at regular intervals over a period and dividing the result by the number of observations (see Box 3.B). (Chapter 3)

multidimensional scaling (MDSCAL) A multivariate statistical technique that aims to develop spatial structure from numerical data by estimating the differences and similarities between analytical units. (Chapters 8, 21)

multinational corporations A firm with the power to coordinate and control operations in several countries. Control may be exercised without the necessity for legal ownership. (Chapter 19)

multiplier effect A term used in *systems thinking* to describe the process by which changes in one field of human activity (subsystem) sometimes act to promote changes in other fields (subsystems) and in turn act on the original subsystem itself. An instance of *positive feedback*, it is thought by some to be one of the primary mechanisms of societal change. (Chapter 8)

multispectral scanning The recording of different portions of the electromagnetic spectrum by one or more sensors. (Chapter 22)

N

National Center for Geographic Information and Analysis (NGIA) Created in 1988 by the National Science Foundation to establish GIS in the United States. Its main centres are in universities in Santa Barbara, Buffalo, and Maine. (Chapter 23)

nation-states In political geography, a community of people of mainly common descent, history, language, etc. forming a state. (Chapter 17)

natural change The net change in the total population of an area due to the balance of births and deaths but ignoring migration. (Chapter 6)

natural resources All the resources that are produced by the earth's natural processes including mineral deposits, fossil fuels, soil, air, water, plants, and animals, and are used by people for agriculture, industry, and other purposes. (Chapter 10)

nearest-neighbour index A measure of spatial pattern first developed by plant ecologists and used by geographers in settlement studies. Values of the index around 1.0 indicate a random pattern. Dispersed or scattered patterns of settlement have values greater than unity, and clustered patterns of settlement have values less than unity. (Chapter 14)

negative feedback In *systems thinking*, this is a process that acts to counter or 'dampen' the potentially disruptive effects of external inputs; it acts as a stabilizing mechanism (see *homeostasis*). (Chapter 1)

neolithic revolution A term coined by V.G. Childe in 1941 to describe the origin and consequences of farming (i.e. the development of stock raising and agriculture), allowing the widespread development of settled village life. (Chapter 5)

nested sampling A type of sampling design (also called hierarchic, or multistage sampling) which divides the study area into a hierarchy of sampling units that 'nest' within one another. Nested sampling is useful in reducing field costs and in investigating variations at different spatial levels within the same area but it entails higher sampling errors. (Chapter 22)

new geography An approach advocated in the 1960s which argued for an explicitly scientific framework of geographical method and theory, with hypotheses rigorously tested, as the proper basis for explanation rather than simply description. (Chapter 24)

nitrogen cycle The natural circulation of nitrogen through the air, soil, and living organisms. Atmospheric nitrogen is converted by soil bacteria, certain algae, and organisms in the root nodules of leguminous plants into organic nitrogen compounds (nitrates). These are absorbed by green plants and synthesized into more complex compounds. When the plants (or animals feeding on them) die they are broken down again by decomposers so that nitrogen gas is released into the atmosphere. (Chapter 4)

nodal region A region whose defining characteristic is the links between its various parts. It is usually defined by flow data (e.g. commuting). Unlike a uniform region it may overlap with other nodal regions and does not need to be spatially contiguous. (Chapters 12, 13)

nominal specification Use of place-names to record a location on the earth's surface (Chapter 21).

non-Euclidean space The familiar Euclidean space with its 'normal' geometry is only one of a set of conceivable spaces. Some geographical methods (e.g. multidimensional scaling) use geometries of other non-Euclidean spaces to solve their locational problems. (Chapter 21)

non-renewable resource Natural resources that are present in the Earth's makeup in finite amounts and cannot be replaced once reserves are exhausted. (Chapter 10)

normative Describes approaches to geography that establish a norm or standard. Thus, they are largely concerned with establishing some optimum condition (e.g. the 'best' location or the 'best' settlement pattern). (Chapter 15)

North Atlantic Treaty Organization (NATO) A military alliance formed in 1949 under US leadership and consisting of North American and western European states. Its original prime purpose was to counter the perceived threat to western Europe from the Soviet Bloc. (Chapter 17)

north–south divide A term used to denote the huge material differences between the rich parts of the world and the three 'southern continents', Latin America, Africa, and Asia. (Chapters 17, 18)

nuclear power The term used to describe the electricity generated from the heat energy released when atoms, usually uranium-28 or plutonium-29, are split. This is called a nuclear reaction. (Chapter 10)

nutrient Any substance or compound, derived from the environment that contributes to the survival, growth, and reproduction of plants or animals. (Chapter 4)

offshore territorial waters A maritime division of the waters bordering a country extending outward from the coastal baseline. The varying claims are made by different countries are partly standardized by the United Nations. (Chapter 17)

orders of geographic magnitude Geographers have to deal with objects that vary considerably in size from the surface of the earth itself down to the level of an individual. Typical orders of geographic magnitude are defined and illustrated in Box 1.C. (Chapter 1)

Organization for Economic Cooperation and Development (OECD) A 'rich countries' club established in 1961 to foster economic growth and the expansion of world trade. Paris-based, it now has 29 members. (Chapter 17)

Organization of Petroleum Exporting Countries (OPEC) Oil-producing countries concerned with pricing and production issues. Vienna-based, it was set up in 1960 and currently has thirteen members. (Chapter 17)

orthophotos A scale-correct photomap created by geometrically correcting aerial photographs or satellite images. (Chapter 23)

overgrazing The result of grazing too many animals on an area of pasture so that the carrying capacity is reduced. The stripping away of vegetation exposes the soil to erosion by wind and water, with risk of desertification. (Chapter 9)

overlay The process in *GIS* of stacking digital representations of various spatial data on top of each other so that each position in the area covered can be analyzed in terms of these data. (Chapter 23)

ozone An enriched oxygen gas that exists naturally in the upper atmosphere (the ozone layer) but is also formed when certain pollutants such as hydrocarbons and nitrogen oxides react with sunlight. It is a major contributor to smog. Even in small concentrations it causes respiratory problems and hinders plant growth. Ozone in the upper atmosphere is being depleted by the build-up of chlorofluorocarbons (CFCs), and seasonal 'holes' have been detected in the layer over both Antarctica and the Arctic. (Chapter 9)

P

paddyfield A flooded field in which rice is planted and cultivated. The fields are later drained as the rice grains ripen. (Chapter 9)

palynology The analysis of fossil pollen as an aid to reconstruction of past vegetation and climates. (Chapter 3)

pandemic An *epidemic* of infectious disease that affects several continents or the whole world. The major outbreaks of cholera in the nineteenth century and of AIDS late in the twentieth century illustrate this phenomenon. (Chapter 20)

Pangaea The former landmass that was postulated once to comprise all the present continents. (Chapter 2)

paradigm The prevailing pattern of thought in a discipline or subdiscipline. (Chapter 1)

paradigm shifts An approach to the history of science developed by Thomas Kuhn, which holds that science develops from a set of common assumptions (a paradigm). Inconsistencies lead to a period of 'revolutionary' science, which ends with the acceptance of a new paradigm, ushering in a period of 'normal' science. The last half century have seen at least two abrupt paradigm shifts in human geography: the move in the middle 1950s towards a logical positivist view and the move in the middle 1970s towards a more phenomenological approach to the field. (Chapter 24)

parallels Imaginary lines of *latitude* on the earth that are parallel to the equator and pass through all places the same distance to the north or south of it. (Chapter 21)

particulates Tiny particles of solid or liquid matter released into the atmosphere through air pollution. (Chapter 9)

pastoralism A way of life based on tending of herding animals such as sheep, cattle, goats, or camels; often nomadic, it involves moving herds according to the natural availability of pasture and water. (Chapter 9)

perception See *mental maps*.

periodic market cycles A variant on *central-place theory* which suggests that activities that cannot be sustained at a single location may persist by cycling over time through a series of locations. (Chapter 14)

permafrost Ground that remains frozen, typically in the polar regions. A layer of soil at the surface may melt in summer, but the water that is released is unable to drain through the frozen subsoil. (Chapter 2)

pH Measurement of the acidity or alkalinity of air, water or soil. Neutral levels are shown as pH7, less that pH7 are acidic and more than pH7 are alkaline. The scale is logarithmic, hence pH5 is ten times more acid than pH6. (Chapter 4)

phenomenology An existential philosophical school which admits that introspective or intuitive attempts to gain geographic knowledge are valid. (Chapter 24)

photochemical smog Air pollution caused by the reactions between sunlight and particulates that produce toxic and irritating compounds. (Chapter 9)

photogrammetry A series of techniques for measuring position and altitude from aerial photographs or images using a stereoscope or stereoplotter. (Chapter 22)

photomosaic A collection of aerial photographs which are joined to form a contiguous view of an area. (Chapter 22)

physical determinism Holds that the environment largely controls human development. It is associated with the German geographer Friedrich Ratzel (1844–1904) and his American disciple Ellen Churchill Semple (1863–1932). (Chapter 11)

physical geography The branch of *geography* which deals with the physical earth: its atmosphere, lithosphere, hydrosphere, and biosphere. (Chapter 2)

phytoplankton Small, drifting plants, mostly algae, that are found in marine and freshwater ecosystems. (Chapter 4)

pixel A term in *remote sensing* to describe the contraction of picture element; smallest unit of information in a grid cell map or scanner image. (Chapters 22, 23)

place A particular point on the earth's surface imbued with human connotations or values. (Chapter 1)

place-names The identification of locations on maps by a nominal scale which names particular points, lines, or areas. (Chapter 21)

place-name evidence Use of the names of settlements or other landscape features to reconstruct the period of occupation or earlier land use. (Chapters 11, 21)

planet's total stock The totality of materials that make up the potential stock from which resources can be won. (Chapter 10)

planimeter Instrument for directly measuring area from maps. (Chapter 23)

plantation settlements A period in European overseas expansion associated with large-scale estates on which monocultural crops were grown. The period is associated largely with subtropical crops (e.g. cotton, tobacco, sugar) and with imported labour using either slavery or indentured labour. (Chapter 5)

plate tectonics A theory of the formation of the earth's surface introduced in the 1960s based on the interaction of rigid plates in the lithosphere which move slowly on the underlying mantle. The theory throws light on a series of earth-building events from mountain building to volcanic and earthquake activity. (Chapter 2)

platforms Objects on which a remote sensor is mounted, usually an aircraft or satellite. (Chapter 22)

Pleistocene epoch The first epoch of the quaternary period of earth history, its name deriving from the Greek '*pleistos*' = most and '*kainos*' = new. It is marked by great fluctuations in temperature with glacial periods followed by warm interglacial periods. (Chapter 3).

plural societies A term introduced by J.S. Furnivall (1948) to describe Southeast Asian societies in which an alien minority ruled over an indigenous majority. It has since been widened to include cultural diversity of race, language, or religion. (Chapter 7)

podzols Soils formed under cool, moist conditions where the natural vegetation is coniferous forest or heath. They are poor and very heavily leached where the soils are sandy. (Chapter 4)

polar coordinate system A spatial reference system for locating objects in terms of the distance and angle from a single centre. (Chapter 21)

polder An area of level at or below sea level that has been reclaimed from the sea or a lake. It is normally used for agriculture. (Chapter 13)

poles The two ends (North Pole and South Pole) of the axis around which the earth spins or rotates. (Chapter 21)

political geography One of the three major divisions of human geography (along with economic and social geography). Political geography covers the spatial political studies carried out by geographers: (a) the mapping of political phenomena (e.g. electoral geography), (b) studying the impact of geography on political events (e.g. boundary drawing), and (c) studying the impact of politics on geography (e.g. boundary effects on trade). (Chapter 17)

pollen analysis. See *palynology*.

pollution A word that derives from the Latin term to make foul or filthy. In ecological studies it refers to a substance in the environment (air, water, soil) which is in the wrong concentrations. (Chapter 9)

polygons In *GIS* the bounded area within which spatial data are captured. Also referred to as patches or *tiles*. (Chapter 23)

population pyramid A device in demography to show the age and sex distribution of a population. (Chapter 6)

positive feedback A term used in *systems thinking* to describe a response in which changing output conditions in the system stimulate further growth in the input; one of the principal factors in generating system change or morphogenesis (see also *multiplier effect*). (Chapter 1)

positivism A philosophical school that holds that human sensory experiences are the exclusive source of valid geographic information about the world. Its theoretical position is that explanations must be empirically verifiable, that there are universal laws in the structure and transformation of human institutions, and that theories that incorporate individualistic elements are not verifiable. (Chapter 24)

possibilism Argues that the environment offers sets of possibilities but that the choice between them is determined by human beings. The French historian Lucien Febvre (1878–1956) was one of the strongest proponents of this view. (Chapter 11)

post-modernism A movement reacting against modern tendencies which began in architecture, spread into literature, then via the social sciences into human geography. It is sceptical of previous theory. Interpretations are regarded as contingent and partial, with a stress on open interpretations and plural views. (Chapter 24)

predator–prey relations The links between the population of one set of animals (the prey) that are hunted for food by another set (the predator). (Chapters 4, 6)

primary forest Any original, virgin forest that has not been cut and may contain massive tress that can be hundreds or even thousands of years old. Also known as oldgrowth forest. (Chapter 4)

primary sector In economic geography, the proportion of the worforce in the winning of natural resources (e.g. agricuture, forestry, mining, fishing). Historically the proportion of employment in this sector has been declining over the last two centuries. (Chapter 8)

primate patterns A country's leading city which is disproportionately larger and functionally more complex than any other. A city dominating a settlement hierarchy in which there is an absence of middle-sized towns between the layer of small rural towns and the top. (Chapter 14)

prime meridians Meridians used as a baseline for the measurement of the east–west position of places on the earth's surface in terms of longitude. The most commonly used prime meridian crosses Greenwich in London. (Chapter 21)

principal components analysis (PCA) A method of multivariate data analysis which expresses the original variation in terms of a minimum number of principal components. These components are linear combinations of the original, partially correlated variables. It has been widely used in human geography to tackle problems where a single concept (e.g. development) can be measured in several ways. (Chapter 18)

probabilism A compromise position between environmentalism and possibilism that assigns different probabilities to alternative patterns of geographic behaviour in a particular location or environment. (Chapter 11)

probabilistic sampling Sampling method, employing probability theory, designed to draw reliable general conclusions about a site or region, based on small sample areas. Four types of sampling strategies are recognized: (1) *simple random sampling*; (2) *stratified random sampling*; (3) *systematic sampling*; (4) stratified systematic sampling. (Chapters 3, 22)

proven reserves Resources that are known to exist, whose size can be estimated, but have not yet been exploited. (Chapter 10)

purchasing power parity (PPP) Statistics that adjust for cost of living in each country based on a basket of goods and services. These are used to obtain PPP estimates of gross domestic product per head. Normally shown on a scale with the United States set at 100. (Chapter 18)

Q

quadtree A data structure in *GIS* for thematic information in a raster database that seeks to minimize data storage. (Chapter 23)

quantitative revolution A period in human geography extending from the 1950s to the late 1970s associated with the widespread adoption of mathematical models and statistical techniques. (Chapter 24)

quarantine Isolation imposed on people or animals that have arrived from outside an area and have been exposed to, and might spread, infectious diseases. The name derives from the Italian for 'forty' referring to the length of the earliest isolation periods imposed by Venice in the fifteenth century. Quarantine laws usually apply to countries but exceptionally may be levied by states within a larger country (e.g. California within the United States). (Chapters 16, 20)

quaternary sector The service industries that supply manufacturers and businesses with legal and financial advice, education and training, information and quality control, catering, and a range of other support activities. (Chapter 8)

R

race Biological varaitions within the human population assocated with minor and secondary variations in blood groups, skin colour, etc. (Chapter 7)

radar images Images in remote sensing derived from a sensor which directs energy at an object and records the rebounded energy as radio waves. (Chapter 22)

radioactive waste The waste products from nuclear power plants, medicine, research, atomic weapons or other processes that have involved nuclear reactions. (Chapter 9)

radio-carbon dating An absolute dating method that measure the decay of the radioactive isotope of carbon (^{4}C) in organic material (see *half-life*). (Chapter 3)

rainforest Usually known at tropical rainforest and found in the equatorial belt where there is heavy rain and no marked dry season. Growth is very lush and rapid. Rainforests probably contain half of all the world's plants and animal species. (Chapter 4)

raised beaches These are remnants of former coastlines, usually the result of processes such as *isostatic uplift* or *tectonic movements*. (Chapter 3)

random-access map (RAM) Storage of geographic data within a *GIS* system in a form from which a large number of different maps can be plotted. (Chapter 23)

random numbers Used in statistics to refer to numbers drawn without bias or conscious choice to ensure equal chances for each item drawn. An ideal random number sequence is one that can only be described by itself (i.e. is pattern free). (Chapter 16)

random sampling Forms of sampling in which the selection of geographical locations for investigation is drawn by random numbers. (Chapter 22)

rank–size rule In its simplest form, the rank–size rule states that within a set of cities, the population of a given city tends to be equal to the population of the largest city divided by the rank of the given city. The basic formula is often modified by a constant to allow variations from the strict rank–size rule. (Chapter 14)

raster In *GIS* a regular grid of cells covering an area. The cells provide a database containing all mapped, spatial information in the form of regular grid cells. It provides a device for displaying information in the form of *pixels* on a computer screen or VDU. Rasterization is the process of converting an image of lines and polygons from vector representation to a gridded representation. (Chapter 23)

read-only map (ROM) Map output from a *GIS* system in the form of a final graphic or printed form. (Chapter 23)

Recent epoch The second half of the Quaternary geological period. (Chapter 3)

recreational yield The amount of use that can be gained from a recreational resource before overcrowding starts to diminish the yield. Technically, the product of the number of users multiplied by the return for each user. Falling recreational yields pose major mangement problems for controlling the overuse of such areas as national parks. (Chapter 10)

refutationist view Approach that holds that science consists of theories about the empirical world, that its goal is to develop better theories, which is achieved by finding mistakes in existing theories, so that it is crucial that theories be falsifiable (vulnerable to error and open to testing). The approach, developed by Karl Popper, emphasizes the importance of testability as a component of scientific theories. (Chapter 24)

region An area of the earth's surface having definable boundaries or characteristics. (Chapters 1, 12)

regional complementarity An element in the *Ullman model* which refers to supply surplus in one region and a demand deficit in another. (Chapter 13)

regional complex analysis One of the three integral characteristics of modern *geography* (along with *ecological analysis* and *spatial analysis*) that lays stress on the character of regions. (Chapter 24)

regional geography The branch of *geography* concerned with the analysis and synthesis of the earth's surface on a place by place basis. (Chapters 12, 24)

relocation diffusion A type of spatial diffusion in which the centre from which the object being spread also changes its location. (Chapter 16)

remote sensing Observation and measurement of the earth's surface using aircraft and satellites as platforms. The sensors includes both photographic images, thermal images, multispectral scanners, and radar images. (Chapter 22)

renewable resource A natural resource that is normally replenished through natural processes. Examples are oxygen in the air and water in lakes. A renewable resource may become non-renewable if it is used up at a quicker rate that it is replenished; for example, forest trees or fish stocks in the oceans. (Chapter 10)

replacement rates Estimates in *demography* of the extent to which a given population is producing enough offspring to replace itself. To calculate the replacement rate, we have to take into account several factors; e.g. (a) the human species produces about 106 male births for every 100 female births; (b) the proportion of female offspring who may be expected to die before they themselves reach the reproductive age. Taking such factors into account, a family size greater than two is needed for replacement reproduction. The actual figure will vary slightly from one country to another, but 2.3 is reasonably representative. (Chapter 6)

reproduction rates The number of girls born to females in the childbearing age groups (roughly 15 to 45 years) in a population. (Chapter 6)

research design Systematic planning of geographic research, usually including (1) the formulation of a strategy to resolve a particular question; (2) the collection and recording of the evidence; (3) the processing and analysis of these data and their interpretation; and (4) the publication of results. (Chapter 24)

reserves That proportion of natural resources (especially stock resources) that is capable of being used with existing technology. (Chapter 10)

resistance curves In diffusion studies, the S-shaped curve which measures the resistance to the take-up of innovations over time. (Chapter 16)

resolution The ability of a *remote-sensing* system to distinguish signals that are close to each other, in time, space, or wavelength. (Chapter 22)

resource conservation. See *conservation.*

resource orientation The tendency for certain industries to locate near their source of localized raw materials. Typical are the mineral-processing industries where a great deal of waste material in the ore can be removed before the refined material is shipped. (Chapter 15)

resources See natural resources. (Chapters 10, 16)

return periods The average time between geophysical events such as flooding at a particular level. The information is also given as the floods that can be expected in a given time period (say, a hundred years). The inverse of the return period is the probability of an event occurring in any individual year. (Chapter 3)

rim settlements A stage in European overseas settlement with the occupation of the maritime fringe of the continents. These are sometimes described as limpet settlements. (Chapter 5)

ring-control stategies A technique for containing epidemic outbreaks by imposing a cordon sanitaire around the source area of the outbreak. (Chapter 16)

Rossby waves Macroscale wave motions in the earth's atmosphere recognized by the Swedish meteorologist,

Carl-Gustav Rossby. They take the form of vast meanders of airflow around a hemisphere and are most clearly identified at higher levels of the atmosphere. Rossby wave theory can predict both the stationarity and drift in the waves and throws important light on suddden switches in historical climatic patterns in the northern hemisphere. (Chapter 3)

Rostow–Taaffe model Spatial version of the economist Rostow's take-off model by the geographer Taaffe. It shows economic development through several stages, each with a distinctive spatial structure. (Chapter 18)

Royal Geographical Society (RGS) Britain's oldest geographical society founded in 1830 for the 'advancement of geographical science'. Its history and present role are described in Box 24.A. (Chapter 24)

rubber-sheeting Technique of spatial transformation used in *GIS* to force the distorted positions of data points from aerial photographs into their correct spatial position on a grid. (Chapter 23)

runoff Water produced by rainfall or melting snow that flows across the land surface into streams and rivers. Delayed runoff is water that soaks into the ground and later emerges on the surface as springs. (Chapter 11)

S

Sahel The region at the southern margin of the Sahara desert in Africa which has suffered a number of prolonged droughts in the last half-century. Some geographers confine the term to the area west of Chad; others include the whole margin east to incude Sudan and Ethiopia. (Chapter 3)

salinization The accumulation of soluble salts near or at the surface of the soil. This occurs naturally in arid and semi-arid areas through evaporation, but can also result from the incorrect application of water for irrigation. Eventually the land becomes worthless for cultivation as plants cannot cope with the high levels of salt. (Chapter 9)

saltwater intrusion The movement of saltwater into freshwater aquifers in coastal and inland areas that occurs when groundwater is withdrawn faster than it is recharged by precipitation. (Chapter 9)

saturation The level at which the population of an area exactly equals its carrying capacity. (Chapter 6)

Sauer hypothesis A hypothesis developed by C.O. Sauer in his *Agricultural Origins and Dispersals* (see Box 5.B) that human agricultural hearths lay outside the great river valleys of the fertile crescent that were conventionally regarded as the likely source areas. (Chapter 5)

savannah A habitat of open grassland with scattered trees in tropical areas. Also known as tropical grassland, it covers areas between tropical rainforest and hot deserts. There is a marked dry season each year and too little rain to support large areas of forest. (Chapter 4)

scale The relation between the size of an object on a map and its size in the real world. (Chapter 1)

scanners A hardware device in GIS to capture map data in numerical form for further analysis. They are available in a wide range of formats from low-resolution portables to large flatbed scanners. (Chapter 23)

scientific determinism A variant of physical determinism in which the argument proceeds from the statistical analysis of sets of data rather than from individual case studies. Yale geographer Ellsworth Huntington (1876–1947) was the leader of this school of thought. (Chapter 11)

sea-level changes Coastlines around the world provide abundant evidence that sea level has not always occupied its present position. Currently there is great interest in the impact of *global warming* on raising sea levels around the world. (Chapters 2, 19)

secondary sector In economic geography, the proportion of the workforce employed in the processing of primary products into manufactured goods. Also termed the industrial sector. (Chapter 8)

sector models In economic geography, the sequence of changes in the proportion of primary, secondary, tertiary, and quaternary employment in the workforce. (Chapter 8)

self-organization The product of a theory derived from thermodynamics which demonstrates that order can arise spontaneously when systems are pushed far from an equilibrium state. The emergence of new structure arises at bifurcation points, or thresholds of instability (cf. *catastrophe* theory). (Chapter 24)

sensors Instruments used to detect the electromagnetic energy associated with a particular object on the earth's surface. (Chapter 22)

sequent waves In settlement geography, an orderly series of phases by which a region is occupied. The series of frontiers (e.g. hunting, cattle ranching, wheat farming) by which the Great Plains of North America was settled by European populations is a classic case. (Chapter 5)

seres The orderly sequence of change in the vegetation of an area over time as it passes through transition stages (seres) toward an equilibrium. (Chapter 4)

settlement hierarchy A classification of settlements by their size and range of services. A hamlet or village is a low-order settlement and a city a high-order settlement. (Chapter 14)

shifting cultivation A method of farming prevalent to tropical areas in which a piece of land is cleared and cultivated until its fertility is diminished. The land is then left to restore itself naturally. (Chapter 9)

sideways-looking airborne radar (SLAR) A *remote-sensing* technique that involves the recording in radar images of the return pulses of electromagnetic radiation sent out from aircraft. (Chapter 22)

Sierra Club An American environmental organization founded by John Muir (see Box 10.C) in 1892. One of the Club's first successes was the preservation of the Yosemite Valley in the Californian Sierra Nevada and its pressure led to Congress passing the Wilderness Act. Today it remains a major environmental pressure group. (Chapter 10)

sieve mapping Use of series of overlapping maps used to locate sub-areas which meet a specified set of specific requirements. (Chapter 23)

Silicon Valley The Santa Clara valley in California where many computer and other high-tech industries using microchip technology have grown up close to a concentration of research-based institutions. (Chapter 19)

simple random sampling A type of *probabilistic sampling* where the areas to be sampled are chosen using a table to random numbers. Drawbacks include (1) defining the site's boundaries beforehand; (2) the nature of random tables results in some areas being allotted clusters of sample squares, while others remain untouched. (Chapter 22)

simulation model A type of *model* that reproduce complex, real-world situations usually in terms of a computer program which reproduces changes through time. Simulation is a useful heuristic device, and can be of considerable help in the development of explanation. Literally, simulation is the art or science of 'pretending'. (Chapter 16)

slash-and-burn farming A method of farming in tropical areas where the vegetation cover is cut and burned to fertilize the land before crops are planted. (Chapter 9)

smog Originally referring to a thick mixture of smoke and fog, the term is now used to describe the haze in many cities caused by a variety of pollutants. Photochemical smog is a complex mixture of air pollutants produced in the lower atmosphere by the reaction of hydrocarbons and nitrogen oxides under the influence of sunlight. Ozone is one of the harmful components. (Chapter 9)

soil erosion The loss of the top layers of soil from an area as a result of erosion by wind or water. In the case of water it may be caused by rill erosion, sheet erosion, or *gullying*. (Chapter 9)

soil horizons Soil is the unconsolidated, weathered layer of material that lies at and immediately below the earth's surface and is able to support plant growth. The A-horizon is the upper layer of a soil, often rich in organic material and subject to leaching as moisture seeps downward. The B-horizon is the next layer where some of the chemical elements (notably iron) leached from the top soil accumulate. Finally the lowest C-horizon is made up of decomposed rock from which soil has not yet begun to form. (Chapter 2)

solar energy The radiant energy produced by the sun, which powers all the earth's natural processes. It can be captured and used to provide domestic heating or converted to produce electrical energy. (Chapter 2)

solar radiation budget In physical geography, an accounting system that measures what happens to radiant energy from the sun when it is intercepted by the earth. (Chapter 2)

solstice Times (in June in the northern hemisphere and December in the southern) when the sun is furthest from the equator. (Chapter 3)

source pricing In this form of *spatial pricing,* the price is established at the production point and the customer pays the transfer costs of moving the product. This system is also termed f.o.b. (free-on-board) pricing. Customers may be charged freight rates on the basis of the actual distance covered, blanket zones, or a uniform postage stamp rate in which the charge levied is unrelated to the distance involved. (Chapter 13)

space In *geography* space is a multipurpose term that refers to a continuous unlimited area or expanse that may extend from local spaces up to the size of the global surface itself. Space may be *isotropic* or *anisotropic* and may be either a reference grid (such as latitude and longitude) or a forced field as revealed by multidimensional mapping. In *remote sensing*, space is also used to describe the physical universe outside the earth's atmosphere as in the NASA space programme. (Chapter 1)

space–cost curves In industrial location, the analysis of optimal sites by plotting both transport and non-transport costs. (Chapter 15)

space–time geography see *time–space geography*

spatial analysis One of the three integral characteristics of modern *geography* (along with *ecological analysis* and *regional complex analysis*) which lays stress on spatial structures. (Chapter 24)

spatial autocorrelation tests A method of interpreting distribution maps by using *autocorrelation* to estimate the statistical significance of clustering and spacing. (Chapter 21)

spatial interaction models A series of ideas that provide understanding of the interdependence between geographical areas. Thus *gravity models* and *Wilson models* throw light on transport flows, while the *Ullman model* is concerned with migration and interregional exchanges. (Chapter 13)

spatial language A broad term introduced by M.F. Dacey to describe the translation of geographic features into map terms using the rules of spatial grammar and syntax. (Chapter 21)

spatial pricing In *economic geography* three alternative pricing strategies are commonly used to recover the costs involved in transporting a product: source pricing, uniform-delivered pricing, and basing point pricing. (Chapter 13)

special economic zone One of the four areas in the People's Republic of China where there is investment to encourage the growth of foreign trade. Joint ventures with foreign companies area feature of commercial developments there. (Chapter 18)

spectral signature In remote sensing, a distinctive set of values on the electromagnetic spectrum which allows the identity of a surface feature to be identified. (Chapter 22)

sphere Thirdmost region within Meinig's four-layer culture region. (Chapter 7)

sphere of influence The area over which a town or city has influence. People living in the area look to that settlement to provide employment, goods and services. (Chapter 13)

spherical coordinates Systems of *latitude* and *longitude*. (Chapter 21)

SPOT French satellites (Systéme Probatoire d'Observation de la Terre) from 1986 providing high-resolution data on the earth's surface. (Chapter 22)

spread effects In *economic geography* the spatial transmission of capital, resources, and skills from the core area to the periphery as proposed in the Myrdal model. (Chapter 18)

Standard Metropolitan Statistical Area (SMSA) In the United States, the name given by the Census Bureau for statistical purposes to an area containing at least one city or urban area with a population of more than 50,000 within a total population of 100,000. Consolidated metropolitan statistical areas are SMSAs with a population of at least 1 million. (Chapter 14)

standardized rates The vital rates in *demography* which are adjusted for such factors as the age and sex structure of a population. (Chapter 6)

state A term used to describe a social formation defined by distinct territorial boundedness, and characterized by strong central government in which the operation of political power is sanctioned by legitimate force. In cultural evolutionist models, it ranks second only to the empire as the most complex societal development stage. (Chapter 17)

stereoscope In *remote sensing*, an instrument that allows height differences to be determined from overlapping aerial photographs. (Chapter 22)

stock resources Non-renewable resources of fossil fuels and minerals used by industry to support modern life including coal, oil, and iron ore. (Chapter 10)

stop-and-go determinism Holds that people determine the rate but not the direction of an area's development. The term was coined by Australian geographer Griffith Taylor (1880–1963). (Chapter 11)

strategic withdrawal A type of regional planning associated with an orderly withdrawal of population and services from an area whose main industry (e.g. coal mining) is in severe decline. (Chapter 18)

stratified random sampling A form of *probabilistic sampling* in which the region or site is divided into natural zones or strata such as cultivated land and forest; units are then chosen by a random number procedure so as to give each zone a number of square proportional to its area, thus overcoming the inherent bias in *simple random sampling*. Stratification can be used when major divisions within the area are already known to the investigator. (Chapter 22)

subsequent boundaries In *political geography* the term for national boundaries that are drawn after an area has been settled and developed. Compare with *antecedent boundaries*. (Chapter 17)

subtropical The climatic zone between the tropics and temperate zones. There are marked seasonal changes of temperature but it is never very cold. (Chapter 2)

superimposed boundaries In political geography, those international boundaries that are established *after* an area has been settled. (Chapter 17)

survivorship curve The proportion of a given population surviving to a particular age. (Chapter 6)

sustainability The concept of using the earth's natural resources to improve people's lives without diminishing the ability of the earth to support life today and in the future. (Chapter 19)

sustained-yield forestry A planned cycle of planting and harvesting that aims to produce a constant flow of timbers from a forest over time. Originally developed for the long-term management of German hardwood forests, the concept has recently been extended to tropical forests. (Chapter 10)

Swedish Society for Anthropology and Geography (SSAG) Sweden's oldest geographical society founded in the 1870s. It publishes *Ymer* and *Geografiska Annaler*. (Chapter 24)

swidden A temporary agricultural plot produced by cutting back and burning the vegetation. (Chapter 9)

synagraphic mapping program (SYMAP) The original grid-cell mapping program developed by Fisher in the Harvard Computer Graphics Unit. It was the pioneer computer cartography method from which modern methods have been derived. (Chapter 23)

synchronic Refers to phenomena considered at a single point in time; i.e. an approach that is not primarily concerned with change (cf. diachronic). (Chapter 1)

synoptic images Those giving a general view of a part of the earth's surface, usually from high-altitude satellites. (Chapter 22)

system A system is a group of things or parts that work together through a regular set of relations. Thus we can regard the beach as a system in which its various parts – shingle, sand, mudbanks, and the like – are linked together through a set of relations involving the energy of waves, tides, and winds. Geographers are particularly interested in systems that link together people and environment. (Chapter 1)

systematic sampling Employs a grid of equally spaced locations to define the sampling points. The origin of the grid is decided by the random location of the first sampling point. The grid also can be randomly oriented. Systematic location gives an even spatial coverage for mapping purposes. This method of regular spacing runs the risk of missing (or hitting) every single example if the distribution itself is regularly spaced. (Chapter 22)

systems thinking A method of formal analysis in which the object of study is viewed as comprising distinct analytical subunits. Thus in archaeology, it comprises a form of explanation in which a society or culture is seen through the interaction and interdependence of its component parts; these are referred to as system parameters, and may include such things as population size, settlement pattern, crop production, technology, etc. (Chapters 1, 2)

T

taiga Russian name given to the coniferous forest and peatland belt that stretches around the world in the northern hemisphere, south of the tundra and north of the deciduous forests and grasslands. (Chapter 4)

tectonic movements Displacements in the plates that make up the earth's crust, often responsible for the occurrence of *raised beaches*. (Chapter 2)

temperature inversion The situation that occurs in particular climatic conditions when a layer of cool, dense air becomes trapped under a layer of warmer, less dense air. Air pollutants in the surface layer are unable to disperse and can build up to levels harmful to living things. (Chapter 9)

terrace A level area of land cut out of a slope allowing crops such as wet rice to be grown. (Chapter 5)

tertiary sector A subdivision of the service industries which are sometimes divided into tertiary (services supplied directly to consumers) and quaternary (services such as auditing or marketing supplied to producers). (Chapter 8)

tessellation In *GIS* the process of dividing an area into smaller, contiguous tiles or *polygons* with no gaps in between them. (Chapter 23)

thematic map Map displaying selected kinds of information relating to specific themes, such as soil, land use, population density, suitability for arable crops, and so on. Many thematic maps are also choropleth maps, but when the attribute is modelled by a continuous field, representation by isolines or colour scales is more appropriate. (Chapter 21)

thermal infrared Records the thermal energy emitted by objects of different temperatures on the earth's surface. (Chapter 22)

thermal pollution The discharge of heat into waterways causing reduced oxygen levels and disrupting natural biological cycles. (Chapter 9)

thermography A non-photographic technique that uses thermal or heat sensors in aircraft to record the temperature of the soil surface. Variations in soil temperature can be the result of the presence of buried structures. (Chapter 22)

Thiessen polygon A *tessellation* in which polygons are created by drawing straight lines between pairs of neighbouring sites, then at the midpoint along each of these lines, a second series of lines are drawn at right angles to the first. Linking the second series of lines creates the Thiessen polygons. Also known as Dirichlet tessellation or Voronoi polygons, they are important in settlement and GIS studies. (Chapter 23)

Third World Describes the countries of Latin America, Africa, Asia, and parts of the Middle East that are not aligned with western countries, or the Soviet Union, and that have no advanced industralization. (Chapter 18)

Thünen model A locational model for explaining spatial variations in agricultural land use. It was put forward by the German land economist Johann von Thünen (see Box 15.C) in his *Isolated State* (1826). (Chapter 15)

tidal energy An alternative energy source in which tidal movement (the fall between high and low tide) is used to turn a turbine to generate electricity. (Chapter 10)

tile A part of the database in a GIS representing a contiguous part of the earth's surface. By splitting a study area into tiles, considerable savings in access times and improvements in system performance can be achieved. (Chapter 23)

time–space geography A branch of human *geography* developed by Torsten Hägerstrand and colleagues in Sweden which lays stress on the role of time constraints on the shaping of human spatial activity. The different kinds of constraints are outlined in Box 1.B. (Chapter 1)

time series Data gathered over a period of time, usually at regular time intervals. (Chapter 3)

T-O map A class of medieval maps in which the world is set within a circle (the letter 'O') and the Black Sea, Mediterranean Sea, and Red Sea form a distinctive shape (the letter 'T'). The Hereford world map is a classic example of a T-O map. (Chapter 21)

topology The mathematical study of continuities of space and spatial properties, such as connectivity, that are unaffected by continuous distortion. Important applications have been made in geography both to transport networks and in GIS data structures. (Chapter 13)

township and range A rectilinear system for surveying the central and western United States based on square one mile by one mile sections. The pattern has left a legacy in settlement patterns and field divisions in rural America today. (Chapter 21)

toxic waste Any form of hazardous waste capable of causing death, serious injury, or illness. (Chapter 9)

trade winds The tropical easterly winds which blow from either side of the equator towards the intertropical convergence zone. Direction tends to be northeasterly in the northern hemisphere and southeasterly in the southern hemisphere but such regularities are modified by monsoonal effects around continental areas. (Chapter 2)

transferability Term used in the *Ullman model* to describe one of the bases of spatial interaction. Covers both (a) transport costs which reflect both the transport system and the commodity being moved and (b) the ability of the commodity to bear these costs. (Chapter 13)

transhumance A type of pastoralism in which the livestock is moved seasonally between one area of pasture and another. The farmers usually move as well to stay with their animals. (Chapter 9)

transpiration The transference of water, drawn up from the soil through the roots and stems of living plants, to the atmosphere as water vapour through pores in their leaves. (Chapter 4)

transport flows In spatial interaction studies, flows of people or freight as distinct from *communication flows*. (Chapter 13)

treeline The limit of tree growth beyond which the growing season each year is not long enough for trees to grow. Treelines may show both latitudinal and altitudinal limits. Their historical movements may be a useful indicator of climatic and other environmental change. (Chapter 4)

trend surface maps A mapping technique whose aim is to highlight the main features of a geographic distribution by smoothing over some of the local irregularities. In this way, important trends can be isolated from the background 'noise' more clearly. Trend surface maps are like filters which cut out 'short-wave' irregularities but allow 'long-wave' regularities to pass through. (Chapter 21)

triangular irregular network (TIN) A vector data structure for representing geographical information that is modelled as a continuous field (usually elevation) which uses tessellated triangles of irregular shape. (Chapter 23)

trophic level The level at which living things are positioned in a food chain. Hence green plants occupying the first level, plant-eaters the second, animal eaters the third, and decomposers the fourth. (Chapter 4)

tropical cyclone A collective term for intense storms that are termed hurricanes (in the Caribbean), typhoons (northwest Pacific), and cyclones (in the Indian Ocean). They originate in the tropical oceans where sea-surface temperatures are 27 °C or warmer (July to October in the northern hemisphere, January to March in the southern) and follow distinctive comma-like paths from lower to higher latitudes. (Chapter 3)

tropics The area lying between the Tropic of Cancer and the Tropic of Capricorn. They mark the latitude farthest from the equator where the sun is directly overhead at midday in midsummer. (Chapter 2)

tropopause The upper limit of the earth's troposphere extending up to 16 km above the equator and 9 km at the poles. (Chapter 2)

tundra The level, treeless land lying in the very cold northern regions of Europe, Asia, and North America, where winters are long and cold and the ground beneath the surface is permanently frozen. (Chapter 4)

turning points Positions in a *time series* when economic activity in a region or country changes direction. When activity is falling during a recession and then swings upward again, the precise time at which it begins to swing is the turning point. (Chapter 12)

U

ubiquitous materials In location theory, those which can be found in any location and therefore play a minor role in locational decision making. (Chapter 15)

Ullman model A broad-based spatial interaction model developed by Edward Ullman (1980) and applied to a very wide range of phenomena from migration to financial flows. The three bases for interaction were seen as regional *complementarity*, *intervening opportunity*, and transferability. (Chapter 13)

uniform delivered pricing In this form of *spatial pricing* the price is the same for all customers regardless of their location. The producers pay all the transport costs involved in shipping their product but recover this by taking the average transport costs into consideration when deciding on the price at which they offer the product. This system is also termed c.i.f. (cost–insurance–freight) pricing. (Chapter 13)

uniformitarianism The principle that the stratification of rocks is due to processes still going on in seas, rivers, and lakes; i.e. that geologically ancient conditions were in essence similar to or 'uniform with' those of our own time. (Chapter 2)

uniform region A region with homogeneous characteristics which are uniform over space. (Chapter 12)

United Nations Environment Programme (UNEP) Founded after the Stockholm Conference in 1972 (see Box 11.B) as a UN agency to coordinate international measures for environmental protection. Its main centre is at Nairobi, Kenya. (Chapter 19)

urban implosion An inward collapse of the relative positions of the world's major cities caused by the reductions in time–space costs. A by-product of such a collapse is to make smaller or poorly connected centres relatively less accessible. (Chapter 8)

urbanization curves Trend lines that measure the historical transformation of a population from rural to urban status through the process of city formation and growth. (Chapter 8)

utility mapping A class of *GIS* applications for managing information about public infrastructure such as water pipes, sewerage, telephone, electricity, and gas networks. (Chapter 23)

V

Vance mercantile model An alternative to static *central-place theory* proposed by J.E. Vance which lays stress on the historical evolution of settlements, particularly as part of Euopean overseas settlement. (Chapter 14)

vector In *GIS* the representation of spatial data by points, lines, and polygons. Vectorization is the conversion of point, line, and area data from a grid to a vector representation. (Chapter 23).

vertex A term used in *graph theory* to describe a node linked by *edges* to other nodes within a network. It is used in determining measures of connectivity. (Chapter 13)

virgin soil epidemics Epidemics that arise from the initial introduction of a disease into an area with a population with no acquired resistance from earlier exposure. Characteristically such epidemics have an unusually high morbidity and mortality. (Chapter 20)

W

Wallace's line A boundary in *biogeography* put forward by Alfred Russel Wallace (see Box 4.C) in 1858. It runs through the Malay archipelago and separates the fauna and flora of Australia and Asia. (Chapter 4)

water table The uppermost level of underground rock that is permanently saturated with groundwater. (Chapter 2)

watershed An imaginary line dividing the headwaters of two separate river systems, also known as a water-parting or divide. The term watershed may also be used to describe the whole area, or basin, drained by a river and its tributaries. (Chapters 2, 4)

Weber model An idealized model of industrial location introduced by Weber (1909) based on reducing transport costs and taking advantage of labour cost differences and agglomeration economies. (Chapter 15)

weight-loss location index A general term for models such as those of *Weber* and von *Thünen* in which locations of industrial or agricultural production are influenced by the need to reduce transport costs on bulky commodities. (Chapter 15)

westerlies Belts of wind with an average position between 35° and 60° of latitude which are major features of the general circulation of the earth's atmosphere. In the northern hemisphere the movement is dominantly northwesterly and in the southern hemisphere southwesterly. Both belts move poleward in summer and equatorward in winter. (Chapter 2)

wetland Any area of low-lying land where the water table is at or near the surface for most of the year, resulting in a flooded or waterlogged landscape. (Chapter 19)

Wilson model A family of *spatial interaction models* developed by A.G. Wilson (see Box 13.B) using *entropy* maximizing methods. It provides a powerful alternative to gravity-type models and has been widely used in regional and local planning applications. (Chapter 13).

world cities A term coined by Patrick Geddes (1915) for 'certain great cities in which a quite disproportionate part of the world's most important business is transacted'. Eight such centres were identified by Hall (1984) in his *World Cities*. (Chapter 19)

world financial centres Major cities that have been identified by a series of stock-market and money-handling as pivots of world financial transactions. New York, London, and Tokyo forms the leading global centres; Chicago, Frankfurt, and Hong Kong are examples from the second tier. (Chapter 19)

World Health Organization (WHO) An agency of the United Nations, established in 1948, responsible for coordinating international health activities. Its main headquarters are in Geneva, Switzerland, but worldwide it has six regional offices. (Chapter 20)

world system A term coined by the historian Wallerstein to designate an economic unit, articulated by trade networks extending far beyond the boundaries of individual political units (nation states), and linking them together in a larger functioning unit. (Chapter 17)

World-Wide Fund for Nature (WWF) A private international organization originally founded in the 1960s primarily concerned with endangered animal and plant species. It merged with the Conservation Foundation in 1990 and now has five million supporters worldwide. (Chapter 19)

XTENT modelling A method of generating settlement hierarchy, that overcomes the limitations of both *central-place theory* and *Thiessen polygons*; it assigns territories to centres based on their scale, assuming that the size of each centre is directly proportional to its area of influence. Hypothetical political maps may thus be constructed from survey data. (Chapter 14)

years of potential life lost (YPLL) In the study of mortality, a measure that weights each death by the number of years which might otherwise have been lived under some 'normal' mortality expectation (see Box 20.A). (Chapter 20)

Z

zero economic growth (ZEG) In conservation, a Utopian condition in which a stable population (see ZPG) is maintained without the need for irrecoverable depletion of the earth's natural resources. (Chapter 18)

zero population growth (ZEP) The ending of population growth when birth and death rates are equal. This would require an average number of around 2.3 children per family. (Chapter 6)

APPENDIX B

Geography on the World-Wide Web

The World-Wide Web (WWW), usually shortened to 'the web', now provides a major new source of information for geographers. Originally invented in 1989 by scientists working at CERN (an international physics laboratory based in Geneva, Switzerland), the web allowed rapid and standardized communication between computers around the world. Within a decade, the World-Wide Web had tens of millions of active users and it continues to grow exponentially. A web site is a set of pages maintained by a university, government agency, company or other organization that stores information (e.g. text, pictures, videos) that you can view. Pages on the web are connected so that you can easily jump from one page to another even though the page may be on a different computer or in another country.

To find pages of interest you make use of a 'search engine'. These are enormous and ever-growing lists of web pages organized into useful search categories and subcategories. Common search engines include Alta Vista, Google, Gopher, Lycos, Webcrawler, and Yahoo. As you explore the web, so you can keep a list of the most useful web pages you find. Netscape calls these 'bookmarks', Mosaic calls them 'hotlists', and Internet Explorer calls them 'favorites'.

The list that follows gives just a sample of the major sources that may be of use to readers of this book. They are arranged alphabetically under general headings. By the time you read this page, the web will have grown and diversified, so do remember that this gives only a fragment of the sites available to you. Readers should note that web sites can be renamed and discontinued, and it may be useful to check out the item using keywords and a good browser. If you fail to raise a site using the reference given here, repeat the search using the relevant key words. An updated list is given on the web site linked to this book (**www.booksites.net/haggett**).

Aerial Photographs Most countries will have agencies from where aerial photographs can be obtained. For the United States the *US Geological Survey (USGS)* is the main source (**www.usgs.gov/photos.html**) with the EROS Data Center near Sioux Falls, South Dakota, housing more than six million photographs, some dating back to the 1940s. Other government departments also have extensive holdings: for example, the *US Department of Agriculture (USDA)* has extensive photographs of the rural landscapes (**www.fsa.usda.gov/pas/airfoto.htm**). For the United Kingdom see the Cambridge University *Department of Aerial Photography* (**www.aerial.cam.ac.uk**) or the commercial company *Aerofilms* (**www.aerofilms.com**). Both indicate the range of material available and show how you can order photos. Australia is typical of countries which maintain a central government archive of aerial photographs (**www.auslig.gov.au/photos**). See also 'satellite photos' in this list.

Biosphere The state of forests and land cover around the world is monitored by the *World Resources Institute* including deforestation information (**www.igc.org**) and by the *UN Food and Agriculture Organization (FAO)* at Rome (**www.fao.org**). The results of the UN Rio de Janeiro conference on biodiversity and follow-up action is given in **www.biodiv.org.** The *Rainforest Workshop* site helps students and teachers learn more about the rainforest (**mh.osd.wednet.edu**) as does the Woods Hole Tropical Research Center with updates on the destruction of the Amazonian rainforest (**www.whrc.org**). Michigan State University maintains a *Tropical Rain Forest Information Center (TRFIC)* with links to NASA satellite data (**www.brsi.msu.edu/trfic**). Check also the pages of the UN *Man and Biosphere Programme (MAB)* which describes regional networks of research on biosphere change (**www.unesco.edu/mab**).

Cartography In addition to the sites giving collections of maps (see below) the *United States Geological Survey* has a site for teachers and students on 'What do maps do?' (**info.er.usgs.gov/eduction/teacher/what-do-maps-show**). This includes map packets, which you can print out for use with lessons. The *Mapmaker* site shows how maps are made and explains some of the terms used in mapmaking (**loki.ur.utk.edu/ut2kids/maps**). The *National Geographic* has a site devoted to map projections and the quest to portray a spherical planet on a flat paper surface (**www.nationalgeographic.com/features/2000/exploration/projections**). It includes sections on how projections are tailored to different parts of the globe and the way distortions are minimized.

Climate and Weather In the United States the *National Oceanic and Atmospheric Administration (NOAA)* maintains a *National Climatic Data Center* which provides a major source for weather and climate data (**www.ncdc.noaa.gov**). Climatic change in general and global warming in particular are covered by the *Intergovernmental Panel on Climate Change (IPCC)* (**www.ipcc.ch/**) and the *US Environmental Protection Agency* (**www.epa.gov/globalwarming/**). Up-to-date weather reports are given in the University of Michigan's *WeatherNet* which also gives cross-links to 300 other meteorology sites on the web (**cirrus.sprl.umich.edu/wxnet**). The Department of Atmospheric Sciences at the University of Illinois at Urbana–Champaign provides an overview of what El Niño is, where it comes from, and what it does (**ww2010.atmos.uiuc.edu/(Gh)/guides/mtr/eln/home.rxml**). El Niño is also well covered in the introduction at NASA's Goddard Space Flight Center (**nsipp.gsfc.nasa.gov/primer/**). The home page for the *Weather Channel* provides a wide range of teacher and student material on weather, climate, and its study from understanding weather charts to storm chasing or backyard observations (**www.weather.com/twc/homepage.twc**). The *American Meteorological Society* provides an introductory college course in weather studies (**www.ametsoc.org/amsedu/online**). In the UK the *Meteorological Office* has a useful site with guides to developing weather situations (**www.meteo.govt.uk**).

Countries and Regions Most countries maintain government web sites with links to a wide range of sources describing the economy of the state. Thus for the United States the *Statistical Abstract* provides a goldmine of regularly updated information on that country (**www.census.gov/statab**) with state and metropolitan area profiles. The UK's *Central Office of Information* (**www.coi.gov.uk/coi**) is an important source. In some countries, such information has a strong spatial and map content. For Canada, the *National Atlas Information Service* provides a guide to Canadian geography (**www-nais.ccm.emr.ca**). For Brazil, the *Instituto Brasileiro de Geografia e Estatistica (IBGE)* provides an abundance of links describing the physical and human geography of that country (**www.ibge.gov.br**). In Sweden the *National Atlas of Sweden* maintains sites describing the geography of that country (**www.sna.se**). Sites with references to a specific geographic region are less common but a good example is the site on Antarctica maintained by New Zealand (**icair.iac.org.nz**).

Development Data The main source for global data is that provided by the *United Nations Statistics Division* which publishes annual abstracts of statistics for many aspects of its member countries (**www.un.org/Depts/unsd**). Information on economic development around the world is given by the *World Bank* (the UN International Bank for Reconstruction and Redevelopment) (**www.worldbank.org**) and by specialist agencies such as the *Food and Agriculture Organization (FAO)* (**www.fao.org**). The UK weekly journal *The Economist* provides both regular updates of economic activity around the world with periodic features on individual countries (**www.economist.com**).

Ecological and Environmental Themes The *United Nations Environment Programme (UNEP)* at Geneva maintains and provides free environmental data sets which are cross-referenced to a GIS (**www.grid.unep.ch**). This provides environmental assessments and early-warning systems. All the leading ecological organizations have web sites of interest. In the United States the *Sierra Club* (**www.sierraclub.org**) has a splendid site describing its work; students may find the Planet Newsletter and John Muir exhibit (see Box 10.C) of interest. One of the most active groups at the international level is *Greenpeace* (**www.greenpeace.org**).

Field Studies Geographers are great travellers and the web contains many thousands of pages with travel information. The *WWW Travel Guide* provides travel information for destinations around the world (**www.infohub.com**). The *Travel Advisory* is a site that provides travel information for countries around the world but also lets you know which countries to avoid due to disease, war, and natural disasters (**www.stolaf.edu/network/travel-advisories.html**).

Geographers and Geography Departments Many leading geographers have their own web sites which tell you about their work, give a brief CV, and give the e-mail addresses where they can be located. If you fail to turn these up on a name search, try searching the university where they are based and look at the home page for the Department of Geography there. Examples of three excellent web sites are those for the Departments of Geography at *Pennsylvania State University* in the USA (**www.geog.psu.edu**), at *Edinburgh University* in Scotland (**www.geo.ed.ac.uk**), and at *Stockholm University* in Sweden (**www.geo.su.se**). But there are now many hundreds of Department sites around the world to choose from and the quality of presentation is increasing rapidly. Interviews with leading geographers around the world on film have been made by Wes and Nancy Dow from their New England base. See their web site (**oz.plymouth.edu/~gof**) for a list of geographers on film and an indication of how you can view these.

Geographic Information Systems (GIS) All the major GIS software companies maintain sites describing their products. The leading company is the *Environmental Sciences Research Institute (ESRI)* of Redlands, California, which has an outstanding example (**www.esri.com**). See also the site maintained at Clark University for its *IDRISI* GIS system (**www.clarklabs.org**) and that of the *Intergraph Corporation* for its GIS software (**www.intergraph.com/gis**). You can also find out about the *National Center for Geographic Information and Analysis (NCGIA)* (**www.ncgia.ucsb.edu**). A Canadian site allows you to use your browser to instruct GIS to build many different maps of Canada (from grizzly-bear distribution to wetlands) and learn about its geography in the process (**atlas.gc.ca/schoolnet/issuemap**). Find out about the *National Center for Geographic Information and Analysis (NCGIA)* (**www.ncgia.ucsb.edu**).

Geographic Societies Each major country in the world has one or more geographic organizations that serve as a focus for geographic information and activities. For example, in the United Kingdom the leading society is the *Royal Geographical Society* (now merged with the Institute of British Geographers) based in London which has a useful web site describing its history, purpose, and programme (**www.rgs.org**). In the United States the leading professional societies are the *Association of American Geographers* (AAG) based in Washington, D.C. (**www.aag.org**) and the *American Geographical Society (AGS)* based in New York (**www.geoplace.com**). The AGS web site includes links to its massive map collection at the University of Wisconsin at Milwaukee (**leardo.lib.uwm.edu**). Other important geographic societies in the United States are the *National Geographic Society* at Washington, D.C. (**www.nationalgeographic.com**) and the *National Council for Geographic Education (NCGE)* at the Indiana University of Pennsylvania (**www.ncge.org**). Links between national societies are provided by the *International Geographical Union* from its headquarters in Bonn, Germany (**www.helsinki.fi/science/igu**).

Health and Disease Data The key organization is the *World Health Organization (WHO)* of Geneva, Switzerland (**www.who.ch**), which acts as a focus for national health agencies around the world. In the United States the main federal agency is the *Centers for Disease Control (CDC)* based at Atlanta (**www.cdc.gov**). Some sites specialize in reporting individual diseases. For example, updated information on the HIV/AIDS pandemic is given by the *United Nations AIDS Agency* (**www.unaids.org**).

Historical Geography A number of the world's leading collections of historical maps have excellent web sites such as the *United States Library of*

Congress (**www.loc.gov**).One of the finest collections of historic maps on the Internet at this *University of Georgia* web site, including material on Colonial and Revolutionary America (**www.libs.uga.edu/darchive/hargrett/maps**). At Oxford University the map room of the *Bodleian Library* contains one of the largest collections in the world (**www.rsl.ox.ac.uk/nnj/mapcase.htm**). It contains a strong collection of British maps but also good examples from Europe and North America. In the UK a *Historical GIS Programme* is developing a historical database for landscape evolution studies (**www.www.geog.qmw.ac.uk/hgns**).

Jobs in Geography Web pages contain a very wide range of job information for geographers that are regularly updated. At the university and college level, the Association of American Geographers maintains its *Jobs in Geography* file (**www.aag.org/Jig**) but access is only through AAG membership. For a more general and open source for geography, GIS, geoscience, and related environmental subjects, see *Jobs with Earthworks* (**ourworld.compuserve.com/homepages/eworks**). General guides to job opportunities with cross-links to a large number of lists are provided by several universities: *Colorado University* (**www.colorado.edu/geography**) and the *University of South Carolina* (**www.cla.sc.edu/geog**) maintain particularly helpful sites. In the UK, jobs at the college level are advertised in the *Times Higher Education Supplement (THES)* (**www.thesjobs.co.uk/**) and at the school level in the *Times Education Supplement (TES)* (**www.jobs.tex.co.uk/**).

Landforms The *US Geological Survey* site provides information to help you better understand several aspects of the earth sciences (**www.usgs.gov**). A rotating earth image with global relief both on the continents and on the seafloor is given in **agcwww.bio.ns.ca/earth/.** The *Color Landform Atlas of the United States* provides colour physical map of each state together with a satellite photo (**fermi.jhuapl.edu/states**). There are also cross-links to other environmental information from watershed maps to toxic dumps.

Map Collections One of the strongest areas of the web is its handling of maps. There a number of outstanding sites and only a sample can be given here. The *PCL Map Collection*, available from the University of Texas Library (**www.lib.utexas.edu/Libs/PCL/Map_collection**), includes not only maps from around the world but links to some of the best map collections on the Internet. The *Map Viewer* site offers maps of the world (**pubweb.parc.xerox.com/map**). The National Geographic Society offers a *MapMachine* site which give online versions of many of the Society's distinctive maps (**www.nationalgeographic.com/resources/ngo/maps/**). This allows the location of maps by clicking on to a spinning globe to get both map coverage and satellite images of the chosen area. Xerox *PARC Map Viewer* gives public domain, copyright-free maps on demand, for the United States and world regions, using several different projections (**mapweb.parc.xerox.com/map/**). The maps show only coastlines, rivers, and borders; they do not have marked cities or roads. For more detailed maps see the *MapQuest* site, which provides customized maps for places all over the world using an interactive atlas (**www.mapquest.com**). This allows users to zoom down to street-level information covering 78 countries and 300 international travel destinations. For the United States the *TIGER Mapping Service* provides a digital map base which can be searched by ZIP code, latitude–longitude, as well as city name (**tiger.census/gov/**). You can mark your maps with a variety of symbols. It is also linked to the Census Bureau's *US Gazetteer*, with information on population. You can save your map as an image and print it.

Natural Hazards Most national environmental agencies have web sites that provide information on current hazards. For the United States the *National Oceanic and Atmospheric Administration (NOAA)* has a good coverage from ozone layer readings to the status of endangered species (**www.noaa.gov**). The *Environmental Protection Agency EPA)* also reports on national measures to curb hazards (**www.epa.gov**). The *Natural Hazards Center* of the University of Colorado at Boulder, Colorado, includes annotated lists on major hazards (**www.colorado.edu/hazards**). The *National Earthquake Information Center* reports on earthquake tremors recorded (**gldfs.cr.usgs.gov**), as does the seismology centre at the *University of California at Berkeley* (**www.seismo.berkeley.edu/seismo/resource**). The Dartmouth College *Flood Monitoring Project* monitors worldwide flood events (**www.dartmouth.edu/artsci/geog/floods**). In Australia the *Centre for Resources and Environmental Studies (CRES)* at the Australian National University is the foremost centre in the southern hemisphere for flood hazard studies (**www.cres.anu.edu.au**).

Population Data The *Population Division* of the United Nations publishes annual volumes of world population statistics and trends (**gopher.undp.org:70/11/ungophers/popin/wdtrends**). The *Population Reference Bureau* has been providing the public with solid information on trends in world and US population and demographics since 1929. This non-profit organization has put together a useful and lively home page, which enables visitors to query the extensive World Population Data Sheet and read current and back issues of the magazine *Population Today*. There are also links to many other online population resources (**www.prb.org/prb/**). The United States *Bureau of the Census* site gives information on past and present demographic statistics for that country (**www.census.gov**).

Satellite Photographs The main sites are maintained by the *US National Space Agency (NASA)*. Its *Earth Observing System* home page allows questions to be followed (**eospso.gsfc.nasa/gov**). See also the photo gallery of the *National Space Science Data Center* (**nssdc.gsfc.nasa.gov/photo-gallery**) and the clearing-house site for aerial and satellite photographs (**nsdi.usgs.gov/products/aerial.html**). In Europe the SPOT Image programme provides high-resolution satellite images (**www.spotimage.fr**). *GlobeXplorer* is the world's largest archive of earth imagery available over the Internet (**www.globexplorer.com**).

Urbanization Much information on urbanization trends are given on the population sites mentioned above. The *World Bank* provides both comparative data on urbanization and a special site for students on leading urban issues (**www.worldbank.org/html/schools/issues/urban.htm**). For individual cities there is a plenitude of sites of variable quality. For the United States the *USA CityLink* gives well-organized sites with atlas and gazeteer elements (**banzai.neosoft.com/citylinl/**). There is an international version with wide coverage of other cities around the world (**www.city.net/countries/**).

Using the Book in Introductory Courses

Introductory courses in geography vary hugely: both from one country to another, and between universities and colleges within the same country, and from department to department at the same level. This book has been designed as a course on *Introduction to Geography* (labelled Geography 100 in many college catalogues). As the preface to the book showed it was written (a) for students who had little background in geography and (b) to cover, at least in outline, the wide range of traditions and approaches within the field.

But the teaching strategy used with this text will need to be tailored to individual circumstances. Some incoming students will have a strong background from high school geography. Some college courses will start with introductory courses already split and with titles such as 'Introduction to Human Geography'. The chart outline ways in which this book may be modified from its original purpose – to serve as the basis of a full one-semester introduction to geography – in order to meet the needs of shorter and more specialized introductory courses. Four such courses are considered here.

Course G100: Introduction to Geography The book is designed to provide material for a year-long (24 week) or semester-long (20 or 16 week) course introducing beginning students at the college level to the richness and range of geography. The basic philosophy behind this approach is discussed in the Preface. The chapter sequences below suggest how the book can be adapted for use in shorter introductory courses.

Course G101: Elements of Human Geography In many colleges the introductory courses are split into two halves – an environmental sciences course (an 'Introduction to Physical Geography') and a social sciences course (an 'Introduction to Human Geography' or an 'Introduction to Cultural Geography). The table suggest ways in which the book might be used in such courses. How many of the chapters not directly included should be assigned for collateral reading or skipped entirely is a matter for the instructor to judge in the light of time constraints and the background of members of the class.

Course G102: Introduction to Spatial Analysis/Economic Geography One common variant on Course G101 (above) is to introduce students to

Length of teaching period	Title of course	Pro-logue	Part I The Earth as a Planet			Part II The Human Population			
		1	2	3	4	5	6	7	8
UK: Year (24 weeks)[a]	Introduction to Geography	●	●	●	●	●	●	●	●
US/Canada: Semester (20 weeks)[b]	Introduction to Geography	●	●●			●●●			
	Elements of Human (Cultural) Geography	◗	■			■	■	■	■
	Elements of Economic Geography	◗	■			●●			●
US/Canada: Semester (16 weeks)[c]	Introduction to Geography	◗	●●			●●			
	Elements of Human (Cultural) Geography	◗	●			●	●	●	●
	Elements of Economic Geography	◗	●			■			●
US/Canada/UK: Term (10 weeks)[d]	Introduction to Geography	◗	●			■			
	Elements of Human (Cultural) Geography	♣	♣			●●●			
	Elements of Economic Geography	♣	♣			●●			
	Geographic Methods (Seminar)	♣	●			♣	●	♣	●
	Geographic Concepts (Seminar)	●	♣		●	♣			●
		1	2	3	4	5	6	7	8

Teaching period: ■ 1½ weeks; ● 1 week; ◗ ½ week.
♣ Chapter skipped or assigned as collateral reading.

[a] Conventionally examination periods follow end of course so all 24 weeks allocated for teaching.
[b] Revisions and examinations in final week so 19 weeks allocated for teaching.
[c] Revisions and examinations in final week so 15 weeks allocated for teaching.
[d] Revisions and examinations in final week so 9 weeks allocated for teaching.

geography through the medium of spatial analysis. Two schemes for using the book in introductory courses that take this approach – usually listed in the catalogue as 'Introduction to Spatial Analysis' or 'Introduction to Economic Geography' – are outlined below. Again, the emphasis is on introducing students to the most elementary principles of the discipline, and the course should be regarded as a broadly based forerunner of more detailed specialized courses.

Other Specialized One-quarter or Term-long (10 week) Introductory Courses A fourth posssible use of this book is in brief introductory courses in particular areas. Five examples of such use are shown in the table. Only limited aspects of regional structure and organization are treated in this book, and supplementary material. More advanced topics in each chapter can be pursued further in the 'One step further ...' section at the end of the chapter.

Part III Resources and Landscapes				Part IV Geographic Structures				Part V Geographic Tensions				Part VI The Geographer's Toolbox			Epi-logue
9	10	11	12	13	14	15	16	17	18	19	20	21	22	23	24
●	●	●	●	●	●	●	●	●	●	●	●	●	●	●	●
●	●	●	●	●●●				●●●				●●			●
●	●	●	●	●●●				●	■		●	♣			◗
●	●	●		■	■	■	■	●	■	■	◗	♣			◗
●●●				●●●				●●●				●			◗
●●●				●●●				●	●●			♣			◗
●●●				●	●	●	●	■		●	●	♣			◗
■				■				■				●			◗
●●				●●				●●				♣			♣
●●				●●●				●●				♣			♣
♣	●	♣		●	♣			♣	●	♣		●	●	●	♣
♣			●	●	♣		●	♣			●	♣		●	●
9	10	11	12	13	14	15	16	17	18	19	20	21	22	23	24

In using the courses outlined below, instructors may want to note the following points:

1 *Chapter sequences.* The sequences charted here are only suggestions. In any book, chapters have to be arranged in a linear (and hopefully logical) sequence but instructors may wish to vary the sequence followed in the book. It is worth recalling that the sequence adopted represents only one of many possible compromises. Indeeed, the twenty-four chapters could be arranged in more than 100 million different ways! The text has been redesigned into more modules so that instructors can more readily rearrange and structure to meet individual needs.

2 *Length.* Each sequence assumes that one week of each period (either a semester or a quarter) will be used for reviews, tests, and the like. The symbols used in the sequence give the suggested length to be devoted to each chapter or whether a chapter is assigned as collateral reading, or skipped, or assigned as optional reading.

3 *Choice of themes.* An attempt has been made to include the majority of concepts that are likely to be of use to the novice geographer. Although I have tried to cover the field in a catholic and impartial way, space is a prime constraint in any volume of this length. Thus, in the last analysis, there must be a heavy personal bias. Although authors cannot eliminate this bias, they should at least identify their own predilections so instructors can correct and modify the text as they choose. My own partiality, where it presents itself, is fairly evident. I have leaned toward systematic theory and hypothesis rather than toward the elaboration of many regional case studies; I have generally adopted contemporary rather than classical statements of such theory; and I have chosen to restrict the discussion of physical geography to those aspects that contribute most directly to the other subfields of geography. Advanced physical geography is taught in most schools in Britain and the Commonwealth, and students there may find this treatment somewhat oversimplified. But many students in the United States and elsewhere may be meeting these topics for the first time. Compromise has therefore been inevitable.

4 *Regional coverage.* Although the emphasis throughout is on concepts and methods, it would be inconceivable to write an introduction to geography without numerous regional case studies. These range widely to stress the global variations in the environment and its exploitation as well as to illustrate differences in the temporal and spatial scales of operation of the forces discussed. Thus, although the bulk of the regional examples is drawn from the world as structured in the second half of the twentieth century and the start of the new century, some cases are deliberately drawn from earlier times. Half the figures are concerned with general relationships, and the other half present specific regional cases. Of these, slightly over a third is drawn from North America, another third from Europe, and the remainder from the rest of the world. This balance reflects partly the pattern of research results and partly the probable distribution of those who may use this book. Therefore, even though care has been taken to diversify the regional case studies, a minority of readers may feel themselves deprived of locally relevant examples and instructors are urged to substitute locally relevant cases. (This was splendidly done by Robert Geipel in the German translation of this book's precursor, *Geographie: Eine moderne Synthese.*) As for the scale of examples, one in six is drawn on the world scale, and one in three deals with an area no larger than a single city. Between these extremes, cases are well distributed along a size continuum.

5 *Quantitative level.* Modern geography lays strong emphasis on the ways of analyzing research problems, and many of these ways are quantitative. Most quantitative methods can be left to later courses, however, and advanced techniques have been eliminated from the text. Those that it seems appropriate to mention are usually described in separate 'marginal' discussions for readers to explore or ignore depending on their own inclinations or those of their instructors. Fuller guides to these techniques are listed in the suggestions for further reading. Each chapter is accompanied by a list of references to guide readers in their further studies. Most are standard and widely available texts found in most college libraries. Journal references, because they change from year to year, are left for more advanced courses.

INDEX

Concepts and terms which are set for review in the 'Reflections' section at the end of each chapter and those that occur in the 'Glossary' (Appendix A) are printed in bold type in this index. The term *passim* is used where more than ten continuous pages of text refer discontinuously to the topic indexed.

The Arts & Crafts Companion

Pamela Todd

First published in the United Kingdom in 2004 by Thames & Hudson Ltd, 181A High Holborn, London WC1V 7QX

www.thamesandhudson.com

First published in paperback in 2008

Created and produced for Thames & Hudson by Palazzo Editions Ltd, 2 Wood Street, Bath, BA1 2JQ

Book design by David Fordham

British Library Cataloguing-in-Publication Data
A catalogue record for this book is available from the British Library

ISBN 978-0-500-28759-0

Printed and bound in Singapore

Page 1: *"Weaver Birds in Foliage," design for a tile by William De Morgan.*

Pages 2–3: *Three stained-glass panels probably designed by Selwyn Image and manufactured by Shrigley and Hunt,* c. *1886.*

Page 4: *This group of four tiles depicting The Flower Fairies was designed by Walter Crane, possibly for Pilkington & Co.,* c. *1900.*

CONTENTS

Above: *Front, back, and end elevation plans for a house at Frinton-on-sea, Essex by C.F.A. Voysey.*

midst bitten mead and
re shorn
the world without is waste and worn
but here
within
our orchar
the guerdon
of our
labour sh

INTRODUCTION

Left: *Detail from William Morris's "Orchard" tapestry, c. 1863. (For complete reproduction of tapestry see page 313.)*

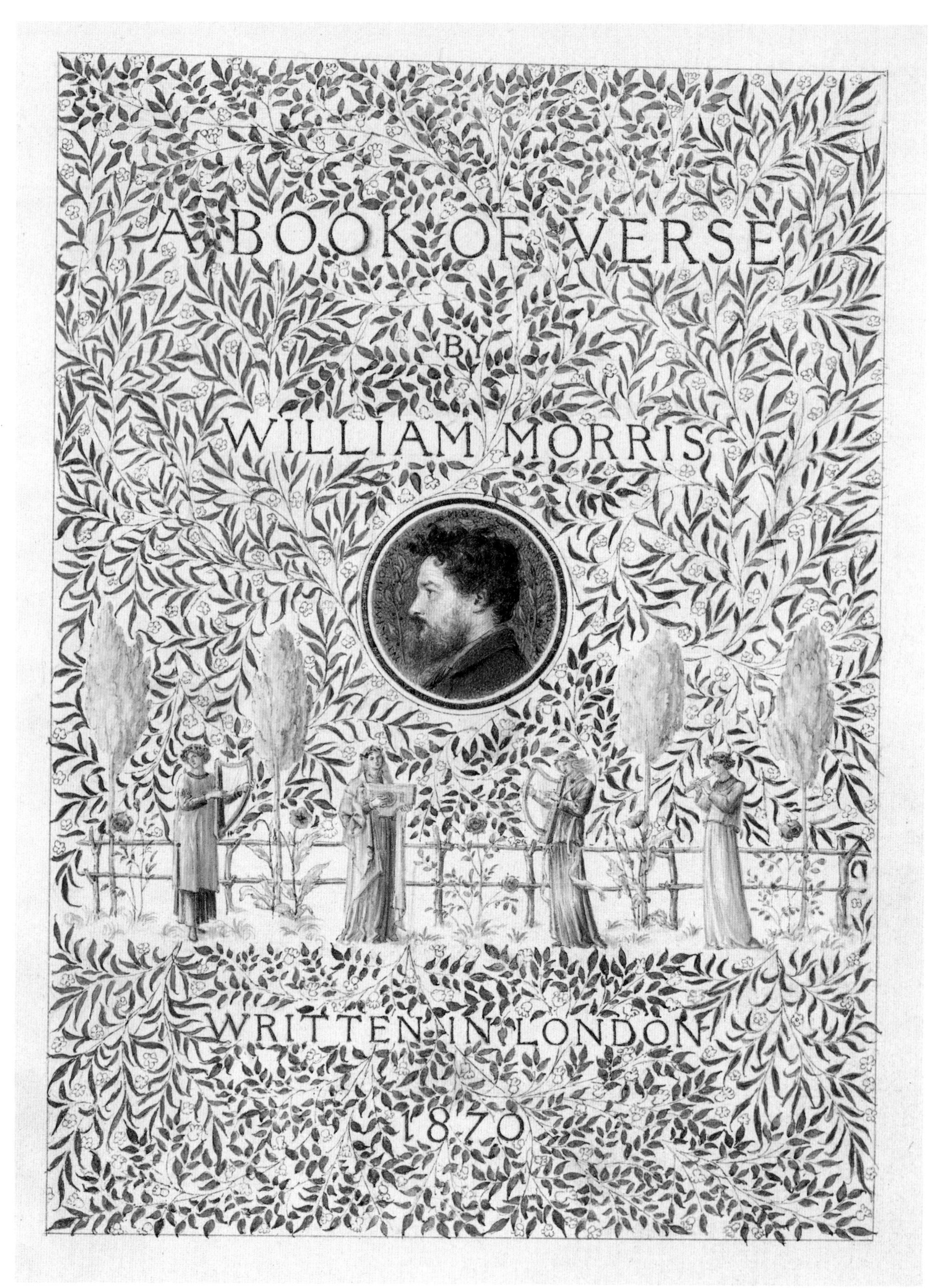
A BOOK OF VERSE
BY
WILLIAM MORRIS
WRITTEN IN LONDON
1870

PHILOSOPHY & BACKGROUND

"What business have we with art at all, if we all cannot share it?"

WILLIAM MORRIS

THE ARTS AND CRAFTS MOVEMENT, founded in the late nineteenth century by a group of British artists and social reformers inspired by John Ruskin, A.W.N. Pugin, and William Morris, sought to stem the tide of Victorian mass production, which its adherents believed degraded the worker and resulted in "shoddy wares." The movement redefined the role of art and craftsmanship, sought to restore dignity to labor, created opportunities for women, and underpinned many social reforms. At its most romantic and intense it offered a complete, if rigorous, model for living, as mapped out in Morris's Utopian novel *News from Nowhere*, and put into practice by, among others, Charles Ashbee in the Cotswolds in England and Elbert Hubbard in East Aurora, New York. The movement was attractive to many who were disenchanted by rapid industrialization; its influences—both social and aesthetic—were felt in Europe, where the ideas cross-fertilized and returned to Britain enriched and enlivened, and in the USA, where its ideals were enthusiastically embraced and adapted. Despite faltering after World War I, it has emerged as a major force in the history of design during the last hundred years and is enduringly popular today.

The movement acquired its name in 1888, when William Morris's tapestries, William De Morgan's tiles, Walter Crane's wallpapers, and Edward Burne-Jones's stained-glass designs went on show on October 4 at the New Gallery in London, along with other work by the newly formed Arts and Crafts Exhibition Society. The show was a success and drew enthusiastic reviews. *The Builder* reported that it was "full of things which seem to have been done because the designer enjoyed doing them…" even if it found some of the exhibits a little "outré and eccentric." Walter Crane set out the Society's mission statement—"to turn our artists into craftsmen and craftsmen into artists"—in the catalog to that first exhibition.

Above: *A ceramic bread plate designed by A.W.N. Pugin and made by Minton's in the encaustic process, c. 1850. The ethical message is made clear in the proverb "Waste Not, Want Not" around the rim.*

Far left: *Title page of* A Book of Verse *by William Morris. Morris put together this illustrated collection of fifty-one of his own poems as a thirtieth birthday present for his dear friend Georgiana Burne-Jones. He was responsible for all the calligraphy and most of the border decorations, though various friends helped, and Charles Fairfax Murray was responsible for the portrait of Morris at the center of this page.*

However, to trace the roots of the movement it is necessary to go back to the Great Exhibition held at Crystal Palace in London in 1851 and to the dismayed and disgusted reaction of the young William Morris and his friends to the "wonderful ugliness" of the mass-produced exhibits, which seemed to them devoid of any soul, dehumanizing the workers who made them. What began as a return to the styles and manners of the medieval period—its cathedrals, furnishing, costumes, and, crucially, its workers' guilds—developed, under the inspiration and guidance of William Morris, to embrace more day-to-day handmade crafts. Unashamedly "artistic," the homes, gardens, and products of the Arts and Crafts movement celebrated the individuality of the craftsmen and craftswomen who made them. For them it was not merely a style, it was a way of life, based on frankly Utopian models. Humble, plain, honest furniture, like that made by Ford Madox Brown for Morris, Marshall, Faulkner & Co., sold alongside more ornate "Arthurian" pieces, but all had in common the individuality of craftsmanship and the vision behind it of beauty and harmony that harked back to the rules and methods of the medieval guild system. Arts and Crafts practitioners truly believed that everyone's quality of life would be improved if only integrity could be restored to objects in daily use.

Above: *"Morris in the Home Mead"—a carving by George Jack, from a sketch by Philip Webb, on the pair of semidetached cottages designed by Webb and built in Kelmscott village for Jane Morris in 1902 as a memorial to Morris.*

The Arts and Crafts movement elevated and ennobled the artisan, and many followed William Morris's moral crusade along the path of socialism and political activism, aiming to restore dignity and satisfaction to workers who had been robbed of all joy in their work by the Victorian doctrine of progress and industrialization. The movement's inclusivity also created opportunities for women, who found in its revival of traditional techniques—such as embroidery, weaving, and enameling—an outlet for their creativity and a respectable way of earning a living. Gender still determined the division of labor, however, with men dominating fields such as metalwork, furniture, and architecture and women working hard to create a niche for themselves in areas like needlework, bookbinding, and pottery.

In the chapters that follow, each dealing with a different aspect of the movement, William Morris emerges as the towering personality and inspiration: a father-figure whose writings, lectures, and practical example had a tremendous and lasting influence on a younger generation of architects, craftsmen, and decorators. Along with men like W.R. Lethaby, A.H. Mackmurdo, Philip Webb, Walter Crane, Ernest Gimson, and the Barnsley brothers, Morris wanted to revitalize native English

Above: *"Summer" and "Winter" from a set of tiles depicting the Four Seasons designed by Walter Crane and produced by Maw & Co., c. 1880.*

traditions, always stressing the importance of honesty to function and local materials. Appalled by Sir Giles Gilbert Scott's recent restoration of Tewkesbury Abbey in Gloucestershire, Morris founded the Society for the Protection of Ancient Buildings, known affectionately as "Anti-Scrape," on March 22, 1877, "for the purposes of watching over and protecting these relics." Webb, Gimson, and the Barnsleys were all active members of the Society, which championed protection rather than restoration. Morris was prepared to prop up a tumbling church but not to tamper with the fabric or ornament of the building as it stood: a principled standpoint that cost him money, for it meant his firm had to stop accepting lucrative stained-glass commissions from churches undergoing restoration.

Morris's defense of medieval craftsmanship was part of his romanticized concept of the rural idyll. "Suppose," he wrote to Edward Burne-Jones's sister-in-law, "people lived in little communities among gardens and green fields, so that you could be in the country in five minutes' walk, and had few wants, almost no furniture for instance, and no servants, and studied the (difficult) arts of enjoying life, and finding out what they really wanted: then I think that one might hope civilisation had really begun." In the face of encroaching urbanization, a vigorous "back to the land" movement, linked to a revival of English folk music and entertainments, had begun to emerge and identify itself strongly with the Arts and Crafts movement. A number of Arts and Crafts guilds were founded in the 1880s, with varying degrees of success. Ruskin's experimental St. George's Guild, designed as a Utopian society—"in essence a kind of enlightened feudalism"—was intended to implement his ideal of a just society, though it was perhaps a little too medieval in scope to offer a serious alternative to the rising tide of capitalism. Other communities were longer-lived if not, ultimately, entirely successful. In his influential book *The Stones of Venice* (1851–3), Ruskin had called for sweeping social change and an end to the illogicality of a situation in which "we want one man to be always thinking and another to be always working, and we call one a gentleman, and the other an operative; whereas the workman ought often to be thinking, and the thinker often to be working, and both should be gentlemen in the best sense.... In each several profession, no master should be too proud to do its hardest work. The painter should grind his own colours; the architect work in the mason's yard with his men; the master-manufacturer be himself a more skilful operative than any man in his mills." Such progressive notions became the cornerstone and creed by which the new guilds operated.

Below: *"Peacock" printed cotton designed for the Century Guild by Arthur Heygate Mackmurdo in 1883.*

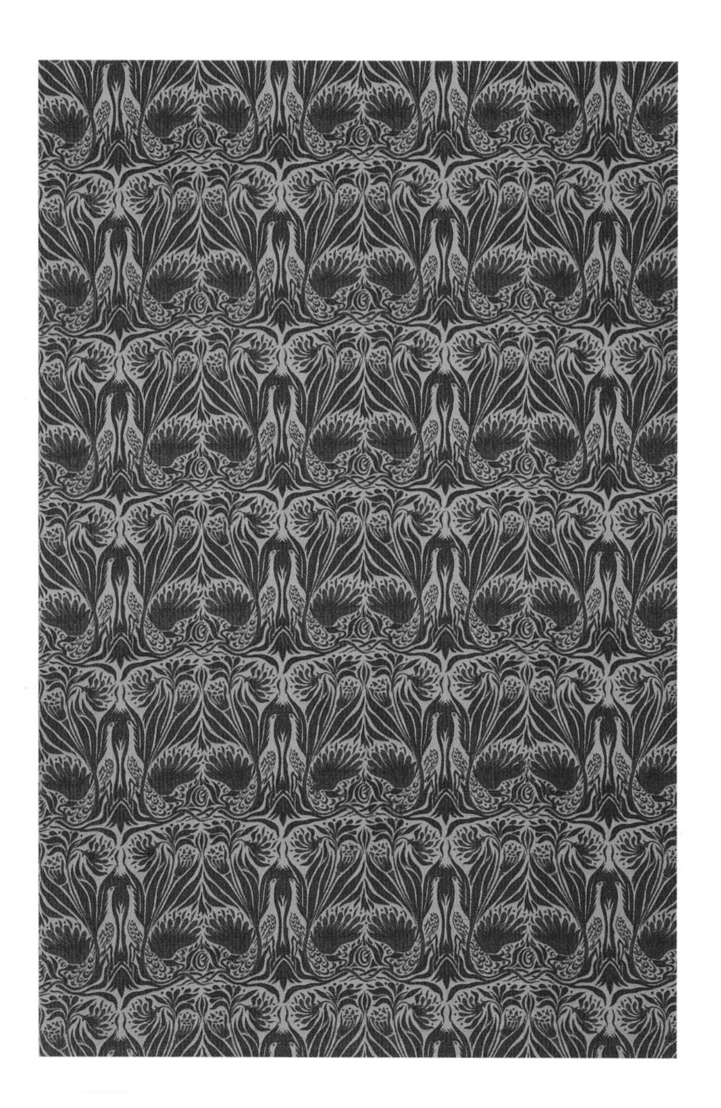

In 1882 Ruskin's ex-pupil A.H. Mackmurdo collaborated with the artist Selwyn Image to form a loose collective of designers into one of the first of the new craft guilds—the Century Guild. It was set up with the aim of supplying all the furniture and furnishings necessary for a house and, most importantly, involving artists in areas of work previously considered the realm of mere tradesmen. Mackmurdo had trained as an architect in the London office of the Gothic revivalist James Brooks, who had inspired him to try to master several crafts himself, including brasswork, embroidery, and cabinet-making, and instilled in him a reforming passion. Mackmurdo saw the Arts and Crafts movement "not as an aesthetic excursion; but as a mighty upheaval of man's spiritual nature" and differed from Morris in this important respect: Morris had aimed to level the disciplines of painting and sculpture to the rank of democratic handicrafts; Mackmurdo wanted to raise the status of crafts to that of "fine art." Century Guild members, including William De Morgan,

W.A.S. Benson, Herbert Horne, Clement Heaton, Benjamin Creswick, and Heywood Sumner, carried out decorative work of various kinds, much of it to Mackmurdo's design but in a cooperative spirit, collaborating on the design of a home and its contents. They also produced the influential journal *The Hobby Horse*, illustrated with woodcut decorations by Selwyn Image and Herbert Horne. Members socialized and met regularly at Mackmurdo's house at 20 Fitzroy Street, London, to hear early English music performed on the viol, lute, and harpsichord.

Further impetus to the craft revival movement came from the Art Workers' Guild (which is still in existence), founded in 1884 by a group of architects and designers including William Lethaby, a pivotal figure in the Arts and Crafts movement. Once again the stated aim of the Guild was to break down barriers between artists and architects, designers and craftsmen, leading to a goal of decorative unity. At their meetings in Queen Square, London, Guild members sat on traditional rush-seated ladderback chairs made by a sixty-seven-year-old carpenter called Philip Clissett, who never joined the movement but whose influence can be seen in the later work of Ernest Gimson, the Barnsley brothers, and Ambrose Heal, all of whom would have sat on Clissett's chairs as they attended lectures.

Above: *Frontispiece from* The Hobby Horse *by Selwyn Image, 1888.*

Then, in 1888, a splinter group from the Art Workers' Guild, including Walter Crane, W.A.S. Benson, Lewis F. Day, and T.J. Cobden-Sanderson, came together to arrange a large exhibition initially to be called "The Combined Arts," a title dropped in favor of "Arts and Crafts." And so the Arts and Crafts Exhibition Society, with Walter Crane as its first chairman, and a committee including William Morris and Edward Burne-Jones, came into being. The Society organized annual exhibitions in the New Gallery in Regent Street, London, providing a useful shop window for the products of the Arts and Craft movement. These included not just the designs of its own members but contributions from Morris & Co., the Century Guild, and C.R. Ashbee's newly formed Guild and School of Handicraft. All exhibitors had to satisfy strict criteria, the most important of which was that their work was entirely handmade. The Society also presented practical demonstrations and accompanying lectures from luminaries like William Morris

Above: *A photograph from* The Studio *showing the Arts & Crafts Exhibition Society stand at the Turin Exhibition in 1902.*

(on tapestry weaving), T.J. Cobden-Sanderson (on bookbinding), Halsey Ricardo (on furniture), Selwyn Image (on designing for the art of embroidery), and Walter Crane (on design).

It is apparent that the figures who did most to shape the fortunes of the Arts and Crafts movement, after Ruskin and Morris, had their roots in architecture: designers such as Mackmurdo, Horne, Philip Webb, Charles Voysey, Norman Shaw, J.D. Sedding, M.H. Baillie Scott, Charles Rennie Mackintosh, Reginald Blomfield, and Lethaby. One architect, however, who felt that the Arts and Crafts movement had taken a wrong turn and was in danger of becoming "a nursery for luxuries, a hothouse for the production of mere trivialities and useless things for the rich," was C.R. Ashbee.

Ashbee had begun with the highest ideals, setting up his Guild of Handicraft in the underprivileged East End of London in 1888 with a starting capital of only £50. By 1902 business was flourishing, and he decided to implement his ambitious plans to relocate to rural Chipping Campden in the Cotswolds, not far from the small but thriving Arts and Crafts group composed of the families of Ernest Gimson and the Barnsley brothers. The town, formerly a center for the wool and silk trade, had fallen on hard times, and the impact of such a large enterprise was tremendous. Ashbee accommodated his Guild members in the local cottages, many of which lay empty, though, controversially, many that did not were emptied of their occupants to make way for the incomers, who could be charged a higher rent. The old silk mill was pressed into service to provide the Guild with silver and jewelry studios, and workshops for woodcarving and cabinet-making. In 1904 Ashbee opened the Campden School of Arts and Crafts at Elm Tree House, where practical lessons in gardening, cookery, and laundry work were given as well as crafts, literature, music, and history. Walter Crane and the poets John Masefield and Edward Carpenter were among the invited lecturers. In keeping with William Morris's Utopian ideals, Ashbee ran the Guild on democratic principles and aimed to promote "a higher standard of craftsmanship, but at the same time ... protect the status of the craftsman." He had explained in a pamphlet published in 1902: "The Guild endeavours to steer a mean between the independence of the artist which is individualistic and often parasitical and the trade shop where the workman is bound to purely commercial and antiquated traditions, and has as a rule neither stake in the business nor any interest beyond his weekly wage."

Devoted to social reform, Ashbee provided his craftsmen with not just homes, but gardens, a swimming pool, and communal activities including amateur theatricals and musical evenings. The impact of a hundred and fifty Londoners arriving *en masse* in a small rural community caused some tensions but gradually, especially after the school was established, the two sides accepted one

Below: *Craftsmen at work in the Guild of Handicraft's metalwork shop at Essex House, before their move to Chipping Campden in 1902.*

Above: *Men and women attending a Guild of Handicraft carving class at Chipping Campden in 1906.*

another and co-existed harmoniously until Ashbee's democratic notions were sabotaged by the market forces that forced him into voluntary liquidation in 1907.

In an article in *House Beautiful* Ashbee defined Arts and Crafts as "occupations or pursuits in the practice of which the individual comes into direct contact with his material, and is enabled to give expression to his own fancy, invention, imagination." That he failed to translate this to a wider public was a source of great sadness to him. Writing to his old friend, the illustrator and etcher F.L. Griggs, he admitted, "We both wanted a better world and were both quite out of touch with the one provided us, while the beauty of life—expressed in that Gloucestershire village—was lost all in all to us."

By 1907, however, the aesthetic ideas behind the movement had taken hold. Many important Arts and Crafts practitioners had taken up positions in the new art colleges being set up around this time and were able to preach their Arts and Crafts message from these prominent platforms, as well as through the established art schools, including the Royal College of Art, which moved swiftly in step with the progressive times. W.R. Lethaby's inspirational philosophy reached a wider audience through his work as co-director of the new Central School of Arts and Crafts in London, of which

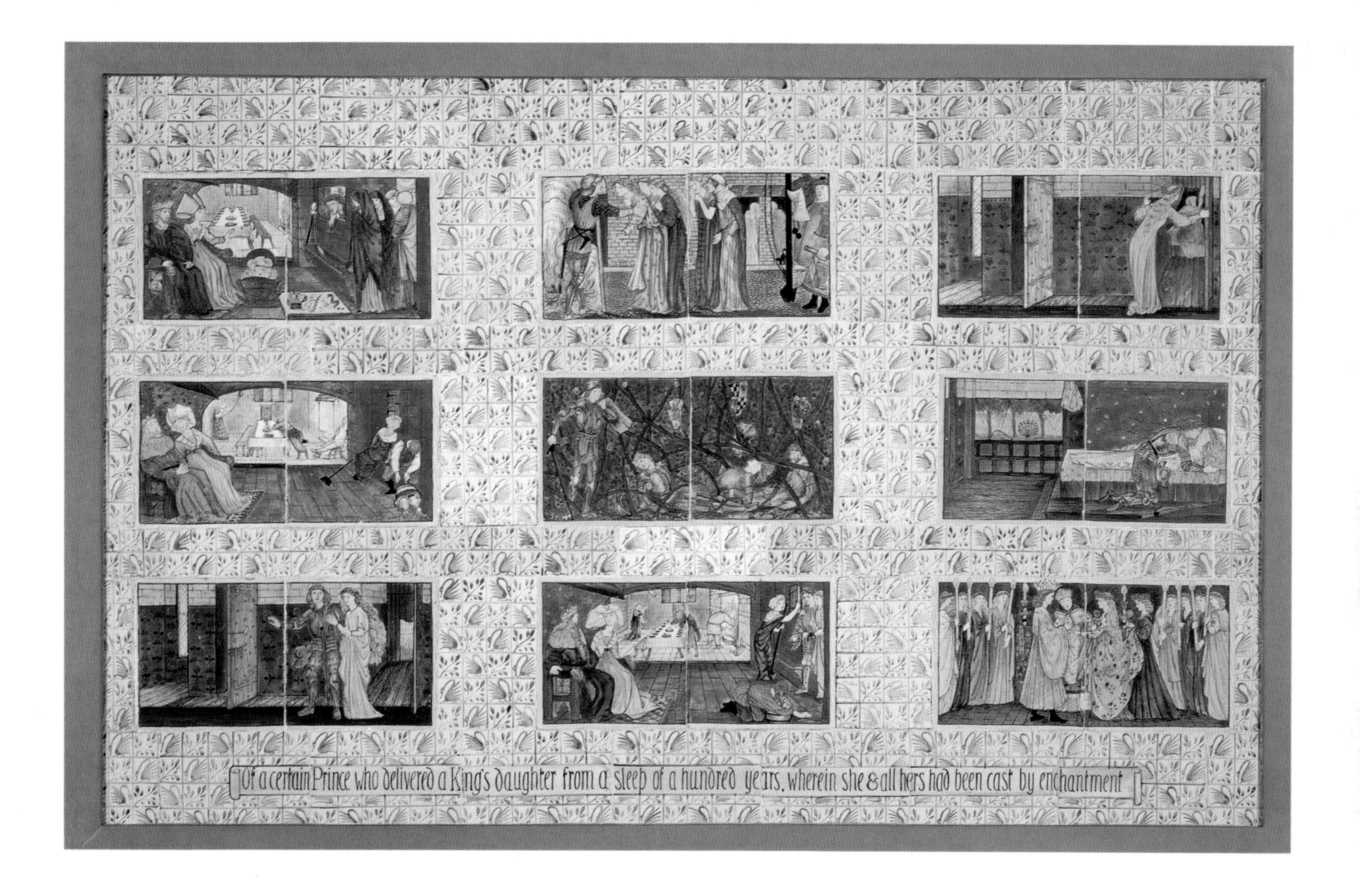

Above: *A tile panel by Edward Burne-Jones based on one of his favorite themes—Sleeping Beauty—for Morris, Marshall, Faulkner & Co., 1894.*

he eventually became principal in 1911. He wanted to raise the standard of the crafts while creating a training ground for designers able to meet the demands of commercial companies like Wedgwood, and he made a practice of placing highly skilled practitioners, rather than teachers, in key posts. For example, he made Halsey Ricardo head of architecture at the school, Christopher Whall head of stained glass, and the sculptor and silversmith Alexander Fisher head of enameling. Meanwhile, in Birmingham and Glasgow, other Arts and Crafts groups bloomed, once again around the focus of an art school sympathetic to their ideals that quickly became the home of the local style. The Birmingham Group included artists and designers such as Arthur Gaskin, Henry Payne, Charles March Gere, Bernard Creswick, the enamelists Sidney Meteyard and his wife Kate Eadie, and Mary Newill, a designer of embroideries and stained glass. All were teaching at the Birmingham School of Art's "Art Laboratories" in the mid-1890s. Inspired by the writings of Ruskin and the work of the English Pre-Raphaelites, especially the Birmingham-born Burne-Jones, they drew on late-romantic poetry and medieval romances for the subject-matter of their mural decorations, book illustrations, embroideries, and lovely Limoges enamel plaques, experimenting with the Renaissance technique

Left: *The impressive western façade of the Glasgow School of Art, designed by Charles Rennie Mackintosh between 1897 and 1899.*

of tempera. Like the Century Guild, Ashbee's craftsmen, and Morris's circle, decorative unity was a paramount concern for the newly formed Birmingham Guild of Handicraft.

In Scotland an artistic revolution was taking place. At Edinburgh the new School of Applied Art offered classes for engravers, plasterers, and cabinet-makers as well as architects, but Glasgow was the real epicenter. Charles Rennie Mackintosh was the most famous member of the circle of artists and architects to emerge from the cutting-edge Glasgow School of Art, which had, in Francis Newbery, a progressive and enlightened director. Newbery introduced technical art studios where artist-craftsmen gave artisans a "technical artistic education" in commercial crafts such as stained glass, bookbinding, metalwork, illumination, and ceramics—even lead plumbing. His wife, Jessie Newbery, taught embroidery at the school and through her teaching, and her own highly creative designs, popularized Glasgow embroidery and brought on some richly talented pupils. Among them were Ann Macbeth and the Macdonald sisters, Frances and Margaret; the Macdonalds, together with

Herbert MacNair and Mackintosh, became known as the "Glasgow Four." The Four pioneered Art Nouveau designs and formed strong links with the Viennese Secessionists, who admired and emulated their extreme modernism. They found an active champion in the German architect and writer Hermann Muthesius, who claimed they were "a seminal influence on the emerging new vocabulary of forms, especially and continuously in Vienna, where an unbreakable bond was forged between them and the leaders of the Vienna Movement." Other leading figures to come out of the Scottish school were the architect George Walton, the designer and illustrator Jessie M. King, and her husband Ernest A. Taylor, the furniture and stained-glass designer.

A similar revival was taking place in the USA, for, although predominantly British, the ideas and ideals of the Arts and Crafts movement soon crossed the Atlantic and found fertile soil in which to take root. An American Arts Workers' Guild was founded in 1885 by the painters Sidney Richmond Burleigh and Charles Walter Stetson, together with the industrialist John Aldrich. Its members designed buildings, silverware, and other metalwork, along with furniture with panel paintings similar in style to that of Morris. The Society of Arts and Crafts, modeled on the British Arts and Crafts Society, was founded in Boston in 1897 by the architects Ralph Adams Cram and Bertram Grosvenor Goodhue, together with Charles Eliot Norton, an art critic of enormous influence—the American equivalent of John Ruskin—and the first professor of fine arts at Harvard. The Society offered lectures, classes, and social gatherings designed to bring together craftsmen, architects, and enthusiastic amateurs. It published a journal, *Handicraft*, and ran a saleroom where work—after a jury selection process—could be sold. New Arts and Crafts academies, such as the Manual Training High School in St. Louis run by Calvin Milton Woodward, which produced such quintessential Arts and Crafts designers as Charles and Henry Greene, were starting to spring up, leading to a rapid expansion of the craft movement.

Just as, in Britain, the Great Exhibition of 1851 had acted as a catalyst for Morris and Ruskin, so the Centennial Exposition in Philadelphia in 1876 revealed the lamentable lack of any truly American style. The industrial arts were dominated by shoddy, machine-produced goods in Empire style and, with a few notable exceptions like handcrafted Shaker furniture, lacked any American cultural identity of their own. The work of Arts and Crafts practitioners like William Morris and Walter Crane on show at the Exposition seemed to offer a way forward.

As in Britain, amateur groups were encouraged to flourish in American mission halls, school classrooms, and at evening classes. Books and manuals were published to satisfy this new consumer market, and magazines and journals such as *House Beautiful* and *Art Workers' Quarterly* appeared, offering practical advice to a largely middle-class market. Pioneering figures such as Elbert Hubbard and Gustav Stickley promoted the movement with a passion, and Arts and Crafts communities

Above: *A color print of the Roycroft shops at Elbert Hubbard's crafts community in East Aurora, near Buffalo, New York.*

similar to those founded in Britain were started in America. The best known of these was undoubtedly the Roycroft craft community in East Aurora, near Buffalo, founded by Hubbard in 1892 after a trip to England and a meeting with William Morris.

Hubbard's vision was of a return to the simple community life of pre-industrial America. Paralleling Ashbee's Chipping Campden experiment, the Roycrofters set up their own shops, farm, bank, smithy, and printing plant, and built houses on communal living lines, attracting such interest that they also had to build a Roycroft Inn to house the throng of curious visitors. Hubbard provided a library and playgrounds and set up a lecture series for his workers, who spent their eight-hour working days in good, well-ventilated conditions, able and encouraged to try their hands at different tasks, and so relieve boredom and explore possibilities: "When a pressman or typesetter found his work monotonous ... he felt at liberty to leave it and turn stonemason or carpenter." In 1901 Hubbard extended the operation, adding furniture to the leather and metalwork gifts his designers were already producing, along with ceramics and household items. Other Utopian craft communities sprang up, including the Rose Valley Arts and Crafts Colony, set up in Philadelphia by the architect William Lightfoot Price, and the Byrdcliffe Colony, started by Ralph Radcliffe Whitehead and his wife Jane Byrd McCall in rural Woodstock, New York State (which is still a thriving summer home to artists, writers, and composers). In Chicago in 1889, Jane Addams and Ellen Gates Starr founded Hull-House which, like Toynbee Hall (the settlement house established in Whitechapel, London, in 1884) provided immigrants with craft skills to lift them out of poverty.

Gustav Stickley, who might be said to have done the most to disseminate the Arts and Crafts message throughout the USA, was, like Hubbard, inspired by a trip to England in 1898, and established his own guild of artisans, United Crafts. Like the British model, he set up a profit-sharing scheme for his craftsmen (though this was soon abandoned) and held regular workshop meetings "to secure harmony and unity of effort." His magazine *The Craftsman* publicized both his products and his philosophy. The first issue contained a monograph on William Morris, the second an appreciation of Ruskin, and the next hailed Morris as "a household name throughout America." Stickley used machinery to liberate his workers from tedious and unnecessary labor but also to maximize production. "As a matter of fact," he insisted, "given the real need for production and the fundamental desire for honest self-expression, the machine can be put to all its legitimate uses as an aid to, and a preparation for, the work of the hand, and the result be quite as vital and satisfying as the best work of the hand alone ... the modern trouble lies not with the use of machinery, but the abuse of it." By taking this position, of course, he ran counter to the central tenet of the British Arts and Crafts movement that work should be handmade, yet his aim of "a simple, democratic art" was undeniably realized in his determinedly simple furniture, which drew on the Shaker ideal of functional beauty.

Above: *Front cover of* The Craftsman, *published by Gustav Stickley's United Crafts in May 1911.*

Both sides of the Atlantic were experiencing an unprecedented rise in the middle classes, which led to a building boom and created a huge new pool of consumers interested in interior decoration and the statement and status "artistic" choices could confer. Designers such as Walter Crane, Lewis F. Day, Christopher Dresser, A.H. Mackmurdo, Louis Comfort Tiffany, and Candace Wheeler met this need by redefining and revolutionizing interiors. Manufacturing firms and department stores were quick to respond to the demand, employing leading Arts and Crafts architects like Voysey and Baillie Scott in Britain, and Frank Lloyd Wright and Charles and Henry Greene in America, who

Above: *Tower House, Bedford Park, west London, designed by C.F.A. Voysey in 1891.*

designed not just the house but the furniture and furnishings to go inside it. Many Arts and Crafts architects also worked as freelance designers for firms such as Liberty's, producing textile patterns and specially tailored ranges of furniture.

The buoyant economy provided a steady stream of commissions for Arts and Crafts architects for large or small homes for the better-off (many of which, like Philip Webb's Standen in West Sussex, Mackintosh's Hill House in Helensburgh, Baillie Scott's Blackwell in Cumbria, and the Greene brothers' Gamble House in Pasadena, have been preserved and stand as living monuments to the Arts and Crafts movement). As home ownership became an achievable aim for the burgeoning middle class in Britain, it fueled a building boom that spread beyond the towns and encouraged an enlightened and innovative group of architects to envision and implement plans for "Garden Cities" in the suburbs surrounding London. The prototype for these had been the development at Bedford Park, just north of Chiswick in west London, near the then-new underground railway station at Turnham Green. It was the brainchild of an idealistic entrepreneur named Jonathan Carr, who, in 1875, started to develop the open fields around his father-in-law's eighteenth-century house, creating a suburban village of small detached and semidetached villas in the Queen Anne style for the artistic middle class. His first choice of architect, E.W. Godwin, was replaced by Richard Norman Shaw, who improved on Godwin's designs and supervised the building of five hundred houses, each with "A Garden and a Bath Room with Hot and Cold Water." The 1881 prospectus declared Bedford Park to be "The Healthiest Place in the World" and it attracted an artistic—if somewhat bourgeois—clientele of actors, artists, and Russian émigrés, who sent their children to the two specially built schools and availed themselves of the daringly unisex tennis club and the cooperative store.

Later Garden City developments combined social principles with early 1900s Arts and Crafts design. The intention was to dissolve the differences between town and country and create an ideal setting in which people could live and work in harmony, happiness, and a healthy environment. These housing experiments were based on Ruskin's model of the fourteenth-century village settlement where working class misery, unemployment, and bad housing could be things of the past. In *Sesame and Lilies*

Above: *A chromolithograph showing the houses on the Bedford Park Estate in London, 1882.*

(1865) Ruskin had called for the building of new homes "strongly, beautifully, and in groups of limited extent, kept in proportion to their streams and walled round, so that there may be no festering and wretched suburb anywhere, but clean and busy street within and the open country without."

Ebenezer Howard, who coined the term "Garden City," was a non-architect theorist influenced by Walt Whitman, Ralph Waldo Emerson, and Ruskin, who wanted to create "a new civilization based on service to the community." In 1898 he wrote an influential book called *Tomorrow: a Peaceful Path to Real Reform*, which was revised as *Garden Cities of Tomorrow* in 1902. The architects Raymond Unwin and Barry Parker, whose extensive work in the the Garden City movement made them enormously influential pioneers of town planning as social engineering, won the competition to plan the first Garden City: Letchworth in Hertfordshire. Ruskin's vision of a "belt of beautiful garden and orchard round the walls, so that from any part of the city perfectly fresh air and grass and sight of far horizon might be reachable in a few minutes' walk" was realized at Letchworth, which was swiftly followed by Dame Henrietta Barnett's Hampstead Garden Suburb in north London and Gidea Park in Romford, Essex. Parker and Unwin were involved in all of them, as were Halsey Ricardo, Edwin Lutyens, and Baillie Scott, who in 1910 designed a pair of show houses at Gidea Park that were filled with furniture and fabrics from Heal's and the Deutsche Werkstätte, made to his designs.

Right: *A picturesque example of the cottages the Quaker Cadbury family provided for their factory workers at Bournville, a low-density housing development designed as a model village.*

A few philanthropic industrialists in northern England had blazed the trail, including men and women like Robert Owen, Sir Titus Salt (whose model estate Saltaire, near Bradford, had been planned and started as early as 1850), and Quakers George and Elizabeth Cadbury (who built the model garden village of Bournville near Birmingham for the workers in Cadbury's chocolate factory in 1895). William Hesketh Lever (later Lord Leverhulme) built Port Sunlight on fifty-two acres on the south bank of the River Mersey in 1889 to provide both a soap factory and a model village of houses "in which our workpeople will be able to live and be comfortable … in which they will be able to know more about the science of life than they can in a back slum and in which they will learn that there is more enjoyment in life than in mere going to and returning from work." Lever wanted three-bedroomed houses, with parkland, playing fields, bath houses, and almshouses for the elderly. One thousand red brick houses were built around a town center laid out with wide avenues and parks. They had all the Arts and Crafts details of white-painted casement windows, tile-hangings, and gables.

Above: *New Earswick, York—the model factory village by Parker and Unwin,* c. *1902, houses "artistic in appearance, sanitary, well-built."*

These model villages were often arranged around a "village green," with perfectly scaled public buildings, schools, a Friends' Meeting House, bandstands, and churches, though rarely inns, for their philanthropic supporters were often active members of the Temperance movement. Though small in scale, the houses tended to be immensely varied in design, involving characteristics of sixteenth-, seventeenth- and eighteenth-century domestic style in brick, timber, and stone. Individual gardens, shared recreation spaces, and plenty of allotments were important features.

Joseph Rowntree, another Quaker chocolate manufacturer, founded the village of New Earswick in Yorkshire and employed Parker and Unwin to lay out and design each house from 1902 onward. Rowntree wanted houses that were "artistic in appearance, sanitary, well built, and yet within the means of men earning about twenty-five shillings a week." And Parker and Unwin supplied them, grouping a variety of cottages, some of brick, others with roughcast exteriors, and all with gardens, around a triangular village green.

Above: *European Arts and Crafts in this interior by the German designer and founder of the Dresden and Munich Werkstätten, Richard Riemerschmid, dating from 1904.*

A similar experiment in workers' housing was unfolding in Europe in the Voysey-influenced rising terraces of Richard Riemerschmid's Hellerau and Hermann Jansen's Berlin projects. They incorporated laundries, a training school, a theater, and workshops for the newly amalgamated Dresden and Munich Werkstätten. In the USA the garden suburb idea was implemented in projects such as Forest Hills Gardens in Queens, New York (planned in 1909 by Frederick Law Olmstead Jr. and Grosvenor Atterbury), Philadelphia's Germantown, and Virginia's Hilton Village (designed in 1918 by Henry Vincent Hubbard and Joseph D. Leland III, to accommodate shipyard workers, and clearly inspired by the published works of Baillie Scott and Parker and Unwin for Letchworth Garden City and Hampstead Garden Suburb).

In Britain, firms like Liberty's and Heal & Son responded to these new homes by creating affordable "cottage" furniture alongside their more expensive and elaborate ranges. They capitalized on the visual appeal of the Arts and Crafts movement without necessarily subscribing to its social aims. Ambrose Heal was a shrewd businessman and talented furniture designer who may have been in tune with the ideals of Arts and Crafts, but was unafraid of employing modern methods. He maintained that "the machine relieves the workman of a good deal of drudgery, and legitimately cheapens production." This was an important point for—despite their best intentions—the insistence on the handcrafting of artefacts meant that most pieces by Arts and Crafts designers were beyond the means of the very people they hoped would benefit from their "improving" presence in their homes.

Above: *Advertisement from* The Studio *yearbook of 1907 promoting Heal & Son's "Country Cottage" look.*

Liberty's introduced the work of some of the period's best designers, including Voysey, Baillie Scott, Arthur Gaskin, Bernard Cuzner, Oliver Baker, and Jessie M. King, to a mass market. However, because of their strict rule that artists should not mark their work, some, like Archibald Knox, their most prolific and innovative designer, remained largely and unfairly obscure for more than half a century. The firm's founder, Arthur Lazenby Liberty, set out to satisfy the seemingly insatiable appetite of the middle classes for fashionable "Aesthetic" interiors. He did sell one-off items but much of the work was machine-made with the odd finishing touch by hand, leading Ashbee to call them "Messrs. Nobody, Novelty and Co." and to blame them for the failure of his Guild of Handicraft. Certainly, Liberty's "Cymric" venture undercut his more expensive, handmade Guild work and accelerated the demise of the silver workshop. Liberty, a clever businessman, completely ignored the social and ethical side of Arts and Crafts while trading on its spirit, visual appeal, and the current fashion for Aesthetic style. While he succeeded in reaching a wider public with his fabrics, metalwork, and furniture than Morris & Co. or any of the Guilds, his policy of anonymity for his designers succeeded in stifling or suppressing the impulse toward celebrating the individual that had been at the heart of the movement at its outset.

As the twentieth century progressed, cracks appeared in the Arts and Crafts movement prompted, once again, by the debate over machines. Some, like Gimson and Ernest Barnsley, argued passionately against any involvement whatsoever with machines or industry, while Lethaby steered a middle course, urging designers to be motivated by a sense of social purpose and responsibility, not just commercial success. In 1913, in his essay on "Art and Workmanship" in *The Imprint*, he wrote: "Although a machine-made thing can never be a work of art in the proper sense, there is no reason why it should not be good in a secondary order—shapely, smooth, strong, well fitting, useful; in fact, like a machine itself. Machine work should show quite frankly that it is the child of the machine; it is the pretence and subterfuge of most machine-made things which make them disgusting." European designers were never as troubled by this dilemma. The essential difference between the British and European Arts and Crafts designers was the latter's readiness to embrace industrialization and harness the machine. By applying

Arts and Crafts values to machine production, preserving integrity in the use of materials, revering the crafts ethic and the ideal of design unity, they sought to reform industrial design. The founders of the Deutsche Werkbund—Hermann Muthesius, Friedrich Naumann, and Henry van de Velde—stressed the need to restore the dignity of labor but in alliance with, rather than in opposition to, the machine. In common with Josef Hoffman and Koloman Moser at the Wiener Werkstätte, they shared the social vision of the founding members of the Arts and Crafts movement and insisted that art and architecture should be put in the service of society in order to create a nobler environment for contemporary humanity. They rejected the *l'art pour l'art* philosophy of the late nineteenth century, turned their backs on the past, and set out to create a new ornament and style appropriate to the machine age. (In this, revolutionary figures such as Christopher Dresser were already ahead of them.)

The central differences between the British and American movements are threefold. First, the USA did not have quite the same class issues as Britain, so there was less of an ideological component; second, the foremost American Arts and Crafts exponents were unembarrassed about promoting, marketing, indeed, commercializing their product; third, their attitude to the machine was in marked contrast to that of their British counterparts. First Stickley and then Frank Lloyd Wright openly designed for machinery and delighted in the new possibilities it provided. "The machine is capable of carrying to fruition high ideals in art, higher than the world has yet seen!" Wright proclaimed in a landmark lecture entitled "The Art and Craft of the Machine." Inevitably this meant that Morris's ideas were diluted, yet paradoxically they reached a wider audience and were made to work in a way that he himself had failed to achieve because of his insistence on expensive handcrafting and his unwillingness to market himself.

By 1915, the wider aims of the Arts and Crafts movement had shrunk. The socialist impulse to change society was no longer driving the movement forward, and what remained had been distilled to individual expression and creativity. For Ashbee the dream had been unraveling for some time. "We have made of a great social movement, a narrow and tiresome little aristocracy working with great skill for the very rich," he wrote despondently in his private journal.

In Britain the movement never really recovered from the impact of World War I, and in the USA it was left reeling from the effects of the Wall Street Crash in 1929. European designers were already responding to Modernism's siren call, although in Scandinavia the Arts and Crafts dream was successfully and perhaps more enduringly accomplished by using the region's rich natural resources to realize Morris's idea of a "decorative, noble, popular art." Gradually, however, the Arts and Crafts movement atrophied in Britain. Its insistence on handcrafting had made its products desirable but prohibitively expensive. Commercial concerns like Liberty's, Pilkington's, Minton, and Wedgwood had been quick to commission Arts and Crafts artists to come up with designs they could mass-produce

cheaply to satisfy the public appetite; they were quicker still to dance to the Modernist tune when the quaint cosy cottageyness of the Arts and Crafts movement began to appear outdated.

Now, Arts and Crafts artefacts are highly prized and often very expensive. Sotheby's recently sold a paneled oak double bed designed by Edwin Lutyens for a sum that would, when it was made, have secured the architect's services and covered the building costs of a sizable country house, with a Gertrude Jekyll garden thrown into the bargain. It is perhaps ironic that today's price for a single Stickley settle sold at auction in America would have cleared its creator's debts and saved him from sliding into bankruptcy. The revival in popularity of the Arts and Crafts style is easily accounted for. Besides being beautiful and simple in form, much of the work celebrates domesticity and we find it easy to identify with its lack of pretension and robust solidity. Yet, as this book sets out to show, there is a great deal more to the movement than the beguiling warmth and homeliness of familiar details of design.

Above: *Light oak four-poster bed with polished steel posts, designed by Edwin Lutyens in 1897 and sold by Sotheby's one hundred years later for £18,000.*

a
Rd
MORRIS
AND
COMPANY,
DECORATORS LTD.
449,
OXFORD STREET.
No 6871 Daffodil
Colours
Width 36
Price 4/2
T.624-1919

THE MAKERS OF THE MOVEMENT

Left: *J.H. Dearle designed this "Daffodil" printed cotton as a furnishing fabric for Morris & Co. in 1891.*

1

WHO WAS WHO

ADDAMS, Jane (1860–1935), American philanthropist. With Ellen Gates Starr (1859–1940), she established the Chicago settlement house Hull-House, modeled on Toynbee Hall, in 1889. She was elected first president of the Women's International League for Peace and Freedom in 1919, and was awarded the Nobel Prize for Peace in 1931.

ASHBEE, Charles Robert (1863–1942), English architect, romantic socialist, and Utopian. Much influenced by William Morris, John Ruskin, and Walt Whitman, Ashbee devoted himself to social reform through craft while still an architecture student (articled to G.F. Bodley) in London. Having been brought up in "discreet luxury," the experience of living in the pioneering settlement house Toynbee Hall, in London's East End, where he taught art to some of London's poorer residents, led him to form the Guild of Handicraft in 1888. The combined school and workshop was devised along the lines of the medieval guild system. In 1890 the Guild expanded and moved to Essex House in the Mile End Road, and in 1902 moved out of London to rural Chipping Campden in the Cotswolds, where Ashbee hoped that by giving his workers their own homes and gardens, with recreational and intellectual facilities (a swimming pool as well as a library), they would live healthier, happier lives. Guild craftsmen were renowned for their work in copper, brass, and iron; they made silver and jewelry, often enameled; they excelled as cabinet-makers and carried out restoration as well as decorative schemes. In 1904 Ashbee opened the School of Arts and Crafts, which operated until 1916 under the sponsorship of Gloucestershire County Council, where Guild members and local people were able to mix in a wide variety of

Far left: *The Morris and Burne-Jones families photographed at the Grange, Fulham, in 1874. The gentleman on the left is Burne-Jones's father.*

Above: *C.R. Ashbee, photographed by Frank Lloyd Wright in 1910.*

BAILLIE SCOTT, Mackay Hugh (1864–1945), influential English architect. He was the eldest of fourteen children, and attended the Royal Agricultural College at Cirencester with a view to taking over the management of the family's Australian sheep farm, but abandoned these plans to become an architect and was articled to Major Charles Davis, the Bath city architect, in 1886. Following his marriage to Florence Nash in 1889, he moved to Douglas on the Isle of Man where he met the designer Archibald Knox. The two collaborated on the design of stained glass, iron grates, and copper fireplace hoods, which were installed in the houses built by Baillie Scott on the island. He became known for his suburban houses in red brick or stucco, planned around spacious living areas with highly decorated interiors, often using elm and oak. His "House for an Art Lover" design won the hightest prize in a competition organized by the Zeitschrift für Innedekoration in 1901, though it was not built. From 1898, Baillie Scott designed furniture for John P. White at the Pyghtle Works, Bedford, whose 1901 catalog included 120 of his pieces. The furniture was also sold through Liberty's. He and his family left the Isle of Man in 1901 and moved to Bedford, where his architectural practice flourished. Seemingly never short of clients, he did not retire until 1939. He divided his remaining years between Ockhams, a Kentish farmhouse

classes aimed at meeting both their practical and spiritual needs. For over two decades Ashbee's democratically run Guild of Handicraft provided a model of communal living, profit sharing, and joyful labor (though with a bias toward craftsmen rather than craftswomen) that was an inspiration to Arts and Crafts idealists worldwide. Sadly, the dream of the simple life in idyllic surroundings came to an end in 1907 when the Guild, which had been experiencing financial difficulties for some time, went into liquidation. Ashbee stayed on in Chipping Campden with his wife Janet and their four daughters until 1919, after which they moved to Kent. Ashbee's furniture and metalwork were exhibited and much admired in Vienna, Munich, and Brussels. His influence extended to the USA, where he had laid the groundwork in 1900 when he visited fourteen American states on a lecture tour, meeting Frank Lloyd Wright in his home on his last day in Chicago. "The real thing is the life," he wrote—and lived up to it.

Above left: *M.H. Baillie Scott was destined to make a major contribution to the Arts and Crafts movement. and his work gained international acclaim.*

Above right: *A bedroom at Upper Dorvel House, the home of Ernest and Alice Barnsley, photographed by Ernest's brother Herbert, c. 1905.*

near Edenbridge, and rooms at the Kensington Palace Mansions Hotel in London.

BARNSLEY, Ernest (1863–1926), English architect and furniture designer. After training with J.D. Sedding in London, he established his own architectural practice in Birmingham in 1892, prior to joining his shyer brother Sidney Barnsley and Ernest Gimson in the Cotswolds in 1893. Affable and enthusiastic by nature, he set up a craft workshop in 1900 but concentrated mostly on his architectural work, much of which was for the Society for the Protection of Ancient Buildings founded by William Morris.

BARNSLEY, Sidney (1865–1926), English furniture designer and maker. Born in Birmingham, he trained as an architect in the London office of Richard Norman Shaw. With Ernest Gimson, Reginald Blomfield, Mervyn Macartney, and W.R. Lethaby, he established the London firm of Kenton & Co., designing furniture according to basic construction principles. Decoration, if used at all, was inlaid, using materials such as mother-of-pearl or ivory. After the collapse of the company, he moved to Gloucestershire and set up a workshop with Gimson and his brother Ernest Barnsley. Sidney married his cousin, Lucy Morley, in 1895. Deaf from the age of nineteen, she was nevertheless practical and efficient, and the Barnsleys lived a rural life following their avowed principles; they baked their own bread, chopped their own wood, kept hens and goats, brewed cider, and made sloe gin. Sidney made his own furniture, working alone, because he enjoyed the making as much as the designing. He sent his two children, Grace and Edward, to Bedales in Hampshire, and in 1910 the school commissioned a new assembly hall from Ernest Gimson and ten years later a library, which was built under the supervision of Sidney Barnsley after Gimson's death. Sidney died in 1926 within months of his brother

and the two are buried, alongside Gimson, at the village of Sapperton, where they lived and worked for so many years. Their graves are marked with plain granite slabs.

BATCHELDER, Ernest Allan (1875–1957), American ceramic artist, design theorist, and teacher. After a period of study at the School of Arts and Crafts in Birmingham, England, he established his own school and tile factory in Pasadena in 1909, producing molded tiles with strong relief designs. The factory produced major ceramic installations for churches and commercial buildings, as well as architectural tiles for bungalow fireplaces that were sold throughout the USA.

Above: *This group photograph taken outside Gimson's Cotswold cottage around 1895 shows, from the left, Sidney Barnsley, his fiancée Lucy Morley, Ernest Gimson, Alice and Ernest Barnsley with their two daughters, Mary and Ethel.*

Right: *An elaborate mahogany and boxwood music cabinet and stand with wax inlaid scene of Orpheus charming the animals, designed by W.A.S. Benson and G.H. Sumner, c. 1889.*

BEHRENS, Peter (1868–1940), German architect and industrial designer. He was a founder of the Dresden and Munich Werkstätten, and later came to be influential in the development of Modernism.

BENSON, William Arthur Smith (1854–1924), English architect, metalworker, and furniture designer. He also designed wallpapers and the silver mounts and hinges of cabinets for Morris & Co., of which he became chairman in 1896 after William Morris's death. He had his own metal workshop in St. Peter's Square, Hammersmith, London, and opened showrooms in Bond Street in 1887. He was a founder member of the Art Workers' Guild in 1884 and the Design and Industries Association in 1915. He retired in 1920.

BIDLAKE, William Henry (1861–1938), English architect and honorary director of the Birmingham Guild of Handicraft. He taught at Birmingham's architecture school for ten years and was, according to The Studio, responsible for influencing and guiding a whole generation of younger architects.

BING, Samuel (formerly Siegfried) (1838–1905), German writer and entrepreneur whose Paris shop, La Maison de l'Art Nouveau, gave its name to Art Nouveau style and had a profound impact on Arts and Crafts designers.

BLOMFIELD, Reginald Theodore (1856–1942), English architect. He was much involved as a young man with the Arts and Crafts movement, founding Kenton & Co. with W.R. Lethaby and others, but moved away to mature along classical lines, building up a highly successful architectural practice and becoming president of the Royal Institute of British Architects in 1912.

BLOUNT, Godfrey (1859–1937), British artist and craftsman. He set up the Haselmere Peasant Industries in Surrey in 1896, as an artistic community with the same idealist aims as C.R. Ashbee's, to combine work—treadle weaving, appliqué embroidery, and making handknotted carpets—and leisure with the "revival of a true country life where handicrafts and the arts of husbandry shall exercise body and mind and express the relation of man to earth and to the fruits of earth." With his wife, Ethel Hine, Blount was a great advocator of

Left: *A fine example of one of Godfrey Blount's designs for an embroidered appliqué hanging, made by the Haslemere Peasant Industries using silks and applied hand-woven linens on a linen ground.*

Right: *The Winter Smoking Room at Cardiff Castle. The third Marquess of Bute employed William Burges to transform the interiors in 1865; the result was a lavish and opulent mix of rich murals, stained glass, marble, gilding, and elaborate wood carvings.*

the simple life and belonged, as did Ashbee and his wife Janet, to the Healthy and Artistic Dress Union, founded in London in 1890 to promote the wearing of comfortable, loose-fitting, hand-woven clothes.

BLOW, Detmar Jellings (1867–1939), English architect. He began his career by apprenticing himself to a mason in Newcastle-upon-Tyne, eventually becoming assistant to Ernest Gimson and working on a series of Cotswold commissions, including Stoneywell Cottage. In 1900 he designed Happisburgh Manor in Norfolk on a butterfly plan like that of E.S. Prior's architecturally significant house The Barn at Exmouth (1897), though Blow claimed that his inspiration came from an idea seeded by Gimson. Between 1895 and 1914 he had one of the largest country-house practices in Britain and earned over £250,000. In 1910 he married the daughter of Lord Tollemache and bought a thousand-acre estate in Gloucestershire, on which he built his own country house, Hilles. His fortunes declined, along with his commissions, following World War I.

BODLEY, George Frederick (1827–1907), English Gothic architect and one of the first patrons of Morris & Co. In 1874 he started his own firm, Watts & Co., with Thomas Garner and George Gilbert Scott Jr., to make hand-blocked wallpapers and textiles in the Queen Anne revival style. He was important to the Arts and Crafts movement because he produced some of its leading architects, including C.R. Ashbee, who joined him as a pupil in 1886, and the Scottish architect Robert Lorimer, who spent a short time in the Bodley office in 1889.

BRADLEY, Will (1868–1962), American typographer, printer, and illustrator. He founded the Wayside Press in Springfield, Massachusetts, in 1896. His illustrations of the "Bradley House" presented the Arts and Crafts ideal in a series of articles for Ladies' Home Journal (1901–2) intended to promote good domestic architecture throughout the USA. Readers were told that, "Mr. Bradley will design practically everything in the pictures," which included furniture, tapestries, wallpapers, and murals and drew heavily on the influences of M.H. Baillie Scott, C.F.A. Voysey, and Frank Lloyd Wright. None of the furniture he illustrated is known to have been executed although he did design and have built three houses for his family.

BRANGWYN, Frank (1867–1956), British artist and etcher, designer of furniture, pottery, metalwork, jewelry, rugs, embroideries, and fans. Largely self-taught, when in his teens he secured a job at Morris & Co., enlarging the designs for tapestries, through the influence of A.H. Mackmurdo. In 1895 he contributed to the decoration of Samuel Bing's Maison de l'Art Nouveau in Paris,

and in 1899 he designed stained glass for Louis Comfort Tiffany. His work on major decorative commissions included the British Rooms for the Venice Biennale exhibitions of 1905 and 1907 and murals in the Rockefeller Center in New York (in collaboration with Jose Maria Sert and Diego Rivera). He was knighted in 1941.

BROWN, Ford Madox (1821–93), British painter and mentor to the Pre-Raphaelites. Brown was a founding member of the Firm, for whom he designed furniture and stained glass, though he fell out with Morris over the restructuring of the business. Twice married and a committed socialist, he was dogged in his early years by poverty and anxiety, and never gained the recognition he deserved. In later years he toiled over a series of important murals depicting local history in Manchester Town Hall for the paltry sum of £300 a year, suffering a stroke that affected his painting arm, but going on to finish the series with his left hand.

BURDEN, Elizabeth (b. 1842), sister to Morris's wife, Janey, and known as Bessy, Burden lived with the family and became a skilled embroiderer, going on to teach at the Royal School of Needlework and becoming an adviser to schools in the London area.

BURGES, William (1827–81), English architect and designer. A disciple of A.W.N. Pugin, he was a leading Gothic revivalist. He also designed ornate and flamboyant furniture in painted and gilded wood, and was well known for his eccentric and playful silver designs featuring otters, mermaids, mice, and spiders.

BURNE-JONES, Edward (1833–98), English artist and member of the Pre-Raphaelite Brotherhood. Born in Birmingham, he met William Morris when they were both theology students at Oxford. Their lifelong collaboration—through the Firm and the Kelmscott Press, on designs for tapestries, tiles, book illustrations, and decorative schemes—was extraordinarily fruitful. Burne-Jones became one of the most important painters of his age and exerted a strong influence on younger artists.

BUTTERFIELD, William (1814–1900), English architect. He is best known for his ecclesiastical buildings, schools, and colleges, including St. Saviour's Church, Coalpit Heath, near Bristol (1844–5), All Saints Church, Margaret Street, London (1849–59), and Keble College, Oxford (1867–83). Imaginative in his use of materials and Gothic details, he was concerned to integrate the design of the building and its furnishings and gave practical consideration to function.

CARPENTER, Edward (1844–1929), influential English poet, farmer, and romantic socialist. He encouraged "the absence of things" and helped to form Arts and Crafts ideals and practice. He promoted the simple life and the beauty of comradeship, together with the idea of "homogenic love," which transformed his homosexual urges into the pure love of comrades working together for the common good. Exposure to the ideas of Ralph Waldo Emerson and Walt Whitman went hand in hand with a concern for the plight of the working poor.

CHARLES, Ethel Mary (1871–1962), first woman member of the Royal Institute of British Architects (1898). She trained in the office of Ernest George and Harold Peto and submitted an entry to the Letchworth Cheap Cottages Exhibition in 1905, carrying on a flourishing practice in Letchworth with her sister Bessie Ada, who in 1900 was elected the second woman member of the RIBA. Ethel also designed a number of houses and cottages in an understated vernacular style in Falmouth, Cornwall.

CLISSETT, Philip (1817–1913), English carpenter responsible for the West Midlands-style rush-seated ladderback chair, variations of which were made by Ernest Gimson (who learned the craft from Clissett), C.R. Ashbee, and Ambrose Heal. Clissett's traditional chairs were used in the hall of the Art Workers' Guild and were therefore seen by influential people.

COBDEN-SANDERSON, Thomas James (1840–1922), English book-designer and binder. A neighbor and friend of William Morris, he was a member of the Art Workers' Guild and founded the Doves Bindery in 1893. In 1900, together with Emery Walker, he set up the Doves Press in Hammersmith Terrace, London, one of the finest and best known among the many British private presses of the early twentieth century.

COCKERELL, Douglas (1870–1945), English bookbinder. A former Doves Bindery apprentice, he taught bookbinding at the Central School of Arts and Crafts in London from 1896.

CONNICK, Charles (1875–1945), leading American stained-glass artist. He ran his Boston studio on a medieval guild system and produced some of the finest stained glass in America, largely Gothic in style.

COOMARASWAMY, Ananda (1877–1947), Anglo-Sinhalese geologist and anthropologist. He commissioned C.R. Ashbee and the Guild of Handicraft to restore and convert a Norman chapel in Chipping Campden for himself and his wife, Ethel Partridge (later Ethel Mairet). He also bought shares in the failing Guild in 1907 and bought one of Ashbee's Essex House presses on which to print his own book, *Medieval Sinhalese Art* (1909).

COOPER, John Paul (1869–1933), English architect, designer, goldsmith, silversmith, and jeweler. Born in Leicester, he began his working career in the office of J.D. Sedding, where he was apprenticed on the advice of W.R. Lethaby. An interest in ornamental plasterwork led to his working with gesso and precious metals, and in 1901 he took up an appointment as head of the metalwork department at the Birmingham School of Art, a post he held for five years. Extraordinarily talented, he made nearly fourteen hundred pieces of jewelry, gesso-work, silverwork, and other metalwork in the workshop he designed himself, along with his house, at Westerham in Kent.

Far left: Sir Bors Sucoures the Mayde, c. *1919, a panel from the Holy Grail window by Charles Connick Associates for Proctor Hall at Princeton University. The window was designed to represent the quest for wisdom and understanding.*

Right: *Walter Crane has chosen to present himself as a fine artist—albeit one in a stiff Victorian collar—in this self portrait with palette; the irises and exotic tower in the background hint at the motifs that influenced him as a designer.*

CRANE, Walter (1845–1915), English artist, muralist, illustrator of children's books, craftsman, designer of wallpapers, textiles and tiles, frieze painter and mosaicist, socialist, writer, and committee man. A friend and disciple of William Morris, he believed "the true root and basis of all Art lies in the handicrafts." He was the son of the portrait painter Thomas Crane, and first exhibited at the Royal Academy at the age of seventeen, but became well known as a nursery book illustrator with the "Toy Books," which began to appear in 1865 and led to his being approached by Jeffrey & Co. to produce designs for nursery wallpapers. His first pattern, "The Queen of Hearts," went into production in 1875 and was followed by over fifty other designs. He devised decorative schemes for private homes and wrote and lectured extensively on home decoration. A crucial meeting with Morris led to Crane joining the Socialist League in 1883 and later the Fabian Society. He was among the founders of the Art Workers' Guild, and became master of the Guild in 1888 and president of the Arts and Crafts Exhibition Society in the same year. He toured the USA in 1891, accompanying an exhibition of his own work and giving lectures. In 1893 he became director of design at Manchester Municipal College and in 1898 was appointed principal of the Royal College of Art, then the leading school of design in Britain.

CRESWICK, Benjamin (1853–1946), British sculptor, metalworker, and Century Guild designer, who taught at the Birmingham School of Art from 1889–1918.

CUZNER, Bernard (1877–1956), English silversmith. The Redditch-born son of a watchmaker, he served his silversmithing apprenticeship with his father and at the Redditch School of Art before moving to Birmingham where he worked and studied under Arthur Gaskin at the Vittoria Street School, later becoming head of the metalwork department at the Birmingham School of Art. He designed for Liberty's "Cymric" range of silver and jewelry.

CZESCHKA, Carl Otto (1878–1960), Austrian architect, painter, designer, and member of the Wiener Werkstätte.

Left: *Lewis F. Day designed this oak embroidery cabinet in 1888 and it was exhibited at the Arts and Crafts Exhibition that same year. Inlaid with ebony and satinwood, it is decorated with painted panels of signs of the Zodiac executed by George McCulloch.*

Right: *An embroidered wool portière designed by J.H. Dearle for Morris & Co., c. 1910.*

DAWSON, Nelson (1859–1942), English enameler and silversmith. After studying enameling under Alexander Fisher he set up a workshop in Chiswick, London, with his wife Edith Robinson in 1893, carrying out enameled decoration that often took the form of delicate jewel-like flowers and insects. In 1900 the Dawsons showed 125 pieces of jewelry at an exhibition at the Fine Art Society in Bond Street. He set up the Artificers' Guild in 1901, which was later acquired by Montague Fordham, first director of the Birmingham Guild of Handicraft.

DAY, Lewis Foreman (1845–1910), leading English designer of textiles, carpets, wallpapers, stained glass, embroidery, pottery, tiles, book covers, and furniture. He brought together a group of artists interested in design, known as "The Fifteen," who joined with the pupils and assistants of Richard Norman Shaw to form the Art Workers' Guild in 1884. He was a founding member of the Arts and Crafts Exhibition Society in 1888, a regular lecturer, author, and contributor to the *Art Journal*. Day's political and social ideas were very different from those of William Morris, though he was influenced by his example.

DEARLE, John Henry (1860–1932), British textile designer. He was taken on as William Morris's first apprentice tapestry weaver in Queen Square, London, because Morris was "influenced by the evident intelligence and brightness of the boy." Dearle went on to supply his own textile designs for Morris & Co., including "Daffodil," one of the Firm's most popular fabrics, and "Persian Brocatel," a loom-woven silk and wool fabric designed specifically for wall-covering at Stanmore Hall, Middlesex. He became art director of the Firm after Morris's death in 1896.

DE MORGAN, William Frend (1839–1917), influential English potter, designer, and novelist. An exceptional colorist, he trained first as a painter at the Royal Academy Schools, where he became part of the Pre-Raphaelite circle and began designing stained glass and tiles for Morris & Co. He set up a kiln in the basement of his own home in Fitzroy Square, London, experimenting with luster decoration for his pottery and eventually blowing the roof off the house. In 1872 he moved to 30 Cheyne Row, which he used as showroom, studio, and workshop. In 1882 he moved the business to Merton Abbey, close to William Morris's new workshops, but found traveling to and from Chelsea time-consuming and deleterious to his health so transferred the works to Fulham in 1888 and entered a ten-year partnership with the architect Halsey

Ricardo. He married the artist Evelyn Pickering in 1887. Although he was a great friend of Morris and a supplier of tiles to the Firm, De Morgan never joined a "collaborative" and was not a member of the Art Workers' Guild. "I could never work," he wrote, "except by myself and in my own manner." He took his inspiration from Iznik wares of the fifteenth to seventeenth centuries and produced "Persian" ceramics in ruby reds, delicate golds, and bluish grays, decorated with birds or plants. Nikolaus Pevsner called him "the great English potter of his age." Although he sometimes decorated ready-made factory blanks, he despised the machine as used in industry. His assistants, Fred and Charles Passenger, remained with him for thirty years, becoming his partners in 1898 and eventually taking over when he retired from the business in 1905. He spent a considerable amount of time in Florence and in his later years became a bestselling novelist.

DEVEY, George (1820–86), English architect. He designed more than twenty country houses and estate cottages, and provided sympathetic additions to Elizabethan or Jacobean buildings. He had first trained as an artist under the Norwich School watercolorist John Sell Cotman, and was one of the first Victorian architects to design new buildings based on the local vernacular, thus ensuring that his new designs harmonized with older buildings. C.F.A. Voysey was one of his assistants from 1880–2.

DIXON, Arthur Stansfield (1856–1929), English architect and designer. Based in Birmingham, he was responsible for much of the design work at the Birmingham Guild of Handicraft, established along similar lines to C.R. Ashbee's idealistic Whitechapel school, where Dixon had worked for a year in 1890.

DOAT, Taxile (1851–1938), French ceramicist who moved from Sèvres to teach at the University City Pottery, St. Louis, Missouri, in 1909. His designs had a lasting influence on American art pottery.

DRESSER, Christopher (1834–1904), Scottish pioneer of modern product design, which he elevated to the status of art. Born in Glasgow, he trained as a botanist and led a counterattack on John Ruskin in his book *The Art of Decorative Design* (1862) arguing, as a scientist, that design should represent the laws of natural growth, not its appearance. Prolific, radical, and decidedly ahead of his time as a designer, he worked in a wide range of materials, styles, and technique, and designed ceramics, glass, metalwork, furniture, carpets, textiles, wallpapers, and other interior decorations. A great innovator and an enthusiastic advocate of the machine, he came under the influence of William Dyce, Henry Cole, and Matthew Digby Wyatt as a student and later lecturer at the Government School of Design. More an admirer of Owen Jones than William Morris, he was greatly influenced by Japanese art following a visit to Japan in 1877, and Tiffany & Co. commissioned him to design a collection of Japanese-style artefacts. While visiting the USA on his way to Japan, he lectured at the new Pennsylvania School of Industrial Art. On his return to London, he set up Dresser & Holme in Farringdon Road, selling Oriental goods, but the business failed. In 1880 he was appointed art

Above: *The Islamic influence is evident in this bowl from the Fulham period by William De Morgan, c. 1890.*

Right: *A kettle made from spun copper and cast brass with an ebony handle, designed by Christopher Dresser, c. 1880, and made by Benham & Froud.*

manager of the Art Furnishers' Alliance, which had a showroom and shop in New Bond Street where his metalwares, pottery, glass, and fabrics were displayed and sold. This venture also failed. By the 1890s, having been design director for both Minton and the Linthorpe Pottery, Dresser was back in Glasgow designing glass for James Couper & Sons. He married Thirza Perry in 1854 and had thirteen children, five of whom died in childhood. He himself died in relative obscurity in France.

EASTLAKE, Charles Locke (1836–1906), English architect and writer. He was the author of the bestselling and influential *Hints on Household Taste* published in 1868 in Britain and 1872 in the USA, which promoted a single, cohesive style in the home.

ELLIS, Harvey (1852–1904), American architect. An immensely talented designer, he trained under Arthur G. Gilman in New York and his own work in Rochester, New York, was inspired by the Romanesque style of Henry Hobson Richardson. A founding member of one of the earliest Arts and Crafts societies in the USA, he joined Gustav Stickley's Craftsman Workshops in 1903 making, in less than a year, an indelible impression on the American Arts and Crafts movement.

ELMSLIE, George Grant (1869–1952), Scottish-born American architect and interior designer. He entered the Chicago offices of J.S. Silsbee (who was responsible for introducing the Shingle style to the American midwest) in 1887 where he met Frank Lloyd Wright. After twenty years working with Louis Sullivan, he set up his own practice in partnership with William G. Purcell and George Feick in 1909. For his most important interiors he also designed oak furniture, together with leaded glass, terracotta ornaments, carpets, and textiles.

FAULKNER, Charles (1834–1892), highly original mathematician and founder member of Morris, Marshall, Faulkner & Co., who followed Morris in his espousal of socialism and led the Oxford branch of the Socialist League.

FISHER, Alexander (1864–1936), British sculptor, enameler, and silversmith. He is widely acknowledged as the master of the revived jewel-like "Limoges" technique of enameling, which he learned from the French enameler Louis Dalpayrat who lectured at the Royal College of Art in 1885. Fisher became head of the enamel workshop at the newly founded Central School of Arts and Crafts in 1896. The most influential enameler in Britain, he promoted enamel paintings as jewels or art objects set in gold, silver, or steel. In 1904 he set up a school of enameling in his studio in Warwick Gardens, Kensington, attracting aristocratic pupils including the Hon. Mrs. Percy Wyndham, chatelaine of Clouds in Wiltshire, who encouraged her friends to commission enameled pieces and portraits from Fisher. He wrote extensively and his work influenced many artists involved in enameling.

FORDHAM, Montague (1869–1942), first director of the Birmingham Guild of Handicraft. He later acquired the Artificers'

Left: *Montague Fordham advertises his services in* The Studio, *1904.*

Below: *Eric Gill designed this book plate for his friend Ananda Coomaraswamy in 1920.*

Guild, set up by Nelson Dawson in 1901, and established it in his gallery at 9 Maddox Street, London. There he sold and popularized the best contemporary decorative work executed by Arts and Crafts practitioners such as Henry Wilson and J. Paul Cooper—rather as the dealer Samuel Bing encouraged Art Nouveau through his Maison de l'Art Nouveau in Paris.

GASKIN, Arthur Joseph (1862–1928), English painter, illustrator, jewelry designer, and enameler. He studied at the Birmingham School of Art, where he subsequently taught, and provided wood-engravings for a number of books produced by William Morris at the Kelmscott Press, including Edmund Spenser's *The Shepheardes Calender* (1896). He married fellow student Georgina (Georgie) Cave France in 1894 and they worked together designing and making jewelry and silverwork inspired by the natural world. In 1902 he became head of the newly founded Vittoria Street School for Jewellers and Silversmiths in Birmingham. He later worked in the Cotswolds and died in Chipping Campden.

GATES, William Day (1852–1935), American founder of the Gates Pottery of Terra Cotta in Illinois in 1885. From 1900 it produced a line of ceramic vases and garden ornaments called Teco Art Pottery, commissioning designs from, among others, Frank Lloyd Wright.

GILL, Eric (1882–1940), English typographer, calligrapher, letter-cutter, wood-engraver, and sculptor. He began his career in London but, like C.R. Ashbee and Ernest Gimson, abandoned the city for the "sweetness, simplicity, freedom, confidence and light" promised by the English countryside. At Ditchling, Sussex, from 1907 he strove toward a close, classless (though male-dominated) community and in his *Autobiography* (1940) hoped that he had "done something towards re-integrating bed and board, the small farm and the worship, the home and the school, earth and heaven."

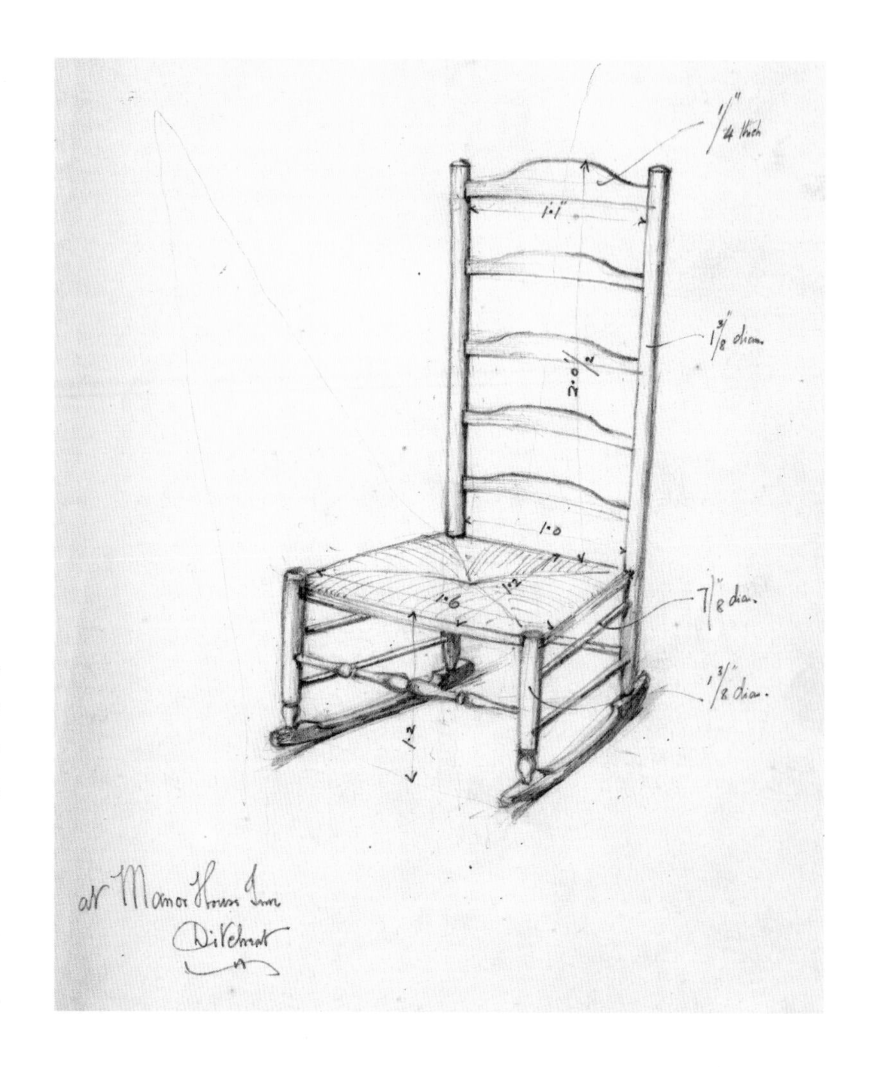

Right: *This page from Ernest Gimson's sketchbook of 1886, drawn at the Manor House Inn, Ditcheat, Somerset, shows his early interest in furniture design.*

GILL, Irving (1870–1936), American architect. In California he pioneered an Arts and Crafts-based style that owed much of its inspiration to the Spanish missions, demonstrated in houses such as the Laughlin (1907) and Dodge houses (1916), both in Los Angeles.

GIMSON, Ernest (1864–1919), English architect and designer of furniture and metalwork. On the personal recommendation of William Morris, a family friend, he moved from his native Leicester to London to enter the office of J.D. Sedding, where he met Ernest Barnsley and through him his brother Sidney, who was working in the office of Richard Norman Shaw. Gimson interested himself in traditional crafts such as decorative plasterwork, woodturning and rush-seated chair-making, and brought these skills with him to Kenton & Co., a youthful furniture-making enterprise he embarked on with W.R. Lethaby, Sidney Barnsley, and two slightly older architects, Reginald Blomfield and Mervyn Macartney, in 1890. Following the collapse of the company two years later he moved with the Barnsleys to Gloucestershire, setting up a workshop in Pinbury Park, which later moved to Daneway House, where he forged his design style, characteristic geometry, and style of surface decoration. He married Emily Ann Thompson, a Yorkshire vicar's daughter, in 1900. He remained committed to the ideal of the simple life in the Cotswolds, working on local architectural commissions and overseeing the craftsmen he employed to implement his furniture and metalwork designs—including Peter van der Waals and Alfred Bucknell, a talented blacksmith. He died of cancer on August 12, 1919, aged only fifty-five, leaving much work unfinished.

GLEESON, Evelyn (1855–1944), founder of the Dun Emer Guild in Dublin in 1902. The Guild was a focus for home industries, including lacemaking, embroidery, and woodcarving. She worked with professionals like Oswald Reeves, an enameler and former pupil of Alexander Fisher, and the stained-glass artists Harry Clarke, Alfred Child (a former apprentice of Christopher Whall), and Sarah Purser. Discussions with Emery Walker and W.B. Yeats led to the foundation of the Dun Emer Press (1903–7) and Dun Emer Industries (1904–8). The latter was run by Yeats's sisters Elizabeth, a bookbinder, and Lily, a former embroidery student of May Morris.

Right: *A watercolor drawing by E.W. Godwin showing a desk and double hanging bookcase. Godwin's impetus, like Morris's, came from needing to furnish his own house and failing to find anything suitable on the market.*

GLESSNER, Frances Macbeth (1848–1932), American philanthropist, silverworker, and needlewoman. Her keen enthusiasm for the Arts and Crafts movement is reflected in Glessner House, Chicago—which she and her husband John commissioned from H.H. Richardson—and its contents.

GODWIN, Edward William (1833–86), English Gothic architect, architectural journalist, and stage and furniture designer. He was born in Bristol, where he established an architectural practice. In 1861 he won the competition for the design of the new Northampton town hall, including its interior decorations, which show the Japanese influence he was coming under. He was retained for several years at £450 a year (in addition to payments for furniture designs) by the London firm of Collinson & Lock, "Art Furnishers," for which he provided designs for furniture, fireplaces, gas brackets, carpets, lockplates, and iron bedsteads. Godwin's light, elegant furniture sometimes incorporated painted panels by friends such as Edward Burne-Jones, Albert Moore, and James McNeill Whistler. He also designed wallpapers with an evident Japanese influence for Jeffrey & Co., as well as tiles for Minton, and was the original architect for Jonathan Carr's Bedford Park estate before being replaced by Richard Norman Shaw in 1877. He lived with the actress Ellen Terry for many years and was described by Max Beerbohm as "the first of the aesthetes."

GREEN, Arthur Romney (1872–1945), self-taught British designer and cabinet-maker. He set up a workshop, first in Surrey and then at Hammersmith, along cooperative lines influenced by the ideas of William Morris, producing furniture very much in keeping with the style of Ernest Gimson, Sidney Barnsley, and other Arts and Crafts designers.

GREENE, Charles Sumner (1868–1957) and **GREENE, Henry Mather** (1870–1954), quintessential American Arts and Crafts architects and designers. They attended the Manual Training High School in St. Louis, one of the first Arts and Crafts academies in the USA, run by Calvin Milton Woodward, followed by architectural training at the Massachusetts Institute of Technology. They established an architectural practice in Pasadena in 1893 and were responsible for iconic buildings such as the Gamble House (1908–9) for which they also designed the garden, furniture, lighting fixtures, and stained glass—as they did for other commissions. The discovery of Japanese prints and Oriental gardens was a revelation to them and they brought some of the same spare elegance and love of the pastoral to their work, which was much admired by C.R. Ashbee. Charles, who

Left and right: *Charles Sumner Greene (left) and his brother Henry Mather Greene (right) were hugely influenced by the Japanese exhibit at the World's Colombian Exposition in Chicago in 1893.*

was generally considered the more artistic of the pair, married Alice Gordon White, an Englishwoman, in 1901, two years after his brother had married Emeline Augusta Dart. After 1909 their work lost much of its Arts and Crafts simplicity, though they continued working until 1923 for rich Californian clients.

GRIGGS, Frederick Landseer Maur (1876–1938), English illustrator, etcher, and conservationist. He joined C.R. Ashbee's Chipping Campden community in 1903, settling there with his wife Nina. From 1902 he illustrated the series Highways and Byways Guides published by Macmillan, a commission that occupied him for the rest of his life.

GROPIUS, Walter (1883–1969), German architect. He studied under Peter Behrens and became leader of one of the most important design schools to follow on from the Arts and Crafts movement: the Bauhaus, founded in Weimar in April 1919, which came to exemplify the Modern movement and the functionalist design ethic—essentially more modern, less machine-averse, and firm in the revolutionary belief that good design served the needs of the ordinary people, the workers. Its early emphasis on guild structure and ideals paralleled C.R. Ashbee's experiment in creating a working community of craftsmen.

GRUEBY, William Henry (1867–1925), American founder of the Grueby Faience Company. Its pottery was recommended through Gustav Stickley's *Craftsman* as "perfectly complementing a Craftsman Home." Grueby began working in the pottery trade at thirteen, for Fiske, Colman & Co. He popularized a matt dark-green glaze on clearly handmade pottery and used simple, restrained leaf and flower motifs.

HEAL, Ambrose (1872–1959), English furniture designer and entrepreneur. He studied at the Slade School of Art and learned his craft as an apprentice cabinet-maker in Warwick, before joining the family firm as a bedding designer in 1893. In tune with the ideals of the Arts and Crafts movement, he was nevertheless one of the first British designers to overcome the movement's dislike of modern methods. He created a strong identity for Heal & Son, using Eric Gill to design the company's brochures and the lettering for its London store. He was a founder member of the Design Club, formed in 1909, and treasurer of the Design and Industries Association, formed in 1915 and inspired by the example of the Deutscher Werkbund to promote cooperation between workers, designers,

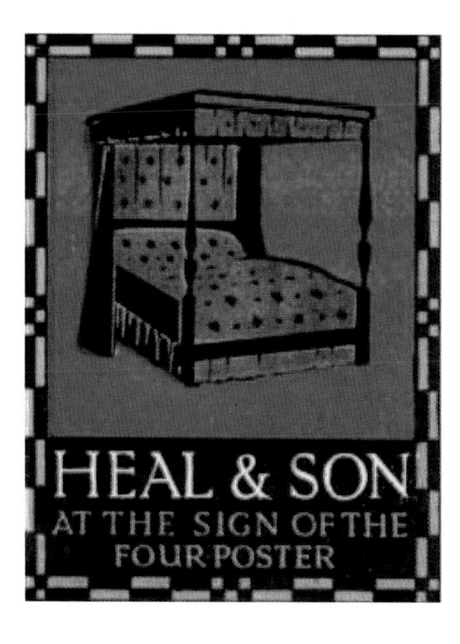

Right: *Heal's trademark "At the Sign of the Four Poster."*

distributors, manufacturers, educationalists, and the public in general, and to raise and improve the standards of design in industry. Although it promoted the best of British design through exhibitions, its acceptance of the necessity of the machine effectively signaled the end of the Arts and Crafts ideal as envisaged by William Morris and John Ruskin.

HEATON, Clement John (1861–1940), English stained-glass artist and metalworker. After training in his father's firm, he became a Century Guild associate. He carried out a number of A.H. Mackmurdo's designs in cloisonné enamels and set up his own firm, Heaton's Cloisonné Mosaics Ltd., in 1885. After 1887 he worked mainly in Switzerland, until moving to the USA in 1912.

HOFFMANN, Josef (1870–1956), Austrian architect, designer, and founder member of the Vienna Secession, along with Koloman Moser, Otto Wagner, Joseph Maria Olbrich, Gustav Klimt, and Egon Schiele. Deeply influenced by William Morris, John Ruskin, and Charles Rennie Mackintosh, Hoffmann visited C.R. Ashbee's Chipping Campden community in 1902 and returned to Austria to found the Wiener Werkstätte in imitation of the Guild of Handicraft, with the aim of creating "an island of tranquility in our own country, which, amid the joyful hum of arts and crafts, would be welcome to anyone who professes faith in Ruskin and Morris." He designed furniture, metalwork, glass, jewelry, and textiles for the Wiener Werkstätte, until the workshops went into liquidation in 1931, and gained the admiration of Le Corbusier.

Below left: *A hammered silvered metal coupe by Josef Hoffmann for the Wiener Werkstätte, c. 1910.*

HOLIDAY, Henry George Alexander (1839–1927), English mural artist and stained-glass designer. He worked for Powells of Whitefriars, the glass manufacturers, as a stained-glass cartoonist and also designed mosaics, enamels, and embroideries, often in collaboration with his wife Catherine, who executed a number of embroidered panels for Morris & Co.

HORNE, Herbert Percy (1864–1916), English architect, designer, poet, and art historian. Articled to A.H. Mackmurdo in 1883, he went on to become a major figure in the Century Guild, editing their journal, *The Hobby Horse*. A prodigious worker, he acted as an agent for designer-craftsmen and for some firms, including Morris & Co. In 1900 he settled in Florence, eventually leaving his considerable collection of Renaissance art to the city, along with the Palazzo Fossi, now the Museo Horne.

HORSLEY, Gerald Callcott (1862–1917), English architect. He was articled to Richard Norman Shaw in 1879 and was a founder member of the Art Workers' Guild in 1884. His commissions, often for churches, are characterized by carved and applied decoration.

Right: *"The Angel with the Trumpet," designed by Herbert Horne for the Century Guild and block printed by Simpson & Godlee from 1884.*

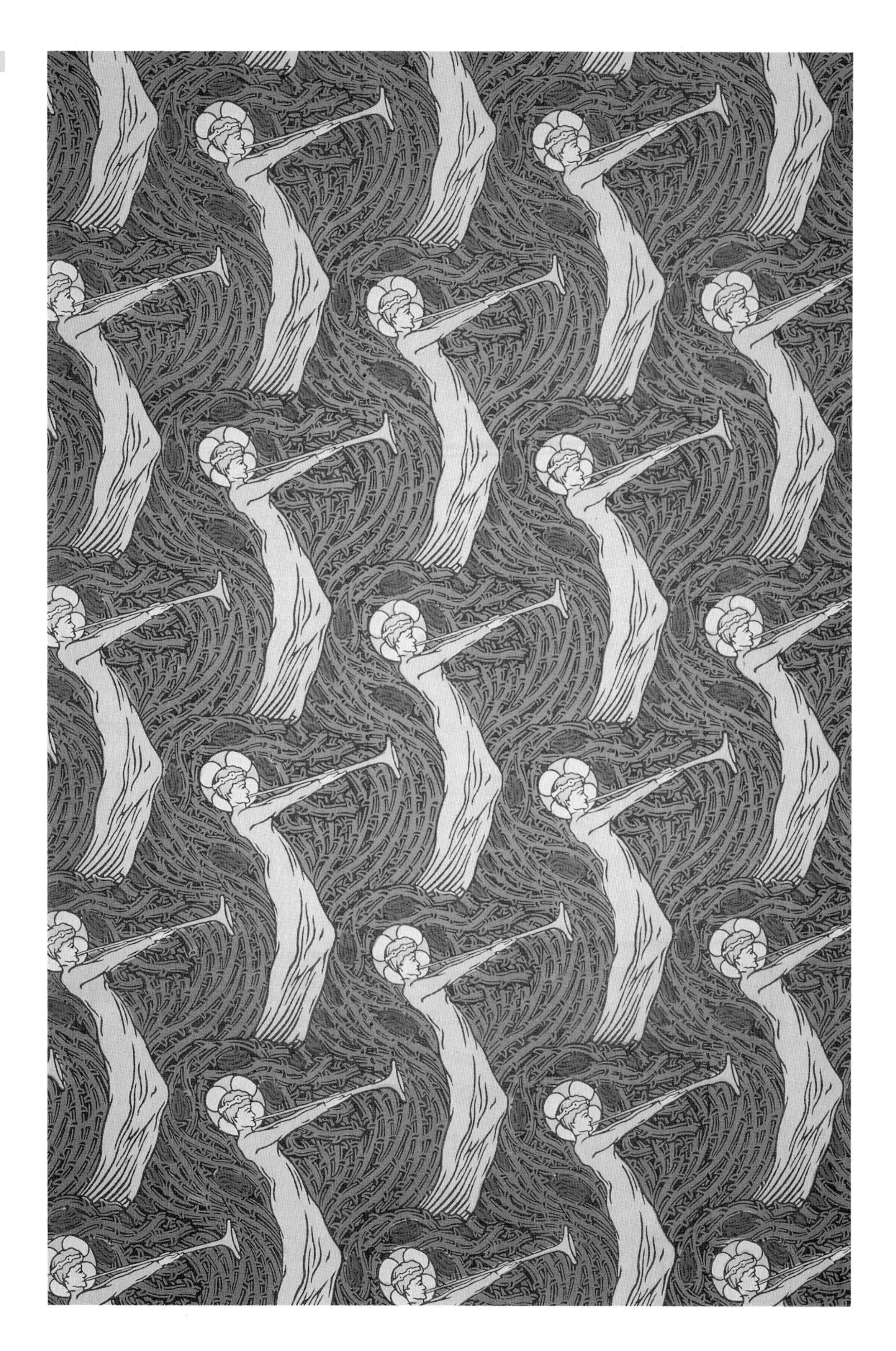

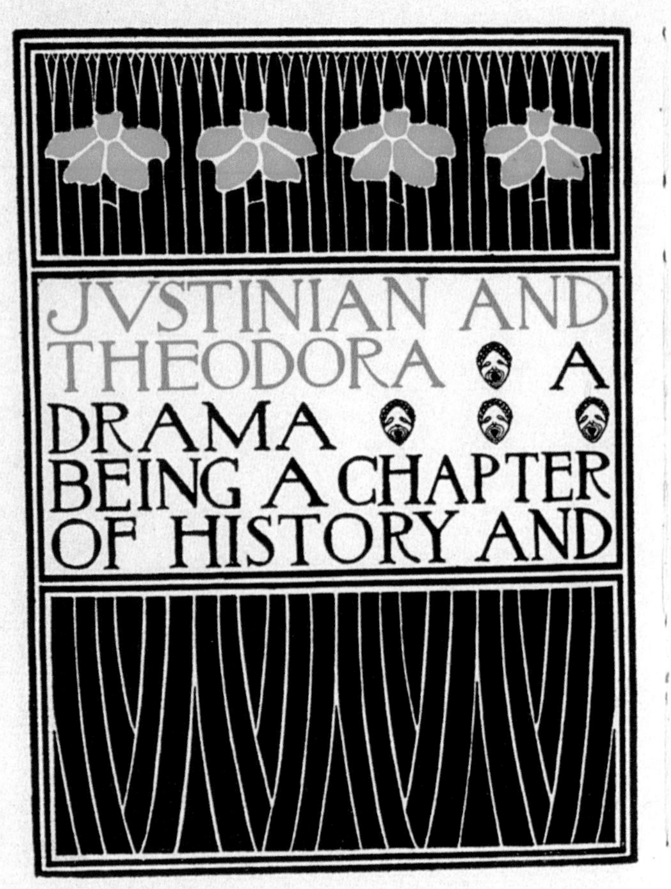

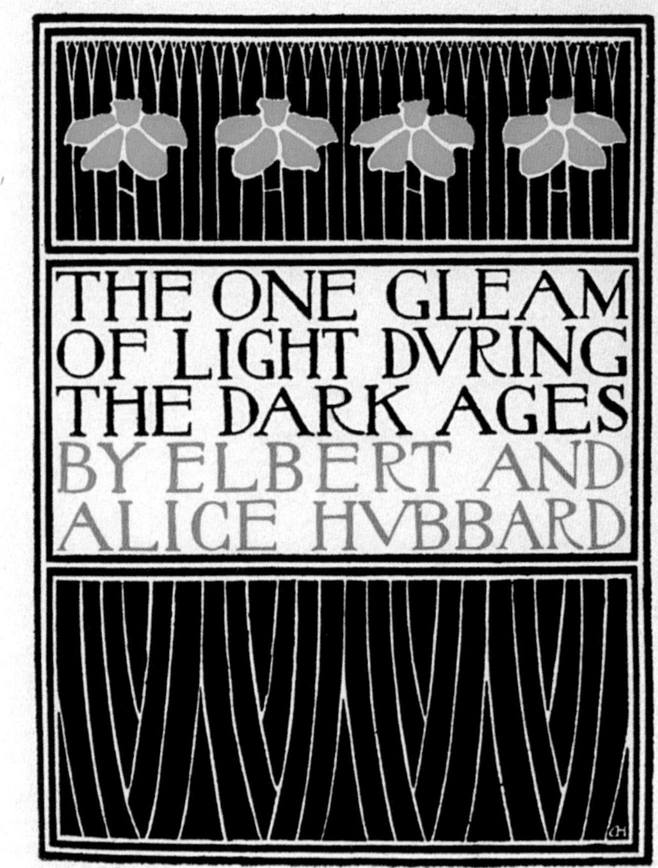

HUBBARD, Elbert Green (1856–1915), American Arts and Crafts practitioner. As a partner in the Larkin soap company in Buffalo, New York, he pioneered mass-marketing techniques, but in 1892, prompted by a meeting with William Morris, he established the Roycroft Press and the quasi-religious Roycroft craft community in nearby East Aurora, seeking a return to the simple community life of pre-industrial America. The Roycrofters—numbering 175 in 1900—built their own shops selling china, glass, ironwork, furniture, and leather goods, as well as their own houses and even a Roycroft Inn for guests, who included C.R. Ashbee and his wife Janet, while the press published Hubbard's books and periodicals. In 1915 Hubbard and his wife were drowned aboard the *Lusitania*, but the shops continued under the management of their son Bert until the community closed in 1938.

HUNTER, Dard (1883–1966), American furniture, lighting, and book designer who studied in Vienna from 1908–9 and worked for Elbert Hubbard, bringing a distinctively European aesthetic to his Roycroft pieces. After 1911 he specialized in papermaking.

IMAGE, Selwyn (1849–1930), important English Century Guild designer along with Herbert Horne and A.H. Mackmurdo. Ordained as a curate in 1872, the influence of John Ruskin led him to leave the church in 1882 to become an artist. He provided graphic designs and woodcuts for the Guild's *The Hobby Horse* and a number of independent publications, as well as designing mosaics, stained glass, and embroideries for the Royal School of Needlework. In 1900 he became master of the Art Workers' Guild and in 1910 Slade Professor of Fine Art at Oxford. Together with Mackmurdo he founded the Fitzroy Picture Society to distribute prints of great paintings to schools.

JACK, George (1855–1932), American-born furniture designer. He came to London in 1875 and became chief assistant to Philip Webb in 1880. As principal furniture designer to Morris & Co. he was responsible for many of the monumental mahogany pieces with inlaid decoration made around the 1890s, including the bookcase in the billiard room at Standen. In 1900 he took over Philip Webb's architectural practice.

Left: *The title-page for Elbert and Alice Hubbard's* Justinian and Theodora *(1906), designed by Dard Hunter for the Roycroft Press.*

JARVIE, Robert Riddle (1865–1941), American metalworker noted for his stylish brass, copper, and bronze candlesticks and lanterns, which he modeled in forms suggestive of tall flowers.

JAUCHEN, Hans W. (1863–1970), German-born founder in 1922 of the art metal shop Old Mission KopperKraft in San Jose, California, in partnership with Fred T. Brosi (d. 1935), an Italian immigrant.

JENSEN, Georg (1866–1935), Danish silver and jewelry designer, with a highly recognizable geometric style. First apprenticed to a goldsmith, he studied sculpture before opening his own silver workshop in Copenhagen in 1904. He employed a number of Arts and Crafts-influenced craftsmen, including Johan Rohde.

JEKYLL, Gertrude (1843–1932), leading English garden designer. Born in London to a wealthy family, she went to the South Kensington School of Art at eighteen, met John Ruskin and William Morris in 1869, and turned to embroidery, designing patterns that were heavily influenced by Morris. She also mastered the art of silver repoussé work. She brought her artistic training and knowledge of color theory to the gardens she started designing in her thirties when her eyesight began to fail and the death of her father occasioned a move, with her mother, to Munstead Heath in Surrey. At the age of forty-six she formed a fruitful partnership with the twenty-year-old Edwin Lutyens, and they worked together on more than one hundred British gardens and houses.

JEWSON, Norman (1884–1975), English architect, wood- and stone-carver, lead and plaster worker. He married Ernest Barnsley's daughter, Mary, after joining Ernest Gimson and the Sapperton group of architects and craftsmen in the Cotswolds, where he lived all his life.

Right: *Edwin Lutyens commissioned Sir William Nicholson to paint this portrait of Gertrude Jekyll, his great friend and collaborator, in 1920.*

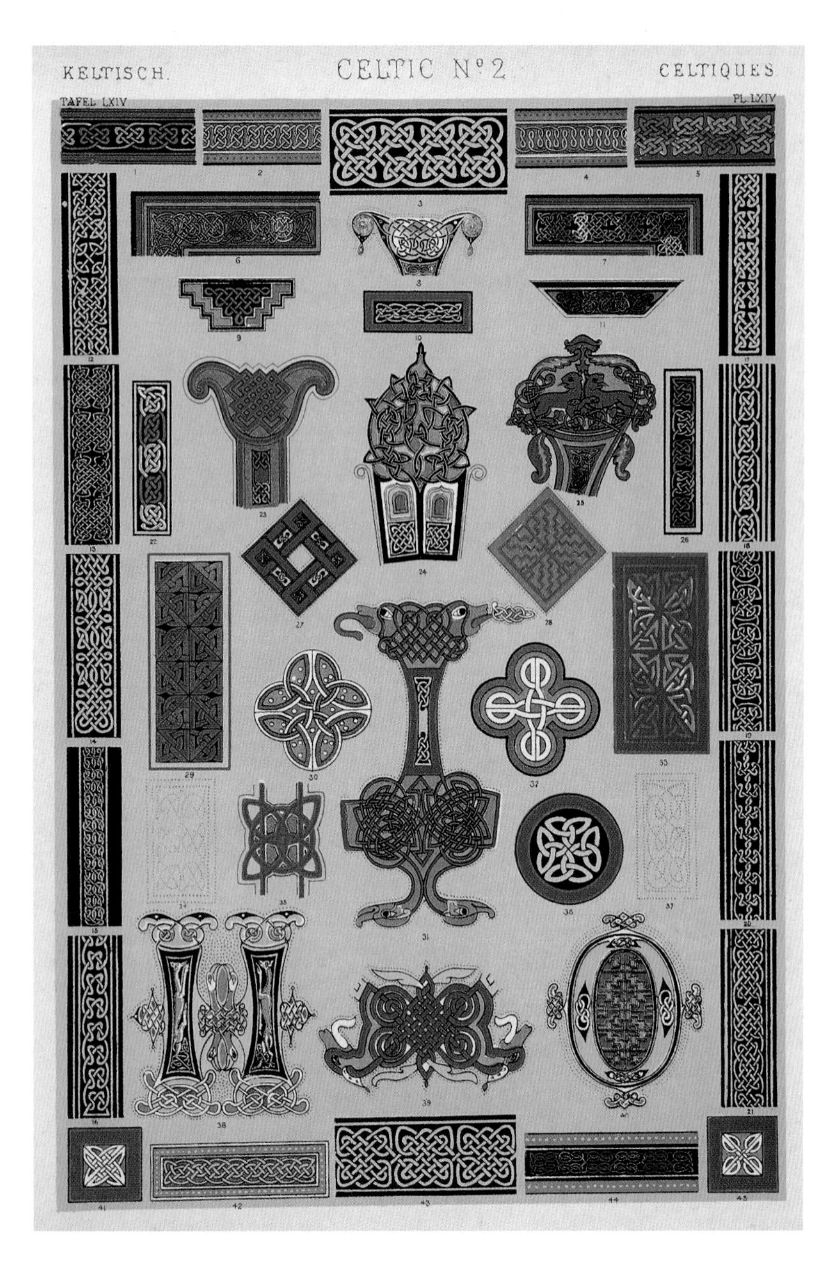

Left: *A page from Owen Jones's* The Grammar of Ornament *showing Celtic knot designs.*

Right: *Clock with enameled dial, designed by Archibald Knox, part of Liberty's Tudric range, c. 1903.*

JOHNSTON, Edward (1872–1944), Scottish calligrapher and typographer who taught lettering and illumination at the Central School of Arts and Crafts (1899–1913), where Eric Gill was one of his students. He later achieved fame as a designer for the London Underground.

JONES, Owen (1809–74), Welsh designer of furniture, wallpaper, textiles, books, and complete interiors, and superintendent of works for the Great Exhibition of 1851. His book *The Grammar of Ornament* (1856)—the first to have full-color plates printed by chromolithography—became a pattern book for British design and taste. William Morris had a copy in his library though Owen's sober, mathematical, geometric patterns were at odds with his own style, which drew freely on nature.

KING, Jessie Marion (1875–1949), Scottish book, textile, and jewelry designer and illustrator. She overcame parental opposition to study at the Glasgow School of Art, where she later became a lecturer in book decoration. She won a gold medal at the Esposizione Internazionale in Turin in 1902. She exhibited with the Scottish Guild of Handicraft (founded in 1905) and also worked for Liberty's, designing for the "Cymric" silver range and providing fabric designs and hand-painted pottery. She married the furniture designer Ernest Taylor and in 1911 they set up a studio in Paris, but they were forced by the outbreak of World War I to return to Scotland, where she continued to work and exhibit until her death.

KIPP, Karl E. (1882–1954), American designer. He ran Elbert Hubbard's Copper Shop in the Roycroft community until 1912, when he left to found his own firm, the Tookay Shop in East Aurora, returning to Roycroft in 1915 after Hubbard's death.

KNOX, Archibald (1864–1933), Isle of Man-born designer of silver, pewter, jewelry, carpets, textiles, and pottery. He studied at Douglas School of Art, specializing in Celtic ornament. Through M.H. Baillie Scott he came into contact with Liberty's and became one of its most prolific, imaginative, and innovative designers (though, like all Liberty designers, anonymous). He produced some of the earliest designs for Liberty's "Cymric" range and was the

inspiration behind the company's "Celtic Revival," enthusiastically supported by its Welsh managing director, John Llewellyn. Knox was influenced by Christopher Dresser.

LA FARGE, John (1835–1910), American painter and stained-glass artist. He was taught by his father, a fresco painter, after which he studied in Paris. Influenced initially by the Pre-Raphaelite painters and William Morris, he identified with the ideals of the Arts and Crafts movement and did much to introduce current European decorative style to the USA. In 1880 he patented a method for producing opalescent glass that became desirable and fashionable among wealthy American collectors, including the Vanderbilts and the Whitneys.

LARSSON, Carl (1853–1919), Swedish artist. His country house, Sundborn, came to epitomize Swedish domestic ideals and set the Arts and Crafts style for a generation through the publication of his book *Ett Hem* ("A Home") in 1898.

LEACH, Bernard (1887–1979), English potter born and brought up in the Far East. On his return from studying pottery in Japan in 1920, he established the Leach Pottery in St. Ives, Cornwall, where he used local clays and natural glazes.

LESSORE, Ada Louise (1882–1956), English ceramic artist and embroiderer. She studied art at the Slade and was a pupil of Edward Johnston at the Central School of Arts and Crafts, collaborating with the calligrapher Graily Hewitt on Virgil's *Aeneid*. She also studied embroidery at the Central School and established herself as a needlewoman, becoming closely associated with May Morris and the Royal School of Needlework. She married Alfred Powell in 1906 and the couple became successful designers for Wedgwood.

Left: *Birds and flower-laden boughs are silhouetted against the moon in this fine leaded-glass window designed by John La Farge in 1884.*

Right: *W.R. Lethaby was responsible for the fireplaces and paneling in the interior of Stanmore Hall, decorated for William Knox D'Arcy by Morris & Co. The Kenton & Co. table in the hall was also probably designed by Lethaby.*

Her younger sister, Therese, who also designed for Wedgwood, was on the fringe of more avant-garde art movements and married two painters: first Bernard Adeney, a founder member of the London Group, and then, in 1926, Walter Sickert.

LETHABY, William Richard (1857–1931), English architect, designer, writer, socialist, educationalist, and pivotal figure in the Arts and Crafts movement. The son of a framemaker in Barnstaple, Devon, he was apprenticed to a local architect and won the RIBA Soane medallion and the Pugin traveling scholarship. He made his way to London in 1879 and became chief clerk to Richard Norman Shaw, working alongside E.S. Prior and Mervyn Macartney, with whom he formed the short-lived furniture company Kenton & Co. He set up his own architectural office in 1889. Lethaby helped to establish the Art Workers' Guild in 1884 (becoming its master in 1911), and embarked on a public promotion of the crafts through exhibitions run by its offshoot, the Arts and Crafts Exhibition Society. He joined the Society for the Protection of Ancient Buildings in 1893 and became the co-director and principal of the Central School of Arts and Crafts, which was dominated by his inspirational philosophy, when it opened in 1896. He appointed highly skilled craftsmen rather than professional teachers, and aimed to raise the standards of the crafts while creating a training ground for designers able to meet the needs of industry. He was a founder member of the Design and Industries Association formed in 1915 and Philip Webb's first biographer. "A work of art is a well-made boot," he once claimed. When he died in 1931 his gravestone in Hampshire was inscribed "Love and labour are all."

LIBERTY, Arthur Lazenby (1843–1917), English founder of the Regent Street, London, store that bears his name. He was the

Left: *An advertisement for Liberty & Co., which appeared on the back cover of a special number of* The Studio *in 1901.*

Right: *A stunningly simple oval table designed and made in oak by Charles P. Limbert's Company of Grand Rapids (and later Holland) Michigan, 1905–10, which incorporates parallel planes and pierced rectangles in the cross braces.*

eldest of eight children born to a draper in Chesham, Buckinghamshire. His interest in design was stimulated by the International Exhibition of 1862 and the following year, aged twenty, he joined Farmer & Rogers Great Shawl and Cloak Emporium in Regent Street as manager of their "Oriental warehouse." He divorced his first wife, Martha Cotham, and became engaged to Emma Louise Blackmore in 1874. With funding from her father, a tailor from Devon, he leased a half-shop directly opposite Farmer & Rogers at 218A Regent Street, which he named East India House. Liberty & Co. opened on May 15, 1875, with a staff of three. It specialized in Oriental art and artefacts—silver kettles, cutlery, and smaller pieces of furniture—which were sold overstamped with the appropriate Liberty & Co. mark and label, a company policy that continued until his death in 1917. He repaid his father-in-law within eighteen months and acquired the second half of the premises at 218 Regent Street. In 1890 Maison Liberty opened in Paris at 38 avenue de l'Opéra, moving to grander premises at 3 boulevard des Capucines fairly swiftly (it closed in 1932). Despite employing some of the period's best designers, among them C.F.A. Voysey, Archibald Knox, Arthur Gaskin, Bernard Cuzner, Olive Baker, and Jessie M. King, much of the work Liberty's sold was machine-made with the odd finishing touch by hand. The company catered to an affluent clientele who, after 1910, included the residents of the new garden cities as well as the London suburbs. It reached a wide public and introduced the work of some important Arts and Crafts designers (though anonymously, for Liberty policy was that most designers and craftworkers did not mark their work), but it undercut the guilds and contributed to their demise, trading on the visual appeal of Arts and Crafts style while ignoring the social and ethical message of the movement.

LIMBERT, Charles, P. (1854–1923), American furniture manufacturer. Having joined the industry as a salesman, he started a manufacturing company in Grand Rapids, Michigan, in 1894, and launched a range of "Dutch Arts and Crafts" furniture in 1902. In 1906 he relocated to Holland, Michigan, and opened his Holland Dutch Arts and Crafts factory in a scenic lakeside location, to provide pleasant and healthy conditions for his workers: "attractive summer cottages and quaint houses with fertile gardens and well kept lawns." He drew inspiration from Japan, the Austrian

Secessionists, and Britain, particularly the geometric furniture of Charles Rennie Mackintosh.

LOGAN, George (1866–1939), Scottish furniture designer who worked with E.A. Taylor at the long-established firm of Glasgow cabinet-makers Wylie & Lochhead.

LOOS, Adolf (1870–1933), innovative Austrian architect and theorist of modern functionalism.

LORIMER, Robert Stodart (1864–1929), Scottish architect who devoted a substantial proportion of his career to restoration and alteration and remained faithful to traditional seventeenth- and eighteenth-century styles.

LUMMIS, Charles Fletcher (1859–1928), American writer who founded the Landmarks Club with the aim of restoring the Spanish missions in California and promoting the appreciation of Spanish, Mexican, and Native Indian culture.

LUTYENS, Edwin Landseer (1869–1944), English architect. Named after his sporting artist father's hero, Lutyens was an almost exact contemporary of Frank Lloyd Wright. Ambitious and precocious, he spent a scant year in the office of Sir Ernest George

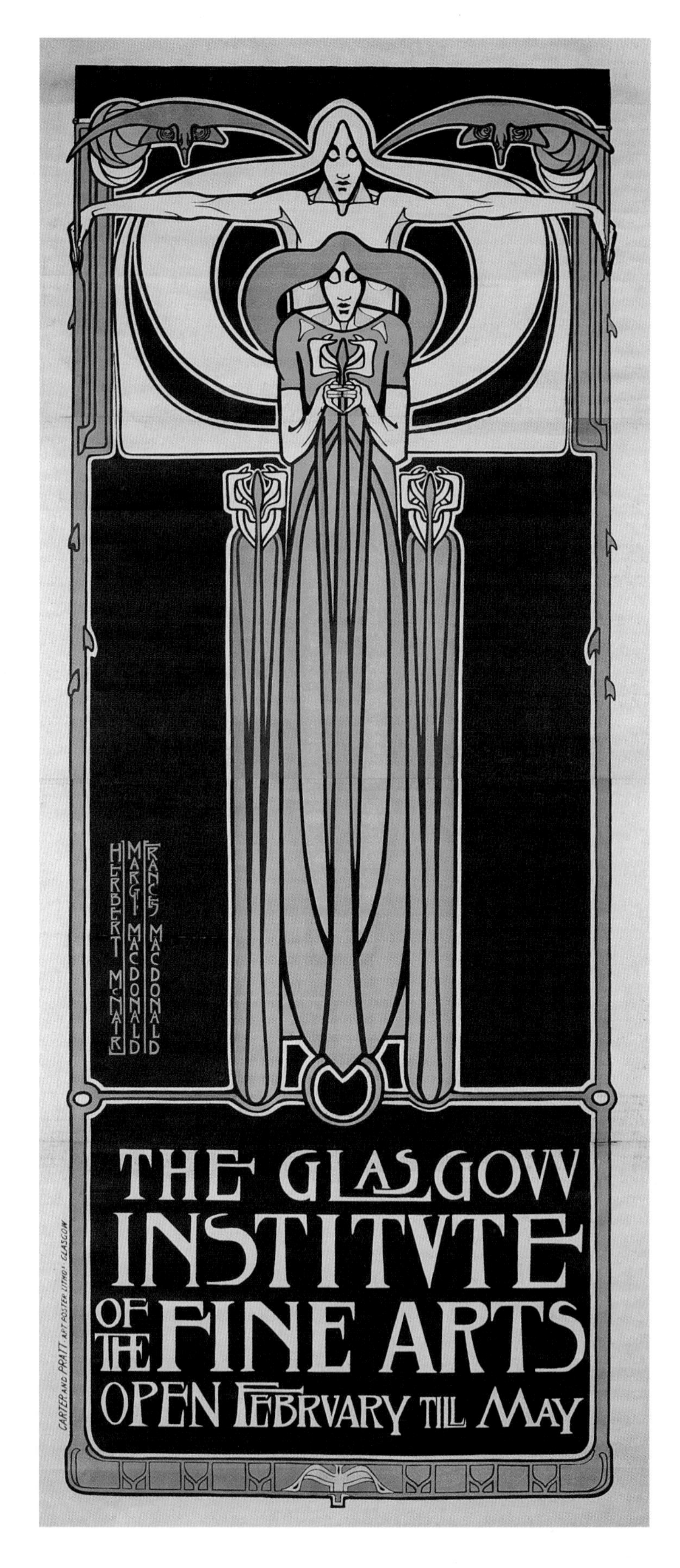

Right: *This poster for the Glasgow Institute of the Fine Arts, 1896, was a collaborative effort by Margaret Macdonald, her sister Frances, and Herbert MacNair, who, together with Charles Rennie Mackintosh, developed a whole new vocabulary of decorative arts.*

when he was eighteen, before setting up in practice on his own. A meeting in 1889 with Gertrude Jekyll led to a long and close collaboration on over one hundred country houses and gardens. He designed the Viceroy's House in New Delhi and, after World War I, became one of the principal architects to the Imperial War Graves Commission, designing both the Cenotaph in Whitehall and the Memorial to the Missing of the Somme at Thiepval in France, among other monuments.

MACARTNEY, Mervyn Edmund (1853–1932), English architect and co-founder of the Art Workers' Guild. He met W.R. Lethaby in Richard Norman Shaw's office, where they both worked, and became involved with him in Kenton & Co., after which he designed furniture for Morris & Co., working with George Jack and W.A.S. Benson. In 1906 he was appointed surveyor to St. Paul's Cathedral and took over the editorship of *The Architectural Review.*

MACBETH, Ann (1875–1948), Scottish designer of bold embroideries and a member of the Glasgow School. She taught at the Glasgow School of Art for many years and executed a number of ecclesiastical commissions for embroidery. She published, with Margaret Swanson, the influential instructional manual *Educational Needlecraft* (1911) and established an embroidery class for children at the School of Art.

MACDONALD, Frances (1873–1921), Scottish designer of furniture, textiles, enamel jewelry, and silverwork. She studied at the Glasgow School of Art with her sister Margaret Macdonald and

Right: *Margaret Mackintosh, photographed by James Craig Annan in the drawing room at 120 Mains Street, Glasgow, c. 1903.*
Far right: *Charles Rennie Mackintosh—"every bit the turn of the century dandy"—photographed by James Craig Annan, 1893.*

they, together with Charles Rennie Mackintosh and Herbert MacNair, formed the group known as the Glasgow Four. The sisters set up a studio in 1894, embracing all manner of crafts, with contributions from Mackintosh and MacNair, whom Frances married in 1899. The Glasgow Four made extensive and valuable contacts abroad—particularly in Europe—where their sparsely furnished white interiors and the ethereal quality of their furnishings was much admired, resulting in a rich cross-fertilization of ideas between them and the Secessionists in Vienna.

MACDONALD Margaret (1864–1933), Scottish designer, artist, and embroiderer. She studied at the Glasgow School of Art where she met, and in 1900 married, Charles Rennie Mackintosh. Their long artistic collaboration appears to have been as harmonious as their marriage. Hermann Muthesius, who visited the newlyweds in their studio-flat home at 120 Mains Street, Glasgow, described them as the *kunstlerpaar* ("the artist-couple"). They had no children. Together they collaborated on the designs of several interiors, including Hill House, Helensburgh, for which Margaret designed the textiles and gesso panels.

MACKINTOSH, Charles Rennie (1868–1928), Scottish architect and designer. The second son in a family of eleven, his father, a police superintendent, instilled in him a deep appreciation of Scotland's cultural heritage and fostered his love of horticulture. He was articled at sixteen to John Hutchison, and left when his apprenticeship ended in 1889 to join the newly founded firm of Honeyman & Keppie as a draftsman. He attended evening classes at the Glasgow School of Art from 1884 where he met the Macdonald sisters—his future wife, Margaret, and Frances—who, together with Mackintosh and Herbert MacNair, formed the Glasgow Four. They pioneered Art Nouveau designs, drawing loosely on ancient Celtic ornament and the economy of Japanese art, and were invited to send their work to the Arts and Crafts Exhibition Society in 1896. That same year Mackintosh won the competition for the design of the new Glasgow School of Art building which, together with his ambitious interior designs for the Glasgow tea-room entrepreneur, Catherine Cranston, became his best-known work in the city. He was extraordinarily influential in Europe and the USA, though recognition in his home town was slow to come. Despite his startling originality, a reputation for

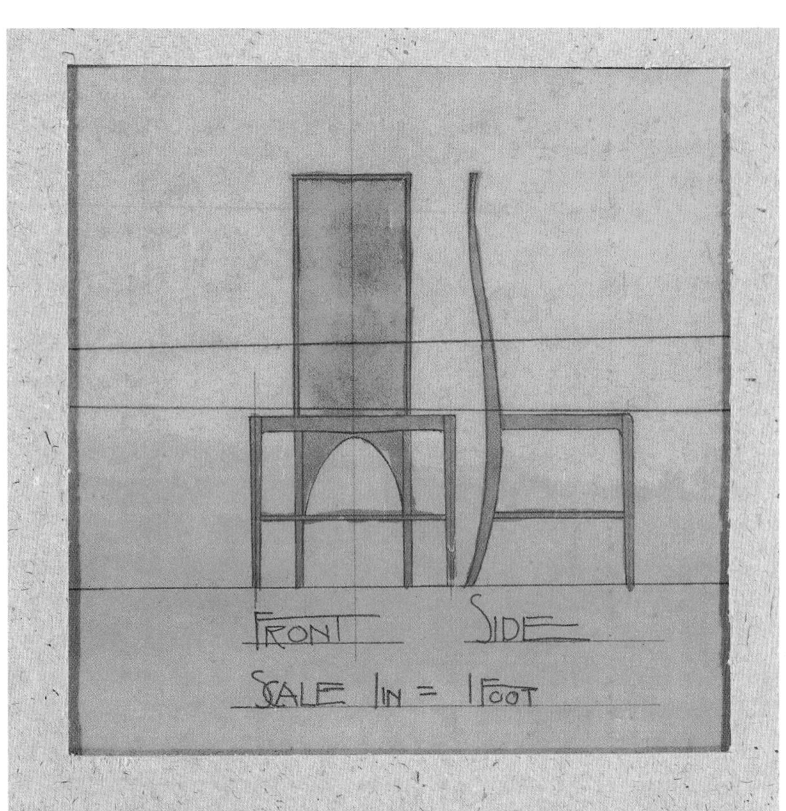

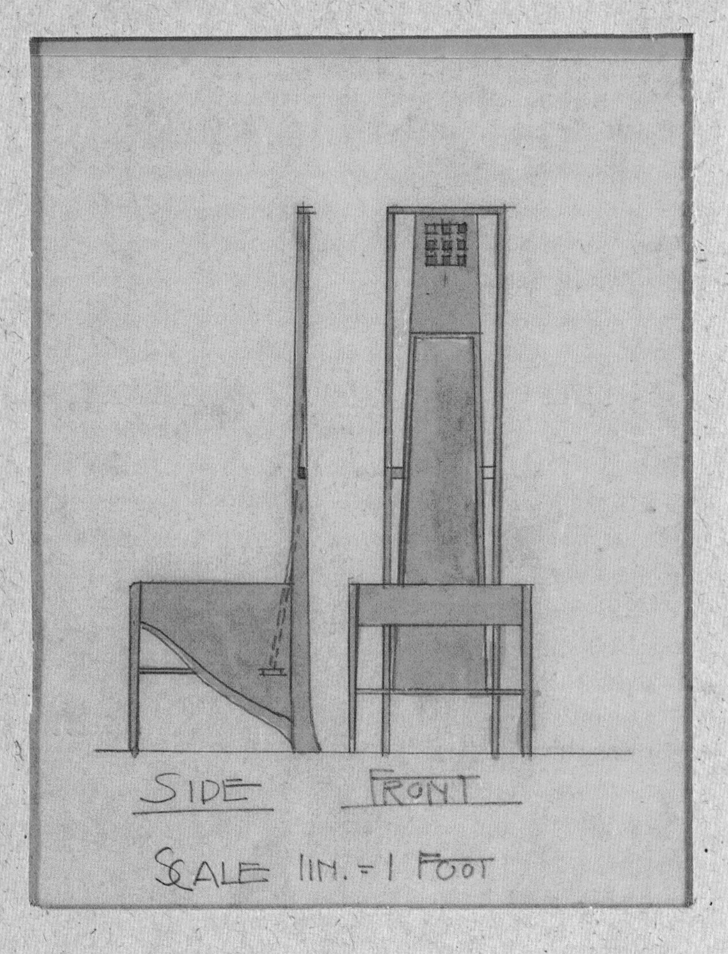

Left: *Charles Rennie Mackintosh's designs for chairs, shown in front and side elevation, for the Room de Luxe in Miss Cranston's Willow Tea Rooms, 1903.*

unreliability, eccentricity, and drinking sparked his professional decline. In 1914 the Mackintoshes left Glasgow. Before settling in London they lived for a year in Walberswick, Suffolk, where their "foreign" accents and correspondence with Viennese and German nationals drew the attention of the police. A continuing versatile creativity is evident in his few later commissions, but it was to be stifled by desperate financial insecurity and illness. Mackintosh never regained his professional status after World War I. Evicted from his Hampstead lodgings, he died in a London nursing home in December 1928, aged sixty, and his death was little remarked in the profession or the press. His architectural drawings, furniture, sketches, watercolors, and flower paintings were deemed practically worthless, valued at £88 16s 2d by the assessor of his Glebe Place studio. Once spurned by his native city, he is now a valuable figurehead for Glasgow's renaissance, his familiar motifs adorning a wide range of profitable merchandise.

MACKMURDO, Arthur Heygate (1851–1942), English architect and designer and prominent figure in the Arts and Crafts movement. He was instrumental in the setting up in 1882 of the Century Guild, a loose collective of designers "to render all branches of art the sphere no longer of the tradesman but of the artist" and to restore building, decoration, glass painting, woodcarving, metalwork, and pottery "to their rightful places beside painting and sculpture." Mackmurdo began his architectural training in London, in the office of the Gothic revivalist James Brooks, an architect he described as a craftsman who designed "every detail to door hinge and prayerbook marker." He taught with John Ruskin at the Working Men's College in London and went on to design

Right: *This MacNair table was exhibited at the Turin Exhibition of 1902 and illustrated in* The Studio *the same year.*

furniture, metalwork, and textiles, employing sinuous motifs and interlaced curves that influenced European Art Nouveau and C.F.A. Voysey, to whom he taught the rudiments of drawing up a repeat pattern for wallpaper or fabric. He was responsible for a number of extremely striking buildings, including the Savoy Hotel (1889).

MACNAIR, James Herbert (1868–1955), Scottish furniture designer and painter, and member of the Glasgow Four. He met Charles Rennie Mackintosh when working in the offices of Honeyman & Keppie, and his future wife Frances Macdonald and her sister Margaret at the Glasgow School of Art, where he was attending evening classes as part of his architectural training. He opened his own business at 227 West George Street in Glasgow, offering furniture design, book illustration, and posters, but in 1897 moved to Liverpool to take up the post of instructor in decorative design at the School of Architecture and Applied Art. He continued with his interior decorating work, in collaboration with his wife, and designed the Writing Room for the Scottish pavilion at the Esposizione Internazionale in Turin in 1902. The Studio published illustrations of the interior of his own house at 54 Oxford Street, Liverpool. In 1908 he returned to Glasgow but, like Mackintosh, found work difficult to come by and eventually died in complete obscurity.

MAHONY, Marion Lucy (1871–1961), American architect. Born in Chicago, she was the second woman to graduate in architecture from Massachusetts Institute of Technology (1894). She worked with Frank Lloyd Wright at his Oak Park office, alongside Walter Burley Griffin, whom she married in 1911, emigrating with him to Australia after his prize-winning plan for Canberra was accepted in 1912.

MAIRET, Ethel (1872–1952), née Partridge, English textile designer. She married the Anglo-Sinhalese geologist and anthropologist Ananda Coomaraswamy and lived in Ceylon and India from 1903–7. On their return to England she and her husband moved into the Norman chapel at Chipping Campden that he had commissioned C.R. Ashbee and the Guild of Handicraft to restore, convert, and decorate. After her divorce she married the Guildsman Philippe Mairet and set up a weaving workshop at her home, Gospels, at Ditchling in Sussex, where she revolutionized the craft of handloom weaving and researched and popularized natural vegetable dyes, becoming a role model for the women she trained there. Her home became the nucleus of a new Arts and Crafts community, which included her brother, the

jeweler Fred Partridge, and was described by Ashbee as "a very alive and happy community with a lot of old Campden and Essex House spirit in it."

MAIRET, Philippe (1886–1975), French illustrator, stained-glass craftsman, actor, author, and translator. He worked with C.R. Ashbee in London and Chipping Campden.

MALMSTEN, Carl (1888–1972), Swedish craftsman, designer of textiles, fabric, and furniture, and teacher. Trained as a cabinet-maker, he subsequently established his own workshop and school for designers and craftsmen in Stockholm.

MARTIN, Robert Wallace (1843–1923), English potter. With his brothers Charles Douglas (1846–1910), Walter Frazier (1857–1912), and Edwin Bruce (1860–1915) he set up the Martin Brothers pottery in Fulham, London, in 1873, moving to Southall in Middlesex in 1877. They became famous for their grotesque but fashionable salt-glazed stoneware.

MATHEWS, Arthur Frank (1860–1945), American muralist and designer. The Wisconsin-born architecture and art student spent five years studying in Paris at the Acadèmie Julien and taught in the Mark Hopkins Institute of Art, San Francisco, on his return to the USA, becoming dean of the School of Design. There he met and married his pupil Lucia Kleinhans and together they opened the Furniture Shop at 1717 California Street, San Francisco, after the great earthquake and fire of 1906. They employed a workforce of between twenty and fifty craftsmen to meet the huge demand for furnishings from wealthy San Francisco residents during the rebuilding of the city. They also founded the Philopolis Press, which published works on city planning and art, including the monthly magazine *Philopolis*, which ran for ten years from 1906. The Furniture Shop closed in 1920.

MAYBECK, Bernard (1862–1957), American architect. He studied architecture at the École des Beaux-Arts in Paris, graduating in 1886, and was concerned to integrate all his buildings with their sites in design and materials. He is best known for the First Church

Left: *A trio of pieces from the Martin Brothers' workshop, featuring, on the vases, handles modeled as grotesque hounds and serpents and, in the center, an example of the birds, for which they became best known.*

Right: *A detail from William Morris's bed curtains at Kelmscott Manor, embroidered by May Morris and helpers, 1891–4.*

of Christ Scientist (1910) in Berkeley, California, and the simple wooden chalets and bungalows he built in the Berkeley Hills and San Francisco Bay area.

MCLAUGHLIN, Mary Louise (1847–1939), American ceramicist. She formed the women's Cincinnati Pottery Club in 1879 and exhibited in the USA and Paris. Her work was known as "Cincinnati Limoges."

MORGAN, Julia (1872–1957), American architect. She was mentored by Bernard Maybeck after studying civil engineering at the University of California at Berkeley, and trained as an architect at the École des Beaux-Arts in Paris, becoming its first woman graduate. She was responsible for over eight hundred buildings, including the Williams house in Berkeley, California, and William Randolph Hearst's palatial home at San Simeon, to which she devoted much of her life from 1919–42.

MORRIS, Janey (1840–1914), daughter of a stablehand in Oxford and statuesque beauty who captivated the hearts of Dante Gabriel Rossetti and William Morris, whom she married in 1858. A skilled embroideress, she aided Morris in his work and collaborated with him on decorative schemes for Red House and Kelmscott Manor.

MORRIS, May (1862–1938), English textile and wallpaper designer, lecturer, and writer. The younger daughter of William Morris, from whom she received her artistic education, May took over the management of the embroidery section of Morris & Co. in 1885 and produced many designs herself, usually depicting flowers or trees. Her *Decorative Needlework* was published in 1893. A participant in her father's political activities, she lectured in London, Birmingham, and the USA, and became a founder member of the Women's Guild of Arts in 1907. She devoted much of her time to editing her father's writings, which appeared in twenty-four volumes between 1910 and 1914. Her "mystic betrothal" to George Bernard Shaw did not result in marriage and, though she was briefly married to Henry Halliday Sparling, she spent the last years of her life in devoted companionship with Mary Lobb at Kelmscott Manor, Gloucestershire.

Left: *A photograph of William Morris taken by Frederick Hollyer in 1874.*

Right: *A page from the Morris & Co. furniture catalog of about 1910, showing the range of rush-seated Morris chairs.*

MORRIS, William (1834–1896), great English polymath and, with John Ruskin, founding father of the Arts and Crafts movement. The eldest son and third child of wealthy parents, he attended Marlborough College, where he acquired a love of landscape and medieval architecture, and Exeter College, Oxford, where he made a lifelong friend in Edward Burne-Jones. He abandoned plans to enter the church in favor of architecture (entering the office of G.E. Street in 1856, where he met Philip Webb) then, on meeting Dante Gabriel Rossetti, he abandoned architecture for art. He married Jane Burden in 1858 and collaborated with Webb on the design of Red House at Upton in Kent. His frustration at failing to find suitable decoration for the house, his "palace of art," led directly to his founding Morris, Marshall, Faulkner & Co. (known as "the Firm") in 1861. It sold furniture designed by Philip Webb, Ford Madox Brown, and later, George Jack, stained glass, painted tiles, wall paintings, embroidery, tapestries and woven hangings, carpets, table glass, metalwork, wallpapers, and chintzes, designed by Burne-Jones, William De Morgan, Walter Crane, and others. Morris's first wallpaper, "Trellis," was produced in 1862. In 1865 the Firm moved from 8 Red Lion Square to 26 Queen Square, London. Showrooms were acquired in Oxford Street in 1875 and the Hammersmith workshop for hand-knotted rugs moved to Merton Abbey in Surrey in 1881. However, by 1875, increasingly bitter and overworked, Morris had reorganized the Firm as Morris & Co. with much acrimony and sundered friendships, to reflect what had always been the case—Morris's main role. He founded "Anti-Scrape," or the Society for the Protection of Ancient Buildings, in 1877 and the Kelmscott Press in 1891, turning in later life increasingly to politics, although he continued to write poems, weave tapestries, design wallpapers, and produce and promote some of the most enduringly lovely Arts and Crafts artefacts. He never visited the USA but he had a profound impact on American culture. The Firm continued in business after his death until World War II, going into voluntary liquidation in 1940.

MOSER, Koloman (1868–1918), Austrian painter and designer, closely involved in the founding of both the Vienna Secession and the Werkstätte.

THE SUSSEX RUSH-SEATED CHAIRS

MORRIS AND COMPANY

449 OXFORD STREET, LONDON, W.

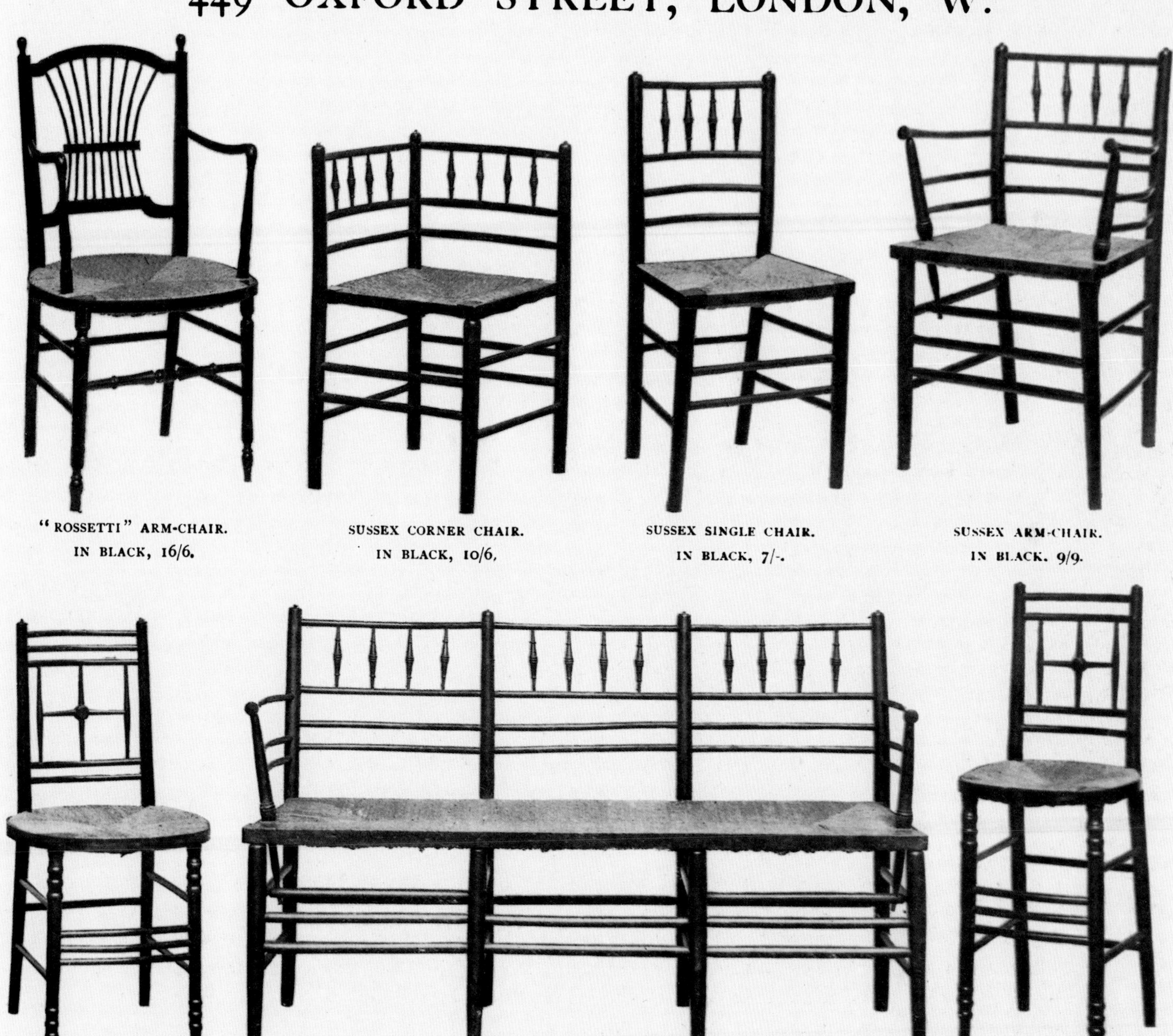

"ROSSETTI" ARM-CHAIR. IN BLACK, 16/6.

SUSSEX CORNER CHAIR. IN BLACK, 10/6.

SUSSEX SINGLE CHAIR. IN BLACK, 7/-.

SUSSEX ARM-CHAIR. IN BLACK. 9/9.

ROUND-SEAT CHAIR. IN BLACK, 10/6.

SUSSEX SETTEE, 4 FT. 6 IN. LONG. IN BLACK, 35/-.

ROUND SEAT PIANO CHAIR. IN BLACK, 10/6.

Left: *One of a pair of stained-glass panels by Mary Newill, probably produced at the Bromsgrove Guild, c. 1905. Newill began with painting and embroidery before designing domestic and ecclesiastical stained glass, winning the first prize in* The Studio *competition in 1897.*

MUNTHE, Gerhard (1849–1929), Norwegian craftsman who pioneered a revival of tapestry weaving, drawing on Norwegian sagas and folk tales, but rendering them in a modern, linear style.

MUTHESIUS, Hermann (1861–1927), German architect and critic. He was cultural attaché to the German embassy in London from 1896–1903 and his book, the monumental *Das englische Haus* ("The English House," 1904–5), provides a definitive contemporary analysis of the domestic scene at a pivotal moment in the history of Arts and Crafts. He wrote about the emergence of "free architecture" in British domestic buildings, admired W.R. Lethaby and the "young generation who now stood upon Morris's shoulders," and actively championed Charles Rennie Mackintosh and the Glasgow Four. Muthesius was a crucial figure in the Deutscher Werkbund, founded in Germany in 1907 and inspired by the British Arts and Crafts movement. It accepted the need to design not just for handcraft but for mass production and its artists, architects, and craftsmen were concerned to refine an object down to its essentials, believing that "machine work may be made beautiful by appropriate handling." The success of the Werkbund contributed to the German national economy and prompted moves toward design reform in Britain, with the formation, a year after a Werkbund exhibition in London, of the Design and Industries Association.

NEWBERY, Francis Henry (1855–1946), the progressive director of the Glasgow School of Art from 1885–1918. In 1892 he introduced technical art studios in which artist-craftsmen gave artisans a "technical artistic education" in commercial crafts.

NEWBERY, Jessie Rowat (1864–1948), Scottish needlework designer and influential and inspiring figure for women. Charles Rennie Mackintosh decorated the family house (her father, William Rowat, was a manufacturer of Paisley shawls) around the time of her marriage to Francis Newbery. She taught embroidery at the Glasgow School of Art from 1894, introducing simple methods and often working with inexpensive fabrics such as calico and flannel. She championed the need for modernity and the availability of quality needlework to all classes and, although her pattern designs were bold and complex, she used the simplest possible stitching.

NEWILL, Mary (1860–1947), versatile and prize-winning British artist, book illustrator, and embroiderer. Part of the Birmingham Group and a member of the Arts and Crafts Exhibition Society, her work shows the strong influence of Edward Burne-Jones.

NEWTON, Ernest (1856–1922), English architect and founder member of the Art Workers' Guild. He studied with Richard

Right: *Women decorators working at the Rookwood Pottery c. 1890, almost ten years after the Cincinnati pottery was founded by Maria Longworth Nichols.*

Norman Shaw and left to found a flourishing country-house practice of his own, building Buller's Wood at Chislehurst, Kent (1889), for the Sanderson Family and a series of stone, brick, and tile cottages for the Lever Brothers at Port Sunlight, Cheshire (1897).

NICHOLS, Maria Longworth (1849–1932), American potter. In 1880 she founded the influential Rookwood Pottery in Cincinnati, set up specifically for constructing and decorating art pottery by hand, usually with naturalistic landscapes and flower designs. Rookwood became commercially successful and received widespread international acclaim. It employed many renowned ceramicists, including Artus Van Briggle, William P. MacDonald, Albert Valentine, and the Japanese ceramicist Kataro Shirayamadani (1865–1948), who became one of the company's principal designers.

NORTON, Charles Eliot (1827–1908), American art critic and Professor of History of Art at Harvard. An American equivalent to John Ruskin, whom he had met, along with William Morris, his influence was enormous. He founded the Boston Society of Arts and Crafts in 1897.

OHR, George Edgar (1857–1918), American potter. He was responsible for some of the most radical and eccentric art pottery of the Arts and Crafts period: he would fashion perilously thin pots and then crumple, twist, or pinch the design into abstraction. He consciously cultivated his image as the "mad potter of Biloxi" in Mississippi and made his shop—called Pot-Ohr-E—into a minor tourist attraction.

OLBRICH, Joseph Maria (1867–1908), Austrian architect and leader of the Secessionists, who aimed to give applied arts the same status as fine art. They staged their first exhibition in 1898 and featured the work of C.R. Ashbee, the Guild of Handicraft, the Glasgow Four, and other leading British Arts and Crafts designers in

their eighth exhibition, in 1900. The Secessionists assimilated the ideals of William Morris and Ashbee but in a more stylized framework of aesthetically acceptable decorative design.

PABST, Daniel (1826–1910), German furniture designer influenced by Christopher Dresser, who settled in Philadelphia in 1849 and opened his own shop and workshop, employing fifty men.

PARKER, Barry (1867–1947), English architect and visionary town planner. In partnership with his second cousin and brother-in-law, Raymond Unwin, he was responsible for much that was good in the Garden City movement in Britain in the early twentieth century. They co-authored *The Art of Building a Home* (1901) and complemented each other as architectural partners. Both socialists (and teetotalers), they wanted to "bring brightness and cheeriness and airiness right into the midst of the house." Their philosophy extended beyond individual house design to embrace the model village of New Earswick in Yorkshire, built in 1902 for the Rowntree family, and garden cities at Letchworth, Hertfordshire (1904–14) and Hampstead Garden Suburb in London (1905–14).

PAYNE, Henry (1868–1940), English landscape, portrait, and stained-glass artist. Based in Birmingham, he was responsible for many important commissions, among them the large five-light Ascension window in E.S. Prior's Arts and Crafts gem, St. Andrew's, Roker, Sunderland, and the seven-light Resurrection window at W.H. Bidlake's St. Agatha's, Birmingham. Payne moved, like many other Arts and Crafts practitioners, to the Cotswolds in 1909 where he built a stained-glass workshop in the garden of his sixteenth-century schoolhouse at Amberley, Gloucestershire, and formed the St. Loe's Guild to promote various Arts and Crafts activities.

PEACH, Harry (1874–1936), enlightened Canadian-born founder, with designer Benjamin Fletcher, of the Leicester-based cane furniture firm Dryad. It produced chairs in modern shapes prompted by the work of Richard Riemerschmid. Peach encouraged his workers to study at art school and take part in social activities. In 1912 he diversified, starting the Dryad Metal Works and Dryad Handicrafts. The latter, supplying materials and tools for a wide variety of craft activities, is still in existence. Peach was a founder member of the Design and Industries Association in 1915.

POWELL, Alfred Hoare (1865–1960), English architect and ceramic artist who studied at the Slade School of Art, joined J.D. Sedding's practice in 1887, and attended W.R. Lethaby's Central School of Arts and Crafts, where he studied calligraphy. He was a close friend of Sidney Barnsley and Ernest Gimson, and joined them in Sapperton, Gloucestershire, in 1901, dividing his time between architectural commissions and making chairs by hand using a primitive pole lathe. A true Arts and Crafts practitioner, he mastered weaving, wood-engraving, and working in metals, stone, and wood before turning to ceramics. He married Louise Lessore in 1906 and together they set up a studio for the painting of ceramics in Red Lion Square, London, working for Wedgwood. Powell's success in this field encouraged Lethaby to invite him to set up a class in china painting at the Central School. He was also closely associated with May Morris and the Royal School of Needlework. Sidney Barnsley's daughter Grace trained with the Powells, going on to work for Wedgwood.

PRICE, William Lightfoot (1861–1916), American architect and founder of the Rose Valley Arts and Crafts Colony, set up in Philadelphia in 1901 to produce furniture in the Gothic revival style.

Left: *The decoration of Madresfield Chapel was commissioned in 1902. All the painting, stained glass, and metalwork was carried out by the Birmingham Group. The tempera-painted frescoes and stained glass are by Henry Payne. The triptych was designed by William Bidlake, painted by Charles Gere, who also designed the altar frontal.*

Right: *A Wedgwood punchbowl by Alfred Powell, decorated with Cotswold scenes and, at the center, Daneway House, c. 1928.*

Left: *A pencil-and-sepia wash drawing showing the interior perspective of E.S. Prior's Church of St. Andrew, Roker, Co. Durham.*

PRIOR, Edward Schroeder (1852–1932), English architect. An ardent disciple of William Morris, Prior was educated at Harrow and Cambridge, joined the office of Richard Norman Shaw in 1874, and set up his own practice in 1880. He was one of the founders of the Art Workers' Guild and held the Slade chair of Fine Art at Cambridge until his death. The "butterfly" plan, which he devised, offered an alternative to the popular L-shaped house: the resulting front elevation, as at The Barn in Exmouth, Devon (1907), formed two arms curving toward each other in a symbolic embrace of welcome.

PUGIN, Augustus Welby Northmore (1812–52), crucial English architect and design theorist. He revitalized the crafts of stained glass, ironwork, and ceramics and wrote a number of books, including *The True Principles of Pointed or Christian Architecture* (1841), which provided the foundation from which the moral aesthetics of Arts and Crafts evolved during the latter half of the nineteenth century. He was married three times, fathered eight children, and designed more than a hundred buildings—mostly churches, with the notable exception of the Houses of Parliament. He converted to Catholicism at twenty-two and died at forty. A passionate advocate of the asymmetry of Gothic architecture, he argued that English architects should abandon classical models in favor of Gothic examples—which he considered more suitable to Christianity—from the late Middle Ages. An ardent sailor, he once

Below: *Armchair, from* Gothic Furniture in the Style of the Fifteenth Century, *designed and etched by A.W.N. Pugin, published in 1830 by Ackermann & Co.*

exclaimed that "There is nothing worth living for but Christian architecture and a boat."

RATHBONE, Richard Llewellyn (1864–1939), leading British metalworker and teacher, who worked on projects with C.F.A. Voysey and A.H. Mackmurdo, and set up a workshop in Liverpool producing metal fittings and utensils.

RHEAD, Frederick Hürten (1880–1942), influential English-born art potter. He emigrated to America in 1902 and was associated with the University City Pottery, St. Louis, Missouri; Roseville in Zanesville, Ohio; and the Arequipa Pottery, located in a sanatarium for working-class tubercular women in California "to give these people interesting work." He proved a better potter than businessman.

RICARDO, Halsey (1854–1928), English architect and designer. He was active in the Art Workers' Guild and was heavily influenced by the theories of John Ruskin. He formed a partnership with William De Morgan in 1888 and often used Persian-style colored glazed tiles in his work—both inside and as facings on the exteriors of his buildings—as in his design for 8 Addison Road, Kensington, London, for Sir Ernest Debenham. He taught architecture at the Central School of Arts and Crafts from 1896.

RICHARDSON, Henry Hobson (1838–86), New England architect and interior designer responsible for introducing romanticism into American architecture. He studied at Harvard University and the École des Beaux Arts, Paris, and spent time in England, where he met William Morris, William Burges, Edward Burne-Jones and William De Morgan. Regarded as America's first "signature" architect, he is best known for the Trinity Church on Copley Square in Boston, the Glessner House in Chicago, and the Marshall Field Wholesale Store, also in Chicago, which sold Morris & Co. products.

RIEMERSCHMID, Richard (1868–1957), German designer and one of the founders of the Dresden and Munich Werkstätten (along with Peter Behrens, Bernhard Pankok, Bruno Paul, and Hermann Obrist). He trained as a painter in Munich and produced sinuous Art Nouveau-inspired designs for metalwork, as well as porcelain, glass, cutlery, light-fittings, carpets, and furnishing textiles. He became a founding member of the Deutscher Werkbund and worked on the plans for Germany's first Garden City at Hellerau near Dresden. Embracing the modern age, he took inspiration from everything from liners to locomotives. "Life, not art, creates style," he wrote, "it is not made, it grows."

ROBERTSON, Hugh Cornwall (1845–1908), American ceramicist. With his father, James, and two brothers, Alexander and George, he ran the Chelsea Keramic Art Works in Massachusetts, making classically inspired vases and art pottery. Following the death of his father and his brothers' move to California in 1884, he abandoned the elaborate decorative format and turned to a more Oriental style, developing his famous Japanese red glaze, known as *sang-de-boeuf* ("bull's blood") and his blue-decorated white Chinese crackle glaze which became very profitable for the now renamed company—Dedham Pottery—which continued in business under his son's care until 1943.

ROBINEAU, Adelaide Alsop (1865–1929), outstanding American ceramicist. China decorator, perfectionist, maker of complicated porcelain pieces, and teacher, she founded, with her French-born husband Samuel, the influential magazine *Keramic Studio* in Syracuse, New York, in 1899, with the aim of providing good designs for other potters. She studied watercolor with William Merritt Chase in New York, but specialized in painting on porcelain, and worked with the celebrated potter Taxile Doat of Sèvres. In 1911 she won the Grand Prix at the Esposizione Internazionale at Turin for her "Scarab Vase," which she made at the University City Pottery, St. Louis, Missouri, founded by the entrepreneur and patron of arts and education Edward Gardner Lewis (who also founded the American Women's League). For four years from 1912 she developed and ran an Arts and Crafts-based summer school specifically for women with children, at her studio, Four Winds.

ROBINSON, William (1838–1935), pioneering Irish garden designer and journalist. Moving to England in his twenties, he wrote *The Wild Garden* (1870) and the extremely popular *The English Flower Garden* (1883), and founded periodicals—*The Garden* (from 1871) and *Gardening Illustrated* (from 1879)—aimed at the new

Left: *A carved oak "Hall Chair" by Charles Rohlfs, c. 1900. "My feeling was," he stated, "to treat my wood well, caress it perhaps, and that desire led to the idea that I must embellish it to evidence my profound regard for a beautiful thing in nature."*

Above: *"Self portrait with blue neckcloth", c. 1873; John Ruskin was one of the giants of the Victorian age and exerted a powerful influence on English art, economics, and the social thinking that underpinned the Arts and Crafts movement.*

type of garden owner then residing in the suburbs. His naturalistic style of planting was taken up by Gertrude Jekyll and other Arts and Crafts gardeners reacting against the excessive artificiality of much of the gardening of the latter half of the nineteenth century. He created his own famous garden at Gravetye Manor, West Sussex, in 1885.

ROHLFS, Charles (1853–1936), American furniture designer. Turning his back on a successful acting career, he set up his workshop in Buffalo, New York, in 1891. His designs combined the sturdy shapes of Arts and Crafts furniture with the sinuousness of Art Nouveau decoration and ornament, and he was commissioned to design furniture for Buckingham Palace.

ROSSETTI, Dante Gabriel (1828–82), precociously talented artistic son of an Italian political exile living in London. Rossetti was a founder member of the Pre-Raphaelite Brotherhood and of the firm of Morris, Marshall, Faulkner & Co., for whom he designed stained glass, furniture, and decorative painted panels. He married Elizabeth Siddal in 1860 and, following her suicide, began an intimate relationship with Janey Morris.

RUSKIN, John (1819–1900), precociously brilliant English writer and social reformer. Advising that "to teach taste is to form character," he became the philosophical inspiration behind the Arts and Crafts movement on both sides of the Atlantic, although he never visited the USA (a country he characterized as full of unbridled competition and relentless ugliness), and his own Utopian experiment, the Guild of St. George, was a failure. He became Slade Professor of Fine Art at Oxford in 1870, and his views on industrialization and the education of working people underpinned the British Labour movement.

SCHULTZ, Robert Weir (1860–1951), Scottish architect who trained in Edinburgh before joining Richard Norman Shaw in 1884, where he met W.R. Lethaby and was introduced to Ernest Gimson and the Barnsley brothers.

SEDDING, John Dando (1838–91), leading English ecclesiastical architect and key figure in the Arts and Crafts movement. In his

Left: *Detail of the door of Swan House, Chelsea Embankment, London, designed by Richard Norman Shaw, 1875–7.*

Far right: *Gustav Stickley, the designer most responsible for the spread of Arts and Crafts principles throughout America and the country's most successful creator of Arts and Crafts furniture, photographed in 1910.*

office at 447 Oxford Street, London, a number of important Arts and Crafts figures first worked, including Alfred Powell, Ernest Gimson, Ernest Barnsley, J. Paul Cooper, Henry Wilson, and W.R. Butler. While articled to G.E. Street (1858), where he worked alongside William Morris, Philip Webb, and Richard Norman Shaw, Sedding developed his understanding of Gothic architecture and ornament, which would later underscore the movement. He also designed wallpapers, embroidery, and church metalwork and was proficient in various other traditional building crafts, including stone-carving and ironwork. He emphasized the interdependence of design and craftsmanship and had a radical vision of "the ideal factory ... where the artist-designer is a handicraftsman and the handicraftsman is an artist in his way." He communicated and promoted his passion for the ideal of medieval craftsmen to others and was instrumental in the revival of weaving, embroidery, and illumination. In 1880 he was appointed diocesan architect of Bath and Wells. In 1885 he designed the richly decorated Holy Trinity Church in Sloane Street, London, and in 1886, in a contrasting severely classical style, the Church of the Holy Redeemer in Clerkenwell. His sudden death from pneumonia in 1891 left a void.

SHAW, Richard Norman (1831–1912), leading British architect. An exponent of the Queen Anne style, and an almost exact contemporary of Philip Webb, Shaw is famous for fostering a number of important Arts and Crafts figures including Mervyn Macartney, E.S. Prior, W.R. Lethaby, Sidney Barnsley, and Robert Weir Schultz. Born in Edinburgh, he trained under the Scottish classicist William Burn and became chief clerk to G.E. Street, where he met Webb in 1859. He went on to set up an office that became a nursery of Arts and Crafts talent, though his own claim to fame is as "the last great Gothic country house architect." Unlike Webb, who shunned publicity and survived on the patronage of a few sympathetic clients, Shaw's output was huge, and he brought his eclectic visual vocabulary to a wide range of commissions. His strong sense of the picturesque is evident in his design for the Bradford Exchange (1864) and the artist Kate Greenaway's house in Frognal, London (1884), which includes a top-floor studio set at an ingenious angle to maximize the north light. In 1877 Jonathan Carr asked him to improve on E.W. Godwin's designs for a development of small detached and semidetached villas at Bedford Park, west London. Alongside his architectural work, Shaw also designed furniture and wallpapers.

SILVER, Arthur (1853–96), English fabric designer. He founded the Silver Studio and supplied Liberty's with some of its most successful and characteristic fabrics (including "Peacock Feather") until the outbreak of World War II. His son Reginald (Rex) Silver (1879–1965) took over the running of the studio on his father's death and, while continuing to supply fabric designs, also designed for Liberty's silver "Cymric" line and "Tudric" pewter range.

SPENCER, Edward (1872–1938), British designer and metalworker. He became a junior designer in the Artificers' Guild, when it was founded in 1901, and chief designer when it transferred to Maddox Street, London. His metalwork was widely publicized through *The Studio*.

SIMMONDS, William George (1876–1968), British painter influenced by Walter Crane. He studied at the Slade School of Art and married Eveline (Eve) Peart (1884–1970) in 1912. Eve Simmonds was a skilled embroiderer and worked closely with Louise Powell (née Lessore), with whom the Simmonds shared a house in Hampstead before moving to the Cotswolds.

STABLER, Harold (1872–1945), British craftsman. He studied woodwork and stone-carving at the Kendal School of Art, helped to found the Design and Industries Association in 1915, and co-founded the Carter, Stabler, Adams pottery in Poole, Dorset, in 1921. He collaborated with his wife, the ceramic and bronze sculptor Phoebe McLeish, on elaborate metalwork and jewelry designs, often embellished with painted enamel panels.

STICKLEY, Gustav (1858–1942), American furniture-maker. With Elbert Hubbard, Stickley was the main promoter of Arts and Crafts style and ideals in America. Born in Wisconsin to German immigrant parents, he trained as a stonemason before going to work at his uncle's furniture factory. He was profoundly influenced by the writings of William Morris and visited England in 1898, where he saw work by and met a range of Arts and Crafts designers, including A.H. Mackmurdo, C.F.A. Voysey, C.R. Ashbee, and M.H. Baillie Scott. In 1899 he established his own guild of artisans, United Crafts, near Syracuse, New York, and soon Americans could order a whole range of products through his mail order catalogs—from comfortable plain oak furniture to textiles, lamps, and carpets—for the home. Before

very long they could even order the home—a bungalow in kit form. In 1902 he opened a metalwork shop. Stickley's magazine *The Craftsman* (published 1901–16) publicized both his products and his philosophy. He attempted to combine the high ideals of British craftsmanship with the more pragmatic necessities of industrial production methods, but he over-extended the company and went bankrupt in April 1915. The Craftsman Workshops were taken over by his brothers' firm, L. & J.G. Stickley (which also made furniture designed by Frank Lloyd Wright) and amalgamated as the Stickley Manufacturing Co.

STICKLEY, Leopold (1869–1957) and STICKLEY, John George (1871–1921), two of Gustav Stickley's brothers who left his United Crafts workshop to set up the L. & J.G. Stickley Furniture Company in 1902, producing machine-aided solid oak furniture.

STREET, George Edmund (1824–81), English architect and designer of exceptional architectural fittings and furniture. A strong supporter of traditional building crafts, he believed that an architect should be multitalented: a builder, a painter, a blacksmith, and a designer of stained glass. He had a profound effect on William

Left: *An oak and leather hexagonal table, with wagon-wheel stretcher, by Gustav Stickley, c. 1912.*

Right: *A painted, wrought iron and bronze grille of an elevator-enclosure cage, designed by Louis Sullivan for the Chicago Stock Exchange and cast by Winslow Brothers' Co., Chicago, 1893–4.*

Morris, who was articled to him for a short while, and Philip Webb, who spent rather longer in his Oxford offices.

SULLIVAN, Louis (1856–1924), American architect. Considered to be the father-figure and spiritual leader of the Prairie school of American architecture, he established his practice in Chicago with Dankmar Adler in 1883. He is best known for commercial rather than domestic buildings, and for mentoring Frank Lloyd Wright.

SUMNER, George Heywood (1853–1940), English wood-engraver and muralist, an associate of the Century Guild (also W.A.S. Benson's brother-in-law) and a member of the Art Workers' Guild.

TAYLOR, Ernest Archibald (1874–1951), Scottish furniture and stained-glass designer. Taylor studied at the Glasgow School of Art before joining the well-established Glasgow cabinet-making firm of Wylie & Lochhead Ltd. He exhibited with the other Glasgow School artists at the 1902 Esposizione Internazionale in Turin, winning a medal and a diploma. He married the textile and jewelry designer Jessie M. King and moved to Manchester where he worked for the decorating firm George Wragge Ltd., becoming very involved in the design of stained glass. In 1911, he and his wife established the Shealing Atelier of Fine Art in Paris, though the outbreak of World War I forced their return to Scotland.

TIFFANY, Louis Comfort (1848–1933), American artist, glass designer and entrepreneur. The son of Charles Tiffany (the owner of New York's Fifth Avenue Tiffany & Co., which specialized in silver and jewelry), he trained first as a painter, then as a glassmaker,

Left: *A "Magnolia" leaded-glass and gilt-bronze Tiffany Studios floor lamp, which sold for almost two million dollars at auction in 1998.*

studying medieval stained-glass panels. He founded Louis Comfort Tiffany & the Associated Artists in New York in 1879, with the cooperation of Lockwood de Forest, Samuel Colman, Candace Wheeler, and the Society of Decorative Art. In 1882–3 they were commissioned to decorate the White House for President Arthur, and other clients included Mark Twain, Cornelius Vanderbilt II, and the British actress Lillie Langtry. Tiffany broke with the Associated Artists in 1885 and set up the independent Tiffany Glass Company. He established Tiffany Studios in 1892 in Corona, New York, as a foundry to supply metal fittings and bases for his glass company, which made lamps, desk sets, candlesticks, metalwares, enamels, and pottery, as well as cut glass. His experiments in glass led him to develop an iridescent satin finish for which he coined the term "Favrile" in 1893. In the same year he won an amazing fifty-four medals at the World's Columbian Exposition held in Chicago. Heavily influenced by William Morris, he wanted his studio—on the top floor of his father's mansion—to be a cross between a European atelier and an Arts and Crafts guild modeled along medieval lines. His trademark electric table, floor, and ceiling lamps sold through Tiffany & Co. in America, Samuel Bing's Maison de l'Art Nouveau in Paris, and the Grafton Gallery in London.

TRAQUAIR, Phoebe Anna (1852–1936), Irish muralist and needlewoman. Born and trained in Dublin, she moved to Edinburgh on her marriage and there expressed the polymath ideal, working as an illuminator, mural decorator, bookbinder, enameler, and embroiderer. An exceptional colorist, she wanted to make her walls "sing" and executed decorations for buildings and furniture by the Scottish Arts and Crafts architect Robert Lorimer.

Right: *An enamel tryptich in silver with abalone shell, designed by Phoebe Anna Traquair, c. 1902.*

UNWIN, Sir Raymond (1863–1940), English architect and father of town planning. In his book *The Art of Building a Home* (1901), co-authored by his partner and brother-in-law Barry Parker, Unwin equated design with problem-solving: "The essence and life of design lies in finding that form for anything which will, with the maximum of conveniences and beauty, fit in for the particular functions it has to perform, and adapt to the special circumstances in which it must be placed." Pre-eminent proponents of the Garden City movement, Parker and Unwin were instrumental in planning Letchworth in Hertfordshire and Hampstead Garden Suburb in London, and also designed twenty-eight houses in the garden village they prepared for Rowntree factory workers at New Earswick, near York, in 1902.

VAN BRIGGLE, Artus (1869–1904), American painter, sculptor, and ceramicist. He worked for Rookwood Pottery from 1887–99 (during which time he was sent to Paris to study for three years from 1893). Ill health forced him to move to Colorado Springs where his style developed to embrace Art Nouveau forms.

VAN ERP, Dirk (1860–1933), Dutch-born American metalworker. While working in a naval shipyard, he began hammering vases out of brass shell casings. He opened the Copper Shop in Oakland, California, in 1908, moving in 1910 to San Francisco, where he produced hammered copper lamps with mica shades and other handcrafted metalwork.

Left: *A chair by Henry van de Velde, one of Europe's foremost peripatetic designers, for Baron von Münchausen, 1904. Van de Velde admired Morris's search for the true principles of design and his socialist beliefs, but found his example invalid because it clung to medievalism.*

VELDE, Henry van de (1863–1957), Belgian architect and designer. An admirer of William Morris, he ran a craft-oriented workshop in Berlin and in 1902 founded a school for the applied arts in Weimar, which became the precursor of the Bauhaus. On his resignation as head of the school in 1914, he recommended Walter Gropius, another admirer of Morris, as his successor, and Gropius founded the Bauhaus in Weimar in 1919.

VOYSEY, Charles Francis Annesley (1857–1941), English architect and one of the most prolific designers of the Arts and Crafts period. He was educated at Dulwich College, after which he worked for the architect J.P. Seddon and then for George Devey. He joined the Art Workers' Guild in 1884 and established himself as a designer of wallpapers and furniture before receiving his first building commissions in the early 1890s. He became a master of "artistic" cottages and smaller country houses and, though he never designed aristocratic mansions like Edwin Lutyens, he was eager to reinstate architecture as the mother of all arts. Voysey was known for the subtlety of proportion and simplicity of form in his work, and for his rigorous attention to detail. He might design every element in a house from the letterplate on the front door to the kitchen fittings, and this rigor also extended to his own appearance, for he liked to wear a special suit of clothes, of his own design, that did away with cuffs and lapels—"dust-traps," he believed—with shirts of "butcher's blue" similar to those worn by William Morris, although Voysey's were set off by a Liberty silk cravat held with a simple gold ring. His designs used plain, good-quality materials made to high standards. His wallpaper designs were sold through Essex & Co. and his fabrics were designed for Alexander Morton & Co. and, of course, Liberty & Co. His work was well known in Europe, where his designs were reproduced and his influence was great.

WAALS, Peter van der (1870–1937), Dutch-born English furniture-maker. Recruited by Ernest Gimson and Ernest Barnsley as foreman for their Sapperton workshop in 1901, Waals had previously worked in Brussels, Berlin, and Vienna. Following the death of Gimson in 1919 he set up his own workshops in an old silk mill at Chalford, Gloucestershire, producing his own designs and some of Gimson's, for clients such as W.A. Cadbury, whose house at King's Norton, near Birmingham, was furnished with many examples of Waals's work.

WALTON, George (1867–1933), Scottish architect and interior designer. He attended evening classes at the Glasgow School of Art before setting up his own business—George Walton & Co.

Above: *"Swallows flying round a tree"—a textile design by C.F.A. Voysey, 1899.*

Right: *C.F.A. Voysey designed and painted the gilded mahogany case for this charming clock in 1895–6. The brass and steel movement was made by Camerer, Cuss & Co.; the case by F. Coote.*

Left: *The Strand showroom of the Kodak photographic company in London, designed by George Walton, c. 1898. Through his friendship with the head of Kodak's European sales organization, George Davison, Walton received the commission to decorate Kodak premises around Europe.*

Right: *Philip Webb's architectural drawing of Red House, without extension, c. 1859.*

Ecclesiastical and House Decorators—at 150–2 Wellington Street, Glasgow, in 1888. He exhibited with the Arts and Crafts Society in 1890 and moved to London (setting up a showroom at 16 Westbourne Park Road) in 1897. He designed the furniture, fittings, and storefronts for branches of the Kodak camera company in Glasgow, London, Milan, Moscow, Vienna, Leningrad, and Brussels and a number of idiosyncratic private houses for directors of Kodak. He also designed the silk hangings used in the Buchanan Street tea-rooms in Glasgow, where Charles Rennie Mackintosh made his name.

WARDLE, Thomas (1831–1908), dyer and printer in Leek, Staffordshire. He helped William Morris with his early experiments in dyeing and worked with Arthur Liberty to introduce "art" or "Liberty" colors to silks and other fabrics from the East.

WATTS, Mary Seton (1849–1938), British painter and craftswoman. The wife of the painter G.F. Watts, she was the director of the Compton Potters' Art Guild in Surrey, a women's guild established in 1896.

WEBB, Philip Speakman (1831–1915), English architect, socialist and designer. He met William Morris when both were briefly working together in the offices of G.E. Street and designed Red House, which has been called the first Arts and Crafts building, for Morris and his wife, Janey, at Upton in Kent in 1859. He designed furniture for Morris's Firm from 1861, as well as glassware and embroidery. Arguably the architectural father-figure of the Arts and Crafts movement, his influence as a designer on those who followed was considerable. A lifelong socialist, modest and shy in public, he had a reputation for highmindedness and high standards. During much of the 1880s and 1890s—during which he was working on a number of country house commissions, including Clouds in Wiltshire (1886) and Standen in Sussex (1891), both decorated by Morris & Co.—he earned less than £320 a year, only slightly more than a master mason. He designed Morris's gravestone and almost his last commission was the memorial cottages to William Morris at Kelmscott, built in the Cotswold tradition, with a relief carved by George Jack, his chief assistant, who after his death wrote: "It is like trying to remember past sunshine—it pleases and it passes, but it also makes things to grow and herein Webb was like the sunshine, and as little recognized and thanked."

WELLES, Clara Barck (1868–1965), American promoter of women's causes, teacher, employer, and owner of the Kalo Shop in Chicago, which she founded in 1900. Run on the lines of C.R. Ashbee's Guild of Handicraft, it made and sold handcrafted silverware and jewelry influenced by Art Nouveau.

WHALL, Christopher Whitworth (1849–1924), leading English designer in the field of architectural glass. He taught stained-glass work at the Central School of Arts and Crafts in London from its foundation in 1896.

WHEELER, Candace (1827–1923), American textile designer and interior decorator, who founded the New York Society of Decorative Art in 1877. She was one of eight children, brought up in a strict

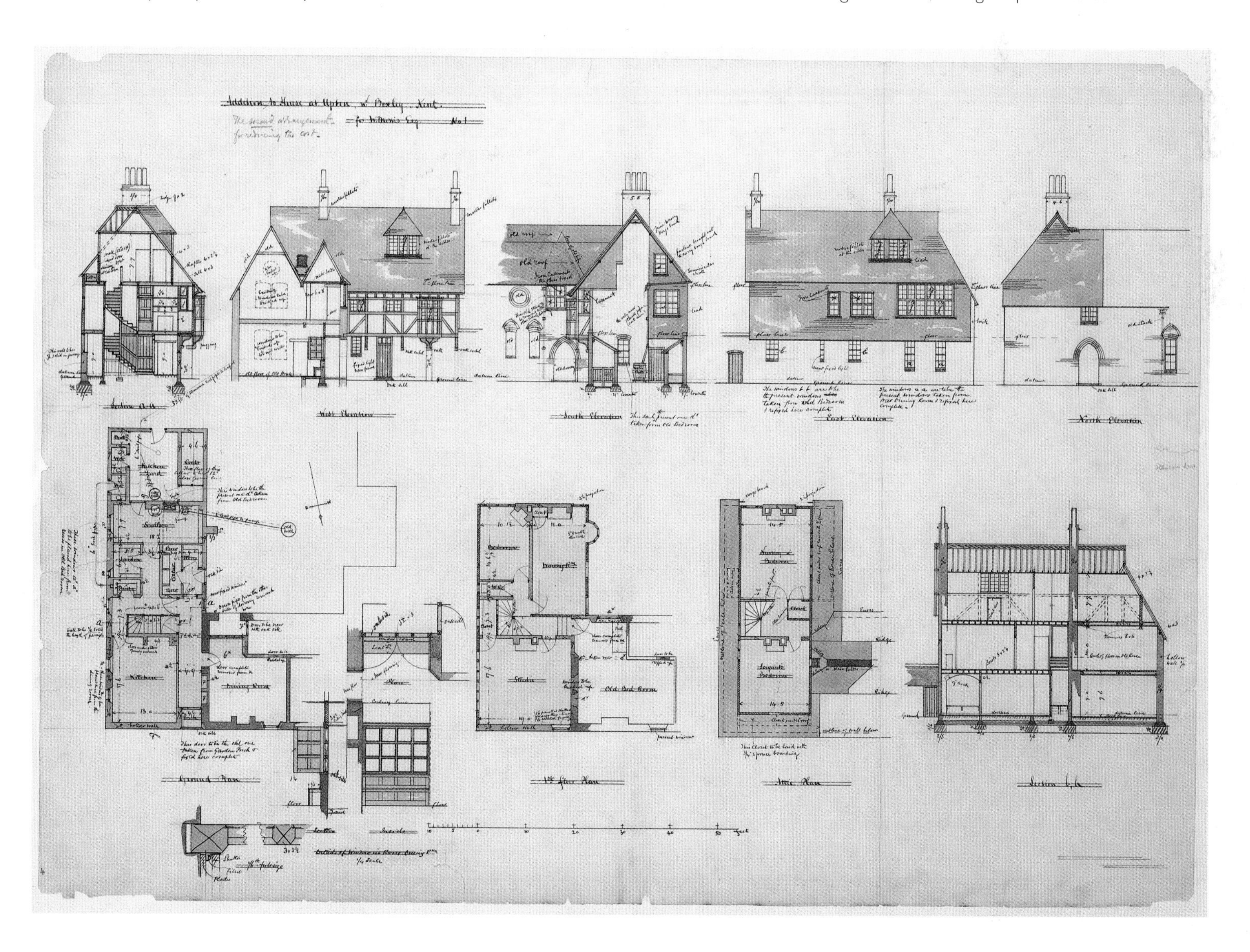

Puritan household in which her mother wove the family's clothes and made their candles. Her marriage at seventeen to Thomas Wheeler, an engineer from New York, introduced her to literature and art. They lived in Brooklyn, then Long Island, and had four children, including a daughter, Dora, who studied at the Acadèmie Julien in Paris and was, like her mother, an admirer of Walter Crane. Both won prizes for their wallpaper designs. When Louis C. Tiffany left Associated Artists, in which Wheeler was a partner, in 1883, she continued the company as an all-female textile design firm until 1907.

WHITEHEAD, Ralph Radcliffe (1854–1929), English-born, Oxford-educated founder, with his wife Jane Byrd McCall, of the American Byrdcliffe Arts Colony in 1902—a Utopian craft community in rural Woodstock, New York, inspired by John Ruskin and devoted to pre-industrial ideas. It produced furniture, metalwork, pottery, and textiles and offered an annual summer school for artists until 1912.

WILDE, Oscar Fingal O'Flahertie Wills (1854–1900), Irish writer and aesthete. In 1882 he embarked on a wildly successful

Far left: *"Consider the Lilies of the Field"—an embroidered portière, designed and made by Candace Wheeler, 1879.*

Left: *Woodcut design for a hair ornament from Henry Wilson's* Silverwork and Jewellery, *1912.*

Right: *Frank Lloyd Wright was already 35 in 1902 when he designed the Susan Lawrence Dana House, followed by other early masterpieces.*

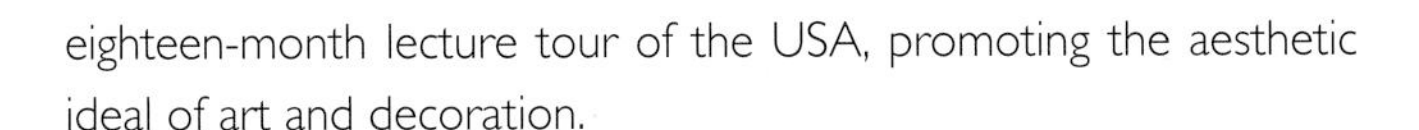

eighteen-month lecture tour of the USA, promoting the aesthetic ideal of art and decoration.

WILSON, Henry (1864–1934), English architect, designer, metalworker, and enameler. He wrote in 1898 that the architect "should be the invisible, inspiring, ever active force animating all the activities necessary for the production of architecture." He inherited J.D. Sedding's practice after Sedding's sudden death in 1891. In 1895 he set up his own metalworking shop in a house at Vicarage Gate in Kensington where for a brief, unsuccessful period he entered into partnership with Alexander Fisher. J. Paul Cooper learnt jewelry-making in Wilson's workshop. From 1896 he taught at the Central School of Arts and Crafts and from 1901 at the Royal College of Art. A member of the Art Workers' Guild from 1892, he became president of the Arts and Crafts Exhibition Society in 1916. His *Silverwork and Jewellery* (1903) was considered one of the best practical manuals on the subject.

WOODROFFE, Paul Vincent (1875–1954), English book illustrator, heraldic artist, and stained-glass designer who set up a stained-glass studio at his house at Westington, Gloucestershire, reconstructed for him by C.R. Ashbee.

WRIGHT, Frank Lloyd (1867–1959), prolific, avant-garde, Wisconsin-born, American architect, furniture, stained-glass, and textile designer. He began work in the office of Dankmar Adler and Louis Sullivan and his long career spanned seven decades and two centuries, during which he received more than a thousand commissions and executed over half of them, most of which still stand today. Much influenced by his mother, Anna Lloyd Wright, a

Unitarian with Welsh roots whom he described as being "in league with the stones of the field," he drew inspiration from nature for what he called his "organic architecture." Wright believed that "the horizontal line is the line of domesticity" and revolutionized the single-family private house by destroying "the box" and opening the house "within itself and out to nature." Nikolaus Pevsner called him "a poet of pure form" in whose "gigantic visions, houses, forests, and hills are all one." The genius of the Prairie school of architecture, his work draws on the economy and grace of Japanese art, and continues to exert enormous influence today. He was a founding member of the Chicago Society of Arts and Crafts at Hull-House in 1897, met C.R. Ashbee in 1900, and was influenced by the ideas of Otto Wagner in Vienna. In his later buildings, such as Fallingwater in Pennsylvania (1935–9), and the Guggenheim Museum in New York, he moved toward a more Modernist style.

YEATS, Lily (1866–1949) sister of the Irish poet William Butler Yeats. She was a close friend of May Morris and was employed from 1866–94 as an embroideress by Morris & Co. Subsequently, she managed the textile section of the Dun Emer Guild, which she co-founded in Dublin in 1902 together with her sister, Elizabeth, and Evelyn Gleeson.

YELLIN, Samuel (1885–1940), Polish-born metalworker who opened a shop in Philadelphia in 1909, where he specialized in hand-hammered repoussé designs for architectural practices.

Left: *The octagonal library in Frank Lloyd Wright's home and studio in Oak Park, Illinois, created an immediate and lasting impression on the clients invited to sit at the oak table and look over presentations for their own houses.*

Right: *This embroidered panel showing an orchard of apple trees in blossom was made by Lily Yeats in 1926, and has a linen label sewn onto the reverse reading "Picture by sister of poet Yeats, Wedding present from Mary Geoffrey Holt (cousin of Yeats)."*

ARCHITECTURE

Left: *The Greene brothers' great Arts and Crafts masterpiece, The Gamble House, built in Pasadena in 1908–9.*

2

EMBRACING THE VERNACULAR

"The Englishman builds his house for himself alone. He feels no urge to impress, has no thought of festive occasions or banquets and the idea of shining in the eyes of the world through lavishness in and of his house simply does not occur to him. Indeed, he even avoids attracting attention to his house by means of striking design or architectonic extravagance, just as he would be loth to appear personally eccentric by wearing a fantastic suit. In particular, the architectonic ostentation, the creation of 'architecture' and 'style' to which we in Germany are still so prone, is no longer to be found in England. It is most instructive to note ... that a movement opposing the imitation of styles and seeking closer ties with simple rural buildings, which began over forty years ago, has had the most gratifying results."

HERMAN MUTHESIUS, *The English House*, 1905

THE ARTS AND CRAFTS MOVEMENT was initially founded upon a strong love of England and all things English, and nowhere is this more readily apparent than in the architecture of the period. A single house—Red House, built at Upton in Kent in 1859 for William Morris by his friend Philip Webb—broke the classical mold, embraced the vernacular, and began a revolution in domestic architecture. For the next half century it pointed the way for succeeding generations of architects who were keen to put function first, to relate their buildings to the landscape, and to build them from carefully selected, often local, materials.

Red House is a deeply pleasing, asymmetrical, L-shaped house, built of warm red brick in a scaled-down Gothic style. It incorporates a great arched entrance porch and steep irregular gabled roofs topped with tall idiosyncratic chimneys and a weathervane ornamented with the initials W M. It was never a grand house, but capacious and comfortable, with a hint of fantasy. Webb's design provided four good bedrooms, another tiny one, and a partitioned dormitory for the cook and two maids at the far end of the western wing. Narrow stairs led down from this to the kitchen and scullery, while a magnificent oak staircase with tapered newelposts guided guests up from the deep expansive

Above: *Lead hopper at Baillie Scott's Blackwell, showing the date 1900, and the initials of his client, Edward Holt.*

Left: *"More a poem than a house"—Red House, designed by Philip Webb in collaboration with William Morris in 1859, helped to change the direction of English domestic building and proved enormously influential.*

entrance hall to the upper rooms, including the drawing room with its high arched ceiling, ribbed with beams, that extended right up into the roof space, and the adjacent light-flooded studio, with its views across orchards and the rolling Kent countryside.

The house represents a stunning collaboration between an architect and an artist and has been called the first Arts and Crafts building, designed by the man credited with having provided "the movement's morality and theory." Webb designed his buildings "as they should be" from the inside out, considering first the functional interior relationships of rooms to corridors and stairwells. He adhered to Pugin's principle of fidelity to place, linking it with the Ruskinian notion of fidelity to function to form the basis for a new national style.

The importance of Augustus Welby Northmore Pugin in tracing the trajectory of Arts and Crafts architecture cannot be overestimated. "But for Pugin," the architect John Dando Sedding admitted in 1888, "we should have had no Morris, no Street, no Burges, no Shaw, no Webb, no Bodley, no Rossetti, no Burne-Jones, no Crane." Though he died at the early age of forty, Pugin designed more than a hundred buildings, most of them churches. A passionate Catholic convert, his vision of Christian architecture, which harked back to the Middle Ages and Gothic asymmetry, came to underpin the Arts and Crafts style. The beauty of a building depended for Pugin on "the fitness of the design to the purpose for which it is intended." His three basic rules for architecture were: structural honesty, originality in design, and the use of regional materials or character.

The writings of Pugin and Ruskin had a profound effect on mid-Victorian architects such as William Butterfield and George Street, in whose Oxford office Philip Webb trained as an architect, forged his friendship with William Morris, and developed the understanding of Gothic architecture and ornament that would characterize his work. Street's office proved an important seed-bed for the movement, fostering not just Webb, but also John Dando Sedding, the leading ecclesiastical architect, and Richard Norman Shaw, the pioneer of the Queen Anne style, whose separate practices provided stimulating starting points for a number of important Arts and Crafts figures. Sedding could boast a roll call at his Oxford Street office that included Henry Wilson, Arthur Grove, W.R. Butler, Alfred H. Powell, and two members of the Cotswold group: Ernest Barnsley and Ernest Gimson. Meanwhile Shaw, who encouraged individuality and was generous about setting his former pupils up with work when they were ready to leave, provided the platform for Mervyn Macartney, William Lethaby, Sidney Barnsley, E.J. May, Ernest Newton, Gerald Horsley, Robert Weir Schultz, and Edward Prior, an ardent disciple of Morris. Prior was responsible for the highly original Arts and Crafts masterpiece of St. Andrew's church at Roker, Sunderland, and for introducing the "X" or "butterfly" plan in his layout of The Barn, a house built on a hill overlooking Exmouth, Devon, in 1897. Prior's curving plan offered an alternative to the popular L-shape and became an important feature of Arts and Crafts

Above: *E.S. Prior used a "butterfly" layout for The Barn at Exmouth in Devon in 1897, curving the front of the house in an open-armed welcoming gesture.*

architecture, adapted most notably by Edwin Lutyens in Papillon Hall in Leicestershire (1903–4). It was also used to great effect by Detmar Blow at Happisburgh Manor in Norfolk (1900).

Shaw—almost an exact contemporary of Webb—had a huge output and often pandered to the tastes and fashions of the *nouveaux riches*, producing dramatic and impressive country houses in a deliberately asymmetrical "Old English" style and elegant London town houses in a broad "Queen Anne" style, with elegant Palladian touches or Dutch gabling. He thought Webb—who shunned publicity and survived on the patronage of a few sympathetic clients—"a very able man indeed, but with a strong liking for the ugly."

Standen, a gracious country house built near East Grinstead in Sussex in 1894 for the successful solicitor James Samuel Beale, is generally considered to be Webb's masterpiece, but before that came Clouds, the "palace of art" he designed as a country seat for the Conservative MP and aristocrat Percy Wyndham and his wife Madeline at East Knoyle in the south-west corner of Wiltshire. Famously conscientious about his work, and busy at the time with Rounton Grange for Sir Isaac Lowthian Bell, Webb was at first reluctant to accept the commission. As his assistant, George Jack, explained, "He would never undertake more work at one time than he could personally supervise in every detail," even though "had he desired he might have built many more houses." Money was never a motivation; his average earnings over forty years, during which time he completed between fifty and sixty buildings, totaled a mere £380 a year. "I do not lay myself out to work for people who do not in any

Above: *Standen in West Sussex, designed by Philip Webb for the solicitor James Beale in 1891–4.*

degree want what I could honestly do for them," he wrote warningly to Wyndham, who knew of his work through the tall red-brick town house at 1 Palace Green, Kensington, which Webb had designed for his friends George and Rosalind Howard. Wyndham believed that in Webb he had found the architect for his "house of the age" and would not accept no for an answer.

The relationship between a house and its setting was a matter of paramount importance to all Arts and Crafts architects. Webb wanted Clouds, despite its monumental size, to appear as if it grew up out of the landscape. He ended up working on it for almost ten years, a period extended by the disaster that befell it soon after its completion when—owing to the carelessness of a maid who had left a lighted candle in a cupboard—the house burned down. It was rebuilt, again by Webb, within two years of the tragedy. His attention to detail was famous. He even made a note of the height of the Wyndhams' head gardener—an exceptionally tall man called Harry Brown—so that he could take this into consideration when designing the doors in the garden walls at Clouds. A committed socialist like William Morris, Webb allowed for coal bunkers on the first- and second-floor landings of the house to make the lives of the servants easier. Owing to the wildly fluctuating fortunes of the Wyndham family after World War I, "the house of the age" has had a rather checkered history, serving at different times as a home for unmarried mothers and a center for the treatment of alcohol and drug dependency. Although much reduced, the ornate plasterwork ceiling, the fireplaces, dadoes, and Della Robbia plaques all survive, incongruous among the institutional furniture.

Standen has fared rather better and, under the careful stewardship of the National Trust, has become a much-visited shrine to the Arts and Crafts movement. More modest in style and form than Clouds, and much less formal, it is built on a long, thin L-shaped plan and assumes an almost "cottagey" simplicity through its use of local materials and techniques, varied coloration, textures, and tidy, weatherboarded, gabled roofs. Webb linked the new house to an existing fifteenth-century farmhouse, Hollybush Farm, and one of its timber barns, using a prominent tower to serve as a transition between the rambling service wing and the symmetrical main block, and planned the gardens and grounds with the same careful attention he gave to the house, preserving as many old trees as possible. "To Webb," wrote W.R. Lethaby, his first biographer and great admirer, "the fields and old buildings of England were a question of quite religious moment."

Below: *Arthur Melville's watercolor of the garden front at Standen, painted in 1896.*

Above: *Melsetter House, Hoy, designed by W.R. Lethaby for the Birmingham businessman Thomas Middlemore in 1898.*

Soft, creamy Sussex stone was quarried from the estate and interspersed with gray Portland stone, which Webb used for the window sills, and the hard red Keymer bricks framing the arches and window surrounds. The result was a house that appeared "to have been built up bit by bit, almost unconsciously."

"Architecture to Webb," Lethaby explained, "was first of all a common tradition of honest building. The great architectures of the past had been noble customary ways of building, naturally developed by the craftsmen engaged in the actual works. Building is a folk art. And all art to Webb meant folk expression embodied and expanding in the several mediums of different materials. Architecture was naturally found out in doing...."

W.R. Lethaby, a pivotal figure in the Arts and Crafts movement who left Shaw's office in 1889 to set up on his own, greatly admired Webb's respect for tradition, functional form, and treatment of materials. He, however, strove to combine romanticism with traditional design and achieved it at Melsetter House, built for a retiring Birmingham businessman, Thomas Middlemore, on Hoy, Orkney, in 1898. Morris's daughter May described Melsetter as "a sort of fairy palace on the edge of the great northern seas, a wonderful place ... remotely and romantically situated with its tapestries

Left: *The exterior of the Church of All Saints, Brockhampton, Herefordshire, by W.R. Lethaby, 1901–2, hailed by Nikolaus Pevsner as "one of the most convincing and most impressive churches of its date in any country";* **Below:** *The uplifting interior of the church.*

and its silken hangings and its carpets ... the embodiment of some of those fairy palaces of which my father wrote with great charm and dignity. But, for all its fineness and dignity, it was a place full of homeliness and the spirit of welcome, a very lovable place. And surely that is the test of an architect's genius: he built for home life as well as dignity."

Hermann Muthesius, the German cultural attaché who made a particular study of Arts and Crafts buildings, was a great admirer of Lethaby. "The number of his houses is not large," he remarked, "but all appear to be masterpieces." Lethaby's last work, the Church of All Saints, Brockhampton, Herefordshire (1901–2) has been called "one of the greatest monuments of the Arts and Crafts movement." It is a deeply romantic, thatched, barnlike building, constructed of local red sandstone with a weatherboarded bell tower and a square stone crossing tower. Inside, the high pointed arches of the concrete vaulted nave spring from just above floor level, creating a scaled-down cathedral-like sense of calm and beauty. Lethaby, like other progressive Arts and Crafts architects, sought and found inspiration in traditional buildings that were not designed by architects—"old work" as Baillie Scott described them, that embodied "a certain aesthetic rightness and beauty expressed in practical ways."

"It is not until we get back to the work of the earlier builders," M.H. Baillie Scott wrote, "that our hearts are touched and thrilled by the strange charm of the building art as then practised." He and other leading Arts and Crafts architects, such as Charles Voysey, C.R. Ashbee, Edwin Lutyens, Raymond Unwin, and Barry Parker, stressed the need for simplicity and aimed to design houses and churches that enhanced rather than dominated their surroundings, combining features from old work found in houses, museums, and churches with contemporary design. They chose to make their buildings from indigenous materials and strove above all to follow local traditions—using stonework in the Cotswolds, pargetted plasterwork in Essex, tile-hanging in Kent, and roughcast in the Lake District—for both economic and aesthetic reasons. They also spearheaded a revival of traditional construction methods, interesting themselves in the details and techniques of crafts such as ornamental plasterwork, ironwork, carving, and sculpting so that their designs could be fully integrated. Unpretentious simplicity and harmony were the keynotes of the Arts and Crafts house.

Plainly, it was an exciting time to be an architect. In Britain, the doubling of the population during the reign of Queen Victoria, rapid strides in industrialization, and the burgeoning of the middle classes had prompted a building boom and ushered in a golden age for architecture. Commissions for all sorts of houses, from gracious country homes to a new form of housing—the semidetached villa—were plentiful. Between 1898 and 1903 an average of 150,000 houses were built each year. Most of these were standard Victorian terraces, much reviled by Baillie Scott who suggested that they should be renamed "The Crimes" for their lack of imagination, light, and cramped conditions. However, the Garden City movement offered a more attractive counterbalance and plenty of work for socially minded progressive young architects.

Architecture was, above all, a respected and valued profession, conferring a privileged status in British society and attracting many able, public-spirited recruits. The usual method of training was the pupilage system, whereby artistically and technically inclined young men (women were very rare, although two—Ethel Mary Charles and her sister Bessie Ada—trained in the office of Ernest George and Harold Peto and became the first women members of the Royal Institute of British Architects in 1898 and 1900 respectively) were placed in the architectural offices, generally in London, of established practitioners. If they were lucky, they were encouraged "to look beyond the confines of the drawing board," but were certainly urged to visit museums and churches, and to spend their weekends and free time touring the Kent, Sussex, Gloucestershire, and Shropshire countryside, gathering ideas and inspiration by filling their sketchbooks with vernacular subjects: traditional brick and tile-hung buildings, hipped-roofed cottages, sundials, oak pegging, and architectural details on old churches. It was common practice for articled pupils to attend lessons three evenings a week at the South Kensington Schools, Royal Academy Schools, or the Architectural Association.

It was William Morris himself who advised the young Ernest Gimson to move to London, giving him a letter of introduction to J.D. Sedding, who accepted him at once. Sedding communicated his enthusiasm for a range of activities—he also designed wallpapers, embroidery, and church metalwork, and was proficient in various other traditional building crafts—to the young Gimson and to Ernest Barnsley (whose brother was learning his trade in that other Arts and Crafts nursery, Richard Norman Shaw's office).

Gimson was always interested in traditional crafts, including decorative plasterwork, woodturning and chair-making. He carried out a complex scheme of plaster decoration in the main rooms of two houses designed by Lethaby: Avon Tyrrell in Hampshire (1891) and The Hurst, near Sutton Coldfield (1893). His friendship with the architect Halsey Ricardo also provided him with plasterwork commissions, including the ceiling decorations for the dazzling new department store of Debenham & Freebody's in Wigmore Street, London, and the exotically decorated Arts and Crafts house at 8 Addison Road, Kensington, for which Gimson provided the plasterwork, William

Above: *A pencil and watercolor sketch by Thomas Hamilton Crawford, 1907, of 8 Addison Road, Kensington, London, designed by Halsey Ricardo for Sir Ernest Debenham.*

De Morgan the ceramic tiles, E.S. Prior the stained glass, and the Birmingham Guild of Handicraft the ironwork. Gimson's sketchbooks from the 1880s show details of traditional farm buildings and furniture as well as measured drawings of churches and country houses. Committed to the ideal of the simple life, Gimson, like Morris, believed in "doing not designing" and his stone cottages in Leicestershire and the Cotswolds demonstrate his commitment to forging a new architecture that used local building traditions. They remain, for some, the most perfect realization of Arts and Crafts theory. In 1915 he built a pair of cottages at Kelmscott in Oxfordshire for May Morris and also designed the village hall—though it was not actually built until 1933. He was also responsible for the library and hall at Bedales School in Hampshire (to which Sidney Barnsley sent his children).

Ernest Barnsley, who also built his own home at Sapperton, is best known architecturally for Rodmarton Manor in Gloucestershire, a honey-colored, multigabled, mullioned, Arts and Crafts masterpiece that he worked on from 1909 until his death in 1926. C.R. Ashbee, who visited the works

Below: *Stoneywell Cottage, designed by Ernest Gimson for his brother Sydney, 1898–9, on an irregular plan, incorporated crooked chimneys, rough, hand-cut stonework, and a thatched roof.*

Right: *Rodmarton Manor, Gloucestershire, designed by Ernest Barnsley, was built by hand in strictest Cotswold tradition over a twenty-year period from 1909 to 1929; the house was finally completed by Norman Jewson after Barnsley's death.*

Above: *A presentation drawing from 1898 by Howard Gaye, depicting Broadleys, a holiday home overlooking Lake Windermere, designed by C.F.A. Voysey for A. Currer Briggs.*

in 1914, called the house "the English Arts and Crafts movement at its best." Barnsley's aim to preserve the original character of typical Cotswold buildings is evident in his design for a pair of cottages bordering Sapperton Common, and it was shared by his brother, Sidney, whose terrace of almshouses and public baths at Painswick showed the same respect for historical buildings and vernacular traditions.

Another Arts and Crafts luminary (now best known for his jewelry) who began work in Sedding's office in 1888 was John Paul Cooper. He was apprenticed there on the advice of W.R. Lethaby, who recommended Sedding to Cooper's father as "the best man in London ... he's quite out of the ordinary run of architects, full of enthusiasm, very clever ... your son cannot do better." That year represented the height of the Arts and Crafts movement in Britain and Cooper found Sedding's office congenial and quite different from other architectural practices. "Sedding was much too original to please the other architects," he wrote in his journal, "& we followed him in excess & looked upon the Academy [Royal Academy of Art] the Institute [RIBA] and the Architectural Association as composed of old fogies, without any life in them." It was a great blow to them all when Sedding, who is best remembered for two London buildings—the richly decorated Holy Trinity Church in Sloane Street and the Church of the Holy Redeemer in Clerkenwell—died suddenly of influenza in 1891.

One important Arts and Crafts practitioner who did not come down through the Street/Sedding/Shaw line was Charles Francis Annesley Voysey. He trained in the offices of J.P. Seddon and George Devey, but established himself first as a designer of wallpapers and furniture in the 1880s—joining the Art Workers' Guild early in 1884—before receiving his first building commissions in the 1890s. His childhood had been turbulent and overshadowed by his curate father's trial for heresy and eventual expulsion from the church, a *cause célèbre* from which Voysey learned some valuable, if uncomfortable, lessons about press attention. In his professional life he was always careful to manipulate the press to his advantage. A clever self-promoter (the antithesis, in fact, of that other great genius Charles Rennie Mackintosh), he commissioned and directed photographs of his houses both in progress and finished, providing them at his own expense to magazine editors, whose subsequent publication of them bestowed on him a sort of celebrity status as an artist-architect. Like Mackintosh, he was ahead of his time and had definite ideas about the nature of the home and the importance of harmony and repose. He became a master of "artistic" cottages and smaller country houses. His houses—such as The Orchard (1899), which he built for himself in Chorleywood, Hertfordshire, and Broadleys (1898), built for Arthur Briggs on a spectacular site overlooking Lake Windermere in Cumbria—with their low roofs, wide eaves, horizontal windows, white roughcast walls, and exposed beams and brickwork, were built to high standards from plain, good-quality materials and were meant to look as if they were part of the natural environment. He was known for his rigorous attention to detail and for the subtlety of

Below: *Sir Edwin Lutyens' Little Thakeham in West Sussex is now a country house hotel.*

Above: *Munstead Wood, designed by Sir Edwin Lutyens for Gertrude Jekyll in 1896.*

proportion and simplicity of form in his work. Candor, comfort, restraint, and clean lines were hallmarks of a Voysey building, and he constantly repeated his view that "we cannot be too simple." He wanted his buildings "to play into the hands of nature" and, although his practice was comparatively small, the notice his work attracted—in *The Architect*, *The British Architect*, *The Studio*, and other Arts and Crafts publications devoted to the modern home and cottage—ensured it was seen by and influenced a generation of younger architects.

Young upcoming men such as Edwin Lutyens saw in Voysey's buildings "the absence of accepted forms … the long, sloping, slate-clad roofs, the white walls clear and clean … and old world made new." Hermann Muthesius also recognized his innovation: "In both interiors and exteriors he strives for a personal style that shall differ from the styles of the past. His means of expression are of the simplest so that there is always an air of primitivism about his houses."

Edwin Lutyens was fortunate in his first client, Gertrude Jekyll. They met in the spring of 1889, "at a tea-table, the silver kettle and the conversation reflecting rhododendrons," as Lutyens described it. She was forty-five, an accomplished painter, silverworker, and embroiderer whose failing eyesight had prompted a move into garden design—an area in which she was beginning to establish quite a reputation. He was twenty and had already, a month before their meeting, precociously set up his own architectural practice. They shared a love of vernacular architecture and spent time together visiting old farms and cottages in Surrey and Sussex around Munstead House, where Jekyll was then living with her mother and just starting to lay out her famous garden on a fifteen-acre plot across the road from her family home. When, a few years later, she decided to build a house to go with the garden, Lutyens, with whom she was already collaborating, was the obvious choice. She wanted a relatively simple house "designed and built in the thorough and honest spirit of the good work of old days"—a house she could "live in and love." Lutyens gave her Munstead Wood, his finest work in a style based on the Surrey vernacular, picturesquely inspired by the cottages and farmhouses they had visited. It was relatively small by the standards of the day, with four main bedrooms upstairs and a living hall, dining room, book room, workshop, and office below. Firmly in the Arts and Crafts tradition, the walls are made of local Bargate stone and the roof of handmade red tiles. The interiors are solid and unpretentious: the ceilings are low and oak-beamed, and the windows are leaded to keep out the sunlight that was so injurious to Jekyll's eyesight. The oak staircase rises in broad, shallow steps to the half-timbered gallery that gives on to the bedrooms above.

Munstead Wood secured many future commissions for Lutyens and became "a mecca for Arts and Crafts architects." Among them was the Scottish architect Robert Lorimer, who visited in 1897 and reported on the newly completed house, "It looks so reasonable, so kindly, so perfectly beautiful that you feel that people might have been making love and living and dying there, and dear little children running about for the last—I was going to say, thousand years, anyway six hundred. They've used old tiles which of course helps but the proportion, the way the thing's built (very long coursed rubble with thick joints and no corners) suggests in fact it has been built by the old people of the old materials in the old 'unhurrying' way but at the same time 'sweet to all modern uses'."

Lutyens was a magpie, collecting and then overlaying his own designs with stylistic quotations from contemporary as well as historical sources, brilliantly synthesizing the ideas he gathered from his contemporaries, while responding to and incorporating the needs and requirements of his affluent clients. Homewood at Knebworth, Hertfordshire, for example, built in 1901 for Lady Lytton, owes much to Philip Webb's Standen period, though in his later work Lutyens moved away from the informality of Arts and Crafts toward a new, though often quirky, grandeur. His houses often have two dissimilar sides, one masking the other: a formal front perhaps with a vernacularized domestic garden side at the back—as in Tigbourne Court (1899) in Witley, Surrey, his "gayest and most elegant building"—or the weatherboarded, gabled cottage-style front at Homewood, which, from the rear, presents as a stuccoed classical villa. One way in which his work is particularly important is the way in which the house relates to the garden. He would often extend the geometric structure of the

Above: *Hill House, Upper Helensburgh, designed by Charles Rennie Mackintosh for the Scottish publisher William Blackie, 1902–4.*

house out into the garden, or he might use the garden elements to fortify and ennoble the house as metaphoric castle. The influence of his early studies, in the company of Gertrude Jekyll, of vernacular buildings around his home at Thursley in Surrey, cannot be overestimated in his development as one of the most important and sought-after country-house architects of the period.

Another young architect upon whose work the influence of Voysey can be felt is Charles Rennie Mackintosh, the great architectural genius to come out of Scotland during this period. His ideas were loudly applauded abroad, especially in Europe, where his avant-garde style was best appreciated. He exhibited with the Vienna Secession artists in their eighth exhibition in 1900 and in 1902 he designed the Scottish pavilion for the Esposizione Internazionale in Turin. Yet recognition in his home town of Glasgow was slow to come. His work was considered eccentric. Some thought the tea-rooms he designed for Catherine Cranston were a joke, and found the Glasgow School of Art (1897–1909)—which has since been hailed as the first truly modern building—too extreme to comprehend. Fortunately the building had impressed the prosperous publisher William Blackie, who commissioned Mackintosh to build him a country house on a site at Upper Helensburgh overlooking the Firth of Clyde. The massive gray walls and circular tower of Hill House, punctuated by irregular windows of varying size and shape, are finished in roughcast rendering topped by roofs of dark blue-gray slate, intended to harmonize, better than any red roof tiles, brick or timber could, with the

Scottish landscape and climate. The monumental effect of the exterior is confounded by the bold geometrical elegance of the interior.

Mackintosh believed that "all great and living architecture has been the direct expression of the needs and beliefs of man at the time of its creation." He wanted to "clothe modern ideas with modern dress" and have "designs by living men for living men." Voysey's simplicity and clear lines exerted an important influence on Mackintosh's early career and the impact of Baillie Scott can also be traced, but his genius resides in his ability to create buildings and spaces that are wholly his own, expressing his own distinctive artistic language.

Tragically, Mackintosh's architectural star was to burn only briefly, eventually extinguished by the neglect that followed his professional decline, as his reputation for unreliability, eccentricity and drinking, put potential clients—already suffering the effects of the economic slump occasioned by the advent of World War I—off commissioning such an uncompromising and individual architect. Flashes of genius are glimpsed in the few commissions he did manage to attract in his later years, but he was to die in poverty and obscurity in London, unremarked by the profession or his native city.

Below: *Blackwell at Bowness on Windermere, designed by M.H. Baillie Scott as a rural holiday retreat for the Manchester brewery owner Sir Edward Holt and his family, 1898–1900.*

Some of the finest Arts and Crafts buildings were small-scale, livable, and affordable homes for a newly prosperous professional class. For the Arts and Crafts architect, as we have seen, beauty resided in line, proportion, texture, workmanship, and most of all in appropriateness. The importance of integrating the designs of furnishings and buildings was paramount, and indulgence in ornament for ornament's sake was anathema. One architect who concerned himself almost exclusively with the smaller country house and its furniture was M.H. Baillie Scott, that fierce critic of town housing, which he lambasted for its endless repetition of identical rooms, "so that an absent-minded occupant of one of them might be excused in entering his neighbour's house in mistake for his own, and would find little in its interior arrangements to undeceive him."

Baillie Scott did not train in an Arts and Crafts practice, he never met William Morris, and he did not belong to the Society for the Protection of Ancient Buildings or the Art Workers' Guild, but the refreshing spirit of the Arts and Crafts ideal is embodied in his country houses, such as Blackwell in Cumbria and White Lodge in Oxfordshire (1898–9), and in his Garden City cottages. Like Voysey, he was at his best when working with small buildings. He sought to achieve "a spirit of repose" in his simple and uncluttered spaces, often leading off a generously proportioned central hall with a corner fireplace inglenook and built-in seating. By reviving the idea of the hall-house, Baillie Scott reignited Morris's notion of the medieval hall "where the house itself was the hall and served for every function of domestic life." Instead of acting as an entry point giving access to the staircase, Baillie Scott made the hall "a general gathering-place with its large fireplace and ample floorspace: no longer a passage ... but a necessary focus to the plan of the house." He pioneered and mastered a new flow of space, breaking down the conventional divisions and making the interior, especially in smaller houses, seem more spacious and flexible. He designed, like Webb, from the inside out, treating the garden as a continuum of the house space and integral to it.

Baillie Scott began his practice on the Isle of Man, where he built a romantic house for himself and his family that he called, in obvious reference to William Morris, the Red House. Here he explored the possibility of opening up the living space. The three main rooms are separated by movable screens that allow for flexibility and transformation. Following the Arts and Crafts model he built in local materials that he knew would mellow with age. Unusually, he was also interested in mass housing—not the terraces he railed against, but more interesting, picturesque groupings, which he explored in his designs for the Garden City movement. These were realized in Hampstead Garden Suburb and at Letchworth, Hertfordshire, where individual houses of great charm, with his signature leaded casement windows and exaggerated fall of the eave line, can still be found. The Garden City movement has been called the ultimate expression of Arts and Crafts and of the revival in English domestic architecture—and Baillie Scott was in the vanguard.

Above: *Rushby Walk, pioneer cul-de-sac terrace layout by Raymond Unwin for the Howard Cottage Society, c. 1906, at Letchworth Garden City in Hertfordshire.*

He was among a number of Arts and Crafts architects employed by Henrietta Barnett, a tireless worker in the East End slums and wife of Canon Samuel Barnett, founder of Toynbee Hall, for her pet project: Hampstead Garden Suburb. Raymond Unwin, whom she instinctively felt was "the man for my beautiful green golden scheme," was made chief planner and Edwin Lutyens was appointed the Trust's consulting architect. Henrietta Barnett was not an easy client, though Unwin dryly reported that in seven years of working together they "reached a good understanding." She summarized her aims for the Suburb in an article in *Contemporary Review*: It was important that no house should spoil another's outlook, that the estate should be planned as a whole (although not in uniform lines), that each house should be surrounded by its own garden, and that every road should be planted with trees. She made ample provision for extra allotments and put in place cooperatives from which garden tools might be borrowed, as well as other communal meeting places: houses of prayer, a library, schools, a lecture hall, clubhouses, shops, baths, washhouses, bakehouses, refreshment rooms, arbors, cooperative stores, playgrounds for smaller children (whose noise was to

Above: *Waterlow Court, an Associated Home for Ladies in Hampstead Garden Suburb, designed by M.H. Baillie Scott in 1908–9, attracted an enthusiastic review from* The British Architect: *"Mr Baillie Scott has shown that our old type of almshouse design, built in quadrangular form, may be dealt with in a sensible modern spirit so as to make economical and artistic housing a possibility."*

be "locally limited"), and "resting places for the aged who cannot walk far." There was to be no public house, because of the strong influence of the Temperance movement, whose involvement was a cornerstone in her scheme to promote the physical wellbeing and moral rectitude of the poor.

Unwin drew inspiration from the Middle Ages and signaled it in the suffixes of the street names—close, way, chase, lea, holm, and garth—as well as in the "Merrie England" arrangement of village greens and in the grander medieval style of the shops with flats above, influenced by ancient German towns like Nuremberg and Rothenburg, that served the Suburb at its edges.

Henrietta Barnett was fortunate in employing some of the most distinguished Arts and Crafts architects of the period. Consequently, her development survives as a scale model of all that was best in the movement: from the serenity of Lutyens' St. Jude's Church in the central square, through the vernacular red brick, steep hipped-roofed, mullioned-windowed semidetached houses of Parker and Unwin to the radical reforming zeal of Baillie Scott's Waterlow Court, a group of two-room flats built around a medieval-style cloistered quad and intended specifically to house "the working lady" (a revolutionary concept in 1909).

Ironically, the Arts and Crafts tradition survived longest—though vulgarized and watered down—in suburban housing schemes, in which thousands of semidetached houses with half-timbered gables, lean-to porches, tile-hanging, and hipped roofs were put up by speculative builders in the interwar period. Such houses still line arterial roads in and out of British cities, leading Nikolaus Pevsner to complain bitterly of being "haunted by miles of semidetached houses mocking at Voysey's work."

A NEW ECLECTIC STYLE

THE ARTS AND CRAFTS IDEAL as exemplified in American architecture is based on an interpretation of the vernacular that varies widely in different areas of the country, drawing on Spanish missions, colonial mansions, and horizontal Prairie houses, and synthesizing timber-based buildings, such as Swiss chalets and Norwegian cabins, with the grace, economy, and apparent simplicity of Japanese landscaping and architecture to create a new, organic style imbued with a deep sense of place. This combination can be traced from Gustav Stickley's modest Craftsman houses, through the Arts and Crafts masterpieces of Greene and Greene, to the Prairie-school architecture of Frank Lloyd Wright, all of which demonstrate the same fidelity to place.

Below: *The Glessner House, designed by H.H. Richardson for John and Frances Glessner in 1885, was completed after the architect's untimely death in 1887.*

The philosophical father of the Prairie school—"the Arts and Crafts movement's manifestation in the Midwest"—was the great American architect Louis Sullivan. Sullivan, Wright, the Greene brothers, and Henry Hobson Richardson were perhaps the most influential American architects of the period. Richardson, a larger-than-life figure, egotistical and eccentric, used natural materials in his work—mostly unadorned stone and wood—and moved toward a new style that identified strongly with nature. He is best known for the towering Trinity Church on Copley Square in Boston, the Marshall Field Wholesale Store in Chicago and the austerely dramatic Glessner House, built for the Arts and Crafts devotees John and Frances Glessner on South Prairie Avenue, Chicago, in 1885–7. Richardson visited Europe and England where, in 1882, he met William Burges, William De Morgan, Edward Burne-Jones, and William Morris on two separate occasions: once in Hammersmith and later at the Firm's Merton Abbey works in Surrey. Morris called him "one of the few modern architects anywhere who have produced a distinctively original style." Richardson, pleased with his reception, wrote back to his wife praising Morris's "straightforward manner."

The quintessential Arts and Crafts men, though, were two brothers, Charles Sumner Greene and Henry Mather Greene, who absorbed the ideals of the movement on two visits to England in 1901 and 1909 and translated them into a series of stunning buildings in California. They had trained as architects at the Massachusetts Institute of Technology, but not before an innovative period of schooling at the new Manual Training High School in St. Louis run by Calvin

Milton Woodward, where, after a morning of academic studies, they spent the second half of each day learning metalworking, carpentry, and other trades. This stood them in good stead as both furniture designers and architects and is amply demonstrated in the visibly high quality and finish of their work. They set up their joint practice in Pasadena in 1903 at the precociously young ages of twenty-five and twenty-three respectively. By the time C.R. Ashbee paid them a visit six years later, business was flourishing and they were in a position to choose, from among a wealthy pool of queuing customers, the clients they felt would most respect their artistic vision and conviction. Ashbee thought they were "among the best there is in this country. Like Lloyd Wright," he observed in his diary, "the spell of Japan is upon him [Charles], he feels the beauty and makes magic out of the horizontal line, but there is in his work more tenderness, more subtlety, more self-effacement than in Wright's work. It is more refined and has more repose."

This serenity shines out from the "ultimate bungalows" designed between 1907–9 and in particular the Blacker (1907) and Gamble (1908) houses. There is a poetry about a Greene and Greene house: the romantic asymmetrical design, utilizing shingle, stone, and—notably—timber to create lovely buildings that sit low in the landscape, signaling their love of the pastoral. The refined simplicity of a Greene and Greene house offered its wealthy occupants sanctuary from the crude realities of industrialization. The broad overhanging roofs and generously proportioned balconies provided seclusion, shade, and subtle shadows from the projecting beams and rafters.

Above: *The architects Charles and Henry Greene worked closely with their clients Mary and David Gamble on the design of the Gamble House, which remained in the family until 1966.*

The New England architect Ralph Adams Cram was an enthusiast: "One must see the real and revolutionary thing in its native haunts of Berkeley and Pasadena," he wrote, "to appreciate it in all its varied charm and its striking beauty. Where it comes from heaven alone knows, but we are glad it arrived, for it gives a new zest to life, a new object for admiration. There are things in it Japanese; things that are Scandinavian; things that hint of Sikkim, Bhutan, and the vastness of Tibet, and yet it all hangs together, it is beautiful, it is contemporary, and for some reason or other it seems to fit California.... It is a wooden style built woodenly, and it has the force and integrity of Japanese

Above: *The drawings for the Gamble House were finalized in February 1908, work began a month later, and the house was completed in a little under a year.*

architecture. Added to this is the elusive element of charm that comes only from the personality of the creator, and charm in a degree hardly matched in other modern work."

The Gamble House, the great icon of American Arts and Crafts, was presented to the City of Pasadena and University of Southern California in 1966 and is now open to the public. Built for David Gamble (of the Proctor & Gamble Company) and his wife Mary on a secluded site on the outskirts of town, it was finished to the highest specifications. As with other commissions, the brothers also designed the garden, furniture, lighting fixtures, and stained glass, creating an astonishing and exciting monument to the Arts and Crafts movement. In March 1908 the *Pasadena Daily News* raved about the mahogany dining room, the teakwood living room, the oak used in the den, and the white cedar elsewhere. It detailed the five bathrooms, three luxurious sleeping porches, two enormous terraces, the garden, and the immense billiard room on the third floor, surrounded by windows on four sides. It salivated over "all the supplementary closets and cupboards known to modern convenience" and boasted about the large kitchen and butler's pantry.

The Greenes' houses were designed with the Californian climate in mind: cross-ventilated, with broad overhanging roofs to provide shade and wide balconies or porches for outdoor sleeping. "The idea," Henry Mather Greene explained in 1912, "was to eliminate everything unnecessary, to make the whole as direct and simple as possible, but always with the beautiful in mind as the first goal." They were fortunate in having wealthy clients, though occasionally Charles was obliged to defend his artistic decisions. "It is too much," he wrote to one client who had questioned the cost of construction of a stone wall, "to expect that anyone may see the excellence of this kind of thing in a few days. The work itself took months to execute and the best years of my life went to develop this style.... Into your busy life I have sought to bring what lay in my power of the best that I could do for Art and for you."

Bernard Maybeck was another noted Californian Arts and Crafts architect, best known for the flamboyant First Church of Christ Scientist in Berkeley (1910), although his domestic buildings—Craftsman-like rustic dwellings influenced by Swiss chalets and English half-timbering—are notable examples of Arts and Crafts style. One such is Grayoaks (1906), the house he built for J.H. Hopps, a wealthy timber merchant who wanted a country house built on a heavily wooded stretch of land in Marin County. The house is relatively modest in size, with warm redwood as the dominant motif, used to panel the interior and clad the exterior. It was Maybeck, in whose office she worked, who encouraged Julia Morgan to study architecture, as he had, at the École des Beaux-Arts in Paris. One of only a handful of female American architects at the time, her version of Arts and Crafts architecture drew from a wide range of references, including Craftsman bungalows, Mediterranean influences, and Spanish colonialism.

Another Californian architect whose work owed much inspiration to the Spanish mission was Irving Gill, who was responsible for the Laughlin house (1907) and the very successful Dodge house (1916). "There is something very restful and satisfying to my mind," he wrote in *The Craftsman* in 1916, "in the simple cube house with creamy walls, sheer and plain, rising boldly into the sky, unrelieved by cornices or overhang of roof, unornamented save for the vines that soften a line or creepers that wreathe a pillar or flowers that inlay color more sentiently than any tile could do. I like the bare honesty of these houses, the childlike frankness, and the chaste simplicity of them."

In New England the Arts and Crafts message was translated by architects such as John Calvin Stevens and Albert Winslow Cobb into a free and democratized interpretation of the colonial styles of the earliest settlers. Describing a house they built in 1899 on the shore of Cape Elizabeth, near Portland, they talk of "the weathered field stone, the very color of the ledges out of which the building grows. The walls above the stone work are of shingles, untouched by paint, but toned a silvery gray by the weather." This harmonizing of a building with its surroundings is a central tenet of the Arts and Crafts creed and was taken up in Philadelphia by architects such as Wilson Eyre of the T-Square Club, who took their inspiration from Pennsylvanian farmhouses to create a new style of domestic architecture best exemplified by the William Turner house of 1907. Eyre initiated the publication of the periodical *House and Garden* in 1901 and acted as its editor for five years, featuring articles on the work of British architects including Lutyens, Lorimer, Prior, and Voysey, as well as showcasing his own work.

The Prairie school of architecture shared with the Arts and Crafts movement the idea of simplicity and respect for materials. A number of important architects are identified with the movement, which spanned the period from 1900 to World War I, including George Grant Elmslie, Walter

Burley Griffin, Francis Barry Byrne, Richard Ernest Schmidt, William Gray Purcell, Robert Closson Spenser Jr., William Eugene Drummond, Marion Mahony, Dwight Heald Perkins, John Shellette Van Bergen, Vernon Spencer Watson, Francis C. Sullivan, Parker Noble Berry, and Percy Dwight Bentley. Louis Sullivan, who believed in an "architecture of democracy", was its spiritual leader but Frank Lloyd Wright, the most versatile of American architects, would emerge as the genius and foremost exponent of the Prairie school. As an architect and designer he was a major influence on American Arts and Crafts architecture. Houses such as the Ward W. Willits house (1902), the Dana-Thomas house (completed in 1904 for the wealthy heiress Susan Lawrence Dana, bought by the Thomas Publishing Company in 1944, and now owned and restored by the state of Illinois), with their imaginative geometry, strong horizontals, long, ground-hugging profiles, and low overhanging roofs, reflect the flat geography of Illinois. Wright described his inspiration for the style: "We of the Middle West are living on the prairie. The prairie has a beauty of its own and we should recognize and accentuate this natural beauty, its quiet level. Hence, gently sloping roofs, low proportions, quiet sky lines, suppressed heavy-set chimneys, and sheltering overhangs, low terraces and out-reaching walls sequestering private gardens."

Above: *The Thomas de Caro house, 1900–6, typifies Frank Lloyd Wright's Prairie style.*

Above: *Frank Lloyd Wright's finished drawing of the east elevation of Susan Lawrence Dana's House. Construction began in the late summer of 1902, and the House was completed before Christmas 1904.*

Right: *The Dana-Thomas House, designed by Frank Lloyd Wright for Susan Lawrence Dana in 1904.*

He identified with the message of Morris—"All artists love and honor William Morris," he wrote—and, in 1897, helped to found the Chicago Arts and Crafts Society. John Ruskin's *The Seven Lamps of Architecture* was one of the first books he owned on the subject and he transmuted the Morris ideal into a twentieth-century idiom, employing a style that was starkly functional, relying on the geometry and juxtaposition of shape and form to create stunning buildings with revolutionary open-plan interiors, for which he would provide the furniture, light fittings, rugs, and patterned leaded-glass windows. Where he departed from the ideals of the British Arts and Crafts movement most markedly was in his attitude to the machine. "My God," he wrote in 1900, "is Machinery, and the art of the future will be the expression of the individual artist through the thousand powers of the machine—the machine doing all those things that the individual workman cannot do. The creative artist is the man who controls all this and understands it."

There is a masculine energy in Wright's work. Nikolaus Pevsner called him "a poet of pure form" in whose "gigantic visions, houses, forests, and hills are all one." His own home, built in Oak Park, Illinois, was a six-room Shingle-style bungalow that he designed for himself and his teenage bride Catherine Tobin in 1889, when he was just twenty-one years old, using $5,000 borrowed from his employer Louis Sullivan. By 1895, the addition of four active children encouraged him to extend the house and, along with a barrel-vaulted playroom, he added a new dining room and, three years later, a new studio complex. Designed as a showpiece to impress new clients, it had an octagonal double-

height drafting room with a balcony suspended by chains above the workspace, an office lit from above and to one side by natural light filtered through art-glass panels, an octagonal library, and a reception hall stretching along the entire façade and unifying the whole. The result was revolutionary and stunning, and provided Wright with a stimulating base from which to work on over a hundred and fifty building designs, surrounded by eager apprentices, for the following thirteen years. During this time Wright's office operated, much as Shaw or Sedding's had decades before, as a nursery for talented architects—among them Walter Burley Griffin and Marion Mahony, one of the first female architects in the USA. (The restored building is now owned by the National Trust for Historic Preservation and operated by the Frank Lloyd Wright Home and Studio Foundation.)

In 1909, four years after a visit to Japan that had a profound influence on his work, Wright's personal life suffered an upheaval when he eloped—scandalously at the time—to Europe with Mamah Borthwick Cheney, the wife of one of his clients, abandoning his family, his country, and his practice. It was not until 1911 that he returned and began plans for Taliesin, a new house to be built overlooking the Wisconsin River on property owned by his mother at Spring Green, Wisconsin.

Below: *The first house Frank Lloyd Wright designed was for himself and his family at Oak Park, Illinois in 1889. In later alterations he added a playroom, a new dining room, and a studio complex*

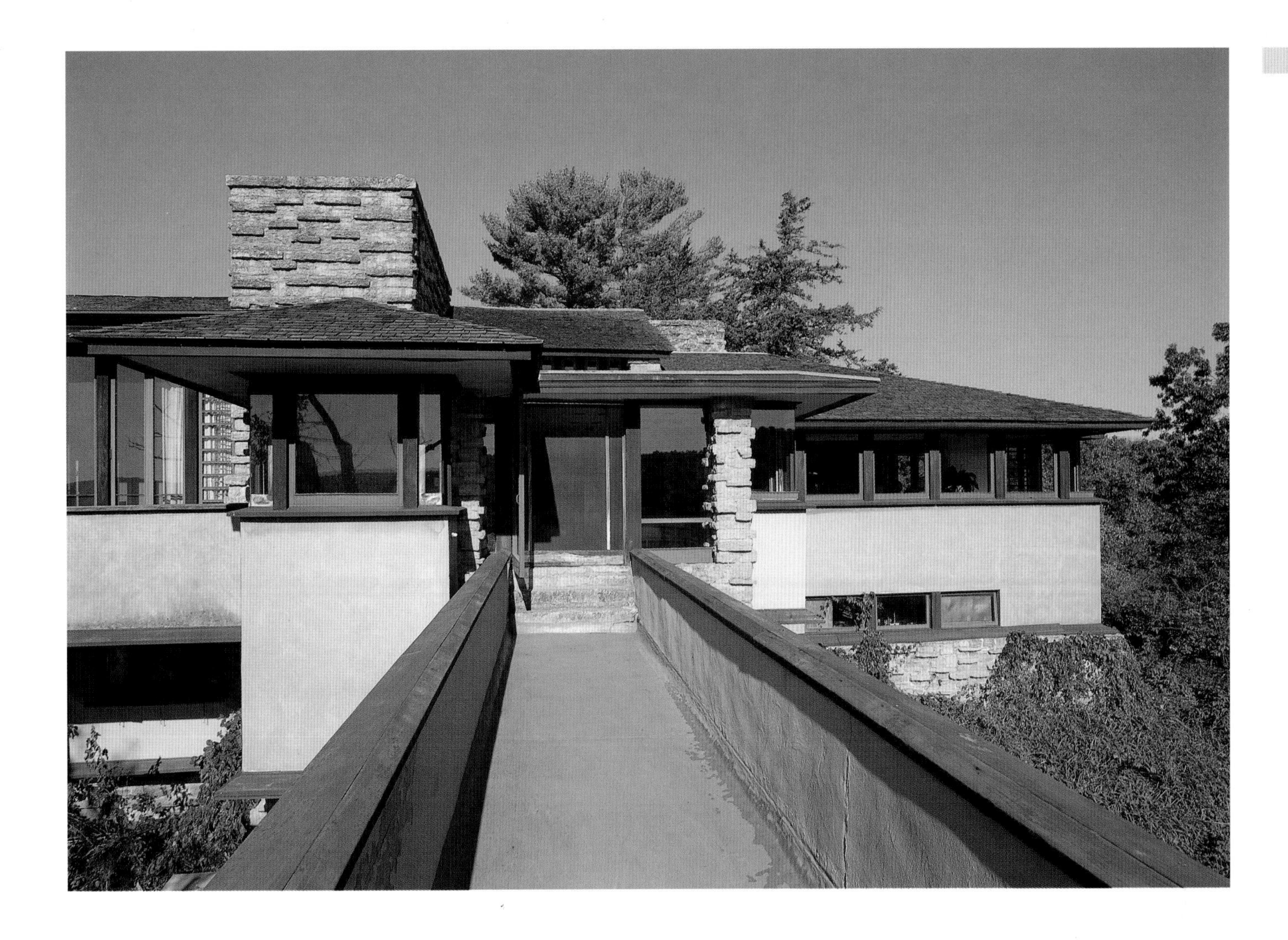

Above: *Frank Lloyd Wright's home Taliesin North in Spring Green, Wisconsin. From its windows, Wright could see across the river valley.*

Taliesin, named after a mythical Welsh bard and whose name means "shining brow", was built from native limestone carried from a nearby quarry and intended by Wright to sit within the landscape—to be "of the hill, not on the hill." The house evolved over a long period to accommodate not just Wright and his family but the Taliesin Fellowship, a community based on craft teaching and agriculture established in the early 1930s. A series of connected buildings, linked by limestone and sand-colored stucco walled passageways, enclosing farm buildings, studio spaces, and courtyards, Taliesin stretched low across the brow of the hill. In the late 1920s it was complemented by Taliesin West, a complex of buildings that evolved over a period of years near Phoenix, Arizona, built by Wright to provide a winter base for himself and the Taliesin Fellowship.

His Prairie school period exactly paralleled the height of the Vienna Secession, when architects like Otto Wagner, Joseph Maria Olbrich, and Josef Hoffmann were making their mark in Europe. They would have a direct influence on the architecture of the 1920s, but the inspiration of Wright, who lived to be ninety-two, colored and went beyond the Modernism of Mies van der Rohe and Walter Gropius into the twenty-first century, where it continues to have relevance.

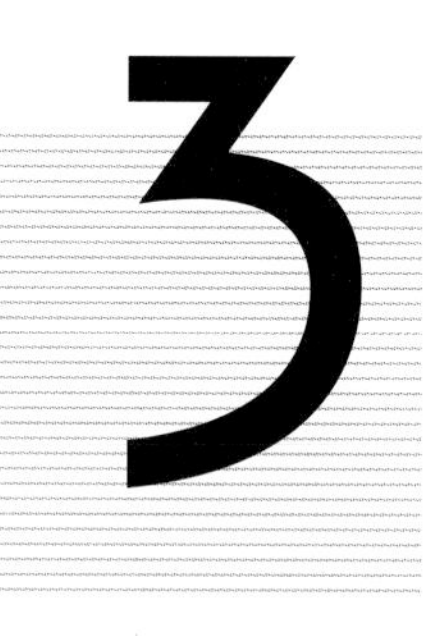

3

ARCHITECTURAL INTERIORS

Right: *The White Drawing Room at Blackwell in Cumbria, designed by M.H. Baillie Scott, incorporates his trademark inglenook and is considered to be one of his finest interiors.*

TOWARDS THE PERFECT LIVING SPACE

Above: *This carved oak ceiling boss can be found in the Main Hall at Baillie Scott's masterpiece, Blackwell.*

"Avoid all things which have no real use or meaning and make those which have especially significant, for there is no one part of your building that may not be made a thing of beauty in itself as related to the whole."

FRANK LLOYD WRIGHT, *The Architect and the Machine*, 1894

ILLIAM MORRIS'S FAMOUS MAXIM, "Have nothing in your houses that you do not know to be useful or do not believe to be beautiful"—delivered during a lecture entitled "Hopes and Fears for Art" in 1882—spoke directly to Arts and Crafts adherents and could be used to sum up a perfect Arts and Crafts interior. The range of perfection is wide, however, stretching from one of Morris's own white-paneled rooms, set off by hand-painted murals, tapestries, rich carpets, and highly patterned chintzes, to the almost spartan simplicity of Sidney Barnsley's scoured and swept spaces, dominated by a sturdy oak table, ranged about by high ladderback chairs with—against a far wall—a vast dresser filled with serviceable china.

In 1882, Morris had long since left his "palace of art" at Red House in Kent and was dividing his time between his two Kelmscotts—Kelmscott House on the banks of the Thames at Hammersmith in London, and Kelmscott Manor, a hundred and thirty miles upriver in Oxfordshire—each decorated in his highly personal style. The Manor—"a beautiful and strangely naif house, Elizabethan in appearance," boasted a romantic upstairs room filled with faded seventeenth-century tapestries and a big parlor lined with "some pleasing paneling," which he had painted white to provide a backdrop for the patterned fabric used on the armchairs and curtains. The interiors of the house in Hammersmith also expressed the extraordinary integrity of Morris's artistic taste. The long first-floor drawing room was covered in his "Bird" hangings and the dining room was papered in "Pimpernel," with a beautiful Persian carpet suspended against one wall from the high ceiling. The rooms were furnished with a vast settle that had been made for Red House, along with cabinets and wardrobes

Left: *A view through to one of the attic bedrooms at Morris's Kelmscott Manor.*

Above: *There is a modest elegance to be found in the spartan simplicity of Sidney and Lucy Barnsley's living room at Beechanger, their house in Sapperton, built in 1904.*

painted with medieval scenes by Dante Gabriel Rossetti and Edward Burne-Jones, but the long dining table was left scrubbed and bare of any tablecloth in a proper Arts and Crafts manner.

The Arts and Crafts movement fueled the new middle-class fashion for interior decoration and the last decades of the nineteenth century witnessed a conscious move away from the clutter and comparative gloom of the traditional Victorian interior toward a lighter, more rational scheme, often unified by recurring decorative motifs in the fabric, fittings, and structural decoration of a room. This was most successfully achieved in the work of innovative architects such as Charles Rennie Mackintosh, C.F.A. Voysey, M.H. Baillie Scott, and Frank Lloyd Wright, who designed from the inside out, rethinking and revolutionizing the use of space, leading to a greater sense of informality, flexibility, and ease of flow. They designed "organically," conceiving the ornamentation in the very ground plan, and were concerned to free the bulk of interior space from the cumbersome mass of furniture and the tyranny of bric-a-brac and occasional tables. They did this by creating fixed or built-in furniture, merging the horizontals and verticals created by the windows, fireplace, bookcases, and seating to create homogeneous and harmonized interiors. They were a new breed of architect-designers, who

Above: *The interior of Stoneywell is kept determinedly plain, almost primitive, with low beams and simple whitewashed walls, though the sinuous curve of the narrow staircase draws the eye, and the jugs and china ranged about the walls and beams create a warm feeling of cozy domesticity.*

had taught themselves the rudiments of traditional crafts, and knew how to work with wood, metal, glass, and fabric to create an integrated design and, crucially, how to organize the interior space. They used local materials and local skills and concerned themselves with every aspect of the design of a house, creating coherent and deeply satisfying interiors. These ranged from the dramatic set pieces of large country houses such as Baillie Scott's Blackwell to the quieter serenity to be found in a more modest Voysey interior. In all the Arts and Crafts interiors, love of wood was on show: wooden paneling, high-backed settles made of oak built into inglenooks, distinctive freestanding pieces of furniture, and, everywhere, evidence of painstaking craftsmanship in the exposed details on wooden staircases, ceiling beams, carved panels, and studded and strap-hinged plain plank doors.

There is a warmth and welcome in an Arts and Crafts interior, a celebration of the domestic and the practical. For Baillie Scott the fireplace was crucial. "In the house the fire is practically a substitute for the sun," he wrote, "and it bears the same relation to the household as the sun does to the landscape. The cheerfulness we experience from the fire is akin to the delight which sunlight brings." He accentuated the fireplace as a symbolic focus by creating deep inglenooks lined with built-in

Left: *The restored Drawing Room at Standen boasts lighting fixtures by Webb and Benson, Morris's "Tulip and Rose" woven wool hangings, a particularly fine example of a hand-knotted Merton Abbey carpet (designed by J.H. Dearle), and Morris's "Sunflower" wallpaper and, though empty, seems to await the arrival of the Beale family, stylishly dressed for dinner.*

Above: *Wood—in the warm paneling, beams, and furniture—is the dominating decorative principle in the interiors at Rodmarton Manor. The pair of walnut writing tables seen here in the Drawing Room were the work of Peter Waals; the oak daybed was made by Sidney Barnsley's son Edward.*

settles. They rapidly became a key feature of Arts and Crafts interiors, both in Britain and the USA, where the archetypal Arts and Crafts room—long with a low-beamed ceiling and leaded windows—combined folk and colonial American motifs to evoke the simple life, as recommended by Gustav Stickley. Stickley believed the living room to be "the executive chamber of the household where the family life centers and from which radiates that indefinable home influence that shapes at last the character of the nation and the age." Images of ideal living rooms, furnished with a Craftsman rocking chair, standing on a Navajo rug, beside a roaring fire, with a Tiffany lamp shedding its buttery light across a sturdy oak table set with art pottery, appeared in his magazine *The Craftsman*.

British architects also used magazines to put across their message. In an article written for *The Studio* in January 1895, Baillie Scott guided his readers round "An Ideal Suburban Villa," which featured a double-height hall, with built-in ingle seating before a wide brick hearth adorned by copper fire dogs: "On entering by the front door, we find ourselves in a wide and low porch from which, through an archway to the right, we catch a glimpse of the staircase which rises from a wide corridor leading to the kitchen," he wrote. "It is difficult for me to picture to you the vista-like effect of the broad corridor, but to get some idea of its general effect I must transport you to some old Cheshire farmhouse, somewhere in the country where people have not yet grown to be ashamed of plain bricks and whitewash." He enlarged and expanded his ideas for ideal living in later articles devoted to "An Artist's House" (October 1896), "A Small Country House" (December 1897), and "A Country House" (February 1900).

Baillie Scott wanted his buildings inside and out to be the product of a single mind—he designed furniture for most of his houses and developed the idea of the integrated interior, searching for simplicity and a sense of repose. He replaced the Victorian entrance hall with an Elizabethan-style dwelling hall (double-height where possible), half-timbered or paneled to evoke the simplicity of the medieval barn. He pioneered and mastered a new flow of space, breaking down the conventional divisions and making interiors, especially in smaller houses, seem more spacious and flexible. "The house rationally planned should primarily consist of at least one good-sized apartment," he wrote, "which, containing no furniture,

Left: *A hand-colored photo-lithograph of Hall, M.H. Baillie Scott's entry for the House for an Art Lover Competition, 1901.*

Left: *Ernest Gimson's plasterwork is shown off to great effect in the hall at Avon Tyrrell, which also boasts W.R. Lethaby's table of 1892–6.*

but that which is really required, leaves an ample floor space at the disposal of its occupants.... In this way, even the laborer's cottage retains its hall, which has now become the kitchen, dining room and parlour." He wanted to move away from the idea of the smaller kind of house being sub-divided to the greatest possible extent into tiny compartments and designed instead around a central space, made flexible by sliding doors and alcove-like spaces for activities such as sewing, reading or dining. These were furnished with built-in, often multipurpose, units combining bookcases, window seats, and cabinets. In his ground-breaking pair of semidetached Elmwood Cottages at Letchworth, he eliminated corridors and connected rooms laterally, increasing the space and connecting the light, airy rooms to the landscape of both the road and the garden in a manner that bears comparison with the work of Frank Lloyd Wright. Baillie Scott explored the frontiers of privacy and sociability in his book *Houses and Gardens* (1906), giving careful thought to the use of space and differentiating between rooms traditionally associated with men or women by using plain or stenciled "feminine" white walls and furniture in the bedrooms, drawing rooms, kitchen, and bathroom, and sturdy oak paneling for the masculine areas—the hall, stairwell, drinking room, smoking, and billiard rooms.

At Blackwell at Bowness, overlooking Lake Windermere in Cumbria, built in 1898 as a holiday home for Edward Holt, lord mayor of Manchester, Baillie Scott used whitewash as a background to the timbered hall and exposed joists and rafters in the dining room. Allowing his imagination full rein, he designed an extravagant double-height "hall-living-room," which contained a massive

Above: *The impressive Main Hall at Blackwell, with its romantic Minstrel's Gallery above the inglenook.*

inglenook fireplace beneath a romantic half-timbered minstrel's gallery. He made the interiors sing, decorating them with exuberant plasterwork, richly carved paneling, and stone corbels, all lit by the glowing colors of stained-glass panels inspired by local wild flowers and peacocks. "Let it be vital, local and modern," he urged. The spatial play of the hall at Blackwell was so innovative that it allowed for the room to be used in multiple ways. (The house, recently restored, now boasts one of the finest Arts and Crafts interiors remaining in existence.) The German writer and architect Hermann Muthesius was in raptures, describing Baillie Scott as a poet whose "ravishing ideas of spatial organization" ensured that every part of the house "down to the smallest corner, is thought out as a place to be lived in."

Baillie Scott's reputation soared in Europe, and he has been credited with making an important contribution to the beginnings of the Arts and Crafts movement in Germany, where his elaborate and highly decorative interior designs—for which he designed not only the spaces but every element within them, including fabrics, stained glass, furniture, carpets, and light fittings—were enthusiastically reviewed in German magazines until the outbreak of the war put an end to his commissions.

Right: *A glimpse of how European followers of Arts and Crafts lived is offered in this interior of Joseph Maria Olbrich's home in the artists' colony at Matildenhohe in Darmstadt, for which he was the chief architect in 1901.*

Above: *Voysey acts upon his belief that houses should have "light, bright, cheerful rooms, easily cleaned and inexpensive to keep" in his design for The Homestead, a green slate roofed house at Frinton-on-Sea, which he designed in 1905–6 for a bachelor client.*

Right: *A period room within Liberty's department store with original wood-paneling, furniture and metalwork from Liberty's early twentieth-century Arts and Crafts ranges.*

Above: *The living hall at Munstead Wood as it was in 1907. The staircase leading up to an oak-beamed first-floor gallery was one of Gertrude Jeykll's favorite features: "It felt firm and solid," she said, "the steps low and broad."*

His attention to detail was legendary. At Blackwell, for example, the door handles and iron window latches are differentiated by some small and often subtle detail in each of the many rooms. As Muthesius remarked, "In Baillie Scott's work each room is an individual creation, the elements of which do not just happen to be available but spring from the overall idea. Baillie Scott is the first to have realized the interior as an autonomous work of art."

C.F.A. Voysey, who described his ideal interior as "a well-proportioned room, with white-washed walls, plain carpet and simple oak furniture," ornamented only by "a simple vase of flowers," also took control of every element of an interior and employed a number of favorite decorative motifs and symbols, such as a stylized heart. His whitewashed rooms boasted large, welcoming, tiled fireplaces, with white or natural oak-beamed ceilings. He believed that homes should have "light, bright, cheerful rooms, easily cleaned and inexpensive to keep." He dismissed the once-fashionable sepias and sludgy greens as "mud and mourning" and in his own house, The Orchard (1899), used his favorite color scheme of green, red, and white, with easily cleaned, durable, slate tiles on the hall and kitchen floor, green cork tile flooring throughout the first floor, and curtains of bright red.

A Voysey house has distinctive qualities of honesty, candor, and simplicity; it provides both physical and spiritual shelter. Voysey's own definition of comfort was "Repose, Cheerfulness, Simplicity, Breadth, Warmth, Quietness in a storm, Economy of upkeep, Evidence of Protection, Harmony with surroundings, Absence of dark passages, even-ness of temperature and making the house a frame to its inmates. Rich and Poor alike will appreciate its qualities."

Charles Rennie Mackintosh took the principle of the integrated interior further, designing friezes, cutlery, silverware, hall-chimes, carpets, and light fixtures for his houses. Like most Arts and Crafts

Below: *Charles Rennie Mackintosh's beautifully designed White Bedroom at Hill House (1902–4) still looks strikingly modern.*

buildings, his interiors reflected his clients' pattern of living, but his geometric precision stamped each commission with his own mark. Hill House near Glasgow is a shining example of his fierce commitment to total stylistic unity. The somewhat severe exterior contrasts completely with the delightful interior spaces—particularly the hall, the drawing room, the study-library, and the main bedroom, over which Mackintosh had complete control. The visionary decorative schemes are the result of an inspired collaboration with his wife Margaret. The elongated lines and delicate geometry in the white-painted principal bedroom made a stunning modern statement. He recessed the bed demurely beneath a barrel-vaulted ceiling, creating an area designed to be screened off from the rest of the room. Downstairs in the drawing room, strikingly decorated with a pale pattern of abstract roses, Mackintosh filled one of the gloriously light bays with a long, low window seat (under which he installed heating) flanked by fitted racks for books and magazines; in the other he created a separate space to house the piano. The effect is magical.

The Greene brothers designed their houses inside and out, and are justly celebrated for creating rooms rich with beautifully crafted wood and glowing with art-glass windows and light fittings. They created a serene and dignified interior at the Gamble House in Pasadena, taking a coordinated

Above: *This bedroom in Carl Larsson's house in Sweden is a fine example of the Scandinavian interpretation of the Arts and Crafts ideal.*

Above: *Wood paneling predominates in the entrance hall of the David Berry Gamble House at Pasadena, looking towards the front doors designed by Charles Greene and made by glass artist Emil Lange.*

approach to the layout and furnishing of the rooms and designing all the furniture, art-glass windows, and mahogany-framed light fittings themselves. Though they actively involved their client, Mary Gamble, in decisions and took her interests and lifestyle into account when planning the interiors, the final result is very much their own creation. The warm tones of wood—Oregon pine, American white oak, redwood, white and red cedar, Honduras mahogany—dominate. The hall and drawing room are paneled with Burma teak, which covers the walls to frieze height, and the rooms open into each other, or to the outside, in a way that recalls the connecting pavilions of a Japanese villa. The whole house glows. In the drawing room a long timber-framed settee sits in the window bay, and an emphatically wide inglenook, with a curved Oriental wooden truss, spreads across one wall, embracing a pair of settles, glass-fronted cabinets, and a table to create a private "room within a room." David Gamble's study, off the entry hall, is furnished with a Morris chair and the desk from his study in Cincinnati, and sturdy Craftsman furniture was ordered for the bedrooms of the Gambles' teenage sons.

The Greenes' father was a doctor who specialized in respiratory diseases; consequently they always paid careful attention to the free flow of air in bedrooms. At the Gamble House, the cross-ventilated family bedrooms each opened onto a sheltered outdoor room, or sleeping porch,

Right: *The dining table in the Gamble House is bathed in glowing light, while the bright Californian sun is softened by the colored glass panels in the windows.*

furnished with rattan armchairs and recliners for use during the day or on particularly warm Californian nights. A generous provision of five bathrooms was made to serve the ground and first-floor bedrooms and, although dressing rooms were not provided, the master bedroom had a large walk-in closet, and maple vanity units were built in to the guest bedroom. Mary Gamble's fine collection of Rookwood vases was displayed on cedar shelves in the master bedroom, which was decorated in dark, earth colors.

Frank Lloyd Wright sought to define and design the perfect living space for contemporary life and in doing so revolutionized the single-family private house. He created interpenetrating spaces by abolishing corners between dining and living areas, making a single L-shaped space that pivoted around a dominant central fireplace, designed "to give a sense of shelter in the look of a building." In a long career, he designed more than three hundred residences, creating not just the outer shell of the building but its decorative interiors as well: art-glass windows and skylights, light fixtures, furniture, carpets and textiles, wall murals—indeed all integral ornament. His carefully conceived built-in furniture and tall spindlebacked chairs, which clearly owe a great debt to Charles Rennie Mackintosh, simplified his functional, open-plan interiors while also unifying the whole design.

Above: *The interior of the Ward W. Willits House, Highland Park, Illinois, designed by Frank Lloyd Wright in 1902.*

Built-in furniture made efficient use of the space. It was orderly and economical and had the added advantage of discouraging his clients from cluttering his spaces with any furniture from previous homes that they might be tempted to introduce. "I tried to make my clients see that furniture and furnishings ... should be seen as minor parts of the building itself, even if detached or kept aside to be used on occasion." In the Robie house in Chicago, for example, the shape of the sofa, with its wide table arms, echoes the ceiling above, and the trim on the furniture matches the moldings on the walls. Wright's interiors were designed to echo the overall design of the house and open freely, one to another, with ingenious transitions, marked by art-glass windows, subtle changes of level or changes of texture on the wall or floor. From the drama of the barrel-vaulted ceiling in the dining room of the Dana-Thomas house to the soaring multilevel spaces of the living room at Taliesin, they are breathtakingly modern.

Wright used geometry to link the different elements and glass to break open boxlike spaces, and he sought ways to dissolve corners. As early as 1900 C.R. Ashbee had recognized his genius. "Wright is to my thinking," he confided in his journal, "far and away the ablest man in our line of work that I have come across in Chicago, perhaps in America. He not only has ideas but the power of expressing them, and his Husser House, over which he took me, showing me every

detail with the keenest delight, is one of the most beautiful and individual of creations that I have been in in America."

"What I call integral ornament," Frank Lloyd Wright explained in a special edition of *House Beautiful* published in 1955, when he was eighty-eight, "is founded upon the same organic simplicities as Beethoven's Fifth Symphony, that amazing revolution in tumult and splendor of sound built upon four tones, based upon a rhythm a child could play on the piano with one finger. Supreme imagination reared the four repeated tones, simple rhythms, into a great symphonic poem that is probably the noblest thought-built edifice in our world. And architecture is like music in this capacity for the symphony."

Above: *The dining-room furniture Frank Lloyd Wright designed for the Robie House in Chicago is among his most famous ensembles. Photograph by Henry Fuermann, c. 1910.*

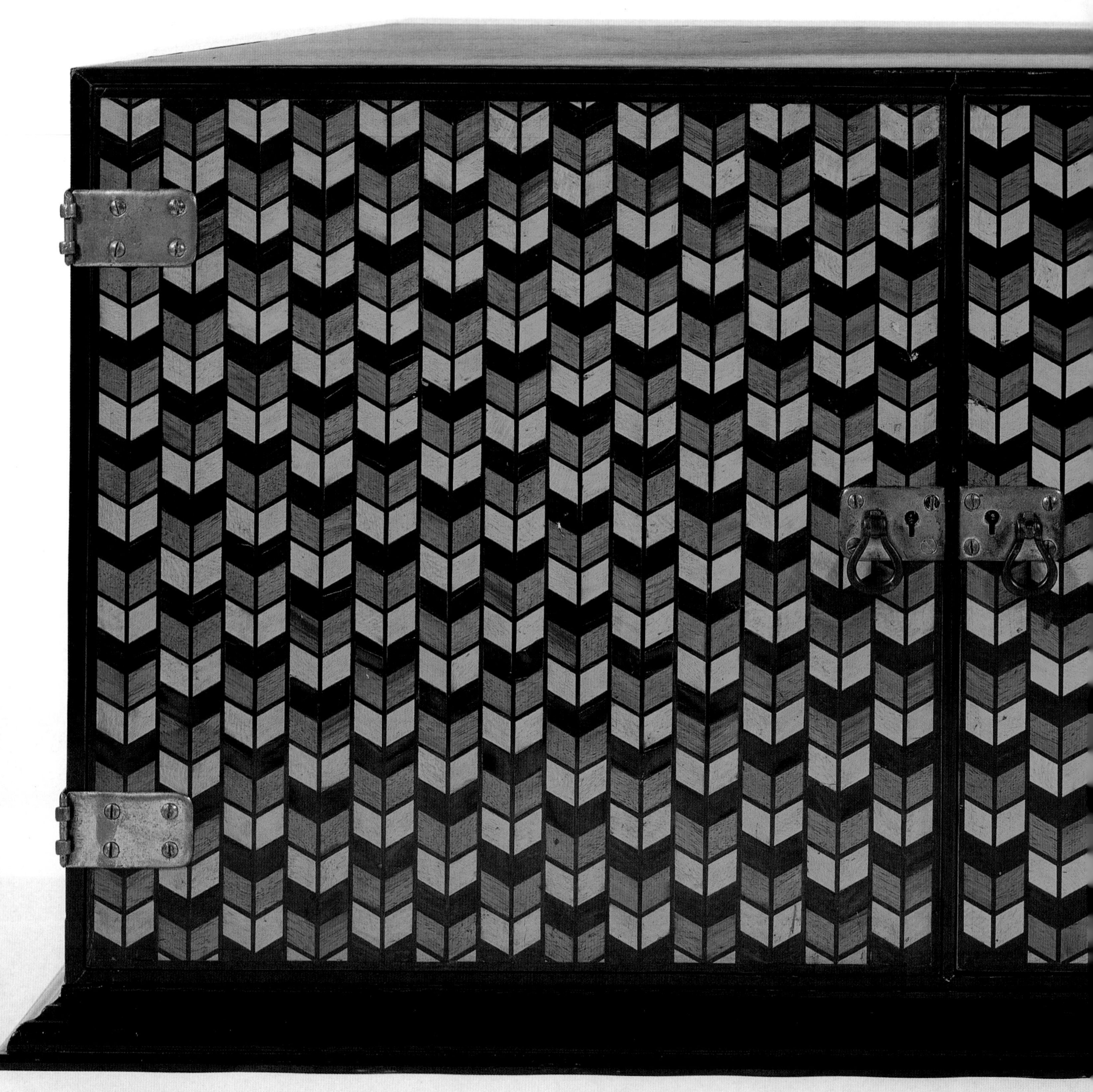

4

FURNITURE

Left: *A Gimson cabinet, made at the Daneway Workshops by Ernest Smith, c. 1903–7. Alfred Bucknell made the handles on the outside, John Paul Cooper those on the inside.*

MODERN ART
Early Prints of Ceylon
BRITAIN
HOW WAS IT DONE?

THE EXPRESSION OF GOOD FEELING

"So I say our furniture should be good citizen's furniture, solid and well made in workmanship, and in design should have nothing about it that is not easily defensible, no monstrosities or extravagances, not even of beauty, lest we weary of it.... Moreover I must needs think of furniture as of two kinds ... one part of it ... the necessary workaday furniture ... simple to the last degree.... But besides this ... there is the other kind of what I shall call state-furniture; I mean sideboards, cabinets and the like ... we need not spare ornament on these but may make them as elegant and elaborate as we can with carving or inlaying or painting; these are the blossoms of the art of furniture."

WILLIAM MORRIS, *The Lesser Arts of Life*, 1882

Above: *Detail of marquetry from a cabinet by Charles Robert Ashbee.*

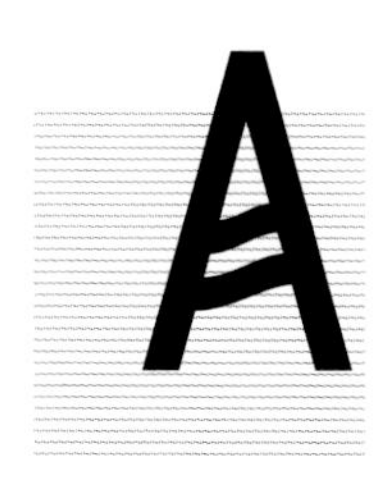

ARTS AND CRAFTS FURNITURE can be found at each end of the spectrum described by William Morris: from the familiar plain and solid oak pieces, perhaps with beaten copper handles and hinges, or distinctive panels of stained glass, to the elaborately decorated sideboards and cabinets that Morris considered the "blossoms" to be "used architecturally to dignify important chambers and important places."

Left: *Morris's enormous white settle was originally designed for his rooms in Red Lion Square, but Philip Webb adapted it and made it the centerpiece of the first-floor drawing room at Red House. He added a canopy, creating a miniature "minstrel's gallery," which also, rather more prosaically, provided access to the doors leading into the roof-space behind.*

THE FIRM, OR MORRIS, MARSHALL, FAULKNER & CO.

IT ALL BEGAN, like so much in the Arts and Crafts movement, with William Morris, who, disenchanted with the furniture he found readily to hand when he began looking for pieces to furnish his student rooms in Red Lion Square, London, designed and made some for himself along monumental medieval lines. There was a vast settle (which proved too large to transport up the stairs to the first-floor rooms he shared with Edward Burne-Jones, and had to be

Above: *King René's Honeymoon Cabinet, built 1860–2, was designed by John Pollard Seddon with panels by Ford Madox Brown, Dante Gabriel Rossetti, and William Morris.*

winched in through a window), some colossal chairs, and a round table "as firm and heavy as a rock," according to Dante Gabriel Rossetti. A few years later, in April 1861, married now and with a beautiful new house in Kent to furnish, Morris banded together with these and other friends to found the firm of Morris, Marshall, Faulkner & Co.—"Fine Art Workmen in Painting, Carving, Furniture and the Metals." The furniture designs were provided by Philip Webb, the architect of Red House, and the painters Dante Gabriel Rossetti and Ford Madox Brown. The latter already had some experience of designing simple and robust furniture for Charles Seddon & Co. and is credited with originating the green stain for oak that was so generally used for art furniture. Webb's first recorded furniture design, dated 1858, was rather more elaborate. Heavily influenced by Augustus Pugin (whose *Gothic Furniture in the Style of the Fifteenth Century* had been published in 1835), he designed a massive wardrobe, which was painted by Burne-Jones with a scene from Chaucer's *Prioress's Tale*, as a wedding present for Morris and Janey. Architects such as Pugin and G.E. Street, in whose offices Webb had trained, had already pioneered the use of plain rectangular forms, solidly executed in unvarnished oak along the lines developed by medieval joiners of rails and posts, then elaborately ornamented. Pugin's Gothic-style furniture was made up for his clients by commercial firms such as J.G. Crace Limited of Wigmore Street, London, and Webb would have been familiar with many of these pieces.

Madox Brown took a rather different approach, designing simple, plain pieces that concentrated on "adaption to need, solidity, a kind of homely beauty and above all absolute dissociation from all false display, veneering and the like." Rossetti, meanwhile, found romance in English country designs of the mid-eighteenth century and adapted them into one of the staples of the Firm, the rush-seated "Sussex chair."

The Firm began life at 8 Red Lion Square, with a workroom referred to by Rossetti as "the Topsaic laboratory" on one floor and a shadowy showroom filled with "bewildering treasures" on another. Here Morris hoped to evoke the spirit of a medieval workshop, where there was pride in work and joy in working together. His lofty aims for simplifying life are exemplified in his much-quoted exhortation: "Have nothing in your houses that you do not know to be useful or believe to

Below: *A pair of enduringly popular chairs, designed around 1865, which sold through the Firm and came to epitomize the Morris look: on the left the "Sussex" chair, in ebonized beech with a rush seat, and, on the right, the "Rossetti" chair, also in ebonized wood, though some, with red painted details on the turning, were made to special order.*

Above: *The elaborate "St George Cabinet" was one of the Firm's showpieces at the International Exhibition of Art and Industry in 1862. Philip Webb designed the cabinet, using mahogany, oak, and pine, but it was Morris himself who decorated the gilded surface with scenes from the legend of St. George. It was priced at 50 guineas but, despite attracting some favorable notice, failed to sell.*

be beautiful." Morris was appalled by the shoddiness of the mass-produced furniture then on offer and felt sure he could offer a better, handmade alternative. By drawing on his hobbies of woodcarving and embroidery and capitalizing on his own experience of decorating Red House, he was rapidly able to combine his genius for design with his talent for business and thus turn himself into a creative shopkeeper who made and sold the things he himself would like to buy.

The Firm's first chance to shine came at the International Exhibition at South Kensington in 1862, where a sofa by Rossetti and half a dozen other pieces, including a cabinet painted with scenes from the life of St. George, as well as chests, chairs, an inlaid escritoire, and bookcases, were on display. The press had a field day: one reviewer claimed the work of the Firm "would be all very well as curiosities in a museum, but they are fit for nothing else."

Despite this reaction, and Philip Webb's assertion that business was conducted "like a picnic," the Firm flourished. In 1865, the workshops were moved to 26 Queen Square and showrooms opened in Oxford Street—right next door, in fact, to John Dando Sedding's offices, where two young architectural students, Ernest Barnsley and Ernest Gimson, would later embark on their own Arts and Crafts path. Morris knew Gimson and had introduced him to Sedding, and it is perfectly possible that this proximity exerted a powerful influence on the two Cotswolds architects, for their own short-lived furniture venture—Kenton & Co.—had strong parallels with the Firm.

The famous "Morris chair," which was to become such a signature piece of the Arts and Crafts movement, first appeared in the showroom in 1866. The initial design was based on a chair found in a Sussex carpenter's shop and adapted by Philip Webb from a sketch made by the Firm's manager, Warington Taylor. Webb incorporated a movable back that could be set at different angles and Morris sold the easy chair in two versions—ebonized or plain wood—with a cushioned seat and back, covered either in chintz or "Utrecht velvet." The Morris chair was still being made in 1913, advertised through the Firm's catalog at 10 guineas, or £8 with the cheaper cotton covers.

The Firm successfully mixed traditional and new ideas, methods, and materials. Much of the early furniture was made in Great Ormond Yard, just around the corner from Queen Square, although after 1881 furniture manufacture—along with tapestry and wallpaper production—moved to larger workshops at Merton Abbey in Surrey. Nine years later a new furniture factory was acquired from Holland & Son in Pimlico, and this flourished under the management of George Jack, one-time

Above: *This adjustable oak swing toilet mirror was made for Morris, Marshall, Faulkner & Co. around 1860 and may have been designed by Philip Webb, though Ford Madox Brown, who was responsible for several sturdy domestic pieces, is equally likely to be responsible.*

Left: *This example of the Morris Adjustable Chair, designed by Philip Webb for the Firm in 1866, is covered with "Bird" design upholstery.*

assistant to Philip Webb and well known for his work as a carver, who now became chief designer for Morris & Co. Jack was responsible for many of the elaborate and highly finished monumental mahogany pieces, often with inlaid decoration, now associated with the Firm. These sat alongside some of Webb's simpler and more functional pieces, which remained in the Firm's catalog well into the twentieth century, though it would be true to say that the furniture side of the business lost some of its impetus even before the death of Morris in 1896.

William Morris never visited the USA, although Morris & Co. products were being sold by American agents from the late 1870s and had a profound impact on design in many fields, including furniture. By 1901 Gustav Stickley was offering seven Morris chair models, with minor differences, through *The Craftsman*, and Morris's ideas and influences can be seen in the work of leading American Arts and Crafts designers from Charles Rohlfs to Frank Lloyd Wright.

Morris & Co. was bought in 1905 by F.C. Marillier and Mrs. Wormald. The firm survived World War I and the interwar period, finally closing for business in 1940.

MORRIS'S INFLUENCE

IT HAS BEEN SUGGESTED that the Arts and Crafts movement had more real influence on twentieth-century furniture design than it had on the great mass of furniture buyers in its own day. Certainly it is true to say that it has emerged as the major force in the history of British design during the last one hundred and thirty years. The insistence of its father-figure, William Morris, on the dignity and joy of labor, his emphasis on handcrafted methods and of honesty to function and material meant that, however well-designed, most of his work and that of the other leading exponents of the Arts and Crafts movement was necessarily expensive and—less comfortably for the socialist Morris—exclusive. His writings, lectures, and practical example, however, had a tremendous influence on a younger generation of architects and designers in the last quarter of the nineteenth century, who developed his ideas and continued the enthusiasm for Arts and Crafts-inspired work.

M.H. Baillie Scott, for example, looked to "simple furniture" to create clear and open space, particularly in smaller homes, where he designed built-in window seats, settles, and dressers, aiming to eliminate clutter. His freestanding furniture was available through John P. White's Pyghtle Works in Bedford. In 1901 its catalog included a hundred and twenty solid, simple, starkly masculine pieces by Baillie Scott. "The furniture," ran the catalog sales pitch, "has been designed and made to meet the

requirements of those who, appreciating soundness of workmanship and simplicity and reasonableness in design, have not found their wants supplied by the furniture on offer to the public in the modern cabinet-maker's shop." Frank Lloyd Wright was pursuing a similar line in the USA, seeking to integrate the furnishings with the architecture of his houses. To create continuity he used similar materials for both and echoed the architectural grammar of the house in built-in cabinets and seating.

THE GUILDS

THE ARTS AND CRAFTS MOVEMENT was, primarily, an attempt at social reform with an emphasis on group work in guilds of craftsmen and designers. The first of these—A.H. Mackmurdo's Century Guild—was formed in 1882. It carried out a number of decorative schemes between 1882 and 1888 and showed furniture at exhibitions both in London and the provinces. Because of its policy of cooperative work, it is difficult to attribute designs with certainty to individual

Above: *Oak settle, with inlay of Macassar ebony, cherry, chestnut, and pewter, designed by M. H. Baillie Scott and made at J. H. White's Pyghtle Works in Bedford in 1901.*

Above: *Oak writing desk by the founder of the Century Guild, A.H. Mackmurdo, c. 1886.*

members of the Guild, but it is likely that most of the furniture—which included chairs, desks, sofas, and cabinets—was designed by Mackmurdo himself with contributions by other Guild members: hinges by Bernard Creswick, for example, or a painted panel by Selwyn Image, or carved ornament designed by Herbert Horne. Century Guild furniture found favor among the critics. *The Builder* spoke of a piano as "a good unpretending piece of work in excellent taste." That taste was simple, depending on proportion, balance, and a sharp contrast of verticals and horizontals, with a classical note prompted by Mackmurdo's early studies of Italian Renaissance architecture, though it could shade off into eccentric stylization, anticipating the sinuous forms of Art Nouveau.

Two years later, in 1884, the Art Workers' Guild was founded by a group of artists and architects led by William Lethaby, and in 1888 a splinter group from the Art Workers' Guild formed the Arts and Crafts Exhibition Society, which organized annual exhibitions in the New Gallery in Regent Street with accompanying lectures and demonstrations. In a talk entitled "Furniture and the Room" E.S. Prior began: "The art of furnishing runs on two wheels—the room and the furniture. As in the bicycle, the inordinate development of one wheel at the expense of its colleague has not been without some great feats, yet too often has provoked catastrophe; so furnishing makes safest progression when, with a juster proportion, its two wheels are kept to moderate and uniform diameters. The room should be for the furniture just as much as the furniture for the room." Across the Atlantic, the members of the American Arts Workers' Guild designed buildings, silverware, and metalwork, along with furniture complete with panel paintings similar in style to that of Morris.

What each guild provided was a platform for displaying and, at a time when there were few retail outlets, for selling, Arts and Crafts pieces. In the first few years of its existence the Arts and Crafts Exhibition Society exhibited furniture by C.R. Ashbee (who later opened his own shop in Brook Street, London, displaying the work of the Guild of Handicraft), Sidney Barnsley, Ernest Gimson, Reginald Blomfield, Ford Madox Brown, George Jack, and W.R. Lethaby. It also, most unusually, attributed each piece to the individual designer rather than the firm that produced it. The critics were intrigued:

> "If we want to produce a nice chest of drawers—say—we must begin at the bare boards, and not at the surface ornament. Nay, perhaps the most valuable lesson of all, to learn, is that it is not *prettiness* that endues a thing with highest charm, but character. A *striking* proof of this is Mr. Madox Brown's delightful deal 'Workman's chest of drawers and glass,' at the Arts and Crafts Exhibition the other day—quite a plain thing with only a jolly, depressed carved shell above the glass and chamfered to the edges of the drawers—made in deal and stained green—that is all! It was just a commonplace thing handled imaginatively, and it gave me as much pleasure as anything in the exhibition. It made me feel that it takes a big man to do a simple thing: for the big artist takes broad views, he gives use its proportionate place, he knows the virtue of restraint, *and he has character to impart.*"

The furniture exhibited was unlike anything commercially produced at the time. "The exhibition is full of things which seem to have been done because the designer and maker enjoyed doing them," wrote one critic in *The Builder*, "not because they were calculated to sell well." And yet, despite being expensive, they did sell well.

KENTON & CO.

THE FIRM OF KENTON & CO., named after a street in Bloomsbury just around the corner from its premises, was set up in 1890 by the Barnsley brothers—the affable Ernest and the shyer Sidney—with Ernest Gimson, W.R. Lethaby, and the slightly older architects Reginald Blomfield and Mervyn Macartney, "with the object of supplying furniture of good design and good workmanship." All of them had trained in the two most notable nurseries of Arts and Crafts architecture, the offices of John Dando Sedding and Richard Norman Shaw. Though Kenton & Co. proved to be a short-lived experiment it was a valuable one, for it provided direct experience both of designing furniture and of managing men in workshop conditions—experience Gimson and Barnsley were able to put into practice when they set up on their own in rural Gloucestershire.

Each of the architects contributed capital of £100 to the venture, with a further £200 being provided by a sixth sleeping partner, a retired cavalry officer called Colonel Mallett. Macartney

Below: *"Wonderful furniture of a commonplace kind,"—an impressive oak dresser by W.R. Lethaby, c. 1900.*

already had some experience supplying furniture designs to Morris & Co., and the new firm was set up very much in the same spirit as Morris's venture. There was no uniform or house style at Kenton & Co. All five architects produced their own designs without meddling: "We made no attempt to interfere with each other's idiosyncrasies," Blomfield recalled in his *Memoirs of an Architect*, "with the result that each followed his own inclination."

It was through his work for Kenton & Co. that Gimson forged his characteristically geometric style and method of surface decoration, so similar to Shaker designs. Lethaby's pieces were also very simple, often of box form—dressers, cupboards, blanket chests, and cabinets in walnut or oak, which he had scrubbed to give a pale, silky finish. They were meant to be placed on stands in the manner of the seventeenth century and were inlaid with designs of plants, boats, animals, or abstract geometric patterns. Gimson was impressed and called this "wonderful furniture of a commonplace kind." It was strikingly modern, revolutionary in its simplicity, and radical in its rethinking of both form

Right: *A mahogany cabinet, veneered with ebony, walnut, and holly, designed by Ernest Gimson for Kenton & Co. in 1890–1, now in the Musée d'Orsay in Paris. The first of a series of elaborately decorated pieces partly inspired by Spanish varguenos and seventeenth-century spice cabinets with small drawers.*

and material. The great sideboard, now in the Victoria and Albert Museum, that Lethaby designed in 1900—of unpolished oak inlaid with unpolished ebony, sycamore, and bleached mahogany—is typical of his furniture design. The twisting foliage of the inlay on the doors springs directly from Morris's flowing patterns, and contrasts dramatically with the more classical work of Reginald Blomfield and Mervyn Macartney, who followed "the elegant motives of the eighteenth century." Sidney Barnsley's bold geometric patterns of shape, color, and texture betrayed a Byzantine influence. He designed his pieces according to basic construction principles, relying on inlays of mother-of-pearl or ivory for decoration.

The Kenton & Co. furniture was made up by a team of professional cabinet-makers. In keeping with the Arts and Crafts belief that a workman inevitably produced better work and gained more personal satisfaction from seeing a single job through from start to finish, each piece was made by a single craftsman. *The Builder* wrote approvingly "all the work is designed by members of the company and made under their personal guidance by their own workmen and each piece of furniture is made entirely by one man and stamped with the initials of the designer and workman. The recognition of the workman was of course a step that has our entire sympathy."

The youthful venture lasted only eighteen months and its high point was undoubtedly the exhibition held at Barnard's Inn in the London Inns of Court in December 1891. C.R. Ashbee attended and

described it as "one of the most beautiful of modern exhibitions of furniture ... where the pieces shown, many of them simple, straightforward & useful pieces, bore the names of Lethaby ... Barnsley, Gimson, and others with whom the Arts and Crafts movement is identified." Despite the critical and financial success of the exhibition—which resulted in sales worth over £700—the firm was wound up the following year and the unsold furniture divided among the partners. Lethaby tells how "to my share fell what we still call 'the Gimson Cabinet' of walnut 'left clean' and unpolished but now mellow and glossy from use; another cabinet which we call 'Blomfield,' 'Barnsley's table,' 'my Oak Chair' and a little revolving bookcase designed by Macartney. After all, these five pieces were not a bad return for £100 down." The Kenton & Co. experience proved invaluable and fueled Lethaby's desire to design his own furniture to complement the houses he was designing.

Following the collapse of the company most of the craftsmen moved with Ernest Gimson and the Barnsley brothers to the Cotswolds, where they produced work that firmly yoked the English rural traditions of craftsmanship to the ideas of William Morris and the ideals of the Arts and Crafts movement.

Above: *A ladderback armchair designed by Ernest Gimson in 1885–90 and made from ash with a rush seat.*

GIMSON & THE BARNSLEYS IN THE COTSWOLDS

THE SOLID OAK FURNITURE made by the Barnsleys and Gimson in the Cotswolds is in the direct tradition of William Morris, Philip Webb, and Ford Madox Brown: good citizen's furniture, certainly. Gimson had met Morris personally a number of times in his youth and valued his advice. Morris had come, at the invitation of Gimson's father—a businessman with a keen interest in the arts and good handiwork—to address the Leicester Secular Society, and had stayed at the Gimsons' house. In *Random Recollections of the Leicester Secular Society*, Ernest's elder brother Sydney recalled how nervous they both were on meeting the great man:

> "Ernest and I went to the station, and two minutes after his train had come in, we were at home with him and captured by his personality. His was a delightfully breezy, virile personality. In his conversations, if they touched on subjects which he felt deeply, came little bursts of temper which subsided as quickly as they arose and left no bad feeling behind them ... his lectures were wonderful in substance and full of arresting thoughts and apt illustrations."

Morris took a close interest in the young architect's career, opening up possibilities for him in London, lending encouragement, and fostering the ideal of a Utopian existence in the country.

While still an architectural student, Gimson had briefly apprenticed himself in 1890 to Philip Clissett, a country chairmaker at Bosbury near Ledbury in Herefordshire. There he had learned how to turn a simple rush-seated ash chair, and he applied this knowledge to the business of design. Though he did not actually make any more furniture with his own hands, he worked closely with his craftsmen and would often modify a design during the course of the work. Sidney Barnsley differed from Gimson in this important respect, for he enjoyed the physical business of making his own furniture by hand just as much as the designing. Gimson and Barnsley did, however, share a dream of establishing a community in the country with themselves as a nucleus around which, they hoped, other craftsmen would gather.

Following the collapse of Kenton & Co., Sidney Barnsley persuaded his brother Ernest to leave his architectural practice in Birmingham and join him and Gimson in Ewen, near Cirencester. It was quite an upheaval for Ernest, who was married with two young daughters, but he persuaded his unenthusiastic wife, Alice,

Above: *A lovely sturdy oak bench chest made by Ernest Gimson.*

to leave her newly built home in Birmingham and take up residence at Pinbury Park, an Elizabethan house on the Gloucestershire estate of Lord Bathurst. He had arranged to lease it at £75 a year on the understanding that he would carry out repairs. He and his family occupied the main house, while Gimson and Sidney Barnsley moved into adjacent cottages converted from former farm buildings.

In a shared workshop the three men began to make solidly constructed domestic furniture, eschewing machines and making a feature of the joinery by showing off the wooden pins, dovetails, and mortise-and-tenon joints. It wasn't for everyone. *The Builder*, in an issue dated October 1899, condemned the work as clumsy and inelegant, attacking the Arts and Crafts style:

> "In the reaction which is taking place against display and over-lavish ornamentation, the new school of designers appears to be losing the sense of style, and of the dignity of design which accompanies it, altogether. The object now seems to be to make a thing as square, as plain, as devoid of any beauty of line as is possible and to call this art."

They worked with local woods—oak, elm, ash, deal, and fruitwoods—which they left unadorned or chose to decorate simply by chamfering the edges of the pieces (a technique that had originated with wheelwrights and became popular with Gimson and the Barnsleys).

In 1895 Sidney married his cousin, Lucy Morley, who proved a strong and supportive wife to the plain-living Sidney, who not only made all his own furniture but chopped his own wood so that he could be warmed twice: once in the chopping and again when he burned the wood. They sought to get close to the land and to live a true country life. "It was wonderful [after] old smoky London to find yourself in those fresh clean rooms," recalled the designer Alfred Powell, "furnished with good oak furniture and a trestle table that at seasonable hours surrendered its drawing-boards to a good English meal, in which figured, if I remember right, at least on guest nights, a great stone jar of best ale."

In 1902 their landlord Lord Bathurst returned to take up residence at Pinbury Park but agreed to build three cottages on his land at the nearby village of Sapperton (twenty miles from Chipping Campden, where C.R. Ashbee was installing his Guild of Handicraft) and offered the farm buildings at Daneway for conversion into a spacious workshop and showrooms. Ernest Gimson and Ernest Barnsley had entered into a short-lived formal partnership, designing and manufacturing furniture, but Sidney chose to work alone, in an outbuilding adjacent to his new cottage at Sapperton which he converted into a workshop. He selected all his own timber and his designs have a rugged quality that reflects the rigor and dedication of his solitary way of working. Because he employed no one to assist him, his output was inevitably smaller. He outlined the plan to his old friend Philip Webb in a letter dated July 1902:

Right: *A walnut and ebony cabinet designed by Ernest Gimson c. 1905 with alternating panels on the drawers to form a geometric pattern. The cabinet was illustrated by Walter Shaw in* The Modern Home, *a showcase for Arts and Crafts furniture.*

> "My brother and Gimson have already started workshops at Daneway having four or five cabinet makers and boys so far, with the hope of chairmakers and modellers in the near future. I am remaining an outsider from this movement and still going on making furniture by myself and handing over to them any orders I cannot undertake, and orders seem to come in too quickly now as we are getting known."

None of the three ever issued a catalog of designs. Instead Gimson sent photographs—which he expected to be returned—to prospective clients. Both fretted about the possibility of their work being copied.

The Daneway workshop went from strength to strength under the foremanship of Peter van der Waals, a Dutchman recruited by Gimson and Ernest Barnsley in 1901. Waals firmly believed that the form, quality, and color of the woods used—oak, occasionally walnut, and chestnut—should dictate the design of the furniture. Sir George Trevelyan, who worked with Waals after Gimson's death, asserted:

> "Gimson would be the first to acknowledge the immense debt he owed to [Waals] as a colleague. Though Gimson was, of course the inspiration and genius, he used Waals from the outset in close co-operation. The association of these two men was an essential factor in the evolving of the Cotswold tradition."

Things were not so harmonious between Gimson and Ernest Barnsley. They had fallen out amid some bitterness, with Barnsley complaining that Gimson "wouldn't compromise in any sort of way." The partnership collapsed and Ernest Barnsley concentrated his efforts on his architectural work, particularly at Rodmarton Manor, while Gimson carried on and expanded the Daneway workshop. By 1914 Peter Waals was in charge of nine woodworkers, four metalworkers, and three apprentices at Daneway. Gimson had leased Hill House Farm to house his workers and paid them an average of 10d an hour for cabinet makers (although Waals received 1s 2d an hour) and the apprentices 3d an hour. By 1918 Waals was receiving £5 for a fifty-hour week. Their costs and the quality of their materials inevitably put their furniture—an average prewar piece cost £12—beyond the reach of local people.

The Kenton & Co. practice of stamping the maker's initials on a piece of furniture was discontinued, with the result that it is very difficult to tell the work of Sidney Barnsley and Ernest Gimson apart. Both men used traditional woodworking techniques such as chamfering or gouging (which Sidney called "tickling" the wood) to add visual interest to their plainer pieces, but they also produced more highly decorated cabinets, sideboards, and church furniture, using mother-of-pearl or ivory inlays. Alfred Powell, who, with his wife Louise, decorated some of Gimson's pieces with oil paintings, claimed:

> "Much insight into his inner life was to be had at Daneway House among all his furniture. At first glance all was of extraordinary interest. Then one saw the beauty of the work, the substance, the development of the various woods, of the ivory, the silver, the brass, of inlays of coloured woods & shell. It was inevitable that you should find in the work now and then a humorous use of peculiar materials, an enjoyment of surprise."

In 1906 (and again in 1916) Gimson shared a stand with May Morris at the Arts and Crafts Exhibition held at Burlington House. He exhibited furniture, some of which she had decorated, in a specially designed bedroom setting. But his real chance to shine came in 1907, when Ernest Debenham offered him the chance to exhibit over eighty items (forty-five of which were furniture) for sale in his new London department store.

For Nikolaus Pevsner, Gimson was the "greatest artist" among his generation of furniture designers: "His chairs give an impression of his honesty, his feeling for the nature of wood, and his unrevolutionary spirit." Gimson died, aged only fifty-five, in 1919. Following his death, Peter Waals set up his own workshops at Chalford, Gloucestershire, producing his own designs and some of Gimson's.

Left: *The furniture in the dining room at Rodmarton Manor was designed by Sidney Barnsley and Peter Waals. The Barnsley built-in dresser is a complicated piece that employed one of his favorite decorative techniques: gouging on the horizontal and vertical edges.*

GUILD OF HANDICRAFT

IN THE SAME YEAR that the Arts and Crafts Exhibition Society was founded, Charles Robert Ashbee, a twenty-five-year-old architect, romantic, and socialist, opened the Guild and School of Handicraft in the East End of London with a capital of £50. Ashbee had been living in the pioneer University Settlement house at Toynbee Hall, an educational institution set up for the working man, where he taught classes on the writings of Ruskin for some years in the 1880s. His readings on the virtues of manual labor inspired his students to decorate the dining room at Toynbee Hall, and it was from these students that he drew the nucleus of his Guild of Handicraft, basing it on the model of the medieval guild system. Life was bleak in the East End and Ashbee wanted to offer something that would improve the lot of the working man. In the event, the School lasted only a few years, but the Guild flourished. In 1890 its expansion encouraged Ashbee to find new premises in a dilapidated Georgian building, Essex House, in the Mile End Road, and in 1902 he ambitiously relocated his Guild to the rural market town of Chipping Campden in the Cotswolds, with the aim of providing his seventy or so workers and their families with better conditions and happier, healthier, simpler lives. By the end of the century Guild craftsmen in the field of furniture were carving, modeling, and cabinet-making, as well as carrying out restoration and undertaking complete decorative schemes. The work of the designer, painter, and craftsman was acknowledged on each piece.

Ashbee's own most original designs were for silver and jewelry, but he made some interesting experiments in furniture, some of the pieces combining a variety of techniques such as woodcarving, metal engraving, and tooled leather. The earliest pieces made by the Guild of Handicraft, such as the oak cabinet painted in red and gilt with a quotation from William Blake's "Auguries of Innocence," are extremely simple in form. In addition to furniture from Ashbee's own designs the Guild made for other designers, notably M.H. Baillie Scott. As well as elegant cabinets and superb piano cases, they made a range of much plainer furniture, including bedsteads and washstands, with handmade handles and modestly carved decoration, that was close in style and price to the furniture commercially produced by Heal & Son.

Large commercial concerns such as Heal's and Liberty's were not slow to imitate the Guild's innovative style, and their undercutting of its prices eventually brought about its demise. The Guild made successive financial losses in 1905 and 1906 and, clearly in trouble, the business was formally wound up at the end of 1907, although some Guildsmen—including Jim Pyment who continued to make furniture alongside the carvers Alec and Fred Miller in the old silk mill where the Guild had set up its workshops—chose to remain in Chipping Campden.

Right: *This writing cabinet by C.R. Ashbee and the Guild of Handicraft was first exhibited at the Woodbury Gallery, London, in fall 1902.*

Above: *Liberty offered this sideboard in their "New Studio" range from 1898.*

LIBERTY'S

ARTHUR LAZENBY LIBERTY was a shrewd businessman with his finger on the fashionable pulse. He had worked his way up from apprentice draper to become the owner of a Regent Street department store in a mere sixteen years. In 1883, recognizing the potential in art furnishings, he opened a decoration studio under the management of the designer Leonard F. Wyburd at Chesham House, 140–50 Regent Street, and set out to provide furniture and furnishings to meet the demand for fashionable Aesthetic interiors, evolving a style that combined commercial Art Nouveau with the design vocabulary of the Arts and Crafts movement. Liberty both "borrowed" from Arts and Crafts designers (for his own range, much of which was designed by Wyburd, but clearly influenced by William Morris and C.F.A. Voysey) and commissioned them as well. The firm stocked designs by Voysey and George Walton and carried a line of eighty-one pieces of furniture designed by

M.H. Baillie Scott. It also stocked and sold Richard Riemerschmid's elegant oak chair, designed in 1899, as well as a very similar, somewhat cheaper, chair produced in the Liberty workshops with chamfered decoration on the legs and side supports.

From 1887 the firm's cabinet-making workshop was turning out simple side chairs, carvers, stools, and country-style furniture in oak with metal handles, strap hinges, and occasionally inset tiles by William De Morgan, and sub-contracting more complicated pieces to a series of manufacturers. Cheaper than the guilds, Liberty's posed a very real threat to them and contributed to their commercial demise in the 1900s. The store traded on the Arts and Crafts ethic—peppering its illustrated catalog with a series of quotations from Ruskin and applying mottoes to furniture much as Morris had—and its visual appeal, while dispensing with the central (and expensive) concept of the artist-craftsman and completely ignoring the movement's social message. And it proved hugely popular. Ironically Liberty's can be credited with both introducing a wider public to the (watered-down) ideal of the Arts and Crafts movement and hastening its end. For the Arts and Crafts style had to compete with its eclectic range—the catalogs romped through historical periods, offering "Elizabethan," "English Renaissance," "Domestic Gothic," and "Tudor" styles (even a "Cromwellian" dining room) alongside the "Quaint" furniture that most closely approximated to Arts and Crafts style and drew on its motifs.

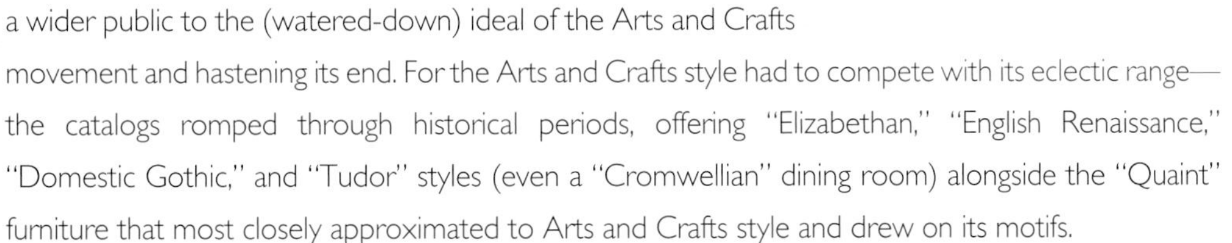

Above: *Leonard F. Wyburd designed this oak washstand for the Liberty Furniture Studio, c. 1894. The tiles are by William De Morgan.*

HEAL & SONS

THE LONG-ESTABLISHED FIRM of Heal & Son produced conventional Victorian furniture, largely beds and bedding, though a new department for sitting-room furniture had been opened in the 1880s. Ambrose Heal, the great-grandson of the store's founder who joined the firm in 1893, changed the course of the whole enterprise. Like Arthur Liberty, he was a shrewd businessman, and though he was largely in tune with the ideals of the Arts and Crafts movement, he was unembarrassed about using modern methods. As William Morris had quickly discovered, the maker of handcrafted furniture spent a very great deal of his time "ministering to the needs of the

Left: *The influence of Ford Madox Brown's stained joiner furniture in sturdy forms is apparent in this painted chest of drawers from Heal & Son, 1900.*

Right: *This high-back oak chair, c. 1898–9, is a classic Charles Rennie Mackintosh.*

swinish rich," but Heal & Son's "Plain Oak Furniture" and "Simple Bedroom Furniture" ranges (which owed a heavy debt to Gimson and the Barnsleys) enabled earnest young couples to fill their homes with the kind of pieces their reading of Morris's *News From Nowhere* had made them yearn for. Heal displayed his furniture in novel "room sets" within the store, and mounted a showcase housing exhibition at Letchworth in 1905 containing cottage furniture of his own design. Available in a range of prices, the pieces were made from cherry, walnut, chestnut, or oak, characterized by simple, stylish design and crafted details—ebonized bandings and chip-carved plain handles. To underscore further the association with Morris and the Utopian dream of a better, simpler life, he called these ranges "Kelmscott" and "Lechlade."

Nikolaus Pevsner, who praised Heal's for its "pleasant bedsteads," detected a new modernism in its style and drew parallels between its plain waxed oak surfaces and the lightness of Voysey's wallpapers. "The close atmosphere of medievalism has vanished," he wrote, "living amongst such objects, we breathe a healthier air."

CHARLES RENNIE MACKINTOSH

The Scottish architect Charles Rennie Mackintosh had an exceptional client in Catherine Cranston, who gave him a creative and a financial free hand in designing her four Glasgow tea rooms—at 114 Argyle Street, 205–9 Ingram Street, 91–3 Buchanan Street, and the Willow Tea Rooms in Sauchiehall Street. Mackintosh was not a commercial furniture designer: each of his pieces was designed to integrate with the space and function of the specific room for which it was intended. It was for Miss Cranston that he designed his now famous high-back chair with pierced oval back, his ubiquitous ladderback chair, and his novel curved lattice-back chair with colored glass inserts.

Beauty mattered to Mackintosh as well as—perhaps more than—function, and his extreme modernism is evident in the exaggeratedly high backs of his dining chairs, which are based on simple geometric forms. His pioneering designs drew

on Art Nouveau, Celtic ornament, and the economy of Japanese art. He championed aesthetic unity in interiors and elevated common objects to the status of art, finding support and praise in Europe, where his work was highly regarded and extremely influential, and in America. In 1900, designs by Mackintosh and his wife Margaret, a fellow member of the Glasgow Four, were rapturously received at the eighth Secession Exhibition in Vienna and he was commissioned by the director of the Wiener Werkstätte to design a music salon for his house.

The couple set up home in Florentine Terrace, Glasgow, in 1906, and together made the interior glow with an ethereal, jewel-like quality. "Margaret has genius, I have only talent," Mackintosh generously said of his wife. They collaborated on the designs of several other house interiors, including Hill House for the publisher William Blackie.

Above: *A Charles Rennie Mackintosh cabinet made for 14 Kingsborough Gardens, Glasgow, in 1902.*

C.F.A. VOYSEY

CHARLES FRANCIS ANNESLEY VOYSEY began designing furniture in earnest in the 1890s. An elegant reticence characterizes his best work, along with an almost playful simplicity and careful proportioning. The wood he favored was untreated oak, planed and left—as he would meticulously note on his drawing—"free from all stain or polish." The freshness and individuality of the pieces comes across in the play of form and shape rather than in any ornament and, although it is not unusual to find some low-relief carving on a Voysey piece, applied metal decoration, sparingly used, such as brass strap hinges and brass and leather panels (as on the Kelmscott cabinet) was a more common form of decoration.

He wanted his furniture to suit the everyday needs of the people who used it and as an architect and a designer constantly stressed the importance of honesty, originality, and simplicity. "We cannot be too simple," he stated in

Left: *The design of this oak dining chair, made in 1902, is distinguished by a broad splat pierced with a single heart shape in characteristic Voysey style.*

Above: *A Voysey oak cabinet with double doors and strap hinges made especially to house and display Morris's Kelmscott Chaucer. Made by F. Coote in 1899.*

1909. "We are too apt to furnish our rooms as if we regarded our wallpapers, furniture and fabrics as far more attractive than our friends." He wanted his houses, inside and out, to exude a feeling of warmth, welcome and friendship, if not always deep repose (his very first design for the "Swan" chair uses a lovely free curving line in its shape but would not prove a very comfortable seat for a long period). If some of his chairs offered slender support, his tables tended towards sturdy functionalism. "Good craftsmanship," he wrote, "is the expression of good feeling, as good feeling leads to right and honest workmanship." His furniture was made for him by a select number of craftsman firms, including F.C. Nielsen and A.W. Simpson of Kendal. Voysey designed a house, Littleholme in Kendal, for Simpson in 1909 and the two men profited from a mutual exchange of ideas.

FURNITURE IN EUROPE

IN 1902, INSPIRED BY A VISIT to Ashbee's Guild of Handicraft, Josef Hoffmann returned to Austria and, together with Koloman Moser, a fellow member of the Vienna Secession, founded the Wiener Werkstätte. By 1905 the workshop employed over a hundred craftworkers and produced everything for the artistic interior, from metalwork, ceramics, glass, bookbinding, wallpapers, textiles, and enamelwork to elegant elongated furniture that owed much to the work of the Glasgow Four, and to Charles Rennie Mackintosh in particular. The furniture they produced was uncompromisingly modern and was condemned as "decadent" by the traditionalists of the British Arts and Crafts movement.

Mackintosh's influence can be felt in the designs that emerged from the Deutscher Werkbund, established in Munich by Hermann Muthesius, Friedrich Naumann, and Henry van de Velde in 1907 and inspired in part by the British Arts and Crafts movement. The Deutscher Werkbund accepted the need to design not just for handcraft but for mass production and produced furniture that was readily recognizable for its clean lines, blunt forms, and prime utilitarian considerations. A Werkbund exhibition in London in 1914 inspired the formation of the Design

Above: *Liberty sold this strikingly modern walnut chair by Richard Riemerschmid, 1899.*

Far right: *An oak chair with checker-pattern rushed seat by Peter Behrens, who went on to become a founder of the Deutscher Werkbund in 1907.*

and Industries Association, which was motivated by British government concerns for design reform with the declared aim of encouraging a public demand for "what is best and soundest in design."

Richard Riemerschmid, the principal designer and one of the founders of the Dresden and Munich Werkstätten, produced sinuous Art Nouveau-inspired designs for furniture that was largely handcrafted, although in 1906 he produced a line of suites called *Maschinenmobel* ("machine-made furniture") that was a huge commercial success. Riemerschmid was a member of the Deutscher Werkbund and enthusiastically embraced the modern age. The cross-fertilization of British and European design ideas is demonstrated by his influence on Harry Peach, a founder member of the Design and Industries Association, who, at his Leicester-based furniture firm of Dryad, made strong cane chairs in modern shapes prompted by the work of Riemerschmid. The chairs were popular in hotels, hospitals, and on ships, though they suffered from the competition of Lloyd Loom in the 1930s.

M.H. Baillie Scott attracted a number of European commissions for which he provided furniture: a music room for A. Wertheim of Berlin, a living-room interior executed by the Werkstätten, and a bedroom-boudoir with furniture made from alder and pear wood stained black and enlivened by mother-of-pearl and ivory inlay, which was exhibited at the Werkstätten exhibition in 1903. Carl Malmsten, the prize-winning Swedish designer of textiles, fabrics and furniture, visited Ernest Gimson's and Sidney Barnsley's furniture workshops in the early years of the century and shared their passion for handcrafted work, as did other contemporary Scandinavian furniture designers, including Alf Sture and Arne Halvosen in Norway. It was the Swedish painter Carl Larsson, however, whose illustrations of his country house, Sundborn, in his book *Ett Hem* (1898) came to epitomize Swedish domestic ideals and set the Arts and Crafts style for a generation.

GUSTAV STICKLEY & THE CRAFTSMAN WORKSHOPS

In America, Gustav Stickley was one of the main promoters of Arts and Crafts style and ideals. Best known as a furniture maker, he was profoundly influenced by the work of William Morris and the ideals of Ruskin. A trip to England in 1898, during which he met Voysey, Ashbee, Samuel Bing, and others, sent him home a passionate Arts and Crafts convert. "I felt that the badly constructed, over-ornate, meaningless furniture that was turned out in such quantities by the factories was not only bad in itself," he wrote, "but that its presence in the homes of the people was an influence that led directly away from the sound qualities which make an honest man and a good citizen."

A year after his visit to England Stickley established his own guild of artisans, United Crafts, in Eastwood, New York, adapting the theories and philosophies of English Arts and Crafts to the American industry and market, and sold a whole range of home products through his mail-order catalogs, including seven chairs based, with minor

Left: *An oak "drop-arm" spindle "Morris" chair by Gustav Stickley, 1905. Stickley offered reasonably priced chairs, tables, desks, beds, and cabinets, usually made of white oak, quarter-sawn to show the grain, to the American middle classes through his Craftsman catalogs.*

Right: *A fine oak dropfront desk by Gustav Stickley, c. 1903.*

differences, on the Morris model, one of which was a lady's Morris chair. He patented the first of these in October 1901. The largest model he described as "a big, deep chair that means comfort to a tired man when he comes home after the day's work." His magazine, *The Craftsman*, publicized both his products and his philosophy. His stated aim was "to do away with all needless ornamentation, returning to plain principles of construction and applying them to the making of simple, strong, comfortable furniture."

Unrepentant about his businesslike approach, Stickley tried to combine high ideals of craftsmanship with the latest machinery to make "furniture that shows plainly what it is, and in which the design and construction harmonize with the wood." He employed hundreds of workers using modern machines, and through *The Craftsman* he declared his intention to produce "a simple, democratic art" that would provide Americans with "material surroundings conducive to plain living and high thinking." His designs included little decoration and were in many ways crude—fidelity to the wood, more often than not oak, was most important. He based his simple, squat forms on seventeenth-century colonial furniture, intending to evoke the "simple life" of the early American settlers, and it proved very successful.

In 1903 Stickley employed the immensely talented architect Harvey Ellis to provide new designs for furniture and interiors. Ellis introduced ornamental inlays of stylized floral patterns in

Left: *A rare and important inlaid oak chair by Harvey Ellis for Gustav Stickley. Ellis, a brilliant but unstable architect, designed for Stickley for only a year in 1903, during which he introduced a more sophisticated European style reminiscent of Olbrich and Van de Velde.*

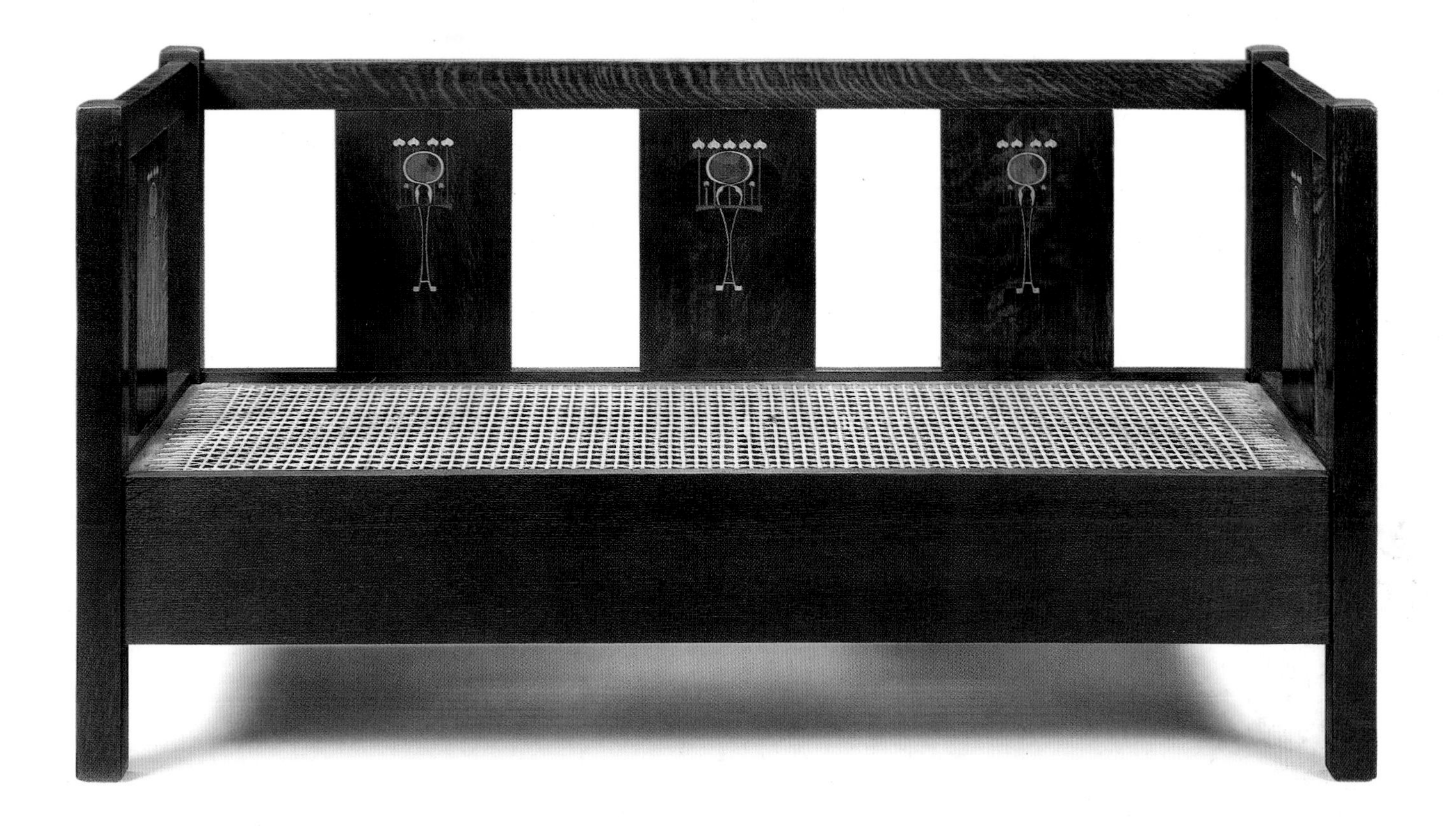

copper, pewter, and colored woods and took the furniture beyond the English styles, elongating and lightening pieces with taller proportions, thinner boards, broad overhangs, and arching curves. His work brought the influence of Charles Rennie Mackintosh and Europe to the American Arts and Crafts movement. Unfortunately, the successful collaboration between the two men would last less than a year, cut short in 1904 by Ellis's untimely death, aged only fifty-two. Stickley, meanwhile, forged ahead, expanding production to include metalwork accessories, lighting, and textiles, and in 1904 he changed the business name to Gustav Stickley's Craftsman Workshops. From 1905–9 he made more attenuated "spindle" furniture, influenced by Frank Lloyd Wright. It was lighter and more elegant but more easily copied and was soon sold by other furniture manufacturers (including his own brothers), leading to his bankruptcy in 1915 and the amalgamating of his firm with his brothers' rival company, L. & J.G. Stickley, to form the Stickley Manufacturing Co.

Today, Gustav Stickley is best known for his "Mission" furniture, so-called because it was associated with the furnishings of eighteenth-century Spanish colonial churches. "A chair, a table, a bookcase or bed [must] fill its mission of usefulness as well as it possibly can," he asserted. "The only decoration that seems in keeping with structural forms lies in the emphasizing of certain features of the construction, such as the mortise, tenon, key, and dovetail."

Above: *An oak settle, inlaid with copper, pewter, and exotic woods, designed by Harvey Ellis for Gustav Stickley, c. 1903. Stickley produced different versions of this form, including a cube armchair and a smaller settee.*

Left: *This sober-looking oak and copper cellarette, c. 1904–5, made and sold through the Roycroft Shop opens to reveal an interior fitted on one door with a tray for bottles and glasses and, on the other, with an ice bucket.*

ELBERT HUBBARD & THE ROYCROFTERS

ANOTHER IMPORTANT OUTLET for American Arts and Crafts furniture in the same period was provided by the craftsmen at Elbert Hubbard's Roycroft Arts and Crafts community in East Aurora, New York, founded in 1892. The furniture the Roycrofters produced was heavy and solid, not unlike Stickley's Mission-style pieces though more Gothic in manner, an influence underscored by the Gothic-style carved letters of the Roycroft name or the orb and double-barred cross symbol with which the pieces were marked. In 1901 *House Beautiful* described a Roycroft chair as "honest and simple enough in construction, but somewhat too austere in design, and altogether too massive to be pleasing." Hubbard sold his furniture range—available in ash, mahogany, or dark, fumed oak—through his shop and catalogs and for some represented the extreme commercial wing of the American Arts and Crafts movement.

CHARLES ROHLFS

AN IMAGINATIVE AMERICAN FURNITURE DESIGNER, Charles Rohlfs opened his workshops in Buffalo, New York, in 1891. His early Gothic-style furniture was exhibited at Marshall Field & Co. in Chicago in 1899 and attracted this praise from Charlotte Moffitt, writing in *House Beautiful*:

> "In the collection shown in Chicago of this queer, dark, crude, mediaeval furniture is a desk, a very marvel of complexity, with endless delights in the way of doors, pigeon-holes, shelves, and drawers.... When closed—that is when the writing shelf is raised and fastened with its rough hasp of dark steel and crude wooden pin—all the drawers and shelves in place, and the doors closed, it looks like nothing so much as a miniature Swiss cottage."

Rohlfs himself described his furniture as having "the spirit of today blended with the poetry of the medieval ages." Later, he combined the sturdy shapes of Arts and Crafts furniture with the sinuousness of Art Nouveau decoration and ornament and, following the international recognition he won at the Esposizione Internazionale in Turin in 1902, he was commissioned to design a set of chairs for Buckingham Palace. His work is too original and too playful to slot neatly under the Arts and Crafts banner, but the solid handcrafting and loving attention to detail earn him a place at its sturdy table.

Right: *A characteristically elongated and elaborately carved oak rocking chair, designed by Charles Rohlfs in 1901.*

CHARLES P. LIMBERT

THE FORMER FURNITURE SALESMAN Charles Limbert started his manufacturing company in Grand Rapids, Michigan in 1894, and in 1902 began making Charles Rennie Mackintosh-inspired geometric furniture out of oak. He used quality materials and produced fine furniture—with the help of the machine, a debt he played down in his company catalogs, which assert: "Limbert's Holland Dutch Arts and Crafts furniture is essentially the result of hand labor, machinery being used where it can be employed to the advantage of the finished article." Limbert drew inspiration from Japan, Austria, and England and the firm became known for incorporating decorative cutouts in the backs and canted sides of a number of its pieces and for its use of angled supports.

Below: *A two-door cabinet with poppy design by Arthur and Lucia Kleinhans Mathews for the Furniture Shop in San Francisco, 1906–20.*

ARTHUR & LUCIA KLEINHANS MATHEWS

THE HUSBAND-AND-WIFE TEAM of Arthur and Lucia Kleinhans Mathews started the Furniture Shop of San Francisco in 1906, offering a complete interior design service to a city rebuilding itself after the huge earthquake and fire of that year. Arthur was responsible for the design and Lucia for the color schemes and carvings on the decorated furniture they supplied, along with wood paneling, murals, paintings, picture frames, and decorative accessories. Both had studied in Paris and their inspiration can be traced back to French classicism, to produce an opulent style far removed from the sturdy functionalism of Stickley and Hubbard. Work was brisk after the earthquake, and they and their team of skilled cabinet-makers, carvers, and decorators executed both large interior schemes and single pieces of furniture, which were often decorated with flowers or Californian landscapes. Their work anticipates Roger Fry's Omega Workshops, a British experiment that followed the Arts and Crafts movement, and produced work that was often of a very high standard.

Above: *Two Greene & Greene rocking chairs on either side of the fireplace in the drawing room of the Gamble House.*

GREENE & GREENE

SOME OF THE MOST EXQUISITE American Arts and Crafts furniture was produced by the Greene brothers to complement the houses they were designing in California. They were much admired by C.R. Ashbee who, following a visit to Charles Greene in his workshop in 1909, wrote of how "they were making without exception the best and most characteristic furniture I have seen in this country. There were beautiful cabinets and chairs of walnut and lignum vitae, exquisite dowelling and pegging, and in all a supreme feeling for the material, quite up to our best English craftsmanship.... I have not felt so at home in any workshop on this side of the Atlantic."

Charles Sumner and Henry Mather Greene had absorbed the Arts and Crafts ideals on two visits to England, in 1901 and 1909. They translated them into the "ultimate bungalows" they designed and the elegant, exciting custom-made furniture that went inside them. Highly accomplished woodworkers themselves, they worked closely with another pair of brothers, the cabinet-makers Peter and John Hall, who executed the Greenes' designs from 1906, with Charles supervising production. Rounded corners and softened edges were a typical feature of the Greenes' furniture, along with a painstaking attention to detail and the use of the highest quality materials.

FRANK LLOYD WRIGHT

Above left: *Frank Lloyd Wright designed this oak chair for the Darwin R. Martin house in Buffalo. He may have taken his inspiration from a chair by Leopold Bauer that was published in* Das Interieur, *1900.*

In the 1880s and 1890s Frank Lloyd Wright began to design his own furniture to fit in with his concept of organic architecture, building in cabinets and shelving, and seating units around fireplaces. He designed his first dining-room table for his own Oak Park home, with tall spindle-backed chairs inspired by Voysey, Mackintosh, and Gimson (later copied by Gustav Stickley and others). He made his furniture from the same materials and with the same finishes as the buildings themselves, in keeping with the "grammar" of the house, relating everything to the whole so that "all are speaking the same language."

A founder member in 1897 of the Chicago Arts and Crafts Society, he took the Morris ideal and reinvented it in a twentieth-century idiom, which certainly did not preclude the aid of machines. Indeed, he celebrated rather than demonized the machine, believing it to be "capable of carrying to fruition high ideals in art—higher than the world has yet seen!" Furthermore he praised the way "the machine ... has made it possible to so use [wood] without waste that the poor as well as the rich may enjoy today beautiful surface treatments of clean, strong forms." His starkly functional style made few concessions to comfort. He relied on the juxtaposition of shape, form, and space rather than applied ornament and used new constructional techniques and new materials to great effect. His furniture tends to be composed of strong horizontals and severe verticals: it is starkly geometric and exceedingly beautiful.

His styles evolved to match his architectural vocabulary, becoming simpler, more cylindrical, and more minimal in keeping with the economy of the low-cost "Usonian" houses he began to design in the 1930s. In 1955 Wright created his own line of furniture for the Heritage-Henredon Furniture Company. Modular pieces made of mahogany in basic circular, rectilinear, and triangular shapes, they were simple and functional and made his work available to people who did not live in a Wright-designed home.

Opposite right: *An oak spindle side chair designed by Frank Lloyd Wright for the Lawrence Memorial Library, Springfield, Illinois, c. 1905–10.*

Right: *A typically elegant high-backed oak chair designed by Frank Lloyd Wright in 1902.*

5

TEXTILES & WALLPAPER

Left: *"Fool's Parsley," a wallpaper design by C.F.A. Voysey who, though primarily an architect, was one of the most prolific and distinguished English pattern designers, influencing the direction of design in England and on the Continent. The bird motif, together with the heart, became in effect the Voysey trademarks.*

SATISFYING THE MIND & THE EYE

"In the Decorative Arts, nothing is finally successful which does not satisfy the mind as well as the eye. A pattern may have beautiful parts and be good in certain relations; but unless it is suitable for the purpose assigned it will not be a decoration."

William Morris, 1877

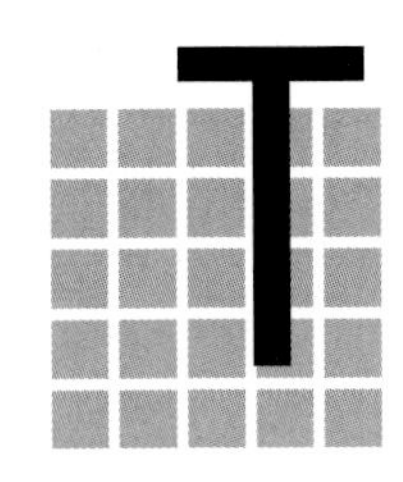

THE ARTS AND CRAFTS MOVEMENT fueled a new middle-class fashion for interior decoration and firms like Morris & Co. and Liberty's—using a range of designers from Walter Crane to C.F.A. Voysey—provided the woven and printed textiles, carpets, wallpapers, and tapestries needed to furnish the "Artistic" home. The Victorian penchant for heavily draping and swathing furniture and windows gave way to a lighter, fresher fashion for less elaborate effect. William Morris's own and other Morris-inspired lightweight chintzes replaced the heavy velvet upholstery; window coverings were simplified and curtains were now arranged on simple wooden or brass poles; while a single oriental or Morris-made carpet was spread over a polished or stained wooden floor. Morris's genius for rich pattern-making and his experiments with natural animal and vegetable dyestuffs created distinctive fabrics and wallpapers that became the keynote of an Arts and Craft home on both sides of the Atlantic. "Aesthetic" colors like peacock blues, russet browns, madder reds, subtle yellows, and sage greens predominated in his luxuriously dense patterns based on natural motifs, while Voysey popularized a lighter palette in designs of almost stencil-like simplicity featuring stylized versions of flowers, foliage, birds, and animals on pale backgrounds.

Above: *Edward Burne-Jones designed the figure for "Pomona" and J.H. Dearle the background for this silk and wool tapestry for Morris & Co., c. 1900.*

Left: *Morris's "Pomegranate" wallpaper design, 1862, was available from the Firm in a choice of three different background colors.*

Above: *An embroidered linen cushion cover with linen appliqué, embroidered with colored silks and needleweaving borders, designed and made by Jessie Newbery, c. 1900.*

EMBROIDERY

A BURGEONING CRAFT in the last quarter of the nineteenth century, embroidery was enthusiastically taken up by women encouraged to make their home surroundings ever more beautiful. The Royal School of Art Needlework came into being and firms such as Morris & Co. and Liberty's in England, the Maison de l'Art Nouveau in Paris, and Gustav Stickley's Craftsman outlets in the USA sold specially dyed silks, wools, linens, and canvas with marked-out designs, which could be made up into curtains, cushions, or table runners. "Simplicity is characteristic of all Craftsman needlework which is bold and plain to a degree," an article in *The Craftsman* reminded its female readers, underscoring the Stickley message that textiles should be used to complement a decorative scheme, not overwhelm it.

What began with a simple daisy design worked by Janey Morris on a piece of cheap blue serge grew into a successful section of Morris & Co., which, by 1885, when May Morris took over the running of the department, produced floral designs for everything from tablecloths to cot quilts. The Morris connection also extended to the Royal School of Art Needlework in South Kensington, for Janey's sister, Bessie Burden, became chief instructress there. In Scotland, Jessie Newbery, wife of the principal of the Glasgow School of Art, established embroidery classes in 1894 and encouraged her students to use embroidery as a means of self-expression.

Once again, the revival started with William Morris. In the late 1850s while living with Edward Burne-Jones in Red Lion Square, Morris had a wooden embroidery frame made—a copy of an old example—and began the painstaking process of teaching himself the art of embroidery by unpicking old pieces to discover how they had been made. He favored wool over silk and eschewed chemical dyes. He wanted to realize a romantic ideal of recreating an English medieval interior and soon enlisted the aid of his housekeeper, Mary Nicolson (known as Red Lion Mary), teaching her the stitches he had only just mastered himself. A year later, after his marriage to Janey, he sketched out a simple daisy design for his wife to work on a piece of indigo-dyed blue serge she had found by chance in a London shop. "He was delighted with it," Janey wrote, "and set to work at once designing flowers—these we worked in bright colors in a simple, rough way—the work went quickly and when we finished we covered the walls of the bedroom at Red House to our great joy." Georgina Burne-Jones, who helped as well, reported that "Morris was a pleased man when he found that his wife could embroider any design that he made, and he did not allow her talent to remain idle." He taught her and her sister Bessie his embroidery techniques and set them to work on an ambitious scheme for a dozen embroidered and appliquéd hangings he had devised for the drawing room of Red House. These depicted female figures—"illustrious women" drawn from the works of Chaucer—that had bold black outlines like stained glass and could almost have passed for cartoons. Other female friends, notably Lucy and Kate Faulkner, were taught "what stitches to use and how to place them" and put to work producing "embroidery of all kinds" using yarns Morris had dyed in subtle shades specifically for the Firm. Many commissions for altar frontals and medieval-style hangings came from churches that were also ordering stained-glass windows from the Firm. Designs for both were provided by Edward Burne-Jones, and Catherine Holiday, wife of the painter and stained-glass artist Henry Holiday, worked many of Burne-Jones's designs in outline stitch.

Below: *"Sunflower" hanging, c. 1876, designed by William Morris and embroidered in silks on a linen ground by Catherine Holiday, one of the most accomplished craftswomen of the period.*

It was around this time that Gertrude Jekyll, recently a student of the South Kensington School of Art, met Morris, who encouraged her interest in embroidery. She went on to design her own patterns, heavily influenced by him, featuring pomegranates, periwinkles, dandelions, and mistletoe. Her designs attracted commissions from the Duke of Westminster and Sir Frederick Leighton, though she was eventually forced by her failing eyesight to abandon such close work and turned instead—and with great success—to garden design.

In 1872 the Royal School of Art Needlework was founded and run by a committee of philanthropically minded aristocratic ladies, led by Lady Victoria Welby and Mrs. Anastasia Dolby, under the presidency of Princess Christian of Schleswig-Holstein, "to give amateurs and gentlewomen instructions in fine needlework and to reproduce from good designs the old English needlework on handmade linen, so often spoken of as Jacobean work." They wanted to provide suitable employment for needy gentlewomen and generally improve the standard of ornamental needlework. Twenty young ladies were chosen from the "impoverished genteel class" to occupy the small apartment above a hat shop in Sloane Street that provided the School with its first premises. Within three years it had expanded, moved to permanent purpose-built premises in Exhibition Road, South Kensington, and had a new patron—the Queen. There were separate workrooms devoted to embroidery, appliqué, and goldwork, as well as an "artistic room" where crewel embroideries from the designs of the artists employed by the School were carried out. The female workers put in an eight-hour day, later reduced to seven, and were paid on a sliding scale from 10 pennies an hour for the most skilled work to 4 pennies an hour for the least skilled. Both William Morris and Walter Crane—who referred to textiles as "the most intimate of the arts of design" because of their close association with daily life—were closely involved with the School, as, glancingly, were Edward Burne-Jones, Val Prinsep, G.F. Bodley, Sir Frederick Leighton, Alexander Fisher, and Selwyn Image. Crane's wife Mary was, in common with most women of her class, a skilled amateur embroiderer and it is probable that his introduction to textiles came from her. He was soon enthused and produced a number of striking designs, including the large and distinguished embroideries that hung at the entrance to the School's stand at the Philadelphia Centennial Exposition of 1876 and had such a powerful impact on Candace Wheeler and Louis Comfort Tiffany.

In 1875 Morris teamed up with Thomas Wardle, a silk dyer with works in Leek in Staffordshire. The two men collaborated on patterns and color schemes, consulting old herbals and reviving Elizabethan recipes for vegetable dyestuffs. Soon the Firm was offering a printed embroidery "kit" of ready-traced patterns for sale and commissions for wall-hangings mounted steadily. Morris's increasingly sophisticated embroideries were designed to be worked by the ladies of a household. Lady Margaret Bell and her daughters Florence and Ada spent eight years executing his designs for

Right: *"Artichoke," designed by William Morris and embroidered in silk by Margaret Beale and her three eldest daughters, c. 1896, for the north bedroom at Standen,*

the wall-hangings he provided for Rounton Grange in Yorkshire, which was designed and built by his old friend Philip Webb in 1872–6. Similarly, Madeline Wyndham, whose husband Percy had just commissioned Philip Webb to design Clouds House for them, executed an elaborate embroidered curtain for a bookcase that was also exhibited at the Philadelphia Centennial Exposition. Mrs. Alexander Ionides and her female relatives embroidered curtains to Morris's design for her London home at 1 Holland Park. Her neighbor Margaret Beale, whose husband James commissioned a weekend country house from Philip Webb—Standen—that would enshrine all that was best in Arts and Crafts, was a fine needlewoman and embroidered one of Morris's best early hangings, a repeating pattern of lotus blossom in subtle Morris shades of peach and brown. Over a number of years, Mrs. Beale and her daughters filled Standen with embroidered versions of several Morris designs, including "Pomegranate," "Vine," and "Artichoke" (which now hangs in the north bedroom).

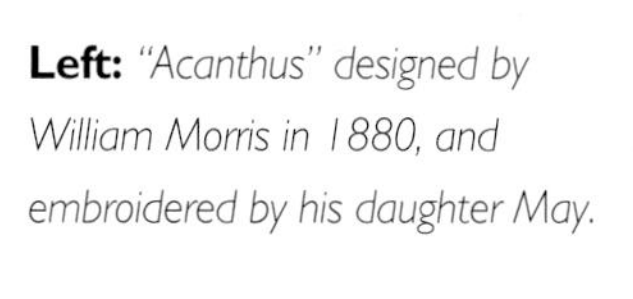

Left: *"Acanthus" designed by William Morris in 1880, and embroidered by his daughter May.*

Above: *The bed hangings surrounding William Morris's bed at Kelmscott Manor were designed and worked by his daughter May, assisted by Ellen Wright and Lily Yeats, c. 1893. Morris's own poem "On the Bed at Kelmscott" appears as an inscription on the valance. His wife, Jane, embroidered the bed cover.*

In 1885 Morris handed over the management of the embroidery section to his twenty-three-year-old daughter May, whom Henry James had discovered embroidering at Queen Square, alongside her sister Jenny, when the two girls were just seven and eight years of age. May recruited girls from the local Hammersmith school as apprentices and was also assisted by W.B. Yeats's sister Lily, Mary De Morgan (sister of the potter), and George Jack's wife. The work was carried out in May's drawing room at 8 Hammersmith Terrace, a short walk away from her parents' Kelmscott House, and was displayed and sold through the Firm's store in Oxford Street. In 1894 an "Orchard" portière, executed for the firm by Mrs. Theodosia Middlemore, was sold for £48, which represented about fourteen-and-a-half weeks' work. The previous year May had published her seminal *Decorative Needlework* and in 1910 she undertook a lecture tour of the USA.

The success of the Royal School of Art Needlework paved the way for pioneering regional organizations. Thomas Wardle's wife, Elizabeth, founded the Leek Embroidery Society in Staffordshire in 1879, which soon gained a reputation for the quality of its rich ecclesiastical embroideries on tussore silk, brocade, and velvet. In 1885 thirty-five members of the Society collaborated in an ambitious scheme to make a full-sized facsimile of the Bayeux Tapestry, which, once completed, went on tour and was eventually bought by the Reading Museum and Art Gallery. In contrast, in Haslemere in Surrey, the members of Godfrey Blount's Peasant Art Society were

using hand-woven linens, vegetable-dyed and appliquéd, to create simple but effective hangings in the traditional Arts and Crafts manner. The focus of the Arts and Crafts movement on rural and folk crafts precipitated a revival of interest in a number of traditional techniques, including smocking, patchwork, and quilting. Walter Crane, a leading practitioner, also designed clothes for himself and his family. His children were dressed according to the principles of "utility, simplicity, picturesqueness" in a sort of peasanty artistic costume. This form of "aesthetic dress" had become popular among an enlightened group of women who refused to wear the tight uncomfortable clothes of the period and favored instead loosely flowing gowns, embroidered in crewel wools and smocked at the yoke, waist, and sleeves. The Rational Dress Society was founded in 1881, encouraging women literally to loosen their stays and adopt "a style of dress based upon considerations of health, comfort, and beauty."

The art of embroidery was raised to such a high standard that a number of leading designers, architects, and artists prominent in the Arts and Crafts movement designed for the medium at some point in their careers. C.F.A. Voysey and M.H. Baillie Scott designed textiles and wrote on the techniques of embroidery. Baillie Scott employed the innovatory technique of using different colors and textures together in his designs, often for appliqué, with satin stitches and metal threads. Walter Crane, Christopher Dresser, and Lewis F. Day all wrote books on repeating design. From 1881 Lewis F. Day was artistic director of the Lancashire furnishing fabric firm of Turnbull & Stockdale and in *Art in Needlework* (1900) he wrote: "Embroidery is often thought of as an idle accomplishment. It is more than that. At the very least it is a handicraft; at the best an art." Ernest Gimson's varied activities also included embroidery designs, drawn from nature in rhythmic flowing patterns. His designs for washstand runners and samplers were often executed in white silk on white linen by his wife, sister, and sisters-in-law.

In Dublin, Evelyn Gleeson founded the Dun Emer Guild in 1902, as a focus for home industries, including lacemaking and embroidery, employing indigenous traditions and rich colors and textures.

Meanwhile, in Birmingham, Mary Newill was teaching embroidery in the new Art School's "art laboratories" alongside painters and illustrators such as Arthur Gaskin, Henry Payne, and Charles Gere. Inspired by the writings of Ruskin and the work of the Pre-Raphaelites, the Birmingham Group, as they came to be known, illustrated late-romantic poetry and medieval romances with their mural decorations, book illustrations, enameling and embroideries, these last often employing simple, homely materials. Like Morris and the Guild of Handicraft they were concerned with decorative unity.

Real innovation, however, was to be found in the work of Scottish and Irish designers. Women such as Phoebe Traquair, a skilled needlewoman who was also an enameler, bookbinder, and muralist, and Jessie Newbery, a teacher of embroidery at the Glasgow School of Art, led the field. Phoebe Traquair's ambitious "Denys" series of four gold and silk embroidered screens depicting allegorical figures was shown at the Arts and Crafts Exhibition Society in 1903. Other pieces, reflecting her interest in myth and legend, nature and history, were exhibited in Europe and at the St. Louis International Exhibition in America. Jessie Newbery's teaching style emphasized design over technique. "I specifically aim at beautifully shaped spaces," she wrote, "and try to make them as important as the patterns." She had three particularly gifted students in the Macdonald sisters, Frances and Margaret, and Ann Macbeth. The young students designed their own distinctive clothes

Below: *An embroidered cream silk panel from a lampshade with beads, ribbons, and braids, by Margaret Macdonald Mackintosh, wife of Charles Rennie Mackintosh, c. 1903.*

Left: *St. George and the Dragon textile, 1905, by the extraordinarily talented Phoebe Traquair, a notable example of a woman attaining recognition and prestige in the Arts and Crafts movement.*

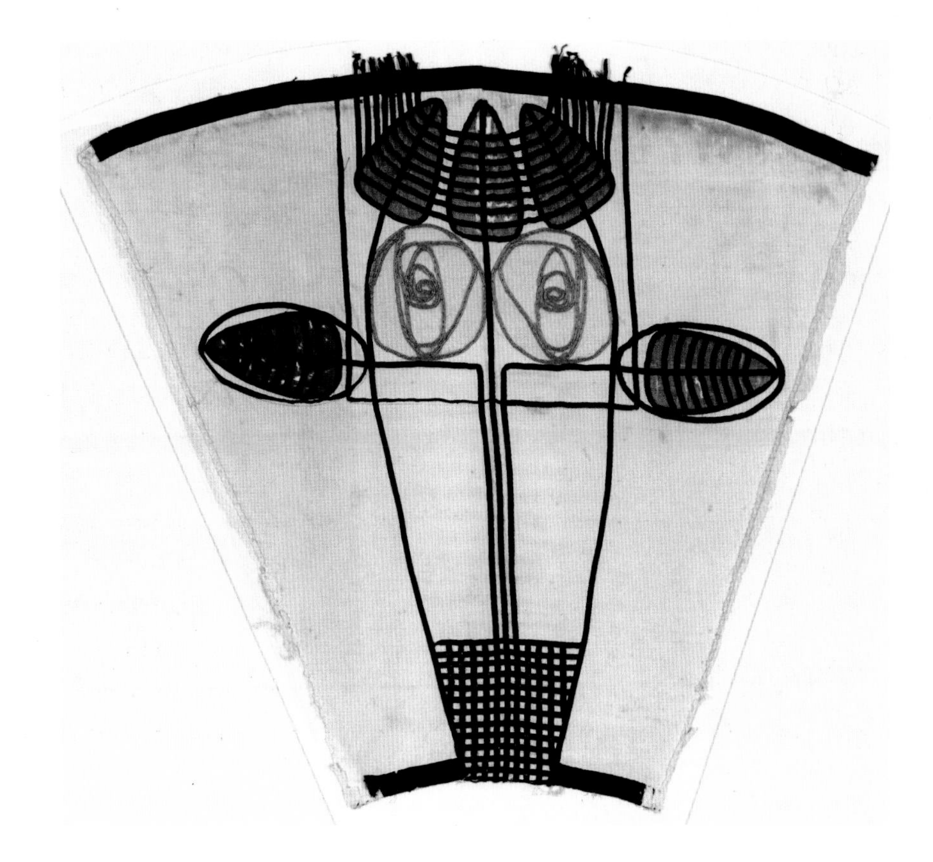

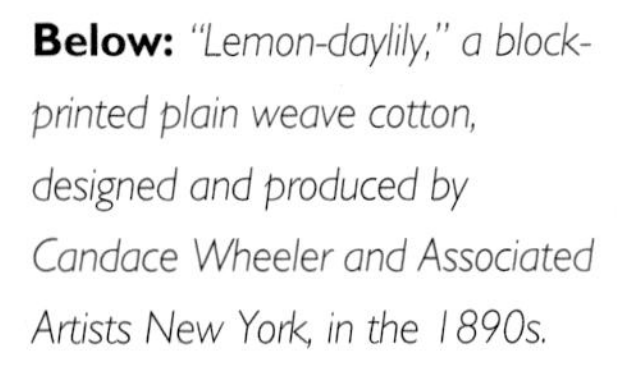

Below: *"Lemon-daylily," a block-printed plain weave cotton, designed and produced by Candace Wheeler and Associated Artists New York, in the 1890s.*

and worked with homely inexpensive fabrics such as burlap, flannel, linen, and unbleached calico to produce practical but stunning items—bags, belts, collars, casement curtains, and cushion covers—in what became known as the "Glasgow" style. They popularized the use of appliqué to fill in large areas of color and used soft silvers, pearly grays, pinks, and lilacs for their bold floral motifs, achieved with the simplest possible stitching. Ann Macbeth went on to teach at the Glasgow School of Art, where she established an embroidery class for children, with the aim of instilling in them a feeling for color and design from an early age. She executed a number of ecclesiastical commissions, including an embroidered panel depicting St. Elizabeth and the altar frontal for St. Mary's Cathedral, Glasgow. She also designed for Liberty's and, in 1911, published the influential instructional manual *Educational Needlecraft*. Another important Scottish school of embroidery was started by Lady Lilian Wemyss and Miss Wemyss to give occupation to the poor in the East Fife area. The Wemyss Castle School carried out all types of embroidery and became a flourishing concern.

The work of Louise Lessore attracted particular notice. A year after marrying Sidney Barnsley's friend the ceramicist Alfred Powell, she designed the altar frontal for E.S. Prior's St. Andrew's church at Roker in Sunderland. The *Art Journal* reviewed the work in 1907, stating: "What she has done beautifully with her needle is part of a unity of design and execution in which the crafts of the weaver, the embroiderer, the wood-carver, metal-worker, and artist in stained glass have been freely employed.... In the work of the loom and of the needle, beauty and grace of the living flowers of the earth are translated into fair and happy art."

The impact of the Royal School of Art Needlework exhibit at the Philadelphia Centennial Exposition in 1876 sparked an interest in "art needlework" in the USA and prompted a revival of embroidery as a fine art. An ever-growing privileged and leisured class of American women—sometimes referred to as the "picnic generation"—with time and money on their hands, traveled to England on vacation and took classes at the Royal School in Kensington, where they also studied the embroideries in the South Kensington Museum. There was a growing interest in the development of the Aesthetic movement and the first comprehensive exhibition of Japanese art in the USA in 1876 was one of the prompts that led Candace Wheeler to found the Society of Decorative Art in 1877. The first branch, established in New York City as an "American Kensington School," was quickly followed by others in Boston, Philadelphia, and Chicago. Mrs. Oliver Wendell Holmes Jr., the daughter-in-law of the poet, was one of the most prominent needlewomen in Boston, famed for her landscape panels embroidered in silk on silk, which she exhibited through the Decorative Art Society as well as in a picture gallery in Boston. Members of the Society exhibited regularly and offered cash prizes for the best designs submitted for window-hangings, screens, portières, and table covers. In 1881 the total prize money topped $3,000. The Society strove to raise the standards of decorative

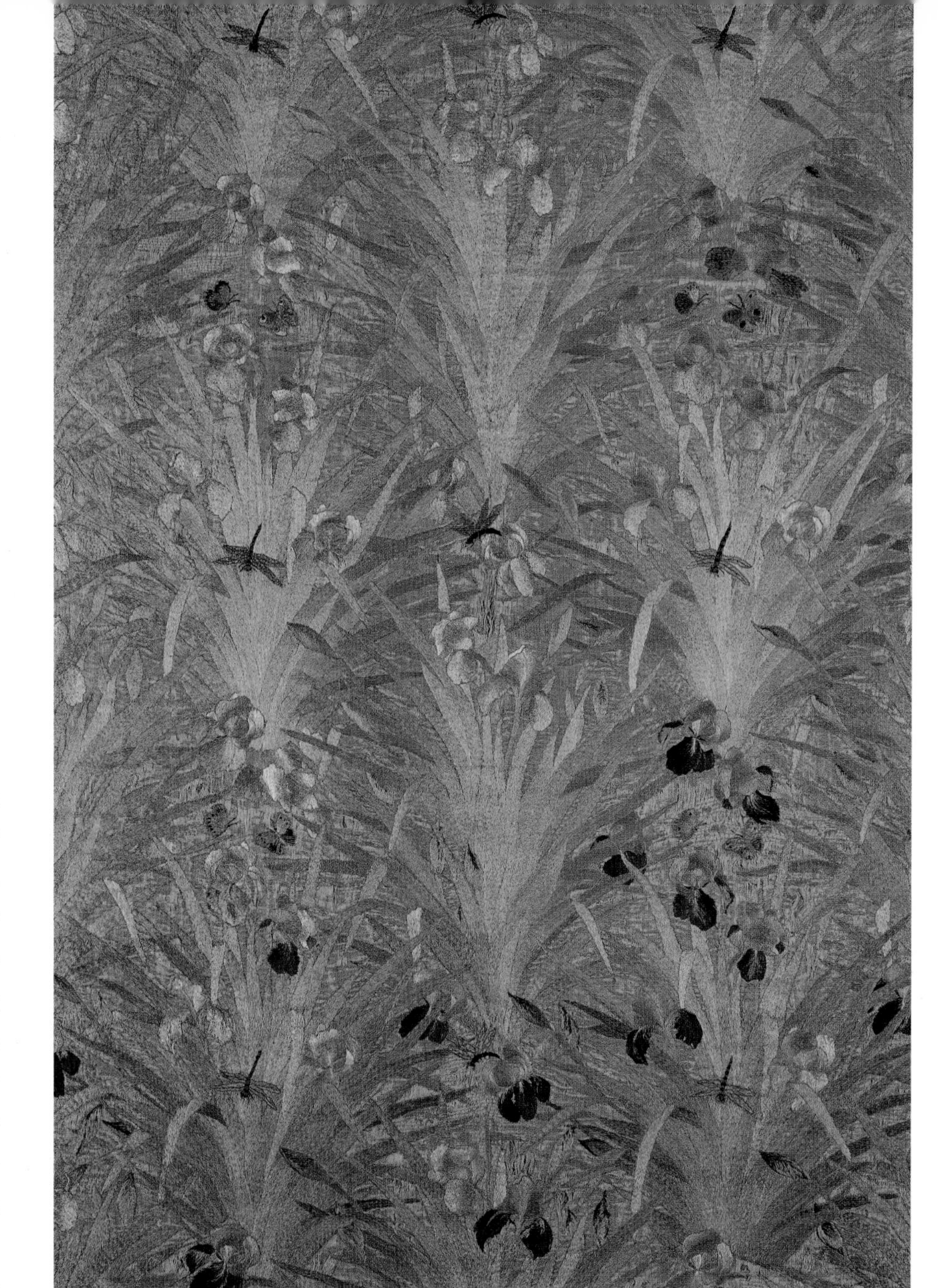

arts and promote a better appreciation of them. It wanted to create a new awareness of the role of interior design and provide middle-class women with both a creative outlet and a respectable means of earning money. "A woman who painted pictures, or even china, or who made artistic embroideries, might sell them without being absolutely shut out from the circle in which she was born and had been reared," wrote Candace Wheeler, in her autobiography *Yesterdays in a Busy Life*, adding the proviso however that "she must not supply things of utility—that was a Brahmanical law."

In 1879 Candace Wheeler joined forces with Louis Tiffany, George Coleman, and Lockwood de Forest to form Louis Comfort Tiffany & the Associated Artists, in which she was responsible for the execution of textiles. "We are going after the money there is in art, but the art is there all the same," Tiffany wrote. In 1882–3 they were commissioned to decorate the White House for President Chester Arthur. For a time they became very fashionable and their Fourth Avenue studios were visited by the great and the good, including Oscar Wilde, Ellen Terry, and Henry Irving. Other clients included Samuel Clemens (Mark Twain) and the British actress Lillie Langtry, for whom they made a silken bed canopy "with loops of full-blown, sunset colored roses" and a coverlet of "the delicatest shade of rose-pink satin, sprinkled plentifully with rose petals fallen from the wreaths above."

Above: *An embroidered silk portière, with glass beads and sequins, designed by Candace Wheeler, c. 1884, and made by Cheney Brothers at Hartford, Connecticut.*

In 1896 the Society of Blue and White Needlework was founded in Deerfield, Massachusetts, to preserve early embroidery techniques. Local women were trained by Ellen Miller and Margaret Whiting to execute designs adapted from examples in the local historical society on tablecloths, bed curtains, and covers. Though most of their yarn was blue and white they experimented with natural dyes (chemical dyes were never used) to produce further colors and exhibited their work at Arts and Crafts Society exhibitions in New York, Chicago, and Boston. They even experimented with eastern influences: "Rub Oriental art through the Puritan sieve," explained Margaret Whiting, "and how odd is the result; how charming and how individual."

Above: *"Seaweed and Dragonflies" embroidered table cover designed and made by the Society of Blue & White Needlework, Deerfield, Massachusetts, c. 1915.*

In keeping with his unified approach to design, Frank Lloyd Wright took a keen interest in the different textiles used in his houses. He favored simple, natural finishes—linens, cottons, and wools in flat weaves or fine velvets and leather for covering chairs. Pattern, if used at all, was geometric. His first wife was an excellent needlewoman and he included custom-designed linens, geometric in design, in several of his early Prairie-style houses. Carpets and floor coverings would be similarly simple and made from natural fibers.

In Austria and Germany hand-printed textiles, among them silks designed by Lotte Frömmel-Fochler and Carl Otto Czeschka, brocades by Josef Hoffmann, and hand-printed linens by Josef Zotti, along with bead bags produced at the workshops of the Wiener Werkstätte, were considered the height of fashion. Meanwhile, in Scandinavia, a revival of folk art and weaving techniques was in full swing. The designer Frieda Hansen, best known for her large woven wall-hangings depicting stylized flowers and motifs taken from Norwegian sagas, founded the Norwegian Tapestry Weaving Studio in Oslo in 1897, and Fanny Churberg set up the Friends of Finnish Handicrafts along Morrisian lines with the aim of preserving peasant traditions in textiles and embroidery.

WOVEN TEXTILES

IN 1881 WILLIAM MORRIS acquired the Merton Abbey Tapestry Works in Surrey and opened his own factory on the banks of the River Wandle, which supplied the copious amounts of water required for the business of madder and indigo dyeing. There he reinstated and perfected the technique of indigo discharge printing, creating some of his best-known block-printed textiles, including "Strawberry Thief," "Evenlode," and "Bird and Anemone." He also experimented with dyes. A young American visitor, Emma Lazarus, captured the atmosphere of the dye-house at Merton in an article entitled "A Day in Surrey with William Morris," which appeared in *The Century Illustrated Magazine* in July 1886:

Below: *William Morris enlisted the help of Philip Webb, who drew the birds in this design for a furnishing fabric entitled "Strawberry Thief." First produced in 1883, it went on to become one of Morris's most successful creations.*

"In the first out-house that we entered stood great vats of liquid dye into which some skeins of unbleached wool were dipped for our amusement; as they were brought dripping forth, they appeared of a sea-green color, but after a few minutes' exposure to the air, they settled into a fast, dusky blue. Scrupulous neatness and order reigned everywhere in the establishment; pleasant smells of dried herbs exhaled from clean vegetable dyes, blent with the wholesome odors of grass and flowers and sunny summer warmth that freely circulated through open doors and windows."

At Merton Morris was able to continue production of his woven and less expensive printed textiles. Designs such as "Small Stem," "Large Stem," and "Coiling Trail" were printed on a fine wool and used for curtains, while the glazed cotton design "Jasmine Trellis" was suitable for lightweight curtains and loose chair covers. In fine weather the brightly colored cottons were laid out to dry in the meadow behind the workshop. Morris spent long hours in the South Kensington Museum (now the Victoria and Albert Museum) sifting through its growing collection of historic textiles and

Left: *Morris & Co. employees printing chintz with hand blocks at Merton Abbey, where the Firm's workshop relocated in 1881.*

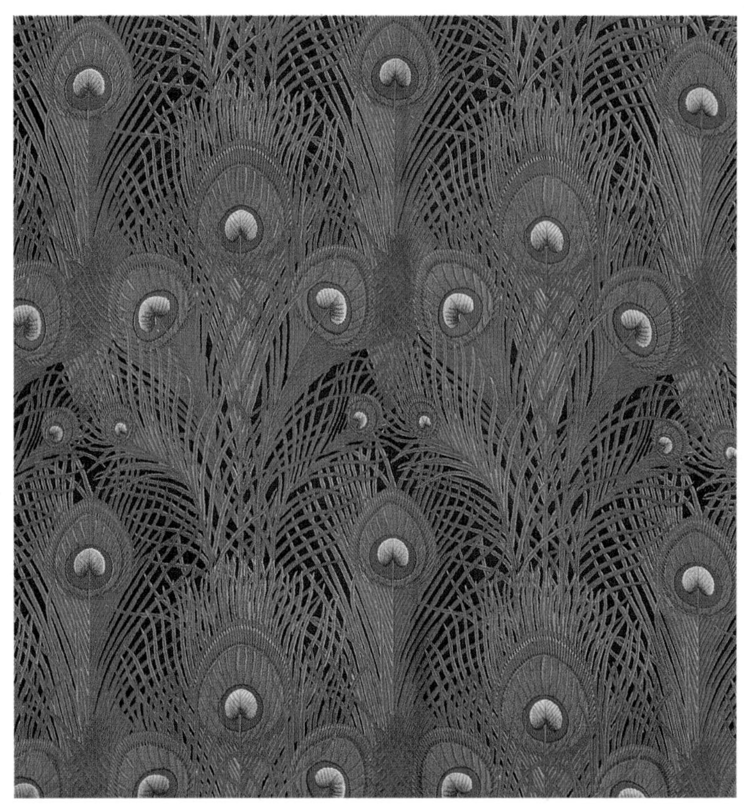

drawing inspiration from early Rhenish examples and, in particular, those from Persia, Turkey, and Italy. His great skill lay in the repeating pattern and early designs such as "Acanthus," "African Marigold," and "Honeysuckle" make clever use of a turnover or "mirror" repeat. Some fine designs were provided for the Firm by Henry Dearle, who had joined Morris as a boy of eighteen and so immersed himself in his master's style that for a while some important panels, portières, and screens by him were wrongly considered to have been the work of Morris. Dearle's finest tapestry, known as "Greenery," was woven in 1892 and bought by Percy Wyndham for the hall at Clouds. A second version of it now hangs in the Metropolitan Museum, New York.

Walter Crane's most successful textile designs, both critically and commercially, were for weaving. He enjoyed the challenge posed by the technical limitations and wrote elegantly in *Line and Form* (1900) about the difficulties of producing "curves by small successive angles," admitting that a certain "squareness of mass becomes a desirable and characteristic feature." Morris & Co. reproduced his design for "The Goose Girl" in traditional arras tapestry in 1881.

Above left: *A woven wool hanging designed by C.F.A. Voysey and manufactured by Alexander Morton & Co., c. 1897.*

Above right: *"Peacock Feathers," a roller-printed cotton, designed by Arthur Silver of the Silver Studios and printed by Rossendale Printers for Liberty & Co., c. 1887.*

Above: *The "Bullerswood" carpet by William Morris, c. 1880.*

CARPETS

William Morris embarked on his first hand-tufted carpet on a loom in a back attic at Queen Square in 1878. He was pleased with the result, and the loom moved with him to Kelmscott House on the bank of the Thames at Hammersmith. It was installed in the coach-house for a couple of years before the whole carpet-making operation was moved to "a long cheerful room" in the Merton Abbey works. Taking Persian carpets as a starting point, Morris created designs, as for his other textiles, based on repeated patterns drawn from nature. He had, however, been experimenting with designs for machine-made carpets before this date and had registered his first two designs on Christmas Eve 1875.

Carpets were not fitted right up to the skirting as is normal today but laid over plain or stained floorboards, and Morris's designs were composed in much the same way, with a central pattern or "field" enclosed by a decorative border. His most inexpensive machine-woven carpets (retailing at

Right: *This large Morris carpet was among the many textiles and wallpapers designed by William Morris that were used in the Glessner House in Chicago.*

approximately 4s per square yard, rather than 4 guineas for a hand-knotted carpet) were intended for use on stairs, in bedrooms, and in the homes of the less well off. They were made for Morris & Co. by the Wilton Royal Carpet Factory, who also produced Axminster and Kidderminster carpets in William Morris designs, and they proved very popular. "Wiltons must be classed as the best kind of machine-woven carpets," ran the sales pitch in the Morris & Co. brochure for the Boston Foreign Fair in 1883. "The patterns they bear are somewhat controlled as to size and colour by the capability of the machine.... If well made the material is very durable, and by skilful treatment in the designing, the restrictions as to colour are not noticeable." They were snapped up in the USA.

Below: *"Tulip and Lily" Kidderminster carpet from Morris & Co., c. 1875.*

The hand-knotted "Hammersmith" carpets, on the other hand, marked with a letter "M" with a hammer and waves to represent the Thames, were individual works of arts on a par with the Firm's richest embroidered hangings and tapestries. Owners—who included wealthy industrialists like Sir Isaac Lowthian Bell, aristocrats such as George Howard and the Hon. Percy Wyndham, and wealthy American clients such as John and Frances Glessner—often hung them on walls rather than laying them on the floor. The Hammersmith carpet Morris designed for the drawing room at Clouds was put up for sale in 1932 when Dick Wyndham disposed of the house and its contents. It did not find a buyer immediately, but soon after it was exhibited at the Victoria and Albert Museum it was acquired by Cambridge University. It still graces the floor of the Combination Room in the Old Schools.

In the quest for decorative unity, Morris's carpets, whether rich or simple, "expressed the proportions of the room." They were intended to harmonize with the Morris & Co. fabric used to cover the chairs and for the curtains: the whole effect was a return to Crane's ideal of "rich and suggestive surface decoration."

Handmade carpets were also designed by A.H. Mackmurdo, whose Century Guild wall rugs were sold through Morris & Co. and advertised through the Guild's magazine *The Hobby Horse*. "There is room for the highest qualities in the pattern of a carpet," wrote Walter Crane in 1887, "... the sincere designer and craftsman ... with his invention and skill applied to the

Above: *This "Tree of Life" carpet was designed by Charles Sumner Greene, 1908–9, to go in the fireplace nook of the living room in the Gamble House.*

accessories of everyday life may do more to keep alive the sense of beauty than the greatest painter that ever lived." Liberty's sold carpets machine-made by the Carlisle carpet and textile firm Alexander Morton & Co. from designs by C.F.A. Voysey and Lindsay Butterfield.

In the USA, Arts and Crafts designs were used on carpets produced by the Bigelow Carpet Co. of Massachusetts. Architects such as George Grant Elmslie and Frank Lloyd Wright designed rugs and carpets for a number of their houses, notably in Elmslie's case the Henry B. Babson house, and in Wright's the Dana-Thomas, Robie, May, Coonley, and Hollyhock residences. Elmslie used beige, rust, and brown backgrounds to set off pale green and turquoise Arts and Crafts motifs, while Wright used solid, earthy tones enlivened by small amounts of geometric pattern, chevrons, and linear borders, designed to continue the other motifs and splashes of color already assembled in his interiors. Oriental rugs were used throughout most of the Gamble House but Charles Sumner Greene did design five custom rugs for the living room, including the fireplace nook, which continued the "Tree of Life" motif used in the stained glass of the entrance, though somewhat abstracted into a fork-like pattern. The colors ranged from shades of olive green and brown through blues and ocher with splashes of rose and mauve.

WALLPAPER

THE MASS PRODUCTION of wallpapers had begun in Britain in the 1840s, revolutionizing interior wall decoration. Up to this point, walls were usually painted, sometimes marbled or grained to look like wood or stenciled and enlivened with a decorative border. Fabric such as watered silk or printed damask was used a good deal on the walls of wealthier homes, but once wallpaper began to be commercially produced by firms such as Jeffrey & Co. it was soon seen on the walls of most Victorian households. By 1860, new technological advancements had pushed production up to nineteen million rolls of wallpaper a year. The publication of Owen Jones's influential book *The Grammar of Ornament* in 1856 had had a huge impact. It became a pattern book for British taste and design, and Gothic, Oriental, Moorish, and classical motifs began to appear on wallpapers, along with the diamond, or diaper, motif used most famously by Augustus Pugin on his wallpaper designs for the Houses of Parliament.

It was the publication of Charles Eastlake's influential *Hints on Household Taste* in 1868 (an American bestseller when published there in 1872, running to seven editions), that promoted a single, cohesive style in the home and paved the way for the Arts and Crafts style that was following

Left: *Jeffrey & Co. printed William Morris's "Willow Bough" wallpapers, which were sold through Morris & Co. from 1887.*

fast behind. As ever, Morris & Co. was shaping the public taste. William Morris's flat, rhythmical, handprinted patterns, based on medieval motifs and on nature, were strikingly fresh and modern. "Wallpapers," he explained in a lecture on pattern designing, "must operate within a little depth. There must be a slight illusion—not as to the forms of the motif, but as to relative depth." He demonstrated this in "Daisy," for example, where the pattern is pleasantly balanced and clearly ornamental, with no recession and no imitation of accidental details. This wallpaper, along with other early designs such as "Fruit" and "Trellis," was inspired by the flowers and fruit found in his garden at Red House (just as his later long-leaved "Willow," which became one of his most enduringly popular decorative designs, was inspired by the willow-bordered river that ran through his country property, Kelmscott Manor).

Between 1862 and 1896 the Firm produced a vast range of patterns in warm, subtle colors—sage green, rusts, ochers, plums, peacock blue, and gold—achieved by avoiding the new chemical methods

Right: *William Morris's "Acanthus" wallpaper design, 1875, was hand printed using wood blocks.*

in favor of traditional vegetable dyes. They proved immensely popular with the public: Margaret Beale ordered all the wallpapers for Standen from Morris & Co., which also supplied a fair number of chintzes for hangings and curtains, hanging "Daffodil" in the morning room and "African Marigold" and "Severn" elsewhere in the house, and Morris's papers were eagerly pasted onto the walls of Richard Norman Shaw's Bedford Park estate houses by their artistic occupants. "Do not be afraid of large patterns," Morris counseled in 1888, "if properly designed they are more restful to the eye than small ones: on the whole, a pattern where the structure is large and the details much broken up is the most useful ... very small rooms, as well as very large ones, look better ornamented with large patterns."

Another important contributor to the field was Walter Crane, who provided elaborate wallpaper designs for Jeffrey & Co., the firm Morris had entrusted with the printing of his first wallpaper designs in 1864, and which had, a year before, printed Owen Jones's series of elaborate papers designed for the Viceroy's Palace in Cairo. Crane was commissioned by Jeffrey & Co.'s entrepreneurial director Metford Warner, "a man of taste

Right: *This repeating pattern of heavily stylized irises was designed by Walter Crane as a wallpaper frieze in 1877.*

and judgment, who," according to Crane, spared "no pains to get the proper effect of a pattern." He was certainly innovative and commissioned designs for papers from E.W. Godwin (who took his inspiration from a Japanese silk patterned with a flowering bamboo motif), William Burges, B.J. Talbert (who also contributed stylized Anglo-Japanese designs), Albert Moore, and Charles Eastlake—none of whom had designed wallpaper before. But it was Crane's work, more than that of any other designer, that established the reputation of Jeffrey & Co. as high-quality manufacturers of wallpapers, friezes, nursery papers, and figurative panels. Crane's designs attracted a great deal of attention from both the popular and the trade press and won awards at international exhibitions in London and Paris, though his wallpapers, like Morris's, were sometimes criticized for being too assertive. As Lewis F. Day retorted in 1903: "I am not sure that I want anyone's personality to call out to me from the walls and the floor of my room." Another critic complained that Crane's

wallpapers were "not retiring enough, they dominate the room by their richness and importance." Crane, however, refused to bow to "the fluctuating harlequin of fashion and trade" and produce "vulgar, commercial work" and some of his hand-printed wallpaper designs, such as "Cockatoo" and "Pomegranate," are beautifully balanced, stunningly colored pieces, though a little would certainly go a long way.

Above: *A typically elaborated "Swan, Rush, & Iris" design for dado wallpaper by Walter Crane, 1875.*

In the middle of a long-running debate raging at this time about whether or not to include animals and birds on wallpapers, Walter Crane cheerfully included human shapes in his friezes and nursery papers and animate forms from macaws to peacocks in many of his wallpapers, although even he conceded that it would be "out of the question to hang pictures on a wall papered with the 'Peacock Garden'."

"The introduction of any members of the Animal Kingdom in wallpapers," thundered the *Building News* of 1872, "involving as it necessarily does, such amount of repetition, is always dangerous. Even

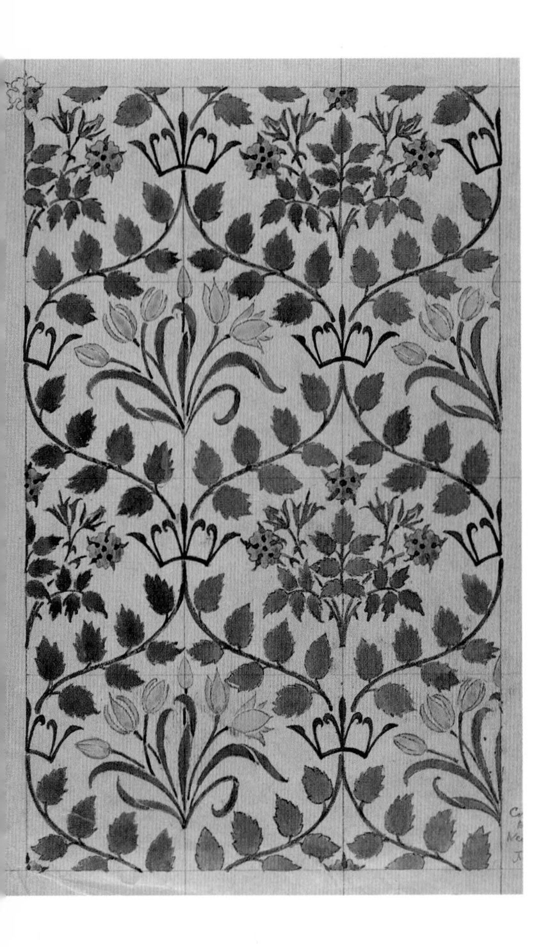

Above: *A watercolor and pencil design by C.F.A. Voysey for a textile or wallpaper, 1919. Exceedingly prolific, Voysey was producing designs for almost fifty years, a period longer than the designing life of William Morris.*

birds, which are the least objectionable, have their difficulties." Morris, with one early exception, omitted birds or small animals from his wallpaper designs, unlike C.F.A. Voysey who, as *The Studio* observed, was "inclined to admit plants and beasts in patterns on condition that they were reduced to mere symbols." This ensured that he arrived at patterns that were happily near to nature and at the same time full of decorative charm. Nikolaus Pevsner, who compared Voysey's wallpapers and chintzes with Morris's "Honeysuckle" to demonstrate the decisive step away from nineteenth-century "historicism," claimed for Voysey a new world of light and youth: "The graceful shapes of birds flying, drifting, or resting, and of tree-tops, with or without leaves, are favorite motives ... and there is an unmistakable kindliness in his childlike stylized trees and affectionately portrayed birds."

In *Hints on Household Taste* Charles Eastlake had popularized a decorative scheme that divided the wall into three horizontal sections: frieze, wallpaper above the dado line, and a darker treatment—including, after 1890, "Anaglypta," an embossed paper devised by Frederick Walton, the inventor of linoleum—to the areas subject to wear and tear below the dado. Walter Crane's wallpaper designs conformed to this pattern initially, but by 1886 he had begun to pioneer the division of the wall into only two parts. "It is usual to accompany the field of the wallpaper with a special frieze and a dado making a complete wall decoration," he wrote in a lecture on applied design. "Although I have made many designs for both, I have come to the conclusion that most rooms look best with the main pattern of the field carried from the skirting to the frieze."

Crane offered general advice on home decoration in his book *Ideals in Art: papers theoretical, practical, critical* (1905) and made specific recommendations for the use of his wallpapers. "Lily" he described as "useful in halls and passages," while "Rose Bush" "would be appropriate for a drawing or living room." But, since the majority of his designs were hand-printed and therefore costly, his advice was presumably directed at the rich and fashionable who would be the only ones able to implement it. It is an irony that the socialist Crane, like Morris, catered to the rich rather than the masses. Prices of wallpapers in 1875 ranged from between 3d and 9d a roll for the servant's quarters, up to 7s 6d to 15s a roll for hand-printed papers for the drawing room.

The first port of call for the artistically minded masses would have been Liberty & Co., which sold designs by Archibald Knox, Harry Napper, M.H. Baillie Scott, Arthur Silver, Lindsay Butterfield, and Voysey. Voysey also designed for Essex & Co., and proved so popular that *The Studio* claimed his name was to wallpaper as "Wellington" was to the boot. He was well known in Europe, where his designs were reproduced and his influence was widespread.

Morris & Co. papers were widely available in the USA (a combination of Morris wallpapers and fabrics can still be seen in the master bedroom of the Glessner House in Chicago), though popular designs were also provided by L.C. Tiffany, Christian Herter, and Candace Wheeler, who, with her

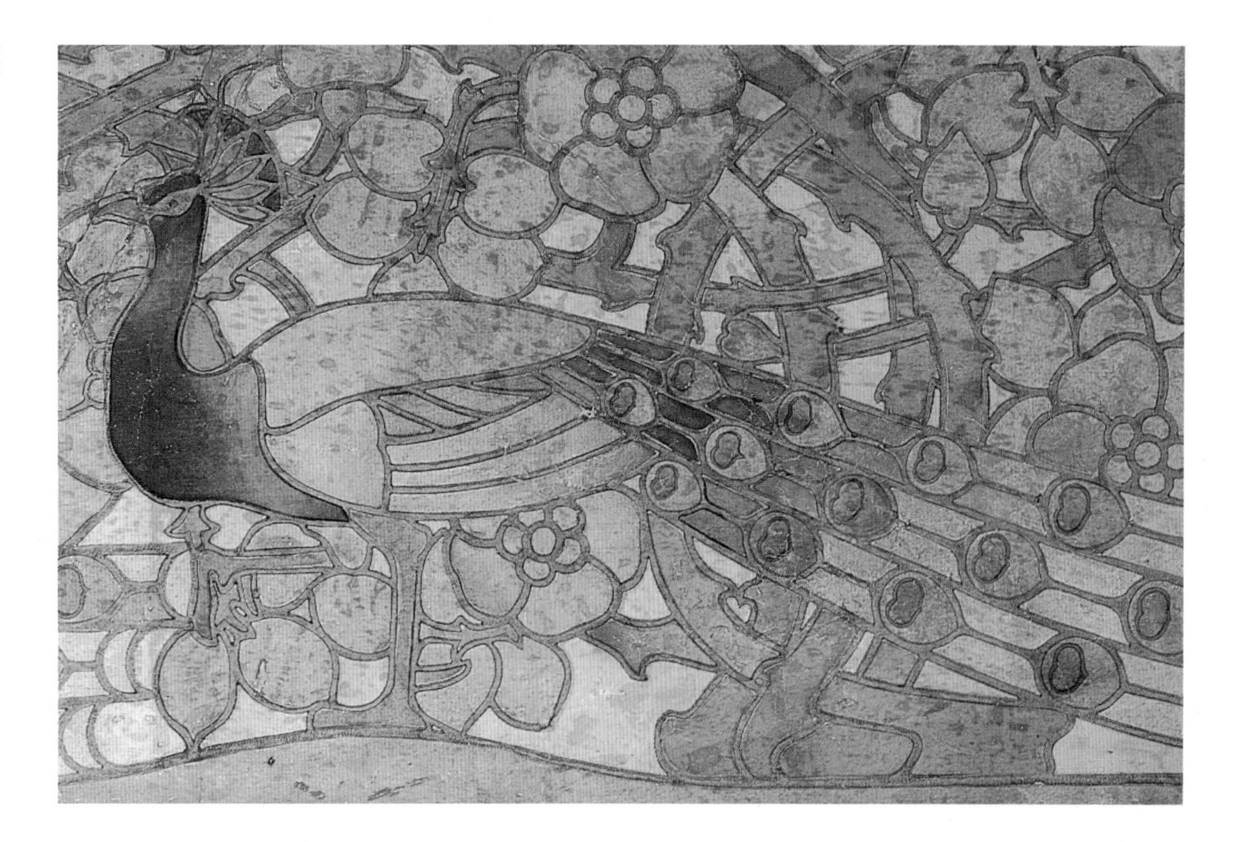

Above: *This "Peacock Frieze" wallpaper, used in the Main Hall at Blackwell, was designed by W. Dennington in 1900 and produced by Shand Kydd Limited.*

daughter Dora, won four prizes in a competition launched by the New York wallpaper manufacturing firm Warren, Fuller, & Lange in 1881. (Dora Wheeler's winning entry, "Peony," owed much to Morris.) A number of manufacturers, including the York Wall Paper Company, marketed patterns based on Morris's designs, as did M.H. Birge & Sons of Buffalo, New York—one of the leading producers of artistic wallpapers, although, as the century turned, the company moved toward a series of Art Nouveau designs with embossed and metallic effects. The craze for Art Nouveau that was sweeping Europe was evidenced in the wallpaper designs of Hector Guimard and Alphonse Mucha, but gradually a lighter, more geometric style prevailed, best exemplified by Josef Hoffmann.

The last decade of the nineteenth century saw a move away from intense patterns to a simpler style, more in keeping with the evolving Arts and Crafts home and best exemplified by Voysey's designs. Flowers, fruit, birds, and foliage were still the main motifs, but the forms were more delicate and the backgrounds lighter. Many Arts and Crafts homes—including Morris's Kelmscott Manor—did not have wallpaper at all, preferring to leave walls or wooden panels painted plain white. This was a decorative scheme favored by M.H. Baillie Scott, who hardly ever used wallpaper (although three elaborate designs by him appeared in *The Studio* in April 1895). Instead he recommended plain colors on the walls, though the ceilings might be stenciled and painted, and the oak boarding, doors, and window sills treated with beeswax polish. "When in doubt," he wrote, "whitewash might well be taken as a maxim to be followed in the decoration of the modern house."

STAINED GLASS & LIGHTING

Right: *The Firm's highly prestigious commission to design The Green Dining Room, 1866–7, for the South Kensington Museum came from the museum's progressive director, Henry Cole. Edward Burne-Jones designed the stained glass and the gilded panels, taking the months of the year as his theme.*

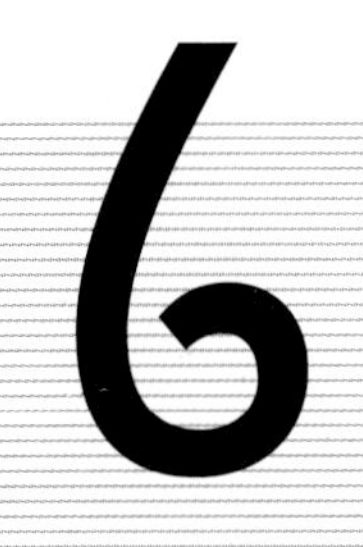

6

MAY DAY MAY DAY

TRANSLUCENT MOSAICS

"The worth of stained glass must mainly depend on the genuineness and spontaneity of the architecture it decorates: if that architecture is less than good, the stained-glass windows in it become a mere congeries of designs without unity of purpose."

William Morris

STAINED GLASS

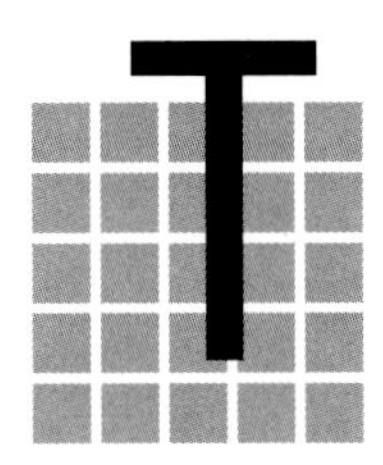

The early Arts and Crafts movement, under the influence of William Morris, did much to revive and popularize the essentially Gothic art of stained glass. Ecclesiastical glass reached its greatest beauty of color, design, and luminosity during the fourteenth century but the combination of the Gothic revival and the Arts and Crafts movement in the nineteenth resulted in a renewal of interest in the use of stained glass both in churches and domestic settings. Practitioners such as William Morris, Edward Burne-Jones, and Walter Crane were great enthusiasts. Indeed the early reputation of the Firm rested on its stained glass, though not its profits, as is made plain by this exasperated letter written by the Firm's manager Warington Taylor to Philip Webb in 1866: "Over £2,000 work in glass done. This should have returned at least 25% therefore over £500 profit. You know well enough there was not £200 profit on glass." Part of the problem was that the jobs were not costed properly and the design fees were not charged separately. The Firm's medal-winning entry at the 1862 International Exhibition in London had ensured a steady stream of commissions from ecclesiastical architects like G.F. Bodley, so Warington Taylor restructured the charging system, recommending a rate of £2 10s to £3 a foot for the glass, plus a separate fee for the original designs. These were most often provided by Edward Burne-Jones, although Dante Gabriel Rossetti provided thirty-six, and Ford

Above: *M.H. Baillie Scott gave careful thought to every detail of the design of Blackwell, 1898–1900, including the stained glass.*

Left: *The Great East Window in John Dando Sedding's great Holy Trinity Church in Sloane Street, London, by Edward Burne-Jones and William Morris, c. 1890.*

Above: *A minstrel angel with a dulcimer by William Morris, for St. Peter and St. Paul Church, Cattistock, Dorset, 1882. Morris employed a number of recurring themes in his work, including the "Legend of Good Women" and musical minstrels, both secular and, as in this case, angelic.*

Madox Brown over a hundred. Morris, William De Morgan, and Charles Faulkner also contributed designs.

The Firm received secular commissions for stained glass as well. The most important and valuable of these were from St. James's Palace for the decoration of the Armoury and Tapestry Room and the South Kensington Museum for the Green Dining Room. The latter provided a powerful and continuing public advertisement for their work.

Morris kept a rigorous control over the production of the Firm's stained glass, personally choosing the colors and vigilantly inspecting every piece of glass before the windows were made up. He was well acquainted with the large traceried windows of York Minster and Merton College Chapel at Oxford and, as a young man, had visited Chartres and the beautiful late thirteenth-century church of St. Urbain at Troyes, which he considered to be "the most satisfactory example of the art of glass-painting." The Firm employed its own glass painters, including George Campfield, a recruit from the Working Men's College who became foreman of the glass works, and a glazier named Holloway, who was taught how to paint directly onto the thick glass keeping as close as possible to the artist's original design. Morris used two types of glass: "pot-metal" glass, in which the coloring matter is fused with the glass when fired, and "flashed" or "ruby" glass, in which a white body is covered with a colored skin, a perilous procedure best performed by professionals like James Powell & Sons of Whitefriars, from whom Morris obtained all his ruby glass.

William Morris's early glass has a touching and haunting quality. It uses bold areas of pure, bright color in its depictions of figures of saints standing before elaborate backgrounds of hedges or screens of fruit-bearing saplings. Morris himself was often responsible for these backgrounds and for the tight clusters of daisies and violets at the feet of the saints. The inspiration is clearly fourteenth-century but it seemed so close to the Gothic ideal that it greatly alarmed the established trade, which set up a petition attacking the 1862 Exhibition windows and accusing Morris of having, in fact, just touched up and remounted genuine old glass. The petition came to nothing, however, and Morris & Co. continued to produce stained glass—in a characteristic free-flowing style, both in lead lines and painting, using strong vibrant colors, Gothic lettering, and naturalistic forms—until the Firm's dissolution in 1940. Of Morris's many

legacies, his stained glass, which can be found in churches and cathedrals up and down Britain and as far afield as India and Australia, is one of the greatest.

Outside the Firm, other important Arts and Crafts names in this field include Henry Holiday, who worked with the architect William Burges and started his own stained-glass studio in 1891, Walter Crane, who collaborated with James Silvester Sparrow, and Christopher Whall. Crane had much to say on the subject. Lead lines, for example, "ought to be fairly complete and agreeable as an arrangement of line even without the color," while color should be used boldly to create a "network of jeweled light," a "translucent mosaic" as he described it in his book *The Bases of Design* (1898), avoiding the use of white, since it led to "holes in the window." Crane's bold color arrangements and juxtaposition of dark and light pieces of glass create a rich overall effect. Crane's glass was manufactured by Messrs. Britten &

Above: *"Chaucer's Good Women: Chaucer Asleep," by Edward Burne-Jones for Morris, Marshall, Faulkner & Co., c. 1864.*

Above: *"Brownies" stained-glass panels from a series depicting fairy characters from a children's story by Mrs. Ewing, designed by Selwyn Image and made by Heaton, Butler & Bayne, early 1890s.*

Gibson, who also made E.S. Prior's "Early English" slab glass, which, unlike normal commercial glass, had an uneven surface. This was because it was blown into a flat mold and then cut into slabs that were naturally thicker at the center than at the edge, which lent a translucent, jewel-like effect, much sought out by Christopher Whall.

Whall, who taught architectural glass at the Central School of Arts and Crafts in London from 1896, was the leading glass designer of his generation and his book *Stained Glass Work* was considered the standard text on the subject. In the final chapter, he pays tribute to Edward Burne-Jones, stating: "To me there is no drapery more beautiful and appropriate for stained glass work in the whole world of art, ancient or modern, as that of Burne-Jones, and especially in his studies and drawings and cartoons for glass." Whall's influence was widespread. Alfred Child, who worked within the Arts and Crafts framework of Evelyn Gleeson's Dublin-based Dun Emer Guild, was a former Whall apprentice, and Henry Payne, though based in Birmingham, benefited from his advice. Whall's work inspired the American stained-glass artist Charles Connick, who was responsible for the windows in Ralph Adam Cram's All Saints Church, Ashmont, Massachusetts. Whall wanted to move away from stylized figures and urged designers to abandon the practice of reusing the same cartoon and instead draw from life and from the model.

As the Arts and Crafts movement developed in the 1890s other painters and designers, many members of the Century Guild, Art Workers' Guild, or other similar societies, undertook work in stained glass. At the Century Guild, stained-glass designs were provided by Clement Heaton and Selwyn Image; at the Art Workers' Guild by Lewis F. Day; in Chipping Campden, Ashbee could rely on Philippe Mairet and Paul Woodroffe, who later set up a stained-glass studio of his own at his house at Westington, Gloucestershire, which was reconstructed for him by Ashbee. Along with Frances Macdonald and Charles Rennie Mackintosh, another husband-and-wife team important in the field of stained glass would emerge from the Glasgow School of Art: Ernest Archibald Taylor, who married Jessie M. King, was very involved in stained-glass design. In Birmingham, historically a

center of stained-glass production, the artist Mary Newill was active. Her work, like that of the others in the Birmingham Group, shows the influence of the Pre-Raphaelites.

In the USA the influence of Morris & Co., which had exhibited at the huge Boston Foreign Fair of 1883, was felt in the work of John La Farge. La Farge was drawn to the work of the Pre-Raphaelites and Morris's flat-pattern-making approach to decoration, although it is the influence of the Japanese prints he collected that can be seen to have had the most lasting impact on the flat, asymmetrical flower designs he incorporated into his stained-glass windows. In 1875 he began to experiment with stained glass and developed a new style that he called American opalescent art glass (reportedly after seeing light pass through an inexpensive tooth-powder bottle). His invention inspired the technical experiments of his more famous successor, Louis Comfort Tiffany. La Farge designed magnificent windows for private clients, including the wealthy Vanderbilts and Whitneys in New York, and Sir

Above left: *An important leaded-glass window depicting peonies blown in the wind, with Kakemono border by John La Farge, 1893–1908. La Farge was influenced by Japanese prints and developed his opalescent glass at about the same time as Tiffany, his chief competitor, whose leaded and plated glass window showing parakeets and gold fish bowl of 1893 is shown* **right.**

Above: *A three-part leaded-glass window designed by Charles and Henry Greene for the Adelaide Tichenor House, Long Beach, California, c. 1904.*

Far right: *Philip Webb, who gave meticulous thought to every aspect of the design of Red House, designed and painted this leaded-glass window, c. 1860.*

Right: *A characteristic stained-glass detail from a cabinet door designed by Charles Rennie Mackintosh for the Glasgow School of Art, 1896–1909.*

Lawrence Alma-Tadema in London, and was responsible for the panels in H.H. Richardson's Trinity Church in Boston, his Crane Library in Quincy, Massachusetts, and the William Watts Sherman house in Newport, Rhode Island.

In Britain, the abolition of window tax in 1851 and the lifting of the duty on glass in 1857 had encouraged the use of larger panes of glass and revived bay and bow windows, although M.H. Baillie Scott still favored old glass over modern sheet glass since "it meets the eye with a friendly twinkle instead of a sullen glare, and the main beauty of all undulations, especially in polished surfaces, is that they given broken reflections instead of glare." But it was the increasing use of stained and art glass in domestic settings that characterized an Arts and Crafts home. There are jeweled windows in Morris's Philip Webb-designed Red House, and architects such as Baillie Scott and Charles Rennie Mackintosh designed front doors and cupboard doors with beautiful glass panels. In the USA, Frank Lloyd Wright and the Greene brothers designed striking leaded glass windows to complement and complete their houses. They were concerned to create not merely the shell of a building but its decorative arts as well. The elaborate stained glass "Tree of Life" triptych that fills the entrance to the Gamble House is typical of their work.

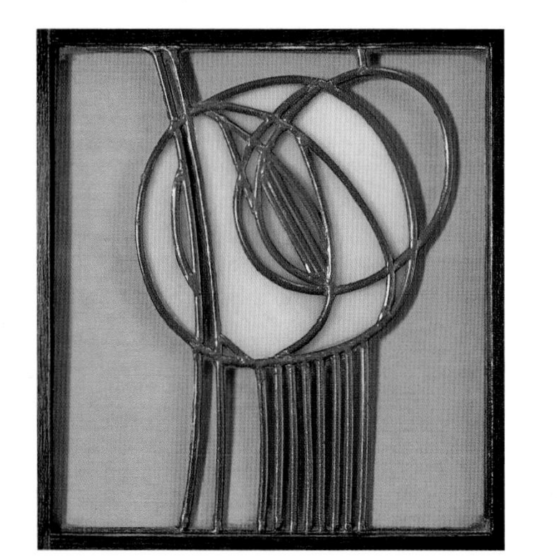

Art-glass windows and light fixtures were important details designed with the same care and attention as the whole building. Charles Rennie Mackintosh's rectilinear glass designs for the Willow Tea Rooms in

Below: *James Powell at his Whitefriars Glassworks executed these designs for four champagne glasses by Philip Webb, c. 1860. All the glasses used by Morris at Red House were designed by Webb. Some were simple and unadorned, the glass pale and greenish, though large decanters were more elaborate.*

Glasgow exerted a direct influence on Frank Lloyd Wright and provided a fruitful interchange with European designers such as Koloman Moser and Josef Hoffmann. Frank Lloyd Wright used skylights and walls made entirely of windows to flood a room with light and break up boxlike interiors—an excellent example can be found in his own Oak Park house and studio, where he installed art-glass skylights.

Arts and Crafts designers also turned their attention to practical domestic items. In an essay entitled "Table Glass" in *Arts and Crafts Essays by Members of the Arts and Crafts Exhibition Society*, George Somers Clarke claimed that "few materials lend themselves more readily to the skill of the craftsman than glass." He makes a plea for "graceful designs" and gives as an example "the old decanter, a massive lump of misshapen material better suited to the purpose of braining a burglar than decorating a table," which "has given place to a light and gracefully formed vessel, covered in many cases with well-designed surface engraving." The table glass designed by Philip Webb for Powell's of Whitefriars conforms to this new model of gracefulness and is powerfully simple and striking. In Europe Richard Riemerschmid and Peter Behrens were designing glassware for Benedikt von Poschinger in Oberzweiselau, and the Wiener Werkstätte was producing wine glasses with matching decanters decorated with colored enamels.

Right: *A decorated Favrile glass vase from Tiffany Studios, c. 1894. Tiffany was one of the most creative and prolific designers of the late nineteenth century and developed his technique for "Favrile" blown-glass vases and bowls in 1893, taking the name from an old English word meaning "hand made."*

Christopher Dresser took glassware to a new plane. Unafraid of machine production and one of the most radical and prolific designers of the nineteenth century, he bridged the divide between Morris and Modernism, anticipating the Bauhaus aesthetic, and made a vital contribution to glassware design. In the 1890s, as he neared retirement, Christopher Dresser designed the "Clutha" glassware range for the Glasgow firm of James Couper & Sons, who sold the blown vases of twisted opaque green glass, often shot with streaks of gold or cream, through Liberty's.

Below: *This hand-beaten brass lamp made at the Birmingham Guild of Handicrafts was probably designed by Arthur Dixon, c. 1893.*

LIGHTING

THE SINGULAR EFFECT ACHIEVED by thoughtful and innovative lighting is a defining aspect of an Arts and Crafts house. Domestic lighting underwent a revolution in the nineteenth century as the developments in paraffin and oil lamps were rapidly superseded by gas and then, in the 1880s, by electricity. Gas-lighting had been an advance on candles and paraffin lamps, but involved pipes, wall brackets, and sconces, all of which had to be fixed to the walls and ceilings and came with a continual soft hissing noise and noxious fumes that were harmful to plants. As early as the 1850s ingenious designers had devised table lamps with long rubber tubes attached to the gas pipe to allow for some mobility, but it was the development by Thomas Edison in October 1879 of the first practical incandescent filament bulb that did away with gas pipes and naked flames, opening up new possibilities for designers.

Early electric lamps were still relatively dim and ceiling pendants were often suspended from pulleys so they might be lowered when in use. Because of its expense and initial inefficiency, electricity was not in common use until the end of the century, but its advent had a radical effect on lighting styles. The gaslit gloom of Victorian interiors was replaced by a fresher, cleaner style, and the new forms of lighting had a profound effect on interior decoration. Entire walls might be composed of continuous, uncurtained art glass in delicate geometric or flowing Art Nouveau designs, while luminous ceiling lamps, positioned above wooden grilles in complex patterns of squares and circles stretched with thin translucent paper, filtered their soft light. The new style of light fitting was seen at its most daring perhaps in the abstract patterned shades designed by Charles Rennie Macintosh for Hill House. A more typical Arts and Crafts light fitting would be a simple wood or metal frame, inset with plain etched glass or stained glass patterned with simple geometric or floral motifs.

Charles Ashbee had electricity installed in his house, the Magpie and Stump in Cheslea, in 1895, hanging translucent enamel shades by strands of twisted wires from roses of beaten metal. Philip Webb was the architect of one of the first houses to be designed for electric lighting from the outset: Standen, in East Grinstead, Sussex. James Beale and his wife Margaret took a keen interest in all the details. On July 7, 1894, Webb wrote to Margaret on the subject of lighting: "I carefully considered as to what might be done for the electric lights in the drawing-room, and concluded that embossed copper sconces standing on the picture rail would be the best form for the finish of these bracket supports—something like this sketch." Margaret Beale approved the design and six sconces were made up by the metalworker John Pearson, each of them slightly different. Webb even added a cautionary note on their cleaning: "The copper plates should not be scoured, but only occasionally rubbed with wash leather." Many of the other fittings were provided by W.A.S. Benson, a founder member of the Art Workers' Guild and leading Arts and Crafts metalworker and cabinet-maker, whose contribution to the field was summed up by Hermann Muthesius in the 1880s when he wrote: "[Benson] created lamps that were to have a revolutionary effect on all our metalware. Benson was the first to develop his design directly out of the purpose and character of the metal as a material. Form was paramount to him. He abandoned ornament at a time when, generally speaking, even the new movement was fond of ornament." After years of upward-reflecting gas flames, Benson created metal shades that deflected the light downward, softening the comparatively harsh electric light. He produced a range of translucent light fittings, plus standard and table lamps using copper, tubular brass, polished steel, and opaque glass.

Above: *W.A.S. Benson: hanging electrolier made of copper and brass, c. 1910.*

In the USA two designers came quickly to the fore. The first was Louis Comfort Tiffany, who had been producing stained-glass windows, door panels, and tiles to bring warmth and light into a wood-paneled interior since 1883. In 1892 he bought his own glass furnace at Corona, near New York, and soon began producing the leaded stained-glass light shades that are now immediately recognizable as icons of the Arts and Crafts interior. These were vastly popular with the public and were soon copied. The Quezal Art Glass and Decorating Company and Steuben Glass Works—both run by former Tiffany employees—produced very similar lines of elaborate glass lampshades; the latter called their

Right: *A lovely example of a hammered copper and stained mica table lamp by Dirk van Erp, c. 1911–12. The conical shade had four vertical leaf-form straps riveted to the top cap and lower rim, enclosing curved and stained mica panels, supported by arms attached to the base.*

range "Aurene," in a conscious echo of Tiffany's "Favrile" iridescent glass. The Grueby Faience Company supplied ceramic bases which, before electrification, housed a fuel canister. But soon Tiffany was using the flowing natural forms of his lamps to disguise electrical wires and colored leaded-glass shades to soften electric light, which—after candles, kerosene, and gas—was considered too harsh. Chemical advances at the turn of the twentieth century meant that glass could be tinted and layered to create new effects, novel surfaces, and textures from matt to a burnished glow. Tiffany experimented enthusiastically with chemical soaks or vapors, perfecting the "Cypriote" finish, which imitated the pitted surface of the ancient excavated Roman glass he had in his own extensive collection, and "Lava," which had thick rivers of gold dripping down a black body. In the studio where he conducted his experiments, he also created magnificent pictorial windows—richly colored landscapes, plant studies, and bible scenes—intended for churches and large public buildings, as well as more modest scenic panels that middle-class Americans could order through manufacturers' pattern books and Tiffany Studios catalogs.

Above: *A fine "Arrowhead" leaded-glass and bronze table lamp from the Tiffany Studios, c. 1900–10. Trained as a painter, Tiffany began studying the techniques of glassmaking when he was twenty-four and enjoyed international acclaim for his work.*

Gustav Stickley and Elbert Hubbard produced plain wooden Arts and Crafts lamps and rugged lanterns, with matte metal frameworks and forged iron chains, at their Craftsman and Roycroft workshops, but a true Arts and Crafts original was Dirk Van Erp, who had emigrated to America from Holland. He began making lamps, candlesticks, and other lighting accessories entirely by hand out of hammered copper shell casings which he brought home from the naval shipyard where he was working. These he fitted with distinctive translucent mica shades which lent a soft amber glow, complementing the copper bases. He began by selling the lamps through craft shops and fairs, where

Far right: *Electric light was still a novelty in many parts of the country in 1904 when Frank Lloyd Wright designed this leaded table lamp. However, in this field, as in many others, he led the way.*

Below right: *One of two leaded-glass and bronze double pedestal lamps by Frank Lloyd Wright, made by the Linden Glass Company for the Dana House, c. 1903, which sold at auction a century later for almost two million dollars.*

Above: *This fine copper and glass lantern was designed by Charles and Henry Greene for the Robert R. Blacker House in Pasadena, c. 1907, and executed in the workshops of Peter Hall and Emil Lange.*

they were received so enthusiastically that he felt emboldened to leave the shipyard and set up the Copper Shop in Oakland, California, in 1908. His pieces are now among the most sought-after and prized Arts and Crafts objects.

American Arts and Crafts architects such as Greene and Greene and Frank Lloyd Wright considered custom lighting an integral part of their designs. In the Greenes' Gamble House, the regularly spaced rectangular lanterns with their leaded-glass shades in warm ochers, oranges, and golds define the separate spaces by the pools of buttery light they shed, while in the Blacker house the light is directed up from six-sided lanterns decorated with art glass patterned with lilies, which echo the pond outside, to reflect off a ceiling decorated with lilies and ripples of water covered in gold leaf. In the Robinson house the dining-room chandelier could be raised or lowered by a system of weights, and in the Thorsen house in Berkeley the light was recessed into the ceiling. Wright's lighting, too, was often indirect. He wanted it to be "made a part of the building. No longer an appliance nor even an appurtenance, but really architecture. This is a new field," he confessed in his essay "In the Cause of Architecture" in 1928. "I touched on it early in my work and can see limitless possibilities of beauty in this one feature … it will soon be a disgrace to an architect to have left anything of a physical nature whatsoever in his building unassimilated in his design as a whole." He did not want "glaring fixtures" but light "incorporated in the wall, which sifts from behind its surface opening appropriately in tremulous

pattern, as sunlight sifts through leaves in the trees." He achieved this effect by creating "decks," or long, deep shelves just below the ceiling, that hid the light fixtures or diffused the light through geometric clear and colored iridescent art-glass skylights as in the May, Dana-Thomas, and Robie houses, among others.

In Europe, Josef Hoffmann and Otto Prutscher at the Wiener Werkstätte and Richard Riemerschmid at the Dresden and Munich Werkstätten produced simple modern designs for electric light fittings and domestic glassware in geometric, almost architectural forms, manufactured by glass companies in Bohemia. In France the work of the highly talented Emile Gallé touched Tiffany with an Art Nouveau wand, while in Sweden Anna Boberg made beautiful glassware at the Reijmyre Glassworks and Gunnar Wennerberg, best known for his ceramic designs, used overlays and different textures to create ravishing new effects in glass at the Kosta factory.

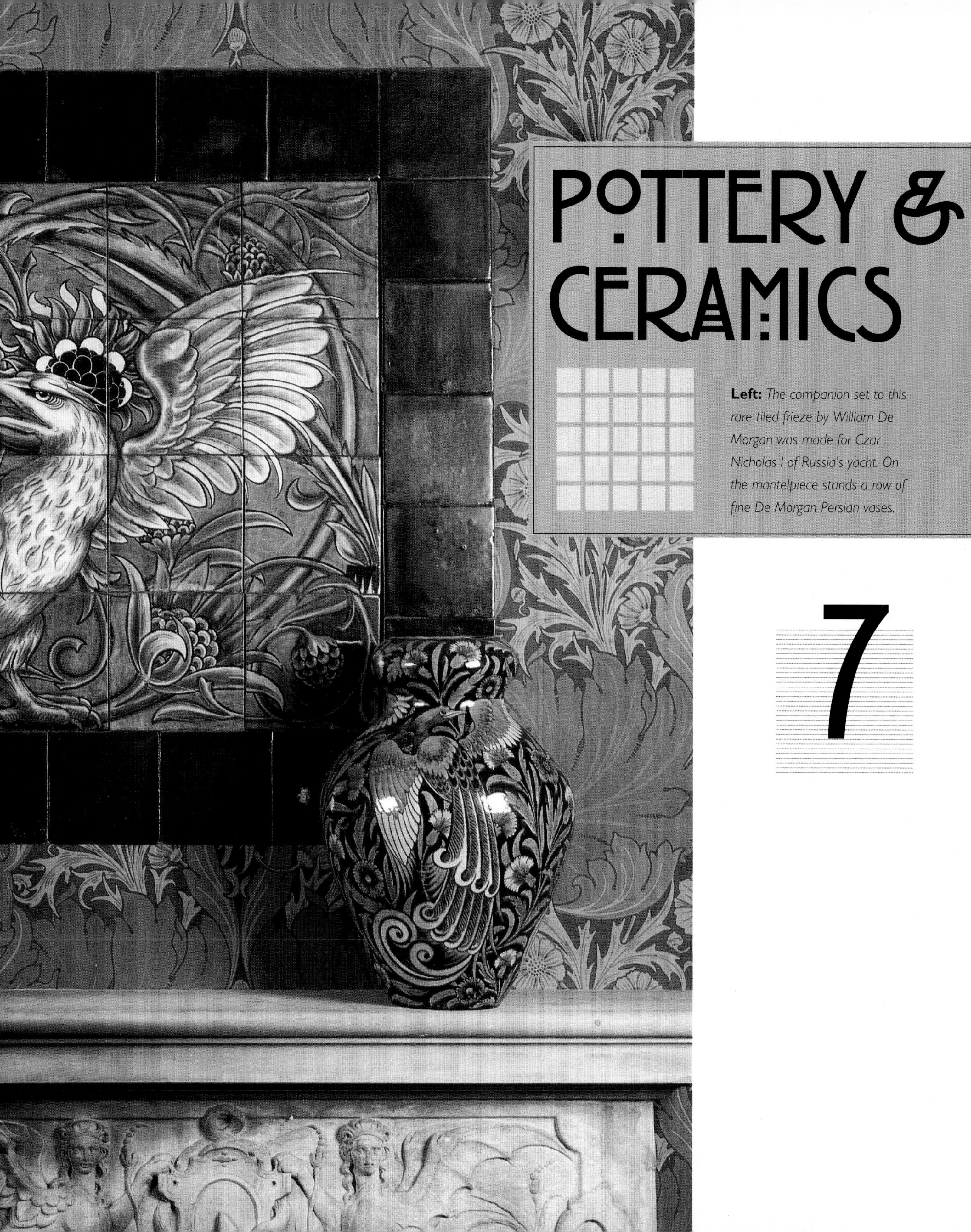

POTTERY & CERAMICS

Left: *The companion set to this rare tiled frieze by William De Morgan was made for Czar Nicholas I of Russia's yacht. On the mantelpiece stands a row of fine De Morgan Persian vases.*

7

UN CHEVALIER SANS PEUR ET SANS REPROCHE

ART FROM THE EARTH

"The pride of the potter is that his clay shall yield to the furnace: flowing and mingling in matchless beauty and endless variety. But the glazes must also acknowledge the artistic restraint by which his whole work is controlled. I endow either porcelain or pottery with brilliant color, pulsing with life and radiance or with tender texture, soft and caressing: color and texture which owe their existence and their quality to the fire—this is art."

CHARLES F. BINNS, writing in *The Craftsman*, 1903

Above: *This "Primrose Tile," dating from 1864, is typical of the simple, floral patterned tiles Morris & Co. offered for sale until the early 1870s.*

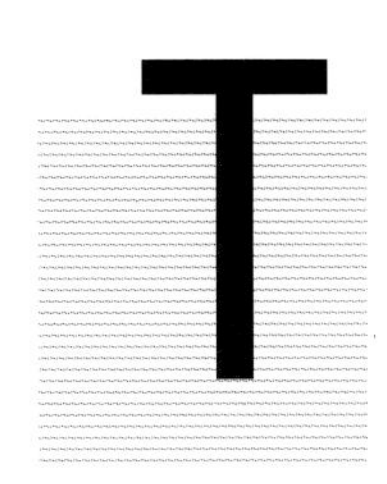

THE ENTHUSIASM FOR ARTS AND CRAFTS PIECES led to the foundation of numerous "studio" potteries and glassworks in both Britain and the USA. Some were set up by individual artist-potters such as William De Morgan, Harold Rathbone, or the eccentric George Edgar Ohr, while others were under the umbrella of well-established and well-respected commercial potteries including Minton, Doulton, and Wedgwood, who started smaller studio departments producing "fine art" pieces from designs by Arts and Crafts luminaries such as C.F.A. Voysey, Lindsay Butterfield, and Walter Crane. The long tradition of making fine pottery and porcelain in Britain—especially in Staffordshire and Derbyshire—was built on by Arts and Crafts designers. In the field of ceramics, makers were able to realize Morris's ideal of designing and making an object from raw material to finished product. Clay was an everyday and inexpensive material, and art potters could work quite simply without reliance on industry or manufacturers.

The closing years of the nineteenth century saw a boom in the American art-pottery market and the opening of a number of new businesses that joined the already established Fulper and Rookwood Potteries: Hugh C. Robertson reopened the family's Chelsea Keramic Art Works as Chelsea Pottery US in Massachusetts in 1891, and William H. Grueby set up the Grueby Faience Company in Boston in 1894. Much of the American art-pottery movement was philanthropic in its ideals. Often enterprises—such as the Marblehead Pottery, the Arequipa Pottery, and the Paul Revere Pottery—were started with the wider social aim of providing education or congenial occupation for poor or convalescing women,

Left: *A charger designed by Walter Crane for Pilkington & Co, 1907, painted by Richard Joyce.*

Above: *A William De Morgan two-handled vase with fish design and Persian coloring, c. 1880.*

and juggled the constraints of profit and philanthropy with varying results. Women designers such as Mary Louise McLaughlin of Cincinnati and Adelaide Alsop Robineau of Syracuse, New York, made outstanding contributions to a field that admitted and encouraged female participation. The standard of American ceramics was extremely high, and many of the emerging potters benefited from the teaching of Charles Fergus Binns—the father of studio pottery—at the New York School of Clayworking and Ceramics. All this activity was welcomed and applauded by the popular press. An article of 1899 in *House Beautiful* congratulated American manufacturers on finding individuals and companies "conducting their work in a spirit that demands first of all that results shall be honest and beautiful.... In the manufacture of pottery, oftener perhaps than in other crafts, one meets with this renascent spirit; possibly because its subtle chemistry offers an opportunity to the scientist as well as to the artist. It is a fascinating and absorbing art, claiming the utmost devotion, but lavishly rewarding the man who can discover its secrets."

The housing boom of the 1880s and 1890s led to an increase in the demand for tiles, which were easily washable, hygienic, and decorative. William De Morgan revived the art of hand-painting tiles while Lewis F. Day and C.F.A. Voysey designed for the leading British manufacturers such as the Pilkington Tile & Pottery Co. and Minton & Co., developers of the encaustic tile in 1840. "This branch of art-manufacture [encaustic tiles] is one of the most hopeful, in regard to taste, now carried on in this country," wrote Charles Eastlake in the late 1870s. "It has not only reached great technical perfection as far as material and color are concerned, but, aided by the designs supplied by many architects of acknowledged skill, it has gradually become a means of decoration which, for beauty of effect, durability, and cheapness, has scarcely a parallel." Both Doulton and Minton supplied blank tiles for fashionable young ladies to decorate. In the USA the father-and-son business of J. & J.G. Low Art Tile Works in Chelsea, Massachusetts, produced high-gloss tiles in the European style from 1878 to around 1900, and Ernest Batchelder became well known for his striking medieval tile designs, but it was William Grueby's company that came to dominate the market.

Above: *A design for a tile panel for an ornamental terrace by Halsey Ricardo for the William De Morgan Tile Company, c. 1890.*

BRITISH POTTERY

WILLIAM FREND DE MORGAN was the most influential British designer working in the field. He had turned to ceramics from stained glass and used exotic colors—ruby reds, delicate golds, and the vivid peacock blues and greens of Iznik and Persian wares, which provided inspiration for his expensive handmade tiles decorated with mythic animals, such as the sea dragon, and with ships, birds, or plants. Originally associated with Morris & Co., he set up his own pottery and showroom in Chelsea in 1872, experimenting with luster glazes, enamel, and decorating ready-made factory blanks. In 1882 he moved to a new purpose-built pottery near Morris at Merton Abbey, where the two men frequently collaborated on tile projects. But it was his ten-year partnership with Halsey Ricardo, beginning in 1888 when he set up a factory at Fulham, that prompted some of his richest work. De Morgan's poor health meant that he often spent extended periods abroad, and he found himself spending so much time in Florence that he set up a studio there and supplied the Italian pottery Cantagalli with designs.

Above: *A twin-handled pottery vase, designed by Walter Crane for Maw & Co. in 1893, decorated with medieval warrior figures drinking from cornucopia.*

Edward Burne-Jones made designs for tiles sold through Morris & Co., such as the "Sleeping Beauty" set commissioned in 1864 by the painter Myles Birket Foster to decorate the overmantels in the bedrooms of his new home, The Hill, at Witley in Surrey. Lucy Faulkner may have copied Burne-Jones's designs onto the tiles and painted them; William Morris was certainly responsible for the design and painting of the surrounding swan pattern.

Walter Crane's move into the field of art pottery provided the opportunity to experiment with different shapes and to work with new luster glazes. His hand-painted designs, such as the striking "Swan" vase in the Victoria and Albert Museum, were, however, expensive "one-offs" made for prestigious displays at international exhibitions and were never meant for a mass market. He was commissioned, along with Day, Voysey, and Frederick Shields to provide designs for the Pilkington Tile & Pottery Co., which was formed in 1891 with William Burton, formerly a chemist at Wedgwood, as manager. Burton was joined by his brother Joseph in 1895 and together they experimented with the wide range of rich glazes that became the hallmark of the company. In 1903 Burton decided to diversify into decorative glazed pottery and launched his "Royal Lancastrian" range, which used "shapes, based on either the forms of the Greek, Persian or Chinese pottery, on some suggestions of natural growth, or on the forms actually evolved from plastic clay in the hands of the potter." He commissioned Crane, Day, and Voysey, giving them complete freedom to interpret the romantic and chivalric themes he wanted for his Lancastrian wares.

Below: *A Della Robbia platter designed by Cassandra Walker, c. 1905.*

The eccentric Martin Brothers—Robert, Charles, Walter, and Edwin—of Southall, London, were responsible for some of the most unusual art pottery to come out of England. Their jugs and vessels, often made in the shape of slyly grinning, grotesque, hybrid creatures, proved popular and became highly fashionable—probably their best-known client was Queen Victoria.

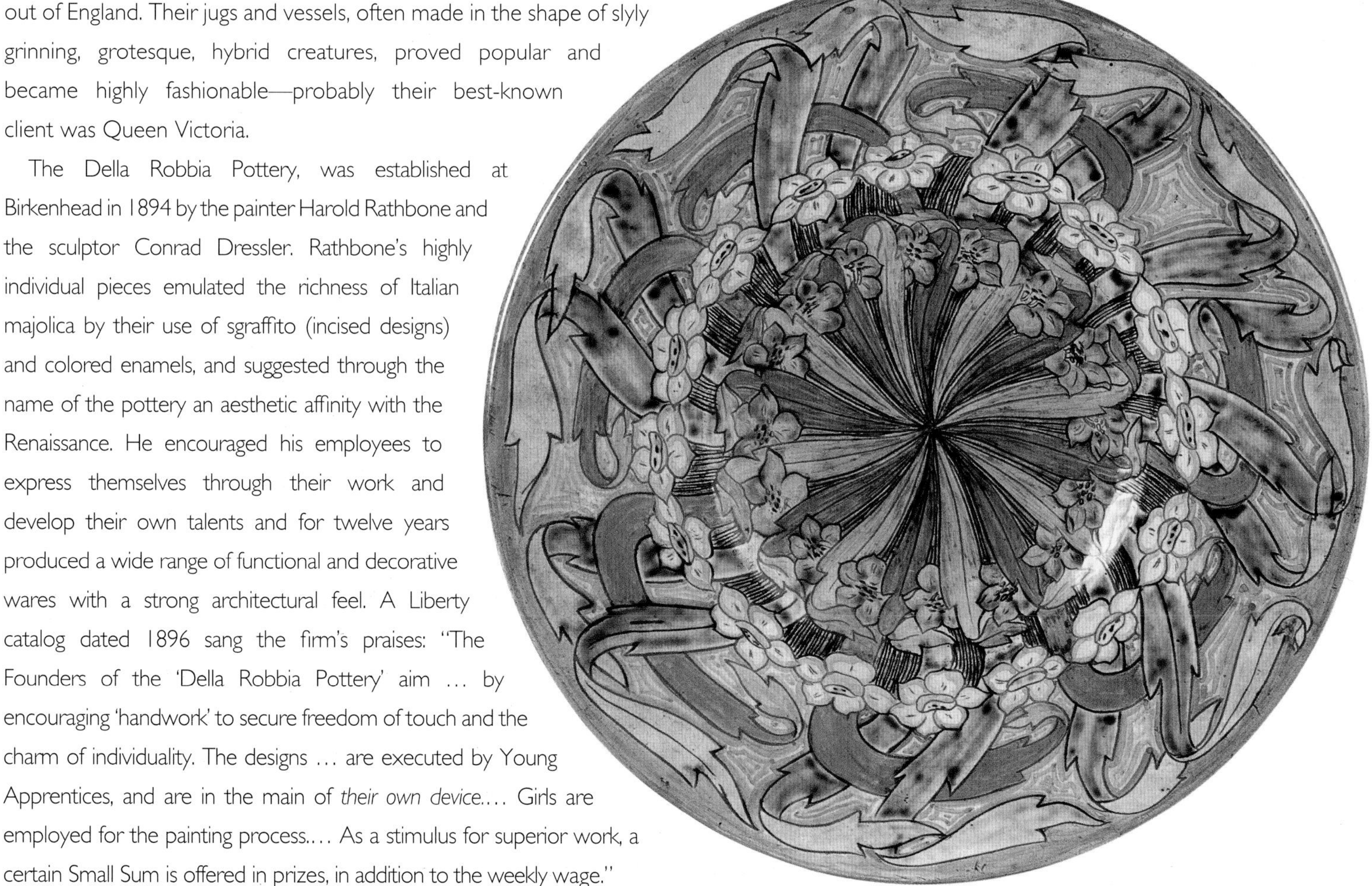

The Della Robbia Pottery, was established at Birkenhead in 1894 by the painter Harold Rathbone and the sculptor Conrad Dressler. Rathbone's highly individual pieces emulated the richness of Italian majolica by their use of sgraffito (incised designs) and colored enamels, and suggested through the name of the pottery an aesthetic affinity with the Renaissance. He encouraged his employees to express themselves through their work and develop their own talents and for twelve years produced a wide range of functional and decorative wares with a strong architectural feel. A Liberty catalog dated 1896 sang the firm's praises: "The Founders of the 'Della Robbia Pottery' aim ... by encouraging 'handwork' to secure freedom of touch and the charm of individuality. The designs ... are executed by Young Apprentices, and are in the main of *their own device*.... Girls are employed for the painting process.... As a stimulus for superior work, a certain Small Sum is offered in prizes, in addition to the weekly wage."

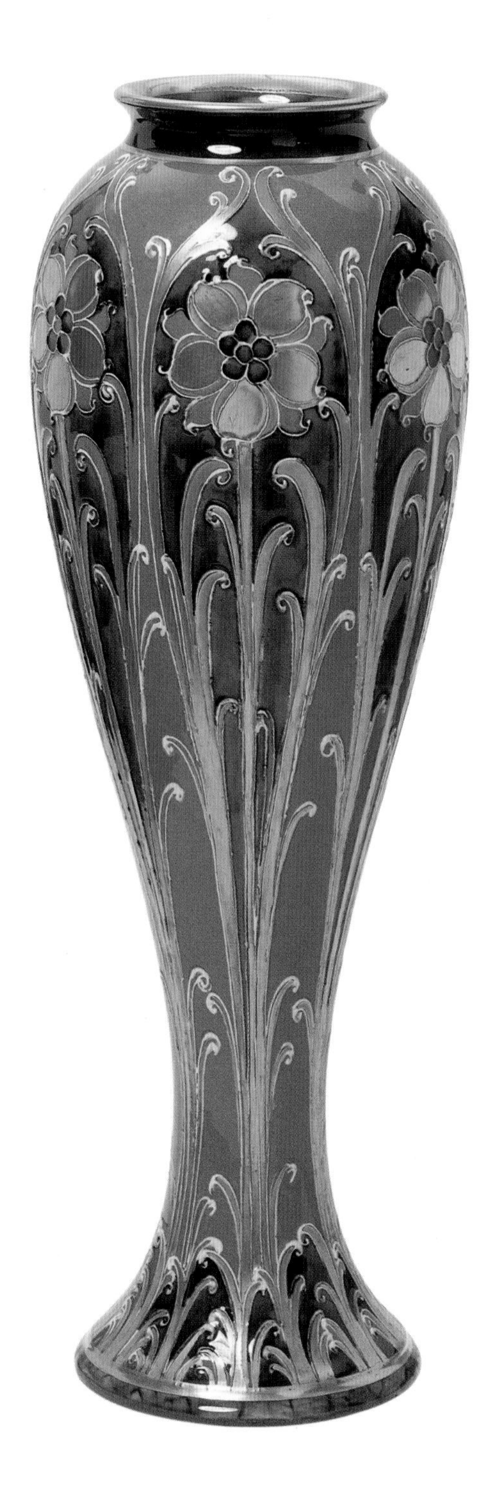

In the village of Compton in Surrey, where the painter George Frederick Watts had settled late in his life, pottery classes organized by his wife Mary proved so successful that the Compton Potters Art Guild, producing ornamental garden wares in porous terracotta along Celtic lines, was set up in 1896. Its pots, sundials, and fountains, including some designed by Archibald Knox, were sold by Liberty's.

The Ruskin Pottery was established in 1898 at West Smethwick near Birmingham by William Howson Taylor, son of the remarkable principal of Birmingham School of Art, Edward Richard Taylor. E.R. Taylor was a friend of Morris and Burne-Jones and a pioneer in the teaching of craft skills, and he provided some of the decorative designs for the pottery. W.H. Taylor was preoccupied with experimental glazes and became famous for his "high-fired" effects and the difficult techniques of "soufflé," luster, and "flambé" glazes.

Sir Edmund Elton was a talented self-taught potter who developed a distinctive crackled metallic finish at the Sunflower Pottery set up at Clevedon Court, his family's Somerset estate, in 1879. Bold and original Sunflower Pottery pieces were sold through Howell & James, a department store in Lower Regent Street, London, which was the leading stockist of modern British pottery until 1883, when Liberty's, which up to then had concentrated on Oriental ceramic imports, began to move into the field.

These small potteries, which drew on local expertise and hired staff from other ceramic studios, had an influence on the bigger factories such as Doulton, Wedgwood, and Minton, which began to emulate them by producing "one-off" decorated products and introducing craft studios. Doulton worked with well-known designers, sculptors, and painters connected with the nearby Lambeth School of Art, including a number of women such as Hannah Barlow, Eliza Simmance, and Mary Mitchell. Wedgwood employed some of the finest modelers, decorators, and craftsmen and, in 1903, took on Louise Powell and her husband Alfred Powell, Birmingham-based associates of the Gimson circle, to shape and decorate luster and tin glaze ware. Tableware patterns by the Powells often featured small repeating flower and leaf designs. They attracted favorable notice from Gordon Forsyth, superintendent of the Stoke-on-Trent Schools of Art, who wrote an official report on "The Position & Tendencies of the Industrial Arts" following the 1925 Paris Exhibition:

> "Wedgwood's make excellent pottery and Mr. and Mrs. Powell are excellent artists. They no doubt design the shapes they decorate, and their work always shows a keen appreciation of suitable treatment of various articles of everyday use. Their best work is found in lordly bowls and plaques.... They are based on the brave and honest pattern work of William Morris, and have a delightfully English style.... English pottery would be very much poorer without their splendid contributions to the artistic side of the craft."

Left: *A green and gold Florian Vase with tube-lined decoration and colors, and gilding designed by the Staffordshire potter William A. Moorcroft and made by James A. Macintyre & Co. at Burslem, c. 1903.*

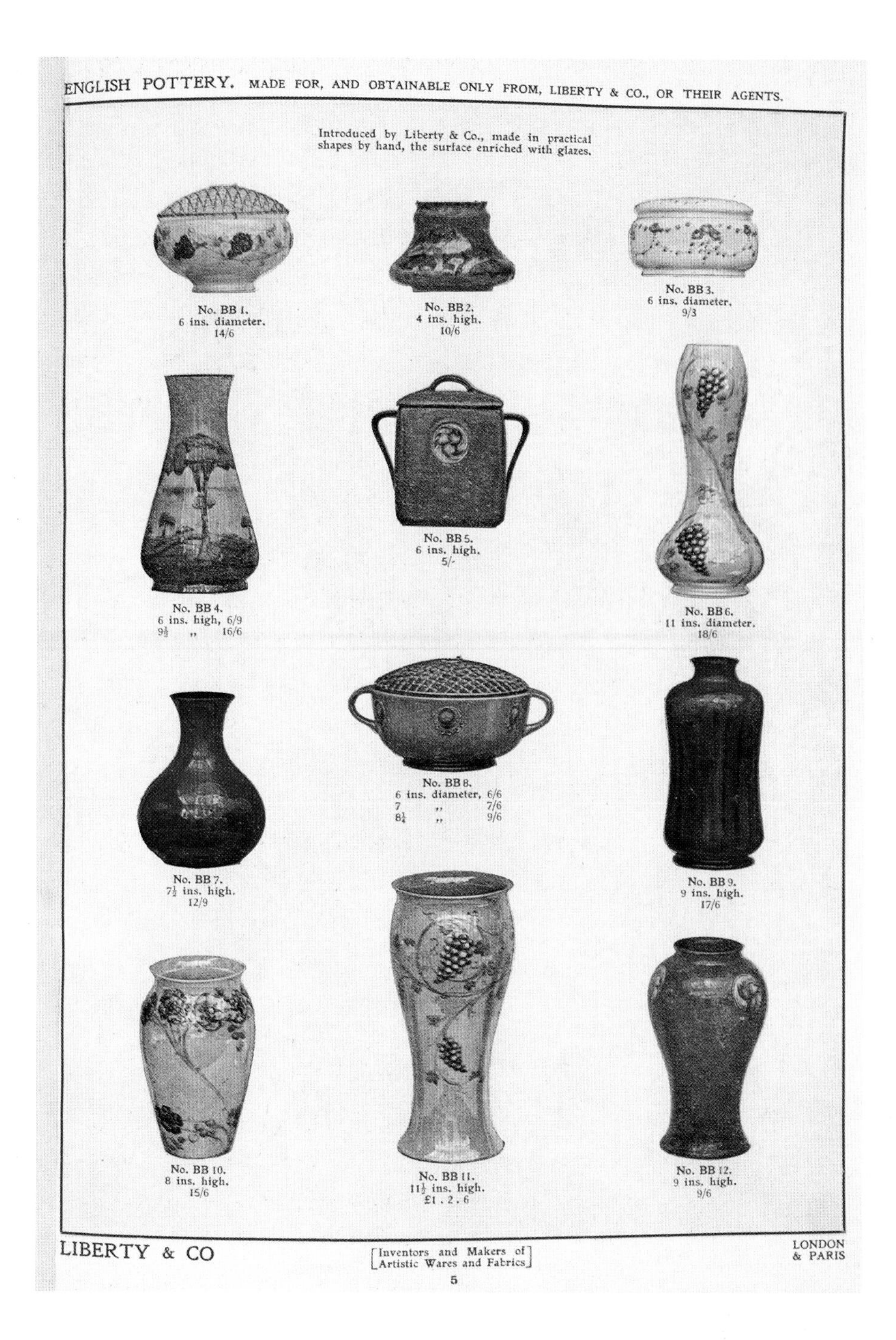

ENGLISH POTTERY. MADE FOR, AND OBTAINABLE ONLY FROM, LIBERTY & CO., OR THEIR AGENTS.

Introduced by Liberty & Co., made in practical shapes by hand, the surface enriched with glazes.

No. BB 1.
6 ins. diameter.
14/6

No. BB 2.
4 ins. high.
10/6

No. BB 3.
6 ins. diameter.
9/3

No. BB 4.
6 ins. high, 6/9
9½ „ 16/6

No. BB 5.
6 ins. high.
5/-

No. BB 6.
11 ins. diameter.
18/6

No. BB 7.
7½ ins. high.
12/9

No. BB 8.
6 ins. diameter, 6/6
7 „ 7/6
8½ „ 9/6

No. BB 9.
9 ins. high.
17/6

No. BB 10.
8 ins. high.
15/6

No. BB 11.
11½ ins. high.
£1 . 2 . 6

No. BB 12.
9 ins. high.
9/6

LIBERTY & CO [Inventors and Makers of Artistic Wares and Fabrics] LONDON & PARIS

5

Above: *A page from the Liberty catalog of 1912 illustrating their range of Moorcroft pottery.*

From 1883 Liberty's began featuring in its catalog English art pottery from dozens of smaller firms, including Aller Vale, Foley, Poole, Moorcroft, Brannum, and Farnham, all of which experimented with hand-throwing, sgraffito work, luster decoration, and unusual glaze effects and came to rely on the department store to retail their wares. Liberty's also provided the London retail outlet for Henry Tooth's Derbyshire-based Bretby Art Pottery. Tooth had first worked with Christopher Dresser—an important influence on a number of potters at this time, including William Ault—at John Harrison's Linthorpe Pottery. Later Liberty's also stocked Royal Doulton, Wedgwood, and Pilkington's Royal Lancastrian.

The Arts and Crafts ideal was not so prevalent in European art pottery, which either embraced Art Nouveau or leaned toward the geometrical Modernism of the Wiener Werkstätte, although Liberty's did sell a range of homely, tin-glazed earthenware made by the Dutch firm Plateelfabriek Zuid-Hollandsche at Gouda, and dark green, blue, and red luster-glazed pottery from Hungary's Zsolnay ceramic works at Pecs.

AMERICAN POTTERY

Below: *A green earthenware vase with a flower decoration by Hattie E. Wilcox, made by the Rookwood Pottery, c. 1900.*

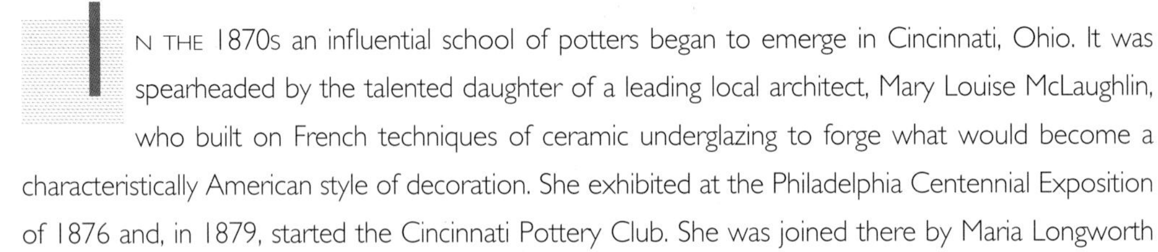

IN THE 1870s an influential school of potters began to emerge in Cincinnati, Ohio. It was spearheaded by the talented daughter of a leading local architect, Mary Louise McLaughlin, who built on French techniques of ceramic underglazing to forge what would become a characteristically American style of decoration. She exhibited at the Philadelphia Centennial Exposition of 1876 and, in 1879, started the Cincinnati Pottery Club. She was joined there by Maria Longworth Nichols, who, a mere year later, with the support of her wealthy art-patron father, had converted her hobby into a thriving concern. Rookwood Pottery, the largest and most influential art pottery in the USA, was founded with the specific aim of constructing and decorating art pottery by hand and enjoyed both commercial and critical success. Rookwood pieces, which included impressive overmantels and fireplace surrounds, concentrated on naturalistic landscapes and flower designs, employing an underglaze decorative technique derived from the French that required mild firing to maintain the warm-colored glaze. Oscar Lovell Triggs, a writer and socialist, called Rookwood "an ideal workshop" in his book *Chapters in the History of the Arts and Crafts Movement* (1902). "The fullest possible freedom is given to the workmen;" he enthused, "they are encouraged to experiment, to express their own individuality, and to increase their culture by study and travel. The spirit of the factory is that of cooperation and good fellowship."

The sizable Japanese display at the Philadelphia Centennial Exposition had a profound effect on Maria Longworth Nichols and, in 1887, she invited the Japanese ceramicist Kataro Shirayamadani to join the firm. He stayed on until his death in 1948, becoming one of the company's principal designers. Rookwood introduced pale cool colors with names like "sea green" and "aerial blue" and, following the award of a Grand Prix at the 1900 Paris Exhibition, became highly fashionable. By moving with the times, the firm continued to flourish and even weathered the Depression, maintaining production until 1960.

Among the many designers who worked for Rookwood, the name Artus Van Briggle stands out. One of Rookwood's most talented decorators, he worked for the company for thirteen years, taking regular sabbaticals to Paris where he drew inspiration from the French Art Nouveau movement and the Oriental collections at the Musée des Arts Decoratifs and Sèvres. In 1902 he moved to Colorado Springs, where he started his own pottery studio, producing matt-

Above: *An earthenware vase by Artus Van Briggle, modeled with a frieze of arrowroot leaves and blooms, in a sheer turquoise glaze, 1906.*

glazed pieces in sculptural Art Nouveau forms until his career was cut short by tuberculosis and an early death in 1904.

William Henry Grueby founded his own business, the Grueby Faience Company, in 1894. At first it produced architectural bricks and tiles, but in 1897 he added an "art wares" section and developed his innovative matt green glaze. The "peculiar texture" of this rich, monotone surface was compared in the company brochure to "the smooth surface of a melon or the bloom of a leaf" and sources of inspiration were cited as "certain common forms in plant life, such as the mullen leaf, the slender marsh grasses, the lotus or tulip, treated in a formal or conventional way." The work won medals and international acclaim at the 1900 International Exposition in Paris. It was also commercially successful, for Grueby adapted handcraft techniques to mass production, standardizing patterns and

Above: *A tulip tile and two vases from the Grueby Pottery, c. 1908.*

Far right: *An early Fulper tapering vase with two cut-out buttressed handles and Flemington Green flambé glaze.*

taking the creative decisions away from the men—many of them art students—who hand-constructed the pieces, which were then decorated with established designs by semi-skilled female students.

A collaboration with Tiffany Studios led to the production of lamps, sold through the Grueby brochure, which boasted: "A fine piece of pottery is essentially an object of utility as well as of decoration. In no way does the Grueby ware fulfill these two purposes more completely than in its lamp forms, whether for oil or electricity. The Grueby-Tiffany lamp combines two recent products of the Applied Arts, the support for the bronze fitting being a Grueby jar made for that special purpose, completed by a leaded or blown-glass shade of Tiffany design and workmanship."

Louis Comfort Tiffany experimented with pottery at his Corona factory and produced a range that was mass-produced (in modest numbers) from hand-thrown originals. It was offered for sale—sometimes unglazed so that customers could choose their own finish—at Tiffany's Fifth Avenue store. It was never as popular as his art glass and lamps, though nowadays Arts and Crafts collectors prize the pieces, which make bold use of organic forms.

Newcomb Pottery, an educational enterprise associated with the H. Sophie Newcomb Memorial College in New Orleans, was founded in 1895. Men were trained to throw pots by hand, and young women—including the gifted Sadie Irvine—to decorate them in an abstract Japanese style. The artists were encouraged to sign their work. The pottery ceased production in 1931.

The Fulper Pottery Company was an old, thriving, and well-established company when the grandson of the founder introduced a new line of ceramic wares—which he called "Vasekraft"—in time for the Christmas market in 1909. The range included lamps, desk and smoking accessories along with vases. Relatively inexpensive (around 1914 the price of a typical table lamp was $35), they were immediately snapped up by an American middle class keen to invest in a piece of well-made art pottery. The shapes were cast in molds, then individually glazed by hand, with a distinctive effect achieved by combining or overlapping different glazes—matt, gloss, or metallic—on the same piece. The company pioneered the use of crystalline glazes. Its pottery lamps, advertised as "Art pottery put to practical uses," were unusual because both the base and the shade were ceramic. An advertisement in *Vogue* in 1913 claimed, "Vase Kraft Lamps and Pottery are much admired for their rich subdued colorings. They lend an air of refined elegance to the surroundings in which they are displayed." Over time many of the ceramic shades (often inset with glass) have become brittle and broken, which explains their relative rarity today. The factory closed in 1929.

William Day Gates started the Gates Pottery of Terra Cotta in Illinois in 1885, producing architectural terra cotta, but in 1900 he introduced a line of ceramic vases and garden ornaments called Teco Art Pottery. "The constant aim," he claimed, "has been to produce an art pottery having originality and true artistic merit, at a comparatively slight cost, and thus make it possible for every lover of art pottery to number among his treasures one or more pieces of this exquisite ware." He commissioned designs from

architects Frank Lloyd Wright and William James Dodd, as well as the sculptor Fritz Wilhelm Albert, and developed a sea-green matt glaze, which he applied to molded (rather than hand-thrown) pots, thus keeping his prices down. Teco ware concentrated on form and color rather than surface decoration and the pieces are distinctively geometric, architectonic, and monumental in design.

The Chelsea Keramic Art Works was a family firm founded by James Robertson outside Boston in 1872, which made classically inspired vases often decorated with applied carved sprigs of flowers, leaves, birds, or bees. His two talented sons, Alexander and Hugh Cornwall Robertson, are names to reckon with in the field of American art pottery. Hugh's pursuit of exotic glazes—he was famous for his Chinese red glaze called *sang-de-boeuf* and his blue-decorated white Chinese crackle glaze—bankrupted the firm in 1889, although he reopened with fresh financial backing two years later, moving the company in 1896 to Dedham, Massachusetts. The Dedham Pottery received honors and recognition at the International Exposition in Paris in 1900 and the St. Louis World's Fair in 1904. It remained in business until 1943.

One of the most outstanding women in the American Arts and Crafts movement was the extraordinarily talented Adelaide Alsop Robineau, a china decorator and maker of complicated porcelain pieces, whose work was sold by Tiffany & Co. In 1911 she won the Grand Prix at the Esposizione Internazionale at Turin for her high-fired porcelain "Scarab Vase," which she made at the University City Pottery, Missouri. Also known as "The Apotheosis of the Toiler," the vase was said to have taken her over a thousand hours to produce.

Frederick Hürten Rhead was born in England but emigrated to the USA in 1902, bringing with him and popularizing the highly ornate sgraffito process he had learned from Harold Rathbone and Conrad Dressler at the Della Robbia Pottery in Birkenhead. He was art director of the Roseville Pottery in Ohio before joining Adelaide Alsop Robineau and the influential French

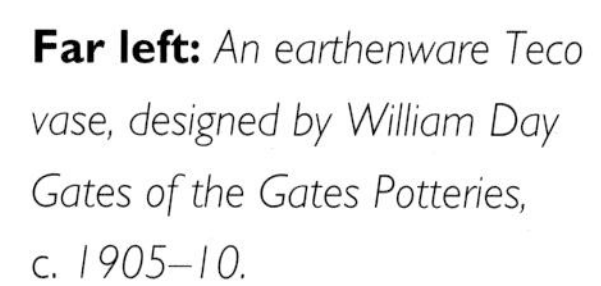

Far left: *An earthenware Teco vase, designed by William Day Gates of the Gates Potteries, c. 1905–10.*

Above: *An urn-shaped vase, decorated with a sgraffito design of wisteria vines in turquoise blue over teal blue, on a high glass blue-black ground, by Arthur Eugene Baggs at the Marblehead Pottery, 1925.*

potter Taxile Doat at the University City Pottery, where he produced some of his finest work. In 1911 he was persuaded to help set up the Arequipa Pottery in California, established in a sanatarium to help rehabilitate nervous or tubercular women. Contemporary photographs show Arequipa Pottery workers seated on cane chairs in the bosky outdoors, decorating pots. Men were hired to throw or mold pots and the patients were trained to decorate and glaze them, often using a design process Rhead called the "raised line," in which slip was trailed onto the surface in decorative patterns and accentuated by other glazes. Rhead stayed on in California and founded his own pottery in 1914 but, although a talented and influential potter, he was not a businessman and the firm folded three years later. "An indefinite idea in the mind of a wealthy person of questionable taste is not easily executed," he moaned, "especially if that person is prepared to pay neither a deposit nor an adequate remuneration for the finished product."

The Marblehead Pottery, like Arequipa, was initially conceived in 1904 as a craft therapy program for patients convalescing in the New England coastal town after which it was named. The work was simple,

elegant, and restrained, following Grueby in its use of glazes, but leaning more toward Frank Lloyd Wright, Charles Rennie Mackintosh, and the Viennese school in its elongated abstractions of natural forms. The chief decorator was a woman called Hannah Tutt and the firm's director was Arthur Eugene Baggs, a former student of Charles Binns at the New York School of Clayworking and Ceramics. In 1919, by which time it was an independent commercial concern, its brochure spoke of "dignity, simplicity, and harmonious color" and claimed: "The aim has been to make the decoration a part of the form, not merely a pretty ornament stuck on at haphazard." It boasted a range of colors from "a beautiful old blue known as Marblehead blue, a warm gray, wisteria, rose, yellow, green, and tobacco brown. All are soft, harmonious tones which lend themselves well to the display of flowers."

In 1911 the Paul Revere Pottery grew out of the Saturday Evening Girls' Club in Boston and gave the daughters of Jewish and Italian immigrants a chance to earn money producing breakfast and tea sets, lamps, vases, and tiles. The girls were trained in throwing, design, and glaze chemistry, and the venture was sustained by a continuing subsidy from its founder, Mrs. James J. Storrow.

Samuel A. Weller's pottery in Zanesville (popularly known as "Clay City"), Ohio, became the American leader in mass-produced low-priced art pottery when it marketed the distinctive metallic glaze developed by the French chemist Jacques Sicard in 1901. The iridescent pottery he produced for Weller—who also employed Frederick H. Rhead—was known as "Sicardo."

The Overbeck Pottery of Cambridge City, Indiana, was started in 1911 by four Overbeck sisters: Elizabeth, who was responsible for the throwing, Mary Frances, who did the decorating, Hannah who was the designer, and Margaret, the driving force, who unfortunately died in the year the firm was established. They worked "in a pleasant old house among the apple trees" producing wheel-thrown pieces decorated with stylized natural motifs. Other important Arts and Crafts concerns included the Grand Feu Art Pottery, established in 1913 by Cornelius Walter Brauckman in Los Angeles, and the California Faience Co., founded by William Victor Bragdon and Chauncey R. Thomas in 1915. Both relied on form and glaze rather than decoration for appeal, and both were influenced by the work of Taxile Doat.

Far left: *"Mission," a glazed matt vase from the Grand Feu Art Pottery, c. 1913, one of four pieces given to the Smithsonian Institution by the firm's founder, Cornelius Walter Brauckman.*

Above: *A small vessel with closed-in rim covered in a fine leathery blue-gray matte glaze from California Faience, c. 1915.*

Final mention must go to Ernest Allan Batchelder, who established his own school and tile factory in Pasadena, in 1909, just as southern California's construction industry was booming and the need for architectural tiles for bungalow fireplaces was at its height. He had studied at the School of Arts and Crafts in Birmingham and was devoted to hand labor and Arts and Crafts principles, shaping his tiles in plaster molds before decorating them with unique glazes. He wrote: "The evil of machinery is largely a question of whether machinery shall use men or men shall use machinery," and his *Principles of Design* (1904) situated both Japanese and Native Indian Art within an Arts and Crafts ethic. He was influenced by medieval and Gothic motifs and sought order in design, achieved through a balance of harmony and rhythm. He also made vessels such as *jardinières* featuring animals, birds, and intertwining floral designs. The boom, which had led to huge demand for his tiles, was followed by the Depression, which finished the firm, though Batchelder survived by scaling down his production to a small home-based operation, which he continued until the early 1950s.

GARDEN·

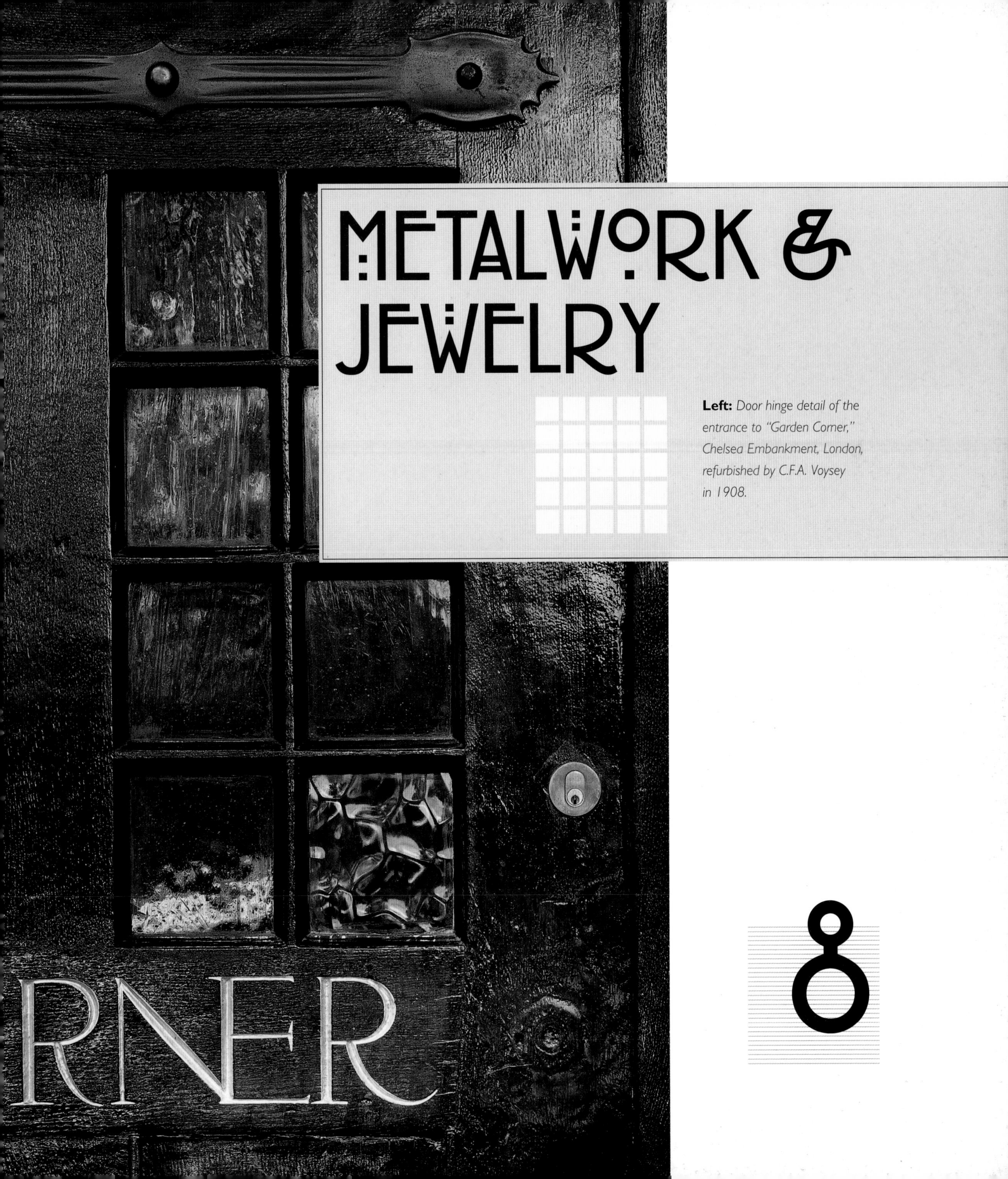

METALWORK & JEWELRY

Left: *Door hinge detail of the entrance to "Garden Corner," Chelsea Embankment, London, refurbished by C.F.A. Voysey in 1908.*

8

BEAUTIFUL, USEFUL, & ENDURING

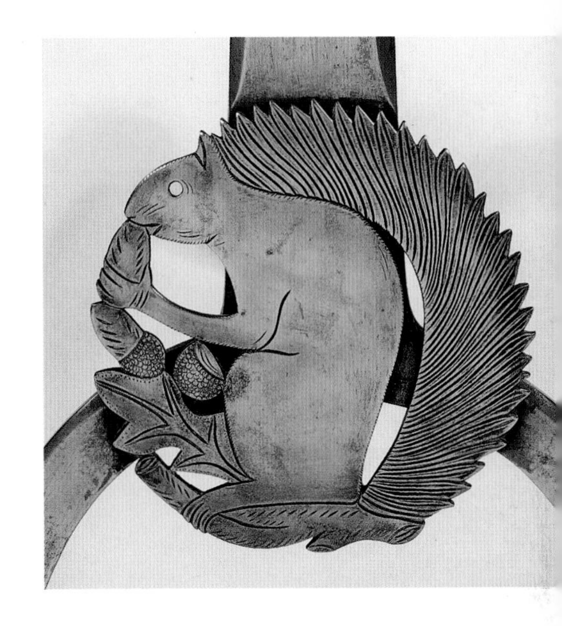

"Metalwork and jewelry in the past were always looked upon as holding a peculiar intermediate position between the Fine and Industrial Arts. The fine arts were those primarily occupied with the expression of ideas, the industrial arts had for their primary object utility, but the expression of ideas was still considered of importance, it was that which gave them their humanizing influence. The Arts and Crafts movement was perhaps above all things a humanizing movement on the part of the artists to give back to the handicrafts the humanitarian aspect which they had lost."

JOHN PAUL COOPER

Above: *Detail of a firedog in polished steel, designed by Ernest Gimson, c. 1905.*

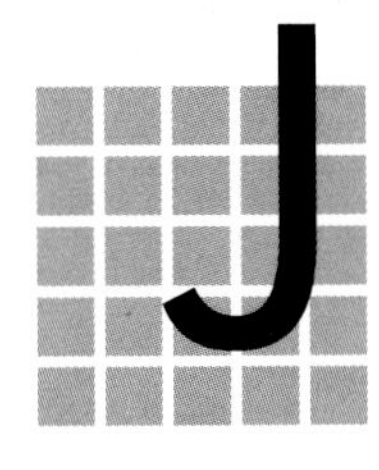

JOHN PAUL COOPER, W.A.S. Benson, Christopher Dresser, Archibald Knox, Charles Rennie Mackintosh, and C.R. Ashbee are among the British luminaries who led the renaissance in metalwork skills and techniques that flourished in England and Scotland at this time. Meanwhile, in the USA, Dirk Van Erp, Gustav Stickley, and Louis Comfort Tiffany were experimenting with traditional materials such as copper, tin, and pewter and using techniques such as enameling on copper to create rich iridescent tones.

The Arts and Crafts ideal of a small workshop producing handmade artefacts was realized by men like Ernest Gimson, the architect, who set up Alfred Bucknell, the son of his local blacksmith, in his own smithy in the wheelwrights' yard at Sapperton in Gloucestershire in 1903. Gimson needed metal fittings for the furniture he made at his Daneway workshops and it naturally followed that he would expect them to be made to the same high standard. Bucknell was in charge of three or four men working in iron, brass, polished steel, and silver producing firedogs, door handles, locks, bolts, window latches, strap hinges, candlesticks and sconces, and other items to Gimson's designs. Charles Voysey also designed his own metal fittings, from letterboxes to keyhole covers, often incorporating motifs such as birds or hearts, both of which were used in his design for the hinges on the cabinet he designed to hold the Kelmscott Chaucer. Architects such as M.H. Baillie Scott and W.R. Lethaby (author of *Leadwork old and ornamental and for the most part English*, published in 1893), who

Far left: *Philip Webb designed the firegrate, fender, fire irons, and paneled surround of the fireplace in the Dining Room at Standen in 1894. The repoussé mild steel cheeks and smoke cowl were made by John Pearson. The grate, plate rack, and fender were created by Thomas Elsley, a London blacksmith often employed by Webb.*

Above: *Philip Webb designed this pair of copper candlesticks around 1860 for Edward Burne-Jones.*

wanted larger pieces such as cast-iron firegrates made to their own design, sought out and found sympathetic manufacturers in Thomas Elsley and H. Longdon & Co., both of London, Yates & Heywood of Rotherham, Yorkshire, and the Coalbrookdale Iron Company. Baillie Scott was good on door details and fireplaces. "In the treatment of wrought-iron work," he wrote, "the best forms will be found to be those which suggest that this cold and hard substance was once, in the heat of the furnace, soft and ductile. And so with brass, and the leadiness of lead." In his work in Hampstead Garden Suburb he fitted simple wrought-iron latches and delightfully shaped tee-hinges to the thick wooden-planked cottage doors he used. For a grander house such as Blackwell in the Lake District he custom-designed all the metalwork, as well as the furniture, stained glass, and fabrics, prompting Hermann Muthesius in *The English House* to praise his "new idea of the interior as an autonomous work of art ... each room is an individual creation, the elements of which spring from the overall idea."

Decorative ironwork was put to bold use by Arts and Crafts architects such as Louis Sullivan, who designed rich and fantastic metal ornament, combining Renaissance and Gothic influences, to adorn

the Carson Pirie Scott department store in Chicago, and Charles Rennie Mackintosh, whose heraldic cast-iron features can be seen on both the interior and exterior of the Glasgow School of Art. The influence of Mackintosh can be felt in Antoni Gaudi's intricate and delicate cast-iron decorations on a number of his buildings in Barcelona and can also be traced in the sinuous Art Nouveau railings of Hector Guimard that still adorn the Paris Metro.

One of the most prolific ironworkers operating in the USA was Samuel Yellin, who opened a metalworking shop in Philadelphia in 1909 specializing in hand-hammered repoussé designs for architectural practices. His monumental gates and grilles were influenced by medieval originals although he used the latest methods and employed two hundred workers in his workshop and showroom.

Smaller workshops and guilds were more the Arts and Crafts norm, however. They were started up by artists working in the field of jewelry, silver, and copperware and many flourished, buoyed up on the wave of the Celtic revival, although they would be swamped by competition from the large commercial manufacturers, including Liberty & Co., which were quick to muscle in on the market, producing machine-made Arts and Crafts silver and copperware, embellished with enamel decoration and finished to look as though it might be hand-hammered, though usually it was not.

It is this hand-hammered finish that generally distinguishes an Arts and Crafts piece. Structural elements, such as rivets or nail heads, were often left visible to provide plain, honest decoration. In an article entitled "Metalwork" written in 1899, W.A.S. Benson—one of the period's foremost metalworkers—discussed the question of surface texture, contrasting "the natural skin of the metal solidified in contact with the mould" with carved or chased work and beaten or wrought work. Benson was a close friend of Edward Burne-Jones and William Morris, who helped him to set up a workshop in Fulham, producing simple turned metalwork, which he expanded two years later to include a foundry, adding a Kensington showroom and eventually an even larger showroom near Morris & Co. on New Bond Street. It was Benson who provided slender pendant light fittings for Standen (one of the first houses to have its own electricity generator) and other Morris & Co. commissioned interiors. He patented his own reflector shades and lanterns and took over as

Above: *A copper and brass hanging light by W.A.S. Benson.*

director of Morris & Co. on the death of Morris in 1896, while continuing to run his own firm until he retired in 1920. Business flourished for Benson even during World War I—an event that marked the end of so much that was good in the Arts and Crafts movement—for his workshop was commissioned to make various war items, including a brass model for torpedo sights that the jeweler J. Paul Cooper also worked on. After the war, Benson was commissioned to execute the presentation sword of honor for Field-Marshal Sir Douglas Haig.

Like William Burges, George Walton, Henry Wilson, and other Arts and Crafts designers, Benson had been influenced by Augustus Pugin's Gothic metalwork designs for goblets, candlesticks, and lanterns, simply decorated with precious metals and jewels, that featured in the two books—*Designs for Gold and Silversmiths* and *Designs for Brass and Ironwork*—published in 1836. However, his later machine-made pieces lean toward the modernity of Christopher Dresser and the angularity of European design. Indeed his work sold well in Paris through Samuel Bing's Maison de l'Art Nouveau.

Christopher Dresser's artistic energies extended to embrace most fields of Arts and Crafts design, but he excelled in his metalwork, from the cast-iron garden and hall furniture—umbrella and coat stands—he designed for manufacture by the Shropshire iron founders Coalbrookdale to the electroplate teapots and coffee services he designed for production by James Dixon & Sons of Sheffield. He had trained as a botanist and brought his almost mathematical interest in plant structure to his designs for silver, metalwork, glass, and pottery. His ideas were ahead of their time and anticipated many of the forms of Bauhaus silverwork, which was itself considered advanced and avant garde in the 1920s. "If the work be beautiful then it is ridiculous to estimate its value as though the material of which it is composed were of greater worth than the amount of life, thought, and painstaking care expended upon its production," he

asserted in 1873. Three years later he visited Japan and returned enthused and inspired. The Japanese were "the only perfect metalworkers which the world has yet produced," he declared, "for they are the only people who do not think of the material, and regard the effect produced as of far greater moment than the material employed." He incorporated Japanese decoration in his own designs and agreed to design a collection for Tiffany & Co. in New York. The cult of Japanese art, which had begun in Europe in the 1850s, was encouraged and taken forward by Arthur Liberty, who stocked highly colored decorative objects such as small silver hinged boxes cloisonné-enameled in an Anglo-Japanese style specially for the export market.

In the regions the best metalwork came out of newly formed schools—institutions such as the Keswick School of Industrial Art in the Lake District, where Harold Stabler taught, Liverpool University Art School which boasted R. Llewellyn Rathbone, and the Newlyn Industrial Class set up

Above: *Christopher Dresser—always ahead of his time—designed this stunning geometric teapot in 1880. It was made by James Dixon & Sons of Sheffield.*

Far left: *One of a pair of polished iron and copper candlesticks known as "The Cawder Candlesticks," made by George Walton, c. 1903.*

Left: *Silver tazza set with chrysoprase designed by C.R.. Ashbee for the Guild of Handicraft, c. 1900.*

by John D. MacKenzie to teach repoussé copper enameling (and embroidery) in Cornwall—and, of course, the guilds.

C.R. Ashbee found work for between twenty and twenty-five metalworkers and jewelers at his Guild of Handicraft in Chipping Campden. As well as the large vases and dishes in copper and brass, embossed with scrolling patterns of plants and fish, that they had produced in London, he expanded production to include jewelry, cutlery, plates, and other domestic ware, often medieval in inspiration, and adorned with simple decoration and stones such as rubies and pearls. Most of the designs were Ashbee's own, although he encouraged Guildsmen to experiment and choose their own stones or enamels.

Right: *A silver-mounted green glass decanter by C.R. Ashbee, c. 1904, marked for the Guild of Handicraft. The bottle was probably made by Powell of Whitefriars.*

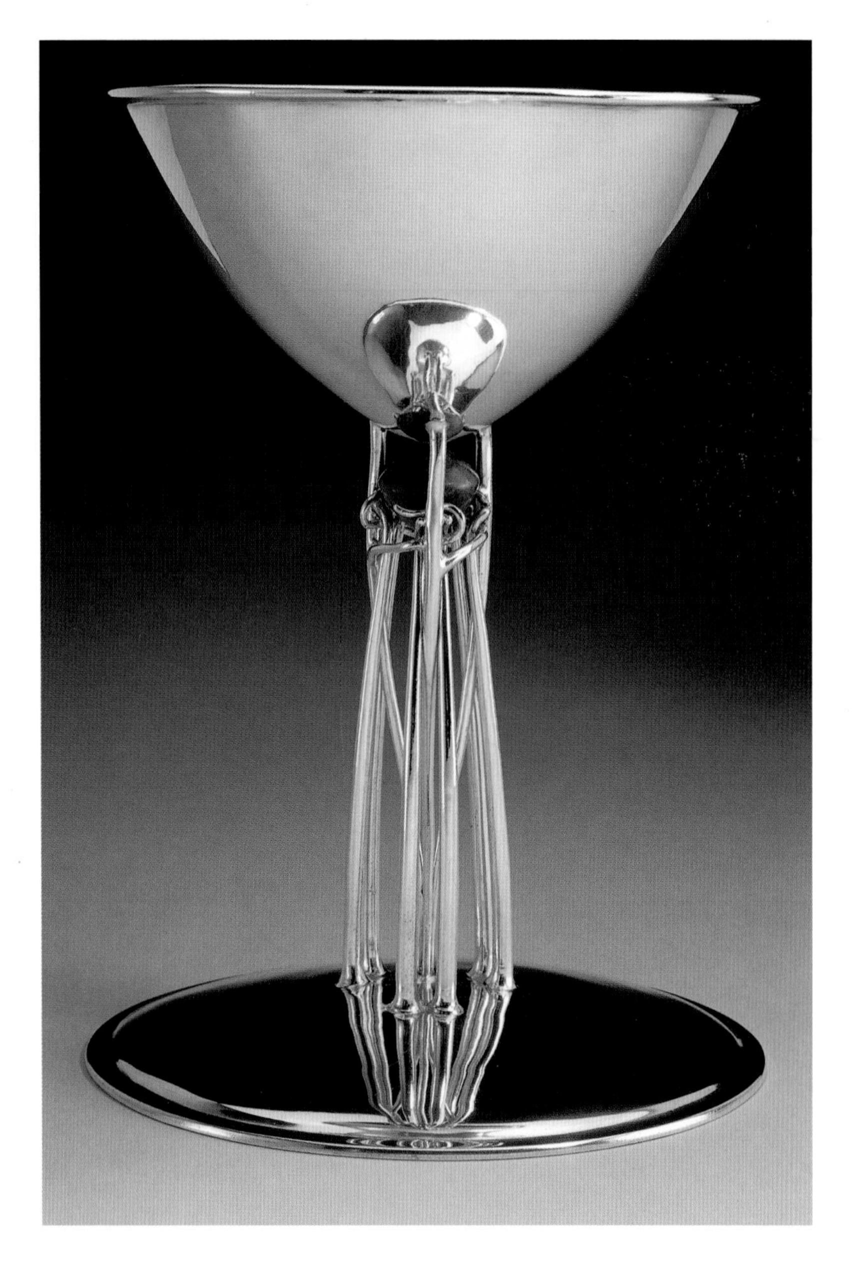

Above: *Silver cup designed by Archibald Knox for Liberty & Co., c. 1900.*

He recruited workers to the Guild very much at random. Will Hardiman, the first of his silversmiths, was a Whitechapel barrow boy; another metalworker, W.A. White, was persuaded to leave his job in a cheap bookshop and join the Guild after he had attended one of Ashbee's Toynbee Hall classes. Few were conventionally trained: they learned skills such as small casting and enameling very much on the job and yet these creative craftsmen were responsible for some of the most elegantly made hammered hollowware, jewelry, and other metalwork of the period. They produced the handmade beaten or hammered doorplates and firedogs that would become such a popular feature of an Arts and Crafts interior and were responsible for the metalwork fittings for the Grand Duke of Hesse's palace at Darmstadt, which they executed to Baillie Scott's designs. Ashbee quickly established the Guild of Handicraft's reputation for innovative design and marketed the fluid and graceful silver, metalwork, and jewelry they produced through the retail premises he had leased at 16a Brook Street in the West End of London. The Guild became famous for four classic silver designs: green-glass, silver-collared decanters; muffin stands; loop-handled bowls; and fruit stands. However, the unique style of the Guild of Handicraft soon found imitators—notably Liberty's, which Ashbee blamed very much for the collapse of his enterprise.

Liberty's sold Arts and Crafts-style gold, silverware, and jewelry in its new "Cymric" collection, which was launched in May 1899, much to the disgust of Ashbee, who firmly believed that the company was undercutting and plagiarizing his handmade items. The Cymric range of spoons, tankards, bowls, salt cellars, clasps, and jewelry proved so popular with women that Liberty's added a pewterware line in the late 1890s under the "Tudric" name. The Cymric range was handmade by the Birmingham-based firm of W.H. Haseler, and Liberty's put the project under "the fostering care" of a new manager, a Welshman called John Llewellyn. He had recently joined the firm from the fashionable Regent Street

Left: *A hot-water jug in "Tudric" pewter, designed by Archibald Knox for Liberty & Co., 1904.*

store Howell & James, and his Celtic roots are reflected in the names chosen for the two ranges. The press reaction was favorable. An article in *The Queen* praised the flower vases, bowls, and tea caddies in the Tudric range, calling them "a delight to those whose artistic instincts have been duly cultivated."

Arthur Liberty cultivated the public's taste by using some of his best designers—including Archibald Knox, Arthur Gaskin, Jessie M. King, the twenty-year-old Rex Silver, Alfred H. Jones, Bernard Cuzner, and Oliver Baker. Together they produced a wide range of jewelry, jewelry boxes, household items including tea sets, jugs, vases, candlesticks, and clocks, and decorative pieces such as cigarette cases and mirror frames decorated with turquoise enamelwork. Knox, who was one of Liberty's most prolific, imaginative, and innovative designers, came from the Isle of Man and, while

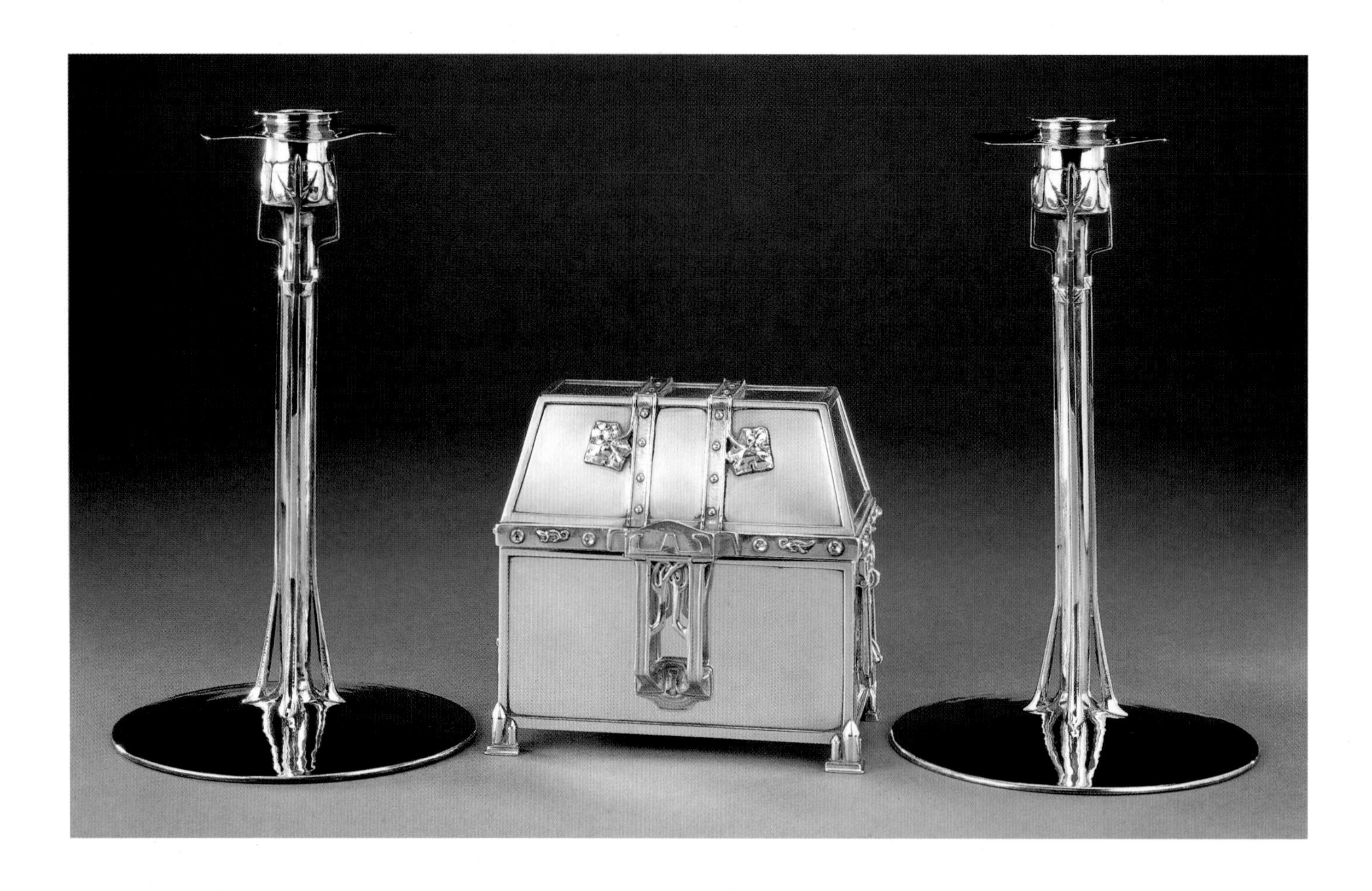

Above: *A silver casket by Alexander Fisher, c. 1900, flanked by a pair of "Conistor" Liberty silver candlesticks, 1906.*

attending the Douglas School of Art, had specialized in the study of Celtic ornament. He was the inspiration behind Liberty's Celtic revival, enthusiastically supported by John Llewellyn and Liberty himself. The extent of his contribution in the field is only just coming to light, unfairly obscured by the Liberty house policy that required all designers to be anonymous.

The Birmingham Guild of Handicraft was established in 1890 along idealistic lines very similar to Ashbee's Guild of Handicraft. Montague Fordham was the Guild's first director and the architect Arthur Dixon, who had worked for a year at Ashbee's School of Handicraft in Whitechapel (and presented him with an elegantly simple brass table lamp as a wedding present), was responsible for much of the design work. The workshop was run along cooperative lines, employing about twenty craftsmen, and produced fresh and distinctive Arts and Crafts pieces, including jewelry and belt buckles, teapots, lampshades, and light fittings, beaten from copper, brass, and silver sheet and fixed with rivets. No machinery, apart from a lathe, was used. Indeed, the members' motto was "By Hammer and Hand" and in this way, working from their premises in the medieval Kyrle Hall in Sheep Street, Birmingham, they made austere rather than ornate items: firedogs, fenders, and chased and

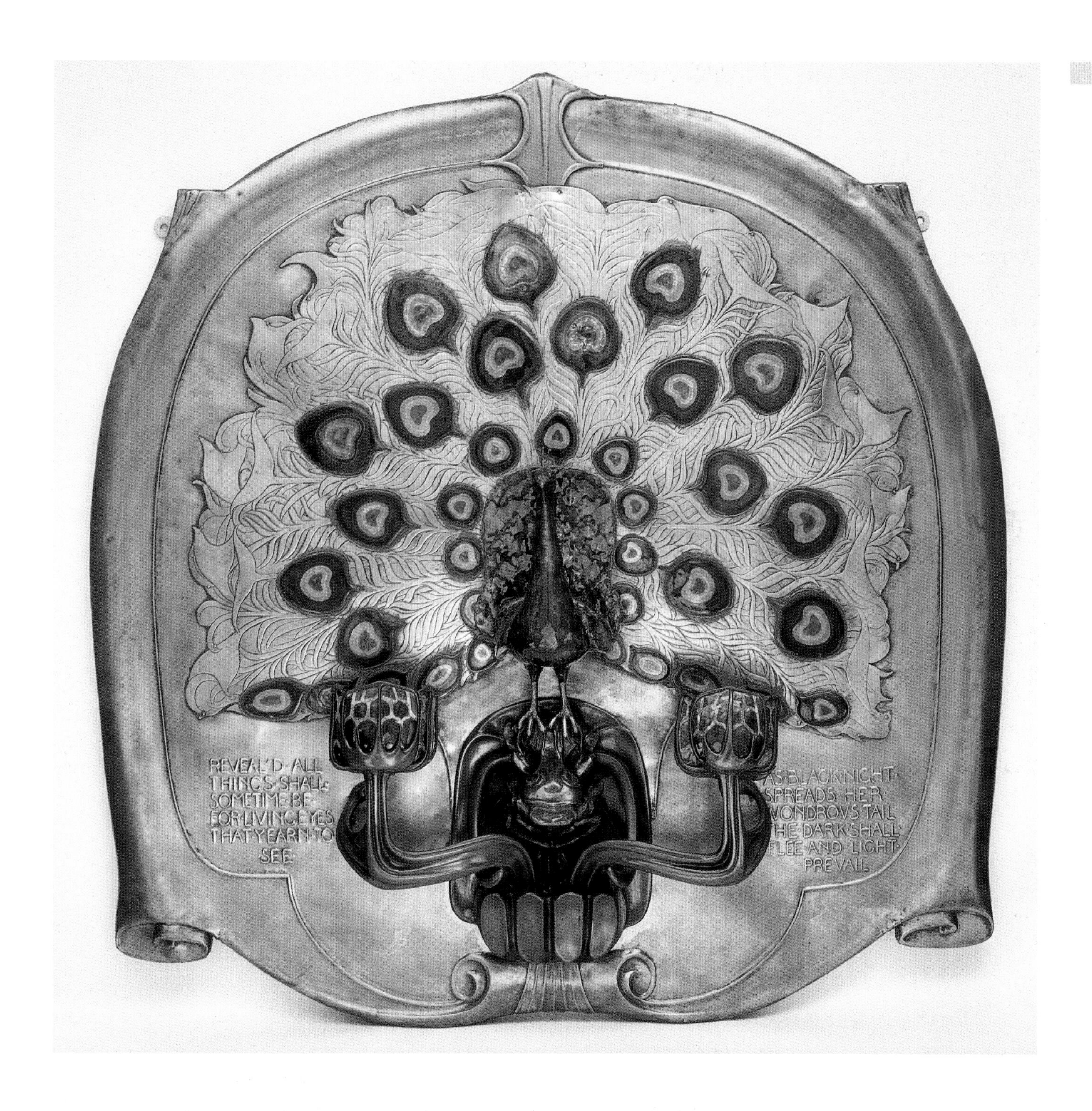

Above: *Peacock sconce in steel, bronze, silver, and brass, designed and made by Alexander Fisher, c. 1899.*

embossed door furniture—fingerplates, lockplates, and hinges made from flat pieces of brass, copper, and gunmetal. The pieces were largely anonymous, hallmarked with the Guild's stamp, and they were exhibited at the Arts and Crafts Exhibition Society in London in 1893, and the following year in Liverpool, Birmingham, and Paris. The Guild published a quarterly, hand-printed magazine entitled *The Quest* and maintained close links with the Birmingham Art School, where Arthur Gaskin

Above: *A pair of copper and wicker table lamps by Gustav Stickley, c. 1909.*

studied and later taught and where Bernard Cuzner headed the metalwork department. In his *First Book of Metalwork* (1931) Cuzner cautioned: "Of the two evils, affected roughness and mechanical smoothness, the first is more deadly by far."

The machine question was debated differently in the USA, where the Arts and Crafts ideal was more loosely interpreted and the emphasis on handmade articles was not so rigorous, although leading practitioners like Gustav Stickley prized craftsmanship above all. Disenchanted with the stamped hardware available, he founded his own metalworking shop. He wanted his hinges, key escutcheons, and handles to have the same structural and simple qualities as his furniture and he set up a smithy to forge them, along with lamp fixtures, chafing dishes, coal buckets, and other fireplace furniture. Stickley's copper, brass, and wrought-iron metalwork is characteristically handcrafted with obvious hammer marks and repoussé designs of simple and stylized floral patterns, English in derivation, made in copper and wrought iron. Later he extended his operation to include larger architectural elements, such as balustrades and fireplace hoods, which brought him closer to achieving his ideal of a fully integrated Arts and Crafts interior provided by Stickley & Co.

Metalwork of all kinds, but lamps in particular, became a popular means of expression and were prolifically produced by many American studios. The Dutch-born Dirk Van Erp opened his Copper Shop in Oakland, California, in 1908, moving in 1910 to San Francisco when he joined with the Canadian craftswoman Elizabeth D'Arcy Gaw. She stayed only a year as his design partner, but introduced a new level of sophistication. Van Erp's expertise was in metalworking, hammering, and the application of subtle patinas. The Copper Shop flourished and until 1929 retailed copper, brass, and iron accessories, along with copper lamps with mica shades. His success found imitators. Another immigrant, the German Hans W. Jauchen, opened a copper showroom in San Francisco in the 1920s in partnership with Fred T. Brosi. He called it Old Mission KopperKraft and produced and sold machine-made lamps that drew heavily on Van Erp's initial designs.

More in keeping with the Arts and Crafts ideal was the metalwork shop set up by Arthur Stone in Gardner, Massachusetts, along the lines of an English guild, with an apprenticeship system. Stone, who was a skillful craftsman well known for his chasing, piercing, fluting, and repoussé work, drew on Celtic, Gothic, Moorish, and Renaissance influences as well as natural forms—the barbed leaves of the arrowhead were a favorite motif—and produced beakers, tankards, flagons, porringers, and punch bowls.

Right: *A copper vase, with a flattened spherical collared rim, designed and executed by Dirk Van Erp, c. 1915.*

Above: *A hammered copper ink well with quatrefoil textured band by Karl Kipp, c. 1908.*

Jewelers such as Lebolt & Company in Chicago opened metalshops to produce hand-hammered coffee and tea services to meet the market's demand. The Heintz Art Metal Shop, established in 1905 in Buffalo, New York, by Otto L. Heintz, supplied decorative accessories and art-metal vases in bronze with handsome sterling silver overlays for over twenty years. Toward the end Heintz substituted copper for silver because it was softer and could be worked more quickly.

One of the most successful ventures was the Roycrofters Copper Shop, established by Elbert Hubbard in 1903, which produced, until 1938, hand-hammered copper vases, trays, bowls, candlesticks, lighting fixtures, and other small, functional objects such as bookends, for middle-class Americans to purchase on their visits to the Roycroft Community, or order from the Copper Shop's successful catalog. Many Roycroft copperware designs were stamped with borders to emulate leather stitching, or pierced with a bold geometric design. Hubbard put Karl E. Kipp, an Austrian former banker, in charge of production from 1908–11,

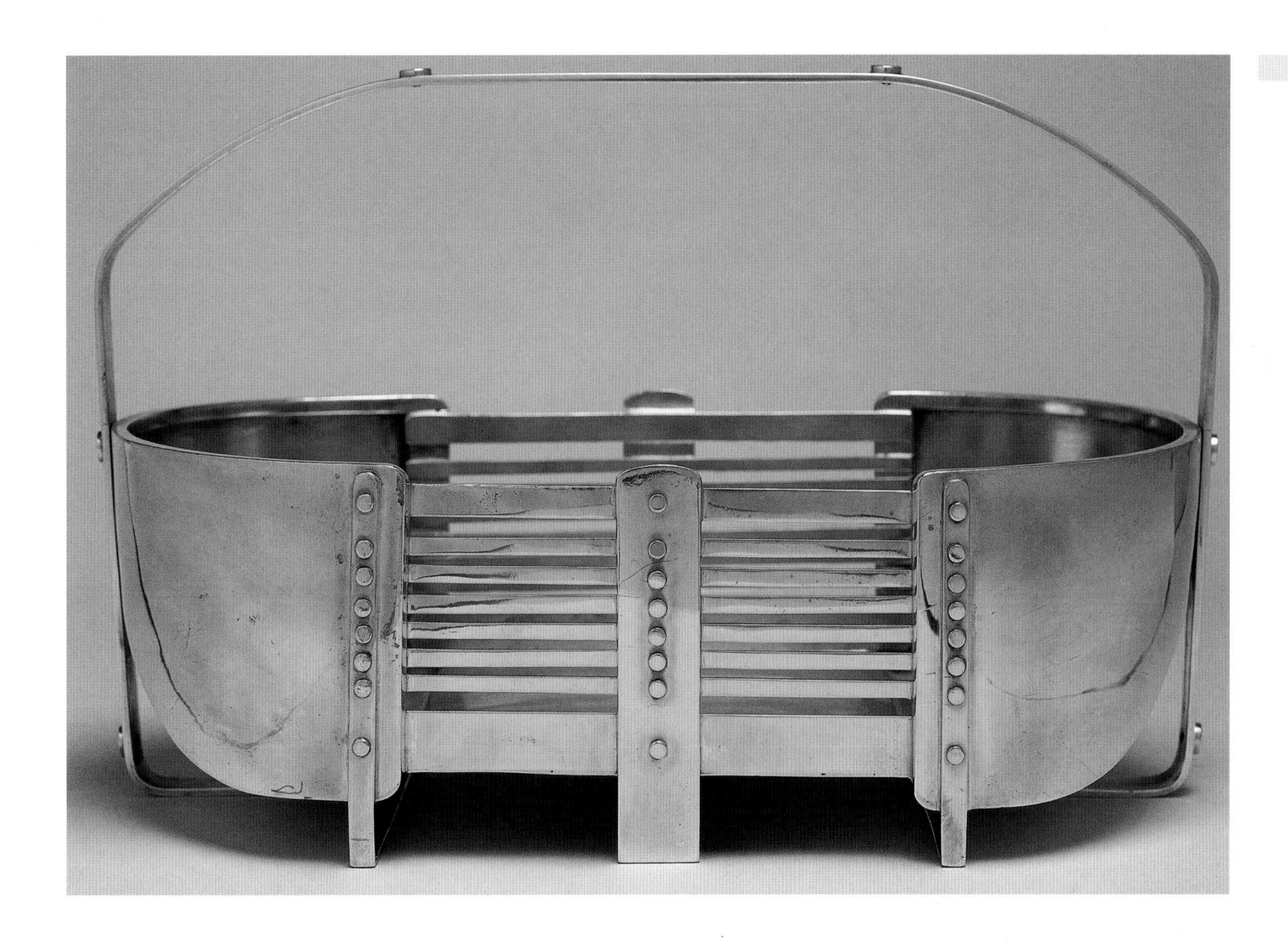

and he was responsible for turning the Copper Shop into a hugely profitable operation, employing thirty-five craftsmen and producing a range of over a hundred and fifty items, many inspired by the Modernist designs of Koloman Moser. Kipp left briefly in 1911 to set up his own studio, the Tookay Shop, where he signed his pieces with an encircled "KK," but was persuaded to return by Hubbard's son after the death of his parents aboard the *Lusitania* in 1915.

Another important Roycroft designer was Dard Hunter, who, in 1908, visited the Wiener Werkstätte and, upon his return, introduced geometric motifs in his copper-mounted table lamps and chandeliers, moving away from the more familiar organic shapes of most American Arts and Crafts metalwork. These European influences owed a lot to Charles Rennie Mackintosh, whose work had greatly informed that of Josef Hoffmann, Koloman Moser, and Carl Otto Czeschka. The members of the Wiener Werkstätte combined fine and applied arts and crafts to convey a whole new outlook on design. No everyday object was too banal, and they experimented with an infinite variety of materials from gold and precious jewels to papier mâché and glass beads, producing extraordinary and ultrafashionable work until 1931, when the workshops went into final liquidation.

Above: *A silver centerpiece by Josef Hoffmann, c. 1903, made by Wurbel and Szkasky, Vienna.*

Far left: *Rare Heintz sterling-on-bronze table lamp with a helmet shade, inset with several amethyst glass jewels and overlaid with a crest in a cross, c. 1910.*

Above: *A matching cream jug and sugar bowl designed by Clara Pauline Barck Welles at the Kalo Shop, c. 1908.*

The cultivated and stylish metalwork designer Robert Riddle Jarvie sold his work, including candlesticks and lanterns, through the Kalo Shop opened in Chicago by Clara Barck Welles. She was an indefatigable promoter of women's causes as well as a teacher, employer, designer, and retailer of handcrafted jewelry influenced by Ashbee and Art Nouveau, trays, jugs, and other classically simple household wares in silver and other metals. The Kalo Shop—the name derives from the Greek word *kalos*, meaning "beautiful," and its motto was "Beautiful, Useful, and Enduring"—continued to operate until 1970.

The best-known American designer in the field was undoubtedly Louis Comfort Tiffany, whose success in the field of art glass had led him to open a foundry in Corona, New York, to supply fittings and bases for his lamps. He also set up a small enamel department in 1898, staffed by a handful of young women and apprentices, who produced small items for as little as $10 and more ambitious and elaborate enameled lamp bases for as much as $900. A critic for *The Commercial Advertiser* commented on the "remarkable achievements in enamel on copper, in line with the other experiments made by Mr. Tiffany, wherein he has secured all the detail, sumptuous coloring and textures of the best of the European workers, retaining a personality quite his own. This enamel is on lamps, plaques, and small boxes, and is most effective."

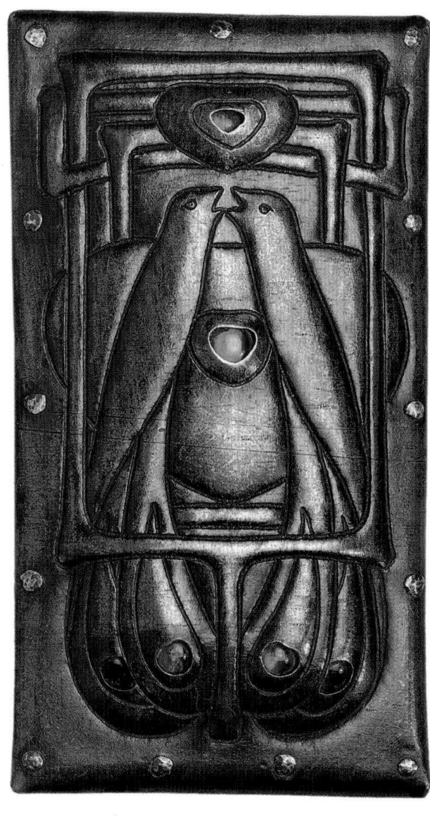

NATURAL, SHIMMERING BEAUTY

ARTS AND CRAFTS JEWELRY is characterized by its workmanship and the simplicity of the semi-precious stones employed. Translucent moonstones, used in combination with mother-of-pearl and amethyst, were popular with Arts and Crafts jewelers, as were pearls, opals, and opaque stones like coral and turquoise, malachite and lapis lazuli. C.R. Ashbee, who was self-taught, loved to use the kind of cheap colorful stones—aquamarines, moonstones, and topaz—despised by the rest of the trade. His designs followed clear, simple lines and reached back for inspiration to Renaissance masters such as Benvenuto Cellini. It was the Guild of Handicraft's metalwork and jewelry that established its

Above: *An aluminium buckle with green pastes and bird designs made around 1901 by the Scottish metalworker Talwyn Morris, who was linked to the Glasgow School.*

Left: *An enameled pendant and chain designed by C.R. Ashbee for the Guild of Handicraft, c. 1905, Above it is a Liberty & Co. necklace attributed to Archibald Knox, c. 1900.*

Above: *Alexander Fisher took the Tristan and Isolde story as his inspiration for this white-metal and enamel belt buckle, 1896.*

reputation for cutting-edge modernity. In *The Studio*, Aymer Vallance recognized that Ashbee's work was new and based on the idea that "The value of a personal ornament consists not in the commercial cost of the materials so much as in the artistic quality of its design and treatment." The Guild's customers were modern women such as Christabel Pankhurst, who wore a silver brooch designed by Ashbee in the form of a stylized flower head, set with an amethyst, a cabochon emerald, and pearls, which together made up the suffragette colors.

This new style of jewelry was worn in conjunction with the novel fashion for artistic dress: long, flowing, softly gathered silk teagowns that allowed the uncorseted wearer a degree of freedom hitherto unimagined. Against the background of these "Aesthetic dresses" with their straight, uncluttered lines, an Arts and Crafts enameled pendant, belt buckle, or cloak clasp could be seen to full advantage, with startling effect.

Alexander Fisher was the widely acknowledged master of the revived "Limoges" technique of enameling. He taught enameling at the Central School of Arts and Crafts in London from 1896, and became its most influential exponent in Britain, experimenting with firing techniques to produce a greater range of colors and pictorial depth, and reviving the medieval chalice and casket. Writing in his

Above: *The Diana diadem, depicting the goddess of hunting in the centre of the diadem, enameled and set with moonstones by Henry Wilson c. 1908.*

book *The Art of Enamelling on Metal* (1906) he maintained that enamel could reflect the "velvet of the purple sea anemone, the jewelled brilliance of sunshine on snow, the hardness greater than that of marble, the flame of sunset, indeed, the very embodiments in colour of the intensity of beauty." He also taught in his London studio, attracting aristocratic pupils for whom enameling became an artistic hobby as well as talented craftsmen and women for whom it became a rigorous professional discipline. For a brief period he entered into a partnership with the architect, designer, metalworker, and enameler Henry Wilson, a former pupil of J.D. Sedding. Wilson's metalwork style was a mix of Byzantine and late Gothic. He experimented with wirework and jewelry, going on to teach at the Central School of Arts and Crafts and from 1901 at the Royal College of Art. The immensely talented John Paul Cooper was one of his pupils. Cooper is best known now as a goldsmith, silversmith, and jeweler but he, too, began his working career in Sedding's congenial architectural office, where he first met Wilson, along with Ernest Gimson, Christopher Whall, and many other Arts and Crafts practitioners.

Cooper was responsible for almost fourteen hundred pieces of jewelry, along with silver frames, brooches, buckles and other metalwork, including domestic plate and hollowware, spoons, teapots, fruit stands, and liturgical silver. In about 1899 he began to work with the material he is most

Right: *A silver brooch set with opals, pink tourmalines, emerald pastes and pearls, part of a set, which includes a pendant necklace, by Arthur and Georgie Gaskin, c. 1914.*

associated with: shagreen, or treated sharkskin—a durable and scratch-resistant material imported from China, which he bought from the London-based firm W.R. Loxley and used on boxes and other pieces mounted in silver. These sold well and became very popular. Cooper made a remarkable gold ring in the form of a castle for the actress Ellen Terry, whose octagonal funeral casket, made to contain her ashes and decorated with repoussé panels of birds and foliage, was also his work. It is now in St. Paul's Church, Covent Garden, London.

In 1901 Cooper took up an appointment as head of the metalwork department at the Birmingham School of Art, an influential post he held for five years until pressure of work forced him to devote more of his time to production, based in his workshop at his house in Westerham in Kent. Between 1902–6 he received numerous commissions from Ernest Gimson for silver and brass handles to be mounted on his furniture.

Birmingham had been a traditional home of the jewelry trade since the eighteenth century although, according to one critic, it was at this time "a locality where a large amount of very deplorable jewelry is produced … and the reason why the vast mass of the trade jewelry manufactured in Birmingham is bad is that in style and outline it is utterly devoid of artistic inspiration." However, the Birmingham School of Art and the recently opened Vittoria Street School for Silversmiths and Jewellers were about to reverse the declining fortunes of the trade and encourage the establishment of a number of important silversmithing workshops in the town, specializing in bold, high-quality work in the Craft Revival or French Art Nouveau style.

The head of the new Vittoria Street school was the painter, enameler, and jewelry designer Arthur Gaskin, who, from 1902 until his retirement to the Cotswolds in 1924, influenced generations of students with his characteristic use of flowers and leaves in silver or gold, set with small colored stones, or decorated with pale opaque enamels. Gaskin formed a highly successful partnership with his wife Georgina Cave France. They favored polished semi-precious stones in their natural state rather than elaborately facet-cut precious stones: in 1915 they made a necklace from stones they collected on a Suffolk beach during a family holiday, including cornelians, agates, and quartz of various colors. Their work attracted notice and prestigious commissions such as the one from the Birmingham City Corporation for an elaborate enameled and jeweled necklace for Queen Alexandra. In an article entitled "The Jewelry of Mr. and Mrs. Gaskin" in *The Studio* the writer praised the originality of their "simple floral designs" and observed how "they did not, and possibly could not, then achieve the mechanical perfection of the trade jeweller."

Other artistic couples also made their mark in the field: Harold and Phoebe Stabler in Liverpool collaborated on setting and enameling, creating lovely gold pieces set with cabochon (rounded) amethysts, turquoise, mother-of pearl, and freshwater pearls. In their workshop in Chiswick, London, Edith and Nelson Dawson worked on their tiny jewel-like enamels mounted in silver, often depicting flowers or insects in minute detail. Both had studied enameling under Alexander Fisher. In 1901 they set up the Artificers' Guild, which was later acquired by Montague Fordham. One of Dawson's workshop employees, Edward Spencer, stayed with the Guild when Fordham bought it and he was responsible for ambitious and elaborate pieces of jewelry including a gold, silver, diamond, and opal "Tree of Life" necklace, complete with a gold phoenix rising from opal flames.

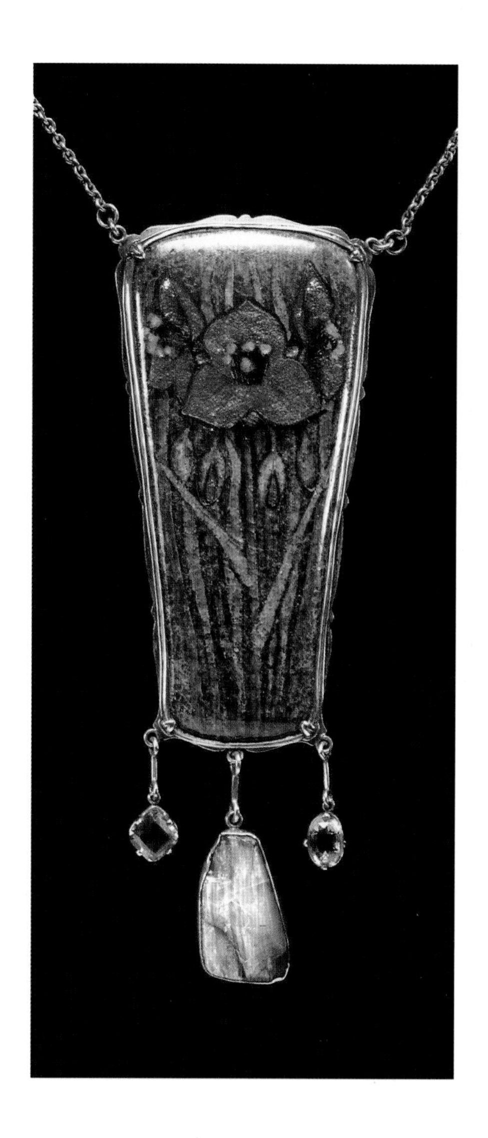

Above: *A gold and silver enameled pendant and chain by Edith and Nelson Dawson, 1900.*

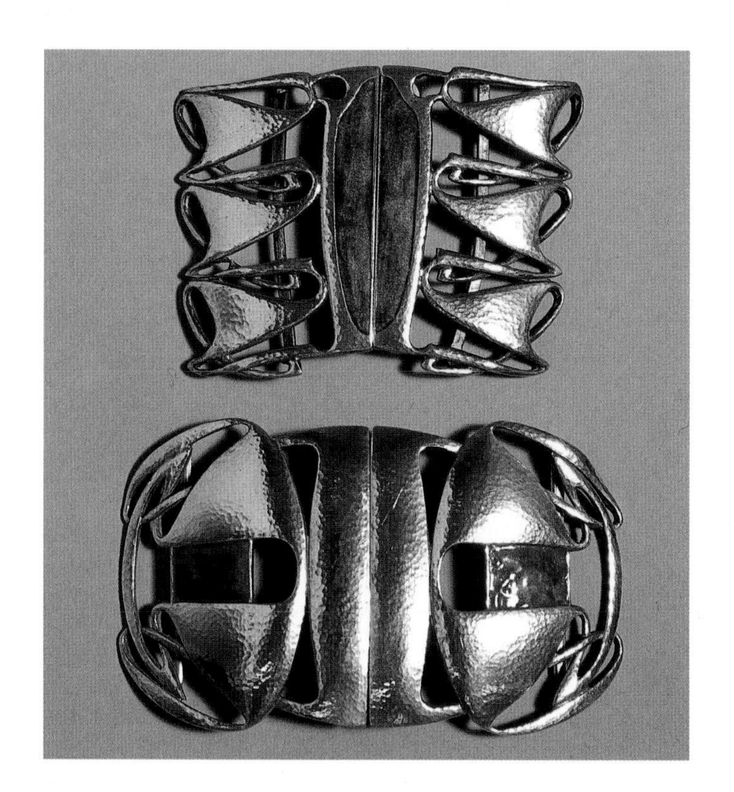

Left: *A pair of silver and enamel buckles designed by Archibald Knox for Liberty & Co., c. 1902.*

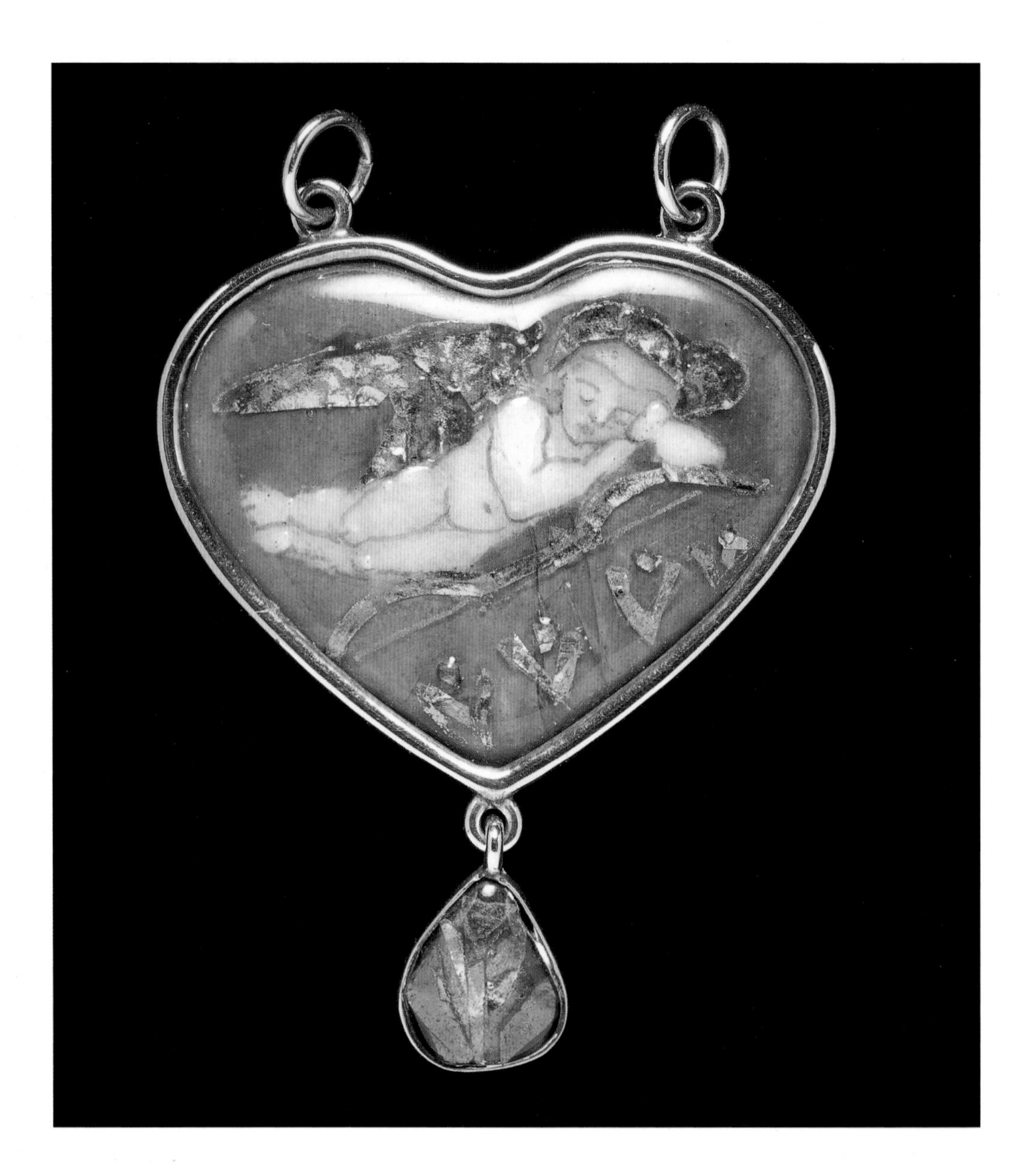

Right: *"Cupid the Earth Upholder," a gold and enamel pendant by Phoebe Traquair, 1902.*

The Arts and Crafts style of translucent enameling and heavily hammered silverwork was once again widely copied by commercial firms, which evolved a type of silverwork, often with a spurious "hand-worked" hammered finish, that was decorated with panels of shaded enamel. Liberty's was quick to board the bandwagon, but other firms including Charles Horner of Sheffield, William Hutton of Birmingham, and Murrle, Bennett & Co. of London mass-produced pieces of indifferent quality but outstanding design, thanks to the work of unsung figures such as Archibald Knox at Liberty, Kate Harris at William Hutton, and F. Rico and R. Win at Murrle, Bennett.

Jewelry was a field in which women excelled. To the roll call of Edith Dawson, Georgie Gaskin, Phoebe Stabler, and Kate Harris can be added the Scottish jewelry designers Jessie M. King, Frances and Margaret Macdonald, and the Dublin-born Phoebe Traquair, an exceptional colorist whose

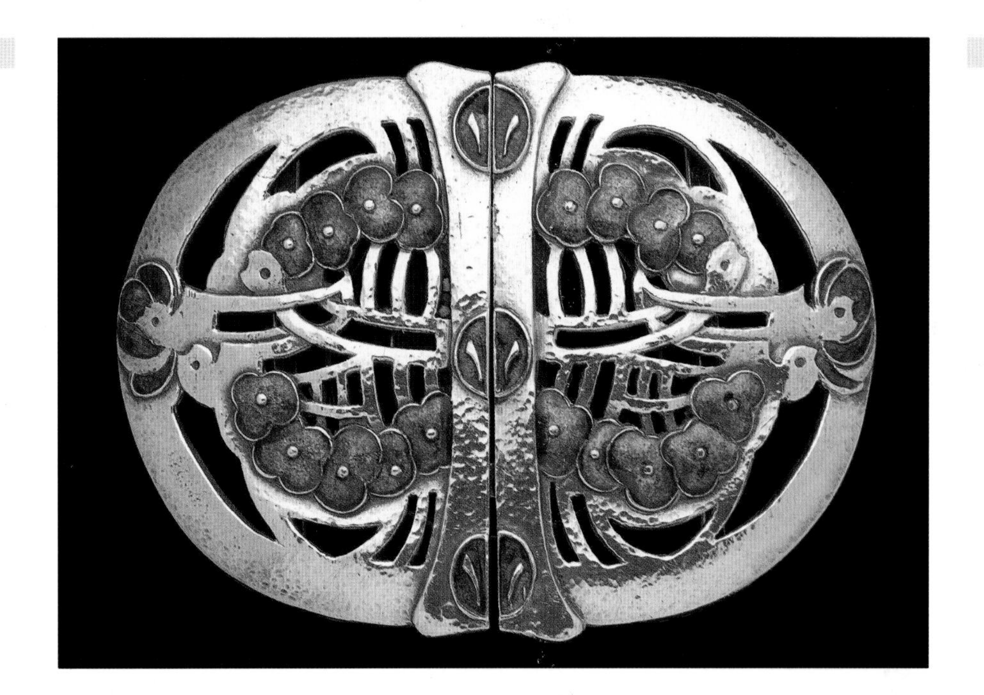

Above: *Silver and enamel two-piece buckle, pierced with a design of stylized Art Nouveau flowers, designed by Jessie M. King for Liberty & Co., marked "Birmingham, 1905–1906."*

lovely enameled jewelry drew on romantic Celtic legend. They designed for a new market, interested less in flashy ostentation and more in subtlety and sophisticated good taste, for the new Aesthetic woman—for themselves.

By 1902 Louis Comfort Tiffany's Tiffany Studios were producing a good deal of remarkable jewelry, with decided Art Nouveau leanings. Indeed, Alphonse Mucha—whose designs were heavily pirated in the USA — is thought to have collaborated with Tiffany on a number of pieces of jewelry. The Unger Brothers and William Kerr, both based in Newark, New Jersey, also produced a range of Art Nouveau jewelry designs. In Europe, jewelry designers such as Joseph Maria Olbrich, the leader of the Secessionists, and Richard Riemerschmid of the Dresden and Munich Werkstätten, produced work that, in its geometrical restraint and linear qualities, showed a growing appreciation of the Glasgow Four and the work of Charles Rennie Mackintosh in particular. The jewelry produced by the Wiener Werkstätte at the turn of the century incorporated formalized motifs taken from the natural world and included designs by Carl Otto Czeschka and Dagobert Peche, which were executed by trained craftsmen employed in the goldsmiths' department. The painter Emil Nolde designed a number of pieces of jewelry, mainly in silver or non-precious metal and semi-precious stones, and the catalogs of the Deutscher Werkbund—founded to bring order, unity, and commercial support to scattered groups—showed exquisite gold and silverwork by the Berlin-based goldsmith Emil Lettré. The Europeans signaled their modernity—and their move away from the Arts and Crafts ethic—not just in their designs but also in their materials, as in the chromium plating used by the Bauhaus designer Naum Slutksy, and in their unembarrassed embrace of the machine and all technological advances. From now on the emphasis would be on functionality and liberating design from the suffocating strictures of superfluous ornamentation.

9

THE PRINTED WORD

Right: *"A Book of Verse," William Morris's gift to Georgiana Burne-Jones on the occasion of her thirtieth birthday.*

32

A GARDEN BY THE S

For which I let slip all delight,
Whereby I grow both deaf and blind
Careless to win, unskilled to find,
And quick to lose what all men se

Yet tottering as I am and weak
Still have I left a little breath
To seek within the jaws of death
An entrance to that happy place
To seek the unforgotten face
Once seen once kissed, once reft f
Anigh the murmuring of the sea

THE BALLAD OF CHRISTINE.

Of silk my gown was shapen,
Scarlet they did on me
Then to the sea-strand was I borne
And laid in a bark of the sea.
O well would I from the World away

But on the sea I might not drown,
To me was God so good,
The billows bore me up aland
Where grew the fair green-wood

There came a knight a riding by
With three swains along the way
And took me up, the little-one
On the sea-strand as I lay

He took me up, and bore me home
To the house that was his own,
And there so long I bode with him
That I was his love alone.

But the very first night we lay abed
Befell this sorrow and harm
That thither came the king's ill men,

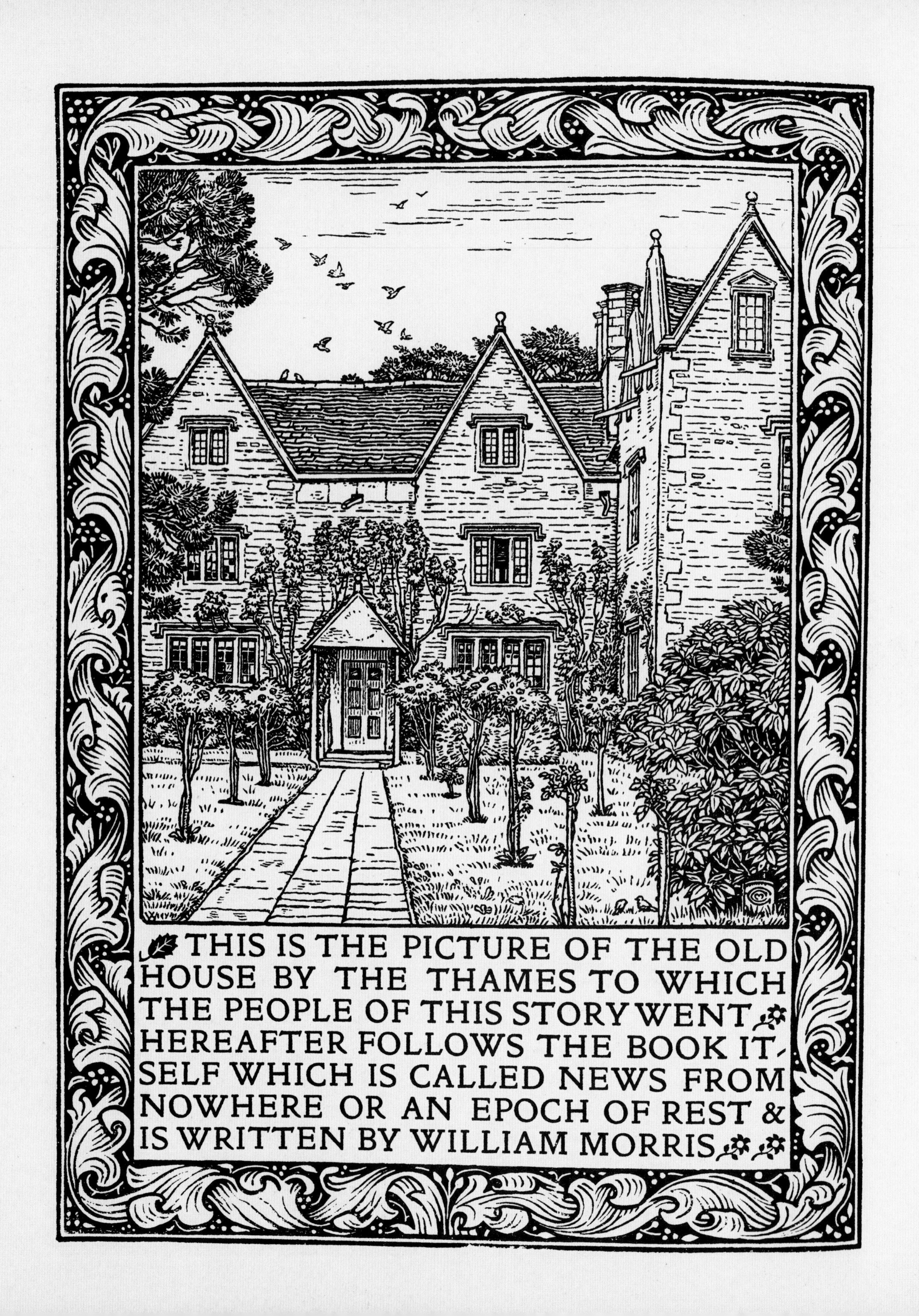
THIS IS THE PICTURE OF THE OLD
HOUSE BY THE THAMES TO WHICH
THE PEOPLE OF THIS STORY WENT
HEREAFTER FOLLOWS THE BOOK IT-
SELF WHICH IS CALLED NEWS FROM
NOWHERE OR AN EPOCH OF REST &
IS WRITTEN BY WILLIAM MORRIS

PRINTING & THE "IDEAL BOOK"

Above: *Initial letter from Felix Shay's posthumous biography of* Elbert Hubbard of East Aurora, *published in 1926, using Hubbard's original artwork.*

"The picture-book is not, perhaps, absolutely necessary to man's life, but it gives us such endless pleasure, and is so intimately connected with the other absolutely necessary art of imaginative literature that it must remain one of the very worthiest things towards the production of which reasonable men should strive."

William Morris, 1893

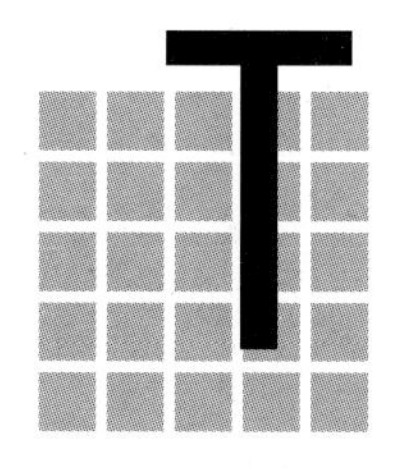

The Arts and Crafts period produced a number of fine private presses. Among them, in Britain, were the Doves Press (1900–17), set up by Emery Walker and William Morris's Hammersmith neighbor T.J. Cobden-Sanderson, for whom art was "every man's duty carried one stage further into beauty," and the Ashendene Press (1894–1935) set up by C.H. St. John Hornby, who would later become a partner in the firm of W.H. Smith. St. John Hornby was responsible for a stunning folio edition of Dante set in a typeface he devised himself, which he called Subiaco. Camille Pissarro's son, Lucien, who settled in England and was himself a painter and illustrator, ran the Eragny Press (1894–1914). Writing in 1900 in *The Ideal Book,* or *Book Beautiful*, Cobden-Sanderson claimed that these small presses contributed to "the wholeness, symmetry, harmony, beauty, without stress or strain" of life.

Importantly, women were involved. Middle-class women, still tied to the home, were able to learn the craft—from male professionals in the trade—of hand-tooling bindings. In 1898 a group of women, including Elizabeth MacColl, Florence and Edith de Rheims, and Constance Karslake formed a federation called the Guild of Women-Binders. In Edinburgh, Sarah Prideaux and Katherine Adams had their own bookbinding workshops, and W.B. Yeats's sisters Elizabeth and Lily, a former embroidery student of May Morris, founded the Dun Emer Press (1903–7) in Dublin.

Writing rather floridly in his essay on bookbinding in *Arts and Crafts Essays by Members of the Arts and Crafts Exhibition Society* Cobden-Sanderson put the case for "A well-bound beautiful book," claiming that it "is individual ... instinct with the hand of him who made it; it is pleasant to feel, to handle, and to see; it is the original work of an original mind working in freedom simultaneously with

Left: *The frontispiece for William Morris's Utopian* News from Nowhere, *published in 1893, which shows Morris's beloved country home, Kelmscott Manor, near Lechlade, Gloucestershire. This drawing of the east front was by C.M. Gere.*

Above: *A bookbinding by the craftswoman, Phoebe Anna Traquair, of embossed leather with spine entitled* Biblia Innocentum *and front cover decorated with scenes of the Creation of the World, Edinburgh, 1897–8.*

hand and heart and brain to produce a thing of use, which all time shall agree ever more and more also call 'a thing of beauty'."

The most important press of the period, however, the one that would have the most seismic effect on the future standards of book design, was undoubtedly William Morris's Kelmscott Press, set up in a rented cottage at 16 Upper Mall, Hammersmith, London, in 1891. "The mere handling of a beautiful thing seemed to give him intense physical pleasure," wrote Morris's first biographer J.W. Mackail, who, as the son-in-law of his closest friend Edward Burne-Jones, had ample opportunity to observe his subject. William Morris had always found books beautiful. As undergraduates at Oxford in the 1850s, he and Burne-Jones had spent long hours poring over "the painted books in the Bodleian." One of his favorite pastimes had been to scour antiquarian booksellers for finds and he bestowed generous gifts on friends, giving the impecunious Burne-Jones a fine edition of Malory's *Morte d'Arthur*. Thirty years later, dissatisfied with the poor quality of modern printing and design, he founded the Kelmscott Press with a view to reviving early Renaissance methods of book production and type design. It was a new passion at a relatively late stage in his life and represented an extraordinary labor of love. As with all his previous enterprises, he researched his subject thoroughly and poured all his energies into producing the best possible finished book. He had ink imported from Germany, paper specially made by hand from a fifteenth-century Venetian model, and he designed three of his own typefaces—Golden, Troy, and Chaucer—based on a Roman typeface designed by Nicholas Jenson, a fifteenth-century Venetian printer. Morris even designed the watermarks for the handmade paper and for a while experimented with making his own ink.

In 1895 the American publisher W. Irving Way persuaded Morris to print an American edition of Dante Gabriel Rossetti's *Hand and Soul*, for distribution by Way & Williams. It was the only Kelmscott Press book printed for the American market but it opened the way for the aesthetic revolution that followed in the field of American typography and printing. It also inspired the book designer Bruce Rogers and the printer Daniel Berkeley Updike, both of whom had worked at one

Below: *Daniel Berkeley Updike,* The Altar Book, *published in only 350 copies by the Merrymount Press in Boston in 1896.*

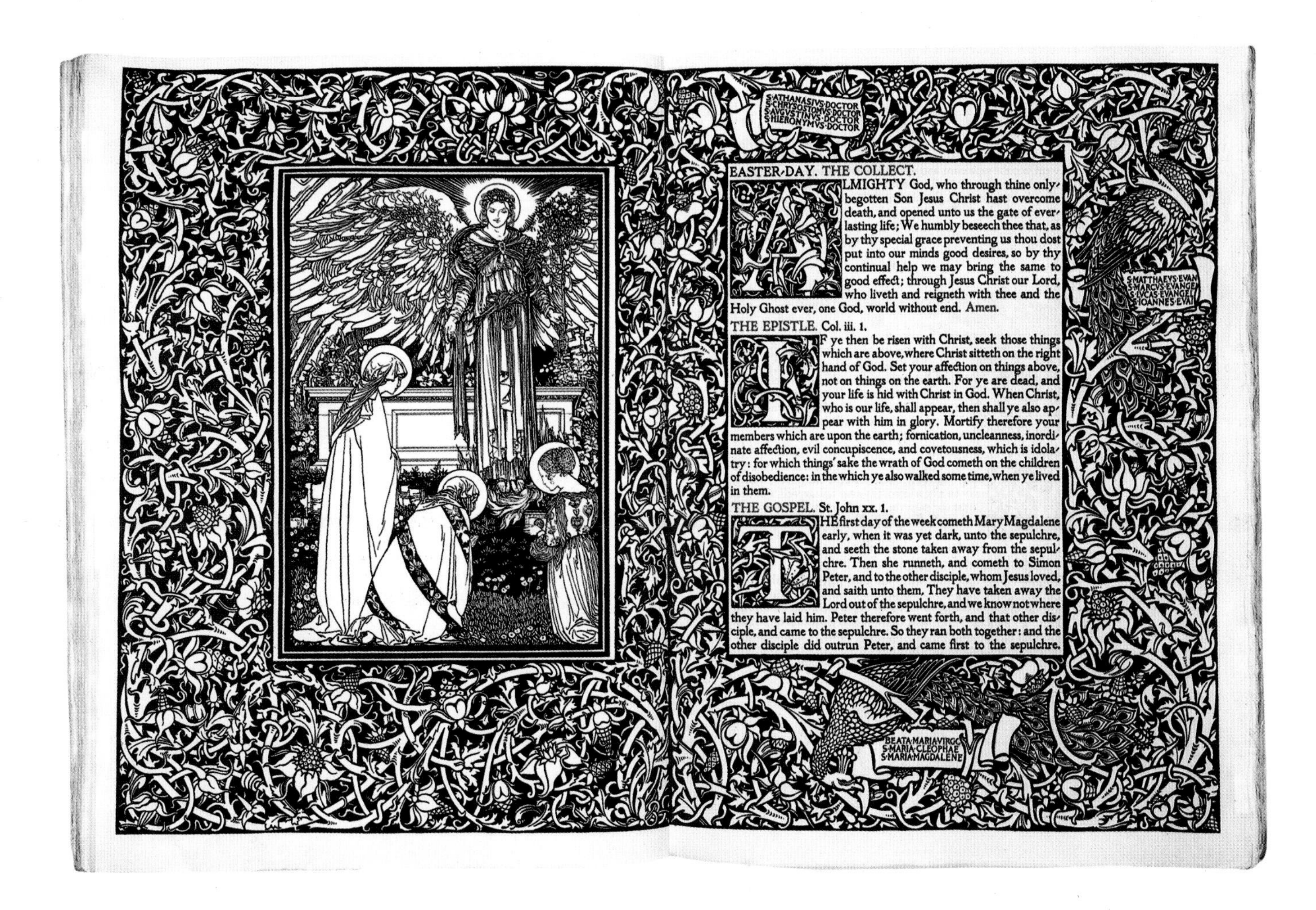

S·ATHANASIVS·DOCTOR
S·CHRYSOSTOMVS·DOCTOR
S·AVGVSTINVS·DOCTOR
S·HIERONYMVS·DOCTOR

EASTER-DAY. THE COLLECT.

ALMIGHTY God, who through thine only-begotten Son Jesus Christ hast overcome death, and opened unto us the gate of everlasting life; We humbly beseech thee that, as by thy special grace preventing us thou dost put into our minds good desires, so by thy continual help we may bring the same to good effect; through Jesus Christ our Lord, who liveth and reigneth with thee and the Holy Ghost ever, one God, world without end. Amen.

THE EPISTLE. Col. iii. 1.

IF ye then be risen with Christ, seek those things which are above, where Christ sitteth on the right hand of God. Set your affection on things above, not on things on the earth. For ye are dead, and your life is hid with Christ in God. When Christ, who is our life, shall appear, then shall ye also appear with him in glory. Mortify therefore your members which are upon the earth; fornication, uncleanness, inordinate affection, evil concupiscence, and covetousness, which is idolatry: for which things' sake the wrath of God cometh on the children of disobedience: in the which ye also walked some time, when ye lived in them.

THE GOSPEL. St. John xx. 1.

THE first day of the week cometh Mary Magdalene early, when it was yet dark, unto the sepulchre, and seeth the stone taken away from the sepulchre. Then she runneth, and cometh to Simon Peter, and to the other disciple, whom Jesus loved, and saith unto them, They have taken away the Lord out of the sepulchre, and we know not where they have laid him. Peter therefore went forth, and that other disciple, and came to the sepulchre. So they ran both together: and the other disciple did outrun Peter, and came first to the sepulchre.

S·MATTHAEVS·EVAN
S·MARCVS·EVANGE
S·LVCAS·EVANGEL
S·IOANNES·EVA

BEATA·MARIA·VIRGO
S·MARIA·CLEOPHAE
S·MARIA·MAGDALENE

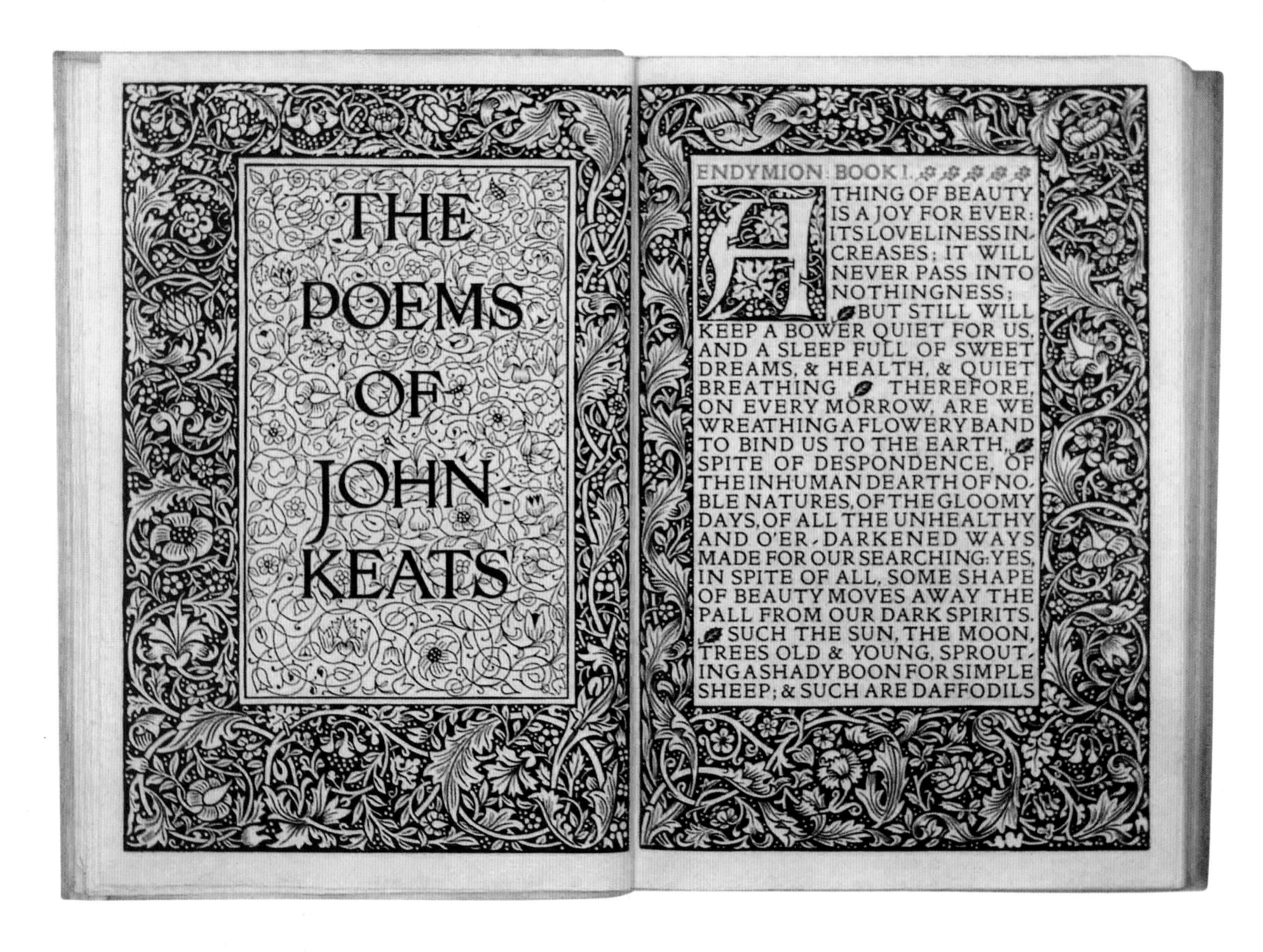
THE POEMS OF JOHN KEATS

ENDYMION: BOOK I.
A THING OF BEAUTY
IS A JOY FOR EVER:
ITS LOVELINESS IN-
CREASES; IT WILL
NEVER PASS INTO
NOTHINGNESS;
BUT STILL WILL
KEEP A BOWER QUIET FOR US,
AND A SLEEP FULL OF SWEET
DREAMS, & HEALTH, & QUIET
BREATHING. THEREFORE,
ON EVERY MORROW, ARE WE
WREATHING A FLOWERY BAND
TO BIND US TO THE EARTH,
SPITE OF DESPONDENCE, OF
THE INHUMAN DEARTH OF NO-
BLE NATURES, OF THE GLOOMY
DAYS, OF ALL THE UNHEALTHY
AND O'ER-DARKENED WAYS
MADE FOR OUR SEARCHING: YES,
IN SPITE OF ALL, SOME SHAPE
OF BEAUTY MOVES AWAY THE
PALL FROM OUR DARK SPIRITS.
SUCH THE SUN, THE MOON,
TREES OLD & YOUNG, SPROUT-
ING A SHADY BOON FOR SIMPLE
SHEEP; & SUCH ARE DAFFODILS

Above: *Title page of the Kelmscott Press Edition of* The Poems of John Keats, *published in 1894.*

time for Houghton, Mifflin & Co. and their subsidiary Riverside Press. Updike's *The Altar Book*, with illustrations by Robert Anning Bell, three years in the preparation and published in 1896, bears comparison with Morris's Kelmscott Chaucer.

"I began printing books," Morris wrote to an American friend in 1895, "with the hope of producing some which would have a definite claim to beauty, while at the same time they should be easy to read and should not dazzle the eye, or trouble the intellect of the reader by eccentricity of form in the letters.... As to the fifteenth-century books, I had noticed that they were always beautiful by force of the mere typography, even without the added ornament, with which many of them are so lavishly supplied. And it was the essence of my undertaking to produce books which it would be a pleasure to look upon as pieces of printing and arrangement of type."

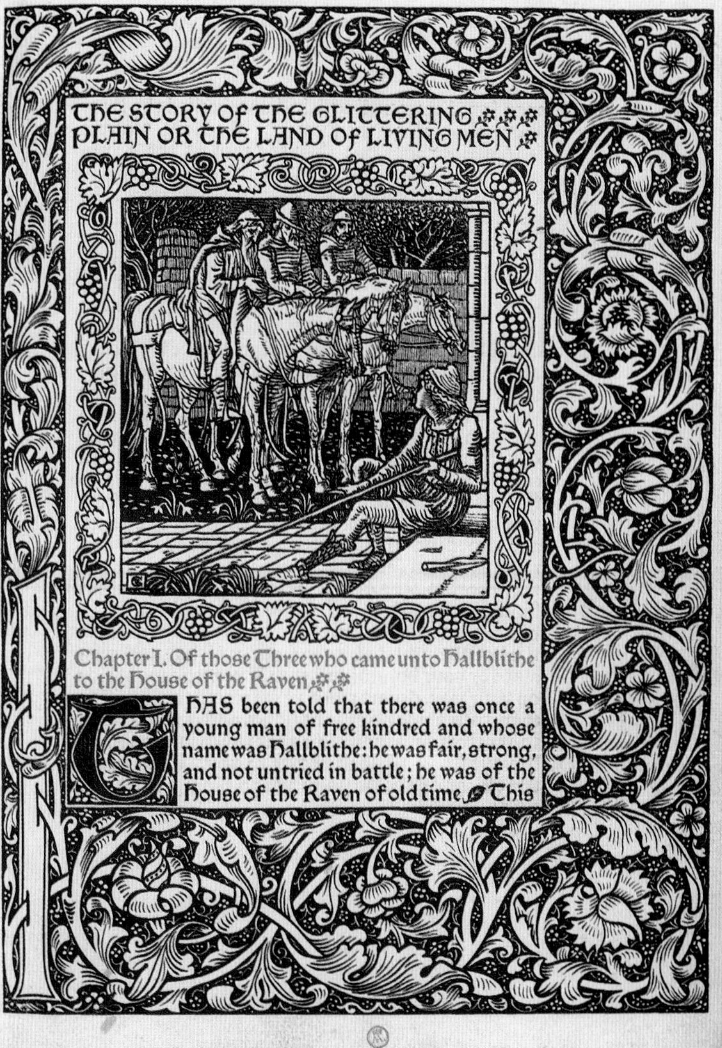
THE STORY OF THE GLITTERING PLAIN OR THE LAND OF LIVING MEN

Chapter I. Of those Three who came unto Hallblithe to the House of the Raven

T HAS been told that there was once a young man of free kindred and whose name was Hallblithe: he was fair, strong, and not untried in battle; he was of the House of the Raven of old time. This

Among the fifty-three titles produced by the Kelmscott Press were Morris's own writings, including *A Dream of John Ball*, Ruskin's *On the Nature of Gothic*, *The Poems of John Keats*, and, most famously, an illustrated edition of Chaucer. The Kelmscott Chaucer has been described as one of the greatest English books ever produced. It is certainly one of the most beautiful. Containing eighty-seven illustrations by Burne-Jones as well as Morris's beautifully decorated initials and intricately patterned borders, it took four years to produce. Thirteen copies were printed on vellum and forty-eight were bound in white pigskin with silver clasps; a special cabinet was designed by Charles Voysey to hold the book. Of the 425 copies that were finally printed, Morris only just lived to see one, for in 1896, after years of robust energy and astonishingly hard work, his health began to decline dramatically and, on October 3, at the comparatively young age of sixty-two, he died. His doctor, Sir William Broadbent,

Above: *Walter Crane designed the title page for William Morris's* The Story of the Glittering Plain or the Land of Living Men, *which was published by the Kelmscott Press in 1896, the year Morris died.*

Left: *A proof of Edward Burne-Jones's illustration for the "Tale of the Clerk of Oxenford," included in the Kelmscott Chaucer, with a hand-painted border by William Morris, 1896.*

famously declared the cause of death to be "simply being William Morris and having done more work than most ten men."

The Kelmscott Press did not long outlive Morris. After its closure, C.R. Ashbee, who had already published a number of books through the Guild of Handicraft, bought one of the Kelmscott's Albion presses, took on some of Morris's craftsmen, and set up the Essex House Press, printing beautiful books such as *The Treatises of Benvenuto Cellini on Goldsmithing and Sculpture* (1898) on handmade paper using Caslon Old Face to achieve the same powerful effect as Morris had with his Chaucer. However, competition from American small presses, and from machine-printed books in a hand-press style, put the Essex House Press under serious financial pressure from the outset. (The second of the Kelmscott Albion presses was bought by Theodore Low de Vinne and Bruce Rogers, who shipped it to the USA and started up one of the many hand-press workshops that mopped up Ashbee's American market.) In 1904 Ashbee cut the hours of his Chipping Campden printers, putting them on half-time working, but soon there was another blow: the order for his two-volume edition of the Bible was cut from three hundred to a mere one hundred copies. Despite having asked William Strang to produce sixty woodcuts, Ashbee had to abandon the edition and was soon dismantling the Press altogether.

He recorded "rather a sorrowful afternoon," in his journal as he described taking stock of "the goods of the Essex House Press" with "old Binning," his chief compositor, who had worked with Morris before joining the Guild and been in the business for over twenty years. Presciently he wrote: "I think the book collectors of a day to come will probably prize some of the books—probably the wrong ones but certainly some," and so it proved. Books such as the Essex House Press edition of Shakespeare's *Poems*, bound in red morocco and outlined in gilt, and Ashbee's own *Endeavour towards the teaching of John Ruskin and William Morris*, bound in green morocco with an onlaid and tooled pattern, which then he could hardly give away, now fetch substantial prices when they come up for auction.

One notable figure who was, for a while, on the fringes of the Arts and Crafts movement, was the artist-craftsman Eric Gill, a typographer and lettercutter of genius, who also mastered the arts of wood-engraving and sculpture. He was interested in the movement and went several times to the Arts and Crafts Exhibition Society show in 1903, exhibiting with them himself in 1906. Commissions came in from Arts and Crafts architects including Ernest Prior and Charles Harrison Townsend (who was responsible for the Whitechapel Gallery and the Horniman Museum), for the lettering on the new Medical School in Cambridge and the lettering of the lichgate for Townsend's church of St. Mary at Great Warley in Essex. Gill found himself part of a brotherhood of architectural craftsmen who saw lettering not as an afterthought but as intrinsic to a building. In 1903 St. John Hornby invited him to paint the name W.H. Smith & Son on the fascia of his first Paris bookshop and then to paint all Smith's signs over a two-year period, creating one of the most recognizable "brands" of the period. Ambrose Heal also valued the "recognition factor" that Gill's plain but beautifully proportioned lettering gave to Heal's, employing him to design the distinctive lettering for the shop, its catalogs, and headed paper.

Above: *An illustration from* Conradin: A Philosophical Ballad, *published by C.R. Ashbee's Essex House Press in a limited edition of 250 printed on paper and this single example on vellum in 1908.*

Gill was in tune with the wider aims of the Arts & Crafts movement—the striving for a simple existence, living in the country, where one made useful things by hand that improved both one's own life and the wider world. He shared the Ruskinian optimism about the creativity of the worker and the social aims of improving the lot of working people. From 1905 he lived for two years in Black Lion Lane, Hammersmith, in the center of an Arts and Crafts community that included May Morris, the metalworker Edward Spencer of the Artificers' Guild, and the Emery Walkers, who all lived in Hammersmith Terrace. Walker's Doves Press was, at the time, completing its most

Above: *Design for the coat of arms of Thomas à Becket by Eric Gill for Messrs. Burns & Oates Limited, London, 1914, showing three Cornish choughs.*

ambitious project, *The English Bible*, printed in five volumes bound in white vellum. The Hammersmith years and the ideas he gleaned there enriched Gill's life in the country, first at Ditchling in Sussex and later at Capel-y-ffin in the Black Mountains of Wales, where he could implement his "idea that life and work and love and the bringing up of a family and clothes and social virtues and food and houses and games and songs and books should all be in the soup together."

The Arts and Crafts period encompassed a golden age in children's book illustration, as figures like Kate Greenaway and Randolph Caldecott came to prominence alongside Walter Crane, whose highly decorative illustrations for nursery rhymes and fairy tales provided idealized glimpses of model Aesthetic interiors and helped to shape the taste of the artistic middle class. Greenaway worked in a studio at the top of a romantic house in Hampstead designed for her by her near neighbor, the Arts and Crafts luminary Richard Norman Shaw, producing illustrations of tidy children in Queen Anne costumes that proved enormously popular, attracting the praise even of Ruskin. Meanwhile, new illustrators including Aubrey Beardsley, Charles Ricketts, and, in the USA, Will Bradley, were incorporating the sinuous lines of Art Nouveau into their book and poster illustration, producing a distinctive, decadent style.

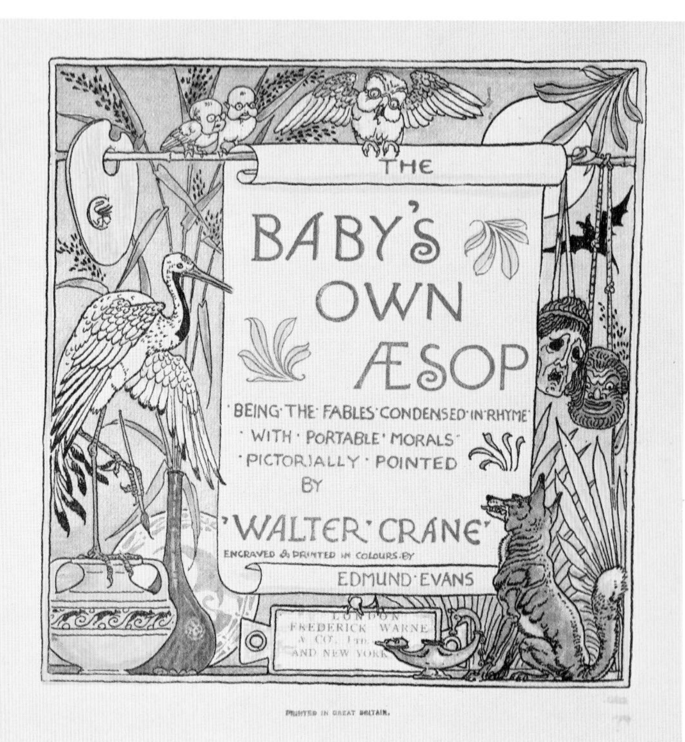

Left: *Frontispiece and title page from Walter Crane's* The Baby's Own Aesop, *published by Frederick Warne, London, c. 1877.*

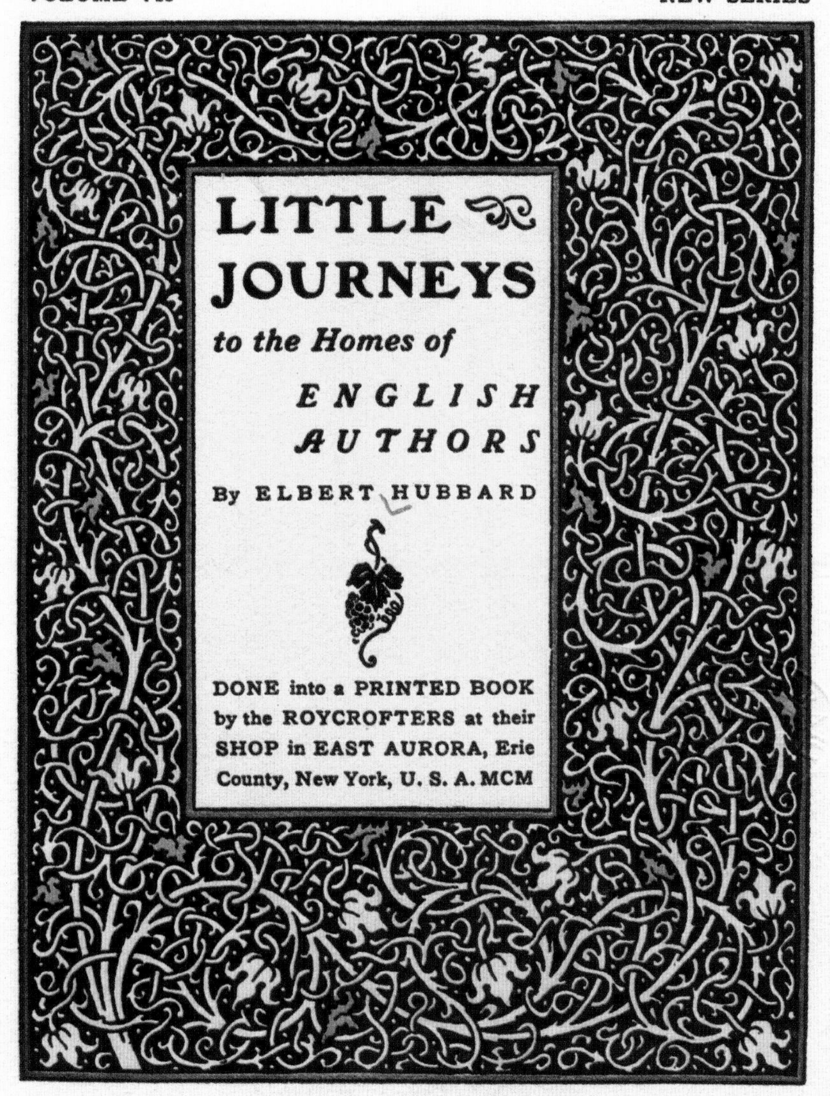

THE EXAMPLE OF MORRIS'S KELMSCOTT PRESS inspired not just Elbert Hubbard's Roycroft Press in East Aurora, New York, but more than fifty small presses that came into existence in the United States from 1895–1910, including the Riverside Press, Copeland & Day, the Merrymount Press of Boston, the University Press of Cambridge, Massachusetts, the De Vinne Press in New York, the Blue Sky Press in Chicago, and Will Bradley's Wayside Press in Springfield, Massachusetts. All reflected the influences of the Arts and Crafts movement in their emphasis on craftsmanship, their use of fine handmade materials and the dense medieval appearance of the type. Editions were limited, signed, and illustrated with woodcuts or wood engravings.

Above: *Title page, illuminated by hand by Maud Baker, of the Roycroft edition of* Little Journeys to the Homes of English Authors *by Elbert Hubbard, published in a limited edition of less than 150 in 1900.*

SO THIS THEN IS YE

RIME

of ye

ANCIENT MARINER

WHEREIN

Is told Whilom on a Day an Ancient Sea-Faring Man Detaineth a Wedding-Guest & Telleth him a Grewsome Tale.

Written by SAMVEL TAYLOR COLE-RIDGE

For ye better Understanding of ye Gentle Reader, Various Pictures are here Inserted by one William W. Denslow

Ye First Edition Corrected and Improved

Done into a Booke by ye merrie ROYCROFTERS at ye *ROYCROFT SHOP*, at ye Sign of ye *Hippocampus*, adjacent to ye Deestrick Academy for ye Younge, which is in *East Aurora*, New York, United States of America. *1899*

Right: *Title page from Elbert Hubbard's edition of Coleridge's "The Rime of the Ancient Mariner," published in East Aurora, New York, 1899.*

Above: *Design on the signature page of* Little Journeys to the Homes of English Authors, *by Elbert Hubbard, 1900.*

True to Arts and Crafts principles, Roycroft books used handmade paper, chamois or vellum bindings, hand-illuminated initials, bordered title pages, and Gothic typefaces, to convey a medieval impression. The first Roycroft book was *The Song of Songs: Which Is Solomon's*, printed in January 1896. The hand tooling on the binding was done in the Roycroft leather shop (which also produced wallets, manicure cases, photograph frames, and bookends and was headed by a German craftsman with Art Nouveau leanings called Frederick C. Kranz, who joined Roycroft in 1903); a young woman called Lucy Edwards, who joined in 1898, oversaw the hand-painted illuminations.

The typographical revolution extended to Europe, where designers such as Rudolf Koch, Peter Behrens, and Henry van de Velde turned the British revival of calligraphy to great effect, creating a new typography based on Fraktur types, and publishing books, posters, and magazines in the new *Jugendstil*, or Art Nouveau style.

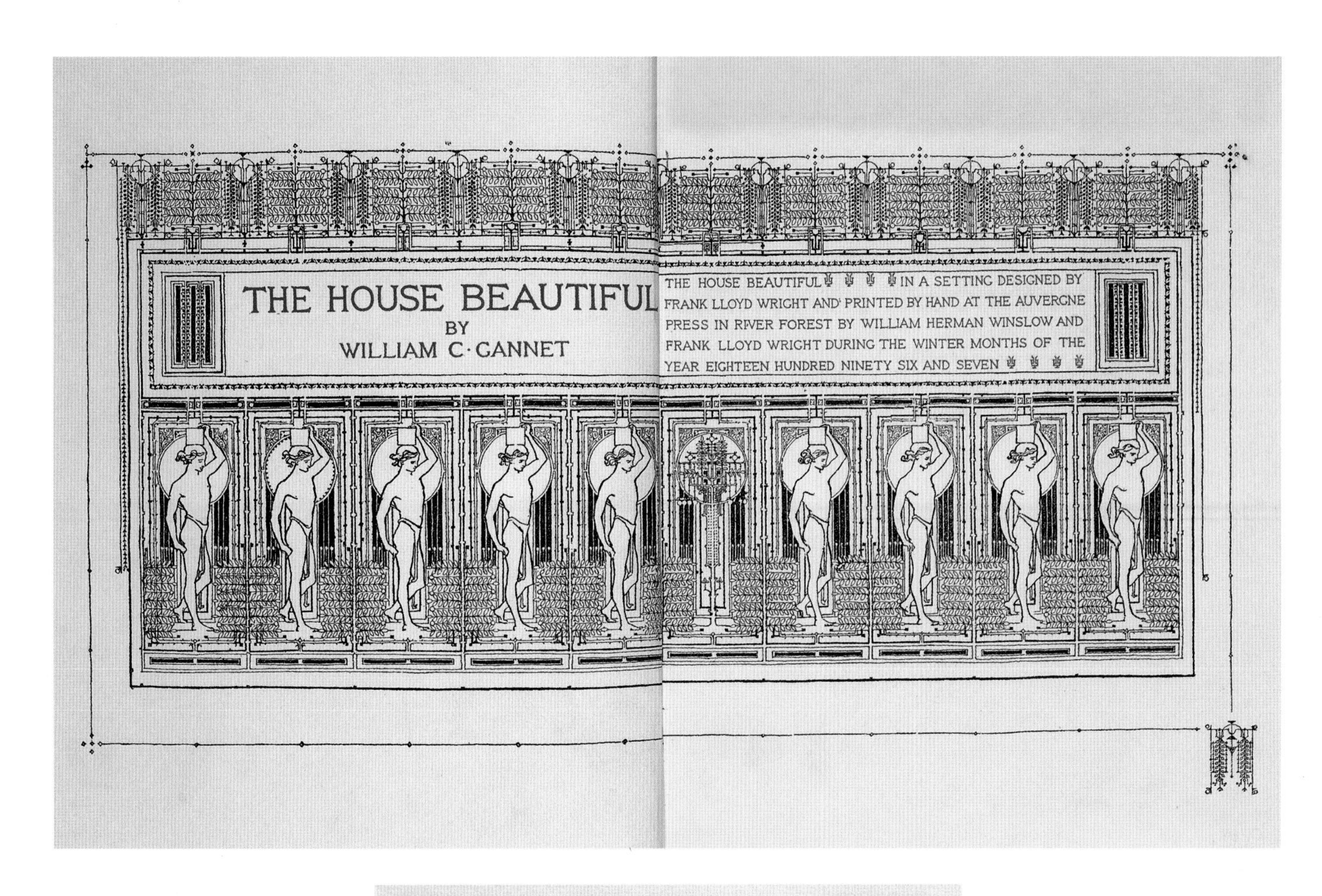

Above: *Title page of* The House Beautiful *by William Channing Gannet, Auvergne Press, River Forest, Illinois, designed by Frank Lloyd Wright, who was commissioned by the publisher, William Herman Winslow, after he had designed a house for him in 1893.*

Left: *Morris's tailpiece to the* Poems of John Keats, *1894.*

GARDENS

Left: *The borders looking toward the stone summer house in the garden at Rodmarton Manor. Well-protected by the yew hedge, tall wall, and the topiary, the herbaceous plants can achieve ten or so feet in a season.*

10

MORE OR LESS WILD

"A garden scheme should have a backbone—a central idea beautifully phrased. Thus the house wall should spring out of a briar bush—with always the best effect, and every wall, path, stone and flower bed has its similar problem and a relative value to the central idea."

EDWIN LUTYENS, writing to his wife Lady Emily, 1908

THE IDEAL ARTS AND CRAFTS GARDEN found inspiration in a rustically romantic old England, turning away from the hard-edged geometry of the rigidly planned Victorian garden to embrace the tumbling profusion of old-fashioned flowers such as poppies, foxgloves, roses, lavender, lupins, irises, delphiniums, phlox, sunflowers, and pinks. While the poor used their gardens to grow food for subsistence, the rich employed teams of gardeners to groom what was essentially a status symbol. Now a burgeoning new middle class saw the garden as an extension of the home, something to be decorated and a place that provided a respectable channel for the creative energies of the lady of the house.

The pioneering garden designers William Robinson and Gertrude Jekyll were already steering fashion away from the elaborate formality and extreme symmetry of the high Victorian garden, dependent on vast seasonal bedding schemes, toward natural groupings and an idealized rural profusion of sweet-smelling flowers such as hollyhocks and wallflowers. "I believe," Robinson wrote, "that the best results can only be got by the owner who knows and loves his ground. The great evil is the stereotyped plan...." He enjoyed much popular and financial success with his books *The Wild Garden* (1870) and *The English Flower Garden* (1883), which promoted the kind of indigenous British plants that inspired William Morris's textile and wallpaper designs: roses on trellis, hollyhocks, sunflowers, and fiery nasturtiums. Passionate and persuasive, Robinson put his ideas into practice in his own garden at Gravetye Manor, a handsome Elizabethan gabled mansion house that he bought in 1885, at the age of forty-seven, with the proceeds of his writing. The house was exceptionally situated in rolling country in West Sussex and it was there, over a period of fifty years, that he

Above: *Gertrude Jekyll's main border at Munstead in 1900, depicted here in a watercolor by Helen Allingham.*

Left: *Gertrude Jekyll's gloriously crowded borders tumble across the aster walks leading up to the north face of Munstead Wood.*

Right: *Engraving from William Robinson's book* The English Flower Garden, *which popularized indigenous British plants, like hollyhocks, sunflowers, and nasturtiums, along with roses on trellises,when published in 1883.*

created his famous garden. He planted tea roses in the flower garden, cutting them back to the ground in severe winters so that he would be rewarded the following season with magnificent flowers right through to October. He planted pansies in "colonies and bold groups ... never in lines and never dotted about singly" and favored mixed plantings of roses, agapanthus, forget-me-nots, campanulas, and carnations. A vast oak pergola, covered in a froth of white wisteria, connected the formal garden to Robinson's "alpine meadow" situated on a steep slope to the south of the house, where he naturalized anemones, scillas, erythroniums, fritillaries, and great drifts of daffodils, which ran down to the fringe of the lake where he had planted white willow. He held strong views and was not interested in discoursing in Latin, insisting that every plant should have an English name. He instructed his head gardener Ernest Markham not to cut the grass until late into the summer to allow the seed to ripen and encouraged the village children to come and play on a sloping bank on the east side of the house. Every year on his birthday they would come and dance around the maypole for him—even though his birthday fell in July—and at ninety-five he was drawing up ambitious plans to move the old orchard and plant a new one.

Robinson's ideas were enthusiastically embraced in Britain, where nostalgia for a simple country life had been fanned by picturesque images of plumply thatched cottages with open rose-framed doors by painters like Helen Allingham. Only a generation earlier, such dwellings would have been looked upon with suspicion and distaste as insanitary hovels, but now they became desirable, even fashionable, and were purchased as weekend retreats by better-off members of the middle class and newly wealthy industrialists, who eagerly embraced the Arts and Crafts ethos of a return to the natural garden. Robinson was also widely read in the USA, where his ideas were put into practice by the Arts and Crafts proselytizer Charles Fletcher Lummis in his wild flower garden at El Alisal in southern California, which he liked to refer to as "the Carpet of God's Country."

"On my own little place there are, today," he boasted in 1905, "at least forty million wild blossoms by calculation. Short of the wandering and unconventional foot-paths, which are almost choked with the urgent plant life beside them, you cannot step anywhere without trampling flowers—maybe ten to a step, as a minimum."

Above: *The timber bridge designed by Philip Webb to link the library wing with the garden beyond the moat at Great Tangley Manor.*

As architects began to take a greater interest in landscaping their sites, a fierce contemporary debate arose concerning the relative merits of the formal versus the natural garden. John Dando Sedding's *Garden Craft Old and New* (published posthumously in 1891) and Reginald Blomfield's *The Formal Garden in England* (1892) championed the case for the architect as overall garden designer, arguing for a return to first principles and the reintroduction of early seventeenth- and eighteenth-century architectural details such as sundials, fountains, and gazebos. Topiaries and trellised walks became a feature of Arts and Crafts gardens, and medieval flowers such as pinks and old roses became fashionable. Architects used masonry and wisteria-covered pergolas to create outdoor rooms: Edwin Lutyens floored them simply with lawn, while E.S. Prior paved them with brick or stone. Prior rhapsodized in 1901 about providing "a sunny wall, a pleasant shade, a seat for rest, and all around the sense of the flowers, their brightness, their fragrance." Ponds and fountains became a feature. Plain oak and elm benches, terracotta flowerpots, and stone troughs were also frequently to be found on terraces that "settled" the house into its site and provided important social space for the taking of tea out of doors. The crevices in lovely old stone terrace walls could be crammed with rock plants such as gypsophila and cerastium, which would cascade down and soften the hard edges of retaining walls.

In 1901 three articles on "Garden-Making" written by E.S. Prior were published in *The Studio*. These included plans for the kind of modest "oblong garden" that corresponded to the reality of the thin suburban plot owned by most middle-class householders. The rise of the middle class had seen a corresponding rise in hobby gardening, encouraged and supported by numerous periodicals carefully aimed at the lady amateur.

The architect Baillie Scott considered a well-designed garden to be "almost as important as a well-designed house.... We can hardly do better," he wrote, "than to try and reproduce some of the beauties of the old English gardens, with their terraces and courts and dusky yew hedges, which make such a splendid background to the bright colors of flowers." This was exactly what William Morris and his architect Philip Webb had sought to achieve at Red House, and later at Kelmscott Manor, where Morris created romantic gardens of topiary hedges, grass walks, and sweeping lawns, with wattled trellises for clambering roses and carefully preserved fruit trees.

Below: *Plants grown in the Conservatory at Standen included bougainvillea, oleander, and plumbago. Mimosa was another particular favorite with the Beales.*

Conservatories, which were associated with the artificial style of hothouse gardening, were rare in Arts and Crafts gardens, which promoted the idea of natural not forced abundance, although Standen in Sussex is a notable exception. James Beale commissioned the London garden designer G.B. Simpson to lay out the grounds surrounding the Standen site in 1890 before he had even commissioned his architect, Philip Webb. Beale's wife Margaret was a keen gardener and corresponded regularly with her Sussex neighbor William Robinson. Following the Arts and Crafts model as described by Herman Muthesius, her garden operated as an extension of the house and was divided into a series of "outdoor rooms, each of which contained and performs a separate function. Thus the garden extends the house into the midst of nature.... This means that the regularly laid out garden must not extend merely to one side of the house, but all the way round it, so that the house appears from all angles to rest on an adequate base." This effect was often achieved by terracing, if the site sloped, with parapets created by box hedging, which like that other Arts and Crafts favorite, yew, lent itself to topiary. For Muthesius, "the point about topiary work is its orderly architectonic form.... Clipped hedges are the walls by means of which the garden-designer delimits his areas. They also lend themselves to rhythmic repetition," establishing "certain points in the geometric composition of the garden. Topiary work is ... the

Above: *June borders of lupin and iris in the garden at Munstead Wood enlivened the pages of* Country Life *in 1912.*

indispensable means of establishing form." Needless to say William Robinson despised such "vegetable sculpture." "Nothing is more miserable for the gardener," he wrote, "or uglier in the landscape, than a garden laid out with clipped Yews."

Gertrude Jekyll was greatly influenced by William Robinson. For her the purpose of a garden was "to give delight and to give refreshment of mind, to soothe, to refine and to lift up the heart." She too abhorred Victorian formality. Born into a comfortably wealthy family, her idea of "a small garden," as set out in her *Gardens for Small Country Houses* (1912), was not perhaps one that a Garden Suburb dweller might recognize as he surveyed the view from his back windows. The elaborate scheme she includes for the "small" garden at Millmead in Bramley, Surrey, measuring 77ft across and 400ft in length, would have been hugely labor-intensive, but she did do much to encourage gardening, especially among children, and, with her artist"s eye for color and form, lifted it to an art form. "When the eye is trained to perceive pictorial effect," she wrote in *Colour Schemes for the Flower Garden* (1908), "it is frequently struck by something—some combination of grouping, lighting, and colour—that is seen to have that complete aspect of unity and beauty that to the artist's

Above: *A profusion of plants growing out of the walls above the curved steps in the garden at Hestercombe.*

eye forms a picture. Such are the impressions that the artist-gardener endeavours to produce in every portion of the garden."

Gertrude Jekyll had trained at the South Kensington School of Art, knew William Morris and John Ruskin, and had it not been for failing eyesight she might have continued her career as an artist, embroiderer, and silverworker. Instead, after her father's death and upon her move to Surrey with her mother, she embarked on a new career, bringing her artistic training and knowledge of color theory to her rich but controlled planting schemes. She proved to be a phenomenally hard worker. Over the course of her career, she wrote over two thousand articles for periodicals, fifteen books (ten of them in the space of nine years), took and developed her own garden photographs, designed or advised on over one hundred and fifty gardens, and, when she found no flower vase suited her particular purpose, designed her own "Munstead glasses."

She is perhaps best known for her professional association with the architect Edwin Lutyens. Together they collaborated on some of the finest garden designs of their time for great houses including Deanery Garden and Folly Farm, both in Berkshire, and Hestercombe House, beautifully situated with wide views of the Vale of Taunton in Somerset. Lutyens provided the architectural shape and Jekyll the planting. The commission for Hestercombe came from Lord Portman, who gave the estate to his

grandson, the Hon. E.W.B. Portman, as a wedding present. In 1903 Lord Portman invited Lutyens and Jekyll to make a garden on a new site, comprising three giant terraces, to the south of the house. Lutyens gave the garden a new architectural symmetry using pools, ponds, a pergola, an orangery, and a rotunda, the three entrances to which were planted with fragrant roses. As her eyesight began to fail, so Gertrude Jekyll's sense of smell became more acute and she liked to use fragrant plants such as rosemary and myrtle at entrances, so that people brushed against them as they passed. For the same reason, she planted lavender, catmint, and roses in the paved Dutch garden. In 1908 *Country Life* claimed the garden at Hestercombe proved "that an architect can be in unison with Nature, that a formal garden can form part of a landscape." At its peak between World Wars I and II, eighteen gardeners were employed to tend the garden, though it fell into serious neglect during World War II when the house was occupied by the US army. Serendipitously, Gertrude Jekyll's own handwritten planting plans were discovered pinned up in a potting shed in the 1970s and the garden has now been restored to its former glory thanks, in part, to a £200,000 grant from English Heritage.

The commission for The Deanery at Sonning in Berkshire came, like so many early Lutyens commissions, from a friend and near neighbor of Gertrude Jekyll's. Edward Hudson was the proprietor of *Country Life* and by publishing rapturous descriptions of their work in his magazine had

Above: *The garden at The Deanery, Sonning, Berks, the house Edwin Lutyens designed for Edward Hudson, proprietor of* Country Life*.*

ensured a wide audience of further potential clients for the pair. The house and garden they created for Hudson has been called a masterpiece of understatement. Jekyll's naturalistic planting softens Lutyens' geometric precision, creating a charming peaceful garden on several levels. In particular, her habit of planting alpines and drought-tolerant plants on steps and in drystone walls created drifts of soft whites, pinks and purples against the soft gray of the walls, in which Lutyens would leave deliberate spaces for her skillful planting.

In 1897 C.F.A. Voysey was commissioned to design a house and garden by the publisher Sir Algernon Methuen in Haslemere, Surrey. New Place fits around the contours of a steeply sloping site with a garden arranged as a series of rooms, terraced to include a formal walled garden, a tennis lawn, a brick arcade covered over with *Azara microphylla*, and a bowling green, complete with thatched arbor to shelter the players awaiting their turn to play.

Voysey's talent was for devising and sculpting space. His horticultural instructions for the garden at New Place are sketchy to say the least. "Flowers big and tall," he indicated for one border on his garden plan. "More or less wild," he instructed elsewhere, although he did have strong views on color. Green he always recognized as the most soothing: "nature never allows her colors to quarrel," he said in a lecture published in *The Arts Connected with Building* (1909). "Her purple trees, with their gossamer of delicate spring green, dwell lovingly with the blue carpet of hyacinths. Harmony is everywhere." In 1904 Methuen, a keen gardener and an expert on alpines, invited Gertrude Jekyll to refine the planting and she added a rose garden, used pale mauve *Abutilon vitifolium* to complement Voysey's gray stone, and grouped white peonies with blue delphiniums in the borders. Green trellis was planted with hydrangeas, escallonias, *Robinia hispida*, *Sophora tetraptera*, and *Fremontodendron californicum*.

Voysey's contemporary, M.H. Baillie Scott, who also worked with Gertrude Jekyll, was, of all the Arts and Crafts architects, the one who gave most serious thought to the small suburban garden. He had always been concerned with the idea of creating a unity between the house and the garden. "One may note," he wrote in an article published in the *Builder's Journal* about Springcot, a holiday cottage and garden he created in 1903, "first of all the importance attached to vistas—vistas arranged with definite terminal effects. One may also observe the usefulness of shade in the garden as well as light, and how embowered paths may be contrasted with the brightness of open spaces." He organized these vistas on a criss-cross pattern. "In passing through these enclosed ways, one loses all conception of the garden's scheme till, at the intersection of a path, one suddenly perceives through vistas of roses and orchard trees some distant garden ornament, or perhaps a seat or a summer house: and so one becomes conscious of a scheme arranged and of well-considered effects. As in a dramatic entertainment, parts of the garden full of tragic shade are followed by open spaces where flowers laugh in the sun."

Above: *An illustration of M.H. Baillie Scott's Undershaw in Guildford.*

Baillie Scott collaborated on two schemes with Gertrude Jekyll and, like her, rebelled against the prevailing Victorian schemes for bedding out gaudy plants at regular intervals. He "composed" his plantings rather as an artist does a picture, relying on color and mass and a profusion of cottage garden flowers—delphiniums, daisies, lavender, hollyhocks, pinks, asters, campanulas, and clambering roses draped across pergolas and trellises. He blurred the distinctions between kitchen and formal gardens, proposing the "scarlet runner bean" as an alternative to red geraniums in a planting scheme and extolling the beauty of "the grey-green foliage and great thistle-like heads of the globe artichokes, the mimic forest of the asparagus bed, and the quaint inflorescence of the onion."

Baillie Scott's early training at the Royal Agricultural College at Cirencester gave him a practical approach to the subject and he took his inspiration from old cottage gardens, writing in his book *House and Gardens*: "Many old cottage gardens, which are to be seen in our villages, show the possibilities of homely beauty which belong to such a union of use and beauty in the garden, and such a garden, worked in the spare time of its owner with a rough and ready love which is his traditional inheritance, will be profitable as well as pleasant." He did not see the need to distinguish between pleasure gardens and kitchen gardens, rather he promoted the idea of uniting use and beauty, responding to nature rather than "striving to mold her to an artificial ideal."

For Baillie Scott the purpose of a small garden was "to grow fruit and vegetables for the household, and also to provide outdoor apartments for the use of the family in fine weather."

Above: *The Well Court at Snowshill Manor takes its name from the central feature, an ancient Venetian well-head. Wade designed the garden as a series of separate courts, sunny ones contrasting with shady ones and different courts for varying moods. "The plan of the garden," he stated, "is much more important than the flowers in it. Walls, steps, and alley ways give a permanent setting, so that it is pleasant and orderly in both summer and winter."*

Larger plots, however, were a different matter. Here he allowed lawns for tennis, croquet, or bowls, took in orchards, and made a kitchen garden, separate rose and flower gardens, all connected by straight paths with perhaps a pergola.

One such garden is to be found in Gloucestershire. The garden at Snowshill Manor was created by Baillie Scott for the architect and collector Charles Paget Wade in 1920. Wade wanted his garden to be an extension of the house, and had architectural rather than horticultural priorities. "A delightful garden can be made in which flowers play a very small part," he asserted, "by using effects of light and shade, vistas, steps to changing levels, terraces, walls, fountains, running water, an old well head or a statue in the right place."

Baillie Scott's true interests lay in smaller country houses, however, and it was natural that he should become increasingly involved in the Garden City movement, designing cottages at Letchworth and houses and gardens for Hampstead Garden Suburb, the brainchild of Henrietta Barnett. It was essential to her that each house in the Suburb should have its own plot of land and that horticulture should be energetically practiced. She herself always "felt perfectly happy when her Canon [her husband Canon Samuel Barnett] was occupied taking plantains out of the lawn."

Summarizing her aims for the Suburb in an article in *Contemporary Review*, she concluded that "each house be surrounded by its own garden and that there be agencies for fostering interest in gardens and allotments and for the co-operative lending of tools." It pleased her enormously that one of the first of many clubs set up within the Suburb was a horticultural society, formed in May 1909. It used the newly built Hampstead Garden Suburb Institute as its headquarters and flourishes still.

Another notable Arts and Crafts garden is to be found at Rodmarton Manor, near Cirencester, built over twenty years from 1909 for the Hon. Claud Biddulph by Ernest Barnsley, "using only local materials and without any kind of mechanical assistance," in defiance of industrialization. Barnsley

Below: *A view of Rodmarton Manor as seen from the White Borders. Work commenced on the garden as the house was being built and, over the years, was subject to alteration and redesign by Margaret Biddulph.*

designed the garden—almost eight acres in all—in collaboration with William Scrubey, the head gardener, and his client's wife Margaret Biddulph, a trained horticulturalist, as a sequence of outdoor rooms, made intimate and secret by long straight walls, tall yew hedges, and screens of pleached limes. The design provided a crisp contrast between the soft informality of the Winter Garden and the architectural shapes of the Topiary Lawn and delightful surprises as concealed areas opened up to view, including the Long Garden with its picturesque summerhouse and secluded stone benches from which visitors could admire the four long herbaceous beds planted with asters, roses, phlox, sedum, and daylilies. Adjacent to this an enormous walled kitchen garden, originally lined with espaliered fruit trees, enclosed vegetable and flower beds. The lovely stone front of the house was planted with climbing roses, clematis, and wisteria. In 1931 *Country Life* described the newly completed house and garden as almost "a village in itself.... In fact, the whole place bears an astonishing testimony to the life and vigor which Mr. and Mrs. Biddulph have given to a tiny village by their enthusiastic encouragement of the arts and crafts."

Gertrude Jekyll designed three gardens for American clients who visited her at Munstead and asked her to provide plans to suit their different plots and needs. The first, for Mr. and Mrs. Glendinning B. Groesbeck in Perintown, Cincinnati (1914), is sited on a steeply sloping valley and takes its inspiration from Italy. Edward Hudson brought the second pair of American clients to Munstead in 1924: Mr. and Mrs. Stanley Resor traveled each summer and wanted a garden for their "Cotswold Cottage" in Greenwich, Connecticut, that they could enjoy in the spring and on their return in the fall. Finally, Miss Annie Burr Jennings sought Gertrude Jekyll out in 1926 and asked her to provide plans for an old-fashioned garden to surround the Old Glebe House garden in Woodbury, Connecticut. Jekyll designed borders of hollyhocks, dahlias, antirrhinums, irises, and peonies.

Far from the cozy cottage profusion of England, however, American Arts and Crafts architects and garden designers, considered and were inspired by the natural world on their doorstep. They incorporated indigenous plants like cacti, and planted native species such as California poppies, California wild oats, and Spanish lily, seeking to create a specifically American garden style. In his *Craftsman Houses* (1913) Gustav Stickley explored the universal need for a garden, writing, "In practically all of us is a deep, distinctive longing to possess a little corner of that green Eden from which our modern and materialistic ways of living have made us exiles."

Frank Lloyd Wright's interest in the overall relationship between the house and its landscape naturally extended to the garden, which he sought to include in his overall view of "organic" architecture and make into a working part of the house. "Let your home appear to grow easily from its site," he said in a public lecture delivered in 1894, "and shape it to sympathize with the surroundings if Nature is manifest there, and if not, try and be as quiet, substantial, and organic as

Far Left: *The garden at Rodmarton, looking toward the house from the Cherry Orchard.*

she would have been if she had the chance." He put these ideas into practice most dramatically at Fallingwater in Pennsylvania (1935–9), where the house juts out over a natural waterfall on Bear Run, mature trees grow through the building and a huge bedrock boulder breaks through the floor of the main room just in front of the hearth. In earlier projects, such as the Robie house in Chicago (1906), and the Avery Coonley house in Riverside, Illinois (1906–9), Wright made the garden a working part of the house. In the Robie house, the children's playroom and the billiard room open directly onto a compact sunken garden court, while the garden enfolds and surrounds the Avery Coonley house. Wright wanted to dissolve the distinctions between outside and in, and he did this with entire walls composed of glazed screens. Living rooms—and even bedrooms—opened onto verandas and terraces, landscaped with simple troughs and urns, cement-rendered to echo the building.

Below: *The Greenes were particularly concerned to place their houses within the landscape. The Japanese influence is evident in the garden of the Gamble House.*

Above: *Black and white photograph of Villa Marcault from* American Gardens, *1902.*

Oriental gardens were an important source of inspiration for Charles and Henry Greene, who used stepping stones and quiet pools, delicate lanterns, and formal tubs as motifs in the garden of the Gamble House (1907–8). Inside, the natural imagery is signaled in the richly colored "Tree of Life" art-glass panel that fills the massive front door, and echoes of the real vines outside resonate in the stylized depiction in the stained glass windows. The Greenes were always concerned to place their buildings sympathetically within the landscape and to incorporate existing trees in the overall design. They used the garden to link the house to the site, anchoring their low-built bungalows with vine-covered pergolas and connecting the outside and in with trellised gateways or loggias and curving Japanese paths. Inspired by the American novelist Edith Wharton's *Italian Villas and Their Gardens* (1904) they created an Italianate water garden in the garden of the Cordelia A. Culbertson house, built in Pasadena in 1911, and Charles laid out the formal gardens of the Fleishhacker family estate in Woodside with a Roman pond and water garden in 1920.

Edith Wharton's book enjoyed huge success in the USA, as did Louise Beebe Wilder's highly influential *Color in my Garden* (1918). Written in a lyrical style, it appealed to the pioneering spirit of Americans. "Much of my youthful life," she wrote, "was spent in old gardens, blossomy inclosures with generations of bloom and sweetness behind them, eloquent of long years of happy human occupancy; and no one is more alive to their charm than I; but during the past twenty years it has fallen to my lot to make three gardens on wholly unimproved ground, and I am ready to testify that there is a deal to be said for a new garden—at least from the standpoint of the owner. It is the fair page, the fresh canvas—opportunity. It affords scope for the age-old joy of creating something—beauty we hope—out of raw materials and the stuff of our dreams."

SOURCE BOOK

UNITED KINGDOM

Birmingham

Birmingham Museum and Art Gallery
Chamberlain Square
Birmingham, B3 3DH
Tel: +44 (0)121 303 2834
Fax: +44 (0)121 303 1394
www.bmag.org.uk
Open: Monday to Thursday and Saturday 10–5, Friday 10–5, Sunday 12.30–5
Houses some fine examples of Arts and Crafts art and artifacts.

Cumbria

Blackwell the Arts and Crafts House
Bowness-on-Windermere
Cumbria, LA23 3JR
Tel: +44 (0)1539 446139
Fax: +44 (0)1539 488486
www.blackwell.org.uk
Open: Seven days a week 10–5, February to December
Built by M.H. Baillie Scott (1897–1900); recently restored and now open to the public.

Brantwood
Coniston, Cumbria, LA21 8AD
Tel: +44 (0)1539 441396
Fax: +44 (0)1539 441263
www.brantwood.org.uk
Open: Daily 11–5.30, all year round
Important as the "lakeland landscape" garden of John Ruskin (1872–1900).

Devon

The Barn
Fox Holes Hill, Exmouth
Devon, EX8 2DF
Tel: +44 (0)1395 224411
Fax: +44 (0)1395 225445
Built by Edward Prior (1897), this is now a country house hotel.

Gloucestershire

The Cheltenham Art Gallery and Museum
Clarence Street, Cheltenham
Gloucestershire, GL50 3JT
Tel: +44 (0)1242 237431
Fax: +44 (0)1242 262334
www.cheltenhammuseum.org.uk
Open: Monday to Saturday 10–5, Sunday 2–4, all year round
Has a gallery devoted to Arts and Crafts furniture and holds regular events featuring William Morris and others.

The Guild of Handicrafts Trust
The Silk Mill, Sheep Street
Chipping Campden, Gloucestershire
Tel: +44 (0)1386 841100
Founded by C.R. Ashbee, having its headquarters in the Silk Mill from 1902 to 1908.

Kelmscott Manor
Kelmscott, Lechlade
Gloucestershire, GL7 3HJ
Tel: +44 (0)1367 252486
Fax: +44 (0)1367 253754
www.kelmscottmanor.co.uk
Open: Wednesdays 11–5 and third Saturday in each month 2–5, April to September
Country home and beautiful garden of William Morris (1871–96), now owned by the Society of Antiquaries.

Owlpen Manor
Uley, Nr Dursley
Gloucestershire, GL11 5BZ
Tel: +44 (0)1453 860261
Fax: +44 (0)1453 860819
www.owlpen.com
Open: Daily (except Mondays) 2–5, April to September
Tudor manor restored by Norman Jewson in 1926 housing important Arts and Crafts collections.

Rodmarton Manor
Cirencester
Gloucestershire, GL7 6PF
Tel: +44 (0)1285 841253
Fax: +44 (0)1285 841298
www.rodmarton-manor.co.uk
Open: Wednesday, Saturday and Bank Holiday Mondays 2–5, May to August
Built by Ernest Barnsley and his Cotswold group of craftsmen (1909 onward).

Herefordshire

All Saints Church
Brockhampton, Near Fownhope
Herefordshire, HR1 4PS
Open: Daily 9–5
W.R. Lethaby's architectural masterpiece.

Hertfordshire

The Garden City Museum
296 Norton Way South, Letchworth
Hertfordshire, SG6 1SU
Tel: +44 (0)1462 482710
Fax: +44 (0)1462 486056
Open: Monday to Saturday 10–5
Once the medieval Hall House designed and built for himself by Barry Parker and now the Garden City Museum.

KENT

Great Maytham Hall
Rolvenden, Cranbrook
Kent, TN17 4NE
Tel: +44 (0)1580 241346
Fax: +44 (0)1580 241038
Open: Wednesday and Thursday 2–5, May to September. *Designed by Sir Edwin Lutyens (1910). This was the house and garden that inspired Frances Hodgson Burnett's* The Secret Garden.

Red House
Red House Lane, Bexleyheath
Kent, DA6 8JF
Tel: +44 (0)20 8304 9878
www.nationaltrust.org.uk
Open: Wednesday, Thursday, Friday, Saturday and Sunday 1 March to 21 December. Admission by booked guided tour with limited free flow entry after 3.30pm. *Designed by Morris and the architect Philip Webb (1859); acquired by The National Trust in 2003.*

LONDON

Geffrye Museum
Kingsland Road London, E2 8EA
Tel: +44 (0)20 7739 9893
www.geffrye-museum.org.uk
Room interiors showing English domestic styles from 1600 to the present day.

Kelmscott House
26 Upper Mall, Hammersmith
London, W6 9TA
Tel: +44 (0)20 8741 3735
Fax: +44 (0)20 8748 5207
www.morrissociety.org/kelmscott_house.html
Open: Basement only Thursday and Saturday 2–5. *Morris's home (1878–96) and now headquarters of the William Morris Society.*

Holy Trinity Church
Sloane Street, Chelsea
London, SW1X 1DF
Tel: +44 (0)20 7730 7270
Fax: +44 (0)20 7730 9287
www.holytrinitysloanestreet.org
Richly decorated with stained glass by Edward Burne-Jones, William Morris and Christopher Whall, and examples of arts and crafts metalwork, sculpture, and other elaborate decorative details.

Victoria and Albert Museum
Cromwell Road, South Kensington
London, SW7 2RL
Tel: +44 (0)20 7942 2000
www.vam.ac.uk
Open: Daily 10–5, Wednesdays and the last Friday of the month 10–10
Holds important examples of the work of William Morris, Edward Burne-Jones and leading members of the Arts and Crafts movement. See especially the Green Dining Room and the recreated rooms in the new English Galleries.

William Morris Gallery
Water House, Lloyd Park, Forest Road
London, E17 4PP
Tel: +44 (0)20 8527 3782
Fax: +44 (0)20 8527 7070
www.lbwf.gov.uk/wmg/home.htm
Open: Tuesday to Saturday 10–5 and the first Sunday in the month 10–12
Morris's home (1848–56), now housing some stunning collections illustrating Morris's life and work.

Bedford Park Chiswick, *and* Hampstead Garden Suburb *are both worth strolling around.*

NORTHUMBERLAND

Cragside
Rothbury, Morpeth
Northumberland, NE65 7PX
Tel: +44 (0)1669 620150
Fax: +44 (0)1669 620066
Open: Daily (except Mondays), April to October
Built by Richard Norman Shaw (1864–95), now managed by the National Trust.

Lindisfarne Castle
Holy Island, Berwick-upon-Tweed
Northumberland, TD15 2SH
Tel: +44 (0)1289 389244
Fax: +44 (0)1289 389349
Open: Saturday to Thursday, March to November
Restoration and conversion by Sir Edwin Lutyens (1903); now managed by the National Trust.

SOMERSET

Hestercombe Gardens
Cheddon Fitzpaine, Nr Taunton
Somerset, TA2 8LG
Tel: +44 (0)1823 413923
Fax: +44 (0)1823 413747
www.hestercombegardens.com
Open: Daily 10–6 (excluding Christmas)
Jekyll and Lutyens' finest surviving garden, recently restored.

SUNDERLAND

Church of St Andrew
Roker, Co Durham
Arts and Crafts masterpiece built by E.S. Prior (1906–07), with choir stalls and paneling by Ernest Gimson, tapestry reredos by Edward Burne-Jones.

Surrey

Goddards
Abinger Lane, Abinger Common
Dorking, Surrey
Tel: +44 (0)1628 825925
Open: Wednesday afternoons by appointment only, April to October
Built by Sir Edwin Lutyens (1898–99); available for rental through the Landmark Trust.

Munstead Wood
Heath Lane, Busbridge
Godalming, Surrey
Open: Gardens (from which exterior views of the house can be seen) 2–6, at the end of April, May, June
Lutyens' first important architectural commission, designed for its owner Gertrude Jekyll, who created the gardens, open occasionally as part of the National Gardens Scheme.

Sussex

Ditchling Museum
Church Lane, Ditchling
Sussex, BN6 8TB
Tel: +44 (0)1273 844744
www.ditchling-museum.com
Open: Tuesday to Saturday 10.30–5, Sunday 2-5, mid-February to mid-December
Small museum with a good Arts and Crafts display, especially related to Edward Johnson, Eric Gill, and the Ditchling community, lighting and furniture by Brangwyn.

Gravetye Manor
East Grinstead
Sussex, RH19 4LJ
Tel: +44 (0)1342 810567
Fax: +44 (0)1342 810080
www.gravetyemanor.co.uk
House once owned by William Robsinson, pioneer of the natural garden, where many of his ideas were realized. Now a "Relais et Chateaux" country house hotel.

Great Dixter
Northiam, Rye
East Sussex, TN31 6PH
Tel: +44 (0)1797 252878
Fax: +44 (0)1797 252879
www.greatdixter.co.uk
Open: Tuesday to Sunday 2–5 all year round
The family home of Christopher Lloyd, with additions by Sir Edwin Lutyens (1910).

Little Thakeham
Merrywood Lane, Storrington
West Sussex
Tel: +44 (0)1903 744416
Designed by Sir Edwin Lutyens (1902); now a country house hotel.

Standen
West Hoathly Road, East Grinstead
Sussex, RH19 4NE
Tel: +44 (0)1342 323029
Fax: +44 (0)1342 316424
Open: Wednesday to Sunday and Bank Holiday Mondays 11–5, March to November
Designed by Philip Webb (1891) and decorated throughout with Morris carpets, fabrics and wallpapers; now owned by the National Trust. For details of the holiday flat ring +44 (0)1225 791199.

West Midlands

Wightwick Manor
Wightwick Bank, Wolverhampton
West Midlands, WV6 8EE
Tel: +44 (0)1902 761108
Fax: +44 (0)1902 764663
Open: Thursday and Saturday 1.30–5, March to December
Built by Edward Ould (1887–93), with Morris & Co. interiors; now owned by the National Trust.

SCOTLAND

East Lothian

Greywalls
Muirfield, Gullane, East Lothian, EH31 2EF
Tel: +44 (0)1620 842144
Fax: +44 (0)1620 842241
www.greywalls.co.uk
Designed by Sir Edwin Lutyens (1901); now a country house hotel.

Fife

Earlshall Castle
Leuchars, Fife
Tel: +44 (0)1334 839205
Restoration by Sir Robert Lorimer (1892) of this sixteenth-century building and gardens; now privately owned.

Glasgow

Glasgow School of Art
167 Renfrew Street,Glasgow, G3 6RQ
Tel: +44 (0)141 332 0521
Charles Rennie Mackintosh's masterwork.

The Hill House
Upper Colquhoun Street, Helensburgh
Glasgow, G84 9AJ
Tel: +44 (0)1436 673900
Fax: +44 (0)1436 674685
www.nts.org.uk

Open: Daily 1.30–5.30, April to October
Designed by Charles Rennie Mackintosh (1902); now owned by the National Trust for Scotland.

The Hunterian Museum
University of Glasgow,Glasgow, G12 8QQ
Tel: +44 (0)141 330 4221
Fax: +44 (0)141 330 3617
www.hunterian.gla.ac.uk
Open: Monday to Saturday 9.30–5
Housing the largest single collection of works by Charles Rennie Mackintosh, including drawings, designs, furniture, and decorative art.

Charles Rennie Mackintosh Society
Queen's Cross Church,
870 Garscube Road, Glasgow G20 7EL
Tel: +44 (0)141 946 6600
Fax: +44 (0)141 945 2321
www.crmsociety.com

The Willow Tea Rooms
217 Sauchiehall Street, Glasgow, G2 3EX
Tel/Fax: +44 (0)141 332 0521
www.willowtearooms.co.uk
Open: Monday to Saturday 9–5, Sunday 12–4, all year round
Designed in 1904 by Charles Rennie Mackintosh, in every detail, from exterior and interoir down to furniture and teaspoons; restored in 1996 and re-opened as tearooms. Also at:
97 Buchanan Street, Glasgow, G1 3HF
Tel/Fax: +44 (0)141 204 5242
Recreation of Mackintosh's White Dining Room and Chinese Room.

Orkney

Melsetter
Isle of Hoy, Orkney, KW16 3M2
Tel: +44 (0)1856 791352
Open: Private visits by arrangement only Thursday, Saturday, and Sunday
Built by W.R. Lethaby (1898) and boasting one of the oldest gardens in Orkney.

UNITED STATES OF AMERICA

Arizona

Frank Lloyd Wright Foundation
PO Box 4430
Scottsdale, AZ 85261-4430
Tel: +1 480 860 2700
Fax: +1 480 391 4009
www.franklloydwright.org
For information on houses designed by Wright open to the public, state by state.

California

The Gamble House
4 Westmoreland Place,
Pasadena, CA 91103
Tel: +1 626 793 3334
www.gamblehouse.org
Open: Thursday to Sunday for guided tours
The only Greene and Greene house (built 1909) regularly open to the public.

Los Angeles County Museum of Art
5905 Wiltshire Boulevard
Los Angeles, CA 90036
Tel: +1 323 857 6000
www.lacma.org
One of the largest displays of Arts and Crafts objects, housing a particularly fine exhibition of Greene and Greene furniture.

Marston House
3525 Seventh Avenue, Balboa Park
San Diego, CA
Tel: +1 619 298 3142
Open: Friday to Sunday 10–4.30
Designed by architect Irving Gill (1905) and furnished in the style of American Arts and Crafts.

Mission Inn
3649 Mission Inn Avenue
Riverside, CA 92501
Tel: +1 909 784 0300
www.missioninn.com
Demonstrates the Spanish mode of the Arts and Crafts style.

Connecticut

Mark Twain House
351 Farmington Avenue
Hartford, CT 06051
Tel: +1 860 247 0998
Fax: +1 860 278 8148
www.marktwainhouse.org
Former home of the writer Mark Twain, decorated by Associated Artists, including work by Louis Comfort Tiffany and Candace Wheeler.

Delaware

Delaware Art Museum
2301 Kentmere Parkway
Wilmington, DE 19806
Tel: +1 303 571 9590
www.delart.org
Open: Tuesday to Friday 10–4, Saturday 10–5, Sunday 1–5
Housing collections of British Arts and Crafts, including some Morris chairs.

Florida

Morse Museum of American Art
445 North Park Avenue, Winter Park
Florida, FA 32789
Tel: +1 407 645 5311
www.morsemuseum.org
Open: Tuesday to Saturday 9.30–4, Sunday 1–4, all year round
Contains the most comprehensive collection of pieces by Louis Comfort Tiffany anywhere in the world.

Illinois

Crabtree Farm
PO Box 218
Lake Bluff, IL 60044
Tel: +1 312 391 8565
Museum and conference center with Stickley furniture and Stickley-infuenced decorative scheme. American and British Arts and Crafts movement collection.

Dana-Thomas House Foundation
300 East Lawrence Avenue
Springfield, IL 62702
Tel: +1 217 782 6776
www.dana-thomas.org
Open: Wednesday to Sunday 9–4
Designed by Frank Lloyd Wright in 1902 for Susan Lawrence Dana; contains the largest collection of site-specific, original Wright art glass and furniture. Now restored and managed by Illinois State Historic Preservation Agency.

Frank Lloyd Wright Home and Studio
951 Chicago Avenue
Oak Park, IL 60302
Tel: +1 708 848 1978
www.wrightplus.org/homestudio
Open: Daily 9–5
Designed by Frank Lloyd Wright (1889–98) and demonstrating his representations of the Arts and Crafts ideals.

Glessner House
1800 South Prairie Avenue
Chicago, IL 60616
Tel: +1 312 326 1480
Fax: +1 312 326 1397
www.glessnerhouse.org
Open: Wednesday to Sunday with three tours daily
Built by H.H. Richardson (1887) and reflecting the William Morris style in the bedroom.

Robie House
5757 Woodrow Avenue, Chicago, IL 60637
Tel: +1 312 702 8374
Designed by Frank Lloyd Wright for Frederick C. Robie; now run by the university administration.

New Jersey

Craftsman Farms
2352 Route 10 West, Manor Lane
Parsippany-Troy Hills, NJ 07950
Tel: +1 973 540 1165
Fax: +1 973 540 1167
www.stickleymuseum.org
Open: Wednesday to Sunday by appointment only, April to November
Created by Gustav Stickley (1911). The log home built by Stickley that acted as the heart of the crafts community and remains one of the most significant landmarks of the American Arts and Crafts movement.

New York

Byrdcliffe Arts and Crafts Colony
The Woodstock Guild, 34 Tinker Street
Woodstock, NY 12498
Tel: +1 845 679 2079
Fax: +1 845 679 4529
www.woodstockguild.org
America's oldest continuing art colony, founded by Ralph Whitehead in 1903.

Metropolitan Museum of Art
Fifth Avenue, New York, NY 10028
Tel: +1 212 879 5500
www.metmuseum.org
Open: Tuesday to Thursday, Sunday 9.30–5.30, Friday and Saturday 9.30–9
Containing a Frank Lloyd Wright room and fine examples of American and European Arts and Crafts pieces.

Roycroft Community
Main and South Grove Streets
East Aurora, NY 14052
Tel: +1 716 655 0571
Established by Elbert Hubbard (1894) as his own craft community based on William Morris's Kelmscott; the fourteen remaining buildings contain many original artefacts.

Stickley Museum
300 Orchard Street
Fayetteville, NY 13104
Tel: +1 315 682 5500
Open: Tuesdays
Stickley's former factory, now home to some fine examples of period furniture.

North Carolina

Grove Park Inn
290 Macon Avenue
Asheville, NC 28804
Tel: +1 800 438 5800
The Arts and Crafts ideal from exterior to Mission furnishings, now a resort home.

BIBLIOGRAPHY

Anscombe, Isabelle and Charlotte Gere, *Arts & Crafts in Britain and America*, Academy Editions, 1978

Anscombe, Isabelle, *Arts and Crafts Style*, Phaidon Press, 1991

Arts and Crafts Essays by members of the Arts and Crafts Exhibition Society, with a Preface by William Morris, Longmans Green and Co., 1899

Ashbee, C.R., *A Book of Cottages and Little Houses*, Batsford, 1906

Ashbee, C.R., *Modern English Silverwork*, Essex House Press, 1909

Ashbee, C.R., *The Guild of Handicraft*, Essex House Press, 1909

Aslin, Elizabeth, *Nineteenth Century English Furniture*, Faber & Faber, 1962

Backemeyer, Sylvia, *W.R. Lethaby 1857–1931*, Lund Humphries, 1984

Benson W.A.S., *Elements of Handicraft and Design*, London, 1893

Bisgrove, Richard, *The Gardens of Gertrude Jekyll*, Frances Lincoln, 1992

Blomfield, Reginald, *The Formal Garden In England*, Macmillan, 1892

Blomfield, Reginald, *Memoirs of an Architect*, Macmillan, 1932

Boris, Eileen, *Art and Labor, Ruskin, Morris and the Craftsman Ideal in America*, Temple University Press, 1986

Bowe, Nicola Gordon and Elizabeth Cumming, *The Arts and Crafts Movements in Dublin & Edinburgh, 1885–1925*, Irish Academic Press, 1998

Bowman, Leslie Greene, *American Arts & Crafts Virtue in Design*, Los Angeles County Museum of Art in association with Little, Brown and Co., 1990

Brandon-Jones, John, *C.F.A. Voysey: Architect and Designer 1857–1941*, Lund Humphries, 1978

Brooks, H. Allen, *The Prairie School: Frank Lloyd Wright and his Midwest Contemporaries*, University of Toronto Press, 1972

Buchanan, William (ed.), *Mackintosh's Masterwork the Glasgow School of Art*, Richard Drew Publishing, 1989

Burchard, John E. and Albert Bush-Brown, *The Architecture of America: A Social and Cultural History*, Victor Gollancz, 1967

Burne-Jones, Georgiana, *Memorials of Edward Burne-Jones* (2 vols), Macmillan, 1904

Calloway, Stephen, *The House of Liberty, Masters of Style & Decoration*, Thames and Hudson, 1992

Campbell, Joan, *The German Werkbund, The Politics of Reform in the Applied Arts*, Princeton University Press, 1978

Carruthers, Annette and Mary Greensted, *Good Citizen's Furniture, The Arts and Crafts Collections at Cheltenham*, Cheltenham Art Gallery and Museums in association with Lund Humphries, 1994

Carruthers, Annette, *Ernest Gimson and the Cotswold Group of Craftsmen*, Leicestershire Arts, Museums and Records Service, 1978

Clark, Robert Judson, *The Arts and Crafts Movement in America 1876–1916*, Princeton University Press, 1972

Cobden-Sanderson, T.J., *The Arts and Crafts Movement*, Hammersmith Publishing Society, 1905

Comino, Mary, *Gimson and the Barnsleys 'Wonderful furniture of a commonplace kind'*, Evans Brothers Limited, 1980

Cook, E.T. and A. Wedderburn (ed.), *The Complete Works of John Ruskin* (39 vols), George Allen, 1903–12

Crane, Walter, *The Bases of Design*, G. Bell & Sons, 1898

Crawford, Alan (ed.), *By Hammer and Hand, The Arts and Crafts Movement in Birmingham*, Birmingham Museums and Art Gallery, 1984

Cross, A.J., *Pilkington's Royal Lancastrian Pottery and Tiles*, Richard Dennis Publications, 1980

Cumming, Elizabeth and Wendy Kaplan, *The Arts and Crafts Movement*, Thames and Hudson, 1991

Davey, Peter, *Arts and Crafts Architecture*, Phaidon Press, 1995

Davison, T. Raffles, *Port Sunlight,: A Record of Its Artistic & Pictorial Aspect*, Batsford, 1916

Dawber, G., *Old Cottages, Farmhouses and other Stone Buildings of the Cotswold Region*, Batsford, 1905

Day, Lewis F., *Nature and Ornament*, Batsford, 1909

Dresser, Chrsitopher, *Principles of Decorative Design*, Cassell, Petter & Galpin, 1873

Eastlake, Sir Charles, *Hints on Household Taste*, London, 1865

Gere, Charlotte and Geoffrey C. Munn, *Artists' Jewellery, Pre-Raphaelite to Arts and Crafts*, Antique Collectors' Club, 1989

Gere, Charlotte and Michael Whiteway, *Nineteenth Century Design From Pugin to Mackintosh*, Weidenfeld & Nicolson, 1993

Grafton Green, Bridget, *Hampstead Garden Suburb 1907–1977 A History*, Hampstead Garden Suburb Residents Association, 1977

Greensted, Mary, *The Arts and Crafts Movement in the Cotswolds*, Alan Sutton, 1993

Greenwood, Martin, *The Designs of William De Morgan*, Richard Dennis Publications, 1989

Haigh, Diane, *Baillie Scott, the Artistic House*, Academy Editions, 1995

Halen, Widar, *Christopher Dresser*, Phaidon Press, 1990

Harrison, Martin, and Bill Waters, *Burne-Jones*, Barrie and Jenkins, 1973

Harrod, Tanya, *The Crafts in Britain in the 20th Century*, Yale University Press, 1999

Haslam, Malcolm, *The Martin Brothers, Potters*, Richard Dennis Publications, 1978

Heal's Catalogue 1853–1934, Middle Class Furnishings, David and Charles, 1972

Henderson, Philip, *The Letters of William Morris to his Family and Friends*, Longmans Green & Co., 1950

Hitchmough, Wendy, *C.F.A. Voysey*, Phaidon Press, 1995

Hitchmough, Wendy, *Arts and Crafts Gardens*, Pavilion, 1997

Hollamby, Edward, *Red House*, Architecture Design and Technology Press, 1991

Howard, Constance, *Twentieth-Century Embroidery in Great Britain to 1939*, Batsford, 1981

Inskip, Peter, *Edwin Lutyens*, Academy Editions, 1986

Jackson, Lesley (ed.), *Whitefriars Glass, The Art of James Powell & Sons*, Richard Dennis Publications, 1996

Jeffrey, Michael, *Christie's Arts and Crafts Style*, Pavilion, 2001
Jekyll, Gertrude with Lawrence Weaver, *Gardens for Small Country Houses*, Country Life Limited, 1912
Jekyll, Gertrude, *Wood and Garden*, Longmans, Green & Co., 1899
Jones, Owen, *The Grammar of Ornament*, Day & Son, 1856
Kelmscott Manor: An Illustrated Guide, Society of Antiquaries of London, 1996
Koch, Robert, *Louis C. Tiffany, Rebel in Glass*, Crown Publishers, 1964
Kornwolf, James D., *M.H. Baillie Scott and the Arts and Crafts Movement*, The John Hopkins Press, 1972
Kuzmanovic, Natasha, *John Paul Cooper, Designer and Craftsman of the Arts and Crafts Movement*, Sutton Publishing, 1999
Lester, Alfred W., *Hampstead Garden Suburb: The Care and Appreciation of its Architectural Heritage*, with a foreword by Sir Nikolaus Pevsner, HGS Design Study Group, 1977
Lethaby, William R., *Philip Webb and his Work*, Oxford University Press, 1935
Lethaby, William R., A.H. Powell and F.L. Griggs, *Ernest Gimson, His Work and Life*, Shakespeare Head Press, 1924
Lind, Carla, *The Wright Style: The Interiors of Frank Lloyd Wright*, Thames and Hudson, 1992
Lind, Carla, *Frank Lloyd Wright's Life and Homes*, Archetype Press Book, 1994
Lubbock, P. (ed.), *Letters of Henry James*, London, 1920
MacCarthy, Fiona, *The Simple Life: C.R. Ashbee in the Cotswolds*, Lund Humphries, 1981
MacCarthy, Fiona, *Eric Gill*, Faber & Faber, 1989
MacCarthy, Fiona, *William Morris*, Faber & Faber, 1994
Mackail, J.W., *The Life of William Morris* (2 vols), Longmans Green, 1899
Masse, H.J.L.J., *The Art-Workers' Guild*, Shakespeare Head Press, 1935
Meister, Maureen, *H.H. Richardson, The Architect, His Peers and Their Era*, The MIT Press, 1999
Moorcroft, A Guide to Moorcroft Pottery 1897–1993, Richard Dennis Publications, 1993
Morris, Barbara, *Victorian Embroidery*, Herbert Jenkins, 1962
Morris, May (ed.), *The Collected Works of William Morris* (24 vols), Longmans Green, 1910–15
Morris, William, *Arts and Crafts Essays*, Longmans Green & Co, 1899
Muthesius, Hermann, *The English House*, edited and with an introduction by Denis Sharp and translated by Janet Seligman, Crosby, Lockwood Staples, 1979
Myers, Richard and Hilary, *William Morris Tiles*, Richard Dennis Publications, 1996
Naylor, Gillian, *The Arts and Crafts Movement, a study of its sources, ideals and influence on design theory*, Studio Vista, 1971
Ohr, Clarissa Campbell (ed.), *Women in the Victorian Art World*, Manchester University Press, 1995
Parry, Linda, *Textiles of the Arts and Crafts Movement*, The Viking Press, 1983
Parry, Linda, *William Morris and the Arts & Crafts Movement, A Source Book*, Studio Editions, 1989
Parry, Linda, *William Morris*, Philip Wilson Publishers in association with the V&A Museum, 1996
Pevsner, Nikolaus, *Pioneers of the Modern Movement from William Morris to Walter Gropius*, Faber & Faber, 1936
Poulson, Christine, *William Morris on Art and Design*, Sheffield Academic Press, 1996
Pugin, A.W.N., *The True Principles of Pointed or Christian Architecture*, London, 1843, reprinted by Academy Editions, 1973
Read, Sir Herbert, *Art and Industry*, Faber & Faber, 1934
Richardson, Margaret, *Architects of the Arts and Crafts Movement*, Trefoil Books, 1983 (published in association with the Royal Institute of British Architects Drawings Collection)
Robinson, William, *The English Flower Garden*, John Murray, 1883
Rothenstein, Sir William, *Men and Memoires* (2 vols), Faber & Faber, 1931
Ruskin, John, *Seven Lamps of Architecture*, George Allen, 1894
Ruskin, John, *The Stones of Venice* (2 vols), Allen and Unwin, 1904
Sedding, John D., *Art and Handicraft*, Kegan Paul, Trench, Trubner & Co., 1893
Sedding, John Dando, *Garden Craft Old and New*, Kegan Paul, Trench, Trubner & Co., 1891
Sheldon, Alexandra, *American Arts and Crafts from the Collection of Alexandra and Sidney Sheldon*, Palm Springs Desert Museum, 1993
Smith, Bruce and Alexander Vertikoff, *Greene & Greene Masterworks*, Chronicle, 1998
Smith, Greg and Sarah Hyde, *Walter Crane 1845–1915, Artist, Designer and Socialist*, Lund Humphries, 1989
Smith, Kathryn, *Frank Lloyd Wright America's Master Architect*, Abbeville Press, 1998
Stanksy, Peter, *Redesigning the World: William Morris, the 1880s and the Arts and Crafts*, Princeton University Press, 1985
Stickley, Gustav, *Craftsman Homes: Architecture and Furnishings of the American Arts and Crafts Movement*, 1909, reprinted Dover Publications, 1979
Surtees, Victoria (ed.), *The Diary of Ford Madox Brown*, Yale University Press, 1981
Tilbrook, A.J., *the Designs of Archibald Knox for Liberty & Co.*, Richard Dennis Publications, 1995
Tinniswood, Adrian, *The Arts and Crafts House*, Mitchell Beazley, 1999
Triggs, Oscar Lovell, *Chapters in the History of the Arts & Crafts Movement*, Chicago: Bohemia Guild of the Industrial Art League, 1902
Volpe, Tod M., *Treasures of the American Arts and Crafts Movement 1890–1920*, Thames and Hudson, 1988
Wallace, Ann, *Arts and Crafts Textiles, The Movement in America* Gibbs Smith, 1999
Watkinson, Ray, *William Morris as Designer*, Studio Vista, 1967
Wilson, H., *Silverwork and Jewellery*, John Hogg, 1912

INDEX

Page numbers in *italic* refer to the illustrations

PICTURE ACKNOWLEDGMENTS

The publishers would like to thank the following for their especial help concerning illustrations for this book: Richard Dennis, John Jesse, John Scott, Alexandra Sheldon and photographers Philip de Bay, Magnus Dennis and Roger Vlitos. Grateful thanks are also due to Helen Brown and Mary Greensted at Cheltenham Museum, Sandy Kitching and team at Blackwell, Simon Biddulph at Rodmarton Manor, Bobbi Mapstone at the Gamble House, Pasadena and all listed sources for access to their collections.

Albright Knox Art Gallery, gift of Darwin R. Martin: 182(l)

Annan Gallery, Glasgow: 63 (l&r)

The Antique Trader at the Millinery Works and Jefferson Smith/Heal & Son Ltd: 29, 51(b)

The Antique Trader at the Millinery Works and Jefferson Smith: 166

Arcaid: Richard Bryant 107, 108, 116, 139, 141, 248-9, Jeremy Cockayne 126, Mark Fiennes 26, 110, 220(b), Lucinda Lambton 297, Alan Weintraub 90, 121, 122

The Art Archive: Eileen Tweedy 15

The Art Institute of Chicago, Gift of Mrs Charles F. Batchelder, 1974.524: 203

Avery Library, Department of Drawings, Columbia University: 287(t)

Birkenhead Collection: photo Magnus Dennis 46, 237, Haslam & Whiteway, London 226

Bridgeman Art Library: Fine Art Society 161, 170, Mallett's 291, V&A 156

Cardiff City Council: 41

Collection of the Carnegie Museum of Art, Pittsburgh, Pennsylvania, Du Puy Fund 1984: 179

Martin Charles: 94, 98, 100, 101, 162, 215, 250, 290, 293, 294, 296

Cheltenham Art Gallery and Museum: 17, 18, 37, 38, 49, 128, 133, CAGM/© Felicity Ashbee 35, CAGM/Bridgeman Art Library 39 73, CAGM/Philip de Bay/Bridgeman Art Library 144-5, 153, 171(r), 251, CAGM/photo Woodley & Quick 165

Chicago Historical Society: 266

Christie's Images: 45, 52, 57, 58, 64, 66, 70, 76, 80, 82, 159, 169, 174, 175, 176, 177, 182, 219(l&r), 220(t), 223, 226, 227, 228, 229(tr&b), 236, 241, 242, 244, 253, 260, 262, 267(bl), 268, 283

Cincinnati Museum Center – Cincinnati Historical Society Library: 71

Courtesy of The Charles J. Connick Stained Glass Foundation, Ltd.: Ian Justice 42

Country Life Picture Library: 104, 129, 136(t), 137, 295

Craftsman Auctions, Lambertville, New Jersey: 243, 247, 264(l&r)

The Craftsman Farms Foundation, Parsippany, New Jersey: 23, 79

Edifice: 24, 105, © Darley 27, 114, © Lewis 78, 113, 119

The Gamble House, Pasadena, CA: 51(t), 117, Interfoto, USA/Oggy Borosov 205, Tim Street-Porter 140, 304

Jenny de Gex: 48(l), 75, 89(l), 292, 305

Glasgow Museums: Art Gallery & Museum, Kelvingrove: 62

Glasgow Picture Library: Eric Thorburn 20

Sonia Halliday Photographs: 216

Haslam & Whiteway, London: 147, 151(r), 168, 171(l), 173

Hedrich-Blessing Photography, Chicago: 115

High Museum of Art, Atlanta, Georgia, Virginia Carroll Crawford Collection 1982.291: 81

By kind permission of the Parochial Church Council, Holy Trinity Church, Sloane Street/photo Courtauld Institute: 214

Angelo Hornak: 97

© Hunterian Art Gallery, University of Glasgow, Mackintosh Collection: 132

Illinois State Preservation Agency: © Frank Lloyd Wright Preservation Trust 89(r), 120

Interior Archive, London/Fritz von der Schulenburg: 131, 136(b), 138

John Jesse Ltd, London: Magnus Dennis 65, 84, 265, 271 (bl), Gate Studios (Mike Bruce) 256, 258, 267(tr), 269

Kelmscott Manor, reproduced with the permission of the Society of Antiquaries of London: photo Nigel Fisher 67, photo A.F. Kersting 193

Christian Korab, Minneapolis, courtesy the Willits-Robinson Preservation Foundation: 142

Lakeland Arts Trust: 111, 124-5, 127, 211, Paul Barker/Country Life 95, Jonathan Lynch 149(r), © Charlotte Wood 129

Andrew Lawson: 302

Los Angeles County Museum of Art, Gift of Max Palevsky/photo © 2003 Museum Associates/LACMA: 178

Matildenhohe/Hessisches Landesmuseum, Darmstadt: 135

Media Union Library, Special Collections, University of Michigan: 143

Metropolitan Museum of Art, New York, Gift of Family of Mrs Candace Wheeler, 1928 (28.34.1): photograph © 2000 197

William Morris Gallery, London: 69, photo Angelo Hornak 154, 200, 218, 276

National Monuments Record: 59

National Museum of American History, Smithsonian Institution, Washington: 246

National Museums of Scotland, reproduced by permission of Dumfries and Galloway Council and the National Trust for Scotland: 273

National Museums of Scotland © the Estate of Phoebe Anna Traquair: 83, 194, 272, 278

© reserved; collection National Portrait Gallery, London: 55

National Trust Photographic Library: John Hammond 99, Michael Caldwell 130, Nadia Mackenzie 146, L&M Gayton 191, Nick Meers 300

The Collection of the Newark Museum, Museum Purchase, 1926: 245

Oakland Museum of California, Gift of Concours d'Antiques, the Art Guild: photo M. Lee Fatheree 180

Pocumtuck Valley Memorial Association, Memorial Hall Museum, Deerfield, Massachusetts: 198

Princeton University Library, Graphic Arts Collection. Department of Rare Books and Special Collections: 279

Private Collections: 277, 285, 286(l), photo Magnus Dennis 11, 13, 230-1, 232, 234, photo Roy Macadam, Norman Mays Photography, Bramsford, Worcester 72

© Réunion des Musées Nationaux/Jean Schormans: 157

Royal Institute of British Architects, British Drawings Collection: 36, 74, 86, 102, 106

Scala, Florence/Pierpont Morgan Library, New York/Art Resource: Gift of the Fellows, 1959.23 77, 284(b)

Scala/Uffizi Gallery © 2000/courtesy of the Ministry for Cultural Heritage, Florence: 43

Alexandra and Sydney Sheldon Collection, Palm Springs: photos Taylor Sherrill 61, 263

Sotheby's, London: 31

Tadema Gallery, London: 270

The Mark Twain House, Hartford, Connecticut: 88, 196

V&A Photo Library, London: 1, 2-3, 4, 7, 8-9, 10, 14, 19, 25, 28, 32-33, 34, 40, 44, 47, 48(r), 50, 53, 56, 68, 85(l&r), 87, 142, 148, 149(l), 150, 151(l), 158, 167, 172, 183, 184-5, 186, 187, 188, 189, 192, 195, 199, 201, 202, 204, 206, 207, 208, 209, 210, 212-13, 217, 221, 222, 224, 233, 235, 238, 239, 240, 252, 254, 255, 257, 259, 261, 271(tr), 274-5, 281, 282, 284(t), 286(r), 313

Roger Vlitos: 12, 280, 287(b), 288-9, 301

Elizabeth Whiting Associates/Tim Street-Porter: 92-3, 181

The Winterthur Library, Printed Book and Periodical Collection: 54

Frank Lloyd Wright © ARS, NY and DACS, London 2003: 90, 120, 142, 182(l&r), 183, 229(b&tr), 287(t)

Picture Researcher: Jenny de Gex

Managing Editor: Sonya Newland